A BRIEF GUIDE TO GETTING THE MOST FROM THIS BOOK

① Read the Book

Feature	Description	Benefit	Page
Section-Opening Scenarios	Every section opens with a scenario presenting a unique application of algebra or trigonometry in your life outside the classroom.	Realizing that algebra and trigonometry are everywhere will help motivate your learning.	461
Detailed Worked-Out Examples	Examples are clearly written and provide step-by-step solutions. No steps are omitted, and each step is thoroughly explained to the right of the mathematics.	The blue annotations will help you to understand the solutions by providing the reason why every step is true.	517
Applications Using Real-World Data	Interesting applications from nearly every discipline, supported by up-to-date real-world data, are included in every section.	Ever wondered how you'll use algebra and trigonometry? This feature will show you how algebra and trigonometry can solve real problems.	519
Study Tips	The book's study tip boxes offer suggestions for problem solving, point out common errors to avoid, and provide informal hints and suggestions.	By seeing common mistakes, you'll be able to avoid them.	498
Explanatory Voice Balloons	Voice balloons help to demystify algebra and trigonometry. They translate mathematical language into plain English, clarify problem-solving procedures, and present alternative ways of understanding.	Does math ever look foreign to you? This feature often translates math into everyday English.	508
Learning Objectives	Every section begins with a list of objectives. Each objective is restated in the margin where the objective is covered.	The objectives focus your reading by emphasizing what's most important and where to find it.	534
Using Technology	The screens displayed in the technology boxes show how graphing utilities verify and visualize algebraic results.	Even if you are not using a graphing utility in the course, this feature will help you understand different approaches to problems.	517

② Work the Problems

Feature	Description	Benefit	Page
Check Point Examples	Each example is followed by a similar matched problem, called a Check Point, that offers you the opportunity to work a similar exercise. The answers to the Check Points are provided in the answer section.	You learn best by doing. You'll solidify your understanding of worked examples if you try a similar problem right away to be sure you understand what you've just read.	482

Extensive and Varied Exercise Sets	An abundant collection of exercises is included in an exercise set at the end of each section. Exercises are organized within categories. Your instructor will usually provide guidance on which exercises to work. The exercises in the first category, Practice Exercises, follow the same order as the section's worked examples.	The parallel order of the Practice Exercises lets you refer to the worked examples and use them as models for solving these problems.	518-521
Practice Plus Problems	This category of exercises contains more challenging problems that often require you to combine several skills or concepts.	It is important to dig in and develop your problem-solving skills. Practice Plus Exercises provide you with ample opportunity to do so.	532

❸ Review for Quizzes and Tests

Feature	Description	Benefit	Page
Mid-Chapter Check Points	At approximately the midway point in the chapter, an integrated set of review exercises allows you to review the skills and concepts you learned separately over several sections.	By combining exercises from the first half of the chapter, the Mid-Chapter Check Points give a comprehensive review before you move on to the material in the remainder of the chapter.	500
Chapter Review Grids	Each chapter contains a review chart that summarizes the definitions and concepts in every section of the chapter. Examples that illustrate these key concepts are referred to by page number in the chart.	Review this chart and you'll know the most important material in the chapter!	560-563
Chapter Review Exercises	A comprehensive collection of review exercises for each of the chapter's sections follows the review grid.	Practice makes perfect. These exercises contain the most significant problems for each of the chapter's sections.	563-567
Chapter Tests	Each chapter contains a practice test with approximately 25 problems that cover the important concepts in the chapter. Take the practice test, check your answers, and then watch the Chapter Test Prep Video CD to see worked-out solutions for any exercises you miss.	You can use the chapter test to determine whether you have mastered the material covered in the chapter.	567
Chapter Test Prep Video CDs	These video CDs found at the back of your text contain worked-out solutions to every exercise in each chapter test.	The videos let you review any exercises you miss on the chapter test.	567
Cumulative Review Exercises	Beginning with Chapter 2, each chapter concludes with a comprehensive collection of mixed cumulative review exercises. These exercises combine problems from previous chapters and the present chapter, providing an ongoing cumulative review.	Ever forget what you've learned? These exercises ensure that you are not forgetting anything as you move forward.	568

MORE TOOLS FOR YOUR MATHEMATICS SUCCESS

Student Study Pack

Get math help when YOU need it! The Student Study Pack provides you with the ultimate set of support resources to go along with your text. The Student Study Pack can be packaged with your textbook, and contains these invaluable tools:

 Student Solutions Manual

A printed manual containing full solutions to odd-numbered textbook exercises.

 Prentice Hall Math Tutor Center

Tutors provide one-on-one tutoring for any problem with an answer at the back of the book. You can contact the Tutor Center via a toll-free phone number, fax, or email.

 CD-ROM Lecture Series

A comprehensive set of textbook-specific CD-ROMs containing short video clips of the textbook objectives being reviewed and key examples being solved.

Tutorial and Homework Options

MYMATHLAB

MyMathLab can be packaged with your textbook. It is a complete online multimedia resource to help you succeed in learning. MyMathLab features:

 The entire textbook online.

 Problem-solving video clips and practice exercises, correlated to the examples and exercises in the text.

 Online tutorial exercises with guided examples.

 Online homework and tests.

Generates a personalized study plan based on your test results.

Tracks all online homework, tests, and tutorial work you do in MyMathLab gradebook.

Precalculus

THIRD EDITION

Robert Blitzer
Miami Dade College

PEARSON

Prentice
Hall

Upper Saddle River, NJ 07458

Library of Congress Cataloging-in-Publication Data

Blitzer, Robert.
    Precalculus / Robert Blitzer.—3rd ed.
        p.   cm.
    Includes index.
    ISBN 0-13-187479-9 (alk. paper)
    1. Algebra—Textbooks.   I. Title.

QA154.3.B586 2007
512—dc22   2005057979

Acquisitions Editor: *Adam Jaworski*
Editor-in-Chief: *Sally Yagan*
Project Manager: *Dawn Murrin*
Production Editor: *Prepare, Inc.*
Assistant Managing Editor: *Bayani Mendoza de Leon*
Senior Managing Editor: *Linda Mihatov Behrens*
Executive Managing Editor: *Kathleen Schiaparelli*
Manufacturing Manager: *Alexis Heydt-Long*
Manufacturing Buyer: *Maura Zaldivar*
Director of Marketing: *Patrice Jones*
Marketing Manager: *Halee Dinsey*
Marketing Assistant: *Joon Won Moon*
Managing Editor, Digital Supplements: *Nicole M. Jackson*
Media Production Editor: *John Cassar*
Director of Creative Services: *Paul Belfanti*
Creative Director: *Juan R. López*
Art Director: *Kenny Beck*
Assistant to the Art Director: *Dina Curro*
Art Editor: *Thomas Benfatti*
Interior Design: *ESM Design*
Manager, Cover Visual Research & Permissions: *Karen Sanatar*
Director, Image Resource Center: *Melinda Reo*
Manager, Rights and Permissions: *Zina Arabia*
Manager, Visual Research: *Beth Brenzel*
Image Permission Coordinator: *Debbie Hewitson*
Photo Researcher: *Melinda Alexander*
Composition: *Prepare, Inc.*
Art Studio: *Scientific Illustrators/Laserwords*
Cover Designers: Cowboy boot pepper—*Suzanne Behnke*; Tie dye pepper—*Dianne Densberger*; Pink pop pepper, Tattoo pepper—*Stacey Abraham*;
        Basketball shoe pepper: *Geoffrey Cassar*; Literary pepper—*Kristine Carney*
Cover Photos: Jalapeño pepper, *photographed by John E. Kelly*, © Getty images. Cowboy boot pepper—Cowboy boots, *photographed by Jules Frazier*,
        © Getty Images; Old wood, *photographed by Siede Preis*, © Getty images; Basketball shoe pepper—Basketball gym floor, *photographed
        by John Giustina*, © Getty images; Basketball shoe, *photographed by David Mager*, © Pearson Education.

© 2007, 2004, 2001 by Prentice-Hall, Inc.
Pearson Prentice Hall
Pearson Education, Inc.
Upper Saddle River, New Jersey 07458

Printed in the United States of America

10   9   8

ISBN 0-13-187479-9 (College Edition)
ISBN 0-13-195993-X (School Edition)

Pearson Education LTD., *London*
Pearson Education Australia PTY, Limited, *Sydney*
Pearson Education Singapore, Pte. Ltd
Pearson Education North Asia Ltd, *Hong Kong*
Pearson Education Canada, Ltd., *Toronto*
Pearson Educación de Mexico, S.A. de C.V.
Pearson Education—Japan, *Tokyo*
Pearson Education Malaysia, Pte. Ltd

Contents

Preface ix
Acknowledgments xiv
To the Student xvii
About the Author xix
Applications Index xx

chapter 5

Analytic Trigonometry   569

chapter 6

Additional Topics in Trigonometry   627

chapter 7

Systems of Equations and Inequalities   709

Chapter 8

Chapter 9

Chapter 10

Chapter 11

Appendix

Preface

Precalculus, **Third Edition** is designed and written to help students make the transition from intermediate algebra into calculus. The book has three fundamental goals:

1. To help students acquire a solid foundation in algebra and trigonometry, preparing them for other courses such as calculus, business calculus, and finite mathematics.
2. To show students how algebra and trigonometry can model and solve authentic real-world problems.
3. To enable students to develop problem-solving skills, while fostering critical thinking, within an interesting setting.

One major obstacle in the way of achieving these goals is the fact that very few students actually read their textbook. This has been a regular source of frustration for me and for my colleagues in the classroom. Anectodal evidence gathered over years highlights two basic reasons that students do not take advantage of their textbook:

- "I'll never use this information."
- "I can't follow the explanations."

I've written every page of the Third Edition with the intent of eliminating these two objections. The ideas and tools I've used to do so are described for the student in "A Brief Guide to Getting the Most from This Book" which appears inside the front cover.

How Does Precalculus Differ from Algebra and Trigonometry?

Precalculus is not simply a condensed version of my Algebra and Trigonometry book. Precalculus students are different from algebra and trigonometry students, and this text reflects those differences. Here are a few examples:

- **Algebra and Trigonometry** devotes an entire chapter to linear equations, rational equations, quadratic equations, radical equations, linear inequalities, and developing models involving these equations and inequalities. **Precalculus** reviews these topics in three sections of the prerequisites chapter (P.7: Equations; P.8 Modeling with Equations; P.9: Linear Inequalities and Absolute Value Inequalities). Functions, the core of any precalculus course, are then introduced in Chapter 1.

- **Precalculus** contains a section on constructing functions from verbal descriptions and formulas (1.10: Modeling with Functions) that is not included in **Algebra and Trigonometry**. Modeling skills are applied to situations that students are likely to see in calculus when solving applied problems involving maximum or minimum values.

- **Precalculus** develolps trigonometry from the perspective of the unit circle (4.2: Trigonometric Functions: The Unit Circle). In **Algebra and Trigonometry**, trigonometry is developed using right triangles.

- **Precalculus** contains a chapter (Chapter 11: Introduction to Calculus) that takes the student into calculus with discussions of limits, continuity, and derivatives. This chapter is not included in **Algebra and Trigonometry**.

- Many of the liberal arts applications in **Algebra and Trigonometry** are replaced by more scientific or higher level applications in **Precalculus**. Some examples:
  → Black Holes in Space (P.2: Exponents and Scientific Notation)
  → Average Velocity (1.5: More on Slope)
  → Newton's Law of Cooling (3.5: Exponential Growth and Decay; Modeling Data)
  → Modeling Involving Mixtures and Uniform Motion (7.1: Systems of Linear Equations in Two Variables)

What's New in the Third Edition?

- **Practice Plus Exercises.** More challenging practice exercises that often require students to combine several skills or concepts have been added to the exercise sets. The 674 Practice Plus Exercises in the Third Edition, averaging 10 of these exercises per exercise set, provide instructors with the option of creating assignments that take practice exercises to a more challenging level than in the previous edition.

- **Mid-Chapter Check Points.** At approximately the midway point in each chapter, an integrated set of review exercises allows students to review and assimilate the skills and concepts they learned separately over several sections. The 314 exercises that make up the Mid-Chapter Check Points, averaging 30 exercises per check point, are of a mixed nature, requiring students to discriminate which concepts or skills to apply. The Mid-Chapter Check Points should help students bring together the different objectives covered in the first half of the chapter before they move on to the material in the remainder of the chapter.

- **New Applications and Real-World Data.** I researched hundreds of books, magazines, almanacs, and online data sites to prepare the Third Edition. The Third Edition contains more new, innovative, applications, supported by data that extend as far up to the present as possible, than any previous revisions of this book.

- **Approximately 2000 New Examples and Exercises.** In addition to the 674 Practice Plus Exercises and the 314 Mid-Chapter Check Points, the Third Edition contains more than 1015 new exercises that appear primarily in the practice and application categories of the exercise sets.

- **Integration of Technology Using Graphical and Numerical Approaches to Problems.** New side-by-side features in the technology boxes connect algebraic solutions to graphical and numerical approaches to problems. Although the use of graphing utilites is optional, students can use the explanatory voice balloons to understand different approaches to problems even if they are not using a graphing utility in the course.

- **Increased Study Tip Boxes.** The book's Study Tip boxes offer suggestions for problem solving, point out common errors to avoid, and provide informal hints and suggestions. These invaluable hints appear in greater abundance in the Third Edition.

- **Chapter Test Prep Video CD.** Packaged at the back of the text, this video CD provides students with step-by-step solutions for each of the exercises in the book's chapter tests.

What Content and Organizational Changes Have Been Made to the Third Edition?

- **Section P.1 (Algebraic Expressions and Real Numbers)** now includes an early discussion of mathematical modeling and mathematical models, central themes of the book. Mathematical models now appear throughout Chapter P. A new discussion of intersection and union of sets paves the way for this notation to be used throughout the book.

- **Section P.2 (Exponents and Scientific Notation)** includes negative numbers in scientific notation, as well as an expanded discussion of converting from decimal to scientific notation.
- **Section P.7 (Equations)** now reviews equations in one section that includes linear, rational, absolute value, quadratic, and radical equations.
- **Section P.8 (Modeling with Equations)** is new to the Third Edition and reviews modeling within the context of equations. The inclusion of this review section is intended to ease the transition into modeling with functions, discussed in Chapter 1.
- **Section P.7 (Equations)** and **Section P.9 (Linear Inequalities and Absolute Value Inequalities)** take the discussion of absolute value to a slightly higher level, including solutions of equations and inequalities such as

$$5|1 - 4x| - 15 = 0 \quad \text{and} \quad -2|3x + 5| + 7 > -13.$$

  Section P.9 contains a new discussion of intersections and unions of intervals.
- **Chapter 1 (Functions and Graphs)** has been reorganized around the book's central idea: functions. Functions are introduced in the second section. All subsequent topics are viewed from the perspective of functions and relations. New multipart exercises that require students to bring together their knowledge of functions appear throughout the chapter.
- **Section 1.2 (Basics of Functions and Their Graphs)** contains a more detailed discussion, including new graphics, of identifying domain and range from a function's graph.
- **Section 1.4 (Linear Functions and Slope)** and **Section 1.5 (More on Slope)** develop lines and slope from the perspective of functions. Section 1.4 contains a new example on using intercepts to graph the general form of a line's equation. Section 1.5 contains a more thoroughly developed example on writing equations of a line perpendicular to a given line.
- **Section 1.6 (Transformations of Functions)** adds the graph of the cube root function, $f(x) = \sqrt[3]{x}$, to the table of common graphs, using this graph, as well as the other six graphs in the table, in the discussion of transformations. New graphics with clarifying voice balloons illustrate transformations. A new discussion of horizontally stretching and shrinking a graph is included among the transformations.
- **Section 1.8 (Inverse Functions)** takes finding the inverse of a function to a slightly higher level, with examples showing how to find f^{-1} for

$$f(x) = \frac{5}{x} + 4 \quad \text{and} \quad f(x) = x^2 - 1, x \geq 0.$$

  In the latter case, both f and f^{-1} are graphed on the same axes, using transformations to obtain the graph of f^{-1}.
- **Section 2.6 (Rational Functions and Their Graphs)** includes a new discussion on using transformations of $f(x) = \frac{1}{x}$ and $f(x) = \frac{1}{x^2}$ to graph rational functions.
- **Section 3.1 (Exponential Functions)** now defines e as the value that

$$\left(1 + \frac{1}{n}\right)^n$$ approaches as $n \to \infty$, using this definition to develop the formula

  for compound interest subject to continuous compounding. New graphics illustrate transformations of exponential functions.
- **Section 3.4 (Exponential and Logarithmic Equations)** has been reorganized into four categories:
  → Solving exponential equations using like bases
  → Solving exponential equations using logarithms and logarithmic properties
  → Solving logarithmic equations using the definition of a logarithm
  → Solving logarithmic equations using the one-to-one property of logarithms

  New examples appear throughout the section to ensure adequate coverage of each category.

- **Section 3.5 (Exponential Growth and Decay; Modeling Data)** contains new examples involving choosing models for data before technology is used to obtain these models.

- **Section 4.1 (Angles and Radian Measure)** contains new examples that help students "think in radians" by drawing angles of known radian measure in standard position without converting to degrees. The theme of "thinking in radians" is extended to new examples involving finding positive angles less than 2π coterminal with angles in standard position measuring $\frac{17\pi}{6}, -\frac{\pi}{12}, \frac{22\pi}{3},$ and $-\frac{17\pi}{6}$.

- **Section 4.2 (Trigonometric Functions: The Unit Circle)** contains new graphics and more developed discussions on using the unit circle to determine the domain and the range of the sine and cosine functions.

- **Section 4.3 (Right Triangle Trigonometry)** contains more examples on radicals in fractions and rationalizing denominators, a problem area for some students.

- **Section 4.4 (Trigonometric Functions of Any Angle)** contains new examples and a reference sheet (in the form of a Study Tip) that combines concepts in the first sections of the chapter, including finding coterminal angles and reference angles, locating special angles, determining the signs of the trigonometric functions in specific quadrants, and finding the trigonometric functions of special angles. Students are reminded that to be successful in trigonometry, it is frequently necessary to connect concepts.

- **Section 4.5 (Graphs of Sine and Cosine Functions) and Section 4.6 (Graphs of Other Trigonometric Functions)** use inequalities to locate one complete cycle for the graphs of trigonometric functions.

- **Section 5.5 (Trigonometric Equations)** contains a new objective on using a calculator to solve trigonometric equations, supported by new examples, including an equation whose solution involves the quadratic formula. Methods for solving quadratic equations—factoring, the square root property, and the quadratic formula— are reviewed in the context of trigonometric equations. A new collection of exercises contains a mixture of all types of trigonometric equations discussed in the section, requiring students to use the most appropriate techinque to solve each equation.

- **Section 6.4 (Graphs of Polar Equations)** is connected to trigonometric equations by solving trigonometric equations to confirm graphical observations.

- **Section 8.3 (Matrix Operations and Their Applications)** contains a new application, related to earlier work with transformations of functions, on using matrix operations to transform and manipulate computer graphics.

- **Section 9.2 (The Hyperbola)** contains a new detailed example on converting the equation of a hyperbola to standard form by completing the square on x and y.

- **Chapter 11 (Introduction to Calculus)** integrates the discussion of one-sided limits into the first two sections of the chapter. **Section 11.1 (Finding Limits Using Tables and Graphs)** contains a new example on finding one-sided limits from a graph. **Section 11.2 (Finding Limits Using Properties of Limits)** contains a new example on finding one-sided limits using limit properties. As a result of these discussions, **Section 11.3 (Limits and Continuity)** is now focused strictly on continuity.

I hope that my love for learning, as well as my respect for the diversity of students I have taught and learned from over the years, is apparent throughout this new edition. By connecting algebra and trigonometry to the whole spectrum of learning, it is my intent to show students that their world is profoundly mathematical, and indeed, π is in the sky.

Robert Blitzer

STUDENT RESOURCES

Student Study Pack

Everything a student needs to succeed in one place. It is available to qualified adopters. Study Pack contains:

- *Student Solutions Manual*

  Fully worked solutions to odd-numbered exercises.

- *CD Lecture Series*

  A comprehensive set of CD-ROMs, tied to the textbook, containing short video clips of an instructor working key book examples.

INSTRUCTOR RESOURCES

Instructor Resource Distribution

All instructor resources can be downloaded from the web site, www.prenhall.com. Select "Browse our catalog," then, click on "Mathematics"; select your course and choose your text. Under "Resources," on the left side, select "instructor" and choose the supplement you need to download. You will be required to run through a one time registration before you can complete this process.

- *TestGen*

  Easily create tests from textbook section objectives. Questions are algorithmically generated allowing for unlimited versions. Edit problems or create your own.

- *Test Item File*

  A printed test bank derived from TestGen.

- *PowerPoint Lecture Slides*

  Fully editable slides that follow the textbook. Project in class or post to a website in an online course.

- *Instructor Solutions Manual*

  Fully worked solutions to all textbook exercises and chapter projects.

Instructor's Edition

Provides answers to *all* exercises in the back of the text.

MathXL®

MathXL® is a powerful online homework, tutorial, and assessment system that accompanies your textbook. Instructors can create edit, and assign online homework and tests using algorithmically generated exercises correlated at the objective level to the textbook. Student work is tracked in an online gradebook. Students can take chapter tests and receive personalized study plans based on their results. The study plan diagnoses weaknesses and links students to tutorial exercises for objectives they need to study. Students can also access video clips from selected exercises. MathXL® is available to qualified adopters. For more information, visit our website at *www.mathxl.com*, or contact your Prentice Hall sales representative for a demonstration.

MyMathLab

MyMathLab is a text-specific, customizable online course for your textbooks. MyMathLab is powered by CourseCompass™—Pearson Education's online teaching and learning environment—and by MathXL®—our online homework, tutorial, and assessment system. MyMathLab gives you the tools you need to deliver all or a portion of your course online, whether your students are in a lab setting or working from home.

MyMathLab provides a rich and flexible set of course materials, featuring free-response exercises that are algorithmically generated for unlimited practice. Students can use online tools such as video lectures and a multimedia textbook to improve their performance. Instructors can use MyMathLab's homework and test managers to select and assign online exercises correlated to the textbook, and can import TestGen tests for added flexibility. The only gradebook—designed specifically for mathematics—automatically tracks students' homework and test results and gives the instructor control over how to calculate final grades. MyMathLab is available to qualified adopters. For more information, visit our website at *www.mymathlab.com* or contact your Prentice Hall sales representative for a product demonstration.

Acknowledgments

I wish to express my appreciation to all the reviewers of my precalculus series for their helpful feedback, frequently transmitted with wit, humor, and intelligence. Every change to this edition is the result of their thoughtful comments and suggestions. In particular, I would like to thank the following people for reviewing **College Algebra**, **Algebra and Trigonometry**, and **Precalculus**.

Barnhill, Kayoko Yates, *Clark College*
Beaver, Timothy, *Isothermal Community College*
Best, Lloyd, *Pacific Union College*
Burgin, Bill, *Gaston College*
Chang, Jimmy, *St. Petersburg College*
Colt, Diana, *University of Minnesota-Duluth*
Densmore, Donna, *Bossier Parish Community College*
Enegren, Disa, *Rose State College*
Fisher, Nancy, *University of Alabama*
Glickman, Cynthia, *Community College of Southern Nevada*
Goel, Sudhir Kumar, *Valdosta State University*
Gordon, Donald, *Manatee Community College*
Gross, David L., *University of Connecticut*
Haack, Joelx K., *University of Northern Iowa*
Haefner, Jeremy, *University of Colorado*
Hague, Joyce, *University of Wisconsin at River Falls*
Hall, Mike, *Univeristy of Mississippi*
Hay-Jahans, Christopher N., *University of South Dakota*
Hernandez, Celeste, *Richland College*
Ihlow, Winfield A., *SUNY College at Oswego*
Johnson, Nancy Raye, *Manatee Community College*
Leesburg, Mary, *Manatee Community College*
Lehmann, Christine Heinecke, *Purdue University North Central*
Levichev, Alexander, *Boston University*
Lin, Zongzhu, *Kansas State University*
Marlin, Benjamin, *Northwestern Oklahoma State University*
Massey, Marilyn, *Collin County Community College*
McCarthy-Germain, Yvelyne, *University of New Orleans*
Miller, James, *West Virginia University*
Pharo, Debra A., *Northwestern Michigan College*
Phoenix, Gloria, *North Carolina Agricultural and Technical State University*
Platt, David, *Front Range Community College*
Pohjanpelto, Juha, *Oregon State University*
Rech, Janice, *University of Nebraska at Omaha*
Salmon, Judith, *Fitchburg State College*
Schultz, Cynthia, *Illinois Valley Community College*
Stump, Chris, *Bethel College*
Trim, Pamela, *Southwest Tennessee Community College*
Turner, Chris, *Arkansas State University*
Van Lommel, Richard E., *California State University-Sacramento*
Van Peursem, Dan, *University of South Dakota*
Van Veldhuizen, Philip, *University of Nevada at Reno*
White, David, *The Victoria College*
Wienckowski, Tracy, *Univesity of Buffalo*

Special thanks to Professor Phoebe Rousse at Louisiana State University for your detailed notebooks of suggestions, including examples and exercise descriptions, as well as actual problems, that helped increase the rigor of the text without affecting its tone or diversity of applications.

Additional acknowledgments are extended to Dan Miller, for the Herculean task of preparing the solutions manuals, Brad Davis, for preparing the answer section and serving as accuracy checker, the Preparè Inc. formatting and production team, including Frank Weihenig and Linda Martino, for the book's brilliant paging, as well as keeping this complex project moving through its many stages, Aaron Darnall at Scientific Illustrators, for superbly illustrating the book, Melinda Alexander, photo researcher, for obtaining the book's new photographs, Kirk Trigsted and Scott Satake, for preparing the videotape series, including the chapter test prep video CDs, and Bayani Mendoza de Leon, assistant managing editor, for orchestrating the entire production process.

I would like to thank my editor at Prentice Hall, Adam Jaworski, and Project Manager, Dawn Murrin, who guided and coordinated the book from manuscript through production. Thanks to Kenny Beck for the beautiful covers and interior design. Finally, thanks to Halee Dinsey and Patrice Jones, for your innovative marketing efforts, to Sally Yagan, for your continuing support, and to the entire Prentice Hall sales force, for your confidence and enthusiasm about the book.

To the Student

I've written this book so that you can learn about the power of algebra and trigonometry and how they relate directly to your life outside the classroom. All concepts are carefully explained, important definitions and procedures are set off in boxes, and worked-out examples that present solutions in a step-by-step manner appear in every section. Each example is followed by a similar matched problem, called a Check Point, for you to try so that you can actively participate in the learning process as you read the book. (Answers to all Check Points appear in the back of the book.) Study Tips offer hints and suggestions and often point out common errors to avoid. A great deal of attention has been given to applying algebra and trigonometry to your life to make your learning experience both interesting and relevant.

As you begin your studies, I would like to offer some specific suggestions for using this book and for being successful in this course:

1. **Read the book.** Read each section with pen (or pencil) in hand. Move through the worked-out examples with great care. These examples provide a model for doing exercises in the exercise sets. As you proceed through the reading, do not give up if you do not understand every single word. Things will become clearer as you read on and see how various procedures are applied to specific worked-out examples.

2. **Work problems every day and check your answers.** The way to learn mathematics is by doing mathematics, which means working the Check Points and assigned exercises in the exercise sets. The more exercises you work, the better you will understand the material.

3. **Review for quizzes and tests.** After completing a chapter, study the chapter summary, work the exercises in the Chapter Review, and work the exercises in the Chapter Test. Answers to all these exercises are given in the back of the book.

> The methods that I've used to help you read the book, work the problems, and review for tests are described in "A Brief Guide to Getting the Most from This Book," which appears inside the front cover. Spend a few minutes reviewing the guide to familiarize yourself with the book's features and their benefits.

4. **Use the resources available with this book.** Additional resources to aid your study may have been purchased by your school. These resources include a Solutions Manual; a Chapter Test Prep Video CD; MyMathLab, an online version of the book with links to multimedia resources; MathXL®, an online homework, tutorial, and assessment system of the text.

I wrote this book in Point Reyes National Seashore, 40 miles north of San Francisco. The park consists of 75,000 acres with miles of pristine surf-washed beaches, forested ridges, and bays bordered by white cliffs. It was my hope to convey the beauty and excitement of mathematics using nature's unspoiled beauty as a source of inspiration and creativity. Enjoy the pages that follow as you empower yourself with the algebra and trigonometry needed to succeed in college, your career, and in your life.

Regards,

Bob

Robert Blitzer

About the Author

Bob Blitzer is a native of Manhattan and received a Bachelor of Arts degree with dual majors in mathematics and psychology (minor: English literature) from the City College of New York. His unusual combination of academic interests led him toward a Master of Arts in mathematics from the University of Miami and a doctorate in behavioral sciences from Nova University. Bob is most energized by teaching mathematics and has taught a variety of mathematics courses at Miami Dade College for nearly 30 years. He has received numerous teaching awards, including Innovator of the Year from the League for Innovations in the Community College, and was among the first group of recipients at Miami Dade College for an endowed chair based on excellence in the classroom. In addition to *Precalculus*, Bob has written textbooks covering introductory algebra, intermediate algebra, college algebra, algebra and trigonometry, and liberal arts mathematics, all published by Prentice Hall.

Applications Index

Prerequisites: Fundamental Concepts of Algebra

HIS CHAPTER REVIEWS fundamental concepts of algebra that are prerequisites for the study of college algebra. Algebra, like all of mathematics, provides the tools to help you recognize, classify, and explore the hidden patterns of your world, revealing its underlying structure. You will see how the special language of algebra describes phenomena as diverse as life expectancy, windchill, costs of reducing environmental pollution, the amount Americans spend on online dating, and, as described in the photo caption, your return to a world where the people you knew have long since departed. In many ways, algebra will provide you with a new way of looking at our world.

THE FUTURE IS NOW: YOU HAVE the opportunity to explore the cosmos in a starship traveling near the speed of light. The experience will enable you to understand the mysteries of the universe first hand, transporting you to unimagined levels of knowing and being. The down side: According to Einstein's theory of relativity, close to the speed of light, your aging rate relative to friends on Earth is nearly zero. You will return from your two-year journey to a futuristic world in which friends and loved ones are long dead. Do you explore space or stay here on Earth?

This discussion is developed algebraically in the essay on page 32 and in Exercise 120 in Exercise Set P. 3.

SECTION P.1 *Algebraic Expressions and Real Numbers*

Objectives

① Evaluate algebraic expressions.

② Use mathematical models.

③ Find the intersection of two sets.

④ Find the union of two sets.

⑤ Recognize subsets of the real numbers.

⑥ Use inequality symbols.

⑦ Evaluate absolute value.

⑧ Use absolute value to express distance.

⑨ Identify properties of the real numbers.

⑩ Simplify algebraic expressions.

Insatiable killer. That's the reputation the gray wolf acquired in the United States in the nineteenth and early twentieth centuries. Although the label was undeserved, an estimated 2 million wolves were shot, trapped, or poisoned. By 1960, the population was reduced to 800 wolves. In this section, you will learn how the special language of algebra describes your world, including the increasing wolf population in the continental United States following the Endangered Species Act of 1973.

Algebraic Expressions

Algebra uses letters, such as x and y, to represent numbers. If a letter is used to represent various numbers, it is called a **variable**. For example, imagine that you are basking in the sun on the beach. We can let x represent the number of minutes that you can stay in the sun without burning with no sunscreen. With a number 6 sunscreen, exposure time without burning is six times as long, or 6 times x. This can be written $6 \cdot x$, but it is usually expressed as $6x$. Placing a number and a letter next to one another indicates multiplication.

Notice that $6x$ combines the number 6 and the variable x using the operation of multiplication. A combination of variables and numbers using the operations of addition, subtraction, multiplication, or division, as well as powers or roots, is called an **algebraic expression**. Here are some examples of algebraic expressions:

$$x + 6, \quad x - 6, \quad 6x, \quad \frac{x}{6}, \quad 3x + 5, \quad x^2 - 3, \quad \sqrt{x} + 7.$$

Many algebraic expressions involve *exponents*. For example, the algebraic expression

$$0.72x^2 + 9.4x + 783$$

approximates the gray wolf population in the United States x years after 1960. The expression x^2 means $x \cdot x$, and is read "x to the second power" or "x squared." The exponent, 2, indicates that the base, x, appears as a factor two times.

Exponential Notation

If n is a counting number (1, 2, 3, and so on),

Exponent or Power

$$b^n = \underbrace{b \cdot b \cdot b \cdot \cdots \cdot b}_{\substack{b \text{ appears as a} \\ \text{factor } n \text{ times.}}}$$

Base

b^n is read "the nth power of b" or "b to the nth power." Thus, the nth power of b is defined as the product of n factors of b. The expression b^n is called an **exponential expression**. Furthermore, $b^1 = b$.

For example,

$$8^2 = 8 \cdot 8 = 64, \quad 5^3 = 5 \cdot 5 \cdot 5 = 125, \quad \text{and} \quad 2^4 = 2 \cdot 2 \cdot 2 \cdot 2 = 16.$$

 Evaluate algebraic expressions.

Evaluating Algebraic Expressions

Evaluating an algebraic expression means to find the value of the expression for a given value of the variable. For example, we can evaluate $6x$ (from the sunscreen example) when $x = 15$. We substitute 15 for x. We obtain $6 \cdot 15$, or 90. This means that if you can stay in the sun for 15 minutes without burning when you don't put on any lotion, then with a number 6 lotion, you can "cook" for 90 minutes without burning.

Many algebraic expressions involve more than one operation. Evaluating an algebraic expression without a calculator involves carefully applying the following order of operations agreement:

The Order of Operations Agreement

1. Perform operations within the innermost parentheses and work outward. If the algebraic expression involves a fraction, treat the numerator and the denominator as if they were each enclosed in parentheses.

2. Evaluate all exponential expressions.

3. Perform multiplications and divisions as they occur, working from left to right.

4. Perform additions and subtractions as they occur, working from left to right.

EXAMPLE 1   Evaluating an Algebraic Expression

Evaluate $7 + 5(x - 4)^3$ for $x = 6$.

Solution

$$
\begin{aligned}
7 + 5(x - 4)^3 &= 7 + 5(6 - 4)^3 && \text{Replace } x \text{ with 6.} \\
&= 7 + 5(2)^3 && \text{First work inside parentheses: } 6 - 4 = 2. \\
&= 7 + 5(8) && \text{Evaluate the exponential expression:} \\
&&& 2^3 = 2 \cdot 2 \cdot 2 = 8. \\
&= 7 + 40 && \text{Multiply: } 5(8) = 40. \\
&= 47 && \text{Add.}
\end{aligned}
$$

Check Point **1** Evaluate $8 + 6(x - 3)^2$ for $x = 13$.

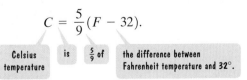 Use mathematical models.

Formulas and Mathematical Models

An **equation** is formed when an equal sign is placed between two algebraic expressions. One aim of algebra is to provide a compact, symbolic description of the world. These descriptions involve the use of *formulas*. A **formula** is an equation that uses letters to express a relationship between two or more variables. Here is an example of a formula:

$$C = \frac{5}{9}(F - 32).$$

Celsius temperature | is | $\frac{5}{9}$ of | the difference between Fahrenheit temperature and 32°.

The process of finding formulas to describe real-world phenomena is called **mathematical modeling**. Such formulas, together with the meaning assigned to the variables, are called **mathematical models**. We often say that these formulas model, or describe, the relationships among the variables.

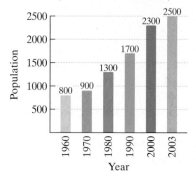

Gray Wolf Population (to the Nearest Hundred)

Year

Figure P.1

Source: U.S. Department of the Interior

EXAMPLE 2   Modeling the Gray Wolf Population

The formula

$$P = 0.72x^2 + 9.4x + 783$$

models the gray wolf population, P, in the United States, x years after 1960. Use the formula to find the population in 1990. How well does the formula model the actual data shown in the bar graph in Figure P.1?

Solution   Because 1990 is 30 years after 1960, we substitute 30 for x in the given formula. Then we use the order of operations to find P, the gray wolf population in 1990.

$P = 0.72x^2 + 9.4x + 783$	This is the given mathematical model.
$P = 0.72(30)^2 + 9.4(30) + 783$	Replace each occurrence of x with 30.
$P = 0.72(900) + 9.4(30) + 783$	Evaluate the exponential expression: $30^2 = 30 \cdot 30 = 900$.
$P = 648 + 282 + 783$	Multiply from left to right: $0.72(900) = 648$ and $9.4(30) = 282$.
$P = 1713$	Add.

The formula indicates that in 1990, the gray wolf population in the United States was 1713. The number given in Figure P.1 is 1700, so the formula models the data quite well.

Check Point 2   Use the formula in Example 2 to find the gray wolf population in 2000. How well does the formula model the data in Figure P.1?

Sometimes a mathematical model gives an estimate that is not a good approximation or is extended to include values of the variable that do not make sense. In these cases, we say that **model breakdown** has occurred.

Sets

Before we describe the set of real numbers, let's be sure you are familiar with some basic ideas about sets. A **set** is a collection of objects whose contents can be clearly determined. The objects in a set are called the **elements** of the set. For example, the set of numbers used for counting can be represented by

$$\{1, 2, 3, 4, 5, \dots\}.$$

The braces, { }, indicate that we are representing a set. This form of representation, called the **roster method**, uses commas to separate the elements of the set. The three dots after the 5, called an *ellipsis*, indicate that there is no final element and that the listing goes on forever.

A set can also be written in **set-builder notation**. In this notation, the elements of the set are described, but not listed. Here is an example:

$$\{x \mid x \text{ is a counting number less than 6}\}.$$

The set of all x such that x is a counting number less than 6.

Study Tip

Grouping symbols such as parentheses, ( ), and square brackets, [ ], are not used to represent sets. Only commas are used to separate the elements of a set. Separators such as colons or semicolons are not used.

The same set written using the roster method is

$$\{1, 2, 3, 4, 5\}.$$

③ Find the intersection of two sets.

If A and B are sets, we can form a new set consisting of all elements that are in both A and B. This set is called the *intersection* of the two sets.

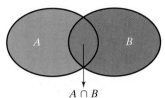

$A \cap B$

Figure P.2 Picturing the intersection of two sets

> **Definition of the Intersection of Sets**
>
> The **intersection** of sets A and B, written $A \cap B$, is the set of elements common to both set A **and** set B. This definition can be expressed in set-builder notation as follows:
>
> $$A \cap B = \{x \mid x \text{ is an element of } A \text{ AND } x \text{ is an element of } B\}.$$

Figure P.2 shows a useful way of picturing the intersection of sets A and B. The figure indicates that $A \cap B$ contains those elements that belong to both A and B at the same time.

EXAMPLE 3   Finding the Intersection of Two Sets

Find the intersection: $\{7, 8, 9, 10, 11\} \cap \{6, 8, 10, 12\}$.

Solution   The elements common to $\{7, 8, 9, 10, 11\}$ and $\{6, 8, 10, 12\}$ are 8 and 10. Thus,

$$\{7, 8, 9, 10, 11\} \cap \{6, 8, 10, 12\} = \{8, 10\}.$$

Check Point 3   Find the intersection: $\{3, 4, 5, 6, 7\} \cap \{3, 7, 8, 9\}$.

If a set has no elements, it is called the **empty set**, or the **null set**, and is represented by the symbol $\emptyset$ (the Greek letter phi). Here is an example that shows how the empty set can result when finding the intersection of two sets:

$$\{2, 4, 6\} \cap \{3, 5, 7\} = \emptyset.$$

These sets have no common elements.

Their intersection has no elements and is the empty set.

④ Find the union of two sets.

Another set that we can form from sets A and B consists of elements that are in A or B or in both sets. This set is called the *union* of the two sets.

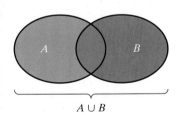

$A \cup B$

Figure P.3 Picturing the union of two sets

> **Definition of the Union of Sets**
>
> The **union** of sets A and B, written $A \cup B$, is the set of elements that are members of set A **or** of set B or of both sets. This definition can be expressed in set-builder notation as follows:
>
> $$A \cup B = \{x \mid x \text{ is an element of } A \text{ OR } x \text{ is an element of } B\}.$$

Figure P.3 shows a useful way of picturing the union of sets A and B. The figure indicates that $A \cup B$ is formed by joining the sets together.

We can find the union of set A and set B by listing the elements of set A. Then, we include any elements of set B that have not already been listed. Enclose all elements that are listed with braces. This shows that the union of two sets is also a set.

Study Tip

When finding the union of two sets, do not list twice any elements that appear in both sets.

EXAMPLE 4   Finding the Union of Two Sets

Find the union: $\{7, 8, 9, 10, 11\} \cup \{6, 8, 10, 12\}$.

Solution   To find $\{7, 8, 9, 10, 11\} \cup \{6, 8, 10, 12\}$, start by listing all the elements from the first set, namely 7, 8, 9, 10, and 11. Now list all the elements from the second set that are not in the first set, namely 6 and 12. The union is the set consisting of all these elements. Thus,

$$\{7, 8, 9, 10, 11\} \cup \{6, 8, 10, 12\} = \{6, 7, 8, 9, 10, 11, 12\}.$$

Check Point 4 Find the union: $\{3, 4, 5, 6, 7\} \cup \{3, 7, 8, 9\}$.

⑤ Recognize subsets of the real numbers.

The Set of Real Numbers

The sets that make up the real numbers are summarized in Table P.1. We refer to these sets as **subsets** of the real numbers, meaning that all elements in each subset are also elements in the set of real numbers.

Notice the use of the symbol $\approx$ in the examples of irrational numbers. The symbol means "is approximately equal to." Thus,

$$\sqrt{2} \approx 1.414214.$$

We can verify that this is only an approximation by multiplying 1.414214 by itself. The product is very close to, but not exactly, 2:

$$1.414214 \times 1.414214 = 2.000001237796.$$

Technology

A calculator with a square root key gives a decimal approximation for $\sqrt{2}$, not the exact value.

Table P.1   Important Subsets of the Real Numbers

Name	Description	Examples
Natural numbers $\mathbb{N}$	$\{1, 2, 3, 4, 5, \dots\}$ These are the numbers that we use for counting.	2, 3, 5, 17
Whole numbers $\mathbb{W}$	$\{0, 1, 2, 3, 4, 5, \dots\}$ The set of whole numbers includes 0 and the natural numbers.	0, 2, 3, 5, 17
Integers $\mathbb{Z}$	$\{\dots, -5, -4, -3, -2, -1, 0, 1, 2, 3, 4, 5, \dots\}$ The set of integers includes the negatives of the natural numbers and the whole numbers.	$-17, -5, -3, -2, 0, 2, 3, 5, 17$
Rational numbers $\mathbb{Q}$	$\left\{\frac{a}{b} \mid a \text{ and } b \text{ are integers and } b \neq 0\right\}$ **This means that b is not equal to zero.** The set of rational numbers is the set of all numbers that can be expressed as a quotient of two integers, with the denominator not 0. Rational numbers can be expressed as terminating or repeating decimals.	$-17 = \frac{-17}{1}, -5 = \frac{-5}{1}, -3, -2,$ $0, 2, 3, 5, 17,$ $\frac{2}{5} = 0.4,$ $\frac{-2}{3} = -0.6666\dots = -0.\overline{6}$
Irrational numbers $\mathbb{I}$	The set of irrational numbers is the set of all numbers whose decimal representations are neither terminating nor repeating. Irrational numbers cannot be expressed as a quotient of integers.	$\sqrt{2} \approx 1.414214$ $-\sqrt{3} \approx -1.73205$ $\pi \approx 3.142$ $-\frac{\pi}{2} \approx -1.571$

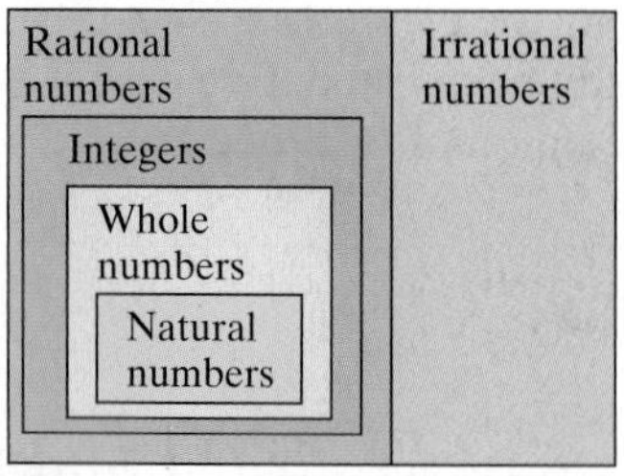

Real numbers

Rational numbers	Irrational numbers
Integers	
Whole numbers	
Natural numbers	

Figure P.4 Every real number is either rational or irrational.

Not all square roots are irrational. For example, $\sqrt{25} = 5$ because $5^2 = 5 \cdot 5 = 25$. Thus, $\sqrt{25}$ is a natural number, a whole number, an integer, and a rational number $\left(\sqrt{25} = \frac{5}{1}\right)$.

The set of *real numbers* is formed by taking the union of the sets of rational numbers and irrational numbers. Thus, every real number is either rational or irrational, as shown in Figure P.4.

Real Numbers

The set of **real numbers** is the set of numbers that are either rational or irrational:

$$\{x \mid x \text{ is rational or } x \text{ is irrational}\}.$$

The symbol $\mathbb{R}$ is used to represent the set of real numbers. Thus,

$$\mathbb{R} = \{x \mid x \text{ is rational}\} \cup \{x \mid x \text{ is irrational}\}.$$

EXAMPLE 5   Recognizing Subsets of the Real Numbers

Consider the following set of numbers:

$$\left\{ -7, -\frac{3}{4}, 0, 0.\overline{6}, \sqrt{5}, \pi, 7.3, \sqrt{81} \right\}.$$

List the numbers in the set that are

    a. natural numbers.   **b.** whole numbers.   **c.** integers.

    d. rational numbers.   **e.** irrational numbers.   **f.** real numbers.

Solution

    a. Natural numbers: The natural numbers are the numbers used for counting. The only natural number in the set is $\sqrt{81}$ because $\sqrt{81} = 9$. (9 multiplied by itself, or 9^2, is 81.)

    b. Whole numbers: The whole numbers consist of the natural numbers and 0. The elements of the set that are whole numbers are 0 and $\sqrt{81}$.

    c. Integers: The integers consist of the natural numbers, 0, and the negatives of the natural numbers. The elements of the set that are integers are $\sqrt{81}$, 0, and -7.

    d. Rational numbers: All numbers in the set that can be expressed as the quotient of integers are rational numbers. These include $-7 \left(-7 = \frac{-7}{1} \right)$, $-\frac{3}{4}$, $0 \left(0 = \frac{0}{1} \right)$, and $\sqrt{81} \left(\sqrt{81} = \frac{9}{1} \right)$. Furthermore, all numbers in the set that are terminating or repeating decimals are also rational numbers. These include $0.\overline{6}$ and 7.3.

    e. Irrational numbers: The irrational numbers in the set are $\sqrt{5} \left(\sqrt{5} \approx 2.236 \right)$ and $\pi (\pi \approx 3.14)$. Both $\sqrt{5}$ and π are only approximately equal to 2.236 and 3.14, respectively. In decimal form, $\sqrt{5}$ and π neither terminate nor have blocks of repeating digits.

    f. Real numbers: All the numbers in the given set are real numbers.

Check Point 5 Consider the following set of numbers:

$$\left\{ -9, -1.3, 0, 0.\overline{3}, \frac{\pi}{2}, \sqrt{9}, \sqrt{10} \right\}.$$

List the numbers in the set that are

    a. natural numbers.   **b.** whole numbers.

    c. integers.   **d.** rational numbers.

    e. irrational numbers.   **f.** real numbers.

The Real Number Line

The **real number line** is a graph used to represent the set of real numbers. An arbitrary point, called the **origin**, is labeled 0. Select a point to the right of 0 and label it 1. The distance from 0 to 1 is called the **unit distance**. Numbers to the right of the origin are **positive** and numbers to the left of the origin are **negative**. The real number line is shown in Figure P.5.

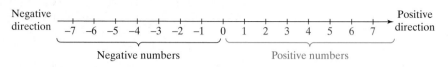

Figure P.5 The real number line

Real numbers are **graphed** on a number line by placing a dot at the correct location for each number. The integers are easiest to locate. In Figure P.6, we've graphed the integers −3, 0, and 4.

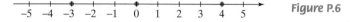

Figure P.6

Every real number corresponds to a point on the number line and every point on the number line corresponds to a real number. We say that there is a **one-to-one correspondence** between all the real numbers and all points on a real number line.

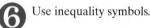 **6** Use inequality symbols.

Ordering the Real Numbers

On the real number line, the real numbers increase from left to right. The lesser of two real numbers is the one farther to the left on a number line. The greater of two real numbers is the one farther to the right on a number line.

Look at the number line in Figure P.7. The integers −4 and −1 are graphed.

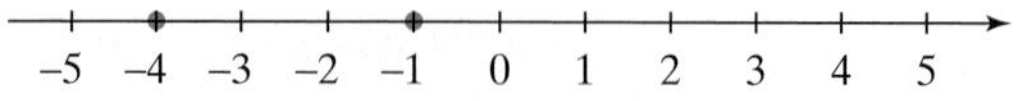

Figure P.7

Observe that −4 is to the left of −1 on the number line. This means that −4 is less than −1.

$$-4 < -1$$

> −4 is less than −1 because −4 is to the **left** of −1 on the number line.

In Figure P.7, we can also observe that −1 is to the right of −4 on the number line. This means that −1 is greater than −4.

$$-1 > -4$$

> −1 is greater than −4 because −1 is to the **right** of −4 on the number line.

The symbols $<$ and $>$ are called **inequality symbols**. These symbols always point to the lesser of the two real numbers when the inequality statement is true.

> −4 is less than −1.

$$-4 < -1$$

The symbol points to −4, the lesser number.

> −1 is greater than −4.

$$-1 > -4$$

The symbol still points to −4, the lesser number.

The symbols $<$ and $>$ may be combined with an equal sign, as shown in the following table:

	Symbols	Meaning	Examples	Explanation
This inequality is true if either the < part or the = part is true.	$a \leq b$	a is less than or equal to b.	$2 \leq 9$ $9 \leq 9$	Because $2 < 9$ Because $9 = 9$
This inequality is true if either the > part or the = part is true.	$b \geq a$	b is greater than or equal to a.	$9 \geq 2$ $2 \geq 2$	Because $9 > 2$ Because $2 = 2$

properties of real numbers listed in Table P.2, without giving it much thought. The properties of the real numbers are especially useful when working with algebraic expressions. For each property listed in Table P.2, a, b, and c represent real numbers, variables, or algebraic expressions.

Table P.2   Properties of the Real Numbers

Name	Meaning	Examples
Commutative Property of Addition	Changing order when adding does not affect the sum. $a + b = b + a$	• $13 + 7 = 7 + 13$ • $13x + 7 = 7 + 13x$
Commutative Property of Multiplication	Changing order when multiplying does not affect the product. $ab = ba$	• $\sqrt{2} \cdot \sqrt{5} = \sqrt{5} \cdot \sqrt{2}$ • $x \cdot 6 = 6x$
Associative Property of Addition	Changing grouping when adding does not affect the sum. $(a + b) + c = a + (b + c)$	• $3 + (8 + x) = (3 + 8) + x$ $= 11 + x$
Associative Property of Multiplication	Changing grouping when multiplying does not affect the product. $(ab)c = a(bc)$	• $-2(3x) = (-2 \cdot 3)x = -6x$
Distributive Property of Multiplication over Addition	Multiplication distributes over addition. $a \cdot (b + c) = a \cdot b + a \cdot c$	• $7(4 + \sqrt{3}) = 7 \cdot 4 + 7 \cdot \sqrt{3}$ $= 28 + 7\sqrt{3}$ • $5(3x + 7) = 5 \cdot 3x + 5 \cdot 7$ $= 15x + 35$
Identity Property of Addition	Zero can be deleted from a sum. $a + 0 = a$ $0 + a = a$	• $\sqrt{3} + 0 = \sqrt{3}$ • $0 + 6x = 6x$
Identity Property of Multiplication	One can be deleted from a product. $a \cdot 1 = a$ $1 \cdot a = a$	• $1 \cdot \pi = \pi$ • $13x \cdot 1 = 13x$
Inverse Property of Addition	The sum of a real number and its additive inverse gives 0, the additive identity. $a + (-a) = 0$ $(-a) + a = 0$	• $\sqrt{5} + (-\sqrt{5}) = 0$ • $-\pi + \pi = 0$ • $6x + (-6x) = 0$ • $(-4y) + 4y = 0$
Inverse Property of Multiplication	The product of a nonzero real number and its multiplicative inverse gives 1, the multiplicative identity. $a \cdot \dfrac{1}{a} = 1, \quad a \neq 0$ $\dfrac{1}{a} \cdot a = 1, \quad a \neq 0$	• $7 \cdot \dfrac{1}{7} = 1$ • $\left(\dfrac{1}{x-3}\right)(x - 3) = 1, \quad x \neq 3$

The Associative Property and the English Language

In the English language, phrases can take on different meanings depending on the way the words are associated with commas. Here are three examples.

- Woman, without her man, is nothing.
  Woman, without her, man is nothing.
- What's the latest dope?
  What's the latest, dope?
- Population of Amsterdam broken down by age and sex
  Population of Amsterdam, broken down by age and sex

Commutative Words and Sentences

The commutative property states that a change in order produces no change in the answer. The words and sentences listed here suggest a characteristic of the commutative property; they read the same from left to right and from right to left!

- dad
- repaper
- never odd or even
- Go deliver a dare, vile dog!
- May a moody baby doom a yam?
- Madam, in Eden I'm Adam.
- Ma is a nun, as I am.
- A man, a plan, a canal: Panama
- Are we not drawn onward, we few, drawn onward to new era?

The properties of the real numbers in Table P.2 on page 11 apply to the operations of addition and multiplication. Subtraction and division are defined in terms of addition and multiplication.

Definitions of Subtraction and Division
Let a and b represent real numbers.

Subtraction: $a - b = a + (-b)$
We call $-b$ the **additive inverse** or **opposite** of b.

Division: $a \div b = a \cdot \frac{1}{b}$, where $b \neq 0$
We call $\frac{1}{b}$ the **multiplicative inverse** or **reciprocal** of b. The quotient of a and b, $a \div b$, can be written in the form $\frac{a}{b}$, where a is the **numerator** and b the **denominator** of the fraction.

Because subtraction is defined in terms of adding an inverse, the distributive property can be applied to subtraction:

$$a(b - c) = ab - ac$$
$$(b - c)a = ba - ca.$$

For example,

$$4(2x - 5) = 4 \cdot 2x - 4 \cdot 5 = 8x - 20.$$

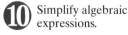 Simplify algebraic expressions.

Simplifying Algebraic Expressions

The **terms** of an algebraic expression are those parts that are separated by addition. For example, consider the algebraic expression

$$7x - 9y + z - 3,$$

which can be expressed as

$$7x + (-9y) + z + (-3).$$

This expression contains four terms, namely $7x$, $-9y$, z, and -3.

The numerical part of a term is called its **coefficient**. In the term $7x$, the 7 is the coefficient. If a term containing one or more variables is written without a coefficient, the coefficient is understood to be 1. Thus, z means $1z$. If a term is a constant, its coefficient is that constant. Thus, the coefficient of the constant term -3 is -3.

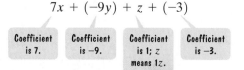

The parts of each term that are multiplied are called the **factors** of the term. The factors of the term $7x$ are 7 and x.

Like terms are terms that have exactly the same variable factors. For example, $3x$ and $7x$ are like terms. The distributive property in the form

$$ba + ca = (b + c)a$$

enables us to add or subtract like terms. For example,

$$3x + 7x = (3 + 7)x = 10x$$
$$7y^2 - y^2 = 7y^2 - 1y^2 = (7 - 1)y^2 = 6y^2.$$

This process is called **combining like terms**.

An algebraic expression is **simplified** when parentheses have been removed and like terms have been combined.

EXAMPLE 8   Simplifying an Algebraic Expression

Simplify: $6(2x^2 + 4x) + 10(4x^2 + 3x)$.

Solution

$$6(2x^2 + 4x) + 10(4x^2 + 3x)$$

$= 6 \cdot 2x^2 + 6 \cdot 4x + 10 \cdot 4x^2 + 10 \cdot 3x$ Use the distributive property to remove the parentheses.

$= 12x^2 + 24x + 40x^2 + 30x$ Multiply.

$= (12x^2 + 40x^2) + (24x + 30x)$ Group like terms.

$= 52x^2 + 54x$ Combine like terms.

$52x^2$ and $54x$ are not like terms. They contain different variable factors, x^2 and x, and cannot be combined.

Check Point 8   Simplify: $7(4x^2 + 3x) + 2(5x^2 + x)$.

Properties of Negatives

The distributive property can be extended to cover more than two terms within parentheses. For example,

This sign represents subtraction.

This sign tells us that the number is negative.

$$-3(4x - 2y + 6) = -3 \cdot 4x - (-3) \cdot 2y - 3 \cdot 6$$
$$= -12x - (-6y) - 18$$
$$= -12x + 6y - 18.$$

The voice balloons illustrate that negative signs can appear side by side. They can represent the operation of subtraction or the fact that a real number is negative. Here is a list of properties of negatives and how they are applied to algebraic expressions:

Properties of Negatives

Let a and b represent real numbers, variables, or algebraic expressions.

Property	Examples
1. $(-1)a = -a$	$(-1)4xy = -4xy$
2. $-(-a) = a$	$-(-6y) = 6y$
3. $(-a)b = -ab$	$(-7)4xy = -7 \cdot 4xy = -28xy$
4. $a(-b) = -ab$	$5x(-3y) = -5x \cdot 3y = -15xy$
5. $-(a + b) = -a - b$	$-(7x + 6y) = -7x - 6y$
6. $-(a - b) = -a + b$	$-(3x - 7y) = -3x + 7y$
$\qquad\qquad = b - a$	$\qquad\qquad = 7y - 3x$

It is not uncommon to see algebraic expressions with parentheses preceded by a negative sign or subtraction. Properties 5 and 6 in the box on the previous page, $-(a + b) = -a - b$ and $-(a - b) = -a + b$, are related to this situation. An expression of the form $-(a + b)$ can be simplified as follows:

$$-(a + b) = -1(a + b) = (-1)a + (-1)b = -a + (-b) = -a - b.$$

Do you see a fast way to obtain the simplified expression on the right? **If a negative sign or a subtraction symbol appears outside parentheses, drop the parentheses and change the sign of every term within the parentheses.** For example,

$$-(3x^2 - 7x - 4) = -3x^2 + 7x + 4.$$

EXAMPLE 9   Simplifying an Algebraic Expression

Simplify: $8x + 2[5 - (x - 3)]$.

Solution

$$8x + 2[5 - (x - 3)]$$

$= 8x + 2[5 - x + 3]$ Drop parentheses and change the sign of each term in parentheses: $-(x - 3) = -x + 3$.

$= 8x + 2[8 - x]$ Simplify inside brackets: $5 + 3 = 8$.

$= 8x + 16 - 2x$ Apply the distributive property:

$$2[8 - x] = 2 \cdot 8 - 2x = 16 - 2x.$$

$= (8x - 2x) + 16$ Group like terms.

$= (8 - 2)x + 16$ Apply the distributive property.

$= 6x + 16$ Simplify.

Check Point 9   Simplify: $6 + 4[7 - (x - 2)]$.

EXERCISE SET P.1

Practice Exercises

In Exercises 1–16, evaluate each algebraic expression for the given value or values of the variable(s).

1. $7 + 5x$, for $x = 10$ **2.** $8 + 6x$, for $x = 5$

3. $6x - y$, for $x = 3$ and $y = 8$

4. $8x - y$, for $x = 3$ and $y = 4$

5. $x^2 + 3x$, for $x = 8$ **6.** $x^2 + 5x$, for $x = 6$

7. $x^2 - 6x + 3$, for $x = 7$ **8.** $x^2 - 7x + 4$, for $x = 8$

9. $4 + 5(x - 7)^3$, for $x = 9$

10. $6 + 5(x - 6)^3$, for $x = 8$

11. $x^2 - 3(x - y)$, for $x = 8$ and $y = 2$

12. $x^2 - 4(x - y)$, for $x = 8$ and $y = 3$

13. $\dfrac{5(x + 2)}{2x - 14}$, for $x = 10$ **14.** $\dfrac{7(x - 3)}{2x - 16}$, for $x = 9$

15. $\dfrac{2x + 3y}{x + 1}$, for $x = -2$ and $y = 4$

16. $\dfrac{2x + y}{xy - 2x}$, for $x = -2$ and $y = 4$

The formula

$$C = \frac{5}{9}(F - 32)$$

expresses the relationship between Fahrenheit temperature, F, and Celsius temperature, C. In Exercises 17–18, use the formula to convert the given Fahrenheit temperature to its equivalent temperature on the Celsius scale.

17. $50°F$ **18.** $86°F$

A football was kicked vertically upward from a height of 4 feet with an initial speed of 60 feet per second. The formula

$$h = 4 + 60t - 16t^2$$

describes the ball's height above the ground, h, in feet, t seconds after it was kicked. Use this formula to solve Exercises 19–20.

19. What was the ball's height 2 seconds after it was kicked?

20. What was the ball's height 3 seconds after it was kicked?

In Exercises 21–28, find the intersection of the sets.

21. $\{1, 2, 3, 4\} \cap \{2, 4, 5\}$ **22.** $\{1, 3, 7\} \cap \{2, 3, 8\}$

23. $\{s, e, t\} \cap \{t, e, s\}$ **24.** $\{r, e, a, l\} \cap \{l, e, a, r\}$

25. $\{1, 3, 5, 7\} \cap \{2, 4, 6, 8, 10\}$

26. $\{0, 1, 3, 5\} \cap \{-5, -3, -1\}$

27. $\{a, b, c, d\} \cap \varnothing$ **28.** $\{w, y, z\} \cap \varnothing$

In Exercises 29–34, find the union of the sets.

29. $\{1, 2, 3, 4\} \cup \{2, 4, 5\}$ **30.** $\{1, 3, 7, 8\} \cup \{2, 3, 8\}$

31. $\{1, 3, 5, 7\} \cup \{2, 4, 6, 8, 10\}$

32. $\{0, 1, 3, 5\} \cup \{2, 4, 6\}$ **33.** $\{a, e, i, o, u\} \cup \varnothing$

34. $\{e, m, p, t, y\} \cup \varnothing$

In Exercises 35–38, list all numbers from the given set that are
a. *natural numbers,* **b.** *whole numbers,* **c.** *integers,* **d.** *rational numbers,* **e.** *irrational numbers,* **f.** *real numbers.*

35. $\left\{-9, -\frac{4}{5}, 0, 0.25, \sqrt{3}, 9.2, \sqrt{100}\right\}$

36. $\left\{-7, -0.\overline{6}, 0, \sqrt{49}, \sqrt{50}\right\}$

37. $\left\{-11, -\frac{5}{6}, 0, 0.75, \sqrt{5}, \pi, \sqrt{64}\right\}$

38. $\left\{-5, -0.\overline{3}, 0, \sqrt{2}, \sqrt{4}\right\}$

39. Give an example of a whole number that is not a natural number.

40. Give an example of a rational number that is not an integer.

41. Give an example of a number that is an integer, a whole number, and a natural number.

42. Give an example of a number that is a rational number, an integer, and a real number.

Determine whether each statement in Exercises 43–50 is true or false.

43. $-13 \le -2$ **44.** $-6 > 2$

45. $4 \ge -7$ **46.** $-13 < -5$

47. $-\pi \ge -\pi$ **48.** $-3 > -13$

49. $0 \ge -6$ **50.** $0 \ge -13$

In Exercises 51–60, rewrite each expression without absolute value bars.

51. $|300|$ **52.** $|-203|$

53. $|12 - \pi|$ **54.** $|7 - \pi|$

55. $|\sqrt{2} - 5|$ **56.** $|\sqrt{5} - 13|$

57. $\dfrac{-3}{|-3|}$ **58.** $\dfrac{-7}{|-7|}$

59. $\||-3| - |-7|\|$ **60.** $\||-5| - |-13|\|$

In Exercises 61–66, evaluate each algebraic expression for $x = 2$ and $y = -5$.

61. $|x + y|$ **62.** $|x - y|$

63. $|x| + |y|$ **64.** $|x| - |y|$

65. $\dfrac{y}{|y|}$ **66.** $\dfrac{|x|}{x} + \dfrac{|y|}{y}$

In Exercises 67–74, express the distance between the given numbers using absolute value. Then find the distance by evaluating the absolute value expression.

67. 2 and 17   **68.** 4 and 15

69. -2 and 5   **70.** -6 and 8

71. -19 and -4 **72.** -26 and -3

73. -3.6 and -1.4 **74.** -5.4 and -1.2

In Exercises 75–84, state the name of the property illustrated.

75. $6 + (-4) = (-4) + 6$

76. $11 \cdot (7 + 4) = 11 \cdot 7 + 11 \cdot 4$

77. $6 + (2 + 7) = (6 + 2) + 7$

78. $6 \cdot (2 \cdot 3) = 6 \cdot (3 \cdot 2)$

79. $(2 + 3) + (4 + 5) = (4 + 5) + (2 + 3)$

80. $7 \cdot (11 \cdot 8) = (11 \cdot 8) \cdot 7$

81. $2(-8 + 6) = -16 + 12$

82. $-8(3 + 11) = -24 + (-88)$

83. $\dfrac{1}{(x + 3)}(x + 3) = 1, x \ne -3$

84. $(x + 4) + [-(x + 4)] = 0$

In Exercises 85–96, simplify each algebraic expression.

85. $5(3x + 4) - 4$ **86.** $2(5x + 4) - 3$

87. $5(3x - 2) + 12x$ **88.** $2(5x - 1) + 14x$

89. $7(3y - 5) + 2(4y + 3)$ **90.** $4(2y - 6) + 3(5y + 10)$

91. $5(3y - 2) - (7y + 2)$ **92.** $4(5y - 3) - (6y + 3)$

93. $7 - 4[3 - (4y - 5)]$ **94.** $6 - 5[8 - (2y - 4)]$

95. $18x^2 + 4 - [6(x^2 - 2) + 5]$

96. $14x^2 + 5 - [7(x^2 - 2) + 4]$

In Exercises 97–102, write each algebraic expression without parentheses.

97. $-(-14x)$ **98.** $-(-17y)$

99. $-(2x - 3y - 6)$ **100.** $-(5x - 13y - 1)$

101. $\frac{1}{3}(3x) + [(4y) + (-4y)]$

102. $\frac{1}{2}(2y) + [(-7x) + 7x]$

Practice Plus

In Exercises 103–110, insert either $<, >,$ or $=$ in the shaded area to make a true statement.

103. $|-6|$ ▨ $|-3|$ **104.** $|-20|$ ▨ $|-50|$

105. $\left|\frac{3}{5}\right|$ ▨ $|-0.6|$ **106.** $\left|\frac{5}{2}\right|$ ▨ $|-2.5|$

107. $\frac{30}{40} - \frac{3}{4}$ ▨ $\frac{14}{15} \cdot \frac{15}{14}$ **108.** $\frac{17}{18} \cdot \frac{18}{17}$ ▨ $\frac{50}{60} - \frac{5}{6}$

109. $\frac{8}{13} \div \frac{8}{13}$ ▨ $|-1|$ **110.** $|-2|$ ▨ $\frac{4}{17} \div \frac{4}{17}$

In Exercises 111–118, write each English phrase as an algebraic expression. Then simplify the expression. Let x represent the number.

111. A number decreased by the sum of the number and four

112. A number decreased by the difference between eight and the number

113. Six times the product of negative five and a number

114. Ten times the product of negative four and a number

115. The difference between the product of five and a number and twice the number

116. The difference between the product of six and a number and negative two times the number

117. The difference between eight times a number and six more than three times the number

118. Eight decreased by three times the sum of a number and six

Application Exercises

The bar graph shows the number of billionaires in the United States from 2000 through 2004.

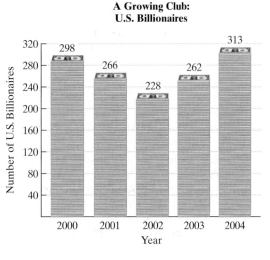

**A Growing Club:
U.S. Billionaires**

Source: Forbes magazine

The formula

$$N = 17x^2 - 65.4x + 302.2$$

models the number of billionaires, N, in the United States, x years after 2000. Use the formula to solve Exercises 119–122.

119. According to the formula, how many U.S. billionaires, to the nearest whole number, were there in 2004? How well does the formula model the actual data shown in the bar graph?

120. According to the formula, how many U.S. billionaires, to the nearest whole number, were there in 2003? How well does the formula model the actual data shown in the bar graph?

121. According to the formula, how many U.S. billionaires, to the nearest whole number, will there be in 2006?

122. According to the formula, how many U.S. billionaires, to the nearest whole number, will there be in 2007?

The bar graph shows that since her blockbuster debut, Britney Spears's album sales have slid downward.

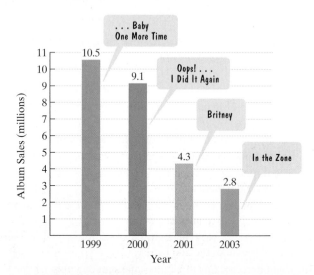

Slipped Discs: Britney Spears's Album Sales

Source: Entertainment Weekly

Here are three mathematical models for the data shown in the graph. In each formula, N represents Britney Spears's album sales, in millions, x years after 1999.

Model 1 $N = -2.04x + 10.24$

Model 2 $N = 0.04x^2 - 3.6x + 11$

Model 3 $N = 0.76x^3 - 4x^2 + 1.8x + 10.5$

Use these models to solve Exercises 123–126.

123. Which of the three models best describes the actual number of album sales in 1999?

124. Which of the three models best describes the actual number of album sales in 2000?

125. Which of the three models best describes the actual number of album sales in 2003?

126. For which formula does model breakdown occur when describing the number of album sales in 2003?

127. You had $10,000 to invest. You put x dollars in a safe, government-insured certificate of deposit paying 5% per year. You invested the remainder of the money in noninsured corporate bonds paying 12% per year. Your total interest earned at the end of the year is given by the algebraic expression

$$0.05x + 0.12(10,000 - x).$$

  a. Simplify the algebraic expression.

  b. Use each form of the algebraic expression to determine your total interest earned at the end of the year if you invested $6000 in the safe, government-insured certificate of deposit.

128. It takes you 50 minutes to get to campus. You spend t minutes walking to the bus stop and the rest of the time riding the bus. Your walking rate is 0.06 miles per minute and the bus travels at a rate of 0.5 miles per minute. The total distance walking and traveling by bus is given by the algebraic expression

$$0.06t + 0.5(50 - t).$$

  a. Simplify the algebraic expression.

  b. Use each form of the algebraic expression to determine the total distance that you travel if you spend 20 minutes walking to the bus stop.

Writing in Mathematics

Writing about mathematics will help you learn mathematics. For all writing exercises in this book, use complete sentences to respond to the question. Some writing exercises can be answered in a sentence; others require a paragraph or two. You can decide how much you need to write as long as your writing clearly and directly answers the question in the exercise. Standard references such as a dictionary and a thesaurus should be helpful.

129. What is an algebraic expression? Give an example with your explanation.

130. If n is a natural number, what does b^n mean? Give an example with your explanation.

131. What does it mean when we say that a formula models real-world phenomena?

132. What is the intersection of sets A and B?

133. What is the union of sets A and B?

134. How do the whole numbers differ from the natural numbers?

135. Can a real number be both rational and irrational? Explain your answer.

136. If you are given two real numbers, explain how to determine which is the lesser.

137. How can $\dfrac{|x|}{x}$ be equal to 1 or -1?

138. Describe the difference between the commutative and the associative properties of addition.

139. Why is $3(x + 7) - 4x$ not simplified? What must be done to simplify the expression?

Critical Thinking Exercises

140. Which one of the following statements is true?
   a. Every rational number is an integer.
   b. Some whole numbers are not integers.
   c. Some rational numbers are not positive.
   d. Irrational numbers cannot be negative.

141. Which of the following is true?
   a. The term x has no coefficient.
   b. $5 + 3(x - 4) = 8(x - 4) = 8x - 32$
   c. $-x - x = -x + (-x) = 0$
   d. $x - 0.02(x + 200) = 0.98x - 4$

In Exercises 142–144, insert either $<$ or $>$ in the shaded area between the numbers to make the statement true.

142. $\sqrt{2}$ ▨ 1.5 **143.** $-\pi$ ▨ -3.5

144. $-\dfrac{3.14}{2}$ ▨ $-\dfrac{\pi}{2}$

145. A business that manufactures small alarm clocks has a weekly fixed cost of $5000. The average cost per clock for the business to manufacture x clocks is described by

$$\frac{0.5x + 5000}{x}.$$

   a. Find the average cost when $x = 100$, 1000, and 10,000.

   b. Like all other businesses, the alarm clock manufacturer must make a profit. To do this, each clock must be sold for at least 50¢ more than what it costs to manufacture. Due to competition from a larger company, the clocks can be sold for $1.50 each and no more. Our small manufacturer can only produce 2000 clocks weekly. Does this business have much of a future? Explain.

SECTION P.2   *Exponents and Scientific Notation*

Objectives

1 Use properties of exponents.

2 Simplify exponential expressions.

3 Use scientific notation.

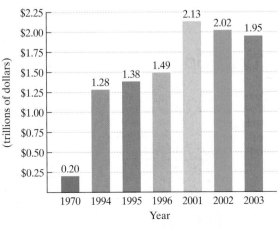

People who complain about paying their income tax can be divided into two types: men and women. Perhaps we can quantify the complaining by examining the data in Figure P.10. The bar graphs show the U.S. population, in millions, and the total amount we paid in federal taxes, in trillions of dollars, for six selected years.

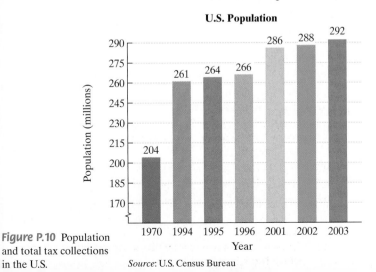

Figure P.10 Population and total tax collections in the U.S.

Source: U.S. Census Bureau

Source: Internal Revenue Service

The bar graph in Figure P.10 shows that in 2003, total tax collections were $1.95 trillion. How can we place this amount in the proper perspective? If the total tax collections were evenly divided among all Americans, how much would each citizen pay in taxes?

In this section, you will learn to use exponents to provide a way of putting large and small numbers in perspective. Using this skill, we will explore the per capita tax for some of the years shown in Figure P.10 on the previous page.

① Use properties of exponents.

Properties of Exponents

The major properties of exponents are summarized in the box that follows.

Properties of Exponents

Property	Examples

The Negative-Exponent Rule

If b is any real number other than 0 and n is a natural number, then

$$b^{-n} = \frac{1}{b^n}.$$

- $5^{-3} = \frac{1}{5^3} = \frac{1}{125}$

- $\frac{1}{4^{-2}} = \frac{1}{\frac{1}{4^2}} = 4^2 = 16$

The Zero-Exponent Rule

If b is any real number other than 0,

$$b^0 = 1.$$

- $7^0 = 1$

- $(-5)^0 = 1$

- $-5^0 = -1$

> Only 5 is raised to the zero power.

The Product Rule

If b is a real number or algebraic expression, and m and n are integers,

$$b^m \cdot b^n = b^{m+n}.$$

- $2^2 \cdot 2^3 = 2^{2+3} = 2^5 = 32$
- $x^{-3} \cdot x^7 = x^{-3+7} = x^4$

When multiplying exponential expressions with the same base, add the exponents. Use this sum as the exponent of the common base.

The Power Rule

If b is a real number or algebraic expression, and m and n are integers,

$$(b^m)^n = b^{mn}.$$

- $(2^2)^3 = 2^{2\cdot3} = 2^6 = 64$

- $(x^{-3})^4 = x^{-3\cdot4} = x^{-12} = \frac{1}{x^{12}}$

When an exponential expression is raised to a power, multiply the exponents. Place the product of the exponents on the base and remove the parentheses.

The Quotient Rule

If b is a nonzero real number or algebraic expression, and m and n are integers,

$$\frac{b^m}{b^n} = b^{m-n}.$$

- $\frac{2^8}{2^4} = 2^{8-4} = 2^4 = 16$

- $\frac{x^3}{x^7} = x^{3-7} = x^{-4} = \frac{1}{x^4}$

When dividing exponential expressions with the same nonzero base, subtract the exponent in the denominator from the exponent in the numerator. Use this difference as the exponent of the common base.

Study Tip

When a negative integer appears as an exponent, switch the position of the base (from numerator to denominator or from denominator to numerator) and make the exponent positive.

Study Tip

$\frac{4^3}{4^5}$ and $\frac{4^5}{4^3}$ represent different numbers:

$$\frac{4^3}{4^5} = 4^{3-5} = 4^{-2} = \frac{1}{4^2} = \frac{1}{16}$$

$$\frac{4^5}{4^3} = 4^{5-3} = 4^2 = 16.$$

Property	Examples
Products Raised to Powers	
If a and b are real numbers or algebraic expressions, and n is an integer, $$(ab)^n = a^n b^n.$$	• $(-2y)^4 = (-2)^4 y^4 = 16y^4$ • $(-2xy)^3 = (-2)^3 x^3 y^3 = -8x^3 y^3$

When a product is raised to a power, raise each factor to that power.

Quotients Raised to Powers	
If a and b are real numbers, $b \neq 0$, or algebraic expressions, and n is an integer, $$\left(\frac{a}{b}\right)^n = \frac{a^n}{b^n}.$$	• $\left(\dfrac{2}{5}\right)^4 = \dfrac{2^4}{5^4} = \dfrac{16}{625}$ • $\left(-\dfrac{3}{x}\right)^3 = \dfrac{(-3)^3}{x^3} = -\dfrac{27}{x^3}$

When a quotient is raised to a power, raise the numerator to that power and divide by the denominator to that power.

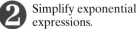 Simplify exponential expressions.

Simplifying Exponential Expressions

Properties of exponents are used to simplify exponential expressions. An exponential expression is **simplified** when

- No parentheses appear.
- No powers are raised to powers.
- Each base occurs only once.
- No negative or zero exponents appear.

Simplifying Exponential Expressions

Example

1. If necessary, remove parentheses by using
$$(ab)^n = a^n b^n \quad \text{or} \quad \left(\frac{a}{b}\right)^n = \frac{a^n}{b^n}.$$
$(xy)^3 = x^3 y^3$

2. If necessary, simplify powers to powers by using
$$(b^m)^n = b^{mn}.$$
$(x^4)^3 = x^{4 \cdot 3} = x^{12}$

3. If necessary, be sure that each base appears only once by using
$$b^m \cdot b^n = b^{m+n} \quad \text{or} \quad \frac{b^m}{b^n} = b^{m-n}.$$
$x^4 \cdot x^3 = x^{4+3} = x^7$

4. If necessary, rewrite exponential expressions with zero powers as 1 $(b^0 = 1)$. Furthermore, write the answer with positive exponents by using
$$b^{-n} = \frac{1}{b^n} \quad \text{or} \quad \frac{1}{b^{-n}} = b^n.$$
$\dfrac{x^5}{x^8} = x^{5-8} = x^{-3} = \dfrac{1}{x^3}$

The following example shows how to simplify exponential expressions. Throughout the example, assume that no variable in a denominator is equal to zero.

EXAMPLE 1   Simplifying Exponential Expressions

Simplify:

a. $(-3x^4 y^5)^3$ **b.** $(-7xy^4)(-2x^5 y^6)$ **c.** $\dfrac{-35x^2 y^4}{5x^6 y^{-8}}$ **d.** $\left(\dfrac{4x^2}{y}\right)^{-3}$.

Solution

a. $(-3x^4y^5)^3 = (-3)^3(x^4)^3(y^5)^3$ Raise each factor inside the parentheses to the third power.

$= (-3)^3x^{4\cdot3}y^{5\cdot3}$ Multiply the exponents when raising powers to powers.

$= -27x^{12}y^{15}$ $(-3)^3 = (-3)(-3)(-3) = -27$

b. $(-7xy^4)(-2x^5y^6) = (-7)(-2)xx^5y^4y^6$ Group factors with the same base.

$= 14x^{1+5}y^{4+6}$ When multiplying expressions with the same base, add the exponents.

$= 14x^6y^{10}$ Simplify.

c. $\dfrac{-35x^2y^4}{5x^6y^{-8}} = \left(\dfrac{-35}{5}\right)\left(\dfrac{x^2}{x^6}\right)\left(\dfrac{y^4}{y^{-8}}\right)$ Group factors with the same base.

$= -7x^{2-6}y^{4-(-8)}$ When dividing expressions with the same base, subtract the exponents.

$= -7x^{-4}y^{12}$ Simplify. Notice that $4 - (-8) = 4 + 8 = 12$.

$= \dfrac{-7y^{12}}{x^4}$ Move the base with the negative exponent, x^{-4}, to the other side of the fraction bar and make the negative exponent positive.

d. $\left(\dfrac{4x^2}{y}\right)^{-3} = \dfrac{(4x^2)^{-3}}{y^{-3}}$ Raise the numerator and the denominator to the -3 power.

$= \dfrac{4^{-3}(x^2)^{-3}}{y^{-3}}$ Raise each factor in the numerator to the -3 power.

$= \dfrac{4^{-3}x^{-6}}{y^{-3}}$ Multiply the exponents when raising a power to a power. $(x^2)^{-3} = x^{2(-3)} = x^{-6}$.

$= \dfrac{y^3}{4^3x^6}$ Move each base with a negative exponent to the other side of the fraction bar and make each negative exponent positive.

$= \dfrac{y^3}{64x^6}$ $4^3 = 4\cdot4\cdot4 = 64$

Check Point 1 Simplify:

a. $(2x^3y^6)^4$ **b.** $(-6x^2y^5)(3xy^3)$ **c.** $\dfrac{100x^{12}y^2}{20x^{16}y^{-4}}$ **d.** $\left(\dfrac{5x}{y^4}\right)^{-2}.$

Study Tip

Try to avoid the following common errors that can occur when simplifying exponential expressions.

Correct	Incorrect	Description of Error
$b^3 \cdot b^4 = b^7$	$b^3 \cdot b^4 = b^{12}$	The exponents should be added, not multiplied.
$3^2 \cdot 3^4 = 3^6$	$3^2 \cdot 3^4 = 9^6$	The common base should be retained, not multiplied.
$\dfrac{5^{16}}{5^4} = 5^{12}$	$\dfrac{5^{16}}{5^4} = 5^4$	The exponents should be subtracted, not divided.
$(4a)^3 = 64a^3$	$(4a)^3 = 4a^3$	Both factors should be cubed.
$b^{-n} = \dfrac{1}{b^n}$	$b^{-n} = -\dfrac{1}{b^n}$	Only the exponent should change sign.
$(a+b)^{-1} = \dfrac{1}{a+b}$	$(a+b)^{-1} = \dfrac{1}{a} + \dfrac{1}{b}$	The exponent applies to the entire expression $a + b$.

③ Use scientific notation.

Scientific Notation

We have seen that in 2003, total tax collections were $1.95 trillion. Because a trillion is 10^{12} (see Table P.3), this amount can be expressed as

$$1.95 \times 10^{12}.$$

The number 1.95×10^{12} is written in a form called *scientific notation*.

Table P.3   Names of Large Numbers

10^2	hundred
10^3	thousand
10^6	million
10^9	billion
10^{12}	trillion
10^{15}	quadrillion
10^{18}	quintillion
10^{21}	sextillion
10^{24}	septillion
10^{27}	octillion
10^{30}	nonillion
10^{100}	googol

Scientific Notation

A number is written in **scientific notation** when it is expressed in the form

$$a \times 10^n,$$

where the absolute value of a is greater than or equal to 1 and less than 10 ($1 \le |a| < 10$), and n is an integer.

It is customary to use the multiplication symbol, $\times$, rather than a dot, when writing a number in scientific notation.

Converting from Scientific to Decimal Notation

Here are two examples of numbers in scientific notation:

$$6.4 \times 10^5 \quad \text{means} \quad 640,000.$$

$$2.17 \times 10^{-3} \quad \text{means} \quad 0.00217.$$

Do you see that the number with the positive exponent is relatively large and the number with the negative exponent is relatively small?

We can use n, the exponent on the 10 in $a \times 10^n$, to change a number in scientific notation to decimal notation. If n is **positive**, move the decimal point in a to the **right** n places. If n is **negative**, move the decimal point in a to the **left** $|n|$ places.

EXAMPLE 2   Converting from Scientific to Decimal Notation

Write each number in decimal notation:

    a. 6.2×10^7 **b.** -6.2×10^7 **c.** 2.019×10^{-3} **d.** -2.019×10^{-3}.

Solution   In each case, we use the exponent on the 10 to move the decimal point. In parts (a) and (b), the exponent is positive, so we move the decimal point to the right. In parts (c) and (d), the exponent is negative, so we move the decimal point to the left.

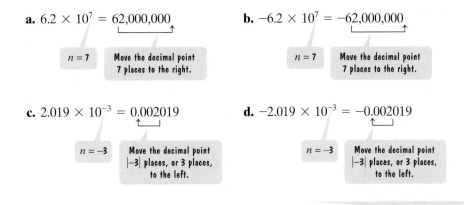

a. $6.2 \times 10^7 = 62,000,000$

$n = 7$ Move the decimal point 7 places to the right.

b. $-6.2 \times 10^7 = -62,000,000$

$n = 7$ Move the decimal point 7 places to the right.

c. $2.019 \times 10^{-3} = 0.002019$

$n = -3$ Move the decimal point $|-3|$ places, or 3 places, to the left.

d. $-2.019 \times 10^{-3} = -0.002019$

$n = -3$ Move the decimal point $|-3|$ places, or 3 places, to the left.

Check Point 2   Write each number in decimal notation:

    a. -2.6×10^9 **b.** 3.017×10^{-6}.

Converting from Decimal to Scientific Notation

To convert from decimal notation to scientific notation, we reverse the procedure of Example 2.

Converting from Decimal to Scientific Notation

Write the number in the form $a \times 10^n$.

- Determine a, the numerical factor. Move the decimal point in the given number to obtain a number whose absolute value is between 1 and 10, including 1.
- Determine n, the exponent on 10^n. The absolute value of n is the number of places the decimal point was moved. The exponent n is positive if the decimal point was moved to the left, negative if the decimal point was moved to the right, and 0 if the decimal point was not moved.

EXAMPLE 3   Converting from Decimal Notation to Scientific Notation

Write each number in scientific notation:

    a. 34,970,000,000,000   **b.** −34,970,000,000,000

    b. 0.0000000000802   **d.** −0.0000000000802.

Solution

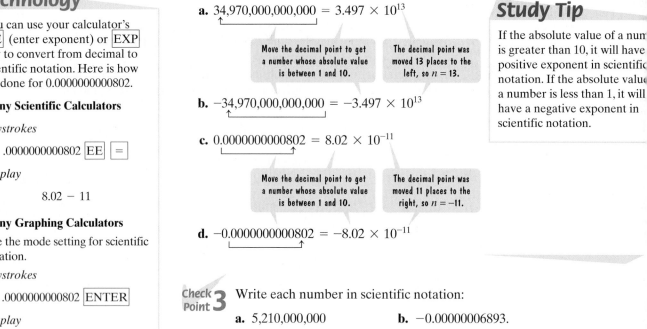

a. $34{,}970{,}000{,}000{,}000 = 3.497 \times 10^{13}$

> Move the decimal point to get a number whose absolute value is between 1 and 10.

> The decimal point was moved 13 places to the left, so $n = 13$.

b. $-34{,}970{,}000{,}000{,}000 = -3.497 \times 10^{13}$

c. $0.0000000000802 = 8.02 \times 10^{-11}$

> Move the decimal point to get a number whose absolute value is between 1 and 10.

> The decimal point was moved 11 places to the right, so $n = -11$.

d. $-0.0000000000802 = -8.02 \times 10^{-11}$

Check Point 3 Write each number in scientific notation:

    a. 5,210,000,000   **b.** −0.00000006893.

EXAMPLE 4   Expressing the U.S. Population in Scientific Notation

In 2003, the population of the United States was approximately 292 million. Express the population in scientific notation.

Solution Because one million is 10^6 (see Table P.3 on page 21), the 2003 population can be expressed as

$$292 \times 10^6.$$

> This factor is not between 1 and 10, so the number is not in scientific notation.

The voice balloon indicates that we need to convert 292 to scientific notation.

$$292 \times 10^6 = (2.92 \times 10^2) \times 10^6 = 2.92 \times 10^{2+6} = 2.92 \times 10^8$$

$$292 = 2.92 \times 10^2$$

In scientific notation, the population is 2.92×10^8.

Check Point 4 Express 410×10^7 in scientific notation.

Computations with Scientific Notation

Properties of exponents are used to perform computations with numbers that are expressed in scientific notation.

Technology

$(6.1 \times 10^5)(4 \times 10^{-9})$
On a Calculator:

Many Scientific Calculators

6.1 $\boxed{\text{EE}}$ 5 $\boxed{\times}$ 4 $\boxed{\text{EE}}$ 9 $\boxed{+/-}$ $\boxed{=}$

Display

$$2.44 - 03$$

Many Graphing Calculators

6.1 $\boxed{\text{EE}}$ 5 $\boxed{\times}$ 4 $\boxed{\text{EE}}$ $\boxed{(-)}$ 9 $\boxed{\text{ENTER}}$

Display (in scientific notation mode)

$$2.44\text{E} - 3$$

EXAMPLE 5   Computations with Scientific Notation

Perform the indicated computations, writing the answers in scientific notation:

a. $(6.1 \times 10^5)(4 \times 10^{-9})$ **b.** $\dfrac{1.8 \times 10^4}{3 \times 10^{-2}}$.

Solution

a. $(6.1 \times 10^5)(4 \times 10^{-9})$

$= (6.1 \times 4) \times (10^5 \times 10^{-9})$ *Regroup factors.*

$= 24.4 \times 10^{5+(-9)}$ *Add the exponents on 10 and multiply the other parts.*

$= 24.4 \times 10^{-4}$ *Simplify.*

$= (2.44 \times 10^1) \times 10^{-4}$ *Convert 24.4 to scientific notation: $24.4 = 2.44 \times 10^1$.*

$= 2.44 \times 10^{-3}$ *$10^1 \times 10^{-4} = 10^{1+(-4)} = 10^{-3}$*

b. $\dfrac{1.8 \times 10^4}{3 \times 10^{-2}} = \left(\dfrac{1.8}{3}\right) \times \left(\dfrac{10^4}{10^{-2}}\right)$ *Regroup factors.*

$= 0.6 \times 10^{4-(-2)}$ *Subtract the exponents on 10 and divide the other parts.*

$= 0.6 \times 10^6$ *Simplify: $4 - (-2) = 4 + 2 = 6$.*

$= (6 \times 10^{-1}) \times 10^6$ *Convert 0.6 to scientific notation: $0.6 = 6 \times 10^{-1}$.*

$= 6 \times 10^5$ *$10^{-1} \times 10^6 = 10^{-1+6} = 10^5$*

Check Point 5 Perform the indicated computations, writing the answers in scientific notation:

a. $(7.1 \times 10^5)(5 \times 10^{-7})$ **b.** $\dfrac{1.2 \times 10^6}{3 \times 10^{-3}}$.

Applications: Putting Numbers in Perspective

We have seen that in 2003, the U.S. government collected $1.95 trillion in taxes. Example 6 shows how we can use scientific notation to comprehend the meaning of a number such as 1.95 trillion.

EXAMPLE 6 Tax per Capita

In 2003, the U.S. government collected 1.95×10^{12} dollars in taxes. At that time, the U.S. population was approximately 292 million, or 2.92×10^8. If the total tax collections were evenly divided among all Americans, how much would each citizen pay? Express the answer in decimal notation, rounded to the nearest dollar.

Solution   The amount that we would each pay, or the tax per capita, is the total amount collected, 1.95×10^{12}, divided by the number of Americans, 2.92×10^8.

$$\frac{1.95 \times 10^{12}}{2.92 \times 10^8} = \left(\frac{1.95}{2.92}\right) \times \left(\frac{10^{12}}{10^8}\right) \approx 0.6678 \times 10^{12-8} = 0.6678 \times 10^4 = 6678$$

> To obtain an answer in decimal notation, it is not necessary to express this number in scientific notation.

> Move the decimal point 4 places to the right.

If total tax collections were evenly divided, we would each pay approximately $6678 in taxes.

Check Point 6   In 2002, the U.S. government collected 2.02×10^{12} dollars in taxes. At that time, the U.S. population was approximately 288 million, or 2.88×10^8. Find the per capita tax, rounded to the nearest dollar, in 2002.

An Application: Black Holes in Space

The concept of a black hole, a region in space where matter appears to vanish, intrigues scientists and nonscientists alike. Scientists theorize that when massive stars run out of nuclear fuel, they begin to collapse under the force of their own gravity. As the star collapses, its density increases. In turn, the force of gravity increases so tremendously that even light cannot escape from the star. Consequently, it appears black.

A mathematical model, called the Schwarzchild formula, describes the critical value to which the radius of a massive body must be reduced for it to become a black hole. This model forms the basis of our next example.

EXAMPLE 7   An Application of Scientific Notation

Use the Schwarzchild formula

$$R_s = \frac{2GM}{c^2}$$

where

R_s = Radius of the star, in meters, that would cause it to become a black hole

M = Mass of the star, in kilograms

G = A constant, called the gravitational constant

$= 6.7 \times 10^{-11} \dfrac{\text{m}^3}{\text{kg} \cdot \text{s}^2}$

c = Speed of light

$= 3 \times 10^8$ meters per second

to determine to what length the radius of the sun must be reduced for it to become a black hole. The sun's mass is approximately 2×10^{30} kilograms.

Solution

$$R_s = \frac{2GM}{c^2}$$ Use the given model.

$$= \frac{2 \times 6.7 \times 10^{-11} \times 2 \times 10^{30}}{(3 \times 10^8)^2}$$ Substitute the given values: $G = 6.7 \times 10^{-11}$, $M = 2 \times 10^{30}$, and $c = 3 \times 10^8$.

$$= \frac{(2 \times 6.7 \times 2) \times (10^{-11} \times 10^{30})}{(3 \times 10^8)^2}$$ Rearrange factors in the numerator.

$$= \frac{26.8 \times 10^{-11+30}}{3^2 \times (10^8)^2}$$ Add exponents in the numerator. Raise each factor in the denominator to the power.

$$= \frac{26.8 \times 10^{19}}{9 \times 10^{16}}$$ Multiply powers to powers: $(10^8)^2 = 10^{8 \cdot 2} = 10^{16}$.

$$= \frac{26.8}{9} \times 10^{19-16}$$ When dividing expressions with the same base, subtract the exponents.

$$\approx 2.978 \times 10^3$$ Simplify.

$$= 2978$$

Although the sun is not massive enough to become a black hole (its radius is approximately 700,000 kilometers), the Schwarzchild model theoretically indicates that if the sun's radius were reduced to approximately 2978 meters, that is, about $\frac{1}{235,000}$ its present size, it would become a black hole.

> **Check Point 7** Pouiseville's law states that the speed of blood, S, in centimeters per second, located r centimeters from the central axis of an artery is
>
> $$S = (1.76 \times 10^5)[(1.44 \times 10^{-2}) - r^2].$$
>
> Find the speed of blood at the central axis of this artery.

EXERCISE SET P.2

Practice Exercises

Evaluate each exponential expression in Exercises 1–22.

1. $5^2 \cdot 2$

2. $6^2 \cdot 2$

3. $(-2)^6$

4. $(-2)^4$

5. -2^6

6. -2^4

7. $(-3)^0$

8. $(-9)^0$

9. -3^0

10. -9^0

11. 4^{-3}

12. 2^{-6}

13. $2^2 \cdot 2^3$

14. $3^3 \cdot 3^2$

15. $(2^2)^3$

16. $(3^3)^2$

17. $\dfrac{2^8}{2^4}$

18. $\dfrac{3^8}{3^4}$

19. $3^{-3} \cdot 3$

20. $2^{-3} \cdot 2$

21. $\dfrac{2^3}{2^7}$

22. $\dfrac{3^4}{3^7}$

Simplify each exponential expression in Exercises 23–64.

23. $x^{-2}y$

24. xy^{-3}

25. $x^0 y^5$

26. $x^7 y^0$

27. $x^3 \cdot x^7$

28. $x^{11} \cdot x^5$

29. $x^{-5} \cdot x^{10}$

30. $x^{-6} \cdot x^{12}$

31. $(x^3)^7$

32. $(x^{11})^5$

33. $(x^{-5})^3$

34. $(x^{-6})^4$

35. $\dfrac{x^{14}}{x^7}$

36. $\dfrac{x^{30}}{x^{10}}$

37. $\dfrac{x^{14}}{x^{-7}}$

38. $\dfrac{x^{30}}{x^{-10}}$

39. $(8x^3)^2$

40. $(6x^4)^2$

41. $\left(-\dfrac{4}{x}\right)^3$

42. $\left(-\dfrac{6}{y}\right)^3$

43. $(-3x^2y^5)^2$

44. $(-3x^4y^6)^3$

45. $(3x^4)(2x^7)$

46. $(11x^5)(9x^{12})$

47. $(-9x^3y)(-2x^6y^4)$

48. $(-5x^4y)(-6x^7y^{11})$

49. $\dfrac{8x^{20}}{2x^4}$

50. $\dfrac{20x^{24}}{10x^6}$

51. $\dfrac{25a^{13}b^4}{-5a^2b^3}$

52. $\dfrac{35a^{14}b^6}{-7a^7b^3}$

53. $\dfrac{14b^7}{7b^{14}}$

54. $\dfrac{20b^{10}}{10b^{20}}$

55. $(4x^3)^{-2}$

56. $(10x^2)^{-3}$

57. $\dfrac{24x^3y^5}{32x^7y^{-9}}$

58. $\dfrac{10x^4y^9}{30x^{12}y^{-3}}$

59. $\left(\dfrac{5x^3}{y}\right)^{-2}$

60. $\left(\dfrac{3x^4}{y}\right)^{-3}$

61. $\left(\dfrac{-15a^4b^2}{5a^{10}b^{-3}}\right)^3$

62. $\left(\dfrac{-30a^{14}b^8}{10a^{17}b^{-2}}\right)^3$

63. $\left(\dfrac{3a^{-5}b^2}{12a^3b^{-4}}\right)^0$

64. $\left(\dfrac{4a^{-5}b^3}{12a^3b^{-5}}\right)^0$

In Exercises 65–76, write each number in decimal notation without the use of exponents.

65. 3.8×10^2

66. 9.2×10^2

67. 6×10^{-4}

68. 7×10^{-5}

69. -7.16×10^6

70. -8.17×10^6

71. 7.9×10^{-1}

72. 6.8×10^{-1}

73. -4.15×10^{-3}

74. -3.14×10^{-3}

75. -6.00001×10^{10}

76. -7.00001×10^{10}

In Exercises 77–86, write each number in scientific notation.

77. 32,000

78. 64,000

79. 638,000,000,000,000,000

80. 579,000,000,000,000,000

81. -5716

82. -3829

83. 0.0027

84. 0.0083

85. -0.00000000504

86. -0.00000000405

In Exercises 87–106, perform the indicated computations. Write the answers in scientific notation. If necessary, round the decimal factor in your scientific notation answer to two decimal places.

87. $(3 \times 10^4)(2.1 \times 10^3)$

88. $(2 \times 10^4)(4.1 \times 10^3)$

89. $(1.6 \times 10^{15})(4 \times 10^{-11})$

90. $(1.4 \times 10^{15})(3 \times 10^{-11})$

91. $(6.1 \times 10^{-8})(2 \times 10^{-4})$

92. $(5.1 \times 10^{-8})(3 \times 10^{-4})$

93. $(4.3 \times 10^8)(6.2 \times 10^4)$

94. $(8.2 \times 10^8)(4.6 \times 10^4)$

95. $\dfrac{8.4 \times 10^8}{4 \times 10^5}$

96. $\dfrac{6.9 \times 10^8}{3 \times 10^5}$

97. $\dfrac{3.6 \times 10^4}{9 \times 10^{-2}}$

98. $\dfrac{1.2 \times 10^4}{2 \times 10^{-2}}$

99. $\dfrac{4.8 \times 10^{-2}}{2.4 \times 10^6}$

100. $\dfrac{7.5 \times 10^{-2}}{2.5 \times 10^6}$

101. $\dfrac{2.4 \times 10^{-2}}{4.8 \times 10^{-6}}$

102. $\dfrac{1.5 \times 10^{-2}}{3 \times 10^{-6}}$

103. $\dfrac{480,000,000,000}{0.00012}$

104. $\dfrac{282,000,000,000}{0.00141}$

105. $\dfrac{0.00072 \times 0.003}{0.00024}$

106. $\dfrac{66,000 \times 0.001}{0.003 \times 0.002}$

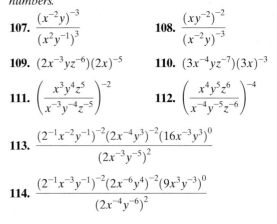

Practice Plus

In Exercises 107–114, simplify each exponential expression. Assume that variables represent nonzero real numbers.

107. $\dfrac{(x^{-2}y)^{-3}}{(x^2y^{-1})^3}$

108. $\dfrac{(xy^{-2})^{-2}}{(x^{-2}y)^{-3}}$

109. $(2x^{-3}yz^{-6})(2x)^{-5}$

110. $(3x^{-4}yz^{-7})(3x)^{-3}$

111. $\left(\dfrac{x^3y^4z^5}{x^{-3}y^{-4}z^{-5}}\right)^{-2}$

112. $\left(\dfrac{x^4y^5z^6}{x^{-4}y^{-5}z^{-6}}\right)^{-4}$

113. $\dfrac{(2^{-1}x^{-2}y^{-1})^{-2}(2x^{-4}y^3)^{-2}(16x^{-3}y^3)^0}{(2x^{-3}y^{-5})^2}$

114. $\dfrac{(2^{-1}x^{-3}y^{-1})^{-2}(2x^{-6}y^4)^{-2}(9x^3y^{-3})^0}{(2x^{-4}y^{-6})^2}$

Application Exercises

The graph shows the number of people in the United States ages 65 and over for the year 2000 and projections beyond. Use 10^6 for one million and the figures shown to solve Exercises 115–118. Express all answers in scientific notation.

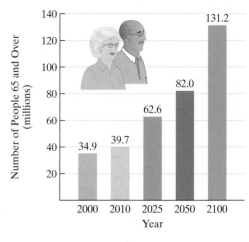

U.S. Population, Ages 65 and Over

Source: U.S. Bureau of the Census

115. How many people 65 and over will there be in 2025?

116. How many people 65 and over will there be in 2050?

117. How many more people 65 and over will there be in 2100 than in 2000?

118. How many more people 65 and over will there be in 2050 than in 2000?

Our ancient ancestors hunted for their meat and expended a great deal of energy chasing it down. Today, our animal protein is raised in cages and on feedlots, delivered in great abundance nearly to our door. Use the numbers shown below to solve Exercises 119–122. Use 10^6 for one million and 10^9 for one billion.

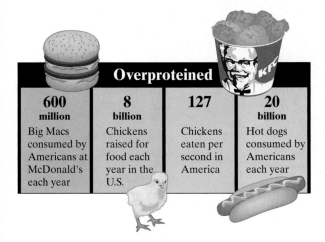

Overproteined

600 million	8 billion	127	20 billion
Big Macs consumed by Americans at McDonald's each year	Chickens raised for food each year in the U.S.	Chickens eaten per second in America	Hot dogs consumed by Americans each year

Source: Time, October 20, 2003

In Exercises 119–120, use 292 million, or 2.92×10^8, for the U.S. population. Express answers in decimal notation, rounded to the nearest whole number.

119. Find the number of hot dogs consumed by each American in a year.

120. If the consumption of Big Macs was divided evenly among all Americans, how many Big Macs would we each consume in a year?

In Exercises 121–122, use the Overproteined table shown above and the fact that there are approximately 3.2×10^7 seconds in a year.

121. How many chickens are raised for food each second in the United States? Express the answer in scientific and decimal notations.

122. How many chickens are eaten per year in the United States? Express the answer in scientific notation.

123. Due to tax cuts and spending increases, the United States began accumulating large deficits in the 1980s. The graph shows the national debt increasing over time.

The National Debt

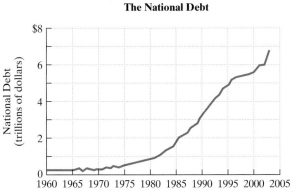

As of November 2003, to finance the deficit, the government had borrowed $6.8 trillion and the national debt was $6.8 trillion, or 6.8×10^{12} dollars. At that time, the U.S. population was approximately 290,000,000 (290 million), or 2.9×10^8. If the national debt was evenly divided among every individual in the United States, how much would each citizen have to pay?

124. In Exercises 121–122, we used 3.2×10^7 as an approximation for the number of seconds in a year. Convert 365 days (one year) to hours, to minutes, and, finally, to seconds, to determine precisely how many seconds there are in a year. Express the answer in scientific notation.

Writing in Mathematics

125. Describe what it means to raise a number to a power. In your description, include a discussion of the difference between -5^2 and $(-5)^2$.

126. Explain the product rule for exponents. Use $2^3 \cdot 2^5$ in your explanation.

127. Explain the power rule for exponents. Use $(3^2)^4$ in your explanation.

128. Explain the quotient rule for exponents. Use $\dfrac{5^8}{5^2}$ in your explanation.

129. Why is $(-3x^2)(2x^{-5})$ not simplified? What must be done to simplify the expression?

130. How do you know if a number is written in scientific notation?

131. Explain how to convert from scientific to decimal notation and give an example.

132. Explain how to convert from decimal to scientific notation and give an example.

Critical Thinking Exercises

133. Which one of the following is true?
   a. $4^{-2} < 4^{-3}$ **b.** $5^{-2} > 2^{-5}$
   c. $(-2)^4 = 2^{-4}$ **d.** $5^2 \cdot 5^{-2} > 2^5 \cdot 2^{-5}$

134. The mad Dr. Frankenstein has gathered enough bits and pieces (so to speak) for $2^{-1} + 2^{-2}$ of his creature-to-be. Write a fraction that represents the amount of his creature that must still be obtained.

135. If $b^A = MN$, $b^C = M$, and $b^D = N$, what is the relationship among A, C, and D?

136. Our hearts beat approximately 70 times per minute. Express in scientific notation how many times the heart beats over a lifetime of 80 years. Round the decimal factor in your scientific notation answer to two decimal places.

Group Exercise

137. Putting Numbers into Perspective. A large number can be put into perspective by comparing it with another number. For example, we put the $1.95 trillion the government collected in taxes (Example 6) and the $6.8 trillion national debt (Exercise 123) by comparing these numbers to the number of U.S. citizens.

For this project, each group member should consult an almanac, a newspaper, or the World Wide Web to find a number greater than one million. Explain to other members of the group the context in which the large number is used. Express the number in scientific notation. Then put the number into perspective by comparing it with another number.

SECTION P.3 *Radicals and Rational Exponents*

Objectives

❶ Evaluate square roots.

❷ Simplify expressions of the form $\sqrt{a^2}$.

❸ Use the product rule to simplify square roots.

❹ Use the quotient rule to simplify square roots.

❺ Add and subtract square roots.

❻ Rationalize denominators.

❼ Evaluate and perform operations with higher roots.

❽ Understand and use rational exponents.

What is the maximum speed at which a racing cyclist can turn a corner without tipping over? The answer, in miles per hour, is given by the algebraic expression $4\sqrt{x}$, where x is the radius of the corner, in feet. Algebraic expressions containing roots describe phenomena as diverse as the distance we can see to the horizon, how we perceive the temperature on a cold day, and Albert Einstein's bizarre concept of how an astronaut moving close to the speed of light would barely age relative to friends watching from Earth. No description of your world can be complete without roots and radicals. In this section, we review the basics of radical expressions and the use of rational exponents to indicate radicals.

❶ Evaluate square roots.

Square Roots

From our earlier work with exponents, we are aware that the square of both 5 and -5 is 25:

$$5^2 = 25 \quad \text{and} \quad (-5)^2 = 25.$$

The reverse operation of squaring a number is finding the *square root* of the number. For example,

- One square root of 25 is 5 because $5^2 = 25$.
- Another square root of 25 is -5 because $(-5)^2 = 25$.

In general, **if $b^2 = a$, then b is a square root of a.**

The symbol $\sqrt{}$ is used to denote the *positive* or *principal square root* of a number. For example,

- $\sqrt{25} = 5$ because $5^2 = 25$ and 5 is positive.
- $\sqrt{100} = 10$ because $10^2 = 100$ and 10 is positive.

The symbol $\sqrt{}$ that we use to denote the principal square root is called a **radical sign**. The number under the radical sign is called the **radicand**. Together we refer to the radical sign and its radicand as a **radical expression**.

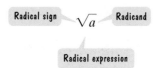

Definition of the Principal Square Root

If a is a nonnegative real number, the nonnegative number b such that $b^2 = a$, denoted by $b = \sqrt{a}$, is the **principal square root** of a.

The symbol $-\sqrt{}$ is used to denote the negative square root of a number. For example,

- $-\sqrt{25} = -5$ because $(-5)^2 = 25$ and -5 is negative.
- $-\sqrt{100} = -10$ because $(-10)^2 = 100$ and -10 is negative.

EXAMPLE 1   Evaluating Square Roots

Evaluate:

    a. $\sqrt{64}$ **b.** $-\sqrt{49}$ **c.** $\sqrt{\dfrac{1}{4}}$ **d.** $\sqrt{9+16}$ **e.** $\sqrt{9} + \sqrt{16}.$

Solution

> **Study Tip**
>
> In Example 1, parts (d) and (e), observe that $\sqrt{9+16}$ is not equal to $\sqrt{9} + \sqrt{16}$. In general,
>
> $$\sqrt{a+b} \neq \sqrt{a} + \sqrt{b}$$
>
> and
>
> $$\sqrt{a-b} \neq \sqrt{a} - \sqrt{b}.$$

    a. $\sqrt{64} = 8$ *The principal square root of 64 is 8. Check: $8^2 = 64$.*

    b. $-\sqrt{49} = -7$ *The negative square root of 49 is -7. Check: $(-7)^2 = 49$.*

    c. $\sqrt{\dfrac{1}{4}} = \dfrac{1}{2}$ *The principal square root of $\frac{1}{4}$ is $\frac{1}{2}$. Check: $\left(\frac{1}{2}\right)^2 = \frac{1}{4}$.*

    d. $\sqrt{9+16} = \sqrt{25}$ *First simplify the expression under the radical sign.*
           $= 5$ *Then take the principal square root of 25, which is 5.*

    e. $\sqrt{9} + \sqrt{16} = 3 + 4$ *$\sqrt{9} = 3$ because $3^2 = 9$. $\sqrt{16} = 4$ because $4^2 = 16$.*
                 $= 7$

Check Point 1   Evaluate:

    a. $\sqrt{81}$ **b.** $-\sqrt{9}$ **c.** $\sqrt{\dfrac{1}{25}}$

    d. $\sqrt{36+64}$ **e.** $\sqrt{36} + \sqrt{64}.$

      A number that is the square of a rational number is called a **perfect square**. All the radicands in Example 1 and Check Point 1 are perfect squares.

- 64 is a perfect square because $64 = 8^2$. Thus, $\sqrt{64} = 8$.

- $\dfrac{1}{4}$ is a perfect square because $\dfrac{1}{4} = \left(\dfrac{1}{2}\right)^2$. Thus, $\sqrt{\dfrac{1}{4}} = \dfrac{1}{2}$.

      Let's see what happens to the radical expression $\sqrt{x}$ if x is a negative number. Is the square root of a negative number a real number? For example, consider $\sqrt{-25}$. Is there a real number whose square is -25? No. Thus, $\sqrt{-25}$ is not a real number. In general, **a square root of a negative number is not a real number**.

      If a number is nonnegative ($a \geq 0$), then $\left(\sqrt{a}\right)^2 = a$. For example,

$$\left(\sqrt{2}\right)^2 = 2, \quad \left(\sqrt{3}\right)^2 = 3, \quad \left(\sqrt{4}\right)^2 = 4, \quad \text{and} \quad \left(\sqrt{5}\right)^2 = 5.$$

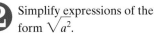 Simplify expressions of the form $\sqrt{a^2}$.

Simplifying Expressions of the Form $\sqrt{a^2}$

You may think that $\sqrt{a^2} = a$. However, this is not necessarily true. Consider the following examples:

$$\sqrt{4^2} = \sqrt{16} = 4$$
$$\sqrt{(-4)^2} = \sqrt{16} = 4.$$

> *The result is not -4, but rather the absolute value of -4, or 4.*

      On the next page is a rule for simplifying expressions of the form $\sqrt{a^2}$.

> **Simplifying $\sqrt{a^2}$**
>
> For any real number a,
> $$\sqrt{a^2} = |a|.$$
> In words, the principal square root of a^2 is the absolute value of a.

For example, $\sqrt{6^2} = |6| = 6$ and $\sqrt{(-6)^2} = |-6| = 6$.

 Use the product rule to simplify square roots.

The Product Rule for Square Roots

A rule for multiplying square roots can be generalized by comparing $\sqrt{25} \cdot \sqrt{4}$ and $\sqrt{25 \cdot 4}$. Notice that

$$\sqrt{25} \cdot \sqrt{4} = 5 \cdot 2 = 10 \quad \text{and} \quad \sqrt{25 \cdot 4} = \sqrt{100} = 10.$$

Because we obtain 10 in both situations, the original radical expressions must be equal. That is,

$$\sqrt{25} \cdot \sqrt{4} = \sqrt{25 \cdot 4}.$$

This result is a special case of the **product rule for square roots** that can be generalized as follows:

> **The Product Rule for Square Roots**
>
> If a and b represent nonnegative real numbers, then
> $$\sqrt{ab} = \sqrt{a} \cdot \sqrt{b} \quad \text{and} \quad \sqrt{a} \cdot \sqrt{b} = \sqrt{ab}.$$
> The square root of a product is the product of the square roots.

A square root is **simplified** when its radicand has no factors other than 1 that are perfect squares. For example, $\sqrt{500}$ is not simplified because it can be expressed as $\sqrt{100 \cdot 5}$ and 100 is a perfect square. Example 2 shows how the product rule is used to remove from the square root any perfect squares that occur as factors.

EXAMPLE 2   Using the Product Rule to Simplify Square Roots

Simplify: **a.** $\sqrt{500}$ **b.** $\sqrt{6x} \cdot \sqrt{3x}$.

Solution

a. $\sqrt{500} = \sqrt{100 \cdot 5}$ Factor 500. 100 is the greatest perfect square factor.

$\phantom{\sqrt{500}} = \sqrt{100}\,\sqrt{5}$ Use the product rule: $\sqrt{ab} = \sqrt{a}\,\sqrt{b}$.

$\phantom{\sqrt{500}} = 10\sqrt{5}$ Write $\sqrt{100}$ as 10. We read $10\sqrt{5}$ as "ten times the square root of 5."

b. We can simplify $\sqrt{6x} \cdot \sqrt{3x}$ using the product rule only if $6x$ and $3x$ represent nonnegative real numbers. Thus, $x \geq 0$.

$\sqrt{6x} \cdot \sqrt{3x} = \sqrt{6x \cdot 3x}$ Use the product rule: $\sqrt{a}\,\sqrt{b} = \sqrt{ab}$.

$\phantom{\sqrt{6x} \cdot \sqrt{3x}} = \sqrt{18x^2}$ Multiply in the radicand.

$\phantom{\sqrt{6x} \cdot \sqrt{3x}} = \sqrt{9x^2 \cdot 2}$ Factor 18. 9 is the greatest perfect square factor.

$\phantom{\sqrt{6x} \cdot \sqrt{3x}} = \sqrt{9x^2}\,\sqrt{2}$ Use the product rule: $\sqrt{ab} = \sqrt{a}\,\sqrt{b}$.

$\phantom{\sqrt{6x} \cdot \sqrt{3x}} = \sqrt{9}\,\sqrt{x^2}\,\sqrt{2}$ Use the product rule to write $\sqrt{9x^2}$ as the product of two square roots.

$\phantom{\sqrt{6x} \cdot \sqrt{3x}} = 3x\sqrt{2}$ $\sqrt{x^2} = |x| = x$ because $x \geq 0$.

Study Tip

When simplifying square roots, always look for the *greatest* perfect square factor possible. The following factorization will lead to further simplification:

$$\sqrt{500} = \sqrt{25 \cdot 20} = \sqrt{25}\sqrt{20} = 5\sqrt{20}.$$

25 is a perfect square factor of 500, but not the greatest perfect square factor.

Because 20 contains a perfect square factor, 4, the simplification is not complete.

$$5\sqrt{20} = 5\sqrt{4 \cdot 5} = 5\sqrt{4}\,\sqrt{5} = 5 \cdot 2\sqrt{5} = 10\sqrt{5}$$

Although the result checks with our simplification using $\sqrt{500} = \sqrt{100 \cdot 5}$, more work is required when the greatest perfect square factor is not used.

Check Point 2 Simplify:

    a. $\sqrt{75}$ **b.** $\sqrt{5x} \cdot \sqrt{10x}$.

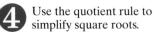 Use the quotient rule to simplify square roots.

The Quotient Rule for Square Roots

Another property for square roots involves division.

The Quotient Rule for Square Roots

If a and b represent nonnegative real numbers and $b \ne 0$, then

$$\sqrt{\frac{a}{b}} = \frac{\sqrt{a}}{\sqrt{b}} \quad \text{and} \quad \frac{\sqrt{a}}{\sqrt{b}} = \sqrt{\frac{a}{b}}.$$

The square root of a quotient is the quotient of the square roots.

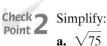 **EXAMPLE 3   Using the Quotient Rule to Simplify Square Roots**

Simplify: **a.** $\sqrt{\dfrac{100}{9}}$ **b.** $\dfrac{\sqrt{48x^3}}{\sqrt{6x}}$.

Solution

    a. $\sqrt{\dfrac{100}{9}} = \dfrac{\sqrt{100}}{\sqrt{9}} = \dfrac{10}{3}$

    b. We can simplify the quotient of $\sqrt{48x^3}$ and $\sqrt{6x}$ using the quotient rule only if $48x^3$ and $6x$ represent nonnegative real numbers and $6x \ne 0$. Thus, $x > 0$.

$$\frac{\sqrt{48x^3}}{\sqrt{6x}} = \sqrt{\frac{48x^3}{6x}} = \sqrt{8x^2} = \sqrt{4x^2}\sqrt{2} = \sqrt{4}\sqrt{x^2}\sqrt{2} = 2x\sqrt{2}$$

$\sqrt{x^2} = |x| = x$ because $x > 0$.

Check Point 3 Simplify: **a.** $\sqrt{\dfrac{25}{16}}$ **b.** $\dfrac{\sqrt{150x^3}}{\sqrt{2x}}$.

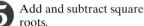 Add and subtract square roots.

Adding and Subtracting Square Roots

Two or more square roots can be combined using the distributive property provided that they have the same radicand. Such radicals are called **like radicals**. For example,

$$7\sqrt{11} + 6\sqrt{11} = (7 + 6)\sqrt{11} = 13\sqrt{11}.$$

7 square roots of 11 plus 6 square roots of 11 result in 13 square roots of 11.

A Radical Idea: Time Is Relative

What does travel in space have to do with radicals? Imagine that in the future we will be able to travel at velocities approaching the speed of light (approximately 186,000 miles per second). According to Einstein's theory of relativity, time would pass more quickly on Earth than it would in the moving spaceship. The radical expression

$$R_f\sqrt{1 - \left(\frac{v}{c}\right)^2}$$

gives the aging rate of an astronaut relative to the aging rate of a friend on Earth, R_f. In the expression, v is the astronaut's speed and c is the speed of light. As the astronaut's speed approaches the speed of light, we can substitute c for v:

$$R_f\sqrt{1 - \left(\frac{v}{c}\right)^2} \quad \text{Let } v = c.$$
$$= R_f\sqrt{1 - \left(\frac{c}{c}\right)^2}$$
$$= R_f\sqrt{1 - 1^2}$$
$$= R_f\sqrt{0} = 0$$

Close to the speed of light, the astronaut's aging rate relative to a friend on Earth is nearly 0. What does this mean? As we age here on Earth, the space traveler would barely get older. The space traveler would return to a futuristic world in which friends and loved ones would be long dead.

EXAMPLE 4   Adding and Subtracting Like Radicals

Add or subtract as indicated:

a. $7\sqrt{2} + 5\sqrt{2}$ **b.** $\sqrt{5x} - 7\sqrt{5x}.$

Solution

a. $7\sqrt{2} + 5\sqrt{2} = (7 + 5)\sqrt{2}$ Apply the distributive property.

$\qquad\qquad\qquad = 12\sqrt{2}$ Simplify.

b. $\sqrt{5x} - 7\sqrt{5x} = 1\sqrt{5x} - 7\sqrt{5x}$ Write $\sqrt{5x}$ as $1\sqrt{5x}$.

$\qquad\qquad\qquad = (1 - 7)\sqrt{5x}$ Apply the distributive property.

$\qquad\qquad\qquad = -6\sqrt{5x}$ Simplify.

Check Point 4   Add or subtract as indicated:

a. $8\sqrt{13} + 9\sqrt{13}$ **b.** $\sqrt{17x} - 20\sqrt{17x}.$

In some cases, radicals can be combined once they have been simplified. For example, to add $\sqrt{2}$ and $\sqrt{8}$, we can write $\sqrt{8}$ as $\sqrt{4 \cdot 2}$ because 4 is a perfect square factor of 8.

$$\sqrt{2} + \sqrt{8} = \sqrt{2} + \sqrt{4 \cdot 2} = 1\sqrt{2} + 2\sqrt{2} = (1 + 2)\sqrt{2} = 3\sqrt{2}$$

EXAMPLE 5   Combining Radicals That First Require Simplification

Add or subtract as indicated:

a. $7\sqrt{3} + \sqrt{12}$ **b.** $4\sqrt{50x} - 6\sqrt{32x}.$

Solution

a. $7\sqrt{3} + \sqrt{12}$

$\quad = 7\sqrt{3} + \sqrt{4 \cdot 3}$ Split 12 into two factors such that one is a perfect square.

$\quad = 7\sqrt{3} + 2\sqrt{3}$ $\sqrt{4 \cdot 3} = \sqrt{4}\sqrt{3} = 2\sqrt{3}$

$\quad = (7 + 2)\sqrt{3}$ Apply the distributive property. You will find that this step is usually done mentally.

$\quad = 9\sqrt{3}$ Simplify.

b. $4\sqrt{50x} - 6\sqrt{32x}$

$\quad = 4\sqrt{25 \cdot 2x} - 6\sqrt{16 \cdot 2x}$ 25 is the greatest perfect square factor of 50x and 16 is the greatest perfect square factor of 32x.

$\quad = 4 \cdot 5\sqrt{2x} - 6 \cdot 4\sqrt{2x}$ $\sqrt{25 \cdot 2x} = \sqrt{25}\sqrt{2x} = 5\sqrt{2x}$ and $\sqrt{16 \cdot 2x} = \sqrt{16}\sqrt{2x} = 4\sqrt{2x}.$

$\quad = 20\sqrt{2x} - 24\sqrt{2x}$ Multiply: $4 \cdot 5 = 20$ and $6 \cdot 4 = 24.$

$\quad = (20 - 24)\sqrt{2x}$ Apply the distributive property.

$\quad = -4\sqrt{2x}$ Simplify.

Check Point 5   Add or subtract as indicated:

a. $5\sqrt{27} + \sqrt{12}$ **b.** $6\sqrt{18x} - 4\sqrt{8x}.$

6 Rationalize denominators.

Rationalizing Denominators

You can use a calculator to compare the approximate values for $\dfrac{1}{\sqrt{3}}$ and $\dfrac{\sqrt{3}}{3}$. The two approximations are the same. This is not a coincidence:

$$\frac{1}{\sqrt{3}} = \frac{1}{\sqrt{3}} \cdot \boxed{\frac{\sqrt{3}}{\sqrt{3}}} = \frac{\sqrt{3}}{\sqrt{9}} = \frac{\sqrt{3}}{3}.$$

> Any number divided by itself is 1. Multiplication by 1 does not change the value of $\frac{1}{\sqrt{3}}$.

This process involves rewriting a radical expression as an equivalent expression in which the denominator no longer contains any radicals. The process is called **rationalizing the denominator**. If the denominator contains the square root of a natural number that is not a perfect square, **multiply the numerator and the denominator by the smallest number that produces the square root of a perfect square in the denominator**.

EXAMPLE 6   Rationalizing Denominators

Rationalize the denominator: **a.** $\dfrac{15}{\sqrt{6}}$ **b.** $\dfrac{12}{\sqrt{8}}$.

Solution

a. If we multiply the numerator and the denominator of $\dfrac{15}{\sqrt{6}}$ by $\sqrt{6}$, the denominator becomes $\sqrt{6} \cdot \sqrt{6} = \sqrt{36} = 6$. Therefore, we multiply by 1, choosing $\dfrac{\sqrt{6}}{\sqrt{6}}$ for 1.

$$\frac{15}{\sqrt{6}} = \frac{15}{\sqrt{6}} \cdot \frac{\sqrt{6}}{\sqrt{6}} = \frac{15\sqrt{6}}{\sqrt{36}} = \frac{15\sqrt{6}}{6} = \frac{5\sqrt{6}}{2}$$

> Multiply by 1.   Simplify: $\frac{15}{6} = \frac{15 \div 3}{6 \div 3} = \frac{5}{2}$.

b. The *smallest* number that will produce a perfect square in the denominator of $\dfrac{12}{\sqrt{8}}$ is $\sqrt{2}$, because $\sqrt{8} \cdot \sqrt{2} = \sqrt{16} = 4$. We multiply by 1, choosing $\dfrac{\sqrt{2}}{\sqrt{2}}$ for 1.

$$\frac{12}{\sqrt{8}} = \frac{12}{\sqrt{8}} \cdot \frac{\sqrt{2}}{\sqrt{2}} = \frac{12\sqrt{2}}{\sqrt{16}} = \frac{12\sqrt{2}}{4} = 3\sqrt{2}$$

Check Point 6   Rationalize the denominator: **a.** $\dfrac{5}{\sqrt{3}}$ **b.** $\dfrac{6}{\sqrt{12}}$.

Radical expressions that involve the sum and difference of the same two terms are called **conjugates**. Thus,

$$\sqrt{a} + \sqrt{b} \quad \text{and} \quad \sqrt{a} - \sqrt{b}$$

are conjugates. Conjugates are used to rationalize denominators because the product of such pairs contains no radicals:

Multiply each term of $\sqrt{a} - \sqrt{b}$
by each term of $\sqrt{a} + \sqrt{b}$.

$$\left(\sqrt{a} + \sqrt{b}\right)\left(\sqrt{a} - \sqrt{b}\right)$$
$$= \sqrt{a}\left(\sqrt{a} - \sqrt{b}\right) + \sqrt{b}\left(\sqrt{a} - \sqrt{b}\right)$$

Distribute $\sqrt{a}$ Distribute $\sqrt{b}$
over $\sqrt{a} - \sqrt{b}$.   over $\sqrt{a} - \sqrt{b}$.

$$= \sqrt{a} \cdot \sqrt{a} - \sqrt{a} \cdot \sqrt{b} + \sqrt{b} \cdot \sqrt{a} - \sqrt{b} \cdot \sqrt{b}$$
$$= \left(\sqrt{a}\right)^2 - \sqrt{ab} + \sqrt{ab} - \left(\sqrt{b}\right)^2$$

$-\sqrt{ab} + \sqrt{ab} = 0$

$$= \left(\sqrt{a}\right)^2 - \left(\sqrt{b}\right)^2$$
$$= a - b.$$

Multiplying Conjugates

$$\left(\sqrt{a} + \sqrt{b}\right)\left(\sqrt{a} - \sqrt{b}\right) = \left(\sqrt{a}\right)^2 - \left(\sqrt{b}\right)^2 = a - b$$

How can we rationalize a denominator if the denominator contains two terms with one or more square roots? **Multiply the numerator and the denominator by the conjugate of the denominator.** Here are three examples of such expressions:

- $\dfrac{7}{5 + \sqrt{3}}$ - $\dfrac{8}{3\sqrt{2} - 4}$ - $\dfrac{h}{\sqrt{x + h} - \sqrt{x}}$

The conjugate of the   The conjugate of the   The conjugate of the
denominator is $5 - \sqrt{3}$.   denominator is $3\sqrt{2} + 4$.   denominator is $\sqrt{x + h} + \sqrt{x}$.

The product of the denominator and its conjugate is found using the formula

$$\left(\sqrt{a} + \sqrt{b}\right)\left(\sqrt{a} - \sqrt{b}\right) = \left(\sqrt{a}\right)^2 - \left(\sqrt{b}\right)^2 = a - b.$$

The simplified product will not contain a radical.

EXAMPLE 7   Rationalizing a Denominator Containing Two Terms

Rationalize the denominator: $\dfrac{7}{5 + \sqrt{3}}$.

Solution   The conjugate of the denominator is $5 - \sqrt{3}$. If we multiply the numerator and denominator by $5 - \sqrt{3}$, the simplified denominator will not contain a radical. Therefore, we multiply by 1, choosing $\dfrac{5 - \sqrt{3}}{5 - \sqrt{3}}$ for 1.

$$\frac{7}{5 + \sqrt{3}} = \frac{7}{5 + \sqrt{3}} \cdot \frac{5 - \sqrt{3}}{5 - \sqrt{3}} = \frac{7\left(5 - \sqrt{3}\right)}{5^2 - \left(\sqrt{3}\right)^2} = \frac{7\left(5 - \sqrt{3}\right)}{25 - 3}$$

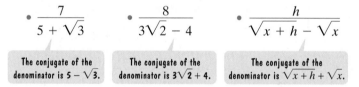

Multiply by 1. $\left(\sqrt{a} + \sqrt{b}\right)\left(\sqrt{a} - \sqrt{b}\right)$
$= \left(\sqrt{a}\right)^2 - \left(\sqrt{b}\right)^2$

$$= \frac{7\left(5 - \sqrt{3}\right)}{22} \text{ or } \frac{35 - 7\sqrt{3}}{22}$$

In either form of the answer, there
is no radical in the denominator.

Check Point 7 Rationalize the denominator: $\dfrac{8}{4 + \sqrt{5}}$.

 Evaluate and perform operations with higher roots.

Other Kinds of Roots

We define the **principal nth root** of a real number a, symbolized by $\sqrt[n]{a}$, as follows:

Definition of the Principal nth Root of a Real Number

$$\sqrt[n]{a} = b \text{ means that } b^n = a.$$

If n, the **index**, is even, then a is nonnegative ($a \geq 0$) and b is also nonnegative ($b \geq 0$). If n is odd, a and b can be any real numbers.

For example,

$$\sqrt[3]{64} = 4 \text{ because } 4^3 = 64 \quad \text{and} \quad \sqrt[5]{-32} = -2 \text{ because } (-2)^5 = -32.$$

The same vocabulary that we learned for square roots applies to nth roots. The symbol $\sqrt[n]{a}$ is called a **radical** and a is called the **radicand**.

A number that is the nth power of a rational number is called a **perfect nth power**. For example, 8 is a perfect third power, or perfect cube, because $8 = 2^3$. Thus, $\sqrt[3]{8} = \sqrt[3]{2^3} = 2$. In general, one of the following rules can be used to find nth roots of perfect nth powers:

Finding nth Roots of Perfect nth Powers

If n is odd, $\sqrt[n]{a^n} = a$.
If n is even, $\sqrt[n]{a^n} = |a|$.

For example,

$$\sqrt[3]{(-2)^3} = -2 \qquad \text{and} \qquad \sqrt[4]{(-2)^4} = |-2| = 2.$$

Absolute value is not needed with odd roots, but is necessary with even roots.

The Product and Quotient Rules for Other Roots

The product and quotient rules apply to cube roots, fourth roots, and all higher roots.

The Product and Quotient Rules for nth Roots

For all real numbers, where the indicated roots represent real numbers,

$$\sqrt[n]{ab} = \sqrt[n]{a} \cdot \sqrt[n]{b} \quad \text{and} \quad \sqrt[n]{\dfrac{a}{b}} = \dfrac{\sqrt[n]{a}}{\sqrt[n]{b}}, \quad b \neq 0.$$

EXAMPLE 8   Simplifying, Multiplying, and Dividing Higher Roots

Simplify: **a.** $\sqrt[3]{24}$ **b.** $\sqrt[4]{8} \cdot \sqrt[4]{4}$ **c.** $\sqrt[4]{\dfrac{81}{16}}$.

Solution

a. $\sqrt[3]{24} = \sqrt[3]{8 \cdot 3}$ *Find the greatest perfect cube that is a factor of 24. $2^3 = 8$, so 8 is a perfect cube and is the greatest perfect cube factor of 24.*

$\phantom{\sqrt[3]{24}} = \sqrt[3]{8} \cdot \sqrt[3]{3}$ $\sqrt[n]{ab} = \sqrt[n]{a}\,\sqrt[n]{b}$

$\phantom{\sqrt[3]{24}} = 2\sqrt[3]{3}$ $\sqrt[3]{8} = 2$

Study Tip

Some higher even and odd roots occur so frequently that you might want to memorize them.

Cube Roots	
$\sqrt[3]{1} = 1$	$\sqrt[3]{125} = 5$
$\sqrt[3]{8} = 2$	$\sqrt[3]{216} = 6$
$\sqrt[3]{27} = 3$	$\sqrt[3]{1000} = 10$
$\sqrt[3]{64} = 4$	

Fourth Roots	Fifth Roots
$\sqrt[4]{1} = 1$	$\sqrt[5]{1} = 1$
$\sqrt[4]{16} = 2$	$\sqrt[5]{32} = 2$
$\sqrt[4]{81} = 3$	$\sqrt[5]{243} = 3$
$\sqrt[4]{256} = 4$	
$\sqrt[4]{625} = 5$	

b. $\sqrt[4]{8} \cdot \sqrt[4]{4} = \sqrt[4]{8 \cdot 4}$ $\sqrt[n]{a} \cdot \sqrt[n]{b} = \sqrt[n]{ab}$

$\qquad\qquad = \sqrt[4]{32}$ Find the greatest perfect fourth power that is a factor of 32

$\qquad\qquad = \sqrt[4]{16 \cdot 2}$ $2^4 = 16$, so 16 is a perfect fourth power and is the greatest perfect fourth power that is a factor of 32.

$\qquad\qquad = \sqrt[4]{16} \cdot \sqrt[4]{2}$ $\sqrt[n]{ab} = \sqrt[n]{a} \cdot \sqrt[n]{b}$

$\qquad\qquad = 2\sqrt[4]{2}$ $\sqrt[4]{16} = 2$

c. $\sqrt[4]{\dfrac{81}{16}} = \dfrac{\sqrt[4]{81}}{\sqrt[4]{16}}$ $\sqrt[n]{\dfrac{a}{b}} = \dfrac{\sqrt[n]{a}}{\sqrt[n]{b}}$

$\qquad\quad\; = \dfrac{3}{2}$ $\sqrt[4]{81} = 3$ because $3^4 = 81$ and $\sqrt[4]{16} = 2$ because $2^4 = 16$.

Check Point 8   Simplify: **a.** $\sqrt[3]{40}$ **b.** $\sqrt[5]{8} \cdot \sqrt[5]{8}$ **c.** $\sqrt[3]{\dfrac{125}{27}}$.

We have seen that adding and subtracting square roots often involves simplifying terms. The same idea applies to adding and subtracting nth roots.

EXAMPLE 9   Combining Cube Roots

Subtract: $5\sqrt[3]{16} - 11\sqrt[3]{2}$.

Solution

$\qquad 5\sqrt[3]{16} - 11\sqrt[3]{2}$

$\qquad = 5\sqrt[3]{8 \cdot 2} - 11\sqrt[3]{2}$ Factor 16. 8 is the greatest perfect cube factor: $2^3 = 8$ and $\sqrt[3]{8} = 2$.

$\qquad = 5 \cdot 2\sqrt[3]{2} - 11\sqrt[3]{2}$ $\sqrt[3]{8 \cdot 2} = \sqrt[3]{8}\,\sqrt[3]{2} = 2\sqrt[3]{2}$

$\qquad = 10\sqrt[3]{2} - 11\sqrt[3]{2}$ Multiply: $5 \cdot 2 = 10$.

$\qquad = (10 - 11)\sqrt[3]{2}$ Apply the distributive property.

$\qquad = -1\sqrt[3]{2}$ or $-\sqrt[3]{2}$ Simplify.

Check Point 9   Subtract: $3\sqrt[3]{81} - 4\sqrt[3]{3}$.

⑧ Understand and use rational exponents.

Rational Exponents

We define rational exponents so that their properties are the same as the properties for integer exponents. For example, we know that exponents are multiplied when an exponential expression is raised to a power. For this to be true,

$$\left(7^{\frac{1}{2}}\right)^2 = 7^{\frac{1}{2} \cdot 2} = 7^1 = 7.$$

We also know that

$$\left(\sqrt{7}\right)^2 = \sqrt{7} \cdot \sqrt{7} = \sqrt{49} = 7.$$

Can you see that the square of both $7^{\frac{1}{2}}$ and $\sqrt{7}$ is 7? It is reasonable to conclude that

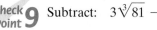

$$7^{\frac{1}{2}} \quad \text{means} \quad \sqrt{7}.$$

We can generalize the fact that $7^{\frac{1}{2}}$ means $\sqrt{7}$ with the following definition:

The Definition of $a^{\frac{1}{n}}$

If $\sqrt[n]{a}$ represents a real number and $n \geq 2$ is an integer, then

$$a^{\frac{1}{n}} = \sqrt[n]{a}.$$

> The denominator of the rational exponent is the radical's index.

Furthermore,

$$a^{-\frac{1}{n}} = \frac{1}{a^{\frac{1}{n}}} = \frac{1}{\sqrt[n]{a}}, \quad a \neq 0.$$

EXAMPLE 10   Using the Definition of $a^{\frac{1}{n}}$

Simplify:

a. $64^{\frac{1}{2}}$ **b.** $125^{\frac{1}{3}}$ **c.** $-16^{\frac{1}{4}}$ **d.** $(-27)^{\frac{1}{3}}$ **e.** $64^{-\frac{1}{3}}$.

Solution

a. $64^{\frac{1}{2}} = \sqrt{64} = 8$

b. $125^{\frac{1}{3}} = \sqrt[3]{125} = 5$

> The denominator is the index.

c. $-16^{\frac{1}{4}} = -(\sqrt[4]{16}) = -2$

> The base is 16 and the negative sign is not affected by the exponent.

d. $(-27)^{\frac{1}{3}} = \sqrt[3]{-27} = -3$

> Parentheses show that the base is −27 and that the negative sign is affected by the exponent.

e. $64^{-\frac{1}{3}} = \dfrac{1}{64^{\frac{1}{3}}} = \dfrac{1}{\sqrt[3]{64}} = \dfrac{1}{4}$

Check Point 10   Simplify:

a. $25^{\frac{1}{2}}$ **b.** $8^{\frac{1}{3}}$ **c.** $-81^{\frac{1}{4}}$ **d.** $(-8)^{\frac{1}{3}}$ **e.** $27^{-\frac{1}{3}}$

In Example 10 and Check Point 10, each rational exponent had a numerator of 1. If the numerator is some other integer, we still want to multiply exponents when raising a power to a power. For this reason,

$$a^{\frac{2}{3}} = \left(a^{\frac{1}{3}}\right)^2 \quad \text{and} \quad a^{\frac{2}{3}} = \left(a^2\right)^{\frac{1}{3}}.$$

> This means $(\sqrt[3]{a})^2$.   This means $\sqrt[3]{a^2}$.

Thus,

$$a^{\frac{2}{3}} = \left(\sqrt[3]{a}\right)^2 = \sqrt[3]{a^2}.$$

Do you see that the denominator, 3, of the rational exponent is the same as the index of the radical? The numerator, 2, of the rational exponent serves as an exponent in each of the two radical forms. We generalize these ideas with the following definition:

The Definition of $a^{\frac{m}{n}}$

If $\sqrt[n]{a}$ represents a real number and $\frac{m}{n}$ is a positive rational number, $n \geq 2$, then

$$a^{\frac{m}{n}} = (\sqrt[n]{a})^m.$$

Also,

$$a^{\frac{m}{n}} = \sqrt[n]{a^m}.$$

Furthermore, if $a^{-\frac{m}{n}}$ is a nonzero real number, then

$$a^{-\frac{m}{n}} = \frac{1}{a^{\frac{m}{n}}}.$$

The first form of the definition of $a^{\frac{m}{n}}$, shown again below, involves taking the root first. This form is often preferable because smaller numbers are involved. Notice that the rational exponent consists of two parts, indicated by the following voice balloons:

The numerator is the exponent.

$$a^{\frac{m}{n}} = (\sqrt[n]{a})^m.$$

The denominator is the radical's index.

EXAMPLE 11   Using the Definition of $a^{\frac{m}{n}}$

Simplify:

a. $27^{\frac{2}{3}}$ **b.** $9^{\frac{3}{2}}$ **c.** $81^{-\frac{3}{4}}$.

Solution

a. $27^{\frac{2}{3}} = (\sqrt[3]{27})^2 = 3^2 = 9$

b. $9^{\frac{3}{2}} = (\sqrt{9})^3 = 3^3 = 27$

c. $81^{-\frac{3}{4}} = \frac{1}{81^{\frac{3}{4}}} = \frac{1}{(\sqrt[4]{81})^3} = \frac{1}{3^3} = \frac{1}{27}$

Technology

Here are the calculator keystroke sequences for $81^{-\frac{3}{4}}$:

Many Scientific Calculators

81 y^x (3 $+/-$ ÷ 4) =

Many Graphing Calculators

81 ∧ ((−) 3 ÷ 4) ENTER .

Check Point 11 Simplify: **a.** $27^{\frac{4}{3}}$ **b.** $4^{\frac{3}{2}}$ **c.** $32^{-\frac{2}{5}}$

Properties of exponents can be applied to expressions containing rational exponents.

EXAMPLE 12   Simplifying Expressions with Rational Exponents

Simplify using properties of exponents:

a. $(5x^{\frac{1}{2}})(7x^{\frac{3}{4}})$ **b.** $\frac{32x^{\frac{5}{3}}}{16x^{\frac{3}{4}}}$.

Solution

a. $(5x^{\frac{1}{2}})(7x^{\frac{3}{4}}) = 5 \cdot 7x^{\frac{1}{2}} \cdot x^{\frac{3}{4}}$ Group factors with the same base.

$= 35x^{\frac{1}{2}+\frac{3}{4}}$ When multiplying expressions with the same base, add the exponents.

$= 35x^{\frac{5}{4}}$ $\frac{1}{2} + \frac{3}{4} = \frac{2}{4} + \frac{3}{4} = \frac{5}{4}$

b. $\dfrac{32x^{\frac{5}{3}}}{16x^{\frac{3}{4}}} = \left(\dfrac{32}{16}\right)\left(\dfrac{x^{\frac{5}{3}}}{x^{\frac{3}{4}}}\right)$ Group factors with the same base.

$= 2x^{\frac{5}{3}-\frac{3}{4}}$ When dividing expressions with the same base, subtract the exponents.

$= 2x^{\frac{11}{12}}$ $\frac{5}{3} - \frac{3}{4} = \frac{20}{12} - \frac{9}{12} = \frac{11}{12}$

Check Point 12 Simplify: **a.** $\left(2x^{\frac{4}{3}}\right)\left(5x^{\frac{8}{3}}\right)$ **b.** $\dfrac{20x^4}{5x^{\frac{3}{2}}}$.

Rational exponents are sometimes useful for simplifying radicals by reducing their index.

EXAMPLE 13   Reducing the Index of a Radical

Simplify: $\sqrt[9]{x^3}$.

Solution $\sqrt[9]{x^3} = x^{\frac{3}{9}} = x^{\frac{1}{3}} = \sqrt[3]{x}$

Check Point 13 Simplify: $\sqrt[6]{x^3}$.

EXERCISE SET P.3

Practice Exercises

Evaluate each expression in Exercises 1–12, or indicate that the root is not a real number.

1. $\sqrt{36}$
2. $\sqrt{25}$
3. $-\sqrt{36}$
4. $-\sqrt{25}$
5. $\sqrt{-36}$
6. $\sqrt{-25}$
7. $\sqrt{25 - 16}$
8. $\sqrt{144 + 25}$
9. $\sqrt{25} - \sqrt{16}$
10. $\sqrt{144} + \sqrt{25}$
11. $\sqrt{(-13)^2}$
12. $\sqrt{(-17)^2}$

Use the product rule to simplify the expressions in Exercises 13–22. In Exercises 17–22, assume that variables represent non-negative real numbers.

13. $\sqrt{50}$
14. $\sqrt{27}$
15. $\sqrt{45x^2}$
16. $\sqrt{125x^2}$
17. $\sqrt{2x} \cdot \sqrt{6x}$
18. $\sqrt{10x} \cdot \sqrt{8x}$
19. $\sqrt{x^3}$
20. $\sqrt{y^3}$
21. $\sqrt{2x^2} \cdot \sqrt{6x}$
22. $\sqrt{6x} \cdot \sqrt{3x^2}$

Use the quotient rule to simplify the expressions in Exercises 23–32. Assume that x > 0.

23. $\sqrt{\dfrac{1}{81}}$
24. $\sqrt{\dfrac{1}{49}}$
25. $\sqrt{\dfrac{49}{16}}$
26. $\sqrt{\dfrac{121}{9}}$
27. $\dfrac{\sqrt{48x^3}}{\sqrt{3x}}$
28. $\dfrac{\sqrt{72x^3}}{\sqrt{8x}}$
29. $\dfrac{\sqrt{150x^4}}{\sqrt{3x}}$
30. $\dfrac{\sqrt{24x^4}}{\sqrt{3x}}$
31. $\dfrac{\sqrt{200x^3}}{\sqrt{10x^{-1}}}$
32. $\dfrac{\sqrt{500x^3}}{\sqrt{10x^{-1}}}$

In Exercises 33–44, add or subtract terms whenever possible.

33. $7\sqrt{3} + 6\sqrt{3}$
34. $8\sqrt{5} + 11\sqrt{5}$
35. $6\sqrt{17x} - 8\sqrt{17x}$
36. $4\sqrt{13x} - 6\sqrt{13x}$
37. $\sqrt{8} + 3\sqrt{2}$
38. $\sqrt{20} + 6\sqrt{5}$
39. $\sqrt{50x} - \sqrt{8x}$
40. $\sqrt{63x} - \sqrt{28x}$
41. $3\sqrt{18} + 5\sqrt{50}$
42. $4\sqrt{12} - 2\sqrt{75}$
43. $3\sqrt{8} - \sqrt{32} + 3\sqrt{72} - \sqrt{75}$
44. $3\sqrt{54} - 2\sqrt{24} - \sqrt{96} + 4\sqrt{63}$

In Exercises 45–54, rationalize the denominator.

45. $\dfrac{1}{\sqrt{7}}$
46. $\dfrac{2}{\sqrt{10}}$
47. $\dfrac{\sqrt{2}}{\sqrt{5}}$
48. $\dfrac{\sqrt{7}}{\sqrt{3}}$
49. $\dfrac{13}{3 + \sqrt{11}}$
50. $\dfrac{3}{3 + \sqrt{7}}$
51. $\dfrac{7}{\sqrt{5} - 2}$
52. $\dfrac{5}{\sqrt{3} - 1}$
53. $\dfrac{6}{\sqrt{5} + \sqrt{3}}$
54. $\dfrac{11}{\sqrt{7} - \sqrt{3}}$

Evaluate each expression in Exercises 55–66, or indicate that the root is not a real number.

55. $\sqrt[3]{125}$
56. $\sqrt[3]{8}$
57. $\sqrt[3]{-8}$
58. $\sqrt[3]{-125}$

59. $\sqrt[4]{-16}$ **60.** $\sqrt[4]{-81}$

61. $\sqrt[4]{(-3)^4}$ **62.** $\sqrt[4]{(-2)^4}$

63. $\sqrt[5]{(-3)^5}$ **64.** $\sqrt[5]{(-2)^5}$

65. $\sqrt[5]{-\frac{1}{32}}$ **66.** $\sqrt[6]{\frac{1}{64}}$

Simplify the radical expressions in Exercises 67–74.

67. $\sqrt[3]{32}$ **68.** $\sqrt[3]{150}$

69. $\sqrt[3]{x^4}$ **70.** $\sqrt[3]{x^5}$

71. $\sqrt[3]{9}\cdot\sqrt[3]{6}$ **72.** $\sqrt[3]{12}\cdot\sqrt[3]{4}$

73. $\dfrac{\sqrt[5]{64x^6}}{\sqrt[5]{2x}}$ **74.** $\dfrac{\sqrt[4]{162x^5}}{\sqrt[4]{2x}}$

In Exercises 75–82, add or subtract terms whenever possible.

75. $4\sqrt[5]{2}+3\sqrt[5]{2}$ **76.** $6\sqrt[5]{3}+2\sqrt[5]{3}$

77. $5\sqrt[3]{16}+\sqrt[3]{54}$ **78.** $3\sqrt[3]{24}+\sqrt[3]{81}$

79. $\sqrt[3]{54xy^3}-y\sqrt[3]{128x}$ **80.** $\sqrt[3]{24xy^3}-y\sqrt[3]{81x}$

81. $\sqrt{2}+\sqrt[3]{8}$ **82.** $\sqrt{3}+\sqrt[3]{15}$

In Exercises 83–90, evaluate each expression without using a calculator.

83. $36^{\frac{1}{2}}$ **84.** $121^{\frac{1}{2}}$

85. $8^{\frac{1}{3}}$ **86.** $27^{\frac{1}{3}}$

87. $125^{\frac{2}{3}}$ **88.** $8^{\frac{2}{3}}$

89. $32^{-\frac{4}{5}}$ **90.** $16^{-\frac{5}{2}}$

In Exercises 91–100, simplify using properties of exponents.

91. $\left(7x^{\frac{1}{3}}\right)\left(2x^{\frac{1}{4}}\right)$ **92.** $\left(3x^{\frac{2}{3}}\right)\left(4x^{\frac{3}{4}}\right)$

93. $\dfrac{20x^{\frac{1}{2}}}{5x^{\frac{1}{4}}}$ **94.** $\dfrac{72x^{\frac{3}{4}}}{9x^{\frac{1}{3}}}$

95. $\left(x^{\frac{2}{3}}\right)^3$ **96.** $\left(x^{\frac{4}{5}}\right)^5$

97. $(25x^4y^6)^{\frac{1}{2}}$ **98.** $(125x^9y^6)^{\frac{1}{3}}$

99. $\dfrac{\left(3y^{\frac{1}{4}}\right)^3}{y^{\frac{1}{12}}}$ **100.** $\dfrac{\left(2y^{\frac{1}{5}}\right)^4}{y^{\frac{3}{10}}}$

In Exercises 101–108, simplify by reducing the index of the radical.

101. $\sqrt[4]{5^2}$ **102.** $\sqrt[4]{7^2}$

103. $\sqrt[3]{x^6}$ **104.** $\sqrt[4]{x^{12}}$

105. $\sqrt[6]{x^4}$ **106.** $\sqrt[9]{x^6}$

107. $\sqrt[9]{x^6y^3}$ **108.** $\sqrt[12]{x^4y^8}$

Practice Plus

In Exercises 109–110, evaluate each expression.

109. $\sqrt[3]{\sqrt[4]{16}+\sqrt{625}}$

110. $\sqrt[3]{\sqrt{\sqrt{169}+\sqrt{9}}+\sqrt{\sqrt[3]{1000}+\sqrt[3]{216}}}$

In Exercises 111–114, simplify each expression. Assume that all variables represent positive numbers.

111. $(49x^{-2}y^4)^{-\frac{1}{2}}\left(xy^{\frac{1}{2}}\right)$ **112.** $(8x^{-6}y^3)^{\frac{1}{3}}\left(x^{\frac{5}{6}}y^{-\frac{1}{3}}\right)^6$

113. $\left(\dfrac{x^{-\frac{5}{4}}y^{\frac{1}{3}}}{x^{-\frac{3}{4}}}\right)^{-6}$ **114.** $\left(\dfrac{x^{\frac{1}{2}}y^{-\frac{7}{4}}}{y^{-\frac{5}{4}}}\right)^{-4}$

Application Exercises

The formula

$$d=\sqrt{\frac{3h}{2}}$$

models the distance, d, in miles, that a person h feet high can see to the horizon. Use this formula to solve Exercises 115–116.

115. The pool deck on a cruise ship is 72 feet above the water. How far can passengers on the pool deck see? Write the answer in simplified radical form. Then use the simplified radical form and a calculator to express the answer to the nearest tenth of a mile.

116. The captain of a cruise ship is on the star deck, which is 120 feet above the water. How far can the captain see? Write the answer in simplified radical form. Then use the simplified radical form and a calculator to express the answer to the nearest tenth of a mile.

Police use the formula $v=2\sqrt{5L}$ to estimate the speed of a car, v, in miles per hour, based on the length, L, in feet, of its skid marks upon sudden braking on a dry asphalt road. Use the formula to solve Exercises 117–118.

117. A motorist is involved in an accident. A police officer measures the car's skid marks to be 245 feet long. Estimate the speed at which the motorist was traveling before braking. If the posted speed limit is 50 miles per hour and the motorist tells the officer he was not speeding, should the officer believe him? Explain.

118. A motorist is involved in an accident. A police officer measures the car's skid marks to be 45 feet long. Estimate the speed at which the motorist was traveling before braking. If the posted speed limit is 35 miles per hour and the motorist tells the officer she was not speeding, should the officer believe her? Explain.

119. In the Peanuts cartoon shown below, Woodstock appears to be working steps mentally. Fill in the missing steps that show how to go from $\dfrac{7\sqrt{2\cdot2\cdot3}}{6}$ to $\dfrac{7}{3}\sqrt{3}$.

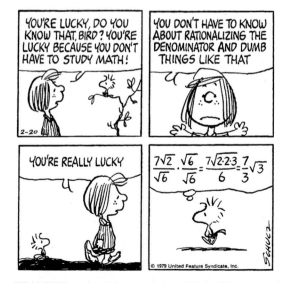

PEANUTS reprinted by permission of United Feature Syndicate, Inc.

120. According to Einstein's theory of relativity, traveling in starships at velocities approaching the speed of light (approximately 186,000 miles per second), time would pass more quickly on Earth than it would in the moving starship. The radical expression

$$R_f \frac{\sqrt{c^2 - v^2}}{\sqrt{c^2}}$$

gives the aging rate of an astronaut relative to the aging rate of a friend, R_f, on Earth. In the expression, v is the astronaut's velocity and c is the speed of light. Use the expression to solve this exercise. Imagine that you are the astronaut on the starship.

a. Use the quotient rule and simplify the expression that shows your aging rate relative to a friend on Earth. Working step-by-step, express your aging rate as

$$R_f \sqrt{1 - \left(\frac{v}{c}\right)^2}.$$

b. You are moving at 90% of the speed of light. Substitute $0.9c$ for v, your velocity, in the simplified expression from part (a). What is your aging rate, correct to two decimal places, relative to a friend on Earth? If you are gone for 44 weeks, approximately how many weeks have passed for your friend?

The way that we perceive the temperature on a cold day depends on both air temperature and wind speed. The windchill is what the air temperature would have to be with no wind to achieve the same chilling effect on the skin. In 2002, the National Weather Service issued new windchill temperatures, shown in the table below. (One reason for this new windchill index is that the wind speed is now calculated at 5 feet, the average height of the human body's face, rather than 33 feet, the height of the standard anemometer, an instrument that calculates wind speed.)

New Windchill Temperature Index

			Air Temperature (°F)									
	30	**25**	**20**	**15**	**10**	**5**	**0**	**−5**	**−10**	**−15**	**−20**	**−25**
5	25	19	13	7	1	−5	−11	−16	−22	−28	−34	−40
10	21	15	9	3	−4	−10	−16	−22	−28	−35	−41	−47
15	19	13	6	0	−7	−13	−19	−26	−32	−39	−45	−51
20	17	11	4	−2	−9	−15	−22	−29	−35	−42	−48	−55
25	16	9	3	−4	−11	−17	−24	−31	−37	−44	−51	−58
30	15	8	1	−5	−12	−19	−26	−33	−39	−46	−53	−60
35	14	7	0	−7	−14	−21	−27	−34	−41	−48	−55	−62
40	13	6	−1	−8	−15	−22	−29	−36	−43	−50	−57	−64
45	12	5	−2	−9	−16	−23	−30	−37	−44	−51	−58	−65
50	12	4	−3	−10	−17	−24	−31	−38	−45	−52	−60	−67
55	11	4	−3	−11	−18	−25	−32	−39	−46	−54	−61	−68
60	10	3	−4	−11	−19	−26	−33	−40	−48	−55	−62	−69

Wind Speed (miles per hour)

▨ Frostbite occurs in 15 minutes or less.

Source: National Weather Service

The windchill temperatures shown in the table can be calculated using

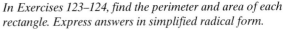

$$C = 35.74 + 0.6215t - 35.74\sqrt[25]{v^4} + 0.4275t\sqrt[25]{v^4},$$

in which C is the windchill, in degrees Fahrenheit, t is the air temperature, in degrees Fahrenheit, and v is the wind speed, in miles per hour. Use the formula to solve Exercises 121–122.

121. a. Rewrite the equation for calculating windchill temperatures using rational exponents.

   b. Use the form of the equation in part (a) and a calculator to find the windchill temperature, to the nearest degree, when the air temperature is 25°F and the wind speed is 30 miles per hour.

122. a. Rewrite the equation for calculating windchill temperatures using rational exponents.

   b. Use the form of the equation in part (a) and a calculator to find the windchill temperature, to the nearest degree, when the air temperature is 35°F and the wind speed is 15 miles per hour.

In Exercises 123–124, find the perimeter and area of each rectangle. Express answers in simplified radical form.

123.

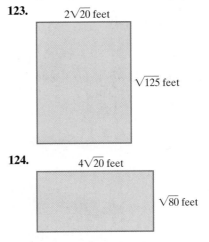

$2\sqrt{20}$ feet

$\sqrt{125}$ feet

124.

$4\sqrt{20}$ feet

$\sqrt{80}$ feet

Writing in Mathematics

125. Explain how to simplify $\sqrt{10} \cdot \sqrt{5}$.

126. Explain how to add $\sqrt{3} + \sqrt{12}$.

127. Describe what it means to rationalize a denominator. Use both $\dfrac{1}{\sqrt{5}}$ and $\dfrac{1}{5 + \sqrt{5}}$ in your explanation.

128. What difference is there in simplifying $\sqrt[3]{(-5)^3}$ and $\sqrt[4]{(-5)^4}$?

129. What does $a^{\frac{m}{n}}$ mean?

130. Describe the kinds of numbers that have rational fifth roots.

131. Why must a and b represent nonnegative numbers when we write $\sqrt{a} \cdot \sqrt{b} = \sqrt{ab}$? Is it necessary to use this restriction in the case of $\sqrt[3]{a} \cdot \sqrt[3]{b} = \sqrt[3]{ab}$? Explain.

132. Answer the question posed in the chapter opener on page 1. What will you do: explore space or stay here on Earth? What are the reasons for your choice?

Critical Thinking Exercises

133. Which one of the following is true?

   a. Neither $(-8)^{\frac{1}{2}}$ nor $(-8)^{\frac{1}{3}}$ represents a real number.

   b. $\sqrt{x^2 + y^2} = x + y$

   c. $8^{-\frac{1}{3}} = -2$

   d. $2^{\frac{1}{2}} \cdot 2^{\frac{1}{2}} = 2$

In Exercises 134–135, fill in each box to make the statement true.

134. $\left(5 + \sqrt{\Box}\right)\left(5 - \sqrt{\Box}\right) = 22$

135. $\sqrt{\Box x^{\Box}} = 5x^7$

136. Find exact value of $\sqrt{13 + \sqrt{2} + \dfrac{7}{3 + \sqrt{2}}}$ without the use of a calculator.

137. Place the correct symbol, $>$ or $<$, in the shaded area between each of the given numbers. *Do not use a calculator.* Then check your result with a calculator.

   a. $3^{\frac{1}{2}} \;\blacksquare\; 3^{\frac{1}{3}}$ **b.** $\sqrt{7} + \sqrt{18} \;\blacksquare\; \sqrt{7 + 18}$

138. a. A mathematics professor recently purchased a birthday cake for her son with the inscription

$$\text{Happy } \left(2^{\frac{5}{2}} \cdot 2^{\frac{3}{4}} \div 2^{\frac{1}{4}}\right)\text{th Birthday.}$$

   How old is the son?

   b. The birthday boy, excited by the inscription on the cake, tried to wolf down the whole thing. Professor Mom, concerned about the possible metamorphosis of her son into a blimp, exclaimed, "Hold on! It is your birthday, so why not take $\dfrac{8^{-\frac{4}{3}} + 2^{-2}}{16^{-\frac{3}{4}} + 2^{-1}}$ of the cake? I'll eat half of what's left over." How much of the cake did the professor eat?

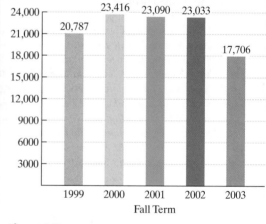

SECTION P.4   *Polynomials*

Objectives

1 Understand the vocabulary of polynomials.

2 Add and subtract polynomials.

3 Multiply polynomials.

4 Use FOIL in polynomial multiplication.

5 Use special products in polynomial multiplication.

6 Perform operations with polynomials in several variables.

Number of Newly-Declared Computer Majors in U.S. and Canadian Colleges

Figure P.11

Source: Computing Research Association

Tech firms might be rebounding from the dot-com bust, but enrollment in college computer programs keeps falling. In the past, a computer degree meant "instant riches, or at least a well-paying, secure job," says San Jose computer science chair David Hayes. "Now, the perception is jobs are going overseas and people are being laid off."

     The bar graph in Figure P.11 shows the number of newly-declared computer science and computer engineering majors for the fall term in U.S. and Canadian colleges from 1999 through 2003. The data can be modeled by the formula

$$N = -365x^4 + 2728x^3 - 7106x^2 + 7372x + 20{,}787,$$

where N is the number of newly-declared computer majors for the fall term x years after 1999.

The algebraic expression on the right side of the equation,

$$-365x^4 + 2728x^3 - 7106x^2 + 7372x + 20{,}787,$$

is an example of a *polynomial*. A **polynomial** is a single term or the sum of two or more terms containing variables in the numerator with whole-number exponents. This particular polynomial contains five terms. Equations containing polynomials are used in such diverse areas as science, business, medicine, psychology, and sociology. In this section, we review basic ideas about polynomials and their operations.

① Understand the vocabulary of polynomials.

How We Describe Polynomials

Consider the polynomial

$$7x^3 - 9x^2 + 13x - 6.$$

We can express this polynomial as

$$7x^3 + (-9x^2) + 13x + (-6).$$

The polynomial contains four terms. It is customary to write the terms in the order of descending powers of the variable. This is the **standard form** of a polynomial.

Some polynomials contain only one variable. Each term of such a polynomial in x is of the form ax^n. If $a \neq 0$, the **degree** of ax^n is n. For example, the degree of the term $7x^3$ is 3.

Study Tip

We can express 0 in many ways, including $0x$, $0x^2$, and $0x^3$. It is impossible to assign a single exponent on the variable. This is why 0 has no defined degree.

The Degree of ax^n

If $a \neq 0$, the degree of ax^n is n. The degree of a nonzero constant is 0. The constant 0 has no defined degree.

Here is an example of a polynomial and the degree of each of its four terms:

$$6x^4 - 3x^3 + 2x - 5.$$

degree 4   degree 3   degree 1   degree of nonzero constant: 0

Notice that the exponent on x for the term $2x$ is understood to be 1: $2x^1$. For this reason, the degree of $2x$ is 1. You can think of -5 as $-5x^0$; thus, its degree is 0.

A polynomial which when simplified has exactly one term is called a **monomial**. A **binomial** is a polynomial that has two terms, each with a different exponent. A **trinomial** is a polynomial with three terms, each with a different exponent. Simplified polynomials with four or more terms have no special names.

The **degree of a polynomial** is the greatest of the degrees of all its terms. For example, $4x^2 + 3x$ is a binomial of degree 2 because the degree of the first term is 2, and the degree of the other term is less than 2. Also, $7x^5 - 2x^2 + 4$ is a trinomial of degree 5 because the degree of the first term is 5, and the degrees of the other terms are less than 5.

Up to now, we have used x to represent the variable in a polynomial. However, any letter can be used. For example,

- $7x^5 - 3x^3 + 8$ is a polynomial (in x) of degree 5. Because there are three terms, the polynomial is a trinomial.
- $6y^3 + 4y^2 - y + 3$ is a polynomial (in y) of degree 3. Because there are four terms, the polynomial has no special name.
- $z^7 + \sqrt{2}$ is a polynomial (in z) of degree 7. Because there are two terms, the polynomial is a binomial.

Not every algebraic expression is a polynomial. Algebraic expressions whose variables do not contain whole number exponents in numerators such as

$$3x^{-2} + 7 \quad \text{and} \quad 5x^{\frac{3}{2}} + 9x^{\frac{1}{2}} + 2$$

are not polynomials. Furthermore, a quotient of polynomials such as

$$\frac{x^2 + 2x + 5}{x^3 - 7x^2 + 9x - 3}$$

is not a polynomial because the form of a polynomial involves only addition and subtraction of terms, not division.

We can tie together the threads of our discussion with the formal definition of a polynomial in one variable. In this definition, the coefficients of the terms are represented by a_n (read "a sub n"), a_{n-1} (read "a sub n minus 1"), a_{n-2}, and so on. The small letters to the lower right of each a are called **subscripts** and are *not* *exponents*. Subscripts are used to distinguish one constant from another when a large and undetermined number of such constants are needed.

Definition of a Polynomial in x

A **polynomial in x** is an algebraic expression of the form

$$a_n x^n + a_{n-1} x^{n-1} + a_{n-2} x^{n-2} + \cdots + a_1 x + a_0,$$

where $a_n, a_{n-1}, a_{n-2}, \ldots, a_1$, and a_0 are real numbers, $a_n \neq 0$, and n is a non-negative integer. The polynomial is of **degree n**, a_n is the **leading coefficient**, and a_0 is the **constant term**.

---

② Add and subtract polynomials.

Adding and Subtracting Polynomials

Polynomials are added and subtracted by combining like terms. For example, we can combine the monomials $-9x^3$ and $13x^3$ using addition as follows:

$$-9x^3 + 13x^3 = (-9 + 13)x^3 = 4x^3.$$

These like terms both contain x to the third power. Add coefficients and keep the same variable factor, x^3.

EXAMPLE 1 Adding and Subtracting Polynomials

Perform the indicated operations and simplify:

a. $(-9x^3 + 7x^2 - 5x + 3) + (13x^3 + 2x^2 - 8x - 6)$
b. $(7x^3 - 8x^2 + 9x - 6) - (2x^3 - 6x^2 - 3x + 9)$.

Solution

a. $(-9x^3 + 7x^2 - 5x + 3) + (13x^3 + 2x^2 - 8x - 6)$

$= (-9x^3 + 13x^3) + (7x^2 + 2x^2) + (-5x - 8x) + (3 - 6)$ Group like terms.

$= 4x^3 + 9x^2 + (-13x) + (-3)$ Combine like terms.

$= 4x^3 + 9x^2 - 13x - 3$ Simplify.

b. $(7x^3 - 8x^2 + 9x - 6) - (2x^3 - 6x^2 - 3x + 9)$

 Change the sign of each coefficient.

$= (7x^3 - 8x^2 + 9x - 6) + (-2x^3 + 6x^2 + 3x - 9)$ Rewrite subtraction as addition of the additive inverse.

$= (7x^3 - 2x^3) + (-8x^2 + 6x^2)$
$\quad + (9x + 3x) + (-6 - 9)$ Group like terms.

$= 5x^3 + (-2x^2) + 12x + (-15)$ Combine like terms.

$= 5x^3 - 2x^2 + 12x - 15$ Simplify.

Study Tip

You can also arrange like terms in columns and combine vertically:

$$
\begin{array}{r}
7x^3 - 8x^2 + 9x - 6 \\
-2x^3 + 6x^2 + 3x - 9 \\
\hline
5x^3 - 2x^2 + 12x - 15
\end{array}
$$

The like terms can be combined by adding their coefficients and keeping the same variable factor.

Check Point 1 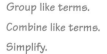 Perform the indicated operations and simplify:

a. $(-17x^3 + 4x^2 - 11x - 5) + (16x^3 - 3x^2 + 3x - 15)$
b. $(13x^3 - 9x^2 - 7x + 1) - (-7x^3 + 2x^2 - 5x + 9)$.

③ Multiply polynomials.

Study Tip

Don't confuse adding and multiplying monomials.

Addition:
$$5x^4 + 6x^4 = 11x^4$$

Multiplication:
$$(5x^4)(6x^4) = (5 \cdot 6)(x^4 \cdot x^4)$$
$$= 30x^{4+4}$$
$$= 30x^8$$

Only like terms can be added or subtracted, but unlike terms may be multiplied.

Addition:
$5x^4 + 3x^2$ cannot be simplified.

Multiplication:
$$(5x^4)(3x^2) = (5 \cdot 3)(x^4 \cdot x^2)$$
$$= 15x^{4+2}$$
$$= 15x^6$$

Multiplying Polynomials

The product of two monomials is obtained by using properties of exponents. For example,

$$(-8x^6)(5x^3) = -8 \cdot 5x^{6+3} = -40x^9.$$

Multiply coefficients and add exponents.

Furthermore, we can use the distributive property to multiply a monomial and a polynomial that is not a monomial. For example,

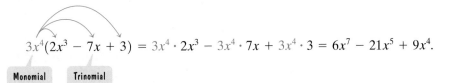

$$3x^4(2x^3 - 7x + 3) = 3x^4 \cdot 2x^3 - 3x^4 \cdot 7x + 3x^4 \cdot 3 = 6x^7 - 21x^5 + 9x^4.$$

Monomial   Trinomial

How do we multiply two polynomials if neither is a monomial? For example, consider

$$(2x + 3)(x^2 + 4x + 5).$$

Binomial   Trinomial

One way to perform this multiplication is to distribute $2x$ throughout the trinomial

$$2x(x^2 + 4x + 5)$$

and 3 throughout the trinomial

$$3(x^2 + 4x + 5).$$

Then combine the like terms that result.

> ### Multiplying Polynomials When Neither Is a Monomial
> Multiply each term of one polynomial by each term of the other polynomial. Then combine like terms.

EXAMPLE 2   Multiplying a Binomial and a Trinomial

Multiply: $(2x + 3)(x^2 + 4x + 5).$

Solution

$(2x + 3)(x^2 + 4x + 5)$

$= 2x(x^2 + 4x + 5) + 3(x^2 + 4x + 5)$ Multiply the trinomial by each term of the binomial.

$= 2x \cdot x^2 + 2x \cdot 4x + 2x \cdot 5 + 3x^2 + 3 \cdot 4x + 3 \cdot 5$ Use the distributive property.

$= 2x^3 + 8x^2 + 10x + 3x^2 + 12x + 15$ Multiply monomials: Multiply coefficients and add exponents.

$= 2x^3 + 11x^2 + 22x + 15$ Combine like terms: $8x^2 + 3x^2 = 11x^2$ and $10x + 12x = 22x.$

Another method for solving Example 2 is to use a vertical format similar to that used for multiplying whole numbers.

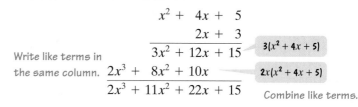

$$
\begin{array}{r}
x^2 + 4x + 5 \\
2x + 3 \\
\hline
3x^2 + 12x + 15 \\
2x^3 + 8x^2 + 10x \\
\hline
2x^3 + 11x^2 + 22x + 15
\end{array}
$$

Write like terms in the same column.

$3(x^2 + 4x + 5)$

$2x(x^2 + 4x + 5)$

Combine like terms.

Check Point 2 Multiply: $(5x - 2)(3x^2 - 5x + 4)$.

④ Use FOIL in polynomial multiplication.

The Product of Two Binomials: FOIL

Frequently, we need to find the product of two binomials. One way to perform this multiplication is to distribute each term in the first binomial through the second binomial. For example, we can find the product of the binomials $3x + 2$ and $4x + 5$ as follows:

$$(3x + 2)(4x + 5) = 3x(4x + 5) + 2(4x + 5)$$

Distribute $3x$ over $4x + 5$.

Distribute 2 over $4x + 5$.

$$= 3x(4x) + 3x(5) + 2(4x) + 2(5)$$

$$= 12x^2 + 15x + 8x + 10.$$

We can also find the product of $3x + 2$ and $4x + 5$ using a method called FOIL, which is based on our work shown above. Any two binomials can be quickly multiplied by using the FOIL method, in which **F** represents the product of the **first** terms in each binomial, **O** represents the product of the **outside** terms, **I** represents the product of the **inside** terms, and **L** represents the product of the **last**, or second terms in each binomial. For example, we can use the FOIL method to find the product of the binomials $3x + 2$ and $4x + 5$ as follows:

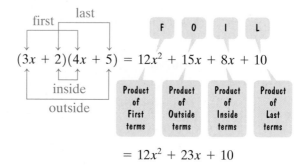

$$(3x + 2)(4x + 5) = 12x^2 + 15x + 8x + 10$$

F — Product of First terms
O — Product of Outside terms
I — Product of Inside terms
L — Product of Last terms

$$= 12x^2 + 23x + 10$$

Combine like terms.

In general, here's how to use the FOIL method to find the product of $ax + b$ and $cx + d$:

Using the FOIL Method to Multiply Binomials

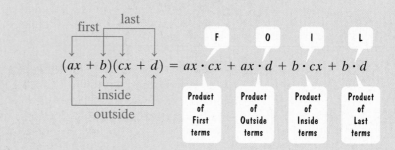

$$(ax + b)(cx + d) = ax \cdot cx + ax \cdot d + b \cdot cx + b \cdot d$$

F — Product of First terms
O — Product of Outside terms
I — Product of Inside terms
L — Product of Last terms

EXAMPLE 3   Using the FOIL Method

Multiply: $(3x + 4)(5x - 3)$.

Solution

$$(3x + 4)(5x - 3) = \underset{F}{3x \cdot 5x} + \underset{O}{3x(-3)} + \underset{I}{4 \cdot 5x} + \underset{L}{4(-3)}$$
$$= 15x^2 - 9x + 20x - 12$$
$$= 15x^2 + 11x - 12 \qquad \textit{Combine like terms.}$$

first   last

inside

outside

Check Point **3** Multiply: $(7x - 5)(4x - 3)$.

⑤ Use special products in polynomial multiplication.

Special Products

There are several products that occur so frequently that it's convenient to memorize the form, or pattern, of these formulas.

Special Products

Let A and B represent real numbers, variables, or algebraic expressions.

Special Product	**Example**
Sum and Difference of Two Terms	
$(A + B)(A - B) = A^2 - B^2$	$(2x + 3)(2x - 3) = (2x)^2 - 3^2$ $= 4x^2 - 9$
Squaring a Binomial	
$(A + B)^2 = A^2 + 2AB + B^2$	$(y + 5)^2 = y^2 + 2 \cdot y \cdot 5 + 5^2$ $= y^2 + 10y + 25$
$(A - B)^2 = A^2 - 2AB + B^2$	$(3x - 4)^4$ $= (3x)^2 - 2 \cdot 3x \cdot 4 + 4^2$ $= 9x^2 - 24x + 16$
Cubing a Binomial	
$(A + B)^3 = A^3 + 3A^2B + 3AB^2 + B^3$	$(x + 4)^3$ $= x^3 + 3x^2(4) + 3x(4)^2 + 4^3$ $= x^3 + 12x^2 + 48x + 64$
$(A - B)^3 = A^3 - 3A^2B + 3AB^2 - B^3$	$(x - 2)^3$ $= x^3 - 3x^2(2) + 3x(2)^2 - 2^3$ $= x^3 - 6x^2 + 12x - 8$

Study Tip

Although it's convenient to memorize these forms, the FOIL method can be used on all five examples in the box. To cube $x + 4$, you can first square $x + 4$ using FOIL and then multiply this result by $x + 4$. In short, you do not necessarily have to utilize these special formulas. What is the advantage of knowing and using these forms?

⑥ Perform operations with polynomials in several variables.

Polynomials in Several Variables

The next time you visit the lumber yard and go rummaging through piles of wood, think *polynomials*, although polynomials a bit different from those we have encountered so far. The forestry industry uses a polynomial in two variables to determine the number of board feet that can be manufactured from a tree with a diameter of x inches and a length of y feet. This polynomial is

$$\tfrac{1}{4}x^2y - 2xy + 4y.$$

In general, a **polynomial in two variables**, x and y, contains the sum of one or more monomials in the form $ax^n y^m$. The constant, a, is the **coefficient**. The exponents, n and m, represent whole numbers. The **degree** of the monomial $ax^n y^m$ is $n + m$. We'll use the polynomial from the forestry industry to illustrate these ideas.

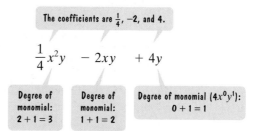

The coefficients are $\frac{1}{4}$, -2, and **4**.

$$\frac{1}{4} x^2 y \quad - 2xy \quad + 4y$$

Degree of monomial: $2 + 1 = 3$

Degree of monomial: $1 + 1 = 2$

Degree of monomial $(4x^0 y^1)$: $0 + 1 = 1$

The **degree of a polynomial in two variables** is the highest degree of all its terms. For the preceding polynomial, the degree is 3.

Polynomials containing two or more variables can be added, subtracted, and multiplied just like polynomials that contain only one variable. For example, we can add the monomials $-7xy^2$ and $13xy^2$ as follows:

$$-7xy^2 + 13xy^2 = (-7 + 13)xy^2 = 6xy^2.$$

These like terms both contain the variable factors x and y^2.

Add coefficients and keep the same variable factors, xy^2.

EXAMPLE 4   Multiplying Polynomials in Two Variables

Multiply: **a.** $(x + 4y)(3x - 5y)$ **b.** $(5x + 3y)^2$.

Solution We will perform the multiplication in part (a) using the FOIL method. We will multiply in part (b) using the formula for the square of a binomial sum, $(A + B)^2$.

a. $(x + 4y)(3x - 5y)$ Multiply these binomials using the FOIL method.

$$\begin{aligned}
&\overset{F}{} \quad \overset{O}{} \quad \overset{I}{} \quad \overset{L}{} \\
&= (x)(3x) + (x)(-5y) + (4y)(3x) + (4y)(-5y) \\
&= 3x^2 - 5xy + 12xy - 20y^2 \\
&= 3x^2 + 7xy - 20y^2 \quad \text{Combine like terms.}
\end{aligned}$$

$$(A + B)^2 \;=\; A^2 \;+\; 2 \cdot A \cdot B \;+\; B^2$$

b. $(5x + 3y)^2 = (5x)^2 + 2(5x)(3y) + (3y)^2$
$$= 25x^2 + 30xy + 9y^2$$

Check Point 4   Multiply:

a. $(7x - 6y)(3x - y)$ **b.** $(2x + 4y)^2$.

Special products can sometimes be used to find the products of certain trinomials, as illustrated in Example 5.

EXAMPLE 5   Using the Special Products

Multiply:

a. $(7x + 5 + 4y)(7x + 5 - 4y)$ **b.** $(3x + y + 1)^2$.

Solution

a. By grouping the first two terms within each of the parentheses, we can find the product using the form for the sum and difference of two terms.

$$(A + B) \cdot (A - B) = A^2 - B^2$$

$$\begin{aligned}[(7x + 5) + 4y] \cdot [(7x + 5) - 4y] &= (7x + 5)^2 - (4y)^2 \\ &= (7x)^2 + 2 \cdot 7x \cdot 5 + 5^2 - (4y)^2 \\ &= 49x^2 + 70x + 25 - 16y^2\end{aligned}$$

b. We can group the terms of $(3x + y + 1)^2$ so that the formula for the square of a binomial can be applied.

$$(A + B)^2 = A^2 + 2 \cdot A \cdot B + B^2$$

$$\begin{aligned}[(3x + y) + 1]^2 &= (3x + y)^2 + 2 \cdot (3x + y) \cdot 1 + 1^2 \\ &= 9x^2 + 6xy + y^2 + 6x + 2y + 1\end{aligned}$$

Check Point 5 Multiply:

a. $(3x + 2 + 5y)(3x + 2 - 5y)$ **b.** $(2x + y + 3)^2$.

EXERCISE SET P.4

Practice Exercises

In Exercises 1–4, is the algebraic expression a polynomial? If it is, write the polynomial in standard form.

1. $2x + 3x^2 - 5$ **2.** $2x + 3x^{-1} - 5$

3. $\dfrac{2x + 3}{x}$ **4.** $x^2 - x^3 + x^4 - 5$

In Exercises 5–8, find the degree of the polynomial.

5. $3x^2 - 5x + 4$ **6.** $-4x^3 + 7x^2 - 11$

7. $x^2 - 4x^3 + 9x - 12x^4 + 63$

8. $x^2 - 8x^3 + 15x^4 + 91$

In Exercises 9–14, perform the indicated operations. Write the resulting polynomial in standard form and indicate its degree.

9. $(-6x^3 + 5x^2 - 8x + 9) + (17x^3 + 2x^2 - 4x - 13)$

10. $(-7x^3 + 6x^2 - 11x + 13) + (19x^3 - 11x^2 + 7x - 17)$

11. $(17x^3 - 5x^2 + 4x - 3) - (5x^3 - 9x^2 - 8x + 11)$

12. $(18x^4 - 2x^3 - 7x + 8) - (9x^4 - 6x^3 - 5x + 7)$

13. $(5x^2 - 7x - 8) + (2x^2 - 3x + 7) - (x^2 - 4x - 3)$

14. $(8x^2 + 7x - 5) - (3x^2 - 4x) - (-6x^3 - 5x^2 + 3)$

In Exercises 15–82, find each product.

15. $(x + 1)(x^2 - x + 1)$ **16.** $(x + 5)(x^2 - 5x + 25)$

17. $(2x - 3)(x^2 - 3x + 5)$ **18.** $(2x - 1)(x^2 - 4x + 3)$

19. $(x + 7)(x + 3)$ **20.** $(x + 8)(x + 5)$

21. $(x - 5)(x + 3)$ **22.** $(x - 1)(x + 2)$

23. $(3x + 5)(2x + 1)$ **24.** $(7x + 4)(3x + 1)$

25. $(2x - 3)(5x + 3)$ **26.** $(2x - 5)(7x + 2)$

27. $(5x^2 - 4)(3x^2 - 7)$ **28.** $(7x^2 - 2)(3x^2 - 5)$

29. $(8x^3 + 3)(x^2 - 5)$ **30.** $(7x^3 + 5)(x^2 - 2)$

31. $(x + 3)(x - 3)$ **32.** $(x + 5)(x - 5)$

33. $(3x + 2)(3x - 2)$ **34.** $(2x + 5)(2x - 5)$

35. $(5 - 7x)(5 + 7x)$ **36.** $(4 - 3x)(4 + 3x)$

37. $(4x^2 + 5x)(4x^2 - 5x)$ **38.** $(3x^2 + 4x)(3x^2 - 4x)$

39. $(1 - y^5)(1 + y^5)$ **40.** $(2 - y^5)(2 + y^5)$

41. $(x + 2)^2$ **42.** $(x + 5)^2$

43. $(2x + 3)^2$ **44.** $(3x + 2)^2$

45. $(x - 3)^2$ **46.** $(x - 4)^2$

47. $(4x^2 - 1)^2$ **48.** $(5x^2 - 3)^2$

49. $(7 - 2x)^2$

50. $(9 - 5x)^2$

51. $(x + 1)^3$

52. $(x + 2)^3$

53. $(2x + 3)^3$

54. $(3x + 4)^3$

55. $(x - 3)^3$

56. $(x - 1)^3$

57. $(3x - 4)^3$

58. $(2x - 3)^3$

59. $(x + 5y)(7x + 3y)$

60. $(x + 9y)(6x + 7y)$

61. $(x - 3y)(2x + 7y)$

62. $(3x - y)(2x + 5y)$

63. $(3xy - 1)(5xy + 2)$

64. $(7x^2y + 1)(2x^2y - 3)$

65. $(7x + 5y)^2$

66. $(9x + 7y)^2$

67. $(x^2y^2 - 3)^2$

68. $(x^2y^2 - 5)^2$

69. $(x - y)(x^2 + xy + y^2)$

70. $(x + y)(x^2 - xy + y^2)$

71. $(3x + 5y)(3x - 5y)$

72. $(7x + 3y)(7x - 3y)$

73. $(x + y + 3)(x + y - 3)$

74. $(x + y + 5)(x + y - 5)$

75. $(3x + 7 - 5y)(3x + 7 + 5y)$

76. $(5x + 7y - 2)(5x + 7y + 2)$

77. $[5y - (2x + 3)][5y + (2x + 3)]$

78. $[8y + (7 - 3x)][8y - (7 - 3x)]$

79. $(x + y + 1)^2$

80. $(x + y + 2)^2$

81. $(2x + y + 1)^2$

82. $(5x + 1 + 6y)^2$

Practice Plus

In Exercises 83–90, perform the indicated operation or operations.

83. $(3x + 4y)^2 - (3x - 4y)^2$ **84.** $(5x + 2y)^2 - (5x - 2y)^2$

85. $(5x - 7)(3x - 2) - (4x - 5)(6x - 1)$

86. $(3x + 5)(2x - 9) - (7x - 2)(x - 1)$

87. $(2x + 5)(2x - 5)(4x^2 + 25)$

88. $(3x + 4)(3x - 4)(9x^2 + 16)$

89. $\dfrac{(2x - 7)^5}{(2x - 7)^3}$

90. $\dfrac{(5x - 3)^6}{(5x - 3)^4}$

Application Exercises

The bar graph shows the number of people in the United States, in millions, who do yoga.

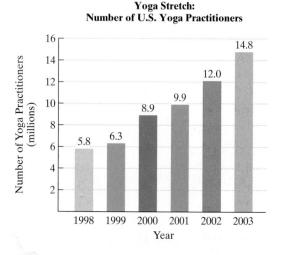

Yoga Stretch:
Number of U.S. Yoga Practitioners

Source: Yoga Journal

Here are four mathematical models for the data shown in the graph. In each formula, N represents the number of U.S. yoga practitioners, in millions, x years after 1998.

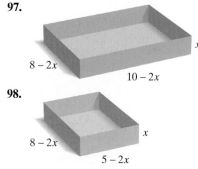

Model 1 $N = 1.8x + 5.1$

Model 2 $N = 5.6(1.2)^x$

Model 3 $N = 0.17x^2 + 0.95x + 5.68$

Model 4 $N = 0.09x^2 + 0.01x^3 + 1.1x + 5.64$

Use these models to solve Exercises 91–96.

91. Which model uses a polynomial that is not in standard form? Rewrite the model in standard form.

92. If x is any real number from 0 to 5, inclusive, which model does not use a polynomial?

93. Which model best describes the data for 2000?

94. Which model best describes the data for 1998?

95. How well does the model of degree 2 describe the data for 2003?

96. How well does the polynomial model that is not in standard form describe the data for 2002?

In Exercises 97–98, write a polynomial in standard form that models, or represents, the volume of the open box.

97.

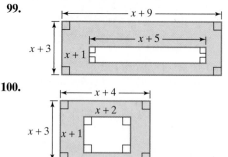

98.

In Exercises 99–100, write a polynomial in standard form that models, or represents, the area of the shaded region.

99.

100.

Writing in Mathematics

101. What is a polynomial in x?

102. Explain how to subtract polynomials.

103. Explain how to multiply two binomials using the FOIL method. Give an example with your explanation.

104. Explain how to find the product of the sum and difference of two terms. Give an example with your explanation.

105. Explain how to square a binomial difference. Give an example with your explanation.

106. Explain how to find the degree of a polynomial in two variables.

107. In the section opener, we used the mathematical model
$$N = -365x^4 + 2728x^3 - 7106x^2 + 7372x + 20{,}787$$
to describe the number of newly-declared computer majors, N, for the fall term x years after 1999. Use a calculator to determine these numbers from 1999 through 2003. Compare your results with the data shown in Figure P.11 on page 42. Describe what you observe.

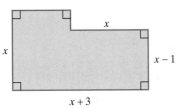

Critical Thinking Exercises

108. Express the area of the plane figure shown as a polynomial in standard form.

In Exercises 109–110, represent the volume of each figure as a polynomial in standard form.

109.

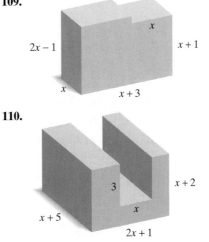

110.

111. Simplify: $(y^n + 2)(y^n - 2) - (y^n - 3)^2$.

SECTION P.5   *Factoring Polynomials*

Objectives

1 Factor out the greatest common factor of a polynomial.

2 Factor by grouping.

3 Factor trinomials.

4 Factor the difference of squares.

5 Factor perfect square trinomials.

6 Factor the sum and difference of two cubes.

7 Use a general strategy for factoring polynomials.

8 Factor algebraic expressions containing fractional and negative exponents.

A two-year-old boy is asked, "Do you have a brother?" He answers, "Yes." "What is your brother's name?" "Tom." Asked if Tom has a brother, the two-year-old replies, "No." The child can go in the direction from self to brother, but he cannot reverse this direction and move from brother back to self.

As our intellects develop, we learn to reverse the direction of our thinking. Reversibility of thought is found throughout algebra. For example, we can multiply polynomials and show that
$$5x(2x + 3) = 10x^2 + 15x.$$
We can also reverse this process and express the resulting polynomial as
$$10x^2 + 15x = 5x(2x + 3).$$

Factoring a polynomial containing the sum of monomials means finding an equivalent expression that is a product.

Factoring $10x^2 + 15x$

Sum of monomials

Equivalent expression that is a product

$$10x^2 + 15x = 5x(2x + 3)$$

The factors of $10x^2 + 15x$ are $5x$ and $2x + 3$.

In this section, we will be **factoring over the set of integers**, meaning that the coefficients in the factors are integers. Polynomials that cannot be factored using integer coefficients are called **irreducible over the integers**, or **prime**.

The goal in factoring a polynomial is to use one or more factoring techniques until each of the polynomial's factors, except possibly for a monomial factor, is prime or irreducible. In this situation, the polynomial is said to be **factored completely**.

We will now discuss basic techniques for factoring polynomials.

 Factor out the greatest common factor of a polynomial.

Common Factors

In any factoring problem, the first step is to look for the *greatest common factor*. The **greatest common factor**, abbreviated GCF, is an expression of the highest degree that divides each term of the polynomial. The distributive property in the reverse direction

$$ab + ac = a(b + c)$$

can be used to factor out the greatest common factor.

Study Tip

The variable part of the greatest common factor always contains the *smallest* power of a variable or algebraic expression that appears in all terms of the polynomial.

EXAMPLE 1 Factoring out the Greatest Common Factor

Factor: **a.** $18x^3 + 27x^2$ **b.** $x^2(x + 3) + 5(x + 3)$.

Solution

a. First, determine the greatest common factor.

> 9 is the greatest integer that divides 18 and 27.

$$18x^3 + 27x^2$$

> x^2 is the greatest expression that divides x^3 and x^2.

The GCF of the two terms of the polynomial is $9x^2$.

$$18x^3 + 27x^2$$
$$= 9x^2(2x) + 9x^2(3) \quad \text{Express each term as the product of the GCF and its other factor.}$$
$$= 9x^2(2x + 3) \quad \text{Factor out the GCF.}$$

b. In this situation, the greatest common factor is the common binomial factor $(x + 3)$. We factor out this common factor as follows:

$$x^2(x + 3) + 5(x + 3) = (x + 3)(x^2 + 5). \quad \text{Factor out the common binomial factor.}$$

Check Point 1 Factor: **a.** $10x^3 - 4x^2$ **b.** $2x(x - 7) + 3(x - 7)$.

 Factor by grouping.

Factoring by Grouping

Some polynomials have only a greatest common factor of 1. However, by a suitable grouping of the terms, it still may be possible to factor. This process, called **factoring by grouping**, is illustrated in Example 2.

EXAMPLE 2 Factoring by Grouping

Factor: $x^3 + 4x^2 + 3x + 12$.

Solution   There is no factor other than 1 common to all terms. However, we can group terms that have a common factor:

$$\boxed{x^3 + 4x^2} + \boxed{3x + 12}.$$

Discovery

In Example 2, group the terms as follows:

$$(x^3 + 3x) + (4x^2 + 12).$$

Factor out the greatest common factor from each group and complete the factoring process. Describe what happens. What can you conclude?

We now factor the given polynomial as follows:

$$
\begin{aligned}
&x^3 + 4x^2 + 3x + 12 \\
&= (x^3 + 4x^2) + (3x + 12) \quad\text{Group terms with common factors.} \\
&= x^2(x + 4) + 3(x + 4) \quad\text{Factor out the greatest common factor from the grouped terms. The remaining two terms have } x + 4 \text{ as a common binomial factor.} \\
&= (x + 4)(x^2 + 3). \quad\text{Factor out the GCF, } x + 4.
\end{aligned}
$$

Thus, $x^3 + 4x^2 + 3x + 12 = (x + 4)(x^2 + 3)$. Check the factorization by multiplying the right side of the equation using the FOIL method. Because the factorization is correct, you should obtain the original polynomial.

Check Point 2 Factor: $x^3 + 5x^2 - 2x - 10$.

③ Factor trinomials.

Factoring Trinomials

To factor a trinomial of the form $ax^2 + bx + c$, a little trial and error may be necessary.

A Strategy for Factoring $ax^2 + bx + c$

Assume, for the moment, that there is no greatest common factor.

1. Find two First terms whose product is ax^2:

$$(\Box x + \quad)(\Box x + \quad) = ax^2 + bx + c.$$

2. Find two Last terms whose product is c:

$$(\Box x + \Box)(\Box x + \Box) = ax^2 + bx + c.$$

3. By trial and error, perform steps 1 and 2 until the sum of the Outside product and the Inside product is bx:

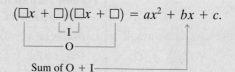

If no such combination exists, the polynomial is prime.

Study Tip

The *error* part of the factoring strategy plays an important role in the process. If you do not get the correct factorization the first time, this is not a bad thing. This error is often helpful in leading you to the correct factorization.

EXAMPLE 3 Factoring Trinomials Whose Leading Coefficients Are 1

Factor: **a.** $x^2 + 6x + 8$ **b.** $x^2 + 3x - 18$.

Solution

a. The factors of the first term are x and x:

$$x^2 + 6x + 8 = (x \quad)(x \quad).$$

To find the second term of each factor, we must find two integers whose product is 8 and whose sum is 6. From the table in the margin, we see that 4 and 2 are the required integers. Thus,

$$x^2 + 6x + 8 = (x + 4)(x + 2) \text{ or } (x + 2)(x + 4).$$

Factors of 8	8, 1	4, 2	−8, −1	−4, −2
Sum of Factors	9	6	−9	−6

This is the desired sum.

We have seen that a polynomial is factored completely when it is written as the product of prime polynomials. To be sure that you have factored completely, check to see whether any factors with more than one term in the factored polynomial can be factored further. If so, continue factoring.

EXAMPLE 7   A Repeated Factorization

Factor completely: $x^4 - 81$.

Solution

$$x^4 - 81 = (x^2)^2 - 9^2$$ Express as the difference of two squares.

$$= (x^2 + 9)(x^2 - 9)$$ The factors are the sum and the difference of the expressions being squared.

$$= (x^2 + 9)(x^2 - 3^2)$$ The factor $x^2 - 9$ is the difference of two squares and can be factored.

$$= (x^2 + 9)(x + 3)(x - 3)$$ The factors of $x^2 - 9$ are the sum and the difference of the expressions being squared.

Are you tempted to further factor $x^2 + 9$, the sum of two squares, in Example 7? Resist the temptation! **The sum of two squares, $A^2 + B^2$, with no common factor other than 1 is a prime polynomial over the integers**.

Check Point 7   Factor completely: $81x^4 - 16$.

 Factor perfect square trinomials.

Factoring Perfect Square Trinomials

Our next factoring technique is obtained by reversing the special products for squaring binomials. The trinomials that are factored using this technique are called **perfect square trinomials**.

Factoring Perfect Square Trinomials

Let A and B be real numbers, variables, or algebraic expressions.

1. $A^2 + 2AB + B^2 = (A + B)^2$

Same sign

2. $A^2 - 2AB + B^2 = (A - B)^2$

Same sign

The two items in the box show that perfect square trinomials come in two forms: one in which the coefficient of the middle term is positive and one in which the coefficient of the middle term is negative. Here's how to recognize a perfect square trinomial:

1. The first and last terms are squares of monomials or integers.
2. The middle term is twice the product of the expressions being squared in the first and last terms.

EXAMPLE 8   Factoring Perfect Square Trinomials

Factor: **a.** $x^2 + 6x + 9$ **b.** $25x^2 - 60x + 36$.

Solution

a. $x^2 + 6x + 9 = x^2 + 2 \cdot x \cdot 3 + 3^2 = (x + 3)^2$ The middle term has a positive sign.

$A^2 \ + \ 2AB \ + \ B^2 \ = \ (A \ + \ B)^2$

b. We suspect that $25x^2 - 60x + 36$ is a perfect square trinomial because $25x^2 = (5x)^2$ and $36 = 6^2$. The middle term can be expressed as twice the product of $5x$ and 6.

$$25x^2 - 60x + 36 = (5x)^2 - 2 \cdot 5x \cdot 6 + 6^2 = (5x - 6)^2$$

$$A^2 \ - \ 2AB \ + \ B^2 \ = \ (A \ - \ B)^2$$

Check Point 8   Factor: **a.** $x^2 + 14x + 49$ **b.** $16x^2 - 56x + 49.$

⑥ Factor the sum and difference of two cubes.

Factoring the Sum and Difference of Two Cubes

We can use the following formulas to factor the sum or the difference of two cubes:

Factoring the Sum and Difference of Two Cubes

1. Factoring the Sum of Two Cubes

$$A^3 + B^3 = (A + B)(A^2 - AB + B^2)$$

Same signs   Opposite signs

2. Factoring the Difference of Two Cubes

$$A^3 - B^3 = (A - B)(A^2 + AB + B^2)$$

Same signs   Opposite signs

EXAMPLE 9   Factoring Sums and Differences of Two Cubes

Factor: **a.** $x^3 + 8$ **b.** $64x^3 - 125.$

Solution

a. To factor $x^3 + 8$, we must express each term as the cube of some monomial. Then we use the formula for factoring $A^3 + B^3$.

$$x^3 + 8 = x^3 + 2^3 = (x + 2)(x^2 - x \cdot 2 + 2^2) = (x + 2)(x^2 - 2x + 4)$$

$$A^3 \ + \ B^3 \ = \ (A \ + \ B) \ (A^2 \ - \ AB \ + \ B^2)$$

b. To factor $64x^3 - 125$, we must express each term as the cube of some monomial. Then use the formula for factoring $A^3 - B^3$.

$$64x^3 - 125 = (4x)^3 - 5^3 = (4x - 5)[(4x)^2 + (4x)(5) + 5^2]$$

$$A^3 \ - \ B^3 \ = \ (A \ - \ B) \ (A^2 \ + \ AB \ + \ B^2)$$

$$= (4x - 5)(16x^2 + 20x + 25)$$

Check Point 9   Factor: **a.** $x^3 + 1$ **b.** $125x^3 - 8.$

Use a general strategy for factoring polynomials.

A Strategy for Factoring Polynomials

It is important to practice factoring a wide variety of polynomials so that you can quickly select the appropriate technique. The polynomial is factored completely when all its polynomial factors, except possibly for monomial factors, are prime. Because of the commutative property, the order of the factors does not matter.

A Strategy for Factoring a Polynomial

1. If there is a common factor, factor out the GCF.

2. Determine the number of terms in the polynomial and try factoring as follows:

  a. If there are two terms, can the binomial be factored by one of the following special forms?

$$\text{Difference of two squares: } A^2 - B^2 = (A + B)(A - B)$$
$$\text{Sum of two cubes: } A^3 + B^3 = (A + B)(A^2 - AB + B^2)$$
$$\text{Difference of two cubes: } A^3 - B^3 = (A - B)(A^2 + AB + B^2)$$

  b. If there are three terms, is the trinomial a perfect square trinomial? If so, factor by one of the following special forms:

$$A^2 + 2AB + B^2 = (A + B)^2$$
$$A^2 - 2AB + B^2 = (A - B)^2$$

  If the trinomial is not a perfect square trinomial, try factoring by trial and error.

  c. If there are four or more terms, try factoring by grouping.

3. Check to see if any factors with more than one term in the factored polynomial can be factored further. If so, factor completely.

EXAMPLE 10   Factoring a Polynomial

Factor: $2x^3 + 8x^2 + 8x$.

Solution

Step 1   If there is a common factor, factor out the GCF. Because $2x$ is common to all terms, we factor it out.

$$2x^3 + 8x^2 + 8x = 2x(x^2 + 4x + 4) \qquad \text{\textit{Factor out the GCF.}}$$

Step 2   Determine the number of terms and factor accordingly. The factor $x^2 + 4x + 4$ has three terms and is a perfect square trinomial. We factor using $A^2 + 2AB + B^2 = (A + B)^2$.

$$2x^3 + 8x^2 + 8x = 2x(x^2 + 4x + 4)$$
$$= 2x(x^2 + 2 \cdot x \cdot 2 + 2^2)$$
$$\underbrace{\qquad}_{A^2 \quad + \quad 2AB \quad + \quad B^2}$$
$$= 2x(x + 2)^2 \qquad A^2 + 2AB + B^2 = (A + B)^2$$

Step 3   Check to see if factors can be factored further. In this problem, they cannot. Thus,

$$2x^3 + 8x^2 + 8x = 2x(x + 2)^2.$$

Check Point **10**   Factor: $3x^3 - 30x^2 + 75x$.

EXAMPLE 11 Factoring a Polynomial

Factor: $x^2 - 25a^2 + 8x + 16$.

Solution

Step 1 If there is a common factor, factor out the GCF. Other than 1 or -1, there is no common factor.

Step 2 Determine the number of terms and factor accordingly. There are four terms. We try factoring by grouping. Grouping into two groups of two terms does not result in a common binomial factor. Let's try grouping as a difference of squares.

$$x^2 - 25a^2 + 8x + 16$$

$$= (x^2 + 8x + 16) - 25a^2$$ Rearrange terms and group as a perfect square trinomial minus $25a^2$ to obtain a difference of squares.

$$= (x + 4)^2 - (5a)^2$$ Factor the perfect square trinomial.

$$= (x + 4 + 5a)(x + 4 - 5a)$$ Factor the difference of squares. The factors are the sum and difference of the expressions being squared.

Step 3 Check to see if factors can be factored further. In this case, they cannot, so we have factored completely.

Check Point 11 Factor: $x^2 - 36a^2 + 20x + 100$.

8 Factor algebraic expressions containing fractional and negative exponents.

Factoring Algebraic Expressions Containing Fractional and Negative Exponents

Although expressions containing fractional and negative exponents are not polynomials, they can be simplified using factoring techniques.

EXAMPLE 12 Factoring Involving Fractional and Negative Exponents

Factor and simplify: $x(x + 1)^{-\frac{3}{4}} + (x + 1)^{\frac{1}{4}}$.

Solution The greatest common factor is $x + 1$ with the *smallest exponent* in the two terms. Thus, the greatest common factor is $(x + 1)^{-\frac{3}{4}}$.

$$x(x + 1)^{-\frac{3}{4}} + (x + 1)^{\frac{1}{4}}$$

$$= (x + 1)^{-\frac{3}{4}}x + (x + 1)^{-\frac{3}{4}}(x + 1)$$ Express each term as the product of the greatest common factor and its other factor.

$$= (x + 1)^{-\frac{3}{4}}[x + (x + 1)]$$ Factor out the greatest common factor.

$$= \frac{2x + 1}{(x + 1)^{\frac{3}{4}}}$$ $b^{-n} = \dfrac{1}{b^n}$

Check Point 12 Factor and simplify: $x(x - 1)^{-\frac{1}{2}} + (x - 1)^{\frac{1}{2}}$.

EXERCISE SET P.5

Practice Exercises

In Exercises 1–10, factor out the greatest common factor.

1. $18x + 27$ **2.** $16x - 24$

3. $3x^2 + 6x$ **4.** $4x^2 - 8x$

5. $9x^4 - 18x^3 + 27x^2$ **6.** $6x^4 - 18x^3 + 12x^2$

7. $x(x + 5) + 3(x + 5)$ **8.** $x(2x + 1) + 4(2x + 1)$

9. $x^2(x - 3) + 12(x - 3)$ **10.** $x^2(2x + 5) + 17(2x + 5)$

In Exercises 11–16, factor by grouping.

11. $x^3 - 2x^2 + 5x - 10$ **12.** $x^3 - 3x^2 + 4x - 12$

13. $x^3 - x^2 + 2x - 2$ **14.** $x^3 + 6x^2 - 2x - 12$

15. $3x^3 - 2x^2 - 6x + 4$ **16.** $x^3 - x^2 - 5x + 5$

In Exercises 17–38, factor each trinomial, or state that the trinomial is prime.

17. $x^2 + 5x + 6$ **18.** $x^2 + 8x + 15$

19. $x^2 - 2x - 15$ **20.** $x^2 - 4x - 5$

21. $x^2 - 8x + 15$ **22.** $x^2 - 14x + 45$

23. $3x^2 - x - 2$ **24.** $2x^2 + 5x - 3$

25. $3x^2 - 25x - 28$ **26.** $3x^2 - 2x - 5$

27. $6x^2 - 11x + 4$ **28.** $6x^2 - 17x + 12$

29. $4x^2 + 16x + 15$ **30.** $8x^2 + 33x + 4$

31. $9x^2 - 9x + 2$ **32.** $9x^2 + 5x - 4$

33. $20x^2 + 27x - 8$ **34.** $15x^2 - 19x + 6$

35. $2x^2 + 3xy + y^2$ **36.** $3x^2 + 4xy + y^2$

37. $6x^2 - 5xy - 6y^2$ **38.** $6x^2 - 7xy - 5y^2$

In Exercises 39–48, factor the difference of two squares.

39. $x^2 - 100$ **40.** $x^2 - 144$

41. $36x^2 - 49$ **42.** $64x^2 - 81$

43. $9x^2 - 25y^2$ **44.** $36x^2 - 49y^2$

45. $x^4 - 16$ **46.** $x^4 - 1$

47. $16x^4 - 81$ **48.** $81x^4 - 1$

In Exercises 49–56, factor each perfect square trinomial.

49. $x^2 + 2x + 1$ **50.** $x^2 + 4x + 4$

51. $x^2 - 14x + 49$ **52.** $x^2 - 10x + 25$

53. $4x^2 + 4x + 1$ **54.** $25x^2 + 10x + 1$

55. $9x^2 - 6x + 1$ **56.** $64x^2 - 16x + 1$

In Exercises 57–64, factor using the formula for the sum or difference of two cubes.

57. $x^3 + 27$ **58.** $x^3 + 64$

59. $x^3 - 64$ **60.** $x^3 - 27$

61. $8x^3 - 1$ **62.** $27x^3 - 1$

63. $64x^3 + 27$ **64.** $8x^3 + 125$

In Exercises 65–92, factor completely, or state that the polynomial is prime.

65. $3x^3 - 3x$ **66.** $5x^3 - 45x$

67. $4x^2 - 4x - 24$ **68.** $6x^2 - 18x - 60$

69. $2x^4 - 162$ **70.** $7x^4 - 7$

71. $x^3 + 2x^2 - 9x - 18$ **72.** $x^3 + 3x^2 - 25x - 75$

73. $2x^2 - 2x - 112$ **74.** $6x^2 - 6x - 12$

75. $x^3 - 4x$ **76.** $9x^3 - 9x$

77. $x^2 + 64$ **78.** $x^2 + 36$

79. $x^3 + 2x^2 - 4x - 8$ **80.** $x^3 + 2x^2 - x - 2$

81. $y^5 - 81y$ **82.** $y^5 - 16y$

83. $20y^4 - 45y^2$ **84.** $48y^4 - 3y^2$

85. $x^2 - 12x + 36 - 49y^2$ **86.** $x^2 - 10x + 25 - 36y^2$

87. $9b^2x - 16y - 16x + 9b^2y$

88. $16a^2x - 25y - 25x + 16a^2y$

89. $x^2y - 16y + 32 - 2x^2$ **90.** $12x^2y - 27y - 4x^2 + 9$

91. $2x^3 - 8a^2x + 24x^2 + 72x$

92. $2x^3 - 98a^2x + 28x^2 + 98x$

In Exercises 93–102, factor and simplify each algebraic expression.

93. $x^{\frac{3}{2}} - x^{\frac{1}{2}}$ **94.** $x^{\frac{3}{4}} - x^{\frac{1}{4}}$

95. $4x^{-\frac{2}{3}} + 8x^{\frac{1}{3}}$ **96.** $12x^{-\frac{3}{4}} + 6x^{\frac{1}{4}}$

97. $(x + 3)^{\frac{1}{2}} - (x + 3)^{\frac{3}{2}}$ **98.** $(x^2 + 4)^{\frac{3}{2}} + (x^2 + 4)^{\frac{7}{2}}$

99. $(x + 5)^{-\frac{1}{2}} - (x + 5)^{-\frac{3}{2}}$

100. $(x^2 + 3)^{-\frac{2}{3}} + (x^2 + 3)^{-\frac{5}{3}}$

101. $(4x - 1)^{\frac{1}{2}} - \frac{1}{3}(4x - 1)^{\frac{3}{2}}$

102. $-8(4x + 3)^{-2} + 10(5x + 1)(4x + 3)^{-1}$

Practice Plus

In Exercises 103–114, factor completely.

103. $10x^2(x + 1) - 7x(x + 1) - 6(x + 1)$

104. $12x^2(x - 1) - 4x(x - 1) - 5(x - 1)$

105. $6x^4 + 35x^2 - 6$ **106.** $7x^4 + 34x^2 - 5$

107. $y^7 + y$ **108.** $(y + 1)^3 + 1$

109. $x^4 - 5x^2y^2 + 4y^4$ **110.** $x^4 - 10x^2y^2 + 9y^4$

111. $(x - y)^4 - 4(x - y)^2$ **112.** $(x + y)^4 - 100(x + y)^2$

113. $2x^2 - 7xy^2 + 3y^4$ **114.** $3x^2 + 5xy^2 + 2y^4$

Application Exercises

115. Your computer store is having an incredible sale. The price on one model is reduced by 40%. Then the sale price is reduced by another 40%. If x is the computer's original price, the sale price can be represented by

$$(x - 0.4x) - 0.4(x - 0.4x).$$

a. Factor out $(x - 0.4x)$ from each term. Then simplify the resulting expression.

b. Use the simplified expression from part (a) to answer these questions. With a 40% reduction followed by a 40% reduction, is the computer selling at 20% of its original price? If not, at what percentage of the original price is it selling?

116. Your local electronics store is having an end-of-the-year sale. The price on a large-screen television had been reduced by 30%. Now the sale price is reduced by another 30%. If x is the television's original price, the sale price can be represented by

$$(x - 0.3x) - 0.3(x - 0.3x).$$

a. Factor out $(x - 0.3x)$ from each term. Then simplify the resulting expression.

b. Use the simplified expression from part (a) to answer these questions. With a 30% reduction followed by a 30% reduction, is the television selling at 40% of its original price? If not, at what percentage of the original price is it selling?

In Exercises 117–120,

   a. *Write an expression for the area of the shaded region.*

   b. *Write the expression in factored form.*

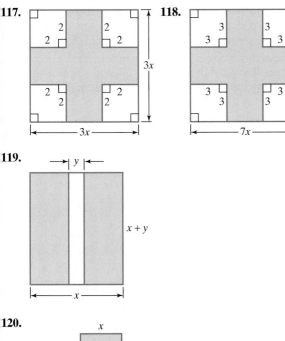

119.

120.

In Exercises 121–122, find the formula for the volume of the region outside the smaller rectangular solid and inside the larger rectangular solid. Then express the volume in factored form.

121.

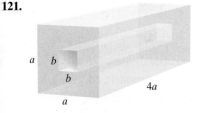

122.

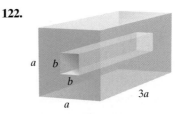

Writing in Mathematics

123. Using an example, explain how to factor out the greatest common factor of a polynomial.

124. Suppose that a polynomial contains four terms. Explain how to use factoring by grouping to factor the polynomial.

125. Explain how to factor $3x^2 + 10x + 8$.

126. Explain how to factor the difference of two squares. Provide an example with your explanation.

127. What is a perfect square trinomial and how is it factored?

128. Explain how to factor $x^3 + 1$.

129. What does it mean to factor completely?

Critical Thinking Exercises

130. Which one of the following is true?

  a. Because $x^2 + 1$ is irreducible over the integers, it follows that $x^3 + 1$ is also irreducible.

  b. One correct factored form for $x^2 - 4x + 3$ is $x(x - 4) + 3$.

  c. $x^3 - 64 = (x - 4)^3$

  d. None of the above is true.

In Exercises 131–134, factor completely.

131. $x^{2n} + 6x^n + 8$ **132.** $-x^2 - 4x + 5$

133. $x^4 - y^4 - 2x^3y + 2xy^3$

134. $(x - 5)^{-\frac{1}{2}}(x + 5)^{-\frac{1}{2}} - (x + 5)^{\frac{1}{2}}(x - 5)^{-\frac{3}{2}}$

In Exercises 135–136, find all integers b so that the trinomial can be factored.

135. $x^2 + bx + 15$ **136.** $x^2 + 4x + b$

Group Exercise

137. Divide the group in half. Without looking at any factoring problems in the book, each group should create five factoring problems. Make sure that some of your problems require at least two factoring strategies. Next, exchange problems with the other half of the group. Work to factor the five problems. After completing the factorizations, evaluate the factoring problems that you were given. Are they too easy? Too difficult? Can the polynomials really be factored? Share your responses with the half of the group that wrote the problems. Finally, grade each other's work in factoring the polynomials. Each factoring problem is worth 20 points. You may award partial credit. If you take off points, explain why points are deducted and how you decided to take off a particular number of points for the error(s) that you found.

CHAPTER P
MID-CHAPTER CHECK POINT

What You Know: We defined the real numbers $[\{x|x \text{ is rational}\} \cup \{x|x \text{ is irrational}\}]$ and graphed them as points on a number line. We reviewed the basic rules of algebra, using these properties to simplify algebraic expressions. We expanded our knowledge of exponents to include exponents other than natural numbers:

$$b^0 = 1; \quad b^{-n} = \frac{1}{b^n}; \quad \frac{1}{b^{-n}} = b^n; \quad b^{\frac{1}{n}} = \sqrt[n]{b};$$

$$b^{\frac{m}{n}} = \left(\sqrt[n]{b}\right)^m = \sqrt[n]{b^m}; \quad b^{-\frac{m}{n}} = \frac{1}{b^{\frac{m}{n}}}.$$

We used properties of exponents to simplify exponential expressions and properties of radicals to simplify radical expressions. We performed operations with polynomials, using a number of fast methods for finding products of polynomials, including the FOIL method for multiplying binomials, a special-product formula for the product of the sum and difference of two terms $[(A + B)(A - B) = A^2 - B^2]$, and special-product formulas for squaring binomials $[(A + B)^2 = A^2 + 2AB = B^2;$ $(A - B)^2 = A^2 - 2AB + B^2]$. We reversed the direction of these formulas and reviewed how to factor polynomials. We used a general strategy, summarized in the box on page 58, for factoring a wide variety of polynomials.

In Exercises 1–27, simplify the given expression or perform the indicated operation (and simplify, if possible), whichever is appropriate.

1. $(3x + 5)(4x - 7)$
2. $(3x + 5) - (4x - 7)$
3. $\sqrt{6} + 9\sqrt{6}$
4. $3\sqrt{12} - \sqrt{27}$
5. $7x + 3[9 - (2x - 6)]$
6. $(8x - 3)^2$
7. $\left(x^{\frac{1}{3}}y^{-\frac{1}{2}}\right)^6$
8. $\left(\frac{2}{7}\right)^0 - 32^{-\frac{2}{5}}$
9. $(2x - 5) - (x^2 - 3x + 1)$
10. $(2x - 5)(x^2 - 3x + 1)$
11. $x^3 + x^3 - x^3 \cdot x^3$
12. $(9a - 10b)(2a + b)$
13. $\{a, c, d, e\} \cup \{c, d, f, h\}$
14. $\{a, c, d, e\} \cap \{c, d, f, h\}$
15. $(3x^2y^3 - xy + 4y^2) - (-2x^2y^3 - 3xy + 5y^2)$
16. $\frac{24x^2y^{13}}{-2x^5y^{-2}}$
17. $\left(\frac{1}{3}x^{-5}y^4\right)(18x^{-2}y^{-1})$
18. $\sqrt[12]{x^4}$
19. $[4y - (3x + 2)][4y + (3x + 2)]$
20. $(x - 2y - 1)^2$
21. $\frac{24 \times 10^3}{2 \times 10^6}$ (Express the answer in scientific notation.)
22. $\frac{\sqrt[3]{32}}{\sqrt[3]{2}}$
23. $(x^3 + 2)(x^3 - 2)$
24. $(x^2 + 2)^2$
25. $\sqrt{50} \cdot \sqrt{6}$
26. $\frac{11}{7 - \sqrt{3}}$
27. $\frac{11}{\sqrt{3}}$

In Exercises 28–34, factor completely, or state that the polynomial is prime.

28. $7x^2 - 22x + 3$
29. $x^2 - 2x + 4$
30. $x^3 + 5x^2 + 3x + 15$
31. $3x^2 - 4xy - 7y^2$
32. $64y - y^4$
33. $50x^3 + 20x^2 + 2x$
34. $x^2 - 6x + 9 - 49y^2$

In Exercises 35–36, factor and simplify each algebraic expression.

35. $x^{-\frac{3}{2}} - 2x^{-\frac{1}{2}} + x^{\frac{1}{2}}$
36. $(x^2 + 1)^{\frac{1}{2}} - 10(x^2 + 1)^{-\frac{1}{2}}$

37. List all the rational numbers in this set:
$$\left\{-11, -\frac{3}{7}, 0, 0.45, \sqrt{23}, \sqrt{25}\right\}.$$

In Exercises 38–39, rewrite each expression without absolute value bars.

38. $|2 - \sqrt{13}|$
39. $x^2|x|$ if $x < 0$

40. If the population of the United States is 2.9×10^8 and each person spends about $120 per year on ice cream, express the total annual spending on ice cream in scientific notation.

41. A human brain contains 3×10^{10} neurons and a gorilla brain contains 7.5×10^9 neurons. How many times as many neurons are in the brain of a human as in the brain of a gorilla?

42. In 2003, 28.5 million U.S. adults browsed Internet personals and 17.4 million posted online personal ads. The bar graph shows the amount spent in the United States, in millions of dollars, on online dating.

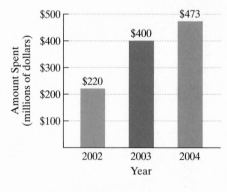

Shopping for a Date: Millions of Dollars Spent in the U.S. on Online Dating

Source: Jupiter Research

Here are three mathematical models for the data shown in the graph. In each formula, D represents the amount spent on online dating, in millions of dollars, x years after 2002.

Model 1 $D = 236(1.5)^x$

Model 2 $D = 127x + 239$

Model 3 $D = -54x^2 + 234x + 220$

a. Which model best describes the data for 2004?

b. According to the polynomial model of degree 1, how much will Americans spend on online dating in 2008?

SECTION P.6   *Rational Expressions*

Objectives

1 Specify numbers that must be excluded from the domain of a rational expression.

2 Simplify rational expressions.

3 Multiply rational expressions.

4 Divide rational expressions.

5 Add and subtract rational expressions.

6 Simplify complex rational expressions.

7 Simplify fractional expressions that occur in calculus.

8 Rationalize numerators.

How can we describe the costs of reducing environmental pollution? We often use algebraic expressions involving quotients of polynomials. For example, the algebraic expression

$$\frac{250x}{100 - x}$$

describes the cost, in millions of dollars, to remove x percent of the pollutants that are discharged into a river. Removing a modest percentage of pollutants, say 40%, is far less costly than removing a substantially greater percentage, such as 95%. We see this by evaluating the algebraic expression for $x = 40$ and $x = 95$.

$$\text{Evaluating } \frac{250x}{100 - x} \quad \text{for}$$

$x = 40:$

$$\text{Cost is } \frac{250(40)}{100 - 40} \approx 167.$$

$x = 95:$

$$\text{Cost is } \frac{250(95)}{100 - 95} = 4750.$$

Discovery

What happens if you try substituting 100 for x in

$$\frac{250x}{100 - x}?$$

What does this tell you about the cost of cleaning up all of the river's pollutants?

The cost increases from approximately $167 million to a possibly prohibitive $4750 million, or $4.75 billion. Costs spiral upward as the percentage of removed pollutants increases.

Many algebraic expressions that describe costs of environmental projects are examples of *rational expressions*. First we will define rational expressions. Then we will review how to perform operations with such expressions.

Rational Expressions

A **rational expression** is the quotient of two polynomials. Some examples are

$$\frac{x - 2}{4}, \quad \frac{4}{x - 2}, \quad \frac{x}{x^2 - 1}, \quad \text{and} \quad \frac{x^2 + 1}{x^2 + 2x - 3}.$$

The set of real numbers for which an algebraic expression is defined is the **domain** of the expression. Because rational expressions indicate division and division by zero is undefined, we must exclude numbers from a rational expression's domain that make the denominator zero.

 Specify numbers that must be excluded from the domain of a rational expression.

EXAMPLE 1   Excluding Numbers from the Domain

Find all the numbers that must be excluded from the domain of each rational expression:

a. $\dfrac{4}{x - 2}$ **b.** $\dfrac{x}{x^2 - 1}$.

Solution   To determine the numbers that must be excluded from each domain, examine the denominators.

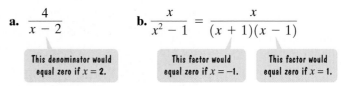

a. $\dfrac{4}{x-2}$ b. $\dfrac{x}{x^2-1} = \dfrac{x}{(x+1)(x-1)}$

This denominator would equal zero if $x = 2$.

This factor would equal zero if $x = -1$.

This factor would equal zero if $x = 1$.

For the rational expression in part (a), we must exclude 2 from the domain. For the rational expression in part (b), we must exclude both -1 and 1 from the domain. These excluded numbers are often written to the right of a rational expression.

$$\dfrac{4}{x-2}, x \neq 2 \qquad \dfrac{x}{x^2-1}, x \neq -1, x \neq 1$$

Check Point 1   Find all the numbers that must be excluded from the domain of each rational expression:

a. $\dfrac{7}{x+5}$ b. $\dfrac{x}{x^2-36}$.

 Simplify rational expressions.

Simplifying Rational Expressions

A rational expression is **simplified** if its numerator and denominator have no common factors other than 1 or -1. The following procedure can be used to simplify rational expressions:

Simplifying Rational Expressions

1. Factor the numerator and the denominator completely.
2. Divide both the numerator and the denominator by any common factors.

EXAMPLE 2   Simplifying Rational Expressions

Simplify: a. $\dfrac{x^3 + x^2}{x + 1}$ b. $\dfrac{x^2 + 6x + 5}{x^2 - 25}$.

Solution

a. $\dfrac{x^3 + x^2}{x + 1} = \dfrac{x^2(x + 1)}{x + 1}$ Factor the numerator. Because the denominator is $x + 1$, $x \neq -1$.

$= \dfrac{x^2\overset{1}{\cancel{(x + 1)}}}{\underset{1}{\cancel{x + 1}}}$ Divide out the common factor, $x + 1$.

$= x^2, x \neq -1$ Denominators of 1 need not be written because $\frac{a}{1} = a$.

b. $\dfrac{x^2 + 6x + 5}{x^2 - 25} = \dfrac{(x + 5)(x + 1)}{(x + 5)(x - 5)}$ Factor the numerator and denominator. Because the denominator is $(x + 5)(x - 5)$, $x \neq -5$ and $x \neq 5$.

$= \dfrac{\overset{1}{\cancel{(x + 5)}}(x + 1)}{\underset{1}{\cancel{(x + 5)}}(x - 5)}$ Divide out the common factor, $x + 5$.

$= \dfrac{x + 1}{x - 5}, \quad x \neq -5, \quad x \neq 5$

Check Point 2   Simplify: a. $\dfrac{x^3 + 3x^2}{x + 3}$ b. $\dfrac{x^2 - 1}{x^2 + 2x + 1}$.

③ Multiply rational expressions.

Multiplying Rational Expressions

The product of two rational expressions is the product of their numerators divided by the product of their denominators. Here is a step-by-step procedure for multiplying rational expressions:

> **Multiplying Rational Expressions**
>
> 1. Factor all numerators and denominators completely.
> 2. Divide numerators and denominators by common factors.
> 3. Multiply the remaining factors in the numerators and multiply the remaining factors in the denominators.

EXAMPLE 3   Multiplying Rational Expressions

Multiply and simplify:

$$\frac{x-7}{x-1} \cdot \frac{x^2-1}{3x-21}.$$

Solution

$$\frac{x-7}{x-1} \cdot \frac{x^2-1}{3x-21}$$

This is the given multiplication problem.

$$= \frac{x-7}{x-1} \cdot \frac{(x+1)(x-1)}{3(x-7)}$$

Factor as many numerators and denominators as possible. Because the denominator has factors of $x - 1$ and $x - 7$, $x \neq 1$ and $x \neq 7$.

$$= \frac{\overset{1}{\cancel{x-7}}}{\underset{1}{\cancel{x-1}}} \cdot \frac{(x+1)\overset{1}{\cancel{(x-1)}}}{3\underset{1}{\cancel{(x-7)}}}$$

Divide numerators and denominators by common factors.

$$= \frac{x+1}{3}, x \neq 1, x \neq 7$$

Multiply the remaining factors in the numerators and denominators.

> These excluded numbers from the domain must also be excluded from the simplified expression's domain.

Check Point 3   Multiply and simplify:

$$\frac{x+3}{x^2-4} \cdot \frac{x^2-x-6}{x^2+6x+9}.$$

④ Divide rational expressions.

Dividing Rational Expressions

The quotient of two rational expressions is the product of the first expression and the multiplicative inverse, or reciprocal, of the second expression. The reciprocal is found by interchanging the numerator and the denominator. Thus, **we find the quotient of two rational expressions by inverting the divisor and multiplying.**

EXAMPLE 4   Dividing Rational Expressions

Divide and simplify:

$$\frac{x^2-2x-8}{x^2-9} \div \frac{x-4}{x+3}.$$

Solution

$$\frac{x^2 - 2x - 8}{x^2 - 9} \div \frac{x - 4}{x + 3}$$

This is the given division problem.

$$= \frac{x^2 - 2x - 8}{x^2 - 9} \cdot \frac{x + 3}{x - 4}$$

Invert the divisor and multiply.

$$= \frac{(x - 4)(x + 2)}{(x + 3)(x - 3)} \cdot \frac{x + 3}{x - 4}$$

Factor as many numerators and denominators as possible. For nonzero denominators, $x \neq -3$, $x \neq 3$, and $x \neq 4$.

$$= \frac{\overset{1}{\cancel{(x - 4)}}(x + 2)}{(x + 3)(x - 3)} \cdot \frac{\overset{1}{\cancel{(x + 3)}}}{\cancel{(x - 4)}}$$

Divide numerators and denominators by common factors.

$$= \frac{x + 2}{x - 3}, x \neq -3, x \neq 3, x \neq 4$$

Multiply the remaining factors in the numerators and the denominators.

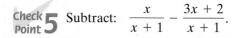 **Check Point 4** Divide and simplify:

$$\frac{x^2 - 2x + 1}{x^3 + x} \div \frac{x^2 + x - 2}{3x^2 + 3}.$$

---

⑤ Add and subtract rational expressions.

Adding and Subtracting Rational Expressions with the Same Denominator

We add or subtract rational expressions with the same denominator by (1) adding or subtracting the numerators, (2) placing this result over the common denominator, and (3) simplifying, if possible.

EXAMPLE 5 Subtracting Rational Expressions with the Same Denominator

Subtract: $\dfrac{5x + 1}{x^2 - 9} - \dfrac{4x - 2}{x^2 - 9}.$

Solution

Study Tip

Example 5 shows that when a numerator is being subtracted, we must subtract every term in that expression.

$$\frac{5x + 1}{x^2 - 9} - \frac{4x - 2}{x^2 - 9} = \frac{5x + 1 - (4x - 2)}{x^2 - 9}$$

Subtract numerators and include parentheses to indicate that both terms are subtracted. Place this difference over the common denominator.

$$= \frac{5x + 1 - 4x + 2}{x^2 - 9}$$

Remove parentheses and then change the sign of each term.

$$= \frac{x + 3}{x^2 - 9}$$

Combine like terms.

$$= \frac{\overset{1}{\cancel{x + 3}}}{\cancel{(x + 3)}(x - 3)}$$

Factor and simplify ($x \neq -3$ and $x \neq 3$).

$$= \frac{1}{x - 3}, x \neq -3, x \neq 3$$

Check Point 5 Subtract: $\dfrac{x}{x + 1} - \dfrac{3x + 2}{x + 1}.$

Adding and Subtracting Rational Expressions with Different Denominators

Rational expressions that have no common factors in their denominators can be added or subtracted using one of the following properties:

$$\frac{a}{b} + \frac{c}{d} = \frac{ad + bc}{bd} \qquad \frac{a}{b} - \frac{c}{d} = \frac{ad - bc}{bd}, b \neq 0, d \neq 0.$$

The denominator, bd, is the product of the factors in the two denominators. Because we are looking at rational expressions that have no common factors in their denominators, the product bd gives the least common denominator.

EXAMPLE 6 Subtracting Rational Expressions Having No Common Factors in Their Denominators

Subtract: $\dfrac{x + 2}{2x - 3} - \dfrac{4}{x + 3}$.

Solution   We need to find the least common denominator. This is the product of the distinct factors in each denominator, namely $(2x - 3)(x + 3)$. We can therefore use the subtraction property given previously as follows:

$$\frac{a}{b} - \frac{c}{d} = \frac{ad - bc}{bd}$$

$$\frac{x + 2}{2x - 3} - \frac{4}{x + 3} = \frac{(x + 2)(x + 3) - (2x - 3)4}{(2x - 3)(x + 3)}$$

Observe that $a = x + 2$, $b = 2x - 3$, $c = 4$, and $d = x + 3$.

$$= \frac{x^2 + 5x + 6 - (8x - 12)}{(2x - 3)(x + 3)}$$

Multiply.

$$= \frac{x^2 + 5x + 6 - 8x + 12}{(2x - 3)(x + 3)}$$

Remove parentheses and then change the sign of each term.

$$= \frac{x^2 - 3x + 18}{(2x - 3)(x + 3)}, x \neq \frac{3}{2}, x \neq -3$$

Combine like terms in the numerator.

Check Point 6   Add: $\dfrac{3}{x + 1} + \dfrac{5}{x - 1}$.

The **least common denominator**, or LCD, of several rational expressions is a polynomial consisting of the product of all prime factors in the denominators, with each factor raised to the greatest power of its occurrence in any denominator. When adding and subtracting rational expressions that have different denominators with one or more common factors in the denominators, it is efficient to find the least common denominator first.

Finding the Least Common Denominator

1. Factor each denominator completely.

2. List the factors of the first denominator.

3. Add to the list in step 2 any factors of the second denominator that do not appear in the list.

4. Form the product of each different factor from the list in step 3. This product is the least common denominator.

EXAMPLE 7   Finding the Least Common Denominator

Find the least common denominator of

$$\frac{7}{5x^2 + 15x} \quad \text{and} \quad \frac{9}{x^2 + 6x + 9}.$$

Solution

Step 1   Factor each denominator completely.

$$5x^2 + 15x = 5x(x + 3)$$
$$x^2 + 6x + 9 = (x + 3)^2$$

Step 2   List the factors of the first denominator.

$$5, x, (x + 3)$$

Step 3   Add any unlisted factors from the second denominator. The second denominator is $(x + 3)^2$ or $(x + 3)(x + 3)$. One factor of $x + 3$ is already in our list, but the other factor is not. We add a second factor of $x + 3$ to the list. We have

$$5, x, (x + 3), (x + 3).$$

Step 4   The least common denominator is the product of all factors in the final list. Thus,

$$5x(x + 3)(x + 3), \quad \text{or} \quad 5x(x + 3)^2$$

is the least common denominator.

 Find the least common denominator of

$$\frac{3}{x^2 - 6x + 9} \quad \text{and} \quad \frac{7}{x^2 - 9}.$$

Finding the least common denominator for two (or more) rational expressions is the first step needed to add or subtract the expressions.

Adding and Subtracting Rational Expressions That Have Different Denominators

1. Find the LCD of the rational expressions.

2. Rewrite each rational expression as an equivalent expression whose denominator is the LCD. To do so, multiply the numerator and the denominator of each rational expression by any factor(s) needed to convert the denominator into the LCD.

3. Add or subtract numerators, placing the resulting expression over the LCD.

4. If possible, simplify the resulting rational expression.

**EXAMPLE 8   Adding Rational Expressions
with Different Denominators**

Add: $\dfrac{x+3}{x^2+x-2}+\dfrac{2}{x^2-1}$.

Solution

Step 1   Find the least common denominator. Start by factoring the denominators.

$$x^2+x-2=(x+2)(x-1)$$
$$x^2-1=(x+1)(x-1)$$

The factors of the first denominator are $x+2$ and $x-1$. The only factor from the second denominator that is not listed is $x+1$. Thus, the least common denominator is

$$(x+2)(x-1)(x+1).$$

Step 2   Write equivalent expressions with the LCD as denominators. We must rewrite each rational expression with a denominator of $(x+2)(x-1)(x+1)$. We do so by multiplying both the numerator and the denominator of each rational expression by any factor(s) needed to convert the expression's denominator into the LCD.

$$\frac{x+3}{(x+2)(x-1)}\cdot\frac{x+1}{x+1}=\frac{(x+3)(x+1)}{(x+2)(x-1)(x+1)}\qquad\frac{2}{(x+1)(x-1)}\cdot\frac{x+2}{x+2}=\frac{2(x+2)}{(x+2)(x+1)(x-1)}$$

Multiply the numerator and denominator by $x+1$ to get $(x+2)(x-1)(x+1)$, the **LCD**.

Multiply the numerator and denominator by $x+2$ to get $(x+2)(x+1)(x-1)$, the **LCD**.

Because $\dfrac{x+1}{x+1}=1$ and $\dfrac{x+2}{x+2}=1$, we are not changing the value of either rational expression, only its appearance.

Now we are ready to perform the indicated addition.

$$\frac{x+3}{x^2+x-2}+\frac{2}{x^2-1}$$

This is the given problem.

$$=\frac{x+3}{(x+2)(x-1)}+\frac{2}{(x+1)(x-1)}$$

Factor the denominators. The LCD is $(x+2)(x+1)(x-1)$.

$$=\frac{(x+3)(x+1)}{(x+2)(x-1)(x+1)}+\frac{2(x+2)}{(x+2)(x-1)(x+1)}$$

Rewrite equivalent expressions with the LCD.

Step 3   Add numerators, putting this sum over the LCD.

$$=\frac{(x+3)(x+1)+2(x+2)}{(x+2)(x-1)(x+1)}$$

$$=\frac{x^2+4x+3+2x+4}{(x+2)(x-1)(x+1)}$$

Perform the multiplications in the numerator.

$$=\frac{x^2+6x+7}{(x+2)(x-1)(x+1)},\ x\neq-2,x\neq1,x\neq-1$$

Combine like terms in the numerator: $4x+2x=6x$ and $3+4=7$.

Step 4   If necessary, simplify. Because the numerator is prime, no further simplification is possible.

Check Point 8   Subtract: $\dfrac{x}{x^2-10x+25}-\dfrac{x-4}{2x-10}$.

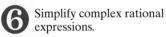

Simplify complex rational expressions.

Complex Rational Expressions

Complex rational expressions, also called **complex fractions**, have numerators or denominators containing one or more rational expressions. Here are two examples of such expressions:

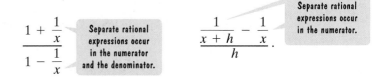

$$\frac{1 + \dfrac{1}{x}}{1 - \dfrac{1}{x}}$$ Separate rational expressions occur in the numerator and the denominator.

$$\frac{\dfrac{1}{x + h} - \dfrac{1}{x}}{h}.$$ Separate rational expressions occur in the numerator.

One method for simplifying a complex rational expression is to combine its numerator into a single expression and combine its denominator into a single expression. Then perform the division by inverting the denominator and multiplying.

EXAMPLE 9   Simplifying a Complex Rational Expression

Simplify: $\dfrac{1 + \dfrac{1}{x}}{1 - \dfrac{1}{x}}.$

Solution

Step 1   **Add to get a single rational expression in the numerator.**

$$1 + \frac{1}{x} = \frac{1}{1} + \frac{1}{x} = \frac{1 \cdot x}{1 \cdot x} + \frac{1}{x} = \frac{x}{x} + \frac{1}{x} = \frac{x + 1}{x}$$

The LCD is $1 \cdot x$, or x.

Step 2   **Subtract to get a single rational expression in the denominator.**

$$1 - \frac{1}{x} = \frac{1}{1} - \frac{1}{x} = \frac{1 \cdot x}{1 \cdot x} - \frac{1}{x} = \frac{x}{x} - \frac{1}{x} = \frac{x - 1}{x}$$

The LCD is $1 \cdot x$, or x.

Step 3   **Perform the division indicated by the main fraction bar: Invert and multiply. If possible, simplify.**

$$\frac{1 + \dfrac{1}{x}}{1 - \dfrac{1}{x}} = \frac{\dfrac{x + 1}{x}}{\dfrac{x - 1}{x}} = \frac{x + 1}{x} \cdot \frac{x}{x - 1} = \frac{x + 1}{\cancel{x}} \cdot \frac{\overset{1}{\cancel{x}}}{x - 1} = \frac{x + 1}{x - 1}$$

Invert and multiply.

Check Point 9   Simplify: $\dfrac{\dfrac{1}{x} - \dfrac{3}{2}}{\dfrac{1}{x} + \dfrac{3}{4}}.$

A second method for simplifying a complex rational expression is to find the least common denominator of all the rational expressions in its numerator and denominator. Then multiply each term in its numerator and denominator by this least common denominator. Because we are multiplying by a form of 1, we will obtain an equivalent expression that does not contain fractions in its numerator or denominator. Here we use this method to simplify the complex rational expression in Example 9.

$$\frac{1+\dfrac{1}{x}}{1-\dfrac{1}{x}} = \frac{\left(1+\dfrac{1}{x}\right)}{\left(1-\dfrac{1}{x}\right)}\cdot\frac{x}{x}$$

The least common denominator of all the rational expressions is x. Multiply the numerator and denominator by x. Because $\frac{x}{x} = 1$, we are not changing the complex fraction ($x \neq 0$).

$$= \frac{1\cdot x + \dfrac{1}{x}\cdot x}{1\cdot x - \dfrac{1}{x}\cdot x}$$

Use the distributive property. Be sure to distribute x to every term.

$$= \frac{x+1}{x-1}, x \neq 0, x \neq 1$$

Multiply. The complex rational expression is now simplified.

EXAMPLE 10   Simplifying a Complex Rational Expression

Simplify: $\dfrac{\dfrac{1}{x+h}-\dfrac{1}{x}}{h}$.

Solution   We will use the method of multiplying each of the three terms, $\dfrac{1}{x+h}, \dfrac{1}{x}$, and h by the least common denominator. The least common denominator is $x(x+h)$.

$$\frac{\dfrac{1}{x+h}-\dfrac{1}{x}}{h}$$

$$= \frac{\left(\dfrac{1}{x+h}-\dfrac{1}{x}\right)x(x+h)}{hx(x+h)}$$

Multiply the numerator and denominator by $x(x+h), h \neq 0, x \neq 0, x \neq -h$.

$$= \frac{\dfrac{1}{x+h}\cdot x(x+h) - \dfrac{1}{x}\cdot x(x+h)}{hx(x+h)}$$

Use the distributive property in the numerator.

$$= \frac{x-(x+h)}{hx(x+h)}$$

Simplify: $\frac{1}{x+h}x(x+h) = x$ and $\frac{1}{x}\cdot x(x+h) = x+h$.

$$= \frac{x-x-h}{hx(x+h)}$$

Subtract in the numerator.

$$= \frac{-h}{hx(x+h)}$$

Simplify: $x-x-h = -h$.

$$= -\frac{1}{x(x+h)}, h \neq 0, x \neq 0, x \neq -h$$

Divide the numerator and denominator by h.

Check Point 10   Simplify: $\dfrac{\dfrac{1}{x+7}-\dfrac{1}{x}}{7}$.

Simplify fractional expressions that occur in calculus.

Fractional Expressions in Calculus

Fractional expressions containing radicals occur frequently in calculus. Because of the radicals, these expressions are not rational expressions. However, they can often be simplified using the procedure for simplifying complex rational expressions.

EXAMPLE 11   Simplifying a Fractional Expression Containing Radicals

Simplify: $\dfrac{\sqrt{9-x^2} + \dfrac{x^2}{\sqrt{9-x^2}}}{9-x^2}$.

Solution

$$\dfrac{\sqrt{9-x^2} + \dfrac{x^2}{\sqrt{9-x^2}}}{9-x^2}$$

The least common denominator of the denominators is $\sqrt{9-x^2}$.

$$= \dfrac{\sqrt{9-x^2} + \dfrac{x^2}{\sqrt{9-x^2}}}{9-x^2} \cdot \dfrac{\sqrt{9-x^2}}{\sqrt{9-x^2}}$$

Multiply the numerator and the denominator by $\sqrt{9-x^2}$.

$$= \dfrac{\sqrt{9-x^2}\,\sqrt{9-x^2} + \dfrac{x^2}{\sqrt{9-x^2}}\,\sqrt{9-x^2}}{(9-x^2)\sqrt{9-x^2}}$$

Use the distributive property in the numerator.

$$= \dfrac{(9-x^2) + x^2}{(9-x^2)^{\frac{3}{2}}}$$

In the denominator:
$$(9-x^2)^1(9-x^2)^{\frac{1}{2}} = (9-x^2)^{1+\frac{1}{2}}$$
$$= (9-x^2)^{\frac{3}{2}}.$$

$$= \dfrac{9}{\sqrt{(9-x^2)^3}}$$

Because the original expression was in radical form, write the denominator in radical form.

Check Point 11   Simplify: $\dfrac{\sqrt{x} + \dfrac{1}{\sqrt{x}}}{x}$.

Rationalize numerators.

Another fractional expression that you will encounter in calculus is

$$\dfrac{\sqrt{x+h} - \sqrt{x}}{h}.$$

Can you see that this expression is not defined if $h = 0$? However, in calculus, you will ask the following question:

What happens to the expression as h takes on values that get closer and closer to 0, such as $h = 0.1$, $h = 0.01$, $h = 0.001$, $h = 0.0001$, and so on?

The question is answered by first **rationalizing the numerator**. This process involves rewriting the fractional expression as an equivalent expression in which the numerator no longer contains any radicals. **To rationalize a numerator, multiply by 1 to eliminate the radical in the *numerator*. Multiply the numerator and the denominator by the conjugate of the numerator.**

EXAMPLE 12   Rationalizing a Numerator

Rationalize the numerator:

$$\frac{\sqrt{x+h}-\sqrt{x}}{h}.$$

Solution   The conjugate of the numerator is $\sqrt{x+h}+\sqrt{x}$. If we multiply the numerator and denominator by $\sqrt{x+h}+\sqrt{x}$, the simplified numerator will not contain a radical. Therefore, we multiply by 1, choosing $\dfrac{\sqrt{x+h}+\sqrt{x}}{\sqrt{x+h}+\sqrt{x}}$ for 1.

$$\frac{\sqrt{x+h}-\sqrt{x}}{h} = \frac{\sqrt{x+h}-\sqrt{x}}{h}\cdot\frac{\sqrt{x+h}+\sqrt{x}}{\sqrt{x+h}+\sqrt{x}}$$ Multiply by 1.

$$= \frac{\left(\sqrt{x+h}\right)^2-\left(\sqrt{x}\right)^2}{h\left(\sqrt{x+h}+\sqrt{x}\right)}$$ $\left(\sqrt{a}-\sqrt{b}\right)\left(\sqrt{a}+\sqrt{b}\right)=$
$\left(\sqrt{a}\right)^2-\left(\sqrt{b}\right)^2$

$$= \frac{x+h-x}{h\left(\sqrt{x+h}+\sqrt{x}\right)}$$ $\left(\sqrt{x+h}\right)^2=x+h$
and $\left(\sqrt{x}\right)^2=x.$

$$= \frac{h}{h\left(\sqrt{x+h}+\sqrt{x}\right)}$$ Simplify; $x+h-x=h.$

$$= \frac{1}{\sqrt{x+h}+\sqrt{x}},\quad h\neq 0$$ Divide both the numerator and denominator by h.

Calculus Preview

In calculus, you will summarize the discussion on the right using the special notation

$$\lim_{h\to 0}\frac{\sqrt{x+h}-\sqrt{x}}{h}=\frac{1}{2\sqrt{x}}.$$

This is read "the limit of $\dfrac{\sqrt{x+h}-\sqrt{x}}{h}$ as h approaches 0 equals $\dfrac{1}{2\sqrt{x}}$." Limits are discussed in Chapter 11, where we present an introduction to calculus.

What happens to $\dfrac{\sqrt{x+h}-\sqrt{x}}{h}$ as h gets closer and closer to 0? In Example 12, we showed that

$$\frac{\sqrt{x+h}-\sqrt{x}}{h}=\frac{1}{\sqrt{x+h}+\sqrt{x}}.$$

As h gets closer to 0, the expression on the right gets closer to $\dfrac{1}{\sqrt{x+0}+\sqrt{x}}=\dfrac{1}{\sqrt{x}+\sqrt{x}}$, or $\dfrac{1}{2\sqrt{x}}$. Thus, the fractional expression $\dfrac{\sqrt{x+h}-\sqrt{x}}{h}$ approaches $\dfrac{1}{2\sqrt{x}}$ as h gets closer to 0.

Check Point 12   Rationalize the numerator: $\dfrac{\sqrt{x+3}-\sqrt{x}}{3}.$

EXERCISE SET P.6

Practice Exercises

In Exercises 1–6, find all numbers that must be excluded from the domain of each rational expression.

1. $\dfrac{7}{x-3}$

2. $\dfrac{13}{x+9}$

3. $\dfrac{x+5}{x^2-25}$

4. $\dfrac{x+7}{x^2-49}$

5. $\dfrac{x-1}{x^2+11x+10}$

6. $\dfrac{x-3}{x^2+4x-45}$

In Exercises 7–14, simplify each rational expression. Find all numbers that must be excluded from the domain of the simplified rational expression.

7. $\dfrac{3x-9}{x^2-6x+9}$

8. $\dfrac{4x-8}{x^2-4x+4}$

9. $\dfrac{x^2-12x+36}{4x-24}$

10. $\dfrac{x^2-8x+16}{3x-12}$

11. $\dfrac{y^2+7y-18}{y^2-3y+2}$

12. $\dfrac{y^2-4y-5}{y^2+5y+4}$

13. $\dfrac{x^2+12x+36}{x^2-36}$

14. $\dfrac{x^2-14x+49}{x^2-49}$

In Exercises 15–32, multiply or divide as indicated.

15. $\dfrac{x-2}{3x+9}\cdot\dfrac{2x+6}{2x-4}$

16. $\dfrac{6x+9}{3x-15}\cdot\dfrac{x-5}{4x+6}$

17. $\dfrac{x^2-9}{x^2}\cdot\dfrac{x^2-3x}{x^2+x-12}$

18. $\dfrac{x^2-4}{x^2-4x+4}\cdot\dfrac{2x-4}{x+2}$

19. $\dfrac{x^2-5x+6}{x^2-2x-3}\cdot\dfrac{x^2-1}{x^2-4}$

20. $\dfrac{x^2+5x+6}{x^2+x-6}\cdot\dfrac{x^2-9}{x^2-x-6}$

21. $\dfrac{x^3-8}{x^2-4}\cdot\dfrac{x+2}{3x}$

22. $\dfrac{x^2+6x+9}{x^3+27}\cdot\dfrac{1}{x+3}$

23. $\dfrac{x+1}{3}\div\dfrac{3x+3}{7}$

24. $\dfrac{x+5}{7}\div\dfrac{4x+20}{9}$

25. $\dfrac{x^2-4}{x}\div\dfrac{x+2}{x-2}$

26. $\dfrac{x^2-4}{x-2}\div\dfrac{x+2}{4x-8}$

27. $\dfrac{4x^2+10}{x-3}\div\dfrac{6x^2+15}{x^2-9}$

28. $\dfrac{x^2+x}{x^2-4}\div\dfrac{x^2-1}{x^2+5x+6}$

29. $\dfrac{x^2-25}{2x-2}\div\dfrac{x^2+10x+25}{x^2+4x-5}$

30. $\dfrac{x^2-4}{x^2+3x-10}\div\dfrac{x^2+5x+6}{x^2+8x+15}$

31. $\dfrac{x^2+x-12}{x^2+x-30}\cdot\dfrac{x^2+5x+6}{x^2-2x-3}\div\dfrac{x+3}{x^2+7x+6}$

32. $\dfrac{x^3-25x}{4x^2}\cdot\dfrac{2x^2-2}{x^2-6x+5}\div\dfrac{x^2+5x}{7x+7}$

In Exercises 33–54, add or subtract as indicated.

33. $\dfrac{4x+1}{6x+5}+\dfrac{8x+9}{6x+5}$

34. $\dfrac{3x+2}{3x+4}+\dfrac{3x+6}{3x+4}$

35. $\dfrac{x^2-2x}{x^2+3x}+\dfrac{x^2+x}{x^2+3x}$

36. $\dfrac{x^2-4x}{x^2-x-6}+\dfrac{4x-4}{x^2-x-6}$

37. $\dfrac{4x-10}{x-2}-\dfrac{x-4}{x-2}$

38. $\dfrac{2x+3}{3x-6}-\dfrac{3-x}{3x-6}$

39. $\dfrac{x^2+3x}{x^2+x-12}-\dfrac{x^2-12}{x^2+x-12}$

40. $\dfrac{x^2-4x}{x^2-x-6}-\dfrac{x-6}{x^2-x-6}$

41. $\dfrac{3}{x+4}+\dfrac{6}{x+5}$

42. $\dfrac{8}{x-2}+\dfrac{2}{x-3}$

43. $\dfrac{3}{x+1}-\dfrac{3}{x}$

44. $\dfrac{4}{x}-\dfrac{3}{x+3}$

45. $\dfrac{2x}{x+2}+\dfrac{x+2}{x-2}$

46. $\dfrac{3x}{x-3}-\dfrac{x+4}{x+2}$

47. $\dfrac{x+5}{x-5}+\dfrac{x-5}{x+5}$

48. $\dfrac{x+3}{x-3}+\dfrac{x-3}{x+3}$

49. $\dfrac{4}{x^2+6x+9}+\dfrac{4}{x+3}$

50. $\dfrac{3}{5x+2}+\dfrac{5x}{25x^2-4}$

51. $\dfrac{3x}{x^2+3x-10}-\dfrac{2x}{x^2+x-6}$

52. $\dfrac{x}{x^2-2x-24}-\dfrac{x}{x^2-7x+6}$

53. $\dfrac{4x^2+x-6}{x^2+3x+2}-\dfrac{3x}{x+1}+\dfrac{5}{x+2}$

54. $\dfrac{6x^2+17x-40}{x^2+x-20}+\dfrac{3}{x-4}-\dfrac{5x}{x+5}$

In Exercises 55–68, simplify each complex rational expression.

55. $\dfrac{\dfrac{x}{3}-1}{x-3}$

56. $\dfrac{\dfrac{x}{4}-1}{x-4}$

57. $\dfrac{1+\dfrac{1}{x}}{3-\dfrac{1}{x}}$

58. $\dfrac{8+\dfrac{1}{x}}{4-\dfrac{1}{x}}$

59. $\dfrac{\dfrac{1}{x}+\dfrac{1}{y}}{x+y}$

60. $\dfrac{1-\dfrac{1}{x}}{xy}$

61. $\dfrac{x-\dfrac{x}{x+3}}{x+2}$

62. $\dfrac{x-3}{x-\dfrac{3}{x-2}}$

63. $\dfrac{\dfrac{3}{x-2}-\dfrac{4}{x+2}}{\dfrac{7}{x^2-4}}$

64. $\dfrac{\dfrac{x}{x-2}+1}{\dfrac{3}{x^2-4}+1}$

65. $$\dfrac{\dfrac{1}{x+1}}{\dfrac{1}{x^2-2x-3}+\dfrac{1}{x-3}}$$

66. $$\dfrac{\dfrac{6}{x^2+2x-15}-\dfrac{1}{x-3}}{\dfrac{1}{x+5}+1}$$

67. $$\dfrac{\dfrac{1}{(x+h)^2}-\dfrac{1}{x^2}}{h}$$

68. $$\dfrac{\dfrac{x+h}{x+h+1}-\dfrac{x}{x+1}}{h}$$

Exercises 69–74 contain fractional expressions that occur frequently in calculus. Simplify each expresson.

69. $$\dfrac{\sqrt{x}-\dfrac{1}{3\sqrt{x}}}{\sqrt{x}}$$

70. $$\dfrac{\sqrt{x}-\dfrac{1}{4\sqrt{x}}}{\sqrt{x}}$$

71. $$\dfrac{\dfrac{x^2}{\sqrt{x^2+2}}-\sqrt{x^2+2}}{x^2}$$

72. $$\dfrac{\sqrt{5-x^2}+\dfrac{x^2}{\sqrt{5-x^2}}}{5-x^2}$$

73. $$\dfrac{\dfrac{1}{\sqrt{x+h}}-\dfrac{1}{\sqrt{x}}}{h}$$

74. $$\dfrac{\dfrac{1}{\sqrt{x+3}}-\dfrac{1}{\sqrt{x}}}{3}$$

In Exercises 75–78, rationalize the numerator.

75. $$\dfrac{\sqrt{x+5}-\sqrt{x}}{5}$$

76. $$\dfrac{\sqrt{x+7}-\sqrt{x}}{7}$$

77. $$\dfrac{\sqrt{x}+\sqrt{y}}{x^2-y^2}$$

78. $$\dfrac{\sqrt{x}+\sqrt{y}}{x^2-y^2}$$

Practice Plus

In Exercises 79–86, perform the indicated operations. Simplify the result, if possible.

79. $$\left(\dfrac{2x+3}{x+1}\cdot\dfrac{x^2+4x-5}{2x^2+x-3}\right)-\dfrac{2}{x+2}$$

80. $$\dfrac{1}{x^2-2x-8}\div\left(\dfrac{1}{x-4}-\dfrac{1}{x+2}\right)$$

81. $$\left(2-\dfrac{6}{x+1}\right)\left(1+\dfrac{3}{x-2}\right)$$

82. $$\left(4-\dfrac{3}{x+2}\right)\left(1+\dfrac{5}{x-1}\right)$$

83. $$\dfrac{y^{-1}-(y+5)^{-1}}{5}$$

84. $$\dfrac{y^{-1}-(y+2)^{-1}}{2}$$

85. $$\left(\dfrac{1}{a^3-b^3}\cdot\dfrac{ac+ad-bc-bd}{1}\right)-\dfrac{c-d}{a^2+ab+b^2}$$

86. $$\dfrac{ab}{a^2+ab+b^2}+\left(\dfrac{ac-ad-bc+bd}{ac-ad+bc-bd}\div\dfrac{a^3-b^3}{a^3+b^3}\right)$$

Application Exercises

87. The rational expression

$$\dfrac{130x}{100-x}$$

describes the cost, in millions of dollars, to inoculate x percent of the population against a particular strain of flu.

a. Evaluate the expression for $x=40$, $x=80$, and $x=90$. Describe the meaning of each evaluation in terms of percentage inoculated and cost.

b. For what value of x is the expression undefined?

c. What happens to the cost as x approaches 100%? How can you interpret this observation?

88. Doctors use the rational expression

$$\dfrac{DA}{A+12}$$

to determine the dosage of a drug prescribed for children. In this expression, A = child's age and D = adult dosage. What is the difference in the child's dosage for a 7-year-old child and a 3-year-old child? Express the answer as a single rational expression in terms of D. Then describe what your answer means in terms of the variables in the rational expression.

89. The bar graph shows the total number of crimes in the United States, in millions, from 1995 through 2002.

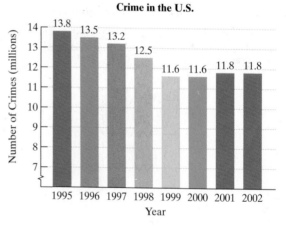

Crime in the U.S.

Source: FBI

The polynomial $3.6t+260$ describes the U.S. population, in millions, t years after 1994. The polynomial $-0.3t+14$ describes the number of crimes in the United States, in millions, t years after 1994.

a. Write a rational expression that describes the crime rate in the United States t years after 1994.

b. According to the rational expression in part (a), what was the crime rate in 2002? Round to two decimal places. How many crimes does this indicate per 100,000 inhabitants?

c. According to the FBI, there were 4119 crimes per 100,000 U.S. inhabitants in 2002. How well does the rational expression that you evaluated in part (b) model this number?

90. The average rate on a round-trip commute having a one-way distance d is given by the complex rational expression

$$\frac{2d}{\dfrac{d}{r_1} + \dfrac{d}{r_2}},$$

in which r_1 and r_2 are the average rates on the outgoing and return trips, respectively. Simplify the expression. Then find your average rate if you drive to campus averaging 40 miles per hour and return home on the same route averaging 30 miles per hour. Explain why the answer is not 25 miles per hour.

In Exercises 91–92, express the perimeter of each rectangle as a single rational expression.

91.

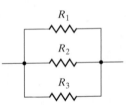

92.

93. If three resistors with resistances R_1, R_2, and R_3 are connected in parallel, their combined resistance is given by the expression

$$\frac{1}{\dfrac{1}{R_1} + \dfrac{1}{R_2} + \dfrac{1}{R_3}}.$$

Simplify the complex rational expression. Then find the combined resistance when R_1 is 4 ohms, R_2 is 8 ohms, and R_3 is 12 ohms.

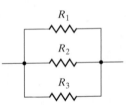

Writing in Mathematics

94. What is a rational expression?

95. Explain how to determine which numbers must be excluded from the domain of a rational expression.

96. Explain how to simplify a rational expression.

97. Explain how to multiply rational expressions.

98. Explain how to divide rational expressions.

99. Explain how to add or subtract rational expressions with the same denominators.

100. Explain how to add rational expressions having no common factors in their denominators. Use $\dfrac{3}{x+5} + \dfrac{7}{x+2}$ in your explanation.

101. Explain how to find the least common denominator for denominators of $x^2 - 100$ and $x^2 - 20x + 100$.

102. Describe two ways to simplify $\dfrac{\dfrac{3}{x} + \dfrac{2}{x^2}}{\dfrac{1}{x^2} + \dfrac{2}{x}}$.

Explain the error in Exercises 103–105. Then rewrite the right side of the equation to correct the error that now exists.

103. $\dfrac{1}{a} + \dfrac{1}{b} = \dfrac{1}{a+b}$

104. $\dfrac{1}{x} + 7 = \dfrac{1}{x+7}$

105. $\dfrac{a}{x} + \dfrac{a}{b} = \dfrac{a}{x+b}$

106. A politician claims that each year the crime rate in the United States is decreasing. Explain how to use the polynomials in Exercise 89 to verify this claim.

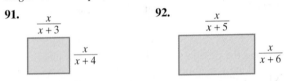 **Critical Thinking Exercises**

107. Which one of the following is true?

a. $\dfrac{a}{b} + \dfrac{a}{c} = \dfrac{a}{b+c}$

b. $6 + \dfrac{1}{x} = \dfrac{7}{x}$

c. $\dfrac{1}{x+3} + \dfrac{x+3}{2} = \dfrac{1}{\cancel{(x+3)}} + \dfrac{\cancel{(x+3)}}{2} = 1 + \dfrac{1}{2} = \dfrac{3}{2}$

d. $\dfrac{x^2 - 25}{x - 5} = x - 5$

e. None of the above is true.

In Exercises 108–110, perform the indicated operations.

108. $\dfrac{1}{x^n - 1} - \dfrac{1}{x^n + 1} - \dfrac{1}{x^{2n} - 1}$

109. $\left(1 - \dfrac{1}{x}\right)\left(1 - \dfrac{1}{x+1}\right)\left(1 - \dfrac{1}{x+2}\right)\left(1 - \dfrac{1}{x+3}\right)$

110. $(x - y)^{-1} + (x - y)^{-2}$

111. In one short sentence, five words or less, explain what

$$\frac{\dfrac{1}{x} + \dfrac{1}{x^2} + \dfrac{1}{x^3}}{\dfrac{1}{x^4} + \dfrac{1}{x^5} + \dfrac{1}{x^6}}$$

does to each number x.

SECTION P.7 *Equations*

Objectives

1. Solve linear equations in one variable.
2. Solve linear equations containing fractions.
3. Solve rational equations with variables in the denominators.
4. Solve a formula for a variable.
5. Solve equations involving absolute value.
6. Solve quadratic equations by factoring.
7. Solve quadratic equations by the square root property.
8. Solve quadratic equations by completing the square.
9. Solve quadratic equations using the quadratic formula.
10. Use the discriminant to determine the number and type of solutions of quadratic equations.
11. Determine the most efficient method to use when solving a quadratic equation.
12. Solve radical equations.

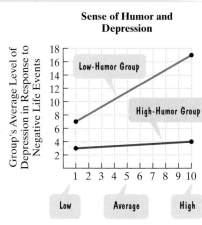

Sense of Humor and Depression

Group's Average Level of Depression in Response to Negative Life Events

Low-Humor Group

High-Humor Group

Low   Average   High

Intensity of Negative Life Event

Figure P.12
Source: Steven Davis and Joseph Palladino, *Psychology,* 3rd Edition, Prentice Hall, 2003.

The belief that humor and laughter can have positive benefits on our lives is not new. The graphs in Figure P.12 indicate that persons with a low sense of humor have higher levels of depression in response to negative life events than those with a high sense of humor. These graphs can be modeled by the following formulas:

Low-Humor Group   High-Humor Group

$$D = \frac{10}{9}x + \frac{53}{9} \qquad D = \frac{1}{9}x + \frac{26}{9}.$$

In each formula, x represents the intensity of a negative life event (from 1, low, to 10, high) and D is the level of depression in response to that event.

Suppose that the low-humor group averages a level of depression of 10 in response to a negative life event. We can determine the intensity of that event by substituting 10 for D in the low-humor model:

$$10 = \frac{10}{9}x + \frac{53}{9}.$$

The two sides of an equation can be reversed. So, we can also express this equation as

$$\frac{10}{9}x + \frac{53}{9} = 10.$$

Notice that the highest exponent on the variable is 1. Such an equation is called a *linear equation in one variable.* In this section, we will review how to solve a variety of equations, including linear equations, quadratic equations, and radical equations.

Linear Equations in One Variable

1 Solve linear equations in one variable.

We begin with a general definition of a linear equation in one variable.

> **Definition of a Linear Equation**
> A **linear equation in one variable** x is an equation that can be written in the form
> $$ax + b = 0,$$
> where a and b are real numbers, and $a \neq 0$.

An example of a linear equation in one variable is

$$4x + 12 = 0.$$

Solving an equation in x involves determining all values of x that result in a true statement when substituted into the equation. Such values are **solutions**, or **roots**, of the equation. For example, substitute -3 for x in $4x + 12 = 0$. We obtain

$$4(-3) + 12 = 0, \quad \text{or} \quad -12 + 12 = 0.$$

This simplifies to the true statement $0 = 0$. Thus, -3 is a solution of the equation $4x + 12 = 0$. We also say that -3 **satisfies** the equation $4x + 12 = 0$, because when we substitute -3 for x, a true statement results. The set of all such solutions is called the equation's **solution set**. For example, the solution set of the equation $4x + 12 = 0$ is $\{-3\}$.

Two or more equations that have the same solution set are called **equivalent equations**. For example, the equations

$$4x + 12 = 0 \quad \text{and} \quad 4x = -12 \quad \text{and} \quad x = -3$$

are equivalent equations because the solution set for each is $\{-3\}$. To solve a linear equation in x, we transform the equation into an equivalent equation one or more times. Our final equivalent equation should be of the form

$$x = \text{a number}.$$

The solution set of this equation is the set consisting of the number.

To generate equivalent equations, we will use the following principles:

Generating Equivalent Equations

An equation can be transformed into an equivalent equation by one or more of the following operations:

Example

1. Simplify an expression by removing grouping symbols and combining like terms.

$$3(x - 6) = 6x - x$$
$$3x - 18 = 5x$$

2. Add (or subtract) the same real number or variable expression on *both* sides of the equation.

$$3x - 18 = 5x$$

Subtract $3x$ from both sides of the equation.

$$3x - 18 - 3x = 5x - 3x$$
$$-18 = 2x$$

3. Multiply (or divide) on *both* sides of the equation by the same *nonzero* quantity.

$$-18 = 2x$$

Divide both sides of the equation by 2.

$$\frac{-18}{2} = \frac{2x}{2}$$
$$-9 = x$$

4. Interchange the two sides of the equation.

$$-9 = x$$
$$x = -9$$

If you look closely at the equations in the box, you will notice that we have solved the equation $3(x - 6) = 6x - x$. The final equation, $x = -9$, with x isolated by itself on the left side, shows that $\{-9\}$ is the solution set. The idea in solving a linear equation is to get the variable by itself on one side of the equal sign and a number by itself on the other side.

Here is a step-by-step procedure for solving a linear equation in one variable. Not all of these steps are necessary to solve every equation.

Solving a Linear Equation

1. Simplify the algebraic expression on each side by removing grouping symbols and combining like terms.
2. Collect all the variable terms on one side and all the numbers, or constant terms, on the other side.
3. Isolate the variable and solve.
4. Check the proposed solution in the original equation.

EXAMPLE 1   Solving a Linear Equation

Solve and check: $2(x - 3) - 17 = 13 - 3(x + 2)$.

Solution

Step 1   Simplify the algebraic expression on each side.

> Do not begin with 13 − 3. Multiplication (the distributive property) is applied before subtraction.

$$2(x - 3) - 17 = 13 - 3(x + 2)$$ This is the given equation.

$$2x - 6 - 17 = 13 - 3x - 6$$ Use the distributive property.

$$2x - 23 = -3x + 7$$ Combine like terms.

Step 2   Collect variable terms on one side and constant terms on the other side. We will collect variable terms on the left by adding $3x$ to both sides. We will collect the numbers on the right by adding 23 to both sides.

$$2x - 23 + 3x = -3x + 7 + 3x$$ Add 3x to both sides.

$$5x - 23 = 7$$ Simplify: 2x + 3x = 5x.

$$5x - 23 + 23 = 7 + 23$$ Add 23 to both sides.

$$5x = 30$$ Simplify.

Step 3   Isolate the variable and solve. We isolate the variable, x, by dividing both sides of $5x = 30$ by 5.

$$\frac{5x}{5} = \frac{30}{5}$$ Divide both sides by 5.

$$x = 6$$ Simplify.

Step 4   Check the proposed solution in the original equation. Substitute 6 for x in the original equation.

$$2(x - 3) - 17 = 13 - 3(x + 2)$$ This is the original equation.

$$2(6 - 3) - 17 \overset{?}{=} 13 - 3(6 + 2)$$ Substitute 6 for x. The question mark indicates that we do not yet know if the two sides are equal.

$$2(3) - 17 \overset{?}{=} 13 - 3(8)$$ Simplify inside parentheses.

$$6 - 17 \overset{?}{=} 13 - 24$$ Multiply.

$$-11 = -11$$ Subtract.

The true statement $-11 = -11$ verifies that the solution set is $\{6\}$.

Discovery

Solve the equation in Example 1 by collecting terms with the variable on the right and numerical terms on the left. What do you observe?

Check Point 1   Solve and check: $4(2x + 1) = 29 + 3(2x - 5)$.

② Solve linear equations containing fractions.

Linear Equations with Fractions

Equations are easier to solve when they do not contain fractions. How do we remove fractions from an equation? We begin by multiplying both sides of the equation by the least common denominator of any fractions in the equation. The least common denominator is the smallest number that all denominators will divide into. Multiplying every term on both sides of the equation by the least common denominator will eliminate the fractions in the equation. Example 2 shows how we "clear an equation of fractions."

EXAMPLE 2   Solving a Linear Equation Involving Fractions

Solve and check: $\dfrac{x+2}{4} - \dfrac{x-1}{3} = 2$.

Solution   The fractional terms have denominators of 4 and 3. The smallest number that is divisible by 4 and 3 is 12. We begin by multiplying both sides of the equation by 12, the least common denominator.

$$\frac{x+2}{4} - \frac{x-1}{3} = 2$$
This is the given equation.

$$12\left(\frac{x+2}{4} - \frac{x-1}{3}\right) = 12 \cdot 2$$
Multiply both sides by 12.

$$12\left(\frac{x+2}{4}\right) - 12\left(\frac{x-1}{3}\right) = 24$$
Use the distributive property and multiply each term on the left by 12.

$$\overset{3}{\cancel{12}}\left(\frac{x+2}{\cancel{4}}\right) - \overset{4}{\cancel{12}}\left(\frac{x-1}{\cancel{3}}\right) = 24$$
Divide out common factors in each multiplication on the left.

$$3(x+2) - 4(x-1) = 24$$
The fractions are now cleared.

$$3x + 6 - 4x + 4 = 24$$
Use the distributive property.

$$-x + 10 = 24$$
Combine like terms: $3x - 4x = -x$ and $6 + 4 = 10$.

$$-x + 10 - 10 = 24 - 10$$
Subtract 10 from both sides.

$$-x = 14$$
Simplify.

> We're not finished. A negative sign should not precede the variable.

Isolate x by multiplying or dividing both sides of this equation by -1.

$$\frac{-x}{-1} = \frac{14}{-1}$$
Divide both sides by -1.

$$x = -14$$
Simplify.

Check the proposed solution. Substitute -14 for x in the original equation. You should obtain $2 = 2$. This true statement verifies that the solution set is $\{-14\}$.

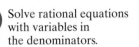 Solve and check: $\dfrac{x-3}{4} = \dfrac{5}{14} - \dfrac{x+5}{7}$.

Rational Equations

③ Solve rational equations with variables in the denominators.

A **rational equation** is an equation containing one or more rational expressions. In Example 2, we solved a rational equation with constants in the denominators. This rational equation was a linear equation. Now, let's consider a rational equation such as

$$\frac{3}{x+6} + \frac{1}{x-2} = \frac{4}{x^2 + 4x - 12}.$$

Can you see how this rational equation differs from the rational equation that we solved earlier? The variable appears in the denominators. Although this rational equation is not a linear equation, the solution procedure still involves multiplying each side by the least common denominator. However, we must avoid any values of the variable that make a denominator zero.

EXAMPLE 3   Solving a Rational Equation

Solve: $\dfrac{3}{x+6} + \dfrac{1}{x-2} = \dfrac{4}{x^2+4x-12}$.

Solution   To identify values of x that make denominators zero, let's factor $x^2 + 4x - 12$, the denominator on the right. This factorization is also necessary in identifying the least common denominator.

$$\frac{3}{x+6} + \frac{1}{x-2} = \frac{4}{(x+6)(x-2)}$$

This denominator is zero if $x = -6$.	This denominator is zero if $x = 2$.	This denominator is zero if $x = -6$ or $x = 2$.

We see that x cannot equal -6 or 2. The least common denominator is $(x+6)(x-2)$.

$\dfrac{3}{x+6} + \dfrac{1}{x-2} = \dfrac{4}{(x+6)(x-2)}, \quad x \neq -6, \quad x \neq 2$ This is the given equation with a denominator factored.

$(x+6)(x-2)\left(\dfrac{3}{x+6} + \dfrac{1}{x-2}\right) = (x+6)(x-2) \cdot \dfrac{4}{(x+6)(x-2)}$ Multiply both sides by $(x+6)(x-2)$, the LCD.

$(x+6)(x-2) \cdot \dfrac{3}{x+6} + (x+6)(x-2) \cdot \dfrac{1}{x-2} = (x+6)(x-2) \cdot \dfrac{4}{(x+6)(x-2)}$ Use the distributive property and divide out common factors.

$3(x-2) + 1(x+6) = 4$ Simplify. This equation is cleared of fractions.

$3x - 6 + x + 6 = 4$ Use the distributive property.

$4x = 4$ Combine like terms.

$\dfrac{4x}{4} = \dfrac{4}{4}$ Divide both sides by 4.

$x = 1$ Simplify. This is not part of the restriction that $x \neq -6$ and $x \neq 2$.

Check the proposed solution. Substitute 1 for x in the original equation. You should obtain $-\frac{4}{7} = -\frac{4}{7}$. This true statement verifies that the solution set is $\{1\}$.

Check Point 3   Solve: $\dfrac{6}{x+3} - \dfrac{5}{x-2} = \dfrac{-20}{x^2+x-6}$.

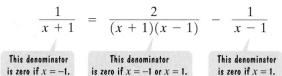

EXAMPLE 4   Solving a Rational Equation

Solve: $\dfrac{1}{x+1} = \dfrac{2}{x^2-1} - \dfrac{1}{x-1}$.

Solution   We begin by factoring $x^2 - 1$.

$$\underbrace{\dfrac{1}{x+1}}_{\substack{\text{This denominator} \\ \text{is zero if } x=-1.}} = \underbrace{\dfrac{2}{(x+1)(x-1)}}_{\substack{\text{This denominator} \\ \text{is zero if } x=-1 \text{ or } x=1.}} - \underbrace{\dfrac{1}{x-1}}_{\substack{\text{This denominator} \\ \text{is zero if } x=1.}}$$

We see that x cannot equal -1 or 1. The least common denominator is $(x+1)(x-1)$.

$$\dfrac{1}{x+1} = \dfrac{2}{(x+1)(x-1)} - \dfrac{1}{(x-1)}, \quad x \neq -1, \quad x \neq 1$$

This is the given equation with a denominator factored.

$$(x+1)(x-1) \cdot \dfrac{1}{x+1} = (x+1)(x-1)\left(\dfrac{2}{(x+1)(x-1)} - \dfrac{1}{x-1}\right)$$

Multiply both sides by $(x+1)(x-1)$, the LCD.

$$\cancel{(x+1)}(x-1) \cdot \dfrac{1}{\cancel{x+1}} = \cancel{(x+1)}\,\cancel{(x-1)} \cdot \dfrac{2}{\cancel{x+1}\,\cancel{(x-1)}} - (x+1)\cancel{(x-1)} \cdot \dfrac{1}{\cancel{(x-1)}}$$

Use the distributive property and divide out common factors.

$$1(x-1) = 2 - (x+1)$$

Simplify. This equation is cleared of fractions.

$$x - 1 = 2 - x - 1$$

Simplify.

$$x - 1 = -x + 1$$

Combine numerical terms.

$$x + x - 1 = -x + x + 1$$

Add x to both sides.

$$2x - 1 = 1$$

Simplify.

$$2x - 1 + 1 = 1 + 1$$

Add 1 to both sides.

$$2x = 2$$

Simplify.

$$\dfrac{2x}{2} = \dfrac{2}{2}$$

Divide both sides by 2.

$$x = 1$$

Simplify.

Study Tip

Reject any proposed solution that causes any denominator in an equation to equal 0.

The proposed solution, 1, is *not* a solution because of the restriction that $x \neq 1$. There is *no solution to this equation*. The solution set for this equation contains no elements. The solution set is $\varnothing$, the empty set.

Check Point 4   Solve: $\dfrac{1}{x+2} = \dfrac{4}{x^2-4} - \dfrac{1}{x-2}$.

 Solve a formula for a variable.

Solving a Formula for One of Its Variables

Solving a formula for a variable means rewriting the formula so that the variable is isolated on one side of the equation. It does not mean obtaining a numerical value for that variable.

To solve a formula for one of its variables, treat that variable as if it were the only variable in the equation. Think of the other variables as if they were numbers.

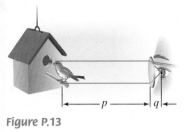

Figure P.13

EXAMPLE 5 Solving a Formula for a Variable

If you wear glasses, did you know that each lens has a measurement called its focal length, f? When an object is in focus, its distance from the lens, p, and the distance from the lens to your retina, q, satisfy the formula

$$\frac{1}{p} + \frac{1}{q} = \frac{1}{f}.$$

(See Figure P.13.) Solve this formula for p.

Solution   Our goal is to isolate the variable p. We begin by multiplying both sides by the least common denominator, pqf, to clear the equation of fractions.

We need to isolate p. $\dfrac{1}{p} + \dfrac{1}{q} = \dfrac{1}{f}$	This is the given formula.
$pqf\left(\dfrac{1}{p} + \dfrac{1}{q}\right) = pqf\left(\dfrac{1}{f}\right)$	Multiply both sides by pqf, the LCD.
$pqf\left(\dfrac{1}{p}\right) + pqf\left(\dfrac{1}{q}\right) = pqf\left(\dfrac{1}{f}\right)$	Use the distributive property on the left side and divide out common factors.
$qf + pf = pq$	Simplify. The formula is cleared of fractions.

We need to isolate p.

To collect terms with p on one side of the equation, subtract pf from both sides. Then factor p from the two resulting terms on the right to convert two occurrences of p into one.

$qf + pf = pq$	This is the equation cleared of fractions.
$qf + pf - pf = pq - pf$	Subtract pf from both sides.
$qf = pq - pf$	Simplify.
$qf = p(q - f)$	Factor out p, the specified variable.
$\dfrac{qf}{q - f} = \dfrac{p(q - f)}{q - f}$	Divide both sides by $q - f$ and solve for p.
$\dfrac{qf}{q - f} = p$	Simplify

Study Tip

You cannot solve $qf + pf = pq$ for p by dividing both sides by q and writing

$$\frac{qf + pf}{q} = p.$$

When a formula is solved for a specified variable, that variable must be isolated on one side. The variable p occurs on both sides of

$$\frac{qf + pf}{q} = p.$$

Check Point **5**   Solve for q: $\dfrac{1}{p} + \dfrac{1}{q} = \dfrac{1}{f}$.

  Solve equations involving absolute value.

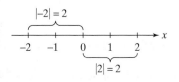

Figure P.14

Equations Involving Absolute Value

We have seen that the absolute value of x, denoted $|x|$, describes the distance of x from zero on a number line. Now consider an **absolute value equation**, such as

$$|x| = 2.$$

This means that we must determine real numbers whose distance from the origin on a number line is 2. Figure P.14 shows that there are two numbers such that $|x| = 2$, namely, 2 and -2. We write $x = 2$ or $x = -2$. This observation can be generalized as shown in the box on the next page.

> **Rewriting an Absolute Value Equation without Absolute Value Bars**
>
> If c is a positive real number and X represents any algebraic expression, then $|X| = c$ is equivalent to $X = c$ or $X = -c$.

EXAMPLE 6   Solving an Equation Involving Absolute Value

Solve: $5|1 - 4x| - 15 = 0$.

Solution

$$5|1 - 4x| - 15 = 0 \qquad \text{This is the given equation.}$$

We need to isolate $|1 - 4x|$, the absolute value expression.

$$5|1 - 4x| = 15 \qquad \text{Add 15 to both sides.}$$

$$|1 - 4x| = 3 \qquad \text{Divide both sides by 5.}$$

$$1 - 4x = 3 \quad \text{or} \quad 1 - 4x = -3 \qquad \text{Rewrite } |X| = c \text{ as } X = c \text{ or } X = -c.$$

$$-4x = 2 \qquad\qquad -4x = -4 \qquad \text{Subtract 1 from both sides of each equation.}$$

$$x = -\tfrac{1}{2} \qquad\qquad x = 1 \qquad \text{Divide both sides of each equation by } -4.$$

Take a moment to check $-\frac{1}{2}$ and 1, the proposed solutions, in the original equation, $5|1 - 4x| - 15 = 0$. In each case, you should obtain the true statement $0 = 0$. The solution set is $\{-\frac{1}{2}, 1\}$.

Check Point **6**   Solve: $4|1 - 2x| - 20 = 0$.

The absolute value of a number is never negative. Thus, if X is an algebraic expression and c is a negative number, then $|X| = c$ has no solution. For example, the equation $|3x - 6| = -2$ has no solution because $|3x - 6|$ cannot be negative. The solution set is $\varnothing$, the empty set.

The absolute value of 0 is 0. Thus, if X is an algebraic expression and $|X| = 0$, the solution is found by solving $X = 0$. For example, the solution of $|x - 2| = 0$ is obtained by solving $x - 2 = 0$. The solution is 2 and the solution set is $\{2\}$.

⑥ Solve quadratic equations by factoring.

Quadratic Equations and Factoring

Linear equations are first-degree polynomial equations of the form $ax + b = 0$. *Quadratic equations* are second-degree polynomial equations and contain an additional term involving the square of the variable.

> **Definition of a Quadratic Equation**
>
> A **quadratic equation** in x is an equation that can be written in the **general form**
> $$ax^2 + bx + c = 0,$$
> where a, b, and c are real numbers, with $a \neq 0$. A quadratic equation in x is also called a **second-degree polynomial equation** in x.

Here are examples of quadratic equations in general form:

$$4x^2 - 2x \quad = 0 \qquad\qquad 2x^2 + 7x - 4 = 0.$$

$a = 4 \quad b = -2 \quad c = 0$ $\qquad\qquad$ $a = 2 \quad b = 7 \quad c = -4$

Some quadratic equations, including the two shown above, can be solved by factoring and using the **zero-product principle**.

The Zero-Product Principle

If the product of two algebraic expressions is zero, then at least one of the factors is equal to zero.

$$\text{If } AB = 0, \text{ then } A = 0 \text{ or } B = 0.$$

The zero-product principle can be applied only when a quadratic equation is in general form, with zero on one side of the equation.

Solving a Quadratic Equation by Factoring

1. If necessary, rewrite the equation in the general form $ax^2 + bx + c = 0$, moving all terms to one side, thereby obtaining zero on the other side.
2. Factor completely.
3. Apply the zero-product principle, setting each factor containing a variable equal to zero.
4. Solve the equations in step 3.
5. Check the solutions in the original equation.

EXAMPLE 7   Solving Quadratic Equations by Factoring

Solve by factoring:

  a $4x^2 - 2x = 0$ **b** $2x^2 + 7x = 4.$

Solution

a. $4x^2 - 2x = 0$ The given equation is in general form, with zero on one side.

$2x(2x - 1) = 0$ Factor.

$2x = 0 \quad \text{or} \quad 2x - 1 = 0$ Use the zero-product principle and set each factor equal to zero.

$x = 0 \qquad\qquad 2x = 1$ Solve the resulting equations.

$$x = \frac{1}{2}$$

Check the proposed solutions, 0 and $\frac{1}{2}$, in the original equation.

Check 0:

$4x^2 - 2x = 0$

$4 \cdot 0^2 - 2 \cdot 0 \stackrel{?}{=} 0$

$0 - 0 \stackrel{?}{=} 0$

$0 = 0, \quad$ true

Check $\frac{1}{2}$:

$4x^2 - 2x = 0$

$4\left(\frac{1}{2}\right)^2 - 2\left(\frac{1}{2}\right) \stackrel{?}{=} 0$

$4\left(\frac{1}{4}\right) - 2\left(\frac{1}{2}\right) \stackrel{?}{=} 0$

$1 - 1 \stackrel{?}{=} 0$

$0 = 0, \quad$ true

The solution set is $\left\{0, \frac{1}{2}\right\}$.

b. $2x^2 + 7x = 4$ This is the given equation.

$2x^2 + 7x - 4 = 4 - 4$ Subtract 4 from both sides and write the quadratic equation in general form.

$2x^2 + 7x - 4 = 0$ Simplify.

$(2x - 1)(x + 4) = 0$ Factor.

$2x - 1 = 0 \quad \text{or} \quad x + 4 = 0$ Use the zero-product principle and set each factor equal to zero.

$2x = 1 \qquad\qquad x = -4$ Solve the resulting equations.

$$x = \frac{1}{2}$$

Check the proposed solutions, $\frac{1}{2}$ and -4, in the original equation.

Check $\frac{1}{2}$:	**Check -4:**
$2x^2 + 7x = 4$	$2x^2 + 7x = 4$
$2\left(\frac{1}{2}\right)^2 + 7\left(\frac{1}{2}\right) \stackrel{?}{=} 4$	$2(-4)^2 + 7(-4) \stackrel{?}{=} 4$
$\frac{1}{2} + \frac{7}{2} \stackrel{?}{=} 4$	$32 + (-28) \stackrel{?}{=} 4$
$4 = 4,$ true	$4 = 4,$ true

The solution set is $\left\{-4, \frac{1}{2}\right\}$.

Check Point 7   Solve by factoring:

    a. $3x^2 - 9x = 0$ **b.** $2x^2 + x = 1$.

⑦ Solve quadratic equations by the square root property.

Quadratic Equations and the Square Root Property

Quadratic equations of the form $u^2 = d$, where u is an algebraic expression, and d is a nonzero real number, can be solved by the *square root property*. First, isolate the squared expression u^2 on one side of the equation and the number d on the other side. Then take the square root of both sides. Remember, there are two numbers whose square is d. One number is $\sqrt{d}$ and one is $-\sqrt{d}$.

    We can use factoring to verify that $u^2 = d$ has these two solutions.

$u^2 = d$	This is the given equation.
$u^2 - d = 0$	Move all terms to one side and obtain zero on the other side.
$\left(u + \sqrt{d}\right)\left(u - \sqrt{d}\right) = 0$	Factor.
$u + \sqrt{d} = 0$ or $u - \sqrt{d} = 0$	Set each factor equal to zero.
$u = -\sqrt{d}$ $u = \sqrt{d}$	Solve the resulting equations.

Because the solutions differ only in sign, we can write them in abbreviated notation as $u = \pm\sqrt{d}$. We read this as "u equals positive or negative the square root of d" or "u equals plus or minus the square root of d."

    Now that we have verified these solutions, we can solve $u^2 = d$ directly by taking square roots. This process is called the **square root property**.

The Square Root Property

If u is an algebraic expression and d is a positive real number, then $u^2 = d$ has exactly two solutions:

$$\text{If } u^2 = d, \text{ then } u = \sqrt{d} \text{ or } u = -\sqrt{d}.$$

Equivalently,

$$\text{If } u^2 = d, \text{ then } u = \pm\sqrt{d}.$$

EXAMPLE 8   Solving Quadratic Equations by the Square Root Property

Solve by the square root property:

    a. $3x^2 - 15 = 0$ **b.** $(x - 2)^2 = 6$.

Solution   To apply the square root property, we need a squared expression by itself on one side of the equation.

$$3x^2 - 15 = 0 \qquad\qquad (x - 2)^2 = 6$$

We want x^2 by itself.

The squared expression is by itself.

a.
$$3x^2 - 15 = 0 \qquad \text{This is the original equation.}$$
$$3x^2 = 15 \qquad \text{Add 15 to both sides.}$$
$$x^2 = 5 \qquad \text{Divide both sides by 3.}$$
$$x = \sqrt{5} \text{ or } x = -\sqrt{5} \qquad \text{Apply the square root property.}$$
$$\text{Equivalently, } x = \pm\sqrt{5}.$$

By checking both proposed solutions in the original equation, we can confirm that the solution set is $\{-\sqrt{5}, \sqrt{5}\}$ or $\{\pm\sqrt{5}\}$.

b.
$$(x - 2)^2 = 6 \qquad \text{This is the original equation.}$$
$$x - 2 = \pm\sqrt{6} \qquad \text{Apply the square root property.}$$
$$x = 2 \pm\sqrt{6} \qquad \text{Add 2 to both sides.}$$

By checking both values in the original equation, we can confirm that the solution set is $\{2 + \sqrt{6}, 2 - \sqrt{6}\}$ or $\{2 \pm \sqrt{6}\}$.

Check Point 8 Solve by the square root property:
a. $3x^2 - 21 = 0$ **b.** $(x + 5)^2 = 11$.

⑧ Solve quadratic equations by completing the square.

Quadratic Equations and Completing the Square

How do we solve an equation in the form $ax^2 + bx + c = 0$ if the trinomial $ax^2 + bx + c$ cannot be factored? We cannot use the zero-product principle in such a case. However, we can convert the equation into an equivalent equation that can be solved using the square root property. This is accomplished by **completing the square**.

Completing the Square

If $x^2 + bx$ is a binomial, then by adding $\left(\dfrac{b}{2}\right)^2$, which is the square of half the coefficient of x, a perfect square trinomial will result. That is,

$$x^2 + bx + \left(\frac{b}{2}\right)^2 = \left(x + \frac{b}{2}\right)^2.$$

We can solve any quadratic equation by completing the square. If the coefficient of the x^2-term is one, we add the square of half the coefficient of x to both sides of the equation. **When you add a constant term to one side of the equation to complete the square, be certain to add the same constant to the other side of the equation.** These ideas are illustrated in Example 9.

EXAMPLE 9 Solving a Quadratic Equation by Completing the Square

Solve by completing the square: $x^2 - 6x + 4 = 0$.

Solution We begin by subtracting 4 from both sides. This is done to isolate the binomial $x^2 - 6x$, so that we can complete the square.

$$x^2 - 6x + 4 = 0 \qquad \text{This is the original equation.}$$
$$x^2 - 6x = -4 \qquad \text{Subtract 4 from both sides.}$$

Next, we work with $x^2 - 6x = -4$ and complete the square. Find half the coefficient of the x-term and square it. The coefficient of the x-term is -6. Half of -6 is -3 and $(-3)^2 = 9$. Thus, we add 9 to both sides of the equation.

$$x^2 - 6x + 9 = -4 + 9$$

Add 9 to both sides of $x^2 - 6x = -4$ to complete the square.

$$(x - 3)^2 = 5$$

Factor and simplify.

$$x - 3 = \sqrt{5} \quad \text{or} \quad x - 3 = -\sqrt{5}$$

Apply the square root property.

$$x = 3 + \sqrt{5} \qquad\qquad x = 3 - \sqrt{5}$$

Add 3 to both sides in each equation.

The solutions are $3 \pm \sqrt{5}$, and the solution set is $\{3 + \sqrt{5}, 3 - \sqrt{5}\}$ or $\{3 \pm \sqrt{5}\}$.

Check Point 9 Solve by completing the square: $x^2 + 4x - 1 = 0$.

Solve quadratic equations using the quadratic formula.

Quadratic Equations and the Quadratic Formula

We can use the method of completing the square to derive a formula that can be used to solve all quadratic equations. The derivation given here also shows a particular quadratic equation, $3x^2 - 2x - 4 = 0$, to specifically illustrate each of the steps.

Notice that if the coefficient of the x^2-term in a quadratic equation is not one, you must divide each side of the equation by this coefficient before completing the square.

Deriving the Quadratic Formula

General Form of a Quadratic Equation	Comment	A Specific Example
$ax^2 + bx + c = 0, a > 0$	This is the given equation.	$3x^2 - 2x - 4 = 0$
$x^2 + \dfrac{b}{a}x + \dfrac{c}{a} = 0$	Divide both sides by a so that the coefficient of x^2 is 1.	$x^2 - \dfrac{2}{3}x - \dfrac{4}{3} = 0$
$x^2 + \dfrac{b}{a}x = -\dfrac{c}{a}$	Isolate the binomial by adding $-\dfrac{c}{a}$ on both sides of the equation.	$x^2 - \dfrac{2}{3}x = \dfrac{4}{3}$
$x^2 + \dfrac{b}{a}x + \left(\dfrac{b}{2a}\right)^2 = -\dfrac{c}{a} + \left(\dfrac{b}{2a}\right)^2$ \n (half)^2	Complete the square. Add the square of half the coefficient of x to both sides.	$x^2 - \dfrac{2}{3}x + \left(-\dfrac{1}{3}\right)^2 = \dfrac{4}{3} + \left(-\dfrac{1}{3}\right)^2$ \n (half)^2
$x^2 + \dfrac{b}{a}x + \dfrac{b^2}{4a^2} = -\dfrac{c}{a} + \dfrac{b^2}{4a^2}$		$x^2 - \dfrac{2}{3}x + \dfrac{1}{9} = \dfrac{4}{3} + \dfrac{1}{9}$
$\left(x + \dfrac{b}{2a}\right)^2 = -\dfrac{c}{a} \cdot \dfrac{4a}{4a} + \dfrac{b^2}{4a^2}$	Factor on the left side and obtain a common denominator on the right side.	$\left(x - \dfrac{1}{3}\right)^2 = \dfrac{4}{3} \cdot \dfrac{3}{3} + \dfrac{1}{9}$
$\left(x + \dfrac{b}{2a}\right)^2 = \dfrac{-4ac + b^2}{4a^2}$	Add fractions on the right side.	$\left(x - \dfrac{1}{3}\right)^2 = \dfrac{12 + 1}{9}$
$\left(x + \dfrac{b}{2a}\right)^2 = \dfrac{b^2 - 4ac}{4a^2}$		$\left(x - \dfrac{1}{3}\right)^2 = \dfrac{13}{9}$
$x + \dfrac{b}{2a} = \pm\sqrt{\dfrac{b^2 - 4ac}{4a^2}}$	Apply the square root property.	$x - \dfrac{1}{3} = \pm\sqrt{\dfrac{13}{9}}$
$x + \dfrac{b}{2a} = \pm\dfrac{\sqrt{b^2 - 4ac}}{2a}$	Take the square root of the quotient, simplifying the denominator.	$x - \dfrac{1}{3} = \pm\dfrac{\sqrt{13}}{3}$
$x = \dfrac{-b}{2a} \pm \dfrac{\sqrt{b^2 - 4ac}}{2a}$	Solve for x by subtracting $\dfrac{b}{2a}$ from both sides.	$x = \dfrac{1}{3} \pm \dfrac{\sqrt{13}}{3}$
$x = \dfrac{-b \pm \sqrt{b^2 - 4ac}}{2a}$	Combine fractions on the right side.	$x = \dfrac{1 \pm \sqrt{13}}{3}$

The formula shown at the bottom of the left column in the table is called the *quadratic formula*. A similar proof shows that the same formula can be used to solve quadratic equations if a, the coefficient of the x^2-term, is negative.

The Quadratic Formula

The solutions of a quadratic equation in general form $ax^2 + bx + c = 0$, with $a \neq 0$, are given by the **quadratic formula**

$$x = \frac{-b \pm \sqrt{b^2 - 4ac}}{2a}.$$

x equals negative *b* plus or minus the square root of $b^2 - 4ac$, all divided by 2*a*.

To use the quadratic formula, write the quadratic equation in general form if necessary. Then determine the numerical values for a (the coefficient of the x^2-term), b (the coefficient of the x-term), and c (the constant term). Substitute the values of a, b, and c into the quadratic formula and evaluate the expression. The $\pm$ sign indicates that there are two solutions of the equation.

EXAMPLE 10   Solving a Quadratic Equation Using the Quadratic Formula

Solve using the quadratic formula: $2x^2 - 6x + 1 = 0$.

Solution   The given equation is in general form. Begin by identifying the values for a, b, and c.

$$4x^2 - 8x + 1 = 0$$

$a = 4$ $b = -8$ $c = 1$

Substituting these values into the quadratic formula and simplifying gives the equation's solutions.

$$x = \frac{-b \pm \sqrt{b^2 - 4ac}}{2a}$$ Use the quadratic formula.

$$= \frac{-(-6) \pm \sqrt{(-6)^2 - 4(2)(1)}}{2 \cdot 2}$$ Substitute the values for a, b, and c: $a = 2$, $b = -6$, and $c = 1$.

$$= \frac{6 \pm \sqrt{36 - 8}}{4}$$ $-(-6) = 6$, $(-6)^2 = (-6)(-6) = 36$, and $4(2)(1) = 8$.

$$= \frac{6 \pm \sqrt{28}}{4}$$ Complete the subtraction under the radical.

$$= \frac{6 \pm 2\sqrt{7}}{4}$$ $\sqrt{28} = \sqrt{4 \cdot 7} = \sqrt{4}\sqrt{7} = 2\sqrt{7}$

$$= \frac{2(3 \pm \sqrt{7})}{4}$$ Factor out 2 from the numerator.

$$= \frac{3 \pm \sqrt{7}}{2}$$ Divide the numerator and denominator by 2.

The solution set is $\left\{ \dfrac{3 + \sqrt{7}}{2}, \dfrac{3 - \sqrt{7}}{2} \right\}$ or $\left\{ \dfrac{3 \pm \sqrt{7}}{2} \right\}$.

To Die at Twenty

Can the equations
$$7x^5 + 12x^3 - 9x + 4 = 0$$
and
$$8x^6 - 7x^5 + 4x^3 - 19 = 0$$
be solved using a formula similar to the quadratic formula? The first equation has five solutions and the second has six solutions, but they cannot be found using a formula. How do we know? In 1832, a 20-year-old Frenchman, Evariste Galois, wrote down a proof showing that there is no general formula to solve equations when the exponent on the variable is 5 or greater. Galois was jailed as a political activist several times while still a teenager. The day after his brilliant proof he fought a duel over a woman. The duel was a political setup. As he lay dying, Galois told his brother, Alfred, of the manuscript that contained his proof: "Mathematical manuscripts are in my room. On the table. Take care of my work. Make it known. Important. Don't cry, Alfred. I need all my courage—to die at twenty." (Our source is Leopold Infeld's biography of Galois, *Whom the Gods Love*. Some historians, however, dispute the story of Galois's ironic death the very day after his algebraic proof. Mathematical truths seem more reliable than historical ones!)

Study Tip

Checking irrational solutions can be time-consuming. The solutions given by the quadratic formula are always correct, unless you have made a careless error. Checking for computational errors or errors in simplification is sufficient.

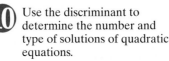

Check Point **10**   Solve using the quadratic formula:

$$2x^2 + 2x - 1 = 0.$$

⑩ Use the discriminant to determine the number and type of solutions of quadratic equations.

Quadratic Equations and the Discriminant

The quantity $b^2 - 4ac$, which appears under the radical sign in the quadratic formula, is called the **discriminant**. Table P.4 shows how the discriminant of the quadratic equation $ax^2 + bx + c = 0$ determines the number and type of solutions.

Table P.4   The Discriminant and the Kinds of Solutions to $ax^2 + bx + c = 0$

Discriminant $b^2 - 4ac$	Kinds of Solutions to $ax^2 + bx + c = 0$
$b^2 - 4ac > 0$	**Two unequal real solutions;** If a, b, and c are rational numbers and the discriminant is a perfect square, the solutions are rational. If the discriminant is not a perfect square, the solutions are irrational.
$b^2 - 4ac = 0$	**One solution (a repeated solution) that is a real number;** If a, b, and c are rational numbers, the repeated solution is also a rational number.
$b^2 - 4ac < 0$	**No real solutions**

EXAMPLE 11   Using the Discriminant

Compute the discriminant of $4x^2 - 8x + 1 = 0$. What does the discriminant indicate about the number and type of solutions?

Solution   Begin by identifying the values for a, b, and c.

$$4x^2 - 8x + 1 = 0$$

$a = 4$ $b = -8$ $c = 1$

Substitute and compute the discriminant:

$$b^2 - 4ac = (-8)^2 - 4 \cdot 4 \cdot 1 = 64 - 16 = 48.$$

The discriminant is 48. Because the discriminant is positive, the equation $4x^2 - 8x + 1 = 0$ has two unequal real solutions.

Check Point **11**   Compute the discriminant of $3x^2 - 2x + 5 = 0$. What does the discriminant indicate about the number and type of solutions?

⑪ Determine the most efficient method to use when solving a quadratic equation.

Determining Which Method to Use

All quadratic equations can be solved by the quadratic formula. However, if an equation is in the form $u^2 = d$, such as $x^2 = 5$ or $(2x + 3)^2 = 8$, it is faster to use the square root property, taking the square root of both sides. If the equation is not in the form $u^2 = d$, write the quadratic equation in general form ($ax^2 + bx + c = 0$). Try to solve the equation by factoring. If $ax^2 + bx + c$ cannot be factored, then solve the quadratic equation by the quadratic formula.

Because we used the method of completing the square to derive the quadratic formula, we no longer need it for solving quadratic equations. However, we will use completing the square later in the book to help graph circles and other kinds of equations.

Table P.5 summarizes our observations about which technique to use when solving a quadratic equation.

Table P.5   Determining the Most Efficient Technique to Use When Solving a Quadratic Equation

Description and Form of the Quadratic Equation	Most Efficient Solution Method	Example
$ax^2 + bx + c = 0$ and $ax^2 + bx + c$ can be factored easily.	Factor and use the zero-product principle.	$3x^2 + 5x - 2 = 0$ $(3x - 1)(x + 2) = 0$ $3x - 1 = 0$ or $x + 2 = 0$ $x = \dfrac{1}{3}$ $x = -2$
$ax^2 + bx = 0$ The quadratic equation has no constant term. $(c = 0)$	Factor and use the zero-product principle.	$6x^2 + 9x = 0$ $3x(2x + 3) = 0$ $3x = 0$ or $2x + 3 = 0$ $x = 0$ $2x = -3$ $x = -\frac{3}{2}$
$ax^2 + c = 0$ The quadratic equation has no x-term. $(b = 0)$	Solve for x^2 and apply the square root property.	$7x^2 - 4 = 0$ $7x^2 = 4$ $x^2 = \dfrac{4}{7}$ $x = \pm\dfrac{2}{\sqrt{7}} = \pm\dfrac{2}{\sqrt{7}}\cdot\dfrac{\sqrt{7}}{\sqrt{7}} = \pm\dfrac{2\sqrt{7}}{7}$
$u^2 = d$; u is a first-degree polynomial.	Use the square root property.	$(x + 4)^2 = 5$ $x + 4 = \pm\sqrt{5}$ $x = -4 \pm \sqrt{5}$
$ax^2 + bx + c = 0$ and $ax^2 + bx + c$ cannot be factored or the factoring is too difficult.	Use the quadratic formula: $x = \dfrac{-b \pm \sqrt{b^2 - 4ac}}{2a}$.	$x^2 - 2x - 6 = 0$ $a = 1$ $b = -2$ $c = -6$ $x = \dfrac{-(-2) \pm \sqrt{(-2)^2 - 4(1)(-6)}}{2}$ $= \dfrac{2 \pm \sqrt{4 - (-24)}}{2}$ $= \dfrac{2 \pm \sqrt{28}}{2} = \dfrac{2 \pm \sqrt{4}\sqrt{7}}{2}$ $= \dfrac{2 \pm 2\sqrt{7}}{2} = \dfrac{2(1 \pm \sqrt{7})}{2}$ $= 1 \pm \sqrt{7}$

⑫ Solve radical equations.

Radical Equations

A **radical equation** is an equation in which the variable occurs in a square root, cube root, or any higher root. An example of a radical equation is

$$\sqrt{x} = 9.$$

We solve the equation by squaring both sides:

$$\left(\sqrt{x}\right)^2 = 9^2$$

Squaring both sides eliminates the square root.

$$x = 81.$$

The proposed solution, 81, can be checked in the original equation, $\sqrt{x} = 9$. Because $\sqrt{81} = 9$, the solution is 81 and the solution set is $\{81\}$.

In general, we solve radical equations with square roots by squaring both sides of the equation. We solve radical equations with nth roots by raising both sides of the equation to the nth power. Unfortunately, if n is even, all the solutions of the equation raised to the even power may not be solutions of the original equation. Consider, for example, the equation

$$x = 4.$$

If we square both sides, we obtain

$$x^2 = 16.$$
$$x = \pm\sqrt{16} = \pm 4.$$

This new equation has two solutions, -4 and 4. By contrast, only 4 is a solution of the original equation, $x = 4$. For this reason, **when raising both sides of an equation to an even power, always check proposed solutions in the original equation**.

Here is a general method for solving radical equations with nth roots:

Solving Radical Equations Containing nth Roots

1. If necessary, arrange terms so that one radical is isolated on one side of the equation.
2. Raise both sides of the equation to the nth power to eliminate the nth root.
3. Solve the resulting equation. If this equation still contains radicals, repeat steps 1 and 2.
4. Check all proposed solutions in the original equation.

Extra solutions may be introduced when you raise both sides of a radical equation to an even power. Such solutions, which are not solutions of the given equation, are called **extraneous solutions** or **extraneous roots**.

EXAMPLE 12 Solving a Radical Equation

Solve: $\sqrt{2x - 1} + 2 = x$.

Solution

Step 1 Isolate a radical on one side. We isolate the radical, $\sqrt{2x - 1}$, by subtracting 2 from both sides.

$$\sqrt{2x - 1} + 2 = x \qquad \text{This is the given equation.}$$
$$\sqrt{2x - 1} = x - 2 \qquad \text{Subtract 2 from both sides.}$$

Step 2 Raise both sides to the nth power. Because n, the index, is 2, we square both sides.

$$\left(\sqrt{2x - 1}\right)^2 = (x - 2)^2$$
$$2x - 1 = x^2 - 4x + 4 \qquad \begin{array}{l}\text{Simplify. Use the formula}\\ (A - B)^2 = A^2 - 2AB + B^2\\ \text{on the right side.}\end{array}$$

Step 3 Solve the resulting equation. Because of the x^2-term, the resulting equation is a quadratic equation. We can obtain 0 on the left side by subtracting $2x$ and adding 1 on both sides.

$$2x - 1 = x^2 - 4x + 4 \qquad \text{The resulting equation is quadratic.}$$
$$0 = x^2 - 6x + 5 \qquad \begin{array}{l}\text{Write in general form, subtracting } 2x \text{ and}\\ \text{adding 1 on both sides.}\end{array}$$
$$0 = (x - 1)(x - 5) \qquad \text{Factor.}$$
$$x - 1 = 0 \quad \text{or} \quad x - 5 = 0 \qquad \text{Set each factor equal to 0.}$$
$$x = 1 \qquad\qquad x = 5 \qquad \text{Solve the resulting equations.}$$

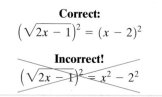

Study Tip

Be sure to square *both sides* of an equation. Do *not* square each term.

Correct:
$$\left(\sqrt{2x - 1}\right)^2 = (x - 2)^2$$

Incorrect!
$$\left(\sqrt{2x - 1}\right)^2 = x^2 - 2^2$$

Step 4 Check the proposed solutions in the original equation.

Check 1:

$\sqrt{2x-1}+2=x$

$\sqrt{2\cdot1-1}+2\overset{?}{=}1$

$\sqrt{1}+2\overset{?}{=}1$

$1+2\overset{?}{=}1$

$3=1,$ false

Check 5:

$\sqrt{2x-1}+2=x$

$\sqrt{2\cdot5-1}+2\overset{?}{=}5$

$\sqrt{9}+2\overset{?}{=}5$

$3+2\overset{?}{=}5$

$5=5,$ true

Thus, 1 is an extraneous solution. The only solution is 5, and the solution set is $\{5\}$.

Check Point 12 Solve: $\sqrt{x+3}+3=x$.

EXERCISE SET P.7

Practice Exercises

In Exercises 1–16, solve each linear equation.

1. $7x-5=72$

2. $6x-3=63$

3. $11x-(6x-5)=40$

4. $5x-(2x-10)=35$

5. $2x-7=6+x$

6. $3x+5=2x+13$

7. $7x+4=x+16$

8. $13x+14=12x-5$

9. $3(x-2)+7=2(x+5)$

10. $2(x-1)+3=x-3(x+1)$

11. $\dfrac{x+3}{6}=\dfrac{3}{8}+\dfrac{x-5}{4}$

12. $\dfrac{x+1}{4}=\dfrac{1}{6}+\dfrac{2-x}{3}$

13. $\dfrac{x}{4}=2+\dfrac{x-3}{3}$

14. $5+\dfrac{x-2}{3}=\dfrac{x+3}{8}$

15. $\dfrac{x+1}{3}=5-\dfrac{x+2}{7}$

16. $\dfrac{3x}{5}-\dfrac{x-3}{2}=\dfrac{x+2}{3}$

*Exercises 17–26 contain rational equations with variables in denominators. For each equation, **a.** Write the value or values of the variable that make a denominator zero. These are the restrictions on the variable. **b.** Keeping the restrictions in mind, solve the equation.*

17. $\dfrac{1}{x-1}+5=\dfrac{11}{x-1}$

18. $\dfrac{3}{x+4}-7=\dfrac{-4}{x+4}$

19. $\dfrac{8x}{x+1}=4-\dfrac{8}{x+1}$

20. $\dfrac{2}{x-2}=\dfrac{x}{x-2}-2$

21. $\dfrac{3}{2x-2}+\dfrac{1}{2}=\dfrac{2}{x-1}$

22. $\dfrac{3}{x+3}=\dfrac{5}{2x+6}+\dfrac{1}{x-2}$

23. $\dfrac{2}{x+1}-\dfrac{1}{x-1}=\dfrac{2x}{x^2-1}$

24. $\dfrac{4}{x+5}+\dfrac{2}{x-5}=\dfrac{32}{x^2-25}$

25. $\dfrac{1}{x-4}-\dfrac{5}{x+2}=\dfrac{6}{x^2-2x-8}$

26. $\dfrac{1}{x-3}-\dfrac{2}{x+1}=\dfrac{8}{x^2-2x-3}$

In Exercises 27–42, solve each formula for the specified variable. Do you recognize the formula? If so, what does it describe?

27. $I=Prt$ for P

28. $C=2\pi r$ for r

29. $T=D+pm$ for p

30. $P=C+MC$ for M

31. $A=\frac12 h(a+b)$ for a

32. $A=\frac12 h(a+b)$ for b

33. $S=P+Prt$ for r

34. $S=P+Prt$ for t

35. $B=\dfrac{F}{S-V}$ for S

36. $S=\dfrac{C}{1-r}$ for r

37. $IR+Ir=E$ for I

38. $A=2lw+2lh+2wh$ for h

39. $\dfrac{1}{p}+\dfrac{1}{q}=\dfrac{1}{f}$ for f

40. $\dfrac{1}{R}=\dfrac{1}{R_1}+\dfrac{1}{R_2}$ for R_1

41. $f=\dfrac{f_1f_2}{f_1+f_2}$ for f_1

42. $f=\dfrac{f_1f_2}{f_1+f_2}$ for f_2

In Exercises 43–54, solve each absolute value equation or indicate the equation has no solution.

43. $|x-2|=7$

44. $|x+1|=5$

45. $|2x-1|=5$

46. $|2x-3|=11$

47. $2|3x-2|=14$

48. $3|2x-1|=21$

49. $2\left|4-\dfrac{5}{2}x\right|+6=18$

50. $4\left|1-\dfrac{3}{4}x\right|+7=10$

51. $|x+1|+5=3$

52. $|x+1|+6=2$

53. $|2x-1|+3=3$

54. $|3x-2|+4=4$

In Exercises 55–60, solve each quadratic equation by factoring.

55. $x^2-3x-10=0$

56. $x^2-13x+36=0$

57. $x^2=8x-15$

58. $x^2=-11x-10$

59. $5x^2=20x$

60. $3x^2=12x$

In Exercises 61–66, solve each quadratic equation by the square root property.

61. $3x^2 = 27$

62. $5x^2 = 45$

63. $5x^2 + 1 = 51$

64. $3x^2 - 1 = 47$

65. $3(x - 4)^2 = 15$

66. $3(x + 4)^2 = 21$

In Exercises 67–74, solve each quadratic equation by completing the square.

67. $x^2 + 6x = 7$

68. $x^2 + 6x = -8$

69. $x^2 - 2x = 2$

70. $x^2 + 4x = 12$

71. $x^2 - 6x - 11 = 0$

72. $x^2 - 2x - 5 = 0$

73. $x^2 + 4x + 1 = 0$

74. $x^2 + 6x - 5 = 0$

In Exercises 75–82, solve each quadratic equation using the quadratic formula.

75. $x^2 + 8x + 15 = 0$

76. $x^2 + 8x + 12 = 0$

77. $x^2 + 5x + 3 = 0$

78. $x^2 + 5x + 2 = 0$

79. $3x^2 - 3x - 4 = 0$

80. $5x^2 + x - 2 = 0$

81. $4x^2 = 2x + 7$

82. $3x^2 = 6x - 1$

Compute the discriminant of each equation in Exercises 83–90. What does the discriminant indicate about the number and type of solutions?

83. $x^2 - 4x - 5 = 0$

84. $4x^2 - 2x + 3 = 0$

85. $2x^2 - 11x + 3 = 0$

86. $2x^2 + 11x - 6 = 0$

87. $x^2 = 2x - 1$

88. $3x^2 = 2x - 1$

89. $x^2 - 3x - 7 = 0$

90. $3x^2 + 4x - 2 = 0$

In Exercises 91–114, solve each quadratic equation by the method of your choice.

91. $2x^2 - x = 1$

92. $3x^2 - 4x = 4$

93. $5x^2 + 2 = 11x$

94. $5x^2 = 6 - 13x$

95. $3x^2 = 60$

96. $2x^2 = 250$

97. $x^2 - 2x = 1$

98. $2x^2 + 3x = 1$

99. $(2x + 3)(x + 4) = 1$

100. $(2x - 5)(x + 1) = 2$

101. $(3x - 4)^2 = 16$

102. $(2x + 7)^2 = 25$

103. $3x^2 - 12x + 12 = 0$

104. $9 - 6x + x^2 = 0$

105. $4x^2 - 16 = 0$

106. $3x^2 - 27 = 0$

107. $x^2 = 4x - 2$

108. $x^2 = 6x - 7$

109. $2x^2 - 7x = 0$

110. $2x^2 + 5x = 3$

111. $\dfrac{1}{x} + \dfrac{1}{x + 2} = \dfrac{1}{3}$

112. $\dfrac{1}{x} + \dfrac{1}{x + 3} = \dfrac{1}{4}$

113. $\dfrac{2x}{x - 3} + \dfrac{6}{x + 3} = -\dfrac{28}{x^2 - 9}$

114. $\dfrac{3}{x - 3} + \dfrac{5}{x - 4} = \dfrac{x^2 - 20}{x^2 - 7x + 12}$

In Exercises 115–124, solve each radical equation. Check all proposed solutions.

115. $\sqrt{3x + 18} = x$

116. $\sqrt{20 - 8x} = x$

117. $\sqrt{x + 3} = x - 3$

118. $\sqrt{x + 10} = x - 2$

119. $\sqrt{2x + 13} = x + 7$

120. $\sqrt{6x + 1} = x - 1$

121. $x - \sqrt{2x + 5} = 5$

122. $x - \sqrt{x + 11} = 1$

123. $\sqrt{2x + 19} - 8 = x$

124. $\sqrt{2x + 15} - 6 = x$

Practice Plus

In Exercises 125–134, solve each equation.

125. $25 - [2 + 5x - 3(x + 2)] =$
$$-3(2x - 5) - [5(x - 1) - 3x + 3$$

126. $45 - [4 - 2x - 4(x + 7)] =$
$$-4(1 + 3x) - [4 - 3(x + 2) - 2(2x - 5)$$

127. $7 - 7x = (3x + 2)(x - 1)$

128. $10x - 1 = (2x + 1)^2$

129. $|x^2 + 2x - 36| = 12$

130. $|x^2 + 6x + 1| = 8$

131. $\dfrac{1}{x^2 - 3x + 2} = \dfrac{1}{x + 2} + \dfrac{5}{x^2 - 4}$

132. $\dfrac{x - 1}{x - 2} + \dfrac{x}{x - 3} = \dfrac{1}{x^2 - 5x + 6}$

133. $\sqrt{x + 8} - \sqrt{x - 4} = 2$

134. $\sqrt{x + 5} - \sqrt{x - 3} = 2$

Exercises 135–136, list all number that must be excluded from the domain of each rational expression.

135. $\dfrac{3}{2x^2 + 4x - 9}$

136. $\dfrac{7}{2x^2 - 8x + 5}$

Application Exercises

In the section opener, we used two formulas to model the level of depression, D, in response to the intensity of a negative life event, x, from 1, low, to 10, high:

Low-Humor Group	High-Humor Group
$D = \dfrac{10}{9}x + \dfrac{53}{9}$	$D = \dfrac{1}{9}x + \dfrac{26}{9}.$

Use these formulas to solve Exercises 137–138.

137. If the high-humor group averages a level of depression of 3.5, or $\dfrac{7}{2}$, in response to a negative life event, what is the intensity of that event? How is the solution shown on the line graph in Figure P.12 on page 77?

138. If the low-humor group averages a level of depression of 10 in response to a negative life event, what is the intensity of that event? How is the solution shown on the line graph in Figure P.12 on page 77?

139. A company wants to increase the 10% peroxide content of its product by adding pure peroxide (100% peroxide). If x liters of pure peroxide are added to 500 liters of its 10% solution, the concentration, C, of the new mixture is given by

$$C = \frac{x + 0.1(500)}{x + 500}.$$

How many liters of pure peroxide should be added to produce a new product that is 28% peroxide?

140. Suppose that x liters of pure acid are added to 200 liters of a 35% acid solution.

    a. Write a formula that gives the concentration, C, of the new mixture. (*Hint*: See Exercise 139.)

    b. How many liters of pure acid should be added to produce a new mixture that is 74% acid?

A driver's age has something to do with his or her chance of getting into a fatal car crash. The bar graph shows the number of fatal vehicle crashes per 100 million miles driven for drivers of various age groups. For example, 25-year-old drivers are involved in 4.1 fatal crashes per 100 million miles driven. Thus, when a group of 25-year-old Americans have driven a total of 100 million miles, approximately 4 have been in accidents in which someone died.

Age of U.S. Drivers and Fatal Crashes

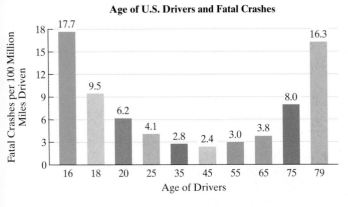

Source: Insurance Institute for Highway Safety

The number of fatal vehicle crashes per 100 million miles, y, for drivers of age x can be modeled by the formula

$$y = 0.013x^2 - 1.19x + 28.24.$$

Use the formula to solve Exercises 141–142.

141. What age groups are expected to be involved in 3 fatal crashes per 100 million miles driven? How well does the formula model the trend in the actual data shown in the bar graph?

142. What age groups are expected to be involved in 10 fatal crashes per 100 million miles driven? How well does the formula model the trend in the actual data shown in the bar graph ?

In 2002, the average surface temperature on Earth was 57.9°F, approximately 1.4° higher than it was one hundred years ago. Worldwide temperatures have risen only 9°F since the end of the last ice age 12,000 years ago. Most climatologists are convinced that over the next one hundred years, global temperatures will continue to increase, possibly setting off a chain of devastating events beginning with a rise in sea levels worldwide and ending with the destruction of water supplies, forests, and agriculture in many parts of the world. The graph shows global annual average temperatures from 1880 through 2002, with projections from 2002 through 2100.

Global Annual Average Temperatures and Projections through 2100

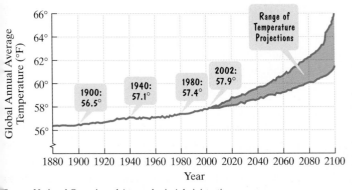

Source: National Oceanic and Atmospheric Administration

The temperature projections shown in the graph can be modeled by two equations:

$$H = 0.083x + 57.9$$

> Models temperatures at the high end of the range

$$L = 0.36\sqrt{x} + 57.9.$$

> Models temperatures at the low end of the range

In these equations, H and L describe projected global annual average temperatures, in degrees Fahrenheit, x years after 2002, where $0 \le x \le 98$. Use the models to solve Exercises 143–146.

143. Use H and L to determine the temperatures at the high and low end of the range of projected global average temperatures for 2100. Round to the nearest tenth of a degree.

144. Use H and L to determine the temperatures at the high and low end of the range of projected global average temperatures for 2080. Round to the nearest tenth of a degree.

145. Use H and L to determine by which year the projected global average temperature will exceed the 2002 average of 57.9° by one degree. Round to the nearest year.

146. Use H and L to determine by which year the projected global average temperature will exceed the 2002 average of 57.9° by two degrees.

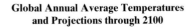

Writing in Mathematics

147. What is a linear equation in one variable? Give an example of this type of equation.

148. Explain how to determine the restrictions on the variable for the equation

$$\frac{3}{x + 5} + \frac{4}{x - 2} = \frac{7}{x^2 + 3x - 6}.$$

149. What does it mean to solve a formula for a variable?

150. Explain how to solve an equation involving absolute value.

151. Why does the procedure that you explained in Exercise 150 not apply to the equation $|x - 2| = -3$? What is the solution set for this equation?

152. What is a quadratic equation?

153. Explain how to solve $x^2 + 6x + 8 = 0$ using factoring and the zero-product principle.

154. Explain how to solve $x^2 + 6x + 8 = 0$ by completing the square.

155. Explain how to solve $x^2 + 6x + 8 = 0$ using the quadratic formula.

156. How is the quadratic formula derived?

157. What is the discriminant and what information does it provide about a quadratic equation?

158. If you are given a quadratic equation, how do you determine which method to use to solve it?

159. In solving $\sqrt{2x - 1} + 2 = x$, why is it a good idea to isolate the radical term? What if we don't do this and simply square each side? Describe what happens.

160. What is an extraneous solution to a radical equation?

Critical Thinking Exercises

161. Which one of the following is true?

    a. The equation $(2x - 3)^2 = 25$ is equivalent to $2x - 3 = 5$.

    b. Every quadratic equation has two distinct numbers in its solution set.

    c. The equation $3y - 1 = 11$ and $3y - 7 = 5$ are equivalent.

    d. The equation $ax^2 + c = 0$, $a \neq 0$, cannot be solved by the quadratic formula.

162. Find b such that $\dfrac{7x + 4}{b} + 13 = x$ will have a solution set given by $\{-6\}$.

163. Write a quadratic equation in general form whose solution set is $\{-3, 5\}$.

165. Solve for C: $V = C - \dfrac{C - S}{L}N$.

165. Solve for t: $s = -16t^2 + v_0 t$.

SECTION P.8   *Modeling with Equations*

Objective

1 Use equations to solve problems.

The human race is undeniably becoming a faster race. Since the beginning of the past century, track-and-field records have fallen in everything from sprints to miles to marathons. The performance arc is clearly rising, but no one knows how much higher it can climb. At some point, even the best-trained body simply has to up and quit. The question is, just where is that point, and is it possible for athletes, trainers, and genetic engineers to push it higher? In this section, you will learn a problem-solving strategy that uses linear equations to determine if anyone will ever run a 3-minute mile.

 Use equations to solve problems.

Problem Solving with Linear Equations

We have seen that a model is a mathematical representation of a real-world situation. In this section, we will be solving problems that are presented in English. This means that we must obtain models by translating from the ordinary language of English into the language of algebraic equations. To translate, however, we must understand the English prose and be familiar with the forms of algebraic language. Here are some general steps we will follow in solving word problems:

Study Tip

When solving word problems, particularly problems involving geometric figures, drawing a picture of the situation is often helpful. Label x on your drawing and, where appropriate, label other parts of the drawing in terms of x.

Strategy for Solving Word Problems

Step 1   Read the problem carefully. Attempt to state the problem in your own words and state what the problem is looking for. Let x (or any variable) represent one of the quantities in the problem.

Step 2   If necessary, write expressions for any other unknown quantities in the problem in terms of x.

Step 3   Write an equation in x that models the verbal conditions of the problem.

Step 4   Solve the equation and answer the problem's question.

Step 5   Check the solution *in the original wording* of the problem, not in the equation obtained from the words.

EXAMPLE 1   Walk It Off

Experts concerned with fitness and health suggest that we should walk 10,000 steps per day, about 5 miles. Depending on stride length, each mile ranges between 2000 and 2500 steps. The graph in Figure P.15 shows the number of steps it takes to burn off various foods. (The data are based on a body weight of 150 to 165 pounds.)

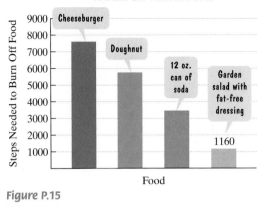

Number of Steps It Takes to Burn Off Various Foods

Figure P.15

Source: The Step Diet Book

The number of steps needed to burn off a cheeseburger exceeds the number needed to burn off a 12-ounce soda by 4140. The number needed to burn off a doughnut exceeds the number needed to burn off a 12-ounce soda by 2300. If you chow down a cheeseburger, doughnut, and 12-ounce soda, a 16,790-step walk is needed to burn off the calories (and perhaps alleviate the guilt). Determine the number of steps it takes to burn off a cheeseburger, a doughnut, and a 12-ounce soda.

Solution

Step 1   Let x represent one of the quantities. We know something about the number of steps needed to burn off a cheeseburger and a doughnut: The numbers exceed that of a 12-ounce soda by 4140 and 2300, respectively. We will let

x = the number of steps needed to burn off a 12-ounce soda.

Step 2   Represent other quantities in terms of x. Because the number of steps needed to burn off a cheeseburger exceeds the number needed to burn off a 12-ounce soda by 4140, let

$x + 4140$ = the number of steps needed to burn off a cheeseburger.

Because the number of steps needed to burn off a doughnut exceeds the number needed to burn off a 12-ounce soda by 2300, let

$x + 2300$ = the number of steps needed to burn off a doughnut.

Step 3   Write an equation in x that models the conditions. A 16,790-step walk is needed to burn off a "meal" consisting of a cheeseburger, a doughnut, and a 12-ounce soda.

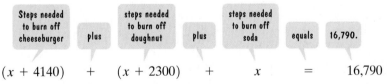

$$(x + 4140) \quad + \quad (x + 2300) \quad + \quad x \quad = \quad 16{,}790$$

Step 4   Solve the equation and answer the question.

$$(x + 4140) + (x + 2300) + x = 16{,}790 \qquad \text{This is the equation that models the problem's conditions.}$$

$$3x + 6440 = 16{,}790 \qquad \text{Remove parentheses, regroup, and combine like terms.}$$

$$3x = 10{,}350 \qquad \text{Subtract 6440 from both sides.}$$

$$x = 3450 \qquad \text{Divide both sides by 3.}$$

Thus,

the number of steps needed to burn off a 12-ounce soda = x = 3450.

the number of steps needed to burn off a cheeseburger
= $x + 4140 = 3450 + 4140 = 7590.$

the number of steps needed to burn off a doughnut
= $x + 2300 = 3450 + 2300 = 5750.$

It takes 7590 steps to burn off a cheeseburger, 5750 steps to burn off a doughnut, and 3450 steps to burn off a 12-ounce soda.

Step 5   Check the proposed solution in the original wording of the problem. The problem states that a 16,790-step walk is needed to burn off the calories in the three foods combined. By adding 7590, 5750, and 3450, the numbers that we found for each of the foods, we obtain

$$7590 + 5750 + 3450 = 16,790,$$

as specified by the problem's conditions.

Study Tip

Modeling with the word "exceeds" can be a bit tricky. It's helpful to identify the smaller quantity. Then add to this quantity to represent the larger quantity. For example, suppose that Tim's height exceeds Tom's height by a inches. Tom is the shorter person. If Tom's height is represented by x, then Tim's height is represented by $x + a$.

Check Point 1   Basketball, bicycle riding, and football are the three sports and recreational activities in the United States with the greatest number of medically treated injuries. In 2004, the number of injuries from basketball exceeded those from football by 0.6 million. The number of injuries from bicycling exceeded those from football by 0.3 million. Combined, basketball, bicycling, and football accounted for 3.9 million injuries. Determine the number of medically treated injuries from each of these recreational activities in 2004.

(*Source:* U.S. Consumer Product Safety Commission)

Mile Records			
1886	4:12.3	1958	3:54.5
1923	4:10.4	1966	3:51.3
1933	4:07.6	1979	3:48.9
1945	4:01.3	1985	3:46.3
1954	3:59.4	1999	3:43.1

Source: U.S.A. Track and Field

EXAMPLE 2   Will Anyone Ever Run a Three-Minute Mile?

One yardstick for measuring how steadily—if slowly—athletic performance has improved is the mile run. In 1923, the record for the mile was a comparatively sleepy 4 minutes, 10.4 seconds. In 1954, Roger Bannister of Britain cracked the 4-minute mark, coming in at 3 minutes, 59.4 seconds. In the half-century since, about 0.3 second per year has been shaved off Bannister's record. If this trend continues, by which year will someone run a 3-minute mile?

Solution   In solving this problem, we will express time for the mile run in seconds. Our interest is in a time of 3 minutes, or 180 seconds.

Step 1   Let x represent one of the quantities. Here is the critical information in the problem:

• In 1954, the record was 3 minutes, 59.4 seconds, or 239.4 seconds.
• The record has decreased by 0.3 second per year since then.

We are interested in when the record will be 180 seconds. Let

$x =$ the number of years after 1954 when someone will run a 3-minute mile.

Step 2   Represent other quantities in terms of x. There are no other unknown quantities to find, so we can skip this step.

Step 3   Write an equation in x that models the conditions.

The 1954 record time	decreased by	0.3 second per year for x years	equals	the 3-minute, or 180-second, mile.
239.4	−	0.3x	=	180

A Poky Species

For a species that prides itself on its athletic prowess, human beings are a pretty poky group. Lions can sprint at up to 50 miles per hour; cheetahs move even faster, flooring it to a sizzling 70 miles per hour. But most humans—with our willowy spines and awkward, upright gait—would have trouble cracking 20 miles per hour with a tail wind, a flat track, and a good pair of running shoes.

Step 4   Solve the equation and answer the question.

$$239.4 - 0.3x = 180 \qquad \text{This is the equation that models the problem's conditions.}$$

$$239.4 - 239.4 - 0.3x = 180 - 239.4 \qquad \text{Subtract 239.4 from both sides.}$$

$$-0.3x = -59.4 \qquad \text{Simplify.}$$

$$\frac{-0.3x}{-0.3} = \frac{-59.4}{-0.3} \qquad \text{Divide both sides by } -0.3.$$

$$x = 198 \qquad \text{Simplify.}$$

Using current trends, by 198 years (gasp!) after 1954, or in 2152, someone will run a 3-minute mile.

Step 5   Check the proposed solution in the original wording of the problem. The problem states that the record time should be 180 seconds. Do we obtain 180 seconds if we decrease the 1954 record time, 239.4 seconds, by 0.3 second per year for 198 years, our proposed solution?

$$239.4 - 0.3(198) = 239.4 - 59.4 = 180$$

This verifies that, using current trends, the 3-minute mile will be run 198 years after 1954.

Check Point 2 Got organic milk? Although organic milk accounts for only 1.2% of the market, consumption is increasing. In 2004, Americans purchased 40.7 million gallons of organic milk, increasing at a rate of 5.6 million gallons per year. If this trend continues, when will Americans purchase 79.9 million gallons of organic milk?

(*Source*: National Dairy Council)

EXAMPLE 3   Education Pays Off

The graph in Figure P.16 shows that for the period from 1982 through 2002, those with the most education had the fastest growth in wages. In 2002, the median annual income for people with a college degree was $52,000. This is a 160% increase over the median income in 1982. What were people with a college degree earning in 1982?

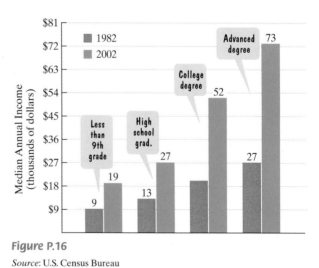

Median Annual Income by Highest Educational Attainment

Figure P.16

Source: U.S. Census Bureau

Solution

Step 1   Let x represent one of the quantities. We will let

x = the median income of people with a college degree in 1982.

Step 2   Represent other quantities in terms of x. There are no other unknown quantities to find, so we can skip this step.

Step 3 **Write an equation in x that models the conditions.** The median income in 1982 plus the 160% increase is the median income in 2002, $52,000.

Median 1982 income	plus	the increase (160% of the median 1982 income)	is	the median 2002 income, $52,000.
x	$+$	$1.6x$	$=$	$52,000$

Step 4 **Solve the equation and answer the question.**

$$x + 1.6x = 52{,}000$$ This is the equation that models the problem's conditions.

$$2.6x = 52{,}000$$ Combine like terms: $x + 1.6x = 1x + 1.6x = 2.6x$.

$$\frac{2.6x}{2.6} = \frac{52{,}000}{2.6}$$ Divide both sides by 2.6.

$$x = 20{,}000$$

In 1982, people with a college degree were earning $20,000.

Step 5 **Check the proposed solution in the original wording of the problem.** The 1982 income, $20,000, plus the 160% increase should equal the 2002 income given in the original wording, $52,000:

$$20{,}000 + 160\% \text{ of } 20{,}000 = 20{,}000 + 1.6(20{,}000) = 20{,}000 + 32{,}000 = 52{,}000.$$

This verifies that in 1982, college graduates were earning $20,000.

Check Point 3 After a 30% price reduction, you purchase a new computer for $840. What was the computer's price before the reduction?

Solving geometry problems usually requires a knowledge of basic geometric ideas and formulas. Formulas for area, perimeter, and volume are given in Table P.6.

Table P.6 Common Formulas for Area, Perimeter, and Volume

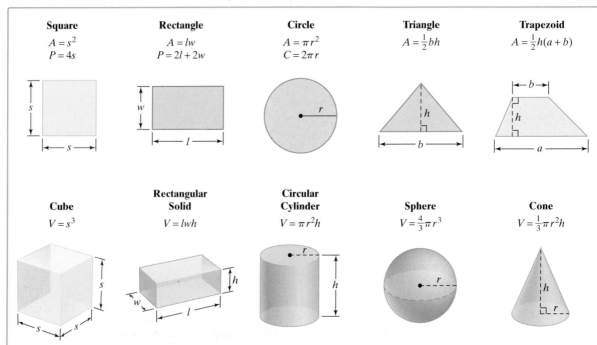

We will be using the formula for the perimeter of a rectangle, $P = 2l + 2w$, in our next example. The formula states that a rectangle's perimeter is the sum of twice its length and twice its width.

EXAMPLE 4 Finding the Dimensions of an American Football Field

The length of an American football field is 200 feet more than the width. If the perimeter of the field is 1040 feet, what are its dimensions?

Solution

Step 1 Let x represent one of the quantities. We know something about the length; the length is 200 feet more than the width. We will let

$$x = \text{the width.}$$

Step 2 Represent other quantities in terms of x. Because the length is 200 feet more than the width, we add 200 to the width to represent the length. Thus,

$$x + 200 = \text{the length.}$$

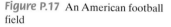

Figure P.17 An American football field

Figure P.17 illustrates an American football field and its dimensions.

Step 3 Write an equation in x that models the conditions. Because the perimeter of the field is 1040 feet,

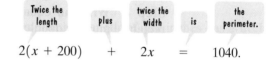

$$2(x + 200) \quad + \quad 2x \quad = \quad 1040.$$

Step 4 Solve the equation and answer the question.

$$2(x + 200) + 2x = 1040 \qquad \text{This is the equation that models the problem's conditions.}$$

$$2x + 400 + 2x = 1040 \qquad \text{Apply the distributive property.}$$

$$4x + 400 = 1040 \qquad \text{Combine like terms: } 2x + 2x = 4x.$$

$$4x = 640 \qquad \text{Subtract 400 from both sides.}$$

$$x = 160 \qquad \text{Divide both sides by 4.}$$

Thus,

$$\text{width} = x = 160.$$
$$\text{length} = x + 200 = 160 + 200 = 360.$$

The dimensions of an American football field are 160 feet by 360 feet. (The 360-foot length is usually described as 120 yards.)

Step 5 Check the proposed solution in the original wording of the problem. The perimeter of the football field using the dimensions that we found is

$$2(160 \text{ feet}) + 2(360 \text{ feet}) = 320 \text{ feet} + 720 \text{ feet} = 1040 \text{ feet.}$$

Because the problem's wording tells us that the perimeter is 1040 feet, our dimensions are correct.

 Check Point 4 The length of a rectangular basketball court is 44 feet more than the width. If the perimeter of the basketball court is 288 feet, what are its dimensions?

According to one mathematical model, the average life expectancy for American men born in 1900 was 55 years. Life expectancy has increased by about 0.2 year for each birth year after 1900. Use this information to solve Exercises 3–4.

3. If this trend continues, for which birth year will the average life expectancy be 85 years?

4. If this trend continues, for which birth year will the average life expectancy be 91 years?

The graph shows the number of Americans without health insurance from 2000 through 2003.

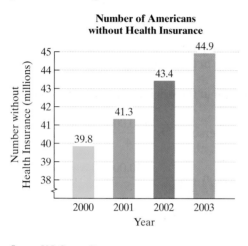

Number of Americans without Health Insurance

Source: U.S. Census Bureau

In 2000, there were 39.8 million Americans without health insurance. This number has increased at an average rate of 1.7 million people per year. Use this description to solve Exercises 5–6.

5. Determine when the number of Americans without health insurance will exceed the number in 2003 by 8.5 million.

6. Determine when the number of Americans without health insurance will exceed the number in 2003 by 10.2 million.

7. In 2005, there were 13,300 students at college A, with a projected enrollment increase of 1000 students per year. In the same year, there were 26,800 students at college B, with a projected enrollment decline of 500 students per year. According to these projections, when will the colleges have the same enrollment? What will be the enrollment in each college at that time?

8. In 2000, the population of Greece was 10,600,000, with projections of a population decrease of 28,000 people per year. In the same year, the population of Belgium was 10,200,000, with projections of a population decrease of 12,000 people per year. (*Source*: United Nations) According to these projections, when will the two countries have the same population? What will be the population at that time?

9. After a 20% reduction, you purchase a television for $336. What was the television's price before the reduction?

10. After a 30% reduction, you purchase a dictionary for $30.80. What was the dictionary's price before the reduction?

11. Including 8% sales tax, an inn charges $162 per night. Find the inn's nightly cost before the tax is added.

12. Including 5% sales tax, an inn charges $252 per night. Find the inn's nightly cost before the tax is added.

The graph shows average yearly earnings in the United States by highest educational attainment. Use the relevant information shown in the graph to solve Exercises 13–14.

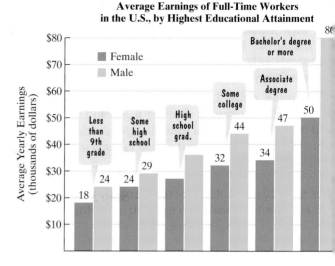

Average Earnings of Full-Time Workers in the U.S., by Highest Educational Attainment

Source: U.S. Census Bureau

13. The annual salary for men with some college is an increase of 22% over the annual salary for men whose highest educational attainment is a high school degree. What is the annual salary, to the nearest thousand dollars, for men whose highest educational attainment is a high school degree?

14. The annual salary for women with an associate degree is an increase of 26% over the annual salary for women whose highest educational attainment is a high school degree. What is the annual salary, to the nearest thousand dollars, for women whose highest educational attainment is a high school degree?

Exercises 15–16 involve markup, the amount added to the dealer's cost of an item to arrive at the selling price of that item.

15. The selling price of a refrigerator is $584. If the markup is 25% of the dealer's cost, what is the dealer's cost of the refrigerator?

16 The selling price of a scientific calculator is $15. If the markup is 25% of the dealer's cost, what is the dealer's cost of the calculator?

17. A rectangular soccer field is twice as long as it is wide. If the perimeter of the soccer field is 300 yards, what are its dimensions?

18. A rectangular swimming pool is three times as long as it is wide. If the perimeter of the pool is 320 feet, what are its dimensions?

19. The length of the rectangular tennis court at Wimbledon is 6 feet longer than twice the width. If the court's perimeter is 228 feet, what are the court's dimensions?

20. The length of a rectangular pool is 6 meters less than twice the width. If the pool's perimeter is 126 meters, what are its dimensions?

1. The rectangular painting in the figure shown measures 12 inches by 16 inches and contains a frame of uniform width around the four edges. The perimeter of the rectangle formed by the painting and its frame is 72 inches. Determine the width of the frame.

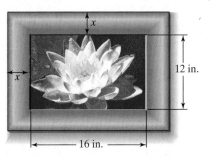

2. The rectangular swimming pool in the figure shown measures 40 feet by 60 feet and contains a path of uniform width around the four edges. The perimeter of the rectangle formed by the pool and the surrounding path is 248 feet. Determine the width of the path.

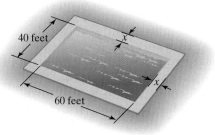

3. The length of a rectangular sign is 3 feet longer than the width. If the sign's area is 54 square feet, find its length and width.

4. A rectangular parking lot has a length that is 3 yards greater than the width. The area of the parking lot is 180 square yards. Find the length and the width.

5. Each side of a square is lengthened by 3 inches. The area of this new, larger square is 64 square inches. Find the length of a side of the original square.

6. Each side of a square is lengthened by 2 inches. The area of this new, larger square is 36 square inches. Find the length of a side of the original square.

7. A pool measuring 10 meters by 20 meters is surrounded by a path of uniform width. If the area of the pool and the path combined is 600 square meters, what is the width of the path?

8. A vacant rectangular lot is being turned into a community vegetable garden measuring 15 meters by 12 meters. A path of uniform width is to surround the garden. If the area of the lot is 378 square meters, find the width of the path surrounding the garden.

9. As part of a landscaping project, you put in a flower bed measuring 20 feet by 30 feet. To finish off the project, you are putting in a uniform border of pine bark around the outside of the rectangular garden. You have enough pine bark to cover 336 square feet. How wide should the border be?

10. As part of a landscaping project, you put in a flower bed measuring 10 feet by 12 feet. You plan to surround the bed with a uniform border of low-growing plants that require

1 square foot each when mature. If you have 168 of these plants, how wide a strip around the flower bed should you prepare for the border?

31. A 20-foot ladder is 15 feet from a house. How far up the house, to the nearest tenth of a foot, does the ladder reach?

32. The base of a 30-foot ladder is 10 feet from a building. If the ladder reaches the flat roof, how tall, to the nearest tenth of a foot, is the building?

33. A tree is supported by a wire anchored in the ground 5 feet from its base. The wire is 1 foot longer than the height that it reaches on the tree. Find the length of the wire.

34. A tree is supported by a wire anchored in the ground 15 feet from its base. The wire is 4 feet longer than the height that it reaches on the tree. Find the length of the wire.

35. A rectangular piece of land whose length its twice its width has a diagonal distance of 64 yards. How many yards, to the nearest tenth of a yard, does a person save by walking diagonally across the land instead of walking its length and its width?

36. A rectangular piece of land whose length is three times its width has a diagonal distance of 92 yards. How many yards, to the nearest tenth of a yard, does a person save by walking diagonally across the land instead of walking its length and its width?

37. A group of people share equally in a $20,000,000 lottery. Before the money is divided, two more winning ticket holders are declared. As a result, each person's share is reduced by $500,000. How many people were in the original group of winners?

38. A group of friends agrees to equally share the cost of a $480,000 vacation condominium. Before the purchase is made, four more people join the group and enter the agreement. As a result, each person's share is reduced by $32,000. How many people were in the original group?

In Exercises 39–42, use the formula

$$\text{Time traveled} = \frac{\text{Distance traveled}}{\text{Average velocity}}.$$

39. A car can travel 300 miles in the same amount of time it takes a bus to travel 180 miles. If the average velocity of the bus is 20 miles per hour slower than the average velocity of the car, find the average velocity for each.

40. A passenger train can travel 240 miles in the same amount of time it takes a freight train to travel 160 miles. If the average velocity of the freight train is 20 miles per hour slower than the average velocity of the passenger train, find the average velocity of each.

41. You ride your bike to campus a distance of 5 miles and return home on the same route. Going to campus, you ride mostly downhill and average 9 miles per hour faster than on your return trip home. If the round trip takes one hour and ten minutes—that is $\frac{7}{6}$ hours—what is your average velocity on the return trip?

42. An engine pulls a train 140 miles. Then a second engine, whose average velocity is 5 miles per hour faster than the first engine, takes over and pulls the train 200 miles. The total time required for both engines is 9 hours. Find the average velocity of each engine.

43. An automobile repair shop charged a customer $448, listing $63 for parts and the remainder for labor. If the cost of labor is $35 per hour, how many hours of labor did it take to repair the car?

44. A repair bill on a sailboat came to $1603, including $532 for parts and the remainder for labor. If the cost of labor is $63 per hour, how many hours of labor did it take to repair the sailboat?

45. An HMO pamphlet contains the following recommended weight for women: "Give yourself 100 pounds for the first 5 feet plus 5 pounds for every inch over 5 feet tall." Using this description, what height corresponds to a recommended weight of 135 pounds?

46. A job pays an annual salary of $33,150, which includes a holiday bonus of $750. If paychecks are issued twice a month, what is the gross amount for each paycheck?

47. You have 35 hits in 140 times at bat. Your batting average is $\frac{35}{140}$, or 0.25. How many consecutive hits must you get to increase your batting average to 0.30?

48. You have 30 hits in 120 times at bat. Your batting average is $\frac{30}{120}$, or 0.25. How many consecutive hits must you get to increase your batting average to 0.28?

Writing in Mathematics

49. In your own words, describe a step-by-step approach for solving algebraic word problems.

50. Write an original word problem that can be solved using an equation. Then solve the problem.

51. The mile records in Example 2 on page 98 are a yardstick for measuring how athletes are getting better and better. Do you think that there is a limit to human performance? Explain your answer. If so, when might we reach it?

52. The bar graph in Exercises 13–14 shows average earnings of U.S. men and women, by highest educational attainment. Describe the trend shown by the graph. Discuss any aspects of the data that surprised you.

53. In your own words, state the Pythagorean Theorem.

54. In the 1939 movie *The Wizard of Oz*, upon being presented with a Th.D. (Doctor of Thinkology), the Scarecrow proudly exclaims, "The sum of the square roots of any two sides of an isosceles triangle is equal to the square root of the remaining side." Did the Scarecrow get the Pythagorean Theorem right? In particular, describe four errors in the Scarecrow's statement.

Critical Thinking Exercises

55. The perimeter of a plot of land in the shape of a right triangle is 12 miles. If one leg of the triangle exceeds the other leg by 1 mile, find the length of each boundary of the land.

56. The price of a dress is reduced by 40%. When the dress still does not sell, it is reduced by 40% of the reduced price. If the price of the dress after both reductions is $72, what was the original price?

57. In a film, the actor Charles Coburn plays an elderly "uncle" character criticized for marrying a woman when he is 3 times her age. He wittily replies, "Ah, but in 20 years time I shall only be twice her age." How old is the "uncle" and the woman?

58. Suppose that we agree to pay you 8¢ for every problem in this chapter that you solve correctly and fine you 5¢ for every problem done incorrectly. If at the end of 26 problems we do not owe each other any money, how many problems did you solve correctly?

59. It was wartime when the Ricardos found out Mrs. Ricardo was pregnant. Ricky Ricardo was drafted and made out a will, deciding that $14,000 in a savings account was to be divided between his wife and his child-to-be. Rather strangely, and certainly with gender bias, Ricky stipulated that if the child were a boy, he would get twice the amount of the mother's portion. If it were a girl, the mother would get twice the amount the girl was to receive. We'll never know what Ricky was thinking of, for (as fate would have it) he did not return from war. Mrs. Ricardo gave birth to twins—a boy and a girl. How was the money divided?

60. A thief steals a number of rare plants from a nursery. On the way out, the thief meets three security guards, one after another. To each security guard, the thief is forced to give one-half the plants that he still has, plus 2 more. Finally, the thief leaves the nursery with 1 lone palm. How many plants were originally stolen?

Group Exercise

61. One of the best ways to learn how to *solve* a word problem in algebra is to *design* word problems of your own. Creating a word problem makes you very aware of precisely how much information is needed to solve the problem. You must also focus on the best way to present information to a reader and on how much information to give. As you write your problem, you gain skills that will help you solve problems created by others.

   The group should design five different word problems that can be solved using equations. All of the problems should be on different topics. For example, the group should not have more than one problem on the perimeter of a rectangle. The group should turn in both the problems and their algebraic solutions.

Linear Inequalities and Absolute Value Inequalities

Objectives

❶ Use interval notation.

❷ Find intersections and unions of intervals.

❸ Solve linear inequalities.

❹ Solve compound inequalities.

❺ Solve absolute value inequalities.

Rent-a-Heap, a car rental company, charges $125 per week plus $0.20 per mile to rent one of their cars. Suppose you are limited by how much money you can spend for the week: You can spend at most $335. If we let x represent the number of miles you drive the heap in a week, we can write an inequality that models the given conditions:

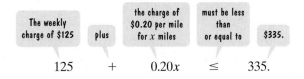

$$125 \quad + \quad 0.20x \quad \leq \quad 335.$$

Placing an inequality symbol between a polynomial of degee 1 and a constant results in a *linear inequality in one variable*. In this section, we will study how to solve linear inequalities such as the one shown above. **Solving an inequality** is the process of finding the set of numbers that make the inequality a true statement. These numbers are called the **solutions** of the inequality and we say that they **satisfy** the inequality. The set of all solutions is called the **solution set** of the inequality. Set-builder notation and a new notation, called *interval notation*, are used to represent solution sets. We begin this section by looking at interval notation.

❶ Use interval notation.

Interval Notation

Subsets of real numbers can be represented using **interval notation**. Suppose that a and b are two real numbers such that $a < b$.

Interval Notation	Graph
The **open interval** (a, b) represents the set of real numbers between, but not including, a and b. $$(a, b) = \{x \mid a < x < b\}$$ *x is greater than a (a < x) and x is less than b (x < b).*	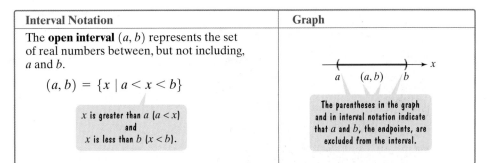 The parentheses in the graph and in interval notation indicate that a and b, the endpoints, are excluded from the interval.

(continued)

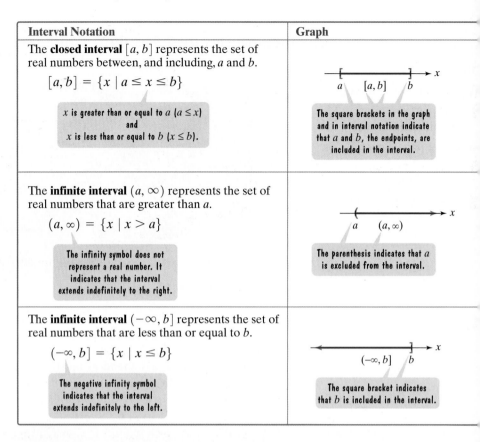

Interval Notation	Graph
The **closed interval** $[a, b]$ represents the set of real numbers between, and including, a and b. $[a, b] = \{x \mid a \le x \le b\}$ *x is greater than or equal to a ($a \le x$)* **and** *x is less than or equal to b ($x \le b$).*	The square brackets in the graph and in interval notation indicate that a and b, the endpoints, are included in the interval.
The **infinite interval** (a, ∞) represents the set of real numbers that are greater than a. $(a, \infty) = \{x \mid x > a\}$ *The infinity symbol does not represent a real number. It indicates that the interval extends indefinitely to the right.*	The parenthesis indicates that a is excluded from the interval.
The **infinite interval** $(-\infty, b]$ represents the set of real numbers that are less than or equal to b. $(-\infty, b] = \{x \mid x \le b\}$ *The negative infinity symbol indicates that the interval extends indefinitely to the left.*	The square bracket indicates that b is included in the interval.

Parentheses and Brackets in Interval Notation

Parentheses indicate endpoints that are not included in an interval. Square brackets indicate endpoints that are included in an interval.

Table P.7 lists nine possible types of intervals used to describe subsets of real numbers.

Table P.7   Intervals on the Real Number Line

Let a and b be real numbers such that $a < b$.

Interval Notation	Set-Builder Notation	Graph
(a, b)	$\{x \mid a < x < b\}$	
$[a, b]$	$\{x \mid a \le x \le b\}$	
$[a, b)$	$\{x \mid a \le x < b\}$	
$(a, b]$	$\{x \mid a < x \le b\}$	
(a, ∞)	$\{x \mid x > a\}$	
$[a, \infty)$	$\{x \mid x \ge a\}$	
$(-\infty, b)$	$\{x \mid x < b\}$	
$(-\infty, b]$	$\{x \mid x \le b\}$	
$(-\infty, \infty)$	$\{x \mid x \text{ is a real number}\}$ or $\mathbb{R}$ (set of all real numbers)	

EXAMPLE 1   Using Interval Notation

Express each interval in set-builder notation and graph:

   a. $(-1, 4]$ **b.** $[2.5, 4]$ **c.** $(-4, \infty)$.

Solution

   a. $(-1, 4] = \{x \mid -1 < x \le 4\}$

   b. $[2.5, 4] = \{x \mid 2.5 \le x \le 4\}$

   c. $(-4, \infty) = \{x \mid x > -4\}$

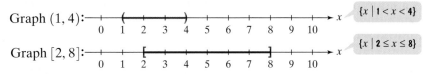

Check Point 1   Express each interval in set-builder notation and graph:

   a. $[-2, 5)$ **b.** $[1, 3.5]$ **c.** $(-\infty, -1)$.

② Find intersections and unions of intervals.

Intersections and Unions of Intervals

In Section P.1, we learned how to find intersections and unions of sets. Recall that $A \cap B$ (A intersection B) is the set of elements common to both set A and set B. By contrast, $A \cup B$ (A union B) is the set of elements in set A or in set B or in both sets.

   Because intervals represent sets, it is possible to find their intersections and unions. Graphs are helpful in this process.

> **Finding Intersections and Unions of Intervals**
>
>   **1.** Graph each interval on a number line.
>
>   **2. a.** To find the intersection, take the portion of the number line that the two graphs have in common.
>
>   **b.** To find the union, take the portion of the number line representing the total collection of numbers in the two graphs.

EXAMPLE 2   Finding Intersections and Unions of Intervals

Use graphs to find each set:

   a. $(1, 4) \cap [2, 8]$ **b.** $(1, 4) \cup [2, 8]$.

Solution

   a. $(1, 4) \cap [2, 8]$, the intersection of the intervals $(1, 4)$ and $[2, 8]$, consists of the numbers that are in both intervals.

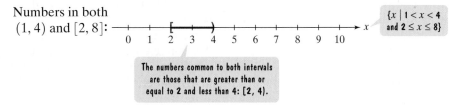

To find $(1, 4) \cap [2, 8]$, take the portion of the number line that the two graphs have in common.

Numbers in both
$(1, 4)$ and $[2, 8]$:

The numbers common to both intervals are those that are greater than or equal to 2 and less than 4: **[2, 4)**.

   Thus, $(1, 4) \cap [2, 8] = [2, 4)$.

b. $(1, 4) \cup [2, 8]$, the union of the intervals $(1, 4)$ and $[2, 8]$, consists of the numbers that are in either one interval or the other (or both).

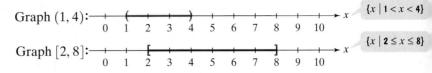

Graph $(1, 4)$: $\qquad$ $\{x \mid 1 < x < 4\}$

Graph $[2, 8]$: $\qquad$ $\{x \mid 2 \le x \le 8\}$

To find $(1, 4) \cup [2, 8]$, take the portion of the number line representing the total collection of numbers in the two graphs.

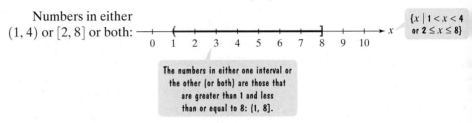

Numbers in either
$(1, 4)$ or $[2, 8]$ or both: $\qquad$ $\{x \mid 1 < x < 4$ or $2 \le x \le 8\}$

The numbers in either one interval or the other (or both) are those that are greater than 1 and less than or equal to 8: $(1, 8]$.

Thus, $(1, 4) \cup [2, 8] = (1, 8]$.

Check Point 2 Use graphs to find each set:

a. $[1, 3] \cap (2, 6)$ **b.** $[1, 3] \cup (2, 6)$.

❸ Solve linear inequalities.

Solving Linear Inequalities in One Variable

We know that a linear equation in x can be expressed as $ax + b = 0$. A **linear inequality in** x can be written in one of the following forms: $ax + b < 0$, $ax + b \le 0$, $ax + b > 0$, $ax + b \ge 0$. In each form, $a \ne 0$. Back to our question that opened this section: How many miles can you drive your Rent-a-Heap car if you can spend at most \$335 per week? We answer the question by solving

$$0.20x + 125 \le 335$$

for x. The solution procedure is nearly identical to that for solving

$$0.20x + 125 = 335.$$

Our goal is to get x by itself on the left side. We do this by subtracting 125 from both sides to isolate $0.20x$:

$0.20x + 125 \le 335$	This is the given inequality.
$0.20x + 125 - 125 \le 335 - 125$	Subtract 125 from both sides
$0.20x \le 210.$	Simplify.

Study Tip

English phrases such as "at least" and "at most" can be modeled by inequalities.

English Sentence	Inequality
x is at least 5.	$x \ge 5$
x is at most 5.	$x \le 5$
x is between 5 and 7.	$5 < x < 7$
x is no more than 5.	$x \le 5$
x is no less than 5.	$x \ge 5$

Finally, we isolate x from $0.20x$ by dividing both sides of the inequality by 0.20:

$\dfrac{0.20x}{0.20} \le \dfrac{210}{0.20}$	Divide both sides by 0.20.
$x \le 1050.$	Simplify.

With at most \$335 per week to spend, you can travel at most 1050 miles.

We started with the inequality $0.20x + 125 \le 335$ and obtained the inequality $x \le 1050$ in the final step. Both of these inequalities have the same solution set, namely $\{x \mid x \le 1050\}$. Inequalities such as these, with the same solution set, are said to be **equivalent**.

We isolated x from $0.20x$ by dividing both sides of $0.20x \le 210$ by 0.20, a positive number. Let's see what happens if we divide both sides of an inequality by a negative number. Consider the inequality $10 < 14$. Divide 10 and 14 by -2:

$$\frac{10}{-2} = -5 \quad \text{and} \quad \frac{14}{-2} = -7.$$

Because -5 lies to the right of -7 on the number line, -5 is greater than -7:
$$-5 > -7.$$
Notice that the direction of the inequality symbol is reversed:

$10 < 14$ Dividing by -2 changes the direction of the inequality symbol.

$-5 > -7.$

In general, **when we multiply or divide both sides of an inequality by a negative number, the direction of the inequality symbol is reversed**. When we reverse the direction of the inequality symbol, we say that we change the *sense* of the inequality.

We can isolate a variable in a linear inequality the same way we can isolate a variable in a linear equation. The following properties are used to create equivalent inequalities:

Properties of Inequalities

Property	The Property in Words	Example
The Addition Property of Inequality If $a < b$, then $a + c < b + c$. If $a < b$, then $a - c < b - c$.	If the same quantity is added to or subtracted from both sides of an inequality, the resulting inequality is equivalent to the original one.	$2x + 3 < 7$ Subtract 3: $2x + 3 - 3 < 7 - 3.$ Simplify: $2x < 4.$
The Positive Multiplication Property of Inequality If $a < b$ and c is positive, then $ac < bc$. If $a < b$ and c is positive, then $\frac{a}{c} < \frac{b}{c}$.	If we multiply or divide both sides of an inequality by the same positive quantity, the resulting inequality is equivalent to the original one.	$2x < 4$ Divide by 2: $\frac{2x}{2} < \frac{4}{2}.$ Simplify: $x < 2.$
The Negative Multiplication Property of Inequality If $a < b$ and c is negative, then $ac > bc$. If $a < b$ and c is negative, $\frac{a}{c} > \frac{b}{c}$.	If we multiply or divide both sides of an inequality by the same negative quantity and reverse the direction of the inequality symbol, the resulting inequality is equivalent to the original one.	$-4x < 20$ Divide by -4 and reverse the sense of the inequality: $\frac{-4x}{-4} > \frac{20}{-4}.$ Simplify: $x > -5.$

EXAMPLE 3   Solving a Linear Inequality

Solve and graph the solution set on a number line:
$$3 - 2x \le 11.$$

Solution

$3 - 2x \le 11$	This is the given inequality.
$3 - 2x - 3 \le 11 - 3$	Subtract 3 from both sides.
$-2x \le 8$	Simplify.
$\dfrac{-2x}{-2} \ge \dfrac{8}{-2}$	Divide both sides by -2 and change the sense of the inequality.
$x \ge -4$	Simplify.

The solution set consists of all real numbers that are greater than or equal to -4, expressed as $\{x \mid x \ge -4\}$ in set-builder notation. The interval notation for this solution set is $[-4, \infty)$. The graph of the solution set is shown as follows:

Discovery

As a partial check, select one number from the solution set for the inequality in Example 3. Substitute that number into the original inequality. Perform the resulting computations. You should obtain a true statement.

Is it possible to perform a partial check using a number that is not in the solution set? What should happen in this case? Try doing this.

Check Point **3**   Solve and graph the solution set on a number line:
$$2 - 3x \le 5.$$

EXAMPLE 4   Solving a Linear Inequality

Solve and graph the solution set on a number line:

$$-2x - 4 > x + 5.$$

Solution

Step 1   Simplify each side. Because each side is already simplified, we can skip this step.

Step 2   Collect variable terms on one side and constant terms on the other side. We will collect variable terms on the left and constant terms on the right.

$-2x - 4 > x + 5$	This is the given inequality.
$-2x - 4 - x > x + 5 - x$	Subtract x from both sides.
$-3x - 4 > 5$	Simplify.
$-3x - 4 + 4 > 5 + 4$	Add 4 to both sides.
$-3x > 9$	Simplify.

Step 3   Isolate the variable and solve. We isolate the variable, x, by dividing both sides by -3. Because we are dividing by a negative number, we must reverse the inequality symbol.

$\dfrac{-3x}{-3} < \dfrac{9}{-3}$	Divide both sides by -3 and change the sense of the inequality.
$x < -3$	Simplify.

Step 4   Express the solution set in set-builder or interval notation and graph the set on a number line. The solution set consists of all real numbers that are less than -3, expressed in set-builder notation as $\{x \mid x < -3\}$. The interval notation for this solution set is $(-\infty, -3)$. The graph of the solution set is shown as follows:

Check Point 4   Solve and graph the solution set on a number line: $3x + 1 > 7x - 15$

④ Solve compound inequalities.

Solving Compound Inequalities

We now consider two inequalities such as

$$-3 < 2x + 1 \quad \text{and} \quad 2x + 1 \leq 3$$

expressed as a **compound inequality**

$$-3 < 2x + 1 \leq 3.$$

The word "and" does not appear when the inequality is written in the shorter form, although intersection is implied. The shorter form enables us to solve both inequalities at once. By performing the same operation on all three parts of the inequality, our goal is to **isolate x in the middle**.

EXAMPLE 5   Solving a Compound Inequality

Solve and graph the solution set on a number line:

$$-3 < 2x + 1 \leq 3.$$

Solution   We would like to isolate x in the middle. We can do this by first subtracting 1 from all three parts of the compound inequality. Then we isolate x from $2x$ by dividing all three parts of the inequality by 2.

$-3 < 2x + 1 \leq 3$	This is the given inequality.
$-3 - 1 < 2x + 1 - 1 \leq 3 - 1$	Subtract 1 from all three parts.
$-4 < 2x \leq 2$	Simplify.
$\dfrac{-4}{2} < \dfrac{2x}{2} \leq \dfrac{2}{2}$	Divide each part by 2.
$-2 < x \leq 1$	Simplify.

Study Tip

You can solve

$$-2x - 4 > x + 5$$

by isolating x on the right side. Add $2x$ to both sides.

$$-2x - 4 + 2x > x + 5 + 2x$$
$$-4 > 3x + 5$$

Now subtract 5 from both sides.

$$-4 - 5 > 3x + 5 - 5$$
$$-9 > 3x$$

Finally, divide both sides by 3.

$$\frac{-9}{3} > \frac{3x}{3}$$
$$-3 > x$$

This last inequality means the same thing as $x < -3$.

The solution set consists of all real numbers greater than -2 and less than or equal to 1, represented by $\{x|-2 < x \le 1\}$ in set-builder notation and $(-2, 1]$ in interval notation. The graph is shown as follows:

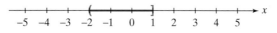

Check Point 5 Solve and graph the solution set on a number line: $1 \le 2x + 3 < 11$.

⑤ Solve absolute value inequalities.

Solving Inequalities with Absolute Value

We know that $|x|$ describes the distance of x from zero on a real number line. We can use this geometric interpretation to solve an inequality such as

$$|x| < 2.$$

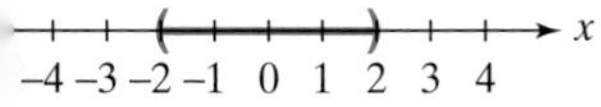

Figure P.20 $|x| < 2$, so $-2 < x < 2$.

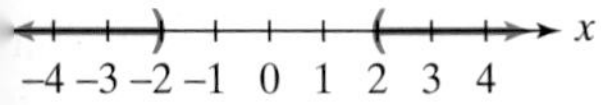

This means that the distance of x from 0 is *less than* 2, as shown in Figure P.20. The interval shows values of x that lie less than 2 units from 0. Thus, x can lie between -2 and 2. That is, x is greater than -2 and less than 2. We write $(-2, 2)$ or $\{x|-2 < x < 2.\}$

Figure P.21 $|x| > 2$, so $x < -2$ or $x > 2$.

Some absolute value inequalities use the "greater than" symbol. For example, $|x| > 2$ means that the distance of x from 0 is *greater than* 2, as shown in Figure P.21. Thus, x can be less than -2 *or* greater than 2. We write $x < -2$ or $x > 2$.

These observations suggest the following principles for solving inequalities with absolute value.

Study Tip

In the $|X| < c$ case, we have one compound inequality to solve. In the $|X| > c$ case, we have two separate inequalities to solve.

Solving an Absolute Value Inequality

If X is an algebraic expression and c is a positive number,

1. The solutions of $|X| < c$ are the numbers that satisfy $-c < X < c$.
2. The solutions of $|X| > c$ are the numbers that satisfy $X < -c$ or $X > c$.

These rules are valid if $<$ is replaced by $\le$ and $>$ is replaced by $\ge$.

EXAMPLE 6   Solving an Absolute Value Inequality

Solve and graph the solution set on a number line: $|x - 4| < 3$.

Solution   We rewrite the inequality without absolute value bars.

$$|X| < c \text{ means } -c < X < c.$$

$$|x - 4| < 3 \text{ means } -3 < x - 4 < 3.$$

We solve the compound inequality by adding 4 to all three parts.

$$-3 < x - 4 < 3$$
$$-3 + 4 < x - 4 + 4 < 3 + 4$$
$$1 < x < 7$$

The solution set is all real numbers greater than 1 and less than 7, denoted by $\{x|1 < x < 7\}$ or $(1, 7)$. The graph of the solution set is shown as follows:

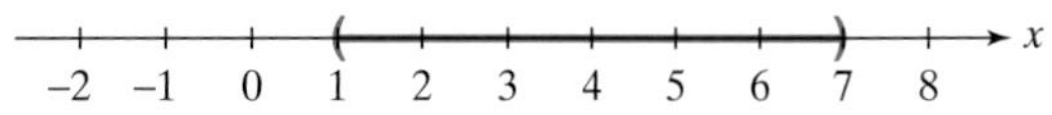

Check Point 6 Solve and graph the solution set on a number line: $|x - 2| < 5$.

EXAMPLE 7   Solving an Absolute Value Inequality

Solve and graph the solution set on a number line: $-2|3x + 5| + 7 \geq -13$.

Solution

$$-2|3x + 5| + 7 \geq -13$$ This is the given inequality.

We need to isolate $|3x + 5|$, the absolute value expression.

$$-2|3x + 5| + 7 - 7 \geq -13 - 7$$ Subtract 7 from both sides.

$$-2|3x + 5| \geq -20$$ Simplify.

$$\frac{-2|3x + 5|}{-2} \leq \frac{-20}{-2}$$ Divide both sides by -2 and change the sense of the inequality.

$$|3x + 5| \leq 10$$ Simplify.

$$-10 \leq 3x + 5 \leq 10$$ Rewrite without absolute value bars: $|X| \leq c$ means $-c \leq X \leq c$.

Now we need to isolate x in the middle.

$$-10 - 5 \leq 3x + 5 - 5 \leq 10 - 5$$ Subtract 5 from all three parts.

$$-15 \leq 3x \leq 5$$ Simplify.

$$\frac{-15}{3} \leq \frac{3x}{3} \leq \frac{5}{3}$$ Divide each part by 3.

$$-5 \leq x \leq \frac{5}{3}$$ Simplify.

The solution set is $\left\{x \mid -5 \leq x \leq \frac{5}{3}\right\}$ in set-builder notation and $\left[-5, \frac{5}{3}\right]$ in interval notation. The graph is shown as follows:

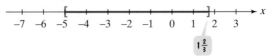

$1\frac{2}{3}$

Check Point 7   Solve and graph the solution set on a number line:
$-3|5x - 2| + 20 \geq -19$.

EXAMPLE 8   Solving an Absolute Value Inequality

Solve and graph the solution set on a number line: $7 < |5 - 2x|$.

Solution   We begin by expressing the inequality with the absolute value expression on the left side:

$$|5 - 2x| > 7.$$ $c < |X|$ means the same thing as $|X| > c$. In both cases, the inequality symbol points to c.

We rewrite this inequality without absolute value bars.

$$|X| > c \quad \text{means} \quad X < -c \quad \text{or} \quad X > c.$$

$|5 - 2x| > 7$ means $5 - 2x < -7$ or $5 - 2x > 7$.

We solve each of these inequalities separately. Then we take the union of their solution sets.

$$5 - 2x < -7 \qquad \text{or} \qquad 5 - 2x > 7$$

These are the inequalities without absolute value bars.

$$5 - 5 - 2x < -7 - 5 \qquad 5 - 5 - 2x > 7 - 5$$

Subtract 5 from both sides.

$$-2x < -12 \qquad\qquad -2x > 2$$

Simplify.

$$\frac{-2x}{-2} > \frac{-12}{-2} \qquad\qquad \frac{-2x}{-2} < \frac{2}{-2}$$

Divide both sides by -2 and change the sense of the inequality.

$$x > 6 \qquad\qquad x < -1$$

Simplify.

The solution set consists of all numbers that are less than -1 or greater than 6. The solution set is $\{x \mid x < -1 \text{ or } x > 6\}$, or, in interval notation $(-\infty, -1) \cup (6, \infty)$. The graph of the solution set is shown as follows:

```
  ◄———)——+——+——+——+——+——+——(——►  x
    -3  -2  -1  0  1  2  3  4  5  6  7  8
```

Study Tip

The graph of the solution set for $|X| > c$ will be divided into two intervals whose union cannot be represented as a single interval. The graph of the solution set for $|X| < c$ will be a single interval. Avoid the common error of rewriting $|X| > c$ as $-c < X > c$.

Check Point 8 Solve and graph the solution set on a number line: $18 < |6 - 3x|$.

Applications

Our next example shows how to use an inequality to select the better deal between two pricing options. We use our strategy for solving word problems, translating from the verbal conditions of the problem to a linear inequality.

EXAMPLE 9   Selecting the Better Deal

Acme Car rental agency charges $4 a day plus $0.15 per mile, whereas Interstate rental agency charges $20 a day and $0.05 per mile. How many miles must be driven to make the daily cost of an Acme rental a better deal than an Interstate rental?

Solution

Step 1   Let x represent one of the quantities. We are looking for the number of miles that must be driven in a day to make Acme the better deal. Thus,

let x = the number of miles driven in a day.

Step 2   Represent other quantities in terms of x. We are not asked to find another quantity, so we can skip this step.

Step 3   Write an inequality in x that models the conditions. Acme is a better deal than Interstate if the daily cost of Acme is less than the daily cost of Interstate.

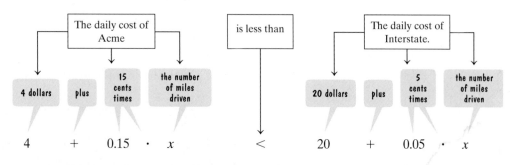

Step 4   Solve the inequality and answer the question.

$$4 + 0.15x < 20 + 0.05x$$ This is the inequality that models the verbal conditions.

$$4 + 0.15x - 0.05x < 20 + 0.05x - 0.05x$$ Subtract 0.05x from both sides.

$$4 + 0.1x < 20$$ Simplify.

$$4 + 0.1x - 4 < 20 - 4$$ Subtract 4 from both sides.

$$0.1x < 16$$ Simplify.

$$\frac{0.1x}{0.1} < \frac{16}{0.1}$$ Divide both sides by 0.1.

$$x < 160$$ Simplify.

Thus, driving fewer than 160 miles per day makes Acme the better deal.

Step 5   Check the proposed solution in the original wording of the problem. One way to do this is to take a mileage less than 160 miles per day to see if Acme is the better deal. Suppose that 150 miles are driven in a day.

$$\text{Cost for Acme} = 4 + 0.15(150) = 26.50$$
$$\text{Cost for Interstate} = 20 + 0.05(150) = 27.50$$

Acme has a lower daily cost, making Acme the better deal.

Check Point 9   A car can be rented from Basic Rental for $260 per week with no extra charge for mileage. Continental charges $80 per week plus 25 cents for each mile driven to rent the same car. How many miles must be driven in a week to make the rental cost for Basic Rental a better deal than Continental's?

EXERCISE SET P.9

Practice Exercises

In Exercises 1–14, express each interval in set-builder notation and graph the interval on a number line.

1. $(1, 6]$ **2.** $(-2, 4]$

3. $[-5, 2)$ **4.** $[-4, 3)$

5. $[-3, 1]$ **6.** $[-2, 5]$

7. $(2, \infty)$ **8.** $(3, \infty)$

9. $[-3, \infty)$ **10.** $[-5, \infty)$

11. $(-\infty, 3)$ **12.** $(-\infty, 2)$

13. $(-\infty, 5.5)$ **14.** $(-\infty, 3.5]$

In Exercises 15–26, use graphs to find each set.

15. $(-3, 0) \cap [-1, 2]$ **16.** $(-4, 0) \cap [-2, 1]$

17. $(-3, 0) \cup [-1, 2]$ **18.** $(-4, 0) \cup [-2, 1]$

19. $(-\infty, 5) \cap [1, 8)$ **20.** $(-\infty, 6) \cap [2, 9)$

21. $(-\infty, 5) \cup [1, 8)$ **22.** $(-\infty, 6) \cup [2, 9)$

23. $[3, \infty) \cap (6, \infty)$ **24.** $[2, \infty) \cap (4, \infty)$

25. $[3, \infty) \cup (6, \infty)$ **26.** $[2, \infty) \cup (4, \infty)$

In all exercises, use interval notation to express solution sets and graph each solution set on a number line.

In Exercises 27–48, solve each linear inequality.

27. $5x + 11 < 26$ **28.** $2x + 5 < 17$

29. $3x - 7 \geq 13$ **30.** $8x - 2 \geq 14$

31. $-9x \geq 36$ **32.** $-5x \leq 30$

33. $8x - 11 \leq 3x - 13$ **34.** $18x + 45 \leq 12x - 8$

35. $4(x + 1) + 2 \geq 3x + 6$

36. $8x + 3 > 3(2x + 1) + x + 5$

37. $2x - 11 < -3(x + 2)$ **38.** $-4(x + 2) > 3x + 20$

39. $1 - (x + 3) \geq 4 - 2x$ **40.** $5(3 - x) \leq 3x - 1$

41. $\frac{x}{4} - \frac{3}{2} \leq \frac{x}{2} + 1$ **42.** $\frac{3x}{10} + 1 \geq \frac{1}{5} - \frac{x}{10}$

43. $1 - \frac{x}{2} > 4$ **44.** $7 - \frac{4}{5}x < \frac{3}{5}$

45. $\frac{x - 4}{6} \geq \frac{x - 2}{9} + \frac{5}{18}$ **46.** $\frac{4x - 3}{6} + 2 \geq \frac{2x - 1}{12}$

47. $3[3(x + 5) + 8x + 7] + 5[3(x - 6)$
$\quad - 2(3x - 5)] < 2(4x + 3)$

48. $5[3(2 - 3x) - 2(5 - x)] - 6[5(x - 2)$
$\quad - 2(4x - 3)] < 3x + 19$

In Exercises 49–56, solve each compound inequality.

49. $6 < x + 3 < 8$ **50.** $7 < x + 5 < 11$

51. $-3 \leq x - 2 < 1$ **52.** $-6 < x - 4 \leq 1$

53. $-11 < 2x - 1 \leq -5$ **54.** $3 \leq 4x - 3 < 19$

55. $-3 \leq \frac{2}{3}x - 5 < -1$ **56.** $-6 \leq \frac{1}{2}x - 4 < -3$

In Exercises 57–92, solve each absolute value inequality.

57. $|x| < 3$ **58.** $|x| < 5$

59. $|x - 1| \leq 2$ **60.** $|x + 3| \leq 4$

61. $|2x - 6| < 8$ **62.** $|3x + 5| < 17$

63. $|2(x - 1) + 4| \leq 8$ **64.** $|3(x - 1) + 2| \leq 20$

65. $\left|\dfrac{2x + 6}{3}\right| < 2$ **66.** $\left|\dfrac{3(x - 1)}{4}\right| < 6$

67. $|x| > 3$ **68.** $|x| > 5$

69. $|x - 1| \ge 2$ **70.** $|x + 3| \ge 4$

71. $|3x - 8| > 7$ **72.** $|5x - 2| > 13$

73. $\left|\dfrac{2x + 2}{4}\right| \ge 2$ **74.** $\left|\dfrac{3x - 3}{9}\right| \ge 1$

75. $\left|3 - \dfrac{2}{3}x\right| > 5$ **76.** $\left|3 - \dfrac{3}{4}x\right| > 9$

77. $3|x - 1| + 2 \ge 8$ **78.** $5|2x + 1| - 3 \ge 9$

79. $-2|x - 4| \ge -4$ **80.** $-3|x + 7| \ge -27$

81. $-4|1 - x| < -16$ **82.** $-2|5 - x| < -6$

83. $3 \le |2x - 1|$ **84.** $9 \le |4x + 7|$

85. $5 > |4 - x|$ **86.** $2 > |11 - x|$

87. $1 < |2 - 3x|$ **88.** $4 < |2 - x|$

89. $12 < \left|-2x + \dfrac{6}{7}\right| + \dfrac{3}{7}$ **90.** $1 < \left|x - \dfrac{11}{3}\right| + \dfrac{7}{3}$

91. $4 + \left|3 - \dfrac{x}{3}\right| \ge 9$ **92.** $\left|2 - \dfrac{x}{2}\right| - 1 \le 1$

Practice Plus

In Exercises 93–96, use interval notation to represent all values of x satisfying the given conditions.

93. $y = 1 - (x + 3) + 2x$ and y is at least 4.

94. $y = 2x - 11 + 3(x + 2)$ and y is at most 0.

95. $y = 7 - \left|\dfrac{x}{2} + 2\right|$ and y is at most 4.

96. $y = 8 - |5x + 3|$ and y is at least 6.

97. When 3 times a number is subtracted from 4, the absolute value of the difference is at least 5. Use interval notation to express the set of all numbers that satisfy this condition.

98. When 4 times a number is subtracted from 5, the absolute value of the difference is at most 13. Use interval notation to express the set of all numbers that satisfy this condition.

Application Exercises

The graphs show that the three components of love, namely passion, intimacy, and commitment, progress differently over time. Passion peaks early in a relationship and then declines. By contrast, intimacy and commitment build gradually. Use the graphs to solve Exercises 99–106.

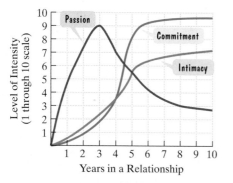

The Course of Love Over Time

Source: R. J. Sternberg. A Triangular Theory of Love, *Psychological Review*, 93, 119–135.

99. Use interval notation to write an inequality that expresses for which years in a relationship intimacy is greater than commitment.

100. Use interval notation to write an inequality that expresses for which years in a relationship passion is greater than or equal to intimacy.

101. What is the relationship between passion and intimacy on for years $[5, 7)$?

102. What is the relationship between intimacy and commitment for years $[4, 7)$?

103. What is the relationship between passion and commitment for years $(6, 8)$?

104. What is the relationship between passion and commitment for years $(7, 9)$?

105. What is the maximum level of intensity for passion? After how many years in a relationship does this occur?

106. After approximately how many years do levels of intensity for commitment exceed the maximum level of intensity for passion?

107. The percentage, P, of U.S. voters who used electronic voting systems, such as optical scans, in national elections can be modeled by the formula
$$P = 3.1x + 25.8,$$
where x is the number of years after 1994. In which years will more than 63% of U.S. voters use electronic systems?

108. The percentage, P, of U.S. voters who used punch cards or lever machines in national elections can be modeled by the formula
$$P = -2.5x + 63.1,$$
where x is the number of years after 1994. In which years will fewer than 38.1% of U.S. voters use punch cards or lever machines?

109. A basic cellular phone plan costs $20 per month for 60 calling minutes. Additional time costs $0.40 per minute. The formula
$$C = 20 + 0.40(x - 60)$$
gives the monthly cost for this plan, C, for x calling minutes, where $x > 60$. How many calling minutes are possible for a monthly cost of at least $28 and at most $40?

110. The formula for converting Fahrenheit temperature, F, to Celsius temperature, C, is
$$C = \dfrac{5}{9}(F - 32).$$
If Celsius temperature ranges from 15° to 35°, inclusive, what is the range for the Fahrenheit temperature? Use interval notation to express this range.

111. If a coin is tossed 100 times, we would expect approximately 50 of the outcomes to be heads. It can be demonstrated that a coin is unfair if h, the number of outcomes that result in heads, satisfies $\left|\dfrac{h - 50}{5}\right| \ge 1.645$. Describe the number of outcomes that determine an unfair coin that is tossed 100 times.

In Exercises 112–123, use the strategy for solving word problems, translating from the verbal conditions of the problem to a linear inequality.

112. A truck can be rented from Basic Rental for $50 per day plus $0.20 per mile. Continental charges $20 per day plus $0.50 per mile to rent the same truck. How many miles must be driven in a day to make the rental cost for Basic Rental a better deal than Continental's?

113. You are choosing between two long-distance telephone plans. Plan A has a monthly fee of $15 with a charge of $0.08 per minute for all long-distance calls. Plan B has a monthly fee of $3 with a charge of $0.12 per minute for all long-distance calls. How many minutes of long-distance calls in a month make plan A the better deal?

114. A city commission has proposed two tax bills. The first bill requires that a homeowner pay $1800 plus 3% of the assessed home value in taxes. The second bill requires taxes of $200 plus 8% of the assessed home value. What price range of home assessment would make the first bill a better deal?

115. A local bank charges $8 per month plus 5¢ per check. The credit union charges $2 per month plus 8¢ per check. How many checks should be written each month to make the credit union a better deal?

116. A company manufactures and sells blank audiocassette tapes. The weekly fixed cost is $10,000 and it costs $0.40 to produce each tape. The selling price is $2.00 per tape. How many tapes must be produced and sold each week for the company to generate a profit?

117. A company manufactures and sells personalized stationery. The weekly fixed cost is $3000 and it costs $3.00 to produce each package of stationery. The selling price is $5.50 per package. How many packages of stationery must be produced and sold each week for the company to generate a profit?

118. An elevator at a construction site has a maximum capacity of 2800 pounds. If the elevator operator weighs 265 pounds and each cement bag weighs 65 pounds, how many bags of cement can be safely lifted on the elevator in one trip?

119. An elevator at a construction site has a maximum capacity of 3000 pounds. If the elevator operator weighs 245 pounds and each cement bag weighs 95 pounds, how many bags of cement can be safely lifted on the elevator in one trip?

120. To earn an A in a course, you must have a final average of at least 90%. On the first four examinations, you have grades of 86%, 88%, 92%, and 84%. If the final examination counts as two grades, what must you get on the final to earn an A in the course?

121. On two examinations, you have grades of 86 and 88. There is an optional final examination, which counts as one grade. You decide to take the final in order to get a course grade of A, meaning a final average of at least 90.
    a. What must you get on the final to earn an A in the course?
    b. By taking the final, if you do poorly, you might risk the B that you have in the course based on the first two exam grades. If your final average is less than 80, you will lose your B in the course. Describe the grades on the final that will cause this to happen.

122. Parts for an automobile repair cost $175. The mechanic charges $34 per hour. If you receive an estimate for at least $226 and at most $294 for fixing the car, what is the time interval that the mechanic will be working on the job?

123. The toll to a bridge is $3.00. A three-month pass costs $7.50 and reduces the toll to $0.50. A six-month pass costs $30 and permits crossing the bridge for no additional fee. How many crossings per three-month period does it take for the three-month pass to be the best deal?

Writing in Mathematics

124. When graphing the solutions of an inequality, what does a parenthesis signify? What does a bracket signify?

125. Describe ways in which solving a linear inequality is similar to solving a linear equation.

126. Describe ways in which solving a linear inequality is different than solving a linear equation.

127. What is a compound inequality and how is it solved?

128. Describe how to solve an absolute value inequality involving the symbol $<$. Give an example.

129. Describe how to solve an absolute value inequality involving the symbol $>$. Give an example.

130. Explain why $|x| < -4$ has no solution.

131. Describe the solution set of $|x| > -4$.

Critical Thinking Exercises

132. Which one of the following is true?
    a. The first step in solving $|2x - 3| > -7$ is to rewrite the inequality as $2x - 3 > -7$ or $2x - 3 < 7$.
    b. The smallest real number in the solution set of $2x > 6$ is
    c. All irrational numbers satisfy $|x - 4| > 0$.
    d. None of these statements is true.

133. What's wrong with this argument? Suppose x and y represent two real numbers, where $x > y$.

$2 > 1$	This is a true statement.
$2(y - x) > 1(y - x)$	Multiply both sides by $y - x$
$2y - 2x > y - x$	Use the distributive property
$y - 2x > -x$	Subtract y from both sides.
$y > x$	Add $2x$ to both sides.

    The final inequality, $y > x$, is impossible because we were initially given $x > y$.

134. Write an absolute value inequality for which the interval shown is the solution.

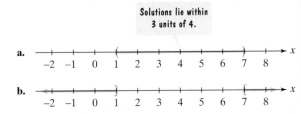

135. Here are two inequalities that describe the range of monthly average temperatures, T, in degrees Fahrenheit for two American cities:

    Model 1: $|T - 57| < 7$
    Model 2: $|T - 50| < 22$.

    Which model describes Albany, New York, and which model describes San Francisco, California?

Group Exercise

136. Each group member should research one situation that provides two different pricing options. These can involve areas such as public transportation options (with or without coupon books), cell phone plans, long-distance telephone plans, or anything of interest. Be sure to bring in all the details for each option. At a second group meeting, select the two pricing situations that are most interesting and relevant. Using each situation, write a word problem about selecting the better of the two options. The word problem should be one that can be solved using a linear inequality. The group should turn in the two problems and their solutions.

Chapter P
Summary, Review, and Test

Summary: Basic Formulas

Definition of Absolute Value

$$|x| = \begin{cases} x & \text{if } x \geq 0 \\ -x & \text{if } x < 0 \end{cases}$$

Distance between Points a and b on a Number Line

$$|a - b| \quad \text{or} \quad |b - a|$$

Properties of Algebra

Commutative	$a + b = b + a$
	$ab = ba$
Associative	$(a + b) + c = a + (b + c)$
	$(ab)c = a(bc)$
Distributive	$a(b + c) = ab + ac$
Identity	$a + 0 = a$
	$a \cdot 1 = a$
Inverse	$a + (-a) = 0$
	$a \cdot \dfrac{1}{a} = 1, a \neq 0$

Properties of Exponents

$$b^{-n} = \frac{1}{b^n}, \quad b^0 = 1, \quad b^m \cdot b^n = b^{m+n},$$

$$(b^m)^n = b^{mn}, \quad \frac{b^m}{b^n} = b^{m-n}, \quad (ab)^n = a^n b^n, \quad \left(\frac{a}{b}\right)^n = \frac{a^n}{b^n}$$

Product and Quotient Rules for nth Roots

$$\sqrt[n]{ab} = \sqrt[n]{a} \cdot \sqrt[n]{b}, \qquad \sqrt[n]{\frac{a}{b}} = \frac{\sqrt[n]{a}}{\sqrt[n]{b}}$$

Rational Exponents

$$a^{\frac{1}{n}} = \sqrt[n]{a}, \quad a^{-\frac{1}{n}} = \frac{1}{a^{\frac{1}{n}}} = \frac{1}{\sqrt[n]{a}},$$

$$a^{\frac{m}{n}} = \left(\sqrt[n]{a}\right)^m = \sqrt[n]{a^m}, \quad a^{-\frac{m}{n}} = \frac{1}{a^{\frac{m}{n}}}$$

Special Products

$$(A + B)(A - B) = A^2 - B^2$$
$$(A + B)^2 = A^2 + 2AB + B^2$$
$$(A - B)^2 = A^2 - 2AB + B^2$$
$$(A + B)^3 = A^3 + 3A^2B + 3AB^2 + B^3$$
$$(A - B)^3 = A^3 - 3A^2B + 3AB^2 - B^3$$

Factoring Formulas

$$A^2 - B^2 = (A + B)(A - B)$$
$$A^2 + 2AB + B^2 = (A + B)^2$$
$$A^2 - 2AB + B^2 = (A - B)^2$$
$$A^3 + B^3 = (A + B)(A^2 - AB + B^2)$$
$$A^3 - B^3 = (A - B)(A^2 + AB + B^2)$$

Absolute Value Equations and Inequalities

1. If $c > 0$, then $|X| = c$ is equivalent to $X = c$ or $X = -c$.

2. If $c > 0$, then $|X| < c$ is equivalent to $-c < X < c$.

3. If $c > 0$, then $|X| > c$ is equivalent to $X < -c$ or $X > c$.

The Quadratic Formula

All quadratic equations

$$ax^2 + bx + c = 0, \quad a \neq 0$$

can be solved by the quadratic formula

$$x = \frac{-b \pm \sqrt{b^2 - 4ac}}{2a}.$$

Review Exercises

You can use these review exercises, like the review exercises at the end of each chapter, to test your understanding of the chapter's topics. However, you can also use these exercises as a prerequisite test to check your mastery of the fundamental algebra skills needed in this book.

P.1

In Exercises 1–2, evaluate each algebraic expression for the given value or values of the variable(s).

1. $3 + 6(x - 2)^3$ for $x = 4$

2. $x^2 - 5(x - y)$ for $x = 6$ and $y = 2$

3. You are riding along an expressway traveling x miles per hour. The formula

$$S = 0.015x^2 + x + 10$$

models the recommended safe distance, S, in feet, between your car and other cars on the expressway. What is the recommended safe distance when your speed is 60 miles per hour?

In Exercises 4–7, let $A = \{a, b, c\}, B = \{a, c, d, e\}$, and $C = \{a, d, f, g\}$. Find the indicated set.

4. $A \cap B$ **5.** $A \cup B$

6. $A \cup C$ **7.** $C \cap A$

8. Consider the set:

$$\left\{-17, -\tfrac{9}{13}, 0, 0.75, \sqrt{2}, \pi, \sqrt{81}\right\}.$$

List all numbers from the set that are **a.** natural numbers, **b.** whole numbers, **c.** integers, **d.** rational numbers, **e.** irrational numbers, **f.** real numbers.

In Exercises 9–11, rewrite each expression without absolute value bars.

9. $|-103|$

10. $|\sqrt{2} - 1|$

11. $|3 - \sqrt{17}|$

12. Express the distance between the numbers -17 and 4 using absolute value. Then evaluate the absolute value.

In Exercises 13–18, state the name of the property illustrated.

13. $3 + 17 = 17 + 3$

14. $(6 \cdot 3) \cdot 9 = 6 \cdot (3 \cdot 9)$

15. $\sqrt{3}\left(\sqrt{5} + \sqrt{3}\right) = \sqrt{15} + 3$

16. $(6 \cdot 9) \cdot 2 = 2 \cdot (6 \cdot 9)$

17. $\sqrt{3}\left(\sqrt{5} + \sqrt{3}\right) = \left(\sqrt{5} + \sqrt{3}\right)\sqrt{3}$

18. $(3 \cdot 7) + (4 \cdot 7) = (4 \cdot 7) + (3 \cdot 7)$

In Exercises 19–22, simplify each algebraic expression.

19. $5(2x - 3) + 7x$

20. $\tfrac{1}{5}(5x) + [(3y) + (-3y)] - (-x)$

21. $3(4y - 5) - (7y + 2)$ **22.** $8 - 2[3 - (5x - 1)]$

23. The bar graph shows the number of endangered animal species in the United States for six selected years. The data can be modeled by the formulas $E = 10x + 166$ and $E = 0.04x^2 + 9.2x + 169$, in which E represents the number of endangered species x years after 1980. Which formula best describes the actual number of endangered animal species in 2000?

Endangered Animal Species in the U.S.

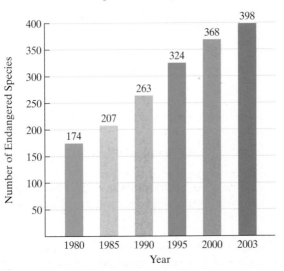

Source: U.S. Fish and Wildlife Service

P.2

Evaluate each exponential expression in Exercises 24–27.

24. $(-3)^3(-2)^2$

25. $2^{-4} + 4^{-1}$

26. $5^{-3} \cdot 5$

27. $\dfrac{3^3}{3^6}$

Simplify each exponential expression in Exercises 28–31.

28. $(-2x^4y^3)^3$

29. $(-5x^3y^2)(-2x^{-11}y^{-2})$

30. $(2x^3)^{-4}$

31. $\dfrac{7x^5y^6}{28x^{15}y^{-2}}$

In Exercises 32–33, write each number in decimal notation.

32. 3.74×10^4

33. 7.45×10^{-5}

In Exercises 34–35, write each number in scientific notation.

34. 3,590,000

35. 0.00725

In Exercises 36–37, perform the indicated operation and write the answer in decimal notation.

36. $(3 \times 10^3)(1.3 \times 10^2)$

37. $\dfrac{6.9 \times 10^3}{3 \times 10^5}$

38. If you earned \$1 million per year (\$$10^6$), how long would it take to accumulate \$1 billion (\$$10^9$)?

39. If the population of the United States is 2.9×10^8 and each person spends about \$150 per year going to the movies (or renting movies), express the total annual spending on movies in scientific notation.

P.3

Use the product rule to simplify the expressions in Exercises 40–43. In Exercises 42–43, assume that variables represent non-negative real numbers.

40. $\sqrt{300}$

41. $\sqrt{12x^2}$

42. $\sqrt{10x} \cdot \sqrt{2x}$

43. $\sqrt{r^3}$

Use the quotient rule to simplify the expressions in Exercises 44–45.

44. $\sqrt{\dfrac{121}{4}}$

45. $\dfrac{\sqrt{96x^3}}{\sqrt{2x}}$ (Assume that $x > 0$.)

In Exercises 46–48, add or subtract terms whenever possible.

46. $7\sqrt{5} + 13\sqrt{5}$

47. $2\sqrt{50} + 3\sqrt{8}$

48. $4\sqrt{72} - 2\sqrt{48}$

In Exercises 49–52, rationalize the denominator.

49. $\dfrac{30}{\sqrt{5}}$

50. $\dfrac{\sqrt{2}}{\sqrt{3}}$

51. $\dfrac{5}{6 + \sqrt{3}}$

52. $\dfrac{14}{\sqrt{7} - \sqrt{5}}$

Evaluate each expression in Exercises 53–56 or indicate that the root is not a real number.

53. $\sqrt[3]{125}$

54. $\sqrt[5]{-32}$

55. $\sqrt[4]{-125}$

56. $\sqrt[4]{(-5)^4}$

Simplify the radical expressions in Exercises 57–61.

57. $\sqrt[3]{81}$

58. $\sqrt[3]{y^5}$

59. $\sqrt[4]{8} \cdot \sqrt[4]{10}$ **60.** $4\sqrt[3]{16} + 5\sqrt[3]{2}$

61. $\dfrac{\sqrt[4]{32x^5}}{\sqrt[4]{16x}}$ (Assume that $x > 0$.)

In Exercises 62–67, evaluate each expression.

62. $16^{\frac{1}{2}}$ **63.** $25^{-\frac{1}{2}}$

64. $125^{\frac{1}{3}}$ **65.** $27^{-\frac{1}{3}}$

66. $64^{\frac{2}{3}}$ **67.** $27^{-\frac{4}{3}}$

In Exercises 68–70, simplify using properties of exponents.

68. $\left(5x^{\frac{2}{3}}\right)\left(4x^{\frac{1}{4}}\right)$ **69.** $\dfrac{15x^{\frac{3}{4}}}{5x^{\frac{1}{2}}}$

70. $(125x^6)^{\frac{2}{3}}$

71. Simplify by reducing the index of the radical: $\sqrt[6]{y^3}$.

P.4

In Exercises 72–73, perform the indicated operations. Write the resulting polynomial in standard form and indicate its degree.

72. $(-6x^3 + 7x^2 - 9x + 3) + (14x^3 + 3x^2 - 11x - 7)$

73. $(13x^4 - 8x^3 + 2x^2) - (5x^4 - 3x^3 + 2x^2 - 6)$

In Exercises 74–80, find each product.

74. $(3x - 2)(4x^2 + 3x - 5)$ **75.** $(3x - 5)(2x + 1)$

76. $(4x + 5)(4x - 5)$ **77.** $(2x + 5)^2$

78. $(3x - 4)^2$ **79.** $(2x + 1)^3$

80. $(5x - 2)^3$

In Exercises 81–87, find each product.

81. $(x + 7y)(3x - 5y)$ **82.** $(3x - 5y)^2$

83. $(3x^2 + 2y)^2$ **84.** $(7x + 4y)(7x - 4y)$

85. $(a - b)(a^2 + ab + b^2)$

86. $[5y - (2x + 1)][5y + (2x + 1)]$

87. $(x + 2y + 4)^2$

P.5

In Exercises 88–104, factor completely, or state that the polynomial is prime.

88. $15x^3 + 3x^2$ **89.** $x^2 - 11x + 28$

90. $15x^2 - x - 2$ **91.** $64 - x^2$

92. $x^2 + 16$ **93.** $3x^4 - 9x^3 - 30x^2$

94. $20x^7 - 36x^3$ **95.** $x^3 - 3x^2 - 9x + 27$

96. $16x^2 - 40x + 25$ **97.** $x^4 - 16$

98. $y^3 - 8$ **99.** $x^3 + 64$

100. $3x^4 - 12x^2$ **101.** $27x^3 - 125$

102. $x^5 - x$ **103.** $x^3 + 5x^2 - 2x - 10$

104. $x^2 + 18x + 81 - y^2$

In Exercises 105–107, factor and simplify each algebraic expression.

105. $16x^{-\frac{3}{4}} + 32x^{\frac{1}{4}}$

106. $(x^2 - 4)(x^2 + 3)^{\frac{1}{2}} - (x^2 - 4)^2(x^2 + 3)^{\frac{3}{2}}$

107. $12x^{-\frac{1}{2}} + 6x^{-\frac{3}{2}}$

P.6

In Exercises 108–110, simplify each rational expression. Also, list all numbers that must be excluded from the domain.

108. $\dfrac{x^3 + 2x^2}{x + 2}$ **109.** $\dfrac{x^2 + 3x - 18}{x^2 - 36}$

110. $\dfrac{x^2 + 2x}{x^2 + 4x + 4}$

In Exercises 111–113, multiply or divide as indicated.

111. $\dfrac{x^2 + 6x + 9}{x^2 - 4} \cdot \dfrac{x + 3}{x - 2}$ **112.** $\dfrac{6x + 2}{x^2 - 1} \div \dfrac{3x^2 + x}{x - 1}$

113. $\dfrac{x^2 - 5x - 24}{x^2 - x - 12} \div \dfrac{x^2 - 10x + 16}{x^2 + x - 6}$

In Exercises 114–117, add or subtract as indicated.

114. $\dfrac{2x - 7}{x^2 - 9} - \dfrac{x - 10}{x^2 - 9}$ **115.** $\dfrac{3x}{x + 2} + \dfrac{x}{x - 2}$

116. $\dfrac{x}{x^2 - 9} + \dfrac{x - 1}{x^2 - 5x + 6}$ **117.** $\dfrac{4x - 1}{2x^2 + 5x - 3} - \dfrac{x + 3}{6x^2 + x - 2}$

In Exercises 118–120, simplify each expression.

118. $\dfrac{\frac{1}{x} - \frac{1}{2}}{\frac{1}{3} - \frac{x}{6}}$ **119.** $\dfrac{3 + \frac{12}{x}}{1 - \frac{16}{x^2}}$ **120.** $\dfrac{3 - \frac{1}{x + 3}}{3 + \frac{1}{x + 3}}$

121. $\dfrac{\sqrt{25 - x^2} + \frac{x^2}{\sqrt{25 - x^2}}}{25 - x^2}$

P.7

In Exercises 122–135, solve each equation.

122. $1 - 2(6 - x) = 3x + 2$

123. $2(x - 4) + 3(x + 5) = 2x - 2$

124. $2x - 4(5x + 1) = 3x + 17$

125. $\dfrac{1}{x - 1} - \dfrac{1}{x + 1} = \dfrac{2}{x^2 - 1}$

126. $\dfrac{4}{x + 2} + \dfrac{2}{x - 4} = \dfrac{30}{x^2 - 2x - 8}$

127. $-4|2x + 1| + 12 = 0$ **128.** $2x^2 - 11x + 5 = 0$

129. $(3x + 5)(x - 3) = 5$ **130.** $3x^2 - 7x + 1 = 0$

131. $x^2 - 9 = 0$ **132.** $(x - 3)^2 - 24 = 0$

133. $\dfrac{2x}{x^2 + 6x + 8} = \dfrac{x}{x + 4} - \dfrac{2}{x + 2}$

134. $\sqrt{8 - 2x} - x = 0$ **135.** $\sqrt{2x - 3} + x = 3$

In Exercises 136–137, solve each formula for the specified variable.

136. $vt + gt^2 = s$ for g **137.** $T = \dfrac{A - P}{Pr}$ for P

In Exercises 138–139, without solving the given quadratic equation, determine the number and type of solutions.

138. $x^2 = 2x - 19$ **139.** $9x^2 - 30x + 25 = 0$

P.8

In Exercises 140–149, use the five-step strategy for solving word problems.

140. The fast-food chains may be touting their "new and improved" salads, but how do they measure up in terms of calories?

Burger King	Taco Bell	Wendy's
Chicken Caesar	Express Taco Salad	Mandarin Chicken Salad

Number of calories exceeds the Chicken Caesar by 125.

Number of calories exceeds the Chicken Caesar by 95.

Source: Newsweek

Combined, the three salads contain 1705 calories. Determine the number of calories in each salad.

141. The bar graph shows that in 1970, 37.4% of U.S. adults smoked cigarettes. For the period from 1970 through 2002, the percentage of smokers among U.S. adults decreased at an average rate of 0.5% per year. If this trend continues, when will only 18.4% of U.S. adults smoke cigarettes?

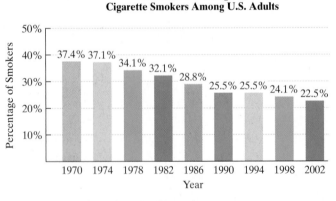

Butt Out: Percentage of Cigarette Smokers Among U.S. Adults

Source: Centers for Disease Control and Prevention

142. After a 20% price reduction, a cordless phone sold for $48. What was the phone's price before the reduction?

143. A salesperson earns $300 per week plus 5% commission of sales. How much must be sold to earn $800 in a week?

144. The length of a rectangular field is 6 yards less than triple the width. If the perimeter of the field is 340 yards, what are its dimensions?

145. In 2007, there were 14,100 students at college A, with a projected enrollment increase of 1500 students per year. In the same year, there were 41,700 students at college B, with a projected enrollment decline of 800 students per year. In which year will the colleges have the same enrollment? What will be the enrollment in each college at that time?

146 An architect is allowed 15 square yards of floor space to add a small bedroom to a house. Because of the room's design in relationship to the existing structure, the width of the rectangular floor must be 7 yards less than two times the length. Find the length and width of the rectangular floor that the architect is permitted.

147. A building casts a shadow that is double the length of its height. If the distance from the end of the shadow to the top of the building is 300 meters, how high is the building? Round to the nearest meter.

148. A painting measuring 10 inches by 16 inches is surrounded by a frame of uniform width. If the combined area of the painting and frame is 280 square inches, determine the width of the frame.

149. Club members equally share the cost of $1500 to charter a fishing boat. Shortly before the boat is to leave, four people decide not to go due to rough seas. As a result, the cost per person is increased by $100. How many people originally intended to go on the fishing trip?

P.9

In Exercises 150–152, express each interval in set-builder notation and graph the interval on a number line.

150. $[-3, 5)$ **151.** $(-2, \infty)$ **152.** $(-\infty, 0]$

In Exercises 153–156, use graphs to find each set.

153. $(-2, 1] \cap [-1, 3)$ **154.** $(-2, 1] \cup [-1, 3)$

155. $[1, 3) \cap (0, 4)$ **156.** $[1, 3) \cup (0, 4)$

In Exercises 157–166, solve each inequality. Use interval notation to express solution sets and graph each solution set on a number line.

157. $-6x + 3 \le 15$ **158.** $6x - 9 \ge -4x - 3$

159. $\dfrac{x}{3} - \dfrac{3}{4} - 1 > \dfrac{x}{2}$ **160.** $6x + 5 > -2(x - 3) - 25$

161. $3(2x - 1) - 2(x - 4) \ge 7 + 2(3 + 4x)$

162. $7 < 2x + 3 \le 9$ **163.** $|2x + 3| \le 15$

164. $\left|\dfrac{2x + 6}{3}\right| > 2$ **165.** $|2x + 5| - 7 \ge -6$

166. $-4|x + 2| + 5 \le -7$

167. A car rental agency rents a certain car for $40 per day with unlimited mileage or $24 per day plus $0.20 per mile. How far can a customer drive this car per day for the $24 option to cost no more than the unlimited mileage option?

168. To receive a B in a course, you must have an average of at least 80% but less than 90% on five exams. Your grades on the first four exams were 95%, 79%, 91%, and 86%. What range of grades on the fifth exam will result in a B for the course?

Chapter P Test

In Exercises 1–18, simplify the given expression or perform the indicated operation (and simplify, if possible), whichever is appropriate.

1. $5(2x^2 - 6x) - (4x^2 - 3x)$

2. $7 + 2[3(x + 1) - 2(3x - 1)]$

3. $\{1, 2, 5\} \cap \{5, a\}$ 4. $\{1, 2, 5\} \cup \{5, a\}$

5. $\dfrac{30x^3y^4}{6x^9y^{-4}}$

6. $\sqrt{6r}\sqrt{3r}$ (Assume that $r \geq 0$.)

7. $4\sqrt{50} - 3\sqrt{18}$

8. $\dfrac{3}{5 + \sqrt{2}}$ 9. $\sqrt[3]{16x^4}$

10. $\dfrac{x^2 + 2x - 3}{x^2 - 3x + 2}$

11. $\dfrac{5 \times 10^{-6}}{20 \times 10^{-8}}$ (Express the answer in scientific notation.)

12. $(2x - 5)(x^2 - 4x + 3)$ 13. $(5x + 3y)^2$

14. $\dfrac{2x + 8}{x - 3} \div \dfrac{x^2 + 5x + 4}{x^2 - 9}$ 15. $\dfrac{x}{x + 3} + \dfrac{5}{x - 3}$

16. $\dfrac{2x + 3}{x^2 - 7x + 12} - \dfrac{2}{x - 3}$

17. $\dfrac{1 - \dfrac{x}{x + 2}}{1 + \dfrac{1}{x}}$ 18. $\dfrac{2x\sqrt{x^2 + 5} - \dfrac{2x^3}{\sqrt{x^2 + 5}}}{x^2 + 5}$

In Exercises 19–24, factor completely, or state that the polynomial is prime.

19. $x^2 - 9x + 18$

20. $x^3 + 2x^2 + 3x + 6$

21. $25x^2 - 9$

22. $36x^2 - 84x + 49$

23. $y^3 - 125$

24. $x^2 + 10x + 25 - 9y^2$

25. Factor and simplify:

$$x(x + 3)^{-\frac{3}{5}} + (x + 3)^{\frac{2}{5}}.$$

26. List all the rational numbers in this set:

$$\left\{-7, -\tfrac{4}{5}, 0, 0.25, \sqrt{3}, \sqrt{4}, \tfrac{22}{7}, \pi\right\}.$$

In Exercises 27–28, state the name of the property illustrated.

27. $3(2 + 5) = 3(5 + 2)$ 28. $6(7 + 4) = 6 \cdot 7 + 6 \cdot 4$

29. Express in scientific notation: 0.00076.

30. Evaluate: $27^{-\frac{5}{3}}$.

31. In 2003, world population was approximately 6.3×10^9. By some projections, world population will double by 2040. Express the population at that time in scientific notation.

32. Your life expectancy is related to the year when you were born. The bar graph shows life expectancy in the United States by year of birth.

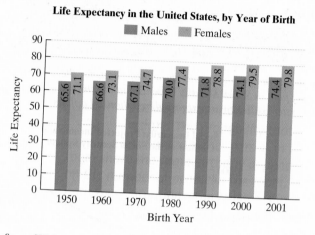

Life Expectancy in the United States, by Year of Birth

Source: U.S. Bureau of the Census

Here are two mathematical models for the data shown in the graph. In each formula, E represents life expectancy for Americans born t years after 1950.

Model 1 $E = 0.17t + 71$

Model 2 $E = 0.18t + 65$

a. Which model describes the data for the men and which model describes the data for the women?

b. According to the model that describes the data for the men, what is the life expectancy for U.S. men born in 2000? How well does the model describe the life expectancy shown in the graph?

c. Use the formula that models the data for the women to determine the year of birth for which U.S. women can expect to live 88 years.

In Exercises 33–47, solve each equation or inequality. Use interval notation to express solution sets of inequalities and graph these solution sets on a number line.

33. $7(x - 2) = 4(x + 1) - 21$

34. $\dfrac{2x - 3}{4} = \dfrac{x - 4}{2} - \dfrac{x + 1}{4}$

35. $\dfrac{2}{x - 3} - \dfrac{4}{x + 3} = \dfrac{8}{x^2 - 9}$

36. $2x^2 - 3x - 2 = 0$ 37. $(3x - 1)^2 = 75$

38. $x(x - 2) = 4$ 39. $\sqrt{x - 3} + 5 = x$

40. $\sqrt{8 - 2x} - x = 0$ 41. $\left|\dfrac{2}{3}x - 6\right| = 2$

42. $-3|4x - 7| + 15 = 0$

43. $\dfrac{2x}{x^2 + 6x + 8} + \dfrac{2}{x + 2} = \dfrac{x}{x + 4}$

EXAMPLE 8   Identifying the Domain and Range of a Function from Its Graph

Use the graph of each function to identify its domain and its range.

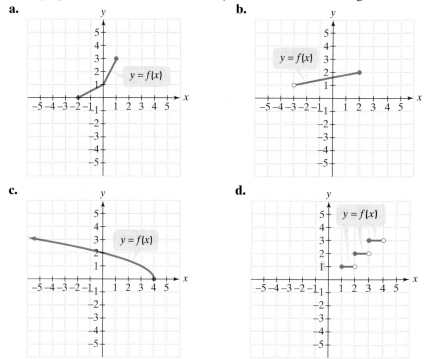

Solution   For the graph of each function, the domain is highlighted in blue on the x-axis and the range is highlighted in green on the y-axis.

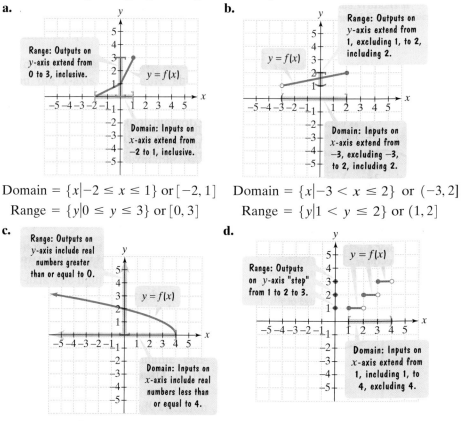

a.

Range: Outputs on y-axis extend from 0 to 3, inclusive.

Domain: Inputs on x-axis extend from −2 to 1, inclusive.

Domain = $\{x|-2 \le x \le 1\}$ or $[-2, 1]$
Range = $\{y|0 \le y \le 3\}$ or $[0, 3]$

b.

Range: Outputs on y-axis extend from 1, excluding 1, to 2, including 2.

Domain: Inputs on x-axis extend from −3, excluding −3, to 2, including 2.

Domain = $\{x|-3 < x \le 2\}$ or $(-3, 2]$
Range = $\{y|1 < y \le 2\}$ or $(1, 2]$

c.

Range: Outputs on y-axis include real numbers greater than or equal to 0.

Domain: Inputs on x-axis include real numbers less than or equal to 4.

Domain = $\{x|x \le 4\}$ or $(-\infty, 4]$
Range = $\{y|y \ge 0\}$ or $[0, \infty)$

d.

Range: Outputs on y-axis "step" from 1 to 2 to 3.

Domain: Inputs on x-axis extend from 1, including 1, to 4, excluding 4.

Domain = $\{x|1 \le x < 4\}$ or $[1, 4)$
Range = $\{y|y = 1, 2, 3\}$

Check Point 8 Use the graph of each function to identify its domain and its range.

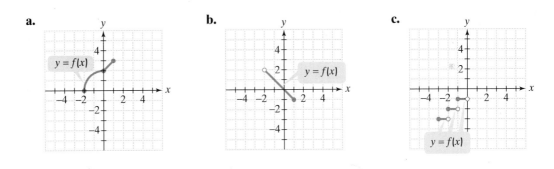

a.   b.   c.

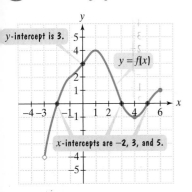

9 Identify intercepts from a function's graph.

Identifying Intercepts from a Function's Graph

Figure 1.23 illustrates how we can identify intercepts from a function's graph. To find the x-intercepts, look for the points at which the graph crosses the x-axis. There are three such points: $(-2, 0)$, $(3, 0)$, and $(5, 0)$. Thus, the x-intercepts are -2, 3, and 5. We express this in function notation by writing $f(-2) = 0$, $f(3) = 0$, and $f(5) = 0$. We say that -2, 3, and 5 are the *zeros of the function*. The **zeros of a function**, f, are the x-values for which $f(x) = 0$.

To find the y-intercept, look for the point at which the graph crosses the y-axis. This occurs at $(0, 3)$. Thus, the y-intercept is 3. We express this in function notation by writing $f(0) = 3$.

By the definition of a function, for each value of x we can have at most one value for y. What does this mean in terms of intercepts? **A function can have more than one x-intercept but at most one y-intercept.**

Figure 1.23 Identifying intercepts

EXERCISE SET 1.2

Practice Exercises

In Exercises 1–10, determine whether each relation is a function. Give the domain and range for each relation.

1. $\{(1, 2), (3, 4), (5, 5)\}$ 2. $\{(4, 5), (6, 7), (8, 8)\}$
3. $\{(3, 4), (3, 5), (4, 4), (4, 5)\}$
4. $\{(5, 6), (5, 7), (6, 6), (6, 7)\}$
5. $\{(3, -2), (5, -2), (7, 1), (4, 9)\}$
6. $\{(10, 4), (-2, 4), (-1, 1), (5, 6)\}$
7. $\{(-3, -3), (-2, -2), (-1, -1), (0, 0)\}$
8. $\{(-7, -7), (-5, -5), (-3, -3), (0, 0)\}$
9. $\{(1, 4), (1, 5), (1, 6)\}$
10. $\{(4, 1), (5, 1), (6, 1)\}$

In Exercises 11–26, determine whether each equation defines y as a function of x.

11. $x + y = 16$ 12. $x + y = 25$
13. $x^2 + y = 16$ 14. $x^2 + y = 25$
15. $x^2 + y^2 = 16$ 16. $x^2 + y^2 = 25$
17. $x = y^2$ 18. $4x = y^2$
19. $y = \sqrt{x + 4}$ 20. $y = -\sqrt{x + 4}$
21. $x + y^3 = 8$ 22. $x + y^3 = 27$

23. $xy + 2y = 1$ 24. $xy - 5y = 1$
25. $|x| - y = 2$ 26. $|x| - y = 5$

In Exercises 27–38, evaluate each function at the given values of the independent variable and simplify.

27. $f(x) = 4x + 5$
  a. $f(6)$ b. $f(x + 1)$ c. $f(-x)$
28. $f(x) = 3x + 7$
  a. $f(4)$ b. $f(x + 1)$ c. $f(-x)$
29. $g(x) = x^2 + 2x + 3$
  a. $g(-1)$ b. $g(x + 5)$ c. $g(-x)$
30. $g(x) = x^2 - 10x - 3$
  a. $g(-1)$ b. $g(x + 2)$ c. $g(-x)$
31. $h(x) = x^4 - x^2 + 1$
  a. $h(2)$ b. $h(-1)$
  c. $h(-x)$ d. $h(3a)$
32. $h(x) = x^3 - x + 1$
  a. $h(3)$ b. $h(-2)$
  c. $h(-x)$ d. $h(3a)$
33. $f(r) = \sqrt{r + 6} + 3$
  a. $f(-6)$ b. $f(10)$ c. $f(x - 6)$

34. $f(r) = \sqrt{25 - r} - 6$

   a. $f(16)$ **b.** $f(-24)$ **c.** $f(25 - 2x)$

35. $f(x) = \dfrac{4x^2 - 1}{x^2}$

   a. $f(2)$ **b.** $f(-2)$ **c.** $f(-x)$

36. $f(x) = \dfrac{4x^3 + 1}{x^3}$

   a. $f(2)$ **b.** $f(-2)$ **c.** $f(-x)$

37. $f(x) = \dfrac{x}{|x|}$

   a. $f(6)$ **b.** $f(-6)$ **c.** $f(r^2)$

38. $f(x) = \dfrac{|x + 3|}{x + 3}$

   a. $f(5)$ **b.** $f(-5)$ **c.** $f(-9 - x)$

In Exercises 39–50, graph the given functions, f and g, in the same rectangular coordinate system. Select integers for x, starting with −2 and ending with 2. Once you have obtained your graphs, describe how the graph of g is related to the graph of f.

39. $f(x) = x, g(x) = x + 3$

40. $f(x) = x, g(x) = x - 4$

41. $f(x) = -2x, g(x) = -2x - 1$

42. $f(x) = -2x, g(x) = -2x + 3$

43. $f(x) = x^2, g(x) = x^2 + 1$

44. $f(x) = x^2, g(x) = x^2 - 2$

45. $f(x) = |x|, g(x) = |x| - 2$

46. $f(x) = |x|, g(x) = |x| + 1$

47. $f(x) = x^3, g(x) = x^3 + 2$

48. $f(x) = x^3, g(x) = x^3 - 1$

49. $f(x) = 3, g(x) = 5$

50. $f(x) = -1, g(x) = 4$

In Exercises 51–54, graph the given square root functions, f and g, in the same rectangular coordinate system. Use the integer values of x given to the right of each function to obtain ordered pairs. Because only nonnegative numbers have square roots that are real numbers, be sure that each graph appears only for values of x that cause the expression under the radical sign to be greater than or equal to zero. Once you have obtained your graphs, describe how the graph of g is related to the graph of f.

51. $f(x) = \sqrt{x}$ $(x = 0, 1, 4, 9)$ and
     $g(x) = \sqrt{x} - 1$ $(x = 0, 1, 4, 9)$

52. $f(x) = \sqrt{x}$ $(x = 0, 1, 4, 9)$ and
     $g(x) = \sqrt{x} + 2$ $(x = 0, 1, 4, 9)$

53. $f(x) = \sqrt{x}$ $(x = 0, 1, 4, 9)$ and
     $g(x) = \sqrt{x - 1}$ $(x = 1, 2, 5, 10)$

54. $f(x) = \sqrt{x}$ $(x = 0, 1, 4, 9)$ and
     $g(x) = \sqrt{x + 2}$ $(x = -2, -1, 2, 7)$

In Exercises 55–64, use the vertical line test to identify graphs in which y is a function of x.

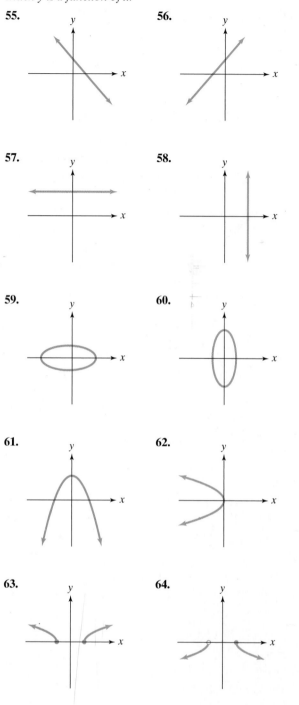

55.

56.

57.

58.

59.

60.

61.

62.

63.

64.

In Exercises 65–70, use the graph of f to find each indicated function value.

65. $f(-2)$

66. $f(2)$

67. $f(4)$

68. $f(-4)$

69. $f(-3)$

70. $f(-1)$

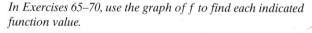
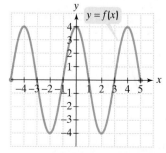

Use the graph of g to solve Exercises 71–76.

71. Find $g(-4)$.

72. Find $g(2)$.

73. Find $g(-10)$.

74. Find $g(10)$.

75. For what value of x is $g(x) = 1$?

76. For what value of x is $g(x) = -1$?

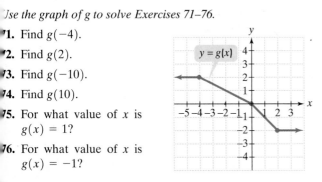

*In Exercises 77–92, use the graph to determine **a.** the function's domain; **b.** the function's range; **c.** the x-intercepts, if any; **d.** the y-intercept, if any; and **e.** the function values indicated below the graphs.*

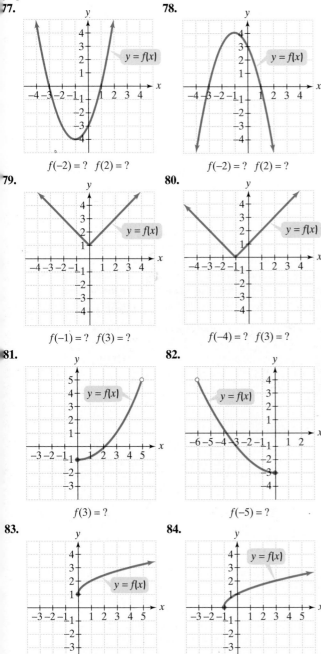

77.

$f(-2) = ?$ $f(2) = ?$

78.

$f(-2) = ?$ $f(2) = ?$

79.

$f(-1) = ?$ $f(3) = ?$

80.

$f(-4) = ?$ $f(3) = ?$

81.

$f(3) = ?$

82.

$f(-5) = ?$

83.

$f(4) = ?$

84.

$f(3) = ?$

85.

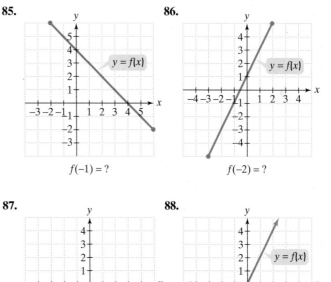

$f(-1) = ?$

86.

$f(-2) = ?$

87.

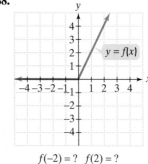

$f(-4) = ?$ $f(4) = ?$

88.

$f(-2) = ?$ $f(2) = ?$

89.

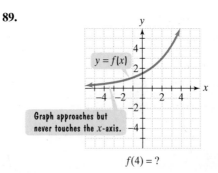

Graph approaches but never touches the x-axis.

$f(4) = ?$

90.

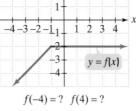

Graph approaches but never touches the dashed vertical line.

On both sides, graph never touches the x-axis.

$y = f(x)$

$f(2) = ?$

91.

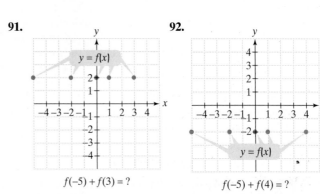

$y = f(x)$

$f(-5) + f(3) = ?$

92.

$y = f(x)$

$f(-5) + f(4) = ?$

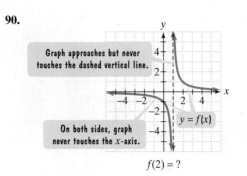

Practice Plus

In Exercises 93–94, let $f(x) = x^2 - x + 4$ *and* $g(x) = 3x - 5.$

93. Find $g(1)$ and $f(g(1))$.   **94.** Find $g(-1)$ and $f(g(-1))$.

In Exercises 95–96, let f and g be defined by the following table:

x	$f(x)$	$g(x)$
-2	6	0
-1	3	4
0	-1	1
1	-4	-3
2	0	-6

95. Find $\sqrt{f(-1) - f(0)} - [g(2)]^2 + f(-2) \div g(2) \cdot g(-1)$.

96. Find $|f(1) - f(0)| - [g(1)]^2 + g(1) \div f(-1) \cdot g(2)$,

In Exercises 97–98, find $f(-x) - f(x)$ *for the given function f. Then simplify the expression.*

97. $f(x) = x^3 + x - 5$ **98.** $f(x) = x^2 - 3x + 7$

Application Exercises

99. The bar graph shows the percentage of children in the world's leading industrial countries who daydream about being rich.

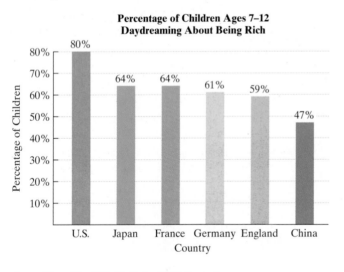

Percentage of Children Ages 7–12 Daydreaming About Being Rich

Source: Roper Starch Worldwide for A.B.C. Research

a. Write a set of six ordered pairs in which countries correspond to the percentage of children daydreaming about being rich. Each ordered pair should be in the form

(country, percent).

b. Is the relation in part (a) a function? Explain your answer.

c. Write a set of six ordered pairs in which the percentage of children daydreaming about being rich corresponds to countries. Each ordered pair should be in the form

(percent, country).

d. Is the relation in part (c) a function? Explain your answer.

100. The bar graph shows the breakdown of political ideologie in the United States.

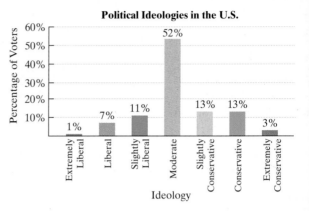

Political Ideologies in the U.S.

Source: Center for Political Studies, University of Michigan

a. Write a set of seven ordered pairs in which politica ideologies correspond to percentages. Each ordered pai should be in the form

(ideology, percent).

Use EL, L, SL, M, SC, C, and EC to represent the respec tive ideologies from left to right.

b. Is the relation in part (a) a function? Explain your answer

c. Write a set of seven ordered pairs in which percentages correspond to political ideologies. Each ordered pai should be in the form

(percent, ideology).

d. Is the relation in part (c) a function? Explain your answer

The male minority? The graphs show enrollment in U.S. colleges, with projections through 2009. The trend indicated by the graphs is among the hottest topics of debate among college-admissions offi- cers. Some private liberal arts colleges have quietly begun special efforts to recruit men—including admissions preferences for them.

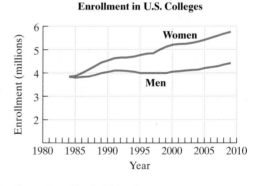

Enrollment in U.S. Colleges

Source: Department of Education

The function

$$W(x) = 0.07x + 4.1$$

models the number of women, $W(x)$, *in millions, enrolled in U.S. colleges x years after 1984. The function*

$$M(x) = 0.01x + 3.9$$

models the number of men, $M(x)$, *in millions, enrolled in U.S. colleges x years after 1984. Use these functions to solve Exercises 101–104.*

101. Find and interpret $W(16)$. Identify this information as a point on the graph for women.

102. Find and interpret $M(16)$. Identify this information as a point on the graph for men.

103. Find and interpret $W(20) - M(20)$.

104. Find and interpret $W(25) - M(25)$.

The wage gap is used to compare the status of women's earnings relative to men's. The wage gap is expressed as a percent and is calculated by dividing the median, or middlemost, annual earnings for women by the median annual earnings for men. The line graph shows the wage gap for selected years from 1960 through 2003.

Median Women's Earnings as a Percentage of Median Men's Earnings in the U.S.

Source: U.S. Women's Bureau

The function

$$P(x) = 0.012x^2 - 0.16x + 60$$

models median women's earnings as a percentage of median men's earnings, $P(x)$, x years after 1960. Use the graph and this function to solve Exercises 105–106.

105. a. Use the graph to estimate, to the nearest percent, women's earnings as a percentage of men's in 2000.

   b. Use the function to find women's earnings as a percentage of men's in 2000.

   c. In 2000, median annual earnings for U.S. women and men were $27,355 and $37,339, respectively. What were women's earnings as a percentage of men's? Use a calculator and round to the nearest tenth of a percent. How well do your answers in parts (a) and (b) model the actual data?

106. a. Use the graph to estimate, to the nearest percent, women's earnings as a percentage of men's in 2003.

   b. Use the function to find women's earnings as a percentage of men's in 2003. Round to the nearest tenth of a percent.

   c. In 2003, median annual earnings for U.S. women and men were $30,724 and $40,668, respectively. What were women's earnings as a percentage of men's? Use a calculator and round to the nearest tenth of a percent. How well do your answers in parts (a) and (b) model the actual data?

In Exercises 107–110, you will be developing functions that model given conditions.

107. A company that manufactures bicycles has a fixed cost of $100,000. It costs $100 to produce each bicycle. The total cost for the company is the sum of its fixed cost and variable costs. Write the total cost, C, as a function of the number of bicycles produced, x. Then find and interpret $C(90)$.

108. A car was purchased for $22,500. The value of the car decreases by $3200 per year for the first six years. Write a function that describes the value of the car, V, after x years, where $0 \le x \le 7$. Then find and interpret $V(3)$.

109. You commute to work a distance of 40 miles and return on the same route at the end of the day. Your average rate on the return trip is 30 miles per hour faster than your average rate on the outgoing trip. Write the total time, T, in hours, devoted to your outgoing and return trips as a function of your rate on the outgoing trip, x. Then find and interpret $T(30)$. Hint:

$$\text{Time traveled} = \frac{\text{Distance traveled}}{\text{Rate of travel}}.$$

110. A chemist working on a flu vaccine needs to mix a 10% sodium-iodine solution with a 60% sodium-iodine solution to obtain a 50-milliliter mixture. Write the amount of sodium iodine in the mixture, S, in milliliters, as a function of the number of milliliters of the 10% solution used, x. Then find and interpret $S(30)$.

Writing in Mathematics

111. What is a relation? Describe what is meant by its domain and its range.

112. Explain how to determine whether a relation is a function. What is a function?

113. How do you determine if an equation in x and y defines y as a function of x?

114. Does $f(x)$ mean f times x when referring to a function f? If not, what does $f(x)$ mean? Provide an example with your explanation.

115. What is the graph of a function?

116. Explain how the vertical line test is used to determine whether a graph represents a function.

117. Explain how to identify the domain and range of a function from its graph.

118. For people filing a single return, federal income tax is a function of adjusted gross income because for each value of adjusted gross income there is a specific tax to be paid. By contrast, the price of a house is not a function of the lot size on which the house sits because houses on same-sized lots can sell for many different prices.

   a. Describe an everyday situation between variables that is a function.

   b. Describe an everyday situation between variables that is not a function.

119. Do you believe that the trend shown by the graphs for Exercises 101–104 should be reversed by providing admissions preferences for men? Explain your position on this issue.

 **Technology Exercise**

120. Use a graphing utility to verify any five pairs of graphs that you drew by hand in Exercises 39–54.

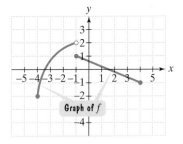

Critical Thinking Exercises

121. Which one of the following is true based on the graph of f in the figure?

Graph of f

    a. The domain of f is $[-4, 1) \cup (1, 4]$.

    b. The range of f is $[-2, 2]$.

    c. $f(-1) - f(4) = 2$

    d. $f(0) = 2.1$

122. If $f(x) = 3x + 7$, find $\dfrac{f(a + h) - f(a)}{h}$.

123. Give an example of a relation with the following characteristics: The relation is a function containing two ordered pairs. Reversing the components in each ordered pair results in a relation that is not a function.

124. If $f(x + y) = f(x) + f(y)$ and $f(1) = 3$, find $f(2)$, $f(3)$, and $f(4)$. Is $f(x + y) = f(x) + f(y)$ for all functions?

SECTION 1.3   *More on Functions and Their Graphs*

Objectives

❶ Find and simplify a function's difference quotient.

❷ Understand and use piecewise functions.

❸ Identify intervals on which a function increases, decreases, or is constant.

❹ Use graphs to locate relative maxima or minima.

❺ Identify even or odd functions and recognize their symmetries.

❻ Graph step functions.

"Our relationship is just going through a phase." When that phase reaches a point where married couples cannot agree on anything else, they agree on a divorce, frequently arranged so that lawyers can live happily ever after. The graph in Figure 1.24 shows the percent distribution of divorces in the United States by number of years of marriage.

Figure 1.24

Source: Divorce Center

You are probably familiar with the words and phrases used to describe the graph in Figure 1.24:

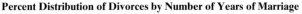

increasing   decreasing   maximum   slowing rate of decrease

In this section, you will enhance your intuitive understanding of ways of describing graphs by viewing these descriptions from the perspective of functions.

① Find and simplify a function's difference quotient.

Functions and Difference Quotients

In the next section, we will be studying the average rate of change of a function. A ratio, called the *difference quotient*, plays an important role in understanding the rate at which functions change.

Definition of a Difference Quotient

The expression

$$\frac{f(x + h) - f(x)}{h}$$

for $h \neq 0$ is called the **difference quotient**.

EXAMPLE 1   Evaluating and Simplifying a Difference Quotient

If $f(x) = 2x^2 - x + 3$, find and simplify each expression:

    a. $f(x + h)$ **b.** $\dfrac{f(x + h) - f(x)}{h}, h \neq 0.$

Solution

a. We find $f(x + h)$ by replacing x with $x + h$ each time that x appears in the equation.

$$f(x) = 2x^2 \quad - \quad x \quad\quad + \quad 3$$

| Replace x with $x + h$. | Replace x with $x + h$. | Replace x with $x + h$. | Copy the 3. There is no x in this term. |

$$f(x + h) = 2(x + h)^2 - (x + h) \quad + \quad 3$$
$$= 2(x^2 + 2xh + h^2) - x - h \ + \ 3$$
$$= 2x^2 + 4xh + 2h^2 - x - h \ + \ 3$$

b. Using our result from part (a), we obtain the following:

This is $f(x + h)$ from part (a).   *This is $f(x)$ from the given equation.*

$$\frac{f(x + h) - f(x)}{h} = \frac{\boxed{2x^2 + 4xh + 2h^2 - x - h + 3} - (2x^2 - x + 3)}{h}$$

$$= \frac{2x^2 + 4xh + 2h^2 - x - h + 3 - 2x^2 + x - 3}{h}$$
Remove parentheses and change the sign of each term in the parentheses.

$$= \frac{(2x^2 - 2x^2) + (-x + x) + (3 - 3) + 4xh + 2h^2 - h}{h}$$
Group like terms.

$$= \frac{4xh + 2h^2 - 1h}{h}$$
Simplify.

We wrote $-h$ as $-1h$ to avoid possible errors in the next factoring step.

$$= \frac{h(4x + 2h - 1)}{h}$$
Factor h from the numerator.

$$= 4x + 2h - 1$$
Divide out identical factors of h in the numerator and denominator.

Check Point 1   If $f(x) = -2x^2 + x + 5$, find and simplify each expression:

    a. $f(x + h)$ **b.** $\dfrac{f(x + h) - f(x)}{h}, h \neq 0.$

② Understand and use piecewise functions.

Piecewise Functions

A cellular phone company offers the following plan:

- $20 per month buys 60 minutes.
- Additional time costs $0.40 per minute.

We can represent this plan mathematically by writing the total monthly cost, C, as a function of the number of calling minutes, t.

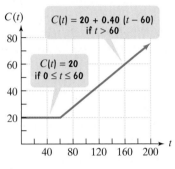

The cost is $20 for up to and including 60 calling minutes.

$$C(t) = \begin{cases} 20 & \text{if } 0 \le t \le 60 \\ 20 + 0.40(t - 60) & \text{if } t > 60 \end{cases}$$

The cost is $20 plus $0.40 per minute for additional time for more than 60 calling minutes.

$20 for first 60 minutes

$0.40 per minute

times the number of calling minutes exceeding 60

Figure 1.25

A function that is defined by two (or more) equations over a specified domain is called a **piecewise function**. Many cellular phone plans can be represented with piecewise functions. The graph of the piecewise function described above is shown in Figure 1.25.

EXAMPLE 2 Evaluating a Piecewise Function

Use the function that describes the cellular phone plan

$$C(t) = \begin{cases} 20 & \text{if } 0 \le t \le 60 \\ 20 + 0.40(t - 60) & \text{if } t > 60 \end{cases}$$

to find and interpret each of the following:
   a. $C(30)$ **b.** $C(100)$.

Solution

a. To find $C(30)$, we let $t = 30$. Because 30 lies between 0 and 60, we use the first line of the piecewise function.

$$C(t) = 20 \qquad \text{This is the function's equation for } 0 \le t \le 60.$$

$$C(30) = 20 \qquad \text{Replace } t \text{ with 30. Regardless of this function's input, the constant output is 20.}$$

This means that with 30 calling minutes, the monthly cost is $20. This can be visually represented by the point (30, 20) on the first piece of the graph in Figure 1.25.

b. To find $C(100)$, we let $t = 100$. Because 100 is greater than 60, we use the second line of the piecewise function.

$$C(t) = 20 + 0.40(t - 60) \qquad \text{This is the function's equation for } t > 60.$$

$$C(100) = 20 + 0.40(100 - 60) \qquad \text{Replace } t \text{ with 100.}$$

$$= 20 + 0.40(40) \qquad \text{Subtract within parentheses: } 100 - 60 = 40.$$

$$= 20 + 16 \qquad \text{Multiply: } 0.40(40) = 16.$$

$$= 36 \qquad \text{Add: } 20 + 16 = 36.$$

Thus, $C(100) = 36$. This means that with 100 calling minutes, the monthly cost is $36. This can be visually represented by the point (100, 36) on the second piece of the graph in Figure 1.25.

Check Point 2 Use the function in Example 2 to find and interpret each of the following:

a. $C(40)$ **b.** $C(80)$.

Identify solutions on the graph in Figure 1.25.

③ Identify intervals on which a function increases, decreases, or is constant.

Increasing and Decreasing Functions

Too late for that flu shot now! It's only 8 A.M. and you're feeling lousy. Your temperature is 101°F. Fascinated by the way that algebra models the world (your author is projecting a bit here), you decide to construct graphs showing your body temperature as a function of the time of day. You decide to let x represent the number of hours after 8 A.M. and $f(x)$ your temperature at time x.

At 8 A.M. your temperature is 101°F and you are not feeling well. However, your temperature starts to decrease. It reaches normal (98.6°F) by 11 A.M. Feeling energized, you construct the graph shown on the right, indicating decreasing temperature for $\{x|0 < x < 3\}$, or on the interval $(0, 3)$.

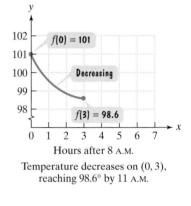

Temperature decreases on $(0, 3)$, reaching 98.6° by 11 A.M.

Did creating that first graph drain you of your energy? Your temperature starts to rise after 11 A.M. By 1 P.M., 5 hours after 8 A.M., your temperature reaches 100°F. However, you keep plotting points on your graph. At the right, we can see that your temperature increases for $\{x|3 < x < 5\}$, or on the interval $(3, 5)$.

The graph of f is decreasing to the left of $x = 3$ and increasing to the right of $x = 3$. Thus, your temperature 3 hours after 8 A.M. was at its lowest point. Your relative minimum temperature was 98.6°.

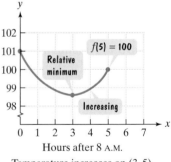

Temperature increases on $(3, 5)$.

By 3 P.M., your temperature is no worse than it was at 1 P.M.: It is still 100°F. (Of course, it's no better, either.) Your temperature remained the same, or constant, for $\{x|5 < x < 7\}$, or on the interval $(5, 7)$.

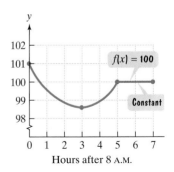

Hours after 8 A.M.

The time-temperature flu scenario illustrates that a function f is increasing when its graph rises from left to right, decreasing when its graph falls from left to right, and remains constant when it neither rises nor falls. Let's now provide a more precise algebraic description for these intuitive concepts.

Did you know that graphs of some equations exhibit exactly the kind of sym metry shown by the attractive face in Figure 1.28? The word *symmetry* comes from the Greek *symmetria*, meaning "the same measure." We can identify graphs with symmetry by looking at a function's equation and determining if the function is *even* or *odd*.

Definition of Even and Odd Functions

The function f is an **even function** if
$$f(-x) = f(x) \quad \text{for all } x \text{ in the domain of } f.$$
The right side of the equation of an even function does not change if x is replaced with $-x$.

The function f is an **odd function** if
$$f(-x) = -f(x) \quad \text{for all } x \text{ in the domain of } f.$$
Every term in the right side of the equation of an odd function changes its sign if x is replaced with $-x$.

EXAMPLE 4 Identifying Even or Odd Functions

Determine whether each of the following functions is even, odd, or neither:

a. $f(x) = x^3 - 6x$ **b.** $g(x) = x^4 - 2x^2$ **c.** $h(x) = x^2 + 2x + 1.$

Solution In each case, replace x with $-x$ and simplify. If the right side of the equation stays the same, the function is even. If every term on the right changes sign, the function is odd.

a. We use the given function's equation, $f(x) = x^3 - 6x$, to find $f(-x)$.

Use $f(x) = x^3 - 6x$.

Replace x with $-x$.

$$f(-x) = (-x)^3 - 6(-x) = (-x)(-x)(-x) - 6(-x) = -x^3 + 6x$$

There are two terms on the right side of the given equation, $f(x) = x^3 - 6x$ and each term changed its sign when we replaced x with $-x$. Because $f(-x) = -f(x)$, f is an odd function.

b. We use the given function's equation, $g(x) = x^4 - 2x^2$, to find $g(-x)$.

Use $g(x) = x^4 - 2x^2$.

Replace x with $-x$.

$$g(-x) = (-x)^4 - 2(-x)^2 = (-x)(-x)(-x)(-x) - 2(-x)(-x)$$
$$= x^4 - 2x^2$$

The right side of the equation of the given function, $g(x) = x^4 - 2x^2$, did not change when we replaced x with $-x$. Because $g(-x) = g(x)$, g is an even function.

c. We use the given function's equation, $h(x) = x^2 + 2x + 1$, to find $h(-x)$.

Use $h(x) = x^2 + 2x + 1$.

Replace x with $-x$.

$$h(-x) = (-x)^2 + 2(-x) + 1 = x^2 - 2x + 1$$

The right side of the equation of the given function, $h(x) = x^2 + 2x + 1$, changed when we replaced x with $-x$. Thus, $h(-x) \neq h(x)$, so h is not an even function. The sign of *each* of the three terms in the equation for $h(x)$ did not change when we replaced x with $-x$. Only the second term changed signs. Thus, $h(-x) \neq -h(x)$, so h is not an odd function. We conclude that h is neither an even nor an odd function.

Check Point 4 Determine whether each of the following functions is even, odd, or neither:

a. $f(x) = x^2 + 6$ **b.** $g(x) = 7x^3 - x$ **c.** $h(x) = x^5 + 1.$

Now, let's see what even and odd functions tell us about a function's graph. Begin with the even function $f(x) = x^2 - 4$, shown in Figure 1.29. The function is even because

$$f(-x) = (-x)^2 - 4 = x^2 - 4 = f(x).$$

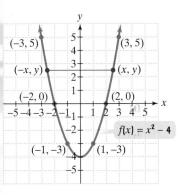

Examine the pairs of points shown, such as $(3, 5)$ and $(-3, 5)$. Notice that we obtain the same y-coordinate whenever we evaluate the function at a value of x and the value of its opposite, $-x$. Like the attractive face, each half of the graph is a mirror image of the other half through the y-axis. If we were to fold the paper along the y-axis, the two halves of the graph would coincide. This causes the graph to be *symmetric with respect to the y-axis*. A graph is **symmetric with respect to the y-axis** if, for every point (x, y) on the graph, the point $(-x, y)$ is also on the graph. All even functions have graphs with this kind of symmetry.

Figure 1.29 y-axis symmetry with $f(-x) = f(x)$

Even Functions and y-Axis Symmetry

The graph of an even function in which $f(-x) = f(x)$ is symmetric with respect to the y-axis.

Now, consider the graph of the function $f(x) = x^3$, shown in Figure 1.30. The function is odd because

$$f(-x) = (-x)^3 = (-x)(-x)(-x) = -x^3 = -f(x).$$

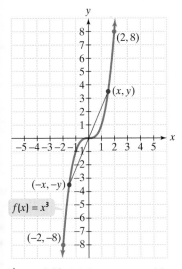

Although the graph in Figure 1.30 is not symmetric with respect to the y-axis, it is symmetric in another way. Look at the pairs of points, such as $(2, 8)$ and $(-2, -8)$. For each point (x, y) on the graph, the point $(-x, -y)$ is also on the graph. The points $(2, 8)$ and $(-2, -8)$ are reflections of one another about the origin. This means that

• the points are the same distance from the origin, and
• the points lie on a line through the origin.

Figure 1.30 Origin symmetry with $f(-x) = -f(x)$

A graph is **symmetric with respect to the origin** if, for every point (x, y) on the graph, the point $(-x, -y)$ is also on the graph. Observe that the first- and third-quadrant portions of $f(x) = x^3$ are reflections of one another with respect to the origin. Notice that $f(x)$ and $f(-x)$ have opposite signs, so that $f(-x) = -f(x)$. All odd functions have graphs with origin symmetry.

Odd Functions and Origin Symmetry

The graph of an odd function in which $f(-x) = -f(x)$ is symmetric with respect to the origin.

Graph step functions.

Table 1.2   Cost of First-Class Mail (Effective June 30, 2002)

Weight Not Over	Cost
1 ounce	$0.37
2 ounces	0.60
3 ounces	0.83
4 ounces	1.06
5 ounces	1.29

Source: U.S. Postal Service

Step Functions

Have you ever mailed a letter that seemed heavier than usual? Perhaps you worried that the letter would not have enough postage. Costs for mailing a letter weighing up to 5 ounces are given in Table 1.2. If your letter weighs an ounce or less, the cost is $0.37. If your letter weighs 1.05 ounces, 1.50 ounces, 1.90 ounces, or 2.00 ounces, the cost "steps" to $0.60. The cost does not take on any value between $0.37 and $0.60. If your letter weighs 2.05 ounces, 2.50 ounces, 2.90 ounces, or 3 ounces, the cost "steps" to $0.83. Cost increases are $0.23 per step.

Now, let's see what the graph of the function that models this situation looks like. Let

$$x = \text{the weight of the letter, in ounces, and}$$
$$y = f(x) = \text{the cost of mailing a letter weighing } x \text{ ounces.}$$

The graph is shown in Figure 1.31. Notice how it consists of a series of steps that jump vertically 0.23 unit at each integer. The graph is constant between each pair of consecutive integers.

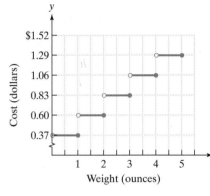

Figure 1.31

Mathematicians have defined functions that describe situations where function values graphically form discontinuous steps. One such function is called the **greatest integer function**, symbolized by int(x) or $[\![x]\!]$. And what is int(x)?

$$\text{int}(x) = \text{the greatest integer that is less than or equal to } x$$

For example,

$$\text{int}(1) = 1, \quad \text{int}(1.3) = 1, \quad \text{int}(1.5) = 1, \quad \text{int}(1.9) = 1.$$

> 1 is the greatest integer that is less than or equal to 1, 1.3, 1.5, and 1.9.

Here are some additional examples:

$$\text{int}(2) = 2, \quad \text{int}(2.3) = 2, \quad \text{int}(2.5) = 2, \quad \text{int}(2.9) = 2.$$

> 2 is the greatest integer that is less than or equal to 2, 2.3, 2.5, and 2.9.

Notice how we jumped from 1 to 2 in the function values for int(x). In particular,

$$\text{If } 1 \le x < 2, \quad \text{then} \quad \text{int}(x) = 1.$$
$$\text{If } 2 \le x < 3, \quad \text{then} \quad \text{int}(x) = 2.$$

The graph of $f(x) = \text{int}(x)$ is shown in Figure 1.32. The graph of the greatest integer function jumps vertically one unit at each integer. However, the graph is constant between each pair of consecutive integers. The rightmost horizontal step shown in the graph illustrates that

$$\text{If } 5 \le x < 6, \quad \text{then} \quad \text{int}(x) = 5.$$

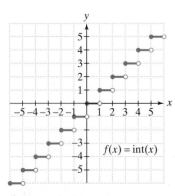

Figure 1.32 The graph of the greatest integer function

In general,

$$\text{If } n \le x < n + 1, \text{ where } n \text{ is an integer,} \quad \text{then} \quad \text{int}(x) = n.$$

By contrast to the graph for the cost of first-class mail, the graph of the greatest integer function includes the point on the left of each horizontal step, but does not include the point on the right. The domain of $f(x) = \text{int}(x)$ is the set of all real numbers, $(-\infty, \infty)$. The range is the set of all integers.

Technology

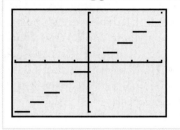

The graph of $f(x) = \text{int}(x)$, shown on the left, was obtained with a graphing utility. By graphing in "dot" mode, we can see the discontinuities at the integers. By looking at the graph, it is impossible to tell that, for each step, the point on the left is included and the point on the right is not. We must trace along the graph to obtain such information.

EXERCISE SET 1.3

Practice Exercises

In Exercises 1–22, find and simplify the difference quotient

$$\frac{f(x + h) - f(x)}{h}, h \neq 0$$

for the given function.

1. $f(x) = 4x$
2. $f(x) = 7x$
3. $f(x) = 3x + 7$
4. $f(x) = 6x + 1$
5. $f(x) = x^2$
6. $f(x) = 2x^2$
7. $f(x) = x^2 - 4x + 3$
8. $f(x) = x^2 - 5x + 8$
9. $f(x) = 2x^2 + x - 1$
10. $f(x) = 3x^2 + x + 5$
11. $f(x) = -x^2 + 2x + 4$
12. $f(x) = -x^2 - 3x + 1$
13. $f(x) = -2x^2 + 5x + 7$
14. $f(x) = -3x^2 + 2x - 1$
15. $f(x) = -2x^2 - x + 3$
16. $f(x) = -3x^2 + x - 1$
17. $f(x) = 6$
18. $f(x) = 7$
19. $f(x) = \dfrac{1}{x}$
20. $f(x) = \dfrac{1}{2x}$
21. $f(x) = \sqrt{x}$
22. $f(x) = \sqrt{x - 1}$

In Exercises 23–28, evaluate each piecewise function at the given values of the independent variable.

23. $f(x) = \begin{cases} 3x + 5 & \text{if } x < 0 \\ 4x + 7 & \text{if } x \geq 0 \end{cases}$

    a. $f(-2)$ **b.** $f(0)$ **c.** $f(3)$

24. $f(x) = \begin{cases} 6x - 1 & \text{if } x < 0 \\ 7x + 3 & \text{if } x \geq 0 \end{cases}$

    a. $f(-3)$ **b.** $f(0)$ **c.** $f(4)$

25. $g(x) = \begin{cases} x + 3 & \text{if } x \geq -3 \\ -(x + 3) & \text{if } x < -3 \end{cases}$

    a. $g(0)$ **b.** $g(-6)$ **c.** $g(-3)$

26. $g(x) = \begin{cases} x + 5 & \text{if } x \geq -5 \\ -(x + 5) & \text{if } x < -5 \end{cases}$

    a. $g(0)$ **b.** $g(-6)$ **c.** $g(-5)$

27. $h(x) = \begin{cases} \dfrac{x^2 - 9}{x - 3} & \text{if } x \neq 3 \\ 6 & \text{if } x = 3 \end{cases}$

    a. $h(5)$ **b.** $h(0)$ **c.** $h(3)$

28. $h(x) = \begin{cases} \dfrac{x^2 - 25}{x - 5} & \text{if } x \neq 5 \\ 10 & \text{if } x = 5 \end{cases}$

    a. $h(7)$ **b.** $h(0)$ **c.** $h(5)$

In Exercises 29–40, use the graph to determine

    a. intervals on which the function is increasing, if any.
    b. intervals on which the function is decreasing, if any.
    c. intervals on which the function is constant, if any.

29.   **30.**

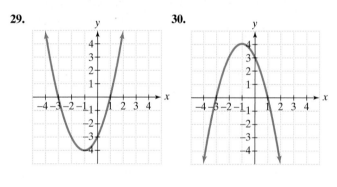

31.   **32.**

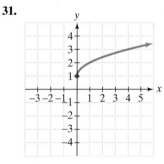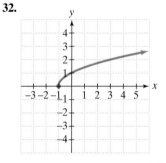

33.   **34.**

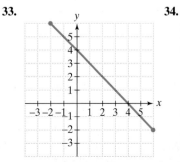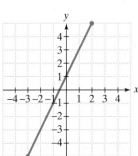

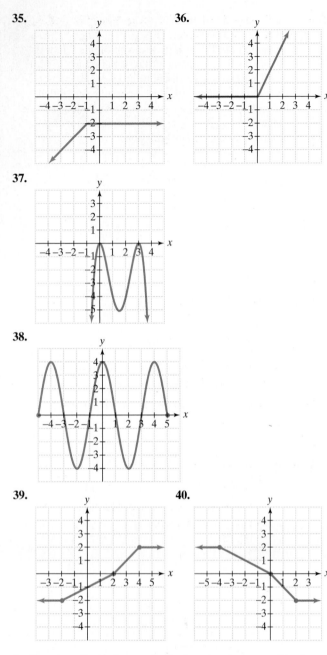

35.

36.

37.

38.

39.

40.

In Exercises 41–44, the graph of a function f is given. Use the graph to find each of the following:

  a. *The numbers, if any, at which f has a relative maximum. What are these relative maxima?*

  b. *The numbers, if any, at which f has a relative minimum. What are these relative minima?*

41.

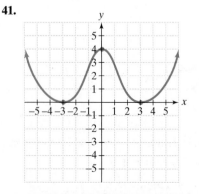

42.

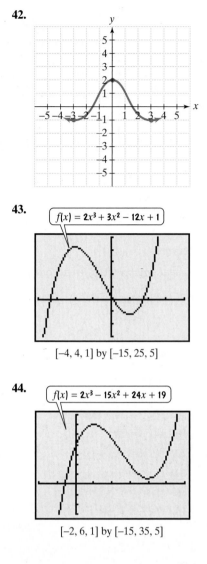

43.

$f(x) = 2x^3 + 3x^2 - 12x + 1$

$[-4, 4, 1]$ by $[-15, 25, 5]$

44.

$f(x) = 2x^3 - 15x^2 + 24x + 19$

$[-2, 6, 1]$ by $[-15, 35, 5]$

In Exercises 45–56, determine whether each function is even, odd, or neither.

45. $f(x) = x^3 + x$ **46.** $f(x) = x^3 - x$

47. $g(x) = x^2 + x$ **48.** $g(x) = x^2 - x$

49. $h(x) = x^2 - x^4$ **50.** $h(x) = 2x^2 + x^4$

51. $f(x) = x^2 - x^4 + 1$ **52.** $f(x) = 2x^2 + x^4 + 1$

53. $f(x) = \frac{1}{5}x^6 - 3x^2$ **54.** $f(x) = 2x^3 - 6x^5$

55. $f(x) = x\sqrt{1 - x^2}$ **56.** $f(x) = x^2\sqrt{1 - x^2}$

In Exercises 57–60, use possible symmetry to determine whether each graph is the graph of an even function, an odd function, or a function that is neither even nor odd.

57.

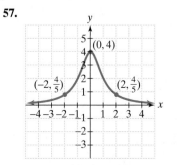

8.

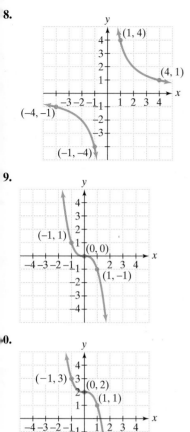

9.

0.

62. Use the graph of f to determine each of the following. Where applicable, use interval notation.

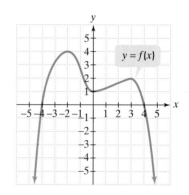

a. the domain of f

b. the range of f

c. the x-intercepts

d. the y-intercept

e. intervals on which f is increasing

f. intervals on which f is decreasing

g. values of x for which $f(x) \le 0$

h. the numbers at which f has a relative maximum

i. the relative maxima of f

j. $f(-2)$

k. the values of x for which $f(x) = 0$

l. Is f even, odd, or neither?

1. Use the graph of f to determine each of the following. Where applicable, use interval notation.

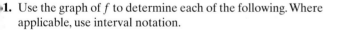
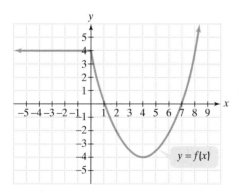

a. the domain of f

b. the range of f

c. the x-intercepts

d. the y-intercept

e. intervals on which f is increasing

f. intervals on which f is decreasing

g. intervals on which f is constant

h. the number at which f has a relative minimum

i. the relative minimum of f

j. $f(-3)$

k. the values of x for which $f(x) = -2$

l. Is f even, odd, or neither?

63. Use the graph of f to determine each of the following. Where applicable, use interval notation.

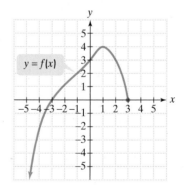

a. the domain of f

b. the range of f

c. the zeros of f

d. $f(0)$

e. intervals on which f is increasing

f. intervals on which f is decreasing

g. values of x for which $f(x) \le 0$

h. any relative maxima and the numbers at which they occur

i. the value of x for which $f(x) = 4$

j. Is $f(-1)$ positive or negative?

64. Use the graph of f to determine each of the following. Where applicable, use interval notation.

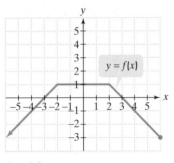

a. the domain of f
b. the range of f
c. the zeros of f
d. $f(0)$
e. intervals on which f is increasing
f. intervals on which f is decreasing
g. intervals on which f is constant
h. values of x for which $f(x) > 0$
i. values of x for which $f(x) = -2$
j. Is $f(4)$ positive or negative?
k. Is f even, odd, or neither?
l. Is $f(2)$ a relative maximum?

In Exercises 65–70, if $f(x) = int(x)$, find each function value.

65. $f(1.06)$ **66.** $f(2.99)$ **67.** $f\left(\frac{1}{3}\right)$

68. $f(-1.5)$ **69.** $f(-2.3)$ **70.** $f(-99.001)$

Practice Plus

In Exercises 71–72, let f be defined by the following graph:

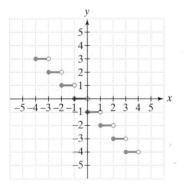

71. Find
$$\sqrt{f(-1.5) + f(-0.9)} - [f(\pi)]^2 + f(-3) \div f(1) \cdot f(-\pi).$$

72. Find
$$\sqrt{f(-2.5) - f(1.9)} - [f(-\pi)]^2 + f(-3) \div f(1) \cdot f(\pi).$$

A cellular phone company offers the following plans. Also given are the piecewise functions that describe these plans. Use this information to solve Exercises 73–74.

Plan A
- $30 per month buys 120 minutes.
- Additional time costs $0.30 per minute.

$$C(t) = \begin{cases} 30 & \text{if } 0 \le t \le 120 \\ 30 + 0.30(t - 120) & \text{if } t > 120 \end{cases}$$

Plan B
- $40 per month buys 200 minutes.
- Additional time costs $0.30 per minute.

$$C(t) = \begin{cases} 40 & \text{if } 0 \le t \le 200 \\ 40 + 0.30(t - 200) & \text{if } t > 200 \end{cases}$$

73. Simplify the algebraic expression in the second line of the piecewise function for plan A. Then use point-plotting to graph the function.

74. Simplify the algebraic expression in the second line of the piecewise function for plan B. Then use point-plotting to graph the function.

In Exercises 75–76, write a piecewise function that describes each cellular phone billing plan. Then graph the function.

75. $50 per month buys 400 minutes. Additional time costs $0.30 per minute.

76. $60 per month buys 450 minutes. Additional time costs $0.35 per minute.

Application Exercises

The figure shows the percentage of Jewish Americans in the U.S. population, $f(x)$, x years after 1900. Use the graph to solve Exercises 77–84.

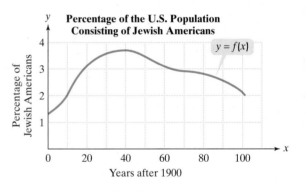

Source: American Jewish Yearbook

77. Use the graph to find a reasonable estimate of $f(60)$. What does this mean in terms of the variables in this situation?

78. Use the graph to find a reasonable estimate of $f(100)$. What does this mean in terms of the variables in this situation?

79. For what value or values of x is $f(x) = 3$? Round to the nearest year. What does this mean in terms of the variables in this situation?

80. For what value or values of x is $f(x) = 2.5$? Round to the nearest year. What does this mean in terms of the variables in this situation?

81. In which year did the percentage of Jewish Americans in the U.S. population reach a maximum? What is a reasonable estimate of the percentage for that year?

82. In which year was the percentage of Jewish Americans in the U.S. population at a minimum? What is a reasonable estimate of the percentage for that year?

83. Explain why f represents the graph of a function.

84. Describe the general trend shown by the graph.

The function

$$f(x) = 0.4x^2 - 36x + 1000$$

models the number of accidents, $f(x)$, per 50 million miles driven as a function of a driver's age, x, in years, where x includes drivers from ages 16 through 74, inclusive. The graph of f is shown. Use the graph of f, and possibly the equation, to solve Exercises 85–88.

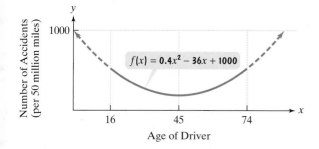

85. State the intervals on which the function is increasing and decreasing. Describe what this means in terms of the variables modeled by the function.

86. For what value of x does the graph reach its lowest point? Use the equation for f to find the minimum value of y. Describe the practical significance of this minimum value.

87. Use the graph to identify two different ages for which drivers have the same number of accidents. Use the equation for f to find the number of accidents for drivers at each of these ages.

88. Use the equation for f to find and interpret $f(50)$. Identify this information as a point on the graph of f.

The graph shows cigarette consumption per U.S. adult from 1910 through 2003. The data can be modeled by the piecewise function

$$f(x) = \begin{cases} 61.9x + 132 & \text{if } 0 \le x \le 30 \\ -2.2x^2 + 256x - 3503 & \text{if } 30 < x \le 93, \end{cases}$$

where x represents years after 1910 and $f(x)$ represents cigarette consumption per U.S. adult. Use this information to solve Exercises 89–92.

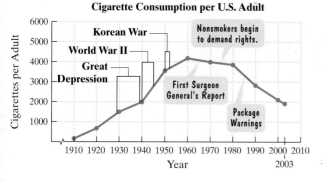

Cigarette Consumption per U.S. Adult

Source: U.S. Department of Health and Human Services

89. Use the piecewise function that models the data to find cigarette consumption in 1940. How well does the function describe the actual consumption for that year shown by the line graph?

90. Use the piecewise function that models the data to find cigarette consumption in 1990. How well does the function describe the actual consumption for that year shown by the line graph?

91. For the period shown, in which year was cigarette consumption at a maximum? Use the graph to find a reasonable estimate of consumption for that year. How well does the piecewise function model this estimate?

92. For the period shown, in which year was cigarette consumption at a minimum? Use the graph to find a reasonable estimate of consumption for that year. How well does the piecewise function model this estimate?

93. The cost of a telephone call between two cities is $0.10 for the first minute and $0.05 for each additional minute or portion of a minute. Draw a graph of the cost, C, in dollars, of the phone call as a function of time, t, in minutes, on the interval $(0, 5]$.

94. A cargo service charges a flat fee of $4 plus $1 for each pound or fraction of a pound to mail a package. Let $C(x)$ represent the cost to mail a package that weighs x pounds. Graph the cost function on the interval $(0, 5]$.

Writing in Mathematics

95. Explain how to find the difference quotient,
$$\frac{f(x + h) - f(x)}{h},$$
if a function's equation is given.

96. What is a piecewise function?

97. What does it mean if a function f is increasing on an interval?

98. Suppose that a function f is increasing on (a, b), decreasing on (b, c), and defined at b. Describe what occurs at $x = b$. What does the function value $f(b)$ represent?

99. If you are given a function's equation, how do you determine if the function is even, odd, or neither?

100. If you are given a function's graph, how do you determine if the function is even, odd, or neither?

101. What is a step function? Give an example of an everyday situation that can be modeled using such a function. Do not use the cost-of-mail example.

102. Explain how to find $\text{int}(-3.000004)$.

Technology Exercises

103. The function

$$f(x) = -0.00002x^3 + 0.008x^2 - 0.3x + 6.95$$

models the number of annual physician visits, $f(x)$, by a person of age x. Graph the function in a $[0, 100, 5]$ by $[0, 40, 2]$ viewing rectangle. What does the shape of the graph indicate about the relationship between one's age and the number of annual physician visits? Use the TRACE or minimum function capability to find the coordinates of the minimum point on the graph of the function. What does this mean?

In Exercises 104–109, use a graphing utility to graph each function. Use a $[-5, 5, 1]$ by $[-5, 5, 1]$ viewing rectangle. Then find the intervals on which the function is increasing, decreasing, or constant.

104. $f(x) = x^3 - 6x^2 + 9x + 1$ 105. $g(x) = |4 - x^2|$

106. $h(x) = |x - 2| + |x + 2|$ 107. $f(x) = x^{\frac{1}{3}}(x - 4)$

108. $g(x) = x^{\frac{2}{3}}$ **109.** $h(x) = 2 - x^{\frac{2}{5}}$

110. a. Graph the functions $f(x) = x^n$ for $n = 2, 4,$ and 6 in a $[-2, 2, 1]$ by $[-1, 3, 1]$ viewing rectangle.

b. Graph the functions $f(x) = x^n$ for $n = 1, 3,$ and 5 in a $[-2, 2, 1]$ by $[-2, 2, 1]$ viewing rectangle.

c. If n is even, where is the graph of $f(x) = x^n$ increasing and where is it decreasing?

d. If n is odd, what can you conclude about the graph of $f(x) = x^n$ in terms of increasing or decreasing behavior?

e. Graph all six functions in a $[-1, 3, 1]$ by $[-1, 3, 1]$ viewing rectangle. What do you observe about the graphs in terms of how flat or how steep they are?

Critical Thinking Exercises

111. Sketch the graph of f using the following properties. (More than one correct graph is possible.) f is a piecewise function that is decreasing on $(-\infty, 2), f(2) = 0, f$ is increasing on $(2, \infty)$, and the range of f is $[0, \infty)$.

112. Define a piecewise function on the intervals $(-\infty, 2]$, $(2, 5)$, and $[5, \infty)$ that does not "jump" at 2 or 5 such that one piece is a constant function, another piece is an increasing function, and the third piece is a decreasing function.

113. Suppose that $h(x) = \dfrac{f(x)}{g(x)}$. The function f can be even, odd, or neither. The same is true for the function g.

a. Under what conditions is h definitely an even function?

b. Under what conditions is h definitely an odd function?

114. Take another look at the cost of first-class mail and its graph (Table 1.2 and Figure 1.31 on page 162. Change the description of the heading in the left column of Table 1.2 so that the graph includes the point on the left of each horizontal step, but does not include the point on the right.

Group Exercise

115. (For assistance with this exercise, refer to the discussion of piecewise functions on page 156, as well as to Exercises 73–74.)

Group members who have cellular phone plans should describe the total monthly cost of the plan as follows:

$\_\_\_\_$ per month buys $\_\_\_\_$ minutes. Additional time costs $ ____ per minute.

(For simplicity, ignore off-peak rates, roaming charges, etc. The group should select any three plans, from "basic" to "premier." For each plan selected, write a piecewise function that describes the plan and graph the function. Graph the three functions in the same rectangular coordinate system. Now examine the graphs. For any given number of calling minutes, the best plan is the one whose graph is lowest at that point. Compare the three calling plans. Over how many minutes does one plan become better than another? (You can check out cellular phone plans by visiting www.point.com.

SECTION 1.4 *Linear Functions and Slope*

Objectives

① Calculate a line's slope.

② Write the point-slope form of the equation of a line.

③ Write and graph the slope-intercept form of the equation of a line.

④ Graph horizontal or vertical lines.

⑤ Recognize and use the general form of a line's equation.

⑥ Use intercepts to graph the general form of a line's equation.

⑦ Model data with linear functions and make predictions.

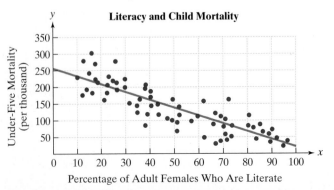

Is there a relationship between literacy and child mortality? As the percentage of adult females who are literate increases, does the mortality of children under five decrease? Figure 1.33, based on data from the United Nations, indicates that this is indeed, the case. Each point in the figure represents one country.

Figure 1.33

Source: United Nations

Data presented in a visual form as a set of points is called a **scatter plot**. Also shown in Figure 1.33 is a line that passes through or near the points. A line that best fits the data points in a scatter plot is called a **regression line**. By writing the equation of this line, we can obtain a model for the data and make predictions about child mortality based on the percentage of literate adult females in a country.

Data often fall on or near a line. In this section, we will use functions to model such data and make predictions. We begin with a discussion of a line's steepness.

① Calculate a line's slope.

The Slope of a Line

Mathematicians have developed a useful measure of the steepness of a line, called the *slope* of the line. Slope compares the vertical change (the **rise**) to the horizontal change (the **run**) when moving from one fixed point to another along the line. To calculate the slope of a line, we use a ratio that compares the change in y (the rise) to the corresponding change in x (the run).

Slope and the Streets of San Francisco

San Francisco's Filbert Street has a slope of 0.613, meaning that for every horizontal distance of 100 feet, the street ascends 61.3 feet vertically. With its 31.5° angle of inclination, the street is too steep to pave and is only accessible by wooden stairs.

Definition of Slope

The **slope** of the line through the distinct points (x_1, y_1) and (x_2, y_2) is

$$\frac{\text{Change in } y}{\text{Change in } x} = \frac{\text{Rise}}{\text{Run}}$$

$$= \frac{y_2 - y_1}{x_2 - x_1}$$

where $x_2 - x_1 \neq 0$.

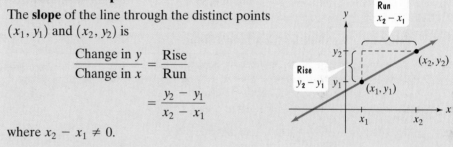

It is common notation to let the letter m represent the slope of a line. The letter m is used because it is the first letter of the French verb *monter*, meaning to rise, or to ascend.

EXAMPLE 1   Using the Definition of Slope

Find the slope of the line passing through each pair of points:

    a. $(-3, -1)$ and $(-2, 4)$ **b.** $(-3, 4)$ and $(2, -2)$.

Solution

    a. Let $(x_1, y_1) = (-3, -1)$ and $(x_2, y_2) = (-2, 4)$. We obtain the slope as follows:

$$m = \frac{\text{Change in } y}{\text{Change in } x} = \frac{y_2 - y_1}{x_2 - x_1} = \frac{4 - (-1)}{-2 - (-3)} = \frac{5}{1} = 5.$$

The situation is illustrated in Figure 1.34(a). The slope of the line is 5, indicating that there is a vertical change, a rise, of 5 units for each horizontal change, a run, of 1 unit. The slope is positive, and the line rises from left to right.

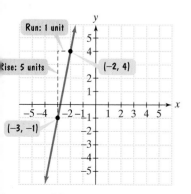

gure 1.34(a) Visualizing slope

Study Tip

When computing slope, it makes no difference which point you call (x_1, y_1) and which point you call (x_2, y_2). If we let $(x_1, y_1) = (-2, 4)$ and $(x_2, y_2) = (-3, -1)$, the slope is still 5:

$$m = \frac{\text{Change in } y}{\text{Change in } x} = \frac{y_2 - y_1}{x_2 - x_1} = \frac{-1 - 4}{-3 - (-2)} = \frac{-5}{-1} = 5.$$

However, you should not subtract in one order in the numerator $(y_2 - y_1)$ and then in a different order in the denominator $(x_1 - x_2)$. The slope is *not* -5:

$$\frac{-1 - 4}{-2 - (-3)} = \frac{-5}{1} = -5. \quad \textit{Incorrect}$$

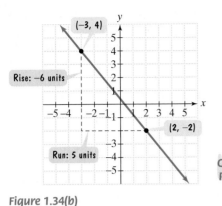

Figure 1.34(b)

b. We can let $(x_1, y_1) = (-3, 4)$ and $(x_2, y_2) = (2, -2)$. The slope of the line shown in Figure 1.34(b) is computed as follows:

$$m = \frac{\text{Change in } y}{\text{Change in } x} = \frac{y_2 - y_1}{x_2 - x_1} = \frac{-2 - 4}{2 - (-3)} = \frac{-6}{5} = -\frac{6}{5}.$$

The slope of the line is $-\frac{6}{5}$. For every vertical change of -6 units (6 units down), there is a corresponding horizontal change of 5 units. The slope is negative and the line falls from left to right.

Check Point **1** Find the slope of the line passing through each pair of points:

a. $(-3, 4)$ and $(-4, -2)$ **b.** $(4, -2)$ and $(-1, 5)$.

Example 1 illustrates that a line with a positive slope is rising from left to right, and a line with a negative slope is falling from left to right. By contrast, a horizontal line neither rises nor falls and has a slope of zero. A vertical line has no horizontal change, so $x_2 - x_1 = 0$ in the formula for slope. Because we cannot divide by zero, the slope of a vertical line is undefined. This discussion is summarized in Table 1.3.

Table 1.3   Possibilities for a Line's Slope

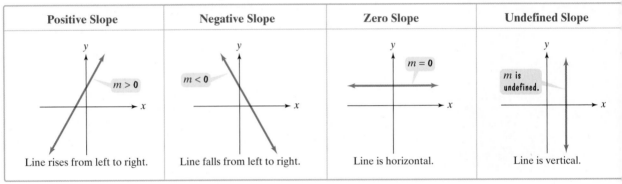

Positive Slope	Negative Slope	Zero Slope	Undefined Slope
$m > 0$	$m < 0$	$m = 0$	m is undefined.
Line rises from left to right.	Line falls from left to right.	Line is horizontal.	Line is vertical.

Study Tip

Always be clear in the way you use language, especially in mathematics. For example, it's not a good idea to say that a line has "no slope." This could mean that the slope is zero or that the slope is undefined.

2 Write the point-slope form of the equation of a line.

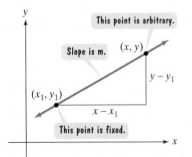

Figure 1.35 A line passing through (x_1, y_1) with slope m

The Point-Slope Form of the Equation of a Line

We can use the slope of a line to obtain various forms of the line's equation. For example, consider a nonvertical line that has slope m and that contains the point (x_1, y_1). Now, let (x, y) represent any other point on the line, shown in Figure 1.35. Keep in mind that the point (x, y) is arbitrary and is not in one fixed position. By contrast, the point (x_1, y_1) is fixed.

Regardless of where the point (x, y) is located, the steepness of the line in Figure 1.35 remains the same. Thus, the ratio for the slope stays a constant m. This means that for all points along the line

$$m = \frac{\text{Change in } y}{\text{Change in } x} = \frac{y - y_1}{x - x_1}.$$

We can clear the fraction by multiplying both sides by $x - x_1$, the least common denominator.

$$m = \frac{y - y_1}{x - x_1} \qquad \text{This is the slope of the line in Figure 1.35.}$$

$$m(x - x_1) = \frac{y - y_1}{x - x_1} \cdot x - x_1 \qquad \text{Multiply both sides by } x - x_1.$$

$$m(x - x_1) = y - y_1 \qquad \text{Simplify: } \frac{y - y_1}{\cancel{x - x_1}} \cdot \cancel{x - x_1} = y - y_1.$$

Now, if we reverse the two sides, we obtain the *point-slope form* of the equation of a line.

> ### Point-Slope Form of the Equation of a Line
>
> The **point-slope form of the equation** of a nonvertical line with slope m that passes through the point (x_1, y_1) is
>
> $$y - y_1 = m(x - x_1).$$

For example, the point-slope form of the equation of the line passing through $(1, 5)$ with slope 2; $(m = 2)$ is

$$y - 5 = 2(x - 1).$$

We will soon be expressing the equation of a nonvertical line in function notation. To do so, we need to solve the point-slope form of a line's equation for y. Example 2 illustrates how to isolate y on one side of the equal sign.

EXAMPLE 2   Writing the Point-Slope Form of the Equation of a Line

Write the point-slope form of the equation of the line with slope 4 that passes through the point $(-1, 3)$. Then solve the equation for y.

Solution   We use the point-slope form of the equation of a line with $m = 4$, $x_1 = -1$, and $y_1 = 3$.

$$y - y_1 = m(x - x_1) \qquad \text{This is the point-slope form of the equation.}$$
$$y - 3 = 4[x - (-1)] \qquad \text{Substitute the given values.}$$
$$y - 3 = 4(x + 1) \qquad \text{We now have the point-slope form of the equation of the given line.}$$

We can solve this equation for y by first applying the distributive property on the right side.

$$y - 3 = 4x + 4$$

Finally, we add 3 to both sides.

$$y = 4x + 7$$

Check Point 2   Write the point-slope form of the equation of the line with slope 6 that passes through the point $(2, -5)$. Then solve the equation for y.

EXAMPLE 3   Writing the Point-Slope Form of the Equation of a Line

Write the point-slope form of the equation of the line passing through the points $(4, -3)$ and $(-2, 6)$. (See Figure 1.36.) Then solve the equation for y.

Solution   To use the point-slope form, we need to find the slope. The slope is the change in the y-coordinates divided by the corresponding change in the x-coordinates.

$$m = \frac{6 - (-3)}{-2 - 4} = \frac{9}{-6} = -\frac{3}{2} \qquad \begin{array}{l} \text{This is the definition of slope using } (4, -3) \\ \text{and } (-2, 6). \end{array}$$

We can take either point on the line to be (x_1, y_1). Let's use $(x_1, y_1) = (4, -3)$. Now, we are ready to write the point-slope form of the equation.

$$y - y_1 = m(x - x_1) \qquad \text{This is the point-slope form of the equation.}$$
$$y - (-3) = -\tfrac{3}{2}(x - 4) \qquad \text{Substitute: } (x_1, y_1) = (4, -3) \text{ and } m = -\tfrac{3}{2}.$$
$$y + 3 = -\tfrac{3}{2}(x - 4) \qquad \text{Simplify.}$$

We now have the point-slope form of the equation of the line shown in Figure 1.36. Now, we solve this equation for y.

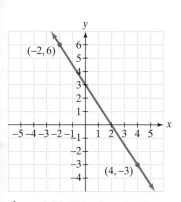

Figure 1.36 Write the point-slope form of the equation of this line.

Discovery

You can use either point for (x_1, y_1) when you write a line's point-slope equation. Rework Example 3 using $(-2, 6)$ for (x_1, y_1). Once you solve for y, you should still obtain

$$y = -\frac{3}{2}x + 3.$$

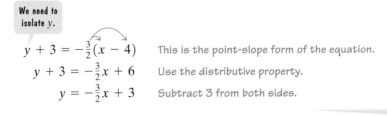

We need to isolate y.

$y + 3 = -\frac{3}{2}(x - 4)$ This is the point-slope form of the equation.

$y + 3 = -\frac{3}{2}x + 6$ Use the distributive property.

$y = -\frac{3}{2}x + 3$ Subtract 3 from both sides.

Check Point 3 Write the point-slope form of the equation of the line passing through the points $(-2, -1)$ and $(-1, -6)$. Then solve the equation for y.

The Slope-Intercept Form of the Equation of a Line

Let's write the point-slope form of the equation of a nonvertical line with slope m and y-intercept b. The line is shown in Figure 1.37. Because the y-intercept is b, the line passes through $(0, b)$. We use the point-slope form with $x_1 = 0$ and $y_1 = b$.

$$y - y_1 = m(x - x_1)$$

Let $y_1 = b$.   Let $x_1 = 0$.

We obtain

$$y - b = m(x - 0).$$

Simplifying on the right side gives us

$$y - b = mx.$$

Finally, we solve for y by adding b to both sides.

$$y = mx + b$$

Thus, if a line's equation is written with y isolated on one side, the x-coefficient is the line's slope and the constant term is the y-intercept. This form of a line's equation is called the *slope-intercept form* of the line.

③ Write and graph the slope-intercept form of the equation of a line.

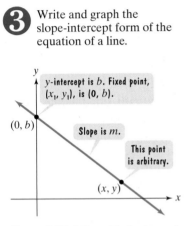

y-intercept is b. Fixed point, (x_1, y_1), is $(0, b)$.

$(0, b)$

Slope is m.

This point is arbitrary.

(x, y)

Figure 1.37 A line with slope m and y-intercept b

> ### Slope-Intercept Form of the Equation of a Line
>
> The **slope-intercept form of the equation** of a nonvertical line with slope m and y-intercept b is
>
> $$y = mx + b.$$

The slope-intercept form of a line's equation, $y = mx + b$, can be expressed in function notation by replacing y with $f(x)$:

$$f(x) = mx + b.$$

We have seen that functions in this form are called **linear functions**. Thus, in the equation of a linear function, the x-coefficient is the line's slope and the constant term is the y-intercept. Here are two examples:

$$y = 2x - 4 \qquad\qquad f(x) = \frac{1}{2}x + 2.$$

The slope is 2.   The y-intercept is −4.   The slope is $\frac{1}{2}$.   The y-intercept is 2.

If a linear function's equation is in slope-intercept form, we can use the y-intercept and the slope to obtain its graph.

> ### Graphing $y = mx + b$ Using the Slope and y-Intercept
>
> **1.** Plot the point containing the y-intercept on the y-axis. This is the point $(0, b)$.
> **2.** Obtain a second point using the slope, m. Write m as a fraction, and use rise over run, starting at the point containing the y-intercept, to plot this point.
> **3.** Use a straightedge to draw a line through the two points. Draw arrowheads at the ends of the line to show that the line continues indefinitely in both directions.

EXAMPLE 4   Graphing Using the Slope and *y*-Intercept

Graph the linear function: $f(x) = -\dfrac{3}{2}x + 2.$

Solution   The equation of the line is in the form $f(x) = mx + b$. We can find the slope, m, by identifying the coefficient of x. We can find the y-intercept, b, by identifying the constant term.

$$f(x) = -\frac{3}{2}x + 2$$

> The slope is $-\frac{3}{2}$.

> The y-intercept is **2.**

Now that we have identified the slope and the y-intercept, we use the three-step procedure to graph the equation.

Step 1   Plot the point containing the *y*-intercept on the *y*-axis. The y-intercept is 2. We plot $(0, 2)$, shown in Figure 1.38.

Step 2   Obtain a second point using the slope, *m*. Write *m* as a fraction, and use rise over run, starting at the point containing the *y*-intercept, to plot this point. The slope, $-\frac{3}{2}$, is already written as a fraction.

$$m = -\frac{3}{2} = \frac{-3}{2} = \frac{\text{Rise}}{\text{Run}}$$

We plot the second point on the line by starting at $(0, 2)$, the first point. Based on the slope, we move 3 units *down* (the rise) and 2 units to the *right* (the run). This puts us at a second point on the line, $(2, -1)$, shown in Figure 1.38.

Step 3   Use a straightedge to draw a line through the two points. The graph of the linear function $f(x) = -\frac{3}{2}x + 2$ is shown as a blue line in Figure 1.38.

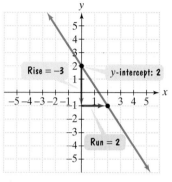

Figure 1.38 The graph of $f(x) = -\frac{3}{2}x + 2$

> **Check Point** **4**   Graph the linear function: $f(x) = \frac{3}{5}x + 1.$

4 Graph horizontal or vertical lines.

Equations of Horizontal and Vertical Lines

Some things change very little. For example, from 1997 through 2003, the federal minimum wage remained constant at $5.15 per hour, indicated by the green bars in Figure 1.39. These bars show the minimum wage before it was adjusted for inflation. Also shown in the figure is a blue horizontal line segment that passes through the tops of the seven green bars.

Figure 1.39

Source: www.dol.gov/esa/public/minwage

We can use $y = mx + b$, the slope-intercept form of a line's equation, to obtain an equation that models the federal minimum wage, y, in current dollars, in year x, where x is between 1997 and 2003, inclusive. The horizontal blue line segment in Figure 1.39 on the previous page provides the values for m and b:

$$y = mx + b.$$

Because the line segment is horizontal, $m = 0$.

The y-intercept is shown as \$5.15, so $b = 5.15$.

Thus, an equation that models the federal minimum wage between 1997 and 2003, inclusive, is

$$y = 0x + 5.15, \quad \text{or } y = 5.15.$$

The federal minimum wage remained constant at \$5.15 per hour. Using function notation, we can write

$$f(x) = 5.15.$$

For all year inputs, x, in the domain from 1997 through 2003, inclusive,

the output is a constant 5.15.

In general, if a line is horizontal, its slope is zero: $m = 0$. Thus, the equation $y = mx + b$ becomes $y = b$, where b is the y-intercept. For example, the graph of $y = -4$ is a horizontal line with a y-intercept of -4. The graph is shown in Figure 1.40. Three of the points along the line are shown and labeled.

x	$y = -4$	(x, y)
-2	-4	$(-2, -4)$
0	-4	$(0, -4)$
3	-4	$(3, -4)$

For all choices of x,

y is a constant -4.

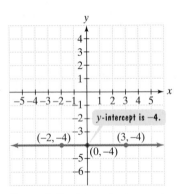

Figure 1.40 The graph of $y = -4$ or $f(x) = -4$

No matter what the x-coordinate is, the corresponding y-coordinate for every point on the line in Figure 1.40 is -4.

Equation of a Horizontal Line

A horizontal line is given by an equation of the form

$$y = b,$$

where b is the y-intercept.

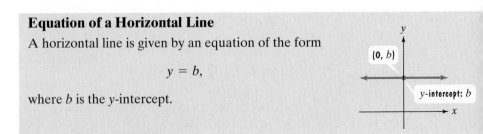

Because any vertical line can intersect the graph of a horizontal line $y = b$ only once, a horizontal line is the graph of a function. Thus, we can express the equation $y = b$ as $f(x) = b$. This linear function is often called a **constant function**. The function modeling the federal minimum wage from 1997 through 2003, namely $f(x) = 5.15$, is an example of a constant function.

Next, let's see what we can discover about the graph of an equation of the form $x = a$ by looking at an example.

EXAMPLE 5 Graphing a Vertical Line

Graph the linear equation: $x = 2$.

Solution   All ordered pairs that are solutions of $x = 2$ have a value of x that is always 2. Any value can be used for y. Let's select three of the possible values for y: $-2, 0$, and 3.

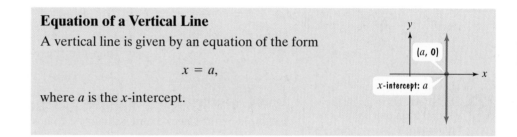

$x = 2$	y	(x, y)
2	-2	$(2, -2)$
2	0	$(2, 0)$
2	3	$(2, 3)$

The table shows that three ordered pairs that are solutions of $x = 2$ are $(2, -2)$, $(2, 0)$, and $(2, 3)$. Drawing a line that passes through the three points gives the vertical line shown in Figure 1.41.

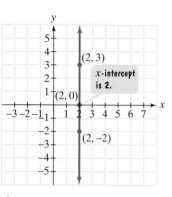

Figure 1.41 The graph of $x = 2$

Equation of a Vertical Line

A vertical line is given by an equation of the form

$$x = a,$$

where a is the x-intercept.

Does a vertical line represent the graph of a linear function? No. Look at the graph of $x = 2$ in Figure 1.41. A vertical line drawn through $(2, 0)$ intersects the graph infinitely many times. This shows that infinitely many outputs are associated with the input 2. **No vertical line is a linear function.**

Check Point 5 Graph the linear equation: $x = -3$.

⑤ Recognize and use the general form of a line's equation.

The General Form of the Equation of a Line

The vertical line whose equation is $x = 5$ cannot be written in slope-intercept form, $y = mx + b$, because its slope is undefined. However, every line has an equation that can be expressed in the form $Ax + By + C = 0$. For example, $x = 5$ can be expressed as $1x + 0y - 5 = 0$, or $x - 5 = 0$. The equation $Ax + By + C = 0$ is called the *general form* of the equation of a line.

General Form of the Equation of a Line

Every line has an equation that can be written in the **general form**

$$Ax + By + C = 0,$$

where A, B, and C are real numbers, and A and B are not both zero.

If the equation of a line is given in general form, it is possible to find the slope, m, and the y-intercept, b, for the line. We solve the equation for y, transforming it into the slope–intercept form $y = mx + b$. In this form, the coefficient of x is the slope of the line and the constant term is its y-intercept.

EXAMPLE 6 Finding the Slope and the y-Intercept

Find the slope and the y-intercept of the line whose equation is $3x + 2y - 4 = 0$

Solution The equation is given in general form. We begin by rewriting it in the form $y = mx + b$. We need to solve for y.

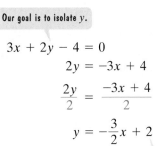

Our goal is to isolate y.

$$3x + 2y - 4 = 0$$ This is the given equation.

$$2y = -3x + 4$$ Isolate the term containing y by adding $-3x + 4$ to both sides.

$$\frac{2y}{2} = \frac{-3x + 4}{2}$$ Divide both sides by 2.

$$y = -\frac{3}{2}x + 2$$ On the right, divide each term in the numerator by 2 to obtain slope-intercept form.

slope   y-intercept

The coefficient of x, $-\frac{3}{2}$, is the slope and the constant term, 2, is the y-intercept. This is the form of the equation that we graphed in Figure 1.38 on page 173.

Check Point 6 Find the slope and the y-intercept of the line whose equation is $3x + 6y - 12 = 0$. Then use the y-intercept and the slope to graph the equation.

6 Use intercepts to graph the general form of a line's equation.

Using Intercepts to Graph $Ax + By + C = 0$

Example 6 and Check Point 6 illustrate that one way to graph the general form of a line's equation is to convert to slope-intercept form, $y = mx + b$. Then use the slope and the y-intercept to obtain the graph.

A second method for graphing $Ax + By + C = 0$ uses intercepts. This method does not require rewriting the general form in a different form.

> ### Using Intercepts to Graph $Ax + By + C = 0$
>
> 1. Find the x-intercept. Let $y = 0$ and solve for x. Plot the point containing the x-intercept on the x-axis.
> 2. Find the y-intercept. Let $x = 0$ and solve for y. Plot the point containing the y-intercept on the y-axis.
> 3. Use a straightedge to draw a line through the two points containing the intercepts. Draw arrowheads at the ends of the line to show that the line continues indefinitely in both directions.

EXAMPLE 7 Using Intercepts to Graph a Linear Equation

Graph using intercepts: $4x - 3y - 6 = 0$.

Solution

Step 1 Find the x-intercept. Let $y = 0$ and solve for x.

$$4x - 3 \cdot 0 - 6 = 0$$ Replace y with 0 in $4x - 3y - 6 = 0$

$$4x - 6 = 0$$ Simplify.

$$4x = 6$$ Add 6 to both sides.

$$x = \frac{6}{4} = \frac{3}{2}$$ Divide both sides by 4.

The x-intercept is $\frac{3}{2}$, so the line passes through $\left(\frac{3}{2}, 0\right)$ or $(1.5, 0)$, as shown in Figure 1.42.

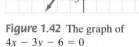

Figure 1.42 The graph of $4x - 3y - 6 = 0$

Step 2   Find the *y*-intercept. Let *x* = 0 and solve for *y*.

$$4 \cdot 0 - 3y - 6 = 0 \qquad \text{Replace x with 0 in } 4x - 3y - 6 = 0.$$
$$-3y - 6 = 0 \qquad \text{Simplify.}$$
$$-3y = 6 \qquad \text{Add 6 to both sides.}$$
$$y = -2 \qquad \text{Divide both sides by } -3.$$

The *y*-intercept is -2, so the line passes through $(0, -2)$, as shown in Figure 1.42.

Step 3   Graph the equation by drawing a line through the two points containing the intercepts. The graph of $4x - 3y - 6 = 0$ is shown in Figure 1.42.

Check Point 7 Graph using intercepts: $3x - 2y - 6 = 0.$

We've covered a lot of territory. Let's take a moment to summarize the various forms for equations of lines.

> ## Equations of Lines
>
> 1. Point-slope form: $y - y_1 = m(x - x_1)$
> 2. Slope-intercept form: $y = mx + b$ or $f(x) = mx + b$
> 3. Horizontal line: $y = b$
> 4. Vertical line: $x = a$
> 5. General form: $Ax + By + C = 0$

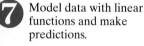

7 Model data with linear functions and make predictions.

Technology

You can use a graphing utility to obtain a model for a scatter plot in which the data points fall on or near a straight line. After entering the data in Figure 1.43(b), a graphing utility displays a scatter plot of the data and the regression line, that is, the line that best fits the data.

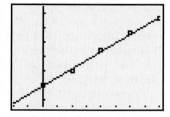

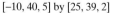
[−10, 40, 5] by [25, 39, 2]

Also displayed is the regression line's equation.

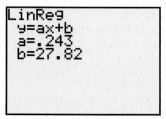
```
LinReg
y=ax+b
a=.243
b=27.82
```

Applications

Linear functions are useful for modeling data that fall on or near a line. For example, the bar graph in Figure 1.43(a) gives the median age of the U.S. population in the indicated year. (The median age is the age in the middle when all the ages of the U.S. population are arranged from youngest to oldest.) The data are displayed as a set of five points in a rectangular coordinate system in Figure 1.43(b).

The Graying of America: Median Age of the U.S. Population

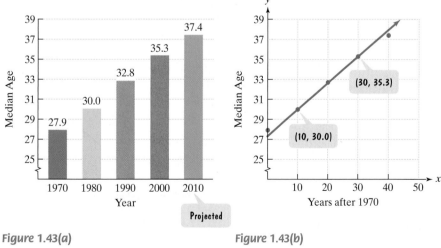

Figure 1.43(a)

Source: U.S. Census Bureau

Figure 1.43(b)

Also shown on the scatter plot in Figure 1.43(b) is a line that passes through or near the five points. By writing the equation of this line, we can obtain a model of the data and make predictions about the median age of the U.S. population in the future.

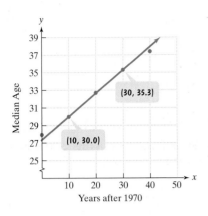

Figure 1.43(b) (repeated)

EXAMPLE 8   Modeling the Graying of America

Write the slope-intercept equation of the line shown in Figure 1.43(b). Use the equation to predict the median age of the U.S. population in 2020.

Solution   The line in Figure 1.43(b) passes through (10, 30.0) and (30, 35.3). We start by finding its slope.

$$m = \frac{\text{Change in } y}{\text{Change in } x} = \frac{35.3 - 30.0}{30 - 10} = \frac{5.3}{20} = 0.265$$

The slope indicates that each year the median age of the U.S. population is increasing by 0.265 years.

Now, we write the line's slope-intercept equation.

$y - y_1 = m(x - x_1)$	Begin with the point-slope form.
$y - 30.0 = 0.265(x - 10)$	Either ordered pair can be (x_1, y_1). Let $(x_1, y_1) = (10, 30.0)$. From above, $m = 0.265$
$y - 30.0 = 0.265x - 2.65$	Apply the distributive property.
$y = 0.265x + 27.35$	Add 30 to both sides and solve for y.

A linear function that models the median age of the U.S. population, $f(x)$, x years after 1970 is

$$f(x) = 0.265x + 27.35.$$

Now, let's use this function to predict the median age in 2020. Because 2020 is 50 years after 1970, we substitute 50 for x and evaluate the function at 50.

$$f(50) = 0.265(50) + 27.35 = 40.6$$

Our model predicts that the median age of the U.S. population in 2020 will be 40.6.

Check Point 8   Use the data points (10, 30.0) and (20, 32.8) from Figure 1.43(b) to write slope-intercept equation that models the median age of the U.S. population x years after 1970. Use this model to predict the median age in 2020.

Cigarettes and Lung Cancer

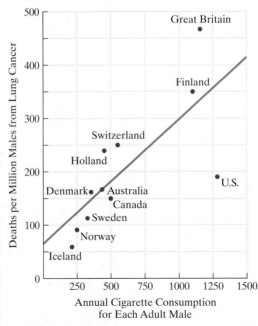

This scatter plot shows a relationship between cigarette consumption among males and deaths due to lung cancer per million males. The data are from 11 countries and date back to a 1964 report by the U.S. Surgeon General. The scatter plot can be modeled by a line whose slope indicates an increasing death rate from lung cancer with increased cigarette consumption. At that time, the tobacco industry argued that in spite of this regression line, tobacco use is not the cause of cancer. Recent data do, indeed, show a causal effect between tobacco use and numerous diseases.

Source: Smoking and Health, Washington, D.C., 1964

EXERCISE SET 1.4

Practice Exercises

In Exercises 1–10, find the slope of the line passing through each pair of points or state that the slope is undefined. Then indicate whether the line through the points rises, falls, is horizontal, or is vertical.

1. $(4, 7)$ and $(8, 10)$ **2.** $(2, 1)$ and $(3, 4)$

3. $(-2, 1)$ and $(2, 2)$ **4.** $(-1, 3)$ and $(2, 4)$

5. $(4, -2)$ and $(3, -2)$ **6.** $(4, -1)$ and $(3, -1)$

7. $(-2, 4)$ and $(-1, -1)$ **8.** $(6, -4)$ and $(4, -2)$

9. $(5, 3)$ and $(5, -2)$ **10.** $(3, -4)$ and $(3, 5)$

In Exercises 11–38, use the given conditions to write an equation for each line in point-slope form and slope-intercept form.

11. Slope $= 2$, passing through $(3, 5)$

12. Slope $= 4$, passing through $(1, 3)$

13. Slope $= 6$, passing through $(-2, 5)$

14. Slope $= 8$, passing through $(4, -1)$

15. Slope $= -3$, passing through $(-2, -3)$

16. Slope $= -5$, passing through $(-4, -2)$

17. Slope $= -4$, passing through $(-4, 0)$

18. Slope $= -2$, passing through $(0, -3)$

19. Slope $= -1$, passing through $\left(-\frac{1}{2}, -2\right)$

20. Slope $= -1$, passing through $\left(-4, -\frac{1}{4}\right)$

21. Slope $= \frac{1}{2}$, passing through the origin

22. Slope $= \frac{1}{3}$, passing through the origin

23. Slope $= -\frac{2}{3}$, passing through $(6, -2)$

24. Slope $= -\frac{3}{5}$, passing through $(10, -4)$

25. Passing through $(1, 2)$ and $(5, 10)$

26. Passing through $(3, 5)$ and $(8, 15)$

27. Passing through $(-3, 0)$ and $(0, 3)$

28. Passing through $(-2, 0)$ and $(0, 2)$

29. Passing through $(-3, -1)$ and $(2, 4)$

30. Passing through $(-2, -4)$ and $(1, -1)$

31. Passing through $(-3, -2)$ and $(3, 6)$

32. Passing through $(-3, 6)$ and $(3, -2)$

33. Passing through $(-3, -1)$ and $(4, -1)$

34. Passing through $(-2, -5)$ and $(6, -5)$

35. Passing through $(2, 4)$ with x-intercept $= -2$

36. Passing through $(1, -3)$ with x-intercept $= -1$

37. x-intercept $= -\frac{1}{2}$ and y-intercept $= 4$

38. x-intercept $= 4$ and y-intercept $= -2$

In Exercises 39–48, give the slope and y-intercept of each line whose equation is given. Then graph the linear function.

39. $y = 2x + 1$ **40.** $y = 3x + 2$

41. $f(x) = -2x + 1$ **42.** $f(x) = -3x + 2$

43. $f(x) = \frac{3}{4}x - 2$ **44.** $f(x) = \frac{3}{4}x - 3$

45. $y = -\frac{3}{5}x + 7$ **46.** $y = -\frac{2}{5}x + 6$

47. $g(x) = -\frac{1}{2}x$ **48.** $g(x) = -\frac{1}{3}x$

In Exercises 49–58, graph each equation in a rectangular coordinate system.

49. $y = -2$ **50.** $y = 4$

51. $x = -3$ **52.** $x = 5$

53. $y = 0$ **54.** $x = 0$

55. $f(x) = 1$ **56.** $f(x) = 3$

57. $3x - 18 = 0$ **58.** $3x + 12 = 0$

In Exercises 59–66,
   a. *Rewrite the given equation in slope-intercept form.*
   b. *Give the slope and y-intercept.*
   c. *Use the slope and y-intercept to graph the linear function.*

59. $3x + y - 5 = 0$ **60.** $4x + y - 6 = 0$

61. $2x + 3y - 18 = 0$ **62.** $4x + 6y + 12 = 0$

63. $8x - 4y - 12 = 0$ **64.** $6x - 5y - 20 = 0$

65. $3y - 9 = 0$ **66.** $4y + 28 = 0$

In Exercises 67–72, use intercepts to graph each equation.

67. $6x - 2y - 12 = 0$ **68.** $6x - 9y - 18 = 0$

69. $2x + 3y + 6 = 0$ **70.** $3x + 5y + 15 = 0$

71. $8x - 2y + 12 = 0$ **72.** $6x - 3y + 15 = 0$

Practice Plus

In Exercises 73–76, find the slope of the line passing through each pair of points or state that the slope is undefined. Assume that all variables represent positive real numbers. Then indicate whether the line through the points rises, falls, is horizontal, or is vertical.

73. $(0, a)$ and $(b, 0)$ **74.** $(-a, 0)$ and $(0, -b)$

75. (a, b) and $(a, b + c)$ **76.** $(a - b, c)$ and $(a, a + c)$

In Exercises 77–78, give the slope and y-intercept of each line whose equation is given. Assume that B ≠ 0.

77. $Ax + By = C$ **78.** $Ax = By - C$

In Exercises 79–80, find the value of y if the line through the two given points is to have the indicated slope.

79. $(3, y)$ and $(1, 4)$, $m = -3$

80. $(-2, y)$ and $(4, -4)$, $m = \frac{1}{3}$

In Exercises 81–82, graph each linear function.

81. $3x - 4f(x) - 6 = 0$ **82.** $6x - 5f(x) - 20 = 0$

83. If one point on a line is $(3, -1)$ and the line's slope is -2, find the y-intercept.

84. If one point on a line is $(2, -6)$ and the line's slope is $-\frac{3}{2}$, find the y-intercept.

Use the figure to make the lists in Exercises 85–86.

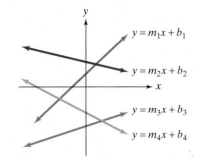

85. List the slopes m_1, m_2, m_3, and m_4 in order of decreasing size.

86. List the y-intercepts b_1, b_2, b_3, and b_4 in order of decreasing size.

 Application Exercises

Though increasing numbers of Americans are obese, fewer are trimming down by watching what they eat. The bar graph shows the percentage of American adults on weight-loss diets for four selected years. The data are displayed as two sets of four points each, one scatter plot for the percentage of dieting women and one for the percentage of dieting men. Also shown in each scatter plot is a line that passes through or near the four points. Use these lines to solve Exercises 87–88.

Percentage of American Adults on Weight-Loss Diets

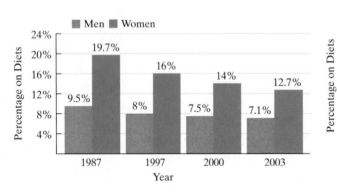

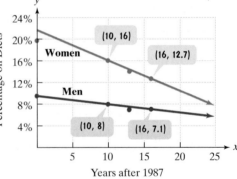

Source: Mediamark Research, American Demographics

87. In this exercise, you will use the blue line for the women shown on the scatter plot to develop a model for the percentage of dieting American women.

  a. Use the two points whose coordinates are shown by the voice balloons to find the point-slope form of the equation of the line that models the percentage of adult women on diets, y, x years after 1987.

  b. Write the equation in part (a) in slope-intercept form. Use function notation.

  c. Use the linear function to predict the percentage of adult women on weight-loss diets in 2007.

88. In this exercise, you will use the red line for men shown on the scatter plot to develop a model for the percentage of dieting American men.

  a. Use the two points whose coordinates are shown by the voice balloons to find the point-slope form of the equation of the line that models the percentage of adult men on diets, y, x years after 1987.

  b. Write the equation in part (a) in slope-intercept form. Use function notation.

  c. Use the linear function to predict the percentage of adult men on weight-loss diets in 2007.

89. The bar graph shows life expectancies for Americans born in seven selected years.

Life Expectancy in the U.S. by Birth Year

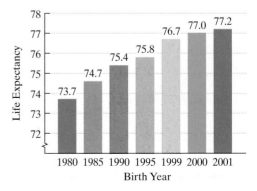

Source: National Center for Health Statistics

a. Let *x* represent the number of birth years after 1980 and let *y* represent life expectancy. Create a scatter plot that displays the data as a set of seven points in a rectangular coordinate system.

b. Draw a line through the two points that show life expectancies for 1985 and 2000. Use the coordinates of these points to write the line's equation in point-slope form and slope-intercept form. Round the slope to two decimal places.

c. Write a linear function that models life expectancy, $E(x)$, for Americans born *x* years after 1980. Then use this function to predict the life expectancy of an American born in 2020.

90. The bar graph shows the number of global HIV/AIDS cases, in millions, from 1999 through 2003.

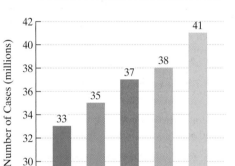

Millions of Worldwide HIV/AIDS Cases

Source: UNAIDS

a. Let *x* represent the number of years after 1999 and let *y* represent the number of HIV/AIDS cases worldwide, in millions. Create a scatter plot that displays the data as a set of five points in a rectangular coordinate system.

b. Draw a line through the two points that show the number of cases in 2000 and 2003. Use the coordinates of these points to write the line's equation in point-slope form and slope-intercept form.

c. Write a linear function that models the number of HIV/AIDS cases worldwide, $A(x)$, in millions, *x* years after 1999. Then use this function to predict the number of cases in 2010.

91. Shown, again, is the scatter plot that indicates a relationship between the percentage of adult females in a country who are literate and the mortality of children under five. Also

shown is a line that passes through or near the points. Find a linear function that models the data by finding the slope-intercept form of the line's equation. Use the function to make a prediction about child mortality based on the percentage of adult females in a country who are literate.

92. Just as money doesn't buy happiness for individuals, the two don't necessarily go together for countries either. However, the scatter plot does show a relationship between a country's annual per capita income and the percentage of people in that country who call themselves "happy."

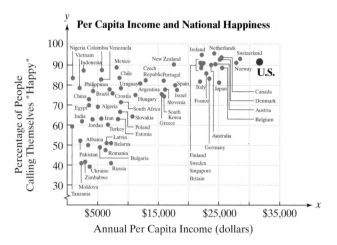

Per Capita Income and National Happiness

Source: Richard Layard, *Happiness: Lessons from a New Science*, Penguin, 2005

Draw a line that fits the data so that the spread of the data points around the line is as small as possible. Use the coordinates of two points along your line to write the slope-intercept form of its equation. Express the equation in function notation and use the linear function to make a prediction about national happiness based on per capita income.

Writing in Mathematics

93. What is the slope of a line and how is it found?

94. Describe how to write the equation of a line if two points along the line are known.

95. Explain how to derive the slope-intercept form of a line's equation, $y = mx + b$, from the point-slope form

$$y - y_1 = m(x - x_1).$$

96. Explain how to graph the equation $x = 2$. Can this equation be expressed in slope-intercept form? Explain.

97. Explain how to use the general form of a line's equation to find the line's slope and *y*-intercept.

98. Explain how to use intercepts to graph the general form of a line's equation.

99. Take another look at the scatter plot in Exercise 91. Although there is a relationship between literacy and child mortality, we cannot conclude that increased literacy causes child mortality to decrease. Offer two or more possible explanations for the data in the scatter plot.

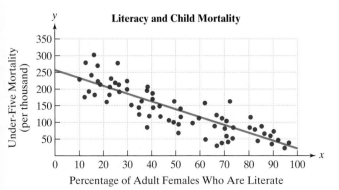

Literacy and Child Mortality

Source: United Nations

Technology Exercises

Use a graphing utility to graph each equation in

Exercises 100–103. Then use the TRACE *feature to trace along the line and find the coordinates of two points. Use these points to compute the line's slope. Check your result by using the coefficient of x in the line's equation.*

100. $y = 2x + 4$ **101.** $y = -3x + 6$

102. $y = -\frac{1}{2}x - 5$ **103.** $y = \frac{3}{4}x - 2$

104. Is there a relationship between alcohol from moderate wine consumption and heart disease death rate? The table gives data from 19 developed countries.

France

Country	A	B	C	D	E	F	G
Liters of alcohol from drinking wine, per person per year (x)	2.5	3.9	2.9	2.4	2.9	0.8	9.1
Deaths from heart disease, per 100,000 people per year (y)	211	167	131	191	220	297	71

U.S.

Country	H	I	J	K	L	M	N	O	P	Q	R	S
(x)	0.8	0.7	7.9	1.8	1.9	0.8	6.5	1.6	5.8	1.3	1.2	2.7
(y)	211	300	107	167	266	227	86	207	115	285	199	172

Source: New York Times, December 28, 1994

a. Use the statistical menu of your graphing utility to enter the 19 ordered pairs of data items shown in the table.

b. Use the DRAW menu and the scatter plot capability to draw a scatter plot of the data.

c. Select the linear regression option. Use your utility to obtain values for a and b for the equation of the regression line, $y = ax + b$. You may also be given a **correlation coefficient**, r. Values of r close to 1 indicate that the points can be described by a linear relationship and the regression line has a positive slope. Values of r close to -1 indicate that the points can be described by a linear relationship and the regression line has a negative slope. Values of r close to 0 indicate no linear relationship between the variables. In this case, a linear model does not accurately describe the data.

d. Use the appropriate sequence (consult your manual) to graph the regression equation on top of the points in the scatter plot.

Critical Thinking Exercises

105. Which one of the following is true?

a. A linear function with nonnegative slope has a graph that rises from left to right.

b. Every line in the rectangular coordinate system has an equation that can be expressed in slope-intercept form.

c. The graph of the linear function $5x + 6y = 30$ is a line passing through the point $(6, 0)$ with slope $-\frac{5}{6}$.

d. The graph of $x = 7$ in the rectangular coordinate system is the single point $(7, 0)$.

In Exercises 106–107, find the coefficients that must be placed in each shaded area so that the function's graph will be a line satisfying the specified conditions.

106. ▢$x +$ ▢$y = 12$; x-intercept $= -2$; y-intercept $= 4$

107. ▢$x +$ ▢$y = 12$; y-intercept $= -6$; slope $= \dfrac{1}{2}$

108. Prove that the equation of a line passing through $(a, 0)$ and $(0, b)$ ($a \neq 0, b \neq 0$) can be written in the form $\dfrac{x}{a} + \dfrac{y}{b} = 1$. Why is this called the *intercept form* of a line?

109. Excited about the success of celebrity stamps, post office officials were rumored to have put forth a plan to institute two new types of thermometers. On these new scales, $°E$ represents degrees Elvis and $°M$ represents degrees Madonna. If it is known that $40°E = 25°M$, $280°E = 125°M$, and degrees Elvis is linearly related to degrees Madonna, write an equation expressing E in terms of M.

Group Exercise

110. In Example 8 on page 178, we used the data in Figure 1.43 on page 177 to develop a linear function that modeled the graying of America. For this group exercise, you might find it helpful to pattern your work after Figure 1.43 and the solution to Example 8. Group members should begin by consulting an almanac, newspaper, magazine, or the Internet to find data that appear to lie approximately on or near a line. Working by hand or using a graphing utility, group members should construct scatter plots for the data that were assembled. If working by hand, draw a line that approximately fits the data in each scatter plot and then write its equation as a function in slope-intercept form. If using a graphing utility, obtain the equation of each regression line. Then use each linear function's equation to make predictions about what might occur in the future. Are there circumstances that might affect the accuracy of the prediction? List some of these circumstances.

SECTION 1.5 *More on Slope*

Objectives

1 Find slopes and equations of parallel and perpendicular lines.

2 Interpret slope as rate of change.

3 Find a function's average rate of change.

Number of People in the U.S. Living Alone

Number Living Alone (millions)

Women

Men

1995 2000 2005 2010

Year

Figure 1.44

Source: Forrester Research

1 Find slopes and equations of parallel and perpendicular lines.

A best guess at the look of our nation in the next decades indicates that the number of men and women living alone will increase each year. Figure 1.44 shows that by 2010, approximately 12 million men and 17 million women will be living alone.

By looking at Figure 1.44, can you tell that the green graph representing women has a greater slope than the blue graph representing men? This indicates a greater yearly rate of change in the millions of women living alone than in the millions of men living alone. In this section, you will learn to interpret slope as a rate of change. You will also explore the relationships between slopes of parallel and perpendicular lines.

Parallel and Perpendicular Lines

Two nonintersecting lines that lie in the same plane are **parallel**. If two lines do not intersect, the ratio of the vertical change to the horizontal change is the same for each line. Because two parallel lines have the same "steepness," they must have the same slope.

Slope and Parallel Lines

1. If two nonvertical lines are parallel, then they have the same slope.

2. If two distinct nonvertical lines have the same slope, then they are parallel.

3. Two distinct vertical lines, both with undefined slopes, are parallel.

EXAMPLE 1   Writing Equations of a Line Parallel to a Given Line

Write an equation of the line passing through $(-3, 1)$ and parallel to the line whose equation is $y = 2x + 1$. Express the equation in point-slope form and slope-intercept form.

Solution The situation is illustrated in Figure 1.45. We are looking for the equation of the red line shown on the left. How do we obtain this equation? Notice that the line passes through the point $(-3, 1)$. Using the point-slope form of the line's equation, we have $x_1 = -3$ and $y_1 = 1$.

$$y - y_1 = m(x - x_1)$$

$y_1 = 1$ $x_1 = -3$

The equation of this line is given: $y = 2x + 1$.

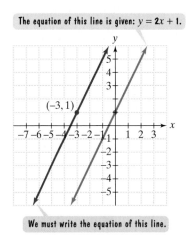

(−3, 1)

We must write the equation of this line.

Figure 1.45

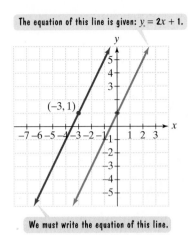

The equation of this line is given: $y = 2x + 1$.

$(-3, 1)$

We must write the equation of this line.

Figure 1.45 (repeated)

With $(x_1, y_1) = (-3, 1)$, the only thing missing from the equation of the red line is m, the slope. Do we know anything about the slope of either line in Figure 1.45? The answer is yes; we know the slope of the blue line on the right, whose equation is given

$$y = 2x + 1$$

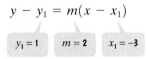

The slope of the blue line on the right in Figure 1.45 is 2.

Parallel lines have the same slope. Because the slope of the blue line is 2, the slope of the red line, the line whose equation we must write, is also 2: $m = 2$. We now have values for x_1, y_1, and m for the red line.

$$y - y_1 = m(x - x_1)$$

$y_1 = 1$ $m = 2$ $x_1 = -3$

The point-slope form of the red line's equation is

$$y - 1 = 2[x - (-3)] \text{ or}$$
$$y - 1 = 2(x + 3).$$

Solving for y, we obtain the slope-intercept form of the equation.

$y - 1 = 2x + 6$ Apply the distributive property.

$y = 2x + 7$ Add 1 to both sides. This is the slope-intercept form, $y = mx + b$, of the equation. Using function notation, the equation is $f(x) = 2x + 7$.

Check Point 1 Write an equation of the line passing through $(-2, 5)$ and parallel to the line whose equation is $y = 3x + 1$. Express the equation in point-slope form and slope-intercept form.

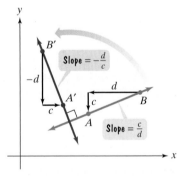

Figure 1.46 Slopes of perpendicular lines

Two lines that intersect at a right angle ($90°$) are said to be **perpendicular**, shown in Figure 1.46. The relationship between the slopes of perpendicular lines is not as obvious as the relationship between parallel lines. Figure 1.46 shows line AB, with slope $\frac{c}{d}$. Rotate line AB through $90°$ to the left to obtain line $A'B'$, perpendicular to line AB. The figure indicates that the rise and the run of the new line are reversed from the original line, but the rise is now negative. This means that the slope of the new line is $-\frac{d}{c}$. Notice that the product of the slopes of the two perpendicular lines is -1:

$$\left(\frac{c}{d}\right)\left(-\frac{d}{c}\right) = -1.$$

This relationship holds for all perpendicular lines and is summarized in the following box:

Slope and Perpendicular Lines

1. If two nonvertical lines are perpendicular, then the product of their slopes is -1.
2. If the product of the slopes of two lines is -1, then the lines are perpendicular.
3. A horizontal line having zero slope is perpendicular to a vertical line having undefined slope.

An equivalent way of stating this relationship is to say that **one line is perpendicular to another line if its slope is the *negative reciprocal* of the slope of the other line**. For example, if a line has slope 5, any line having slope $-\frac{1}{5}$ is perpendicular to it. Similarly, if a line has slope $-\frac{3}{4}$, any line having slope $\frac{4}{3}$ is perpendicular to it.

**EXAMPLE 2   Writing Equations of a Line Perpendicular
to a Given Line**

a. Find the slope of any line that is perpendicular to the line whose equation is
$x + 4y - 8 = 0$.

b. Write the equation of the line passing through $(3, -5)$ and perpendicular to the
line whose equation is $x + 4y - 8 = 0$. Express the equation in general form.

Solution

a. We begin by writing the equation of the given line, $x + 4y - 8 = 0$, in slope-
intercept form. Solve for y.

$x + 4y - 8 = 0$ This is the given equation.

       $4y = -x + 8$ To isolate the y-term, subtract x and add 8 on both sides.

       $y = -\dfrac{1}{4}x + 2$ Divide both sides by 4.

Slope is $-\frac{1}{4}$.

The given line has slope $-\frac{1}{4}$. Any line perpendicular to this line has a slope
that is the negative reciprocal of $-\frac{1}{4}$. Thus, the slope of any perpendicular
line is 4.

b. Let's begin by writing the point-slope form of the perpendicular line's equation. Because the line passes through the point $(3, -5)$, we have $x_1 = 3$ and
$y_1 = -5$. In part (a), we determined that the slope of any line perpendicular to
$x + 4y - 8 = 0$ is 4, so the slope of this particular perpendicular line must
also be 4: $m = 4$.

$$y - y_1 = m(x - x_1)$$

$y_1 = -5$ $m = 4$ $x_1 = 3$

The point-slope form of the perpendicular line's equation is

$$y - (-5) = 4(x - 3) \text{ or}$$
$$y + 5 = 4(x - 3).$$

How can we express this equation in general form $(Ax + By + C = 0)$? We
need to obtain zero on one side of the equation. Let's do this and keep A, the
coefficient of x, positive.

$y + 5 = 4(x - 3)$ This is the point-slope form of the line's equation.

$y + 5 = 4x - 12$ Apply the distributive property.

$y - y + 5 - 5 = 4x - y - 12 - 5$ To obtain 0 on the left, subtract y and subtract 5 on both sides.

$0 = 4x - y - 17$ Simplify.

The general form of the perpendicular line's equation is $4x - y - 17 = 0$.

Check Point **2**   **a.** Find the slope of any line that is perpendicular to the line whose equation is $x + 3y - 12 = 0$.

      b. Write the equation of the line passing through $(-2, -6)$ and perpen-
dicular to the line whose equation is $x + 3y - 12 = 0$. Express the
equation in general form.

Interpret slope as rate of change.

Slope as Rate of Change

Slope is defined as the ratio of a change in y to a corresponding change in x. I describes how fast y is changing with respect to x. For a linear function, slope may be interpreted as the rate of change of the dependent variable per unit change in the independent variable.

Our next example shows how slope can be interpreted as a rate of change in an applied situation. When calculating slope in applied problems, keep track of the units in the numerator and the denominator.

EXAMPLE 3 **Slope as a Rate of Change**

The line graphs for the number of women and men living alone are shown again in Figure 1.47. Find the slope of the line segment for the women. Describe what this slope represents.

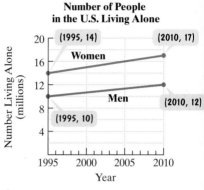

Number of People in the U.S. Living Alone

Figure 1.47
Source: Forrester Research

Solution We let x represent a year and y the number of women living alone in that year. The two points shown on the line segment for women have the following coordinates:

$$(1995, 14) \quad \text{and} \quad (2010, 17).$$

| In 1995, 14 million U.S. women lived alone. | In 2010, 17 million U.S. women are projected to live alone. |

Now we compute the slope:

$$m = \frac{\text{Change in } y}{\text{Change in } x} = \frac{17 - 14}{2010 - 1995}$$

The unit in the numerator is *million women*.

$$= \frac{3}{15} = \frac{1}{5} = \frac{0.2 \text{ million women}}{\text{year}}.$$

The unit in the denominator is *year*.

The slope indicates that the number of U.S. women living alone is projected to increase by 0.2 million each year. The rate of change is 0.2 million women per year

 Check Point 3 Use the graph in Figure 1.47 to find the slope of the line segment for the men. Express the slope correct to two decimal places and describe what it represents.

In Check Point 3 did you find that the slope of the line segment for the men is different from that of the women? The rate of change for men living alone is not equal to the rate of change for women living alone. Because of these different slopes, if you extend the line segments in Figure 1.47, the resulting lines will inter- sect. They are not parallel.

 Find a function's average rate of change.

The Average Rate of Change of a Function

If the graph of a function is not a straight line, the **average rate of change** between any two points is the slope of the line containing the two points. This line is called a **secant line**. For example, Figure 1.48 shows the graph of a particular man's height in inches, as a function of his age, in years. Two points on the graph are labeled: $(13, 57)$ and $(18, 76)$. At age 13, this man was 57 inches tall and at age 18, he was 76 inches tall.

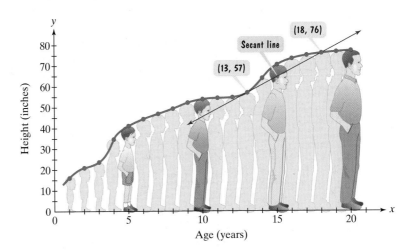

Figure 1.48 Height as a function of age

The man's average growth rate between ages 13 and 18 is the slope of the secant line containing $(13, 57)$ and $(18, 76)$:

$$m = \frac{\text{Change in } y}{\text{Change in } x} = \frac{76 - 57}{18 - 13} = \frac{19}{5} = 3\frac{4}{5}.$$

This man's average rate of change, or average growth rate, from age 13 to age 18 was $3\frac{4}{5}$, or 3.8, inches per year.

The Average Rate of Change of a Function

Let $(x_1, f(x_1))$ and $(x_2, f(x_2))$ be distinct points on the graph of a function f. (See Figure 1.49.) The **average rate of change of f** from x_1 to x_2, denoted by $\frac{\Delta y}{\Delta x}$ (read "delta y divided by delta x" or "change in y divided by change in x"), is

$$\frac{\Delta y}{\Delta x} = \frac{f(x_2) - f(x_1)}{x_2 - x_1}.$$

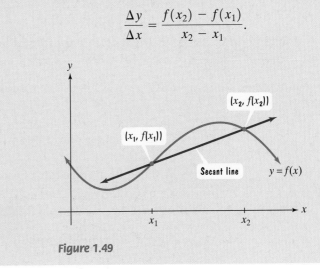

Figure 1.49

EXAMPLE 4   Finding the Average Rate of Change

Find the average rate of change of $f(x) = x^2$ from

a. $x_1 = 0$ to $x_2 = 1$ **b.** $x_1 = 1$ to $x_2 = 2$ **c.** $x_1 = -2$ to $x_2 = 0$.

Solution

a. The average rate of change of $f(x) = x^2$ from $x_1 = 0$ to $x_2 = 1$ is

$$\frac{\Delta y}{\Delta x} = \frac{f(x_2) - f(x_1)}{x_2 - x_1} = \frac{f(1) - f(0)}{1 - 0} = \frac{1^2 - 0^2}{1} = 1.$$

Figure 1.50(a) shows the secant line of $f(x) = x^2$ from $x_1 = 0$ to $x_2 = 1$. The average rate of change is positive and the function is increasing on the interval $(0, 1)$.

b. The average rate of change of $f(x) = x^2$ from $x_1 = 1$ to $x_2 = 2$ is

$$\frac{\Delta y}{\Delta x} = \frac{f(x_2) - f(x_1)}{x_2 - x_1} = \frac{f(2) - f(1)}{2 - 1} = \frac{2^2 - 1^2}{1} = 3.$$

Figure 1.50(b) shows the secant line of $f(x) = x^2$ from $x_1 = 1$ to $x_2 = 2$. The average rate of change is positive and the function is increasing on the interval $(1, 2)$. Can you see that the graph rises more steeply on the interval $(1, 2)$ than on $(0, 1)$? This is because the average rate of change from $x_1 = 1$ to $x_2 = 2$ is greater than the average rate of change from $x_1 = 0$ to $x_2 = 1$.

c. The average rate of change of $f(x) = x^2$ from $x_1 = -2$ to $x_2 = 0$ is

$$\frac{\Delta y}{\Delta x} = \frac{f(x_2) - f(x_1)}{x_2 - x_1} = \frac{f(0) - f(-2)}{0 - (-2)} = \frac{0^2 - (-2)^2}{2} = \frac{-4}{2} = -2.$$

Figure 1.50(c) shows the secant line of $f(x) = x^2$ from $x_1 = -2$ to $x_2 = 0$. The average rate of change is negative and the function is decreasing on the interval $(-2, 0)$.

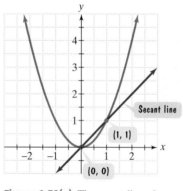
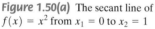

Figure 1.50(a) The secant line of $f(x) = x^2$ from $x_1 = 0$ to $x_2 = 1$

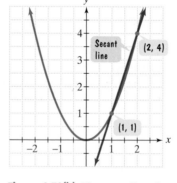
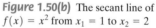

Figure 1.50(b) The secant line of $f(x) = x^2$ from $x_1 = 1$ to $x_2 = 2$

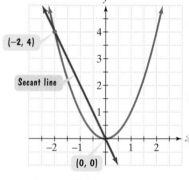
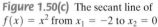

Figure 1.50(c) The secant line of $f(x) = x^2$ from $x_1 = -2$ to $x_2 = 0$

Check Point 4 Find the average rate of change of $f(x) = x^3$ from

a. $x_1 = 0$ to $x_2 = 1$ **b.** $x_1 = 1$ to $x_2 = 2$ **c.** $x_1 = -2$ to $x_2 = 0$

Suppose we are interested in the average rate of change of f from $x_1 = x$ to $x_2 = x + h$. In this case, the average rate of change is

$$\frac{\Delta y}{\Delta x} = \frac{f(x_2) - f(x_1)}{x_2 - x_1} = \frac{f(x + h) - f(x)}{x + h - x} = \frac{f(x + h) - f(x)}{h}.$$

Do you recognize the last expression? It is the difference quotient that you used in Section 1.3. Thus, the difference quotient gives the average rate of change of a function from x to $x + h$. In the difference quotient, h is thought of as a number very close to 0. In this way, the average rate of change can be found for a very short interval.

EXAMPLE 5   Finding the Average Rate of Change

When a person receives a drug injected into a muscle, the concentration of the drug in the body, measured in milligrams per 100 milliliters, is a function of the time elapsed after the injection, measured in hours. Figure 1.51 shows the graph of such a function, where x represents hours after the injection and $f(x)$ is the drug's concentration at time x. Find the average rate of change in the drug's concentration between 3 and 7 hours.

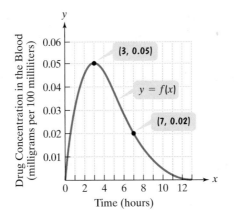

Figure 1.51 Concentration of a drug as a function of time

Solution   At 3 hours, the drug's concentration is 0.05 and at 7 hours, the concentration is 0.02. The average rate of change in its concentration between 3 and 7 hours is

$$\frac{\Delta y}{\Delta x} = \frac{f(x_2) - f(x_1)}{x_2 - x_1} = \frac{f(7) - f(3)}{7 - 3} = \frac{0.02 - 0.05}{7 - 3} = \frac{-0.03}{4} = -0.0075.$$

The average rate of change is -0.0075. This means that the drug's concentration is decreasing at an average rate of 0.0075 milligrams per 100 milliliters per hour.

Study Tip

Units used to describe x and y tend to "pile up" when expressing the rate of change of y with respect to x. The unit used to express the rate of change of y with respect to x is

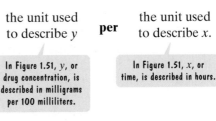

In Figure 1.51, the rate of change is described in terms of milligrams per 100 milliliters per hour.

Check Point 5   Use Figure 1.51 to find the average rate of change in the drug's concentration between 1 hour and 3 hours.

The **average velocity** of an object is its change in position divided by the change in time between the starting and ending positions. If a function expresses an object's position in terms of time, the function's average rate of change describes the object's average velocity.

> ## Average Velocity of an Object
>
> Suppose that a function expresses an object's position, $s(t)$, in terms of time, t. The **average velocity** of the object from t_1 to t_2 is
>
> $$\frac{\Delta s}{\Delta t} = \frac{s(t_2) - s(t_1)}{t_2 - t_1}.$$

EXAMPLE 6 Finding Average Velocity

The distance, $s(t)$, in feet, traveled by a ball rolling down a ramp is given by the function

$$s(t) = 5t^2,$$

where t is the time, in seconds, after the ball is released. Find the ball's average velocity from

a. $t_1 = 2$ seconds to $t_2 = 3$ seconds.

b. $t_1 = 2$ seconds to $t_2 = 2.5$ seconds.

c. $t_1 = 2$ seconds to $t_2 = 2.01$ seconds.

Solution

a. The ball's average velocity between 2 and 3 seconds is

$$\frac{\Delta s}{\Delta t} = \frac{s(3) - s(2)}{3 \text{ sec} - 2 \text{ sec}} = \frac{5 \cdot 3^2 - 5 \cdot 2^2}{1 \text{ sec}} = \frac{45 \text{ ft} - 20 \text{ ft}}{1 \text{ sec}} = 25 \text{ ft/sec.}$$

b. The ball's average velocity between 2 and 2.5 seconds is

$$\frac{\Delta s}{\Delta t} = \frac{s(2.5) - s(2)}{2.5 \text{ sec} - 2 \text{ sec}} = \frac{5(2.5)^2 - 5 \cdot 2^2}{0.5 \text{ sec}} = \frac{31.25 \text{ ft} - 20 \text{ ft}}{0.5 \text{ sec}} = 22.5 \text{ ft/sec.}$$

c. The ball's average velocity between 2 and 2.01 seconds is

$$\frac{\Delta s}{\Delta t} = \frac{s(2.01) - s(2)}{2.01 \text{ sec} - 2 \text{ sec}} = \frac{5(2.01)^2 - 5 \cdot 2^2}{0.01 \text{ sec}} = \frac{20.2005 \text{ ft} - 20 \text{ ft}}{0.01 \text{ sec}} = 20.05 \text{ ft/sec.}$$

In Example 6, observe that each calculation begins at 2 seconds and involves shorter and shorter time intervals. In calculus, this procedure leads to the concept of *instantaneous*, as opposed to *average*, velocity.

Check Point 6 The distance, $s(t)$, in feet, traveled by a ball rolling down a ramp is given by the function

$$s(t) = 4t^2,$$

where t is the time, in seconds, after the ball is released. Find the ball's average velocity from

a. $t_1 = 1$ second to $t_2 = 2$ seconds.

b. $t_1 = 1$ second to $t_2 = 1.5$ seconds.

c. $t_1 = 1$ second to $t_2 = 1.01$ seconds.

EXERCISE SET 1.5

Practice Exercises

In Exercises 1–4, write an equation for line L in point-slope form and slope-intercept form.

1.

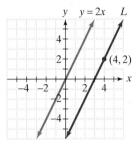

L is parallel to $y = 2x$.

2.

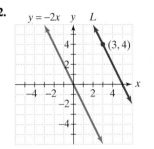

L is parallel to $y = -2x$.

3.

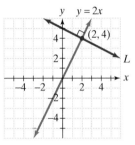

L is perpendicular to $y = 2x$.

4.

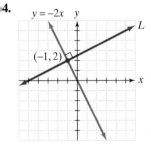

L is perpendicular to $y = -2x$.

In Exercises 5–8, use the given conditions to write an equation for each line in point-slope form and slope-intercept form.

5. Passing through $(-8, -10)$ and parallel to the line whose equation is $y = -4x + 3$

6. Passing through $(-2, -7)$ and parallel to the line whose equation is $y = -5x + 4$

7. Passing through $(2, -3)$ and perpendicular to the line whose equation is $y = \frac{1}{5}x + 6$

8. Passing through $(-4, 2)$ and perpendicular to the line whose equation is $y = \frac{1}{3}x + 7$

In Exercises 9–12, use the given conditions to write an equation for each line in point-slope form and general form.

9. Passing through $(-2, 2)$ and parallel to the line whose equation is $2x - 3y - 7 = 0$

10. Passing through $(-1, 3)$ and parallel to the line whose equation is $3x - 2y - 5 = 0$

11. Passing through $(4, -7)$ and perpendicular to the line whose equation is $x - 2y - 3 = 0$

12. Passing through $(5, -9)$ and perpendicular to the line whose equation is $x + 7y - 12 = 0$

In Exercises 13–18, find the average rate of change of the function from x_1 to x_2.

13. $f(x) = 3x$ from $x_1 = 0$ to $x_2 = 5$

14. $f(x) = 6x$ from $x_1 = 0$ to $x_2 = 4$

15. $f(x) = x^2 + 2x$ from $x_1 = 3$ to $x_2 = 5$

16. $f(x) = x^2 - 2x$ from $x_2 = 3$ to $x_2 = 6$

17. $f(x) = \sqrt{x}$ from $x_1 = 4$ to $x_2 = 9$

18. $f(x) = \sqrt{x}$ from $x_1 = 9$ to $x_2 = 16$

In Exercises 19–20, suppose that a ball is rolling down a ramp. The distance traveled by the ball is given by the function in each exercise, where t is the time, in seconds, after the ball is released, and s(t) is measured in feet. For each given function, find the ball's average velocity from

a. $t_1 = 3$ to $t_2 = 4$. **b.** $t_1 = 3$ to $t_2 = 3.5$.

c. $t_1 = 3$ to $t_2 = 3.01$. **d.** $t_1 = 3$ to $t_2 = 3.001$.

19. $s(t) = 10t^2$ **20.** $s(t) = 12t^2$

Practice Plus

In Exercises 21–26, write the slope-intercept equation of a function f whose graph satisfies the given conditions.

21. The graph of f passes through $(-1, 5)$ and is perpendicular to the line whose equation is $x = 6$.

22. The graph of f passes through $(-2, 6)$ and is perpendicular to the line whose equation is $x = -4$.

23. The graph of f passes through $(-6, 4)$ and is perpendicular to the line that has an x-intercept of 2 and a y-intercept of -4.

24. The graph of f passes through $(-5, 6)$ and is perpendicular to the line that has an x-intercept of 3 and a y-intercept of -9.

25. The graph of f is perpendicular to the line whose equation is $3x - 2y - 4 = 0$ and has the same y-intercept as this line.

26. The graph of f is perpendicular to the line whose equation is $4x - y - 6 = 0$ and has the same y-intercept as this line.

Application Exercises

In Exercises 27–30, a linear function that models data is described. Find the slope of each model. Then describe what this means in terms of the rate of change of the dependent variable per unit change in the independent variable.

27. The linear function $f(x) = 0.01x + 57.7$ models the global average temperature of Earth, $f(x)$, in degrees Fahrenheit, x years after 1995.

28. The linear function $f(x) = 2x + 10$ models the amount, $f(x)$, in billions of dollars, that the drug industry spent on marketing information about drugs to doctors x years after 2000. (*Source:* IMS Health)

29. The linear function $f(x) = -0.52x + 24.7$ models the percentage of U.S. adults who smoked cigarettes, $f(x)$, x years after 1997. (*Source:* National Center for Health Statistics)

30. The linear function $f(x) = -0.28x + 1.7$ models the percentage of U.S. taxpayers who were audited by the IRS, $f(x)$, x years after 1996. (*Source:* IRS)

The bar graph shows the average amount that U.S. consumers spent on four pieces of the entertainment pie from 2002 through 2004.

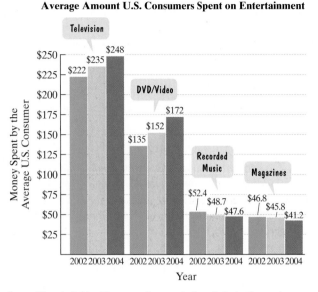

Average Amount U.S. Consumers Spent on Entertainment

Source: Veronis, Suhler, Stevenson, Communications Industry Forecast and Report

In Exercises 31–34, find a linear function in slope-intercept form that models the given description. Each function should model the average amount, $f(x)$, that U.S. consumers spent on the mode of entertainment x years after 2002.

31. In 2002, the average U.S. consumer spent $222 on television (broadcast, cable, and satellite) and this amount has increased at an average rate of $13 per year since then.

32. In 2002, the average U.S. consumer spent $135 on DVDs and videos, and this amount has increased at an average rate of $18.50 per year since then.

33. In 2002, the average U.S. consumer spent $52.40 on recorded music and this amount has decreased at an average rate of $2.40 per year since then.

34. In 2002, the average U.S. consumer spent $46.80 on magazines and this amount has decreased at an average rate of $2.80 per year since then.

The graph shows the percentage of sales of recorded music in the United States for rock and rap/hip-hop from 1997 through 2003. Use the information shown to solve Exercises 35–36.

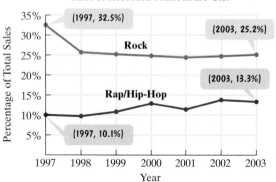

Sales of Recorded Music in the U.S.

Source: RIAA

35. Find the average rate of change in the percentage of total sales of rock from 1997 through 2003. Round to the nearest hundredth of a percent.

36. Find the average rate of change in the percentage of total sales of rap/hip-hop from 1997 through 2003. Round to the nearest hundredth of a percent.

Writing in Mathematics

37. If two lines are parallel, describe the relationship between their slopes.

38. If two lines are perpendicular, describe the relationship between their slopes.

39. If you know a point on a line and you know the equation of a line perpendicular to this line, explain how to write the line's equation.

40. A formula in the form $y = mx + b$ models the cost, y, of a four-year college x years after 2005. Would you expect m to be positive, negative, or zero? Explain your answer.

41. What is a secant line?

42. What is the average rate of change of a function?

Technology Exercise

43. a. Why are the lines whose equations are $y = \frac{1}{3}x + 1$ and $y = -3x - 2$ perpendicular?

b. Use a graphing utility to graph the equations in a $[-10, 10, 1]$ by $[-10, 10, 1]$ viewing rectangle. Do the lines appear to be perpendicular?

c. Now use the zoom square feature of your utility. Describe what happens to the graphs. Explain why this is so.

Critical Thinking Exercises

In Exercises 44–45, draw a graph that illustrates each data description.

44. From 1971 through 1980, the percentage of Americans who were obese held constant at 15%. After 1980, this percentage increased at a rate of 0.8% per year.

(*Source:* National Center for Health Statistics)

45. In 1970, the daily calories in the U.S. food supply was 3300 per person. From 1970 through 1975, daily calories in the food supply decreased at a rate of 20 calories per person per year. From 1975 through the present, this number has increased at a rate of 28 calories per person per year.

(*Source:* USDA, Economic Research Service)

46 What is the slope of a line that is perpendicular to the line whose equation is $Ax + By + C = 0$, $A \neq 0$ and $B \neq 0$?

47. Determine the value of A so that the line whose equation is $Ax + y - 2 = 0$ is perpendicular to the line containing the points $(1, -3)$ and $(-2, 4)$.

CHAPTER 1
MID-CHAPTER CHECK POINT

What You Know: We learned that a function is a relation in which no two ordered pairs have the same first component and different second components. We represented functions as equations and used function notation. We graphed functions and applied the vertical line test to identify graphs of functions. We determined the domain and range of a function from its graph, using inputs on the x-axis for the domain and outputs on the y-axis for the range. We used graphs to identify intervals on which functions increase, decrease, or are constant, as well as to locate relative maxima or minima. We identified even functions [$f(-x) = f(x)$: y-axis symmetry] and odd functions [$f(-x) = -f(x)$: origin symmetry]. Finally, we studied linear functions and slope, using slope (change in y divided by change in x) to develop various forms for equations of lines:

Point-slope form	Slope-intercept form	Horizontal line	Vertical line	General form
$y - y_1 = m(x - x_1)$	$y = f(x) = mx + b$	$y = f(x) = b$	$x = a$	$Ax + By + C = 0$

We saw that parallel lines have the same slope and that perpendicular lines have slopes that are negative reciprocals. For linear functions, slope was interpreted as the rate of change of the dependent variable per unit change in the independent variable. For nonlinear functions, the slope of the secant line between $(x_1, f(x_1))$ and $(x_2, f(x_2))$ described the average rate of change of f from x_1 to x_2: $\dfrac{f(x_2) - f(x_1)}{x_2 - x_1}$.

In Exercises 1–6, determine whether each relation is a function. Give the domain and range for each relation.

1. $\{(2, 6), (1, 4), (2, -6)\}$ **2.** $\{(0, 1), (2, 1), (3, 4)\}$

3.

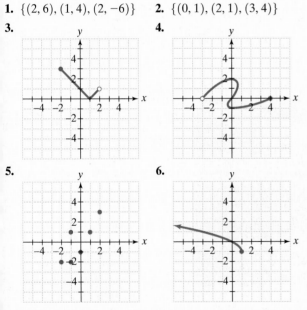

4.

5.

6.

In Exercises 7–8, determine whether each equation defines y as a function of x.

7. $x^2 + y = 5$ **8.** $x + y^2 = 5$

Use the graph of f to solve Exercises 9–24. Where applicable, use interval notation.

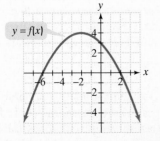

$y = f(x)$

9. Explain why f represents the graph of a function.

10. Find the domain of f.

11. Find the range of f.

12. Find the x-intercept(s).

13. Find the y-intercept.

14. Find the interval(s) on which f is increasing.

15. Find the interval(s) on which f is decreasing.

16. At what number does f have a relative maximum?

17. What is the relative maximum of f?

18. Find $f(-4)$.

19. For what value or values of x is $f(x) = -2$?

20. For what value or values of x is $f(x) = 0$?

21. For what values of x is $f(x) > 0$?

22. Is $f(100)$ positive or negative?

23. Is f even, odd, or neither?

24. Find the average rate of change of f from $x_1 = -4$ to $x_2 = 4$.

In Exercises 25–36, graph each equation in a rectangular coordinate system.

25. $y = -2x$

26. $y = -2$

27. $x + y = -2$

28. $y = \frac{1}{3}x - 2$

29. $x = 3.5$

30. $4x - 2y = 8$

31. $f(x) = x^2 - 4$

32. $f(x) = x - 4$

33. $f(x) = |x| - 4$

34. $5y = -3x$

35. $5y = 20$

36. $f(x) = \begin{cases} -1 & \text{if} \quad x \le 0 \\ 1 & \text{if} \quad x > 0 \end{cases}$

37. Let $f(x) = -2x^2 + x - 5$.

    a. Find $f(-x)$. Is f even, odd, or neither?

    b. Find $\dfrac{f(x + h) - f(x)}{h}$, $h \ne 0$.

38. Let $C(x) = \begin{cases} 30 & \text{if} & 0 \le t \le 200 \\ 30 + 0.40(t - 200) & \text{if} & t > 200 \end{cases}$.

    a. Find $C(150)$.   **b.** Find $C(250)$.

In Exercises 39–42, write a function in slope-intercept form whose graph satisfies the given conditions.

39. Slope $= -2$, passing through $(-4, 3)$

40. Passing through $(-1, -5)$ and $(2, 1)$

41. Passing through $(3, -4)$ and parallel to the line whose equation is $3x - y - 5 = 0$

42. Passing through $(-4, -3)$ and perpendicular to the line whose equation is $2x - 5y - 10 = 0$

43. Determine whether the line through $(2, -4)$ and $(7, 0)$ is parallel to a second line through $(-4, 2)$ and $(1, 6)$.

44. The graph shows the percentage of U.S. colleges that offered distance learning by computer for selected years from 1995 through 2002.

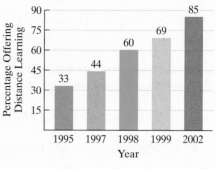

Percentage of U.S. Colleges Offering Distance Learning by Computer

Source : International Data Corporation

The data can be modeled by the linear function $f(x) = 7.8x + 33$, where x is the number of years after 1995 and $f(x)$ is the percentage of U.S. colleges offering distance learning. Find the slope of this function and describe its meaning as a rate of change.

45. Find the average rate of change of $f(x) = 3x^2 - x$ from $x_1 = -1$ to $x_2 = 2$.

Objectives

❶ Recognize graphs of common functions.

❷ Use vertical shifts to graph functions.

❸ Use horizontal shifts to graph functions.

❹ Use reflections to graph functions.

❺ Use vertical stretching and shrinking to graph functions.

❻ Use horizontal stretching and shrinking to graph functions.

❼ Graph functions involving a sequence of transformations.

Have you seen *Terminator 2, The Mask*, or *The Matrix*? These were among the first films to use spectacular effects in which a character or object having one shape was transformed in a fluid fashion into a quite different shape. The name for such a transformation is **morphing**. The effect allows a real actor to be seamlessly transformed into a computer-generated animation. The animation can be made to perform impossible feats before it is morphed back to the conventionally filmed image.

    Like transformed movie images, the graph of one function can be turned into the graph of a different function. To do this, we need to rely on a function's equation. Knowing that a graph is a transformation of a familiar graph makes graphing easier.

 Recognize graphs of common functions.

Graphs of Common Functions

Table 1.4 on the next page gives names to seven frequently encountered functions in algebra. The table shows each function's graph and lists characteristics of the function. Study the shape of each graph and take a few minutes to verify the function's characteristics from its graph. Knowing these graphs is essential for analyzing their transformations into more complicated graphs.

le 1.4   Algebra's Common Graphs

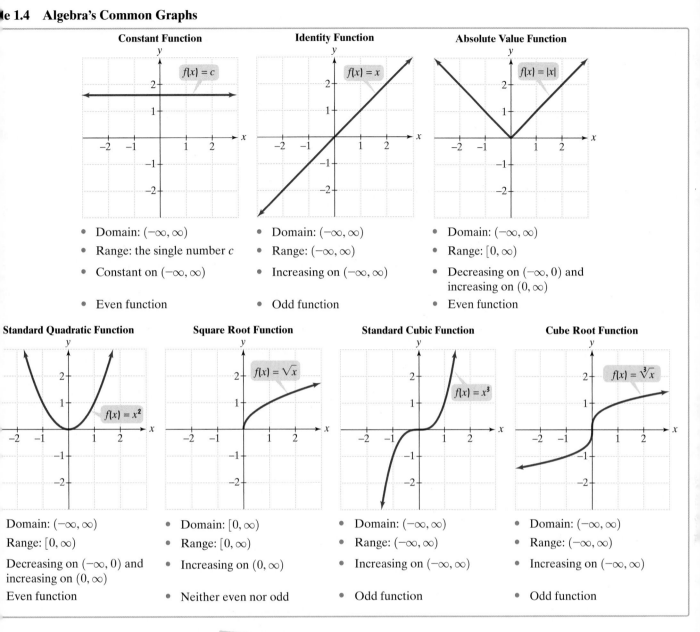

Constant Function

$f(x) = c$

- Domain: $(-\infty, \infty)$
- Range: the single number c
- Constant on $(-\infty, \infty)$
- Even function

Identity Function

$f(x) = x$

- Domain: $(-\infty, \infty)$
- Range: $(-\infty, \infty)$
- Increasing on $(-\infty, \infty)$
- Odd function

Absolute Value Function

$f(x) = |x|$

- Domain: $(-\infty, \infty)$
- Range: $[0, \infty)$
- Decreasing on $(-\infty, 0)$ and increasing on $(0, \infty)$
- Even function

Standard Quadratic Function

$f(x) = x^2$

- Domain: $(-\infty, \infty)$
- Range: $[0, \infty)$
- Decreasing on $(-\infty, 0)$ and increasing on $(0, \infty)$
- Even function

Square Root Function

$f(x) = \sqrt{x}$

- Domain: $[0, \infty)$
- Range: $[0, \infty)$
- Increasing on $(0, \infty)$
- Neither even nor odd

Standard Cubic Function

$f(x) = x^3$

- Domain: $(-\infty, \infty)$
- Range: $(-\infty, \infty)$
- Increasing on $(-\infty, \infty)$
- Odd function

Cube Root Function

$f(x) = \sqrt[3]{x}$

- Domain: $(-\infty, \infty)$
- Range: $(-\infty, \infty)$
- Increasing on $(-\infty, \infty)$
- Odd function

Discovery

The study of how changing a function's equation can affect its graph can be explored with a graphing utility. Use your graphing utility to verify the hand-drawn graphs as you read this section.

② Use vertical shifts to graph functions.

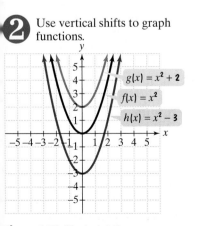

$g(x) = x^2 + 2$

$f(x) = x^2$

$h(x) = x^2 - 3$

Vertical Shifts

Let's begin by looking at three graphs whose shapes are the same. Figure 1.52 shows the graphs. The black graph in the middle is the standard quadratic function, $f(x) = x^2$. Now, look at the blue graph on the top. The equation of this graph, $g(x) = x^2 + 2$, adds 2 to the right side of $f(x) = x^2$. The y-coordinate of each point of g is 2 more than the corresponding y-coordinate of each point of f. What effect does this have on the graph of f? It shifts the graph vertically up by 2 units.

$$g(x) = x^2 + 2 = f(x) + 2$$

The graph of g shifts the graph of f up 2 units.

Finally, look at the red graph on the bottom in Figure 1.52. The equation of this graph, $h(x) = x^2 - 3$, subtracts 3 from the right side of $f(x) = x^2$. The y-coordinate of each

Figure 1.52 Vertical shifts

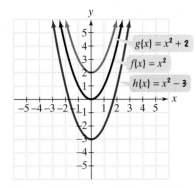

Figure 1.52 (repeated) Vertical shifts

point of h is 3 less than the corresponding y-coordinate of each point of f. What effect does this have on the graph of f? It shifts the graph vertically down by 3 units.

$$h(x) = x^2 - 3 = f(x) - 3$$

The graph of h · · · shifts the graph of f down 3 units.

In general, if c is positive, $y = f(x) + c$ shifts the graph of f upward c units and $y = f(x) - c$ shifts the graph of f downward c units. These are called **vertical shifts** of the graph of f.

Vertical Shifts

Let f be a function and c a positive real number.

- The graph of $y = f(x) + c$ is the graph of $y = f(x)$ shifted c units vertically upward.
- The graph of $y = f(x) - c$ is the graph of $y = f(x)$ shifted c units vertically downward.

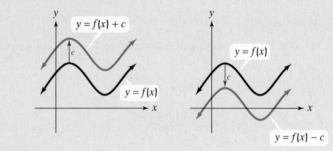

Study Tip

To keep track of transformations, identify a number of points on the given function's graph. Then analyze what happens to the coordinates of these points with each transformation.

EXAMPLE 1 Vertical Shift Down

Use the graph of $f(x) = |x|$ to obtain the graph of $g(x) = |x| - 4$.

Solution The graph of $g(x) = |x| - 4$ has the same shape as the graph of $f(x) = |x|$. However, it is shifted down vertically 4 units.

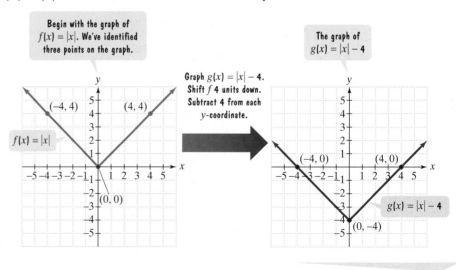

Check Point 1 Use the graph of $f(x) = |x|$ to obtain the graph of $g(x) = |x| + 3$.

3 Use horizontal shifts to graph functions.

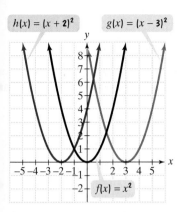

Figure 1.53 Horizontal shifts

Horizontal Shifts

We return to the graph of $f(x) = x^2$, the standard quadratic function. In Figure 1.53, the graph of function f is in the middle of the three graphs. By contrast to the vertical shift situation, this time there are graphs to the left and to the right of the graph of f. Look at the blue graph on the right. The equation of this graph, $g(x) = (x - 3)^2$, subtracts 3 from each value of x before squaring it. What effect does this have on the graph of $f(x) = x^2$? It shifts the graph horizontally to the right by 3 units.

$$g(x) = (x - 3)^2 = f(x - 3)$$

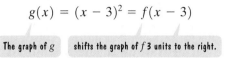

Does it seem strange that *subtracting* 3 in the domain causes a shift of 3 units to the *right*? Perhaps a partial table of coordinates for each function will numerically convince you of this shift.

x	$f(x) = x^2$
-2	$(-2)^2 = 4$
-1	$(-1)^2 = 1$
0	$0^2 = 0$
1	$1^2 = 1$
2	$2^2 = 4$

x	$g(x) = (x - 3)^2$
1	$(1 - 3)^2 = (-2)^2 = 4$
2	$(2 - 3)^2 = (-1)^2 = 1$
3	$(3 - 3)^2 = 0^2 = 0$
4	$(4 - 3)^2 = 1^2 = 1$
5	$(5 - 3)^2 = 2^2 = 4$

Notice that for the values of $f(x)$ and $g(x)$ to be the same, the values of x used in graphing g must each be 3 units greater than those used to graph f. For this reason, the graph of g is the graph of f shifted 3 units to the right.

Now, look at the red graph on the left in Figure 1.53. The equation of this graph, $h(x) = (x + 2)^2$, adds 2 to each value of x before squaring it. What effect does this have on the graph of $f(x) = x^2$? It shifts the graph horizontally to the left by 2 units.

$$h(x) = (x + 2)^2 = f(x + 2)$$

The graph of h shifts the graph of f **2** units to the left.

In general, if c is positive, $y = f(x + c)$ shifts the graph of f to the left c units and $y = f(x - c)$ shifts the graph of f to the right c units. These are called **horizontal shifts** of the graph of f.

Study Tip

On a number line, if x represents a number and c is positive, then $x + c$ lies c units to the right of x and $x - c$ lies c units to the left of x. This orientation does not apply to horizontal shifts: $f(x + c)$ causes a shift of c units to the left and $f(x - c)$ causes a shift of c units to the right.

Horizontal Shifts

Let f be a function and c a positive real number.

• The graph of $y = f(x + c)$ is the graph of $y = f(x)$ shifted to the left c units.

• The graph of $y = f(x - c)$ is the graph of $y = f(x)$ shifted to the right c units.

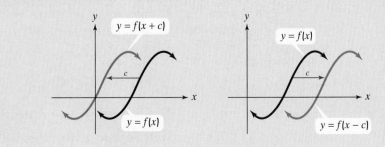

EXAMPLE 2   Horizontal Shift to the Left

Use the graph of $f(x) = \sqrt{x}$ to obtain the graph of $g(x) = \sqrt{x + 5}$.

Solution   Compare the equations for $f(x) = \sqrt{x}$ and $g(x) = \sqrt{x + 5}$. The equation for g adds 5 to each value of x before taking the square root.

$$y = g(x) = \sqrt{x + 5} = f(x + 5)$$

The graph of g shifts the graph of f 5 units to the left.

The graph of $g(x) = \sqrt{x + 5}$ has the same shape as the graph of $f(x) = \sqrt{x}$. However, it is shifted horizontally to the left 5 units.

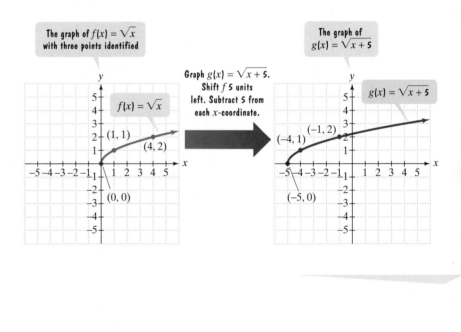

Check Point **2**   Use the graph of $f(x) = \sqrt{x}$ to obtain the graph of $g(x) = \sqrt{x - 4}$.

Some functions can be graphed by combining horizontal and vertical shifts. These functions will be variations of a function whose equation you know how to graph, such as the standard quadratic function, the standard cubic function, the square root function, the cube root function, or the absolute value function.

In our next example, we will use the graph of the standard quadratic function, $f(x) = x^2$, to obtain the graph of $h(x) = (x + 1)^2 - 3$. We will graph three functions:

$$f(x) = x^2 \qquad g(x) = (x + 1)^2 \qquad h(x) = (x + 1)^2 - 3.$$

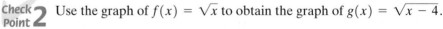

Start by graphing the standard quadratic function.   Shift the graph of f horizontally one unit to the left.   Shift the graph of g vertically down 3 units.

EXAMPLE 3   Combining Horizontal and Vertical Shifts

Use the graph of $f(x) = x^2$ to obtain the graph of $h(x) = (x + 1)^2 - 3$.

Solution

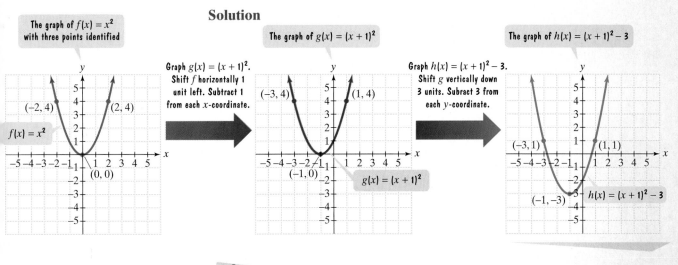

The graph of $f(x) = x^2$ with three points identified

$(-2, 4)$ $(2, 4)$

$f(x) = x^2$

$(0, 0)$

Graph $g(x) = (x + 1)^2$. Shift f horizontally 1 unit left. Subtract 1 from each x-coordinate.

The graph of $g(x) = (x + 1)^2$

$(-3, 4)$ $(1, 4)$

$(-1, 0)$

$g(x) = (x + 1)^2$

Graph $h(x) = (x + 1)^2 - 3$. Shift g vertically down 3 units. Subtract 3 from each y-coordinate.

The graph of $h(x) = (x + 1)^2 - 3$

$(-3, 1)$ $(1, 1)$

$(-1, -3)$ $h(x) = (x + 1)^2 - 3$

Discovery

Work Example 3 by first shifting the graph of $f(x) = x^2$ three units down, graphing $g(x) = x^2 - 3$. Now, shift this graph one unit left to graph $h(x) = (x + 1)^2 - 3$. Did you obtain the last graph shown in the solution of Example 3? What can you conclude?

Check Point 3   Use the graph of $f(x) = \sqrt{x}$ to obtain the graph of $h(x) = \sqrt{x - 1} - 2$.

 Use reflections to graph functions.

Reflections of Graphs

This photograph shows a reflection of an old bridge in a Maryland river. This perfect reflection occurs because the surface of the water is absolutely still. A mild breeze rippling the water's surface would distort the reflection.

Is it possible for graphs to have mirror-like qualities? Yes. Figure 1.54 shows the graphs of $f(x) = x^2$ and $g(x) = -x^2$. The graph of g is a **reflection about the x-axis** of the graph of f. For corresponding values of x, the y-coordinates of g are the opposites of the y-coordinates of f. In general, the graph of $y = -f(x)$ reflects the graph of f about the x-axis. Thus, the graph of g is a reflection of the graph of f about the x-axis because

$$g(x) = -x^2 = -f(x).$$

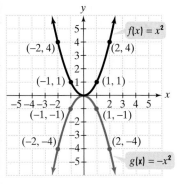

$(-2, 4)$ $f(x) = x^2$ $(2, 4)$

$(-1, 1)$ $(1, 1)$

$(-1, -1)$ $(1, -1)$

$(-2, -4)$ $(2, -4)$

$g(x) = -x^2$

Figure 1.54 Reflection about the x-axis

Reflection about the x-Axis

The graph of $y = -f(x)$ is the graph of $y = f(x)$ reflected about the x-axis.

EXAMPLE 4   Reflection about the *x*-Axis

Use the graph of $f(x) = \sqrt[3]{x}$ to obtain the graph of $g(x) = -\sqrt[3]{x}$.

Solution   Compare the equations for $f(x) = \sqrt[3]{x}$ and $g(x) = -\sqrt[3]{x}$. The graph of g is a reflection about the *x*-axis of the graph of f because

$$g(x) = -\sqrt[3]{x} = -f(x).$$

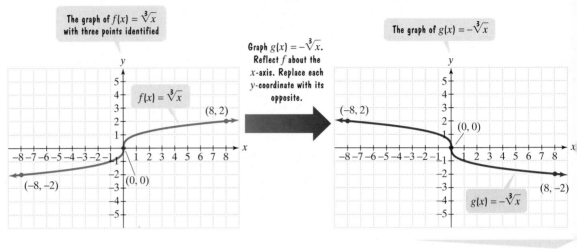

The graph of $f(x) = \sqrt[3]{x}$ with three points identified

Graph $g(x) = -\sqrt[3]{x}$. Reflect f about the *x*-axis. Replace each *y*-coordinate with its opposite.

The graph of $g(x) = -\sqrt[3]{x}$

Check Point 4   Use the graph of $f(x) = |x|$ to obtain the graph of $g(x) = -|x|$.

It is also possible to reflect graphs about the *y*-axis.

Reflection about the *y*-Axis

The graph of $y = f(-x)$ is the graph of $y = f(x)$ reflected about the *y*-axis.

For corresponding values of y, the *x*-coordinates of $y = f(-x)$ are the opposite of those of $y = f(x)$.

EXAMPLE 5   Reflection about the *y*-Axis

Use the graph of $f(x) = \sqrt{x}$ to obtain the graph of $h(x) = \sqrt{-x}$.

Solution   Compare the equations for $f(x) = \sqrt{x}$ and $h(x) = \sqrt{-x}$. The graph of h is a reflection about the *y*-axis of the graph of f because

$$h(x) = \sqrt{-x} = f(-x).$$

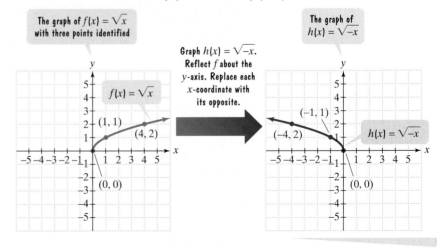

The graph of $f(x) = \sqrt{x}$ with three points identified

Graph $h(x) = \sqrt{-x}$. Reflect f about the *y*-axis. Replace each *x*-coordinate with its opposite.

The graph of $h(x) = \sqrt{-x}$

Check Point 5   Use the graph of $f(x) = \sqrt[3]{x}$ to obtain the graph of $h(x) = \sqrt[3]{-x}$.

⑤ Use vertical stretching and shrinking to graph functions.

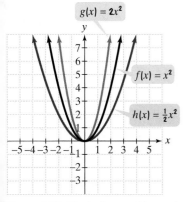

Figure 1.55 Vertically stretching and shrinking $f(x) = x^2$

Vertical Stretching and Shrinking

Morphing does much more than move an image horizontally, vertically, or about an axis. An object having one shape is transformed into a different shape. Horizontal shifts, vertical shifts, and reflections do not change the basic shape of a graph. Graphs remain rigid and proportionally the same when they undergo these transformations. How can we shrink and stretch graphs, thereby altering their basic shapes?

Look at the three graphs in Figure 1.55. The black graph in the middle is the graph of the standard quadratic function, $f(x) = x^2$. Now, look at the blue graph on the top. The equation of this graph is $g(x) = 2x^2$, or $g(x) = 2f(x)$. Thus, for each x, the y-coordinate of g is 2 times as large as the corresponding y-coordinate on the graph of f. The result is a narrower graph because the values of y are rising faster. We say that the graph of g is obtained by vertically *stretching* the graph of f. Now, look at the red graph on the bottom. The equation of this graph is $h(x) = \frac{1}{2}x^2$, or $h(x) = \frac{1}{2}f(x)$. Thus, for each x, the y-coordinate of h is one-half as large as the corresponding y-coordinate on the graph of f. The result is a wider graph because the values of y are rising more slowly. We say that the graph of h is obtained by vertically *shrinking* the graph of f.

These observations can be summarized as follows:

Vertically Stretching and Shrinking Graphs

Let f be a function and c a positive real number.
- If $c > 1$, the graph of $y = cf(x)$ is the graph of $y = f(x)$ vertically stretched by multiplying each of its y-coordinates by c.
- If $0 < c < 1$, the graph of $y = cf(x)$ is the graph of $y = f(x)$ vertically shrunk by multiplying each of its y-coordinates by c.

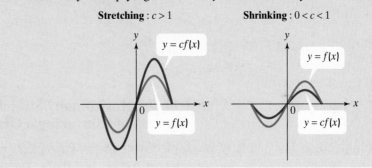

EXAMPLE 6 Vertically Shrinking a Graph

Use the graph of $f(x) = x^3$ to obtain the graph of $h(x) = \frac{1}{2}x^3$.

Solution The graph of $h(x) = \frac{1}{2}x^3$ is obtained by vertically shrinking the graph of $f(x) = x^3$.

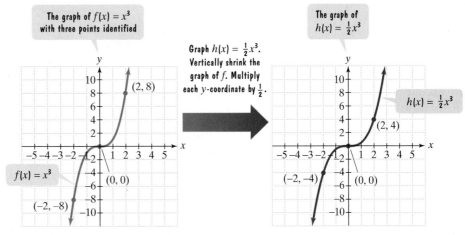

Check Point 6 Use the graph of $f(x) = |x|$ to obtain the graph of $g(x) = 2|x|$.

⑥ Use horizontal stretching and shrinking to graph functions.

Horizontal Stretching and Shrinking

It is also possible to horizontally stretch and shrink graphs.

Horizontally Stretching and Shrinking Graphs

Let f be a function and c a positive real number.

- If $c > 1$, the graph of $y = f(cx)$ is the graph of $y = f(x)$ horizontally shrunk by dividing each of its x-coordinates by c.

- If $0 < c < 1$, the graph of $y = f(cx)$ is the graph of $y = f(x)$ horizontally stretched by dividing each of its x-coordinates by c.

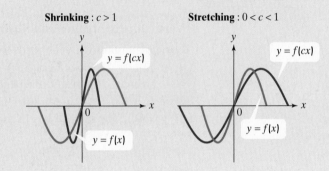

Shrinking : $c > 1$ **Stretching** : $0 < c < 1$

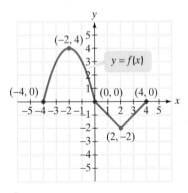

Figure 1.56

EXAMPLE 7 Horizontally Stretching and Shrinking a Graph

Use the graph of $y = f(x)$ in Figure 1.56 to obtain each of the following graphs:

a. $g(x) = f(2x)$ **b.** $h(x) = f\left(\frac{1}{2}x\right)$.

Solution

a. The graph of $g(x) = f(2x)$ is obtained by horizontally shrinking the graph of $y = f(x)$.

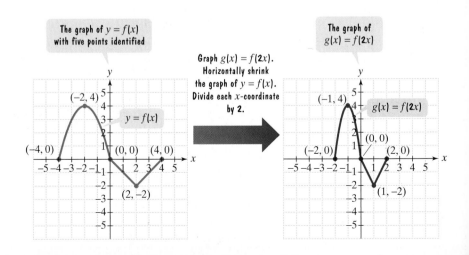

The graph of $y = f(x)$ with five points identified

Graph $g(x) = f(2x)$. Horizontally shrink the graph of $y = f(x)$. Divide each x-coordinate by 2.

The graph of $g(x) = f(2x)$

b. The graph of $h(x) = f\left(\frac{1}{2}x\right)$ is obtained by horizontally stretching the graph of $y = f(x)$.

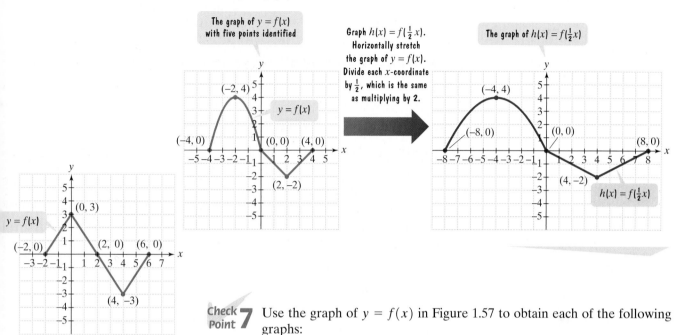

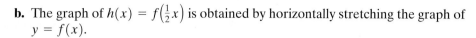

Figure 1.57

Check Point 7 Use the graph of $y = f(x)$ in Figure 1.57 to obtain each of the following graphs:

a. $g(x) = f(2x)$ **b.** $h(x) = f\left(\frac{1}{2}x\right)$.

7 Graph functions involving a sequence of transformations.

Sequences of Transformations

Table 1.5 summarizes the procedures for transforming the graph of $y = f(x)$.

Table 1.5   Summary of Transformations
In each case, c represents a positive real number.

To Graph:	Draw the Graph of f and:	Changes in the Equation of $y = f(x)$
Vertical shifts $y = f(x) + c$ $y = f(x) - c$	Raise the graph of f by c units. Lower the graph of f by c units.	c is added to $f(x)$. c is subtracted from $f(x)$.
Horizontal shifts $y = f(x + c)$ $y = f(x - c)$	Shift the graph of f to the left c units. Shift the graph of f to the right c units.	x is replaced with $x + c$. x is replaced with $x - c$.
Reflection about the x-axis $y = -f(x)$	Reflect the graph of f about the x-axis.	$f(x)$ is multiplied by -1.
Reflection about the y-axis $y = f(-x)$	Reflect the graph of f about the y-axis.	x is replaced with $-x$.
Vertical stretching or shrinking $y = cf(x), c > 1$ $y = cf(x), 0 < c < 1$	Multiply each y-coordinate of $y = f(x)$ by c, vertically stretching the graph of f. Multiply each y-coordinate of $y = f(x)$ by c, vertically shrinking the graph of f.	$f(x)$ is multiplied by $c, c > 1$. $f(x)$ is multiplied by $c, 0 < c < 1$.
Horizontal stretching or shrinking $y = f(cx), c > 1$ $y = f(cx), 0 < c < 1$	Divide each x-coordinate of $y = f(x)$ by c, horizontally shrinking the graph of f. Divide each x-coordinate of $y = f(x)$ by c, horizontally stretching the graph of f.	x is replaced with $cx, c > 1$. x is replaced with $cx, 0 < c < 1$.

A function involving more than one transformation can be graphed by per forming transformations in the following order:

1. Horizontal shifting   **2.** Stretching or shrinking

3. Reflecting   **4.** Vertical shifting

EXAMPLE 8   Graphing Using a Sequence of Transformations

Use the graph of $y = f(x)$ given in Figure 1.56 of Example 7 on page 202, and repeated below, to graph $y = -\frac{1}{2}f(x - 1) + 3$.

Solution   Our graphs will evolve in the following order:

1. Horizontal shifting: Graph $y = f(x - 1)$ by shifting the graph of $y = f(x)$ 1 unit to the right.

2. Shrinking: Graph $y = \frac{1}{2}f(x - 1)$ by shrinking the previous graph by a factor of $\frac{1}{2}$.

3. Reflecting: Graph $y = -\frac{1}{2}f(x - 1)$ by reflecting the previous graph about the x-axis.

4. Vertical shifting: Graph $y = -\frac{1}{2}f(x - 1) + 3$ by shifting the previous graph up 3 units.

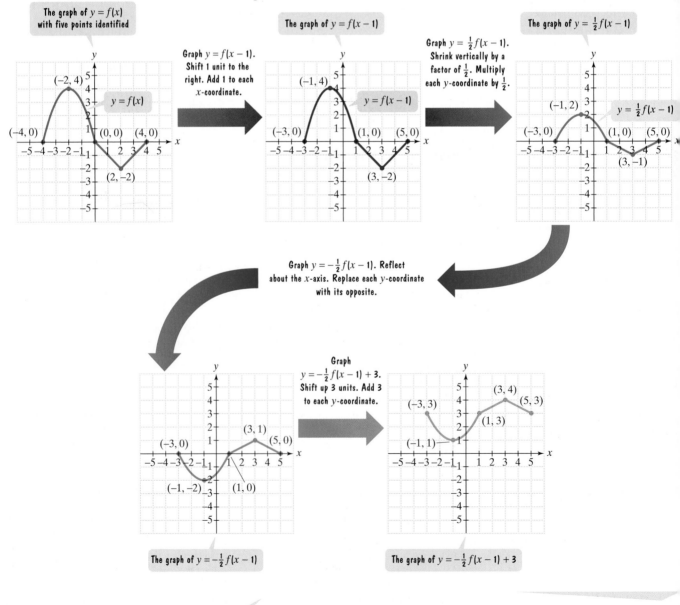

Check Point **8**   Use the graph of $y = f(x)$ given in Figure 1.57 of Check Point 7 on page 203 to graph $y = -\frac{1}{3}f(x + 1) - 2$.

EXAMPLE 9   Graphing Using a Sequence of Transformations

Use the graph of $f(x) = x^2$ to graph $g(x) = 2(x + 3)^2 - 1$.

Solution   Our graphs will evolve in the following order:

1. Horizontal shifting: Graph $y = (x + 3)^2$ by shifting the graph of $f(x) = x^2$ three units to the left.

2 Stretching: Graph $y = 2(x + 3)^2$ by stretching the previous graph by a factor of 2.

3. Vertical shifting: Graph $g(x) = 2(x + 3)^2 - 1$ by shifting the previous graph down 1 unit.

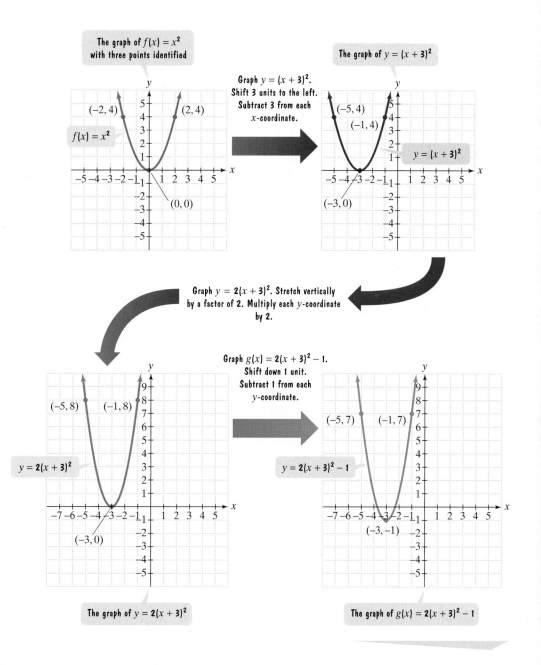

Check Point **9**   Use the graph of $f(x) = x^2$ to graph $g(x) = 2(x - 1)^2 + 3$.

EXERCISE SET 1.6

Practice Exercises

In Exercises 1–16, use the graph of $y = f(x)$ to graph each function g.

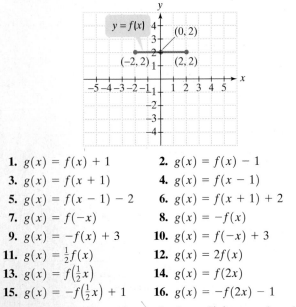

1. $g(x) = f(x) + 1$
2. $g(x) = f(x) - 1$
3. $g(x) = f(x + 1)$
4. $g(x) = f(x - 1)$
5. $g(x) = f(x - 1) - 2$
6. $g(x) = f(x + 1) + 2$
7. $g(x) = f(-x)$
8. $g(x) = -f(x)$
9. $g(x) = -f(x) + 3$
10. $g(x) = f(-x) + 3$
11. $g(x) = \frac{1}{2}f(x)$
12. $g(x) = 2f(x)$
13. $g(x) = f\left(\frac{1}{2}x\right)$
14. $g(x) = f(2x)$
15. $g(x) = -f\left(\frac{1}{2}x\right) + 1$
16. $g(x) = -f(2x) - 1$

In Exercises 17–32, use the graph of $y = f(x)$ to graph each function g.

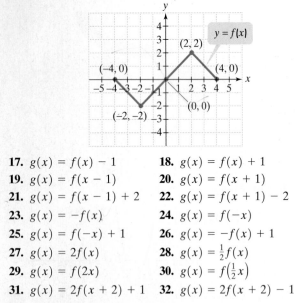

17. $g(x) = f(x) - 1$
18. $g(x) = f(x) + 1$
19. $g(x) = f(x - 1)$
20. $g(x) = f(x + 1)$
21. $g(x) = f(x - 1) + 2$
22. $g(x) = f(x + 1) - 2$
23. $g(x) = -f(x)$
24. $g(x) = f(-x)$
25. $g(x) = f(-x) + 1$
26. $g(x) = -f(x) + 1$
27. $g(x) = 2f(x)$
28. $g(x) = \frac{1}{2}f(x)$
29. $g(x) = f(2x)$
30. $g(x) = f\left(\frac{1}{2}x\right)$
31. $g(x) = 2f(x + 2) + 1$
32. $g(x) = 2f(x + 2) - 1$

In Exercises 33–44, use the graph of $y = f(x)$ to graph each function g.

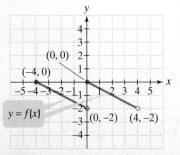

33. $g(x) = f(x) + 2$
34. $g(x) = f(x) - 2$
35. $g(x) = f(x + 2)$
36. $g(x) = f(x - 2)$
37. $g(x) = -f(x + 2)$
38. $g(x) = -f(x - 2)$
39. $g(x) = -\frac{1}{2}f(x + 2)$
40. $g(x) = -\frac{1}{2}f(x - 2)$
41. $g(x) = -\frac{1}{2}f(x + 2) - 2$
42. $g(x) = -\frac{1}{2}f(x - 2) + 2$
43. $g(x) = \frac{1}{2}f(2x)$
44. $g(x) = 2f\left(\frac{1}{2}x\right)$

In Exercises 45–52, use the graph of $y = f(x)$ to graph each function g.

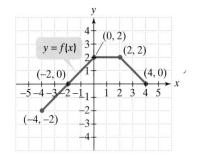

45. $g(x) = f(x - 1) - 1$
46. $g(x) = f(x + 1) + 1$
47. $g(x) = -f(x - 1) + 1$
48. $g(x) = -f(x + 1) - 1$
49. $g(x) = 2f\left(\frac{1}{2}x\right)$
50. $g(x) = \frac{1}{2}f(2x)$
51. $g(x) = \frac{1}{2}f(x + 1)$
52. $g(x) = 2f(x - 1)$

In Exercises 53–66, begin by graphing the standard quadratic function, $f(x) = x^2$. Then use transformations of this graph to graph the given function.

53. $g(x) = x^2 - 2$
54. $g(x) = x^2 - 1$
55. $g(x) = (x - 2)^2$
56. $g(x) = (x - 1)^2$
57. $h(x) = -(x - 2)^2$
58. $h(x) = -(x - 1)^2$
59. $h(x) = (x - 2)^2 + 1$
60. $h(x) = (x - 1)^2 + 2$
61. $g(x) = 2(x - 2)^2$
62. $g(x) = \frac{1}{2}(x - 1)^2$
63. $h(x) = 2(x - 2)^2 - 1$
64. $h(x) = \frac{1}{2}(x - 1)^2 - 1$
65. $h(x) = -2(x + 1)^2 + 1$
66. $h(x) = -2(x + 2)^2 + 1$

In Exercises 67–80, begin by graphing the square root function, $f(x) = \sqrt{x}$. Then use transformations of this graph to graph the given function.

67. $g(x) = \sqrt{x} + 2$
68. $g(x) = \sqrt{x} + 1$
69. $g(x) = \sqrt{x + 2}$
70. $g(x) = \sqrt{x + 1}$
71. $h(x) = -\sqrt{x + 2}$
72. $h(x) = -\sqrt{x + 1}$
73. $h(x) = \sqrt{-x + 2}$
74. $h(x) = \sqrt{-x + 1}$
75. $g(x) = \frac{1}{2}\sqrt{x + 2}$
76. $g(x) = 2\sqrt{x + 1}$
77. $h(x) = \sqrt{x + 2} - 2$
78. $h(x) = \sqrt{x + 1} - 1$
79. $g(x) = 2\sqrt{x + 2} - 2$
80. $g(x) = 2\sqrt{x + 1} - 1$

In Exercises 81–94, begin by graphing the absolute value function, $f(x) = |x|$. Then use transformations of this graph to graph the given function.

81. $g(x) = |x| + 4$
82. $g(x) = |x| + 3$
83. $g(x) = |x + 4|$
84. $g(x) = |x + 3|$
85. $h(x) = |x + 4| - 2$
86. $h(x) = |x + 3| - 2$
87. $h(x) = -|x + 4|$
88. $h(x) = -|x + 3|$

89. $g(x) = -|x + 4| + 1$ **90.** $g(x) = -|x + 4| + 2$

91. $h(x) = 2|x + 4|$ **92.** $h(x) = 2|x + 3|$

93. $g(x) = -2|x + 4| + 1$ **94.** $g(x) = -2|x + 3| + 2$

In Exercises 95–106, begin by graphing the standard cubic function, $f(x) = x^3$. Then use transformations of this graph to graph the given function.

95. $g(x) = x^3 - 3$ **96.** $g(x) = x^3 - 2$

97. $g(x) = (x - 3)^3$ **98.** $g(x) = (x - 2)^3$

99. $h(x) = -x^3$ **100.** $h(x) = -(x - 2)^3$

101. $h(x) = \frac{1}{2}x^3$ **102.** $h(x) = \frac{1}{4}x^3$

103. $r(x) = (x - 3)^3 + 2$ **104.** $r(x) = (x - 2)^3 + 1$

105. $h(x) = \frac{1}{2}(x - 3)^3 - 2$ **106.** $h(x) = \frac{1}{2}(x - 2)^3 - 1$

In Exercises 107–118, begin by graphing the cube root function, $f(x) = \sqrt[3]{x}$. Then use transformations of this graph to graph the given function.

107. $g(x) = \sqrt[3]{x} + 2$ **108.** $g(x) = \sqrt[3]{x} - 2$

109. $g(x) = \sqrt[3]{x + 2}$ **110.** $g(x) = \sqrt[3]{x - 2}$

111. $h(x) = \frac{1}{2}\sqrt[3]{x + 2}$ **112.** $h(x) = \frac{1}{2}\sqrt[3]{x - 2}$

113. $r(x) = \frac{1}{2}\sqrt[3]{x + 2} - 2$ **114.** $r(x) = \frac{1}{2}\sqrt[3]{x - 2} + 2$

115. $h(x) = -\sqrt[3]{x + 2}$ **116.** $h(x) = -\sqrt[3]{x - 2}$

117. $g(x) = \sqrt[3]{-x - 2}$ **118.** $g(x) = \sqrt[3]{-x + 2}$

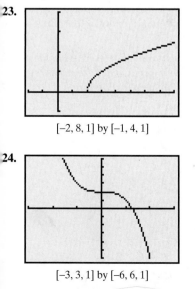

Practice Plus

In Exercises 119–122, use transformations of the graph of the greatest integer function, $f(x) = \mathrm{int}(x)$, to graph each function. (The graph of $f(x) = \mathrm{int}(x)$ is shown in Figure 1.32 on page 162.)

119. $g(x) = 2\,\mathrm{int}\,(x + 1)$ **120.** $g(x) = 3\,\mathrm{int}\,(x - 1)$

121. $h(x) = \mathrm{int}(-x) + 1$ **122.** $h(x) = \mathrm{int}(-x) - 1$

In Exercises 123–126, write a possible equation for the function whose graph is shown. Each graph shows a transformation of a common function.

123.

[−2, 8, 1] by [−1, 4, 1]

124.

[−3, 3, 1] by [−6, 6, 1]

125.

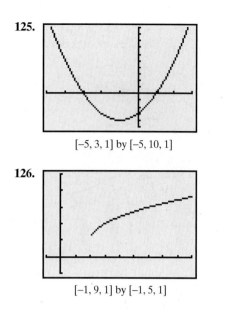

[−5, 3, 1] by [−5, 10, 1]

126.

[−1, 9, 1] by [−1, 5, 1]

Application Exercises

127. The function $f(x) = 2.9\sqrt{x} + 20.1$ models the median height, $f(x)$, in inches, of boys who are x months of age. The graph of f is shown.

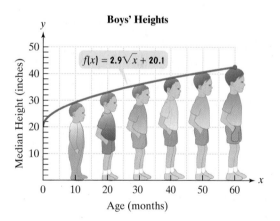

Boys' Heights

$f(x) = 2.9\sqrt{x} + 20.1$

Median Height (inches)

Age (months)

Source: Laura Walther Nathanson, *The Portable Pediatrician for Parents*

a. Describe how the graph can be obtained using transformations of the square root function $f(x) = \sqrt{x}$.

b. According to the model, what is the median height of boys who are 48 months, or four years, old? Use a calculator and round to the nearest tenth of an inch. The actual median height for boys at 48 months is 40.8 inches. How well does the model describe the actual height?

c. Use the model to find the average rate of change, in inches per month, between birth and 10 months. Round to the nearest tenth.

d. Use the model to find the average rate of change, in inches per month, between 50 and 60 months. Round to the nearest tenth. How does this compare with your answer in part (c)? How is this difference shown by the graph?

128. The function $f(x) = 3.1\sqrt{x} + 19$ models the median height, $f(x)$, in inches, of girls who are x months of age. The graph of f is shown.

Girls' Heights

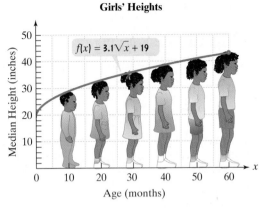

$f(x) = 3.1\sqrt{x} + 19$

Source: Laura Walther Nathanson, *The Portable Pediatrician for Parents*

a. Describe how the graph can be obtained using transformations of the square root function $f(x) = \sqrt{x}$.

b. According to the model, what is the median height of girls who are 48 months, or four years, old? Use a calculator and round to the nearest tenth of an inch. The actual median height for girls at 48 months is 40.2 inches. How well does the model describe the actual height?

c. Use the model to find the average rate of change, in inches per month, between birth and 10 months. Round to the nearest tenth.

d. Use the model to find the average rate of change, in inches per month, between 50 and 60 months. Round to the nearest tenth. How does this compare with your answer in part (c)? How is this difference shown by the graph?

Writing in Mathematics

129. What must be done to a function's equation so that its graph is shifted vertically upward?

130. What must be done to a function's equation so that its graph is shifted horizontally to the right?

131. What must be done to a function's equation so that its graph is reflected about the x-axis?

132. What must be done to a function's equation so that its graph is reflected about the y-axis?

133. What must be done to a function's equation so that its graph is stretched vertically?

134. What must be done to a function's equation so that its graph is shrunk horizontally?

Technology Exercises

135. a. Use a graphing utility to graph $f(x) = x^2 + 1$.

b. Graph $f(x) = x^2 + 1$, $g(x) = f(2x)$, $h(x) = f(3x)$, and $k(x) = f(4x)$ in the same viewing rectangle.

c. Describe the relationship among the graphs of f, g, h and k, with emphasis on different values of x for point on all four graphs that give the same y-coordinate.

d. Generalize by describing the relationship between the graph of f and the graph of g, where $g(x) = f(cx)$ for $c > 1$.

e. Try out your generalization by sketching the graphs of $f(cx)$ for $c = 1$, $c = 2$, $c = 3$, and $c = 4$ for a function of your choice.

136. a. Use a graphing utility to graph $f(x) = x^2 + 1$.

b. Graph $f(x) = x^2 + 1$, $g(x) = f\left(\frac{1}{2}x\right)$, and $h(x) = f\left(\frac{1}{4}x\right)$ in the same viewing rectangle.

c. Describe the relationship among the graphs of f, g, and h with emphasis on different values of x for points on all three graphs that give the same y-coordinate.

d. Generalize by describing the relationship between the graph of f and the graph of g, where $g(x) = f(cx)$ for $0 < c < 1$.

e. Try out your generalization by sketching the graphs of $f(cx)$ for $c = 1$, and $c = \frac{1}{2}$, and $c = \frac{1}{4}$ for a function of your choice.

Critical Thinking Exercises

137. Which one of the following is true?

a. If $f(x) = |x|$ and $g(x) = |x + 3| + 3$, then the graph of g is a translation of the graph of f three units to the right and three units upward.

b. If $f(x) = -\sqrt{x}$ and $g(x) = \sqrt{-x}$, then f and g have identical graphs.

c. If $f(x) = x^2$ and $g(x) = 5(x^2 - 2)$, then the graph of g can be obtained from the graph of f by stretching f five units followed by a downward shift of two units.

d. If $f(x) = x^3$ and $g(x) = -(x - 3)^3 - 4$, then the graph of g can be obtained from the graph of f by moving it three units to the right, reflecting about the x-axis, and then moving the resulting graph down four units.

In Exercises 138–141, functions f and g are graphed in the same rectangular coordinate system. If g is obtained from f through a sequence of transformations, find an equation for g.

138.

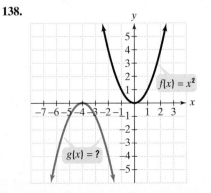

$f(x) = x^2$

$g(x) = ?$

139.

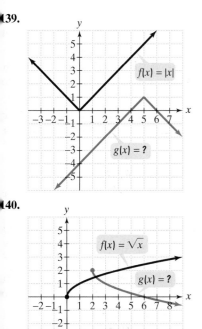

141.

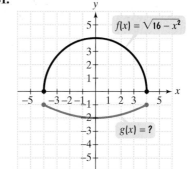

140.

For Exercises 142–145, assume that (a, b) is a point on the graph of f. What is the corresponding point on the graph of each of the following functions?

142. $y = f(-x)$ **143.** $y = 2f(x)$

144. $y = f(x - 3)$ **145.** $y = f(x) - 3$

SECTION 1.7 *Combinations of Functions; Composite Functions*

Objectives

❶ Find the domain of a function.

❷ Combine functions using the algebra of functions, specifying domains.

❸ Form composite functions.

❹ Determine domains for composite functions.

❺ Write functions as compositions.

America's big three automakers, GM, Ford, and Chrysler, know that consumers have always been willing to pay more for cool design and a hot car. In 2004, Detroit gave consumers that opportunity, unleashing 40 new cars or updated models. The Big Three's discovery of the automobile in 2004 might seem odd, but from 1990 through 2003, Detroit focused much of its energy and money on SUVs and trucks, which commanded high prices and high profits, and saw less competition from imports. The line graphs in Figure 1.58 show the number, in millions, of cars and SUVs sold by the Big Three from 1990 through 2003. In this section, we will look at these data from the perspective of functions. By considering total sales of cars and SUVs, you will see that functions can be combined using procedures that will remind you of combining algebraic expressions.

Figure 1.58 Sales by America's Big Three automakers

Study Tip

Throughout this section, we will be using the intersection of sets expressed in interval notation. Recall that the intersection of sets A and B, written $A \cap B$, is the set of elements common to both set A and set B. When sets A and B are in interval notation, to find the intersection, graph each interval and take the portion of the number line that the two graphs have in common. We will also be using notation involving the union of sets A and B, $A \cup B$, meaning the set of elements in A or in B or in both. For more detail, see Section P.1, pages 5–6, and Section P.9, pages 111–112.

1 Find the domain of a function.

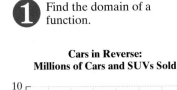

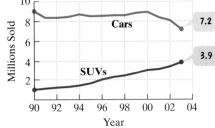

Cars in Reverse:
Millions of Cars and SUVs Sold

Figure 1.58 (repeated)

The Domain of a Function

We begin with two functions that model the data in Figure 1.58.

$$C(x) = -0.14x + 9 \qquad S(x) = 0.22x + 1$$

Car sales, $C(x)$, in millions, x years after 1990

SUV sales, $S(x)$, in millions, x years after 1990

How far beyond 1990 should we extend these models? The trend in Figure 1.58 shows decreasing car sales and increasing SUV sales. Because the three automakers focused on SUVs from 1990 through 2003 and launched a fleet of new cars in 2004, the trend shown by the data changed in 2004. Thus, we should not extend the models beyond 2003. Because x represents the number of years after 1990,

$$\text{Domain of } C = \{x \mid x = 0, 1, 2, 3, \ldots, 13\}$$

and

$$\text{Domain of } S = \{x \mid x = 0, 1, 2, 3, \ldots, 13\}.$$

Functions that model data often have their domains explicitly given with the function's equation. However, for most functions, only an equation is given and the domain is not specified. In cases like this, the domain of a function f is the largest set of real numbers for which the value of $f(x)$ is a real number. For example, consider the function

$$f(x) = \frac{1}{x - 3}.$$

Because division by 0 is undefined (and not a real number), the denominator, $x - 3$, cannot be 0. Thus, x cannot equal 3. The domain of the function consists of all real numbers other than 3, represented by

$$\text{Domain of } f = \{x \mid x \text{ is a real number and } x \neq 3\}.$$

Using interval notation,

$$\text{Domain of } f = (-\infty, 3) \cup (3, \infty).$$

All real numbers less than 3   or   All real numbers greater than 3

Now consider a function involving a square root:

$$g(x) = \sqrt{x - 3}.$$

Because only nonnegative numbers have square roots that are real numbers, the expression under the square root sign, $x - 3$, must be nonnegative. We can use inspection to see that $x - 3 \geq 0$ if $x \geq 3$. The domain of g consists of all real numbers that are greater than or equal to 3:

$$\text{Domain of } g = \{x \mid x \geq 3\} \text{ or } [3, \infty).$$

Finding a Function's Domain

If a function f does not model data or verbal conditions, its domain is the largest set of real numbers for which the value of $f(x)$ is a real number. Exclude from a function's domain real numbers that cause division by zero and real numbers that result in a square root of a negative number.

EXAMPLE 1   Finding the Domain of a Function

Find the domain of each function:

a. $f(x) = x^2 - 7x$ **b.** $g(x) = \dfrac{3x + 2}{x^2 - 2x - 3}$ **c.** $h(x) = \sqrt{3x + 12}$.

Solution

a. The function $f(x) = x^2 - 7x$ contains neither division nor a square root. For every real number, x, the algebraic expression $x^2 - 7x$ represents a real number. Thus, the domain of f is the set of all real numbers.

$$\text{Domain of } f = (-\infty, \infty)$$

b. The function $g(x) = \dfrac{3x + 2}{x^2 - 2x - 3}$ contains division. Because division by 0 is undefined, we must exclude from the domain the values of x that cause the denominator, $x^2 - 2x - 3$, to be 0. We can identify these values by setting $x^2 - 2x - 3$ equal to 0.

$$x^2 - 2x - 3 = 0 \qquad \text{Set the function's denominator equal to 0.}$$
$$(x + 1)(x - 3) = 0 \qquad \text{Factor.}$$
$$x + 1 = 0 \quad \text{or} \quad x - 3 = 0 \qquad \text{Set each factor equal to 0.}$$
$$x = -1 \qquad\qquad x = 3 \qquad \text{Solve the resulting equations.}$$

We must exclude -1 and 3 from the domain of g.

$$\text{Domain of } g = (-\infty, -1) \cup (-1, 3) \cup (3, \infty)$$

c. The function $h(x) = \sqrt{3x + 12}$ contains an even root. Because only nonnegative numbers have real square roots, the quantity under the radical sign, $3x + 12$, must be greater than or equal to 0.

$$3x + 12 \geq 0 \qquad \text{Set the function's radicand greater than} \\ \text{or equal to 0.}$$
$$3x \geq -12 \qquad \text{Subtract 12 from both sides.}$$
$$x \geq -4 \qquad \text{Divide both sides by 3. Division by a positive} \\ \text{number preserves the sense of the inequality.}$$

The domain of h consists of all real numbers greater than or equal to -4.

$$\text{Domain of } h = [-4, \infty)$$

The domain is highlighted on the x-axis in Figure 1.59.

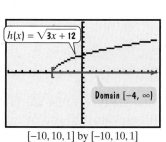

$h(x) = \sqrt{3x + 12}$

Domain $[-4, \infty)$

$[-10, 10, 1]$ by $[-10, 10, 1]$

Figure 1.59

Check Point **1**   Find the domain of each function:

a. $f(x) = x^2 + 3x - 17$ **b.** $g(x) = \dfrac{5x}{x^2 - 49}$

c. $h(x) = \sqrt{9x - 27}$.

The Algebra of Functions

We return to the functions that model millions of car and SUV sales from 1990 through 2003:

$$C(x) = -0.14x + 9 \qquad S(x) = 0.22x + 1.$$

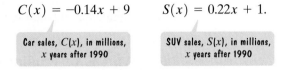

Car sales, $C(x)$, in millions, x years after 1990

SUV sales, $S(x)$, in millions, x years after 1990

How can we use these functions to find total sales of cars and SUVs in 2003? Because 2003 is 13 years after 1990 and 13 is in the domain of each function, we need to find the sum of two function values:

$$C(13) + S(13).$$

Here is how it's done:

$$C(13) = -0.14(13) + 9 = 7.18 \qquad S(13) = 0.22(13) + 1 = 3.86$$

Substitute 13 for x in $C(x) = -0.14x + 9$.

7.18 million cars were sold in 2003.

Substitute 13 for x in $S(x) = 0.22x + 1$.

3.86 million SUVs were sold in 2003.

$$C(13) + S(13) = 7.18 + 3.86 = 11.04.$$

Thus, a total of 11.04 million cars and SUVs were sold in 2003.

There is a second way that we can obtain this number. We can first add the functions C and S to obtain a new function, $C + S$. To do so, we add the terms to the right of the equal sign for $C(x)$ to the terms to the right of the equal sign for $S(x)$:

$$
\begin{aligned}
(C + S)(x) &= C(x) + S(x) \\
&= (-0.14x + 9) + (0.22x + 1) \qquad \text{Add terms for } C(x) \text{ and } S(x). \\
&= 0.08x + 10. \qquad\qquad\qquad\quad \text{Combine like terms.}
\end{aligned}
$$

Thus,

$$(C + S)(x) = 0.08x + 10.$$

Total car and SUV sales, in millions, x years after 1990

Do you see how we can use this new function to find total car and SUV sales in 2003? Substitute 13 for x in the equation for $C + S$:

$$(C + S)(13) = 0.08(13) + 10 = 11.04.$$

Substitute 13 for x in $(C + S)(x) = 0.08x + 10$.

As we found above, a total of 11.04 million cars and SUVs were sold in 2003.

The domain of the new function, $C + S$, consists of the numbers x that are in the domain of C **and** in the domain of S. If D_c represents the domain of C and D_s represents the domain of S, the domain of $C + S$ is $D_c \cap D_s$. Because both C and S model data from 1990 through 2003,

$$\text{Domain of } C + S = \{0, 1, 2, 3, \ldots, 13\}.$$

The function that models total car and SUV sales illustrates that functions can be added algebraically. We can also combine functions using subtraction, multiplication, and division by performing operations with the algebraic expressions that appear on the right side of the equations. The domain for each of these functions consists of all real numbers that are common to the domains of the functions being combined. Furthermore, when combining functions using division, values that make the divisor zero must be excluded from the domain.

The following definitions summarize our discussion:

Combine functions using the algebra of functions, specifying domains.

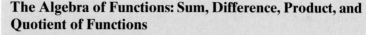

The Algebra of Functions: Sum, Difference, Product, and Quotient of Functions

Let f and g be two functions. The **sum** $f + g$, the **difference** $f - g$, the **product** fg, and the **quotient** $\dfrac{f}{g}$ are functions whose domains are the set of all real numbers common to the domains of f and g $(D_f \cap D_g)$, defined as follows:

1. Sum: $(f + g)(x) = f(x) + g(x)$
2. Difference: $(f - g)(x) = f(x) - g(x)$
3. Product: $(fg)(x) = f(x) \cdot g(x)$
4. Quotient: $\left(\dfrac{f}{g}\right)(x) = \dfrac{f(x)}{g(x)}$, provided $g(x) \neq 0$.

EXAMPLE 2 Combining Functions

Let $f(x) = 2x - 1$ and $g(x) = x^2 + x - 2$. Find each of the following functions:

a. $(f + g)(x)$ **b.** $(f - g)(x)$ **c.** $(fg)(x)$ **d.** $\left(\dfrac{f}{g}\right)(x)$.

Determine the domain for each function.

Solution

a. $(f + g)(x) = f(x) + g(x)$ This is the definition of the sum $f + g$.

$\qquad\qquad\quad = (2x - 1) + (x^2 + x - 2)$ Substitute the given functions.

$\qquad\qquad\quad = x^2 + 3x - 3$ Remove parentheses and combine like terms.

b. $(f - g)(x) = f(x) - g(x)$ This is the definition of the difference $f - g$.

$\qquad\qquad\quad = (2x - 1) - (x^2 + x - 2)$ Substitute the given functions.

$\qquad\qquad\quad = 2x - 1 - x^2 - x + 2$ Remove parentheses and change the sign of each term in the second set of parentheses.

$\qquad\qquad\quad = -x^2 + x + 1$ Combine like terms and arrange terms in descending powers of x.

c. $(fg)(x) = f(x) \cdot g(x)$ This is the definition of the product fg.

$\qquad\qquad = (2x - 1)(x^2 + x - 2)$ Substitute the given functions.

$\qquad\qquad = 2x(x^2 + x - 2) - 1(x^2 + x - 2)$ Multiply each term in the second factor by 2x and -1, respectively.

$\qquad\qquad = 2x^3 + 2x^2 - 4x - x^2 - x + 2$ Use the distributive property.

$\qquad\qquad = 2x^3 + (2x^2 - x^2) + (-4x - x) + 2$ Rearrange terms so that like terms are adjacent.

$\qquad\qquad = 2x^3 + x^2 - 5x + 2$ Combine like terms.

d. $\left(\dfrac{f}{g}\right)(x) = \dfrac{f(x)}{g(x)}$ This is the definition of the quotient $\dfrac{f}{g}$.

$\qquad\qquad = \dfrac{2x - 1}{x^2 + x - 2}$ Substitute the given functions. This rational expression cannot be simplified.

Study Tip

If the function $\frac{f}{g}$ can be simplified, determine the domain *before* simplifying.

Example:

$$f(x) = x^2 - 4 \text{ and}$$
$$g(x) = x - 2$$

$$\left(\frac{f}{g}\right)(x) = \frac{x^2 - 4}{x - 2}$$

$x \neq 2$. The domain of $\frac{f}{g}$ is $(-\infty, 2) \cup (2, \infty)$.

$$= \frac{\overset{1}{\cancel{(x+2)}}(x-2)}{\underset{1}{\cancel{(x-2)}}} = x + 2$$

Because the equations for f and g do not involve division or contain even roots, the domain of both f and g is the set of all real numbers. Thus, the domain of $f + g$, $f - g$, and fg is the set of all real numbers, $(-\infty, \infty)$.

The function $\frac{f}{g}$ contains division. We must exclude from its domain values of x that cause the denominator, $x^2 + x - 2$, to be 0. Let's identify these values.

$$x^2 + x - 2 = 0 \qquad \text{Set the denominator of } \tfrac{f}{g} \text{ equal to 0.}$$
$$(x + 2)(x - 1) = 0 \qquad \text{Factor.}$$
$$x + 2 = 0 \quad \text{or} \quad x - 1 = 0 \qquad \text{Set each factor equal to 0.}$$
$$x = -2 \qquad\qquad x = 1 \qquad \text{Solve the resulting equations.}$$

We must exclude -2 and 1 from the domain of $\frac{f}{g}$.

$$\text{Domain of } \frac{f}{g} = (-\infty, -2) \cup (-2, 1) \cup (1, \infty)$$

Check Point 2 Let $f(x) = x - 5$ and $g(x) = x^2 - 1$. Find each of the following functions:

a. $(f + g)(x)$ **b.** $(f - g)(x)$ **c.** $(fg)(x)$ **d.** $\left(\dfrac{f}{g}\right)(x)$.

Determine the domain for each function.

EXAMPLE 3 Adding Functions and Determining the Domain

Let $f(x) = \sqrt{x + 3}$ and $g(x) = \sqrt{x - 2}$. Find each of the following:

a. $(f + g)(x)$ **b.** the domain of $f + g$.

Solution

a. $(f + g)(x) = f(x) + g(x) = \sqrt{x + 3} + \sqrt{x - 2}$

b. The domain of $f + g$ is the set of all real numbers that are common to the domain of f and the domain of g. Thus, we must find the domains of f and before finding their intersection. We will do so for f first.

Note that $f(x) = \sqrt{x + 3}$ is a function involving the square root $x + 3$. Because the square root of a negative quantity is not a real number, the value of $x + 3$ must be nonnegative. Thus, the domain of f is all x such that $x + 3 \geq 0$. Equivalently, the domain is $\{x \mid x \geq -3\}$, or $[-3, \infty)$.

Likewise, $g(x) = \sqrt{x - 2}$ is also a square root function. Because the square root of a negative quantity is not a real number, the value of $x - $ must be nonnegative. Thus, the domain of g is all x such that $x - 2 \geq 0$. Equivalently, the domain is $\{x \mid x \geq 2\}$, or $[2, \infty)$.

Now, we can use a number line to determine $D_f \cap D_g$, the domain of $f + g$. Figure 1.60 shows the domain of f in blue and the domain of g in red. Can you see that all real numbers greater than or equal to 2 are common to both domains? This is shown in purple on the number line. Thus, the domain of $f + g$ is $[2, \infty)$.

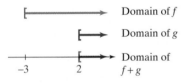

Domain of f
Domain of g
Domain of $f + g$

Figure 1.60 Finding the domain of the sum $f + g$

Technology

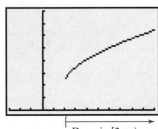

Domain $[2, \infty)$

The graph on the left is the graph of

$$y = \sqrt{x + 3} + \sqrt{x - 2}$$

in a $[-3, 10, 1]$ by $[0, 8, 1]$ viewing rectangle. The graph reveals what we discovered algebraically in Example 3(b). The domain of this function is $[2, \infty)$.

Check Point 3   Let $f(x) = \sqrt{x - 3}$ and $g(x) = \sqrt{x + 1}$. Find each of the following:

  a. $(f + g)(x)$ **b.** the domain of $f + g$.

③ Form composite functions.

Composite Functions

There is another way of combining two functions. To help understand this new combination, suppose that your local computer store is having a sale. The models that are on sale cost either $300 less than the regular price or 85% of the regular price. If x represents the computer's regular price, the discounts can be described with the following functions:

$$f(x) = x - 300 \qquad g(x) = 0.85x.$$

| The computer is on sale for $300 less than its regular price. | The computer is on sale for 85% of its regular price. |

At the store, you bargain with the salesperson. Eventually, she makes an offer you can't refuse. The sale price will be 85% of the regular price followed by a $300 reduction:

$$0.85x - 300.$$

| 85% of the regular price | followed by a $300 reduction |

In terms of the functions f and g, this offer can be obtained by taking the output of $g(x) = 0.85x$, namely $0.85x$, and using it as the input of f:

$$f(x) = x - 300$$

Replace x with 0.85x, the output of g(x) = 0.85x.

$$f(0.85x) = 0.85x - 300.$$

Because $0.85x$ is $g(x)$, we can write this last equation as

$$f(g(x)) = 0.85x - 300.$$

We read this equation as "f of g of x is equal to $0.85x - 300$." We call $f(g(x))$ the **composition of the function f with g**, or a **composite function**. This composite function is written $f \circ g$. Thus,

$$(f \circ g)(x) = f(g(x)) = 0.85x - 300.$$

This can be read "f of g of x" or "f composed with g of x."

Like all functions, we can evaluate $f \circ g$ for a specified value of x in the function's domain. For example, here's how to find the value of the composite function describing the offer you cannot refuse at 1400:

$$(f \circ g)(x) = 0.85x - 300$$

Replace x with 1400.

$$(f \circ g)(1400) = 0.85(1400) - 300 = 1190 - 300 = 890.$$

Because $(f \circ g)(1400) = 890$, this means that a computer that regularly sells fo $1400 is on sale for $890 subject to both discounts. We can use a partial table o coordinates for each of the discount functions, g and f, to numerically verify thi result.

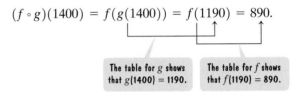

	Computer's regular price	85% of the regular price		85% of the regular price	$300 reduction
	x	$g(x) = 0.85x$		x	$f(x) = x - 300$
	1200	1020		1020	720
	1300	1105		1105	805
	1400	1190		1190	890

Using these tables, we can find $(f \circ g)(1400)$:

$$(f \circ g)(1400) = f(g(1400)) = f(1190) = 890.$$

The table for g shows that $g(1400) = 1190$.

The table for f shows that $f(1190) = 890$.

This verifies that a computer that regularly sells for $1400 is on sale for $890 subjec to both discounts.

Before you run out to buy a computer, let's generalize our discussion of the computer's double discount and define the composition of any two functions.

The Composition of Functions

The **composition of the function f with g** is denoted by $f \circ g$ and is defined by the equation

$$(f \circ g)(x) = f(g(x)).$$

The **domain of the composite function $f \circ g$** is the set of all x such that

1. x is in the domain of g and
2. $g(x)$ is in the domain of f.

The composition of f with g, $f \circ g$, is pictured as a machine with inputs an outputs in Figure 1.61. The diagram indicates that the output of g, or $g(x)$, become the input for "machine" f. If $g(x)$ is not in the domain of f, it cannot be input int machine f, and so $g(x)$ must be discarded.

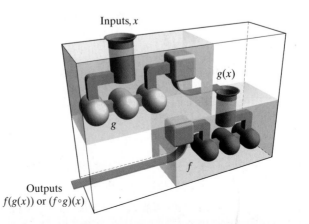

Inputs, x

$g(x)$

g

f

Outputs
$f(g(x))$ or $(f \circ g)(x)$

Figure 1.61 Inputting one function into a second function

EXAMPLE 4   Forming Composite Functions

Given $f(x) = 3x - 4$ and $g(x) = x^2 - 2x + 6$, find each of the following composite functions:

a. $(f \circ g)(x)$ **b.** $(g \circ f)(x)$.

Solution

a. We begin with $(f \circ g)(x)$, the composition of f with g. Because $(f \circ g)(x)$ means $f(g(x))$, we must replace each occurrence of x in the equation for f with $g(x)$.

$$f(x) = 3x - 4 \qquad \text{This is the given equation for f.}$$

Replace x with $g(x)$.

$$(f \circ g)(x) = f(g(x)) = 3g(x) - 4$$

$$= 3(x^2 - 2x + 6) - 4 \qquad \text{Because } g(x) = x^2 - 2x + 6, \text{ replace } g(x) \text{ with } x^2 - 2x + 6.$$

$$= 3x^2 - 6x + 18 - 4 \qquad \text{Use the distributive property.}$$

$$= 3x^2 - 6x + 14 \qquad \text{Simplify.}$$

Thus, $(f \circ g)(x) = 3x^2 - 6x + 14$.

b. Next, we find $(g \circ f)(x)$, the composition of g with f. Because $(g \circ f)(x)$ means $g(f(x))$, we must replace each occurrence of x in the equation for g with $f(x)$.

$$g(x) = x^2 - 2x + 6 \qquad \text{This is the equation for g.}$$

Replace x with $f(x)$.

$$(g \circ f)(x) = g(f(x)) = (f(x))^2 - 2f(x) + 6$$

$$= (3x - 4)^2 - 2(3x - 4) + 6 \qquad \text{Because } f(x) = 3x - 4, \text{ replace } f(x) \text{ with } 3x - 4.$$

$$= 9x^2 - 24x + 16 - 6x + 8 + 6 \qquad \text{Use } (A - B)^2 = A^2 - 2AB + B^2 \text{ to square } 3x - 4.$$

$$= 9x^2 - 30x + 30 \qquad \text{Simplify:} -24x - 6x = -30x \text{ and } 16 + 8 + 6 = 30.$$

Thus, $(g \circ f)(x) = 9x^2 - 30x + 30$. **Notice that $(f \circ g)(x)$ is not the same function as $(g \circ f)(x)$.**

Check Point 4 Given $f(x) = 5x + 6$ and $g(x) = 2x^2 - x - 1$, find each of the following composite functions:

a. $(f \circ g)(x)$ **b.** $(g \circ f)(x)$.

Determine domains for composite functions.

We need to be careful in determining the domain for a composite function.

Excluding Values from the Domain of $(f \circ g)(x) = f(g(x))$

The following values must be excluded from the input x:

- If x is not in the domain of g, it must not be in the domain of $f \circ g$.
- Any x for which $g(x)$ is not in the domain of f must not be in the domain of $f \circ g$.

EXAMPLE 5   Forming a Composite Function and Finding Its Domain

Given $f(x) = \dfrac{2}{x-1}$ and $g(x) = \dfrac{3}{x}$, find each of the following:

a. $(f \circ g)(x)$ **b.** the domain of $f \circ g$.

Solution

a. Because $(f \circ g)(x)$ means $f(g(x))$, we must replace x in $f(x) = \dfrac{2}{x-1}$ with $g(x)$.

$$(f \circ g)(x) = f(g(x)) = \frac{2}{g(x) - 1} = \frac{2}{\frac{3}{x} - 1} = \frac{2}{\frac{3}{x} - 1} \cdot \frac{x}{x} = \frac{2x}{3 - x}$$

$g(x) = \frac{3}{x}$

Simplify the complex fraction by multiplying by $\frac{x}{x}$, or 1.

Study Tip

The procedure for simplifying complex fractions can be found in Section P.6, pages 70–71.

Thus, $(f \circ g)(x) = \dfrac{2x}{3 - x}$.

b. We determine values to exclude from the domain of $(f \circ g)(x)$ in two steps

Rules for Excluding Numbers from the Domain of $(f \circ g)(x) = f(g(x))$	Applying the Rules to $f(x) = \dfrac{2}{x-1}$ and $g(x) = \dfrac{3}{x}$
If x is not in the domain of g, it must not be in the domain of $f \circ g$.	Because $g(x) = \dfrac{3}{x}$, 0 is not in the domain of g. Thus, 0 must be excluded from the domain of $f \circ g$.
Any x for which $g(x)$ is not in the domain of f must not be in the domain of $f \circ g$.	Because $f(g(x)) = \dfrac{2}{g(x) - 1}$, we must exclude from the domain of $f \circ g$ any x for which $g(x) = 1$. $\dfrac{3}{x} = 1$ Set $g(x)$ equal to 1. $3 = x$ Multiply both sides by x. 3 must be excluded from the domain of $f \circ g$.

We see that 0 and 3 must be excluded from the domain of $f \circ g$. The domain of $f \circ g$ is

$$(-\infty, 0) \cup (0, 3) \cup (3, \infty).$$

Check Point 5 Given $f(x) = \dfrac{4}{x + 2}$ and $g(x) = \dfrac{1}{x}$, find each of the following:

a. $(f \circ g)(x)$ **b.** the domain of $f \circ g$.

⑤ Write functions as compositions.

Decomposing Functions

When you form a composite function, you "compose" two functions to form a new function. It is also possible to reverse this process. That is, you can "decompose" a given function and express it as a composition of two functions. Although there is more than one way to do this, there is often a "natural" selection that comes to mind first. For example, consider the function h defined by

$$h(x) = (3x^2 - 4x + 1)^5.$$

The function h takes $3x^2 - 4x + 1$ and raises it to the power 5. A natural way to write h as a composition of two functions is to raise the function $g(x) = 3x^2 - 4x + 1$ to the power 5. Thus, if we let

$$f(x) = x^5 \text{ and } g(x) = 3x^2 - 4x + 1, \text{ then}$$
$$(f \circ g)(x) = f(g(x)) = f(3x^2 - 4x + 1) = (3x^2 - 4x + 1)^5.$$

EXAMPLE 6   Writing a Function as a Composition

Express $h(x)$ as a composition of two functions:

$$h(x) = \sqrt[3]{x^2 + 1}.$$

Study Tip

Suppose the form of function h is $h(x) = (\text{algebraic expression})^{\text{power}}$. Function h can be expressed as a composition, $f \circ g$, using

$f(x) = x^{\text{power}}$
$g(x) = \text{algebraic expression}$.

Solution   The function h takes $x^2 + 1$ and takes its cube root. A natural way to write h as a composition of two functions is to take the cube root of the function $g(x) = x^2 + 1$. Thus, we let

$$f(x) = \sqrt[3]{x} \text{ and } g(x) = x^2 + 1.$$

We can check this composition by finding $(f \circ g)(x)$. This should give the original function, namely $h(x) = \sqrt[3]{x^2 + 1}$.

$$(f \circ g)(x) = f(g(x)) = f(x^2 + 1) = \sqrt[3]{x^2 + 1} = h(x)$$

Check Point **6**   Express $h(x)$ as a composition of two functions:

$$h(x) = \sqrt{x^2 + 5}.$$

EXERCISE SET 1.7

Practice Exercises

In Exercises 1–30, find the domain of each function.

1. $f(x) = 3(x - 4)$

2. $f(x) = 2(x + 5)$

3. $g(x) = \dfrac{3}{x - 4}$

4. $g(x) = \dfrac{2}{x + 5}$

5. $f(x) = x^2 - 2x - 15$

6. $f(x) = x^2 + x - 12$

7. $g(x) = \dfrac{3}{x^2 - 2x - 15}$

8. $g(x) = \dfrac{2}{x^2 + x - 12}$

9. $f(x) = \dfrac{1}{x + 7} + \dfrac{3}{x - 9}$

10. $f(x) = \dfrac{1}{x + 8} + \dfrac{3}{x - 10}$

11. $g(x) = \dfrac{1}{x^2 + 1} - \dfrac{1}{x^2 - 1}$

12. $g(x) = \dfrac{1}{x^2 + 4} - \dfrac{1}{x^2 - 4}$

13. $h(x) = \dfrac{4}{\dfrac{3}{x} - 1}$

14. $h(x) = \dfrac{5}{\dfrac{4}{x} - 1}$

15. $f(x) = \dfrac{1}{\dfrac{4}{x - 1} - 2}$

16. $f(x) = \dfrac{1}{\dfrac{4}{x - 2} - 3}$

17. $f(x) = \sqrt{x - 3}$

18. $f(x) = \sqrt{x + 2}$

19. $g(x) = \dfrac{1}{\sqrt{x - 3}}$

20. $g(x) = \dfrac{1}{\sqrt{x + 2}}$

21. $g(x) = \sqrt{5x + 35}$

22. $g(x) = \sqrt{7x - 70}$

23. $f(x) = \sqrt{24 - 2x}$

24. $f(x) = \sqrt{84 - 6x}$

25. $h(x) = \sqrt{x - 2} + \sqrt{x + 3}$

26. $h(x) = \sqrt{x - 3} + \sqrt{x + 4}$

27. $g(x) = \dfrac{\sqrt{x - 2}}{x - 5}$

28. $g(x) = \dfrac{\sqrt{x - 3}}{x - 6}$

29. $f(x) = \dfrac{2x + 7}{x^3 - 5x^2 - 4x + 20}$

30. $f(x) = \dfrac{7x + 2}{x^3 - 2x^2 - 9x + 18}$

In Exercises 31–48, find $f + g$, $f - g$, fg, and $\frac{f}{g}$. Determine the domain for each function.

31. $f(x) = 2x + 3$, $g(x) = x - 1$

32. $f(x) = 3x - 4$, $g(x) = x + 2$

33. $f(x) = x - 5$, $g(x) = 3x^2$

34. $f(x) = x - 6$, $g(x) = 5x^2$

35. $f(x) = 2x^2 - x - 3$, $g(x) = x + 1$

36. $f(x) = 6x^2 - x - 1$, $g(x) = x - 1$

37. $f(x) = 3 - x^2$, $g(x) = x^2 + 2x - 15$

38. $f(x) = 5 - x^2$, $g(x) = x^2 + 4x - 12$

39. $f(x) = \sqrt{x}$, $g(x) = x - 4$

40. $f(x) = \sqrt{x}$, $g(x) = x - 5$

41. $f(x) = 2 + \dfrac{1}{x}$, $g(x) = \dfrac{1}{x}$

42. $f(x) = 6 - \dfrac{1}{x}$, $g(x) = \dfrac{1}{x}$

43. $f(x) = \dfrac{5x + 1}{x^2 - 9}$, $g(x) = \dfrac{4x - 2}{x^2 - 9}$

44. $f(x) = \dfrac{3x + 1}{x^2 - 25}$, $g(x) = \dfrac{2x - 4}{x^2 - 25}$

45. $f(x) = \sqrt{x + 4}$, $g(x) = \sqrt{x - 1}$

46. $f(x) = \sqrt{x + 6}$, $g(x) = \sqrt{x - 3}$

47. $f(x) = \sqrt{x - 2}$, $g(x) = \sqrt{2 - x}$

48. $f(x) = \sqrt{x - 5}$, $g(x) = \sqrt{5 - x}$

In Exercises 49–64, find

   a. $(f \circ g)(x)$;   **b.** $(g \circ f)(x)$;   **c.** $(f \circ g)(2)$.

49. $f(x) = 2x$, $g(x) = x + 7$

50. $f(x) = 3x$, $g(x) = x - 5$

51. $f(x) = x + 4$, $g(x) = 2x + 1$

52. $f(x) = 5x + 2$, $g(x) = 3x - 4$

53. $f(x) = 4x - 3$, $g(x) = 5x^2 - 2$

54. $f(x) = 7x + 1$, $g(x) = 2x^2 - 9$

55. $f(x) = x^2 + 2$, $g(x) = x^2 - 2$

56. $f(x) = x^2 + 1$, $g(x) = x^2 - 3$

57. $f(x) = 4 - x$, $g(x) = 2x^2 + x + 5$

58. $f(x) = 5x - 2$, $g(x) = -x^2 + 4x - 1$

59. $f(x) = \sqrt{x}$, $g(x) = x - 1$

60. $f(x) = \sqrt{x}$, $g(x) = x + 2$

61. $f(x) = 2x - 3$, $g(x) = \dfrac{x + 3}{2}$

62. $f(x) = 6x - 3$, $g(x) = \dfrac{x + 3}{6}$

63. $f(x) = \dfrac{1}{x}$, $g(x) = \dfrac{1}{x}$

64. $f(x) = \dfrac{2}{x}$, $g(x) = \dfrac{2}{x}$

In Exercises 65–72, find

   a. $(f \circ g)(x)$;   **b.** *the domain of $f \circ g$.*

65. $f(x) = \dfrac{2}{x + 3}$, $g(x) = \dfrac{1}{x}$

66. $f(x) = \dfrac{5}{x + 4}$, $g(x) = \dfrac{1}{x}$

67. $f(x) = \dfrac{x}{x + 1}$, $g(x) = \dfrac{4}{x}$

68. $f(x) = \dfrac{x}{x + 5}$, $g(x) = \dfrac{6}{x}$

69. $f(x) = \sqrt{x}$, $g(x) = x - 2$

70. $f(x) = \sqrt{x}$, $g(x) = x - 3$

71. $f(x) = x^2 + 4$, $g(x) = \sqrt{1 - x}$

72. $f(x) = x^2 + 1$, $g(x) = \sqrt{2 - x}$

In Exercises 73–80, express the given function h as a composition of two functions f and g so that $h(x) = (f \circ g)(x)$.

73. $h(x) = (3x - 1)^4$ **74.** $h(x) = (2x - 5)^3$

75. $h(x) = \sqrt[3]{x^2 - 9}$ **76.** $h(x) = \sqrt{5x^2 + 3}$

77. $h(x) = |2x - 5|$ **78.** $h(x) = |3x - 4|$

79. $h(x) = \dfrac{1}{2x - 3}$ **80.** $h(x) = \dfrac{1}{4x + 5}$

Practice Plus

Use the graphs of f and g to solve Exercises 81–88.

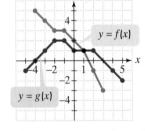

81. Find $(f + g)(-3)$.   **82.** Find $(g - f)(-2)$.

83. Find $(fg)(2)$.   **84.** Find $\left(\dfrac{g}{f}\right)(3)$.

85. Find the domain of $f + g$. **86.** Find the domain of $\dfrac{f}{g}$.

87. Graph $f + g$.   **88.** Graph $f - g$.

In Exercises 89–92, use the graphs of f and g to evaluate each composite function.

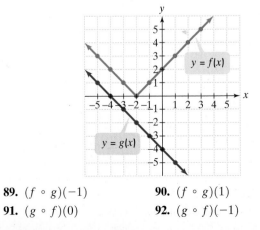

89. $(f \circ g)(-1)$ **90.** $(f \circ g)(1)$

91. $(g \circ f)(0)$ **92.** $(g \circ f)(-1)$

In Exercises 93–94, find all values of x satisfying the given conditions.

93. $f(x) = 2x - 5, g(x) = x^2 - 3x + 8$, and $(f \circ g)(x) = 7$.

94. $f(x) = 1 - 2x, g(x) = 3x^2 + x - 1$, and $(f \circ g)(x) = -5$.

Application Exercises

The table shows the total number of births and the total number of deaths in the United States from 1995 through 2003.

Births and Deaths in the U.S.

Year	Births	Deaths
1995	3,899,589	2,312,132
1996	3,891,494	2,314,690
1997	3,880,894	2,314,245
1998	3,941,553	2,337,256
1999	3,959,417	2,391,399
2000	4,058,814	2,403,351
2001	4,025,933	2,416,425
2002	4,022,000	2,436,000
2003	4,093,000	2,423,000

Source: Department of Health and Human Services

The data can be modeled by the following functions:

Number of births → $B(x) = 26,208x + 3,869,910$

Number of deaths → $D(x) = 17,964x + 2,300,198$.

In each function, x represents the number of years after 1995. Assume that the functions apply only to the years shown in the table. Use these functions to solve Exercises 95–98.

95. Find the domain of B.

96. Find the domain of D.

97. a. Find $(B - D)(x)$. What does this function represent?

b. Use the function in part (a) to find $(B - D)(8)$. What does this mean in terms of the U.S. population and to which year does this apply?

c. Use the data shown in the table to find $(B - D)(8)$. How well does the difference of functions used in part (b) model this number?

98. a. Find $(B - D)(x)$. What does this function represent?

b. Use the function in part (a) to find $(B - D)(6)$. What does this mean in terms of the U.S. population and to which year does this apply?

c. Use the data shown in the table to find $(B - D)(6)$. How well does the difference of functions used in part (b) model this number?

Consider the following functions:

$f(x) = $ *population of the world's more developed regions in year x*

$g(x) = $ *population of the world's less developed regions in year x*

$h(x) = $ *total world population in year x.*

Use these functions and the graphs shown to answer Exercises 99–102.

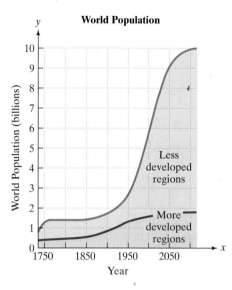

World Population

Source: Population Reference Bureau

99. What does the function $f + g$ represent?

100. What does the function $h - g$ represent?

101. Use the graph to estimate $(f + g)(2000)$.

102. Use the graph to estimate $(h - g)(2000)$.

103. A company that sells radios has yearly fixed costs of $600,000. It costs the company $45 to produce each radio. Each radio will sell for $65. The company's costs and revenue are modeled by the following functions:

$C(x) = 600,000 + 45x$ This function models the company's costs.

$R(x) = 65x$.   This function models the company's revenue.

Find and interpret $(R - C)(20,000)$, $(R - C)(30,000)$, and $(R - C)(40,000)$.

104. A department store has two locations in a city. From 2000 through 2004, the profits for each of the store's two branches are modeled by the functions $f(x) = -0.44x + 13.62$ and $g(x) = 0.51x + 11.14$. In each model, x represents the number of years after 2000, and f and g represent the profit, in millions of dollars.

a. What is the slope of f? Describe what this means.

b. What is the slope of g? Describe what this means.

c. Find $f + g$. What is the slope of this function? What does this mean?

105. The regular price of a computer is x dollars. Let $f(x) = x - 400$ and $g(x) = 0.75x$.

a. Describe what the functions f and g model in terms of the price of the computer.

b. Find $(f \circ g)(x)$ and describe what this models in terms of the price of the computer.

c. Repeat part (b) for $(g \circ f)(x)$.

d. Which composite function models the greater discount on the computer, $f \circ g$ or $g \circ f$? Explain.

106. The regular price of a pair of jeans is x dollars. Let $f(x) = x - 5$ and $g(x) = 0.6x$.

    a. Describe what functions f and g model in terms of the price of the jeans.

    b. Find $(f \circ g)(x)$ and describe what this models in terms of the price of the jeans.

    c. Repeat part (b) for $(g \circ f)(x)$.

    d. Which composite function models the greater discount on the jeans, $f \circ g$ or $g \circ f$? Explain.

Writing in Mathematics

107. If a function is defined by an equation, explain how to find its domain.

108. If equations for f and g are given, explain how to find $f - g$.

109. If equations for two functions are given, explain how to obtain the quotient function and its domain.

110. Describe a procedure for finding $(f \circ g)(x)$. What is the name of this function?

111. Describe the values of x that must be excluded from the domain of $(f \circ g)(x)$.

112. We opened the section with two functions that modeled the data in Figure 1.58 on page 209. Car sales, $C(x)$, in millions, x years after 1990 were modeled by $C(x) = -0.14x + 9$. SUV sales, $S(x)$, in millions, x years after 1990 were modeled by $S(x) = 0.22x + 1$. Explain how these models were obtained from Figure 1.58.

Technology Exercises

113. The function $f(t) = -0.14t^2 + 0.51t + 31.6$ models the U.S. population ages 65 and older, $f(t)$, in millions, t years after 1990. The function $g(t) = 0.54t^2 + 12.64t + 107.1$ models the total yearly cost of Medicare, $g(t)$, in billions of dollars t years after 1990. Graph the function $\frac{g}{f}$ in a $[0, 15, 1]$ by $[0, 60, 10]$ viewing rectangle. What does the shape of the graph indicate about the per capita costs of Medicare for the U.S. population ages 65 and over with increasing time?

114. Graph $y_1 = x^2 - 2x$, $y_2 = x$, and $y_3 = y_1 \div y_2$ in the same $[-10, 10, 1]$ by $[-10, 10, 1]$ viewing rectangle. Then use the [TRACE] feature to trace along y_3. What happens at $x = 0$? Explain why this occurs.

115. Graph $y_1 = \sqrt{2 - x}$, $y_2 = \sqrt{x}$, and $y_3 = \sqrt{2 - y_2}$ in the same $[-4, 4, 1]$ by $[0, 2, 1]$ viewing rectangle. If y_1 represents f and y_2 represents g, use the graph of y_3 to find the domain of $f \circ g$. Then verify your observation algebraically.

Critical Thinking Exercises

116. Which one of the following is true? _____

    a. If $f(x) = x^2 - 4$ and $g(x) = \sqrt{x^2 - 4}$, then $(f \circ g)(x) = -x^2$ and $(f \circ g)(5) = -25$.

    b. There can never be two functions f and g, where $f \neq g$ for which $(f \circ g)(x) = (g \circ f)(x)$.

    c. If $f(7) = 5$ and $g(4) = 7$, then $(f \circ g)(4) = 35$.

    d. If $f(x) = \sqrt{x}$ and $g(x) = 2x - 1$, then $(f \circ g)(5) = g(2)$.

117. Prove that if f and g are even functions, then fg is also an even function.

118. Define two functions f and g so that $f \circ g = g \circ f$.

119. Use the graphs given in Exercises 99–102 to create a graph that shows the population, in billions, of less developed regions from 1950 through 2050.

SECTION 1.8 *Inverse Functions*

Objectives

❶ Verify inverse functions.

❷ Find the inverse of a function.

❸ Use the horizontal line test to determine if a function has an inverse function.

❹ Use the graph of a one-to-one function to graph its inverse function.

❺ Find the inverse of a function and graph both functions on the same axes.

In most societies, women say they prefer to marry men who are older than themselves, whereas men say they prefer women who are younger. Evolutionary psychologists attribute these preferences to female concern with a partner's material resources and male concern with a partner's fertility (*Source:* David M. Buss *Psychological Inquiry*, 6, 1–30). When the man is considerably older than the woman, people rarely comment. However, when the woman is older, as in the relationship between actors Ashton Kutcher and Demi Moore, people take notice.

Figure 1.62 shows the preferred age in a mate in five selected countries. We can focus on the data for the women and define a function.

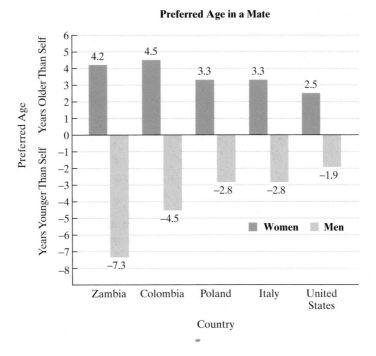

Preferred Age in a Mate

Figure 1.62
Source: Carole Wade and Carol Tavris, *Psychology*, 6th Edition, Prentice Hall, 2000

Let the domain of the function be the set of the five countries shown in the graph. Let the range be the set of the average number of years women in each of the respective countries prefer men who are older than themselves. The function can be written as follows:

f: {(Zambia, 4.2), (Colombia, 4.5), (Poland, 3.3), (Italy, 3.3), (U.S., 2.5)}.

Now let's "undo" f by interchanging the first and second components in each of its ordered pairs. Switching the inputs and outputs of f, we obtain the following relation:

Same first component

Undoing f: {(4.2, Zambia), (4.5, Colombia), (3.3, Poland), (3.3, Italy), (2.5, U.S.)}.

Different second components

Can you see that this relation is not a function? Two of its ordered pairs have the same first component and different second components. This violates the definition of a function.

If a function f is a set of ordered pairs, (x, y), then the changes produced by f can be "undone" by reversing the components of all the ordered pairs. The resulting relation, (y, x), may or may not be a function. In this section, we will develop these ideas by studying functions whose compositions have a special "undoing" relationship.

Inverse Functions

Here are two functions that describe situations related to the price of a computer, x:

$$f(x) = x - 300 \qquad g(x) = x + 300.$$

The function f, $f(x) = x - 300$, subtracts \$300 from the computer's price and the function g, $g(x) = x + 300$, adds \$300 to the computer's price. Let's see what $f(g(x))$ does. Put $g(x)$ into f:

$$f(x) = x - 300 \qquad \text{This is the given equation for f.}$$

Replace x with $g(x)$.

$$f(g(x)) = g(x) - 300$$
$$= x + 300 - 300 \qquad \text{Because } g(x) = x + 300,$$
$$\text{replace } g(x) \text{ with } x + 300.$$
$$= x.$$

This is the computer's original price.

By putting $g(x)$ into f and finding $f(g(x))$, we see that the computer's price x, went through two changes: the first, an increase; the second, a decrease:

$$x + 300 - 300.$$

The final price of the computer, x, is identical to its starting price, x.

In general, if the changes made to x by a function g are undone by the changes made by a function f, then

$$f(g(x)) = x.$$

Assume, also, that this "undoing" takes place in the other direction:

$$g(f(x)) = x.$$

Under these conditions, we say that each function is the *inverse function* of the other. The fact that g is the inverse of f is expressed by renaming g as f^{-1}, read "f-inverse." For example, the inverse functions

$$f(x) = x - 300 \qquad g(x) = x + 300$$

are usually named as follows:

$$f(x) = x - 300 \qquad f^{-1}(x) = x + 300.$$

We can use partial tables of coordinates for f and f^{-1} to gain numerical insight into the relationship between a function and its inverse function.

Computer's regular price	\$300 reduction		Price with \$300 reduction	\$300 price increase
x	$f(x) = x - 300$		x	$f^{-1}(x) = x + 300$
1200	900		900	1200
1300	1000		1000	1300
1400	1100		1100	1400

Ordered pairs for f:
(1200, 900), (1300, 1000), (1400, 1100)

Ordered pairs for f^{-1}:
(900, 1200), (1000, 1300), (1100, 1400)

The tables illustrate that if a function f is the set of ordered pairs (x, y), then its inverse, f^{-1}, is the set of ordered pairs (y, x). Using these tables, we can see how one function's changes to x are undone by the other function:

$$(f^{-1} \circ f)(1300) = f^{-1}(f(1300)) = f^{-1}(1000) = 1300.$$

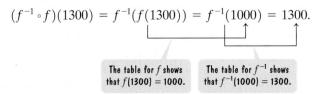

The table for f shows that $f(1300) = 1000$.

The table for f^{-1} shows that $f^{-1}(1000) = 1300$.

The final price of the computer, \$1300, is identical to its starting price, \$1300.

With these ideas in mind, we present the formal definition of the inverse of a function:

Definition of the Inverse of a Function

Let f and g be two functions such that

$$f(g(x)) = x \qquad \text{for every } x \text{ in the domain of } g$$

and

$$g(f(x)) = x \qquad \text{for every } x \text{ in the domain of } f.$$

The function g is the **inverse of the function** f and is denoted by f^{-1} (read "f-inverse"). Thus, $f(f^{-1}(x)) = x$ and $f^{-1}(f(x)) = x$. The domain of f is equal to the range of f^{-1}, and vice versa.

 Verify inverse functions.

EXAMPLE 1   Verifying Inverse Functions

Show that each function is the inverse of the other:

$$f(x) = 3x + 2 \quad \text{and} \quad g(x) = \frac{x-2}{3}.$$

Solution To show that f and g are inverses of each other, we must show that $f(g(x)) = x$ and $g(f(x)) = x$. We begin with $f(g(x))$.

$$f(x) = 3x + 2 \qquad \text{This is the equation for } f.$$

Replace x with $g(x)$.

$$f(g(x)) = 3g(x) + 2 = 3\left(\frac{x-2}{3}\right) + 2 = x - 2 + 2 = x$$

$$g(x) = \frac{x-2}{3}$$

Next, we find $g(f(x))$.

$$g(x) = \frac{x-2}{3} \qquad \text{This is the equation for } g.$$

Replace x with $f(x)$.

$$g(f(x)) = \frac{f(x) - 2}{3} = \frac{(3x+2) - 2}{3} = \frac{3x}{3} = x$$

$$f(x) = 3x + 2$$

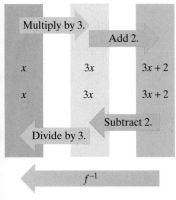

Figure 1.63 f^{-1} undoes the changes produced by f.

Because g is the inverse of f (and vice versa), we can use inverse notation and write

$$f(x) = 3x + 2 \quad \text{and} \quad f^{-1}(x) = \frac{x-2}{3}.$$

Notice how f^{-1} undoes the changes produced by f: f changes x by *multiplying* by 3 and *adding* 2, and f^{-1} undoes this by *subtracting* 2 and *dividing* by 3. This "undoing" process is illustrated in Figure 1.63.

Check Point 1 Show that each function is the inverse of the other:

$$f(x) = 4x - 7 \quad \text{and} \quad g(x) = \frac{x+7}{4}.$$

② Find the inverse of a function.

Finding the Inverse of a Function

The definition of the inverse of a function tells us that the domain of f is equal to the range of f^{-1}, and vice versa. This means that if the function f is the set of ordered pairs (x, y), then the inverse of f is the set of ordered pairs (y, x). If a function is defined by an equation, we can obtain the equation for f^{-1}, the inverse of f, by interchanging the role of x and y in the equation for the function f.

Finding the Inverse of a Function

The equation for the inverse of a function f can be found as follows:

1. Replace $f(x)$ with y in the equation for $f(x)$.
2. Interchange x and y.
3. Solve for y. If this equation does not define y as a function of x, the function f does not have an inverse function and this procedure ends. If this equation does define y as a function of x, the function f has an inverse function.
4. If f has an inverse function, replace y in step 3 by $f^{-1}(x)$. We can verify our result by showing that $f(f^{-1}(x)) = x$ and $f^{-1}(f(x)) = x$.

Study Tip

The procedure for finding a function's inverse uses a *switch-and-solve* strategy. Switch x and y, and then solve for y.

EXAMPLE 2 Finding the Inverse of a Function

Find the inverse of $f(x) = 7x - 5$.

Solution

Step 1 Replace $f(x)$ with y:
$$y = 7x - 5.$$

Step 2 Interchange x and y:
$$x = 7y - 5. \qquad \text{This is the inverse function.}$$

Step 3 Solve for y:
$$x + 5 = 7y \qquad \text{Add 5 to both sides.}$$
$$\frac{x + 5}{7} = y. \qquad \text{Divide both sides by 7.}$$

Discovery

In Example 2, we found that if $f(x) = 7x - 5$, then
$$f^{-1}(x) = \frac{x + 5}{7}.$$
Verify this result by showing that
$$f(f^{-1}(x)) = x$$
and
$$f^{-1}(f(x)) = x.$$

Step 4 Replace y with $f^{-1}(x)$:
$$f^{-1}(x) = \frac{x + 5}{7}. \qquad \text{The equation is written with } f^{-1} \text{ on the left.}$$

Thus, the inverse of $f(x) = 7x - 5$ is $f^{-1}(x) = \dfrac{x + 5}{7}$.

The inverse function, f^{-1}, undoes the changes produced by f. f changes x b multiplying by 7 and subtracting 5. f^{-1} undoes this by adding 5 and dividing by 7.

Check Point **2** Find the inverse of $f(x) = 2x + 7$.

EXAMPLE 3 Finding the Inverse of a Function

Find the inverse of $f(x) = x^3 + 1$.

Solution

Step 1 Replace $f(x)$ with y: $y = x^3 + 1$.

Step 2 Interchange x and y: $x = y^3 + 1$.

Step 3 Solve for y:

> Our goal is to isolate y. Because $\sqrt[3]{y^3} = y$, we will take the cube root of both sides of the equation.

$$x - 1 = y^3 \qquad \text{Subtract 1 from both sides.}$$
$$\sqrt[3]{x - 1} = \sqrt[3]{y^3} \qquad \text{Take the cube root on both sides.}$$
$$\sqrt[3]{x - 1} = y. \qquad \text{Simplify.}$$

Step 4 Replace y with $f^{-1}(x)$: $f^{-1}(x) = \sqrt[3]{x - 1}$.

Thus, the inverse of $f(x) = x^3 + 1$ is $f^{-1}(x) = \sqrt[3]{x - 1}$.

Check Point 3   Find the inverse of $f(x) = 4x^3 - 1$.

EXAMPLE 4   Finding the Inverse of a Function

Find the inverse of $f(x) = \dfrac{5}{x} + 4$.

Solution

Step 1   Replace $f(x)$ with y:

$$y = \frac{5}{x} + 4.$$

Step 2   Interchange x and y:

$$x = \frac{5}{y} + 4.$$

Our goal is to isolate y. To get y out of the denominator, we will multiply both sides of the equation by y, $y \neq 0$.

Step 3   Solve for y:

$$x = \frac{5}{y} + 4 \qquad \text{This is the equation from step 2.}$$

$$xy = \left(\frac{5}{y} + 4\right)y \qquad \text{Multiply both sides by } y, y \neq 0.$$

$$xy = \frac{5}{y} \cdot y + 4y \qquad \text{Use the distributive property.}$$

$$xy = 5 + 4y \qquad \text{Simplify: } \frac{5}{y} \cdot y = 5.$$

$$xy - 4y = 5 \qquad \text{Subtract } 4y \text{ from both sides.}$$

$$y(x - 4) = 5 \qquad \text{Factor out } y \text{ from } xy - 4y \text{ to obtain a single occurrence of } y.$$

$$\frac{y(x - 4)}{x - 4} = \frac{5}{x - 4} \qquad \text{Divide both sides by } x - 4.$$

$$y = \frac{5}{x - 4}. \qquad \text{Simplify.}$$

Step 4   Replace y with $f^{-1}(x)$:

$$f^{-1}(x) = \frac{5}{x - 4}.$$

Thus, the inverse of $f(x) = \dfrac{5}{x} + 4$ is $f^{-1}(x) = \dfrac{5}{x - 4}$.

Check Point 4   Find the inverse of $f(x) = \dfrac{3}{x} - 1$.

Use the horizontal line test to determine if a function has an inverse function.

The Horizontal Line Test and One-to-One Functions

Let's see what happens if we try to find the inverse of the standard quadratic function $f(x) = x^2$.

Step 1 **Replace $f(x)$ with y:** $y = x^2$.

Step 2 **Interchange x and y:** $x = y^2$.

Step 3 **Solve for y:** We apply the square root property to solve $y^2 = x$ for y. We obtain

$$y = \pm\sqrt{x}.$$

The $\pm$ in this last equation shows that for certain values of x (all positive real numbers), there are two values of y. Because this equation does not represent y as a function of x, the standard quadratic function $f(x) = x^2$ does not have an inverse function.

We can use a few of the solutions of $y = x^2$ to illustrate numerically that this function does not have an inverse:

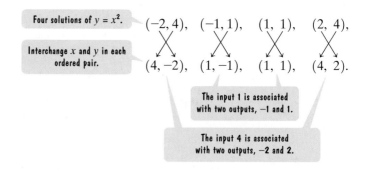

Four solutions of $y = x^2$.

$(-2, 4), \quad (-1, 1), \quad (1,\ 1), \quad (2,\ 4),$

Interchange x and y in each ordered pair.

$(4, -2), \quad (1, -1), \quad (1,\ 1), \quad (4,\ 2).$

The input 1 is associated with two outputs, −1 and 1.

The input 4 is associated with two outputs, −2 and 2.

A function provides exactly one output for each input. Thus, the ordered pairs in the bottom row do not define a function.

Can we look at the graph of a function and tell if it represents a function with an inverse? Yes. The graph of the standard quadratic function $f(x) = x^2$ is shown in Figure 1.64. Four units above the x-axis, a horizontal line is drawn. This line intersects the graph at two of its points, $(-2, 4)$ and $(2, 4)$. Inverse functions have ordered pairs with the coordinates reversed. We just saw what happened when we interchanged x and y. We obtained $(4, -2)$ and $(4, 2)$, and these ordered pairs do not define a function.

If any horizontal line, such as the one in Figure 1.64, intersects a graph at two or more points, the set of these points will not define a function when their coordinates are reversed. This suggests the **horizontal line test** for inverse functions.

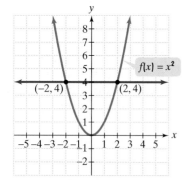

Figure 1.64 The horizontal line intersects the graph twice.

Discovery

How might you restrict the domain of $f(x) = x^2$, graphed in Figure 1.64, so that the remaining portion of the graph passes the horizontal line test?

The Horizontal Line Test For Inverse Functions

A function f has an inverse that is a function, f^{-1}, if there is no horizontal line that intersects the graph of the function f at more than one point.

EXAMPLE 5 Applying the Horizontal Line Test

Which of the following graphs represent functions that have inverse functions?

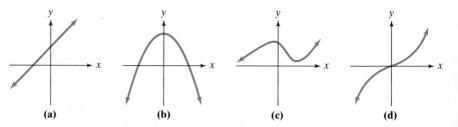

(a)   (b)   (c)   (d)

Solution   Notice that horizontal lines can be drawn in graphs (b) and (c) that intersect the graphs more than once. These graphs do not pass the horizontal line test. These are not the graphs of functions with inverse functions. By contrast, no horizontal line can be drawn in graphs (a) and (d) that intersects the graphs more than once. These graphs pass the horizontal line test. Thus, the graphs in parts (a) and (d) represent functions that have inverse functions.

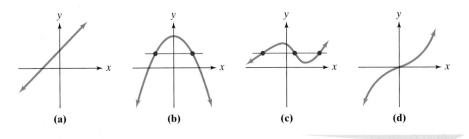

(a)   (b)   (c)   (d)

Check Point 5   Which of the following graphs represent functions that have inverse functions?

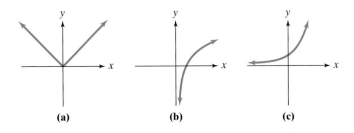

(a)   (b)   (c)

A function passes the horizontal line test when no two different ordered pairs have the same second component. This means that if $x_1 \neq x_2$, then $f(x_1) \neq f(x_2)$. Such a function is called a **one-to-one function**. Thus, **a one-to-one function is a function in which no two different ordered pairs have the same second component**. Only one-to-one functions have inverse functions. Any function that passes the horizontal line test is a one-to-one function. Any one-to-one function has a graph that passes the horizontal line test.

④ Use the graph of a one-to-one function to graph its inverse function.

Graphs of f and f^{-1}

There is a relationship between the graph of a one-to-one function, f, and its inverse, f^{-1}. Because inverse functions have ordered pairs with the coordinates interchanged, if the point (a, b) is on the graph of f then the point (b, a) is on the graph of f^{-1}. The points (a, b) and (b, a) are symmetric with respect to the line $y = x$. Thus, **the graph of f^{-1} is a reflection of the graph of f about the line $y = x$.** This is illustrated in Figure 1.65.

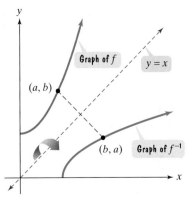

Figure 1.65 The graph of f^{-1} is a reflection of the graph of f about $y = x$.

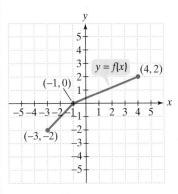

Figure 1.66

EXAMPLE 6   Graphing the Inverse Function

Use the graph of f in Figure 1.66 to draw the graph of its inverse function.

Solution   We begin by noting that no horizontal line intersects the graph of f at more than one point, so f does have an inverse function. Because the

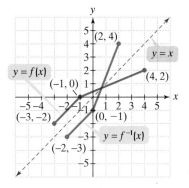

$y = f(x)$

$(2, 4)$

$y = x$

$(-1, 0)$

$(4, 2)$

$(-3, -2)$

$(0, -1)$

$(-2, -3)$

$y = f^{-1}(x)$

Figure 1.67 The graphs of f and f^{-1}

5 Find the inverse of a function and graph both functions on the same axes.

points $(-3, -2), (-1, 0)$, and $(4, 2)$ are on the graph of f, the graph of the inverse function, f^{-1}, has points with these ordered pairs reversed. Thus, $(-2, -3), (0, -1)$, and $(2, 4)$ are on the graph of f^{-1}. We can use these points to graph f^{-1}. The graph of f^{-1} is shown in green in Figure 1.67. Note that the green graph of f^{-1} is the reflection of the blue graph of f about the line $y = x$.

Check Point 6 The graph of function f consists of two line segments, one segment from $(-2, -2)$ to $(-1, 0)$ and a second segment from $(-1, 0)$ to $(1, 2)$. Graph f and use the graph to draw the graph of its inverse function.

In our final example, we will first find f^{-1}. Then we will graph f and f^{-1} in the same rectangular coordinate system.

EXAMPLE 7 Finding the Inverse of a Domain-Restricted Function

Find the inverse of $f(x) = x^2 - 1$ if $x \geq 0$. Graph f and f^{-1} in the same rectangular coordinate system.

Solution The graph of $f(x) = x^2 - 1$ is the graph of the standard quadratic function shifted vertically down 1 unit. Figure 1.68 shows the function's graph. This graph fails the horizontal line test, so the function $f(x) = x^2 - 1$ does not have an inverse function. By restricting the domain to $x \geq 0$, as given, we obtain a new function whose graph is shown in red in Figure 1.69. This red portion of the graph is increasing on the interval $(0, \infty)$ and passes the horizontal line test. This tells us that $f(x) = x^2 - 1$ has an inverse function if we restrict its domain to $x \geq 0$. We use our four-step procedure to find this inverse function. Begin with $f(x) = x^2 - 1, x \geq 0$.

Step 1 Replace $f(x)$ with y: $y = x^2 - 1, x \geq 0$.

Step 2 Interchange x and y: $x = y^2 - 1, y \geq 0$.

Step 3 Solve for y:

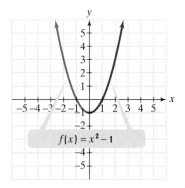

$f(x) = x^2 - 1$

Figure 1.68

$$x = y^2 - 1, y \geq 0 \qquad \text{This is the equation from step 2.}$$

$$x + 1 = y^2 \qquad \text{Add 1 to both sides.}$$

$$\sqrt{x + 1} = y \qquad \text{Apply the square root property.}$$

Because $y \geq 0$, take only the principal square root and not the negative square root.

Step 4 Replace y with $f^{-1}(x)$: $f^{-1}(x) = \sqrt{x + 1}$.

Thus, the inverse of $f(x) = x^2 - 1, x \geq 0$, is $f^{-1}(x) = \sqrt{x + 1}$. The graphs of f and f^{-1} are shown in Figure 1.69. We obtained the graph of $f^{-1}(x) = \sqrt{x + 1}$ by shifting the graph of the square root function, $y = \sqrt{x}$, horizontally to the left 1 unit. Note that the green graph of f^{-1} is the reflection of the red graph of f about the line $y = x$.

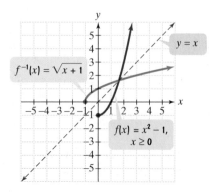

$f^{-1}(x) = \sqrt{x + 1}$

$y = x$

$f(x) = x^2 - 1,$
$x \geq 0$

Figure 1.69

Check Point 7 Find the inverse of $f(x) = x^2 + 1$ if $x \geq 0$. Graph f and f^{-1} in the same rectangular coordinate system.

EXERCISE SET 1.8

Practice Exercises

In Exercises 1–10, find $f(g(x))$ and $g(f(x))$ and determine whether each pair of functions f and g are inverses of each other.

1. $f(x) = 4x$ and $g(x) = \dfrac{x}{4}$

2. $f(x) = 6x$ and $g(x) = \dfrac{x}{6}$

3. $f(x) = 3x + 8$ and $g(x) = \dfrac{x - 8}{3}$

4. $f(x) = 4x + 9$ and $g(x) = \dfrac{x - 9}{4}$

5. $f(x) = 5x - 9$ and $g(x) = \dfrac{x + 5}{9}$

6. $f(x) = 3x - 7$ and $g(x) = \dfrac{x + 3}{7}$

7. $f(x) = \dfrac{3}{x - 4}$ and $g(x) = \dfrac{3}{x} + 4$

8. $f(x) = \dfrac{2}{x - 5}$ and $g(x) = \dfrac{2}{x} + 5$

9. $f(x) = -x$ and $g(x) = -x$

10. $f(x) = \sqrt[3]{x - 4}$ and $g(x) = x^3 + 4$

The functions in Exercises 11–28 are all one-to-one. For each function,

a. *Find an equation for $f^{-1}(x)$, the inverse function.*

b. *Verify that your equation is correct by showing that $f(f^{-1}(x)) = x$ and $f^{-1}(f(x)) = x$.*

11. $f(x) = x + 3$ **12.** $f(x) = x + 5$

13. $f(x) = 2x$ **14.** $f(x) = 4x$

15. $f(x) = 2x + 3$ **16.** $f(x) = 3x - 1$

17. $f(x) = x^3 + 2$ **18.** $f(x) = x^3 - 1$

19. $f(x) = (x + 2)^3$ **20.** $f(x) = (x - 1)^3$

21. $f(x) = \dfrac{1}{x}$ **22.** $f(x) = \dfrac{2}{x}$

23. $f(x) = \sqrt{x}$ **24.** $f(x) = \sqrt[3]{x}$

25. $f(x) = \dfrac{7}{x} - 3$ **26.** $f(x) = \dfrac{4}{x} + 9$

27. $f(x) = \dfrac{2x + 1}{x - 3}$ **28.** $f(x) = \dfrac{2x - 3}{x + 1}$

Which graphs in Exercises 29–34 represent functions that have inverse functions?

29.

30.

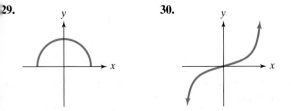

31.

32.

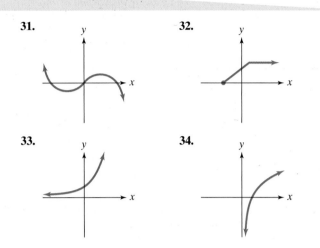

33.

34.

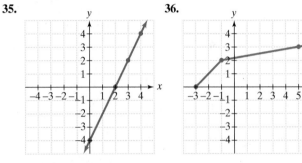

In Exercises 35–38, use the graph of f to draw the graph of its inverse function.

35.

36.

37.

38.

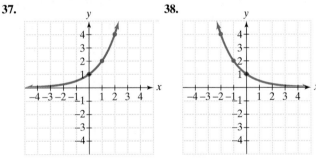

In Exercises 39–52,

a. *Find an equation for $f^{-1}(x)$.*

b. *Graph f and f^{-1} in the same rectangular coordinate system.*

c. *Use interval notation to give the domain and the range of f and f^{-1}.*

39. $f(x) = 2x - 1$ **40.** $f(x) = 2x - 3$

41. $f(x) = x^2 - 4, x \geq 0$ **42.** $f(x) = x^2 - 1, x \leq 0$

43. $f(x) = (x - 1)^2, x \leq 1$ **44.** $f(x) = (x - 1)^2, x \geq 1$

45. $f(x) = x^3 - 1$ **46.** $f(x) = x^3 + 1$

47. $f(x) = (x + 2)^3$ **48.** $f(x) = (x - 2)^3$

(Hint for Exercises 49–52: To solve for a variable involving an nth root, raise both sides of the equation to the nth power: $\left(\sqrt[n]{y}\right)^n = y$.)

49. $f(x) = \sqrt{x - 1}$ **50.** $f(x) = \sqrt{x} + 2$

51. $f(x) = \sqrt[3]{x} + 1$ **52.** $f(x) = \sqrt[3]{x - 1}$

Practice Plus

In Exercises 53–58, f and g are defined by the following tables. Use the tables to evaluate each composite function.

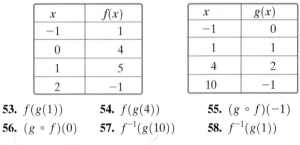

x	$f(x)$
-1	1
0	4
1	5
2	-1

x	$g(x)$
-1	0
1	1
4	2
10	-1

53. $f(g(1))$ 54. $f(g(4))$ 55. $(g \circ f)(-1)$
56. $(g \circ f)(0)$ 57. $f^{-1}(g(10))$ 58. $f^{-1}(g(1))$

In Exercises 59–64, let

$$f(x) = 2x - 5$$
$$g(x) = 4x - 1$$
$$h(x) = x^2 + x + 2.$$

Evaluate the indicated function without finding an equation for the function.

59. $(f \circ g)(0)$ 60. $(g \circ f)(0)$ 61. $f^{-1}(1)$
62. $g^{-1}(7)$ 63. $g(f[h(1)])$ 64. $f(g[h(1)])$

Application Exercises

65. Refer to Figure 1.62 on page 223. Recall that the bar graphs in the figure show the preferred age in a mate in five selected countries.
   a. Consider a function, f, whose domain is the set of the five countries shown in the graph. Let the range be the set of the average number of years men in each of the respective countries prefer women who are younger than themselves. Write the function f as a set of ordered pairs.
   b. Write the relation that is the inverse of f as a set of ordered pairs. Is this relation a function? Explain your answer.

66. The bar graph shows the number of days in a school year for the six countries with the longest school years. (To put these numbers in perspective, a school year in the United States is 180 days.)

Countries with the Longest School Years

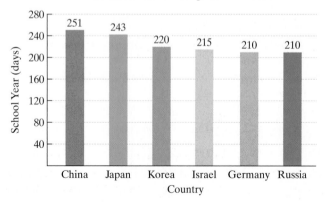

Source: UNESCO

   a. Consider a function, f, whose domain is the set of six countries shown. Let the range be the number of days in the school year for each respective country. Write the function f as a set of ordered pairs.

   b. Write the relation that is the inverse of f as a set of ordered pairs. Is this relation a function? Explain your answer.

67. The graph represents the probability of two people in the same room sharing a birthday as a function of the number of people in the room. Call the function f.

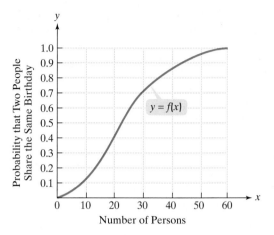

   a. Explain why f has an inverse that is a function.
   b. Describe in practical terms the meaning of $f^{-1}(0.25), f^{-1}(0.5)$, and $f^{-1}(0.7)$.

68. A study of 900 working women in Texas showed that their feelings changed throughout the day. As the graph indicates, the women felt better as time passed, except for a blip (that's slang for relative maximum) at lunchtime.

Average Level of Happiness at Different Times of Day

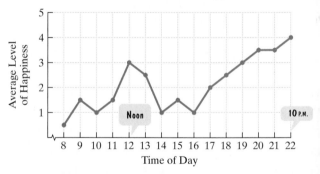

Source: D. Kahneman et. al., "A Survey Method for Characterizing Daily Life Experience," *Science.*

   a. Does the graph have an inverse that is a function? Explain your answer.
   b. Identify two or more times of day when the average happiness level is 3. Express your answers as ordered pairs.
   c. Do the ordered pairs in part (b) indicate that the graph represents a one-to-one function? Explain your answer.

69. The formula

$$y = f(x) = \frac{9}{5}x + 32$$

is used to convert from x degrees Celsius to y degrees Fahrenheit. The formula

$$y = g(x) = \frac{5}{9}(x - 32)$$

is used to convert from x degrees Fahrenheit to y degrees Celsius. Show that f and g are inverse functions.

Writing in Mathematics

70. Explain how to determine if two functions are inverses of each other.

71. Describe how to find the inverse of a one-to-one function.

72. What is the horizontal line test and what does it indicate?

73. Describe how to use the graph of a one-to-one function to draw the graph of its inverse function.

74. How can a graphing utility be used to visually determine if two functions are inverses of each other?

75. What explanations can you offer for the trends shown by the graph in Exercise 68?

Technology Exercises

In Exercises 76–83, use a graphing utility to graph the function. Use the graph to determine whether the function has an inverse that is a function (that is, whether the function is one-to-one).

76. $f(x) = x^2 - 1$

77. $f(x) = \sqrt[3]{2 - x}$

78. $f(x) = \dfrac{x^3}{2}$

79. $f(x) = \dfrac{x^4}{4}$

80. $f(x) = \text{int}(x - 2)$

81. $f(x) = |x - 2|$

82. $f(x) = (x - 1)^3$

83. $f(x) = -\sqrt{16 - x^2}$

In Exercises 84–86, use a graphing utility to graph f and g in the same viewing rectangle. In addition, graph the line $y = x$ and visually determine if f and g are inverses.

84. $f(x) = 4x + 4,\ g(x) = 0.25x - 1$

85. $f(x) = \dfrac{1}{x} + 2,\ g(x) = \dfrac{1}{x - 2}$

86. $f(x) = \sqrt[3]{x} - 2,\ g(x) = (x + 2)^3$

Critical Thinking Exercises

87. Which one of the following is true?

  a. The inverse of $\{(1, 4), (2, 7)\}$ is $\{(2, 7), (1, 4)\}$.

  b. The function $f(x) = 5$ is one-to-one.

  c. If $f(x) = 3x$, then $f^{-1}(x) = \dfrac{1}{3x}$.

  d. The domain of f is the same as the range of f^{-1}.

88. If $f(x) = 3x$ and $g(x) = x + 5$, find $(f \circ g)^{-1}(x)$ and $(g^{-1} \circ f^{-1})(x)$.

89. Show that
$$f(x) = \frac{3x - 2}{5x - 3}$$
is its own inverse.

90. *Freedom 7* was the spacecraft that carried the first American into space in 1961. Total flight time was 15 minutes and the spacecraft reached a maximum height of 116 miles. Consider a function, s, that expresses *Freedom 7*'s height, $s(t)$, in miles, after t minutes. Is s a one-to-one function? Explain your answer.

91. If $f(2) = 6$, and f is one-to-one, find x satisfying $8 + f^{-1}(x - 1) = 10$.

Group Exercise

92. In Tom Stoppard's play *Arcadia*, the characters dream and talk about mathematics, including ideas involving graphing, composite functions, symmetry, and lack of symmetry in things that are tangled, mysterious, and unpredictable. Group members should read the play. Present a report on the ideas discussed by the characters that are related to concepts that we studied in this chapter. Bring in a copy of the play and read appropriate excerpts.

SECTION 1.9   Distance and Midpoint Formulas; Circles

Objectives

1 Find the distance between two points.

2 Find the midpoint of a line segment.

3 Write the standard form of a circle's equation.

4 Give the center and radius of a circle whose equation is in standard form.

5 Convert the general form of a circle's equation to standard form.

It's a good idea to know your way around a circle. Clocks, angles, maps, and compasses are based on circles. Circles occur everywhere in nature: in ripples on water, patterns on a moth's wings, and cross sections of trees. Some consider the circle to be the most pleasing of all shapes.

The rectangular coordinate system gives us a unique way of knowing a circle. It enables us to translate a circle's geometric definition into an algebraic equation. To do this, we must first develop a formula for the distance between any two points in rectangular coordinates.

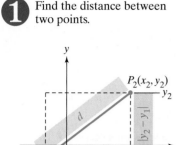

1 Find the distance between two points.

Figure 1.70

The Distance Formula

Using the Pythagorean Theorem, we can find the distance between the two point $P_1(x_1, y_1)$ and $P_2(x_2, y_2)$ in the rectangular coordinate system. The two points are illustrated in Figure 1.70.

The distance that we need to find is represented by d and shown in blue. Notice that the distance between two points on the dashed horizontal line is the absolute value of the difference between the x-coordinates of the two points. This distance, $|x_2 - x_1|$, is shown in pink. Similarly, the distance between two points on the dashed vertical line is the absolute value of the difference between the y-coordinates of the two points. This distance, $|y_2 - y_1|$, is also shown in pink.

Because the dashed lines are horizontal and vertical, a right triangle is formed. Thus, we can use the Pythagorean Theorem to find the distance d. Squaring the lengths of the triangle's sides results in positive numbers, so absolute value notation is not necessary.

$$d^2 = (x_2 - x_1)^2 + (y_2 - y_1)^2$$ Apply the Pythagorean Theorem to the right triangle in Figure 1.70.

$$d = \pm\sqrt{(x_2 - x_1)^2 + (y_2 - y_1)^2}$$ Apply the square root property.

$$d = \sqrt{(x_2 - x_1)^2 + (y_2 - y_1)^2}$$ Because distance is nonnegative, write only the principal square root.

This result is called the **distance formula**.

The Distance Formula

The distance, d, between the points (x_1, y_1) and (x_2, y_2) in the rectangular coordinate system is

$$d = \sqrt{(x_2 - x_1)^2 + (y_2 - y_1)^2}.$$

To compute the distance between two points, find the square of the difference between the x-coordinates plus the square of the difference between the y-coordinates. The principal square root of this sum is the distance.

When using the distance formula, it does not matter which point you call (x_1, y_1) and which you call (x_2, y_2).

EXAMPLE 1   Using the Distance Formula

Find the distance between $(-1, 4)$ and $(3, -2)$.

Solution   We will let $(x_1, y_1) = (-1, 4)$ and $(x_2, y_2) = (3, -2)$.

$$d = \sqrt{(x_2 - x_1)^2 + (y_2 - y_1)^2}$$ Use the distance formula.

$$= \sqrt{(-2 - 4)^2 + [3 - (-1)]^2}$$ Substitute the given values.

$$= \sqrt{(-6)^2 + 4^2}$$ Perform operations inside grouping symbols: $-2 - 4 = -6$ and $3 - (-1) = 3 + 1 = 4$.

$$= \sqrt{36 + 16}$$ Caution: This does not equal $\sqrt{36} + \sqrt{16}$.   Square -6 and 4.

$$= \sqrt{52}$$ Add.

$$= \sqrt{4 \cdot 13} = 2\sqrt{13} \approx 7.21$$ $\sqrt{52} = \sqrt{4 \cdot 13} = \sqrt{4}\sqrt{13} = 2\sqrt{13}$

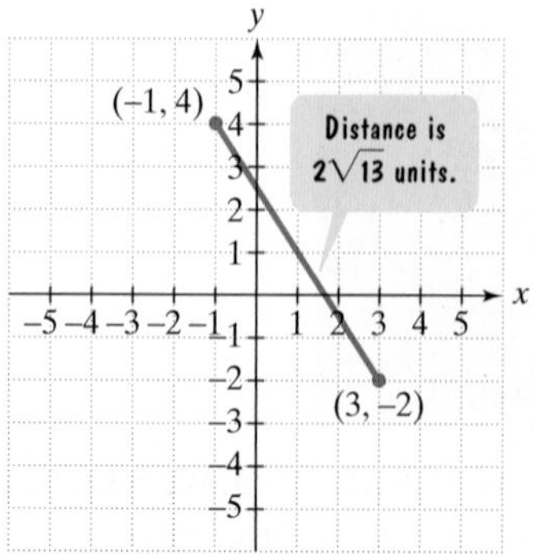

The distance between the given points is $2\sqrt{13}$ units, or approximately 7.21 units. The situation is illustrated in Figure 1.71.

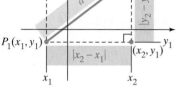

Figure 1.71 Finding the distance between two points

Check Point 1   Find the distance between $(-4, 9)$ and $(1, -3)$.

② Find the midpoint of a line segment.

The Midpoint Formula

The distance formula can be used to derive a formula for finding the midpoint of a line segment between two given points. The formula is given as follows:

The Midpoint Formula

Consider a line segment whose endpoints are (x_1, y_1) and (x_2, y_2). The coordinates of the segment's midpoint are

$$\left(\frac{x_1 + x_2}{2}, \frac{y_1 + y_2}{2} \right).$$

To find the midpoint, take the average of the two x-coordinates and the average of the two y-coordinates.

Study Tip

The midpoint formula requires finding the *sum* of coordinates. By contrast, the distance formula requires finding the *difference* of coordinates:

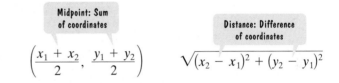

It's easy to confuse the two formulas. Be sure to use addition, not subtraction, when applying the midpoint formula.

EXAMPLE 2   Using the Midpoint Formula

Find the midpoint of the line segment with endpoints $(1, -6)$ and $(-8, -4)$.

Solution   To find the coordinates of the midpoint, we average the coordinates of the endpoints.

$$\text{Midpoint} = \left(\frac{1 + (-8)}{2}, \frac{-6 + (-4)}{2} \right) = \left(\frac{-7}{2}, \frac{-10}{2} \right) = \left(-\frac{7}{2}, -5 \right)$$

Average the x-coordinates.   Average the y-coordinates.

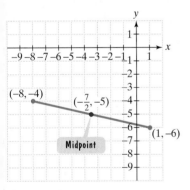

Figure 1.72 Finding a line segment's midpoint

Figure 1.72 illustrates that the point $\left(-\frac{7}{2}, -5 \right)$ is midway between the points $(1, -6)$ and $(-8, -4)$.

Check Point 2 Find the midpoint of the line segment with endpoints $(1, 2)$ and $(7, -3)$.

Circles

Our goal is to translate a circle's geometric definition into an equation. We begin with this geometric definition.

Definition of a Circle

A **circle** is the set of all points in a plane that are equidistant from a fixed point, called the **center**. The fixed distance from the circle's center to any point on the circle is called the **radius**.

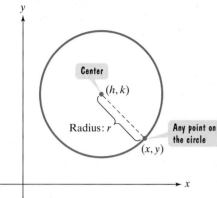

Figure 1.73 is our starting point for obtaining a circle's equation. We've placed the circle into a rectangular coordinate system. The circle's center is (h, k) and its radius is r. We let (x, y) represent the coordinates of any point on the circle.

What does the geometric definition of a circle tell us about point (x, y) in Figure 1.73? The point is on the circle if and only if its distance from the center is r. We can use the distance formula to express this idea algebraically:

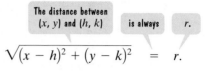

$$\sqrt{(x - h)^2 + (y - k)^2} = r.$$

Figure 1.73 A circle centered at (h, k) with radius r

Squaring both sides of $\sqrt{(x - h)^2 + (y - k)^2} = r$ yields the *standard form of the equation of a circle*.

 Write the standard form of a circle's equation.

The Standard Form of the Equation of a Circle

The **standard form of the equation of a circle** with center (h, k) and radius r is
$$(x - h)^2 + (y - k)^2 = r^2.$$

EXAMPLE 3 Finding the Standard Form of a Circle's Equation

Write the standard form of the equation of the circle with center $(0, 0)$ and radius 2. Graph the circle.

Solution   The center is $(0, 0)$. Because the center is represented as (h, k) in the standard form of the equation, $h = 0$ and $k = 0$. The radius is 2, so we will let $r = 2$ in the equation.

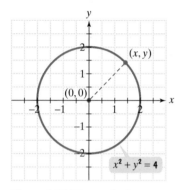

$(x - h)^2 + (y - k)^2 = r^2$	This is the standard form of a circle's equation.
$(x - 0)^2 + (y - 0)^2 = 2^2$	Substitute 0 for h, 0 for k, and 2 for r.
$x^2 + y^2 = 4$	Simplify.

Figure 1.74 The graph of $x^2 + y^2 = 4$

The standard form of the equation of the circle is $x^2 + y^2 = 4$. Figure 1.74 shows the graph.

 Check Point 3   Write the standard form of the equation of the circle with center $(0, 0)$ and radius 4.

Technology

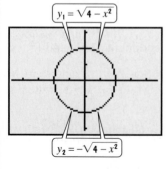

To graph a circle with a graphing utility, first solve the equation for y.
$$x^2 + y^2 = 4$$
$$y^2 = 4 - x^2$$
$$y = \pm\sqrt{4 - x^2}$$

y is not a function of x.

Graph the two equations
$$y_1 = \sqrt{4 - x^2} \quad \text{and} \quad y_2 = -\sqrt{4 - x^2}$$
in the same viewing rectangle. The graph of $y_1 = \sqrt{4 - x^2}$ is the top semicircle because y is always positive. The graph of $y_2 = -\sqrt{4 - x^2}$ is the bottom semicircle because y is always negative. Use a $\boxed{\text{ZOOM SQUARE}}$ setting so that the circle looks like a circle. (Many graphing utilities have problems connecting the two semicircles because the segments directly across horizontally from the center become nearly vertical.)

Example 3 and Check Point 3 involved circles centered at the origin. The standard form of the equation of all such circles is $x^2 + y^2 = r^2$, where r is the circle's radius. Now, let's consider a circle whose center is not at the origin.

EXAMPLE 4   Finding the Standard Form of a Circle's Equation

Write the standard form of the equation of the circle with center $(-2, 3)$ and radius 4.

Solution   The center is $(-2, 3)$. Because the center is represented as (h, k) in the standard form of the equation, $h = -2$ and $k = 3$. The radius is 4, so we will let $r = 4$ in the equation.

$$(x - h)^2 + (y - k)^2 = r^2 \qquad \text{This is the standard form of a circle's equation.}$$
$$[x - (-2)]^2 + (y - 3)^2 = 4^2 \qquad \text{Substitute } -2 \text{ for } h, 3 \text{ for } k, \text{ and } 4 \text{ for } r.$$
$$(x + 2)^2 + (y - 3)^2 = 16 \qquad \text{Simplify.}$$

The standard form of the equation of the circle is $(x + 2)^2 + (y - 3)^2 = 16$.

Check Point 4   Write the standard form of the equation of the circle with center $(5, -6)$ and radius 10.

④ Give the center and radius of a circle whose equation is in standard form.

EXAMPLE 5   Using the Standard Form of a Circle's Equation to Graph the Circle

a. Find the center and radius of the circle whose equation is

$$(x - 2)^2 + (y + 4)^2 = 9.$$

b. Graph the equation.

c. Use the graph to identify the relation's domain and range.

Solution

a. We begin by finding the circle's center, (h, k), and its radius, r. We can find the values for h, k, and r by comparing the given equation to the standard form of the equation of a circle, $(x - h)^2 + (y - k)^2 = r^2$.

$$(x - 2)^2 + (y + 4)^2 = 9$$

$$(x - 2)^2 + (y - (-4))^2 = 3^2$$

This is $(x - h)^2$, with $h = 2$.   This is $(y - k)^2$, with $k = -4$.   This is r^2, with $r = 3$.

We see that $h = 2$, $k = -4$, and $r = 3$. Thus, the circle has center $(h, k) = (2, -4)$ and a radius of 3 units.

b. To graph this circle, first plot the center $(2, -4)$. Because the radius is 3, you can locate at least four points on the circle by going out three units to the right, to the left, up, and down from the center.

The points three units to the right and to the left of $(2, -4)$ are $(5, -4)$ and $(-1, -4)$, respectively. The points three units up and down from $(2, -4)$ are $(2, -1)$ and $(2, -7)$, respectively.

Using these points, we obtain the graph in Figure 1.75.

c. The four points that we located on the circle can be used to determine the relation's domain and range. The points $(-1, -4)$ and $(5, -4)$ show that values of x extend from -1 to 5, inclusive:

$$\text{Domain} = [-1, 5].$$

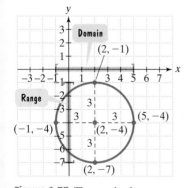

Figure 1.75 The graph of $(x - 2)^2 + (y + 4)^2 = 9$

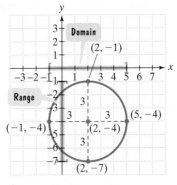

Figure 1.75 (repeated)

The points $(2, -7)$ and $(2, -1)$ in Figure 1.75 show that values of y extend from -7 to -1, inclusive:

$$\text{Range} = [-7, -1].$$

Check Point 5   **a.** Find the center and radius of the circle whose equation is
$$(x + 3)^2 + (y - 1)^2 = 4.$$

      b. Graph the equation.

      c. Use the graph to identify the relation's domain and range.

If we square $x - 2$ and $y + 4$ in the standard form of the equation from Example 5, we obtain another form for the circle's equation.

$(x - 2)^2 + (y + 4)^2 = 9$	This is the standard form of the equation from Example 5.
$x^2 - 4x + 4 + y^2 + 8y + 16 = 9$	Square $x - 2$ and $y + 4$.
$x^2 + y^2 - 4x + 8y + 20 = 9$	Combine constants and rearrange terms.
$x^2 + y^2 - 4x + 8y + 11 = 0$	Subtract 9 from both sides.

This result suggests that an equation in the form $x^2 + y^2 + Dx + Ey + F = 0$ can represent a circle. This is called the *general form of the equation of a circle.*

The General Form of the Equation of a Circle

The **general form of the equation of a circle** is
$$x^2 + y^2 + Dx + Ey + F = 0,$$
where D, E, and F are real numbers.

---

⑤ Convert the general form of a circle's equation to standard form.

We can convert the general form of the equation of a circle to the standard form $(x - h)^2 + (y - k)^2 = r^2$. We do so by completing the square on x and y. Let's see how this is done.

EXAMPLE 6   **Converting the General Form of a Circle's Equation to Standard Form and Graphing the Circle**

Study Tip

To review completing the square, see Section P.7, pages 87–88.

Write in standard form and graph: $x^2 + y^2 + 4x - 6y - 23 = 0$.

Solution   Because we plan to complete the square on both x and y, let's rearrange the terms so that x-terms are arranged in descending order, y-terms are arranged in descending order, and the constant term appears on the right.

$x^2 + y^2 + 4x - 6y - 23 = 0$	This is the given equation.
$(x^2 + 4x \quad\;) + (y^2 - 6y \quad\;) = 23$	Rewrite in anticipation of completing the square.
$(x^2 + 4x + 4) + (y^2 - 6y + 9) = 23 + 4 + 9$	Complete the square on x: $\frac{1}{2} \cdot 4 = 2$ and $2^2 = 4$, so add 4 to both sides. Complete the square on y: $\frac{1}{2}(-6) = -3$ and $(-3)^2 = 9$, so add 9 to both sides.

> Remember that numbers added on the left side must also be added on the right side.

$(x + 2)^2 + (y - 3)^2 = 36$	Factor on the left and add on the right.

9.

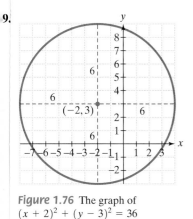

Figure 1.76 The graph of
$(x + 2)^2 + (y - 3)^2 = 36$

This last equation, $(x + 2)^2 + (y - 3)^2 = 36$, is in standard form. We can identify the circle's center and radius by comparing this equation to the standard form of the equation of a circle, $(x - h)^2 + (y - k)^2 = r^2$.

$$(x + 2)^2 + (y - 3)^2 = 36$$

$$(x - (-2))^2 + (y - 3)^2 = 6^2$$

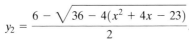

This is $(x - h)^2$, with $h = -2$.   This is $(y - k)^2$, with $k = 3$.   This is r^2, with $r = 6$.

We use the center, $(h, k) = (-2, 3)$, and the radius, $r = 6$, to graph the circle. The graph is shown in Figure 1.76.

Technology

To graph $x^2 + y^2 + 4x - 6y - 23 = 0$, rewrite the equation as a quadratic equation in y.
$$y^2 - 6y + (x^2 + 4x - 23) = 0$$

Now solve for y using the quadratic formula, with $a = 1$, $b = -6$, and $c = x^2 + 4x - 23$.

$$y = \frac{-b \pm \sqrt{b^2 - 4ac}}{2a} = \frac{-(-6) \pm \sqrt{(-6)^2 - 4 \cdot 1(x^2 + 4x - 23)}}{2 \cdot 1} = \frac{6 \pm \sqrt{36 - 4(x^2 + 4x - 23)}}{2}$$

Because we will enter these equations, there is no need to simplify. Enter

$$y_1 = \frac{6 + \sqrt{36 - 4(x^2 + 4x - 23)}}{2}$$

and

$$y_2 = \frac{6 - \sqrt{36 - 4(x^2 + 4x - 23)}}{2}.$$

Use a ZOOM SQUARE setting. The graph is shown on the right.

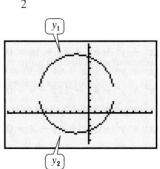

Check Point 6 Write in standard form and graph:

$$x^2 + y^2 + 4x - 4y - 1 = 0.$$

EXERCISE SET 1.9

Practice Exercises

In Exercises 1–18, find the distance between each pair of points. If necessary, round answers to two decimals places.

1. $(2, 3)$ and $(14, 8)$ **2.** $(5, 1)$ and $(8, 5)$

3. $(4, -1)$ and $(-6, 3)$ **4.** $(2, -3)$ and $(-1, 5)$

5. $(0, 0)$ and $(-3, 4)$ **6.** $(0, 0)$ and $(3, -4)$

7. $(-2, -6)$ and $(3, -4)$ **8.** $(-4, -1)$ and $(2, -3)$

9. $(0, -3)$ and $(4, 1)$ **10.** $(0, -2)$ and $(4, 3)$

11. $(3.5, 8.2)$ and $(-0.5, 6.2)$ **12.** $(2.6, 1.3)$ and $(1.6, -5.7)$

13. $\left(0, -\sqrt{3}\right)$ and $\left(\sqrt{5}, 0\right)$ **14.** $\left(0, -\sqrt{2}\right)$ and $\left(\sqrt{7}, 0\right)$

15. $\left(3\sqrt{3}, \sqrt{5}\right)$ and $\left(-\sqrt{3}, 4\sqrt{5}\right)$

16. $\left(2\sqrt{3}, \sqrt{6}\right)$ and $\left(-\sqrt{3}, 5\sqrt{6}\right)$

17. $\left(\frac{7}{3}, \frac{1}{5}\right)$ and $\left(\frac{1}{3}, \frac{6}{5}\right)$ **18.** $\left(-\frac{1}{4}, -\frac{1}{7}\right)$ and $\left(\frac{3}{4}, \frac{6}{7}\right)$

In Exercises 19–30, find the midpoint of each line segment with the given endpoints.

19. $(6, 8)$ and $(2, 4)$ **20.** $(10, 4)$ and $(2, 6)$

21. $(-2, -8)$ and $(-6, -2)$ **22.** $(-4, -7)$ and $(-1, -3)$

23. $(-3, -4)$ and $(6, -8)$ **24.** $(-2, -1)$ and $(-8, 6)$

25. $\left(-\frac{7}{2}, \frac{3}{2}\right)$ and $\left(-\frac{5}{2}, -\frac{11}{2}\right)$

26. $\left(-\frac{2}{5}, \frac{7}{15}\right)$ and $\left(-\frac{2}{5}, -\frac{4}{15}\right)$

27. $\left(8, 3\sqrt{5}\right)$ and $\left(-6, 7\sqrt{5}\right)$ **28.** $\left(7\sqrt{3}, -6\right)$ and $\left(3\sqrt{3}, -2\right)$

29. $\left(\sqrt{18}, -4\right)$ and $\left(\sqrt{2}, 4\right)$ **30.** $\left(\sqrt{50}, -6\right)$ and $\left(\sqrt{2}, 6\right)$

SECTION 1.10 Modeling with Functions

Objectives

❶ Construct functions from verbal descriptions.

❷ Construct functions from formulas.

A can of Coca-Cola is sold every six seconds throughout the world.

Study Tip

In calculus, you will solve problems involving maximum or minimum values of functions. Such problems often require creating the function that is to be maximized or minimized. Quite often, the calculus is fairly routine. It is the algebraic setting up of the function that causes difficulty. That is why the material in this section is so important.

In 2005, to curb consumption of sugared soda, the Center for Science in the Public Interest (CSPI) urged the FDA to slap cigarette-style warning labels on these drinks, citing statistics like this: In 2004, the average American drank 37 gallons—60,000 calories—of what CSPI calls "liquid candy." Despite the variety of its reputations throughout the world, the soft drink industry has spent far more time reducing the amount of aluminum in its cylindrical cans than addressing the problems of the nutritional disaster floating within its packaging. In the 1960s, one pound of aluminum made fewer than 20 cans; today, almost 30 cans come out of the same amount. The thickness of the can wall is less than five-thousandths of an inch, about the same as a magazine cover.

Many real-world problems involve constructing mathematical models that are functions. The problem of minimizing the amount of aluminum needed to manufacture a 12-ounce soft-drink can first requires that we express the surface area of all such cans as a function of their radius. In constructing such a function, we must be able to translate a verbal description into a mathematical representation—that is, a mathematical model.

In Chapter P, Section P.8, we reviewed how to obtain equations that modeled problems' verbal conditions. Earlier in this chapter, we used real-world data to obtain models that were linear functions. In this section, you will learn to use verbal descriptions and formulas to obtain models involving both linear and nonlinear functions.

❶ Construct functions from verbal descriptions.

Functions from Verbal Descriptions

There is no rigid step-by-step procedure that can be used to construct a function from a verbal description. Read the problem carefully. Attempt to write a critical sentence that describes the function's conditions in terms of its independent variable, x. In the following examples, we will use voice balloons that show these critical sentences, or **verbal models**. Then translate the verbal model into the algebraic notation used to represent a function's equation.

EXAMPLE 1 Modeling Costs of Long-Distance Carriers

You are choosing between two long-distance telephone plans. Plan A has a monthly fee of $20 with a charge of $0.05 per minute for all long-distance calls. Plan B has a monthly fee of $5 with a charge of $0.10 per minute for all long-distance calls.

    a. Express the monthly cost for plan A, f, as a function of the number of minutes of long-distance calls in a month, x.

    b. Express the monthly cost for plan B, g, as a function of the number of minutes of long-distance calls in a month, x.

    c. For how many minutes of long-distance calls will the costs for the two plans be the same?

Solution

a. The monthly cost for plan A is the monthly fee, $20, plus the per-minute charge, $0.05, times the number of minutes of long-distance calls, x.

Monthly cost for plan A	equals	monthly fee	plus	per-minute charge	times	the number of minutes of long-distance calls.
$f(x)$	$=$	20	$+$	0.05	$\cdot$	x

The function $f(x) = 0.05x + 20$, expressed in slope-intercept form, models the monthly cost, f, in dollars, in terms of the number of minutes of long-distance calls, x.

b. The monthly cost for plan B is the monthly fee, $5, plus the per-minute charge, $0.10, times the number of minutes of long-distance calls, x.

Monthly cost for plan B	equals	monthly fee	plus	per-minute charge	times	the number of minutes of long-distance calls.
$g(x)$	$=$	5	$+$	0.10	$\cdot$	x

The function $g(x) = 0.10x + 5$, expressed in slope-intercept form, models the monthly cost, g, in dollars, in terms of the number of minutes of long-distance calls, x.

c. We are interested in how many minutes of long-distance calls, x, result in the same monthly costs, f and g, for the two plans. Thus, we must set the equations for f and g equal to each other. We then solve the resulting linear equation for x.

The monthly cost for plan A	must equal	the monthly cost for plan B.
$20 + 0.05x$	$=$	$5 + 0.10x$

$$20 + 0.05x = 5 + 0.10x \qquad \text{This equation models equal monthly costs.}$$
$$20 = 5 + 0.05x \qquad \text{Subtract 0.05x from both sides.}$$
$$15 = 0.05x \qquad \text{Subtract 5 from both sides.}$$
$$300 = x \qquad \text{Divide both sides by 0.05: } \tfrac{15}{0.05} = 300.$$

The costs for the two plans will be the same with 300 minutes of long-distance calls. Take a moment to verify that $f(300) = g(300) = 35$. Thus, the cost for each plan will be $35.

In Example 1, the functions modeling the costs for the two plans are both linear functions of the form $f(x) = mx + b$. Based on the meaning of the functions' variables, we can interpret slope and y-intercept as follows:

Plan A	**Plan B**
$f(x) = 0.05x + 20$	$g(x) = 0.10x + 5.$

| The slope indicates that the rate of change in the plan's cost is $0.05 per minute. | The y-intercept indicates the starting cost with no long-distance calls is $20. | The slope indicates that the rate of change in the plan's cost is $0.10 per minute. | The y-intercept indicates the starting cost with no long-distance calls is $5. |

Technology

We can use a graphing utility to numerically or graphically verify our work in Example 1(c). Enter the linear functions that model the costs for the two plans.

The monthly cost for plan A	must equal	the monthly cost for plan B.

$$20 + 0.05x \quad = \quad 5 + 0.10x$$

Enter $y_1 = 20 + .05x$.   Enter $y_2 = 5 + .10x$.

Numeric Check
Display a table for y_1 and y_2.

Graphic Check
Display graphs for y_1 and y_2. Use the intersection feature.

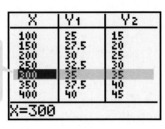

When $x = 300$, y_1 and y_2 have the same value, 35. With 300 minutes of calls, costs are the same, $35, for both plans.

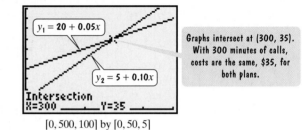

Graphs intersect at (300, 35). With 300 minutes of calls, costs are the same, $35, for both plans.

[0, 500, 100] by [0, 50, 5]

Check Point 1 You are choosing between two long-distance telephone plans. Plan A has a monthly fee of $15 with a charge of $0.08 per minute for all long-distance calls. Plan B has a monthly fee of $3 with a charge of $0.12 per minute for all long-distance calls.

    a. Express the monthly cost for plan A, f, as a function of the number of minutes of long-distance calls in a month, x.

    b. Express the monthly cost for plan B, g, as a function of the number of minutes of long-distance calls in a month, x.

    c. For how many minutes of long-distance calls will the costs for the two plans be the same?

EXAMPLE 2   Modeling the Number of Customers and Revenue

On a certain route, an airline carries 6000 passengers per month, each paying $200. A market survey indicates that for each $1 increase in the ticket price, the airline will lose 100 passengers.

    a. Express the number of passengers per month, N, as a function of the ticket price, x.

    b. The airline's monthly revenue for the route is the product of the number of passengers and the ticket price. Express the monthly revenue, R, as a function of the ticket price, x.

Solution

    a. The number of passengers, N, depends on the ticket price, x. In particular, the number of passengers is the original number, 6000, minus the number lost to the fare increase. The following table shows how to find the number lost to the fare increase:

English Phrase	Algebraic Translation
Ticket price	x
Amount of fare increase: ticket price minus original ticket price	$x - 200$ *The original ticket price was $200.*
Decrease in passengers due to the fare increase: 100 times the dollar amount of the fare increase	$100(x - 200)$ *100 passengers are lost for each dollar of fare increase.*

The number of passengers per month, N, is the original number of passengers, 6000, minus the decrease due to the fare increase.

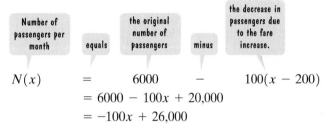

$$N(x) = 6000 - 100(x - 200)$$
$$= 6000 - 100x + 20{,}000$$
$$= -100x + 26{,}000$$

The linear function $N(x) = -100x + 26{,}000$ models the number of passengers per month, N, in terms of the price per ticket, x. The linear function's slope, -100, indicates that the rate of change is a loss of 100 passengers per dollar of fare increase.

b. The monthly revenue for the route is the number of passengers, $-100x + 26{,}000$, times the ticket price, x.

> Monthly revenue   equals   the number of passengers   times   the ticket price.

$$R(x) = (-100x + 26{,}000) \cdot x$$
$$= -100x^2 + 26{,}000x$$

The function $R(x) = -100x^2 + 26{,}000x$ models the airline's monthly revenue for the route, R, in terms of the ticket price, x.

The revenue function in Example 2 is of the form $f(x) = ax^2 + bx + c$. Any function of this form, where $a \neq 0$, is called a **quadratic function**. In this chapter, we used the U-shaped graph of the standard quadratic function, $f(x) = x^2$, to graph various transformations. In the next chapter, you will study quadratic functions in detail, including where maximum or minimum values occur.

Check Point **2** On a certain route, an airline carries 8000 passengers per month, each paying $100. A market survey indicates that for each $1 increase in ticket price, the airline will lose 100 passengers.
  a. Express the number of passengers per month, N, as a function of the ticket price, x.
  b. Express the monthly revenue for the route, R, as a function of the ticket price, x.

Functions from Formulas

In Chapter P, Section P.8, we used basic geometric formulas to obtain equations that modeled geometric situations. Formulas for area, perimeter, and volume are given in Table P.6 on page 100. Obtaining functions that model geometric situations requires a knowledge of these formulas. Take a moment to turn to page 100 and be sure that you are familiar with the 13 formulas given in the table.

Technology

The graph of the function $R(x) = -100x^2 + 26{,}000x$, the model for the airline's revenue, R, in terms of the ticket price, x, is shown in a $[0, 260, 13]$ by $[0, 1{,}700{,}000, 100{,}000]$ viewing rectangle. The graphing utility's maximum function feature indicates that a ticket price of $130 yields the maximum revenue. As the ticket price exceeds $130, the maximum revenue, $1,690,000, starts to decrease.

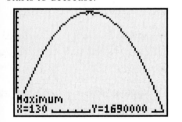

 Construct functions from formulas.

In our next example, we will obtain a function using the formula for the volume of a rectangular solid, $V = lwh$. A rectangular solid's volume is the product of its length, width, and height.

EXAMPLE 3   Obtaining a Function from a Geometric Formula

A machine produces open boxes using square sheets of metal measuring 12 inches on each side. The machine cuts equal-sized squares from each corner. Then it shapes the metal into an open box by turning up the sides.

a. Express the volume of the box, V, in cubic inches, as a function of the length of the side of the square cut from each corner, x, in inches.

b. Find the domain of V.

Solution

a. The situation is illustrated in Figure 1.77. The volume of the box in the lower portion of the figure is the product of its length, width, and height. The height of the box is the same as the side of the square cut from each corner, x. Because the 12-inch square has x inches cut from each corner, the length of the resulting box is $12 - x - x$, or $12 - 2x$. Similarly, the width of the resulting box is also $12 - 2x$.

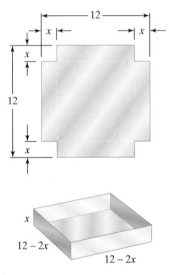

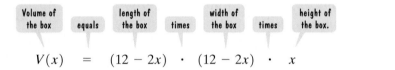

$$V(x) \quad = \quad (12 - 2x) \quad \cdot \quad (12 - 2x) \quad \cdot \quad x$$

The function $V(x) = x(12 - 2x)^2$ models the volume of the box, V, in terms of the length of the side of the square cut from each corner, x.

b. The formula for V involves a polynomial, $x(12 - 2x)^2$, which is defined for any real number, x. However, in the function $V(x) = x(12 - 2x)^2$, x represents the number of inches cut from each corner of the 12-inch square. Thus, $x > 0$. To produce an open box, the machine must cut less than 6 inches from each corner of the 12-inch square. Thus, $x < 6$. The domain of V is $\{x | 0 < x < 6\}$, or, in interval notation, $(0, 6)$.

Figure 1.77 Producing open boxes using square sheets of metal

Technology

The graph of the function $V(x) = x(12 - 2x)^2$, the model for the volume of the box in Figure 1.77, is shown in a $[0, 6, 1]$ by $[0, 130, 13]$ viewing rectangle. The graphing utility's maximum function feature indicates that the volume of the box is a maximum, 128 cubic inches, when the side of the square cut from each corner of the metal sheet is 2 inches.

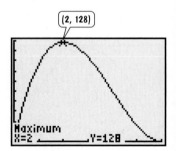

 Check Point 3 A machine produces open boxes using rectangular sheets of metal measuring 15 inches by 8 inches. The machine cuts equal-sized squares from each corner. Then it shapes the metal into an open box by turning up the sides.

a. Express the volume of the box, V, in cubic inches, as a function of the length of the side of the square cut from each corner, x, in inches.

b. Find the domain of V.

In many situations, the conditions of the problem result in a function whose equation contains more than one variable. If this occurs, use the given information to write an equation among these variables. Then use this equation to eliminate all but one of the variables in the function's expression.

EXAMPLE 4   Modeling the Area of a Rectangle with a Fixed Perimeter

You have 140 yards of fencing to enclose a rectangular garden. Express the area of the garden, A, as a function of one of its dimensions, x.

Solution   Figure 1.78 illustrates three of your options for enclosing the garden. In each case, the perimeter of the rectangle, twice the length plus twice the width, is 140 yards. By contrast, the area, length times width, varies according to the length of a side, x.

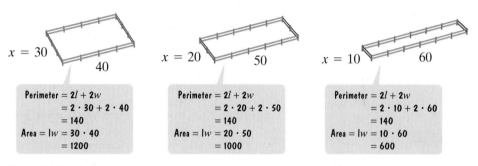

Figure 1.78   Rectangles with a fixed perimeter and varying areas

As specified, x represents one of the dimensions of the rectangle. In particular, let

$$x = \text{the length of the garden}$$
$$y = \text{the width of the garden.}$$

The area, A, of the garden is the product of its length and its width:

$$A = xy.$$

There are two variables in this formula—the garden's length, x, and its width, y. We need to transform this into a function in which A is represented by one variable, x, the garden's length. Thus, we must express the width, y, in terms of the length, x. We do this using the information that you have 140 yards of fencing.

$2x + 2y = 140$	The perimeter, twice the length plus twice the width, is 140 yards.
$2y = 140 - 2x$	Subtract 2x from both sides.
$y = \dfrac{140 - 2x}{2}$	Divide both sides by 2.
$y = 70 - x$	Divide each term in the numerator by 2.

Now we substitute $70 - x$ for y in the formula for area.

$$A = xy = x(70 - x)$$

The rectangle and its dimensions are illustrated in Figure 1.79. Because A is a function of x, we can write

$$A(x) = x(70 - x) \quad \text{or} \quad A(x) = 70x - x^2.$$

This function models the area, A, of a rectangular garden with a perimeter of 140 yards in terms of the length of a side, x.

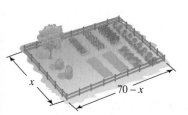

Figure 1.79

Technology

The graph of the function

$$A(x) = x(70 - x),$$

the model for the area of the garden in Example 4, is shown in a $[0, 70, 5]$ by $[0, 1400, 100]$ viewing rectangle. The graph shows that as the length of a side increases, the enclosed area increases, then decreases. The area of the garden is a maximum, 1225 square yards, when the length of one of its sides is 35 yards.

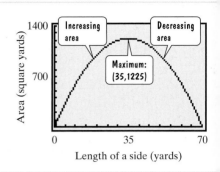

Check Point 4 You have 200 feet of fencing to enclose a rectangular garden. Express the area of the garden, A, as a function of one of its dimensions, x.

In order to save on production costs, manufacturers need to use the least amount of material for containers that are required to hold a specified volume of their product. Using the least amount of material involves minimizing the surface area of the container. Formulas for surface area, A, are given in Table 1.6.

Table 1.6 Common Formulas for Surface Area

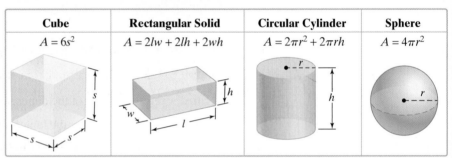

Cube	Rectangular Solid	Circular Cylinder	Sphere
$A = 6s^2$	$A = 2lw + 2lh + 2wh$	$A = 2\pi r^2 + 2\pi rh$	$A = 4\pi r^2$

Technology

The graph of the function

$$A(r) = 2\pi r^2 + \frac{44}{r}, \text{ or}$$

$$y = 2\pi x^2 + \frac{44}{x},$$

the model for the surface area of the soft-drink can in Figure 1.80, was obtained with a graphing utility.

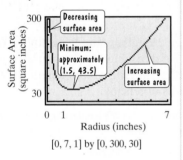

$[0, 7, 1]$ by $[0, 300, 30]$

As the radius increases, the can's surface area decreases, then increases. Using a graphing utility's minimum function feature, it can be shown that when $x \approx 1.5$, the value of y is smallest ($y \approx 43.5$). Thus, the least amount of material needed to manufacture the can, approximately 43.5 square inches, occurs when its radius is approximately 1.5 inches. In calculus, you will learn techniques that give the exact radius needed to minimize the can's surface area.

EXAMPLE 5 Modeling the Surface Area of a Soft-Drink Can with Fixed Volume

Figure 1.80 shows a cylindrical soft-drink can. The can is to have a volume of 12 fluid ounces, approximately 22 cubic inches. Express the surface area of the can, A, in square inches, as a function of its radius, r, in inches.

Solution The surface area, A, of the cylindrical can in Figure 1.80 is given by

$$A = 2\pi r^2 + 2\pi rh.$$

Figure 1.80

There are two variables in this formula—the can's radius, r, and its height, h. We need to transform this into a function in which A is represented by one variable, r, the radius of the can. Thus, we must express the height, h, in terms of the radius, r. We do this using the information that the can's volume, $V = \pi r^2 h$, must be 22 cubic inches.

$$\pi r^2 h = 22 \qquad \text{The volume of the can is 22 cubic inches.}$$

$$h = \frac{22}{\pi r^2} \qquad \text{Divide both sides by } \pi r^2 \text{ and solve for } h.$$

Now we substitute $\frac{22}{\pi r^2}$ for h in the formula for surface area.

$$A = 2\pi r^2 + 2\pi rh = 2\pi r^2 + 2\pi r\left(\frac{22}{\pi r^2}\right) = 2\pi r^2 + \frac{44}{r}$$

Because A is a function of r, the can's radius, we can express the surface area of the can as

$$A(r) = 2\pi r^2 + \frac{44}{r}.$$

Check Point 5 A cylindrical can is to hold 1 liter, or 1000 cubic centimeters, of oil. Express the surface area of the can, A, in square centimeters, as a function of its radius, r, in centimeters.

Our next example involves constructing a function that models simple interest. The annual simple interest that an investment earns is given by the formula

$$I = Pr,$$

where I is the simple interest, P is the principal, and r is the simple interest rate, expressed in decimal form. Suppose, for example, that you deposit \$2000 ($P = 2000$) in an account that has a simple interest rate of 3% ($r = 0.03$). The annual simple interest is computed as follows:

$$I = Pr = (2000)(0.03) = 60.$$

The annual interest is \$60.

EXAMPLE 6 Modeling Simple Interest

You inherit \$16,000 with the stipulation that for the first year the money must be placed in two investments expected to pay 6% and 8% annual interest, respectively. Express the expected interest, I, as a function of the amount of money invested at 6%, x.

Solution As specified, x represents the amount invested at 6%. We will let y represent the amount invested at 8%. The expected interest, I, on the two investments combined is the expected interest on the 6% investment plus the expected interest on the 8% investment.

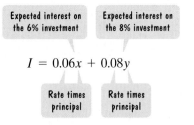

$$I = 0.06x + 0.08y$$

There are two variables in this formula—the amount invested at 6%, x, and the amount invested at 8%, y. We need to transform this into a function in which I is represented by one variable, x, the amount invested at 6%. Thus, we must express the amount invested at 8%, y, in terms of the amount invested at 6%, x. We do this using the information that you have \$16,000 to invest.

$$x + y = 16{,}000 \qquad \text{The sum of the amounts invested at each rate must be \$16,000.}$$

$$y = 16{,}000 - x \qquad \text{Subtract } x \text{ from both sides and solve for } y.$$

Now we substitute $16{,}000 - x$ for y in the formula for interest.

$$I = 0.06x + 0.08y = 0.06x + 0.08(16{,}000 - x).$$

Because I is now a function of x, the amount invested at 6%, the expected interes can be expressed as

$$I(x) = 0.06x + 0.08(16{,}000 - x).$$

Check Point 6 You place $25,000 in two investments expected to pay 7% and 9% annua interest, respectively. Express the expected interest, I, as a function of the amount of money invested at 7%, x.

Our next example involves constructing a function using the distance formula In the previous section, we saw that the distance between two points in the rectangu lar coordinate system is the square root of the difference between their x-coordinate squared plus the difference between their y-coordinates squared.

EXAMPLE 7 Modeling the Distance from the Origin to a Point on a Graph

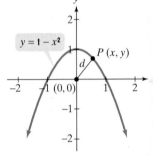

Technology

The graph of the function
$$d(x) = \sqrt{x^4 - x^2 + 1}, \text{ or}$$

$$y = \sqrt{x^4 - x^2 + 1},$$

the model for the distance, d, shown in Figure 1.81, was obtained with a graphing utility.

Minimum: approximately $(-0.71, 0.87)$

Minimum: approximately $(0.71, 0.87)$

$[-2, 2, 1]$ by $[0, 2, 1]$

Using a graphing utility's minimum function feature, it can be shown that when $x \approx -0.71$ and when $x \approx 0.71$, the value of d is smallest ($d \approx 0.87$ is a relative minimum). Examine Figure 1.81. The graph's y-axis symmetry indicates that there are two points on $y = 1 - x^2$ whose distance to the origin is smallest.

Figure 1.81 shows that $P(x, y)$ is a point on the graph of $y = 1 - x^2$. Express the distance, d, from P to the origin as a function of the point's x-coordinate.

Solution We use the distance formula to find the distance, d, from $P(x, y)$ to the origin, $(0, 0)$.

$$d = \sqrt{(x - 0)^2 + (y - 0)^2} = \sqrt{x^2 + y^2}$$

There are two variables in this formula—the point's x-coordinate and its y-coordinate. We need to transform this into a function in which d is represented by one variable, x, the point's x-coordinate. Thus, we must express y in terms of x. We do this using the information shown in Figure 1.81, namely that $P(x, y)$ is a point on the graph of $y = 1 - x^2$ This means that we can replace y with $1 - x^2$ in our formula for d.

Figure 1.81

$$d = \sqrt{x^2 + y^2} = \sqrt{x^2 + (1 - x^2)^2} = \sqrt{x^2 + 1 - 2x^2 + x^4} = \sqrt{x^4 - x^2 + 1}$$

Square $1 - x^2$ using $(A - B)^2 = A^2 - 2AB + B^2$.

The distance, d, from $P(x, y)$ to the origin can be expressed as a function of th point's x-coordinate as

$$d(x) = \sqrt{x^4 - x^2 + 1}.$$

Check Point 7 Let $P(x, y)$ be a point on the graph of $y = x^3$. Express the distance, d from P to the origin as a function of the point's x-coordinate.

EXERCISE SET 1.10

Practice and Application Exercises

1. A car rental agency charges $200 per week plus $0.15 per mile to rent a car.

   a. Express the weekly cost to rent the car, f, as a function of the number of miles driven during the week, x.

   b. How many miles did you drive during the week if the weekly cost to rent the car was $320?

2. A car rental agency charges $180 per week plus $0.25 pe mile to rent a car.

   a. Express the weekly cost to rent the car, f, as a function c the number of miles driven during the week, x.

   b. How many miles did you drive during the week if th weekly cost to rent the car was $395?

3. One yardstick for measuring how steadily—if slowly—athletic performance has improved is the mile run. In 1954, Roger Bannister of Britain cracked the 4-minute mark, setting the record for running a mile in 3 minutes, 59.4 seconds, or 239.4 seconds. In the half-century since then, the record has decreased by 0.3 second per year.

   a. Express the record time for the mile run, M, as a function of the number of years after 1954, x.

   b. If this trend continues, in which year will someone run a 3-minute, or 180-second, mile?

4. According to the National Center for Health Statistics, in 1990, 28% of babies in the United States were born to parents who were not married. Throughout the 1990s, this increased by approximately 0.6% per year.

   a. Express the percentage of babies born out of wedlock, P, as a function of the number of years after 1990, x.

   b. If this trend continues, in which year will 40% of babies be born out of wedlock?

5. The bus fare in a city is $1.25. People who use the bus have the option of purchasing a monthly coupon book for $21.00. With the coupon book, the fare is reduced to $0.50.

   a. Express the total monthly cost to use the bus without a coupon book, f, as a function of the number of times in a month the bus is used, x.

   b. Express the total monthly cost to use the bus with a coupon book, g, as a function of the number of times in a month the bus is used, x.

   c. Determine the number of times in a month the bus must be used so that the total monthly cost without the coupon book is the same as the total monthly cost with the coupon book. What will be the monthly cost for each option?

6. A coupon book for a bridge costs $21 per month. The toll for the bridge is normally $2.50, but it is reduced to $1 for people who have purchased the coupon book.

   a. Express the total monthly cost to use the bridge without a coupon book, f, as a function of the number of times in a month the bridge is crossed, x.

   b. Express the total monthly cost to use the bridge with a coupon book, g, as a function of the number of times in a month the bridge is crossed, x.

   c. Determine the number of times in a month the bridge must be crossed so that the total monthly cost without the coupon book is the same as the total monthly cost with the coupon book. What will be the monthly cost for each option?

7. You are choosing between two plans at a discount warehouse. Plan A offers an annual membership of $100 and you pay 80% of the manufacturer's recommended list price. Plan B offers an annual membership fee of $40 and you pay 90% of the manufacturer's recommended list price.

   a. Express the total yearly amount paid to the warehouse under plan A, f, as a function of the dollars of merchandise purchased during the year, x.

   b. Express the total yearly amount paid to the warehouse under plan B, g, as a function of the dollars of merchandise purchased during the year, x.

   c. How many dollars of merchandise would you have to purchase in a year to pay the same amount under both plans? What will be the total yearly amount paid to the warehouse for each plan?

8. You are choosing between two plans at a discount warehouse. Plan A offers an annual membership fee of $300 and you pay 70% of the manufacturer's recommended list price. Plan B offers an annual membership fee of $40 and you pay 90% of the manufacturer's recommended list price.

   a. Express the total yearly amount paid to the warehouse under plan A, f, as a function of the dollars of merchandise purchased during the year, x.

   b. Express the total yearly amount paid to the warehouse under plan B, g, as a function of the dollars of merchandise purchased during the year, x.

   c. How many dollars of merchandise would you have to purchase in a year to pay the same amount under both plans? What will be the total yearly amount paid to the warehouse for each plan?

9. A football team plays in a large stadium. With a ticket price of $20, the average attendance at recent games has been 30,000. A market survey indicates that for each $1 increase in the ticket price, attendance decreases by 500.

   a. Express the number of spectators at a football game, N, as a function of the ticket price, x.

   b. Express the revenue from a football game, R, as a function of the ticket price, x.

10. A baseball team plays in a large stadium. With a ticket price of $15, the average attendance at recent games has been 20,000. A market survey indicates that for each $1 increase in the ticket price, attendance decreases by 400.

    a. Express the number of spectators at a baseball game, N, as a function of the ticket price, x.

    b. Express the revenue from a baseball game, R, as a function of the ticket price, x.

11. On a certain route, an airline carries 9000 passengers per month, each paying $150. A market survey indicates that for each $1 decrease in the ticket price, the airline will gain 50 passengers.

    a. Express the number of passengers per month, N, as a function of the ticket price, x.

    b. Express the monthly revenue for the route, R, as a function of the ticket price, x.

12. On a certain route, an airline carries 7000 passengers per month, each paying $90. A market survey indicates that for each $1 decrease in the ticket price, the airline will gain 60 passengers.

    a. Express the number of passengers per month, N, as a function of the ticket price, x.

    b. Express the monthly revenue for the route, R, as a function of the ticket price, x.

13. The annual yield per lemon tree is fairly constant at 320 pounds per tree when the number of trees per acre is 50 or fewer. For each additional tree over 50, the annual yield per tree for all trees on the acre decreases by 4 pounds due to overcrowding.

    a. Express the yield per tree, Y, in pounds, as a function of the number of lemon trees per acre, x.

    b. Express the total yield for an acre, T, in pounds, as a function of the number of lemon trees per acre, x.

14. The annual yield per orange tree is fairly constant at 270 pounds per tree when the number of trees per acre is 30 or fewer. For each additional tree over 30, the annual yield per tree for all trees on the acre decreases by 3 pounds due to overcrowding.

    a. Express the yield per tree, Y, in pounds, as a function of the number of orange trees per acre, x.

    b. Express the total yield for an acre, T, in pounds, as a function of the number of orange trees per acre, x.

15. An open box is made from a square piece of cardboard 24 inches on a side by cutting identical squares from the corners and turning up the sides.

    a. Express the volume of the box, V, as a function of the length of the side of the square cut from each corner, x.

    b. Find and interpret $V(2)$, $V(3)$, $V(4)$, $V(5)$, and $V(6)$. What is happening to the volume of the box as the length of the side of the square cut from each corner increases?

    c. Find the domain of V.

16. An open box is made from a square piece of cardboard 30 inches on a side by cutting identical squares from the corners and turning up the sides.

    a. Express the volume of the box, V, as a function of the length of the side of the square cut from each corner, x.

    b. Find and interpret $V(3)$, $V(4)$, $V(5)$, $V(6)$, and $V(7)$. What is happening to the volume of the box as the length of the side of the square cut from each corner increases?

    c. Find the domain of V.

17. A rain gutter is made from sheets of aluminum that are 20 inches wide. As shown in the figure, the edges are turned up to form right angles. Express the cross-sectional area of the gutter, A, as a function of its depth, x.

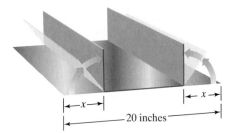

18. A piece of wire is 8 inches long. The wire is cut into two pieces and then each piece is bent into a square. Express the sum of the areas of these squares, A, as a function of the length of the cut, x.

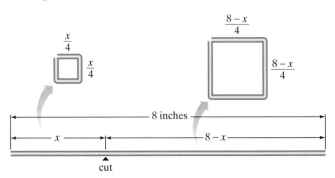

19. The sum of two numbers is 66. Express the product of the numbers, P, as a function of one of the numbers, x.

20. The sum of two numbers is 50. Express the product of the numbers, P, as a function of one of the numbers, x.

21. You have 800 feet of fencing to enclose a rectangular field. Express the area of the field, A, as a function of one of its dimensions, x.

22. You have 600 feet of fencing to enclose a rectangular field. Express the area of the field, A, as a function of one of its dimensions, x.

23. As in Exercise 21, you have 800 feet of fencing to enclose a rectangular field. However, one side of the field lies along a canal and requires no fencing. Express the area of the field, A, as a function of one of its dimensions, x.

24. As in Exercise 22, you have 600 feet of fencing to enclose a rectangular field. However, one side of the field lies along a canal and requires no fencing. Express the area of the field, A, as a function of one of its dimensions, x.

25. You have 1000 feet of fencing to enclose a rectangular playground and subdivide it into two smaller playgrounds by placing the fencing parallel to one of the sides. Express the area of the playground, A, as a function of one of its dimensions, x.

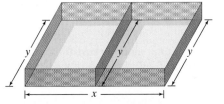

26. You have 1200 feet of fencing to enclose a rectangular region and subdivide it into three smaller rectangular regions by placing two fences parallel to one of the sides. Express the area of the enclosed rectangular region, A, as a function of one of its dimensions, x.

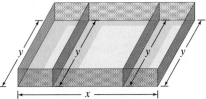

27. A new running track is to be constructed in the shape of a rectangle with semicircles at each end. The track is to be 440 yards long. Express the area of the region enclosed by the track, A, as a function of its radius, r.

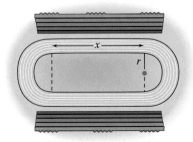

28. Work Exercise 27 if the length of the track is increased to 880 yards.

29. A contractor is to build a warehouse whose rectangular floor will have an area of 4000 square feet. The warehouse will be separated into two rectangular rooms by an interior wall. The cost of the exterior walls is $175 per linear foot and the cost of the interior wall is $125 per linear foot. Express the contractor's cost for building the walls, C, as a function of one of the dimensions of the warehouse's rectangular floor, x.

0. The area of a rectangular garden is 125 square feet. The garden is to be enclosed on three sides by a brick wall costing $20 per foot and on one side by a fence costing $9 per foot. Express the cost to enclose the garden, C, as a function of one of its dimensions, x.

1. The figure shows an open box with a square base. The box is to have a volume of 10 cubic feet. Express the amount of material needed to construct the box, A, as a function of the length of a side of its square base, x.

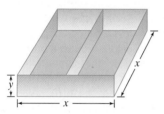

2. The figure shows an open box with a square base and a partition down the middle. The box is to have a volume of 400 cubic inches. Express the amount of material needed to construct the box, A, as a function of the length of a side of its square base, x.

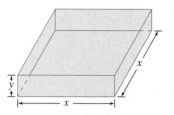

3. The figure shows a package whose front is a square. The length plus girth (the distance around) of the package is 300 inches. (This is the maximum length plus girth permitted by Federal Express for its overnight service.) Express the volume of the package, V, as a function of the length of a side of its square front, x.

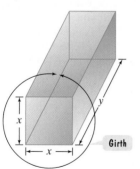

4. Work Exercise 33 if the length plus girth of the package is 108 inches.

5. Your grandmother needs your help. She has $50,000 to invest. Part of this money is to be invested in noninsured bonds paying 15% annual interest. The rest of this money is to be invested in a government-insured certificate of deposit paying 7% annual interest.
  a. Express the interest from both investments, I, as a function of the amount of money invested in noninsured bonds, x.
  b. Your grandmother told you that she requires $6000 per year in extra income from both these investments. How much money should be placed in each investment?

36. You inherit $18,750 with the stipulation that for the first year the money must be placed in two investments expected to pay 10% and 12% annual interest, respectively.
  a. Express the expected interest from both investments, I, as a function of the amount of money invested at 10%, x.
  b. If the total interest earned for the year was $2117, how much money was invested at each rate?

37. You invested $8000, part of it in a stock that paid 12% annual interest. However, the rest of the money suffered a 5% loss. Express the total annual income from both investments, I, as a function of the amount invested in the 12% stock, x.

38. You invested $12,000, part of it in a stock that paid 14% annual interest. However, the rest of the money suffered a 6% loss. Express the total annual income from both investments, I, as a function of the amount invested in the 14% stock, x.

39. Let $P(x, y)$ be a point on the graph of $y = x^2 - 4$. Express the distance, d, from P to the origin as a function of the point's x-coordinate.

40. Let $P(x, y)$ be a point on the graph of $y = x^2 - 8$. Express the distance, d, from P to the origin as a function of the point's x-coordinate.

41. Let $P(x, y)$ be a point on the graph of $y = \sqrt{x}$. Express the distance, d, from P to $(1, 0)$ as a function of the point's x-coordinate.

42. Let $P(x, y)$ be a point on the graph of $y = \sqrt{x}$. Express the distance, d, from P to $(2, 0)$ as a function of the point's x-coordinate.

43. The figure shows a rectangle with two vertices on a semicircle of radius 2 and two vertices on the x-axis. Let $P(x, y)$ be the vertex that lies in the first quadrant.

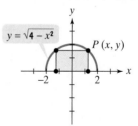

  a. Express the area of the rectangle, A, as a function of x.
  b. Express the perimeter of the rectangle, P, as a function of x.

44. The figure shows a rectangle with two vertices on a semicircle of radius 3 and two vertices on the x-axis. Let $P(x, y)$ be the vertex that lies in the first quadrant.

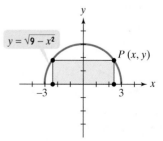

  a. Express the area of the rectangle, A, as a function of x.
  b. Express the perimeter of the rectangle, P, as a function of x.

45. Two vertical poles of length 6 feet and 8 feet, respectively, stand 10 feet apart. A cable reaches from the top of one pole to some point on the ground between the poles and then to the top of the other pole. Express the amount of cable used, f, as a function of the distance of the cable from the 6-foot pole, x.

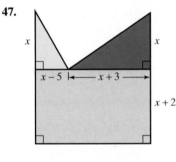

6 ft   8 ft

x $10 - x$

46. Towns A and B are located 6 miles and 3 miles, respectively, from a major expressway. The point on the expressway closest to town A is 12 miles from the point on the expressway closest to town B. Two new roads are to be built from A to the expressway and then to B. Express the combined lengths of the new roads, f, as a function of where the roads are positioned from A, shown as x in the figure.

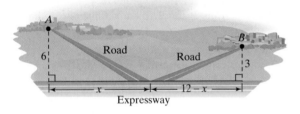

A

6   Road   Road B

3

x $12 - x$

Expressway

Practice Plus

In Exercises 47–48, express the area of each figure, A, as a function of one of its dimensions, x. Write the function's equation as a polynomial in standard form.

47.

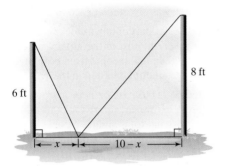

x x

$x - 5$ $x + 3$

$x + 2$

48.

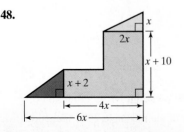

x

$2x$

$x + 10$

$x + 2$

$4x$

$6x$

In Exercises 49–50, express the volume of each figure, V, as a function of one of its dimensions, x. Write the function's equation as a polynomial in standard form.

49.

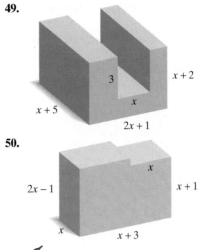

3    $x + 2$

x

$x + 5$

$2x + 1$

50.

x

$2x - 1$ $x + 1$

x $x + 3$

Writing in Mathematics

51. Throughout this section, we started with familiar formulas and created functions by substitution. Describe a specific situation in which we obtained a function using this technique.

52. Describe what should be displayed on the screen of a graphing utility to illustrate the solution that you obtained in Exercise 5(c) or Exercise 6(c).

53. In Exercise 9(b) or Exercise 10(b), describe what important information the team owners could learn from the revenue function.

54. In Exercise 13(b) or 14(b), describe what important information the growers could learn from the total-yield function.

55. In Exercise 31 or 32, describe what important information the box manufacturer could learn from the surface area function.

56. In calculus, you will learn powerful tools that reveal how functions behave. However, before applying these tools, there will be situations in which you are first required to obtain these functions from verbal descriptions. This is why your work in this section is so important. Because there is no rigid step-by-step procedure for modeling from verbal conditions, you might have had some difficulties obtaining functions for the assigned exercises. Discuss what you did if this happened to you. Did your course of action enhance your ability to model with functions?

Technology Exercises

57. Use a graphing utility to graph the function that you obtained in Exercise 1 or Exercise 2. Then use the TRACE or ZOOM feature to verify your answer in part (b) of the exercise.

58. Use a graphing utility to graph the two functions, f and g, that you obtained in any one exercise from Exercises 5–8. Then use the TRACE or INTERSECTION feature to verify your answer in part (c) of the exercise.

59. Use a graphing utility to graph the volume-of-the-box function, V, that you obtained in Exercise 15 or Exercise 16. Then use the TRACE or maximum function feature to find the length of the side of the square that should be cut from each corner of the cardboard to create a box with the greatest possible volume. What is the maximum volume of the open box?

0. Use a graphing utility to graph the area function, A, that you obtained in Exercise 21 or Exercise 22. Then use an appropriate feature on your graphing utility to find the dimensions of the field that result in the greatest possible area. What is the maximum area?

1. Use a graphing utility to graph the area function, A, that you obtained in Exercise 25 or Exercise 26. Then use an appropriate feature on your graphing utility to find the dimensions that result in the greatest possible area. What is the maximum area?

2. Use the maximum or minimum function feature of a graphing utility to provide useful numerical information to any one of the following: the manufacturer of the rain gutters in Exercise 17; the person enclosing the playground in Exercise 25; the contractor in Exercise 29; the manufacturer of the cylindrical cans in Check Point 5 (page 249).

Critical Thinking Exercises

3. You are on an island 2 miles from the nearest point P on a straight shoreline, as shown in the figure. Six miles down the shoreline from point P is a restaurant, shown as point R. To reach the restaurant, you first row from the island to point Q, averaging 2 miles per hour. Then you jog the distance from Q to R, averaging 5 miles per hour. Express the time, T, it takes to go from the island to the restaurant as a function of the distance, x, from P, where you land the boat.

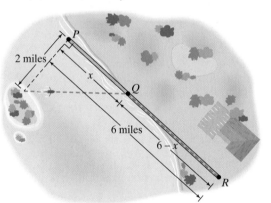

64. A pool measuring 20 meters by 10 meters is surrounded by a path of uniform width, as shown in the figure. Express the area of the path, A, in square meters, as a function of its width, x, in meters.

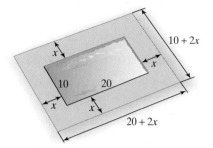

65. The figure shows a Norman window that has the shape of a rectangle with a semicircle attached at the top. The diameter of the semicircle is equal to the width of the rectangle. The window has a perimeter of 12 feet. Express the area of the window, A, as a function of its radius, r.

66. The figure shows water running into a container in the shape of a cone. The radius of the cone is 6 feet and its height is 12 feet. Express the volume of the water in the cone, V, as a function of the height of the water, h.

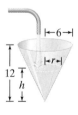

Chapter 1
Summary, Review, and Test

Summary

DEFINITIONS AND CONCEPTS	EXAMPLES

1.1 Graphs and Graphing Utilities

a. The rectangular coordinate system consists of a horizontal number line, the x-axis, and a vertical number line, the y-axis, intersecting at their zero points, the origin. Each point in the system corresponds to an ordered pair of real numbers (x, y). The first number in the pair is the x-coordinate; the second number is the y-coordinate. See Figure 1.1 on page 128.

Ex. 1, p. 129

b. An ordered pair is a solution of an equation in two variables if replacing the variables by the corresponding coordinates results in a true statement. The ordered pair is said to satisfy the equation. The graph of the equation is the set of all points whose coordinates satisfy the equation. One method for graphing an equation is to plot ordered-pair solutions and connect them with a smooth curve or line.

Ex. 2, p. 130;
Ex. 3, p. 130

DEFINITIONS AND CONCEPTS	EXAMPLES
c. An x-intercept of a graph is the x-coordinate of a point where the graph intersects the x-axis. The y-coordinate corresponding to an x-intercept is always zero. A y-intercept of a graph is the y-coordinate of a point where the graph intersects the y-axis. The x-coordinate corresponding to a y-intercept is always zero.	Ex. 5, p. 132

1.2 Basics of Functions and Their Graphs

a. A relation is any set of ordered pairs. The set of first components is the domain of the relation and the set of second components is the range.	Ex. 1, p. 139
b. A function is a correspondence from a first set, called the domain, to a second set, called the range, such that each element in the domain corresponds to exactly one element in the range. If any element in a relation's domain corresponds to more than one element in the range, the relation is not a function.	Ex. 2, p. 141
c. Functions are usually given in terms of equations involving x and y, in which x is the independent variable and y is the dependent variable. If an equation is solved for y and more than one value of y can be obtained for a given x, then the equation does not define y as a function of x. If an equation defines a function, the value of the function at x, $f(x)$, often replaces y.	Ex. 3, p. 142; Ex. 4, p. 143
d. The graph of a function is the graph of its ordered pairs.	Ex. 5, p. 144
e. The vertical line test for functions: If any vertical line intersects a graph in more than one point, the graph does not define y as a function of x.	Ex. 6, p. 145
f. The graph of a function can be used to determine the function's domain and its range. To find the domain, look for all the inputs on the x-axis that correspond to points on the graph. To find the range, look for all the outputs on the y-axis that correspond to points on the graph.	Ex. 8, p. 148
g. The zeros of a function, f, are the values of x for which $f(x) = 0$. At these values, the graph of f has x-intercepts. A function can have more than one x-intercept but at most one y-intercept.	Figure 1.23, p. 149

1.3 More on Functions and Their Graphs

a. The difference quotient is $$\frac{f(x + h) - f(x)}{h}, h \neq 0.$$	Ex. 1, p. 155
b. Piecewise functions are defined by two (or more) equations over a specified domain.	Ex. 2, p. 156
c. A function is increasing on intervals where its graph rises, decreasing on intervals where it falls, and constant on intervals where it neither rises nor falls. Precise definitions are given in the box on page 158.	Ex. 3, p. 158
d. If the graph of a function is given, we can often visually locate the number(s) at which the function has a relative maximum or relative minimum. Precise definitions are given in the box on page 159.	Figure 1.27, p. 159
e. The graph of an even function in which $f(-x) = f(x)$ is symmetric with respect to the y-axis. The graph of an odd function in which $f(-x) = -f(x)$ is symmetric with respect to the origin.	Ex. 4, p. 160
f. The graph of $f(x) = \text{int}(x)$, where $\text{int}(x)$ is the greatest integer that is less than or equal to x, has function values that form discontinuous steps, shown in Figure 1.32 on page 162. If $n \leq x < n + 1$, where n is an integer, then $\text{int}(x) = n$.	

1.4 Linear Functions and Slope

a. The slope, m, of the line through (x_1, y_1) and (x_2, y_2) is $m = \dfrac{y_2 - y_1}{x_2 - x_1}$.	Ex. 1, p. 169
b. Equations of lines include point-slope form, $y - y_1 = m(x - x_1)$, slope-intercept form, $y = mx + b$, and general form, $Ax + By + C = 0$. The equation of a horizontal line is $y = b$; a vertical line is $x = a$. A vertical line is not a linear function.	Ex. 2, p. 171; Ex. 3, p. 171; Ex. 5, p. 175
c. Linear functions in the form $f(x) = mx + b$ can be graphed using the slope, m, and the y-intercept, b. (See the box on page 172.) Linear equations in the general form $Ax + By + C = 0$ can be solved for y and graphed using the slope and the y-intercept. Intercepts can also be used to graph $Ax + By + C = 0$. (See the box on page 176.)	Ex. 4, p. 173; Ex. 6, p. 176; Ex. 7, p. 176

DEFINITIONS AND CONCEPTS	**EXAMPLES**

1.5 *More on Slope*

a. Parallel lines have equal slopes. Perpendicular lines have slopes that are negative reciprocals.
Ex. 1, p. 183;
Ex. 2, p. 185

b. The slope of a linear function is the rate of change of the dependent variable per unit change of the independent variable.
Ex. 3, p. 186

c. The average rate of change of f from x_1 to x_2 is
$$\frac{\Delta y}{\Delta x} = \frac{f(x_2) - f(x_1)}{x_2 - x_1}.$$
Ex. 4, p. 187;
Ex. 5, p. 189

d. If a function expresses an object's position, $s(t)$, in terms of time, t, the average velocity of the object from t_1 to t_2 is
$$\frac{\Delta s}{\Delta t} = \frac{s(t_2) - s(t_1)}{t_2 - t_1}.$$
Ex. 6, p. 190

1.6 *Transformations of Functions*

a. Table 1.4 on page 195 shows the graphs of the constant function, $f(x) = c$, the identity function, $f(x) = x$, the absolute value function, $f(x) = |x|$, the standard quadratic function, $f(x) = x^2$, the square root function, $f(x) = \sqrt{x}$, the standard cubic function, $f(x) = x^3$, and the cube root function, $f(x) = \sqrt[3]{x}$. The table also lists characteristics of each function.

b. Table 1.5 on page 203 summarizes how to graph a function using vertical shifts, $y = f(x) \pm c$, horizontal shifts, $y = f(x \pm c)$, reflections about the x-axis, $y = -f(x)$, reflections about the y-axis, $y = f(-x)$, vertical stretching, $y = cf(x), c > 1$, vertical shrinking, $y = cf(x), 0 < c < 1$, horizontal shrinking, $y = f(cx), c > 1$, and horizontal stretching, $y = f(cx), 0 < c < 1$.
Ex. 1, p. 196;
Ex. 2, p. 198;
Ex. 3, p. 199;
Ex. 4, p. 200;
Ex. 5, p. 200;
Ex. 6, p. 201;
Ex. 7, p. 202

c. A function involving more than one transformation can be graphed in the following order: (1) horizontal shifting; (2) stretching or shrinking; (3) reflecting; (4) vertical shifting.
Ex. 8, p. 204;
Ex. 9, p. 205

1.7 *Combinations of Functions; Composite Functions*

a. If a function f does not model data or verbal conditions, its domain is the largest set of real numbers for which the value of $f(x)$ is a real number. Exclude from a function's domain real numbers that cause division by zero and real numbers that result in a square root of a negative number.
Ex. 1, p. 211

b. When functions are given as equations, they can be added, subtracted, multiplied, or divided by performing operations with the algebraic expressions that appear on the right side of the equations. Definitions for the sum $f + g$, the difference $f - g$, the product fg, and the quotient $\frac{f}{g}$ functions, with domains $D_f \cap D_g$, and $g(x) \neq 0$ for the quotient function, are given in the box on page 213.
Ex. 2, p. 213;
Ex. 3, p. 214

c. The composition of functions f and g, $f \circ g$, is defined by $(f \circ g)(x) = f(g(x))$. The domain of the composite function $f \circ g$ is given in the box on page 216. This composite function is obtained by replacing each occurrence of x in the equation for f with $g(x)$.
Ex. 4, p. 217;
Ex. 5, p. 218

1.8 *Inverse Functions*

a. If $f(g(x)) = x$ and $g(f(x)) = x$, function g is the inverse of function f, denoted f^{-1} and read "f–inverse." Thus, to show that f and g are inverses of each other, one must show that $f(g(x)) = x$ and $g(f(x)) = x$.
Ex. 1, p. 225

b. The procedure for finding a function's inverse uses a switch-and-solve strategy. Switch x and y, and then solve for y. The procedure is given in the box on page 226.
Ex. 2, p. 226;
Ex. 3, p. 226;
Ex. 4, p. 227

c. The horizontal line test for inverse functions: A function f has an inverse that is a function, f^{-1}, if there is no horizontal line that intersects the graph of the function f at more than one point.
Ex. 5, p. 228

d. A one-to-one function is one in which no two different ordered pairs have the same second component. Only one-to-one functions have inverse functions.

e. If the point (a, b) is on the graph of f, then the point (b, a) is on the graph of f^{-1}. The graph of f^{-1} is a reflection of the graph of f about the line $y = x$.
Ex. 6, p. 229;
Ex. 7, p. 230

DEFINITIONS AND CONCEPTS	EXAMPLES

1.9 *Distance and Midpoint Formulas; Circles*

a. The distance, d, between the points (x_1, y_1) and (x_2, y_2) is given by $d = \sqrt{(x_2 - x_1)^2 + (y_2 - y_1)^2}$.   Ex. 1, p. 234

b. The midpoint of the line segment whose endpoints are (x_1, y_1) and (x_2, y_2) is the point with coordinates $\left(\dfrac{x_1 + x_2}{2}, \dfrac{y_1 + y_2}{2} \right)$.   Ex. 2, p. 235

c. The standard form of the equation of a circle with center (h, k) and radius r is $(x - h)^2 + (y - k)^2 = r^2$.   Ex. 3, p. 236;
Ex. 4, p. 237;
Ex. 5, p. 237

d. The general form of the equation of a circle is $x^2 + y^2 + Dx + Ey + F = 0$.

e. To convert from the general form to the standard form of a circle's equation, complete the square on x and y.   Ex. 6, p. 238

1.10 *Modeling with Functions*

a. Verbal models are often helpful in obtaining functions from verbal descriptions.   Ex. 1, p. 242;
Ex. 2, p. 244

b. Functions can be constructed from formulas, such as formulas for area, perimeter, and volume (Table P. 6 on page 100) and formulas for surface area (Table 1.6 on page 248).   Ex. 3, p. 246

c. If a problem's conditions are modeled by a function whose equation contains more than one variable, use the given information to write an equation among these variables. Then use this equation to eliminate all but one of the variables in the function's expression.   Ex. 4, p. 247;
Ex. 5, p. 248;
Ex. 6, p. 249;
Ex. 7, p. 250

Review Exercises

1.1

Graph each equation in Exercises 1–4. Let $x = -3, -2, -1, 0, 1, 2,$ and 3.

1. $y = 2x - 2$ **2.** $y = x^2 - 3$

3. $y = x$ **4.** $y = |x| - 2$

5. What does a $[-20, 40, 10]$ by $[-5, 5, 1]$ viewing rectangle mean? Draw axes with tick marks and label the tick marks to illustrate this viewing rectangle.

In Exercises 6–8, use the graph and determine the x-intercepts, if any, and the y-intercepts, if any. For each graph, tick marks along the axes represent one unit each.

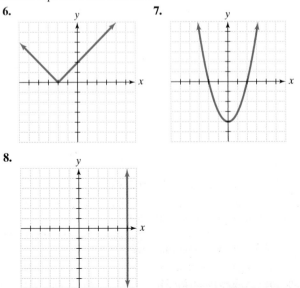

Afghanistan accounts for 76% of the world's illegal opium production. Opium-poppy cultivation nets big money in a country where most people earn less than $1 per day. (Source: Newsweek) The line graph shows opium-poppy cultivation, in thousands of acres, in Afghanistan from 1990 through 2004. Use the graph to solve Exercises 9–14.

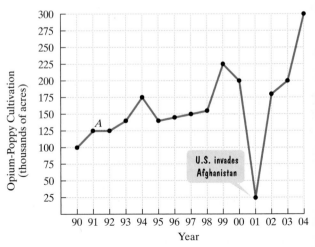

Source: U.N. Office on Drugs and Crime

9. What are the coordinates of point A? What does this mean in terms of the information given by the graph?

10. In which year were 150 thousand acres used for opium poppy cultivation?

1. For the period shown, when did opium cultivation reach a minimum? How many thousands of acres were used to cultivate the illegal crop?

2. For the period shown, when did opium cultivation reach a maximum? How many thousands of acres were used to cultivate the illegal crop?

3. Between which two years did opium cultivation not change?

4. Between which two years did opium cultivation increase at the greatest rate? What is a reasonable estimate of the increase, in thousands of acres, used to cultivate the illegal crop during this period?

.2 and 1.3

In Exercises 15–17, determine whether each relation is a function. Give the domain and range for each relation.

5. $\{(2, 7), (3, 7), (5, 7)\}$ **16.** $\{(1, 10), (2, 500), (13, \pi)\}$

7. $\{(12, 13), (14, 15), (12, 19)\}$

In Exercises 18–20, determine whether each equation defines y as a function of x.

8. $2x + y = 8$ **19.** $3x^2 + y = 14$

0. $2x + y^2 = 6$

In Exercises 21–24, evaluate each function at the given values of the independent variable and simplify.

1. $f(x) = 5 - 7x$

  a. $f(4)$ **b.** $f(x + 3)$ **c.** $f(-x)$

2. $g(x) = 3x^2 - 5x + 2$

  a. $g(0)$ **b.** $g(-2)$
  c. $g(x - 1)$ **d.** $g(-x)$

3. $g(x) = \begin{cases} \sqrt{x - 4} & \text{if } x \geq 4 \\ 4 - x & \text{if } x < 4 \end{cases}$

  a. $g(13)$ **b.** $g(0)$ **c.** $g(-3)$

4. $f(x) = \begin{cases} \dfrac{x^2 - 1}{x - 1} & \text{if } x \neq 1 \\ 12 & \text{if } x = 1 \end{cases}$

  a. $f(-2)$ **b.** $f(1)$ **c.** $f(2)$

In Exercises 25–30, use the vertical line test to identify graphs in which y is a function of x.

5.
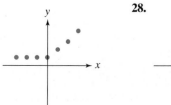

26.

7.
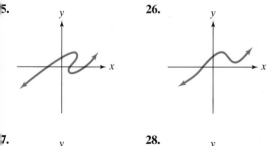

28.
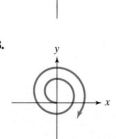

29.   **30.**
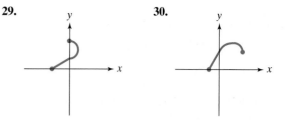

In Exercises 31–32, find and simplify the difference quotient

$$\frac{f(x + h) - f(x)}{h}, \quad h \neq 0$$

for the given function.

31. $f(x) = 8x - 11$ **32.** $f(x) = -2x^2 + x + 10$

*In Exercises 33–35, use the graph to determine **a.** the function's domain; **b.** the function's range; **c.** the x-intercepts, if any; **d.** the y-intercept, if any; **e.** intervals on which the function is increasing, decreasing, or constant; and **f.** the function values indicated below the graphs.*

33.
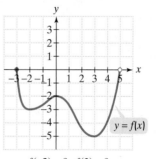

$f(-2) = ?$ $f(3) = ?$

34.
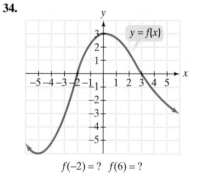

$f(-2) = ?$ $f(6) = ?$

35.
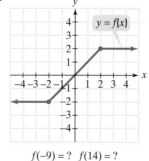

$f(-9) = ?$ $f(14) = ?$

In Exercises 36–37, find each of the following:

  a. *The numbers, if any, at which f has a relative maximum. What are these relative maxima?*

  b. *The numbers, if any, at which f has a relative minimum. What are these relative minima?*

36. Use the graph in Exercise 33.

37. Use the graph in Exercise 34.

In Exercises 38–40, determine whether each function is even, odd, or neither. State each function's symmetry. If you are using a graphing utility, graph the function and verify its possible symmetry.

38. $f(x) = x^3 - 5x$

39. $f(x) = x^4 - 2x^2 + 1$

40. $f(x) = 2x\sqrt{1 - x^2}$

41. The graph shows the height, in meters, of an eagle in terms of its time, in seconds, in flight.

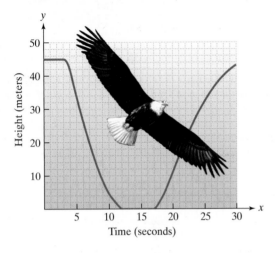

Time (seconds)

a. Is the eagle's height a function of time? Use the graph to explain why or why not.

b. On which interval is the function decreasing? Describe what this means in practical terms.

c. On which intervals is the function constant? What does this mean for each of these intervals?

d. On which interval is the function increasing? What does this mean?

42. A cargo service charges a flat fee of $5 plus $1.50 for each pound or fraction of a pound. Graph shipping cost, $C(x)$, in dollars, as a function of weight, x, in pounds, for $0 < x \le 5$.

1.4 and 1.5

In Exercises 43–46, find the slope of the line passing through each pair of points or state that the slope is undefined. Then indicate whether the line through the points rises, falls, is horizontal, or is vertical.

43. $(3, 2)$ and $(5, 1)$

44. $(-1, -2)$ and $(-3, -4)$

45. $\left(-3, \frac{1}{4}\right)$ and $\left(6, \frac{1}{4}\right)$

46. $(-2, 5)$ and $(-2, 10)$

In Exercises 47–50, use the given conditions to write an equation for each line in point-slope form and slope-intercept form.

47. Passing through $(-3, 2)$ with slope -6

48. Passing through $(1, 6)$ and $(-1, 2)$

49. Passing through $(4, -7)$ and parallel to the line whose equation is $3x + y - 9 = 0$

50. Passing through $(-3, 6)$ and perpendicular to the line whose equation is $y = \frac{1}{3}x + 4$

51. Write the general form of the equation of the line passing through $(-12, -1)$ and perpendicular to the line whose equation is $6x - y - 4 = 0$.

In Exercises 52–55, give the slope and y-intercept of each line whose equation is given. Then graph the line.

52. $y = \frac{2}{5}x - 1$

53. $f(x) = -4x + 5$

54. $2x + 3y + 6 = 0$

55. $2y - 8 = 0$

56. Graph using intercepts: $2x - 5y - 10 = 0$.

57. Graph: $2x - 10 = 0$.

58. You can click a mouse and bet the house. The points in the graph show the dizzying growth of online gambling. With more than 1800 sites, the industry has become the Web' biggest moneymaker.

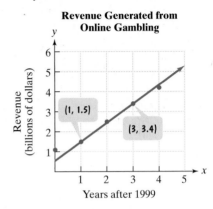

Years after 1999

Source: Newsweek

a. Use the two points whose coordinates are shown by th voice balloons to find the point-slope form of the equa tion of the line that models revenue from onlin gambling, y, in billions of dollars, x years after 1999.

b. Write the equation in part (a) in slope-intercept form.

c. In 2003, nearly $3.5 billion was lost on Internet bet triggering a sharp backlash that threatened to shut dow Internet wagering. If this crackdown on the industry is nc successful, use your slope-intercept model to predict th billions of dollars in revenue from online gambling in 2009

59. The graph shows new AIDS diagnoses among the genera U.S. population, y, for year x, where $1999 \le x \le 2003$.

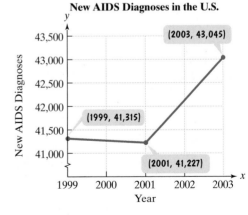

Year

Source: Centers for Disease Control

a. Find the slope of the line passing through (1999, 41,31! and (2001, 41,227). Then express the slope as a rate c change with the proper units attached.

b. Find the slope of the line passing through (2001, 41,22 and (2003, 43,045). Then express the slope as a rate c change.

c. Draw a line passing through (1999, 41,315) and (2003, 43,045) and find its slope. Is the slope the average of the slopes of the lines that you found in parts (a) and (b)? Explain your answer.

0. Find the average rate of change of $f(x) = x^2 - 4x$ from $x_1 = 5$ to $x_2 = 9$.

1. A person standing on the roof of a building throws a ball directly upward. The ball misses the rooftop on its way down and eventually strikes the ground. The function

$$s(t) = -16t^2 + 64t + 80$$

describes the ball's height above the ground, $s(t)$ in feet, t seconds after it was thrown.

  a. Find the ball's average velocity between the time it was thrown and 2 seconds later.

  b. Find the ball's average velocity between 2 and 4 seconds after it was thrown.

  c. What do the signs in your answers to parts (a) and (b) mean in terms of the direction of the ball's motion?

.6

Exercises 62–66, use the graph of $y = f(x)$ to graph each function g.

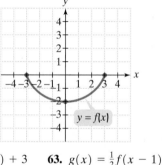

2. $g(x) = f(x + 2) + 3$ **63.** $g(x) = \frac{1}{2}f(x - 1)$

. $g(x) = -f(2x)$ **65.** $g(x) = 2f\left(\frac{1}{2}x\right)$

. $g(x) = -f(-x) - 1$

Exercises 67–70, begin by graphing the standard quadratic function, $f(x) = x^2$. Then use transformations of this graph to graph the given function.

7. $g(x) = x^2 + 2$ **68.** $h(x) = (x + 2)^2$

. $r(x) = -(x + 1)^2$ **70.** $y(x) = \frac{1}{2}(x - 1)^2 + 1$

Exercises 71–73, begin by graphing the square root function, $x) = \sqrt{x}$. Then use transformations of this graph to graph the ven function.

. $g(x) = \sqrt{x + 3}$ **72.** $h(x) = \sqrt{3 - x}$

. $r(x) = 2\sqrt{x + 2}$

Exercises 74–76, begin by graphing the absolute value function, $x) = |x|$. Then use transformations of this graph to graph the ven function.

. $g(x) = |x + 2| - 3$ **75.** $h(x) = -|x - 1| + 1$

. $r(x) = \frac{1}{2}|x + 2|$

Exercises 77–79, begin by graphing the standard cubic function, $f(x) = x^3$. Then use transformations of this graph to graph e given function.

. $g(x) = \frac{1}{2}(x - 1)^3$ **78.** $h(x) = -(x + 1)^3$

. $r(x) = \frac{1}{4}x^3 - 1$

In Exercises 80–82, begin by graphing the cube root function, $f(x) = \sqrt[3]{x}$. Then use transformations of this graph to graph the given function.

80. $g(x) = \sqrt[3]{x + 2} - 1$ **81.** $h(x) = -\sqrt[3]{2x}$

82. $r(x) = -2\sqrt[3]{-x}$

1.7

In Exercises 83–88, find the domain of each function.

83. $f(x) = x^2 + 6x - 3$ **84.** $g(x) = \dfrac{4}{x - 7}$

85. $h(x) = \sqrt{8 - 2x}$ **86.** $f(x) = \dfrac{x}{x^2 + 4x - 21}$

87. $g(x) = \dfrac{\sqrt{x - 2}}{x - 5}$ **88.** $f(x) = \sqrt{x - 1} + \sqrt{x + 5}$

In Exercises 89–91, find $f + g, f - g, fg,$ and $\frac{f}{g}$. Determine the domain for each function.

89. $f(x) = 3x - 1$, $g(x) = x - 5$

90. $f(x) = x^2 + x + 1$, $g(x) = x^2 - 1$

91. $f(x) = \sqrt{x + 7}$, $g(x) = \sqrt{x - 2}$

In Exercises 92–93, find **a.** $(f \circ g)(x)$; **b.** $(g \circ f)(x)$;
c. $(f \circ g)(3)$.

92. $f(x) = x^2 + 3$, $g(x) = 4x - 1$

93. $f(x) = \sqrt{x}$, $g(x) = x + 1$

In Exercises 94–95, find **a.** $(f \circ g)(x)$; **b.** *the domain of* $(f \circ g)$.

94. $f(x) = \dfrac{x + 1}{x - 2}$, $g(x) = \dfrac{1}{x}$

95. $f(x) = \sqrt{x - 1}$, $g(x) = x + 3$

In Exercises 96–97, express the given function h as a composition of two functions f and g so that $h(x) = (f \circ g)(x)$.

96. $h(x) = (x^2 + 2x - 1)^4$

97. $h(x) = \sqrt[3]{7x + 4}$

1.8

In Exercises 98–99, find $f(g(x))$ and $g(f(x))$ and determine whether each pair of functions f and g are inverses of each other.

98. $f(x) = \dfrac{3}{5}x + \dfrac{1}{2}$ and $g(x) = \dfrac{5}{3}x - 2$

99. $f(x) = 2 - 5x$ and $g(x) = \dfrac{2 - x}{5}$

The functions in Exercises 100–102 are all one-to-one. For each function,

  a. *Find an equation for $f^{-1}(x)$, the inverse function.*

  b. *Verify that your equation is correct by showing that $f(f^{-1}(x)) = x$ and $f^{-1}(f(x)) = x$.*

100. $f(x) = 4x - 3$ **101.** $f(x) = 8x^3 + 1$

102. $f(x) = \dfrac{2}{x} + 5$

Which graphs in Exercises 103–106 represent functions that have inverse functions?

103.

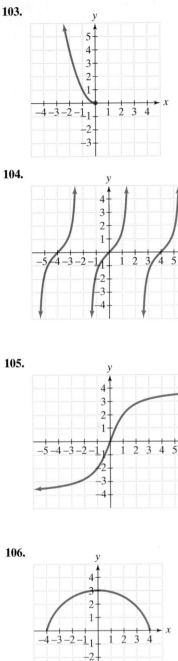

104.

105.

106.

107. Use the graph of f in the figure shown to draw the graph of its inverse function.

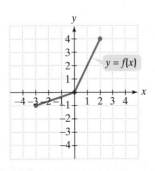

In Exercises 108–109, find an equation for $f^{-1}(x)$. Then graph f and f^{-1} in the same rectangular coordinate system.

108. $f(x) = 1 - x^2, x \geq 0$ **109.** $f(x) = \sqrt{x} + 1$

1.9

In Exercises 110–111, find the distance between each pair of points. If necessary, round answers to two decimal places.

110. $(-2, 3)$ and $(3, -9)$ **111.** $(-4, 3)$ and $(-2, 5)$

In Exercises 112–113, find the midpoint of each line segment with the given endpoints.

112. $(2, 6)$ and $(-12, 4)$ **113.** $(4, -6)$ and $(-15, 2)$

In Exercises 114–115, write the standard form of the equation of the circle with the given center and radius.

114. Center $(0, 0), r = 3$ **115.** Center $(-2, 4), r = 6$

In Exercises 116–118, give the center and radius of each circle and graph its equation. Use the graph to identify the relation's domain and range.

116. $x^2 + y^2 = 1$ **117.** $(x + 2)^2 + (y - 3)^2 = 9$
118. $x^2 + y^2 - 4x + 2y - 4 = 0$

1.10

119. In 2000, the average weekly salary for workers in the United States was $567. This amount has increased by approximately $15 per year.

   a. Express the average weekly salary for U.S. workers, W, as a function of the number of years after 2000, x.

   b. If this trend continues, in which year will the average weekly salary be $702?

120. You are choosing between two long-distance telephone plans. Plan A has a monthly fee of $15 with a charge of $0.0 per minute. Plan B has a monthly fee of $5 with a charge of $0.07 per minute.

   a. Express the monthly cost for plan A, f, as a function of the number of minutes of long-distance calls in a month, x.

   b. Express the monthly cost for plan B, g, as a function of the number of minutes of long-distance calls in a month, x.

   c. For how many minutes of long-distance calls will the costs for the two plans be the same?

121. A 400-room hotel can rent every one of its rooms $120 per room. For each $1 increase in rent, two fewer rooms are rented.

   a. Express the number of rooms rented, N, as a function of the rent, x.

   b. Express the hotel's revenue, R, as a function of the rent,

122. An open box is made by cutting identical squares from the corners of a 16-inch by 24-inch piece of cardboard and then turning up the sides.

   a. Express the volume of the box, V, as a function of the length of the side of the square cut from each corner,

   b. Find the domain of V.

123. You have 400 feet of fencing to enclose a rectangular lot and divide it in two by another fence that is parallel to one side of the lot. Express the area of the rectangular lot, A, as a function of the length of the fence that divides the rectangular lot, x.

124. The figure shows a box with a square base and a square top. The box is to have a volume of 8 cubic feet. Express the surface area of the box, A, as a function of the length of a side of its square base, x.

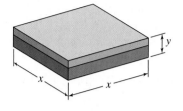

125. You inherit $10,000 with the stipulation that for the first year the money must be placed in two investments expected to earn 8% and 12% annual interest, respectively. Express the expected interest from both investments, I, as a function of the amount of money invested at 8%, x.

Chapter 1 Test

1. List by letter all relations that are not functions.

   a. $\{(7,5), (8,5), (9,5)\}$

   b. $\{(5,7), (5,8), (5,9)\}$

   c.

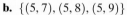

   d. $x^2 + y^2 = 100$

   e.

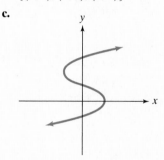

2. Use the graph of $y = f(x)$ to solve this exercise.

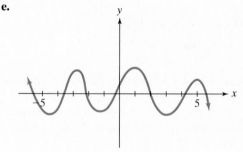

   a. What is $f(4) - f(-3)$?

   b. What is the domain of f?

   c. What is the range of f?

   d. On which interval or intervals is f increasing?

   e. On which interval or intervals is f decreasing?

   f. For what number does f have a relative maximum? What is the relative maximum?

   g. For what number does f have a relative minimum? What is the relative minimum?

   h. What are the x-intercepts?

   i. What is the y-intercept?

3. Use the graph of $y = f(x)$ to solve this exercise.

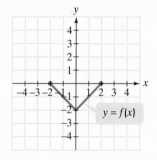

   a. What are the zeros of f?

   b. Find the value(s) of x for which $f(x) = -1$.

   c. Find the value(s) of x for which $f(x) = -2$.

   d. Is f even, odd, or neither?

   e. Does f have an inverse function?

   f. Is $f(0)$ a relative maximum, a relative minimum, or neither?

   g. Graph $g(x) = f(x + 1) - 1$.

   h. Graph $h(x) = \frac{1}{2}f\left(\frac{1}{2}x\right)$.

   i. Graph $r(x) = -f(-x) + 1$.

   j. Find the average rate of change of f from $x_1 = -2$ to $x_2 = 1$.

In Exercises 4–15, graph each equation in a rectangular coordinate system. If two functions are indicated, graph both in the same system. Then use your graphs to identify each relation's domain and range.

   4. $x + y = 4$ **5.** $x^2 + y^2 = 4$

6. $f(x) = 4$

7. $f(x) = -\frac{1}{3}x + 2$

8. $(x + 2)^2 + (y - 1)^2 = 9$

9. $f(x) = \begin{cases} 2 & \text{if } x \le 0 \\ -1 & \text{if } x > 0 \end{cases}$

10. $x^2 + y^2 + 4x - 6y - 3 = 0$

11. $f(x) = |x|$ and $g(x) = \frac{1}{2}|x + 1| - 2$

12. $f(x) = x^2$ and $g(x) = -(x - 1)^2 + 4$

13. $f(x) = 2x - 4$ and f^{-1}

14. $f(x) = x^3 - 1$ and f^{-1}

15. $f(x) = x^2 - 1, x \ge 0$, and f^{-1}

In Exercises 16–23, let $f(x) = x^2 - x - 4$ and $g(x) = 2x - 6$.

16. Find $f(x - 1)$.

17. Find $\dfrac{f(x + h) - f(x)}{h}$.

18. Find $(g - f)(x)$.

19. Find $\left(\dfrac{f}{g}\right)(x)$ and its domain.

20. Find $(f \circ g)(x)$.

21. Find $(g \circ f)(x)$.

22. Find $g(f(-1))$.

23. Find $f(-x)$. Is f even, odd, or neither?

In Exercises 24–25, use the given conditions to write an equation for each line in point-slope form and slope-intercept form.

24. Passing through $(2, 1)$ and $(-1, -8)$

25. Passing through $(-4, 6)$ and perpendicular to the line whose equation is $y = -\frac{1}{4}x + 5$

26. Write the general form of the equation of the line passing through $(-7, -10)$ and parallel to the line whose equation is $4x + 2y - 5 = 0$.

27. When adjusted for inflation, the federal minimum wage from 1997 through 2003 continued to decrease. The points in the scatter plot show the minimum hourly inflation-adjusted wage for this period. Also shown is a line that passes through or near the points.

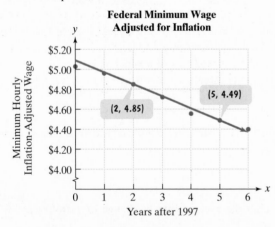

Federal Minimum Wage Adjusted for Inflation

Source: www.dul.gov/esa/pudha/minwage

a. Use the two points whose coordinates are shown by the voice balloons to find the point-slope form of the equation of the line that models the minimum hourly inflation-adjusted wage, y, x years after 1997.

b. Write the equation in part (a) in slope-intercept form. Use function notation.

c. Use the linear function to predict the minimum hourly inflation-adjusted wage in 2007.

28. Find the average rate of change of $f(x) = 3x^2 - 5$ from $x_1 = 6$ to $x_2 = 10$.

29. If $g(x) = \begin{cases} \sqrt{x - 3} & \text{if } x \ge 3 \\ 3 - x & \text{if } x < 3 \end{cases}$, find $g(-1)$ and $g(7)$.

In Exercises 30–31, find the domain of each function.

30. $f(x) = \dfrac{3}{x + 5} + \dfrac{7}{x - 1}$

31. $f(x) = 3\sqrt{x + 5} + 7\sqrt{x - 1}$

32. If $f(x) = \dfrac{7}{x - 4}$ and $g(x) = \dfrac{2}{x}$, find $(f \circ g)(x)$ and the domain of $f \circ g$.

33. Express $h(x) = (2x + 3)^7$ as a composition of two functions f and g so that $h(x) = (f \circ g)(x)$.

34. Find the length and the midpoint of the line segment whose endpoints are $(2, -2)$ and $(5, 2)$.

35. In 1980, the winning time for women in the Olympic 500-meter speed skating event was 41.78 seconds. The average rate of decrease in the winning time has been about 0.1 second per year.

a. Express the winning time, T, in this event as a function of the number of years after 1980, x.

b. According to the function, when will the winning time be 35.7 seconds?

36. The annual yield per walnut tree is fairly constant at 50 pounds per tree when the number of trees per acre is 30 or fewer. For each additional tree over 30, the annual yield per tree for all trees on the acre decreases by 1.5 pounds due to overcrowding.

a. Express the yield per tree, Y, in pounds, as a function of the number of walnut trees per acre, x.

b. Express the total yield for an acre, T, in pounds, as a function of the number of walnut trees per acre, x.

37. You have 600 yards of fencing to enclose a rectangular field. Express the area of the field, A, as a function of one of its dimensions, x.

38. A closed rectangular box with a square base has a volume of 8000 cubic centimeters. Express the surface area of the box, A, as a function of the length of a side of its square base, x.

Polynomial and Rational Functions

HERE IS A FUNCTION THAT *models the age in human years, $H(x)$, of a dog that is x years old:*

$$H(x) = -0.001618x^4 + 0.077326x^3$$
$$-1.2367x^2 + 11.460x + 2.914.$$

The function contains variables to powers that are whole numbers and is an example of a **polynomial function**. *In this chapter, we study polynomial functions and functions that consist of quotients of polynomials, called* **rational functions**.

ONE OF THE JOYS OF YOUR LIFE IS YOUR dog, your very special buddy. Lately, however, you've noticed that your companion is slowing down a bit. He's now 8 years old and you wonder how this translates into human years. You remember something about every year of a dog's life being equal to seven years for a human. Is there a more accurate description?

This problem appears as Exercises 63–64 in Exercise Set 2.5.

SECTION 2.1   *Complex Numbers*

Objectives

❶ Add and subtract complex numbers.

❷ Multiply complex numbers.

❸ Divide complex numbers.

❹ Perform operations with square roots of negative numbers.

❺ Solve quadratic equations with complex imaginary solutions.

THE KID WHO LEARNED ABOUT MATH

ON THE STREET

If you divide 6,973 by 0, you die.

Once, this guy tried to find the square root of -9, and his eyeballs turned black.

This girl my brother knows found out exactly what π equals, but she went nuts.

R. Chast

Who is this kid warning us about ou eyeballs turning black if we attempt t find the square root of −9? Don believe what you hear on the stree Although square roots of negative num bers are not real numbers, they do play significant role in algebra. In this sec tion, we move beyond the real number and discuss square roots with negativ radicands.

The Imaginary Unit *i*

In the next section, we will study equations whose solutions may involve the squar roots of negative numbers. Because the square of a real number is never negative there is no real number x such that $x^2 = -1$. To provide a setting in which such equa tions have solutions, mathematicians invented an expanded system of numbers, th complex numbers. The *imaginary number i*, defined to be a solution of the equatio $x^2 = -1$, is the basis of this new set.

The Imaginary Unit *i*

The **imaginary unit *i*** is defined as

$$i = \sqrt{-1}, \text{ where } i^2 = -1.$$

Using the imaginary unit i, we can express the square root of any negativ number as a real multiple of i. For example,

$$\sqrt{-25} = \sqrt{-1}\sqrt{25} = i\sqrt{25} = 5i.$$

We can check this result by squaring $5i$ and obtaining -25.

$$(5i)^2 = 5^2 i^2 = 25(-1) = -25$$

A new system of numbers, called *complex numbers*, is based on adding mult ples of i, such as $5i$, to the real numbers.

Complex Numbers and Imaginary Numbers

The set of all numbers in the form

$$a + bi$$

with real numbers a and b, and i, the imaginary unit, is called the set of **complex numbers**. The real number a is called the **real part** and the real number b is called the **imaginary part** of the complex number $a + bi$. If $b \neq 0$, then the complex number is called an **imaginary number** (Figure 2.1). An imaginary number in the form bi is called a **pure imaginary number**.

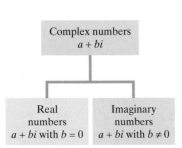

Figure 2.1 The complex number system

Complex numbers
$a + bi$

Real numbers
$a + bi$ with $b = 0$

Imaginary numbers
$a + bi$ with $b \neq 0$

Here are some examples of complex numbers. Each number can be written in the form $a + bi$.

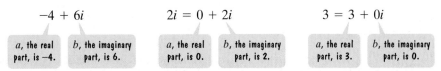

$$-4 + 6i \qquad\qquad 2i = 0 + 2i \qquad\qquad 3 = 3 + 0i$$

| a, the real part, is −4. | b, the imaginary part, is 6. | a, the real part, is 0. | b, the imaginary part, is 2. | a, the real part, is 3. | b, the imaginary part, is 0. |

Can you see that b, the imaginary part, is not zero in the first two complex numbers? Because $b \neq 0$, these complex numbers are imaginary numbers. Furthermore, the imaginary number $2i$ is a pure imaginary number. By contrast, the imaginary part of the complex number on the right is zero. This complex number is not an imaginary number. The number 3, or $3 + 0i$, is a real number.

A complex number is said to be **simplified** if it is expressed in the **standard form** $a + bi$. If b is a radical, we usually write i before b. For example, we write $7 + i\sqrt{5}$ rather than $7 + \sqrt{5}i$, which could easily be confused with $7 + \sqrt{5i}$.

Expressed in standard form, two complex numbers are equal if and only if their real parts are equal and their imaginary parts are equal.

Equality of Complex Numbers

$a + bi = c + di$ if and only if $a = c$ and $b = d$.

① Add and subtract complex numbers.

Operations with Complex Numbers

The form of a complex number $a + bi$ is like the binomial $a + bx$. Consequently, we can add, subtract, and multiply complex numbers using the same methods we used for binomials, remembering that $i^2 = -1$.

Adding and Subtracting Complex Numbers

1. $(a + bi) + (c + di) = (a + c) + (b + d)i$
   In words, this says that you add complex numbers by adding their real parts, adding their imaginary parts, and expressing the sum as a complex number.
2. $(a + bi) - (c + di) = (a - c) + (b - d)i$
   In words, this says that you subtract complex numbers by subtracting their real parts, subtracting their imaginary parts, and expressing the difference as a complex number.

EXAMPLE 1   Adding and Subtracting Complex Numbers

Perform the indicated operations, writing the result in standard form:

    a. $(5 - 11i) + (7 + 4i)$ **b.** $(-5 + i) - (-11 - 6i)$.

Solution

  a. $(5 - 11i) + (7 + 4i)$

    $= 5 - 11i + 7 + 4i$ Remove the parentheses.

    $= 5 + 7 - 11i + 4i$ Group real and imaginary terms.

    $= (5 + 7) + (-11 + 4)i$ Add real parts and add imaginary parts.

    $= 12 - 7i$ Simplify.

  b. $(-5 + i) - (-11 - 6i)$

    $= -5 + i + 11 + 6i$ Remove the parentheses. Change signs of real and imaginary parts in the complex number being subtracted.

    $= -5 + 11 + i + 6i$ Group real and imaginary terms.

    $= (-5 + 11) + (1 + 6)i$ Add real parts and add imaginary parts.

    $= 6 + 7i$ Simplify.

Study Tip

The following examples, using the same integers as in Example 1, show how operations with complex numbers are just like operations with polynomials.

  a. $(5 - 11x) + (7 + 4x)$

    $= 12 - 7x$

  b. $(-5 + x) - (-11 - 6x)$

    $= -5 + x + 11 + 6x$

    $= 6 + 7x$

Multiply complex numbers.

Check
Point **1** Add or subtract as indicated:

    a. $(5 - 2i) + (3 + 3i)$ **b.** $(2 + 6i) - (12 - i)$.

Multiplication of complex numbers is performed the same way as multiplicatio of polynomials, using the distributive property and the FOIL method. After completin the multiplication, we replace any occurrences of i^2 with -1. This idea is illustrated i the next example.

EXAMPLE 2   Multiplying Complex Numbers

Find the products:

    a. $4i(3 - 5i)$ **b.** $(7 - 3i)(-2 - 5i)$.

Solution

    a. $4i(3 - 5i)$

$$= 4i \cdot 3 - 4i \cdot 5i \qquad \text{Distribute } 4i \text{ throughout the parentheses.}$$
$$= 12i - 20i^2 \qquad \text{Multiply.}$$
$$= 12i - 20(-1) \qquad \text{Replace } i^2 \text{ with } -1.$$
$$= 20 + 12i \qquad \text{Simplify to } 12i + 20 \text{ and write in standard form}$$

    b. $(7 - 3i)(-2 - 5i)$

          F   O   I   L

$$= -14 - 35i + 6i + 15i^2 \qquad \text{Use the FOIL method.}$$
$$= -14 - 35i + 6i + 15(-1) \qquad i^2 = -1$$
$$= -14 - 15 - 35i + 6i \qquad \text{Group real and imaginary terms.}$$
$$= -29 - 29i \qquad \text{Combine real and imaginary terms.}$$

Check
Point **2** Find the products:

    a. $7i(2 - 9i)$ **b.** $(5 + 4i)(6 - 7i)$.

 Divide complex numbers.

Complex Conjugates and Division

It is possible to multiply complex numbers and obtain a real number. This occur when we multiply $a + bi$ and $a - bi$.

          F   O   I   L

$$(a + bi)(a - bi) = a^2 - abi + abi - b^2i^2 \qquad \text{Use the FOIL method.}$$
$$= a^2 - b^2(-1) \qquad i^2 = -1$$
$$= a^2 + b^2 \qquad \text{Notice that this product eliminates } i.$$

For the complex number $a + bi$, we define its *complex conjugate* to be $a - b$ The multiplication of complex conjugates results in a real number.

> **Conjugate of a Complex Number**
>
> The **complex conjugate** of the number $a + bi$ is $a - bi$, and the complex conju gate of $a - bi$ is $a + bi$. The multiplication of complex conjugates gives a real number.
>
> $$(a + bi)(a - bi) = a^2 + b^2$$
> $$(a - bi)(a + bi) = a^2 + b^2$$

Complex conjugates are used to divide complex numbers. By multiplying th numerator and the denominator of the division by the complex conjugate of th denominator, you will obtain a real number in the denominator.

EXAMPLE 3   Using Complex Conjugates to Divide Complex Numbers

Divide and express the result in standard form: $\dfrac{7 + 4i}{2 - 5i}$.

Solution   The complex conjugate of the denominator, $2 - 5i$, is $2 + 5i$. Multiplication of both the numerator and the denominator by $2 + 5i$ will eliminate i from the denominator.

$$\frac{7 + 4i}{2 - 5i} = \frac{(7 + 4i)}{(2 - 5i)} \cdot \frac{(2 + 5i)}{(2 + 5i)}$$

Multiply the numerator and the denominator by the complex conjugate of the denominator.

$$= \frac{\overset{F}{14} + \overset{O}{35i} + \overset{I}{8i} + \overset{L}{20i^2}}{2^2 + 5^2}$$

Use the FOIL method in the numerator and $(a - bi)(a + bi) = a^2 + b^2$ in the denominator.

$$= \frac{14 + 43i + 20(-1)}{29}$$

Combine imaginary terms and replace i^2 with -1.

$$= \frac{-6 + 43i}{29}$$

Combine real terms in the numerator:
$14 + 20(-1) = 14 - 20 = -6$.

$$= -\frac{6}{29} + \frac{43}{29}i$$

Express the answer in standard form.

Observe that the quotient is expressed in the standard form $a + bi$, with $a = -\frac{6}{29}$ and $b = \frac{43}{29}$.

Check Point 3   Divide and express the result in standard form: $\dfrac{5 + 4i}{4 - i}$.

❹ Perform operations with square roots of negative numbers.

Roots of Negative Numbers

The square of $4i$ and the square of $-4i$ both result in -16:

$$(4i)^2 = 16i^2 = 16(-1) = -16 \qquad (-4i)^2 = 16i^2 = 16(-1) = -16.$$

Consequently, in the complex number system -16 has two square roots, namely, $4i$ and $-4i$. We call $4i$ the **principal square root** of -16.

Principal Square Root of a Negative Number

For any positive real number b, the **principal square root** of the negative number $-b$ is defined by

$$\sqrt{-b} = i\sqrt{b}.$$

Consider the multiplication problem

$$5i \cdot 2i = 10i^2 = 10(-1) = -10.$$

This problem can also be given in terms of principal square roots of negative numbers:

$$\sqrt{-25} \cdot \sqrt{-4}.$$

Because the product rule for radicals only applies to real numbers, multiplying radicands is incorrect. **When performing operations with square roots of negative numbers, begin by expressing all square roots in terms of i.** Then perform the indicated operation.

Correct: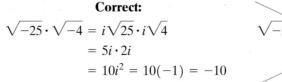

$$\sqrt{-25} \cdot \sqrt{-4} = i\sqrt{25} \cdot i\sqrt{4}$$
$$= 5i \cdot 2i$$
$$= 10i^2 = 10(-1) = -10$$

Incorrect: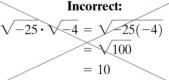

$$\sqrt{-25} \cdot \sqrt{-4} = \sqrt{-25(-4)}$$
$$= \sqrt{100}$$
$$= 10$$

EXAMPLE 4 Operations Involving Square Roots of Negative Numbers

Perform the indicated operations and write the result in standard form:

a. $\sqrt{-18} - \sqrt{-8}$ **b.** $\left(-1 + \sqrt{-5}\right)^2$ **c.** $\dfrac{-25 + \sqrt{-50}}{15}$.

Solution Begin by expressing all square roots of negative numbers in terms of i.

a. $\sqrt{-18} - \sqrt{-8} = i\sqrt{18} - i\sqrt{8} = i\sqrt{9 \cdot 2} - i\sqrt{4 \cdot 2}$
$$= 3i\sqrt{2} - 2i\sqrt{2} = i\sqrt{2}$$

$$(A + B)^2 = A^2 + 2AB + B^2$$

b. $\left(-1 + \sqrt{-5}\right)^2 = \left(-1 + i\sqrt{5}\right)^2 = (-1)^2 + 2(-1)(i\sqrt{5}) + (i\sqrt{5})^2$
$$= 1 - 2i\sqrt{5} + 5i^2$$
$$= 1 - 2i\sqrt{5} + 5(-1)$$
$$= -4 - 2i\sqrt{5}$$

c. $\dfrac{-25 + \sqrt{-50}}{15}$

$$= \dfrac{-25 + i\sqrt{50}}{15} \qquad \sqrt{-b} = i\sqrt{b}$$

$$= \dfrac{-25 + 5i\sqrt{2}}{15} \qquad \sqrt{50} = \sqrt{25 \cdot 2} = 5\sqrt{2}$$

$$= \dfrac{-25}{15} + \dfrac{5i\sqrt{2}}{15} \qquad \text{Write the complex number in standard form.}$$

$$= -\dfrac{5}{3} + i\dfrac{\sqrt{2}}{3} \qquad \text{Simplify.}$$

Check Point 4 Perform the indicated operations and write the result in standard form:

a. $\sqrt{-27} + \sqrt{-48}$ **b.** $\left(-2 + \sqrt{-3}\right)^2$ **c.** $\dfrac{-14 + \sqrt{-12}}{2}$.

⑤ Solve quadratic equations with complex imaginary solutions.

Quadratic Equations with Complex Imaginary Solutions

We have seen that a quadratic equation can be expressed in the general form

$$ax^2 + bx + c = 0, \quad a \neq 0.$$

All quadratic equations can be solved by the quadratic formula:

$$x = \dfrac{-b \pm \sqrt{b^2 - 4ac}}{2a}.$$

Study Tip

If you need to review quadratic equations and how to solve them, read Section P. 7, beginning on page 84.

Recall that the quantity $b^2 - 4ac$, which appears under the radical sign in the quadratic formula, is called the discriminant. If the discriminant is negative, a quadratic equation has no real solutions. However, quadratic equations with negative discriminants do have two solutions. These solutions are imaginary numbers that are complex conjugates.

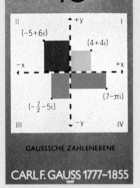

Complex Numbers on a Postage Stamp

DEUTSCHE BUNDESPOST
40
II (-5+6i)
(4+4i)
III $\left(-\frac{7}{2}-5i\right)$
(7-πi)
IV
GAUSSSCHE ZAHLENEBENE
CARL F. GAUSS 1777-1855

This stamp honors the work done by the German mathematician Carl Friedrich Gauss (1777–1855) with complex numbers. Gauss represented complex numbers as points in the plane.

EXAMPLE 5   A Quadratic Equation with Imaginary Solutions

Solve using the quadratic formula: $3x^2 - 2x + 4 = 0$.

Solution   The given equation is in general form. Begin by identifying the values for a, b, and c.

$$3x^2 - 2x + 4 = 0$$

$$a = 3 \qquad b = -2 \qquad c = 4$$

$$x = \frac{-b \pm \sqrt{b^2 - 4ac}}{2a}$$

Use the quadratic formula.

$$= \frac{-(-2) \pm \sqrt{(-2)^2 - 4(3)(4)}}{2(3)}$$

Substitute the values for a, b, and c: $a = 3$, $b = -2$, and $c = 4$,

$$= \frac{2 \pm \sqrt{4 - 48}}{6}$$

$-(-2) = 2$ and $(-2)^2 = (-2)(-2) = 4$.

$$= \frac{2 \pm \sqrt{-44}}{6}$$

Subtract under the radical. Because the number under the radical sign is negative, the solutions will not be real numbers.

$$= \frac{2 \pm 2i\sqrt{11}}{6}$$

$\sqrt{-44} = \sqrt{4(11)(-1)} = 2i\sqrt{11}$

$$= \frac{2\left(1 \pm i\sqrt{11}\right)}{6}$$

Factor 2 from the numerator.

$$= \frac{1 \pm i\sqrt{11}}{3}$$

Divide numerator and denominator by 2.

$$= \frac{1}{3} \pm i\frac{\sqrt{11}}{3}$$

Write the complex numbers in standard form.

The solutions are complex conjugates, and the solution set is $\left\{\dfrac{1}{3} + i\dfrac{\sqrt{11}}{3}, \dfrac{1}{3} - i\dfrac{\sqrt{11}}{3}\right\}$ or $\left\{\dfrac{1}{3} \pm i\dfrac{\sqrt{11}}{3}\right\}$.

Check Point 5   Solve using the quadratic formula:

$$x^2 - 2x + 2 = 0.$$

EXERCISE SET 2.1

Practice Exercises

In Exercises 1–8, add or subtract as indicated and write the result in standard form.

1. $(7 + 2i) + (1 - 4i)$

2. $(-2 + 6i) + (4 - i)$

3. $(3 + 2i) - (5 - 7i)$

4. $(-7 + 5i) - (-9 - 11i)$

5. $6 - (-5 + 4i) - (-13 - i)$

6. $7 - (-9 + 2i) - (-17 - i)$

7. $8i - (14 - 9i)$

8. $15i - (12 - 11i)$

In Exercises 9–20, find each product and write the result in standard form.

9. $-3i(7i - 5)$

10. $-8i(2i - 7)$

11. $(-5 + 4i)(3 + i)$

12. $(-4 - 8i)(3 + i)$

13. $(7 - 5i)(-2 - 3i)$

14. $(8 - 4i)(-3 + 9i)$

15. $(3 + 5i)(3 - 5i)$

16. $(2 + 7i)(2 - 7i)$

17. $(-5 + i)(-5 - i)$

18. $(-7 - i)(-7 + i)$

19. $(2 + 3i)^2$

20. $(5 - 2i)^2$

In Exercises 21–28, divide and express the result in standard form.

21. $\dfrac{2}{3-i}$

22. $\dfrac{3}{4+i}$

23. $\dfrac{2i}{1+i}$

24. $\dfrac{5i}{2-i}$

25. $\dfrac{8i}{4-3i}$

26. $\dfrac{-6i}{3+2i}$

27. $\dfrac{2+3i}{2+i}$

28. $\dfrac{3-4i}{4+3i}$

In Exercises 29–44, perform the indicated operations and write the result in standard form.

29. $\sqrt{-64} - \sqrt{-25}$

30. $\sqrt{-81} - \sqrt{-144}$

31. $5\sqrt{-16} + 3\sqrt{-81}$

32. $5\sqrt{-8} + 3\sqrt{-18}$

33. $\left(-2 + \sqrt{-4}\right)^2$

34. $\left(-5 - \sqrt{-9}\right)^2$

35. $\left(-3 - \sqrt{-7}\right)^2$

36. $\left(-2 + \sqrt{-11}\right)^2$

37. $\dfrac{-8 + \sqrt{-32}}{24}$

38. $\dfrac{-12 + \sqrt{-28}}{32}$

39. $\dfrac{-6 - \sqrt{-12}}{48}$

40. $\dfrac{-15 - \sqrt{-18}}{33}$

41. $\sqrt{-8}\left(\sqrt{-3} - \sqrt{5}\right)$

42. $\sqrt{-12}\left(\sqrt{-4} - \sqrt{2}\right)$

43. $\left(3\sqrt{-5}\right)\left(-4\sqrt{-12}\right)$

44. $\left(3\sqrt{-7}\right)\left(2\sqrt{-8}\right)$

In Exercises 45–50, solve each quadratic equation using the quadratic formula. Express solutions in standard form.

45. $x^2 - 6x + 10 = 0$

46. $x^2 - 2x + 17 = 0$

47. $4x^2 + 8x + 13 = 0$

48. $2x^2 + 2x + 3 = 0$

49. $3x^2 = 8x - 7$

50. $3x^2 = 4x - 6$

Practice Plus

In Exercises 51–56, perform the indicated operation(s) and write the result in standard form.

51. $(2 - 3i)(1 - i) - (3 - i)(3 + i)$

52. $(8 + 9i)(2 - i) - (1 - i)(1 + i)$

53. $(2 + i)^2 - (3 - i)^2$

54. $(4 - i)^2 - (1 + 2i)^2$

55. $5\sqrt{-16} + 3\sqrt{-81}$

56. $5\sqrt{-8} + 3\sqrt{-18}$

57. Evaluate $x^2 - 2x + 2$ for $x = 1 + i$.

58. Evaluate $x^2 - 2x + 5$ for $x = 1 - 2i$.

59. Evaluate $\dfrac{x^2 + 19}{2 - x}$ for $x = 3i$.

60. Evaluate $\dfrac{x^2 + 11}{3 - x}$ for $x = 4i$.

Application Exercises

Complex numbers are used in electronics to describe the current in an electric circuit. Ohm's law relates the current in a circuit, I, in amperes, the voltage of the circuit, E, in volts, and the resistance of the circuit, R, in ohms, by the formula $E = IR$. Use this formula to solve Exercises 61–62.

61. Find E, the voltage of a circuit, if $I = (4 - 5i)$ amperes and $R = (3 + 7i)$ ohms.

62. Find E, the voltage of a circuit, if $I = (2 - 3i)$ amperes and $R = (3 + 5i)$ ohms.

63. The mathematician Girolamo Cardano is credited with the first use (in 1545) of negative square roots in solving the now-famous problem, "Find two numbers whose sum is 10 and whose product is 40." Show that the complex numbers $5 + i\sqrt{15}$ and $5 - i\sqrt{15}$ satisfy the conditions of the problem. (Cardano did not use the symbolism $i\sqrt{15}$ or even $\sqrt{-15}$. He wrote R.m 15 for $\sqrt{-15}$, meaning "radix minus 15." He regarded the numbers $5 + $ R.m 15 and $5 - $ R.m 15 as "fictitious" or "ghost numbers," and considered the problem "manifestly impossible." But in a mathematically adventurous spirit, he exclaimed, "Nevertheless, we will operate.")

Writing in Mathematics

64. What is i?

65. Explain how to add complex numbers. Provide an example with your explanation.

66. Explain how to multiply complex numbers and give an example.

67. What is the complex conjugate of $2 + 3i$? What happens when you multiply this complex number by its complex conjugate?

68. Explain how to divide complex numbers. Provide an example with your explanation.

69. Explain each of the three jokes in the cartoon on page 266.

70. A stand-up comedian uses algebra in some jokes, including one about a telephone recording that announces "You have just reached an imaginary number. Please multiply by i and dial again." Explain the joke.

Explain the error in Exercises 71–72.

71. $\sqrt{-9} + \sqrt{-16} = \sqrt{-25} = i\sqrt{25} = 5i$

72. $\left(\sqrt{-9}\right)^2 = \sqrt{-9} \cdot \sqrt{-9} = \sqrt{81} = 9$

Critical Thinking Exercises

73. Which one of the following is true?

   a. Some irrational numbers are not complex numbers.

   b. $(3 + 7i)(3 - 7i)$ is an imaginary number.

   c. $\dfrac{7 + 3i}{5 + 3i} = \dfrac{7}{5}$

   d. In the complex number system, $x^2 + y^2$ (the sum of two squares) can be factored as $(x + yi)(x - yi)$.

In Exercises 74–76, perform the indicated operations and write the result in standard form.

74. $\dfrac{4}{(2 + i)(3 - i)}$

75. $\dfrac{1 + i}{1 + 2i} + \dfrac{1 - i}{1 - 2i}$

76. $\dfrac{8}{1 + \dfrac{2}{i}}$

SECTION 2.2 *Quadratic Functions*

Objectives

1 Recognize characteristics of parabolas.

2 Graph parabolas.

3 Determine a quadratic function's minimum or maximum value.

4 Solve problems involving a quadratic function's minimum or maximum value.

The Food Stamp Program is the first line of defense against hunger for millions of American families. The program provides benefits for eligible participants to purchase approved food items at approved food stores. Over half of all participants are children; one out of six is a low-income older adult. The function

$$f(x) = 0.22x^2 - 0.50x + 7.68$$

models the number of households, $f(x)$, in millions, participating in the program x years after 1999. For example, to find the number of households receiving food stamps in 2005, substitute 6 for x because 2005 is 6 years after 1999:

$$f(6) = 0.22(6)^2 - 0.50(6) + 7.68 = 12.6.$$

Thus, in 2005, 12.6 million households received food stamps.

The function $f(x) = 0.22x^2 - 0.50x + 7.68$ is an example of a quadratic function. We have seen that a **quadratic function** is any function of the form

$$f(x) = ax^2 + bx + c,$$

where a, b, and c are real numbers, with $a \neq 0$. A quadratic function is a polynomial function whose greatest exponent is 2. In this section, we study quadratic functions and their graphs.

1 Recognize characteristics of parabolas.

Graphs of Quadratic Functions

The graph of any quadratic function is called a **parabola**. Parabolas are shaped like cups, as shown in Figure 2.2. If the coefficient of x^2 (the value of a in $ax^2 + bx + c$) is positive, the parabola opens upward. If the coefficient of x^2 is negative, the graph opens downward. The **vertex** (or turning point) of the parabola is the lowest point on the graph when it opens upward and the highest point on the graph when it opens downward.

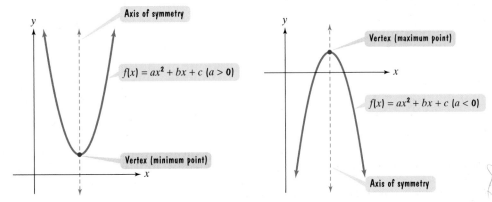

$a > 0$: Parabola opens upward.

$a < 0$: Parabola opens downward.

Figure 2.2 Characteristics of graphs of quadratic functions

 Graph parabolas.

Look at the unusual image of the word "mirror" shown below. The artist, Scott Kim, has created the image so that the two halves of the whole are mirror images of each other. A parabola shares this kind of symmetry, in which a line through the vertex divides the figure in half. Parabolas are symmetric with respect to this line, called the **axis of symmetry**. If a parabola is folded along its axis of symmetry, the two halves match exactly.

Graphing Quadratic Functions in Standard Form

In our earlier work with transformations, we applied a series of transformations to the graph of $f(x) = x^2$. The graph of this function is a parabola. The vertex for this parabola is $(0, 0)$. In Figure 2.3(a), the graph of $f(x) = ax^2$ for $a > 0$ is shown in black; it opens *upward*. In Figure 2.3(b), the graph of $f(x) = ax^2$ for $a < 0$ is shown in black; it opens *downward*.

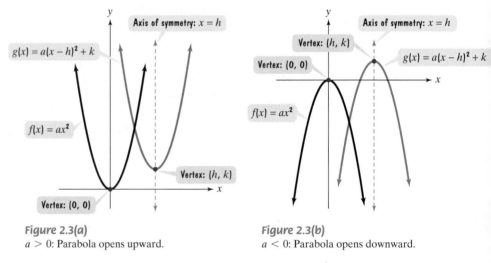

Figure 2.3(a)
$a > 0$: Parabola opens upward.

Figure 2.3(b)
$a < 0$: Parabola opens downward.

Transformations of $f(x) = ax^2$

Figure 2.3(a) and 2.3(b) also show the graph of $g(x) = a(x - h)^2 + k$ in blue. Compare these graphs to those of $f(x) = ax^2$. Observe that h determines the horizontal shift and k determines the vertical shift of the graph of $f(x) = ax^2$:

$$g(x) = a(x - h)^2 + k.$$

If $h > 0$, the graph of $f(x) = ax^2$ is shifted h units to the right.

If $k > 0$, the graph of $y = a(x - h)^2$ is shifted k units up.

Consequently, the vertex $(0, 0)$ on the black graph of $f(x) = ax^2$ moves to the point (h, k) on the blue graph of $g(x) = a(x - h)^2 + k$. The axis of symmetry is the vertical line whose equation is $x = h$.

The form of the expression for g is convenient because it immediately identifies the vertex of the parabola as (h, k). This is the **standard form** of a quadratic function.

The Standard Form of a Quadratic Function

The quadratic function

$$f(x) = a(x - h)^2 + k, \quad a \neq 0$$

is in **standard form**. The graph of f is a parabola whose vertex is the point (h, k). The parabola is symmetric with respect to the line $x = h$. If $a > 0$, the parabola opens upward; if $a < 0$, the parabola opens downward.

The sign of a in $f(x) = a(x - h)^2 + k$ determines whether the parabola opens upward or downward. Furthermore, if $|a|$ is small, the parabola opens more flatly than if $|a|$ is large. Here is a general procedure for graphing parabolas whose equations are in standard form:

Graphing Quadratic Functions with Equations in Standard Form

To graph $f(x) = a(x - h)^2 + k$,

1. Determine whether the parabola opens upward or downward. If $a > 0$, it opens upward. If $a < 0$, it opens downward.
2. Determine the vertex of the parabola. The vertex is (h, k).
3. Find any x-intercepts by solving $f(x) = 0$. The function's real zeros are the x-intercepts.
4. Find the y-intercept by computing $f(0)$.
5. Plot the intercepts, the vertex, and additional points as necessary. Connect these points with a smooth curve that is shaped like a cup.

In the graphs that follow, we will show each axis of symmetry as a dashed vertical line. Because this vertical line passes through the vertex, (h, k), its equation is $x = h$. The line is dashed because it is not part of the parabola.

EXAMPLE 1   Graphing a Quadratic Function in Standard Form

Graph the quadratic function $f(x) = -2(x - 3)^2 + 8$.

Solution   We can graph this function by following the steps in the preceding box. We begin by identifying values for a, h, and k.

Standard form
$$f(x) = a(x - h)^2 + k$$

$a = -2$ $h = 3$ $k = 8$

Given function
$$f(x) = -2(x - 3)^2 + 8$$

Step 1   Determine how the parabola opens.   Note that a, the coefficient of x^2, is -2. Thus, $a < 0$; this negative value tells us that the parabola opens downward.

Step 2   Find the vertex.   The vertex of the parabola is at (h, k). Because $h = 3$ and $k = 8$, the parabola has its vertex at $(3, 8)$.

Step 3   Find the x-intercepts by solving $f(x) = 0$.   Replace $f(x)$ with 0 in $f(x) = -2(x - 3)^2 + 8$.

$0 = -2(x - 3)^2 + 8$	Find x-intercepts, setting f(x) equal to 0.
$2(x - 3)^2 = 8$	Solve for x. Add $2(x - 3)^2$ to both sides of the equation.
$(x - 3)^2 = 4$	Divide both sides by 2.
$x - 3 = \sqrt{4}$ or $x - 3 = -\sqrt{4}$	Apply the square root property.
$x - 3 = 2$ $\qquad$ $x - 3 = -2$	$\sqrt{4} = 2$
$x = 5$ $\qquad\qquad$ $x = 1$	Add 3 to both sides in each equation.

The x-intercepts are 1 and 5. The parabola passes through $(1, 0)$. and $(5, 0)$.

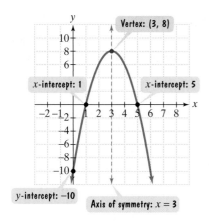

Vertex: (3, 8)

Figure 2.4 The graph of $f(x) = -2(x - 3)^2 + 8$

Step 4   Find the y-intercept by computing $f(0)$. Replace x with 0 in $f(x) = -2(x - 3)^2 + 8$.

$$f(0) = -2(0 - 3)^2 + 8 = -2(-3)^2 + 8 = -2(9) + 8 = -10$$

The y-intercept is -10. The parabola passes through $(0, -10)$.

Step 5   Graph the parabola. With a vertex at $(3, 8)$, x-intercepts at 1 and 5, and y-intercept at -10, the graph of f is shown in Figure 2.4. The axis of symmetry is the vertical line whose equation is $x = 3$.

> **Check Point 1**   Graph the quadratic function $f(x) = -(x - 1)^2 + 4$.

EXAMPLE 2   Graphing a Quadratic Function in Standard Form

Graph the quadratic function $f(x) = (x + 3)^2 + 1$.

Solution   We begin by finding values for a, h, and k.

$$f(x) = a(x - h)^2 + k \quad \text{Standard form of quadratic function}$$
$$f(x) = (x + 3)^2 + 1 \quad \text{Given function}$$
$$f(x) = 1(x - (-3))^2 + 1$$

$$a = 1 \qquad h = -3 \qquad k = 1$$

Step 1   Determine how the parabola opens. Note that a, the coefficient of x^2, is 1. Thus, $a > 0$; this positive value tells us that the parabola opens upward.

Step 2   Find the vertex. The vertex of the parabola is at (h, k). Because $h = -3$ and $k = 1$, the parabola has its vertex at $(-3, 1)$.

Step 3   Find the x-intercepts by solving $f(x) = 0$. Replace $f(x)$ with 0 in $f(x) = (x + 3)^2 + 1$. Because the vertex is at $(-3, 1)$, which lies above the x-axis, and the parabola opens upward, it appears that this parabola has no x-intercepts. We can verify this observation algebraically.

$$0 = (x + 3)^2 + 1 \quad \text{Find possible x-intercepts, setting } f(x) \text{ equal to 0.}$$

$$-1 = (x + 3)^2 \quad \text{Solve for x. Subtract 1 from both sides.}$$

$$x + 3 = \sqrt{-1} \quad \text{or} \quad x + 3 = -\sqrt{-1} \quad \text{Apply the square root property.}$$
$$x + 3 = i \qquad\qquad x + 3 = -i \qquad \sqrt{-1} = i$$
$$x = -3 + i \qquad\qquad x = -3 - i \quad \text{The solutions are } -3 \pm i.$$

Because this equation has no real solutions, the parabola has no x-intercepts.

Step 4   Find the y-intercept by computing $f(0)$. Replace x with 0 in $f(x) = (x + 3)^2 + 1$.

$$f(0) = (0 + 3)^2 + 1 = 3^2 + 1 = 9 + 1 = 10$$

The y-intercept is 10. The parabola passes through $(0, 10)$.

Step 5   Graph the parabola. With a vertex at $(-3, 1)$, no x-intercepts, and a y-intercept at 10, the graph of f is shown in Figure 2.5. The axis of symmetry is the vertical line whose equation is $x = -3$.

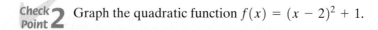

Axis of symmetry: $x = -3$ y-intercept: 10   Vertex: $(-3, 1)$

Figure 2.5 The graph of $f(x) = (x + 3)^2 + 1$

> **Check Point 2**   Graph the quadratic function $f(x) = (x - 2)^2 + 1$.

Graphing Quadratic Functions in the Form $f(x) = ax^2 + bx + c$

Quadratic functions are frequently expressed in the form $f(x) = ax^2 + bx + c$. How can we identify the vertex of a parabola whose equation is in this form? Completing the square provides the answer to this question.

$$f(x) = ax^2 + bx + c$$

$$= a\left(x^2 + \frac{b}{a}x\right) + c \qquad\qquad \text{Factor out } a \text{ from } ax^2 + bx.$$

$$= a\left(x^2 + \frac{b}{a}x + \frac{b^2}{4a^2}\right) + c - a\left(\frac{b^2}{4a^2}\right)$$

> Complete the square by adding the square of half the coefficient of x.

> By completing the square, we added $a \cdot \dfrac{b^2}{4a^2}$. To avoid changing the function's equation, we must subtract this term.

$$= a\left(x + \frac{b}{2a}\right)^2 + c - \frac{b^2}{4a}$$

> Write the trinomial as the square of a binomial and simplify the constant term.

Compare this form of the equation with a quadratic function's **standard form**.

> Standard form

$$f(x) = a(x - h)^2 + k$$

$$h = -\frac{b}{2a} \qquad k = c - \frac{b^2}{4a}$$

> Equation under discussion

$$f(x) = a\left(x - \left(-\frac{b}{2a}\right)\right)^2 + c - \frac{b^2}{4a}$$

The important part of this observation is that h, the x-coordinate of the vertex, is $-\dfrac{b}{2a}$. The y-coordinate can be found by evaluating the function at $-\dfrac{b}{2a}$.

The Vertex of a Parabola Whose Equation Is $f(x) = ax^2 + bx + c$

Consider the parabola defined by the quadratic function $f(x) = ax^2 + bx + c$. The parabola's vertex is $\left(-\dfrac{b}{2a}, f\left(-\dfrac{b}{2a}\right)\right)$.

We can apply our five-step procedure and graph parabolas in the form $f(x) = ax^2 + bx + c$. The only step that is different is how we determine the vertex.

EXAMPLE 3 Graphing a Quadratic Function in the Form $f(x) = ax^2 + bx + c$

Graph the quadratic function $f(x) = -x^2 - 2x + 1$. Use the graph to identify the function's domain and its range.

Solution

Step 1 Determine how the parabola opens. Note that a, the coefficient of x^2, is -1. Thus, $a < 0$; this negative value tells us that the parabola opens downward.

Step 2 Find the vertex. We know that the x-coordinate of the vertex is $x = -\dfrac{b}{2a}$. We identify a, b, and c in $f(x) = ax^2 + bx + c$.

$$f(x) = -x^2 - 2x + 1$$

> $a = -1$ $b = -2$ $c = 1$

Substitute -1 for a and -2 for b into the equation for the x-coordinate:

$$x = -\frac{b}{2a} = -\frac{-2}{2(-1)} = -\left(\frac{-2}{-2}\right) = -1.$$

The x-coordinate of the vertex is -1 and the vertex is at $(-1, f(-1))$. We substitute -1 for x in the equation of the function, $f(x) = -x^2 - 2x + 1$, to find the y-coordinate

$$f(-1) = -(-1)^2 - 2(-1) + 1 = -1 + 2 + 1 = 2.$$

The vertex is at $(-1, 2)$.

Step 3   Find the x-intercepts by solving $f(x) = 0$. Replace $f(x)$ with 0 in $f(x) = -x^2 - 2x + 1$. We obtain $0 = -x^2 - 2x + 1$. This equation cannot be solved by factoring. We will use the quadratic formula to solve it.

$$-x^2 - 2x + 1 = 0$$

$$\boxed{a = -1} \quad \boxed{b = -2} \quad \boxed{c = 1}$$

$$x = \frac{-b \pm \sqrt{b^2 - 4ac}}{2a} = \frac{-(-2) \pm \sqrt{(-2)^2 - 4(-1)(1)}}{2(-1)} = \frac{2 \pm \sqrt{4 - (-4)}}{-2}$$

> To locate the x-intercepts, we need decimal approximations. Thus, there is no need to simplify the radical form of the solutions.

$$x = \frac{2 + \sqrt{8}}{-2} \approx -2.4 \quad \text{or} \quad x = \frac{2 - \sqrt{8}}{-2} \approx 0.4$$

The x-intercepts are approximately -2.4 and 0.4. The parabola passes through $(-2.4, 0)$ and $(0.4, 0)$.

Step 4   Find the y-intercept by computing $f(0)$. Replace x with 0 in $f(x) = -x^2 - 2x + 1$.

$$f(0) = -0^2 - 2 \cdot 0 + 1 = 1$$

The y-intercept is 1. The parabola passes through $(0, 1)$.

Step 5   Graph the parabola. With a vertex at $(-1, 2)$, x-intercepts at approximately -2.4 and 0.4, and a y-intercept at 1, the graph of f is shown in Figure 2.6(a). The axis of symmetry is the vertical line whose equation is $x = -1$.

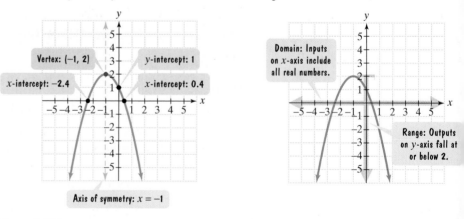

Figure 2.6(a) The graph of $f(x) = -x^2 - 2x + 1$

Figure 2.6(b) Determining the domain and range of $f(x) = -x^2 - 2x + 1$

Study Tip

The domain of any quadratic function includes all real numbers. If the vertex is the graph's highest point, the range includes all real numbers at or below the y-coordinate of the vertex. If the vertex is the graph's lowest point, the range includes all real numbers at or above the y-coordinate of the vertex.

Now we are ready to determine the domain and range of $f(x) = -x^2 - 2x + 1$. We can use the parabola, shown again in Figure 2.6(b), to do so. To find the domain, look for all the inputs on the x-axis that correspond to points on the graph. As the graph widens and continues to fall at both ends, can you see that these inputs include all real numbers?

Domain of f is $\{x | x$ is a real number$\}$ or $(-\infty, \infty)$.

To find the range, look for all the outputs on the y-axis that correspond to points on the graph. Figure 2.6(b) shows that the parabola's vertex, $(-1, 2)$, is the highest point on the graph. Because the y-coordinate of the vertex is 2, outputs on the y-axis fall at or below 2.

$$\text{Range of } f \text{ is } \{y \mid y \le 2\} \text{ or } (-\infty, 2].$$

Check Point 3 Graph the quadratic function $f(x) = -x^2 + 4x + 1$. Use the graph to identify the function's domain and its range.

③ Determine a quadratic function's minimum or maximum value.

Minimum and Maximum Values of Quadratic Functions

Consider the quadratic function $f(x) = ax^2 + bx + c$. If $a > 0$, the parabola opens upward and the vertex is its lowest point. If $a < 0$, the parabola opens downward and the vertex is its highest point. The x-coordinate of the vertex is $-\dfrac{b}{2a}$. Thus, we can find the minimum or maximum value of f by evaluating the quadratic function at $x = -\dfrac{b}{2a}$.

Minimum and Maximum: Quadratic Functions

Consider the quadratic function $f(x) = ax^2 + bx + c$.

1. If $a > 0$, then f has a minimum that occurs at $x = -\dfrac{b}{2a}$. This minimum value is $f\left(-\dfrac{b}{2a}\right)$.

2. If $a < 0$, then f has a maximum that occurs at $x = -\dfrac{b}{2a}$. This maximum value is $f\left(-\dfrac{b}{2a}\right)$.

In each case, the value of x gives the location of the minimum or maximum value. The value of y, or $f\left(-\dfrac{b}{2a}\right)$, gives that minimum or maximum value.

EXAMPLE 4   Obtaining Information about a Quadratic Function from Its Equation

Consider the quadratic function $f(x) = -3x^2 + 6x - 13$.

  a. Determine, without graphing, whether the function has a minimum value or a maximum value.

  b. Find the minimum or maximum value and determine where it occurs.

  c. Identify the function's domain and its range.

Solution   We begin by identifying a, b, and c in the function's equation:

$$f(x) = -3x^2 + 6x - 13.$$

$$a = -3 \qquad b = 6 \qquad c = -13$$

  a. Because $a < 0$, the function has a maximum value.

  b. The maximum value occurs at

$$x = -\frac{b}{2a} = -\frac{6}{2(-3)} = -\frac{6}{-6} = -(-1) = 1.$$

The maximum value occurs at $x = 1$ and the maximum value of $f(x) = -3x^2 + 6x - 13$ is

$$f(1) = -3 \cdot 1^2 + 6 \cdot 1 - 13 = -3 + 6 - 13 = -10.$$

We see that the maximum is -10 at $x = 1$.

c. Like all quadratic functions, the domain is $(-\infty, \infty)$. Because the function's maximum value is -10, the range includes all real numbers at or below -10. The range is $(-\infty, -10]$.

We can use the graph of $f(x) = -3x^2 + 6x - 13$ to visualize the results of Example 4. Figure 2.7 shows the graph in a $[-6, 6, 1]$ by $[-50, 20, 10]$ viewing rectangle. The maximum function feature verifies that the function's maximum is -10 at $x = 1$. Notice that x gives the location of the maximum and y gives the maximum value. Notice, too, that the maximum value is -10 and not the ordered pair $(1, -10)$.

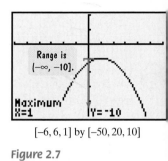

$[-6, 6, 1]$ by $[-50, 20, 10]$

Figure 2.7

> **Check Point 4** Repeat parts (a) through (c) of Example 4 using the quadratic function $f(x) = 4x^2 - 16x + 1000$.

④ Solve problems involving a quadratic function's minimum or maximum value.

Applications of Quadratic Functions

When did the minimum number of households participate in the food stamp program? What is the age of a driver having the least number of car accidents? If you throw a baseball vertically upward, after how many seconds will it reach its maximum height and what is that height? The answers to these questions involve finding the maximum or minimum value of a quadratic function, as well as where this value occurs.

EXAMPLE 5   The Food Stamp Program

Figure 2.8 shows the number of U.S. households, in millions, participating in the Food Stamp Program from 1999 through 2004. The function

$$f(x) = 0.22x^2 - 0.50x + 7.68$$

models the number of households, $f(x)$, in millions, participating in the program x years after 1999. According to this function, in which year was the number of participants at a minimum? How many households received food stamps for that year? How well does this model the data shown in Figure 2.8?

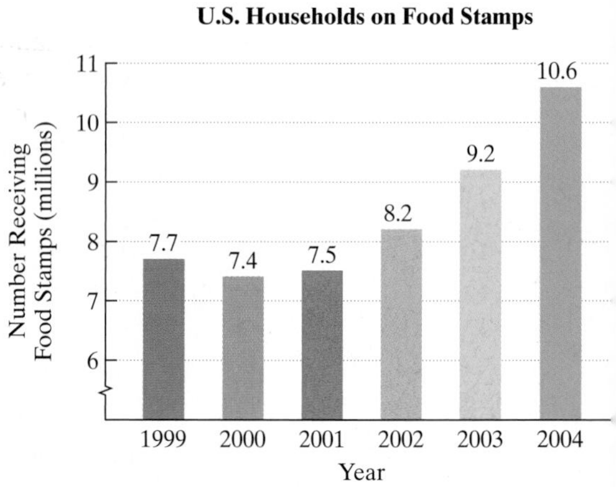

U.S. Households on Food Stamps

Number Receiving Food Stamps (millions)

11 —
10 —   10.6
 9 —   9.2
 8 —   7.7   8.2
 7 —   7.4   7.5
 6 —

    1999   2000   2001   2002   2003   2004
                    Year

Figure 2.8

Source: Food Stamp Program

Solution   We begin by identifying a, b, and c in the function's equation:

$$f(x) = 0.22x^2 - 0.50x + 7.68.$$

$a = 0.22$ $b = -0.50$ $c = 7.68$

Because $a > 0$, the function has a minimum value. The minimum value occurs at

$$x = -\frac{b}{2a} = -\frac{(-0.50)}{2(0.22)} = \frac{0.50}{0.44} \approx 1.$$

This means that the number of households receiving food stamps was at a minimum approximately 1 year after 1999, in 2000. Using the model $f(x) = 0.22x^2 - 0.50x + 7.68$, the number of households, in millions, for that year was

$$f(1) = 0.22(1)^2 - 0.50(1) + 7.68 = 7.4.$$

In 2000, the number of households receiving food stamps was at a minimum of 7.4 million. Because this is precisely what is shown in Figure 2.8 on the previous page, the function models the data extremely well.

Technology

Because of the decreasing-increasing cuplike shape of the data in Figure 2.8, a quadratic function is an appropriate model. We entered the data using

(number of years after 1999, millions of participants).

Upon entering the QUADratic REGression program, we obtain the results shown in the screen on the right. Thus, the quadratic function of best fit is

$$f(x) = 0.22x^2 - 0.50x + 7.68,$$

where x represents the number of years after 1999 and $f(x)$ represents the number of U.S. households, in millions, on food stamps.

Data:
(0, 7.7), (1,7.4), (2, 7.5),
(3, 8.2), (4, 9.2), (5, 10.6)

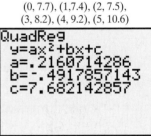

```
QuadReg
 y=ax²+bx+c
 a=.2160714286
 b=-.4917857143
 c=7.682142857
```

Check Point 5 The function $f(x) = 0.4x^2 - 36x + 1000$ models the number of accidents, $f(x)$, per 50 million miles driven, for drivers x years old, where $16 \le x \le 74$. What is the age of a driver having the least number of car accidents? What is the minimum number of car accidents per 50 million miles driven?

Quadratic functions can also be modeled from verbal conditions. Once we have obtained a quadratic function, we can then use the x-coordinate of the vertex to determine its maximum or minimum value. Here is a step-by-step strategy for solving these kinds of problems:

Strategy for Solving Problems Involving Maximizing or Minimizing Quadratic Functions

1. Read the problem carefully and decide which quantity is to be maximized or minimized.
2. Use the conditions of the problem to express the quantity as a function in one variable.
3. Rewrite the function in the form $f(x) = ax^2 + bx + c$.
4. Calculate $-\dfrac{b}{2a}$. If $a > 0$, f has a minimum at $x = -\dfrac{b}{2a}$. This minimum value is $f\left(-\dfrac{b}{2a}\right)$. If $a < 0$, f has a maximum at $x = -\dfrac{b}{2a}$. This maximum value is $f\left(-\dfrac{b}{2a}\right)$.
5. Answer the question posed in the problem.

EXAMPLE 6   Minimizing a Product

Among all pairs of numbers whose difference is 10, find a pair whose product is as small as possible. What is the minimum product?

Solution

Step 1   Decide what must be maximized or minimized. We must minimize th
product of two numbers. Calling the numbers x and y, and calling the product P, w
must minimize

$$P = xy.$$

Step 2   Express this quantity as a function in one variable. In the formula $P = xy$
P is expressed in terms of two variables, x and y. However, because the difference o
the numbers is 10, we can write

$$x - y = 10.$$

We can solve this equation for y in terms of x (or vice versa), substitute the resul
into $P = xy$, and obtain P as a function of one variable.

$$-y = -x + 10 \qquad \text{Subtract } x \text{ from both sides of } x - y = 10.$$

$$y = x - 10 \qquad \text{Multiply both sides of the equation by } -1 \text{ and solve for } y.$$

Now we substitute $x - 10$ for y in $P = xy$.

$$P = xy = x(x - 10).$$

Because P is now a function of x, we can write

$$P(x) = x(x - 10).$$

Step 3   Write the function in the form $f(x) = ax^2 + bx + c$. We apply the
distributive property to obtain

$$P(x) = x(x - 10) = x^2 - 10x.$$

$$\boxed{a = 1} \quad \boxed{b = -10}$$

Step 4   Calculate $-\dfrac{b}{2a}$. If $a > 0$, the function has a minimum at this value. The
voice balloons show that $a = 1$ and $b = -10$.

$$x = -\frac{b}{2a} = -\frac{-10}{2(1)} = -(-5) = 5$$

This means that the product, P, of two numbers whose difference is 10 is a minimum
when one of the numbers, x, is 5.

Step 5   Answer the question posed by the problem. The problem asks for the two
numbers and the minimum product. We found that one of the numbers, x, is 5. Now
we must find the second number, y.

$$y = x - 10 = 5 - 10 = -5$$

The number pair whose difference is 10 and whose product is as small as possible is
5, -5. The minimum product is $5(-5)$, or -25.

Technology

The ⌐TABLE⌐ feature of a graphing utility can be used to verify
our work in Example 6.

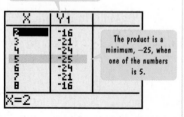

Enter $y_1 = x^2 - 10x$, the
function for the product,
when one of the numbers
is x.

The product is a
minimum, −25, when
one of the numbers
is 5.

**Check
Point 6**   Among all pairs of numbers whose difference is 8, find a pair whose product is as small as possible. What is the minimum product?

EXAMPLE 7 Maximizing Area

You have 100 yards of fencing to enclose a rectangular region. Find the dimensions of the rectangle that maximize the enclosed area. What is the maximum area?

Solution

Step 1 Decide what must be maximized or minimized. We must maximize area. What we do not know are the rectangle's dimensions, x and y.

Step 2 Express this quantity as a function in one variable. Because we must maximize area, we have $A = xy$. We need to transform this into a function in which A is represented by one variable. Because you have 100 yards of fencing, the perimeter of the rectangle is 100 yards. This means that

$$2x + 2y = 100.$$

We can solve this equation for y in terms of x, substitute the result into $A = xy$, and obtain A as a function in one variable. We begin by solving for y.

$$2y = 100 - 2x \quad \text{Subtract 2x from both sides.}$$

$$y = \frac{100 - 2x}{2} \quad \text{Divide both sides by 2.}$$

$$y = 50 - x \quad \text{Divide each term in the numerator by 2.}$$

Now we substitute $50 - x$ for y in $A = xy$.

$$A = xy = x(50 - x)$$

The rectangle and its dimensions are illustrated in Figure 2.9. Because A is now a function of x, we can write

$$A(x) = x(50 - x).$$

This function models the area, $A(x)$, of any rectangle whose perimeter is 100 yards in terms of one of its dimensions, x.

Step 3 Write the function in the form $f(x) = ax^2 + bx + c$. We apply the distributive property to obtain

$$A(x) = x(50 - x) = 50x - x^2 = -x^2 + 50x.$$

$a = -1$ $b = 50$

Step 4 Calculate $-\dfrac{b}{2a}$. If $a < 0$, the function has a maximum at this value. The voice balloons show that $a = -1$ and $b = 50$.

$$x = -\frac{b}{2a} = -\frac{50}{2(-1)} = 25$$

This means that the area, $A(x)$, of a rectangle with perimeter 100 yards is a maximum when one of the rectangle's dimensions, x, is 25 yards.

Step 5 Answer the question posed by the problem. We found that $x = 25$. Figure 2.9 shows that the rectangle's other dimension is $50 - x = 50 - 25 = 25$. The dimensions of the rectangle that maximize the enclosed area are 25 yards by 25 yards. The rectangle that gives the maximum area is actually a square with an area of 25 yards · 25 yards, or 625 square yards.

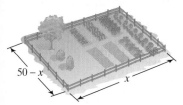

Figure 2.9 What value of x will maximize the rectangle's area?

$50 - x$

x

Technology

The graph of the area function

$$A(x) = x(50 - x)$$

was obtained with a graphing utility using a $[0, 50, 2]$ by $[0, 700, 25]$ viewing rectangle. The maximum function feature verifies that a maximum area of 625 square yards occurs when one of the dimensions is 25 yards.

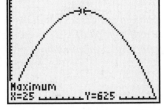

Maximum
X=25   Y=625

Check Point 7 You have 120 feet of fencing to enclose a rectangular region. Find the dimensions of the rectangle that maximize the enclosed area. What is the maximum area?

The ability to express a quantity to be maximized or minimized as a function in one variable plays a critical role in solving max-min problems. In calculus, you will learn a technique for maximizing or minimizing all functions, not only quadratic functions.

EXERCISE SET 2.2

Practice Exercises

In Exercises 1–4, the graph of a quadratic function is given. Write the function's equation, selecting from the following options.

$$f(x) = (x + 1)^2 - 1 \qquad g(x) = (x + 1)^2 + 1$$
$$h(x) = (x - 1)^2 + 1 \qquad j(x) = (x - 1)^2 - 1$$

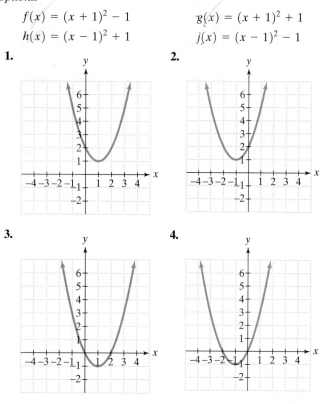

1.

2.

3.

4.

In Exercises 5–8, the graph of a quadratic function is given. Write the function's equation, selecting from the following options.

$$f(x) = x^2 + 2x + 1 \qquad g(x) = x^2 - 2x + 1$$
$$h(x) = x^2 - 1 \qquad j(x) = -x^2 - 1$$

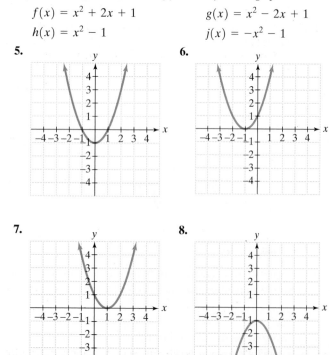

5.

6.

7.

8.

In Exercises 9–16, find the coordinates of the vertex for the parabola defined by the given quadratic function.

9. $f(x) = 2(x - 3)^2 + 1$ **10.** $f(x) = -3(x - 2)^2 + 12$
11. $f(x) = -2(x + 1)^2 + 5$ **12.** $f(x) = -2(x + 4)^2 - 8$
13. $f(x) = 2x^2 - 8x + 3$ **14.** $f(x) = 3x^2 - 12x + 1$
15. $f(x) = -x^2 - 2x + 8$ **16.** $f(x) = -2x^2 + 8x - 1$

In Exercises 17–38, use the vertex and intercepts to sketch the graph of each quadratic function. Give the equation of the parabola's axis of symmetry. Use the graph to determine the function's domain and range.

17. $f(x) = (x - 4)^2 - 1$ **18.** $f(x) = (x - 1)^2 - 2$
19. $f(x) = (x - 1)^2 + 2$ **20.** $f(x) = (x - 3)^2 + 2$
21. $y - 1 = (x - 3)^2$ **22.** $y - 3 = (x - 1)^2$
23. $f(x) = 2(x + 2)^2 - 1$ **24.** $f(x) = \frac{5}{4} - \left(x - \frac{1}{2}\right)^2$
25. $f(x) = 4 - (x - 1)^2$ **26.** $f(x) = 1 - (x - 3)^2$
27. $f(x) = x^2 - 2x - 3$ **28.** $f(x) = x^2 - 2x - 15$
29. $f(x) = x^2 + 3x - 10$ **30.** $f(x) = 2x^2 - 7x - 4$
31. $f(x) = 2x - x^2 + 3$ **32.** $f(x) = 5 - 4x - x^2$
33. $f(x) = x^2 + 6x + 3$ **34.** $f(x) = x^2 + 4x - 1$
35. $f(x) = 2x^2 + 4x - 3$ **36.** $f(x) = 3x^2 - 2x - 4$
37. $f(x) = 2x - x^2 - 2$ **38.** $f(x) = 6 - 4x + x^2$

In Exercises 39–44, an equation of a quadratic function is given.

   a. Determine, without graphing, whether the function has a minimum value or a maximum value.
   b. Find the minimum or maximum value and determine where it occurs.
   c. Identify the function's domain and its range.

39. $f(x) = 3x^2 - 12x - 1$ **40.** $f(x) = 2x^2 - 8x - 3$
41. $f(x) = -4x^2 + 8x - 3$ **42.** $f(x) = -2x^2 - 12x + 3$
43. $f(x) = 5x^2 - 5x$ **44.** $f(x) = 6x^2 - 6x$

Practice Plus

In Exercises 45–48, give the domain and the range of each quadratic function whose graph is described.

45. The vertex is $(-1, -2)$ and the parabola opens up.
46. The vertex is $(-3, -4)$ and the parabola opens down.
47. Maximum $= -6$ at $x = 10$
48. Minimum $= 18$ at $x = -6$

In Exercises 49–52, write an equation in standard form of the parabola that has the same shape as the graph of $f(x) = 2x^2$, but with the given point as the vertex.

49. $(5, 3)$ **50.** $(7, 4)$
51. $(-10, -5)$ **52.** $(-8, -6)$

In Exercises 53–56, write an equation in standard form of the parabola that has the same shape as the graph of $f(x) = 3x^2$ or $g(x) = -3x^2$, but with the given maximum or minimum.

53. Maximum $= 4$ at $x = -2$
54. Maximum $= -7$ at $x = 5$
55. Minimum $= 0$ at $x = 11$
56. Minimum $= 0$ at $x = 9$

Application Exercises

57. The graph shows per capita U.S. adult wine consumption (in gallons per person) for selected years from 1980 through 2003. The function

$$f(x) = 0.005x^2 - 0.104x + 2.626$$

models U.S. wine consumption, $f(x)$, in gallons per person, x years after 1980. According to this function, in which year was wine consumption at a minimum? Round to the nearest year. What does the function give for per capita consumption, to the nearest tenth of a gallon, for that year? How well does this model the data shown in the graph?

Wine Consumption per U.S. Adult

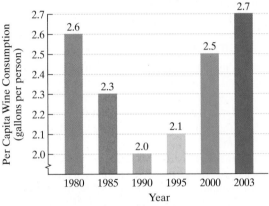

Source: Adams Business Media

58. After declining in the late 1990s, the number of gang-related murders across the United States has increased in recent years. The graph shows the number of gang-related homicides in the United States. The function

$$f(x) = 33x^2 - 255x + 1230$$

models the number of gang-related homicides across the nation, $f(x)$, x years after 1995. According to this function, in which year was the number of homicides at a minimum? Round to the nearest year. What does the function give for the number of gang-related murders for that year? How well does this model the data shown in the graph?

Gang-Related Homicides in the U.S.

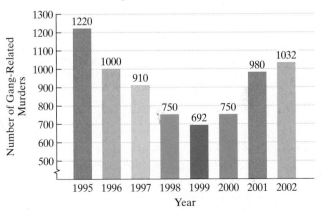

Source: Professor James Alan Fox, Northeastern University

59. A person standing close to the edge on the top of a 200-foot building throws a baseball vertically upward. The quadratic function

$$s(t) = -16t^2 + 64t + 200$$

models the ball's height above the ground, $s(t)$, in feet, t seconds after it was thrown.

   a. After how many seconds does the ball reach its maximum height? What is the maximum height?

   b. How many seconds does it take until the ball finally hits the ground? Round to the nearest tenth of a second.

   c. Find $s(0)$ and describe what this means.

   d. Use your results from parts (a) through (c) to graph the quadratic function. Begin the graph with $t = 0$ and end with the value of t for which the ball hits the ground.

60. A person standing close to the edge on the top of a 160-foot building throws a baseball vertically upward. The quadratic function

$$s(t) = -16t^2 + 64t + 160$$

models the ball's height above the ground, $s(t)$, in feet, t seconds after it was thrown.

   a. After how many seconds does the ball reach its maximum height? What is the maximum height?

   b. How many seconds does it take until the ball finally hits the ground? Round to the nearest tenth of a second.

   c. Find $s(0)$ and describe what this means.

   d. Use your results from parts (a) through (c) to graph the quadratic function. Begin the graph with $t = 0$ and end with the value of t for which the ball hits the ground.

61. Among all pairs of numbers whose sum is 16, find a pair whose product is as large as possible. What is the maximum product?

62. Among all pairs of numbers whose sum is 20, find a pair whose product is as large as possible. What is the maximum product?

63. Among all pairs of numbers whose difference is 16, find a pair whose product is as small as possible. What is the minimum product?

64. Among all pairs of numbers whose difference is 24, find a pair whose product is as small as possible. What is the minimum product?

65. You have 600 feet of fencing to enclose a rectangular plot that borders on a river. If you do not fence the side along the river, find the length and width of the plot that will maximize the area. What is the largest area that can be enclosed?

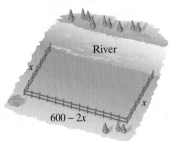

66. You have 200 feet of fencing to enclose a rectangular plot that borders on a river. If you do not fence the side along the river, find the length and width of the plot that will maximize the area. What is the largest area that can be enclosed?

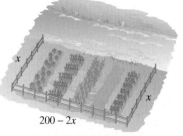

$200 - 2x$

67. You have 50 yards of fencing to enclose a rectangular region. Find the dimensions of the rectangle that maximize the enclosed area. What is the maximum area?

68. You have 80 yards of fencing to enclose a rectangular region. Find the dimensions of the rectangle that maximize the enclosed area. What is the maximum area?

69. A rectangular playground is to be fenced off and divided in two by another fence parallel to one side of the playground. Six hundred feet of fencing is used. Find the dimensions of the playground that maximize the total enclosed area. What is the maximum area?

70. A rectangular playground is to be fenced off and divided in two by another fence parallel to one side of the playground. Four hundred feet of fencing is used. Find the dimensions of the playground that maximize the total enclosed area. What is the maximum area?

71. A rain gutter is made from sheets of aluminum that are 20 inches wide by turning up the edges to form right angles. Determine the depth of the gutter that will maximize its cross-sectional area and allow the greatest amount of water to flow. What is the maximum cross-sectional area?

72. A rain gutter is made from sheets of aluminum that are 12 inches wide by turning up the edges to form right angles. Determine the depth of the gutter that will maximize its cross-sectional area and allow the greatest amount of water to flow. What is the maximum cross-sectional area?

If you have difficulty obtaining the functions to be maximized in Exercises 73–76, read Example 2 in Section 1.10 on pages 244–245.

73. On a certain route, an airline carries 8000 passengers per month, each paying $50. A market survey indicates that for each $1 increase in the ticket price, the airline will lose 100 passengers. Find the ticket price that will maximize the airline's monthly revenue for the route. What is the maximum monthly revenue?

74. A car rental agency can rent every one of its 200 cars at $30 per day. Far each $1 increase in rate, five fewer cars are rented. Find the rental amount that will maximize the agency's daily revenue. What is the maximum daily revenue?

75. The annual yield per walnut tree is fairly constant at 60 pounds per tree when the number of trees per acre is 20 or fewer. For each additional tree over 20, the annual yield per tree for all trees on the acre decreases by 2 pounds due to overcrowding. How many walnut trees should be planed per acre to maximize the annual yield for the acre? What is the maximum number of pounds of walnuts per acre?

76. The annual yield per cherry tree is fairly constant at 50 pounds per tree when the number of trees per acre is 30 or fewer. For each additional tree aver 30, the annual yield per tree for all trees on the acre decreases by 1 pound due to overcrowding. How many cherry trees should be planted per acre to maximize the annual yield for the acre? What is the maximum number of pounds of cherries per acre?

Writing in Mathematics

77. What is a quadratic function?

78. What is a parabola? Describe its shape.

79. Explain how to decide whether a parabola opens upward or downward.

80. Describe how to find a parabola's vertex if its equation is expressed in standard form. Give an example.

81. Describe how to find a parabola's vertex if its equation is in the form $f(x) = ax^2 + bx + c$. Use $f(x) = x^2 - 6x + 8$ as an example.

82. A parabola that opens upward has its vertex at $(1, 2)$. Describe as much as you can about the parabola based on this information. Include in your discussion the number of x-intercepts (if any) for the parabola.

Technology Exercises

83. Use a graphing utility to verify any five of your hand-drawn graphs in Exercises 17–38.

84. a. Use a graphing utility to graph $y = 2x^2 - 82x + 720$ in a standard viewing rectangle. What do you observe?

b. Find the coordinates of the vertex for the given quadratic function.

c. The answer to part (b) is $(20.5, -120.5)$. Because the leading coefficient, 2, of the given function is positive, the vertex is a minimum point on the graph. Use this fact to help find a viewing rectangle that will give a relatively complete picture of the parabola. With an axis of symmetry at $x = 20.5$, the setting for x should extend past this, so try Xmin = 0 and Xmax = 30. The setting for y should include (and probably go below) the y-coordinate of the graph's minimum y-value, so try Ymin = −130. Experiment with Ymax until your utility shows the parabola's major features.

d. In general, explain how knowing the coordinates of a parabola's vertex can help determine a reasonable viewing rectangle on a graphing utility for obtaining a complete picture of the parabola.

In Exercises 85–88, find the vertex for each parabola. Then determine a reasonable viewing rectangle on your graphing utility and use it to graph the quadratic function.

85. $y = -0.25x^2 + 40x$ **86.** $y = -4x^2 + 20x + 160$

87. $y = 5x^2 + 40x + 600$ **88.** $y = 0.01x^2 + 0.6x + 100$

89. The following data show fuel efficiency, in miles per gallon, for all U.S. automobiles in the indicated year.

x (Years after 1940)	y (Average Number of Miles per Gallon for U.S. Automobiles)
1940: 0	14.8
1950: 10	13.9
1960: 20	13.4
1970: 30	13.5
1980: 40	15.9
1990: 50	20.2
2000: 60	22.0

Source: U.S. Department of Transportation

a. Use a graphing utility to draw a scatter plot of the data. Explain why a quadratic function is appropriate for modeling these data.
b. Use the quadratic regression feature to find the quadratic function that best fits the data.
c. Use the model in part (b) to determine the worst year for automobile fuel efficiency. What was the average number of miles per gallon for that year?
d. Use a graphing utility to draw a scatter plot of the data and graph the quadratic function of best fit on the scatter plot.

Critical Thinking Exercises

90. Which one of the following is true?
  a. No quadratic functions have a range of $(-\infty, \infty)$.
  b. The vertex of the parabola described by $f(x) = 2(x - 5)^2 - 1$ is at $(5, 1)$.
  c. The graph of $f(x) = -2(x + 4)^2 - 8$ has one y-intercept and two x-intercepts.
  d. The maximum value of y for the quadratic function $f(x) = -x^2 + x + 1$ is 1.

In Exercises 91–92, find the axis of symmetry for each parabola whose equation is given. Use the axis of symmetry to find a second point on the parabola whose y-coordinate is the same as the given point.

91. $f(x) = 3(x + 2)^2 - 5$; $(-1, -2)$
92. $f(x) = (x - 3)^2 + 2$; $(6, 11)$

In Exercises 93–94, write the equation of each parabola in standard form.

93. Vertex: $(-3, -4)$; The graph passes through the point $(1, 4)$.
94. Vertex: $(-3, -1)$; The graph passes through the point $(-2, -3)$.
95. Find the point on the line whose equation is $2x + y - 2 = 0$ that is closest to the origin. *Hint:* Minimize the distance function by minimizing the expression under the square root.

96. A 300-room hotel can rent every one of its rooms at $80 per room. For each $1 increase in rent, three fewer rooms are rented. Each rented room costs the hotel $10 to service per day. How much should the hotel charge for each room to maximize its daily profit? What is the maximum daily profit?
97. A track and field area is to be constructed in the shape of a rectangle with semicircles at each end. The inside perimeter of the track is to be 440 yards. Find the dimensions of the rectangle that maximize the area of the rectangular portion of the field.

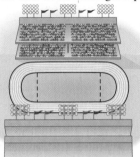

Group Exercise

98. Each group member should consult an almanac, newspaper, magazine, or the Internet to find data that initially increase and then decrease, or vice versa, and therefore can be modeled by a quadratic function. Group members should select the two sets of data that are most interesting and relevant. For each data set selected,
  a. Use the quadratic regression feature of a graphing utility to find the quadratic function that best fits the data.
  b. Use the equation of the quadratic function to make a prediction from the data. What circumstances might affect the accuracy of your prediction?
  c. Use the equation of the quadratic function to write and solve a problem involving maximizing or minimizing the function.

SECTION 2.3 *Polynomial Functions and Their Graphs*

Objectives

1. Identify polynomial functions.
2. Recognize characteristics of graphs of polynomial functions.
3. Determine end behavior.
4. Use factoring to find zeros of polynomial functions.
5. Identify zeros and their multiplicities.
6. Use the Intermediate Value Theorem.
7. Understand the relationship between degree and turning points.
8. Graph polynomial functions.

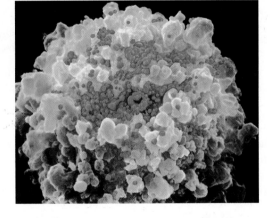

Magnified 6000 times, this color-scanned image shows a T-lymphocyte blood cell (green) infected with the HIV virus (red). Depletion of the number of T-cells causes destruction of the immune system.

In 1980, U.S. doctors diagnosed 41 cases of a rare form of cancer, Kaposi's sarcoma, that involved skin lesions, pneumonia, and severe immunological deficiencies. All cases involved gay men ranging in age from 26 to 51. By the end of 2002, approximately 890,000 Americans, straight and gay, male and female, old and young, were infected with the HIV virus.

Modeling AIDS-related data and making predictions about the epidemic's havoc is serious business. Figure 2.10 shows the number of AIDS cases diagnosed in the United States from 1983 through 2002.

AIDS Cases Diagnosed in the U.S., 1983–2002

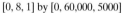

Figure 2.10

Source: Department of Health and Human Services

① Identify polynomial functions.

Changing circumstances and unforeseen events can result in models for AIDS-related data that are not particularly useful over long periods of time. For example, the function

$$f(x) = -49x^3 + 806x^2 + 3776x + 2503$$

models the number of AIDS cases diagnosed in the United States x years after 1983. The model was obtained using a portion of the data shown in Figure 2.10, namely cases diagnosed from 1983 through 1991, inclusive. Figure 2.11 shows the graph of f from 1983 through 1991. This function is an example of a *polynomial function of degree 3.*

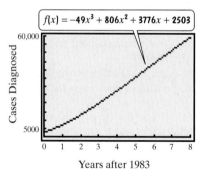

[0, 8, 1] by [0, 60,000, 5000]

Figure 2.11 The graph of a function modeling the number of AIDS cases from 1983 through 1991

Definition of a Polynomial Function

Let n be a nonnegative integer and let $a_n, a_{n-1}, \ldots, a_2, a_1, a_0$ be real numbers, with $a_n \neq 0$. The function defined by

$$f(x) = a_n x^n + a_{n-1} x^{n-1} + \cdots + a_2 x^2 + a_1 x + a_0$$

is called a **polynomial function of degree n**. The number a_n, the coefficient of the variable to the highest power, is called the **leading coefficient**.

A constant function $f(x) = c$, where $c \neq 0$, is a polynomial function of degree 0. A linear function $f(x) = mx + b$, where $m \neq 0$, is a polynomial function of degree 1. A quadratic function $f(x) = ax^2 + bx + c$, where $a \neq 0$, is a polynomial function of degree 2. In this section, we focus on polynomial functions of degree 3 or higher.

② Recognize characteristics of graphs of polynomial functions.

Smooth, Continuous Graphs

Polynomial functions of degree 2 or higher have graphs that are *smooth* and *continuous*. By **smooth**, we mean that the graphs contain only rounded curves with no sharp corners. By **continuous**, we mean that the graphs have no breaks and can be drawn without lifting your pencil from the rectangular coordinate system. These ideas are illustrated in Figure 2.12 on the next page.

Graphs of Polynomial Functions

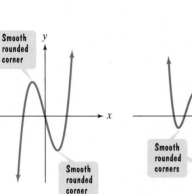

Not Graphs of Polynomial Functions

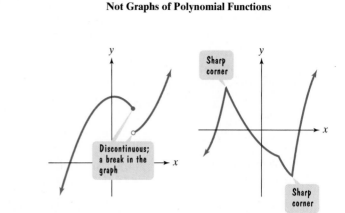

Figure 2.12 Recognizing graphs of polynomial functions

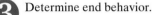

③ Determine end behavior.

End Behavior of Polynomial Functions

Figure 2.13 shows the graph of the function

$$f(x) = -49x^3 + 806x^2 + 3776x + 2503,$$

which models the number of U.S. AIDS cases from 1983 through 1991. Look what happens to the graph when we extend the year up through 2005. By year 21 (2004), the values of y are negative and the function no longer models AIDS cases. We've added an arrow to the graph at the far right to emphasize that it continues to decrease without bound. It is this far-right *end behavior* of the graph that makes it inappropriate for modeling AIDS cases into the future.

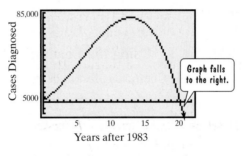

Years after 1983

$[0, 22, 1]$ by $[-10{,}000, 85{,}000, 5000]$

Figure 2.13 By extending the viewing rectangle, we see that y is eventually negative and the function no longer models the number of AIDS cases. Model breakdown occurs by 2004.

The behavior of a graph of a function to the far left or the far right is called its **end behavior**. Although the graph of a polynomial function may have intervals where it increases or decreases, the graph will eventually rise or fall without bound as it moves far to the left or far to the right.

How can you determine whether the graph of a polynomial function goes up or down at each end? The end behavior of a polynomial function

$$f(x) = a_n x^n + a_{n-1} x^{n-1} + \cdots + a_1 x + a_0$$

depends upon the leading term $a_n x^n$, because when $|x|$ is large, the other terms are relatively insignificant in size. In particular, the sign of the leading coefficient, a_n, and the degree, n, of the polynomial function reveal its end behavior. In terms of end behavior, only the term of highest degree counts, as summarized by the **Leading Coefficient Test**.

Study Tip

Odd-degree polynomial functions have graphs with opposite behavior at each end. Even-degree polynomial functions have graphs with the same behavior at each end.

The Leading Coefficient Test

As x increases or decreases without bound, the graph of the polynomial function

$$f(x) = a_nx^n + a_{n-1}x^{n-1} + a_{n-2}x^{n-2} + \cdots + a_1x + a_0 \quad (a_n \neq 0)$$

eventually rises or falls. In particular,

1. For n odd:

If the leading coefficient is positive, the graph falls to the left and rises to the right.

If the leading coefficient is negative, the graph rises to the left and falls to the right.

2. For n even:

If the leading coefficient is positive, the graph rises to the left and to the right.

If the leading coefficient is negative, the graph falls to the left and to the right.

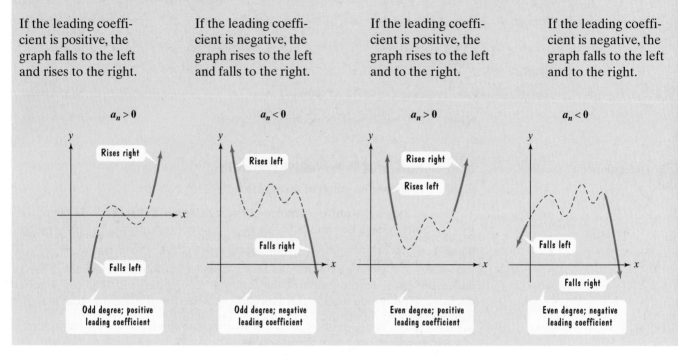

EXAMPLE 1   Using the Leading Coefficient Test

Use the Leading Coefficient Test to determine the end behavior of the graph of

$$f(x) = x^3 + 3x^2 - x - 3.$$

Solution   We begin by identifying the sign of the leading coefficient and the degree of the polynomial.

$$f(x) = x^3 + 3x^2 - x - 3$$

The leading coefficient, 1, is positive.

The degree of the polynomial, 3, is odd.

The degree of the function f is 3, which is odd. Odd-degree polynomial functions have graphs with opposite behavior at each end. The leading coefficient, 1, is positive. Thus, the graph falls to the left and rises to the right. The graph of f is shown in Figure 2.14.

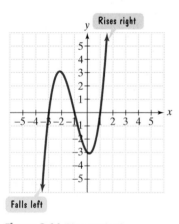

Figure 2.14 The graph of $f(x) = x^3 + 3x^2 - x - 3$

Check Point 1   Use the Leading Coefficient Test to determine the end behavior of the graph of $f(x) = x^4 - 4x^2$.

EXAMPLE 2   Using the Leading Coefficient Test

Use end behavior to explain why

$$f(x) = -49x^3 + 806x^2 + 3776x + 2503$$

is only an appropriate model for AIDS cases for a limited time period.

Solution We begin by identifying the sign of the leading coefficient and the degree of the polynomial.

$$f(x) = -49x^3 + 806x^2 + 3776x + 2503$$

| The leading coefficient, −49, is negative. | The degree of the polynomial, 3, is odd. |

The degree of f is 3, which is odd. Odd-degree polynomial functions have graphs with opposite behavior at each end. The leading coefficient, −49, is negative. Thus, the graph rises to the left and falls to the right. The fact that the graph falls to the right indicates that at some point the number of AIDS cases will be negative, an impossibility. If a function has a graph that decreases without bound over time, it will not be capable of modeling nonnegative phenomena over long time periods. Model breakdown will eventually occur.

Check Point **2** The polynomial function

$$f(x) = -0.27x^3 + 9.2x^2 - 102.9x + 400$$

models the ratio of students to computers in U.S. public schools x years after 1980. Use end behavior to determine whether this function could be an appropriate model for computers in the classroom well into the twenty-first century. Explain your answer.

If you use a graphing utility to graph a polynomial function, it is important to select a viewing rectangle that accurately reveals the graph's end behavior. If the viewing rectangle, or window, is too small, it may not accurately show the end behavior.

EXAMPLE 3   Using the Leading Coefficient Test

The graph of $f(x) = -x^4 + 8x^3 + 4x^2 + 2$ was obtained with a graphing utility using a $[-8, 8, 1]$ by $[-10, 10, 1]$ viewing rectangle. The graph is shown in Figure 2.15(a). Does the graph show the end behavior of the function?

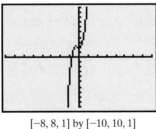   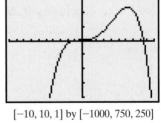

$[-8, 8, 1]$ by $[-10, 10, 1]$ $[-10, 10, 1]$ by $[-1000, 750, 250]$

Figure 2.15(a)   *Figure 2.15(b)*

Solution We begin by identifying the sign of the leading coefficient and the degree of the polynomial.

$$f(x) = -x^4 + 8x^3 + 4x^2 + 2$$

| The leading coefficient, −1, is negative. | The degree of the polynomial, 4, is even. |

The degree of f is 4, which is even. Even-degree polynomial functions have graphs with the same behavior at each end. The leading coefficient, −1, is negative. Thus, the graph should fall to the left and fall to the right. The graph in Figure 2.15(a) is falling to the left, but it is not falling to the right. Therefore, the graph is not complete enough to show end behavior. A more complete graph of the function is shown in a larger viewing rectangle in Figure 2.15(b).

Use factoring to find zeros of polynomial functions.

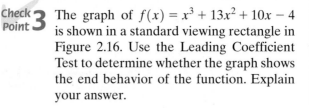

The graph of $f(x) = x^3 + 13x^2 + 10x - 4$ is shown in a standard viewing rectangle in Figure 2.16. Use the Leading Coefficient Test to determine whether the graph shows the end behavior of the function. Explain your answer.

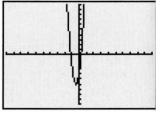

Figure 2.16

Zeros of Polynomial Functions

If f is a polynomial function, then the values of x for which $f(x)$ is equal to 0 are called the **zeros** of f. These values of x are the **roots**, or **solutions**, of the polynomial equation $f(x) = 0$. Each real root of the polynomial equation appears as an x-intercept of the graph of the polynomial function.

EXAMPLE 4   Finding Zeros of a Polynomial Function

Find all zeros of $f(x) = x^3 + 3x^2 - x - 3$.

Solution   By definition, the zeros are the values of x for which $f(x)$ is equal to 0. Thus, we set $f(x)$ equal to 0:

$$f(x) = x^3 + 3x^2 - x - 3 = 0.$$

We solve the polynomial equation $x^3 + 3x^2 - x - 3 = 0$ for x as follows:

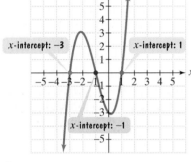

Figure 2.17

$x^3 + 3x^2 - x - 3 = 0$	This is the equation needed to find the function's zeros.
$x^2(x + 3) - 1(x + 3) = 0$	Factor x^2 from the first two terms and -1 from the last two terms.
$(x + 3)(x^2 - 1) = 0$	A common factor of $x + 3$ is factored from the expression.
$x + 3 = 0$ or $x^2 - 1 = 0$	Set each factor equal to 0.
$x = -3$ $x^2 = 1$	Solve for x.
$x = \pm 1$	Remember that if $x^2 = d$, then $x^2 = \pm\sqrt{d}$.

The zeros of f are -3, -1, and 1. The graph of f in Figure 2.17 shows that each zero is an x-intercept. The graph passes through the points $(-3, 0)$, $(-1, 0)$, and $(1, 0)$.

Technology

A graphing utility can be used to verify that -3, -1, and 1 are the three real zeros of $f(x) = x^3 + 3x^2 - x - 3$.

Numeric Check
Display a table for the function.

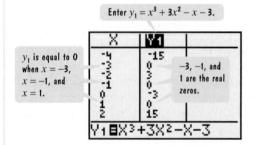

y_1 is equal to 0 when $x = -3$, $x = -1$, and $x = 1$.

Graphic Check
Display a graph for the function. The x-intercepts indicate that -3, -1, and 1 are the real zeros.

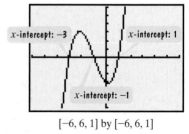

$[-6, 6, 1]$ by $[-6, 6, 1]$

The utility's $\boxed{\text{ZERO}}$ feature on the graph of f also verifies that -3, -1, and 1 are the function's real zeros.

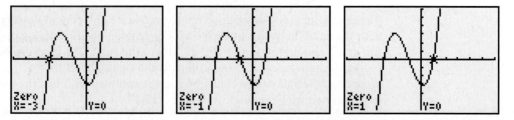

Check Point **4**   Find all zeros of $f(x) = x^3 + 2x^2 - 4x - 8$.

EXAMPLE 5   Finding Zeros of a Polynomial Function

Find all zeros of $f(x) = -x^4 + 4x^3 - 4x^2$.

Solution   We find the zeros of f by setting $f(x)$ equal to 0 and solving the resulting equation.

$$-x^4 + 4x^3 - 4x^2 = 0 \qquad \text{We now have a polynomial equation.}$$

$$x^4 - 4x^3 + 4x^2 = 0 \qquad \text{Multiply both sides by } -1. \text{ This step is optional.}$$

$$x^2(x^2 - 4x + 4) = 0 \qquad \text{Factor out } x^2.$$

$$x^2(x - 2)^2 = 0 \qquad \text{Factor completely.}$$

$$x^2 = 0 \quad \text{or} \quad (x - 2)^2 = 0 \qquad \text{Set each factor equal to 0.}$$

$$x = 0 \qquad\qquad\qquad x = 2 \qquad \text{Solve for x.}$$

The zeros of $f(x) = -x^4 + 4x^3 - 4x^2$ are 0 and 2. The graph of f, shown in Figure 2.18, has x-intercepts at 0 and 2. The graph passes through the points $(0, 0)$ and $(2, 0)$.

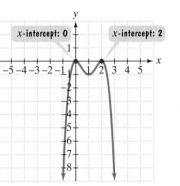

x-intercept: 0 x-intercept: 2

Figure 2.18 The zeros of $f(x) = -x^4 + 4x^3 - 4x^2$, namely 0 and 2, are the x-intercepts for the graph of f.

Check Point **5**   Find all zeros of $f(x) = x^4 - 4x^2$.

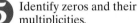

5 Identify zeros and their multiplicities.

Multiplicities of Zeros

We can use the results of factoring to express a polynomial as a product of factors. For instance, in Example 5, we can use our factoring to express the function's equation as follows:

$$f(x) = -x^4 + 4x^3 - 4x^2 = -(x^4 - 4x^3 + 4x^2) = -x^2(x - 2)^2.$$

The factor x occurs twice: $x^2 = x \cdot x$.

The factor $(x - 2)$ occurs twice: $(x - 2)^2 = (x - 2)(x - 2)$.

Notice that each factor occurs twice. In factoring the equation for the polynomial function f, if the same factor $x - r$ occurs k times, but not $k + 1$ times, we call r a **zero with multiplicity k**. For the polynomial function

$$f(x) = -x^2(x - 2)^2,$$

0 and 2 are both zeros with multiplicity 2.

Multiplicity provides another connection between zeros and graphs. The multiplicity of a zero tells us whether the graph of a polynomial function touches the x-axis at the zero and turns around, or if the graph crosses the x-axis at the zero. For example, look again at the graph of $f(x) = -x^4 + 4x^3 - 4x^2$ in Figure 2.18. Each zero, 0 and 2, is a zero with multiplicity 2. The graph of f touches, but does not cross, the x-axis at each of these zeros of even multiplicity. By contrast, a graph crosses the x-axis at zeros of odd multiplicity.

> ### Multiplicity and x-Intercepts
>
> If r is a zero of **even multiplicity**, then the graph **touches** the x-axis **and turns around** at r. If r is a zero of **odd multiplicity**, then the graph **crosses** the x-axis at r. Regardless of whether the multiplicity of a zero is even or odd, graphs tend to flatten out at zeros with multiplicity greater than one.

If a polynomial function's equation is expressed as a product of linear factors, we can quickly identify zeros and their multiplicities.

EXAMPLE 6   Finding Zeros and Their Multiplicities

Find the zeros of $f(x) = (x + 1)(2x - 3)^2$ and give the multiplicity of each zero. State whether the graph crosses the x-axis or touches the x-axis and turns around at each zero.

Solution   We find the zeros of f by setting $f(x)$ equal to 0:

$$(x + 1)(2x - 3)^2 = 0.$$

Set each factor equal to 0.

$x + 1 = 0$
$x = -1$

$2x - 3 = 0$
$x = \frac{3}{2}$

$$(x + 1)^1(2x - 3)^2 = 0$$

This exponent is 1.
Thus, the multiplicity of −1 is 1.

This exponent is 2.
Thus, the multiplicity of $\frac{3}{2}$ is 2.

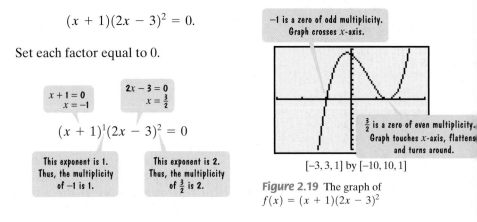

−1 is a zero of odd multiplicity.
Graph crosses x-axis.

$\frac{3}{2}$ is a zero of even multiplicity. Graph touches x-axis, flattens and turns around.

$[-3, 3, 1]$ by $[-10, 10, 1]$

Figure 2.19 The graph of $f(x) = (x + 1)(2x - 3)^2$

The zeros of $f(x) = (x + 1)(2x - 3)^2$ are −1, with multiplicity 1, and $\frac{3}{2}$, with multiplicity 2. Because the multiplicity of −1 is odd, the graph crosses the x-axis at this zero. Because the multiplicity of $\frac{3}{2}$ is even, the graph touches the x-axis and turns around at this zero. These relationships are illustrated by the graph of f in Figure 2.19.

Check Point 6   Find the zeros of $f(x) = -4\left(x + \frac{1}{2}\right)^2(x - 5)^3$ and give the multiplicity of each zero. State whether the graph crosses the x-axis or touches the x-axis and turns around at each zero.

⑥ Use the Intermediate Value Theorem.

The Intermediate Value Theorem

The *Intermediate Value Theorem* tells us of the existence of real zeros. The idea behind the theorem is illustrated in Figure 2.20. The figure shows that if $(a, f(a))$ lies below the x-axis and $(b, f(b))$ lies above the x-axis, the smooth, continuous graph of a polynomial function f must cross the x-axis at some value c between a and b. This value is a real zero for the function.

These observations are summarized in the **Intermediate Value Theorem**.

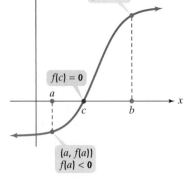

$(b, f(b))$
$f(b) > 0$

$f(c) = 0$
a
c
b

$(a, f(a))$
$f(a) < 0$

Figure 2.20 The graph must cross the x-axis at some value between a and b.

The Intermediate Value Theorem for Polynomials

Let f be a polynomial function with real coefficients. If $f(a)$ and $f(b)$ have opposite signs, then there is at least one value of c between a and b for which $f(c) = 0$. Equivalently, the equation $f(x) = 0$ has at least one real root between a and b.

EXAMPLE 7   Using the Intermediate Value Theorem

Show that the polynomial function $f(x) = x^3 - 2x - 5$ has a real zero between 2 and 3.

Solution   Let us evaluate f at 2 and at 3. If $f(2)$ and $f(3)$ have opposite signs, then there is at least one real zero between 2 and 3. Using $f(x) = x^3 - 2x - 5$, we obtain

$$f(2) = 2^3 - 2 \cdot 2 - 5 = 8 - 4 - 5 = -1$$

$f(2)$ is negative.

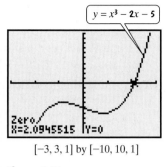

$$y = x^3 - 2x - 5$$

Zero.
X=2.0945515   Y=0

[-3, 3, 1] by [-10, 10, 1]

Figure 2.21

⑦ Understand the relationship between degree and turning points.

⑧ Graph polynomial functions.

Study Tip

Remember that, without calculus, it is often impossible to give the exact location of turning points. However, you can obtain additional points satisfying the function to estimate how high the graph rises or how low it falls. To obtain these points, use values of x between (and to the left and right of) the x-intercepts.

and

$$f(3) = 3^3 - 2 \cdot 3 - 5 = 27 - 6 - 5 = 16.$$

$f(3)$ is positive.

Because $f(2) = -1$ and $f(3) = 16$, the sign change shows that the polynomial function has a real zero between 2 and 3. This zero is actually irrational and is approximated using a graphing utility's $\boxed{\text{ZERO}}$ feature as 2.0945515 in Figure 2.21.

Check Point 7 Show that the polynomial function $f(x) = 3x^3 - 10x + 9$ has a real zero between -3 and -2.

Turning Points of Polynomial Functions

The graph of $f(x) = x^5 - 6x^3 + 8x + 1$ is shown in Figure 2.22. The graph has four smooth **turning points**. At each turning point, the graph changes direction from increasing to decreasing or vice versa. The given equation has 5 as its greatest exponent and is therefore a polynomial function of degree 5. Notice that the graph has four turning points. In general, **if f is a polynomial function of degree n, then the graph of f has at most $n - 1$ turning points**.

Figure 2.22 illustrates that the y-coordinate of each turning point is either a relative maximum or a relative minimum of f. Without the aid of a graphing utility or a knowledge of calculus, it is difficult and often impossible to locate turning points of polynomial functions with degrees greater

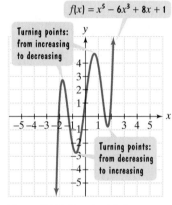

$f(x) = x^5 - 6x^3 + 8x + 1$

Turning points: from increasing to decreasing

Turning points: from decreasing to increasing

Figure 2.22 Graph with four turning points

than 2. If necessary, test values can be taken between the x-intercepts to get a general idea of how high the graph rises or how low the graph falls. For the purpose of graphing in this section, a general estimate is sometimes appropriate and necessary.

A Strategy for Graphing Polynomial Functions

Here's a general strategy for graphing a polynomial function. A graphing utility is a valuable complement, but not a necessary component, to this strategy. If you are using a graphing utility, some of the steps listed in the following box will help you to select a viewing rectangle that shows the important parts of the graph.

Graphing a Polynomial Function

$$f(x) = a_n x^n + a_{n-1} x^{n-1} + a_{n-2} x^{n-2} + \cdots + a_1 x + a_0, \, a_n \neq 0$$

1. Use the Leading Coefficient Test to determine the graph's end behavior.
2. Find x-intercepts by setting $f(x) = 0$ and solving the resulting polynomial equation. If there is an x-intercept at r as a result of $(x - r)^k$ in the complete factorization of $f(x)$, then
   a. If k is even, the graph touches the x-axis at r and turns around.
   b. If k is odd, the graph crosses the x-axis at r.
   c. If $k > 1$, the graph flattens out at $(r, 0)$.
3. Find the y-intercept by computing $f(0)$.
4. Use symmetry, if applicable, to help draw the graph:
   a. y-axis symmetry: $f(-x) = f(x)$
   b. Origin symmetry: $f(-x) = -f(x)$.
5. Use the fact that the maximum number of turning points of the graph is $n - 1$ to check whether it is drawn correctly.

EXAMPLE 8   Graphing a Polynomial Function

Graph: $f(x) = x^4 - 2x^2 + 1$.

Solution

Step 1   Determine end behavior. Identify the sign of a_n, the leading coefficient, and the degree, n, of the polynomial function.

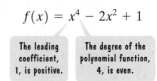

$$f(x) = x^4 - 2x^2 + 1$$

| The leading coefficient, 1, is positive. | The degree of the polynomial function, 4, is even. |

Because the degree, 4, is even, the graph has the same behavior at each end. The leading coefficient, 1, is positive. Thus, the graph rises to the left and rises to the right.

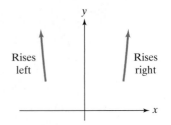

Step 2   Find x-intercepts (zeros of the function) by setting $f(x) = 0$.

$$x^4 - 2x^2 + 1 = 0$$

$(x^2 - 1)(x^2 - 1) = 0$ *Factor.*

$(x + 1)(x - 1)(x + 1)(x - 1) = 0$ *Factor completely.*

$(x + 1)^2(x - 1)^2 = 0$ *Express the factorization in a more compact form.*

$(x + 1)^2 = 0$ or $(x - 1)^2 = 0$ *Set each factorization equal to 0.*

$x = -1$ $x = 1$ *Solve for x.*

We see that -1 and 1 are both repeated zeros with multiplicity 2. Because of the even multiplicity, the graph touches the x-axis at -1 and 1 and turns around. Furthermore, the graph tends to flatten out at these zeros with multiplicity greater than one.

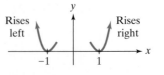

Step 3   Find the y-intercept by computing $f(0)$. We use $f(x) = x^4 - 2x^2 + 1$ and compute $f(0)$.

$$f(0) = 0^4 - 2 \cdot 0^2 + 1 = 1$$

There is a y-intercept at 1, so the graph passes through $(0, 1)$.

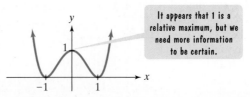

It appears that 1 is a relative maximum, but we need more information to be certain.

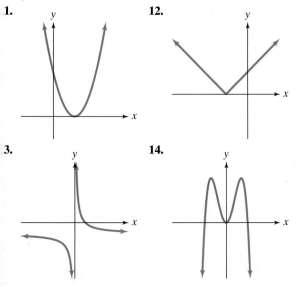

Figure 2.23 The graph of $f(x) = x^4 - 2x^2 + 1$

Step 4   Use possible symmetry to help draw the graph. Our partial graph suggests y-axis symmetry. Let's verify this by finding $f(-x)$.

$$f(x) = x^4 - 2x^2 + 1$$

Replace x with $-x$.

$$f(-x) = (-x)^4 - 2(-x)^2 + 1 = x^4 - 2x^2 + 1$$

Because $f(-x) = f(x)$, the graph of f is symmetric with respect to the y-axis. Figure 2.23 shows the graph of $f(x) = x^4 - 2x^2 + 1$.

Step 5   Use the fact that the maximum number of turning points of the graph is $n - 1$ to check whether it is drawn correctly. Because $n = 4$, the maximum number of turning points is $4 - 1$, or 3. Because the graph in Figure 2.23 has three turning points, we have not violated the maximum number possible. Can you see how this verifies that 1 is indeed a relative maximum and $(0, 1)$ is a turning point? If the graph rose above 1 on either side of $x = 0$, it would have to rise above 1 on the other side as well because of symmetry. This would require additional turning points to smoothly curve back to the x-intercepts. The graph already has three turning points, which is the maximum number for a fourth-degree polynomial function.

Check Point **8**   Use the five-step strategy to graph $f(x) = x^3 - 3x^2$.

EXERCISE SET 2.3

Practice Exercises

In Exercises 1–10, determine which functions are polynomial functions. For those that are, identify the degree.

1. $f(x) = 5x^2 + 6x^3$ **2.** $f(x) = 7x^2 + 9x^4$

3. $g(x) = 7x^5 - \pi x^3 + \frac{1}{5}x$ **4.** $g(x) = 6x^7 + \pi x^5 + \frac{2}{3}x$

5. $h(x) = 7x^3 + 2x^2 + \frac{1}{x}$ **6.** $h(x) = 8x^3 - x^2 + \frac{2}{x}$

7. $f(x) = x^{\frac{1}{2}} - 3x^2 + 5$ **8.** $f(x) = x^{\frac{1}{3}} - 4x^2 + 7$

9. $f(x) = \dfrac{x^2 + 7}{x^3}$ **10.** $f(x) = \dfrac{x^2 + 7}{3}$

In Exercises 11–14, identify which graphs are not those of polynomial functions.

11.

12.

13.

14.

In Exercises 15–18, use the Leading Coefficient Test to determine the end behavior of the graph of the given polynomial function. Then use this end behavior to match the polynomial function with its graph. [The graphs are labeled (a) through (d).]

15. $f(x) = -x^4 + x^2$ **16.** $f(x) = x^3 - 4x^2$

17. $f(x) = (x - 3)^2$

18. $f(x) = -x^3 - x^2 + 5x - 3$

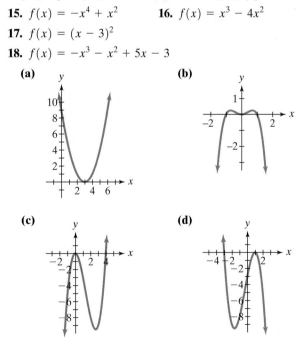

In Exercises 19–24, use the Leading Coefficient Test to determine the end behavior of the graph of the polynomial function.

19. $f(x) = 5x^3 + 7x^2 - x + 9$

20. $f(x) = 11x^3 - 6x^2 + x + 3$

21. $f(x) = 5x^4 + 7x^2 - x + 9$

22. $f(x) = 11x^4 - 6x^2 + x + 3$

23. $f(x) = -5x^4 + 7x^2 - x + 9$

24. $f(x) = -11x^4 - 6x^2 + x + 3$

In Exercises 25–32, find the zeros for each polynomial function and give the multiplicity for each zero. State whether the graph crosses the x-axis, or touches the x-axis and turns around, at each zero.

25. $f(x) = 2(x - 5)(x + 4)^2$

26. $f(x) = 3(x + 5)(x + 2)^2$

27. $f(x) = 4(x - 3)(x + 6)^3$

28. $f(x) = -3\left(x + \frac{1}{2}\right)(x - 4)^3$

29. $f(x) = x^3 - 2x^2 + x$

30. $f(x) = x^3 + 4x^2 + 4x$

31. $f(x) = x^3 + 7x^2 - 4x - 28$

32. $f(x) = x^3 + 5x^2 - 9x - 45$

In Exercises 33–40, use the Intermediate Value Theorem to show that each polynomial has a real zero between the given integers.

33. $f(x) = x^3 - x - 1$; between 1 and 2

34. $f(x) = x^3 - 4x^2 + 2$; between 0 and 1

35. $f(x) = 2x^4 - 4x^2 + 1$; between −1 and 0

36. $f(x) = x^4 + 6x^3 - 18x^2$; between 2 and 3

37. $f(x) = x^3 + x^2 - 2x + 1$; between −3 and −2

38. $f(x) = x^5 - x^3 - 1$; between 1 and 2

39. $f(x) = 3x^3 - 10x + 9$; between −3 and −2

40. $f(x) = 3x^3 - 8x^2 + x + 2$; between 2 and 3

In Exercises 41–64,

    a. *Use the Leading Coefficient Test to determine the graph's end behavior.*

    b. *Find the x-intercepts. State whether the graph crosses the x-axis, or touches the x-axis and turns around, at each intercept.*

    c. *Find the y-intercept.*

    d. *Determine whether the graph has y-axis symmetry, origin symmetry, or neither.*

    e. *If necessary, find a few additional points and graph the function. Use the maximum number of turning points to check whether it is drawn correctly.*

41. $f(x) = x^3 + 2x^2 - x - 2$ **42.** $f(x) = x^3 + x^2 - 4x - 4$

43. $f(x) = x^4 - 9x^2$ **44.** $f(x) = x^4 - x^2$

45. $f(x) = -x^4 + 16x^2$ **46.** $f(x) = -x^4 + 4x^2$

47. $f(x) = x^4 - 2x^3 + x^2$ **48.** $f(x) = x^4 - 6x^3 + 9x^2$

49. $f(x) = -2x^4 + 4x^3$ **50.** $f(x) = -2x^4 + 2x^3$

51. $f(x) = 6x^3 - 9x - x^5$ **52.** $f(x) = 6x - x^3 - x^5$

53. $f(x) = 3x^2 - x^3$ **54.** $f(x) = \frac{1}{2} - \frac{1}{2}x^4$

55. $f(x) = -3(x - 1)^2(x^2 - 4)$

56. $f(x) = -2(x - 4)^2(x^2 - 25)$

57. $f(x) = x^2(x - 1)^3(x + 2)$

58. $f(x) = x^3(x + 2)^2(x + 1)$

59. $f(x) = -x^2(x - 1)(x + 3)$

60. $f(x) = -x^2(x + 2)(x - 2)$

61. $f(x) = -2x^3(x - 1)^2(x + 5)$

62. $f(x) = -3x^3(x - 1)^2(x + 3)$

63. $f(x) = (x - 2)^2(x + 4)(x - 1)$

64. $f(x) = (x + 3)(x + 1)^3(x + 4)$

Practice Plus

In Exercises 65–72, complete graphs of polynomial functions whose zeros are integers are shown.

    a. *Find the zeros and state whether the multiplicity of each zero is even or odd.*

    b. *Write an equation, expressed as the product of factors, of a polynomial function that might have each graph. Use a leading coefficient of 1 or −1, and make the degree of f as small as possible.*

    c. *Use both the equation in part (b) and the graph to find the y-intercept.*

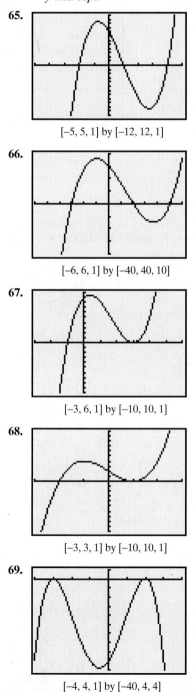

65.

[−5, 5, 1] by [−12, 12, 1]

66.

[−6, 6, 1] by [−40, 40, 10]

67.

[−3, 6, 1] by [−10, 10, 1]

68.

[−3, 3, 1] by [−10, 10, 1]

69.

[−4, 4, 1] by [−40, 4, 4]

70.

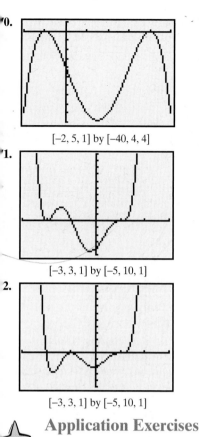

[−2, 5, 1] by [−40, 4, 4]

71.

[−3, 3, 1] by [−5, 10, 1]

72.

[−3, 3, 1] by [−5, 10, 1]

Application Exercises

The bar graph shows the cumulative number of deaths from AIDS in the United States from 1990 through 2002.

Cumulative Number of Deaths from AIDS in the U.S.

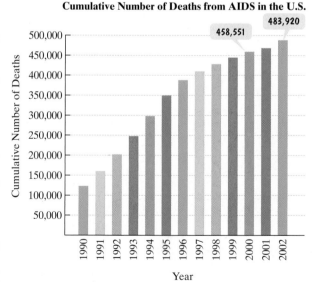

Year

Source: Centers for Disease Control

The data in the bar graph can be modeled by the following second- and third-degree polynomial functions:

Cumulative number of AIDS deaths x years after 1990

$$f(x) = -2212x^2 + 57{,}575x + 107{,}896$$
$$g(x) = -84x^3 - 702x^2 + 50{,}609x + 113{,}435.$$

Use these functions to solve Exercises 73–76.

73. Use both functions to find the cumulative number of AIDS deaths in 2000. Which function provides a better description for the actual number shown in the bar graph?

74. Use both functions to find the cumulative number of AIDS deaths in 2002. Which function provides a better description for the actual number shown in the bar graph?

75. Use the Leading Coefficient Test to determine the end behavior to the right for the graph of f. Will this function be useful in modeling the cumulative number of AIDS deaths over an extended period of time? Explain your answer.

76. Use the Leading Coefficient Test to determine the end behavior to the right for the graph of g. Will this function be useful in modeling the cumulative number of AIDS deaths over an extended period of time? Explain your answer.

77. Although it has been more than 50 years since the Supreme Court ruled against school segregation, data from the Civil Rights Project at Harvard University indicate that integration and academic equality remain elusive. The graph shows the percentage of the average African-American student's classmates who were white for the period from 1970 through 2002.

Percentage of the Average African-American Student's Classmates Who Were White

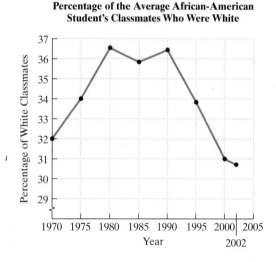

Year
2002

Source: Civil Rights Project, Harvard University

a. For which years was the percentage of white classmates increasing?

b. For which years was the percentage of white classmates decreasing?

c. How many turning points (from increasing to decreasing or from decreasing to increasing) does the graph have for the period shown?

d. Suppose that a polynomial function is used to model the data shown in the graph using

(number of years after 1970, percentage of the average African-American student's classmates who were white).

Use the number of turning points to determine the degree of the polynomial function of best fit.

e. For the model in part (d), should the leading coefficient of the polynomial function be positive or negative? Explain your answer.

78. The graphs show the percentage of husbands and wives with one or more children who said their marriage was going well "all the time" at various stages in their relationships.

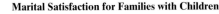

Marital Satisfaction for Families with Children

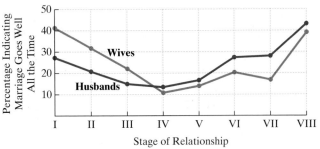

Stage   I: Beginning families
Stage II: Child-bearing families
Stage III: Families with preschool children
Stage IV: Families with school-age children

Stage   V: Families with teenagers
Stage VI: Families with adult children leaving home
Stage VII: Families in the middle years
Stage VIII: Aging families

Source: Rollins, B., & Feldman, H. (1970), Marital satisfaction over the family life cycle. *Journal of Marriage and the Family, 32,* 20–28.

a. Between which stages was marital satisfaction for wives decreasing?

b. Between which stages was marital satisfaction for wives increasing?

c. How many turning points (from decreasing to increasing or from increasing to decreasing) are shown in the graph for wives?

d. Suppose that a polynomial function is used to model the data shown in the graph for wives using

(stage in the relationship, percentage indicating that the marriage was going well all the time).

Use the number of turning points to determine the degree of the polynomial function of best fit.

e. For the model in part (d), should the leading coefficient of the polynomial function be positive or negative? Explain your answer.

Writing in Mathematics

79. What is a polynomial function?

80. What do we mean when we describe the graph of a polynomial function as smooth and continuous?

81. What is meant by the end behavior of a polynomial function?

82. Explain how to use the Leading Coefficient Test to determine the end behavior of a polynomial function.

83. Why is a third-degree polynomial function with a negative leading coefficient not appropriate for modeling nonnegative real-world phenomena over a long period of time?

84. What are the zeros of a polynomial function and how are they found?

85. Explain the relationship between the multiplicity of a zero and whether or not the graph crosses or touches the x-axis at that zero.

86. If f is a polynomial function, and $f(a)$ and $f(b)$ have opposite signs, what must occur between a and b? If $f(a)$ and $f(b)$ have the same sign, does it necessarily mean that this will not occur? Explain your answer.

87. Explain the relationship between the degree of a polynomial function and the number of turning points on its graph.

88. Can the graph of a polynomial function have no x-intercepts? Explain.

89. Can the graph of a polynomial function have no y-intercept? Explain.

90. Describe a strategy for graphing a polynomial function. In your description, mention intercepts, the polynomial's degree, and turning points.

91. The graphs shown in Exercise 78 indicate that marital satisfaction tends to be greatest at the beginning and at the end of the stages in the relationship, with a decline occurring in the middle. What explanations can you offer for this trend?

Technology Exercises

92. Use a graphing utility to verify any five of the graphs that you drew by hand in Exercises 41–64.

Write a polynomial function that imitates the end behavior of each graph in Exercises 93–96. The dashed portions of the graphs indicate that you should focus only on imitating the left and right behavior of the graph and can be flexible about what occurs between the left and right ends. Then use your graphing utility to graph the polynomial function and verify that you imitated the end behavior shown in the given graph.

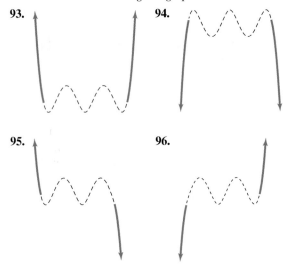

93.

94.

95.

96.

In Exercises 97–100, use a graphing utility with a viewing rectangle large enough to show end behavior to graph each polynomial function.

97. $f(x) = x^3 + 13x^2 + 10x - 4$

98. $f(x) = -2x^3 + 6x^2 + 3x - 1$

99. $f(x) = -x^4 + 8x^3 + 4x^2 + 2$

100. $f(x) = -x^5 + 5x^4 - 6x^3 + 2x + 20$

In Exercises 101–102, use a graphing utility to graph f and g in the same viewing rectangle. Then use the ZOOM OUT *feature to show that f and g have identical end behavior.*

101. $f(x) = x^3 - 6x + 1, \quad g(x) = x^3$

102. $f(x) = -x^4 + 2x^3 - 6x, \quad g(x) = -x^4$

Critical Thinking Exercises

03. Which one of the following is true?

   a. If $f(x) = -x^3 + 4x$, then the graph of f falls to the left and falls to the right.

   b. A mathematical model that is a polynomial of degree n whose leading term is $a_n x^n$, n odd and $a_n < 0$, is ideally suited to describe nonnegative phenomena over unlimited periods of time.

   c. There is more than one third-degree polynomial function with the same three x-intercepts.

   d. The graph of a function with origin symmetry can rise to the left and to the right.

Use the descriptions in Exercises 104–105 to write an equation of a polynomial function with the given characteristics. Use a graphing utility to graph your function to see if you are correct. If not, modify the function's equation and repeat this process.

104. Crosses the x-axis at -4, 0, and 3; lies above the x-axis between -4 and 0; lies below the x-axis between 0 and 3

105. Touches the x-axis at 0 and crosses the x-axis at 2; lies below the x-axis between 0 and 2

SECTION 2.4 Dividing Polynomials; Remainder and Factor Theorems

Objectives

1. Use long division to divide polynomials.
2. Use synthetic division to divide polynomials.
3. Evaluate a polynomial using the Remainder Theorem.
4. Use the Factor Theorem to solve a polynomial equation.

A moth has moved into your closet. She appeared in your bedroom at night, but somehow her relatively stout body escaped your clutches. Within a few weeks, swarms of moths in your tattered wardrobe suggest that Mama Moth was in the family way. There must be at least 200 critters nesting in every crevice of your clothing.

Two hundred plus moth-tykes from one female moth—is this possible? Indeed it is. The number of eggs, $f(x)$, in a female moth is a function of her abdominal width, x, in millimeters, modeled by

$$f(x) = 14x^3 - 17x^2 - 16x + 34, \quad 1.5 \le x \le 3.5.$$

Because there are 200 moths feasting on your favorite sweaters, Mama's abdominal width can be estimated by finding the solutions of the polynomial equation

$$14x^3 - 17x^2 - 16x + 34 = 200.$$

How can we solve such an equation? You might begin by subtracting 200 from both sides to obtain zero on one side. But then what? The factoring that we used in the previous section will not work in this situation.

In Section 2.5, we will present techniques for solving certain kinds of polynomial equations. These techniques will further enhance your ability to manipulate algebraically the polynomial functions that model your world. Because these techniques are based on understanding polynomial division, in this section we look at two methods for dividing polynomials. (We'll return to Mama Moth's abdominal width in the exercise set.)

① Use long division to divide polynomials.

Long Division of Polynomials and the Division Algorithm

We begin by looking at division by a polynomial containing more than one term, such a

$$x + 3\overline{)x^2 + 10x + 21}.$$

Divisor has two terms and is a binomial.

The polynomial dividend has three terms and is a trinomial.

When a divisor has more than one term, the four steps used to divide whole numbers–**divide, multiply, subtract, bring down the next term**—form the repetitive procedure for polynomial long division.

EXAMPLE 1 Long Division of Polynomials

Divide $x^2 + 10x + 21$ by $x + 3$.

Solution   The following steps illustrate how polynomial division is very similar to numerical division.

$$x + 3\overline{)x^2 + 10x + 21}$$

Arrange the terms of the dividend $(x^2 + 10x + 21)$ and the divisor $(x + 3)$ in descending powers of x.

$$\begin{array}{r} x \\ x + 3\overline{)x^2 + 10x + 21} \end{array}$$

Divide x^2 (the first term in the dividend) by x (the first term in the divisor): $\dfrac{x^2}{x} = x$. Align like terms.

$x(x + 3) = x^2 + 3x$

$$\begin{array}{r} x \\ x + 3\overline{)x^2 + 10x + 21} \\ x^2 + 3x \end{array}$$

Multiply each term in the divisor $(x + 3)$ by x, aligning terms of the product under like terms in the dividend.

$$\begin{array}{r} x \\ x + 3\overline{)x^2 + 10x + 21} \\ \ominus x^2 \ominus 3x \\ \hline 7x \end{array}$$

Change signs of the polynomial being subtracted.

Subtract $x^2 + 3x$ from $x^2 + 10x$ by changing the sign of each term in the lower expression and adding.

$$\begin{array}{r} x \\ x + 3\overline{)x^2 + 10x + 21} \\ x^2 + 3x \downarrow \\ \hline 7x + 21 \end{array}$$

Bring down 21 from the original dividend and add algebraically to form a new dividend.

$$\begin{array}{r} x + 7 \\ x + 3\overline{)x^2 + 10x + 21} \\ x^2 + 3x \\ \hline 7x + 21 \end{array}$$

Find the second term of the quotient. Divide the first term of $7x + 21$ by x, the first term of the divisor: $\dfrac{7x}{x} = 7$.

$7(x + 3) = 7x + 21$

$$\begin{array}{r} x + 7 \\ x + 3\overline{)x^2 + 10x + 21} \\ x^2 + 3x \\ \hline 7x + 21 \\ \ominus 7x \ominus 21 \\ \hline 0 \end{array}$$

Remainder

Multiply the divisor $(x + 3)$ by 7, aligning under like terms in the new dividend. Then subtract to obtain the remainder of 0.

The quotient is $x + 7$. Because the remainder is 0, we can conclude that $x + 3$ is a factor of $x^2 + 10x + 21$ and

$$\frac{x^2 + 10x + 21}{x + 3} = x + 7.$$

Check Point 1   Divide $x^2 + 14x + 45$ by $x + 9$.

Before considering additional examples, let's summarize the general procedure for dividing one polynomial by another.

Long Division of Polynomials

1. Arrange the terms of both the dividend and the divisor in descending powers of the variable.
2. **Divide** the first term in the dividend by the first term in the divisor. The result is the first term of the quotient.
3. **Multiply** every term in the divisor by the first term in the quotient. Write the resulting product beneath the dividend with like terms lined up.
4. **Subtract** the product from the dividend.
5. **Bring down** the next term in the original dividend and write it next to the remainder to form a new dividend.
6. Use this new expression as the dividend and repeat this process until the remainder can no longer be divided. This will occur when the degree of the remainder (the highest exponent on a variable in the remainder) is less than the degree of the divisor.

In our next long division, we will obtain a nonzero remainder.

EXAMPLE 2   Long Division of Polynomials

Divide $4 - 5x - x^2 + 6x^3$ by $3x - 2$.

Solution   We begin by writing the dividend in descending powers of x.

$$4 - 5x - x^2 + 6x^3 = 6x^3 - x^2 - 5x + 4$$

$2x^2(3x - 2) = 6x^3 - 4x^2$

$$\begin{array}{r} 2x^2 \\ 3x - 2 \overline{)6x^3 - x^2 - 5x + 4} \\ \underline{6x^3 \oplus 4x^2} \\ 3x^2 - 5x \end{array}$$

Change signs of the polynomial being subtracted.

Divide: $\dfrac{6x^3}{3x} = 2x^2.$

Multiply: $2x^2(3x - 2) = 6x^3 - 4x^2.$
Subtract $6x^3 - 4x^2$ from $6x^3 - x^2$ and bring down $-5x$.

Now we divide $3x^2$ by $3x$ to obtain x, multiply x and the divisor, and subtract.

$x(3x - 2) = 3x^2 - 2x$

$$\begin{array}{r} 2x^2 + x \\ 3x - 2 \overline{)6x^3 - x^2 - 5x + 4} \\ \underline{6x^3 - 4x^2} \\ 3x^2 - 5x \\ \underline{3x^2 \oplus 2x} \\ -3x + 4 \end{array}$$

Change signs of the polynomial being subtracted.

Divide: $\dfrac{3x^2}{3x} = x.$

Multiply: $x(3x - 2) = 3x^2 - 2x.$
Subtract $3x^2 - 2x$ from $3x^2 - 5x$ and bring down 4.

Now we divide $-3x$ by $3x$ to obtain -1, multiply -1 and the divisor, and subtract.

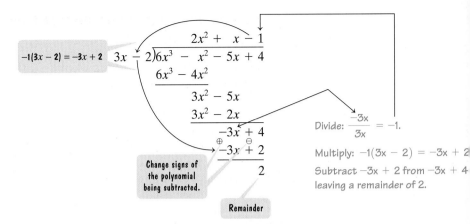

The quotient is $2x^2 + x - 1$ and the remainder is 2. When there is a nonzero remainder, as in this example, list the quotient, plus the remainder above the divisor. Thus,

$$\frac{6x^3 - x^2 - 5x + 4}{3x - 2} = \underbrace{2x^2 + x - 1}_{\text{Quotient}} + \underbrace{\frac{2}{3x - 2}}_{\substack{\text{Remainder} \\ \text{above divisor}}}.$$

An important property of division can be illustrated by clearing fractions in the equation that concluded Example 2. Multiplying both sides of this equation by $3x - 2$ results in the following equation:

$$6x^3 - x^2 - 5x + 4 = (3x - 2)(2x^2 + x - 1) + 2.$$

$$\underbrace{}_{\text{Dividend}} \quad \underbrace{}_{\text{Divisor}} \quad \underbrace{}_{\text{Quotient}} \quad \underbrace{}_{\text{Remainder}}$$

Polynomial long division is checked by multiplying the divisor with the quotient and then adding the remainder. This should give the dividend. The process illustrates the **Division Algorithm**.

> ## The Division Algorithm
> If $f(x)$ and $d(x)$ are polynomials, with $d(x) \neq 0$, and the degree of $d(x)$ is less than or equal to the degree of $f(x)$, then there exist unique polynomials $q(x)$ and $r(x)$ such that
>
> $$f(x) \quad = \quad d(x) \quad \cdot \quad q(x) \quad + \quad r(x).$$
>
> $$\underbrace{}_{\text{Dividend}} \quad \underbrace{}_{\text{Divisor}} \quad \underbrace{}_{\text{Quotient}} \quad \underbrace{}_{\text{Remainder}}$$
>
> The remainder, $r(x)$, equals 0 or it is of degree less than the degree of $d(x)$. If $r(x) = 0$, we say that $d(x)$ **divides evenly** into $f(x)$ and that $d(x)$ and $q(x)$ are **factors** of $f(x)$.

Check Point 2 Divide $7 - 11x - 3x^2 + 2x^3$ by $x - 3$. Express the result in the form quotient, plus remainder divided by divisor.

If a power of x is missing in either a dividend or a divisor, add that power of x with a coefficient of 0 and then divide. In this way, like terms will be aligned as you carry out the long division.

EXAMPLE 3   Long Division of Polynomials

Divide $6x^4 + 5x^3 + 3x - 5$ by $3x^2 - 2x$.

Solution   We write the dividend, $6x^4 + 5x^3 + 3x - 5$, as $6x^4 + 5x^3 + 0x^2 + 3x - 5$ to keep all like terms aligned.

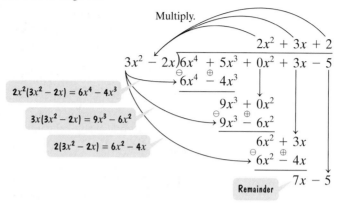

The division process is finished because the degree of $7x - 5$, which is 1, is less than the degree of the divisor $3x^2 - 2x$, which is 2. The answer is

$$\frac{6x^4 + 5x^3 + 3x - 5}{3x^2 - 2x} = 2x^2 + 3x + 2 + \frac{7x - 5}{3x^2 - 2x}.$$

Check Point 3   Divide $2x^4 + 3x^3 - 7x - 10$ by $x^2 - 2x$.

 Use synthetic division to divide polynomials.

Dividing Polynomials Using Synthetic Division

We can use **synthetic division** to divide polynomials if the divisor is of the form $x - c$. This method provides a quotient more quickly than long division. Let's compare the two methods showing $x^3 + 4x^2 - 5x + 5$ divided by $x - 3$.

Long Division

Quotient

Divisor $x - c$; $c = 3$

$$
\begin{array}{r}
x^2 + 7x + 16 \\
x - 3\overline{)\,x^3 + 4x^2 - 5x + 5} \\
\underline{x^3 - 3x^2} \\
7x^2 - 5x \\
\underline{7x^2 - 21x} \\
16x + 5 \\
\underline{16x - 48} \\
53
\end{array}
$$

Dividend

Remainder

Synthetic Division

$$
\begin{array}{r|rrrr}
3 & 1 & 4 & -5 & 5 \\
  &   & 3 & 21 & 48 \\
\hline
  & 1 & 7 & 16 & 53
\end{array}
$$

Notice the relationship between the polynomials in the long division process and the numbers that appear in synthetic division.

These are the coefficients of the dividend $x^3 + 4x^2 - 5x + 5$.

The divisor is $x - 3$. This is 3, or c, in $x - c$.

$$
\begin{array}{r|rrrr}
3 & 1 & 4 & -5 & 5 \\
  &   & 3 & 21 & 48 \\
\hline
  & 1 & 7 & 16 & 53
\end{array}
$$

These are the coefficients of the quotient $x^2 + 7x + 16$.

This is the remainder.

Now let's look at the steps involved in synthetic division.

Synthetic Division

To divide a polynomial by $x - c$:

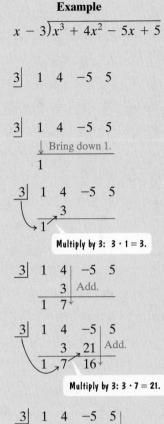

Example

1. Arrange the polynomial in descending powers, with a 0 coefficient for any missing term.

2. Write c for the divisor, $x - c$. To the right, write the coefficients of the dividend.

3. Write the leading coefficient of the dividend on the bottom row.

4. Multiply c (in this case, 3) times the value just written on the bottom row. Write the product in the next column in the second row.

5. Add the values in this new column, writing the sum in the bottom row.

6. Repeat this series of multiplications and additions until all columns are filled in.

7. Use the numbers in the last row to write the quotient, plus the remainder above the divisor. **The degree of the first term of the quotient is one less than the degree of the first term of the dividend.** The final value in this row is the remainder.

EXAMPLE 4   Using Synthetic Division

Use synthetic division to divide $5x^3 + 6x + 8$ by $x + 2$.

Solution   The divisor must be in the form $x - c$. Thus, we write $x + 2$ as $x - (-2)$. This means that $c = -2$. Writing a 0 coefficient for the missing x^2-term in the dividend, we can express the division as follows:

$$x - (-2)\overline{)5x^3 + 0x^2 + 6x + 8}.$$

Now we are ready to set up the problem so that we can use synthetic division.

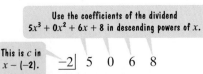

We begin the synthetic division process by bringing down 5. This is followed by a series of multiplications and additions.

1. Bring down 5.

$$\begin{array}{r|rrrr} -2 & 5 & 0 & 6 & 8 \\ & & & & \\ \hline & 5 & & & \end{array}$$

2. Multiply: $-2(5) = -10$.

$$\begin{array}{r|rrrr} -2 & 5 & 0 & 6 & 8 \\ & & -10 & & \\ \hline & 5 & & & \end{array}$$

Multiply by −2.

3. Add: $0 + (-10) = -10$.

$$\begin{array}{r|rrrr} -2 & 5 & 0 & 6 & 8 \\ & & -10 & & \text{Add.} \\ \hline & 5 & -10 & & \end{array}$$

4. Multiply: $-2(-10) = 20$.

$$\begin{array}{r|rrrr} -2 & 5 & 0 & 6 & 8 \\ & & -10 & 20 & \\ \hline & 5 & -10 & & \end{array}$$

Multiply by −2.

5. Add: $6 + 20 = 26$.

$$\begin{array}{r|rrrr} -2 & 5 & 0 & 6 & 8 \\ & & -10 & 20 & \text{Add.} \\ \hline & 5 & -10 & 26 & \end{array}$$

6. Multiply: $-2(26) = -52$.

$$\begin{array}{r|rrrr} -2 & 5 & 0 & 6 & 8 \\ & & -10 & 20 & -52 \\ \hline & 5 & -10 & 26 & \end{array}$$

Multiply by −2.

7. Add: $8 + (-52) = -44$.

$$\begin{array}{r|rrrr} -2 & 5 & 0 & 6 & 8 \\ & & -10 & 20 & -52 & \text{Add.} \\ \hline & 5 & -10 & 26 & -44 \end{array}$$

The numbers in the last row represent the coefficients of the quotient and the remainder. The degree of the first term of the quotient is one less than that of the dividend. Because the degree of the dividend, $5x^3 + 6x + 8$, is 3, the degree of the quotient is 2. This means that the 5 in the last row represents $5x^2$.

$$\begin{array}{r|rrrr} -2 & 5 & 0 & 6 & 8 \\ & & -10 & 20 & -52 \\ \hline & 5 & -10 & 26 & -44 \end{array}$$

The quotient is $5x^2 - 10x + 26$.   The remainder is −44.

Thus,

$$x + 2\overline{)5x^3 + 6x + 8} = 5x^2 - 10x + 26 - \frac{44}{x + 2}$$

Check Point 4 Use synthetic division to divide $x^3 - 7x - 6$ by $x + 2$.

  Evaluate a polynomial using the Remainder Theorem.

The Remainder Theorem

Let's consider the Division Algorithm when the dividend, $f(x)$, is divided by $x - c$. In this case, the remainder must be a constant because its degree is less than one, the degree of $x - c$.

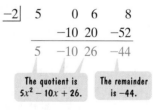

$$f(x) = (x - c)q(x) + r$$

Dividend   Divisor   Quotient   The remainder, r, is a constant when dividing by $x - c$.

Now let's evaluate f at c.

$$f(c) = (c - c)q(c) + r$$ Find $f(c)$ by letting $x = c$ in $f(x) = (x - c)q(x) + r$.
This will give an expression for r.

$$f(c) = 0 \cdot q(c) + r$$ $c - c = 0$

$$f(c) = r$$ $0 \cdot q(c) = 0$ and $0 + r = r$.

What does this last equation mean? If a polynomial is divided by $x - c$, the remain der is the value of the polynomial at c. This result is called the **Remainder Theorem**.

The Remainder Theorem

If the polynomial $f(x)$ is divided by $x - c$, then the remainder is $f(c)$.

Example 5 shows how we can use the Remainder Theorem to evaluate a poly nomial function at 2. Rather than substituting 2 for x, we divide the function b $x - 2$. The remainder is $f(2)$.

EXAMPLE 5   Using the Remainder Theorem to Evaluate a Polynomial Function

Given $f(x) = x^3 - 4x^2 + 5x + 3$, use the Remainder Theorem to find $f(2)$.

Solution   By the Remainder Theorem, if $f(x)$ is divided by $x - 2$, then th remainder is $f(2)$. We'll use synthetic division to divide.

$$
\begin{array}{r|rrrr}
2 & 1 & -4 & 5 & 3 \\
  &   & 2 & -4 & 2 \\
\hline
  & 1 & -2 & 1 & 5 \\
\end{array}
$$ Remainder

The remainder, 5, is the value of $f(2)$. Thus, $f(2) = 5$. We can verify that this i correct by evaluating $f(2)$ directly. Using $f(x) = x^3 - 4x^2 + 5x + 3$, we obtain

$$f(2) = 2^3 - 4 \cdot 2^2 + 5 \cdot 2 + 3 = 8 - 16 + 10 + 3 = 5.$$

Check Point **5**   Given $f(x) = 3x^3 + 4x^2 - 5x + 3$, use the Remainder Theorem to fin $f(-4)$.

4 Use the Factor Theorem to solve a polynomial equation.

The Factor Theorem

Let's look again at the Division Algorithm when the divisor is of the form $x - c$.

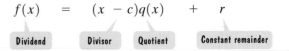

$$f(x) \quad = \quad (x - c)q(x) \quad + \quad r$$

Dividend   Divisor   Quotient   Constant remainder

By the Remainder Theorem, the remainder r is $f(c)$, so we can substitute $f(c)$ for

$$f(x) = (x - c)q(x) + f(c).$$

Notice that if $f(c) = 0$, then

$$f(x) = (x - c)q(x)$$

so that $x - c$ is a factor of $f(x)$. This means that for the polynomial function $f(x)$, $f(c) = 0$, then $x - c$ is a factor of $f(x)$.

Let's reverse directions and see what happens if $x - c$ is a factor of $f(x)$. Th means that

$$f(x) = (x - c)q(x).$$

If we replace x in $f(x) = (x - c)q(x)$ with c, we obtain

$$f(c) = (c - c)q(c) = 0 \cdot q(c) = 0.$$

Thus, if $x - c$ is a factor of $f(x)$, then $f(c) = 0$.

We have proved a result known as the **Factor Theorem**.

The Factor Theorem

Let $f(x)$ be a polynomial.

   a. If $f(c) = 0$, then $x - c$ is a factor of $f(x)$.

   b. If $x - c$ is a factor of $f(x)$, then $f(c) = 0$.

The example that follows shows how the Factor Theorem can be used to solve a polynomial equation.

EXAMPLE 6   Using the Factor Theorem

Solve the equation $2x^3 - 3x^2 - 11x + 6 = 0$ given that 3 is a zero of $f(x) = 2x^3 - 3x^2 - 11x + 6$.

Solution   We are given that 3 is a zero of $f(x) = 2x^3 - 3x^2 - 11x + 6$. This means that $f(3) = 0$. Because $f(3) = 0$, the Factor Theorem tells us that $x - 3$ is a factor of $f(x)$. We'll use synthetic division to divide $f(x)$ by $x - 3$.

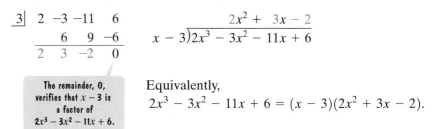

The remainder, 0, verifies that $x - 3$ is a factor of $2x^3 - 3x^2 - 11x + 6$.

Equivalently,

$$2x^3 - 3x^2 - 11x + 6 = (x - 3)(2x^2 + 3x - 2).$$

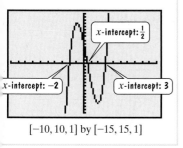
Now we can solve the polynomial equation.

$2x^3 - 3x^2 - 11x + 6 = 0$ This is the given equation.

$(x - 3)(2x^2 + 3x - 2) = 0$ Factor using the result from the synthetic division.

$(x - 3)(2x - 1)(x + 2) = 0$ Factor the trinomial.

$x - 3 = 0$ or $2x - 1 = 0$ or $x + 2 = 0$ Set each factor equal to 0.

$x = 3$ $x = \frac{1}{2}$ $x = -2$ Solve for x.

The solution set is $\left\{-2, \frac{1}{2}, 3\right\}$.

Based on the Factor Theorem, the following statements are useful in solving polynomial equations:

   1. If $f(x)$ is divided by $x - c$ and the remainder is zero, then c is a zero of f and c is a root of the polynomial equation $f(x) = 0$.

   2. If $f(x)$ is divided by $x - c$ and the remainder is zero, then $x - c$ is a factor of $f(x)$.

Check Point 6   Solve the equation $15x^3 + 14x^2 - 3x - 2 = 0$ given that -1 is a zero of $f(x) = 15x^3 + 14x^2 - 3x - 2$.

EXERCISE SET 2.4

Practice Exercises

In Exercises 1–16, divide using long division. State the quotient, $q(x)$, and the remainder, $r(x)$.

1. $(x^2 + 8x + 15) \div (x + 5)$

2. $(x^2 + 3x - 10) \div (x - 2)$

3. $(x^3 + 5x^2 + 7x + 2) \div (x + 2)$

4. $(x^3 - 2x^2 - 5x + 6) \div (x - 3)$

5. $(6x^3 + 7x^2 + 12x - 5) \div (3x - 1)$

6. $(6x^3 + 17x^2 + 27x + 20) \div (3x + 4)$

7. $(12x^2 + x - 4) \div (3x - 2)$

8. $(4x^2 - 8x + 6) \div (2x - 1)$

9. $\dfrac{2x^3 + 7x^2 + 9x - 20}{x + 3}$

10. $\dfrac{3x^2 - 2x + 5}{x - 3}$

11. $\dfrac{4x^4 - 4x^2 + 6x}{x - 4}$

12. $\dfrac{x^4 - 81}{x - 3}$

13. $\dfrac{6x^3 + 13x^2 - 11x - 15}{3x^2 - x - 3}$

14. $\dfrac{x^4 + 2x^3 - 4x^2 - 5x - 6}{x^2 + x - 2}$

15. $\dfrac{18x^4 + 9x^3 + 3x^2}{3x^2 + 1}$

16. $\dfrac{2x^5 - 8x^4 + 2x^3 + x^2}{2x^3 + 1}$

In Exercises 17–32, divide using synthetic division.

17. $(2x^2 + x - 10) \div (x - 2)$

18. $(x^2 + x - 2) \div (x - 1)$

19. $(3x^2 + 7x - 20) \div (x + 5)$

20. $(5x^2 - 12x - 8) \div (x + 3)$

21. $(4x^3 - 3x^2 + 3x - 1) \div (x - 1)$

22. $(5x^3 - 6x^2 + 3x + 11) \div (x - 2)$

23. $(6x^5 - 2x^3 + 4x^2 - 3x + 1) \div (x - 2)$

24. $(x^5 + 4x^4 - 3x^2 + 2x + 3) \div (x - 3)$

25. $(x^2 - 5x - 5x^3 + x^4) \div (5 + x)$

26. $(x^2 - 6x - 6x^3 + x^4) \div (6 + x)$

27. $\dfrac{x^5 + x^3 - 2}{x - 1}$

28. $\dfrac{x^7 + x^5 - 10x^3 + 12}{x + 2}$

29. $\dfrac{x^4 - 256}{x - 4}$

30. $\dfrac{x^7 - 128}{x - 2}$

31. $\dfrac{2x^5 - 3x^4 + x^3 - x^2 + 2x - 1}{x + 2}$

32. $\dfrac{x^5 - 2x^4 - x^3 + 3x^2 - x + 1}{x - 2}$

In Exercises 33–40, use synthetic division and the Remainder Theorem to find the indicated function value.

33. $f(x) = 2x^3 - 11x^2 + 7x - 5;\quad f(4)$

34. $f(x) = x^3 - 7x^2 + 5x - 6;\quad f(3)$

35. $f(x) = 3x^3 - 7x^2 - 2x + 5;\quad f(-3)$

36. $f(x) = 4x^3 + 5x^2 - 6x - 4;\quad f(-2)$

37. $f(x) = x^4 + 5x^3 + 5x^2 - 5x - 6;\quad f(3)$

38. $f(x) = x^4 - 5x^3 + 5x^2 + 5x - 6;\quad f(2)$

39. $f(x) = 2x^4 - 5x^3 - x^2 + 3x + 2;\quad f\left(-\dfrac{1}{2}\right)$

40. $f(x) = 6x^4 + 10x^3 + 5x^2 + x + 1;\quad f\left(-\dfrac{2}{3}\right)$

41. Use synthetic division to divide
$$f(x) = x^3 - 4x^2 + x + 6 \text{ by } x + 1.$$
Use the result to find all zeros of f.

42. Use synthetic division to divide
$$f(x) = x^3 - 2x^2 - x + 2 \text{ by } x + 1.$$
Use the result to find all zeros of f.

43. Solve the equation $2x^3 - 5x^2 + x + 2 = 0$ given that 2 is zero of $f(x) = 2x^3 - 5x^2 + x + 2$.

44. Solve the equation $2x^3 - 3x^2 - 11x + 6 = 0$ given that $-$ is a zero of $f(x) = 2x^3 - 3x^2 - 11x + 6$.

45. Solve the equation $12x^3 + 16x^2 - 5x - 3 = 0$ given that $-$ is a root.

46. Solve the equation $3x^3 + 7x^2 - 22x - 8 = 0$ given that $-$ is a root.

Practice Plus

In Exercises 47–50, use the graph or the table to determine a solution of each equation. Use synthetic division to verify that this number is a solution of the equation. Then solve the polynomial equation.

47. $x^3 + 2x^2 - 5x - 6 = 0$

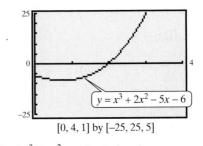

$[0, 4, 1]$ by $[-25, 25, 5]$

48. $2x^3 + x^2 - 13x + 6 = 0$

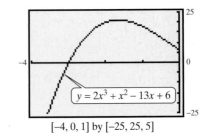

$[-4, 0, 1]$ by $[-25, 25, 5]$

49. $6x^3 - 11x^2 + 6x - 1 = 0$

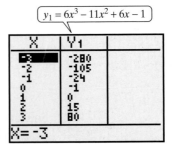

50. $2x^3 + 11x^2 - 7x - 6 = 0$

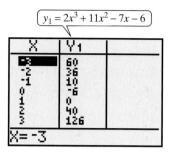

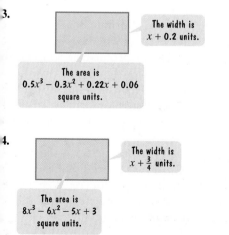

Application Exercises

51. a. Use synthetic division to show that 3 is a solution of the polynomial equation

$$14x^3 - 17x^2 - 16x - 177 = 0.$$

b. Use the solution from part (a) to solve this problem. The number of eggs, $f(x)$, in a female moth is a function of her abdominal width, x, in millimeters, modeled by

$$f(x) = 14x^3 - 17x^2 - 16x + 34.$$

What is the abdominal width when there are 211 eggs?

52. a. Use synthetic division to show that 2 is a solution of the polynomial equation

$$2h^3 + 14h^2 - 72 = 0.$$

b. Use the solution from part (a) to solve this problem. The width of a rectangular box is twice the height and the length is 7 inches more than the height. If the volume is 72 cubic inches, find the dimensions of the box.

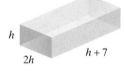

In Exercises 53–54, write a polynomial that represents the length of each rectangle.

53.

The width is $x + 0.2$ units.

The area is $0.5x^3 - 0.3x^2 + 0.22x + 0.06$ square units.

54.

The width is $x + \frac{3}{4}$ units.

The area is $8x^3 - 6x^2 - 5x + 3$ square units.

During the 1980s, the controversial economist Arthur Laffer promoted the idea that tax increases lead to a reduction in government revenue. Called supply-side economics, the theory uses functions such as

$$f(x) = \frac{80x - 8000}{x - 110}, \ 30 \le x \le 100.$$

This function models the government tax revenue, $f(x)$, in tens of billions of dollars, in terms of the tax rate, x. The graph of the function is shown. It illustrates tax revenue decreasing quite dramatically as the tax rate increases. At a tax rate of (gasp) 100%, the government takes all our money and no one has an incentive to work. With no income earned, zero dollars in tax revenue is generated.

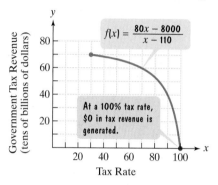

Use function f and its graph to solve Exercises 55–56.

55. a. Find and interpret $f(30)$. Identify the solution as a point on the graph of the function.

b. Rewrite the function by using long division to perform

$$(80x - 8000) \div (x - 110).$$

Then use this new form of the function to find $f(30)$. Do you obtain the same answer as you did in part (a)?

c. Is f a polynomial function? Explain your answer.

56. a. Find and interpret $f(40)$. Identify the solution as a point on the graph of the function.

b. Rewrite the function by using long division to perform

$$(80x - 8000) \div (x - 110).$$

Then use this new form of the function to find $f(40)$. Do you obtain the same answer as you did in part (a)?

c. Is f a polynomial function? Explain your answer.

Writing in Mathematics

57. Explain how to perform long division of polynomials. Use $2x^3 - 3x^2 - 11x + 7$ divided by $x - 3$ in your explanation.

58. In your own words, state the Division Algorithm.

59. How can the Division Algorithm be used to check the quotient and remainder in a long division problem?

60. Explain how to perform synthetic division. Use the division problem in Exercise 57 to support your explanation.

61. State the Remainder Theorem.

62. Explain how the Remainder Theorem can be used to find $f(-6)$ if $f(x) = x^4 + 7x^3 + 8x^2 + 11x + 5$. What advantage is there to using the Remainder Theorem in this situation rather than evaluating $f(-6)$ directly?

63. How can the Factor Theorem be used to determine if $x - 1$ is a factor of $x^3 - 2x^2 - 11x + 12$?

64. If you know that -2 is a zero of
$$f(x) = x^3 + 7x^2 + 4x - 12,$$
explain how to solve the equation
$$x^3 + 7x^2 + 4x - 12 = 0.$$

Technology Exercise

65. For each equation that you solved in Exercises 43–46, use a graphing utility to graph the polynomial function defined by the left side of the equation. Use end behavior to obtain a complete graph. Then use the graph's x-intercepts to verify your solutions.

Critical Thinking Exercises

66. Which one of the following is true?
   a. If a trinomial in x of degree 6 is divided by a trinomial in x of degree 3, the degree of the quotient is 2.

   b. Synthetic division could not be used to find the quotient of $10x^3 - 6x^2 + 4x - 1$ and $x - \frac{1}{2}$.

   c. Any problem that can be done by synthetic division can also be done by the method for long division of polynomials.

   d. If a polynomial long-division problem results in a remainder that is a whole number, then the divisor is a factor of the dividend.

67. Find k so that $4x + 3$ is a factor of
$$20x^3 + 23x^2 - 10x + k.$$

68. When $2x^2 - 7x + 9$ is divided by a polynomial, the quotient is $2x - 3$ and the remainder is 3. Find the polynomial.

69. Find the quotient of $x^{3n} + 1$ and $x^n + 1$.

70. Synthetic division is a process for dividing a polynomial by $x - c$. The coefficient of x is 1. How might synthetic division be used if you are dividing by $2x - 4$?

71. Use synthetic division to show that 5 is a solution of
$$x^4 - 4x^3 - 9x^2 + 16x + 20 = 0.$$
Then solve the polynomial equation.

SECTION 2.5 Zeros of Polynomial Functions

Objectives

1 Use the Rational Zero Theorem to find possible rational zeros.

2 Find zeros of a polynomial function.

3 Solve polynomial equations.

4 Use the Linear Factorization Theorem to find polynomials with given zeros.

5 Use Descartes's Rule of Signs.

You stole my formula!

Tartaglia's Secret Formula for One Solution of $x^3 + mx = n$

$$x = \sqrt[3]{\sqrt{\left(\frac{n}{2}\right)^2 + \left(\frac{m}{3}\right)^3} + \frac{n}{2}} - \sqrt[3]{\sqrt{\left(\frac{n}{2}\right)^2 + \left(\frac{m}{3}\right)^3} - \frac{n}{2}}$$

Popularizers of mathematics are sharing bizarre stories that are giving math a secure place in popular culture. One episode, able to compete with the wildest fare served up by television talk shows and the tabloids, involves three Italian mathematicians and, of all things, zeros of polynomial functions.

Tartaglia (1499–1557), poor and starving, has found a formula that gives a root for a third-degree polynomial equation. Cardano (1501–1576) begs Tartaglia to reveal the secret formula, wheedling it from him with the promise he will find the impoverished Tartaglia a patron. Then Cardano publishes his famous work *Ars Magna*, in which he presents Tartaglia's formula as his own. Cardano uses his most talented student, Ferrari (1522–1565), who derived a formula for a root of a fourth-degree polynomial equation, to falsely accuse Tartaglia of plagiarism. The dispute becomes violent and Tartaglia is fortunate to escape alive.

The noise from this "You Stole My Formula" episode is quieted by the work of French mathematician Evariste Galois (1811–1832). Galois proved that there is no general formula for finding roots of polynomial equations of degree 5 or higher. There are, however, methods for finding roots. In this section, we study methods for finding zeros of polynomial functions. We begin with a theorem that plays an important role in this process.

Study Tip

Be sure you are familiar with the various kinds of zeros of polynomial functions. Here's a quick example:

$$f(x) = (x + 3)(2x - 1)(x + \sqrt{2})(x - \sqrt{2})(x - 4 + 5i)(x - 4 - 5i).$$

Zeros: $-3,$ $\dfrac{1}{2},$ $-\sqrt{2},$ $\sqrt{2},$ $4 - 5i,$ $4 + 5i$

Rational zeros   Irrational zeros   Complex imaginary zeros

Real zeros   Nonreal zeros

① Use the Rational Zero Theorem to find possible rational zeros.

The Rational Zero Theorem

The Rational Zero Theorem provides us with a tool that we can use to make a list of all possible rational zeros of a polynomial function. Equivalently, the theorem gives all possible rational roots of a polynomial equation. Not every number in the list will be a zero of the function, but every rational zero of the polynomial function will appear somewhere in the list.

The Rational Zero Theorem

If $f(x) = a_n x^n + a_{n-1}x^{n-1} + \cdots + a_1 x + a_0$ has *integer* coefficients and $\dfrac{p}{q}$ (where $\dfrac{p}{q}$ is reduced to lowest terms) is a rational zero of f, then p is a factor of the constant term, a_0, and q is a factor of the leading coefficient, a_n.

You can explore the "why" behind the Rational Zero Theorem in Exercise 90 of Exercise Set 2.5. For now, let's see if we can figure out what the theorem tells us about possible rational zeros. To use the theorem, list all the integers that are factors of the constant term, a_0. Then list all the integers that are factors of the leading coefficient, a_n. Finally list all possible rational zeros:

$$\text{Possible rational zeros} = \frac{\text{Factors of the constant term}}{\text{Factors of the leading coefficient}}.$$

EXAMPLE 1   Using the Rational Zero Theorem

List all possible rational zeros of $f(x) = -x^4 + 3x^2 + 4$.

Solution   The constant term is 4. We list all of its factors: $\pm 1, \pm 2, \pm 4$. The leading coefficient is -1. Its factors are ± 1.

Factors of the constant term, 4: $\pm 1, \pm 2, \pm 4$
Factors of the leading coefficient, -1: ± 1

Because

$$\text{Possible rational zeros} = \frac{\text{Factors of the constant term}}{\text{Factors of the leading coefficient}},$$

we must take each number in the first row, $\pm 1, \pm 2, \pm 4$, and divide by each number in the second row, ± 1.

$$\text{Possible rational zeros} = \frac{\text{Factors of } 4}{\text{Factors of } -1} = \frac{\pm 1, \pm 2, \pm 4}{\pm 1} = \pm 1, \quad \pm 2, \quad \pm 4$$

Divide ± 1 by ± 1.   Divide ± 2 by ± 1.   Divide ± 4 by ± 1.

Study Tip

Always keep in mind the relationship among zeros, roots, and x-intercepts. The zeros of a function f are the roots, or solutions, of the equation $f(x) = 0$. Furthermore, the real zeros, or real roots, are the x-intercepts of the graph of f.

There are six possible rational zeros, ± 1, ± 2, and ± 4. The graph of $f(x) = -x^4 + 3x^2 + 4$ is shown in Figure 2.24. The x-intercepts are -2 and 2. Thus, -2 and 2 are the actual rational zeros.

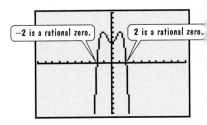

Figure 2.24 The graph of $f(x) = -x^4 + 3x^2 + 4$ shows that -2 and 2 are rational zeros.

Check Point 1 List all possible rational zeros of
$$f(x) = x^3 + 2x^2 - 5x - 6.$$

EXAMPLE 2 Using the Rational Zero Theorem

List all possible rational zeros of $f(x) = 15x^3 + 14x^2 - 3x - 2$.

Solution The constant term is -2 and the leading coefficient is 15.

$$\text{Possible rational zeros} = \frac{\text{Factors of the constant term, } -2}{\text{Factors of the leading coefficient, } 15} = \frac{\pm 1, \pm 2}{\pm 1, \pm 3, \pm 5, \pm 15}$$

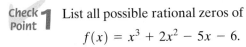

$$= \pm 1, \ \pm 2, \ \pm\frac{1}{3}, \ \pm\frac{2}{3}, \ \pm\frac{1}{5}, \ \pm\frac{2}{5}, \ \pm\frac{1}{15}, \ \pm\frac{2}{15}$$

Divide ± 1 and ± 2 by ± 1.   Divide ± 1 and ± 2 by ± 3.   Divide ± 1 and ± 2 by ± 5.   Divide ± 1 and ± 2 by ± 15.

There are 16 possible rational zeros. The actual solution set of
$$15x^3 + 14x^2 - 3x - 2 = 0$$
is $\left\{ -1, -\frac{1}{3}, \frac{2}{5} \right\}$, which contains three of the 16 possible zeros.

Check Point 2 List all possible rational zeros of
$$f(x) = 4x^5 + 12x^4 - x - 3.$$

 Find zeros of a polynomial function.

How do we determine which (if any) of the possible rational zeros are rational zeros of the polynomial function? To find the first rational zero, we can use a trial-and-error process involving synthetic division: If $f(x)$ is divided by $x -$ and the remainder is zero, then c is a zero of f. After we identify the first rational zero, we use the result of the synthetic division to factor the original polynomial. Then we set each factor equal to zero to identify any additional rational zeros.

EXAMPLE 3 Finding Zeros of a Polynomial Function

Find all zeros of $f(x) = x^3 + 2x^2 - 5x - 6$.

Solution We begin by listing all possible rational zeros.

Possible rational zeros

$$= \frac{\text{Factors of the constant term, } -6}{\text{Factors of the leading coefficient, } 1} = \frac{\pm 1, \pm 2, \pm 3, \pm 6}{\pm 1} = \pm 1, \pm 2, \pm 3, \pm 6$$

Divide the eight numbers in the numerator by ± 1.

Now we will use synthetic division to see if we can find a rational zero among the possible rational zeros ± 1, ± 2, ± 3, ± 6. Keep in mind that if $f(x)$ is divided b

$x - c$ and the remainder is zero, then c is a zero of f. Let's start by testing 1. If 1 is not a rational zero, then we will test other possible rational zeros.

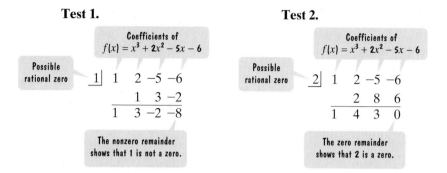

Test 1.

Possible rational zero

$$\underline{1|}\ \ \begin{array}{rrrr} 1 & 2 & -5 & -6 \\ & 1 & 3 & -2 \\ \hline 1 & 3 & -2 & -8 \end{array}$$

Coefficients of $f(x) = x^3 + 2x^2 - 5x - 6$

The nonzero remainder shows that 1 is not a zero.

Test 2.

Possible rational zero

$$\underline{2|}\ \ \begin{array}{rrrr} 1 & 2 & -5 & -6 \\ & 2 & 8 & 6 \\ \hline 1 & 4 & 3 & 0 \end{array}$$

Coefficients of $f(x) = x^3 + 2x^2 - 5x - 6$

The zero remainder shows that 2 is a zero.

The zero remainder tells us that 2 is a zero of the polynomial function $f(x) = x^3 + 2x^2 - 5x - 6$. Equivalently, 2 is a solution, or root, of the polynomial equation $x^3 + 2x^2 - 5x - 6 = 0$. Thus, $x - 2$ is a factor of the polynomial. The first three numbers in the bottom row of the synthetic division on the right, 1, 4, and 3, give the coefficients of the other factor. This factor is $x^2 + 4x + 3$.

$$x^3 + 2x^2 - 5x - 6 = 0 \qquad \text{\small Finding the zeros of } f(x) = x^3 + 2x^2 - 5x - 6 \text{ is}$$
$$\text{\small the same as finding the roots of this equation.}$$

$$(x - 2)(x^2 + 4x + 3) = 0 \qquad \text{\small Factor using the result from the synthetic division.}$$

$$(x - 2)(x + 3)(x + 1) = 0 \qquad \text{\small Factor completely.}$$

$$x - 2 = 0 \ \ \text{or} \ \ x + 3 = 0 \ \ \text{or} \ \ x + 1 = 0 \qquad \text{\small Set each factor equal to zero.}$$

$$x = 2 \qquad\qquad x = -3 \qquad\qquad x = -1 \qquad \text{\small Solve for x.}$$

The solution set is $\{-3, -1, 2\}$. The zeros of f are -3, -1, and 2.

Check Point 3 Find all zeros of

$$f(x) = x^3 + 8x^2 + 11x - 20.$$

Our work in Example 3 involved finding zeros of a third-degree polynomial function. The Rational Zero Theorem is a tool that allows us to rewrite such functions as products of two factors, one linear and one quadratic. Zeros of the quadratic factor are found by factoring, the quadratic formula, or the square root property.

EXAMPLE 4   Finding Zeros of a Polynomial Function

Find all zeros of $f(x) = x^3 + 7x^2 + 11x - 3$.

Solution   We begin by listing all possible rational zeros.

$$\text{Possible rational zeros} = \frac{\text{Factors of the constant term, } -3}{\text{Factors of the leading coefficient, } 1} = \frac{\pm 1, \pm 3}{\pm 1} = \pm 1, \pm 3$$

Now we will use synthetic division to see if we can find a rational zero among the four possible rational zeros.

Test 1.

$$\underline{1|}\ \ \begin{array}{rrrr} 1 & 7 & 11 & -3 \\ & 1 & 8 & 19 \\ \hline 1 & 8 & 19 & 16 \end{array}$$

Test -1.

$$\underline{-1|}\ \ \begin{array}{rrrr} 1 & 7 & 11 & -3 \\ & -1 & -6 & -5 \\ \hline 1 & 6 & 5 & -8 \end{array}$$

Test 3.

$$\underline{3|}\ \ \begin{array}{rrrr} 1 & 7 & 11 & -3 \\ & 3 & 30 & 123 \\ \hline 1 & 10 & 41 & 120 \end{array}$$

Test -3.

$$\underline{-3|}\ \ \begin{array}{rrrr} 1 & 7 & 11 & -3 \\ & -3 & -12 & 3 \\ \hline 1 & 4 & -1 & 0 \end{array}$$

The zero remainder in the final synthetic division on the previous page tells us tha[t] -3 is a zero of the polynomial function $f(x) = x^3 + 7x^2 + 11x - 3$. To find a[ll] zeros of f, we proceed as follows:

$$x^3 + 7x^2 + 11x - 3 = 0$$ Finding the zeros of f is the same thing as finding the roots of f(x) = 0.

$$(x + 3)(x^2 + 4x - 1) = 0$$ This result is from the last synthetic division on the previous page. The first three number[s] in the bottom row, 1, 4, and −1, give the coefficients of the second factor.

$$x + 3 = 0 \quad \text{or} \quad x^2 + 4x - 1 = 0$$ Set each factor equal to 0.
$$x = -3$$ Solve the linear equation.

We can use the quadratic formula to solve $x^2 + 4x - 1 = 0$.

$$x = \frac{-b \pm \sqrt{b^2 - 4ac}}{2a}$$ We use the quadratic formula because $x^2 + 4x -$ cannot be factored.

$$= \frac{-4 \pm \sqrt{4^2 - 4(1)(-1)}}{2(1)}$$ Let a = 1, b = 4, and c = −1.

$$= \frac{-4 \pm \sqrt{20}}{2}$$ Multiply and subtract under the radical: $4^2 - 4(1)(-1) = 16 - (-4) = 16 + 4 = 20$

$$= \frac{-4 \pm 2\sqrt{5}}{2}$$ $\sqrt{20} = \sqrt{4 \cdot 5} = 2\sqrt{5}$

$$= -2 \pm \sqrt{5}$$ Divide the numerator and the denominator by 2.

The solution set is $\left\{-3, -2 - \sqrt{5}, -2 + \sqrt{5}\right\}$. The zeros of f are -3, $-2 - \sqrt{5}$ and $-2 + \sqrt{5}$. Among these three real zeros, one zero is rational and two ar[e] irrational.

 Check Point 4 Find all zeros of $f(x) = x^3 + x^2 - 5x - 2$.

If the degree of a polynomial function or equation is 4 or higher, it is ofte[n] necessary to find more than one linear factor by synthetic division.

One way to speed up the process of finding the first zero is to graph th[e] function. Any x-intercept is a zero.

③ Solve polynomial equations.

EXAMPLE 5   Solving a Polynomial Equation

Solve: $x^4 - 6x^2 - 8x + 24 = 0$.

Solution   Recall that we refer to the *zeros* of a polynomial function and the *roo[ts]* of a polynomial equation. Because we are given an equation, we will use the wor[d] "roots," rather than "zeros," in the solution process. We begin by listing all possib[le] rational roots.

Possible rational roots $= \dfrac{\text{Factors of the constant term, 24}}{\text{Factors of the leading coefficient, 1}}$

$$= \frac{\pm 1, \pm 2, \pm 3, \pm 4, \pm 6, \pm 8, \pm 12, \pm 24}{\pm 1}$$

$$= \pm 1, \pm 2, \pm 3, \pm 4, \pm 6, \pm 8, \pm 12, \pm 24$$

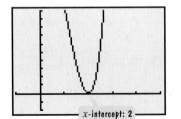

Figure 2.25 The graph of $f(x) = x^4 - 6x^2 - 8x + 24$ in a $[-1, 5, 1]$ by $[-2, 10, 1]$ viewing rectangle

The graph of $f(x) = x^4 - 6x^2 - 8x + 24$ is shown in Figure 2.25. Becaus[e] the x-intercept is 2, we will test 2 by synthetic division and show that it is a ro[ot] of the given equation. Without the graph, the procedure would be to start th[e]

trial-and-error synthetic division with 1 and proceed until a zero remainder is found, as we did in Example 4.

$$
\begin{array}{r|rrrrr}
2 & 1 & 0 & -6 & -8 & 24 \\
  &   & 2 & 4 & -4 & -24 \\
\hline
  & 1 & 2 & -2 & -12 & 0
\end{array}
$$

Careful!
$x^4 - 6x^2 - 8x + 24 = x^4 + 0x^3 - 6x^2 - 8x + 24$

The zero remainder indicates that 2 is a root of $x^4 - 6x^2 - 8x + 24 = 0.$

Now we can rewrite the given equation in factored form.

$$x^4 - 6x^2 - 8x + 24 = 0 \qquad \text{This is the given equation.}$$
$$(x - 2)(x^3 + 2x^2 - 2x - 12) = 0 \qquad \text{This is the result obtained from the synthetic division. The first four numbers in the bottom row, 1, 2, } -2, \text{ and } -12, \text{ give the coefficients of the second factor.}$$

$$x - 2 = 0 \quad \text{or} \quad x^3 + 2x^2 - 2x - 12 = 0 \qquad \text{Set each factor equal to 0.}$$

We can use the same approach to look for rational roots of the polynomial equation $x^3 + 2x^2 - 2x - 12 = 0$, listing all possible rational roots. Without the graph in Figure 2.25, the procedure would be to start testing possible rational roots by trial-and-error synthetic division. However, take a second look at the graph in Figure 2.25. Because the graph turns around at 2, this means that 2 is a root of even multiplicity. Thus, 2 must also be a root of $x^3 + 2x^2 - 2x - 12 = 0$, confirmed by the following synthetic division.

$$
\begin{array}{r|rrrr}
2 & 1 & 2 & -2 & -12 \\
  &   & 2 & 8 & 12 \\
\hline
  & 1 & 4 & 6 & 0
\end{array}
$$

These are the coefficients of $x^3 + 2x^2 - 2x - 12 = 0.$

The zero remainder indicates that 2 is a root of $x^3 + 2x^2 - 2x - 12 = 0.$

Now we can solve the original equation as follows:

$$x^4 - 6x^2 - 8x + 24 = 0 \qquad \text{This is the given equation.}$$
$$(x - 2)(x^3 + 2x^2 - 2x - 12) = 0 \qquad \text{This factorization was obtained from the first synthetic division.}$$
$$(x - 2)(x - 2)(x^2 + 4x + 6) = 0 \qquad \text{This factorization was obtained from the second synthetic division. The first three numbers in the bottom row, 1, 4, and 6, give the coefficients of the third factor.}$$

$$x - 2 = 0 \quad \text{or} \quad x - 2 = 0 \quad \text{or} \quad x^2 + 4x + 6 = 0 \qquad \text{Set each factor equal to 0.}$$
$$x = 2 \qquad\qquad x = 2 \qquad\qquad\qquad\qquad \text{Solve the linear equations.}$$

We can use the quadratic formula to solve $x^2 + 4x + 6 = 0$.

$$x = \frac{-b \pm \sqrt{b^2 - 4ac}}{2a} \qquad \text{We use the quadratic formula because } x^2 + 4x + 6 \text{ cannot be factored.}$$
$$= \frac{-4 \pm \sqrt{4^2 - 4(1)(6)}}{2(1)} \qquad \text{Let } a = 1, b = 4, \text{ and } c = 6.$$
$$= \frac{-4 \pm \sqrt{-8}}{2} \qquad \text{Multiply and subtract under the radical: } 4^2 - 4(1)(6) = 16 - 24 = -8.$$
$$= \frac{-4 \pm 2i\sqrt{2}}{2} \qquad \sqrt{-8} = \sqrt{4(2)(-1)} = 2i\sqrt{2}$$
$$= -2 \pm i\sqrt{2} \qquad \text{Simplify.}$$

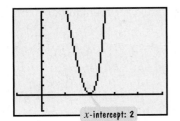

Figure 2.25 (repeated)

The solution set of the original equation, $x^4 - 6x^2 - 8x + 24 = 0$, is $\{2, -2 - i\sqrt{2}, -2 + i\sqrt{2}\}$. The graph of $f(x) = x^4 - 6x^2 - 8x + 24$ in Figure 2.25 illustrates that a graphing utility does not reveal the two imaginary roots.

In Example 5, 2 is a repeated root of the equation with multiplicity 2. The example illustrates two general properties:

Properties of Polynomial Equations

1. If a polynomial equation is of degree n, then counting multiple roots separately, the equation has n roots.

2. If $a + bi$ is a root of a polynomial equation with real coefficients ($b \neq 0$), then the complex imaginary number $a - bi$ is also a root. Complex imaginary roots, if they exist, occur in conjugate pairs.

Check Point 5 Solve: $x^4 - 6x^3 + 22x^2 - 30x + 13 = 0$.

The Fundamental Theorem of Algebra

The fact that a polynomial equation of degree n has n roots is a consequence of a theorem proved in 1799 by a 22-year-old student named Carl Friedrich Gauss in his doctoral dissertation. His result is called the **Fundamental Theorem of Algebra**.

The Fundamental Theorem of Algebra

If $f(x)$ is a polynomial of degree n, where $n \geq 1$, then the equation $f(x) = 0$ has at least one complex root.

Suppose, for example, that $f(x) = 0$ represents a polynomial equation of degree n. By the Fundamental Theorem of Algebra, we know that this equation has at least one complex root; we'll call it c_1. By the Factor Theorem, we know that $x - c_1$ is a factor of $f(x)$. Therefore, we obtain

$$(x - c_1)q_1(x) = 0 \qquad \text{The degree of the polynomial } q_1(x) \text{ is } n - 1.$$
$$x - c_1 = 0 \quad \text{or} \quad q_1(x) = 0. \qquad \text{Set each factor equal to 0.}$$

If the degree of $q_1(x)$ is at least 1, by the Fundamental Theorem of Algebra, the equation $q_1(x) = 0$ has at least one complex root. We'll call it c_2. The Factor Theorem gives us

$$q_1(x) = 0 \qquad \text{The degree of } q_1(x) \text{ is } n - 1.$$
$$(x - c_2)q_2(x) = 0 \qquad \text{The degree of } q_2(x) \text{ is } n - 2.$$
$$x - c_2 = 0 \quad \text{or} \quad q_2(x) = 0. \qquad \text{Set each factor equal to 0.}$$

Let's see what we have up to this point and then continue the process.

$$f(x) = 0 \qquad \text{This is the original polynomial equation of degree } n.$$

$$(x - c_1)q_1(x) = 0 \qquad \text{This is the result from our first application of the Fundamental Theorem.}$$

$$(x - c_1)(x - c_2)q_2(x) = 0 \qquad \text{This is the result from our second application of the Fundamental Theorem.}$$

By continuing this process, we will obtain the product of n linear factors. Setting each of these linear factors equal to zero results in n complex roots. Thus, if $f(x)$ is a polynomial of degree n, where $n \geq 1$, then $f(x) = 0$ has exactly n roots, where roots are counted according to their multiplicity.

④ Use the Linear Factorization Theorem to find polynomials with given zeros.

The Linear Factorization Theorem

In Example 5, we found that $x^4 - 6x^2 - 8x + 24 = 0$ has $\{2, -2 \pm i\sqrt{2}\}$ as a solution set, where 2 is a repeated root with multiplicity 2. The polynomial can be factored over the complex nonreal numbers as follows:

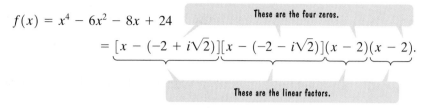

$$f(x) = x^4 - 6x^2 - 8x + 24$$

These are the four zeros.

$$= \underbrace{[x - (-2 + i\sqrt{2})]}\underbrace{[x - (-2 - i\sqrt{2})]}\underbrace{(x - 2)}\underbrace{(x - 2)}.$$

These are the linear factors.

This fourth-degree polynomial has four linear factors. Just as an nth-degree polynomial equation has n roots, an nth-degree polynomial has n linear factors. This is formally stated as the **Linear Factorization Theorem**.

The Linear Factorization Theorem

If $f(x) = a_nx^n + a_{n-1}x^{n-1} + \cdots + a_1x + a_0$, where $n \geq 1$ and $a_n \neq 0$, then

$$f(x) = a_n(x - c_1)(x - c_2) \cdots (x - c_n),$$

where $c_1, c_2, \ldots, c_n$ are complex numbers (possibly real and not necessarily distinct). In words: An nth-degree polynomial can be expressed as the product of a nonzero constant and n linear factors.

Many of our problems involving polynomial functions and polynomial equations dealt with the process of finding zeros and roots. The Linear Factorization Theorem enables us to reverse this process, finding a polynomial function when the zeros are given.

EXAMPLE 6   Finding a Polynomial Function with Given Zeros

Find a fourth-degree polynomial function $f(x)$ with real coefficients that has $-2, 2$, and i as zeros and such that $f(3) = -150$.

Solution   Because i is a zero and the polynomial has real coefficients, the conjugate, $-i$, must also be a zero. We can now use the Linear Factorization Theorem.

$f(x) = a_n(x - c_1)(x - c_2)(x - c_3)(x - c_4)$ — This is the linear factorization for a fourth-degree polynomial.

$= a_n(x + 2)(x - 2)(x - i)(x + i)$ — Use the given zeros: $c_1 = -2, c_2 = 2, c_3 = i$, and, from above, $c_4 = -i$.

$= a_n(x^2 - 4)(x^2 + 1)$ — Multiply: $(x - i)(x + i) = x^2 - i^2 = x^2 - (-1) = x^2 + 1$.

$f(x) = a_n(x^4 - 3x^2 - 4)$ — Complete the multiplication.

$f(3) = a_n(3^4 - 3 \cdot 3^2 - 4) = -150$ — To find a_n, use the fact that $f(3) = -150$.

$a_n(81 - 27 - 4) = -150$ — Solve for a_n.

$50a_n = -150$ — Simplify: $81 - 27 - 4 = 50$.

$a_n = -3$ — Divide both sides by 50.

Substituting -3 for a_n in the formula for $f(x)$, we obtain

$$f(x) = -3(x^4 - 3x^2 - 4).$$

Equivalently,

$$f(x) = -3x^4 + 9x^2 + 12.$$

Technology

The graph of $f(x) = -3x^4 + 9x^2 + 12$, shown in a $[-3, 3, 1]$ by $[-200, 20, 20]$ viewing rectangle, verifies that -2 and 2 are real zeros. By tracing along the curve, we can check that $f(3) = -150$.

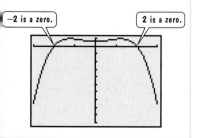

-2 is a zero.   2 is a zero.

⑤ Use Descartes's Rule of Sign.

"An equation can have as many true [positive] roots as it contains changes of sign, from plus to minus or from minus to plus." René Descartes (1596–1650) in *La Géométrie* (1637)

Check Point 6 Find a third-degree polynomial function $f(x)$ with real coefficients that has -3 and i as zeros and such that $f(1) = 8$.

Descartes's Rule of Signs

Because an nth-degree polynomial equation might have roots that are imaginary numbers, we should note that such an equation can have *at most n* real roots. **Descartes's Rule of Signs** provides even more specific information about the number of real zeros that a polynomial can have. The rule is based on considering *variations in sign* between consecutive coefficients. For example, the function $f(x) = 3x^7 - 2x^5 - x^4 + 7x^2 + x - 3$ has three sign changes:

$$f(x) = 3x^7 - 2x^5 - x^4 + 7x^2 + x - 3.$$

sign change   sign change   sign change

> ### Descartes's Rule of Signs
> Let $f(x) = a_n x^n + a_{n-1} x^{n-1} + \cdots + a_2 x^2 + a_1 x + a_0$ be a polynomial with real coefficients.
> **1.** The number of *positive real zeros* of f is either
>   **a.** the same as the number of sign changes of $f(x)$
>   or
>   **b.** less than the number of sign changes of $f(x)$ by a positive even integer.
>   If $f(x)$ has only one variation in sign, then f has exactly one positive real zero.
> **2.** The number of *negative real zeros* of f is either
>   **a.** the same as the number of sign changes of $f(-x)$
>   or
>   **b.** less than the number of sign changes of $f(-x)$ by a positive even integer.
>   If $f(-x)$ has only one variation in sign, then f has exactly one negative real zero.

Study Tip

The number of real zeros given by Descartes's Rule of Signs includes rational zeros from a list of possible rational zeros, as well as irrational zeros not on the list. It does not include any imaginary zeros.

Table 2.1 illustrates what Descartes's Rule of Signs tells us about the positive real zeros of various polynomial functions.

Table 2.1   Descartes's Rule of Signs and Positive Real Zeros

Polynomial Function	Sign Changes	Conclusion
$f(x) = 3x^7 - 2x^5 - x^4 + 7x^2 + x - 3.$ sign change sign change sign change	3	There are 3 positive real zeros. or There is $3 - 2 = 1$ positive real zero.
$f(x) = 4x^5 + 2x^4 - 3x^2 + x + 5$ sign change sign change	2	There are 2 positive real zeros. or There are $2 - 2 = 0$ positive real zeros.
$f(x) = -7x^6 - 5x^4 + x + 9$ sign change	1	There is 1 positive real zero.

EXAMPLE 7   Using Descartes's Rule of Signs

Determine the possible numbers of positive and negative real zeros of $f(x) = x^3 + 2x^2 + 5x + 4$.

Solution

1. To find possibilities for positive real zeros, count the number of sign changes in the equation for $f(x)$. Because all the coefficients are positive, there are no variations in sign. Thus, there are no positive real zeros.

2. To find possibilities for negative real zeros, count the number of sign changes in the equation for $f(-x)$. We obtain this equation by replacing x with $-x$ in the given function.

$$f(x) \; = \; x^3 \; + \; 2x^2 \; + \; 5x + 4$$

Replace x with $-x$.

$$f(-x) = (-x)^3 + 2(-x)^2 + 5(-x) + 4$$
$$= -x^3 + 2x^2 - 5x + 4$$

Now count the sign changes.

$$f(x) = -x^3 + 2x^2 - 5x + 4x$$

sign change   sign change   sign change

There are three variations in sign. The number of negative real zeros of f is either equal to the number of sign changes, 3, or is less than this number by an even integer. This means that either there are 3 negative real zeros or there is $3 - 2 = 1$ negative real zero.

What do the results of Example 7 mean in terms of solving
$$x^3 + 2x^2 + 5x + 4 = 0?$$

Without using Descartes's Rule of Signs, we list the possible rational roots as follows:

Possible rational roots

$$= \frac{\text{Factors of the constant term, 4}}{\text{Factors of the leading coefficient, 1}} = \frac{\pm 1, \pm 2, \pm 4}{\pm 1} = \pm 1, \pm 2, \pm 4$$

However, Descartes's Rule of Signs informed us that $f(x) = x^3 + 2x^2 + 5x + 4$ has no positive real zeros. Thus, the polynomial equation $x^3 + 2x^2 + 5x + 4 = 0$ has no positive real roots. This means that we can eliminate the positive numbers from our list of possible rational roots. Possible rational roots include only -1, -2, and -4. We can use synthetic division and test two of the three possible rational roots of $x^3 + 2x^2 + 5x + 4 = 0$ as follows:

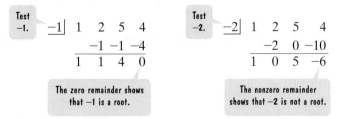

Test -1.

-1	1	2	5	4
		-1	-1	-4
	1	1	4	0

The zero remainder shows that -1 is a root.

Test -2.

-2	1	2	5	4
		-2	0	-10
	1	0	5	-6

The nonzero remainder shows that -2 is not a root.

By solving the equation $x^3 + 2x^2 + 5x + 4 = 0$, you will find that this equation of degree 3 has three roots. One root is -1 and the other two roots are imaginary numbers in a conjugate pair. Verify this by completing the solution process.

Check Point 7   Determine the possible numbers of positive and negative real zeros of $f(x) = x^4 - 14x^3 + 71x^2 - 154x + 120$.

EXERCISE SET 2.5

Practice Exercises

In Exercises 1–8, use the Rational Zero Theorem to list all possible rational zeros for each given function.

1. $f(x) = x^3 + x^2 - 4x - 4$
2. $f(x) = x^3 + 3x^2 - 6x - 8$
3. $f(x) = 3x^4 - 11x^3 - x^2 + 19x + 6$
4. $f(x) = 2x^4 + 3x^3 - 11x^2 - 9x + 15$
5. $f(x) = 4x^4 - x^3 + 5x^2 - 2x - 6$
6. $f(x) = 3x^4 - 11x^3 - 3x^2 - 6x + 8$
7. $f(x) = x^5 - x^4 - 7x^3 + 7x^2 - 12x - 12$
8. $f(x) = 4x^5 - 8x^4 - x + 2$

In Exercises 9–16,
   a. List all possible rational zeros.
   b. Use synthetic division to test the possible rational zeros and find an actual zero.
   c. Use the quotient from part (b) to find the remaining zeros of the polynomial function.

9. $f(x) = x^3 + x^2 - 4x - 4$
10. $f(x) = x^3 - 2x^2 - 11x + 12$
11. $f(x) = 2x^3 - 3x^2 - 11x + 6$
12. $f(x) = 2x^3 - 5x^2 + x + 2$
13. $f(x) = x^3 + 4x^2 - 3x - 6$
14. $f(x) = 2x^3 + x^2 - 3x + 1$
15. $f(x) = 2x^3 + 6x^2 + 5x + 2$
16. $f(x) = x^3 - 4x^2 + 8x - 5$

In Exercises 17–24,
   a. List all possible rational roots.
   b. Use synthetic division to test the possible rational roots and find an actual root.
   c. Use the quotient from part (b) to find the remaining roots and solve the equation.

17. $x^3 - 2x^2 - 11x + 12 = 0$ 18. $x^3 - 2x^2 - 7x - 4 = 0$
19. $x^3 - 10x - 12 = 0$ 20. $x^3 - 5x^2 + 17x - 13 = 0$
21. $6x^3 + 25x^2 - 24x + 5 = 0$
22. $2x^3 - 5x^2 - 6x + 4 = 0$
23. $x^4 - 2x^3 - 5x^2 + 8x + 4 = 0$
24. $x^4 - 2x^2 - 16x - 15 = 0$

In Exercises 25–32, find an nth-degree polynomial function with real coefficients satisfying the given conditions. If you are using a graphing utility, use it to graph the function and verify the real zeros and the given function value.

25. $n = 3$; 1 and $5i$ are zeros; $f(-1) = -104$
26. $n = 3$; 4 and $2i$ are zeros; $f(-1) = -50$
27. $n = 3$; -5 and $4 + 3i$ are zeros; $f(2) = 91$
28. $n = 3$; 6 and $-5 + 2i$ are zeros; $f(2) = -636$
29. $n = 4$; i and $3i$ are zeros; $f(-1) = 20$
30. $n = 4$; $-2, -\frac{1}{2}$, and i are zeros; $f(1) = 18$
31. $n = 4$; $-2, 5$, and $3 + 2i$ are zeros; $f(1) = -96$
32. $n = 4$; $-4, \frac{1}{3}$, and $2 + 3i$ are zeros; $f(1) = 100$

In Exercises 33–38, use Descartes's Rule of Signs to determine the possible number of positive and negative real zeros for each given function.

33. $f(x) = x^3 + 2x^2 + 5x + 4$
34. $f(x) = x^3 + 7x^2 + x + 7$
35. $f(x) = 5x^3 - 3x^2 + 3x - 1$
36. $f(x) = -2x^3 + x^2 - x + 7$
37. $f(x) = 2x^4 - 5x^3 - x^2 - 6x + 4$
38. $f(x) = 4x^4 - x^3 + 5x^2 - 2x - 6$

In Exercises 39–52, find all zeros of the polynomial function or solve the given polynomial equation. Use the Rational Zero Theorem, Descartes's Rule of Signs, and possibly the graph of the polynomial function shown by a graphing utility as an aid in obtaining the first zero or the first root.

39. $f(x) = x^3 - 4x^2 - 7x + 10$
40. $f(x) = x^3 + 12x^2 + 21x + 10$
41. $2x^3 - x^2 - 9x - 4 = 0$
42. $3x^3 - 8x^2 - 8x + 8 = 0$
43. $f(x) = x^4 - 2x^3 + x^2 + 12x + 8$
44. $f(x) = x^4 - 4x^3 - x^2 + 14x + 10$
45. $x^4 - 3x^3 - 20x^2 - 24x - 8 = 0$
46. $x^4 - x^3 + 2x^2 - 4x - 8 = 0$
47. $f(x) = 3x^4 - 11x^3 - x^2 + 19x + 6$
48. $f(x) = 2x^4 + 3x^3 - 11x^2 - 9x + 15$
49. $4x^4 - x^3 + 5x^2 - 2x - 6 = 0$
50. $3x^4 - 11x^3 - 3x^2 - 6x + 8 = 0$
51. $2x^5 + 7x^4 - 18x^2 - 8x + 8 = 0$
52. $4x^5 + 12x^4 - 41x^3 - 99x^2 + 10x + 24 = 0$

Practice Plus

Exercises 53–60, show incomplete graphs of given polynomial functions.
   a. Find all the zeros of each function.
   b. Without using a graphing utility, draw a complete graph of the function.

53. $f(x) = -x^3 + x^2 + 16x - 16$

[−5, 0, 1] by [−40, 25, 5]

54. $f(x) = -x^3 + 3x^2 - 4$

[−2, 0, 1] by [−10, 10, 1]

55. $f(x) = 4x^3 - 8x^2 - 3x + 9$

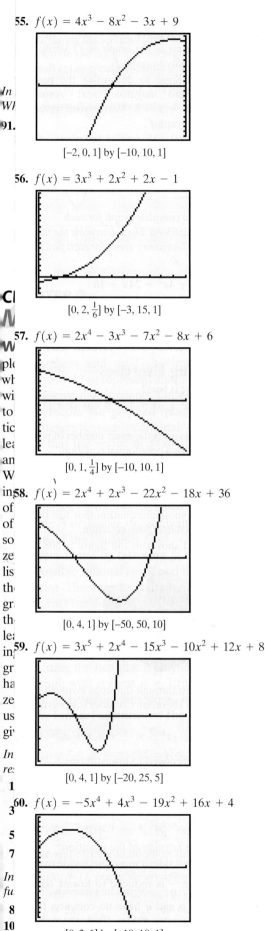

[−2, 0, 1] by [−10, 10, 1]

56. $f(x) = 3x^3 + 2x^2 + 2x - 1$

$[0, 2, \frac{1}{6}]$ by [−3, 15, 1]

57. $f(x) = 2x^4 - 3x^3 - 7x^2 - 8x + 6$

$[0, 1, \frac{1}{4}]$ by [−10, 10, 1]

58. $f(x) = 2x^4 + 2x^3 - 22x^2 - 18x + 36$

[0, 4, 1] by [−50, 50, 10]

59. $f(x) = 3x^5 + 2x^4 - 15x^3 - 10x^2 + 12x + 8$

[0, 4, 1] by [−20, 25, 5]

60. $f(x) = -5x^4 + 4x^3 - 19x^2 + 16x + 4$

[0, 2, 1] by [−10, 10, 1]

Application Exercises

The graphs are based on a study of the percentage of professional works completed in each age decade of life by 738 people who lived to be at least 79. Use the graphs to solve Exercises 61–62.

Age Trends in Professional Productivity

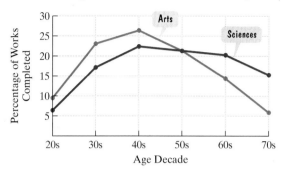

Source: Dennis, W. (1966), Creative productivity between the ages of 20 and 80 years. *Journal of Gerontology*, 21, 1–8.

61. Suppose that a polynomial function f is used to model the data shown in the graph for the arts using

(age decade, percentage of works completed).

a. Use the graph to solve the polynomial equation $f(x) = 27$. Describe what this means in terms of an age decade and productivity.

b. Describe the degree and the leading coefficient of a function f that can be used to model the data in the graph.

62. Suppose that a polynomial function g is used to model the data shown in the graph for the sciences using

(age decade, percentage of works completed).

a. Use the graph to solve the polynomial equation $g(x) = 20$. Find only the meaningful value of x and then describe what this means in terms of an age decade and productivity.

b. Describe the degree and the leading coefficient of a function g that can be used to model the data in the graph.

The polynomial function

$$H(x) = -0.001618x^4 + 0.077326x^3 - 1.2367x^2 + 11.460x + 2.914$$

models the age in human years, $H(x)$, of a dog that is x years old, where $x \geq 1$. Although the coefficients make it difficult to solve equations algebraically using this function, a graph of the function makes approximate solutions possible.

Use the graph shown to solve Exercises 63–64. Round all answers to the nearest year.

Dog's Age in Human Years

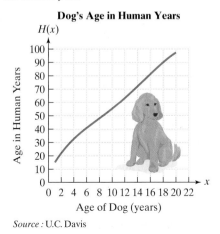

Source: U.C. Davis

As $x \to 0^+$, $f(x) \to \infty$.

> **As x approaches 0 from the right, $f(x)$ approaches infinity (that is, the graph rises).**

Observe that the plus ($+$) superscript on the 0 ($x \to 0^+$) is read "from the right."

Now let's see what happens to the function values, $f(x)$, as x gets farther awa from the origin. The following tables suggest what happens to $f(x)$ as x increases o decreases without bound.

x increases without bound:

x	1	10	100	1000
$f(x) = \dfrac{1}{x}$	1	0.1	0.01	0.001

x decreases without bound:

x	-1	-10	-100	-1000
$f(x) = \dfrac{1}{x}$	-1	-0.1	-0.01	-0.001

It appears that as x increases or decreases without bound, the function values, $f(x)$ are getting progressively closer to 0.

Figure 2.26 illustrates the end behavior of $f(x) = \dfrac{1}{x}$ as x increases or decrease without bound. The graph shows that the function values, $f(x)$, are approaching (This means that as x increases or decreases without bound, the graph of f x approaching the horizontal line $y = 0$ (that is, the x-axis). We use arrow notation t describe this situation:

As $x \to \infty$, $f(x) \to 0$ and as $x \to -\infty$, $f(x) \to 0$.

> **As x approaches infinity (that is, increases without bound), $f(x)$ approaches 0.**

> **As x approaches negative infinity (that is, decreases without bound), $f(x)$ approaches 0.**

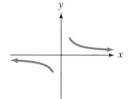

Figure 2.26 $f(x)$ approaches 0 as x increases or decreases without bound.

Thus, as x approaches infinity ($x \to \infty$) or as x approaches negative infinity ($x \to -\infty$), the function values are approaching zero: $f(x) \to 0$.

The graph of the reciprocal function $f(x) = \dfrac{1}{x}$ is shown in Figure 2.27. Unlike the graph of a polynomial function, the graph of the reciprocal function has a break and is composed of two distinct branches.

The arrow notation used throughout our discussion of the reciprocal function is summarized in the following box:

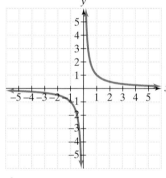

Figure 2.27 The graph of the reciprocal function $f(x) = \dfrac{1}{x}$

Arrow Notation

Symbol	Meaning
$x \to a^+$	x approaches a from the right.
$x \to a^-$	x approaches a from the left.
$x \to \infty$	x approaches infinity; that is, x increases without bound.
$x \to -\infty$	x approaches negative infinity; that is, x decreases without bound.

In calculus, you will use **limits** to convey ideas involving a function's end behavior or its possible asymptotic behavior. For example, examine the graph o $f(x) = \dfrac{1}{x}$ in Figure 2.27 and its end behavior to the right. As $x \to \infty$, the values o $f(x)$ approach 0: $f(x) \to 0$. In calculus, this is symbolized by

$$\lim_{x \to \infty} f(x) = 0.$$

This is read "the limit of $f(x)$ as x approaches infinity equals zero."

Another basic rational function is $f(x) = \dfrac{1}{x^2}$. The graph of this even function, with y-axis symmetry and positive function values, is shown in Figure 2.28. Like the reciprocal function, the graph has a break and is composed of two distinct branches.

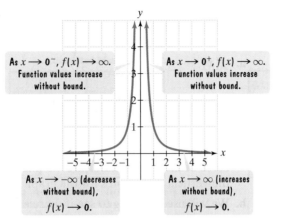

As $x \to 0^-$, $f(x) \to \infty$.
Function values increase
without bound.

As $x \to 0^+$, $f(x) \to \infty$.
Function values increase
without bound.

As $x \to -\infty$ (decreases
without bound),
$f(x) \to 0$.

As $x \to \infty$ (increases
without bound),
$f(x) \to 0$.

Figure 2.28 The graph of $f(x) = \dfrac{1}{x^2}$

 Identify vertical asymptotes.

Vertical Asymptotes of Rational Functions

Look again at the graph of $f(x) = \dfrac{1}{x^2}$ in Figure 2.28. The curve approaches, but does not touch, the y-axis. The y-axis, or $x = 0$, is said to be a *vertical asymptote* of the graph. A rational function may have no vertical asymptotes, one vertical asymptote, or several vertical asymptotes. The graph of a rational function never intersects a vertical asymptote. We will use dashed lines to show asymptotes.

Definition of a Vertical Asymptote

The line $x = a$ is a **vertical asymptote** of the graph of a function f if $f(x)$ increases or decreases without bound as x approaches a.

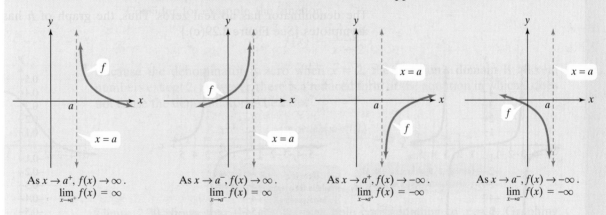

As $x \to a^+$, $f(x) \to \infty$.
$\lim\limits_{x \to a^+} f(x) = \infty$

As $x \to a^-$, $f(x) \to \infty$.
$\lim\limits_{x \to a^-} f(x) = \infty$

As $x \to a^+$, $f(x) \to -\infty$.
$\lim\limits_{x \to a^+} f(x) = -\infty$

As $x \to a^-$, $f(x) \to -\infty$.
$\lim\limits_{x \to a^-} f(x) = -\infty$

Thus, as x approaches a from either the left or the right, $f(x) \to \infty$ or $f(x) \to -\infty$.

If the graph of a rational function has vertical asymptotes, they can be located using the following theorem:

Locating Vertical Asymptotes

If $f(x) = \dfrac{p(x)}{q(x)}$ is a rational function in which $p(x)$ and $q(x)$ have no common factors and a is a zero of $q(x)$, the denominator, then $x = a$ is a vertical asymptote of the graph of f.

EXAMPLE 4   Using Transformations to Graph a Rational Function

Use the graph of $f(x) = \dfrac{1}{x^2}$ to graph $g(x) = \dfrac{1}{(x-2)^2} + 1$.

Solution

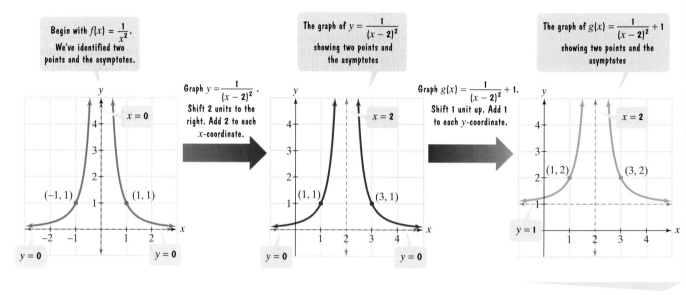

Begin with $f(x) = \dfrac{1}{x^2}$.
We've identified two points and the asymptotes.

Graph $y = \dfrac{1}{(x-2)^2}$.
Shift 2 units to the right. Add 2 to each x-coordinate.

The graph of $y = \dfrac{1}{(x-2)^2}$ showing two points and the asymptotes

Graph $g(x) = \dfrac{1}{(x-2)^2} + 1$.
Shift 1 unit up. Add 1 to each y-coordinate.

The graph of $g(x) = \dfrac{1}{(x-2)^2} + 1$ showing two points and the asymptotes

Check Point 4 Use the graph of $f(x) = \dfrac{1}{x}$ to graph $g(x) = \dfrac{1}{x+2} - 1$.

 Graph rational functions.

Graphing Rational Functions

Rational functions that are not transformations of $f(x) = \dfrac{1}{x}$ or $f(x) = \dfrac{1}{x^2}$ can be graphed using the following suggestions:

Strategy for Graphing a Rational Function

The following strategy can be used to graph

$$f(x) = \frac{p(x)}{q(x)},$$

where p and q are polynomial functions with no common factors.

1. Determine whether the graph of f has symmetry.

$$f(-x) = f(x): y\text{-axis symmetry}$$
$$f(-x) = -f(x): \text{origin symmetry}$$

2. Find the y-intercept (if there is one) by evaluating $f(0)$.

3. Find the x-intercepts (if there are any) by solving the equation $p(x) = 0$.

4. Find any vertical asymptote(s) by solving the equation $q(x) = 0$.

5. Find the horizontal asymptote (if there is one) using the rule for determining the horizontal asymptote of a rational function.

6. Plot at least one point between and beyond each x-intercept and vertical asymptote.

7. Use the information obtained previously to graph the function between and beyond the vertical asymptotes.

EXAMPLE 5   Graphing a Rational Function

Graph: $f(x) = \dfrac{2x}{x-1}$.

Solution

Step 1   Determine symmetry.

$$f(-x) = \frac{2(-x)}{-x-1} = \frac{-2x}{-x-1} = \frac{2x}{x+1}$$

Because $f(-x)$ does not equal $f(x)$ or $-f(x)$, the graph has neither y-axis nor origin symmetry.

Step 2   Find the y-intercept. Evaluate $f(0)$.

$$f(0) = \frac{2 \cdot 0}{0-1} = \frac{0}{-1} = 0$$

The y-intercept is 0, so the graph passes through the origin.

Step 3   Find x-intercept(s). This is done by solving $p(x) = 0$.

$$2x = 0 \qquad \text{\small Set the numerator equal to 0.}$$

$$x = 0 \qquad \text{\small Divide both sides by 2.}$$

There is only one x-intercept. This verifies that the graph passes through the origin.

Step 4   Find the vertical asymptote(s). Solve $q(x) = 0$, thereby finding zeros of the denominator.

$$x - 1 = 0 \qquad \text{\small Set the denominator equal to 0.}$$

$$x = 1 \qquad \text{\small Add 1 to both sides.}$$

The equation of the vertical asymptote is $x = 1$.

Step 5   Find the horizontal asymptote. Because the numerator and denominator of $f(x) = \dfrac{2x}{x-1}$ have the same degree, 1, the leading coefficients of the numerator and denominator, 2 and 1, respectively, are used to obtain the equation of the horizontal asymptote. The equation is

$$y = \frac{2}{1} = 2.$$

The equation of the horizontal asymptote is $y = 2$.

Step 6   Plot points between and beyond each x-intercept and vertical asymptote. With an x-intercept at 0 and a vertical asymptote at $x = 1$, we evaluate the function at $-2, -1, \frac{1}{2}, 2,$ and 4.

x	-2	-1	$\dfrac{1}{2}$	2	4
$f(x) = \dfrac{2x}{x-1}$	$\dfrac{4}{3}$	1	-2	4	$\dfrac{8}{3}$

Figure 2.32 shows these points, the y-intercept, the x-intercept, and the asymptotes.

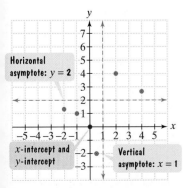

Figure 2.32 Preparing to graph the rational function $f(x) = \dfrac{2x}{x-1}$

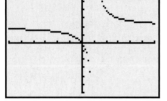

Step 7 Graph the function. The graph of $f(x) = \dfrac{2x}{x - 1}$ is shown in Figure 2.33

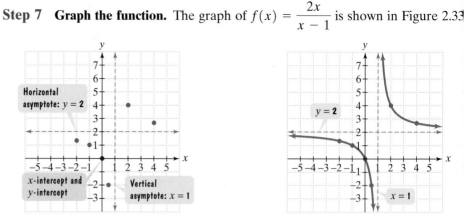

Figure 2.32 (repeated) Preparing to graph the rational function $f(x) = \dfrac{2x}{x - 1}$

Figure 2.33 The graph of $f(x) = \dfrac{2x}{x - 1}$

Check Point 5 Graph: $f(x) = \dfrac{3x}{x - 2}$.

EXAMPLE 6 Graphing a Rational Function

Graph: $f(x) = \dfrac{3x^2}{x^2 - 4}$.

Solution

Step 1 Determine symmetry. $f(-x) = \dfrac{3(-x)^2}{(-x)^2 - 4} = \dfrac{3x^2}{x^2 - 4} = f(x)$: The grap of f is symmetric with respect to the y-axis.

Step 2 Find the y-intercept. $f(0) = \dfrac{3 \cdot 0^2}{0^2 - 4} = \dfrac{0}{-4} = 0$: The y-intercept is 0, s the graph passes through the origin.

Step 3 Find the x-intercept(s). $3x^2 = 0$, so $x = 0$: The x-intercept is 0, verifying that the graph passes through the origin.

Step 4 Find the vertical asymptote(s). Set $q(x) = 0$.

$$x^2 - 4 = 0 \qquad \text{Set the denominator equal to 0.}$$
$$x^2 = 4 \qquad \text{Add 4 to both sides.}$$
$$x = \pm 2 \qquad \text{Use the square root property.}$$

The vertical asymptotes are $x = -2$ and $x = 2$.

Step 5 Find the horizontal asymptote. Because the numerator and denominato of $f(x) = \dfrac{3x^2}{x^2 - 4}$ have the same degree, 2, their leading coefficients, 3 and 1, are used to determine the equation of the horizontal asymptote. The equation i $y = \dfrac{3}{1} = 3$.

Step 6 Plot points between and beyond each x-intercept and vertical asymptote With an x-intercept at 0 and vertical asymptotes at $x = -2$ and $x = 2$, we evaluate the function at -3, -1, 1, 3, and 4.

x	-3	-1	1	3	4
$f(x) = \dfrac{3x^2}{x^2 - 4}$	$\dfrac{27}{5}$	-1	-1	$\dfrac{27}{5}$	4

Figure 2.34 at the top of the next page shows these points, the y-intercept, the x-intercept, and the asymptotes.

Step 7   Graph the function. The graph of $f(x) = \dfrac{3x^2}{x^2 - 4}$ is shown in Figure 2.35. The y-axis symmetry is now obvious.

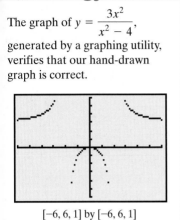
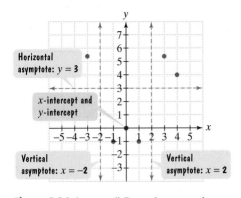

Figure 2.34 (repeated) Preparing to graph $f(x) = \dfrac{3x^2}{x^2 - 4}$

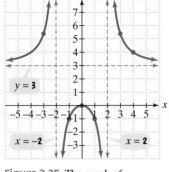

Figure 2.35 The graph of $f(x) = \dfrac{3x^2}{x^2 - 4}$

Check Point 6   Graph: $f(x) = \dfrac{2x^2}{x^2 - 9}$.

Example 7 illustrates that not every rational function has vertical and horizontal asymptotes.

EXAMPLE 7   Graphing a Rational Function

Graph: $f(x) = \dfrac{x^4}{x^2 + 1}$.

Solution

Step 1   Determine symmetry. $f(-x) = \dfrac{(-x)^4}{(-x)^2 + 1} = \dfrac{x^4}{x^2 + 1} = f(x)$: The graph of f is symmetric with respect to the y-axis.

Step 2   Find the y-intercept. $f(0) = \dfrac{0^4}{0^2 + 1} = \dfrac{0}{1} = 0$: The y-intercept is 0.

Step 3   Find the x-intercept(s). $x^4 = 0$, so $x = 0$: The x-intercept is 0.

Step 4   Find the vertical asymptote. Set $q(x) = 0$.

$$x^2 + 1 = 0 \qquad \text{Set the denominator equal to 0.}$$
$$x^2 = -1 \qquad \text{Subtract 1 from both sides.}$$

Although this equation has imaginary roots ($x = \pm i$), there are no real roots. Thus, the graph of f has no vertical asymptotes.

Step 5   Find the horizontal asymptote. Because the degree of the numerator, 4, is greater than the degree of the denominator, 2, there is no horizontal asymptote.

Step 6   Plot points between and beyond each x-intercept and vertical asymptote. With an x-intercept at 0 and no vertical asymptotes, let's look at function values at

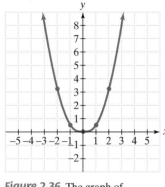

Figure 2.36 The graph of
$$f(x) = \frac{x^4}{x^2 + 1}$$

$-2, -1, 1,$ and $2.$ You can evaluate the function at 1 and 2. Use y-axis symmetry to obtain function values at -1 and -2:

$$f(-1) = f(1) \text{ and } f(-2) = f(2).$$

x	-2	-1	1	2
$f(x) = \dfrac{x^4}{x^2 + 1}$	$\dfrac{16}{5}$	$\dfrac{1}{2}$	$\dfrac{1}{2}$	$\dfrac{16}{5}$

Step 7 Graph the function. Figure 2.36 shows the graph of f using the points obtained from the table and y-axis symmetry. Notice that as x approaches infinity or negative infinity ($x \to \infty$ or $x \to -\infty$), the function values, $f(x)$, are getting larger without bound [$f(x) \to \infty$].

Check Point 7 Graph: $f(x) = \dfrac{x^4}{x^2 + 2}$.

 Identify slant asymptotes.

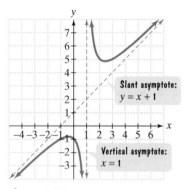

Figure 2.37 The graph of
$$f(x) = \frac{x^2 + 1}{x - 1} \text{ with a slant asymptote}$$

Slant Asymptotes

Examine the graph of

$$f(x) = \frac{x^2 + 1}{x - 1},$$

shown in Figure 2.37. Note that the degree of the numerator, 2, is greater than the degree of the denominator, 1. Thus, the graph of this function has no horizontal asymptote. However, the graph has a **slant asymptote**, $y = x + 1$.

The graph of a rational function has a slant asymptote if the degree of the numerator is one more than the degree of the denominator. The equation of the slant asymptote can be found by division. For example, to find the slant asymptote for the graph of $f(x) = \dfrac{x^2 + 1}{x - 1}$, divide $x - 1$ into $x^2 + 1$:

$$\begin{array}{c|ccc} 1 & 1 & 0 & 1 \\ & & 1 & 1 \\ \hline & 1 & 1 & 2 \end{array}$$

$$\begin{array}{r} 1x + 1 + \dfrac{2}{x-1} \\ x - 1 \overline{)\, x^2 + 0x + 1} \end{array}$$

Remainder

Observe that

$$f(x) = \frac{x^2 + 1}{x - 1} = \underbrace{x + 1} + \frac{2}{x - 1}.$$

The equation of the slant asymptote is $y = x + 1.$

As $|x| \to \infty$, the value of $\dfrac{2}{x - 1}$ is approximately 0. Thus, when $|x|$ is large, the function is very close to $y = x + 1 + 0$. This means that as $x \to \infty$ or as $x \to -\infty$, the graph of f gets closer and closer to the line whose equation is $y = x + 1$. The line $y = x + 1$ is a slant asymptote of the graph.

In general, if $f(x) = \dfrac{p(x)}{q(x)}, p$ and q have no common factors, and the degree of p is one greater than the degree of q, find the slant asymptote by dividing $q(x)$ into $p(x)$. The division will take the form

$$\frac{p(x)}{q(x)} = mx + b + \frac{\text{remainder}}{q(x)}.$$

Slant asymptote:
$y = mx + b$

The equation of the slant asymptote is obtained by dropping the term with the remainder. Thus, the equation of the slant asymptote is $y = mx + b$.

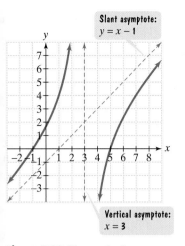

Slant asymptote:
$y = x - 1$

Vertical asymptote:
$x = 3$

Figure 2.38 The graph of
$f(x) = \dfrac{x^2 - 4x - 5}{x - 3}$

EXAMPLE 8   Finding the Slant Asymptote of a Rational Function

Find the slant asymptote of $f(x) = \dfrac{x^2 - 4x - 5}{x - 3}$.

Solution   Because the degree of the numerator, 2, is exactly one more than the degree of the denominator, 1, and $x - 3$ is not a factor of $x^2 - 4x - 5$, the graph of f has a slant asymptote. To find the equation of the slant asymptote, divide $x - 3$ into $x^2 - 4x - 5$:

$$\begin{array}{r|rrr} 3 & 1 & -4 & -5 \\ & & 3 & -3 \\ \hline & 1 & -1 & -8 \end{array}$$

Remainder

$$x - 3 \overline{\smash{\big)}\,x^2 - 4x - 5} \qquad 1x - 1 - \dfrac{8}{x-3}$$

Drop the remainder term and you'll have the equation of the slant asymptote.

The equation of the slant asymptote is $y = x - 1$. Using our strategy for graphing rational functions, the graph of $f(x) = \dfrac{x^2 - 4x - 5}{x - 3}$ is shown in Figure 2.38.

Check Point 8   Find the slant asymptote of $f(x) = \dfrac{2x^2 - 5x + 7}{x - 2}$.

 Solve applied problems involving rational functions.

Applications

There are numerous examples of asymptotic behavior in functions that describe real-world phenomena. Let's consider an example from the business world. The **cost function**, C, for a business is the sum of its fixed and variable costs:

$$C(x) = (\text{fixed cost}) + cx.$$

Cost per unit times the number of units produced, x.

The **average cost** per unit for a company to produce x units is the sum of its fixed and variable costs divided by the number of units produced. The **average cost function** is a rational function that is denoted by $\overline{C}$. Thus,

Cost of producing x units: fixed plus variable costs

$$\overline{C}(x) = \dfrac{(\text{fixed cost}) + cx}{x}.$$

Number of units produced

EXAMPLE 9   Average Cost of Producing a Wheelchair

A company is planning to manufacture wheelchairs that are light, fast, and beautiful. The fixed monthly cost will be $500,000 and it will cost $400 to produce each radically innovative chair.

   a. Write the cost function, C, of producing x wheelchairs.

   b. Write the average cost function, $\overline{C}$, of producing x wheelchairs.

   c. Find and interpret $\overline{C}(1000)$, $\overline{C}(10{,}000)$, and $\overline{C}(100{,}000)$.

   d. What is the horizontal asymptote for the graph of the average cost function, $\overline{C}$? Describe what this represents for the company.

Solution

a. The cost function of producing x wheelchairs, C, is the sum of the fixed cost and the variable costs.

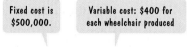

Fixed cost is $500,000.

Variable cost: $400 for each wheelchair produced

$$C(x) = 500,000 + 400x$$

b. The average cost function of producing x wheelchairs, $\overline{C}$, is the sum of the fixed and variable costs divided by the number of wheelchairs produced.

$$\overline{C}(x) = \frac{500,000 + 400x}{x} \quad \text{or} \quad \overline{C}(x) = \frac{400x + 500,000}{x}$$

c. We evaluate $\overline{C}$ at 1000, 10,000, and 100,000, interpreting the results.

$$\overline{C}(1000) = \frac{400(1000) + 500,000}{1000} = 900$$

The average cost per wheelchair of producing 1000 wheelchairs per month is $900.

$$\overline{C}(10,000) = \frac{400(10,000) + 500,000}{10,000} = 450$$

The average cost per wheelchair of producing 10,000 wheelchairs per month is $450.

$$\overline{C}(100,000) = \frac{400(100,000) + 500,000}{100,000} = 405$$

The average cost per wheelchair of producing 100,000 wheelchairs per month is $405. Notice that with higher production levels, the cost of producing each wheelchair decreases.

d. We developed the average cost function

$$\overline{C}(x) = \frac{400x + 500,000}{x}$$

in which the degree of the numerator, 1, is equal to the degree of the denominator, 1. The leading coefficients of the numerator and denominator, 400 and 1, are used to obtain the equation of the horizontal asymptote. The equation of the horizontal asymptote is

$$y = \frac{400}{1} \quad \text{or} \quad y = 400.$$

The horizontal asymptote is shown in Figure 2.39. This means that the more wheelchairs produced per month, the closer the average cost per wheelchair for the company comes to $400. The least possible cost per wheelchair is approaching $400. Competitively low prices take place with high production levels, posing a major problem for small businesses.

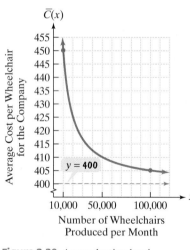

Figure 2.39 As production level increases, the average cost per wheelchair approaches $400:
$$\lim_{x \to \infty} \overline{C}(x) = 400.$$

Check Point 9 The time: the not-too-distant future. A new company is hoping to replace traditional computers and two-dimensional monitors with its virtual reality system. The fixed monthly cost will be $600,000 and it will cost $500 to produce each system.

a. Write the cost function, C, of producing x virtual reality systems.

b. Write the average cost function, $\overline{C}$, of producing x virtual reality systems.

c. Find and interpret $\overline{C}(1000)$, $\overline{C}(10,000)$, and $\overline{C}(100,000)$.

d. What is the horizontal asymptote for the graph of the average cost function, $\overline{C}$? Describe what this represents for the company.

If an object moves at an average velocity v, the distance, s, covered in time t is given by the formula

$$s = vt.$$

Thus, *distance = velocity · time*. Objects that move in accordance with this formula are said to be in **uniform motion**. In Example 10, we use a rational function to model time, t, in uniform motion. Solving the uniform motion formula for t, we obtain

$$t = \frac{s}{v}.$$

Thus, time is the quotient of distance and average velocity.

EXAMPLE 10   Time Involved in Uniform Motion

Two commuters drove to work a distance of 40 miles and then returned again on the same route. The average velocity on the return trip was 30 miles per hour faster than the average velocity on the outgoing trip. Express the total time required to complete the round trip, T, as a function of the average velocity on the outgoing trip, x.

Solution   As specified, the average velocity on the outgoing trip is represented by x. Because the average velocity on the return trip was 30 miles per hour faster than the average velocity on the outgoing trip, let

$$x + 30 = \text{the average velocity on the return trip.}$$

The sentence that we use as a verbal model to write our rational function is

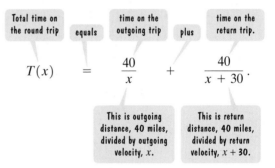

The function that expresses the total time required to complete the round trip is

$$T(x) = \frac{40}{x} + \frac{40}{x + 30}.$$

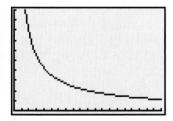

Figure 2.40 The graph of $T(x) = \frac{40}{x} + \frac{40}{x + 30}$. As average velocity increases, time for the trip decreases: $\lim\limits_{x \to \infty} T(x) = 0$.

Once you have modeled a problem's conditions with a function, you can use a graphing utility to explore the function's behavior. For example let's graph the function in Example 10. Because it seems unlikely that an average outgoing velocity exceeds 60 miles per hour with an average return velocity that is 30 miles per hour faster, we graph the function for $0 \le x \le 60$. Figure 2.40 shows the graph of $T(x) = \frac{40}{x} + \frac{40}{x + 30}$ in a $[0, 60, 3]$ by $[0, 10, 1]$ viewing rectangle. Notice that the function is decreasing. This shows decreasing times with increasing average velocities. Can you see that the vertical asymptote is $x = 0$, or the y-axis? This indicates that close to an outgoing average velocity of zero miles per hour, the round trip will take nearly forever: $\lim\limits_{x \to 0} T(x) = \infty$.

Check Point 10   Two commuters drove to work a distance of 20 miles and then returned again on the same route. The average velocity on the return trip was 10 miles per hour slower than the average velocity on the outgoing trip. Express the total time required to complete the round trip, T, as a function of the average velocity on the outgoing trip, x.

EXERCISE SET 2.6

Practice Exercises

In Exercises 1–8, find the domain of each rational function.

1. $f(x) = \dfrac{5x}{x - 4}$

2. $f(x) = \dfrac{7x}{x - 8}$

3. $g(x) = \dfrac{3x^2}{(x - 5)(x + 4)}$

4. $g(x) = \dfrac{2x^2}{(x - 2)(x + 6)}$

5. $h(x) = \dfrac{x + 7}{x^2 - 49}$

6. $h(x) = \dfrac{x + 8}{x^2 - 64}$

7. $f(x) = \dfrac{x + 7}{x^2 + 49}$

8. $f(x) = \dfrac{x + 8}{x^2 + 64}$

Use the graph of the rational function in the figure shown to complete each statement in Exercises 9–14.

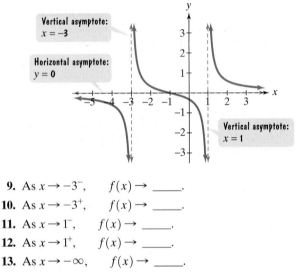

9. As $x \to -3^-$, $\quad f(x) \to$ _____.

10. As $x \to -3^+$, $\quad f(x) \to$ _____.

11. As $x \to 1^-$, $\quad f(x) \to$ _____.

12. As $x \to 1^+$, $\quad f(x) \to$ _____.

13. As $x \to -\infty$, $\quad f(x) \to$ _____.

14. As $x \to \infty$, $\quad f(x) \to$ _____.

Use the graph of the rational function in the figure shown to complete each statement in Exercises 15–20.

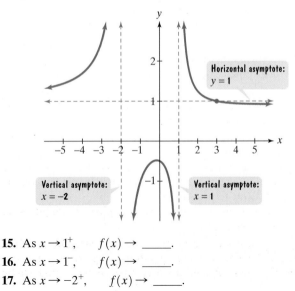

15. As $x \to 1^+$, $\quad f(x) \to$ _____.

16. As $x \to 1^-$, $\quad f(x) \to$ _____.

17. As $x \to -2^+$, $\quad f(x) \to$ _____.

18. As $x \to -2^-$, $\quad f(x) \to$ _____.

19. As $x \to \infty$, $\quad f(x) \to$ _____.

20. As $x \to -\infty$, $\quad f(x) \to$ _____.

In Exercises 21–28, find the vertical asymptotes, if any, of the graph of each rational function.

21. $f(x) = \dfrac{x}{x + 4}$

22. $f(x) = \dfrac{x}{x - 3}$

23. $g(x) = \dfrac{x + 3}{x(x + 4)}$

24. $g(x) = \dfrac{x + 3}{x(x - 3)}$

25. $h(x) = \dfrac{x}{x(x + 4)}$

26. $h(x) = \dfrac{x}{x(x - 3)}$

27. $r(x) = \dfrac{x}{x^2 + 4}$

28. $r(x) = \dfrac{x}{x^2 + 3}$

In Exercises 29–36, find the horizontal asymptote, if any, of the graph of each rational function.

29. $f(x) = \dfrac{12x}{3x^2 + 1}$

30. $f(x) = \dfrac{15x}{3x^2 + 1}$

31. $g(x) = \dfrac{12x^2}{3x^2 + 1}$

32. $g(x) = \dfrac{15x^2}{3x^2 + 1}$

33. $h(x) = \dfrac{12x^3}{3x^2 + 1}$

34. $h(x) = \dfrac{15x^3}{3x^2 + 1}$

35. $f(x) = \dfrac{-2x + 1}{3x + 5}$

36. $f(x) = \dfrac{-3x + 7}{5x - 2}$

In Exercises 37–48, use transformations of $f(x) = \dfrac{1}{x}$ or $f(x) = \dfrac{1}{x^2}$ to graph each rational function.

37. $g(x) = \dfrac{1}{x - 1}$

38. $g(x) = \dfrac{1}{x - 2}$

39. $h(x) = \dfrac{1}{x} + 2$

40. $h(x) = \dfrac{1}{x} + 1$

41. $g(x) = \dfrac{1}{x + 1} - 2$

42. $g(x) = \dfrac{1}{x + 2} - 2$

43. $g(x) = \dfrac{1}{(x + 2)^2}$

44. $g(x) = \dfrac{1}{(x + 1)^2}$

45. $h(x) = \dfrac{1}{x^2} - 4$

46. $h(x) = \dfrac{1}{x^2} - 3$

47. $h(x) = \dfrac{1}{(x - 3)^2} + 1$

48. $h(x) = \dfrac{1}{(x - 3)^2} + 2$

In Exercises 49–70, follow the seven steps on page 334 to graph each rational function.

49. $f(x) = \dfrac{4x}{x - 2}$

50. $f(x) = \dfrac{3x}{x - 1}$

51. $f(x) = \dfrac{2x}{x^2 - 4}$

52. $f(x) = \dfrac{4x}{x^2 - 1}$

53. $f(x) = \dfrac{2x^2}{x^2 - 1}$

54. $f(x) = \dfrac{4x^2}{x^2 - 9}$

5. $f(x) = \dfrac{-x}{x+1}$

56. $f(x) = \dfrac{-3x}{x+2}$

7. $f(x) = -\dfrac{1}{x^2-4}$

58. $f(x) = -\dfrac{2}{x^2-1}$

9. $f(x) = \dfrac{2}{x^2+x-2}$

60. $f(x) = \dfrac{-2}{x^2-x-2}$

1. $f(x) = \dfrac{2x^2}{x^2+4}$

62. $f(x) = \dfrac{4x^2}{x^2+1}$

3. $f(x) = \dfrac{x+2}{x^2+x-6}$

64. $f(x) = \dfrac{x-4}{x^2-x-6}$

5. $f(x) = \dfrac{x^4}{x^2+2}$

66. $f(x) = \dfrac{2x^4}{x^2+1}$

7. $f(x) = \dfrac{x^2+x-12}{x^2-4}$

68. $f(x) = \dfrac{x^2}{x^2+x-6}$

9. $f(x) = \dfrac{3x^2+x-4}{2x^2-5x}$

70. $f(x) = \dfrac{x^2-4x+3}{(x+1)^2}$

n Exercises 71–78, **a.** *Find the slant asymptote of the graph of ach rational function and* **b.** *Follow the seven-step strategy and se the slant asymptote to graph each rational function.*

1. $f(x) = \dfrac{x^2-1}{x}$

72. $f(x) = \dfrac{x^2-4}{x}$

3. $f(x) = \dfrac{x^2+1}{x}$

74. $f(x) = \dfrac{x^2+4}{x}$

5. $f(x) = \dfrac{x^2+x-6}{x-3}$

76. $f(x) = \dfrac{x^2-x+1}{x-1}$

7. $f(x) = \dfrac{x^3+1}{x^2+2x}$

78. $f(x) = \dfrac{x^3-1}{x^2-9}$

Practice Plus

In Exercises 79–84, the equation for f is given by the mplified expression that results after performing the indicated peration. Write the equation for f and then graph the function.

9. $\dfrac{5x^2}{x^2-4} \cdot \dfrac{x^2+4x+4}{10x^3}$

80. $\dfrac{x-5}{10x-2} \div \dfrac{x^2-10x+25}{25x^2-1}$

1. $\dfrac{x}{2x+6} - \dfrac{9}{x^2-9}$

82. $\dfrac{2}{x^2+3x+2} - \dfrac{4}{x^2+4x+3}$

3. $\dfrac{1 - \dfrac{3}{x+2}}{1 + \dfrac{1}{x-2}}$

84. $\dfrac{x - \dfrac{1}{x}}{x + \dfrac{1}{x}}$

Exercises 85–88, use long division to rewrite the equation or g in the form quotient, plus remainder divided by divisor. Then se this form of the function's equation and transformations of $f(x) = \dfrac{1}{x}$ *to graph g.*

5. $g(x) = \dfrac{2x+7}{x+3}$

86. $g(x) = \dfrac{3x+7}{x+2}$

7. $g(x) = \dfrac{3x-7}{x-2}$

88. $g(x) = \dfrac{2x-9}{x-4}$

Application Exercises

89. A company is planning to manufacture mountain bikes. The fixed monthly cost will be $100,000 and it will cost $100 to produce each bicycle.
  a. Write the cost function, C, of producing x mountain bikes.
  b. Write the average cost function, $\overline{C}$, of producing x mountain bikes.
  c. Find and interpret $\overline{C}(500), \overline{C}(1000), \overline{C}(2000),$ and $\overline{C}(4000)$.
  d. What is the horizontal asymptote for the graph of the average cost function, $\overline{C}$? Describe what this means in practical terms.

90. A company that manufactures running shoes has a fixed monthly cost of $300,000. It costs $30 to produce each pair of shoes.
  a. Write the cost function, C, of producing x pairs of shoes.
  b. Write the average cost function, $\overline{C}$, of producing x pairs of shoes.
  c. Find and interpret $\overline{C}(1000), \overline{C}(10,000),$ and $\overline{C}(100,000)$.
  d. What is the horizontal asymptote for the graph of the average cost function, $\overline{C}$? Describe what this represents for the company.

91. The function
$$f(x) = \dfrac{6.5x^2 - 20.4x + 234}{x^2 + 36}$$
models the pH level, $f(x)$, of the human mouth x minutes after a person eats food containing sugar. The graph of this function is shown in the figure.

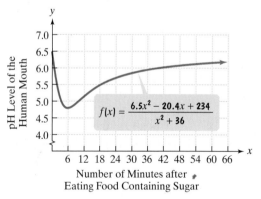

a. Use the graph to obtain a reasonable estimate, to the nearest tenth, of the pH level of the human mouth 42 minutes after a person eats food containing sugar.
b. After eating sugar, when is the pH level the lowest? Use the function's equation to determine the pH level, to the nearest tenth, at this time.
c. According to the graph, what is the normal pH level of the human mouth?
d. What is the equation of the horizontal asymptote associated with this function? Describe what this means in terms of the mouth's pH level over time.
e. Use the graph to describe what happens to the pH level during the first hour.

92. A drug is injected into a patient and the concentration of the drug in the bloodstream is monitored. The drug's concentration, $C(t)$, in milligrams per liter, after t hours is modeled by
$$C(t) = \dfrac{5t}{t^2 + 1}.$$

The graph of this rational function, obtained with a graphing utility, is shown in the figure.

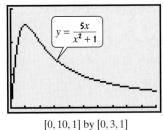

$$y = \frac{5x}{x^2 + 1}$$

[0, 10, 1] by [0, 3, 1]

a. Use the graph to obtain a reasonable estimate of the drug's concentration after 3 hours. Then verify this estimate algebraically.

b. Use the function's equation, $C(t) = \dfrac{5t}{t^2 + 1}$, to find the horizontal asymptote for the graph. Describe what this means about the drug's concentration in the patient's bloodstream as time increases.

*Among all deaths from a particular disease, the percentage that are smoking related (21–39 cigarettes per day) is a function of the disease's **incidence ratio**. The incidence ratio describes the number of times more likely smokers are than nonsmokers to die from the disease. The following table shows the incidence ratios for heart disease and lung cancer for two age groups.*

Incidence Ratios

	Heart Disease	**Lung Cancer**
Ages 55–64	1.9	10
Ages 65–74	1.7	9

Source: Alexander M. Walker, *Observations and Inference*, Epidemiology Resources Inc., 1991.

For example, the incidence ratio of 9 in the table means that smokers between the ages of 65 and 74 are 9 times more likely than nonsmokers in the same age group to die from lung cancer. The rational function

$$P(x) = \frac{100(x - 1)}{x}$$

models the percentage of smoking-related deaths among all deaths from a disease, $P(x)$, in terms of the disease's incidence ratio, x. The graph of the rational function is shown. Use this function to solve Exercises 93–96.

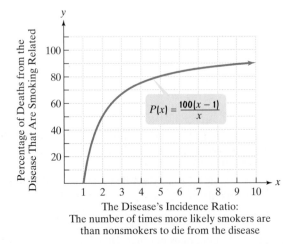

$$P(x) = \frac{100(x-1)}{x}$$

The Disease's Incidence Ratio:
The number of times more likely smokers are than nonsmokers to die from the disease

93. Find $P(10)$. Describe what this means in terms of the incidence ratio, 10, given in the table. Identify your solution as a point on the graph.

94. Find $P(9)$. Round to the nearest percent. Describe what this means in terms of the incidence ratio, 9, given in the table. Identify your solution as a point on the graph.

95. What is the horizontal asymptote of the graph? Describe what this means about the percentage of deaths caused by smoking with increasing incidence ratios.

96. According to the model and its graph, is there a disease for which all deaths are caused by smoking? Explain your answer.

97. The graph shows the U.S. population, by gender, for selected years from 1950 through 2002.

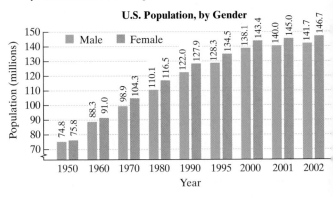

U.S. Population, by Gender

Source: U.S. Census Bureau

a. Write a fraction that shows the ratio of males to females in 1995. Then express the fraction as a decimal, rounded to the nearest thousandth. How many males per 1000 females were there in 1995?

b. How many males per 1000 females were there in 2002?

c. The function $p(x) = 1.256x + 74.2$ models the male U.S. population, $p(x)$, in millions, x years after 1950. The function $q(x) = 1.324x + 76.71$ models the female U.S. population, $q(x)$, in millions, x years after 1950. Write a function that models the ratio of males to females x years after 1950.

d. Use the function that you wrote in part (c) to find the number of males per 1000 females in 1995. How well does the function model the actual number that you determined in part (a)?

e. Use the function that you wrote in part (c) to find the number of males per 1000 females in 2002. How well does the function model the actual number that you determined in part (b)?

f. What is the equation of the horizontal asymptote associated with the function in part (c)? Round to the nearest thousandth. What does this mean about the number of males per 1000 females over time?

Exercises 98–101 involve writing a rational function that models a problem's conditions.

98. You drive from your home to a vacation resort 600 miles away. You return on the same highway. The average velocity on the return trip is 10 miles per hour slower than the average velocity on the outgoing trip. Express the total time required to complete the round trip, T, as a function of the average velocity on the outgoing trip, x.

99. A tourist drives 90 miles along a scenic highway and then takes a 5-mile walk along a hiking trail. The average velocity driving is nine times that while hiking. Express the total time for driving and hiking, T, as a function of the average velocity on the hike, x.

100. A contractor is constructing the house shown in the figure. The cross section up to the roof is in the shape of a rectangle. The area of the rectangular floor of the house is 2500 square feet. Express the perimeter of the rectangular floor, P, as a function of the width of the rectangle, x.

Length

Width: x

101. The figure shows a page with 1-inch margins at the top and the bottom and half-inch side margins. A publishing company is willing to vary the page dimensions subject to the condition that the printed area of the page is 50 square inches. Express the total area of the page, A, as a function of the width of the rectangle containing the print, x.

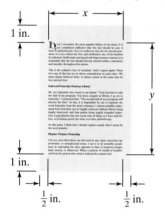

x

1 in.

y

1 in.

$\frac{1}{2}$ in. $\frac{1}{2}$ in.

Writing in Mathematics

102. What is a rational function?

103. Use everyday language to describe the graph of a rational function f such that as $x \rightarrow -\infty$, $f(x) \rightarrow 3$.

104. Use everyday language to describe the behavior of a graph near its vertical asymptote if $f(x) \rightarrow \infty$ as $x \rightarrow -2^-$ and $f(x) \rightarrow -\infty$ as $x \rightarrow -2^+$.

105. If you are given the equation of a rational function, explain how to find the vertical asymptotes, if any, of the function's graph.

106. If you are given the equation of a rational function, explain how to find the horizontal asymptote, if any, of the function's graph.

107. Describe how to graph a rational function.

108. If you are given the equation of a rational function, how can you tell if the graph has a slant asymptote? If it does, how do you find its equation?

109. Is every rational function a polynomial function? Why or why not? Does a true statement result if the two adjectives *rational* and *polynomial* are reversed? Explain.

110. Although your friend has a family history of heart disease, he smokes, on average, 25 cigarettes per day. He sees the table showing incidence ratios for heart disease (see Exercises 93–96) and feels comfortable that they are less than 2, compared to 9 and 10 for lung cancer. He claims that all family deaths have been from heart disease and decides not to give up smoking. Use the given function and its graph to describe some additional information not given in the table that might influence his decision.

Technology Exercises

111. Use a graphing utility to verify any five of your hand-drawn graphs in Exercises 37–78.

112. Use a graphing utility to graph $y = \dfrac{1}{x}$, $y = \dfrac{1}{x^3}$, and $\dfrac{1}{x^5}$ in the same viewing rectangle. For odd values of n, how does changing n affect the graph of $y = \dfrac{1}{x^n}$?

113. Use a graphing utility to graph $y = \dfrac{1}{x^2}$, $y = \dfrac{1}{x^4}$, and $y = \dfrac{1}{x^6}$ in the same viewing rectangle. For even values of n, how does changing n affect the graph of $y = \dfrac{1}{x^n}$?

114. Use a graphing utility to graph

$$f(x) = \frac{x^2 - 4x + 3}{x - 2} \quad \text{and} \quad g(x) = \frac{x^2 - 5x + 6}{x - 2}.$$

What differences do you observe between the graph of f and the graph of g? How do you account for these differences?

115. The rational function

$$f(x) = \frac{27{,}725(x - 14)}{x^2 + 9} - 5x$$

models the number of arrests, $f(x)$, per 100,000 drivers, for driving under the influence of alcohol, as a function of a driver's age, x.

a. Graph the function in a $[0, 70, 5]$ by $[0, 400, 20]$ viewing rectangle.

b. Describe the trend shown by the graph.

c. Use the ZOOM and TRACE features or the maximum function feature of your graphing utility to find the age that corresponds to the greatest number of arrests. How many arrests, per 100,000 drivers, are there for this age group?

Critical Thinking Exercises

116. Which one of the following is true?

a. The graph of a rational function cannot have both a vertical asymptote and a horizontal asymptote.

b. It is not possible to have a rational function whose graph has no y-intercept.

c. The graph of a rational function can have three horizontal asymptotes.

d. The graph of a rational function can never cross a vertical asymptote.

117. Which one of the following is true?

a. The function $f(x) = \dfrac{1}{\sqrt{x} - 3}$ is a rational function.

b. The x-axis is a horizontal asymptote for the graph of $f(x) = \dfrac{4x - 1}{x + 3}$.

c. The number of televisions that a company can produce per week after t weeks of production is given by

$$N(t) = \frac{3000t^2 + 30{,}000t}{t^2 + 10t + 25}.$$

Using this model, the company will eventually be able to produce 30,000 televisions in a single week.

d. None of the given statements is true.

In Exercises 118–121, write the equation of a rational function $f(x) = \dfrac{p(x)}{q(x)}$ *having the indicated properties, in which the degrees of p and q are as small as possible. More than one correct function may be possible. Graph your function using a graphing utility to verify that it has the required properties.*

118. f has a vertical asymptote given by $x = 3$, a horizontal asymptote $y = 0$, y-intercept at -1, and no x-intercept.

119. f has vertical asymptotes given by $x = -2$ and $x = 2$, horizontal asymptote $y = 2$, y-intercept at $\frac{9}{2}$, x-intercepts at -3 and 3, and y-axis symmetry.

120. f has a vertical asymptote given by $x = 1$, a slant asymptote whose equation is $y = x$, y-intercept at 2, and x-intercepts at -1 and 2.

121. f has no vertical, horizontal, or slant asymptotes, and no x-intercepts.

SECTION 2.7 Polynomial and Rational Inequalities

Objectives

1 Solve polynomial inequalities.

2 Solve rational inequalities.

3 Solve problems modeled by polynomial or rational inequalities.

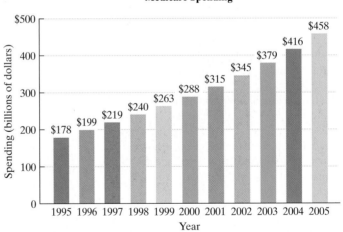

People are going to live longer in the twenty-first century. This will put added pressure on the Social Security and Medicare systems. The bar graph in Figure 2.41 shows the cost of Medicare, in billions of dollars, through 2005.

Medicare Spending

Figure 2.41

Source: Congressional Budget Office

Medicare spending, $f(x)$, in billions of dollars, x years after 1995 can be modeled by the quadratic function

$$f(x) = 1.2x^2 + 15.2x + 181.4.$$

To determine in which years Medicare spending will exceed $500 billion, we must solve the inequality

$$1.2x^2 + 15.2x + 181.4 > 500.$$

Medicare spending   exceeds $500 billion.

We begin by subtracting 500 from both sides. This will give us zero on the right:

$$1.2x^2 + 15.2x + 181.4 - 500 > 500 - 500$$

$$1.2x^2 + 15.2x - 318.6 > 0.$$

The form of $1.2x^2 + 15.2x - 318.6 > 0$ is $ax^2 + bx + c > 0$. Such an inequality is called a *polynomial inequality*.

> ### Definition of a Polynomial Inequality
> A polynomial inequality is any inequality that can be put into one of the forms
> $$f(x) < 0, \quad f(x) > 0, \quad f(x) \le 0, \quad \text{or} \quad f(x) \ge 0,$$
> where f is a polynomial function.

In this section, we establish the basic techniques for solving polynomial inequalities. We will use these techniques to solve inequalities involving rational functions.

Solving Polynomial Inequalities

Graphs can help us visualize the solutions of polynomial inequalities. For example, the graph of $f(x) = x^2 - 7x + 10$ is shown in Figure 2.42. The x-intercepts, 2 and 5, are **boundary points** between where the graph lies above the x-axis, shown in blue, and where the graph lies below the x-axis, shown in red.

Locating the x-intercepts of a polynomial function, f, is an important step in finding the solution set for polynomial inequalities in the form $f(x) < 0$ or $f(x) > 0$. We use the x-intercepts of f as boundary points that divide the real number line into intervals. On each interval, the graph of f is either above the x-axis $[f(x) > 0]$ or below the x-axis $[f(x) < 0]$. For this reason, x-intercepts play a fundamental role in solving polynomial inequalities. The x-intercepts are found by solving the equation $f(x) = 0$.

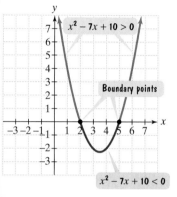

Figure 2.42

 Solve polynomial inequalities.

> ### Procedure for Solving Polynomial Inequalities
> **1.** Express the inequality in the form
> $$f(x) < 0 \quad \text{or} \quad f(x) > 0,$$
> where f is a polynomial function.
> **2.** Solve the equation $f(x) = 0$. The real solutions are the **boundary points**.
> **3.** Locate these boundary points on a number line, thereby dividing the number line into intervals.
> **4.** Choose one representative number, called a **test value**, within each interval and evaluate f at that number.
>   **a.** If the value of f is positive, then $f(x) > 0$ for all numbers, x, in the interval.
>   **b.** If the value of f is negative, then $f(x) < 0$ for all numbers, x, in the interval.
> **5.** Write the solution set, selecting the interval or intervals that satisfy the given inequality.
>
> This procedure is valid if $<$ is replaced by $\le$ or $>$ is replaced by $\ge$. However, if the inequality involves $\le$ or $\ge$, include the boundary points [the solutions of $f(x) = 0$] in the solution set.

EXAMPLE 1   Solving a Polynomial Inequality

Solve and graph the solution set on a real number line: $2x^2 + x > 15$.

Solution

Step 1   Express the inequality in the form $f(x) < 0$ or $f(x) > 0$. We begin by rewriting the inequality so that 0 is on the right side.

$$2x^2 + x > 15 \qquad \text{This is the given inequality.}$$
$$2x^2 + x - 15 > 15 - 15 \qquad \text{Subtract 15 from both sides.}$$
$$2x^2 + x - 15 > 0 \qquad \text{Simplify.}$$

This inequality is equivalent to the one we wish to solve. It is in the form $f(x) > 0$, where $f(x) = 2x^2 + x - 15$.

Step 2   Solve the equation $f(x) = 0$. We find the x-intercepts of $f(x) = 2x^2 + x - 1!$ by solving the equation $2x^2 + x - 15 = 0$.

$$2x^2 + x - 15 = 0$$ This polynomial equation is a quadratic equation.

$$(2x - 5)(x + 3) = 0$$ Factor.

$$2x - 5 = 0 \quad \text{or} \quad x + 3 = 0$$ Set each factor equal to 0.

$$x = \frac{5}{2} \qquad\qquad x = -3$$ Solve for x.

The x-intercepts of f are -3 and $\frac{5}{2}$. We will use these x-intercepts as boundary point on a number line.

Step 3   Locate the boundary points on a number line and separate the line int intervals. The number line with the boundary points is shown as follows:

The boundary points divide the number line into three intervals:

$$(-\infty, \quad -3) \quad \left(-3, \frac{5}{2}\right) \quad \left(\frac{5}{2}, \quad \infty\right).$$

Step 4   Choose one test value within each interval and evaluate f at that number

Interval	Test Value	Substitute into $f(x) = 2x^2 + x - 15$	Conclusion
$(-\infty, -3)$	-4	$f(-4) = 2(-4)^2 + (-4) - 15$ $= 13$, positive	$f(x) > 0$ for all x in $(-\infty, -3)$.
$\left(-3, \frac{5}{2}\right)$	0	$f(0) = 2 \cdot 0^2 + 0 - 15$ $= -15$, negative	$f(x) < 0$ for all x in $\left(-3, \frac{5}{2}\right)$.
$\left(\frac{5}{2}, \infty\right)$	3	$f(3) = 2 \cdot 3^2 + 3 - 15$ $= 6$, positive	$f(x) > 0$ for all x in $\left(\frac{5}{2}, \infty\right)$.

Technology

The solution set for

$$2x^2 + x > 15$$

or, equivalently,

$$2x^2 + x - 15 > 0$$

can be verified with a graphing utility. The graph of $f(x) = 2x^2 + x - 15$ was obtained using a $[-10, 10, 1]$ by $[-16, 6, 1]$ viewing rectangle. The graph lies above the x-axis, representing $>$, for all x in $\left(-\infty, -3\right)$ or $\left(\frac{5}{2}, \infty\right)$.

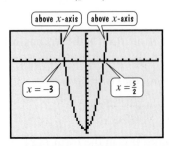

Step 5   Write the solution set, selecting the interval or intervals that satisfy th given inequality. We are interested in solving $2x^2 + x - 15 > 0$, wher $f(x) = 2x^2 + x - 15$. Based on our work in step 4, we see that $f(x) > 0$ for all x i $(-\infty, -3)$ or $\left(\frac{5}{2}, \infty\right)$. Thus, the solution set of the given inequality, $2x^2 + x > 1!$ or, equivalently, $2x^2 + x - 15 > 0$, is

$$(-\infty, -3) \cup \left(\frac{5}{2}, \infty\right) \text{ or } \left\{x \middle| x < -3 \text{ or } x > \frac{5}{2}\right\}.$$

The graph of the solution set on a number line is shown as follows:

Check Point 1   Solve and graph the solution set: $x^2 - x > 20$.

EXAMPLE 2   Solving a Polynomial Inequality

Solve and graph the solution set on a real number line: $x^3 + x^2 \leq 4x + 4$.

Solution

Step 1   Express the inequality in the form $f(x) \leq 0$ or $f(x) \geq 0$. We begin b rewriting the inequality so that 0 is on the right side.

$$x^3 + x^2 \leq 4x + 4 \qquad \text{This is the given inequality.}$$
$$x^3 + x^2 - 4x - 4 \leq 4x + 4 - 4x - 4 \qquad \text{Subtract } 4x + 4 \text{ from both sides.}$$
$$x^3 + x^2 - 4x - 4 \leq 0 \qquad \text{Simplify.}$$

This inequality is equivalent to the one we wish to solve. It is in the form $f(x) \leq 0$, where $f(x) = x^3 + x^2 - 4x - 4$.

Step 2 Solve the equation $f(x) = 0$. We find the x-intercepts of $f(x) = x^3 + x^2 - 4x - 4$ by solving the equation $x^3 + x^2 - 4x - 4 = 0$.

$$x^3 + x^2 - 4x - 4 = 0 \qquad \text{This polynomial equation is of degree 3.}$$
$$x^2(x + 1) - 4(x + 1) = 0 \qquad \begin{array}{l}\text{Factor } x^2 \text{ from the first two terms and } -4 \\ \text{from the last two terms.}\end{array}$$
$$(x + 1)(x^2 - 4) = 0 \qquad \begin{array}{l}\text{A common factor of } x + 1 \text{ is factored from} \\ \text{the expression.}\end{array}$$
$$x + 1 = 0 \quad \text{or} \quad x^2 - 4 = 0 \qquad \text{Set each factor equal to 0.}$$
$$x = -1 \qquad\qquad x^2 = 4 \qquad \text{Solve for x.}$$
$$x = \pm 2 \qquad \text{Use the square root property.}$$

The x-intercepts of f are -2, -1, and 2. We will use these x-intercepts as boundary points on a number line.

Step 3 Locate the boundary points on a number line and separate the line into intervals. The number line with the boundary points is shown as follows:

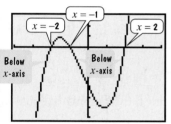

The boundary points divide the number line into four intervals:

$$(-\infty, -2) \quad (-2, -1) \quad (-1, 2) \quad (2, \infty).$$

Step 4 Choose one test value within each interval and evaluate f at that number.

Technology

The solution set for
$$x^3 + x^2 \leq 4x + 4$$
or, equivalently,
$$x^3 + x^2 - 4x - 4 \leq 0$$
can be verified with a graphing utility. The graph of $f(x) = x^3 + x^2 - 4x - 4$ lies on or below the x-axis, representing $\leq$, for all x in $(-\infty, -2]$ or $[-1, 2]$.

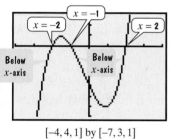

$[-4, 4, 1]$ by $[-7, 3, 1]$

Interval	Test Value	Substitute into $f(x) = x^3 + x^2 - 4x - 4$	Conclusion
$(-\infty, -2)$	-3	$f(-3) = (-3)^3 + (-3)^2 - 4(-3) - 4$ $= -10$, negative	$f(x) < 0$ for all x in $(-\infty, -2)$.
$(-2, -1)$	-1.5	$f(-1.5) = (-1.5)^3 + (-1.5)^2 - 4(-1.5) - 4$ $= 0.875$, positive	$f(x) > 0$ for all x in $(-2, -1)$.
$(-1, 2)$	0	$f(0) = 0^3 + 0^2 - 4 \cdot 0 - 4$ $= -4$, negative	$f(x) < 0$ for all x in $(-1, 2)$.
$(2, \infty)$	3	$f(3) = 3^3 + 3^2 - 4 \cdot 3 - 4$ $= 20$, positive	$f(x) > 0$ for all x in $(2, \infty)$.

Step 5 Write the solution set, selecting the interval or intervals that satisfy the given inequality. We are interested in solving $x^3 + x^2 - 4x - 4 \leq 0$, where $f(x) = x^3 + x^2 - 4x - 4$. Based on our work in step 4, we see that $f(x) < 0$ for all x in $(-\infty, -2)$ or $(-1, 2)$. However, because the inequality involves $\leq$ (less than or *equal to*), we must also include the solutions of $x^3 + x^2 - 4x - 4 = 0$, namely -2, -1, and 2, in the solution set. Thus, the solution set of the given inequality, $x^3 + x^2 \leq 4x + 4$, or, equivalently, $x^3 + x^2 - 4x - 4 \leq 0$, is

$$(-\infty, -2] \cup [-1, 2]$$
$$\text{or} \quad \{x \mid x \leq -2 \text{ or } -1 \leq x \leq 2\}.$$

The graph of the solution set on a number line is shown as follows:

Check Point **2** Solve and graph the solution set on a real number line: $x^3 + 3x^2 \leq x + 3$

② Solve rational inequalities.

Solving Rational Inequalities

A **rational inequality** is any inequality that can be put into one of the forms

$$f(x) < 0, \quad f(x) > 0, \quad f(x) \leq 0, \quad \text{or} \quad f(x) \geq 0,$$

where f is a rational function. An example of a rational inequality is

$$\frac{3x + 3}{2x + 4} > 0.$$

This inequality is in the form $f(x) > 0$, where f is the rational function given by

$$f(x) = \frac{3x + 3}{2x + 4}.$$

The graph of f is shown in Figure 2.43.
We can find the x-intercept of f by setting the numerator equal to 0:

$$3x + 3 = 0$$
$$3x = -3$$
$$x = -1.$$

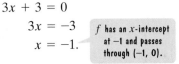
f has an x-intercept at −1 and passes through (−1, 0).

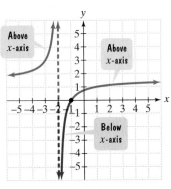

Figure 2.43 The graph of
$$f(x) = \frac{3x + 3}{2x + 4}$$

We can determine where f is undefined by setting the denominator equal to 0:

$$2x + 4 = 0$$
$$2x = -4$$
$$x = -2.$$

f is undefined at −2. Figure 2.43 shows that the function's vertical asymptote is x = −2.

By setting both the numerator and the denominator of f equal to 0, we obtained −2 and −1. These numbers separate the x-axis into three intervals: $(-\infty, -2), (-2, -1)$, and $(-1, \infty)$. On each interval, the graph of f is either above the x-axis $[f(x) > 0]$ or below the x-axis $[f(x) < 0]$.

Examine the graph in Figure 2.43 carefully. Can you see that it is above the x-axis for all x in $(-\infty, -2)$ or $(-1, \infty)$, shown in blue? Thus, the solution set of $\frac{3x + 3}{2x + 4} > 0$ is $(-\infty, -2) \cup (-1, \infty)$. By contrast, the graph of f lies below the x-axis for all x in $(-2, -1)$, shown in red. Thus, the solution set of $\frac{3x + 3}{2x + 4} < 0$ is $(-2, -1)$.

The first step in solving a rational inequality is to bring all terms to one side, obtaining zero on the other side. Then express the rational function on the nonzero side as a single quotient. The second step is to set the numerator and the denominator of f equal to zero. The solutions of these equations serve as boundary points that separate the real number line into intervals. At this point, the procedure is the same as the one we used for solving polynomial inequalities.

Study Tip

Do not begin solving

$$\frac{x + 1}{x + 3} \geq 2$$

by multiplying both sides by $x + 3$. We do not know if $x + 3$ is positive or negative. Thus, we do not know whether or not to change the sense of the inequality.

EXAMPLE 3   Solving a Rational Inequality

Solve and graph the solution set: $\dfrac{x + 1}{x + 3} \geq 2$.

Solution

Step 1   Express the inequality so that one side is zero and the other side is a single quotient. We subtract 2 from both sides to obtain zero on the right.

$$\frac{x+1}{x+3} \geq 2 \qquad \text{This is the given inequality.}$$

$$\frac{x+1}{x+3} - 2 \geq 0 \qquad \text{Subtract 2 from both sides, obtaining 0 on the right.}$$

$$\frac{x+1}{x+3} - \frac{2(x+3)}{x+3} \geq 0 \qquad \text{The least common denominator is } x+3. \text{ Express 2 in terms of this denominator.}$$

$$\frac{x+1 - 2(x+3)}{x+3} \geq 0 \qquad \text{Subtract rational expressions.}$$

$$\frac{x+1 - 2x - 6}{x+3} \geq 0 \qquad \text{Apply the distributive property.}$$

$$\frac{-x-5}{x+3} \geq 0 \qquad \text{Simplify.}$$

This inequality is equivalent to the one we wish to solve. It is in the form $f(x) \geq 0$, where $f(x) = \dfrac{-x-5}{x+3}$.

Step 2   Set the numerator and the denominator of f equal to zero. The real solutions are the boundary points.

$$-x - 5 = 0 \qquad x + 3 = 0 \qquad \text{Set the numerator and denominator equal to 0. These are the values that make the previous quotient zero or undefined.}$$

$$x = -5 \qquad x = -3 \qquad \text{Solve for } x.$$

We will use these solutions as boundary points on a number line.

Step 3   Locate the boundary points on a number line and separate the line into intervals. The number line with the boundary points is shown as follows:

The boundary points divide the number line into three intervals:

$$(-\infty, -5) \quad (-5, -3) \quad (-3, \infty).$$

Step 4   Choose one test value within each interval and evaluate f at that number.

Interval	Test Value	Substitute into $f(x) = \dfrac{-x-5}{x+3}$	Conclusion
$(-\infty, -5)$	-6	$f(-6) = \dfrac{-(-6) - 5}{-6 + 3}$ $= -\dfrac{1}{3}, \text{ negative}$	$f(x) < 0$ for all x in $(-\infty, -5)$.
$(-5, -3)$	-4	$f(-4) = \dfrac{-(-4) - 5}{-4 + 3}$ $= 1, \text{ positive}$	$f(x) > 0$ for all x in $(-5, -3)$.
$(-3, \infty)$	0	$f(0) = \dfrac{-0 - 5}{0 + 3}$ $= -\dfrac{5}{3}, \text{ negative}$	$f(x) < 0$ for all x in $(-3, \infty)$.

Step 5   Write the solution set, selecting the interval or intervals that satisfy the given inequality. We are interested in solving $\dfrac{-x-5}{x+3} \geq 0$, where $f(x) = \dfrac{-x-5}{x+3}$. Based on our work in step 4, we see that $f(x) > 0$ for all x in $(-5, -3)$. However,

Study Tip

Never include the value that causes a rational function's denominator to equal zero in the solution set of a rational inequality. Division by zero is undefined.

Discovery

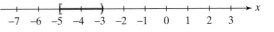

Because $(x + 3)^2$ is positive, it is possible so solve

$$\frac{x + 1}{x + 3} \geq 2$$

by first multiplying both sides by $(x + 3)^2$ (where $x \neq -3$). This will not change the sense of the inequality and will clear the fraction. Try using this solution method and compare it to the solution on pages 350–352.

because the inequality involves $\geq$ (greater than or *equal to*), we must also includ⬛ the solution of $f(x) = 0$, namely the value that we obtained when we set the nu⬛ merator of f equal to zero. Thus, we must include -5 in the solution set. The solutio⬛ set of the given inequality is

$$[-5, -3) \text{ or } \{x | -5 \leq x < -3\}.$$

The graph of the solution set on a number line is shown as follows:

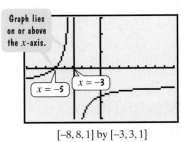

Technology

The solution set for

$$\frac{x + 1}{x + 3} \geq 2$$

or, equivalently,

$$\frac{-x - 5}{x + 3} \geq 0$$

can be verified with a graphing utility. The graph of $f(x) = \dfrac{-x - 5}{x + 3}$ lies on or above the x-axis, representing $\geq$, for all x in $[-5, -3)$.

Graph lies on or above the x-axis.

$x = -5$ $x = -3$

$[-8, 8, 1]$ by $[-3, 3, 1]$

Check Point 3 Solve and graph the solution set: $\dfrac{2x}{x + 1} \geq 1$.

 Solve problems modeled by polynomial or rational inequalities.

Applications

We are surrounded by evidence that the world is profoundly mathematical. Fo⬛ example, did you know that every time you throw an object vertically upward, i⬛ changing height above the ground can be described by a quadratic function? Th⬛ same function can be used to describe objects that are falling, such as sky divers.

The Position Function for a Free-Falling Object Near Earth's Surface

An object that is falling or vertically projected into the air has its height above the ground, $s(t)$, in feet, given by

$$s(t) = -16t^2 + v_0 t + s_0,$$

where v_0 is the original velocity (initial velocity) of the object, in feet per second, t is the time that the object is in motion, in seconds, and s_0 is the original height (initial height) of the object, in feet.

In Example 4, we solve a polynomial inequality in a problem about th⬛ position of a free-falling object.

EXAMPLE 4   Using the Position Function

A ball is thrown vertically upward from the top of the Leaning Tower of Pisa (19⬛ feet high) with an initial velocity of 96 feet per second (Figure 2.44). During whic⬛ time period will the ball's height exceed that of the tower?

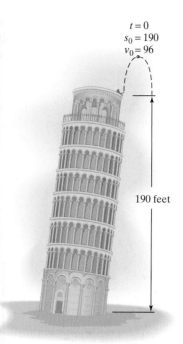

$t = 0$
$s_0 = 190$
$v_0 = 96$

190 feet

Figure 2.44 Throwing a ball from 190 feet with a velocity of 96 feet per second

Solution

$$s(t) = -16t^2 + v_0 t + s_0$$

This is the position function for a free-falling object.

$$s(t) = -16t^2 + 96t + 190$$

Because v_0 (initial velocity) $= 96$ and s_0 (initial position) $= 190$, substitute these values into the formula.

When will $s(t)$, the ball's height,   exceed that   of the tower?

$$-16t^2 \quad + \quad 96t \quad + 190 \quad > \quad 190$$

$$-16t^2 + 96t + 190 > 190$$

This is the inequality that models the problem's question. We must find t.

$$-16t^2 + 96t > 0$$

Subtract 190 from both sides. This inequality is in the form $f(t) > 0$, where $f(t) = -16t^2 + 96t$.

$$-16t^2 + 96t = 0$$

Solve the equation $f(t) = 0$.

$$-16t(t - 6) = 0$$

Factor.

$$-16t = 0 \quad \text{or} \quad t - 6 = 0$$

Set each factor equal to 0.

$$t = 0 \qquad\qquad t = 6$$

Solve for t. The boundary points are 0 and 6.

```
  +--+--+--●--+--+--+--+--+--●--+--+--→ t
 -2 -1  0  1  2  3  4  5  6  7  8
```

Locate these values on a number line, with $t \geq 0$.

The intervals are $(-\infty, 0)$, $(0, 6)$, and $(6, \infty)$. For our purposes, the mathematical model is useful only from $t = 0$ until the ball hits the ground. (By setting $-16t^2 + 96t + 190$ equal to zero, we find $t \approx 7.57$; the ball hits the ground after approximately 7.57 seconds.) Thus, we use $(0, 6)$ and $(6, 7.57)$ for our intervals.

Interval	Test Value	Substitute into $f(t) = -16t^2 + 96t$	Conclusion
$(0, 6)$	1	$f(1) = -16 \cdot 1^2 + 96 \cdot 1$ $= 80$, positive	$f(t) > 0$ for all t in $(0, 6)$.
$(6, 7.57)$	7	$f(7) = -16 \cdot 7^2 + 96 \cdot 7$ $= -112$, negative	$f(t) < 0$ for all t in $(6, 7.57)$.

We are interested in solving $-16t^2 + 96t > 0$, where $f(t) = -16t^2 + 96t$. We see that $f(t) > 0$ for all t in $(0, 6)$. This means that the ball's height exceeds that of the tower between 0 and 6 seconds.

Technology

The graphs of

$$y_1 = -16x^2 + 96x + 190$$

and

$$y_2 = 190$$

are shown in a

$$[0, 8, 1] \text{ by } [0, 360, 36]$$

seconds in motion   height, in feet

viewing rectangle. The graphs show that the ball's height exceeds that of the tower between 0 and 6 seconds.

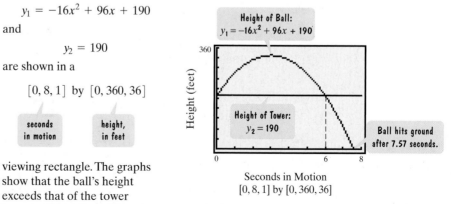

Height of Ball:
$y_1 = -16x^2 + 96x + 190$

Height of Tower:
$y_2 = 190$

Ball hits ground after 7.57 seconds.

Seconds in Motion
$[0, 8, 1]$ by $[0, 360, 36]$

Check Point 4 An object is propelled straight up from ground level with an initial velocity of 80 feet per second. Its height at time t is modeled by

$$s(t) = -16t^2 + 80t,$$

where the height, $s(t)$, is measured in feet and the time, t, is measured in seconds. In which time interval will the object be more than 64 feet above the ground?

EXERCISE SET 2.7

Practice Exercises

Solve each polynomial inequality in Exercises 1–38 and graph the solution set on a real number line. Express each solution set in interval notation.

1. $(x - 4)(x + 2) > 0$
2. $(x + 3)(x - 5) > 0$

3. $(x - 7)(x + 3) \le 0$
4. $(x + 1)(x - 7) \le 0$

5. $x^2 - 5x + 4 > 0$
6. $x^2 - 4x + 3 < 0$

7. $x^2 + 5x + 4 > 0$
8. $x^2 + x - 6 > 0$

9. $x^2 - 6x + 9 < 0$
10. $x^2 - 2x + 1 > 0$

11. $3x^2 + 10x - 8 \le 0$
12. $9x^2 + 3x - 2 \ge 0$

13. $2x^2 + x < 15$
14. $6x^2 + x > 1$

15. $4x^2 + 7x < -3$
16. $3x^2 + 16x < -5$

17. $5x \le 2 - 3x^2$
18. $4x^2 + 1 \ge 4x$

19. $x^2 - 4x \ge 0$
20. $x^2 + 2x < 0$

21. $2x^2 + 3x > 0$
22. $3x^2 - 5x \le 0$

23. $-x^2 + x \ge 0$
24. $-x^2 + 2x \ge 0$

25. $x^2 \le 4x - 2$
26. $x^2 \le 2x + 2$

27. $x^2 - 6x + 9 < 0$
28. $4x^2 - 4x + 1 \ge 0$

29. $(x - 1)(x - 2)(x - 3) \ge 0$

30. $(x + 1)(x + 2)(x + 3) \ge 0$

31. $x^3 + 2x^2 - x - 2 \ge 0$ **32.** $x^3 + 2x^2 - 4x - 8 \ge 0$

33. $x^3 - 3x^2 - 9x + 27 < 0$ **34.** $x^3 + 7x^2 - x - 7 < 0$

35. $x^3 + x^2 + 4x + 4 > 0$ **36.** $x^3 - x^2 + 9x - 9 > 0$

37. $x^3 \ge 9x^2$
38. $x^3 \le 4x^2$

Solve each rational inequality in Exercises 39–56 and graph the solution set on a real number line. Express each solution set in interval notation.

39. $\dfrac{x - 4}{x + 3} > 0$
40. $\dfrac{x + 5}{x - 2} > 0$

41. $\dfrac{x + 3}{x + 4} < 0$
42. $\dfrac{x + 5}{x + 2} < 0$

43. $\dfrac{-x + 2}{x - 4} \ge 0$
44. $\dfrac{-x - 3}{x + 2} \le 0$

45. $\dfrac{4 - 2x}{3x + 4} \le 0$
46. $\dfrac{3x + 5}{6 - 2x} \ge 0$

47. $\dfrac{x}{x - 3} > 0$
48. $\dfrac{x + 4}{x} > 0$

49. $\dfrac{(x + 4)(x - 1)}{x + 2} \le 0$
50. $\dfrac{(x + 3)(x - 2)}{x + 1} \le 0$

51. $\dfrac{x + 1}{x + 3} < 2$
52. $\dfrac{x}{x - 1} > 2$

53. $\dfrac{x + 4}{2x - 1} \le 3$
54. $\dfrac{1}{x - 3} < 1$

55. $\dfrac{x - 2}{x + 2} \le 2$
56. $\dfrac{x}{x + 2} \ge 2$

Practice Plus

In Exercises 57–60, find the domain of each function.

57. $f(x) = \sqrt{2x^2 - 5x + 2}$ **58.** $f(x) = \dfrac{1}{\sqrt{4x^2 - 9x + 2}}$

59. $f(x) = \sqrt{\dfrac{2x}{x + 1} - 1}$ **60.** $f(x) = \sqrt{\dfrac{x}{2x - 1} - 1}$

Solve each inequality in Exercises 61–66 and graph the solution set on a real number line.

61. $|x^2 + 2x - 36| > 12$
62. $|x^2 + 6x + 1| > 8$

63. $\dfrac{3}{x + 3} > \dfrac{3}{x - 2}$
64. $\dfrac{1}{x + 1} > \dfrac{2}{x - 1}$

65. $\dfrac{x^2 - x - 2}{x^2 - 4x + 3} > 0$
66. $\dfrac{x^2 - 3x + 2}{x^2 - 2x - 3} > 0$

In Exercises 67–68, use the graph of the polynomial function to solve each inequality.

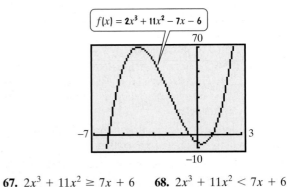

$f(x) = 2x^3 + 11x^2 - 7x - 6$

67. $2x^3 + 11x^2 \ge 7x + 6$ **68.** $2x^3 + 11x^2 < 7x + 6$

n Exercises 69–70, use the graph of the rational function to solve ʌch inequality.

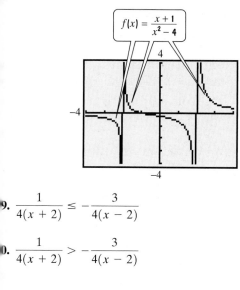

$$f(x) = \frac{x+1}{x^2-4}$$

9. $\dfrac{1}{4(x+2)} \le -\dfrac{3}{4(x-2)}$

0. $\dfrac{1}{4(x+2)} > -\dfrac{3}{4(x-2)}$

Application Exercises

ʼse the position function

$$s(t) = -16t^2 + v_0 t + s_0$$

v_0 = initial velocity, s_0 = initial position, t = time)

answer Exercises 71–72.

1. Divers in Acapulco, Mexico, dive headfirst at 8 feet per second from the top of a cliff 87 feet above the Pacific Ocean. During which time period will the diver's height exceed that of the cliff?

2. You throw a ball straight up from a rooftop 160 feet high with an initial velocity of 48 feet per second. During which time period will the ball's height exceed that of the rooftop?

he bar graph in Figure 2.41 on page 346 shows the cost of ¹edicare, in billions of dollars, through 2005. Using the regression .ature of a graphing utility, these data can be modeled by

a linear function, $f(x) = 27x + 163$;

a quadratic function, $g(x) = 1.2x^2 + 15.2x + 181.4$.

each function, x represents the number of years after 1995. Use ese functions to solve Exercises 73–76.

. The graph indicates that Medicare spending reached $379 billion in 2003. Find the amount predicted by each of the functions, f and g, for that year. How well do the functions model the value in the graph?

. The graph indicates that Medicare spending reached $458 billion in 2005. Find the amount predicted by each of the functions, f and g, for that year. How well do the functions model the value in the graph? Which function serves as a better model for that year?

. For which years does the quadratic model indicate that Medicare spending will exceed $536.6 billion?

. For which years does the quadratic model indicate that Medicare spending will exceed $629.4 billion?

It's vacation time. You drive 90 miles along a scenic highway and then take a 5-mile run along a hiking trail. Your driving rate is nine times that of your running rate. The graph shows the total time you spend driving and running, $f(x)$, as a function of your running rate, x. Use the rational function and its graph to solve Exercises 77–81.

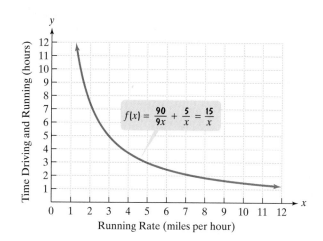

$$f(x) = \frac{90}{9x} + \frac{5}{x} = \frac{15}{x}$$

77. Describe your running rate if you have no more than a total of 3 hours for driving and running. Use a rational inequality to solve the problem. Then explain how your solution is shown on the graph.

78. Describe your running rate if you have no more than a total of 5 hours for driving and running. Use a rational inequality to solve the problem. Then explain how your solution is shown on the graph.

79. Describe the behavior of the graph as $x \to \infty$. What does this show about the time driving and running as a function of your running rate?

80. Describe the behavior of the graph as $x \to 0^+$. What does this show about the time driving and running as a function of your running rate?

81. Describe how to use the formula $t = \dfrac{s}{v}$ and the problem's verbal conditions to obtain the function's equation displayed in the voice balloon.

82. The perimeter of a rectangle is 50 feet. Describe the possible lengths of a side if the area of the rectangle is not to exceed 114 square feet.

83. The perimeter of a rectangle is 180 feet. Describe the possible lengths of a side if the area of the rectangle is not to exceed 800 square feet.

Writing in Mathematics

84. What is a polynomial inequality?

85. What is a rational inequality?

86. If f is a polynomial or rational function, explain how the graph of f can be used to visualize the solution set of the inequality $f(x) < 0$.

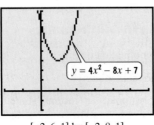

Technology Exercises

87. Use a graphing utility to verify your solution sets to any three of the polynomial inequalities that you solved algebraically in Exercises 1–38.

88. Use a graphing utility to verify your solution sets to any three of the rational inequalities that you solved algebraically in Exercises 39–56.

Solve each inequality in Exercises 89–94 using a graphing utility.

89. $x^2 + 3x - 10 > 0$

90. $2x^2 + 5x - 3 \leq 0$

91. $x^3 + x^2 - 4x - 4 > 0$

92. $\dfrac{x - 4}{x - 1} \leq 0$

93. $\dfrac{x + 2}{x - 3} \leq 2$

94. $\dfrac{1}{x + 1} \leq \dfrac{2}{x + 4}$

Critical Thinking Exercises

95. Which one of the following is true?

   a. The solution set of $x^2 > 25$ is $(5, \infty)$.

   b. The inequality $\dfrac{x - 2}{x + 3} < 2$ can be solved by multiplying both sides by $x + 3$, resulting in the equivalent inequality $x - 2 < 2(x + 3)$.

   c. $(x + 3)(x - 1) \geq 0$ and $\dfrac{x + 3}{x - 1} \geq 0$ have the same solution set.

   d. None of these statements is true.

96. Write a polynomial inequality whose solution set is $[-3, 5]$.

97. Write a rational inequality whose solution set is $(-\infty, -4) \cup [3, \infty)$.

In Exercises 98–101, use inspection to describe each inequality's solution set. Do not solve any of the inequalities.

98. $(x - 2)^2 > 0$ **99.** $(x - 2)^2 \leq 0$

100. $(x - 2)^2 < -1$ **101.** $\dfrac{1}{(x - 2)^2} > 0$

102. The graphing utility screen shows the graph of $y = 4x^2 - 8x + 7$.

$[-2, 6, 1]$ by $[-2, 8, 1]$

a. Use the graph to describe the solution set of $4x^2 - 8x + 7 > 0$.

b. Use the graph to describe the solution set of $4x^2 - 8x + 7 < 0$.

c. Use an algebraic approach to verify each of you descriptions in parts (a) and (b).

103. The graphing utility screen shows the graph of $y = \sqrt{27 - 3x^2}$. Write and solve a quadratic inequalit that explains why the graph only appears for $-3 \leq x \leq $

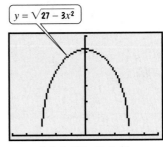

$[-5, 5, 1]$ by $[0, 6, 1]$

Group Exercise

104. This exercise is intended as a group learning experience an is appropriate for groups of three to five people. Befor working on the various parts of the problem, reread th description of the position function on page 352.

   a. Drop a ball from a height of 3 feet, 6 feet, and 12 fee Record the number of seconds it takes for the ball to h the ground.

   b. For each of the three initial positions, use the positio function to determine the time required for the ball t hit the ground.

   c. What factors might result in differences between th times that you recorded and the times indicated by th function?

   d. What appears to be happening to the time required for free-falling object to hit the ground as its initial height doubled? Verify this observation algebraically and wit a graphing utility.

   e. Repeat part (a) using a sheet of paper rather than a ba What differences do you observe? What factor seems be ignored in the position function?

   f. What is meant by the acceleration of gravity and ho does this number appear in the position function for free-falling object?

Objectives

1 Solve direct variation problems.

2 Solve inverse variation problems.

3 Solve combined variation problems.

4 Solve problems involving joint variation.

Have you ever wondered how telecommunication companies estimate the number of phone calls expected per day between two cities? The formula

$$C = \frac{0.02 P_1 P_2}{d^2}$$

shows that the daily number of phone calls, C, increases as the populations of the cities, P_1 and P_2, in thousands, increase and decreases as the distance, d, between the cities increases.

Certain formulas occur so frequently in applied situations that they are given special names. Variation formulas show how one quantity changes in relation to other quantities. Quantities can vary *directly, inversely,* or *jointly.* In this section, we look at situations that can be modeled by each of these kinds of variation. And think of this: The next time you get one of those "all-circuits-are-busy" messages, you will be able to use a variation formula to estimate how many other callers you're competing with for those precious 5-cent minutes.

1 Solve direct variation problems.

Direct Variation

When you swim underwater, the pressure in your ears depends on the depth at which you are swimming. The formula

$$p = 0.43d$$

describes the water pressure, p, in pounds per square inch, at a depth of d feet. We can use this linear function to determine the pressure in your ears at various depths:

If $d = 20$, $p = 0.43(20) = 8.6$. At a depth of 20 feet, water pressure is 8.6 pounds per square inch.

Doubling the depth doubles the pressure.

If $d = 40$, $p = 0.43(40) = 17.2$. At a depth of 40 feet, water pressure is 17.2 pounds per square inch.

Doubling the depth doubles the pressure.

If $d = 80$, $p = 0.43(80) = 34.4$. At a depth of 80 feet, water pressure is 34.4 pounds per square inch.

The formula $p = 0.43d$ illustrates that water pressure is a constant multiple of your underwater depth. If your depth is doubled, the pressure is doubled; if your depth is tripled, the pressure is tripled; and so on. Because of this, the pressure in your ears is said to **vary directly** as your underwater depth. The **equation of variation** is

$$p = 0.43d.$$

Generalizing our discussion of pressure and depth on the previous page, w obtain the following statement:

Direct Variation

If a situation is described by an equation in the form

$$y = kx,$$

where k is a nonzero constant, we say that **y varies directly as x** or **y is directly proportional to x**. The number k is called the **constant of variation** or the **constant of proportionality**.

Can you see that **the direct variation equation, $y = kx$, is a special case of th** **linear function $y = mx + b$**? When $m = k$ and $b = 0$, $y = mx + b$ become $y = kx$. Thus, the slope of a direct variation equation is k, the constant of variation Because b, the y-intercept, is 0, the graph of a direct variation equation is a lin passing through the origin. This is illustrated in Figure 2.45, which shows the grap of $p = 0.43d$: Water pressure varies directly as depth.

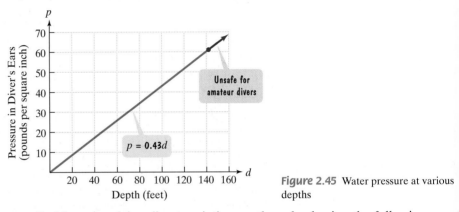

Figure 2.45 Water pressure at various depths

Problems involving direct variation can be solved using the following proce dure. This procedure applies to direct variation problems, as well as to the othe kinds of variation problems that we will discuss.

Solving Variation Problems

1. Write an equation that describes the given English statement.
2. Substitute the given pair of values into the equation in step 1 and solve for k, the constant of variation.
3. Substitute the value of k into the equation in step 1.
4. Use the equation from step 3 to answer the problem's question.

EXAMPLE 1   Solving a Direct Variation Problem

Many areas of Northern California depend on the snowpack of the Sierra Nevac mountain range for their water supply. The volume of water produced from meltir snow varies directly as the volume of snow. Meteorologists have determined th 250 cubic centimeters of snow will melt to 28 cubic centimeters of water. How muc water does 1200 cubic centimeters of melting snow produce?

Solution

Step 1   Write an equation. We know that y varies directly as x is expressed as

$$y = kx.$$

By changing letters, we can write an equation that describes the following Englis statement: Volume of water, W, varies directly as volume of snow, S.

$$W = kS$$

Step 2   Use the given values to find k. We are told that 250 cubic centimeters of snow will melt to 28 cubic centimeters of water. Substitute 250 for S and 28 for W in the direct variation equation. Then solve for k.

$$W = kS \qquad \text{Volume of water varies directly as volume of melting snow.}$$

$$28 = k(250) \qquad \text{250 cubic centimeters of snow melt to 28 cubic centimeters of water.}$$

$$\frac{28}{250} = \frac{k(250)}{250} \qquad \text{Divide both sides by 250.}$$

$$0.112 = k \qquad \text{Simplify.}$$

Step 3   Substitute the value of k into the equation.

$$W = kS \qquad \text{This is the equation from step 1.}$$

$$W = 0.112S \qquad \text{Replace } k \text{, the constant of variation, with 0.112.}$$

Step 4   Answer the problem's question. How much water does 1200 cubic centimeters of melting snow produce? Substitute 1200 for S in $W = 0.112S$ and solve for W.

$$W = 0.112S \qquad \text{Use the equation from step 3.}$$

$$W = 0.112(1200) \qquad \text{Substitute 1200 for } S.$$

$$W = 134.4 \qquad \text{Multiply.}$$

A snowpack measuring 1200 cubic centimeters will produce 134.4 cubic centimeters of water.

Check Point 1   The number of gallons of water, W, used when taking a shower varies directly as the time, t, in minutes, in the shower. A shower lasting 5 minutes uses 30 gallons of water. How much water is used in a shower lasting 11 minutes?

The direct variation equation $y = kx$ is a linear function. If $k > 0$, then the slope of the line is positive. Consequently, as x increases, y also increases.

A direct variation situation can involve variables to higher powers. For example, y can vary directly as $x^2 (y = kx^2)$ or as $x^3 (y = kx^3)$.

Direct Variation with Powers

y **varies directly as the nth power of x** if there exists some nonzero constant k such that

$$y = kx^n.$$

We also say that y **is directly proportional to the nth power of x**.

Direct variation with whole number powers is modeled by polynomial functions. In our next example, the graph of the variation equation is the familiar parabola.

EXAMPLE 2   Solving a Direct Variation Problem

The distance, s, that a body falls from rest varies directly as the square of the time, t, of the fall. If skydivers fall 64 feet in 2 seconds, how far will they fall in 4.5 seconds?

Solution

Step 1   Write an equation. We know that y *varies directly as the square of x* is expressed as

$$y = kx^2.$$

By changing letters, we can write an equation that describes the following Englis▮ statement: Distance, s, varies directly as the square of time, t, of the fall.

$$s = kt^2$$

Step 2   Use the given values to find k. Skydivers fall 64 feet in 2 seconds. Subst▮tute 64 for s and 2 for t in the direct variation equation. Then solve for k.

$s = kt^2$	Distance varies directly as the square of time
$64 = k \cdot 2^2$	Skydivers fall 64 feet in 2 seconds.
$64 = 4k$	Simplify: $2^2 = 4$.
$\dfrac{64}{4} = \dfrac{4k}{4}$	Divide both sides by 4.
$16 = k$	Simplify.

Step 3   Substitute the value of k into the equation.

$s = kt^2$	Use the equation from step 1.
$s = 16t^2$	Replace k, the constant of variation, with 16.

Step 4   Answer the problem's question. How far will the skydivers fall in 4.▮ seconds? Substitute 4.5 for t in $s = 16t^2$ and solve for s.

$$s = 16(4.5)^2 = 16(20.25) = 324$$

Thus, in 4.5 seconds, the skydivers will fall 324 feet.

We can express the variation equation from Example 2 in function notatio▮ writing

$$s(t) = 16t^2.$$

The distance that a body falls from rest is a function of the time, t, of the fall. Th▮ parabola that is the graph of this quadratic function is shown in Figure 2.46. The grap▮ increases rapidly from left to right, showing the effects of the acceleration of gravit▮

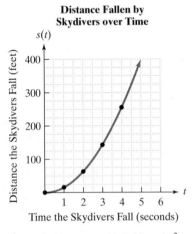

Distance Fallen by Skydivers over Time

Figure 2.46 The graph of $s(t) = 16t^2$

 Solve inverse variation problems.

 Check Point 2 The distance required to stop a car varies directly as the square of i▮ speed. If 200 feet are required to stop a car traveling 60 miles per hou▮ how many feet are required to stop a car traveling 100 miles per hour?

Inverse Variation

The distance from San Francisco to Los Angeles is 420 miles. The time that it take▮ to drive from San Francisco to Los Angeles depends on the rate at which one drive▮ and is given by

$$\text{Time} = \frac{420}{\text{Rate}}.$$

For example, if you average 30 miles per hour, the time for the drive is

$$\text{Time} = \frac{420}{30} = 14,$$

or 14 hours. If you average 50 miles per hour, the time for the drive is

$$\text{Time} = \frac{420}{50} = 8.4,$$

or 8.4 hours. As your rate (or speed) increases, the time for the trip decreases and vice versa. This is illustrated by the graph in Figure 2.47.

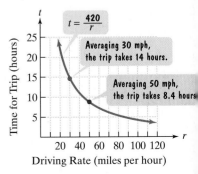

Figure 2.47

We can express the time for the San Francisco–Los Angeles trip using t for time and r for rate:

$$t = \frac{420}{r}.$$

This equation is an example of an **inverse variation** equation. Time, t, **varies inversely** as rate, r. When two quantities vary inversely, one quantity increases as the other decreases and vice versa.

Generalizing, we obtain the following statement:

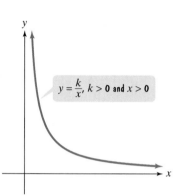

y

$y = \frac{k}{x}, k > 0$ and $x > 0$

x

Inverse Variation

If a situation is described by an equation in the form

$$y = \frac{k}{x},$$

where k is a nonzero constant, we say that **y varies inversely as x** or **y is inversely proportional to x**. The number k is called the **constant of variation**.

Notice that **the inverse variation equation**

$$y = \frac{k}{x}, \quad \text{or} \quad f(x) = \frac{k}{x},$$

Figure 2.48 The graph of the inverse variation equation

is a rational function. For $k > 0$ and $x > 0$, the graph of the function takes on the shape shown in Figure 2.48.

We use the same procedure to solve inverse variation problems as we did to solve direct variation problems. Example 3 illustrates this procedure.

EXAMPLE 3   Solving an Inverse Variation Problem

When you use a spray can and press the valve at the top, you decrease the pressure of the gas in the can. This decrease of pressure causes the volume of the gas in the can to increase. Because the gas needs more room than is provided in the can, it expands in spray form through the small hole near the valve. In general, if the temperature is constant, the pressure, P, of a gas in a container varies inversely as the volume, V, of the container. The pressure of a gas sample in a container whose volume is 8 cubic inches is 12 pounds per square inch. If the sample expands to a volume of 22 cubic inches, what is the new pressure of the gas?

2P

P

2V

V

Doubling the pressure halves the volume.

Solution

Step 1   Write an equation. We know that y *varies inversely as* x is expressed as

$$y = \frac{k}{x}.$$

By changing letters, we can write an equation that describes the following English statement: The pressure, P, of a gas in a container varies inversely as the volume, V.

$$P = \frac{k}{V}.$$

Step 2   Use the given values to find k. The pressure of a gas sample in a container whose volume is 8 cubic inches is 12 pounds per square inch. Substitute 12 for P and 8 for V in the inverse variation equation. Then solve for k.

$$P = \frac{k}{V} \qquad \text{Pressure varies inversely as volume.}$$

$$12 = \frac{k}{8} \qquad \text{The pressure in an 8 cubic-inch container is 12 pounds per square inch.}$$

$$12 \cdot 8 = \frac{k}{8} \cdot 8 \qquad \text{Multiply both sides by 8.}$$

$$96 = k \qquad \text{Simplify.}$$

Step 3   Substitute the value of k into the equation.

$$P = \frac{k}{V}$$ Use the equation from step 1.

$$P = \frac{96}{V}$$ Replace k, the constant of variation, with 96.

Step 4   Answer the problem's question. We need to find the pressure when the volume expands to 22 cubic inches. Substitute 22 for V and solve for P.

$$P = \frac{96}{V} = \frac{96}{22} = 4\frac{4}{11}$$

When the volume is 22 cubic inches, the pressure of the gas is $4\frac{4}{11}$ pounds per square inch.

 Check Point 3   The length of a violin string varies inversely as the frequency of its vibrations. A violin string 8 inches long vibrates at a frequency of 640 cycles per second. What is the frequency of a 10-inch string?

③ Solve combined variation problems.

Combined Variation

In **combined variation**, direct and inverse variation occur at the same time. For example, as the advertising budget, A, of a company increases, its monthly sales, S, also increase. Monthly sales vary directly as the advertising budget:

$$S = kA.$$

By contrast, as the price of the company's product, P, increases, its monthly sales, S, decrease. Monthly sales vary inversely as the price of the product:

$$S = \frac{k}{P}.$$

We can combine these two variation equations into one combined equation:

$$S = \frac{kA}{P}.$$

Monthly sales , S, vary directly as the advertising budget, A, and inversely as the price of the product, P.

The following example illustrates an application of combined variation.

EXAMPLE 4   Solving a Combined Variation Problem

The owners of Rollerblades Plus determine that the monthly sales, S, of its skates vary directly as its advertising budget, A, and inversely as the price of the skates, P. When $60,000 is spent on advertising and the price of the skates is $40, the monthly sales are 12,000 pairs of rollerblades.

    a. Write an equation of variation that describes this situation.

    b. Determine monthly sales if the amount of the advertising budget is increased to $70,000.

Solution

    a. Write an equation.

$$S = \frac{kA}{P}.$$

Translate "sales vary directly as the advertising budget and inversely as the skates' price."

Use the given values to find k.

$$12,000 = \frac{k(60,000)}{40}$$ When \$60,000 is spent on advertising $(A = 60,000)$ and the price is \$40 $(P = 40)$, monthly sales are 12,000 units $(S = 12,000)$.

$$12,000 = k \cdot 1500$$ Divide 60,000 by 40.

$$\frac{12,000}{1500} = \frac{k \cdot 1500}{1500}$$ Divide both sides of the equation by 1500.

$$8 = k$$ Simplify.

Therefore, the equation of variation that describes monthly sales is

$$S = \frac{8A}{P}.$$ Substitute 8 for k in $S = \frac{kA}{P}$.

b. The advertising budget is increased to \$70,000, so $A = 70,000$. The skates' price is still \$40, so $P = 40$.

$$S = \frac{8A}{P}$$ This is the equation from part (a).

$$S = \frac{8(70,000)}{40}$$ Substitute 70,000 for A and 40 for P.

$$S = 14,000$$ Simplify.

With a \$70,000 advertising budget and \$40 price, the company can expect to sell 14,000 pairs of rollerblades in a month (up from 12,000).

Check Point 4 The number of minutes needed to solve an exercise set of variation problems varies directly as the number of problems and inversely as the number of people working to solve the problems. It takes 4 people 32 minutes to solve 16 problems. How many minutes will it take 8 people to solve 24 problems?

④ Solve problems involving joint variation.

Joint Variation

Joint variation is a variation in which a variable varies directly as the product of two or more other variables. Thus, the equation $y = kxz$ is read "y varies jointly as x and z."

Joint variation plays a critical role in Isaac Newton's formula for gravitation:

$$F = G\frac{m_1 m_2}{d^2}.$$

The formula states that the force of gravitation, F, between two bodies varies jointly as the product of their masses, m_1 and m_2, and inversely as the square of the distance between them, d. (G is the gravitational constant.) The formula indicates that gravitational force exists between any two objects in the universe, increasing as the distance between the bodies decreases. One practical result is that the pull of the moon on the oceans is greater on the side of Earth closer to the moon. This gravitational imbalance is what produces tides.

EXAMPLE 5   Modeling Centrifugal Force

The centrifugal force, C, of a body moving in a circle varies jointly with the radius of the circular path, r, and the body's mass, m, and inversely with the square of the time, t, it takes to move about one full circle. A 6-gram body moving in a circle with radius 100 centimeters at a rate of 1 revolution in 2 seconds has a centrifugal force of 6000 dynes. Find the centrifugal force of an 18-gram body moving in a circle with radius 100 centimeters at a rate of 1 revolution in 3 seconds.

Solution

$$C = \frac{krm}{t^2}$$

Translate "Centrifugal force, C, varies jointly with radius, r, and mass, m, and inversely with the square of time, t."

$$6000 = \frac{k(100)(6)}{2^2}$$

A 6-gram body ($m = 6$) moving in a circle with radius 100 centimeters ($r = 100$) at 1 revolution in 2 seconds ($t = 2$) has a centifugal force of 6000 dynes ($C = 6000$).

$$6000 = 150k$$

Simplify.

$$40 = k$$

Divide both sides by 150 and solve for k.

$$C = \frac{40rm}{t^2}$$

Substitute 40 for k in the model for centrifugal force.

$$C = \frac{40(100)(18)}{3^2}$$

Find centifugal force, C, of an 18-gram body ($m = 18$) moving in a circle with radius 100 centimeters ($r = 100$) at 1 revolution in 3 seconds ($t = 3$).

$$= 8000$$

Simplify.

The centrifugal force is 8000 dynes.

 Check Point 5 The volume of a cone, V, varies jointly as its height, h, and the square of it radius, r. A cone with a radius measuring 6 feet and a height measuring 1 feet has a volume of 120π cubic feet. Find the volume of a cone having radius of 12 feet and a height of 2 feet.

EXERCISE SET 2.8

 Practice Exercises

Use the four-step procedure for solving variation problems given on page 358 to solve Exercises 1–10.

1. y varies directly as x. $y = 65$ when $x = 5$. Find y when $x = 12$.

2. y varies directly as x. $y = 45$ when $x = 5$. Find y when $x = 13$.

3. y varies inversely as x. $y = 12$ when $x = 5$. Find y when $x = 2$.

4. y varies inversely as x. $y = 6$ when $x = 3$. Find y when $x = 9$.

5. y varies directly as x and inversely as the square of z. $y = 20$ when $x = 50$ and $z = 5$. Find y when $x = 3$ and $z = 6$.

6. a varies directly as b and inversely as the square of c. $a = 7$ when $b = 9$ and $c = 6$. Find a when $b = 4$ and $c = 8$.

7. y varies jointly as x and z. $y = 25$ when $x = 2$ and $z = 5$. Find y when $x = 8$ and $z = 12$.

8. C varies jointly as A and T. $C = 175$ when $A = 2100$ and $T = 4$. Find C when $A = 2400$ and $T = 6$.

9. y varies jointly as a and b and inversely as the square root of c. $y = 12$ when $a = 3, b = 2$, and $c = 25$. Find y when $a = 5, b = 3$, and $c = 9$.

10. y varies jointly as m and the square of n and inversely as p. $y = 15$ when $m = 2, n = 1$, and $p = 6$. Find y whe $m = 3, n = 4$, and $p = 10$.

Practice Plus

In Exercises 11–20, write an equation that expresses each relationship. Then solve the equation for y.

11. x varies jointly as y and z.

12. x varies jointly as y and the square of z.

13. x varies directly as the cube of z and inversely as y.

14. x varies directly as the cube root of z and inversely as y.

15. x varies jointly as y and z and inversely as the square roc of w.

16. x varies jointly as y and z and inversely as the square of u

17. x varies jointly as z and the sum of y and w.

18. x varies jointly as z and the difference between y and w.

19. x varies directly as z and inversely as the difference betwee y and w.

20. x varies directly as z and inversely as the sum of y and w.

Application Exercises

Use the four-step procedure for solving variation problems given on page 358 to solve Exercises 21–36.

1. An alligator's tail length, T, varies directly as its body length, B. An alligator with a body length of 4 feet has a tail length of 3.6 feet. What is the tail length of an alligator whose body length is 6 feet?

|← — Body length, B — →|← — Tail length, T — →|

2. An object's weight on the moon, M, varies directly as its weight on Earth, E. Neil Armstrong, the first person to step on the moon on July 20, 1969, weighed 360 pounds on Earth (with all of his equipment on) and 60 pounds on the moon. What is the moon weight of a person who weighs 186 pounds on Earth?

3. The height that a ball bounces varies directly as the height from which it was dropped. A tennis ball dropped from 12 inches bounces 8.4 inches. From what height was the tennis ball dropped if it bounces 56 inches?

4. The distance that a spring will stretch varies directly as the force applied to the spring. A force of 12 pounds is needed to stretch a spring 9 inches. What force is required to stretch the spring 15 inches?

5. If all men had identical body types, their weight would vary directly as the cube of their height. Shown below is Robert Wadlow, who reached a record height of 8 feet 11 inches (107 inches) before his death at age 22. If a man who is 5 feet 10 inches tall (70 inches) with the same body type as Mr. Wadlow weighs 170 pounds, what was Robert Wadlow's weight shortly before his death?

6. On a dry asphalt road, a car's stopping distance varies directly as the square of its speed. A car traveling at 45 miles per hour can stop in 67.5 feet. What is the stopping distance for a car traveling at 60 miles per hour?

27. The figure shows that a bicyclist tips the cycle when making a turn. The angle B, formed by the vertical direction and the bicycle, is called the banking angle. The banking angle varies inversely as the cycle's turning radius. When the turning radius is 4 feet, the banking angle is 28°. What is the banking angle when the turning radius is 3.5 feet?

28. The water temperature of the Pacific Ocean varies inversely as the water's depth. At a depth of 1000 meters, the water temperature is 4.4° Celsius. What is the water temperature at a depth of 5000 meters?

29. Radiation machines, used to treat tumors, produce an intensity of radiation that varies inversely as the square of the distance from the machine. At 3 meters, the radiation intensity is 62.5 milliroentgens per hour. What is the intensity at a distance of 2.5 meters?

30. The illumination provided by a car's headlight varies inversely as the square of the distance from the headlight. A car's headlight produces an illumination of 3.75 footcandles at a distance of 40 feet. What is the illumination when the distance is 50 feet?

31. Body-mass index, or BMI, takes both weight and height into account when assessing whether an individual is underweight or overweight. BMI varies directly as one's weight, in pounds, and inversely as the square of one's height, in inches. In adults, normal values for the BMI are between 20 and 25, inclusive. Values below 20 indicate that an individual is underweight and values above 30 indicate that an individual is obese. A person who weighs 180 pounds and is 5 feet, or 60 inches, tall has a BMI of 35.15. What is the BMI, to the nearest tenth, for a 170 pound person who is 5 feet 10 inches tall. Is this person overweight?

32. One's intelligence quotient, or IQ, varies directly as a person's mental age and inversely as that person's chronological age. A person with a mental age of 25 and a chronological age of 20 has an IQ of 125. What is the chronological age of a person with a mental age of 40 and an IQ of 80?

33. The heat loss of a glass window varies jointly as the window's area and the difference between the outside and inside temperatures. A window 3 feet wide by 6 feet long loses 1200 Btu per hour when the temperature outside is 20° colder than the temperature inside. Find the heat loss through a glass window that is 6 feet wide by 9 feet long when the temperature outside is 10° colder than the temperature inside.

34. Kinetic energy varies jointly as the mass and the square of the velocity. A mass of 8 grams and velocity of 3 centimeters per second has a kinetic energy of 36 ergs. Find the kinetic energy for a mass of 4 grams and velocity of 6 centimeters per second.

35. Sound intensity varies inversely as the square of the distance from the sound source. If you are in a movie theater and you change your seat to one that is twice as far from the speakers, how does the new sound intensity compare to that of your original seat?

36. Many people claim that as they get older, time seems to pass more quickly. Suppose that the perceived length of a period of time is inversely proportional to your age. How long will a year seem to be when you are three times as old as you are now?

37. The average number of daily phone calls, C, between two cities varies jointly as the product of their populations, P_1 and P_2, and inversely as the square of the distance, d, between them.

   a. Write an equation that expresses this relationship.

   b. The distance between San Francisco (population: 777,000) and Los Angeles (population: 3,695,000) is 420 miles. If the average number of daily phone calls between the cities is 326,000, find the value of k to two decimal places and write the equation of variation.

   c. Memphis (population: 650,000) is 400 miles from New Orleans (population: 490,000). Find the average number of daily phone calls, to the nearest whole number, between these cities.

38. The force of wind blowing on a window positioned at a right angle to the direction of the wind varies jointly as the area of the window and the square of the wind's speed. It is known that a wind of 30 miles per hour blowing on a window measuring 4 feet by 5 feet exerts a force of 150 pounds. During a storm with winds of 60 miles per hour, should hurricane shutters be placed on a window that measures 3 feet by 4 feet and is capable of withstanding 300 pounds of force?

39. The table shows the values for the current, I, in an electric circuit and the resistance, R, of the circuit.

I (amperes)	0.5	1.0	1.5	2.0	2.5	3.0	4.0	5.0
R (ohms)	12.0	6.0	4.0	3.0	2.4	2.0	1.5	1.2

   a. Graph the ordered pairs in the table of values, with values of I along the x-axis and values of R along the y-axis. Connect the eight points with a smooth curve.

   b. Does current vary directly or inversely as resistance? Use your graph and explain how you arrived at your answer.

   c. Write an equation of variation for I and R, using one of the ordered pairs in the table to find the constant of variation. Then use your variation equation to verify the other seven ordered pairs in the table.

Writing in Mathematics

40. What does it mean if two quantities vary directly?

41. In your own words, explain how to solve a variation problem.

42. What does it mean if two quantities vary inversely?

43. Explain what is meant by combined variation. Give an example with your explanation.

44. Explain what is meant by joint variation. Give an example with your explanation.

In Exercises 45–46, describe in words the variation shown by the given equation.

45. $z = \dfrac{k\sqrt{x}}{y^2}$ **46.** $z = kx^2\sqrt{y}$

47. We have seen that the daily number of phone calls between two cities varies jointly as their populations and inversely as the square of the distance between them. This model, used by telecommunication companies to estimate the line capacities needed among various cities, is called the *gravity model*. Compare the model to Newton's formula for gravitation on page 363 and describe why the name *gravity model* is appropriate.

Technology Exercise

48. Use a graphing utility to graph any three of the variation equations in Exercises 21–30. Then TRACE along each curve and identify the point that corresponds to the problem's solution.

Critical Thinking Exercises

49. In a hurricane, the wind pressure varies directly as the square of the wind velocity. If wind pressure is a measure of a hurricane's destructive capacity, what happens to this destructive power when the wind speed doubles?

50. The illumination from a light source varies inversely as the square of the distance from the light source. If you raise a lamp from 15 inches to 30 inches over your desk, what happens to the illumination?

51. The heat generated by a stove element varies directly as the square of the voltage and inversely as the resistance. If the voltage remains constant, what needs to be done to triple the amount of heat generated?

52. Galileo's telescope brought about revolutionary changes in astronomy. A comparable leap in our ability to observe the universe took place as a result of the Hubble Space Telescope. The space telescope was able to see stars and galaxies whose brightness is $\frac{1}{50}$ of the faintest objects observable using ground-based telescopes. Use the fact that the brightness of a point source, such as a star, varies inversely as the square of its distance from an observer to show that the space telescope was able to see about seven times farther than a ground-based telescope.

Group Exercise

53. Begin by deciding on a product that interests the group because you are now in charge of advertising this product. Members were told that the demand for the product varies directly as the amount spent on advertising and inversely as the price of the product. However, as more money is spent on advertising, the price of your product rises. Under what conditions would members recommend an increased expense in advertising? Once you've determined what your product is, write formulas for the given conditions and experiment with hypothetical numbers. What other factors might you take into consideration in terms of your recommendation? How do these factor affect the demand for your product?

Chapter 2
Summary, Review, and Test

Summary

DEFINITIONS AND CONCEPTS	**EXAMPLES**

2.1 Complex Numbers

a. The imaginary unit i is defined as
$$i = \sqrt{-1}, \text{ where } i^2 = -1.$$
The set of numbers in the form $a + bi$ is called the set of complex numbers; a is the real part and b is the imaginary part. If $b = 0$, the complex number is a real number. If $b \neq 0$, the complex number is an imaginary number. Complex numbers in the form bi are called pure imaginary numbers.

b. Rules for adding and subtracting complex numbers are given in the box on page 267. — Ex. 1, p. 267

c. To multiply complex numbers, multiply as if they are polynomials. After completing the multiplication, replace i^2 with -1 and simplify. — Ex. 2, p. 268

d. The complex conjugate of $a + bi$ is $a - bi$ and vice versa. The multiplication of complex conjugates gives a real number:
$$(a + bi)(a - bi) = a^2 + b^2.$$

e. To divide complex numbers, multiply the numerator and the denominator by the complex conjugate of the denominator. — Ex. 3, p. 269

f. When performing operations with square roots of negative numbers, begin by expressing all square roots in terms of i. The principal square root of $-b$ is defined by
$$\sqrt{-b} = i\sqrt{b}.$$ — Ex. 4, p. 270

g. Quadratic equations $(ax^2 + bx + c = 0, a \neq 0)$ with negative discriminants $(b^2 - 4ac < 0)$ have imaginary solutions that are complex conjugates. — Ex. 5, p. 271

2.2 Quadratic Functions

a. A quadratic function is of the form $f(x) = ax^2 + bx + c, a \neq 0$.

b. The standard form of a quadratic function is $f(x) = a(x - h)^2 + k, a \neq 0$.

c. The graph of a quadratic function is a parabola. The vertex is (h, k) or $\left(-\dfrac{b}{2a}, f\left(-\dfrac{b}{2a}\right)\right)$. A procedure for graphing a quadratic function is given in the box on page 275. — Ex. 1, p. 275; Ex. 2, p. 276; Ex. 3, p. 277

d. See the box on page 279 for minimum or maximum values of quadratic functions. — Ex. 4, p. 279; Ex. 5, p. 280

e. A strategy for solving problems involving maximizing or minimizing quadratic functions is given in the box on page 281. — Ex. 6, p. 281; Ex. 7, p. 283

2.3 Polynomial Functions and Their Graphs

a. Polynomial Function of Degree n: $f(x) = a_n x^n + a_{n-1} x^{n-1} + \cdots + a_2 x^2 + a_1 x + a_0, a_n \neq 0$

b. The graphs of polynomial functions are smooth and continuous. — Fig. 2.12, p. 289

c. The end behavior of the graph of a polynomial function depends on the leading term, given by the Leading Coefficient Test in the box on page 290. — Ex. 1, p. 290; Ex. 2, p. 290; Ex. 3, p. 291

d. The values of x for which $f(x)$ is equal to 0 are the zeros of the polynomial function f. These values are the roots, or solutions, of the polynomial equation $f(x) = 0$. — Ex. 4, p. 292; Ex. 5, p. 293

e. If $x - r$ occurs k times in a polynomial function's factorization, r is a repeated zero with multiplicity k. If k is even, the graph touches the x-axis and turns around at r. If k is odd, the graph crosses the x-axis at r. — Ex. 6, p. 294

f. The Intermediate Value Theorem: If f is a polynomial function and $f(a)$ and $f(b)$ have opposite signs, there is at least one value of c between a and b for which $f(c) = 0$. — Ex. 7, p. 294

g. If f is a polynomial of degree n, the graph of f has at most $n - 1$ turning points. — Fig. 2.22, p. 295

h. A strategy for graphing a polynomial function is given in the box on page 295. — Ex. 8, p. 296

DEFINITIONS AND CONCEPTS	EXAMPLES

2.4 Dividing Polynomials; Remainder and Factor Theorems

a. Long division of polynomials is performed by dividing, multiplying, subtracting, bringing down the next term, and repeating this process until the degree of the remainder is less than the degree of the divisor. The details are given in the box on page 303.
Ex. 1, p. 302; Ex. 2, p. 303; Ex. 3, p. 305

b. The Division Algorithm: $f(x) = d(x)q(x) + r(x)$. The dividend is the product of the divisor and the quotient plus the remainder.

c. Synthetic division is used to divide a polynomial by $x - c$. The details are given in the box on page 306.
Ex. 4, p. 306

d. The Remainder Theorem: If a polynomial $f(x)$ is divided by $x - c$, then the remainder is $f(c)$.
Ex. 5, p. 308

e. The Factor Theorem: If $x - c$ is a factor of a polynomial function $f(x)$, then c is a zero of f and a root of $f(x) = 0$. If c is a zero of f or a root of $f(x) = 0$, then $x - c$ is a factor of $f(x)$.
Ex. 6, p. 309

2.5 Zeros of Polynomial Functions

a. The Rational Zero Theorem states that the possible rational zeros of a polynomial
$$\text{function} = \frac{\text{Factors of the constant term}}{\text{Factors of the leading coefficient}}.$$ The theorem is stated in the box on page 313.
Ex. 1, p. 313; Ex. 2, p. 314; Ex. 3, p. 314; Ex. 4, p. 315; Ex. 5, p. 316

b. Number of roots: If $f(x)$ is a polynomial of degree $n \geq 1$, then, counting multiple roots separately, the equation $f(x) = 0$ has n roots.

c. If $a + bi$ is a root of $f(x) = 0$, then $a - bi$ is also a root.

d. The Linear Factorization Theorem: An nth-degree polynomial can be expressed as the product of n linear factors. Thus, $f(x) = a_n(x - c_1)(x - c_2) \cdots (x - c_n)$.
Ex. 6, p. 319

e. Descartes's Rule of Signs: The number of positive real zeros of f equals the number of sign changes of $f(x)$ or is less than that number by an even integer. The number of negative real zeros of f applies a similar statement to $f(-x)$.
Table 2.1, p. 320; Ex. 7, p. 321

2.6 Rational Functions and Their Graphs

a. Rational function: $f(x) = \dfrac{p(x)}{q(x)}$; p and q are polynomial functions and $q(x) \neq 0$. The domain of f is the set of all real numbers excluding values of x that make $q(x)$ zero.
Ex. 1, p. 326

b. Arrow notation is summarized in the box on page 328.

c. The line $x = a$ is a vertical asymptote of the graph of f if $f(x)$ increases or decreases without bound as x approaches a. Vertical asymptotes are identified using the location theorem in the box on page 329.
Ex. 2, p. 330

d. The line $y = b$ is a horizontal asymptote of the graph of f if $f(x)$ approaches b as x increases or decreases without bound. Horizontal asymptotes are identified using the location theorem in the box on page 332.
Ex. 3, p. 332

e. Table 2.2 on page 333 shows the graphs of $f(x) = \dfrac{1}{x}$ and $f(x) = \dfrac{1}{x^2}$. Some rational functions can be graphed using transformations of these common graphs.
Ex. 4, p. 334

f. A strategy for graphing rational functions is given in the box on page 334.
Ex. 5, p. 335; Ex. 6, p. 336; Ex. 7, p. 337

g. The graph of a rational function has a slant asymptote when the degree of the numerator is one more than the degree of the denominator. The equation of the slant asymptote is found using division and dropping the remainder term.
Ex. 8, p. 339

2.7 Polynomial and Rational Inequalities

a. A polynomial inequality can be expressed as $f(x) < 0, f(x) > 0, f(x) \leq 0$, or $f(x) \geq 0$, where f is a polynomial function. A procedure for solving polynomial inequalities is given in the box on page 347.
Ex. 1, p. 347; Ex. 2, p. 348

DEFINITIONS AND CONCEPTS	EXAMPLES

b. A rational inequality can be expressed as $f(x) < 0$, $f(x) > 0$, $f(x) \le 0$, or $f(x) \ge 0$, where f is a rational function. The procedure for solving such inequalities begins with expressing them so that one side is zero and the other side is a single quotient. Find boundary points by setting the numerator and denominator equal to zero. Then follow a procedure similar to that for solving polynomial inequalities.

Ex. 3, p. 350

2.8 Modeling Using Variation

a. A procedure for solving variation problems is given in the lower box on page 358.

b. English Statement	Equation	
y varies directly as x. y is directly proportional to x.	$y = kx$	Ex. 1, p. 358
y varies directly as x^n. y is directly proportional to x^n.	$y = kx^n$	Ex. 2, p. 359
y varies inversely as x. y is inversely proportional to x.	$y = \dfrac{k}{x}$	Ex. 3, p. 361; Ex. 4, p. 362
y varies inversely as x^n. y is inversely proportional to x^n.	$y = \dfrac{k}{x^n}$	
y varies jointly as x and z.	$y = kxz$	Ex. 5, p. 363

Review Exercises

2.1

In Exercises 1–10 perform the indicated operations and write the result in standard form.

1. $(8 - 3i) - (17 - 7i)$ **2.** $4i(3i - 2)$

3. $(7 - i)(2 + 3i)$ **4.** $(3 - 4i)^2$

5. $(7 + 8i)(7 - 8i)$ **6.** $\dfrac{6}{5 + i}$

7. $\dfrac{3 + 4i}{4 - 2i}$ **8.** $\sqrt{-32} - \sqrt{-18}$

9. $\left(-2 + \sqrt{-100}\right)^2$ **10.** $\dfrac{4 + \sqrt{-8}}{2}$

In Exercises 11–12, solve each quadratic equation using the quadratic formula. Express solutions in standard form.

11. $x^2 - 2x + 4 = 0$ **12.** $2x^2 - 6x + 5 = 0$

2.2

In Exercises 13–16, use the vertex and intercepts to sketch the graph of each quadratic function. Give the equation for the parabola's axis of symmetry. Use the graph to determine the function's domain and range.

13. $f(x) = -(x + 1)^2 + 4$ **14.** $f(x) = (x + 4)^2 - 2$

15. $f(x) = -x^2 + 2x + 3$ **16.** $f(x) = 2x^2 - 4x - 6$

In Exercises 17–18, use the function's equation, and not its graph, to find

   a. *the minimum or maximum value and where it occurs.*

   b. *the function's domain and its range.*

17. $f(x) = -x^2 + 14x - 106$ **18.** $f(x) = 2x^2 + 12x + 703$

19. The function

$$f(x) = -0.02x^2 + x + 1$$

models the yearly growth of a young redwood tree, $f(x)$, in inches, with x inches of rainfall per year. How many inches of rainfall per year result in maximum tree growth? What is the maximum yearly growth?

20. Suppose that a quadratic function is used to model the data shown in the graph using

U.S. Divorce Rate

Source: National Center for Health Statistics

(number of years after 1960, divorce rate per 1000 population).

Determine, without obtaining an actual quadratic function that models the data, the approximate coordinates of the vertex for the function's graph. Describe what this means in practical terms.

21. A field bordering a straight stream is to be enclosed. The side bordering the stream is not to be fenced. If 1000 yards of fencing material is to be used, what are the dimensions of the largest rectangular field that can be fenced? What is the maximum area?

22. Among all pairs of numbers whose difference is 14, find a pair whose product is as small as possible. What is the minimum product?

23. You have 1000 feet of fencing to construct six corrals, as shown in the figure. Find the dimensions that maximize the enclosed area. What is the maximum area?

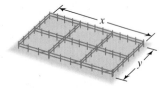

24. The annual yield per fruit tree is fairly constant at 150 pounds per tree when the number of trees per acre is 35 or fewer. For each additional tree over 35, the annual yield per tree for all trees on the acre decreases by 4 pounds due to overcrowding. How many fruit trees should be planted per acre to maximize the annual yield for the acre? What is the maximum number of pounds of fruit per acre?

2.3

In Exercises 25–28, use the Leading Coefficient Test to determine the end behavior of the graph of the given polynomial function. Then use this end behavior to match the polynomial function with its graph. [The graphs are labeled (a) through (d).]

25. $f(x) = -x^3 + x^2 + 2x$ **26.** $f(x) = x^6 - 6x^4 + 9x^2$

27. $f(x) = x^5 - 5x^3 + 4x$ **28.** $f(x) = -x^4 + 1$

a.

b.

c.

d.

29. The polynomial function

$$f(x) = -0.87x^3 + 0.35x^2 + 81.62x + 7684.94$$

models the number of thefts, $f(x)$, in thousands, in the United States x years after 1987. Will this function be useful in modeling the number of thefts over an extended period of time? Explain your answer.

30 A herd of 100 elk is introduced to a small island. The number of elk, $f(x)$, after x years is modeled by the polynomial function

$$f(x) = -x^4 + 21x^2 + 100.$$

Use the Leading Coefficient Test to determine the graph's end behavior to the right. What does this mean about what will eventually happen to the elk population?

In Exercises 31–32, find the zeros for each polynomial function and give the multiplicity of each zero. State whether the graph crosses the x-axis, or touches the x-axis and turns around, at each zero.

31. $f(x) = -2(x - 1)(x + 2)^2(x + 5)^3$

32. $f(x) = x^3 - 5x^2 - 25x + 125$

33. Show that $f(x) = x^3 - 2x - 1$ has a real zero between 1 and 2.

In Exercises 34–39,

> **a.** *Use the Leading Coefficient Test to determine the graph's end behavior.*
>
> **b.** *Determine whether the graph has y-axis symmetry, origin symmetry, or neither.*
>
> **c.** *Graph the function.*

34. $f(x) = x^3 - x^2 - 9x + 9$

35. $f(x) = 4x - x^3$

36. $f(x) = 2x^3 + 3x^2 - 8x - 12$

37. $f(x) = -x^4 + 25x^2$

38. $f(x) = -x^4 + 6x^3 - 9x^2$

39. $f(x) = 3x^4 - 15x^3$

In Exercises 40–41, graph each polynomial function.

40. $f(x) = 2x^2(x - 1)^3(x + 2)$

41. $f(x) = -x^3(x + 4)^2(x - 1)$

2.4

In Exercises 42–44, divide using long division.

42. $(4x^3 - 3x^2 - 2x + 1) \div (x + 1)$

43. $(10x^3 - 26x^2 + 17x - 13) \div (5x - 3)$

44. $(4x^4 + 6x^3 + 3x - 1) \div (2x^2 + 1)$

In Exercises 45–46, divide using synthetic division.

45. $(3x^4 + 11x^3 - 20x^2 + 7x + 35) \div (x + 5)$

46. $(3x^4 - 2x^2 - 10x) \div (x - 2)$

47. Given $f(x) = 2x^3 - 7x^2 + 9x - 3$, use the Remainder Theorem to find $f(-13)$.

48. Use synthetic division to divide $f(x) = 2x^3 + x^2 - 13x + $ by $x - 2$. Use the result to find all zeros of f.

49. Solve the equation $x^3 - 17x + 4 = 0$ given that 4 is a root

2.5

In Exercises 50–51, use the Rational Zero Theorem to list all possible rational zeros for each given function.

50. $f(x) = x^4 - 6x^3 + 14x^2 - 14x + 5$

51. $f(x) = 3x^5 - 2x^4 - 15x^3 + 10x^2 + 12x - 8$

In Exercises 52–53, use Descartes's Rule of Signs to determine the possible number of positive and negative real zeros for each given function.

52. $f(x) = 3x^4 - 2x^3 - 8x + 5$

53. $f(x) = 2x^5 - 3x^3 - 5x^2 + 3x - 1$

54. Use Descartes's Rule of Signs to explain why $2x^4 + 6x^2 + 8 = 0$ has no real roots.

For Exercises 55–61,

    a. List all possible rational roots or rational zeros.

    b. Use Descartes's Rule of Signs to determine the possible number of positive and negative real roots or real zeros.

    c. Use synthetic division to test the possible rational roots or zeros and find an actual root or zero.

    d. Use the quotient from part (c) to find all the remaining zeros or roots.

55. $f(x) = x^3 + 3x^2 - 4$

56. $f(x) = 6x^3 + x^2 - 4x + 1$

57. $8x^3 - 36x^2 + 46x - 15 = 0$

58. $2x^3 + 9x^2 - 7x + 1 = 0$

59. $x^4 - x^3 - 7x^2 + x + 6 = 0$

60. $4x^4 + 7x^2 - 2 = 0$

61. $f(x) = 2x^4 + x^3 - 9x^2 - 4x + 4$

In Exercises 62–63, find an nth-degree polynomial function with real coefficients satisfying the given conditions. If you are using a graphing utility, graph the function and verify the real zeros and the given function value.

62. $n = 3$; 2 and $2 - 3i$ are zeros; $f(1) = -10$

63. $n = 4$; i is a zero; -3 is a zero of multiplicity 2; $f(-1) = 16$

In Exercises 64–65, find all the zeros of each polynomial function and write the polynomial as a product of linear factors.

64. $f(x) = 2x^4 + 3x^3 + 3x - 2$

65. $g(x) = x^4 - 6x^3 + x^2 + 24x + 16$

In Exercises 66–69, graphs of fifth-degree polynomial functions are shown. In each case, specify the number of real zeros and the number of imaginary zeros. Indicate whether there are any real zeros with multiplicity other than 1.

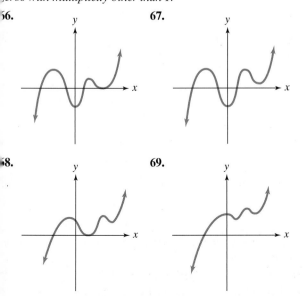

66.

67.

68.

69.

2.6

In Exercises 70–71, use transformations of $f(x) = \dfrac{1}{x}$ or $f(x) = \dfrac{1}{x^2}$ to graph each rational function.

70. $g(x) = \dfrac{1}{(x+2)^2} - 1$ **71.** $h(x) = \dfrac{1}{x-1} + 3$

In Exercises 72–79, find the vertical asymptotes, if any, the horizontal asymptote, if one exists, and the slant asymptote, if there is one, of the graph of each rational function. Then graph the rational function.

72. $f(x) = \dfrac{2x}{x^2 - 9}$ **73.** $g(x) = \dfrac{2x - 4}{x + 3}$

74. $h(x) = \dfrac{x^2 - 3x - 4}{x^2 - x - 6}$ **75.** $r(x) = \dfrac{x^2 + 4x + 3}{(x+2)^2}$

76. $y = \dfrac{x^2}{x + 1}$ **77.** $y = \dfrac{x^2 + 2x - 3}{x - 3}$

78. $f(x) = \dfrac{-2x^3}{x^2 + 1}$ **79.** $g(x) = \dfrac{4x^2 - 16x + 16}{2x - 3}$

80. A company is planning to manufacture affordable graphing calculators. The fixed monthly cost will be $50,000 and it will cost $25 to produce each calculator.

    a. Write the cost function, C, of producing x graphing calculators.

    b. Write the average cost function, $\overline{C}$, of producing x graphing calculators.

    c. Find and interpret $\overline{C}(50), \overline{C}(100), \overline{C}(1000)$, and $\overline{C}(100,000)$.

    d. What is the horizontal asymptote for the graph of this function and what does it represent?

81. In Palo Alto, California, a government agency ordered computer-related companies to contribute to a monetary pool to clean up underground water supplies. (The companies had stored toxic chemicals in leaking underground containers.) The rational function

$$C(x) = \dfrac{200x}{100 - x}$$

models the cost, $C(x)$, in tens of thousands of dollars, for removing x percent of the contaminants.

    a. Find and interpret $C(90) - C(50)$.

    b. What is the equation for the vertical asymptote? What does this mean in terms of the variables given by the function?

Exercises 82–83 involve rational functions that model the given situations. In each case, find the horizontal asymptote as $x \to \infty$ and then describe what this means in practical terms.

82. $f(x) = \dfrac{150x + 120}{0.05x + 1}$; the number of bass, $f(x)$, after x months in a lake that was stocked with 120 bass

83. $P(x) = \dfrac{72,900}{100x^2 + 729}$; the percentage, $P(x)$, of people in the United States with x years of education who are unemployed

84. The function $p(x) = 1.96x + 3.14$ models the number of nonviolent prisoners, $p(x)$, in thousands, in New York State prisons x years after 1980. The function $q(x) = 3.04x + 21.79$ models the total number of prisoners, $q(x)$, in thousands, in New York State prisons x years after 1980.

a. Write a function that models the fraction of nonviolent prisoners in New York State prisons x years after 1980.

b. What is the equation of the horizontal asymptote associated with the function in part (a)? Describe what this means about the percentage, to the nearest tenth of a percent, of nonviolent prisoners in New York State prisons over time.

c. Use your equation in part (b) to explain why, in 1998, New York State implemented a strategy where more nonviolent offenders are granted parole and more violent offenders are denied parole.

85. A jogger ran 4 miles and then walked 2 miles. The average velocity running was 3 miles per hour faster than the average velocity walking. Express the total time for running and walking, T, as a function of the average velocity walking, x.

86. The area of a rectangular floor is 1000 square feet. Express the perimeter of the floor, P, as a function of the width of the rectangle, x.

2.7

In Exercises 87–92, solve each inequality and graph the solution set on a real number line.

87. $2x^2 + 5x - 3 < 0$

88. $2x^2 + 9x + 4 \geq 0$

89. $x^3 + 2x^2 > 3x$

90. $\dfrac{x - 6}{x + 2} > 0$

91. $\dfrac{(x + 1)(x - 2)}{x - 1} \geq 0$

92. $\dfrac{x + 3}{x - 4} \leq 5$

93. Use the position function

$$s(t) = -16t^2 + v_0 t + s_0$$

to solve this problem. A projectile is fired vertically upward from ground level with an initial velocity of 48 feet per second. During which time period will the projectile's height exceed 32 feet?

2.8

Solve the variation problems in Exercises 94–99.

94. An electric bill varies directly as the amount of electricity used. The bill for 1400 kilowatts of electricity is \$98. What is the bill for 2200 kilowatts of electricity?

95. The distance that a body falls from rest is directly proportional to the square of the time of the fall. If skydivers fall 144 feet in 3 seconds, how far will they fall in 10 seconds?

96. The time it takes to drive a certain distance is inversely proportional to the rate of travel. If it takes 4 hours at 50 miles per hour to drive the distance, how long will it take at 40 miles per hour?

97. The loudness of a stereo speaker, measured in decibels, varies inversely as the square of your distance from the speaker. When you are 8 feet from the speaker, the loudness is 28 decibels. What is the loudness when you are 4 feet from the speaker?

98. The time required to assemble computers varies directly as the number of computers assembled and inversely as the number of workers. If 30 computers can be assembled by 6 workers in 10 hours, how long would it take 5 workers to assemble 40 computers?

99. The volume of a pyramid varies jointly as its height and the area of its base. A pyramid with a height of 15 feet and a base with an area of 35 square feet has a volume of 175 cubic feet. Find the volume of a pyramid with a height of 20 feet and a base with an area of 120 square feet.

Chapter 2 Test

In Exercises 1–3, perform the indicated operations and write the result in standard form.

1. $(6 - 7i)(2 + 5i)$

2. $\dfrac{5}{2 - i}$

3. $2\sqrt{-49} + 3\sqrt{-64}$

4. Solve and express solutions in standard form: $x^2 = 4x - 8$.

In Exercises 5–6, use the vertex and intercepts to sketch the graph of each quadratic function. Give the equation for the parabola's axis of symmetry. Use the graph to determine the function's domain and range.

5. $f(x) = (x + 1)^2 + 4$

6. $f(x) = x^2 - 2x - 3$

7. Determine, without graphing, whether the quadratic function $f(x) = -2x^2 + 12x - 16$ has a minimum value or a maximum value. Then find

a. the minimum or maximum value and where it occurs.

b. the function's domain and its range.

8. The function $f(x) = -x^2 + 46x - 360$ models the daily profit, $f(x)$, in hundreds of dollars, for a company that manufactures x computers daily. How many computers should be manufactured each day to maximize profit? What is the maximum daily profit?

9. Among all pairs of numbers whose sum is 14, find a pair whose product is as large as possible. What is the maximum product?

10. Consider the function $f(x) = x^3 - 5x^2 - 4x + 20$.

a. Use factoring to find all zeros of f.

b. Use the Leading Coefficient Test and the zeros of f to graph the function.

11. Use end behavior to explain why the graph shown below cannot be the graph of $f(x) = x^5 - x$. Then use intercepts to explain why the graph cannot represent $f(x) = x^5 - x$.

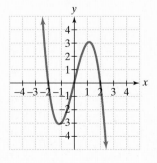

12. The graph of $f(x) = 6x^3 - 19x^2 + 16x - 4$ is shown in the figure.

   a. Based on the graph of f, find the root of the equation $6x^3 - 19x^2 + 16x - 4 = 0$ that is an integer.

   b. Use synthetic division to find the other two roots of $6x^3 - 19x^2 + 16x - 4 = 0$.

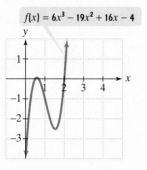

$f(x) = 6x^3 - 19x^2 + 16x - 4$

13. Use the Rational Zero Theorem to list all possible rational zeros of $f(x) = 2x^3 + 11x^2 - 7x - 6$.

14. Use Descartes's Rule of Signs to determine the possible number of positive and negative real zeros of
$$f(x) = 3x^5 - 2x^4 - 2x^2 + x - 1.$$

15. Solve: $x^3 + 9x^2 + 16x - 6 = 0$.

16. Consider the function whose equation is given by $f(x) = 2x^4 - x^3 - 13x^2 + 5x + 15$.

   a. List all possible rational zeros.

   b. Use the graph of f in the figure shown and synthetic division to find all zeros of the function.

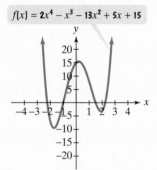

$f(x) = 2x^4 - x^3 - 13x^2 + 5x + 15$

17. Use the graph of $f(x) = x^3 + 3x^2 - 4$ in the figure shown to factor $x^3 + 3x^2 - 4$.

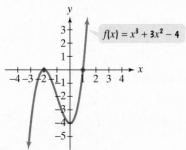

$f(x) = x^3 + 3x^2 - 4$

18. Find a fourth-degree polynomial function $f(x)$ with real coefficients that has -1, 1, and i as zeros and such that $f(3) = 160$.

19. The figure shows an incomplete graph of $f(x) = -3x^3 - 4x^2 + x + 2$. Find all the zeros of the function. Then draw a complete graph.

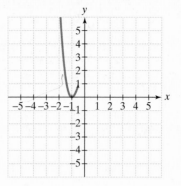

In Exercises 20–25, find the domain of each rational function and graph the function.

20. $f(x) = \dfrac{1}{(x+3)^2}$

21. $f(x) = \dfrac{1}{x-1} + 2$

22. $f(x) = \dfrac{x}{x^2 - 16}$

23. $f(x) = \dfrac{x^2 - 9}{x - 2}$

24. $f(x) = \dfrac{x+1}{x^2 + 2x - 3}$

25. $f(x) = \dfrac{4x^2}{x^2 + 3}$

26. A company is planning to manufacture pocket-sized televisions. The fixed monthly cost will be \$300,000 and it will cost \$10 to produce each television.

   a. Write the average cost function, $\overline{C}$, of producing x televisions.

   b. What is the horizontal asymptote for the graph of this function and what does it represent?

27. Rational functions can be used to model learning. Many of these functions model the proportion of correct responses as a function of the number of trials of a particular task. One such model, called a learning curve, is

$$f(x) = \frac{0.9x - 0.4}{0.9x + 0.1},$$

where $f(x)$ is the proportion of correct responses after x trials. If $f(x) = 0$, there are no correct responses. If $f(x) = 1$, all responses are correct. The graph of the rational function is shown.

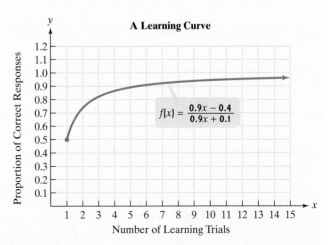

a. According to the graph, what proportion of responses are correct after 5 learning trials?

b. According to the graph, how many learning trials are necessary for 0.95 of the responses to be correct?

c. Use the function's equation to write the equation of the horizontal asymptote. What does this mean in terms of the variables modeled by the learning curve?

Solve each inequality in Exercises 28–29 and graph the solution set on a real number line. Express each solution set in interval notation.

28. $x^2 < x + 12$

29. $\dfrac{2x + 1}{x - 3} \leq 3$

30. The intensity of light received at a source varies inversely as the square of the distance from the source. A particular light has an intensity of 20 foot-candles at 15 feet. What is the light's intensity at 10 feet?

Cumulative Review Exercises (Chapters P–2)

Use the graph of $y = f(x)$ to solve Exercises 1–6.

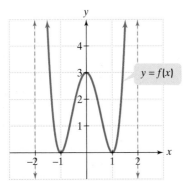

1. Find the domain and the range of f.

2. Find the zeros and the least possible multiplicity of each zero.

3. Where does the relative maximum occur?

4. Find $(f \circ f)(-1)$.

5. Use arrow notation to complete this statement:
$f(x) \to \infty$ as _____ or as _____.

6. Graph $g(x) = f(x + 2) + 1$.

In Exercises 7–12, solve each equation or inequality.

7. $|2x - 1| = 3$

8. $3x^2 - 5x + 1 = 0$

9. $9 + \dfrac{3}{x} = \dfrac{2}{x^2}$

10. $x^3 + 2x^2 - 5x - 6 = 0$

11. $|2x - 5| > 3$

12. $3x^2 > 2x + 5$

In Exercises 13–18, graph each equation in a rectangular coordinate system. If two functions are given, graph both in the same system.

13. $f(x) = x^3 - 4x^2 - x + 4$

14. $f(x) = x^2 + 2x - 8$

15. $f(x) = x^2(x - 3)$

16. $f(x) = \dfrac{x - 1}{x - 2}$

17. $f(x) = |x|$ and $g(x) = -|x| - 1$

18. $x^2 + y^2 - 2x + 4y - 4 = 0$

In Exercises 19–20, let $f(x) = 2x^2 - x - 1$ and $g(x) = 4x - 1$.

19. Find $(f \circ g)(x)$.

20. Find $\dfrac{f(x + h) - f(x)}{h}$.

Exponential and Logarithmic Functions

WHAT WENT WRONG ON THE space shuttle *Challenger*? Will population growth lead to a future without comfort or individual choice? Can I put aside a small amount of money and have millions for early retirement? Why did I feel I was walking too slowly on my visit to New York City? Why are people in California at far greater risk from drunk drivers than from earthquakes? What is the difference between earthquakes measuring 6 and 7 on the Richter scale? And what can I hope to accomplish in weightlifting?

The functions that you will be learning about in this chapter will provide you with the mathematics for answering these questions. You will see how these remarkable functions enable us to predict the future and rediscover the past.

YOU'VE RECENTLY TAKEN UP weightlifting, recording the maximum number of pounds you can lift at the end of each week. At first your weight limit increases rapidly, but now you notice that this growth is beginning to level off. You wonder about a function that would serve as a mathematical model to predict the number of pounds you can lift as you continue the sport.

This problem appears as Exercise 51 in Exercise Set 3.5 and as the group project (Exercise 64) on page 437.

SECTION 3.1 *Exponential Functions*

Objectives

❶ Evaluate exponential functions.

❷ Graph exponential functions.

❸ Evaluate functions with base *e*.

❹ Use compound interest formulas.

The space shuttle *Challenger* exploded approximately 73 seconds into flight on January 28, 1986. The tragedy involved damage to ○-rings, which were used to sea the connections between different sections of the shuttle engines. The number o ○-rings damaged increases dramatically as temperature falls.

The function

$$f(x) = 13.49(0.967)^x - 1$$

models the number of ○-rings expected to fail when the temperature is $x°$F. Can yo see how this function is different from polynomial functions? The variable x is in th exponent. Functions whose equations contain a variable in the exponent are calle **exponential functions**. Many real-life situations, including population growth growth of epidemics, radioactive decay, and other changes that involve rapi increase or decrease, can be described using exponential functions.

Definition of the Exponential Function

The **exponential function** f **with base** b is defined by

$$f(x) = b^x \quad \text{or} \quad y = b^x,$$

where b is a positive constant other than 1 ($b > 0$ and $b \neq 1$) and x is any real number.

Here are some examples of exponential functions:

$$f(x) = 2^x \qquad g(x) = 10^x \qquad h(x) = 3^{x+1} \qquad j(x) = \left(\frac{1}{2}\right)^{x-1}.$$

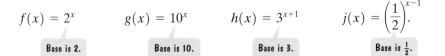

Base is 2. Base is 10. Base is 3. Base is $\frac{1}{2}$.

Each of these functions has a constant base and a variable exponent. By contras the following functions are not exponential functions:

$$F(x) = x^2 \qquad G(x) = 1^x \qquad H(x) = (-1)^x \qquad J(x) = x^x.$$

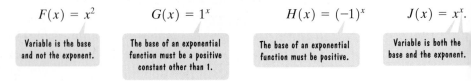

Variable is the base and not the exponent. The base of an exponential function must be a positive constant other than 1. The base of an exponential function must be positive. Variable is both the base and the exponent.

Why is $G(x) = 1^x$ not classified as an exponential function? The number raised to any power is 1. Thus, the function G can be written as $G(x) = 1$, which is constant function.

Why is $H(x) = (-1)^x$ not an exponential function? The base of an exponential function must be positive to avoid having to exclude many values of x from the domain that result in nonreal numbers in the range:

$$H(x) = (-1)^x \qquad H\left(\frac{1}{2}\right) = (-1)^{\frac{1}{2}} = \sqrt{-1} = i.$$

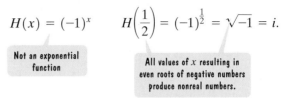

Not an exponential
function

All values of x resulting in
even roots of negative numbers
produce nonreal numbers.

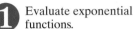

1 Evaluate exponential functions.

You will need a calculator to evaluate exponential expressions. Most scientific calculators have a $\boxed{y^x}$ key. Graphing calculators have a $\boxed{\wedge}$ key. To evaluate expressions of the form b^x, enter the base b, press $\boxed{y^x}$ or $\boxed{\wedge}$, enter the exponent x, and finally press $\boxed{=}$ or $\boxed{\text{ENTER}}$.

EXAMPLE 1   Evaluating an Exponential Function

The exponential function $f(x) = 13.49(0.967)^x - 1$ describes the number of O-rings expected to fail, $f(x)$, when the temperature is $x°F$. On the morning the *Challenger* was launched, the temperature was $31°F$, colder than any previous experience. Find the number of O-rings expected to fail at this temperature.

Solution   Because the temperature was $31°F$, substitute 31 for x and evaluate the function.

$$f(x) = 13.49(0.967)^x - 1 \qquad \text{This is the given function.}$$
$$f(31) = 13.49(0.967)^{31} - 1 \qquad \text{Substitute 31 for x.}$$

Use a scientific or graphing calculator to evaluate $f(31)$. Press the following keys on your calculator to do this:

Scientific calculator: $13.49 \boxed{\times} .967 \boxed{y^x} 31 \boxed{-} 1 \boxed{=}$

Graphing calculator: $13.49 \boxed{\times} .967 \boxed{\wedge} 31 \boxed{-} 1 \boxed{\text{ENTER}}$.

The display should be approximately 3.7668627.

$$f(31) = 13.49(0.967)^{31} - 1 \approx 3.8 \approx 4$$

Thus, four O-rings are expected to fail at a temperature of $31°F$.

Check Point 1   Use the function in Example 1 to find the number of O-rings expected to fail at a temperature of $60°F$. Round to the nearest whole number.

2 Graph exponential functions.

Graphing Exponential Functions

We are familiar with expressions involving b^x, where x is a rational number. For example,

$$b^{1.7} = b^{\frac{17}{10}} = \sqrt[10]{b^{17}} \quad \text{and} \quad b^{1.73} = b^{\frac{173}{100}} = \sqrt[100]{b^{173}}.$$

However, note that the definition of $f(x) = b^x$ includes all real numbers for the domain x. You may wonder what b^x means when x is an irrational number, such as $b^{\sqrt{3}}$ or b^π. Using closer and closer approximations for $\sqrt{3}\,(\sqrt{3} \approx 1.73205)$, we can think of $b^{\sqrt{3}}$ as the value that has the successively closer approximations

$$b^{1.7}, b^{1.73}, b^{1.732}, b^{1.73205}, \ldots .$$

In this way, we can graph exponential functions with no holes, or points of discontinuity, at the irrational domain values.

EXAMPLE 2 Graphing an Exponential Function

Graph: $f(x) = 2^x$.

Solution We begin by setting up a table of coordinates.

x	$f(x) = 2^x$
-3	$f(-3) = 2^{-3} = \dfrac{1}{8}$
-2	$f(-2) = 2^{-2} = \dfrac{1}{4}$
-1	$f(-1) = 2^{-1} = \dfrac{1}{2}$
0	$f(0) = 2^0 = 1$
1	$f(1) = 2^1 = 2$
2	$f(2) = 2^2 = 4$
3	$f(3) = 2^3 = 8$

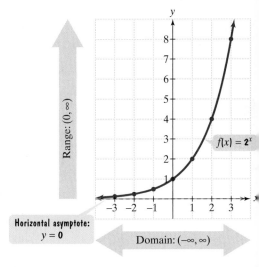

Figure 3.1 The graph of $f(x) = 2^x$

We plot these points, connecting them with a continuous curve. Figure 3.1 shows the graph of $f(x) = 2^x$. Observe that the graph approaches, but never touches, the negative portion of the x-axis. Thus, the x-axis, or $y = 0$, is a horizontal asymptote. The range is the set of all positive real numbers. Although we used integers for x in our table of coordinates, you can use a calculator to find additional points. For example, $f(0.3) = 2^{0.3} \approx 1.231$ and $f(0.95) = 2^{0.95} \approx 1.932$. The points $(0.3, 1.231)$ and $(0.95, 1.932)$ approximately fit the graph.

Check Point 2 Graph: $f(x) = 3^x$.

EXAMPLE 3 Graphing an Exponential Function

Graph: $g(x) = \left(\dfrac{1}{2}\right)^x$.

Solution We begin by setting up a table of coordinates. We compute the function values by noting that

$$g(x) = \left(\frac{1}{2}\right)^x = (2^{-1})^x = 2^{-x}.$$

x	$g(x) = \left(\dfrac{1}{2}\right)^x$ or 2^{-x}
-3	$g(-3) = 2^{-(-3)} = 2^3 = 8$
-2	$g(-2) = 2^{-(-2)} = 2^2 = 4$
-1	$g(-1) = 2^{-(-1)} = 2^1 = 2$
0	$g(0) = 2^{-0} = 1$
1	$g(1) = 2^{-1} = \dfrac{1}{2^1} = \dfrac{1}{2}$
2	$g(2) = 2^{-2} = \dfrac{1}{2^2} = \dfrac{1}{4}$
3	$g(3) = 2^{-3} = \dfrac{1}{2^3} = \dfrac{1}{8}$

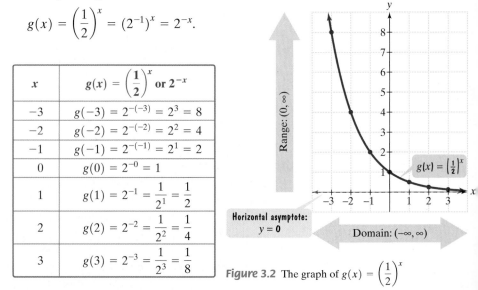

Figure 3.2 The graph of $g(x) = \left(\dfrac{1}{2}\right)^x$

We plot these points, connecting them with a continuous curve. Figure 3.2 shows the graph of $g(x) = \left(\frac{1}{2}\right)^x$. This time the graph approaches, but never touches, the *positive* portion of the *x*-axis. Once again, the *x*-axis, or $y = 0$, is a horizontal asymptote. The range consists of all positive real numbers.

Do you notice a relationship between the graphs of $f(x) = 2^x$ and $g(x) = \left(\frac{1}{2}\right)^x$ in Figures 3.1 and 3.2? The graph of $g(x) = \left(\frac{1}{2}\right)^x$ is the graph of $f(x) = 2^x$ reflected about the *y*-axis:

$$g(x) = \left(\frac{1}{2}\right)^x = 2^{-x} = f(-x)$$

> Recall that the graph of $y = f(-x)$ is the graph of $y = f(x)$ reflected about the *y*-axis.

Check Point 3   Graph: $f(x) = \left(\frac{1}{3}\right)^x$. Note that $f(x) = \left(\frac{1}{3}\right)^x = (3^{-1})^x = 3^{-x}$.

Four exponential functions have been graphed in Figure 3.3. Compare the black and green graphs, where $b > 1$, to those in blue and red, where $b < 1$. When $b > 1$, the value of *y* increases as the value of *x* increases. When $b < 1$, the value of *y* decreases as the value of *x* increases. Notice that all four graphs pass through $(0, 1)$.

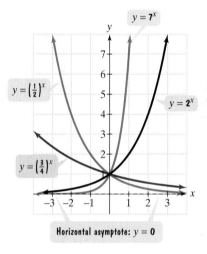

Figure 3.3 Graphs of four exponential functions

These graphs illustrate the following general characteristics of exponential functions:

Characteristics of Exponential Functions of the Form $f(x) = b^x$

1. The domain of $f(x) = b^x$ consists of all real numbers: $(-\infty, \infty)$. The range of $f(x) = b^x$ consists of all positive real numbers: $(0, \infty)$.
2. The graphs of all exponential functions of the form $f(x) = b^x$ pass through the point $(0, 1)$ because $f(0) = b^0 = 1 (b \neq 0)$. The *y*-intercept is 1.
3. If $b > 1$, $f(x) = b^x$ has a graph that goes up to the right and is an increasing function. The greater the value of *b*, the steeper the increase.
4. If $0 < b < 1$, $f(x) = b^x$ has a graph that goes down to the right and is a decreasing function. The smaller the value of *b*, the steeper the decrease.
5. $f(x) = b^x$ is one-to-one and has an inverse that is a function.
6. The graph of $f(x) = b^x$ approaches, but does not touch, the *x*-axis. The *x*-axis, or $y = 0$, is a horizontal asymptote.

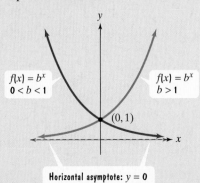

Transformations of Exponential Functions

The graphs of exponential functions can be translated vertically or horizontally, reflected, stretched, or shrunk. These transformations are summarized in Table 3.1.

Table 3.1 Transformations Involving Exponential Functions

In each case, c represents a positive real number.

Transformation	Equation	Description
Vertical translation	$g(x) = b^x + c$ $g(x) = b^x - c$	• Shifts the graph of $f(x) = b^x$ upward c units. • Shifts the graph of $f(x) = b^x$ downward c units.
Horizontal translation	$g(x) = b^{x+c}$ $g(x) = b^{x-c}$	• Shifts the graph of $f(x) = b^x$ to the left c units. • Shifts the graph of $f(x) = b^x$ to the right c units.
Reflection	$g(x) = -b^x$ $g(x) = b^{-x}$	• Reflects the graph of $f(x) = b^x$ about the x-axis. • Reflects the graph of $f(x) = b^x$ about the y-axis.
Vertical stretching or shrinking	$g(x) = cb^x$	• Vertically stretches the graph of $f(x) = b^x$ if $c > 1$. • Vertically shrinks the graph of $f(x) = b^x$ if $0 < c < 1$.
Horizontal stretching or shrinking	$g(x) = b^{cx}$	• Horizontally shrinks the graph of $f(x) = b^x$ if $c > 1$ • Horizontally stretches the graph of $f(x) = b^x$ if $0 < c < 1$.

EXAMPLE 4 Transformations Involving Exponential Functions

Use the graph of $f(x) = 3^x$ to obtain the graph of $g(x) = 3^{x+1}$.

Solution The graph of $g(x) = 3^{x+1}$ is the graph of $f(x) = 3^x$ shifted 1 unit to the left.

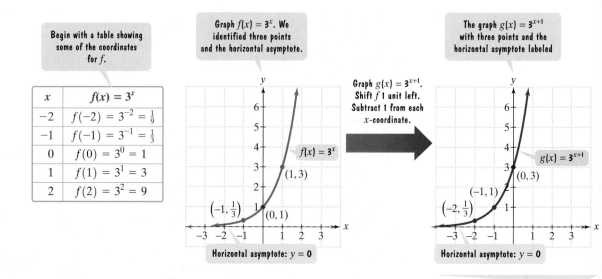

Begin with a table showing some of the coordinates for f.

x	$f(x) = 3^x$
-2	$f(-2) = 3^{-2} = \frac{1}{9}$
-1	$f(-1) = 3^{-1} = \frac{1}{3}$
0	$f(0) = 3^0 = 1$
1	$f(1) = 3^1 = 3$
2	$f(2) = 3^2 = 9$

Graph $f(x) = 3^x$. We identified three points and the horizontal asymptote.

Graph $g(x) = 3^{x+1}$. Shift f 1 unit left. Subtract 1 from each x-coordinate.

The graph $g(x) = 3^{x+1}$ with three points and the horizontal asymptote labeled

Check Point 4 Use the graph of $f(x) = 3^x$ to obtain the graph of $g(x) = 3^{x-1}$.

If an exponential function is translated upward or downward, the horizontal asymptote is shifted by the amount of the vertical shift.

EXAMPLE 5   Transformations Involving Exponential Functions

Use the graph of $f(x) = 2^x$ to obtain the graph of $g(x) = 2^x - 3$.

Solution   The graph of $g(x) = 2^x - 3$ is the graph of $f(x) = 2^x$ shifted down 3 units.

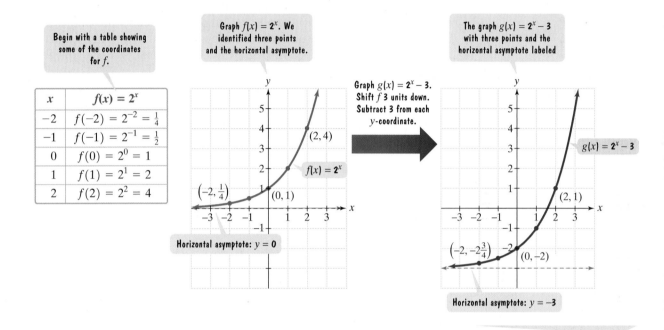

Begin with a table showing some of the coordinates for f.

x	$f(x) = 2^x$
-2	$f(-2) = 2^{-2} = \frac{1}{4}$
-1	$f(-1) = 2^{-1} = \frac{1}{2}$
0	$f(0) = 2^0 = 1$
1	$f(1) = 2^1 = 2$
2	$f(2) = 2^2 = 4$

Graph $f(x) = 2^x$. We identified three points and the horizontal asymptote.

Graph $g(x) = 2^x - 3$. Shift f 3 units down. Subtract 3 from each y-coordinate.

The graph $g(x) = 2^x - 3$ with three points and the horizontal asymptote labeled

Horizontal asymptote: $y = 0$

Horizontal asymptote: $y = -3$

Check Point 5   Use the graph of $f(x) = 2^x$ to obtain the graph of $g(x) = 2^x + 1$.

---

3 Evaluate functions with base e.

The Natural Base e

An irrational number, symbolized by the letter e, appears as the base in many applied exponential functions. The number e is defined as the value that $\left(1 + \frac{1}{n}\right)^n$ approaches as n gets larger and larger. Table 3.2 shows the values of $\left(1 + \frac{1}{n}\right)^n$ for increasingly large values of n. As $n \to \infty$, the approximate value of e to nine decimal places is

$$e \approx 2.718281827.$$

The irrational number e, approximately 2.72, is called the **natural base**. The function $f(x) = e^x$ is called the **natural exponential function**.

Use a scientific or graphing calculator with an $\boxed{e^x}$ key to evaluate e to various powers. For example, to find e^2, press the following keys on most calculators:

Scientific calculator: 2 $\boxed{e^x}$

Graphing calculator: $\boxed{e^x}$ 2 $\boxed{\text{ENTER}}$.

The display should be approximately 7.389.

$$e^2 \approx 7.389$$

Table 3.2

n	$\left(1 + \dfrac{1}{n}\right)^n$
1	2
2	2.25
5	2.48832
10	2.59374246
100	2.704813829
1000	2.716923932
10,000	2.718145927
100,000	2.718268237
1,000,000	2.718280469
1,000,000,000	2.718281827

As $n \to \infty$, $\left(1 + \dfrac{1}{n}\right)^n \to e$.

Technology

As $n \to \infty$, the graph of $y = \left(1 + \frac{1}{n}\right)^n$ approaches the graph of $y = e$.

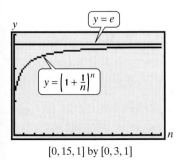

$y = e$

$y = \left(1 + \frac{1}{n}\right)^n$

$[0, 15, 1]$ by $[0, 3, 1]$

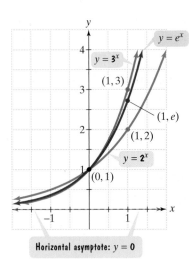

Figure 3.4 Graphs of three exponential functions

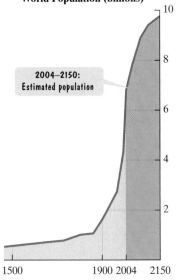

World Population (billions)

2004–2150:
Estimated population

1500   1900 2004   2150

Source: U.N. Population Division

The number e lies between 2 and 3. Because $2^2 = 4$ and $3^2 = 9$, it makes sense tha e^2, approximately 7.389, lies between 4 and 9.

Because $2 < e < 3$, the graph of $y = e^x$ is between the graphs of $y = 2^x$ and $y = 3^x$, shown in Figure 3.4.

EXAMPLE 6   World Population

In a report entitled *Resources and Man*, the U.S. National Academy of Sciences concluded that a world population of 10 billion "is close to (if not above) the maximum that an intensely managed world might hope to support with some degree of comfor and individual choice." At the time the report was issued in 1969, world population was approximately 3.6 billion, with a growth rate of 2% per year. The function

$$f(x) = 3.6e^{0.02x}$$

describes world population, $f(x)$, in billions, x years after 1969. Use the function to find world population in the year 2020. Is there cause for alarm?

Solution   Because 2020 is 51 years after 1969, we substitute 51 for x in $f(x) = 3.6e^{0.02x}$:

$$f(51) = 3.6e^{0.02(51)}.$$

Perform this computation on your calculator.

Scientific calculator: 3.6 $\boxed{\times}$ $\boxed{(}$ $\boxed{.02}$ $\boxed{\times}$ 51 $\boxed{)}$ $\boxed{e^x}$ $\boxed{=}$

Graphing calculator: 3.6 $\boxed{\times}$ $\boxed{e^x}$ $\boxed{(}$ $\boxed{.02}$ $\boxed{\times}$ 51 $\boxed{)}$ $\boxed{\text{ENTER}}$

The display should be approximately 9.9835012. Thus,

$$f(51) = 3.6e^{0.02(51)} \approx 9.98.$$

This indicates that world population in the year 2020 will be approximately 9.98 billion. Because this number is quite close to 10 billion, the given function suggests that there may be cause for alarm.

World population in 2004 was approximately 6.4 billion, but the growth rate was no longer 2%. It had slowed down to 1.23%. Using this current growth rate exponential functions now predict a world population of 7.8 billion in the year 2020 Experts think the population may stabilize at 10 billion after 2200 if the growth rate continues to decline.

 6 The function $f(x) = 6.4e^{0.0123x}$ describes world population, $f(x)$, in billions, x years after 2004 subject to a growth rate of 1.23% annually. Use the function to predict world population in 2050.

 Use compound interest formulas.

Compound Interest

We all want a wonderful life with fulfilling work, good health, and loving relationships And let's be honest: Financial security wouldn't hurt! Achieving this goal depends on understanding how money in savings accounts grows in remarkable ways as a result of *compound interest*. **Compound interest** is interest computed on your origina investment as well as on any accumulated interest.

Suppose a sum of money, called the **principal**, P, is invested at an annua percentage rate r, in decimal form, compounded once per year. Because the interest is added to the principal at year's end, the accumulated value, A, is

$$A = P + Pr = P(1 + r).$$

The accumulated amount of money follows this pattern of multiplying the previous principal by $(1 + r)$ for each successive year, as indicated in Table 3.3.

Table 3.3

Time in Years	Accumulated Value after Each Compounding
0	$A = P$
1	$A = P(1 + r)$
2	$A = P(1 + r)(1 + r) = P(1 + r)^2$
3	$A = P(1 + r)^2(1 + r) = P(1 + r)^3$
4	$A = P(1 + r)^3(1 + r) = P(1 + r)^4$
$\vdots$	$\vdots$
t	$A = P(1 + r)^t$

This formula gives the balance, A, that a principal, P, is worth after t years at interest rate r, compounded once a year.

Most savings institutions have plans in which interest is paid more than once a year. If compound interest is paid twice a year, the compounding period is six months. We say that the interest is **compounded semiannually**. When compound interest is paid four times a year, the compounding period is three months and the interest is said to be **compounded quarterly**. Some plans allow for monthly compounding or daily compounding.

In general, when compound interest is paid n times a year, we say that there are **n compounding periods per year**. The formula $A = P(1 + r)^t$ can be adjusted to take into account the number of compounding periods in a year. If there are n compounding periods per year, in each time period the interest rate is $i = \frac{r}{n}$ and there are nt time periods in t years. This results in the following formula for the balance, A, after t years:

$$A = P\left(1 + \frac{r}{n}\right)^{nt}.$$

Some banks use **continuous compounding**, where the number of compounding periods increases infinitely (compounding interest every trillionth of a second, every quadrillionth of a second, etc.). Let's see what happens to the balance, A, as $n \to \infty$.

$$\frac{n}{r} \cdot rt = nt$$

$$A = P\left(1 + \frac{r}{n}\right)^{nt} = P\left[\left(1 + \frac{1}{\frac{n}{r}}\right)^{\frac{n}{r}}\right]^{rt} = P\left[\left(1 + \frac{1}{h}\right)^{h}\right]^{rt} = Pe^{rt}$$

Let $h = \frac{n}{r}$.
As $n \to \infty$, $h \to \infty$.

As $h \to \infty$, by definition $\left(1 + \frac{1}{h}\right)^{h} \to e$.

We see that the formula for continuous compounding is $A = Pe^{rt}$. Although continuous compounding sounds terrific, it yields only a fraction of a percent more interest over a year than daily compounding.

Formulas for Compound Interest

After t years, the balance, A, in an account with principal P and annual interest rate r (in decimal form) is given by the following formulas:

1. For n compoundings per year: $A = P\left(1 + \frac{r}{n}\right)^{nt}$

2. For continuous compounding: $A = Pe^{rt}$.

EXAMPLE 7   Choosing between Investments

You decide to invest $8000 for 6 years and you have a choice between two accounts. The first pays 7% per year, compounded monthly. The second pays 6.85% per year, compounded continuously. Which is the better investment?

Solution   The better investment is the one with the greater balance in the account after 6 years. Let's begin with the account with monthly compounding. We use the compound interest model with $P = 8000, r = 7\% = 0.07, n = 12$ (monthly compounding means 12 compoundings per year), and $t = 6$.

$$A = P\left(1 + \frac{r}{n}\right)^{nt} = 8000\left(1 + \frac{0.07}{12}\right)^{12 \cdot 6} \approx 12{,}160.84$$

The balance in this account after 6 years is $12,160.84.

For the second investment option, we use the model for continuous compounding with $P = 8000, r = 6.85\% = 0.0685$, and $t = 6$.

$$A = Pe^{rt} = 8000e^{0.0685(6)} \approx 12{,}066.60$$

The balance in this account after 6 years is $12,066.60, slightly less than the previous amount. Thus, the better investment is the 7% monthly compounding option.

Check Point 7   A sum of $10,000 is invested at an annual rate of 8%. Find the balance in the account after 5 years subject to **a.** quarterly compounding and **b.** continuous compounding.

EXERCISE SET 3.1

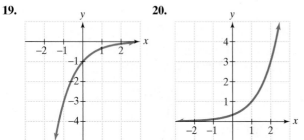

Practice Exercises

In Exercises 1–10, approximate each number using a calculator. Round your answer to three decimal places.

1. $2^{3.4}$ **2.** $3^{2.4}$ **3.** $3^{\sqrt{5}}$ **4.** $5^{\sqrt{3}}$ **5.** $4^{-1.5}$

6. $6^{-1.2}$ **7.** $e^{2.3}$ **8.** $e^{3.4}$ **9.** $e^{-0.95}$ **10.** $e^{-0.75}$

In Exercises 11–18, graph each function by making a table of coordinates. If applicable, use a graphing utility to confirm your hand-drawn graph.

11. $f(x) = 4^x$ **12.** $f(x) = 5^x$

13. $g(x) = \left(\frac{3}{2}\right)^x$ **14.** $g(x) = \left(\frac{4}{3}\right)^x$

15. $h(x) = \left(\frac{1}{2}\right)^x$ **16.** $h(x) = \left(\frac{1}{3}\right)^x$

17. $f(x) = (0.6)^x$ **18.** $f(x) = (0.8)^x$

In Exercises 19–24, the graph of an exponential function is given. Select the function for each graph from the following options:

$$f(x) = 3^x, g(x) = 3^{x-1}, h(x) = 3^x - 1,$$
$$F(x) = -3^x, G(x) = 3^{-x}, H(x) = -3^{-x}.$$

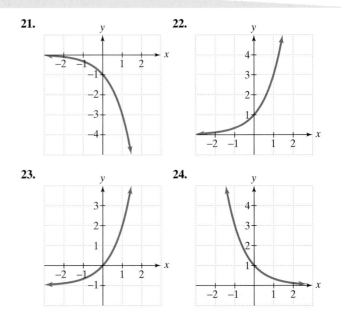

In Exercises 25–34, begin by graphing $f(x) = 2^x$. Then use transformations of this graph to graph the given function. Be sure to graph and give equations of the asymptotes. Use the graphs to determine each function's domain and range. If applicable, use a graphing utility to confirm your hand-drawn graphs.

25. $g(x) = 2^{x+1}$ **26.** $g(x) = 2^{x+2}$

27. $g(x) = 2^x - 1$ **28.** $g(x) = 2^x + 2$

29. $h(x) = 2^{x+1} - 1$ **30.** $h(x) = 2^{x+2} - 1$

1. $g(x) = -2^x$

3. $g(x) = 2 \cdot 2^x$

32. $g(x) = 2^{-x}$

34. $g(x) = \frac{1}{2} \cdot 2^x$

The figure shows the graph of $f(x) = e^x$. In Exercises 35–46, use transformations of this graph to graph each function. Be sure to give equations of the asymptotes. Use the graphs to determine each function's domain and range. If applicable, use a graphing utility to confirm your hand-drawn graphs.

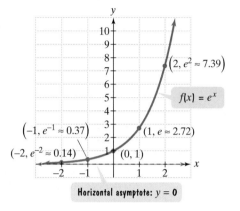

5. $g(x) = e^{x-1}$

7. $g(x) = e^x + 2$

39. $h(x) = e^{x-1} + 2$

41. $h(x) = e^{-x}$

43. $g(x) = 2e^x$

45. $h(x) = e^{2x} + 1$

36. $g(x) = e^{x+1}$

38. $g(x) = e^x - 1$

40. $h(x) = e^{x+1} - 1$

42. $h(x) = -e^x$

44. $g(x) = \frac{1}{2} e^x$

46. $h(x) = e^{\frac{x}{2}} + 2$

In Exercises 47–52, graph functions f and g in the same rectangular coordinate system. Graph and give equations of all asymptotes. If applicable, use a graphing utility to confirm your hand-drawn graphs.

47. $f(x) = 3^x$ and $g(x) = 3^{-x}$

48. $f(x) = 3^x$ and $g(x) = -3^x$

49. $f(x) = 3^x$ and $g(x) = \frac{1}{3} \cdot 3^x$

50. $f(x) = 3^x$ and $g(x) = 3 \cdot 3^x$

51. $f(x) = \left(\frac{1}{2}\right)^x$ and $g(x) = \left(\frac{1}{2}\right)^{x-1} + 1$

52. $f(x) = \left(\frac{1}{2}\right)^x$ and $g(x) = \left(\frac{1}{2}\right)^{x-1} + 2$

Use the compound interest formulas $A = P\left(1 + \dfrac{r}{n}\right)^{nt}$ and $A = Pe^{rt}$ to solve Exercises 53–56. Round answers to the nearest cent.

53. Find the accumulated value of an investment of $10,000 for 5 years at an interest rate of 5.5% if the money is **a.** compounded semiannually; **b.** compounded quarterly; **c.** compounded monthly; **d.** compounded continuously.

54. Find the accumulated value of an investment of $5000 for 10 years at an interest rate of 6.5% if the money is **a.** compounded semiannually; **b.** compounded quarterly; **c.** compounded monthly; **d.** compounded continuously.

55. Suppose that you have $12,000 to invest. Which investment yields the greater return over 3 years: 7% compounded monthly or 6.85% compounded continuously?

56. Suppose that you have $6000 to invest. Which investment yields the greater return over 4 years: 8.25% compounded quarterly or 8.3% compounded semiannually?

Practice Plus

In Exercises 57–58, graph f and g in the same rectangular coordinate system. Then find the point of intersection of the two graphs.

57. $f(x) = 2^x$, $g(x) = 2^{-x}$

58. $f(x) = 2^{x+1}$, $g(x) = 2^{-x+1}$

59. Graph $y = 2^x$ and $x = 2^y$ in the same rectangular coordinate system.

60. Graph $y = 3^x$ and $x = 3^y$ in the same rectangular coordinate system.

In Exercises 61–64, give the equation of each exponential function whose graph is shown.

61.

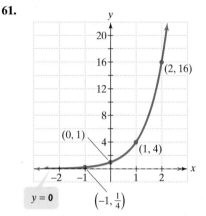

62.

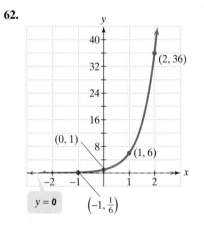

63.

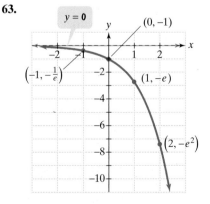

64.

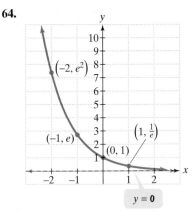

70. A decimal approximation for π is 3.141593. Use a calculator to find $2^3, 2^{3.1}, 2^{3.14}, 2^{3.141}, 2^{3.1415}, 2^{3.14159}$, and $2^{3.141593}$. Now find 2^π. What do you observe?

Use a calculator with an $\boxed{e^x}$ *key to solve Exercises 71–77.*
The graph shows the number of words, in millions, in the U.S. federal tax code for selected years from 1955 through 2000. The data can be modeled by

$$f(x) = 0.16x + 1.43 \quad \text{and} \quad g(x) = 1.8e^{0.04x},$$

in which $f(x)$ and $g(x)$ represent the number of words, in millions, in the federal tax code x years after 1955. Use these functions to solve Exercises 71–72.

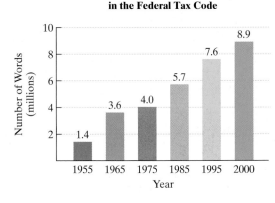

 Application Exercises

Use a calculator with a $\boxed{y^x}$ *key or a* $\boxed{\wedge}$ *key to solve Exercises 65–70.*

65. India is currently one of the world's fastest-growing countries. By 2040, the population of India will be larger than the population of China; by 2050, nearly one-third of the world's population will live in these two countries alone. The exponential function $f(x) = 574(1.026)^x$ models the population of India, $f(x)$, in millions, x years after 1974.

a. Substitute 0 for x and, without using a calculator, find India's population in 1974.

b. Substitute 27 for x and use your calculator to find India's population, to the nearest million, in the year 2001 as modeled by this function.

c. Find India's population, to the nearest million, in the year 2028 as predicted by this function.

d. Find India's population, to the nearest million, in the year 2055 as predicted by this function.

e. What appears to be happening to India's population every 27 years?

66. The 1986 explosion at the Chernobyl nuclear power plant in the former Soviet Union sent about 1000 kilograms of radioactive cesium-137 into the atmosphere. The function $f(x) = 1000(0.5)^{\frac{x}{30}}$ describes the amount, $f(x)$, in kilograms, of cesium-137 remaining in Chernobyl x years after 1986. If even 100 kilograms of cesium-137 remain in Chernobyl's atmosphere, the area is considered unsafe for human habitation. Find $f(80)$ and determine if Chernobyl will be safe for human habitation by 2066.

The formula $S = C(1 + r)^t$ models inflation, where $C =$ the value today, $r =$ the annual inflation rate, and $S =$ the inflated value t years from now. Use this formula to solve Exercises 67–68. Round answers to the nearest dollar.

67. If the inflation rate is 6%, how much will a house now worth $465,000 be worth in 10 years?

68. If the inflation rate is 3%, how much will a house now worth $510,000 be worth in 5 years?

69. A decimal approximation for $\sqrt{3}$ is 1.7320508. Use a calculator to find $2^{1.7}, 2^{1.73}, 2^{1.732}, 2^{1.73205}$, and $2^{1.7320508}$. Now find $2^{\sqrt{3}}$. What do you observe?

71. Which function, the linear or the exponential, is a better model for the data in 2000?

72. Which function, the linear or the exponential, is a better model for the data in 1985?

73. In college, we study large volumes of information—information that, unfortunately, we do not often retain for very long. The function

$$f(x) = 80e^{-0.5x} + 20$$

describes the percentage of information, $f(x)$, that a particular person remembers x weeks after learning the information.

a. Substitute 0 for x and, without using a calculator, find the percentage of information remembered at the moment it is first learned.

b. Substitute 1 for x and find the percentage of information that is remembered after 1 week.

c. Find the percentage of information that is remembered after 4 weeks.

d. Find the percentage of information that is remembered after one year (52 weeks).

74. In 1626, Peter Minuit convinced the Wappinger Indians to sell him Manhattan Island for $24. If the Native Americans had put the $24 into a bank account paying 5% interest, how much would the investment have been worth in the year 2005 if interest were compounded

a. monthly? **b.** continuously?

The bar graph shows the number of identity theft complaints to the Federal Trade Commission from 2000 through 2004. (The problem is much worse: The graph shows only the complaints. According to an FCC survey, 9.9 million Americans—about 1 in 30—were victims of identity theft from spring 2002 to spring 2003.)

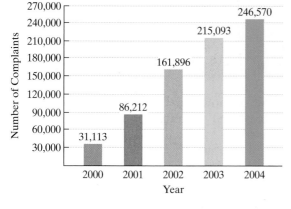

Number of Identity Theft Complaints to the Federal Trade Commission

Source: Federal Trade Commission

The functions

$$f(x) = \frac{258,051}{1 + 6.78e^{-1.21x}} \quad \text{and} \quad g(x) = 55,979.5x + 36,217.8$$

model the number of identity theft complaints to the FCC, $f(x)$ or $g(x)$, x years after 2000. Use these functions to solve Exercises 75–76.

75. Which function is a better model for the number of complaints in 2004?

76. Which function is a better model for the number of complaints in 2003?

Writing in Mathematics

77. What is an exponential function?

78. What is the natural exponential function?

79. Use a calculator to evaluate $\left(1 + \dfrac{1}{x}\right)^x$ for $x = 10, 100, 1000,$ 10,000, 100,000, and 1,000,000. Describe what happens to the expression as x increases.

80. Describe how you could use the graph of $f(x) = 2^x$ to obtain a decimal approximation for $\sqrt{2}$.

81. The exponential function $y = 2^x$ is one-to-one and has an inverse function. Try finding the inverse function by exchanging x and y and solving for y. Describe the difficulty that you encounter in this process. What is needed to overcome this problem?

82. In 2004, world population was approximately 6.4 billion with an annual growth rate of 1.23%. Discuss two factors that would cause this growth rate to slow down over the next ten years.

Technology Exercises

83. Graph $y = 13.49(0.967)^x - 1$, the function for the number of O-rings expected to fail at $x°$F, in a $[0, 90, 10]$ by $[0, 20, 5]$ view-

ing rectangle. If NASA engineers had used this function and its graph, is it likely they would have allowed the *Challenger* to be launched when the temperature was $31°$F? Explain.

84. You have \$10,000 to invest. One bank pays 5% interest compounded quarterly and the other pays 4.5% interest compounded monthly.

   a. Use the formula for compound interest to write a function for the balance in each account at any time t.

   b. Use a graphing utility to graph both functions in an appropriate viewing rectangle. Based on the graphs, which bank offers the better return on your money?

85. a. Graph $y = e^x$ and $y = 1 + x + \dfrac{x^2}{2}$ in the same viewing rectangle.

   b. Graph $y = e^x$ and $y = 1 + x + \dfrac{x^2}{2} + \dfrac{x^3}{6}$ in the same viewing rectangle.

   c. Graph $y = e^x$ and $y = 1 + x + \dfrac{x^2}{2} + \dfrac{x^3}{6} + \dfrac{x^4}{24}$ in the same viewing rectangle.

   d. Describe what you observe in parts (a)–(c). Try generalizing this observation.

Critical Thinking Exercises

86. Which one of the following is true?

   a. As the number of compounding periods increases on a fixed investment, the amount of money in the account over a fixed interval of time will increase without bound.

   b. The functions $f(x) = 3^{-x}$ and $g(x) = -3^x$ have the same graph.

   c. If $f(x) = 2^x$, then $f(a + b) = f(a) + f(b)$.

   d. The functions $f(x) = \left(\frac{1}{3}\right)^x$ and $g(x) = 3^{-x}$ have the same graph.

87. The graphs labeled (a)–(d) in the figure represent $y = 3^x$, $y = 5^x$, $y = \left(\frac{1}{3}\right)^x$, and $y = \left(\frac{1}{5}\right)^x$, but not necessarily in that order. Which is which? Describe the process that enables you to make this decision.

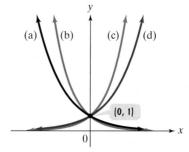

88. Graph $f(x) = 2^x$ and its inverse function in the same rectangular coordinate system.

89. The *hyperbolic cosine* and *hyperbolic sine* functions are defined by

$$\cosh x = \frac{e^x + e^{-x}}{2} \quad \text{and} \quad \sinh x = \frac{e^x - e^{-x}}{2}.$$

   a. Show that $\cosh x$ is an even function.

   b. Show that $\sinh x$ is an odd function.

   c. Prove that $(\cosh x)^2 - (\sinh x)^2 = 1$.

SECTION 3.2 *Logarithmic Functions*

Objectives

❶ Change from logarithmic to exponential form.

❷ Change from exponential to logarithmic form.

❸ Evaluate logarithms.

❹ Use basic logarithmic properties.

❺ Graph logarithmic functions.

❻ Find the domain of a logarithmic function.

❼ Use common logarithms.

❽ Use natural logarithms.

The earthquake that ripped through northern California on October 17, 1989 measured 7.1 on the Richter scale, killed more than 60 people, and injured more than 2400. Shown here is San Francisco's Marina district, where shock waves tossed houses off their foundations and into the street.

A higher measure on the Richter scale is more devastating than it seems because for each increase in one unit on the scale, there is a tenfold increase in the intensity of an earthquake. In this section, our focus is on the inverse of the exponential function, called the logarithmic function. The logarithmic function will help you to understand diverse phenomena, including earthquake intensity, human memory, and the pace of life in large cities.

The Definition of Logarithmic Functions

No horizontal line can be drawn that intersects the graph of an exponential function at more than one point. This means that the exponential function is one-to-one and has an inverse. The inverse function of the exponential function with base b is called the *logarithmic function with base b*.

Definition of the Logarithmic Function

For $x > 0$ and $b > 0, b \neq 1$,

$$y = \log_b x \text{ is equivalent to } b^y = x.$$

The function $f(x) = \log_b x$ is the **logarithmic function with base b**.

The equations

$$y = \log_b x \quad \text{and} \quad b^y = x$$

are different ways of expressing the same thing. The first equation is in **logarithmic form** and the second equivalent equation is in **exponential form**.

Notice that a **logarithm, y, is an exponent**. You should learn the location of the base and exponent in each form.

Location of Base and Exponent in Exponential and Logarithmic Forms

Exponent

Logarithmic Form: $y = \log_b x$

Base

Exponent

Exponential Form: $b^y = x$

Base

Study Tip

To change from logarithmic form to the more familiar exponential form, use this pattern:

$$y = \log_b x \quad \text{means} \quad b^y = x.$$

1 Change from logarithmic to exponential form.

EXAMPLE 1   Changing from Logarithmic to Exponential Form

Write each equation in its equivalent exponential form:

a. $2 = \log_5 x$ **b.** $3 = \log_b 64$ **c.** $\log_3 7 = y$.

Solution   We use the fact that $y = \log_b x$ means $b^y = x$.

a. $2 = \log_5 x$ means $5^2 = x$.   **b.** $3 = \log_b 64$ means $b^3 = 64$.

> Logarithms are exponents.   Logarithms are exponents.

c. $\log_3 7 = y$ or $y = \log_3 7$ means $3^y = 7$.

Check Point 1   Write each equation in its equivalent exponential form:

a. $3 = \log_7 x$ **b.** $2 = \log_b 25$ **c.** $\log_4 26 = y$.

2 Change from exponential to logarithmic form.

EXAMPLE 2   Changing from Exponential to Logarithmic Form

Write each equation in its equivalent logarithmic form:

a. $12^2 = x$ **b.** $b^3 = 8$ **c.** $e^y = 9$.

Solution   We use the fact that $b^y = x$ means $y = \log_b x$.

a. $12^2 = x$ means $2 = \log_{12} x$.   **b.** $b^3 = 8$ means $3 = \log_b 8$.

> Exponents are logarithms.   Exponents are logarithms.

c. $e^y = 9$ means $y = \log_e 9$.

Check Point 2   Write each equation in its equivalent logarithmic form:

a. $2^5 = x$ **b.** $b^3 = 27$ **c.** $e^y = 33$.

3 Evaluate logarithms.

Remembering that logarithms are exponents makes it possible to evaluate some logarithms by inspection. The logarithm of x with base b, $\log_b x$, is the exponent to which b must be raised to get x. For example, suppose we want to evaluate $\log_2 32$. We ask, 2 to what power gives 32? Because $2^5 = 32$, $\log_2 32 = 5$.

EXAMPLE 3   Evaluating Logarithms

Evaluate:

a. $\log_2 16$ **b.** $\log_3 9$ **c.** $\log_{25} 5$.

Solution

Logarithmic Expression	Question Needed for Evaluation	Logarithmic Expression Evaluated
a. $\log_2 16$	2 to what power gives 16?	$\log_2 16 = 4$ because $2^4 = 16$.
b. $\log_3 9$	3 to what power gives 9?	$\log_3 9 = 2$ because $3^2 = 9$.
c. $\log_{25} 5$	25 to what power gives 5?	$\log_{25} 5 = \frac{1}{2}$ because $25^{\frac{1}{2}} = \sqrt{25} = 5$.

Check Point 3   Evaluate:

a. $\log_{10} 100$ **b.** $\log_3 3$ **c.** $\log_{36} 6$.

Figure 3.6 illustrates the relationship between the graph of an exponential function, shown in blue, and its inverse, a logarithmic function, shown in red, for bases greater than 1 and for bases between 0 and 1. Also shown and labeled are the exponential function's horizontal asymptote ($y = 0$) and the logarithmic function's vertical asymptote ($x = 0$).

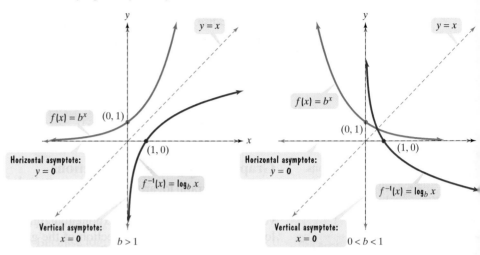

Figure 3.6 Graphs of exponential and logarithmic functions

Characteristics of the Graphs of Logarithmic Functions of the Form $f(x) = \log_b x$

- The x-intercept is 1. There is no y-intercept.
- The y-axis, or $x = 0$, is a vertical asymptote. As $x \to 0^+$, $\log_b x \to -\infty$ or ∞.
- If $b > 1$, the function is increasing. If $0 < b < 1$, the function is decreasing.
- The graph is smooth and continuous. It has no sharp corners or gaps.

The graphs of logarithmic functions can be translated vertically or horizontally, reflected, stretched, or shrunk. These transformations are summarized in Table 3.4.

Table 3.4 Transformations Involving Logarithmic Functions

In each case, c represents a positive real number.

Transformation	Equation	Description
Vertical translation	$g(x) = \log_b x + c$	• Shifts the graph of $f(x) = \log_b x$ upward c units.
	$g(x) = \log_b x - c$	• Shifts the graph of $f(x) = \log_b x$ downward c units.
Horizontal translation	$g(x) = \log_b(x + c)$	• Shifts the graph of $f(x) = \log_b x$ to the left c units. Vertical asymptote: $x = -c$
	$g(x) = \log_b(x - c)$	• Shifts the graph of $f(x) = \log_b x$ to the right c units. Vertical asymptote: $x = c$
Reflection	$g(x) = -\log_b x$	• Reflects the graph of $f(x) = \log_b x$ about the x-axis.
	$g(x) = \log_b(-x)$	• Reflects the graph of $f(x) = \log_b x$ about the y-axis.
Vertical stretching or shrinking	$g(x) = c \log_b x$	• Vertically stretches the graph of $f(x) = \log_b x$ if $c > 1$. • Vertically shrinks the graph of $f(x) = \log_b x$ if $0 < c < 1$.
Horizontal stretching or shrinking	$g(x) = \log_b(cx)$	• Horizontally shrinks the graph of $f(x) = \log_b x$ if $c > 1$. • Horizontally stretches the graph of $f(x) = \log_b x$ if $0 < c < 1$.

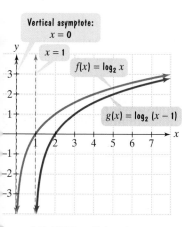

Figure 3.7 Shifting $f(x) = \log_2 x$ one unit to the right

For example, Figure 3.7 illustrates that the graph of $g(x) = \log_2(x - 1)$ is the graph of $f(x) = \log_2 x$ shifted one unit to the right. If a logarithmic function is translated to the left or to the right, both the x-intercept and the vertical asymptote are shifted by the amount of the horizontal shift. In Figure 3.7, the x-intercept of f is 1. Because g is shifted one unit to the right, its x-intercept is 2. Also observe that the vertical asymptote for f, the y-axis, or $x = 0$, is shifted one unit to the right for the vertical asymptote for g. Thus, $x = 1$ is the vertical asymptote for g.

Here are some other examples of transformations of graphs of logarithmic functions:

- The graph of $g(x) = 3 + \log_4 x$ is the graph of $f(x) = \log_4 x$ shifted up three units, shown in Figure 3.8.
- The graph of $h(x) = -\log_2 x$ is the graph of $f(x) = \log_2 x$ reflected about the x-axis, shown in Figure 3.9.
- The graph of $r(x) = \log_2(-x)$ is the graph of $f(x) = \log_2 x$ reflected about the y-axis, shown in Figure 3.10.

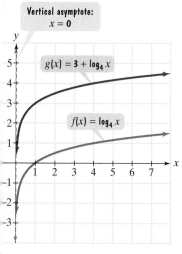

Figure 3.8 Shifting vertically up three units

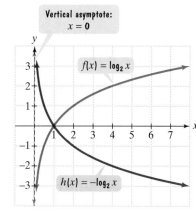

Figure 3.9 Reflection about the x-axis

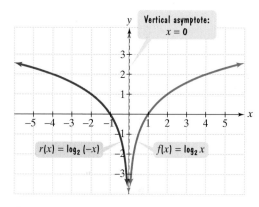

Figure 3.10 Reflection about the y-axis

⑥ Find the domain of a logarithmic function.

The Domain of a Logarithmic Function

In Section 3.1, we learned that the domain of an exponential function of the form $f(x) = b^x$ includes all real numbers and its range is the set of positive real numbers. Because the logarithmic function reverses the domain and the range of the exponential function, the **domain of a logarithmic function of the form $f(x) = \log_b x$ is the set of all positive real numbers.** Thus, $\log_2 8$ is defined because the value of x in the logarithmic expression, 8, is greater than zero and therefore is included in the domain of the logarithmic function $f(x) = \log_2 x$. However, $\log_2 0$ and $\log_2(-8)$ are not defined because 0 and -8 are not positive real numbers and therefore are excluded from the domain of the logarithmic function $f(x) = \log_2 x$. In general, **the domain of $f(x) = \log_b g(x)$ consists of all x for which $g(x) > 0$.**

EXAMPLE 7   Finding the Domain of a Logarithmic Function

Find the domain of $f(x) = \log_4(x + 3)$.

Solution   The domain of f consists of all x for which $x + 3 > 0$. Solving this inequality for x, we obtain $x > -3$. Thus, the domain of f is $(-3, \infty)$. This is illustrated in Figure 3.11. The vertical asymptote is $x = -3$ and all points on the graph of f have x-coordinates that are greater than -3.

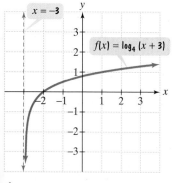

Figure 3.11 The domain of $g(x) = \log_4(x + 3)$ is $(-3, \infty)$.

Check Point 7   Find the domain of $f(x) = \log_4(x - 5)$.

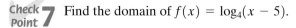

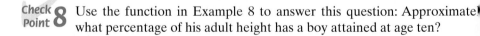

⑦ Use common logarithms.

Common Logarithms

The logarithmic function with base 10 is called the **common logarithmic function**. The function $f(x) = \log_{10} x$ is usually expressed as $f(x) = \log x$. A calculator with a $\boxed{\text{LOG}}$ key can be used to evaluate common logarithms. Here are some examples:

Logarithm	Most Scientific Calculator Keystrokes	Most Graphing Calculator Keystrokes	Display (or Approximate Display)
$\log 1000$	$1000\ \boxed{\text{LOG}}$	$\boxed{\text{LOG}}\ 1000\ \boxed{\text{ENTER}}$	3
$\log \dfrac{5}{2}$	$\boxed{(}\ 5\ \boxed{\div}\ 2\ \boxed{)}\ \boxed{\text{LOG}}$	$\boxed{\text{LOG}}\ \boxed{(}\ 5\ \boxed{\div}\ 2\ \boxed{)}\ \boxed{\text{ENTER}}$	0.39794
$\dfrac{\log 5}{\log 2}$	$5\ \boxed{\text{LOG}}\ \boxed{\div}\ 2\ \boxed{\text{LOG}}\ \boxed{=}$	$\boxed{\text{LOG}}\ 5\ \boxed{\div}\ \boxed{\text{LOG}}\ 2\ \boxed{\text{ENTER}}$	2.32193
$\log(-3)$	$3\ \boxed{+/-}\ \boxed{\text{LOG}}$	$\boxed{\text{LOG}}\ \boxed{(-)}\ 3\ \boxed{\text{ENTER}}$	$\boxed{\text{ERROR}}$

Some graphing calculators display an open parenthesis when the $\boxed{\text{LOG}}$ key is pressed. In this case, remember to close the set of parentheses after entering the function's domain value: $\boxed{\text{LOG}}\ 5\ \boxed{)}\ \boxed{\div}\ \boxed{\text{LOG}}\ 2\ \boxed{)}\ \boxed{\text{ENTER}}$.

The error message given by many calculators for $\log(-3)$ is a reminder that the domain of the common logarithmic function, $f(x) = \log x$, is the set of positive real numbers. In general, the domain of $f(x) = \log g(x)$ consists of all x for which $g(x) > 0$.

Many real-life phenomena start with rapid growth and then the growth begins to level off. This type of behavior can be modeled by logarithmic functions.

EXAMPLE 8 Modeling Height of Children

The percentage of adult height attained by a boy who is x years old can be modeled by

$$f(x) = 29 + 48.8 \log(x + 1),$$

where x represents the boy's age and $f(x)$ represents the percentage of his adult height. Approximately what percentage of his adult height has a boy attained at age eight?

Solution We substitute the boy's age, 8, for x and evaluate the function.

$$f(x) = 29 + 48.8 \log(x + 1) \qquad \text{This is the given function.}$$
$$f(8) = 29 + 48.8 \log(8 + 1) \qquad \text{Substitute 8 for x.}$$
$$= 29 + 48.8 \log 9 \qquad \text{Graphing calculator keystrokes:}$$
$$\qquad\qquad\qquad\qquad\qquad 29\ \boxed{+}\ 48.8\ \boxed{\text{LOG}}\ 9\ \boxed{\text{ENTER}}$$
$$\approx 76$$

Thus, an 8-year-old boy has attained approximately 76% of his adult height.

Check Point **8** Use the function in Example 8 to answer this question: Approximately what percentage of his adult height has a boy attained at age ten?

The basic properties of logarithms that were listed earlier in this section can be applied to common logarithms.

Properties of Common Logarithms

General Properties	Common Logarithm Properties
1. $\log_b 1 = 0$	**1.** $\log 1 = 0$
2. $\log_b b = 1$	**2.** $\log 10 = 1$
3. $\log_b b^x = x$	**3.** $\log 10^x = x$
4. $b^{\log_b x} = x$	**4.** $10^{\log x} = x$

Inverse properties

The property $\log 10^x = x$ can be used to evaluate common logarithms involving powers of 10. For example,

$$\log 100 = \log 10^2 = 2, \quad \log 1000 = \log 10^3 = 3, \quad \text{and} \quad \log 10^{7.1} = 7.1.$$

EXAMPLE 9   Earthquake Intensity

The magnitude, R, on the Richter scale of an earthquake of intensity I is given by

$$R = \log\frac{I}{I_0},$$

where I_0 is the intensity of a barely felt zero-level earthquake. The earthquake that destroyed San Francisco in 1906 was $10^{8.3}$ times as intense as a zero-level earthquake. What was its magnitude on the Richter scale?

Solution   Because the earthquake was $10^{8.3}$ times as intense as a zero-level earthquake, the intensity, I, is $10^{8.3}I_0$.

$$R = \log\frac{I}{I_0} \qquad \text{This is the formula for magnitude on the Richter scale.}$$

$$R = \log\frac{10^{8.3}I_0}{I_0} \qquad \text{Substitute } 10^{8.3}I_0 \text{ for } I.$$

$$= \log 10^{8.3} \qquad \text{Simplify.}$$

$$= 8.3 \qquad \text{Use the property } \log 10^x = x.$$

San Francisco's 1906 earthquake registered 8.3 on the Richter scale.

Check Point 9   Use the formula in Example 9 to solve this problem. If an earthquake is 10,000 times as intense as a zero-level quake ($I = 10,000I_0$), what is its magnitude on the Richter scale?

  Use natural logarithms.

Natural Logarithms

The logarithmic function with base e is called the **natural logarithmic function**. The function $f(x) = \log_e x$ is usually expressed as $f(x) = \ln x$, read "el en of x." A calculator with an $\boxed{\text{LN}}$ key can be used to evaluate natural logarithms. Keystrokes are identical to those shown for common logarithmic evaluations on page 394.

Like the domain of all logarithmic functions, the domain of the natural logarithmic function $f(x) = \ln x$ is the set of all positive real numbers. Thus, the domain of $f(x) = \ln g(x)$ consists of all x for which $g(x) > 0$.

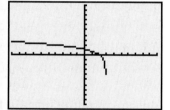

[−10, 10, 1] by [−10, 10, 1]

Figure 3.12 The domain of $f(x) = \ln(3 - x)$ is $(-\infty, 3)$.

EXAMPLE 10   Finding Domains of Natural Logarithmic Functions

Find the domain of each function:

a. $f(x) = \ln(3 - x)$ **b.** $h(x) = \ln(x - 3)^2.$

Solution

a. The domain of f consists of all x for which $3 - x > 0$. Solving this inequality for x, we obtain $x < 3$. Thus, the domain of f is $\{x|x < 3\}$ or $(-\infty, 3)$. This is verified by the graph in Figure 3.12.

b. The domain of h consists of all x for which $(x - 3)^2 > 0$. It follows that the domain of h is all real numbers except 3. Thus, the domain of h is $\{x|x \neq 3\}$ or $(-\infty, 3) \cup (3, \infty)$. This is shown by the graph in Figure 3.13. To make it more obvious that 3 is excluded from the domain, we used a $\boxed{\text{DOT}}$ format.

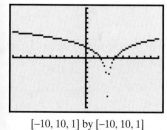

[−10, 10, 1] by [−10, 10, 1]

Figure 3.13 3 is excluded from the domain of $h(x) = \ln(x - 3)^2.$

Check Point 10   Find the domain of each function:

a. $f(x) = \ln(4 - x)$ **b.** $h(x) = \ln x^2.$

The basic properties of logarithms that were listed earlier in this section can be applied to natural logarithms.

Properties of Natural Logarithms

General Properties	**Natural Logarithm Properties**
1. $\log_b 1 = 0$	**1.** $\ln 1 = 0$
2. $\log_b b = 1$	**2.** $\ln e = 1$
3. $\log_b b^x = x$ *Inverse properties*	**3.** $\ln e^x = x$
4. $b^{\log_b x} = x$	**4.** $e^{\ln x} = x$

Examine the inverse properties, $\ln e^x = x$ and $e^{\ln x} = x$. Can you see how ln and e "undo" one another? For example,

$$\ln e^2 = 2, \ \ln e^{7x^2} = 7x^2, \ e^{\ln 2} = 2, \ \text{and} \ e^{\ln 7x^2} = 7x^2.$$

EXAMPLE 11   Dangerous Heat: Temperature in an Enclosed Vehicle

When the outside air temperature is anywhere from 72° to 96° Fahrenheit, the temperature in an enclosed vehicle climbs by 43° in the first hour. The bar graph in Figure 3.14 shows the temperature increase throughout the hour. The function

$$f(x) = 13.4 \ln x - 11.6$$

models the temperature increase, $f(x)$, in degrees Fahrenheit, after x minutes. Use the function to find the temperature increase, to the nearest degree, after 50 minutes. How well does the function model the actual increase shown in Figure 3.14?

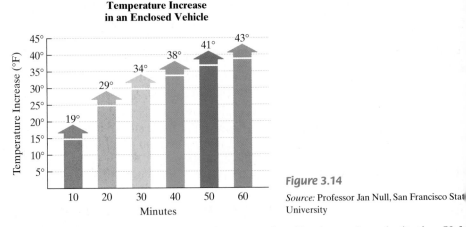

Temperature Increase in an Enclosed Vehicle

Figure 3.14

Source: Professor Jan Null, San Francisco State University

Solution   We find the temperature increase after 50 minutes by substituting 50 for x and evaluating the function at 50.

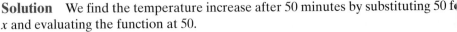

$$f(x) = 13.4 \ln x - 11.6 \qquad \text{This is the given function.}$$
$$f(50) = 13.4 \ln 50 - 11.6 \qquad \text{Substitute 50 for x.}$$
$$\approx 41 \qquad \text{Graphing calculator keystrokes: 13.4 } \boxed{\text{ln}} \ 50 \ \boxed{-} \ 11.}$$

$\boxed{\text{ENTER}}$. On some calculators, a parenthesis is needed after 50.

According to the function, the temperature will increase by approximately 41° after 50 minutes. Because the increase shown in Figure 3.14 is 41°, the function models the actual increase extremely well.

Check Point 11   Use the function in Example 11 to find the temperature increase, to the nearest degree, after 30 minutes. How well does the function model the actual increase shown in Figure 3.14?

The Curious Number e

You will learn more about each curiosity mentioned below when you take calculus.

- The number e was named by the Swiss mathematician Leonhard Euler (1707–1783), who proved that it is the limit as $n \to \infty$ of $\left(1 + \dfrac{1}{n}\right)^n$.

- e features in Euler's remarkable relationship $e^{i\pi} = -1$, in which $i = \sqrt{-1}$.

- The first few decimal places of e are fairly easy to remember:
$e = 2.7\ 1828\ 1828\ 45\ 90\ 45\ldots.$

- The best approximation of e using numbers less than 1000 is also easy to remember:
$e \approx \dfrac{878}{323} \approx 2.71826\ldots.$

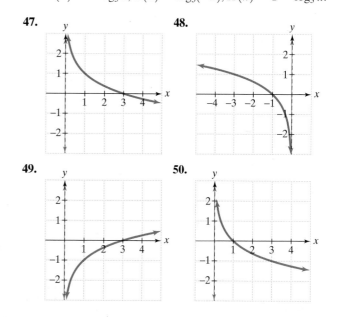

Figure 3.15

- Isaac Newton (1642–1727), one of the cofounders of calculus, showed that
$e^x = 1 + x + \dfrac{x^2}{2!} + \dfrac{x^3}{3!} + \dfrac{x^4}{4!} + \cdots$, from which we obtain $e = 1 + 1 + \dfrac{1}{2!} + \dfrac{1}{3!} + \dfrac{1}{4!} + \cdots$, an infinite sum suitable for calculation because its terms decrease so rapidly. (*Note*: $n!$ (n factorial) is the product of all the consecutive integers from n down to 1:
$n! = n(n-1)(n-2)(n-3) \cdot \cdots \cdot 3 \cdot 2 \cdot 1.$)

- The area of the region bounded by $y = \dfrac{1}{x}$, the x-axis, $x = 1$ and $x = t$ (shaded in Figure 3.15) is a function of t, designated by $A(t)$. Grégoire de Saint-Vincent, a Belgian Jesuit (1584–1667), spent his entire professional life attempting to find a formula for $A(t)$. With his student, he showed that $A(t) = \ln t$, becoming one of the first mathematicians to make use of the logarithmic function for something other than a computational device.

EXERCISE SET 3.2

Practice Exercises

In Exercises 1–8, write each equation in its equivalent exponential form.

1. $4 = \log_2 16$ **2.** $6 = \log_2 64$

3. $2 = \log_3 x$ **4.** $2 = \log_9 x$

5. $5 = \log_b 32$ **6.** $3 = \log_b 27$

7. $\log_6 216 = y$ **8.** $\log_5 125 = y$

Exercises 9–20, write each equation in its equivalent logarithmic form.

9. $2^3 = 8$ **10.** $5^4 = 625$ **11.** $2^{-4} = \frac{1}{16}$ **12.** $5^{-3} = \frac{1}{125}$

13. $\sqrt[3]{8} = 2$ **14.** $\sqrt[3]{64} = 4$ **15.** $13^2 = x$ **16.** $15^2 = x$

17. $b^3 = 1000$ **18.** $b^3 = 343$ **19.** $7^y = 200$ **20.** $8^y = 300$

Exercises 21–42, evaluate each expression without using a calculator.

21. $\log_4 16$ **22.** $\log_7 49$ **23.** $\log_2 64$ **24.** $\log_3 27$

25. $\log_5 \frac{1}{5}$ **26.** $\log_6 \frac{1}{6}$ **27.** $\log_2 \frac{1}{8}$ **28.** $\log_3 \frac{1}{9}$

29. $\log_7 \sqrt{7}$ **30.** $\log_6 \sqrt{6}$ **31.** $\log_2 \frac{1}{\sqrt{2}}$ **32.** $\log_3 \frac{1}{\sqrt{3}}$

33. $\log_{64} 8$ **34.** $\log_{81} 9$ **35.** $\log_5 5$ **36.** $\log_{11} 11$

37. $\log_4 1$ **38.** $\log_6 1$ **39.** $\log_5 5^7$ **40.** $\log_4 4^6$

41. $8^{\log_8 19}$ **42.** $7^{\log_7 23}$

43. Graph $f(x) = 4^x$ and $g(x) = \log_4 x$ in the same rectangular coordinate system.

44. Graph $f(x) = 5^x$ and $g(x) = \log_5 x$ in the same rectangular coordinate system.

45. Graph $f(x) = \left(\frac{1}{2}\right)^x$ and $g(x) = \log_{\frac{1}{2}} x$ in the same rectangular coordinate system.

46. Graph $f(x) = \left(\frac{1}{4}\right)^x$ and $g(x) = \log_{\frac{1}{4}} x$ in the same rectangular coordinate system.

In Exercises 47–52, the graph of a logarithmic function is given. Select the function for each graph from the following options:

$$f(x) = \log_3 x, \; g(x) = \log_3(x-1), \; h(x) = \log_3 x - 1,$$
$$F(x) = -\log_3 x, \; G(x) = \log_3(-x), \; H(x) = 1 - \log_3 x.$$

47.

48.

49.

50.

51.

52.

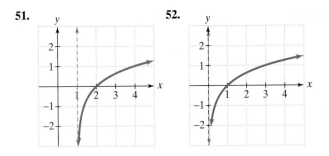

In Exercises 53–58, begin by graphing $f(x) = \log_2 x$. Then use transformations of this graph to graph the given function. What is the vertical asymptote? Use the graphs to determine each function's domain and range.

53. $g(x) = \log_2(x + 1)$ **54.** $g(x) = \log_2(x + 2)$

55. $h(x) = 1 + \log_2 x$ **56.** $h(x) = 2 + \log_2 x$

57. $g(x) = \frac{1}{2} \log_2 x$ **58.** $g(x) = -2 \log_2 x$

The figure shows the graph of $f(x) = \log x$. In Exercises 59–64, use transformations of this graph to graph each function. Graph and give equations of the asymptotes. Use the graphs to determine each function's domain and range.

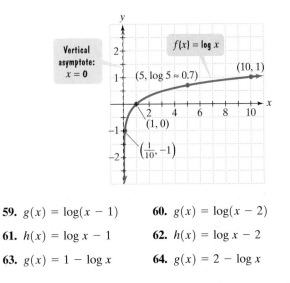

59. $g(x) = \log(x - 1)$ **60.** $g(x) = \log(x - 2)$

61. $h(x) = \log x - 1$ **62.** $h(x) = \log x - 2$

63. $g(x) = 1 - \log x$ **64.** $g(x) = 2 - \log x$

The figure shows the graph of $f(x) = \ln x$. In Exercises 65–74, use transformations of this graph to graph each function. Graph and give equations of the asymptotes. Use the graphs to determine each function's domain and range.

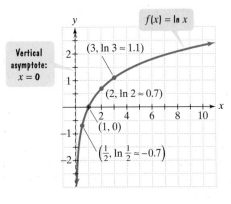

65. $g(x) = \ln(x + 2)$ **66.** $g(x) = \ln(x + 1)$

67. $h(x) = \ln(2x)$ **68.** $h(x) = \ln\left(\frac{1}{2}x\right)$

69. $g(x) = 2 \ln x$ **70.** $g(x) = \frac{1}{2} \ln x$

71. $h(x) = -\ln x$ **72.** $h(x) = \ln(-x)$

73. $g(x) = 2 - \ln x$ **74.** $g(x) = 1 - \ln x$

In Exercises 75–80, find the domain of each logarithmic function

75. $f(x) = \log_5(x + 4)$ **76.** $f(x) = \log_5(x + 6)$

77. $f(x) = \log(2 - x)$ **78.** $f(x) = \log(7 - x)$

79. $f(x) = \ln(x - 2)^2$ **80.** $f(x) = \ln(x - 7)^2$

In Exercises 81–100, evaluate or simplify each expression withou using a calculator.

81. $\log 100$ **82.** $\log 1000$ **83.** $\log 10^7$ **84.** $\log 10^8$

85. $10^{\log 33}$ **86.** $10^{\log 53}$ **87.** $\ln 1$ **88.** $\ln e$

89. $\ln e^6$ **90.** $\ln e^7$ **91.** $\ln \frac{1}{e^6}$ **92.** $\ln \frac{1}{e^7}$

93. $e^{\ln 125}$ **94.** $e^{\ln 300}$ **95.** $\ln e^{9x}$ **96.** $\ln e^{13x}$

97. $e^{\ln 5x^2}$ **98.** $e^{\ln 7x^2}$ **99.** $10^{\log \sqrt{x}}$ **100.** $10^{\log \sqrt[3]{x}}$

 Practice Plus

In Exercises 101–104, write each equation in its equivalent exponential form. Then solve for x.

101. $\log_3(x - 1) = 2$ **102.** $\log_5(x + 4) = 2$

103. $\log_4 x = -3$ **104.** $\log_{64} x = \frac{2}{3}$

In Exercises 105–108, evaluate each expression without using a calculator.

105. $\log_3(\log_7 7)$ **106.** $\log_5(\log_2 32)$

107. $\log_2(\log_3 81)$ **108.** $\log(\ln e)$

In Exercises 109–112, find the domain of each logarithmic function.

109. $f(x) = \ln(x^2 - x - 2)$ **110.** $f(x) = \ln(x^2 - 4x - 12)$

111. $f(x) = \log\left(\frac{x + 1}{x - 5}\right)$ **112.** $f(x) = \log\left(\frac{x - 2}{x + 5}\right)$

Application Exercises

The percentage of adult height attained by a girl who i x years old can be modeled by

$$f(x) = 62 + 35 \log(x - 4),$$

where x represents the girl's age (from 5 to 15) and $f(x)$ represents the percentage of her adult height. Use the function to solve Exercises 113–114. Round answers to the nearest tenth of percent.

113. Approximately what percentage of her adult height has girl attained at age 13?

114. Approximately what percentage of her adult height has girl attained at age ten?

The bar graph shows the percentage of U.S. companies that performed drug tests on employees or job applicants in five selected years from 1998 through 2003. The function

$$f(x) = -4.9 \ln x + 73.8$$

models the percentage of such companies x years after 1997. Use this function to solve Exercises 115–116. Round answers to the nearest percent.

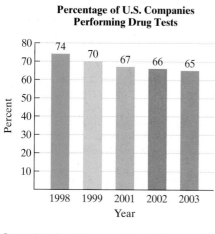

Percentage of U.S. Companies Performing Drug Tests

Source: American Management Association

15. Use the function to find the percentage of U.S. companies that performed drug tests in 2003. How well does this model the actual number shown for that year?

16. Use the function to predict the percentage of U.S. companies that will be performing drug tests in 2008.

The loudness level of a sound, D, in decibels, is given by the formula

$$D = 10 \log(10^{12} I),$$

where I is the intensity of the sound, in watts per meter². Decibel levels range from 0, a barely audible sound, to 160, a sound resulting in a ruptured eardrum. Use the formula to solve Exercises 117–118.

17. The sound of a blue whale can be heard 500 miles away, reaching an intensity of 6.3×10^6 watts per meter². Determine the decibel level of this sound. At close range, can the sound of a blue whale rupture the human eardrum?

18. What is the decibel level of a normal conversation, 3.2×10^{-6} watt per meter²?

19. Students in a psychology class took a final examination. As part of an experiment to see how much of the course content they remembered over time, they took equivalent forms of the exam in monthly intervals thereafter. The average score for the group, $f(t)$, after t months was modeled by the function

$$f(t) = 88 - 15 \ln(t + 1), \qquad 0 \le t \le 12.$$

a. What was the average score on the original exam?

b. What was the average score after 2 months? 4 months? 6 months? 8 months? 10 months? one year?

c. Sketch the graph of f (either by hand or with a graphing utility). Describe what the graph indicates in terms of the material retained by the students.

Writing in Mathematics

120. Describe the relationship between an equation in logarithmic form and an equivalent equation in exponential form.

121. What question can be asked to help evaluate $\log_3 81$?

122. Explain why the logarithm of 1 with base b is 0.

123. Describe the following property using words: $\log_b b^x = x$.

124. Explain how to use the graph of $f(x) = 2^x$ to obtain the graph of $g(x) = \log_2 x$.

125. Explain how to find the domain of a logarithmic function.

126. Logarithmic models are well suited to phenomena in which growth is initially rapid but then begins to level off. Describe something that is changing over time that can be modeled using a logarithmic function.

127. Suppose that a girl is 4 feet 6 inches at age 10. Explain how to use the function in Exercises 113–114 to determine how tall she can expect to be as an adult.

Technology Exercises

In Exercises 128–131, graph f and g in the same viewing rectangle. Then describe the relationship of the graph of g to the graph of f.

128. $f(x) = \ln x, g(x) = \ln(x + 3)$

129. $f(x) = \ln x, g(x) = \ln x + 3$

130. $f(x) = \log x, g(x) = -\log x$

131. $f(x) = \log x, g(x) = \log(x - 2) + 1$

132. Students in a mathematics class took a final examination. They took equivalent forms of the exam in monthly intervals thereafter. The average score, $f(t)$, for the group after t months was modeled by the human memory function $f(t) = 75 - 10 \log(t + 1)$, where $0 \le t \le 12$. Use a graphing utility to graph the function. Then determine how many months will elapse before the average score falls below 65.

133. In parts (a)–(c), graph f and g in the same viewing rectangle.

a. $f(x) = \ln(3x), g(x) = \ln 3 + \ln x$

b. $f(x) = \log(5x^2), g(x) = \log 5 + \log x^2$

c. $f(x) = \ln(2x^3), g(x) = \ln 2 + \ln x^3$

d. Describe what you observe in parts (a)–(c). Generalize this observation by writing an equivalent expression for $\log_b (MN)$, where $M > 0$ and $N > 0$.

e. Complete this statement: The logarithm of a product is equal to _____.

134. Graph each of the following functions in the same viewing rectangle and then place the functions in order from the one that increases most slowly to the one that increases most rapidly.

$$y = x, y = \sqrt{x}, y = e^x, y = \ln x, y = x^x, y = x^2$$

Critical Thinking Exercises

135. Which one of the following is true?

a. $\dfrac{\log_2 8}{\log_2 4} = \dfrac{8}{4}$

b. $\log(-100) = -2$

c. The domain of $f(x) = \log_2 x$ is $(-\infty, \infty)$.

d. $\log_b x$ is the exponent to which b must be raised to obtain x.

136. Without using a calculator, find the exact value of

$$\frac{\log_3 81 - \log_\pi 1}{\log_{2\sqrt{2}} 8 - \log 0.001}.$$

137. Without using a calculator, find the exact value of $\log_4[\log_3(\log_2 8)]$.

138. Without using a calculator, determine which is the greater number: $\log_4 60$ or $\log_3 40$.

Group Exercise

139. This group exercise involves exploring the way we grow. Group members should create a graph for the function that models the percentage of adult height attained by a boy who is x years old, $f(x) = 29 + 48.8 \log(x + 1)$. Let $x = 1, 2, 3, \ldots, 12$, find function values, and connect the resulting points with a smooth curve. Then create a graph for the function that models the percentage of adult height attained by a girl who is x years old, $g(x) = 62 + 35 \log(x - 4)$. Let $x = 5, 6, 7, \ldots, 15$, find function values, and connect the resulting points with a smooth curve. Group members should then discuss similarities and differences in the growth patterns for boys and girls based on the graphs.

SECTION 3.3 *Properties of Logarithms*

Objectives

1 Use the product rule.

2 Use the quotient rule.

3 Use the power rule.

4 Expand logarithmic expressions.

5 Condense logarithmic expressions.

6 Use the change-of-base property.

We all learn new things in different ways. In this section, we consider important properties of logarithms. What would be the most effective way for you to learn about these properties? Would it be helpful to use your graphing utility and discover one of these properties for yourself? To do so, work Exercise 133 in Exercise Set 3.2 before continuing. Would the properties become more meaningful if you could see exactly where they come from? If so, you will find details of the proofs of some of these properties in the appendix. The remainder of our work in this chapter will be based on the properties of logarithms that you learn in this section.

1 Use the product rule.

The Product Rule

Properties of exponents correspond to properties of logarithms. For example, when we multiply with the same base, we add exponents:

$$b^m \cdot b^n = b^{m+n}.$$

This property of exponents, coupled with an awareness that a logarithm is an exponent, suggests the following property, called the **product rule**:

Discovery

We know that log 100,000 = 5. Show that you get the same result by writing 100,000 as 1000 · 100 and then using the product rule. Then verify the product rule by using other numbers whose logarithms are easy to find.

The Product Rule

Let b, M, and N be positive real numbers with $b \neq 1$.
$$\log_b(MN) = \log_b M + \log_b N$$
The logarithm of a product is the sum of the logarithms.

When we use the product rule to write a single logarithm as the sum of two logarithms, we say that we are **expanding a logarithmic expression**. For example, we can use the product rule to expand $\ln(7x)$:

$$\ln(7x) = \ln 7 + \ln x.$$

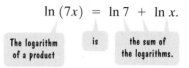

The logarithm of a product | is | the sum of the logarithms.

EXAMPLE 1   Using the Product Rule

Use the product rule to expand each logarithmic expression:

a. $\log_4(7 \cdot 5)$ **b.** $\log(10x)$.

Solution

a. $\log_4(7 \cdot 5) = \log_4 7 + \log_4 5$ The logarithm of a product is the sum of the logarithms.

b. $\log(10x) = \log 10 + \log x$ The logarithm of a product is the sum of the logarithms. These are common logarithms with base 10 understood.

$\qquad\qquad = 1 + \log x$ Because $\log_b b = 1$, then log 10 = 1.

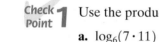 Use the product rule to expand each logarithmic expression:

a. $\log_6(7 \cdot 11)$ **b.** $\log(100x)$.

 Use the quotient rule.

The Quotient Rule

When we divide with the same base, we subtract exponents:

$$\frac{b^m}{b^n} = b^{m-n}.$$

This property suggests the following property of logarithms, called the **quotient rule**:

Discovery

We know that $\log_2 16 = 4$. Show that you get the same result by writing 16 as $\frac{32}{2}$ and then using the quotient rule. Then verify the quotient rule using other numbers whose logarithms are easy to find.

The Quotient Rule

Let b, M, and N be positive real numbers with $b \neq 1$.
$$\log_b\left(\frac{M}{N}\right) = \log_b M - \log_b N$$
The logarithm of a quotient is the difference of the logarithms.

When we use the quotient rule to write a single logarithm as the difference of two logarithms, we say that we are **expanding a logarithmic expression**. For example, we can use the quotient rule to expand $\log\left(\frac{x}{2}\right)$:

$$\log\left(\frac{x}{2}\right) = \log x - \log 2.$$

The logarithm of a quotient | is | the difference of the logarithms.

EXAMPLE 2   Using the Quotient Rule

Use the quotient rule to expand each logarithmic expression:

a. $\log_7\left(\dfrac{19}{x}\right)$ **b.** $\ln\left(\dfrac{e^3}{7}\right)$.

Solution

a. $\log_7\left(\dfrac{19}{x}\right) = \log_7 19 - \log_7 x$ The logarithm of a quotient is the difference of th logarithms.

b. $\ln\left(\dfrac{e^3}{7}\right) = \ln e^3 - \ln 7$ The logarithm of a quotient is the difference of th logarithms. These are natural logarithms wit base e understood.

$= 3 - \ln 7$ Because $\ln e^x = x$, then $\ln e^3 = 3$.

Check Point 2 Use the quotient rule to expand each logarithmic expression:

a. $\log_8\left(\dfrac{23}{x}\right)$ **b.** $\ln\left(\dfrac{e^5}{11}\right)$.

 Use the power rule.

The Power Rule

When an exponential expression is raised to a power, we multiply exponents:

$$(b^m)^n = b^{mn}.$$

This property suggests the following property of logarithms, called the **power rul**

> **The Power Rule**
>
> Let b and M be positive real numbers with $b \neq 1$, and let p be any real number.
>
> $$\log_b M^p = p \log_b M$$
>
> The logarithm of a number with an exponent is the product of the exponent and the logarithm of that number.

When we use the power rule to "pull the exponent to the front," we say that w are **expanding a logarithmic expression**. For example, we can use the power rule t expand $\ln x^2$:

$$\ln x^2 = 2 \ln x.$$

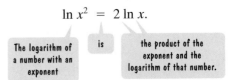

Figure 3.16 shows the graphs of $y = \ln x^2$ and $y = 2 \ln x$ in $[-5, 5, 1]$ t $[-5, 5, 1]$ viewing rectangles. Are $\ln x^2$ and $2 \ln x$ the same? The graphs illustra that $y = \ln x^2$ and $y = 2 \ln x$ have different domains. The graphs are only the san if $x > 0$. Thus, we should write

$$\ln x^2 = 2 \ln x \text{ for } x > 0.$$

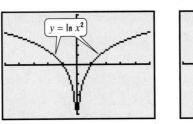

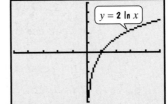

Domain: $(-\infty, 0) \cup (0, \infty)$ Domain: $(0, \infty)$

Figure 3.16 $\ln x^2$ and $2 \ln x$ have different domains.

When expanding a logarithmic expression, you might want to determine whether the rewriting has changed the domain of the expression. For the rest of this section, assume that all variables and variable expressions represent positive numbers.

EXAMPLE 3   Using the Power Rule

Use the power rule to expand each logarithmic expression:

    a. $\log_5 7^4$ **b.** $\ln \sqrt{x}$ **c.** $\log(4x)^5$.

Solution

    a. $\log_5 7^4 = 4 \log_5 7$ The logarithm of a number with an exponent is the exponent times the logarithm of the number.

    b. $\ln \sqrt{x} = \ln x^{\frac{1}{2}}$ Rewrite the radical using a rational exponent.

              $= \frac{1}{2} \ln x$ Use the power rule to bring the exponent to the front.

    c. $\log(4x)^5 = 5 \log(4x)$ We immediately apply the power rule because the entire variable expression, 4x, is raised to the 5th power.

 Use the power rule to expand each logarithmic expression:

    a. $\log_6 3^9$ **b.** $\ln \sqrt[3]{x}$ **c.** $\log(x + 4)^2$.

 Expand logarithmic expressions.

Expanding Logarithmic Expressions

It is sometimes necessary to use more than one property of logarithms when you expand a logarithmic expression. Properties for expanding logarithmic expressions are as follows:

Properties for Expanding Logarithmic Expressions

For $M > 0$ and $N > 0$:

1. $\log_b(MN) = \log_b M + \log_b N$ Product rule

2. $\log_b\left(\dfrac{M}{N}\right) = \log_b M - \log_b N$ Quotient rule

3. $\log_b M^p = p \log_b M$ Power rule

Study Tip

The graphs show that

$$\ln (x + 3) \neq \ln x + \ln 3.$$

$y = \ln x$ shifted 3 units left $y = \ln x$ shifted 3 units up

In general,

$$\log_b(M + N) \neq \log_b M + \log_b N.$$

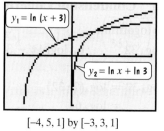

$[-4, 5, 1]$ by $[-3, 3, 1]$

Try to avoid the following errors:

Incorrect!

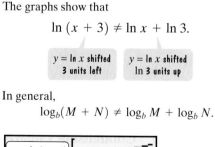

Use the change-of-base property.

The Change-of-Base Property

We have seen that calculators give the values of both common logarithms (base 10) and natural logarithms (base e). To find a logarithm with any other base, we can use the following change-of-base property:

The Change-of-Base Property

For any logarithmic bases a and b, and any positive number M,

$$\log_b M = \frac{\log_a M}{\log_a b}.$$

The logarithm of M with base b is equal to the logarithm of M with any new base divided by the logarithm of b with that new base.

In the change-of-base property, base b is the base of the original logarithm. Base a is a new base that we introduce. Thus, the change-of-base property allows us to change from base b to *any* new base a, as long as the newly introduced base is a positive number not equal to 1.

The change-of-base property is used to write a logarithm in terms of quantities that can be evaluated with a calculator. Because calculators contain keys for common (base 10) and natural (base e) logarithms, we will frequently introduce base 10 or base e.

Change-of-Base Property	Introducing Common Logarithms	Introducing Natural Logarithms
$\log_b M = \dfrac{\log_a M}{\log_a b}$	$\log_b M = \dfrac{\log_{10} M}{\log_{10} b}$	$\log_b M = \dfrac{\log_e M}{\log_e b}$
a is the new introduced base.	10 is the new introduced base.	e is the new introduced base.

Using the notations for common logarithms and natural logarithms, we have the following results:

The Change-of-Base Property: Introducing Common and Natural Logarithms

Introducing Common Logarithms	Introducing Natural Logarithms
$\log_b M = \dfrac{\log M}{\log b}$	$\log_b M = \dfrac{\ln M}{\ln b}$

EXAMPLE 7   Changing Base to Common Logarithms

Use common logarithms to evaluate $\log_5 140$.

Solution Because $\log_b M = \dfrac{\log M}{\log b}$,

$$\log_5 140 = \frac{\log 140}{\log 5}$$

$$\approx 3.07.$$

Use a calculator: 140 [LOG] ÷ 5 [LOG] [=] or [LOG] 140 ÷ [LOG] 5 [ENTER]. On some calculators, parentheses are needed after 140 and 5.

This means that $\log_5 140 \approx 3.07$.

Discovery

Find a reasonable estimate of $\log_5 140$ to the nearest whole number. To what power can you raise 5 in order to get 140? Compare your estimate to the value obtained in Example 7.

 Use common logarithms to evaluate $\log_7 2506$.

EXAMPLE 8   Changing Base to Natural Logarithms

Use natural logarithms to evaluate $\log_5 140$.

Solution   Because $\log_b M = \dfrac{\ln M}{\ln b}$,

$$\log_5 140 = \frac{\ln 140}{\ln 5}$$

$$\approx 3.07.$$

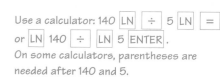

Use a calculator: 140 [LN] [÷] 5 [LN] [=]
or [LN] 140 [÷] [LN] 5 [ENTER].
On some calculators, parentheses are
needed after 140 and 5.

We have again shown that $\log_5 140 \approx 3.07$.

Check Point 8   Use natural logarithms to evaluate $\log_7 2506$.

Technology

We can use the change-of-base property to graph logarithmic functions with bases other than 10 or e on a graphing utility. For example, Figure 3.17 shows the graphs of

$$y = \log_2 x \quad \text{and} \quad y = \log_{20} x$$

in a $[0, 10, 1]$ by $[-3, 3, 1]$ viewing rectangle. Because $\log_2 x = \dfrac{\ln x}{\ln 2}$ and $\log_{20} x = \dfrac{\ln x}{\ln 20}$, the functions are entered as

$$y_1 = \boxed{LN}\, x \;\boxed{÷}\; \boxed{LN}\, 2$$
$$\text{and}\quad y_2 = \boxed{LN}\, x \;\boxed{÷}\; \boxed{LN}\, 20.$$

On some calculators, parentheses
are needed after x, 2, and 20.

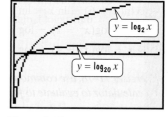

Figure 3.17 Using the change-of-base property to graph logarithmic functions

---

EXERCISE SET 3.3

Practice Exercises

In Exercises 1–40, use properties of logarithms to expand each logarithmic expression as much as possible. Where possible, evaluate logarithmic expressions without using a calculator.

1. $\log_5(7 \cdot 3)$
2. $\log_8(13 \cdot 7)$
3. $\log_7(7x)$

4. $\log_9(9x)$
5. $\log(1000x)$
6. $\log(10{,}000x)$

7. $\log_7\!\left(\dfrac{7}{x}\right)$
8. $\log_9\!\left(\dfrac{9}{x}\right)$
9. $\log\!\left(\dfrac{x}{100}\right)$

10. $\log\!\left(\dfrac{x}{1000}\right)$
11. $\log_4\!\left(\dfrac{64}{y}\right)$
12. $\log_5\!\left(\dfrac{125}{y}\right)$

13. $\ln\!\left(\dfrac{e^2}{5}\right)$
14. $\ln\!\left(\dfrac{e^4}{8}\right)$
15. $\log_b x^3$

16. $\log_b x^7$
17. $\log N^{-6}$
18. $\log M^{-8}$

19. $\ln \sqrt[5]{x}$
20. $\ln \sqrt[7]{x}$
21. $\log_b(x^2 y)$

22. $\log_b(xy^3)$
23. $\log_4\!\left(\dfrac{\sqrt{x}}{64}\right)$
24. $\log_5\!\left(\dfrac{\sqrt{x}}{25}\right)$

25. $\log_6\!\left(\dfrac{36}{\sqrt{x+1}}\right)$
26. $\log_8\!\left(\dfrac{64}{\sqrt{x+1}}\right)$
27. $\log_b\!\left(\dfrac{x^2 y}{z^2}\right)$

28. $\log_b\!\left(\dfrac{x^3 y}{z^2}\right)$
29. $\log\sqrt{100x}$
30. $\ln\sqrt{ex}$

31. $\log\sqrt[3]{\dfrac{x}{y}}$
32. $\log\sqrt[5]{\dfrac{x}{y}}$
33. $\log_b\!\left(\dfrac{\sqrt{x}\,y^3}{z^3}\right)$

34. $\log_b\!\left(\dfrac{\sqrt[3]{x}\,y^4}{z^5}\right)$
35. $\log_5 \sqrt[3]{\dfrac{x^2 y}{25}}$
36. $\log_2 \sqrt[5]{\dfrac{xy^4}{16}}$

37. $\ln\!\left[\dfrac{x^3\sqrt{x^2+1}}{(x+1)^4}\right]$
38. $\ln\!\left[\dfrac{x^4\sqrt{x^2+3}}{(x+3)^5}\right]$

39. $\log\!\left[\dfrac{10x^2\sqrt[3]{1-x}}{7(x+1)^2}\right]$
40. $\log\!\left[\dfrac{100x^3\sqrt[3]{5-x}}{3(x+7)^2}\right]$

SECTION 3.4 Exponential and Logarithmic Equations

Objectives

❶ Use like bases to solve exponential equations.

❷ Use logarithms to solve exponential equations.

❸ Use the definition of a logarithm to solve logarithmic equations.

❹ Use the one-to-one property of logarithms to solve logarithmic equations.

❺ Solve applied problems involving exponential and logarithmic equations.

You inherited $30,000. You would like to put aside $25,000 and eventually have ov half a million dollars for early retirement. Is this possible? In this section, you w see how techniques for solving equations with variable exponents provide an a swer to the question.

Exponential Equations

An **exponential equation** is an equation containing a variable in an exponen Examples of exponential equations include

$$2^{3x-8} = 16, \qquad 4^x = 15, \quad \text{and} \quad 40e^{0.6x} = 240.$$

Some exponential equations can be solved by expressing each side of th equation as a power of the same base. All exponential functions are one-to-one—tha is, no two different ordered pairs have the same second component. Thus, if b is positive number other than 1 and $b^M = b^N$, then $M = N$.

Solving Exponential Equations by Expressing Each Side as a Power of the Same Base

$$\text{If} \quad b^M = b^N, \text{ then } M = N.$$

Express each side as a power of the same base. ⟶ Set the exponents equal to each other.

1. Rewrite the equation in the form $b^M = b^N$.
2. Set $M = N$.
3. Solve for the variable.

---

❶ Use like bases to solve exponential equations.

Technology

The graphs of

$$y_1 = 2^{3x-8}$$
$$\text{and} \quad y_2 = 16$$

have an intersection point whose x-coordinate is 4. This verifies that $\{4\}$ is the solution set of $2^{3x-8} = 16$.

$y_2 = 16$

$y_1 = 2^{3x-8}$

$x = 4$

$[-1, 5, 1]$ by $[0, 20, 1]$

EXAMPLE 1 Solving Exponential Equations

Solve: **a.** $2^{3x-8} = 16$ **b.** $27^{x+3} = 9^{x-1}$.

Solution In each equation, express both sides as a power of the same base. The set the exponents equal to each other and solve for the variable.

a. Because 16 is 2^4, we express each side of $2^{3x-8} = 16$ in terms of base 2.

$2^{3x-8} = 16$	This is the given equation.
$2^{3x-8} = 2^4$	Write each side as a power of the same base
$3x - 8 = 4$	If $b^M = b^N$, $b > 0$ and $b \neq 1$, then $M = N$.
$3x = 12$	Add 8 to both sides.
$x = 4$	Divide both sides by 3.

Substituting 4 for x into the original equation produces the true statemen $16 = 16$. The solution set is $\{4\}$.

b. Because $27 = 3^3$ and $9 = 3^2$, we can express both sides of $27^{x+3} = 9^{x-1}$ in terms of base 3.

$$27^{x+3} = 9^{x-1}$$ This is the given equation.

$$\left(3^3\right)^{x+3} = \left(3^2\right)^{x-1}$$ Write each side as a power of the same base.

$$3^{3(x+3)} = 3^{2(x-1)}$$ When an exponential expression is raised to a power, multiply exponents.

$$3(x + 3) = 2(x - 1)$$ If two powers of the same base are equal, then the exponents are equal.

$$3x + 9 = 2x - 2$$ Apply the distributive property.

$$x + 9 = -2$$ Subtract 2x from both sides.

$$x = -11$$ Subtract 9 from both sides.

Substituting -11 for x into the original equation produces $27^{-8} = 9^{-12}$, which simplifies to the true statement $3^{-24} = 3^{-24}$. The solution set is $\{-11\}$.

Check Point 1   Solve: **a.** $5^{3x-6} = 125$ **b.** $8^{x+2} = 4^{x-3}$.

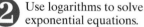

② Use logarithms to solve exponential equations.

Most exponential equations cannot be rewritten so that each side has the same base. Logarithms are extremely useful in solving such equations. The solution begins with isolating the exponential expression and taking the natural logarithm on both sides. Why can we do this? All logarithmic relations are functions. Thus, if M and N are positive real numbers and $M = N$, then $\log_b M = \log_b N$.

Using Natural Logarithms to Solve Exponential Equations

1. Isolate the exponential expression.
2. Take the natural logarithm on both sides of the equation.
3. Simplify using one of the following properties:

$$\ln b^x = x \ln b \quad \text{or} \quad \ln e^x = x.$$

4. Solve for the variable.

EXAMPLE 2   Solving an Exponential Equation

Solve: $4^x = 15$.

Discovery

The base that is used when taking the logarithm on both sides of an equation can be any base at all. Solve $4^x = 15$ by taking the common logarithm on both sides. Solve again, this time taking the logarithm with base 4 on both sides. Use the change-of-base property to show that the solutions are the same as the one obtained in Example 2.

Solution   Because the exponential expression, 4^x, is already isolated on the left, we begin by taking the natural logarithm on both sides of the equation.

$$4^x = 15$$ This is the given equation.

$$\ln 4^x = \ln 15$$ Take the natural logarithm on both sides.

$$x \ln 4 = \ln 15$$ Use the power rule and bring the variable exponent to the front: $\ln b^x = x \ln b$.

$$x = \frac{\ln 15}{\ln 4}$$ Solve for x by dividing both sides by ln 4.

We now have an exact value for x. We use the exact value for x in the equation's solution set. Thus, the equation's solution is $\dfrac{\ln 15}{\ln 4}$ and the solution set is $\left\{\dfrac{\ln 15}{\ln 4}\right\}$.

We can obtain a decimal approximation by using a calculator: $x \approx 1.95$. Because $4^2 = 16$, it seems reasonable that the solution to $4^x = 15$ is approximately 1.95.

Solve: $5^x = 134$. Find the solution set and then use a calculator to obtain a decimal approximation to two decimal places for the solution.

EXAMPLE 3 Solving an Exponential Equation

Solve: $40e^{0.6x} - 3 = 237$.

Solution We begin by adding 3 to both sides and dividing both sides by 40 to isolate the exponential expression, $e^{0.6x}$. Then we take the natural logarithm on both sides of the equation.

$40e^{0.6x} - 3 = 237$	This is the given equation.
$40e^{0.6x} = 240$	Add 3 to both sides.
$e^{0.6x} = 6$	Isolate the exponential factor by dividing both sides by 40.
$\ln e^{0.6x} = \ln 6$	Take the natural logarithm on both sides.
$0.6x = \ln 6$	Use the inverse property $\ln e^x = x$ on the left.
$x = \dfrac{\ln 6}{0.6} \approx 2.99$	Divide both sides by 0.6 and solve for x.

Thus, the solution of the equation is $\dfrac{\ln 6}{0.6} \approx 2.99$. Try checking this approximate solution in the original equation to verify that $\left\{\dfrac{\ln 6}{0.6}\right\}$ is the solution set.

Solve: $7e^{2x} - 5 = 58$. Find the solution set and then use a calculator to obtain a decimal approximation to two decimal places for the solution.

EXAMPLE 4 Solving an Exponential Equation

Solve: $5^{x-2} = 4^{2x+3}$.

Solution Because each exponential expression is isolated on one side of the equation, we begin by taking the natural logarithm on both sides.

$$5^{x-2} = 4^{2x+3}$$ This is the given equation.

$$\ln 5^{x-2} = \ln 4^{2x+3}$$ Take the natural logarithm on both sides.

Be sure to insert parentheses around the binomials.

$$(x - 2)\ln 5 = (2x + 3)\ln 4$$ Use the power rule and bring the variable exponents to the front: $\ln b^x = x \ln b$.

Remember that ln 5 and ln 4 are constants, not variables.

$$x \ln 5 - 2\ln 5 = 2x\ln 4 + 3\ln 4$$ Use the distributive property to distribute ln 5 and ln 4 to both terms in parentheses.

$$x \ln 5 - 2x\ln 4 = 2\ln 5 + 3\ln 4$$ Collect variable terms involving x on the left by subtracting $2x \ln 4$ and adding $2 \ln 5$ on both sides.

$$x(\ln 5 - 2\ln 4) = 2\ln 5 + 3\ln 4$$ Factor out x from the two terms on the left.

$$x = \dfrac{2\ln 5 + 3\ln 4}{\ln 5 - 2\ln 4}$$ Isolate x by dividing both sides by $\ln 5 - 2 \ln 4$.

Discovery

Use properties of logarithms to show that the solution in Example 4 can be expressed as

$$\dfrac{\ln 1600}{\ln\left(\frac{5}{16}\right)}.$$

The solution set is $\left\{\dfrac{2\ln 5 + 3\ln 4}{\ln 5 - 2\ln 4}\right\}$. The solution is approximately -6.34.

Solve: $3^{2x-1} = 7^{x+1}$. Find the solution set and then use a calculator to obtain a decimal approximation to two decimal places for the solution.

EXAMPLE 5   Solving an Exponential Equation

Solve: $e^{2x} - 4e^x + 3 = 0$.

Solution   The given equation is quadratic in form. If $u = e^x$, the equation can be expressed as $u^2 - 4u + 3 = 0$. Because this equation can be solved by factoring, we factor to isolate the exponential term.

$e^{2x} - 4e^x + 3 = 0$	This is the given equation.
$(e^x - 3)(e^x - 1) = 0$	Factor on the left. Notice that if $u = e^x$, $u^2 - 4u + 3 = (u - 3)(u - 1)$.
$e^x - 3 = 0$ or $e^x - 1 = 0$	Set each factor equal to 0.
$e^x = 3$ $e^x = 1$	Solve for e^x.
$\ln e^x = \ln 3$ $x = 0$	Take the natural logarithm on both sides of the first equation. The equation on the right can be solved by inspection.
$x = \ln 3$	$\ln e^x = x$

The solution set is $\{0, \ln 3\}$. The solutions are 0 and $\ln 3$, which is approximately 1.10.

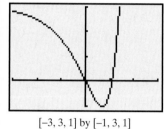
Check Point 5   Solve: $e^{2x} - 8e^x + 7 = 0$. Find the solution set and then use a calculator to obtain a decimal approximation to two decimal places, if necessary, for the solutions.

 Use the definition of a logarithm to solve logarithmic equations.

Logarithmic Equations

A **logarithmic equation** is an equation containing a variable in a logarithmic expression. Examples of logarithmic equations include

$$\log_4(x + 3) = 2 \quad \text{and} \quad \ln(x + 2) - \ln(4x + 3) = \ln\left(\frac{1}{x}\right).$$

Some logarithmic equations can be expressed in the form $\log_b M = c$. We can solve such equations by rewriting them in exponential form.

Using the Definition of a Logarithm to Solve Logarithmic Equations

1. Express the equation in the form $\log_b M = c$.
2. Use the definition of a logarithm to rewrite the equation in exponential form:

$$\log_b M = c \quad \text{means} \quad b^c = M.$$

Logarithms are exponents.

3. Solve for the variable.
4. Check proposed solutions in the original equation. Include in the solution set only values for which $M > 0$.

EXAMPLE 6   Solving Logarithmic Equations

Solve: **a.** $\log_4(x + 3) = 2$ **b.** $3 \ln(2x) = 12$.

Solution   The form $\log_b M = c$ involves a single logarithm whose coefficient is 1 on one side and a constant on the other side. Equation (a) is already in this form. We will need to divide both sides of equation (b) by 3 to obtain this form.

Technology

The graphs of

$$y_1 = \log_4(x + 3) \text{ and } y_2 = 2$$

have an intersection point whose x-coordinate is 13. This verifies that $\{13\}$ is the solution set for $\log_4(x + 3) = 2$.

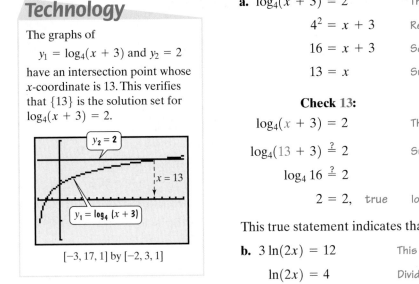

$[-3, 17, 1]$ by $[-2, 3, 1]$

a. $\log_4(x + 3) = 2$ This is the given equation.

$\qquad 4^2 = x + 3$ Rewrite in exponential form: $\log_b M = c$ means $b^c = M$.

$\qquad 16 = x + 3$ Square 4.

$\qquad 13 = x$ Subtract 3 from both sides.

Check 13:

$\log_4(x + 3) = 2$ This is the given logarithmic equation.

$\log_4(13 + 3) \overset{?}{=} 2$ Substitute 13 for x.

$\log_4 16 \overset{?}{=} 2$

$2 = 2, \quad \text{true} \qquad \log_4 16 = 2 \text{ because } 4^2 = 16.$

This true statement indicates that the solution set is $\{13\}$.

b. $3 \ln(2x) = 12$ This is the given equation.

$\ln(2x) = 4$ Divide both sides by 3.

$\log_e(2x) = 4$ Rewrite the natural logarithm showing base e. This step is optional.

$e^4 = 2x$ Rewrite in exponential form: $\log_b M = c$ means $b^c = M$.

$\dfrac{e^4}{2} = x$ Divide both sides by 2.

Check $\dfrac{e^4}{2}$:

$3 \ln(2x) = 12$ This is the given logarithmic equation.

$3 \ln\left[2\left(\dfrac{e^4}{2}\right) \right] \overset{?}{=} 12$ Substitute $\dfrac{e^4}{2}$ for x.

$3 \ln e^4 \overset{?}{=} 12$ Simplify: $\dfrac{\cancel{2}}{1} \cdot \dfrac{e^4}{\cancel{2}} = e^4.$

$3 \cdot 4 \overset{?}{=} 12$ Because $\ln e^x = x$, we conclude $\ln e^4 = 4$.

$12 = 12, \quad \text{true}$

This true statement indicates that the solution set is $\left\{\dfrac{e^4}{2}\right\}$.

Check Point 6 Solve:

a. $\log_2(x - 4) = 3$ **b.** $4 \ln(3x) = 8.$

Logarithmic expressions are defined only for logarithms of positive real numbers. **Always check proposed solutions of a logarithmic equation in the original equation. Exclude from the solution set any proposed solution that produces the logarithm of a negative number or the logarithm of 0.**

To rewrite the logarithmic equation $\log_b M = c$ in the equivalent exponential form $b^c = M$, we need a single logarithm whose coefficient is one. It is sometimes necessary to use properties of logarithms to condense logarithms into a single logarithm. In the next example, we use the product rule for logarithms to obtain a single logarithmic expression on the left side.

EXAMPLE 7   Solving a Logarithmic Equation

Solve: $\log_2 x + \log_2(x - 7) = 3$.

Solution

$\log_2 x + \log_2(x - 7) = 3$	This is the given equation.
$\log_2[x(x - 7)] = 3$	Use the product rule to obtain a single logarithm: $\log_b M + \log_b N = \log_b(MN)$.
$2^3 = x(x - 7)$	Rewrite in exponential form: $\log_b M = c$ means $b^c = M$.
$8 = x^2 - 7x$	Evaluate 2^3 on the left and apply the distributive property on the right.
$0 = x^2 - 7x - 8$	Set the equation equal to 0.
$0 = (x - 8)(x + 1)$	Factor.
$x - 8 = 0$ or $x + 1 = 0$	Set each factor equal to 0.
$x = 8$ $\qquad\qquad$ $x = -1$	Solve for x.

Check 8:

$\log_2 x + \log_2(x - 7) = 3$

$\log_2 8 + \log_2(8 - 7) \overset{?}{=} 3$

$\log_2 8 + \log_2 1 \overset{?}{=} 3$

$3 + 0 \overset{?}{=} 3$

$3 = 3,$ true

Check -1:

$\log_2 x + \log_2(x - 7) = 3$

$\log_2(-1) + \log_2(-1 - 7) \overset{?}{=} 3$

The number -1 does not check. Negative numbers do not have logarithms.

The solution set is $\{8\}$.

Check Point 7   Solve: $\log x + \log(x - 3) = 1$.

④ Use the one-to-one property of logarithms to solve logarithmic equations.

Some logarithmic equations can be expressed in the form $\log_b M = \log_b N$. Because all logarithmic functions are one-to-one, we can conclude that $M = N$.

> **Using the One-to-One Property of Logarithms to Solve Logarithmic Equations**
>
> 1. Express the equation in the form $\log_b M = \log_b N$. This form involves a single logarithm whose coefficient is 1 on each side of the equation.
> 2. Use the one-to-one property to rewrite the equation without logarithms: If $\log_b M = \log_b N$, then $M = N$.
> 3. Solve for the variable.
> 4. Check proposed solutions in the original equation. Include in the solution set only values for which $M > 0$ and $N > 0$.

EXAMPLE 8   Solving a Logarithmic Equation

Solve: $\ln(x + 2) - \ln(4x + 3) = \ln\left(\dfrac{1}{x}\right)$.

Solution   In order to apply the one-to-one property of logarithms, we need a single logarithm whose coefficient is 1 on each side of the equation. The right side is already in this form. We can obtain a single logarithm on the left side by applying the quotient rule.

$$\ln(x + 2) - \ln(4x + 3) = \ln\left(\frac{1}{x}\right)$$ This is the given equation.

$$\ln\left(\frac{x + 2}{4x + 3}\right) = \ln\left(\frac{1}{x}\right)$$ Use the quotient rule to obtain a single logarithm on the left side: $\log_b M - \log_b N = \log_b\left(\dfrac{M}{N}\right).$

$$\frac{x + 2}{4x + 3} = \frac{1}{x}$$ Use the one-to-one property: If $\log_b M = \log_b N$, then $M = N$.

$$x(4x + 3)\left(\frac{x + 2}{4x + 3}\right) = x(4x + 3)\left(\frac{1}{x}\right)$$ Multiply both sides by $x(4x + 3)$, the LCD.

$$x(x + 2) = 4x + 3$$ Simplify.

$$x^2 + 2x = 4x + 3$$ Apply the distributive property.

$$x^2 - 2x - 3 = 0$$ Subtract $4x + 3$ from both sides and set the equation equal to 0.

$$(x - 3)(x + 1) = 0$$ Factor.

$$x - 3 = 0 \quad \text{or} \quad x + 1 = 0$$ Set each factor equal to 0.

$$x = 3 \qquad\qquad x = -1$$ Solve for x.

Substituting 3 for x into the original equation produces the true statemen $\ln\left(\frac{1}{3}\right) = \ln\left(\frac{1}{3}\right)$. However, substituting -1 produces logarithms of negative numbers Thus, -1 is not a solution. The solution set is $\{3\}$.

Technology

A graphing utility's TABLE feature can be used to verify that $\{3\}$ is the solution set of

$$\ln(x + 2) - \ln(4x + 3) = \ln\left(\frac{1}{x}\right).$$

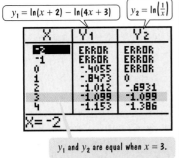

$y_1 = \ln(x + 2) - \ln(4x + 3)$ $y_2 = \ln\left(\frac{1}{x}\right)$

y_1 and y_2 are equal when $x = 3$.

Check Point 8 Solve: $\ln(x - 3) = \ln(7x - 23) - \ln(x + 1)$.

⑤ Solve applied problems involving exponential and logarithmic equations.

Applications

Our first applied example provides a mathematical perspective on the old slogan "Alcoho and driving don't mix." In California, where 38% of fatal traffic crashes involve drinking drivers, it is illegal to drive with a blood alcohol concentration of 0.08 or higher. At these levels, drivers may be arrested and charged with driving under the influence.

EXAMPLE 9   Alcohol and Risk of a Car Accident

Medical research indicates that the risk of having a car accident increases exponentially as the concentration of alcohol in the blood increases. The risk is modeled by

$$R = 6e^{12.77x},$$

where x is the blood alcohol concentration and R, given as a percent, is the risk o having a car accident. What blood alcohol concentration corresponds to a 20% risk of a car accident?

Solution   For a risk of 20% we let $R = 20$ in the equation and solve for x, the blood alcohol concentration.

$$R = 6e^{12.77x}$$ This is the given equation.

$$6e^{12.77x} = 20$$ Substitute 20 for R and (optional) reverse the two sides of the equation.

$$e^{12.77x} = \frac{20}{6}$$ Isolate the exponential factor by dividing both sides by 6.

$$\ln e^{12.77x} = \ln\left(\frac{20}{6}\right)$$ Take the natural logarithm on both sides.

$$12.77x = \ln\left(\frac{20}{6}\right)$$ Use the inverse property $\ln e^x = x$ on the left.

$$x = \frac{\ln\left(\frac{20}{6}\right)}{12.77} \approx 0.09$$ Divide both sides by 12.77 and solve for x.

Visualizing the Relationship between Blood Alcohol Concentration and the Risk of a Car Accident

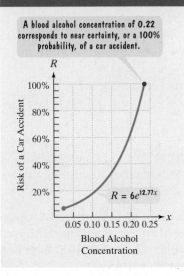

A blood alcohol concentration of 0.22 corresponds to near certainty, or a 100% probability, of a car accident.

Risk of a Car Accident

$R = 6e^{12.77x}$

Blood Alcohol Concentration

For a blood alcohol concentration of 0.09, the risk of a car accident is 20%. In many states, it is illegal to drive at 0.08, which is below this blood alcohol concentration.

Check Point **9** Use the formula in Example 9 to answer this question: What blood alcohol concentration corresponds to a 7% risk of a car accident? (In many states, drivers under the age of 21 can lose their licenses for driving at this level.)

Suppose that you inherit $30,000. Is it possible to invest $25,000 and have over half a million dollars for early retirement? Our next example illustrates the power of compound interest.

EXAMPLE 10   Revisiting the Formula for Compound Interest

The formula

$$A = P\left(1 + \frac{r}{n}\right)^{nt}$$

describes the accumulated value, A, of a sum of money, P, the principal, after t years at annual percentage rate r (in decimal form) compounded n times a year. How long will it take $25,000 to grow to $500,000 at 9% annual interest compounded monthly?

Solution

$$A = P\left(1 + \frac{r}{n}\right)^{nt} \qquad \text{This is the given formula.}$$

$$500,000 = 25,000\left(1 + \frac{0.09}{12}\right)^{12t} \qquad \begin{array}{l} A(\text{the desired accumulated value}) = \$500,000, \\ P(\text{the principal}) = \$25,000, \\ r(\text{the interest rate}) = 9\% = 0.09, \text{ and } n = 12 \\ (\text{monthly compounding}). \end{array}$$

Our goal is to solve the equation for t. Let's reverse the two sides of the equation and then simplify within parentheses.

$$25,000\left(1 + \frac{0.09}{12}\right)^{12t} = 500,000 \qquad \text{Reverse the two sides of the previous equation.}$$

$$25,000(1 + 0.0075)^{12t} = 500,000 \qquad \text{Divide within parentheses: } \frac{0.09}{12} = 0.0075.$$

$$25,000(1.0075)^{12t} = 500,000 \qquad \text{Add within parentheses.}$$

$$(1.0075)^{12t} = 20 \qquad \text{Divide both sides by 25,000.}$$

$$\ln(1.0075)^{12t} = \ln 20 \qquad \text{Take the natural logarithm on both sides.}$$

$$12t \ln(1.0075) = \ln 20 \qquad \begin{array}{l}\text{Use the power rule to bring the exponent to} \\ \text{the front: } \ln M^p = p \ln M. \end{array}$$

$$t = \frac{\ln 20}{12 \ln 1.0075} \qquad \text{Solve for } t, \text{ dividing both sides by } 12 \ln 1.0075.$$

$$\approx 33.4 \qquad \text{Use a calculator.}$$

After approximately 33.4 years, the $25,000 will grow to an accumulated value of $500,000. If you set aside the money at age 20, you can begin enjoying a life of leisure at about age 53.

Check Point **10** How long, to the nearest tenth of a year, will it take $1000 to grow to $3600 at 8% annual interest compounded quarterly?

Playing Doubles: Interest Rates and Doubling Time

One way to calculate what your savings will be worth at some point in the future is to consider doubling time. The following table shows how long it takes for your money to double at different annual interest rates subject to continuous compounding.

Annual Interest Rate	Years to Double
5%	13.9 years
7%	9.9 years
9%	7.7 years
11%	6.3 years

Of course, the first problem is collecting some money to invest. The second problem is finding a reasonably safe investment with a return of 9% or more.

EXAMPLE 11   The Growth in the Number of U.S. Internet Users

The bar graph in Figure 3.18 shows the number, in millions, of Internet users in the United States from 2000 through 2003. The function

$$f(x) = 34.1 \ln x + 117.7$$

models the number of U.S. Internet users, $f(x)$, in millions, x years after 1999. By which year will there be 200 million Internet users in the United States?

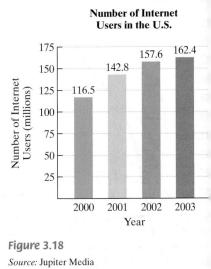

Number of Internet Users in the U.S.

Figure 3.18

Source: Jupiter Media

Solution   We substitute 200 for $f(x)$ and solve for x, the number of years after 1999.

$f(x) = 34.1 \ln x + 117.7$ This is the given function.

$200 = 34.1 \ln x + 117.7$ Substitute 200 for f(x).

Our goal is to isolate $\ln x$ in the equation $200 = 34.1 \ln x + 117.7$. We can then find x by using the definition of a logarithm to rewrite the equation in exponential form.

$34.1 \ln x + 117.7 = 200$ Reverse the two sides of the equation.

$34.1 \ln x = 82.3$ Subtract 117.7 from both sides.

$\ln x = \dfrac{82.3}{34.1}$ Divide both sides by 34.1.

$\log_e x = \dfrac{82.3}{34.1}$ Rewrite the natural logarithm showing base e. This step is optional.

$e^{\frac{82.3}{34.1}} = x$ Rewrite in exponential form: $\log_b M = c$ means $b^c = M$.

$11 \approx x$ Use a calculator.

Approximately 11 years after 1999, in the year 2010, there will be 200 million Internet users in the United States.

Check Point **11**   Use the function in Example 11 to find in which year there will be 210 million Internet users in the United States.

EXERCISE SET 3.4

 Practice Exercises

Solve each exponential equation in Exercises 1–22 by expressing each side as a power of the same base and then equating exponents.

1. $2^x = 64$

2. $3^x = 81$

3. $5^x = 125$

4. $5^x = 625$

5. $2^{2x-1} = 32$

6. $3^{2x+1} = 27$

7. $4^{2x-1} = 64$

8. $5^{3x-1} = 125$

9. $32^x = 8$

10. $4^x = 32$

11. $9^x = 27$

12. $125^x = 625$

13. $3^{1-x} = \frac{1}{27}$

14. $5^{2-x} = \frac{1}{125}$

15. $6^{\frac{x-3}{4}} = \sqrt{6}$

16. $7^{\frac{x-2}{6}} = \sqrt{7}$

17. $4^x = \dfrac{1}{\sqrt{2}}$

18. $9^x = \dfrac{1}{\sqrt[3]{3}}$

19. $8^{x+3} = 16^{x-1}$

20. $8^{1-x} = 4^{x+2}$

21. $e^{x+1} = \dfrac{1}{e}$

22. $e^{x+4} = \dfrac{1}{e^{2x}}$

Solve each exponential equation in Exercises 23–48. Express the solution set in terms of natural logarithms. Then use a calculator to obtain a decimal approximation, correct to two decimal places, for the solution.

23. $10^x = 3.91$

24. $10^x = 8.07$

25. $e^x = 5.7$ **26.** $e^x = 0.83$

27. $5^x = 17$ **28.** $19^x = 143$

29. $5e^x = 23$ **30.** $9e^x = 107$

31. $3e^{5x} = 1977$ **32.** $4e^{7x} = 10{,}273$

33. $e^{1-5x} = 793$ **34.** $e^{1-8x} = 7957$

35. $e^{5x-3} - 2 = 10{,}476$ **36.** $e^{4x-5} - 7 = 11{,}243$

37. $7^{x+2} = 410$ **38.** $5^{x-3} = 137$

39. $7^{0.3x} = 813$ **40.** $3^{\frac{x}{7}} = 0.2$

41. $5^{2x+3} = 3^{x-1}$ **42.** $7^{2x+1} = 3^{x+2}$

43. $e^{2x} - 3e^x + 2 = 0$ **44.** $e^{2x} - 2e^x - 3 = 0$

45. $e^{4x} + 5e^{2x} - 24 = 0$ **46.** $e^{4x} - 3e^{2x} - 18 = 0$

47. $3^{2x} + 3^x - 2 = 0$ **48.** $2^{2x} + 2^x - 12 = 0$

Solve each logarithmic equation in Exercises 49–90. Be sure to reject any value of x that is not in the domain of the original logarithmic expressions. Give the exact answer. Then, where necessary, use a calculator to obtain a decimal approximation, correct to two decimal places, for the solution.

49. $\log_3 x = 4$ **50.** $\log_5 x = 3$

51. $\ln x = 2$ **52.** $\ln x = 3$

53. $\log_4(x + 5) = 3$ **54.** $\log_5(x - 7) = 2$

55. $\log_3(x - 4) = -3$ **56.** $\log_7(x + 2) = -2$

57. $\log_4(3x + 2) = 3$ **58.** $\log_2(4x + 1) = 5$

59. $5 \ln(2x) = 20$ **60.** $6 \ln(2x) = 30$

61. $6 + 2 \ln x = 5$ **62.** $7 + 3 \ln x = 6$

63. $\ln \sqrt{x + 3} = 1$ **64.** $\ln \sqrt{x + 4} = 1$

65. $\log_5 x + \log_5(4x - 1) = 1$

66. $\log_6(x + 5) + \log_6 x = 2$

67. $\log_3(x - 5) + \log_3(x + 3) = 2$

68. $\log_2(x - 1) + \log_2(x + 1) = 3$

69. $\log_2(x + 2) - \log_2(x - 5) = 3$

70. $\log_4(x + 2) - \log_4(x - 1) = 1$

71. $2 \log_3(x + 4) = \log_3 9 + 2$

72. $3 \log_2(x - 1) = 5 - \log_2 4$

73. $\log_2(x - 6) + \log_2(x - 4) - \log_2 x = 2$

74. $\log_2(x - 3) + \log_2 x - \log_2(x + 2) = 2$

75. $\log(x + 4) = \log x + \log 4$

76. $\log(5x + 1) = \log(2x + 3) + \log 2$

77. $\log(3x - 3) = \log(x + 1) + \log 4$

78. $\log(2x - 1) = \log(x + 3) + \log 3$

79. $2 \log x = \log 25$

80. $3 \log x = \log 125$

81. $\log(x + 4) - \log 2 = \log(5x + 1)$

82. $\log(x + 7) - \log 3 = \log(7x + 1)$

83. $2 \log x - \log 7 = \log 112$

84. $\log(x - 2) + \log 5 = \log 100$

85. $\log x + \log(x + 3) = \log 10$

86. $\log(x + 3) + \log(x - 2) = \log 14$

87. $\ln(x - 4) + \ln(x + 1) = \ln(x - 8)$

88. $\log_2(x - 1) - \log_2(x + 3) = \log_2\left(\dfrac{1}{x}\right)$

89. $\ln(x - 2) - \ln(x + 3) = \ln(x - 1) - \ln(x + 7)$

90. $\ln(x - 5) - \ln(x + 4) = \ln(x - 1) - \ln(x + 2)$

Practice Plus

In Exercises 91–100, solve each equation.

91. $5^{2x} \cdot 5^{4x} = 125$ **92.** $3^{x+2} \cdot 3^x = 81$

93. $2\,|\ln x| - 6 = 0$ **94.** $3\,|\log x| - 6 = 0$

95. $3^{x^2} = 45$ **96.** $5^{x^2} = 50$

97. $\ln(2x + 1) + \ln(x - 3) - 2 \ln x = 0$

98. $\ln 3 - \ln(x + 5) - \ln x = 0$

99. $5^{x^2 - 12} = 25^{2x}$ **100.** $3^{x^2 - 12} = 9^{2x}$

Application Exercises

Use the formula $R = 6e^{12.77x}$, where x is the blood alcohol concentration and R, given as a percent, is the risk of having a car accident, to solve Exercises 101–102.

101. What blood alcohol concentration corresponds to a 25% risk of a car accident?

102. What blood alcohol concentration corresponds to a 50% risk of a car accident?

103. The formula $A = 18.9e^{0.0055t}$ models the population of New York State, A, in millions, t years after 2000.

    a. What was the population of New York in 2000?

    b. When will the population of New York reach 19.6 million?

104. The formula $A = 15.9e^{0.0235t}$ models the population of Florida, A, in millions, t years after 2000.

    a. What was the population of Florida in 2000?

    b. When will the population of Florida reach 19.2 million?

In Exercises 105–108, complete the table for a savings account subject to n compoundings yearly $\left[A = P\left(1 + \dfrac{r}{n}\right)^{nt}\right]$. Round answers to one decimal place.

	Amount Invested	Number of Compounding Periods	Annual Interest Rate	Accumulated Amount	Time t in Years
105.	$12,500	4	5.75%	$20,000	
106.	$7250	12	6.5%	$15,000	
107.	$1000	360		$1400	2
108.	$5000	360		$9000	4

In Exercises 109–112, complete the table for a savings account subject to continuous compounding ($A = Pe^{rt}$). Round answers to one decimal place.

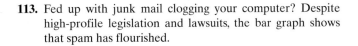

	Amount Invested	Annual Interest Rate	Accumulated Amount	Time t in Years
109.	$8000	8%	Double the amount invested	
110.	$8000		$12,000	2
111.	$2350		Triple the amount invested	7
112.	$17,425	4.25%	$25,000	

113. Fed up with junk mail clogging your computer? Despite high-profile legislation and lawsuits, the bar graph shows that spam has flourished.

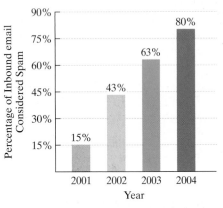

Spam Slam: Percentage of Inbound email in the U.S. Considered Spam

Source: Meta Group

The function $f(x) = 13.4 + 46.3 \ln x$ models the percentage of inbound email in the United States considered spam, $f(x)$, x years after 2000.

a. How well does the function model the data for 2003?

b. If law enforcement against spammers does not change and the model is projected into the future, when will 96% of inbound e-mail be spam? Round to the nearest year.

114. The bar graph shows the number of children under 18 as a percentage of the total U.S. population.

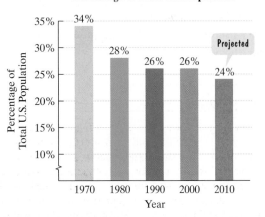

Number of Children Under 18 as a Percentage of Total U.S. Population

Projected

Source: www.childstats.gov

The function $f(x) = 34 - 2.6 \ln x$ models the number of children under 18 as a percentage of the total U.S. population, $f(x)$, x years after 1969.

a. How well does the function model the projected data for 2010?

b. According to the model, when will children under 18 decline to 23% of the total U.S. population? Round to the nearest year.

The function $P(x) = 95 - 30 \log_2 x$ models the percentage, $P(x)$, of students who could recall the important features of a classroom lecture as a function of time, where x represents the number of days that have elapsed since the lecture was given. The figure shows the graph of the function. Use this information to solve Exercises 115–116. Round answers to one decimal place.

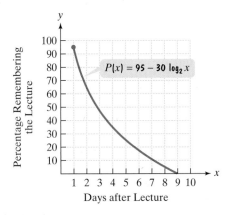

$P(x) = 95 - 30 \log_2 x$

Days after Lecture

115. After how many days do only half the students recall the important features of the classroom lecture? (Let $P(x) = 50$ and solve for x.) Locate the point on the graph that conveys this information.

116. After how many days have all students forgotten the important features of the classroom lecture? (Let $P(x) = 0$ and solve for x.) Locate the point on the graph that conveys this information.

The pH of a solution ranges from 0 to 14. An acid solution has a pH less than 7. Pure water is neutral and has a pH of 7. Normal, unpolluted rain has a pH of about 5.6. The pH of a solution is given by

$$pH = -\log x,$$

where x represents the concentration of the hydrogen ions in the solution, in moles per liter. Use the formula to solve Exercises 117–118.

117. An environmental concern involves the destructive effects of acid rain. The most acidic rainfall ever had a pH of 2.4. What was the hydrogen ion concentration? Express the answer as a power of 10 and then round to the nearest thousandth.

118. The figure shows very acidic rain in the northeast United States. What is the hydrogen ion concentration of rainfall with a pH of 4.2? Express the answer as a power of 10 and then round to the nearest hundred-thousandth.

Acid Rain over Canada and the United States

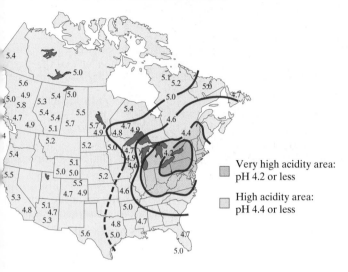

Source: National Atmospheric Program

Writing in Mathematics

119. Explain how to solve an exponential equation when both sides can be written as a power of the same base.

120. Explain how to solve an exponential equation when both sides cannot be written as a power of the same base. Use $3^x = 140$ in your explanation.

121. Explain the differences between solving $\log_3(x - 1) = 4$ and $\log_3(x - 1) = \log_3 4$.

122. In many states, a 17% risk of a car accident with a blood alcohol concentration of 0.08 is the lowest level for charging a motorist with driving under the influence. Do you agree with the 17% risk as a cutoff percentage, or do you feel that the percentage should be lower or higher? Explain your answer. What blood alcohol concentration corresponds to what you believe is an appropriate percentage?

Technology Exercises

In Exercises 123–130, use your graphing utility to graph each side of the equation in the same viewing rectangle. Then use the x-coordinate of the intersection point to find the equation's solution set. Verify this value by direct substitution into the equation.

123. $2^{x+1} = 8$

124. $3^{x+1} = 9$

125. $\log_3(4x - 7) = 2$

126. $\log_3(3x - 2) = 2$

127. $\log(x + 3) + \log x = 1$

128. $\log(x - 15) + \log x = 2$

129. $3^x = 2x + 3$

130. $5^x = 3x + 4$

Hurricanes are one of nature's most destructive forces. These low-pressure areas often have diameters of over 500 miles. The function $f(x) = 0.48 \ln(x + 1) + 27$ models the barometric air pressure, $f(x)$, in inches of mercury, at a distance of x miles from the eye of a hurricane. Use this function to solve Exercises 131–132.

131. Graph the function in a $[0, 500, 50]$ by $[27, 30, 1]$ viewing rectangle. What does the shape of the graph indicate about barometric air pressure as the distance from the eye increases?

132. Use an equation to answer this question: How far from the eye of a hurricane is the barometric air pressure 29 inches of mercury? Use the TRACE and ZOOM features or the intersect command of your graphing utility to verify your answer.

133. The function $P(t) = 145e^{-0.092t}$ models a runner's pulse, $P(t)$, in beats per minute, t minutes after a race, where $0 \le t \le 15$. Graph the function using a graphing utility. TRACE along the graph and determine after how many minutes the runner's pulse will be 70 beats per minute. Round to the nearest tenth of a minute. Verify your observation algebraically.

134. The function $W(t) = 2600(1 - 0.51e^{-0.075t})^3$ models the weight, $W(t)$, in kilograms, of a female African elephant at age t years. (1 kilogram $\approx$ 2.2 pounds) Use a graphing utility to graph the function. Then TRACE along the curve to estimate the age of an adult female elephant weighing 1800 kilograms.

Critical Thinking Exercises

135. Which one of the following is true?
  a. If $\log(x + 3) = 2$, then $e^2 = x + 3$.
  b. If $\log(7x + 3) - \log(2x + 5) = 4$, then the equation in exponential form is $10^4 = (7x + 3) - (2x + 5)$.
  c. If $x = \dfrac{1}{k} \ln y$, then $y = e^{kx}$.
  d. Examples of exponential equations include $10^x = 5.71$, $e^x = 0.72$, and $x^{10} = 5.71$.

136. If $4000 is deposited into an account paying 3% interest compounded annually and at the same time $2000 is deposited into an account paying 5% interest compounded annually, after how long will the two accounts have the same balance? Round to the nearest year.

Solve each equation in Exercises 137–139. Check each proposed solution by direct substitution or with a graphing utility.

137. $(\ln x)^2 = \ln x^2$ **138.** $(\log x)(2 \log x + 1) = 6$

139. $\ln(\ln x) = 0$

Group Exercise

140. Research applications of logarithmic functions as mathematical models and plan a seminar based on your group's research. Each group member should research one of the following areas or any other area of interest: pH (acidity of solutions), intensity of sound (decibels), brightness of stars, human memory, progress over time in a sport, profit over time. For the area that you select, explain how logarithmic functions are used and provide examples.

SECTION 3.5   Exponential Growth and Decay; Modeling Data

Objectives

❶ Model exponential growth and decay.

❷ Use logistic growth models.

❸ Use Newton's Law of Cooling.

❹ Model data with exponential and logarithmic functions.

❺ Express an exponential model in base e.

The most casual cruise on the Internet shows how people disagree when it comes to making predictions about the effects of the world's growing population. Some argue that there is a recent slowdown in the growth rate, economies remain robust, and famines in North Korea and Ethiopia are aberrations rather than signs of the future. Others say that the 6.3 billion people on Earth is twice as many as can be supported in middle-class comfort, and the world is running out of arable land and fresh water. Debates about entities that are growing exponentially can be approached mathematically: We can create functions that model data and use these functions to make predictions. In this section, we will show you how this is done.

❶ Model exponential growth and decay.

Exponential Growth and Decay

One of algebra's many applications is to predict the behavior of variables. This can be done with *exponential growth* and *decay models*. With exponential growth or decay, quantities grow or decay at a rate directly proportional to their size. Populations that are growing exponentially grow extremely rapidly as they get larger because there are more adults to have offspring. For example, the **growth rate** for world population is 1.3%, or 0.013. This means that each year world population is 1.3% more than what it was in the previous year. In 2001, world population was approximately 6.2 billion. Thus, we compute the world population in 2002 as follows:

$$6.2 \text{ billion} + 13\% \text{ of } 6.2 \text{ billion} = 6.2 + (0.013)(6.2) = 6.2806.$$

This computation suggests that 6.2806 billion people populated the world in 2002. The 0.0806 billion represents an increase of 80.6 million people from 2001 to 2002, the equivalent of the population of Germany. Using 1.3% as the annual growth rate, world population for 2003 is found in a similar manner:

$$6.2806 + 1.3\% \text{ of } 6.2806 = 6.2806 + (0.013)(6.2806) \approx 6.3622.$$

This computation suggests that approximately 6.3622 billion people populated the world in 2003.

The explosive growth of world population may remind you of the growth of money in an account subject to compound interest. Just as the growth rate for world population is multiplied by the population plus any increase in the population, a compound interest rate is multiplied by your original investment plus any accumulated interest. The balance in an account subject to continuous compounding and world population are special cases of *exponential growth models*.

Study Tip

You have seen the formula for exponential growth before, but with different letters. It is the formula for compound interest with continous compounding.

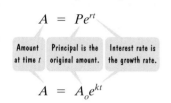

Exponential Growth and Decay Models

The mathematical model for **exponential growth** or **decay** is given by

$$f(t) = A_0 e^{kt} \quad \text{or} \quad A = A_0 e^{kt}.$$

- **If $k > 0$, the function models the amount, or size, of a *growing* entity.** A_0 is the original amount, or size, of the growing entity at time $t = 0$, A is the amount at time t, and k is a constant representing the growth rate.

- **If $k < 0$, the function models the amount, or size, of a *decaying* entity.** A_0 is the original amount, or size, of the decaying entity at time $t = 0$, A is the amount at time t, and k is a constant representing the decay rate.

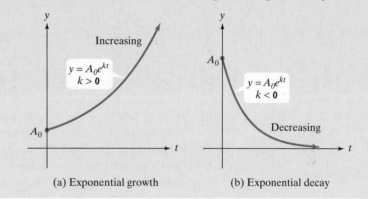

(a) Exponential growth   (b) Exponential decay

Sometimes we need to use given data to determine k, the rate of growth or decay. After we compute the value of k, we can use the formula $A = A_0 e^{kt}$ to make predictions. This idea is illustrated in our first two examples.

EXAMPLE 1   Modeling the Growth of the U.S. Population

The graph in Figure 3.19 shows the U.S. population, in millions, for five selected years from 1970 through 2003. In 1970, the U.S. population was 203.3 million. By 2003, it had grown to 294 million.

a. Find the exponential growth function that models the data for 1970 through 2003.

b. By which year will the U.S. population reach 315 million?

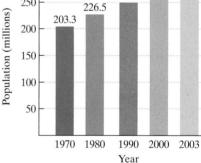

Figure 3.19

Source: Bureau of the Census

Solution

a. We use the exponential growth model

$$A = A_0 e^{kt},$$

in which t is the number of years after 1970. This means that 1970 corresponds to $t = 0$. At that time the U.S. population was 203.3 million, so we substitute 203.3 for A_0 in the growth model:

$$A = 203.3 e^{kt}.$$

We are given that 294 million is the population in 2003. Because 2003 is 33 years after 1970, when $t = 33$ the value of A is 294. Substituting these numbers into the growth model will enable us to find k, the growth rate. We know that $k > 0$ because the problem involves growth.

$$A = 203.3e^{kt}$$ Use the growth model $A = A_0 e^{kt}$ with $A_0 = 203.3$.

$$294 = 203.3e^{k \cdot 33}$$ When $t = 33$, $A = 294$. Substitute these numbers into the model.

$$e^{33k} = \frac{294}{203.3}$$ Isolate the exponential factor by dividing both sides by 203.3. We also reversed the sides.

$$\ln e^{33k} = \ln\left(\frac{294}{203.3}\right)$$ Take the natural logarithm on both sides.

$$33k = \ln\left(\frac{294}{203.3}\right)$$ Simplify the left side using $\ln e^x = x$.

$$k = \frac{\ln\left(\frac{294}{203.3}\right)}{33} \approx 0.011$$ Divide both sides by 33 and solve for k. Then use a calculator.

The value of k, approximately 0.011, indicates a growth rate of about 1.1%. This means that the U.S. population is increasing by approximately 1.1% per year. We substitute 0.011 for k in the growth model, $A = 203.3e^{kt}$, to obtain the exponential growth function for the U.S. population. It is

$$A = 203.3e^{0.011t},$$

where t is measured in years after 1970.

b. To find the year in which the U.S. population will reach 315 million, substitute 315 for A in the model from part (a) and solve for t.

$$A = 203.3e^{0.011t}$$ This is the model from part (a).

$$315 = 203.3e^{0.011t}$$ Substitute 315 for A.

$$e^{0.011t} = \frac{315}{203.3}$$ Divide both sides by 203.3. We also reversed the sides.

$$\ln e^{0.011t} = \ln\left(\frac{315}{203.3}\right)$$ Take the natural logarithm on both sides.

$$0.011t = \ln\left(\frac{315}{203.3}\right)$$ Simplify on the left using $\ln e^x = x$.

$$t = \frac{\ln\left(\frac{315}{203.3}\right)}{0.011} \approx 40$$ Divide both sides by 33 and solve for t. Then use a calculator.

Because t represents the number of years after 1970, the model indicates that the U.S. population will reach 315 million by 1970 + 40, or in the year 2010.

Check Point 1 In 1990, the population of Africa was 643 million and by 2000 it had grown to 813 million.

a. Use the exponential growth model $A = A_0 e^{kt}$, in which t is the number of years after 1990, to find the exponential growth function that models the data.

b. By which year will Africa's population reach 2000 million, or two billion?

Creating an Inaccurate Picture by Leaving Something Out

On Monday, October 19, 1987, the Dow Jones Industrial Average plunged 508 points, losing 22.6% of its value. The graph shown on the left, which appeared in a major newspaper following "Black Monday" (as it was instantly dubbed), creates the impression that the Dow average had been "bullish" from 1972 through 1987, increasing throughout this period. The graph creates this inaccurate picture by leaving something out. The graph on the right illustrates that the stock market rose and fell sharply over these years. The impressively smooth curve on the left was obtained by plotting only three of the data points. By ignoring most of the data, increases and decreases are not accounted for and the actual behavior of the market over the 15 years leading to "Black Monday" is inaccurately conveyed.

In Example 1, we used only two data values, the population for 1970 and the population for 2003, to develop a model for U.S. population growth from 1970 through 2003. By not using data for any other years, have we created a model that inaccurately describes both the

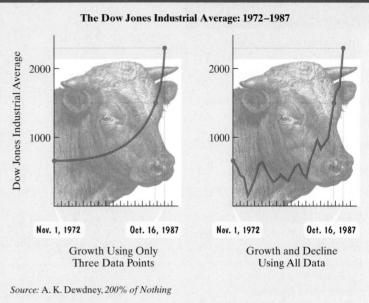

existing data and future population projections given by the U.S. Census Bureau? Something else to think about: Is an exponential model the best choice for describing U.S. population growth, or might a linear model provide a better description? We return to these issues in Exercises 54–58 in the exercise set.

Our next example involves exponential decay and its use in determining the age of fossils and artifacts. The method is based on considering the percentage of carbon-14 remaining in the fossil or artifact. Carbon-14 decays exponentially with a *half-life* of approximately 5715 years. The **half-life** of a substance is the time required for half of a given sample to disintegrate. Thus, after 5715 years a given amount of carbon-14 will have decayed to half the original amount. Carbon dating is useful for artifacts or fossils up to 80,000 years old. Older objects do not have enough carbon-14 left to determine age accurately.

EXAMPLE 2   Carbon-14 Dating: The Dead Sea Scrolls

a. Use the fact that after 5715 years a given amount of carbon-14 will have decayed to half the original amount to find the exponential decay model for carbon-14.

b. In 1947, earthenware jars containing what are known as the Dead Sea Scrolls were found by an Arab Bedouin herdsman. Analysis indicated that the scroll wrappings contained 76% of their original carbon-14. Estimate the age of the Dead Sea Scrolls.

Solution

a. We begin with the exponential decay model $A = A_0 e^{kt}$. We know that $k < 0$ because the problem involves the decay of carbon-14. After 5715 years ($t = 5715$), the amount of carbon-14 present, A, is half the original amount, A_0. Thus, we can substitute $\frac{A_0}{2}$ for A in the exponential decay model. This will enable us to find k, the decay rate.

$$A = A_0 e^{kt}$$ Begin with the exponential decay model.

$$\frac{A_0}{2} = A_0 e^{k(5715)}$$ After 5715 years ($t = 5715$), $A = \frac{A_0}{2}$ (because the amount present, A, is half the original amount, A_0).

$$\frac{1}{2} = e^{5715k}$$ Divide both sides of the equation by A_0.

$$\ln\left(\frac{1}{2}\right) = \ln e^{5715k} \qquad \text{Take the natural logarithm on both sides.}$$

$$\ln\left(\frac{1}{2}\right) = 5715k \qquad \text{Simplify the right side using } \ln e^x = x.$$

$$k = \frac{\ln\left(\frac{1}{2}\right)}{5715} \approx -0.000121 \qquad \text{Divide both sides by 5715 and solve for } k.$$

Substituting for k in the decay model, $A = A_0 e^{kt}$, the model for carbon-14 is

$$A = A_0 e^{-0.000121t}.$$

b. In 1947, the Dead Sea Scrolls contained 76% of their original carbon-14. To find their age in 1947, substitute $0.76A_0$ for A in the model from part (a) and solve for t.

$$A = A_0 e^{-0.000121t} \qquad \text{This is the decay model for carbon-14.}$$

$$0.76A_0 = A_0 e^{-0.000121t} \qquad \begin{array}{l}\text{A, the amount present, is 76\% of the original}\\ \text{amount, so } A = 0.76A_0.\end{array}$$

$$0.76 = e^{-0.000121t} \qquad \text{Divide both sides of the equation by } A_0.$$

$$\ln 0.76 = \ln e^{-0.000121t} \qquad \text{Take the natural logarithm on both sides.}$$

$$\ln 0.76 = -0.000121t \qquad \text{Simplify the right side using } \ln e^x = x.$$

$$t = \frac{\ln 0.76}{-0.000121} \approx 2268 \qquad \text{Divide both sides by } -0.000121 \text{ and solve for } t.$$

The Dead Sea Scrolls are approximately 2268 years old plus the number of years between 1947 and the current year.

Check Point 2 Strontium-90 is a waste product from nuclear reactors. As a consequence of fallout from atmospheric nuclear tests, we all have a measurable amount of strontium-90 in our bones.

a. The half-life of strontium-90 is 28 years, meaning that after 28 years a given amount of the substance will have decayed to half the original amount. Find the exponential decay model for strontium-90.

b. Suppose that a nuclear accident occurs and releases 60 grams of strontium-90 into the atmosphere. How long will it take for strontium-90 to decay to a level of 10 grams?

② Use logistic growth models.

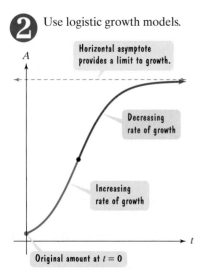

Figure 3.20 The logistic growth curve has a horizontal asymptote that identifies the limit of the growth of A over time.

Logistic Growth Models

From population growth to the spread of an epidemic, nothing on Earth can grow exponentially indefinitely. Growth is always limited. This is shown in Figure 3.20 by the horizontal asymptote. The *logistic growth model* is a function used to model situations of this type.

Logistic Growth Model

The mathematical model for limited logistic growth is given by

$$f(t) = \frac{c}{1 + ae^{-bt}} \quad \text{or} \quad A = \frac{c}{1 + ae^{-bt}},$$

where a, b, and c are constants, with $c > 0$ and $b > 0$.

As time increases ($t \rightarrow \infty$), the expression ae^{-bt} in the model approaches 0 and A gets closer and closer to c. This means that $y = c$ is a horizontal asymptote for the graph of the function. Thus, the value of A can never exceed c and c represents the limiting size that A can attain.

EXAMPLE 3   Modeling the Spread of the Flu

The function

$$f(t) = \frac{30,000}{1 + 20e^{-1.5t}}$$

describes the number of people, $f(t)$, who have become ill with influenza t weeks after its initial outbreak in a town with 30,000 inhabitants.

    a. How many people became ill with the flu when the epidemic began?

    b. How many people were ill by the end of the fourth week?

    c. What is the limiting size of $f(t)$, the population that becomes ill?

Solution

    a. The time at the beginning of the flu epidemic is $t = 0$. Thus, we can find the number of people who were ill at the beginning of the epidemic by substituting 0 for t.

$$f(t) = \frac{30,000}{1 + 20e^{-1.5t}} \quad \text{This is the given logistic growth function.}$$

$$f(0) = \frac{30,000}{1 + 20e^{-1.5(0)}} \quad \text{When the epidemic began, } t = 0.$$

$$= \frac{30,000}{1 + 20} \quad e^{-1.5(0)} = e^0 = 1$$

$$\approx 1429$$

Approximately 1429 people were ill when the epidemic began.

    b. We find the number of people who were ill at the end of the fourth week by substituting 4 for t in the logistic growth function.

$$f(t) = \frac{30,000}{1 + 20e^{-1.5t}} \quad \text{Use the given logistic growth function.}$$

$$f(4) = \frac{30,000}{1 + 20e^{-1.5(4)}} \quad \text{To find the number of people ill by the end of week four, let } t = 4.$$

$$\approx 28,583 \quad \text{Use a calculator.}$$

Approximately 28,583 people were ill by the end of the fourth week. Compared with the number of people who were ill initially, 1429, this illustrates the virulence of the epidemic.

    c. Recall that in the logistic growth model, $f(t) = \dfrac{c}{1 + ae^{-bt}}$, the constant c represents the limiting size that $f(t)$ can attain. Thus, the number in the numerator, 30,000, is the limiting size of the population that becomes ill.

Technology

The graph of the logistic growth function for the flu epidemic

$$y = \frac{30,000}{1 + 20e^{-1.5x}}$$

can be obtained using a graphing utility. We started x at 0 and ended at 10. This takes us to week 10. (In Example 3, we found that by week 4 approximately 28,583 people were ill.) We also know that 30,000 is the limiting size, so we took values of y up to 30,000. Using a [0, 10, 1] by [0, 30,000, 3000] viewing rectangle, the graph of the logistic growth function is shown below.

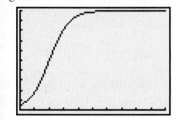

Check Point 3 In a learning theory project, psychologists discovered that

$$f(t) = \frac{0.8}{1 + e^{-0.2t}}$$

is a model for describing the proportion of correct responses, $f(t)$, after t learning trials.

    a. Find the proportion of correct responses prior to learning trials taking place.

    b. Find the proportion of correct responses after 10 learning trials.

    c. What is the limiting size of $f(t)$, the proportion of correct responses, as continued learning trials take place?

Use Newton's Law of Cooling.

Modeling Cooling

Over a period of time, a cup of hot coffee cools to the temperature of the surrounding air. **Newton's Law of Cooling**, named after Sir Isaac Newton, states that the temperature of a heated object decreases exponentially over time toward the temperature of the surrounding medium.

Study Tip

Newton's Law of Cooling applies to any situation in which an object's temperature is different from that of the surrounding medium, Thus, it can be used to model a heated object cooling to room temperature as well as a frozen object thawing to room temperature.

Newton's Law of Cooling

The temperature, T, of a heated object at time t is given by

$$T = C + (T_0 - C)e^{kt},$$

where C is the constant temperature of the surrounding medium, T_0 is the initial temperature of the heated object, and k is a negative constant that is associated with the cooling object.

EXAMPLE 4   Using Newton's Law of Cooling

A cake removed from the oven has a temperature of 210°F. It is left to cool in a room that has a temperature of 70°F. After 30 minutes, the temperature of the cake is 140°F.

a. Use Newton's Law of Cooling to find a model for the temperature of the cake, T, after t minutes.

b. What is the temperature of the cake after 40 minutes?

c. When will the temperature of the cake be 90°F?

Solution

a. We use Newton's Law of Cooling

$$T = C + (T_0 - C)e^{kt}.$$

When the cake is removed from the oven, its temperature is 210°F. This is its initial temperature: $T_0 = 210$. The constant temperature of the room is 70°F: $C = 70$. Substitute these values into Newton's Law of Cooling. Thus, the temperature of the cake, T, in degrees Fahrenheit, at time t, in minutes, is

$$T = 70 + (210 - 70)e^{kt} = 70 + 140e^{kt}.$$

After 30 minutes, the temperature of the cake is 140°F. This means that when $t = 30$, $T = 140$. Substituting these numbers into Newton's Law of Cooling will enable us to find k, a negative constant.

$T = 70 + 140e^{kt}$	Use Newton's Law of Cooling from above.
$140 = 70 + 140e^{k \cdot 30}$	When $t = 30$, $T = 140$. Substitute these numbers into the cooling model.
$70 = 140e^{30k}$	Subtract 70 from both sides.
$e^{30k} = \frac{1}{2}$	Isolate the exponential factor by dividing both sides by 140. We also reversed the sides.
$\ln e^{30k} = \ln\left(\frac{1}{2}\right)$	Take the natural logarithm on both sides.
$30k = \ln\left(\frac{1}{2}\right)$	Simplify the left side using $\ln e^x = x$.
$k = \dfrac{\ln\left(\frac{1}{2}\right)}{30} \approx -0.0231$	Divide both sides by 30 and solve for k.

We substitute -0.0231 for k into Newton's Law of Cooling, $T = 70 + 140e^{kt}$. The temperature of the cake, T, in degrees Fahrenheit, after t minutes is modeled by

$$T = 70 + 140e^{-0.0231t}.$$

b. To find the temperature of the cake after 40 minutes, we substitute 40 for t into the cooling model from part (a) and evaluate to find T.

$$T = 70 + 140e^{-0.0231(40)} \approx 126$$

After 40 minutes, the temperature of the cake will be approximately 126°F.

c. To find when the temperature of the cake will be 90°F, we substitute 90 for T into the cooling model from part (a) and solve for t.

$T = 70 + 140e^{-0.0231t}$	This is the cooling model from part (a).
$90 = 70 + 140e^{-0.0231t}$	Substitute 90 for T.
$20 = 140e^{-0.0231t}$	Subtract 70 from both sides.
$e^{-0.0231t} = \frac{1}{7}$	Divide both sides by 140. We also reversed the sides.
$\ln e^{-0.0231t} = \ln\left(\frac{1}{7}\right)$	Take the natural logarithm on both sides.
$-0.0231t = \ln\left(\frac{1}{7}\right)$	Simplify the left side using $\ln e^x = x$.
$t = \dfrac{\ln\left(\frac{1}{7}\right)}{-0.0231} \approx 84$	Solve for t by dividing both sides by -0.0231.

The temperature of the cake will be 90°F after approximately 84 minutes.

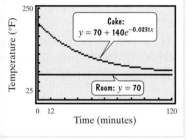

Technology

The graphs illustrate how the temperature of the cake decreases exponentially over time toward the 70°F room temperature.

Cake: $y = 70 + 140e^{-0.0231x}$

Room: $y = 70$

Check Point 4 An object is heated to 100°C. It is left to cool in a room that has a temperature of 30°C. After 5 minutes, the temperature of the object is 80°C.

a. Use Newton's Law of Cooling to find a model for the temperature of the object, T, after t minutes.

b. What is the temperature of the object after 20 minutes?

c. When will the temperature of the object be 35°C?

④ Model data with exponential and logarithmic functions.

The Art of Modeling

Throughout this chapter, we have been working with models that were given. However, we can create functions that model data by observing patterns in scatter plots. Figure 3.21 shows scatter plots for data that are exponential or logarithmic.

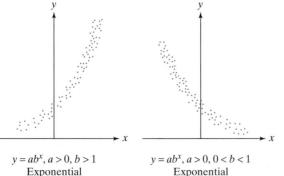
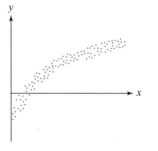
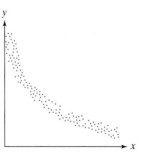

$y = ab^x, a > 0, b > 1$
Exponential

$y = ab^x, a > 0, 0 < b < 1$
Exponential

$y = a + b \ln x, a > 0, b > 0$
Logarithmic

$y = a + b \ln x, a > 0, b < 0$
Logarithmic

Figure 3.21 Scatter plots for exponential or logarithmic models

EXAMPLE 5   Choosing a Model for Data

Figure 3.22(a) shows the percentage of U.S. households with televisions that subscribe to cable television. The data are displayed for five selected years from 1980 through 2002. A scatter plot is shown in Figure 3.22(b). What function would be a good choice for modeling the data?

Percentage of U.S. Households with TVs with Cable Television

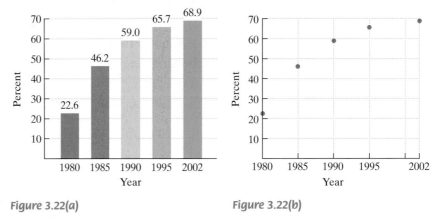

Figure 3.22(a)

Figure 3.22(b)

Source: Nielsen Media Research

Solution   Because the data in the scatter plot increase rapidly at first and then begin to level off a bit, the shape suggests that a logarithmic function is a good choice for modeling the data.

Check Point **5**   Table 3.5 shows the populations of various cities, in thousands, and the average walking speed, in feet per second, of a person living in the city. Create a scatter plot for the data. Based on the scatter plot, what function would be a good choice for modeling the data?

Table 3.5   Population and Walking Speed

Population (thousands)	Walking Speed (feet per second)
5.5	0.6
14	1.0
71	1.6
138	1.9
342	2.2

Source: Mark and Helen Bornstein, "The Pace of Life"

How can we obtain a logarithmic function that models the data for the percentage of U.S. households with cable television shown in Figure 3.22(a)? A graphing utility can be used to obtain a logarithmic model of the form $y = a + b \ln x$. **Because the domain of the logarithmic function is the set of positive numbers, zero must not be a value for x.** What does this mean for our cable television data that begin in the year 1980? We must start values of x after 0. Thus, we'll assign x to represent the number of years after 1979. This gives us the data shown in Table 3.6. Using the Logarithmic REGression option, we obtain the equation in Figure 3.23.

Table 3.6

x, Number of Years after 1979	*y*, Percentage of U.S. Households with Cable TV
1 (1980)	22.6
6 (1985)	46.2
11 (1990)	59.0
16 (1995)	65.7
23 (2002)	68.9

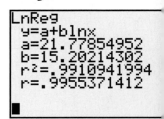

Figure 3.23 A logarithmic model for the data in Table 3.6

From Figure 3.23, we see that the logarithmic model of the data, with numbers rounded to three decimal places, is

$$y = 21.779 + 15.202 \ln x.$$

The number r that appears in Figure 3.23 is called the **correlation coefficient** and is a measure of how well the model fits the data. The value of r is such that $-1 \le r \le 1$. A positive r means that as the x-values increase, so do the y-values. A negative r means that as the x-values increase, the y-values decrease. **The closer that r is to −1 or 1, the better the model fits the data.** Because r is approximately 0.996, the model fits the data very well.

EXAMPLE 6 Choosing a Model for Data

Figure 3.24(a) shows world population, in billions, for seven selected years from 1950 through 2003. A scatter plot is shown in Figure 3.24(b). Suggest two functions that would be good choices for modeling the data.

World Population, 1950–2003

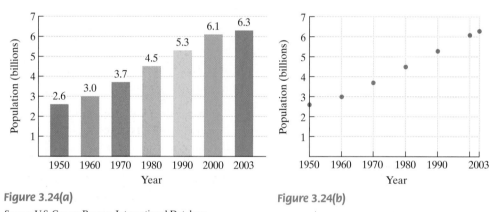

Figure 3.24(a)

Figure 3.24(b)

Source: U.S. Census Bureau, International Database

Solution   Because the data in the scatter plot appear to increase more and more rapidly, the shape suggests that an exponential model might be a good choice. Furthermore, we can probably draw a line that passes through or near the seven points. Thus, a linear function would also be a good choice for modeling the data.

Check Point **6** In 2003, 49.3 million tons of paper were recycled in the United States. Table 3.7 shows the percentage of all paper recycled for five selected years from 1970 through 2003. Create a scatter plot for the data. Based on the scatter plot, what function would be a good choice for modeling the data?

Table 3.7   Percentage of All Paper Recycled in the U.S.

Year	Percent
1970	22.4%
1980	26.2%
1990	33.5%
2000	46.0%
2003	50.3%

Source: The American Forest and Paper Association

World Population, 1950–2003

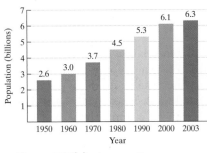

Figure 3.24(a) (repeated)

If we choose to model world population shown in Figure 3.24(a) with an exponential function, a graphing utility's Exponential REGression option can be used to obtain the function's equation. With this feature, a graphing utility fits the data to an exponential model of the form $y = ab^x$.

Although the domain of the exponential function $y = ab^x$ is the set of all real numbers, some graphing utilities only accept positive values for x. What does this mean for our data for world population that starts in the year 1950? We will start values of x after 0. Thus, we'll assign x to represent the number of years after 1949. This gives us the data shown in Table 3.8. Using the Exponential REGression option, we obtain the equation in Figure 3.25.

Table 3.8

x, Numbers of Years after 1949	y, World Population (billions)
1 (1950)	2.6
11 (1960)	3.0
21 (1970)	3.7
31 (1980)	4.5
41 (1990)	5.3
51 (2000)	6.1
54 (2003)	6.3

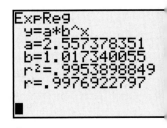

Figure 3.25 An exponential model for the data in Table 3.8

From Figure 3.25, we see that the exponential model of the data for world population x years after 1949, with numbers rounded to three decimal places, is

$$y = 2.557(1.017)^x.$$

The correlation coefficient, r, is close to 1, indicating that the model fits the data very well.

When using a graphing utility to model data, begin with a scatter plot, drawn either by hand or with the graphing utility, to obtain a general picture for the shape of the data. It might be difficult to determine which model best fits the data—linear, logarithmic, exponential, quadratic, or something else. If necessary, use your graphing utility to fit several models to the data. The best model is the one that yields the value of r, the correlation coefficient, closest to 1 or -1. Finding a proper fit for data can be almost as much art as it is mathematics. In this era of technology, the process of creating models that best fit data is one that involves more decision making than computation.

Study Tip

Once you have obtained one or more models for data, you can use a graphing utility's TABLE feature to numerically see how well each model describes the data. Enter the models as y_1, y_2, and so on. Create a table, scroll through the table, and compare the table values given by the models to the actual data.

⑤ Express an exponential model in base e.

Expressing $y = ab^x$ in Base e

Graphing utilities display exponential models in the form $y = ab^x$. However, our discussion of exponential growth involved base e. Because of the inverse property $b = e^{\ln b}$, we can rewrite any model in the form $y = ab^x$ in terms of base e.

Expressing an Exponential Model in Base e

$$y = ab^x \quad \text{is equivalent to} \quad y = ae^{(\ln b)\cdot x}$$

EXAMPLE 7   Rewriting the Model for World Population in Base e

We have seen that the function

$$y = 2.557(1.017)^x$$

models world population, y, in billions, x years after 1949. Rewrite the model in terms of base e.

Solution We use the two equivalent equations shown in the voice balloons to rewrite the model in terms of base e.

$$y = ab^x \qquad\qquad y = ae^{(\ln b)\cdot x}$$

$$y = 2.557(1.017)^x \quad \text{is equivalent to} \quad y = 2.557e^{(\ln 1.017)\cdot x}.$$

Using $\ln 1.017 \approx 0.017$, the exponential growth model for world population, y, in billions, x years after 1949 is

$$y = 2.557e^{0.017x}.$$

In Example 7, we can replace y with A and x with t so that the model has the same letters as those in the exponential growth model $A = A_0e^{kt}$.

$$A = A_0 \ e^{kt} \qquad \text{This is the exponential growth model.}$$

$$A = 2.557e^{0.017t} \qquad \text{This is the model for world population.}$$

The value of k, 0.017, indicates a growth rate of 1.7%. Although this is an excellent model for the data, we must be careful about making projections about world population using this growth function. Why? World population growth rate is now 1.3%, not 1.7%, so our model will overestimate future populations.

Check Point 7 Rewrite $y = 4(7.8)^x$ in terms of base e. Express the answer in terms of a natural logarithm and then round to three decimal places.

EXERCISE SET 3.5

Practice Exercises and Application Exercises

The exponential models describe the population of the indicated country, A, in millions, t years after 2003. Use these models to solve Exercises 1–6.

India	$A = 1049.7e^{0.015t}$
Iraq	$A = 24.7e^{0.028t}$
Japan	$A = 127.2e^{0.001t}$
Russia	$A = 144.5e^{-0.004t}$

1. What was the population of Japan in 2003?

2. What was the population of Iraq in 2003?

3. Which country has the greatest growth rate? By what percentage is the population of that country increasing each year?

4. Which country has a decreasing population? By what percentage is the population of that country decreasing each year?

5. When will India's population be 1238 million?

6. When will India's population be 1416 million?

About the size of New Jersey, Israel has seen its population soar to more than 6 million since it was established. With the help of U.S. aid, the country now has a diversified economy rivaling those of other developed Western nations. By contrast, the Palestinians, living under Israeli occupation and a corrupt regime, endure bleak conditions. The graphs show that by 2050, Palestinians in the West Bank, Gaza Strip, and East Jerusalem will outnumber Israelis. Exercises 7–8, involve the projected growth of these two populations.

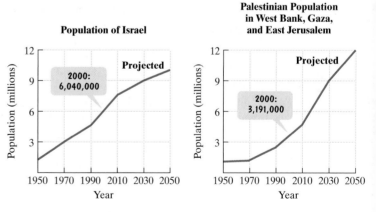

Source: Newsweek

7. a. In 2000, the population of Israel was approximately 6.04 million and by 2050 it is projected to grow to 10 million. Use the exponential growth model $A = A_0e^{kt}$, in which t is the number of years after 2000, to find an exponential growth function that models the data.

    b. In which year will Israel's population be 9 million?

8. a. In 2000, the population of the Palestinians in the West Bank, Gaza Strip, and East Jerusalem was approximately 3.2 million and by 2050 it is projected to grow to 12 million. Use the exponential growth model $A = A_0 e^{kt}$, in which t is the number of years after 2000, to find the exponential growth function that models the data.

  b. In which year will the Palestinian population be 9 million?

An artifact originally had 16 grams of carbon-14 present. The decay model $A = 16e^{-0.000121t}$ describes the amount of carbon-14 present after t years. Use this model to solve Exercises 9–10.

9. How many grams of carbon-14 will be present in 5715 years?

10. How many grams of carbon-14 will be present in 11,430 years?

11. The half-life of the radioactive element krypton-91 is 10 seconds. If 16 grams of krypton-91 are initially present, how many grams are present after 10 seconds? 20 seconds? 30 seconds? 40 seconds? 50 seconds?

12. The half-life of the radioactive element plutonium-239 is 25,000 years. If 16 grams of plutonium-239 are initially present, how many grams are present after 25,000 years? 50,000 years? 75,000 years? 100,000 years? 125,000 years?

Use the exponential decay model for carbon-14, $A = A_0 e^{-0.000121t}$, to solve Exercises 13–14.

13. Prehistoric cave paintings were discovered in a cave in France. The paint contained 15% of the original carbon-14. Estimate the age of the paintings.

14. Skeletons were found at a construction site in San Francisco in 1989. The skeletons contained 88% of the expected amount of carbon-14 found in a living person. In 1989, how old were the skeletons?

15. The August 1978 issue of *National Geographic* described the 1964 find of bones of a newly discovered dinosaur weighing 170 pounds, measuring 9 feet, with a 6-inch claw on one toe of each hind foot. The age of the dinosaur was estimated using potassium-40 dating of rocks surrounding the bones.

  a. Potassium-40 decays exponentially with a half-life of approximately 1.31 billion years. Use the fact that after 1.31 billion years a given amount of potassium-40 will have decayed to half the original amount to show that the decay model for potassium-40 is given by $A = A_0 e^{-0.52912t}$, where t is in billions of years.

  b. Analysis of the rocks surrounding the dinosaur bones indicated that 94.5% of the original amount of potassium-40 was still present. Let $A = 0.945 A_0$ in the model in part (a) and estimate the age of the bones of the dinosaur.

16. A bird species in danger of extinction has a population that is decreasing exponentially ($A = A_0 e^{kt}$). Five years ago the population was at 1400 and today only 1000 of the birds are alive. Once the population drops below 100, the situation will be irreversible. When will this happen?

17. Use the exponential growth model, $A = A_0 e^{kt}$, to show that the time it takes a population to double (to grow from A_0 to $2A_0$) is given by $t = \dfrac{\ln 2}{k}$.

18. Use the exponential growth model, $A = A_0 e^{kt}$, to show that the time it takes a population to triple (to grow from A_0 to $3A_0$) is given by $t = \dfrac{\ln 3}{k}$.

Use the formula $t = \dfrac{\ln 2}{k}$ that gives the time for a population with a growth rate k to double to solve Exercises 19–20. Express each answer to the nearest whole year.

19. The growth model $A = 4e^{0.007t}$ describes New Zealand's population, A, in millions, t years after 2003.

  a. What is New Zealand's growth rate?

  b. How long will it take New Zealand to double its population?

20. The growth model $A = 104.9e^{0.017t}$ describes Mexico's population, A, in millions, t years after 2003.

  a. What is Mexico's growth rate?

  b. How long will it take Mexico to double its population?

21. The logistic growth function

$$f(t) = \frac{100{,}000}{1 + 5000e^{-t}}$$

describes the number of people, $f(t)$, who have become ill with influenza t weeks after its initial outbreak in a particular community.

  a. How many people became ill with the flu when the epidemic began?

  b. How many people were ill by the end of the fourth week?

  c. What is the limiting size of the population that becomes ill?

Shown, again, is world population, in billions, for seven selected years from 1950 through 2003. Using a graphing utility's logistic REGression option, we obtain the equation shown on the screen.

x, Numbers of Years after 1949	y, World Population (billions)
1 (1950)	2.6
11 (1960)	3.0
21 (1970)	3.7
31 (1980)	4.5
41 (1990)	5.3
51 (2000)	6.1
54 (2003)	6.3

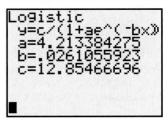

We see that a logistic growth model for world population, $f(x)$, in billions, x years after 1949 is

$$f(x) = \frac{12.85}{1 + 4.21e^{-0.026x}}.$$

Use this function to solve Exercises 22–26.

22. How well does the function model the data for 2000?

23. How well does the function model the data for 2003?

24. When will world population reach 7 billion?

25. When will world population reach 8 billion?

6. According to the model, what is the limiting size of the population that Earth will eventually sustain? What does this mean in terms of the statement made by the U.S. National Academy of Sciences that 10 billion is the maximum that the world can support with some degree of comfort and individual choice?

The logistic growth function

$$P(x) = \frac{90}{1 + 271e^{-0.122x}}$$

models the percentage, $P(x)$, of Americans who are x years old with some coronary heart disease. Use the function to solve Exercises 27–30.

7. What percentage of 20-year-olds have some coronary heart disease?

8. What percentage of 80-year-olds have some coronary heart disease?

9. At what age is the percentage of some coronary heart disease 50%?

0. At what age is the percentage of some coronary heart disease 70%?

Use Newton's Law of Cooling, $T = C + (T_0 - C)e^{kt}$, to solve Exercises 31–34.

1. A bottle of juice initially has a temperature of 70°F. It is left to cool in a refrigerator that has a temperature of 45°F. After 10 minutes, the temperature of the juice is 55°F.
   a. Use Newton's Law of Cooling to find a model for the temperature of the juice, T, after t minutes.
   b. What is the temperature of the juice after 15 minutes?
   c. When will the temperature of the juice be 50°F?

2. A pizza removed from the oven has a temperature of 450°F. It is left sitting in a room that has a temperature of 70°F. After 5 minutes, the temperature of the pizza is 300°F.
   a. Use Newton's Law of Cooling to find a model for the temperature of the pizza, T, after t minutes.
   b. What is the temperature of the pizza after 20 minutes?
   c. When will the temperature of the pizza be 140°F?

3. A frozen steak initially has a temperature of 28°F. It is left to thaw in a room that has a temperature of 75°F. After 10 minutes, the temperature of the steak has risen to 38°F. After how many minutes will the temperature of the steak be 50°F?

4. A frozen steak initially has a temperature of 24°F. It is left to thaw in a room that has a temperature of 65°F. After 10 minutes, the temperature of the steak has risen to 30°F. After how many minutes will the temperature of the steak be 45°F?

Exercises 35–40 present data in the form of tables. For each data set shown by the table,
   a. *Create a scatter plot for the data.*
   b. *Use the scatter plot to determine whether an exponential function, a logarithmic function, or a linear function is the*

best choice for modeling the data. (If applicable, in Exercise 61, you will use your graphing utility to obtain these functions.)

35. Percent of Miscarriages, by Age

Woman's Age	Percent of Miscarriages
22	9%
27	10%
32	13%
37	20%
42	38%
47	52%

Source: Time

36. Number of Countries Connected to the Internet

Year	Number of Countries Connected to the Internet
1985	11
1991	91
1994	146
1997	195
2002	220

Source: Medard Gabel, Global Inc., 2003

37. Number of Illegal Immigrants Living in the U.S.

Year	Number of Illegal Immigrants (millions)
1992	3.4
1996	5.0
2000	7.0
2004	8.0

Source: U.S. Department of Homeland Security

38. Number of U.S. Households with Pets

Year	Number with Pets (millions)
1998	54.0
1999	58.2
2000	61.1
2001	63.0
2002	64.2

Source: American Pet Products Manufacturers Association

39. Alcohol Use by U.S. High School Seniors

Year	Percentage Using Alcohol during 30 Days Preceding the Survey
1980	72.0%
1985	65.9%
1990	57.1%
1995	51.3%
2000	50.0%
2002	48.6%
2003	47.5%

Source: U.S. Department of Health and Human Services

40. U.S. Vehicle Fatality Rates

Year	Deaths per 100 Million Vehicle Miles Traveled
1965	5.50
1970	4.50
1975	3.40
1985	2.50
1990	2.00
1995	1.60
2000	1.50
2003	1.48

Source: National Highway Traffic Safety Administration

In Exercises 41–44, rewrite the equation in terms of base e. Express the answer in terms of a natural logarithm and then round to three decimal places.

41. $y = 100(4.6)^x$
42. $y = 1000(7.3)^x$
43. $y = 2.5(0.7)^x$
44. $y = 4.5(0.6)^x$

Writing in Mathematics

45. Nigeria has a growth rate of 0.025 or 2.5%. Describe what this means.

46. How can you tell whether an exponential model describes exponential growth or exponential decay?

47. Suppose that a population that is growing exponentially increases from 800,000 people in 2003 to 1,000,000 people in 2006. Without showing the details, describe how to obtain the exponential growth function that models the data.

48. What is the half-life of a substance?

49. Describe a difference between exponential growth and logistic growth.

50. Describe the shape of a scatter plot that suggests modeling the data with an exponential function.

51. You take up weightlifting and record the maximum number of pounds you can lift at the end of each week. You start off with rapid growth in terms of the weight you can lift from week to week, but then the growth begins to level off. Describe how to obtain a function that models the number of pounds you can lift at the end of each week. How can you use this function to predict what might happen if you continue the sport?

52. Would you prefer that your salary be modeled exponentially or logarithmically? Explain your answer.

53. One problem with all exponential growth models is that nothing can grow exponentially forever. Describe factors that might limit the size of a population.

Technology Exercises

In Example 1 on page 423, we used two data points and an exponential function to model the population of the United States from 1970 through 2003. The data are shown again in the table. Use all five data points to solve Exercises 54–58.

x, Number of Years after 1969	y, U.S. Population (millions)
1 (1970)	203.3
11 (1980)	226.5
21 (1990)	248.7
31 (2000)	281.4
34 (2003)	294.0

54. a. Use your graphing utility's Exponential REGression option to obtain a model of the form $y = ab^x$ that fits the data. How well does the correlation coefficient, r, indicate that the model fits the data?

b. Rewrite the model in terms of base e. By what percentage is the population of the United States increasing each year?

55. Use your graphing utility's Logarithmic REGression option to obtain a model of the form $y = a + b \ln x$ that fits the data. How well does the correlation coefficient, r, indicate that the model fits the data?

56. Use your graphing utility's Linear REGression option to obtain a model of the form $y = ax + b$ that fits the data. How well does the correlation coefficient, r, indicate that the model fits the data?

57. Use your graphing utility's Power REGression option to obtain a model of the form $y = ax^b$ that fits the data. How well does the correlation coefficient, r, indicate that the model fits the data?

58. Use the values of r in Exercises 54–57 to select the two models of best fit. Use each of these models to predict by which year the U.S. population will reach 315 million. How do these answers compare to the year we found in Example 1, namely 2010? If you obtained different years, how do you account for this difference?

59. In Exercises 27–30, you worked with the logistic growth function

$$P(x) = \frac{90}{1 + 271e^{-0.122x}}$$

which models the percentage, $P(x)$, of Americans who are x years old with some coronary heart disease. Use your graphing utility to graph the function in a $[0, 100, 10]$ by $[0, 100, 10]$ viewing rectangle. Describe as specifically as possible what the logistic curve indicates about aging and the percentage of Americans with coronary heart disease.

60. The figure shows the number of people in the United States age 65 and over, with projected figures for the year 2010 and beyond.

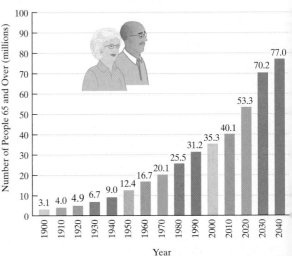

U.S. Population Age 65 and Over

Source: U.S. Bureau of the Census

a. Let x represent the number of years after 1899 and let y represent the U.S. population age 65 and over, in millions. Use your graphing utility to find the model that best fits the data in the bar graph.

b. Rewrite the model in terms of base e. By what percentage is the 65 and over population increasing each year?

1. In Exercises 35–40, you determined the best choice for the kind of function that modeled the data in each table. For each of these exercises that you worked, use a graphing utility to find the actual function that best fits the data. Then use the model to make a reasonable prediction for a value that exceeds those shown in the table's first column.

Critical Thinking Exercises

2. The exponential growth models describe the population of the indicated country, A, in millions, t years after 2003.

Canada $A = 32.2e^{0.003t}$

Uganda $A = 25.6e^{0.03t}$

According to these models, which one of the following is true?

a. In 2003, Uganda's population was ten times that of Canada's.

b. In 2003, Canada's population exceeded Uganda's by 660,000.

c. In 2012, Uganda's population will exceed Canada's.

d. None of these statements is true.

63. Use Newton's Law of Cooling, $T = C + (T_0 - C)e^{-kt}$, to solve this exercise. At 9:00 A.M., a coroner arrived at the home of a person who had died during the night. The temperature of the room was 70°F, and at the time of death the person had a body temperature of 98.6°F. The coroner took the body's temperature at 9:30 A.M., at which time it was 85.6°F, and again at 10:00 A.M., when it was 82.7°F. At what time did the person die?

Group Exercises

64. This activity is intended for three or four people who would like to take up weightlifting. Each person in the group should record the maximum number of pounds that he or she can lift at the end of each week for the first 10 consecutive weeks. Use the Logarithmic REGression option of a graphing utility to obtain a model showing the amount of weight that group members can lift from week 1 through week 10. Graph each of the models in the same viewing rectangle to observe similarities and differences among weight-growth patterns of each member. Use the functions to predict the amount of weight that group members will be able to lift in the future. If the group continues to work out together, check the accuracy of these predictions.

65. Each group member should consult an almanac, newspaper, magazine, or the Internet to find data that can be modeled by exponential or logarithmic functions. Group members should select the two sets of data that are most interesting and relevant. For each set selected, find a model that best fits the data. Each group member should make one prediction based on the model and then discuss a consequence of this prediction. What factors might change the accuracy of the prediction?

Chapter 3
Summary, Review, and Test

Summary

DEFINITIONS AND CONCEPTS	EXAMPLES

3.1 Exponential Functions

a. The exponential function with base b is defined by $f(x) = b^x$, where $b > 0$ and $b \neq 1$. — Ex. 1, p. 377

b. Characteristics of exponential functions and graphs for $0 < b < 1$ and $b > 1$ are shown in the box on page 379. — Ex. 2, p. 378; Ex. 3, p. 378

c. Transformations involving exponential functions are summarized in Table 3.1 on page 380. — Ex. 4, p. 380; Ex. 5, p. 381

d. The natural exponential function is $f(x) = e^x$. The irrational number e is called the natural base, where $e \approx 2.7183$. e is the value that $\left(1 + \dfrac{1}{n}\right)^n$ approaches as $n \to \infty$. — Ex. 6, p. 382

e. Formulas for compound interest: After t years, the balance, A, in an account with principal P and annual interest rate r (in decimal form) is given by one of the following formulas: — Ex. 7, p. 384

1. For n compoundings per year: $A = P\left(1 + \dfrac{r}{n}\right)^{nt}$

2. For continuous compounding: $A = Pe^{rt}$

DEFINITIONS AND CONCEPTS	**EXAMPLES**

3.2 Logarithmic Functions

a. Definition of the logarithmic function: For $x > 0$ and $b > 0, b \neq 1$, $y = \log_b x$ is equivalent to $b^y = x$. The function $f(x) = \log_b x$ is the logarithmic function with base b. This function is the inverse function of the exponential function with base b.

Ex. 1, p. 389;
Ex. 2, p. 389;
Ex. 3, p. 389

b. Graphs of logarithmic functions for $b > 1$ and $0 < b < 1$ are shown in Figure 3.6 on page 392. Characteristics of the graphs are summarized in the box that follows the figure.

Ex. 6, p. 391

c. Transformations involving logarithmic functions are summarized in Table 3.4 on page 392.

Figures 3.7–3.10
p. 393

d. The domain of a logarithmic function of the form $f(x) = \log_b x$ is the set of all positive real numbers. The domain of $f(x) = \log_b g(x)$ consists of all x for which $g(x) > 0$.

Ex. 7, p. 393;
Ex. 10, p. 395

e. Common and natural logarithms: $f(x) = \log x$ means $f(x) = \log_{10} x$ and is the common logarithmic function. $f(x) = \ln x$ means $f(x) = \log_e x$ and is the natural logarithmic function.

Ex. 8, p. 394;
Ex. 9, p. 395;
Ex. 11, p. 396

f. Basic Logarithmic Properties

Base b ($b > 0, b \neq 1$)	**Base 10** (Common Logarithms)	**Base e** (Natural Logarithms)
$\log_b 1 = 0$	$\log 1 = 0$	$\ln 1 = 0$
$\log_b b = 1$	$\log 10 = 1$	$\ln e = 1$
$\log_b b^x = x$	$\log 10^x = x$	$\ln e^x = x$
$b^{\log_b x} = x$	$10^{\log x} = x$	$e^{\ln x} = x$

Ex. 4, p. 390

Ex. 5, p. 391

3.3 Properties of Logarithms

a. *The Product Rule:* $\log_b(MN) = \log_b M + \log_b N$

Ex. 1, p. 401

b. *The Quotient Rule:* $\log_b\left(\dfrac{M}{N}\right) = \log_b M - \log_b N$

Ex. 2, p. 402

c. *The Power Rule:* $\log_b M^p = p \log_b M$

Ex. 3, p. 403

d. *The Change-of-Base Property:*

The General Property	**Introducing Common Logarithms**	**Introducing Natural Logarithms**
$\log_b M = \dfrac{\log_a M}{\log_a b}$	$\log_b M = \dfrac{\log M}{\log b}$	$\log_b M = \dfrac{\ln M}{\ln b}$

Ex. 7, p. 406;
Ex. 8, p. 407

e. Properties for expanding logarithmic expressions are given in the box on page 403.

Ex. 4, p. 404

f. Properties for condensing logarithmic expressions are given in the box on page 404.

Ex. 5, p. 404;
Ex. 6, p. 405

3.4 Exponential and Logarithmic Equations

a. An exponential equation is an equation containing a variable in an exponent. Some exponential equations can be solved by expressing each side as a power of the same base: If $b^M = b^N$, then $M = N$. Details are in the box on page 410.

Ex. 1, p. 410

b. The procedure for using natural logarithms to solve exponential equations is given in the box on page 411. The solution procedure involves isolating the exponential expression and taking the natural logarithm on both sides.

Ex. 2, p. 411;
Ex. 3, p. 412;
Ex. 4, p. 412;
Ex. 5, p. 413

c. A logarithmic equation is an equation containing a variable in a logarithmic expression. Some logarithmic equations can be expressed in the form $\log_b M = c$. The definition of a logarithm is used to rewrite the equation in exponential form: $b^c = M$. See the box on page 413. When checking logarithmic equations, reject proposed solutions that produce the logarithm of a negative number or the logarithm of 0 in the original equation.

Ex. 6, p. 413;
Ex. 7, p. 415

DEFINITIONS AND CONCEPTS	EXAMPLES

d. Some logarithmic equations can be expressed in the form $\log_b M = \log_b N$. Use the one-to-one property to rewrite the equation without logarithms: $M = N$. See the box on page 415. — Ex. 8, p. 415

3.5 Exponential Growth and Decay; Modeling Data

a. Exponential growth and decay models are given by $A = A_0 e^{kt}$ in which t represents time, A_0 is the amount present at $t = 0$, and A is the amount present at time t. If $k > 0$, the model describes growth and k is the growth rate. If $k < 0$, the model describes decay and k is the decay rate. — Ex. 1, p. 423; Ex. 2, p. 425

b. The logistic growth model, given by $A = \dfrac{c}{1 + ae^{-bt}}$, describes situations in which growth is limited. $y = c$ is a horizontal asymptote for the graph, and growth, A, can never exceed c. — Ex. 3, p. 427

c. Newton's Law of Cooling: The temperature, T, of a heated object at time t is given by
$$T = C + (T_0 - C)e^{kt},$$
where C is the constant temperature of the surrounding medium, T_0 is the initial temperature of the heated object, and k is a negative constant. — Ex. 4, p. 428

d. Scatter plots for exponential and logarithmic models are shown in Figure 3.21 on page 429. When using a graphing utility to model data, the closer that the correlation coefficient, r, is to -1 or 1, the better the model fits the data. — Ex. 5, p. 430; Ex. 6, p. 431

e. Expressing an Exponential Model in Base e: $y = ab^x$ is equivalent to $y = ae^{(\ln b)\cdot x}$. — Ex. 7, p. 432

Review Exercises

3.1

In Exercises 1–4, the graph of an exponential function is given. Select the function for each graph from the following options:

$$f(x) = 4^x, g(x) = 4^{-x},$$
$$h(x) = -4^{-x}, r(x) = -4^{-x} + 3.$$

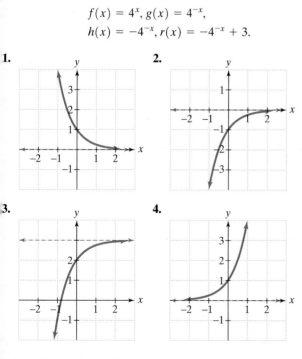

1.

2.

3.

4.

In Exercises 5–9, graph f and g in the same rectangular coordinate system. Use transformations of the graph of f to obtain the graph of g. Graph and give equations of all asymptotes. Use the graphs to determine each function's domain and range.

5. $f(x) = 2^x$ and $g(x) = 2^{x-1}$

6. $f(x) = 3^x$ and $g(x) = 3^x - 1$

7. $f(x) = 3^x$ and $g(x) = -3^x$

8. $f(x) = \left(\frac{1}{2}\right)^x$ and $g(x) = \left(\frac{1}{2}\right)^{-x}$

9. $f(x) = e^x$ and $g(x) = 2e^{\frac{x}{2}}$

Use the compound interest formulas to solve Exercises 10–11.

10. Suppose that you have $5000 to invest. Which investment yields the greater return over 5 years: 5.5% compounded semiannually or 5.25% compounded monthly?

11. Suppose that you have $14,000 to invest. Which investment yields the greater return over 10 years: 7% compounded monthly or 6.85% compounded continuously?

12. A cup of coffee is taken out of a microwave oven and placed in a room. The temperature, T, in degrees Fahrenheit, of the coffee after t minutes is modeled by the function $T = 70 + 130e^{-0.04855t}$. The graph of the function is shown in the figure.

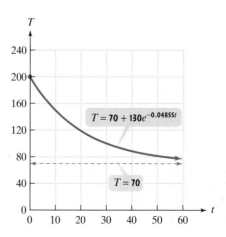

Use the graph shown at the bottom of the previous page to answer each of the following questions.

a. What was the temperature of the coffee when it was first taken out of the microwave?

b. What is a reasonable estimate of the temperature of the coffee after 20 minutes? Use your calculator to verify this estimate.

c. What is the limit of the temperature to which the coffee will cool? What does this tell you about the temperature of the room?

3.2

In Exercises 13–15, write each equation in its equivalent exponential form.

13. $\frac{1}{2} = \log_{49} 7$ **14.** $3 = \log_4 x$ **15.** $\log_3 81 = y$

In Exercises 16–18, write each equation in its equivalent logarithmic form.

16. $6^3 = 216$ **17.** $b^4 = 625$ **18.** $13^y = 874$

In Exercises 19–29, evaluate each expression without using a calculator. If evaluation is not possible, state the reason.

19. $\log_4 64$ **20.** $\log_5 \frac{1}{25}$ **21.** $\log_3(-9)$

22. $\log_{16} 4$ **23.** $\log_{17} 17$ **24.** $\log_3 3^8$

25. $\ln e^5$ **26.** $\log_3 \frac{1}{\sqrt{3}}$ **27.** $\ln \frac{1}{e^2}$

28. $\log \frac{1}{1000}$ **29.** $\log_3(\log_8 8)$

30. Graph $f(x) = 2^x$ and $g(x) = \log_2 x$ in the same rectangular coordinate system.

31. Graph $f(x) = \left(\frac{1}{3}\right)^x$ and $g(x) = \log_{\frac{1}{3}} x$ in the same rectangular coordinate system.

In Exercises 32–35, the graph of a logarithmic function is given. Select the function for each graph from the following options:

$$f(x) = \log x, g(x) = \log(-x),$$
$$h(x) = \log(2 - x), r(x) = 1 + \log(2 - x).$$

32.

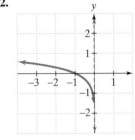

33.

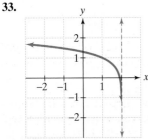

34.

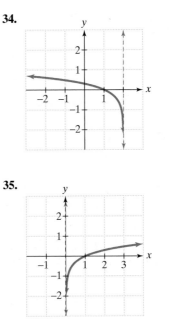

35.

In Exercises 36–38, begin by graphing $f(x) = \log_2 x$. Then use transformations of this graph to graph the given function. What is the graph's x-intercept? What is the vertical asymptote? Use the graphs to determine each function's domain and range.

36. $g(x) = \log_2(x - 2)$ **37.** $h(x) = -1 + \log_2 x$

38. $r(x) = \log_2(-x)$

In Exercises 39–40, graph f and g in the same rectangular coordinate system. Use transformations of the graph of f to obtain the graph of g. Graph and give equations of all asymptotes. Use the graphs to determine each function's domain and range.

39. $f(x) = \log x$ and $g(x) = -\log(x + 3)$

40. $f(x) = \ln x$ and $g(x) = -\ln(2x)$

In Exercises 41–43, find the domain of each logarithmic function.

41. $f(x) = \log_8(x + 5)$ **42.** $f(x) = \log(3 - x)$

43. $f(x) = \ln(x - 1)^2$

In Exercises 44–46, use inverse properties of logarithms to simplify each expression.

44. $\ln e^{6x}$ **45.** $e^{\ln \sqrt{x}}$ **46.** $10^{\log 4x^2}$

47. On the Richter scale, the magnitude, R, of an earthquake of intensity I is given by $R = \log \frac{I}{I_0}$, where I_0 is the intensity of a barely felt zero-level earthquake. If the intensity of an earthquake is $1000I_0$, what is its magnitude on the Richter scale?

48. Students in a psychology class took a final examination. As part of an experiment to see how much of the course content they remembered over time, they took equivalent forms of the exam in monthly intervals thereafter. The average score, $f(t)$, for the group after t months is modeled by the function $f(t) = 76 - 18 \log(t + 1)$, where $0 \le t \le 12$.

a. What was the average score when the exam was first given?

b. What was the average score after 2 months? 4 months? 6 months? 8 months? one year?

c. Use the results from parts (a) and (b) to graph f. Describe what the shape of the graph indicates in terms of the material retained by the students.

9. The formula

$$t = \frac{1}{c}\ln\left(\frac{A}{A - N}\right)$$

describes the time, t, in weeks, that it takes to achieve mastery of a portion of a task. In the formula, A represents maximum learning possible, N is the portion of the learning that is to be achieved, and c is a constant used to measure an individual's learning style. A 50-year-old man decides to start running as a way to maintain good health. He feels that the maximum rate he could ever hope to achieve is 12 miles per hour. How many weeks will it take before the man can run 5 miles per hour if $c = 0.06$ for this person?

.3

n Exercises 50–53, use properties of logarithms to expand each ogarithmic expression as much as possible. Where possible, valuate logarithmic expressions without using a calculator.

0. $\log_6(36x^3)$

51. $\log_4\left(\frac{\sqrt{x}}{64}\right)$

2. $\log_2\left(\frac{xy^2}{64}\right)$

53. $\ln\sqrt[3]{\frac{x}{e}}$

n Exercises 54–57, use properties of logarithms to condense each ogarithmic expression. Write the expression as a single logarithm hose coefficient is 1.

4. $\log_b 7 + \log_b 3$

55. $\log 3 - 3\log x$

6. $3\ln x + 4\ln y$

57. $\frac{1}{2}\ln x - \ln y$

Exercises 58–59, use common logarithms or natural logarithms nd a calculator to evaluate to four decimal places.

8. $\log_6 72,348$

59. $\log_4 0.863$

Exercises 60–63, determine whether each equation is true or lse. Where possible, show work to support your conclusion. If e statement is false, make the necessary change(s) to produce a ue statement.

). $(\ln x)(\ln 1) = 0$

1. $\log(x + 9) - \log(x + 1) = \dfrac{\log(x + 9)}{\log(x + 1)}$

2. $(\log_2 x)^4 = 4\log_2 x$

63. $\ln e^x = x\ln e$

.4

Exercises 64–72, solve each exponential equation. Where ecessary, express the solution set in terms of natural logarithms d use a calculator to obtain a decimal approximation, correct to o decimal places, for the solution.

. $2^{4x-2} = 64$

65. $125^x = 25$

66. $9^{x+2} = 27^{-x}$

67. $8^x = 12,143$

68. $9e^{5x} = 1269$

69. $e^{12-5x} - 7 = 123$

70. $5^{4x+2} = 37,500$

71. $3^{x+4} = 7^{2x-1}$

72. $e^{2x} - e^x - 6 = 0$

In Exercises 73–78, solve each logarithmic equation.

73. $\log_4(3x - 5) = 3$

74. $3 + 4\ln(2x) = 15$

75. $\log_2(x + 3) + \log_2(x - 3) = 4$

76. $\log_3(x - 1) - \log_3(x + 2) = 2$

77. $\ln(x + 4) - \ln(x + 1) = \ln x$

78. $\log_4(2x + 1) = \log_4(x - 3) + \log_4(x + 5)$

79. The function $P(x) = 14.7e^{-0.21x}$ models the average atmospheric pressure, $P(x)$, in pounds per square inch, at an altitude of x miles above sea level. The atmospheric pressure at the peak of Mt. Everest, the world's highest mountain, is 4.6 pounds per square inch. How many miles above sea level, to the nearest tenth of a mile, is the peak of Mt. Everest?

80. The amount of carbon dioxide in the atmosphere, measured in parts per million, has been increasing as a result of the burning of oil and coal. The buildup of gases and particles traps heat and raises the planet's temperature, a phenomenon called the greenhouse effect. Carbon dioxide accounts for about half of the warming. The function $f(t) = 364(1.005)^t$ projects carbon dioxide concentration, $f(t)$, in parts per million, t years after 2000. Using the projections given by the function, when will the carbon dioxide concentration be double the preindustrial level of 280 parts per million?

81. The function $W(x) = 0.37\ln x + 0.05$ models the average walking speed, $W(x)$, in feet per second, of residents in a city whose population is x thousand. Visitors to New York City frequently feel they are moving too slowly to keep pace with New Yorkers' average walking speed of 3.38 feet per second. What is the population of New York City? Round to the nearest thousand.

82. Use the formula for compound interest with n compoundings per year to solve this problem. How long, to the nearest tenth of a year, will it take $12,500 to grow to $20,000 at 6.5% annual interest compounded quarterly?

Use the formula for continuous compounding to solve Exercises 83–84.

83. How long, to the nearest tenth of a year, will it take $50,000 to triple in value at 7.5% annual interest compounded continuously?

84. What interest rate, to the nearest percent, is required for an investment subject to continuous compounding to triple in 5 years?

3.5

85. According to the U.S. Bureau of the Census, in 1990 there were 22.4 million residents of Hispanic origin living in the United States. By 2000, the number had increased to 35.3 million. The exponential growth function $A = 22.4e^{kt}$ describes the U.S. Hispanic population, A, in millions, t years after 1990.

  a. Find k, correct to three decimal places.

  b. Use the resulting model to project the Hispanic resident population in 2010.

  c. In which year will the Hispanic resident population reach 60 million?

86. Use the exponential decay model, $A = A_0e^{kt}$, to solve this exercise. The half-life of polonium-210 is 140 days. How long will it take for a sample of this substance to decay to 20% of its original amount?

87. The function

$$f(t) = \frac{500,000}{1 + 2499e^{-0.92t}}$$

models the number of people, $f(t)$, in a city who have become ill with influenza t weeks after its initial outbreak.

  a. How many people became ill with the flu when the epidemic began?

  b. How many people were ill by the end of the sixth week?

  c. What is the limiting size of $f(t)$, the population that becomes ill?

88. Use Newton's Law of Cooling, $T = C + (T_0 - C)e^{kt}$, to solve this exercise. You are served a cup of coffee that has a temperature of 185°F. The room temperature is 65°F. After 2 minutes, the temperature of the coffee is 155°F.

  a. Write a model for the temperature of the coffee, T, after t minutes.

  b. When will the temperature of the coffee be 105°F?

Exercises 89–90 present data in the form of tables. For each data set shown by the table,

  a. *Create a scatter plot for the data.*

  b. *Use the scatter plot to determine whether an exponential function or a logarithmic function is the better choice for modeling the data.*

89. Percentage of the U.S. Population, Ages 25 or Older, with a College Degree

Year	Percent
1980	17.0%
1991	21.4%
1995	23.0%
2000	25.6%
2001	26.2%
2002	26.7%

Source: U.S. Census Bureau

90. Projection of U.S. Jobs Moving Overseas

Year	Number of Jobs Moving Overseas (millions)
2003	0.3
2008	1.0
2010	1.5
2012	2.5
2015	3.3

Source: Forrester Research, Inc.

In Exercises 91–92, rewrite the equation in terms of base e. Express the answer in terms of a natural logarithm and then round to three decimal places.

91. $y = 73(2.6)^x$

92. $y = 6.5(0.43)^x$

93. The figure shows world population projections through the year 2150. The data are from the United Nations Family Planning Program and are based on optimistic or pessimistic expectations for successful control of human population growth. Suppose that you are interested in modeling these data using exponential, logarithmic, linear, and quadratic functions. Which function would you use to model each of the projections? Explain your choices. For the choice corresponding to a quadratic model, would your formula involve one with a positive or negative leading coefficient? Explain.

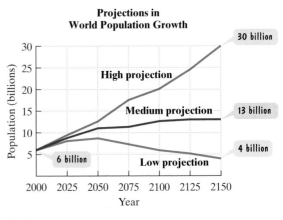

Projections in World Population Growth

Source: U.N.

Chapter 3 Test

1. Graph $f(x) = 2^x$ and $g(x) = 2^{x+1}$ in the same rectangular coordinate system.

2. Graph $f(x) = \log_2 x$ and $g(x) = \log_2(x - 1)$ in the same rectangular coordinate system.

3. Write in exponential form: $\log_5 125 = 3$.

4. Write in logarithmic form: $\sqrt{36} = 6$.

5. Find the domain of $f(x) = \ln(3 - x)$.

In Exercises 6–7, use properties of logarithms to expand each logarithmic expression as much as possible. Where possible, evaluate logarithmic expressions without using a calculator.

6. $\log_4(64x^5)$

7. $\log_3\left(\dfrac{\sqrt[3]{x}}{81}\right)$

In Exercises 8–9, write each expression as a single logarithm.

8. $6 \log x + 2 \log y$

9. $\ln 7 - 3 \ln x$

10. Use a calculator to evaluate $\log_{15} 71$ to four decimal places.

In Exercises 11–18, solve each equation.

11. $3^{x-2} = 9^{x+4}$

12. $5^x = 1.4$

13. $400e^{0.005x} = 1600$

14. $e^{2x} - 6e^x + 5 = 0$

15. $\log_6(4x - 1) = 3$

16. $2 \ln(3x) = 8$

17. $\log x + \log(x + 15) = 2$

18. $\ln(x - 4) - \ln(x + 1) = \ln 6$

19. On the decibel scale, the loudness of a sound, D, in decibels, is given by $D = 10 \log \dfrac{I}{I_0}$, where I is the intensity of the sound, in watts per meter^2, and I_0 is the intensity of a sound barely audible to the human ear. If the intensity of a sound is $10^{12}I_0$, what is its loudness in decibels? (Such a sound is potentially damaging to the ear.)

In Exercises 20–22, simplify each expression.

20. $\ln e^{5x}$

21. $\log_b b$

22. $\log_6 1$

Use the compound interest formulas to solve Exercises 23–25.

23. Suppose you have $3000 to invest. Which investment yields the greater return over 10 years: 6.5% compounded semiannually or 6% compounded continuously? How much more (to the nearest dollar) is yielded by the better investment?

24. How long, to the nearest tenth of a year, will it take $4000 to grow to $8000 at 5% annual interest compounded quarterly?

25. What interest rate, to the nearest tenth of a percent, is required for an investment subject to continuous compounding to double in 10 years?

26. The function
$$A = 82.3e^{-0.002t}$$
models the population of Germany, A, in millions, t years after 2003.
   a. What was the population of Germany in 2003?
   b. Is the population of Germany increasing or decreasing? Explain.
   c. In which year will the population of Germany be 81.5 million?

27. The 1990 population of Europe was 509 million; in 2000, it was 729 million. Write the exponential growth function that describes the population of Europe, in millions, t years after 1990.

28. Use the exponential decay model for carbon-14, $A = A_0 e^{-0.000121t}$, to solve this exercise. Bones of a prehistoric man were discovered and contained 5% of the original amount of carbon-14. How long ago did the man die?

29. The logistic growth function
$$f(t) = \dfrac{140}{1 + 9e^{-0.165t}}$$
describes the population, $f(t)$, of an endangered species of elk t years after they were introduced to a nonthreatening habitat.
   a. How many elk were initially introduced to the habitat?
   b. How many elk are expected in the habitat after 10 years?
   c. What is the limiting size of the elk population that the habitat will sustain?

In Exercises 30–33, determine whether the values in each table belong to an exponential function, a logarithmic function, a linear function, or a quadratic function.

30.

x	y
0	3
1	1
2	-1
3	-3
4	-5

31.

x	y
$\frac{1}{3}$	-1
1	0
3	1
9	2
27	3

32.

x	y
0	1
1	5
2	25
3	125
4	625

33.

x	y
0	12
1	3
2	0
3	3
4	12

34. Rewrite $y = 96(0.38)^x$ in terms of base e. Express the answer in terms of a natural logarithm and then round to three decimal places.

Cumulative Review Exercises (Chapters P–3)

In Exercises 1–8, solve each equation or inequality.

1. $|3x - 4| = 2$

2. $x^2 + 2x + 5 = 0$

3. $x^4 + x^3 - 3x^2 - x + 2 = 0$

4. $e^{5x} - 32 = 96$

5. $\log_2(x + 5) + \log_2(x - 1) = 4$

6. $\ln(x + 4) + \ln(x + 1) = 2\ln(x + 3)$

7. $14 - 5x \geq -6$

8. $|2x - 4| \leq 2$

In Exercises 9–14, graph each equation in a rectangular coordinate system. If two functions are indicated, graph both in the same system.

9. $(x - 3)^2 + (y + 2)^2 = 4$

10. $f(x) = (x - 2)^2 - 1$

11. $f(x) = \dfrac{x^2 - 1}{x^2 - 4}$

12. $f(x) = (x - 2)^2(x + 1)$

13. $f(x) = 2x - 4$ and $f^{-1}(x)$

14. $f(x) = \ln x$ and $g(x) = \ln(x - 2) + 1$

15. Write the point-slope form and the slope-intercept form c the line passing through $(1, 3)$ and $(3, -3)$.

16. If $f(x) = x^2$ and $g(x) = x + 2$, find $(f \circ g)(x)$ an $(g \circ f)(x)$.

17. You discover that the number of hours you sleep each nigh varies inversely as the square of the number of cups of coffe consumed during the early evening. If 2 cups of coffee ar consumed, you get 8 hours of sleep. If the number of cups c coffee is doubled, how many hours should you expect t sleep?

A baseball player hits a pop fly into the air. The function

$$s(t) = -16t^2 + 64t + 5$$

models the ball's height above the ground, s(t), in feet, t seconds after it is hit. Use the function to solve Exercises 18–19.

18. When does the baseball reach its maximum height? What i that height?

19. After how many seconds does the baseball hit the ground Round to the nearest tenth of a second.

20. You are paid time-and-a-half for each hour worked over 4 hours a week. Last week you worked 50 hours and earne $660. What is your normal hourly salary?

Trigonometric Functions

HAVE YOU HAD DAYS WHERE *your physical, intellectual, and emotional potentials were all at their peak? Then there are* those other days when we feel we should not even bother getting out of bed. Do our potentials run in oscillating cycles like the tides? Can they be described mathematically? In this chapter, you will encounter functions that enable us to model phenomena that occur in cycles.

WHAT A DAY! IT STARTED WHEN YOU added two miles to your morning run. You've experienced a feeling of peak physical well-being ever since. College was wonderful: You actually enjoyed two difficult lectures and breezed through a math test that had you worried. Now you're having dinner with a group of old friends. You experience the warmth from bonds of friendship filling the room.

Graphs of functions showing a person's *biorhythms*, the physical, intellectual, and emotional cycles we experience in life, are presented in Exercises 75–82 of Exercise Set 4.5.

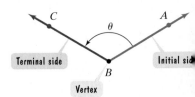

SECTION 4.1 *Angles and Radian Measure*

Objectives

❶ Recognize and use the vocabulary of angles.

❷ Use degree measure.

❸ Use radian measure.

❹ Convert between degrees and radians.

❺ Draw angles in standard position.

❻ Find coterminal angles.

❼ Find the length of a circular arc.

❽ Use linear and angular speed to describe motion on a circular path.

The San Francisco Museum of Modern Art was constructed in 1995 to illustrate how art and architecture can enrich one another. The exterior involves geometric shape symmetry, and unusual facades. Although there are no windows, natural ligh streams in through a truncated cylindrical skylight that crowns the building. Th architect worked with a scale model of the museum at the site and observed ho light hit it during different times of the day. These observations were used to cut th cylindrical skylight at an angle that maximizes sunlight entering the interior.

Angles play a critical role in creating modern architecture. They are als fundamental in trigonometry. In this section, we begin our study of trigonometry looking at angles and methods for measuring them.

 Recognize and use the vocabulary of angles.

Angles

The hour hand of a clock suggests a **ray**, a part of a line that has only one endpoi and extends forever in the opposite direction. An **angle** is formed by two rays tha have a common endpoint. One ray is called the **initial side** and the other th **terminal side**.

A rotating ray is often a useful way to think about angles. The ray in Figure 4 rotates from 12 to 2. The ray pointing to 12 is the **initial side** and the ray pointing to 2 is the **terminal side**. The common endpoint of an angle's initial side and terminal side is the **vertex** of the angle.

Figure 4.2 shows an angle. The arrow near the vertex shows the direction and the amount of rotation from the initial side to the terminal side. Several methods can be used to name an angle. Lowercase Greek letters, such as α (alpha), β (beta), γ (gamma and θ (theta), are often used.

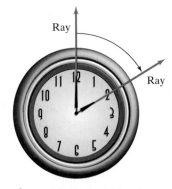

Figure 4.1 Clock with hands forming an angle

Figure 4.2 An angle; two rays with a common endpoint

An angle is in **standard position** if

- its vertex is at the origin of a rectangular coordinate system and
- its initial side lies along the positive *x*-axis.

The angles in Figure 4.3 at the top of the next page are both in standard position.

When we see an initial side and a terminal side in place, there are two kin of rotation that could have generated the angle. The arrow in Figure 4.3(a) indicat that the rotation from the initial side to the terminal side is in the counterclockwi

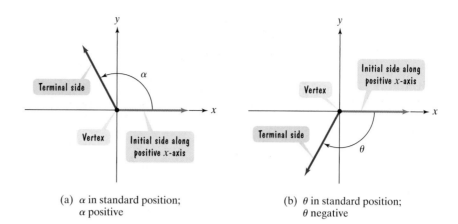

(a) α in standard position;
     α positive

(b) θ in standard position;
     θ negative

Figure 4.3 Two angles in standard position

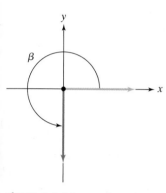

Figure 4.4 β is a quadrantal angle.

direction. **Positive angles** are generated by counterclockwise rotation. Thus, angle α is positive. By contrast, the arrow in Figure 4.3(b) shows that the rotation from the initial side to the terminal side is in the clockwise direction. **Negative angles** are generated by clockwise rotation. Thus, angle θ is negative.

When an angle is in standard position, its terminal side can lie in a quadrant. We say that the angle **lies in that quadrant**. For example, in Figure 4.3(a), the terminal side of angle α lies in quadrant II. Thus, angle α lies in quadrant II. By contrast, in Figure 4.3(b), the terminal side of angle θ lies in quadrant III. Thus, angle θ lies in quadrant III.

Must all angles in standard position lie in a quadrant? The answer is no. The terminal side can lie on the x-axis or the y-axis. For example, angle β in Figure 4.4 has a terminal side that lies on the negative y-axis. An angle is called a **quadrantal angle** if its terminal side lies on the x-axis or the y-axis. Angle β in Figure 4.4 is an example of a quadrantal angle.

② Use degree measure.

A complete 360° rotation

Measuring Angles Using Degrees

Angles are measured by determining the amount of rotation from the initial side to the terminal side. One way to measure angles is in **degrees**, symbolized by a small, raised circle °. Think of the hour hand of a clock. From 12 noon to 12 midnight, the hour hand moves around in a complete circle. By definition, the ray has rotated through 360 degrees, or 360°. Using 360° as the amount of rotation of a ray back onto itself, a degree, 1°, is $\frac{1}{360}$ of a complete rotation.

Figure 4.5 shows that certain angles have special names. An **acute angle** measures less than 90° [see Figure 4.5(a)]. A **right angle**, one quarter of a complete rotation, measures 90° [Figure 4.5(b)]. Examine the right angle—do you see a small square at the vertex? This symbol is used to indicate a right angle. An **obtuse angle** measures more than 90°, but less than 180° [Figure 4.5(c)]. Finally, a **straight angle**, one-half a complete rotation, measures 180° [Figure 4.5(d)].

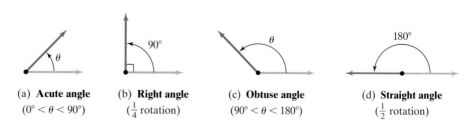

(a) **Acute angle**
$(0° < \theta < 90°)$

(b) **Right angle**
$(\frac{1}{4}$ rotation$)$

(c) **Obtuse angle**
$(90° < \theta < 180°)$

(d) **Straight angle**
$(\frac{1}{2}$ rotation$)$

Figure 4.5 Classifying angles by their degree measurement

We will be using notation such as $\theta = 60°$ to refer to an angle θ whose measure is 60°. We also refer to *an angle of* 60° or a 60° *angle*, rather than using the more precise (but cumbersome) phrase *an angle whose measure is* 60°.

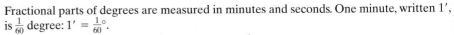

Technology

Fractional parts of degrees are measured in minutes and seconds. One minute, written $1'$, is $\frac{1}{60}$ degree: $1' = \frac{1}{60}°$.

One second, written $1''$, is $\frac{1}{3600}$ degree: $1'' = \frac{1}{3600}°$.
For example,

$$31°47'12''$$

$$= \left(31 + \frac{47}{60} + \frac{12}{3600}\right)°$$

$$\approx 31.787°.$$

Many calculators have keys for changing an angle from degree-minute-second notation (D°M′S″) to a decimal form and vice versa.

③ Use radian measure.

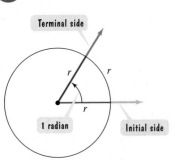

Figure 4.6 For a 1-radian angle, the intercepted arc and the radius are equal.

Measuring Angles Using Radians

Another way to measure angles is in *radians*. Let's first define an angle measuring **1 radian**. We use a circle of radius r. In Figure 4.6, we've constructed an angle whose vertex is at the center of the circle. Such an angle is called a **central angle**. Notice that this central angle intercepts an arc along the circle measuring r units. The radius of the circle is also r units. The measure of such an angle is 1 radian.

Definition of a Radian

One radian is the measure of the central angle of a circle that intercepts an arc equal in length to the radius of the circle.

The **radian measure** of any central angle is the length of the intercepted arc divided by the circle's radius. In Figure 4.7(a), the length of the arc intercepted by angle β is double the radius, r. We find the measure of angle β in radians by dividing the length of the intercepted arc by the radius.

$$\beta = \frac{\text{length of the intercepted arc}}{\text{radius}} = \frac{2r}{r} = 2$$

Thus, angle β measures 2 radians.

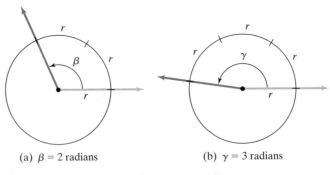

(a) $\beta = 2$ radians   (b) $\gamma = 3$ radians

Figure 4.7 Two central angles measured in radians

In Figure 4.7(b), the length of the intercepted arc is triple the radius, r. Let us find the measure of angle γ:

$$\gamma = \frac{\text{length of the intercepted arc}}{\text{radius}} = \frac{3r}{r} = 3.$$

Thus, angle γ measures 3 radians.

Radian Measure

Consider an arc of length s on a circle of radius r. The measure of the central angle, θ, that intercepts the arc is

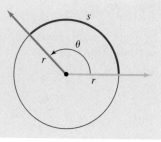

$$\theta = \frac{s}{r} \text{ radians.}$$

EXAMPLE 1   Computing Radian Measure

A central angle, θ, in a circle of radius 6 inches intercepts an arc of length 15 inches. What is the radian measure of θ?

Solution   Angle θ is shown in Figure 4.8. The radian measure of a central angle is the length of the intercepted arc, s, divided by the circle's radius, r. The length of the intercepted arc is 15 inches: $s = 15$ inches. The circle's radius is 6 inches: $r = 6$ inches. Now we use the formula for radian measure to find the radian measure of θ.

$$\theta = \frac{s}{r} = \frac{15 \text{ inches}}{6 \text{ inches}} = 2.5$$

Thus, the radian measure of θ is 2.5.

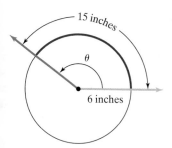

Figure 4.8

Study Tip

Before applying the formula for radian measure, be sure that the same unit of length is used for the intercepted arc, s, and the radius, r.

In Example 1, notice that the units (inches) cancel when we use the formula for radian measure. We are left with a number with no units. Thus, if an angle θ has a measure of 2.5 radians, we can write $\theta = 2.5$ radians or $\theta = 2.5$. We will often include the word *radians* simply for emphasis. There should be no confusion as to whether radian or degree measure is being used. Why is this so? If θ has a degree measure of, say, $2.5°$, we must include the degree symbol and write $\theta = 2.5°$, and *not* $\theta = 2.5$.

Check Point 1   A central angle, θ, in a circle of radius 12 feet intercepts an arc of length 42 feet. What is the radian measure of θ?

Relationship between Degrees and Radians

4 Convert between degrees and radians.

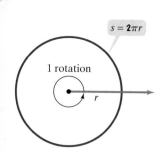

Figure 4.9 A complete rotation

How can we obtain a relationship between degrees and radians? We compare the number of degrees and the number of radians in one complete rotation, shown in Figure 4.9. We know that $360°$ is the amount of rotation of a ray back onto itself. The length of the intercepted arc is equal to the circumference of the circle. Thus, the radian measure of this central angle is the circumference of the circle divided by the circle's radius, r. The circumference of a circle of radius r is $2\pi r$. We use the formula for radian measure to find the radian measure of the $360°$ angle.

$$\theta = \frac{s}{r} = \frac{\text{the circle's circumference}}{r} = \frac{2\pi r}{r} = 2\pi$$

Because one complete rotation measures $360°$ and 2π radians,

$$360° = 2\pi \text{ radians.}$$

Dividing both sides by 2, we have

$$180° = \pi \text{ radians.}$$

Dividing this last equation by $180°$ or π gives the conversion rules in the box on the next page.

Study Tip

The unit you are converting *to* appears in the *numerator* of the conversion factor.

Conversion between Degrees and Radians

Using the basic relationship π radians = $180°$,

1. To convert degrees to radians, multiply degrees by $\dfrac{\pi \text{ radians}}{180°}$.

2. To convert radians to degrees, multiply radians by $\dfrac{180°}{\pi \text{ radians}}$.

Angles that are fractions of a complete rotation are usually expressed in radian measure as fractional multiples of π, rather than as decimal approximation. For example, we write $\theta = \dfrac{\pi}{2}$ rather than using the decimal approximation $\theta \approx 1.5\overline{7}$.

EXAMPLE 2 Converting from Degrees to Radians

Convert each angle in degrees to radians:

a. $30°$ **b.** $90°$ **c.** $-135°$.

Solution To convert degrees to radians, multiply by $\dfrac{\pi \text{ radians}}{180°}$. Observe how the degree units cancel.

a. $30° = 30° \cdot \dfrac{\pi \text{ radians}}{180°} = \dfrac{30\pi}{180} \text{ radians} = \dfrac{\pi}{6} \text{ radians}$

b. $90° = 90° \cdot \dfrac{\pi \text{ radians}}{180°} = \dfrac{90\pi}{180} \text{ radians} = \dfrac{\pi}{2} \text{ radians}$

c. $-135° = -135° \cdot \dfrac{\pi \text{ radians}}{180°} = -\dfrac{135\pi}{180} \text{ radians} = -\dfrac{3\pi}{4} \text{ radians}$

> Divide the numerator and denominator by 45.

Check Point 2 Convert each angle in degrees to radians:

a. $60°$ **b.** $270°$ **c.** $-300°$.

EXAMPLE 3 Converting from Radians to Degrees

Convert each angle in radians to degrees:

a. $\dfrac{\pi}{3}$ radians   **b.** $-\dfrac{5\pi}{3}$ radians   **c.** 1 radian.

Solution To convert radians to degrees, multiply by $\dfrac{180°}{\pi \text{ radians}}$. Observe how the radian units cancel.

a. $\dfrac{\pi}{3} \text{radians} = \dfrac{\pi \text{ radians}}{3} \cdot \dfrac{180°}{\pi \text{ radians}} = \dfrac{180°}{3} = 60°$

b. $-\dfrac{5\pi}{3} \text{radians} = -\dfrac{5\pi \text{ radians}}{3} \cdot \dfrac{180°}{\pi \text{ radians}} = -\dfrac{5 \cdot 180°}{3} = -300°$

c. $1 \text{ radian} = 1 \text{ radian} \cdot \dfrac{180°}{\pi \text{ radians}} = \dfrac{180°}{\pi} \approx 57.3°$

Study Tip

In Example 3(c), we see that 1 radian is approximately 57°. Keep in mind that a radian is much larger than a degree.

Check Point 3 Convert each angle in radians to degrees:

a. $\dfrac{\pi}{4}$ radians   **b.** $-\dfrac{4\pi}{3}$ radians   **c.** 6 radians.

⑤ Draw angles in standard position.

Drawing Angles in Standard Position

Although we can convert angles in radians to degrees, it is helpful to "think in radians" without having to make this conversion. To become comfortable with radian measure, consider angles in standard position: Each vertex is at the origin and each initial side lies along the positive *x*-axis. Think of the terminal side of the angle revolving around the origin. Thinking in radians means determining what part of a complete revolution or how many full revolutions will produce an angle whose radian measure is known. And here's the thing: We want to do this without having to convert from radians to degrees.

Figure 4.10 is a starting point for learning to think in radians. The figure illustrates that when the terminal side makes one full revolution, it forms an angle whose radian measure is 2π. The figure shows the quadrantal angles formed by $\frac{3}{4}$ of a revolution, $\frac{1}{2}$ of a revolution, and $\frac{1}{4}$ of a revolution.

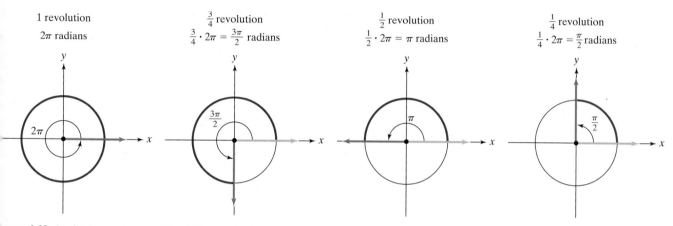

1 revolution
2π radians

$\frac{3}{4}$ revolution
$\frac{3}{4} \cdot 2\pi = \frac{3\pi}{2}$ radians

$\frac{1}{2}$ revolution
$\frac{1}{2} \cdot 2\pi = \pi$ radians

$\frac{1}{4}$ revolution
$\frac{1}{4} \cdot 2\pi = \frac{\pi}{2}$ radians

Figure 4.10 Angles formed by revolutions of terminal sides

EXAMPLE 4   Drawing Angles in Standard Position

Draw and label each angle in standard position:

a. $\theta = \dfrac{\pi}{4}$ **b.** $\alpha = \dfrac{5\pi}{4}$ **c.** $\beta = -\dfrac{3\pi}{4}$ **d.** $\gamma = \dfrac{9\pi}{4}$.

theta   alpha   beta   gamma

Solution   Because we are drawing angles in standard position, each vertex is at the origin and each initial side lies along the positive *x*-axis.

a. An angle of $\dfrac{\pi}{4}$ radians is a positive angle. It is obtained by rotating the terminal side counterclockwise. Because 2π is a full-circle revolution, we can express $\dfrac{\pi}{4}$ as a fractional part of 2π to determine the necessary rotation:

$$\frac{\pi}{4} = \frac{1}{8} \cdot 2\pi$$

$\dfrac{\pi}{4}$ is $\dfrac{1}{8}$ of a complete revolution of 2π radians.

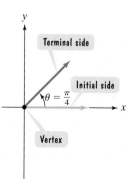

Figure 4.11

We see that $\theta = \dfrac{\pi}{4}$ is obtained by rotating the terminal side counterclockwise for $\dfrac{1}{8}$ of a revolution. The angle lies in quadrant I and is shown in Figure 4.11.

b. An angle of $\dfrac{5\pi}{4}$ radians is a positive angle. It is obtained by rotating the termina[l] side counterclockwise. Here are two ways to determine the necessary rotatio[n].

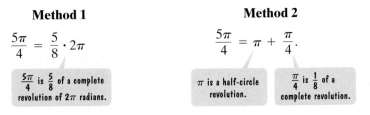

Method 1	**Method 2**

$$\frac{5\pi}{4} = \frac{5}{8} \cdot 2\pi$$

$\frac{5\pi}{4}$ is $\frac{5}{8}$ of a complete revolution of 2π radians.

$$\frac{5\pi}{4} = \pi + \frac{\pi}{4}.$$

π is a half-circle revolution.

$\frac{\pi}{4}$ is $\frac{1}{8}$ of a complete revolution.

Method 1 shows that $\alpha = \dfrac{5\pi}{4}$ is obtained by rotating the terminal sid[e] counterclockwise for $\dfrac{5}{8}$ of a revolution. Method 2 shows that $\alpha = \dfrac{5\pi}{4}$ [is] obtained by rotating the terminal side counterclockwise for half of a revolu[tion followed by a counterclockwise rotation of $\dfrac{1}{4}$ of a revolution. The angl[e] lies in quadrant III and is shown in Figure 4.12.

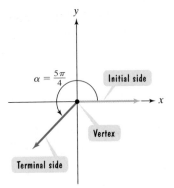

Figure 4.12

c. An angle of $-\dfrac{3\pi}{4}$ is a negative angle. It is obtained by rotating the termina[l] side clockwise. We use $\left| -\dfrac{3\pi}{4} \right|$, or $\dfrac{3\pi}{4}$, to determine the necessary rotation.

Method 1	**Method 2**

$$\frac{3\pi}{4} = \frac{3}{8} \cdot 2\pi$$

$\frac{3\pi}{4}$ is $\frac{3}{8}$ of a complete revolution of 2π radians.

$$\frac{3\pi}{4} = \frac{2\pi}{4} + \frac{\pi}{4} = \frac{\pi}{2} + \frac{\pi}{4}$$

$\frac{\pi}{2}$ is a quarter-circle revolution.

$\frac{\pi}{4}$ is $\frac{1}{8}$ of a complete revolution.

Method 1 shows that $\beta = -\dfrac{3\pi}{4}$ is obtained by rotating the terminal sid[e] clockwise for $\dfrac{3}{8}$ of a revolution. Method 2 shows that $\beta = -\dfrac{3\pi}{4}$ is obtained b[y] rotating the terminal side clockwise for $\dfrac{1}{4}$ of a revolution followed by a cloc[k]wise rotation of $\dfrac{1}{8}$ of a revolution. The angle lies in quadrant III and is show[n] in Figure 4.13.

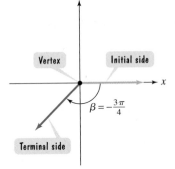

Figure 4.13

d. An angle of $\dfrac{9\pi}{4}$ radians is a positive angle. It is obtained by rotating the termin[al] side counterclockwise. Here are two methods to determine the necessary rotatio[n].

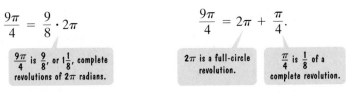

Method 1	**Method 2**

$$\frac{9\pi}{4} = \frac{9}{8} \cdot 2\pi$$

$\frac{9\pi}{4}$ is $\frac{9}{8}$, or $1\frac{1}{8}$, complete revolutions of 2π radians.

$$\frac{9\pi}{4} = 2\pi + \frac{\pi}{4}.$$

2π is a full-circle revolution.

$\frac{\pi}{4}$ is $\frac{1}{8}$ of a complete revolution.

Method 1 shows that $\gamma = \dfrac{9\pi}{4}$ is obtained by rotating the terminal sid[e] counterclockwise for $1\dfrac{1}{8}$ revolutions. Method 2 shows that $\gamma = \dfrac{9\pi}{4}$ is obtaine[d] by rotating the terminal side counterclockwise for a full-circle revolutio[n] followed by a counterclockwise rotation of $\dfrac{1}{8}$ of a revolution. The angle lies [in] quadrant I and is shown in Figure 4.14.

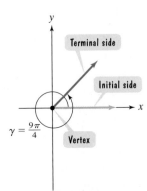

Figure 4.14

Check Point 4 Draw and label each angle in standard position:

a. $\theta = -\dfrac{\pi}{4}$ **b.** $\alpha = \dfrac{3\pi}{4}$ **c.** $\beta = -\dfrac{7\pi}{4}$ **d.** $\gamma = \dfrac{13\pi}{4}$.

Figure 4.15 illustrates the degree and radian measures of angles that you will commonly see in trigonometry. Each angle is in standard position, so that the initial side lies along the positive x-axis. We will be using both degree and radian measure for these angles.

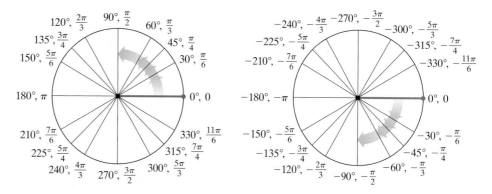

Figure 4.15 Degree and radian measures of selected positive and negative angles

Table 4.1 describes some of the positive angles in Figure 4.15 in terms of revolutions of the angle's terminal side around the origin.

Study Tip

When drawing the angles in Table 4.1 and Figure 4.15, it is helpful to first divide the rectangular coordinate system into eight equal sectors:

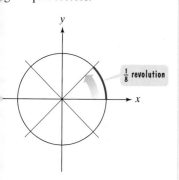

12 equal sectors:

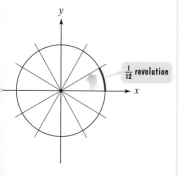

Perhaps we should call this study "Making a Clone of Arc."

Table 4.1

Terminal Side	Radian Measure of Angle	Degree Measure of Angle
$\dfrac{1}{12}$ revolution	$\dfrac{1}{12} \cdot 2\pi = \dfrac{\pi}{6}$	$\dfrac{1}{12} \cdot 360° = 30°$
$\dfrac{1}{8}$ revolution	$\dfrac{1}{8} \cdot 2\pi = \dfrac{\pi}{4}$	$\dfrac{1}{8} \cdot 360° = 45°$
$\dfrac{1}{6}$ revolution	$\dfrac{1}{6} \cdot 2\pi = \dfrac{\pi}{3}$	$\dfrac{1}{6} \cdot 360° = 60°$
$\dfrac{1}{4}$ revolution	$\dfrac{1}{4} \cdot 2\pi = \dfrac{\pi}{2}$	$\dfrac{1}{4} \cdot 360° = 90°$
$\dfrac{1}{3}$ revolution	$\dfrac{1}{3} \cdot 2\pi = \dfrac{2\pi}{3}$	$\dfrac{1}{3} \cdot 360° = 120°$
$\dfrac{1}{2}$ revolution	$\dfrac{1}{2} \cdot 2\pi = \pi$	$\dfrac{1}{2} \cdot 360° = 180°$
$\dfrac{2}{3}$ revolution	$\dfrac{2}{3} \cdot 2\pi = \dfrac{4\pi}{3}$	$\dfrac{2}{3} \cdot 360° = 240°$
$\dfrac{3}{4}$ revolution	$\dfrac{3}{4} \cdot 2\pi = \dfrac{3\pi}{2}$	$\dfrac{3}{4} \cdot 360° = 270°$
$\dfrac{7}{8}$ revolution	$\dfrac{7}{8} \cdot 2\pi = \dfrac{7\pi}{4}$	$\dfrac{7}{8} \cdot 360° = 315°$
1 revolution	$1 \cdot 2\pi = 2\pi$	$1 \cdot 360° = 360°$

⑥ Find coterminal angles.

Coterminal Angles

Two angles with the same initial and terminal sides but possibly different rotation⃗ are called **coterminal angles**.

Every angle has infinitely many coterminal angles. Why? Think of an angle i⃗ standard position. If the rotation of the angle is extended by one or more complet⃗ rotations of 360° or 2π, clockwise or counterclockwise, the result is an angle wit⃗ the same initial and terminal sides as the original angle.

> ### Coterminal Angles
>
> Increasing or decreasing the degree measure of an angle in standard position by an integer multiple of 360° results in a coterminal angle. Thus, an angle of $\theta°$ is coterminal with angles of $\theta° \pm 360°k$, where k is an integer.
>
> Increasing or decreasing the radian measure of an angle by an integer multiple of 2π results in a coterminal angle. Thus, an angle of θ radians is coterminal with angles of $\theta \pm 2\pi k$, where k is an integer.

Two coterminal angles for an angle of $\theta°$ can be found by adding 360° to $\theta°$ an⃗ subtracting 360° from $\theta°$.

EXAMPLE 5   Finding Coterminal Angles

Assume the following angles are in standard position. Find a positive angle less tha⃗ 360° that is coterminal with each of the following:

    a. a 420° angle   **b.** a $-120°$ angle.

Solution   We obtain the coterminal angle by adding or subtracting 360°. Th⃗ requirement to obtain a positive angle less than 360° determines whether we shoul⃗ add or subtract.

    a. For a 420° angle, subtract 360° to find a positive coterminal angle.

$$420° - 360° = 60°$$

A 60° angle is coterminal with a 420° angle. Figure 4.16(a) illustrates that thes⃗ angles have the same initial and terminal sides.

    b. For a $-120°$ angle, add 360° to find a positive coterminal angle.

$$-120° + 360° = 240°$$

A 240° angle is coterminal with a $-120°$ angle. Figure 4.16(b) illustrates tha⃗ these angles have the same initial and terminal sides.

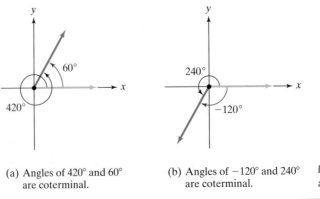

(a) Angles of 420° and 60°   (b) Angles of $-120°$ and 240°   **Figure 4.16** Pairs of cotermin⃗
    are coterminal.   are coterminal.   angles

Check **5** Find a positive angle less than 360° that is coterminal with each of t⃗
Point following:

    a. a 400° angle   **b.** a $-135°$ angle.

Two coterminal angles for an angle of θ radians can be found by adding 2π to θ and subtracting 2π from θ.

EXAMPLE 6   Finding Coterminal Angles

Assume the following angles are in standard position. Find a positive angle less than 2π that is coterminal with each of the following:

a. a $\dfrac{17\pi}{6}$ angle   **b.** a $-\dfrac{\pi}{12}$ angle.

Solution   We obtain the coterminal angle by adding or subtracting 2π. The requirement to obtain a positive angle less than 2π determines whether we should add or subtract.

a. For a $\dfrac{17\pi}{6}$, or $2\dfrac{5}{6}\pi$, angle, subtract 2π to find a positive coterminal angle.

$$\frac{17\pi}{6} - 2\pi = \frac{17\pi}{6} - \frac{12\pi}{6} = \frac{5\pi}{6}$$

A $\dfrac{5\pi}{6}$ angle is coterminal with a $\dfrac{17\pi}{6}$ angle. Figure 4.17(a) illustrates that these angles have the same initial and terminal sides.

b. For a $-\dfrac{\pi}{12}$ angle, add 2π to find a positive coterminal angle.

$$-\frac{\pi}{12} + 2\pi = -\frac{\pi}{12} + \frac{24\pi}{12} = \frac{23\pi}{12}$$

A $\dfrac{23\pi}{12}$ angle is coterminal with a $-\dfrac{\pi}{12}$ angle. Figure 4.17(b) illustrates that these angles have the same initial and terminal sides.

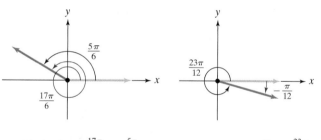

(a) Angles of $\frac{17\pi}{6}$ and $\frac{5\pi}{6}$ are coterminal.   **(b)** Angles of $-\frac{\pi}{12}$ and $\frac{23\pi}{12}$ are coterminal.

Figure 4.17 Pairs of coterminal angles

Check Point 6   Find a positive angle less than 2π that is coterminal with each of the following:

a. a $\dfrac{13\pi}{5}$ angle   **b.** a $-\dfrac{\pi}{15}$ angle.

To find a positive coterminal angle less than $360°$ or 2π, it is sometimes necessary to add or subtract more than one multiple of $360°$ or 2π.

EXAMPLE 7   Finding Coterminal Angles

Find a positive angle less than $360°$ or 2π that is coterminal with each of the following:

a. a $750°$ angle   **b.** a $\dfrac{22\pi}{3}$ angle   **c.** a $-\dfrac{17\pi}{6}$ angle.

EXAMPLE 9   Finding Linear Speed

A wind machine used to generate electricity has blades that are 10 feet in length (see Figure 4.18). The propeller is rotating at four revolutions per second. Find the linear speed, in feet per second, of the tips of the blades.

Solution   We are given ω, the angular speed.

$$\omega = 4 \text{ revolutions per second}$$

We use the formula $v = r\omega$ to find v, the linear speed. Before applying the formula, we must express ω in radians per second.

Figure 4.18

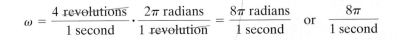

$$\omega = \frac{4 \text{ revolutions}}{1 \text{ second}} \cdot \frac{2\pi \text{ radians}}{1 \text{ revolution}} = \frac{8\pi \text{ radians}}{1 \text{ second}} \quad \text{or} \quad \frac{8\pi}{1 \text{ second}}$$

The angular speed of the propeller is 8π radians per second. The linear speed is

$$v = r\omega = 10 \text{ feet} \cdot \frac{8\pi}{1 \text{ second}} = \frac{80\pi \text{ feet}}{\text{second}}.$$

The linear speed of the tips of the blades is 80π feet per second, which i̇ approximately 251 feet per second.

Check Point 9   Long before iPods that hold thousands of songs and play them with superb audio quality, individual songs were delivered on 75-rpm and 45-rpm circular records. A 45-rpm record has an angular speed of 45 revolution per minute. Find the linear speed, in inches per minute, at the point where the needle is 1.5 inches from the record's center.

EXERCISE SET 4.1

Practice Exercises

In Exercises 1–6, the measure of an angle is given. Classify the angle as acute, right, obtuse, or straight.

1. $135°$ **2.** $177°$ **3.** $83.135°$

4. $87.177°$ **5.** π **6.** $\dfrac{\pi}{2}$

In Exercises 7–12, find the radian measure of the central angle of a circle of radius r that intercepts an arc of length s.

Radius, r	Arc length, s
7. 10 inches	40 inches
8. 5 feet	30 feet
9. 6 yards	8 yards
10. 8 yards	18 yards
11. 1 meter	400 centimeters
12. 1 meter	600 centimeters

In Exercises 13–20, convert each angle in degrees to radians. Express your answer as a multiple of π.

13. $45°$ **14.** $18°$ **15.** $135°$

16. $150°$ **17.** $300°$ **18.** $330°$

19. $-225°$ **20.** $-270°$

In Exercises 21–28, convert each angle in radians to degrees.

21. $\dfrac{\pi}{2}$ **22.** $\dfrac{\pi}{9}$ **23.** $\dfrac{2\pi}{3}$

24. $\dfrac{3\pi}{4}$ **25.** $\dfrac{7\pi}{6}$ **26.** $\dfrac{11\pi}{6}$

27. -3π **28.** -4π

In Exercises 29–34, convert each angle in degrees to radians. Round to two decimal places.

29. $18°$ **30.** $76°$ **31.** $-40°$

32. $-50°$ **33.** $200°$ **34.** $250°$

In Exercises 35–40, convert each angle in radians to degrees. Round to two decimal places.

35. 2 radians   **36.** 3 radians

37. $\dfrac{\pi}{13}$ radians   **38.** $\dfrac{\pi}{17}$ radians

39. -4.8 radians   **40.** -5.2 radians

In Exercises 41–56, use the circle shown in the rectangular coordinate system to draw each angle in standard position. State the quadrant in which the angle lies. When an angle's measure is given in radians, work the exercise without converting to degrees.

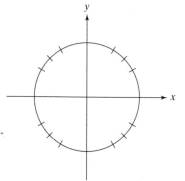

41. $\frac{7\pi}{6}$ **42.** $\frac{4\pi}{3}$ **43.** $\frac{3\pi}{4}$

44. $\frac{7\pi}{4}$ **45.** $-\frac{2\pi}{3}$ **46.** $-\frac{5\pi}{6}$

47. $-\frac{5\pi}{4}$ **48.** $-\frac{7\pi}{4}$ **49.** $\frac{16\pi}{3}$

50. $\frac{14\pi}{3}$ **51.** $120°$ **52.** $150°$

53. $-210°$ **54.** $-240°$ **55.** $420°$

56. $405°$

In Exercises 57–70, find a positive angle less than 360° or 2π that is coterminal with the given angle.

57. $395°$ **58.** $415°$ **59.** $-150°$

60. $-160°$ **61.** $-765°$ **62.** $-760°$

63. $\frac{19\pi}{6}$ **64.** $\frac{17\pi}{5}$ **65.** $\frac{23\pi}{5}$

66. $\frac{25\pi}{6}$ **67.** $-\frac{\pi}{50}$ **68.** $-\frac{\pi}{40}$

69. $-\frac{31\pi}{7}$ **70.** $-\frac{38\pi}{9}$

In Exercises 71–74, find the length of the arc on a circle of radius r intercepted by a central angle θ. Express arc length in terms of π. Then round your answer to two decimal places.

Radius, r	Central angle, θ
71. 12 inches	$\theta = 45°$
72. 16 inches	$\theta = 60°$
73. 8 feet	$\theta = 225°$
74. 9 yards	$\theta = 315°$

In Exercises 75–76, express each angular speed in radians per second.

75. 6 revolutions per second **76.** 20 revolutions per second

Practice Plus

Use the circle shown in the rectangular coordinate system to solve Exercises 77–82. Find two angles, in radians, between −2π and 2π such that each angle's terminal side passes through the origin and the given point.

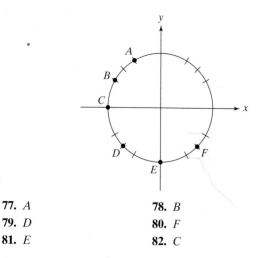

77. A **78.** B

79. D **80.** F

81. E **82.** C

In Exercises 83–86, find the positive radian measure of the angle that the second hand of a clock moves through in the given time.

83. 55 seconds   **84.** 35 seconds

85. 3 minutes and 40 seconds

86. 4 minutes and 25 seconds

 ### Application Exercises

87. The minute hand of a clock moves from 12 to 2 o'clock, or $\frac{1}{6}$ of a complete revolution. Through how many degrees does it move? Through how many radians does it move?

88. The minute hand of a clock moves from 12 to 4 o'clock, or $\frac{1}{3}$ of a complete revolution. Through how many degrees does it move? Through how many radians does it move?

89. The minute hand of a clock is 8 inches long and moves from 12 to 2 o'clock. How far does the tip of the minute hand move? Express your answer in terms of π and then round to two decimal places.

90. The minute hand of a clock is 6 inches long and moves from 12 to 4 o'clock. How far does the tip of the minute hand move? Express your answer in terms of π and then round to two decimal places.

91. The figure shows a highway sign that warns of a railway crossing. The lines that form the cross pass through the circle's center and intersect at right angles. If the radius of the circle is 24 inches, find the length of each of the four arcs formed by the cross. Express your answer in terms of π and then round to two decimal places.

92. The radius of a wheel rolling on the ground is 80 centimeters. If the wheel rotates through an angle of 60°, how many centimeters does it move? Express your answer in terms of π and then round to two decimal places.

How do we measure the distance between two points, A and B, on Earth? We measure along a circle with a center, C, at the center of Earth. The radius of the circle is equal to the distance from C to the surface. Use the fact that Earth is a sphere of radius equal to approximately 4000 miles to solve Exercises 93–96.

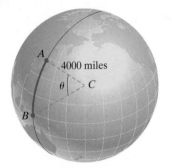

93. If two points, A and B, are 8000 miles apart, express angle θ in radians and in degrees.

94. If two points, A and B, are 10,000 miles apart, express angle θ in radians and in degrees.

95. If $\theta = 30°$, find the distance between A and B to the nearest mile.

96. If $\theta = 10°$, find the distance between A and B to the nearest mile.

97. The angular speed of a point on Earth is $\frac{\pi}{12}$ radians per hour. The Equator lies on a circle of radius approximately 4000 miles. Find the linear velocity, in miles per hour, of a point on the Equator.

98. A Ferris wheel has a radius of 25 feet. The wheel is rotating at two revolutions per minute. Find the linear speed, in feet per minute, of a seat on this Ferris wheel.

99. A water wheel has a radius of 12 feet. The wheel is rotating at 20 revolutions per minute. Find the linear speed, in feet per minute, of the water.

100. On a carousel, the outer row of animals is 20 feet from the center. The inner row of animals is 10 feet from the center. The carousel is rotating at 2.5 revolutions per minute. What is the difference, in feet per minute, in the linear speeds of the animals in the outer and inner rows? Round to the nearest foot per minute.

Writing in Mathematics

101. What is an angle?

102. What determines the size of an angle?

103. Describe an angle in standard position.

104. Explain the difference between positive and negative angles. What are coterminal angles?

105. Explain what is meant by one radian.

106. Explain how to find the radian measure of a central angle.

107. Describe how to convert an angle in degrees to radians.

108. Explain how to convert an angle in radians to degrees.

109. Explain how to find the length of a circular arc.

110. If a carousel is rotating at 2.5 revolutions per minute, explain how to find the linear speed of a child seated on one of the animals.

111. The angular velocity of a point on Earth is $\frac{\pi}{12}$ radians per hour. Describe what happens every 24 hours.

112. Have you ever noticed that we use the vocabulary of angles in everyday speech? Here is an example:

My opinion about art museums took a 180° turn after visiting the San Francisco Museum of Modern Art.

Explain what this means. Then give another example of the vocabulary of angles in everyday use.

Technology Exercises

 In Exercises 113–116, use the keys on your calculator or graphing utility for converting an angle in degrees, minutes, and seconds $(D°M'S'')$ *into decimal form, and vice versa.*

In Exercises 113–114, convert each angle to a decimal in degrees. Round your answer to two decimal places.

113. $30°15'10''$ 114. $65°45'20''$

In Exercises 115–116, convert each angle to $D°M'S''$ *form. Round your answer to the nearest second.*

115. $30.42°$ 116. $50.42°$

Critical Thinking Exercises

117. If $\theta = \frac{3}{2}$, is this angle larger or smaller than a right angle?

118. A railroad curve is laid out on a circle. What radius should be used if the track is to change direction by 20° in a distance of 100 miles? Round your answer to the nearest mile.

119. Assuming Earth to be a sphere of radius 4000 miles, how many miles north of the Equator is Miami, Florida, if it is 26° north from the Equator? Round your answer to the nearest mile.

SECTION 4.2   *Trigonometric Functions: The Unit Circle*

Objectives

❶ Use a unit circle to define trigonometric functions of real numbers.

❷ Recognize the domain and range of sine and cosine functions.

❸ Find exact values of the trigonometric functions at $\frac{\pi}{4}$.

❹ Use even and odd trigonometric functions.

❺ Recognize and use fundamental identities.

❻ Use periodic properties.

❼ Evaluate trigonometric functions with a calculator.

There is something comforting in the repetition of some of nature's patterns. The ocean level at a beach varies between high and low tide approximately every 12 hours. The number of hours of daylight oscillates from a maximum on the summer solstice, June 21, to a minimum on the winter solstice, December 21. Then it increases to the same maximum the following June 21. Some believe that cycles, called biorhythms, represent physical, emotional, and intellectual aspects of our lives. In this chapter, we study six functions, the six *trigonometric functions*, that are used to model phenomena that occur again and again.

Calculus and the Unit Circle

The word *trigonometry* means measurement of triangles. Trigonometric functions, with domains consisting of sets of angles, were first defined using right triangles. By contrast, problems in calculus are solved using functions whose domains are sets of real numbers. Therefore, we introduce the trigonometric functions using unit circles and radians, rather than right triangles and degrees.

A **unit circle** is a circle of radius 1, with its center at the origin of a rectangular coordinate system. The equation of this unit circle is $x^2 + y^2 = 1$. Figure 4.19 shows a unit circle in which the central angle measures t radians.

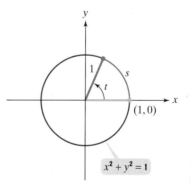

Figure 4.19 Unit circle with a central angle measuring t radians

We can use the formula for the length of a circular arc, $s = r\theta$, to find the length of the intercepted arc.

$$s = r\theta = 1 \cdot t = t$$

| The radius of a unit circle is 1. | The radian measure of the central angle is t. |

Thus, the length of the intercepted arc is t. This is also the radian measure of the central angle. Thus, **in a unit circle, the radian measure of the central angle is equal to the length of the intercepted arc.** Both are given by the same *real number t*.

In Figure 4.20, the radian measure of the angle and the length of the intercepted arc are both shown by t. Let $P = (x, y)$ denote the point on the unit circle that has arc length t from $(1, 0)$. Figure 4.20(a) shows that if t is positive, point P is reached by moving counterclockwise along the unit circle from $(1, 0)$. Figure 4.20(b) shows that if t is negative, point P is reached by moving clockwise along the unit circle from $(1, 0)$. For each real number t, there corresponds a point $P = (x, y)$ on the unit circle.

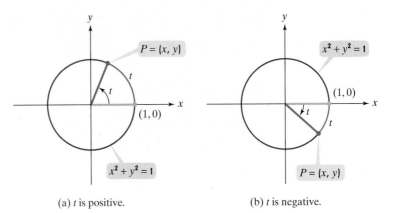

(a) t is positive.   (b) t is negative.

Figure 4.20

<table>
<tr><td>①</td><td>Use a unit circle to define trigonometric functions of real numbers.</td></tr>
</table>

The Six Trigonometric Functions

We begin the study of trigonometry by defining the six trigonometric functions. The inputs of these functions are real numbers, represented by t in Figure 4.20. The outputs involve the point $P = (x, y)$ on the unit circle that corresponds to t and the coordinates of this point.

The trigonometric functions have names that are words, rather than single letters such as f, g, and h. For example, the **sine of t** is the y-coordinate of point P on the unit circle:

$$\sin t = y.$$

Input is the real number t.   Output is the y-coordinate of a point on the unit circle.

The value of y depends on the real number t and thus is a function of t. The expression $\sin t$ really means $\sin(t)$, where sine is the name of the function and t, a real number, is an input.

For example, a point $P = (x, y)$ on the unit circle corresponding to a real number t is shown in Figure 4.21 for $\pi < t < \dfrac{3\pi}{2}$. We see that the coordinates of $P = (x, y)$ are $x = -\dfrac{3}{5}$ and $y = -\dfrac{4}{5}$. Because the sine function is the y-coordinate of P, the value of this trigonometric function at the real number t is

$$\sin t = -\frac{4}{5}.$$

Here are the names of the six trigonometric functions, along with their abbreviations.

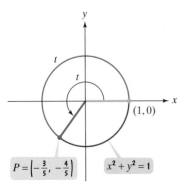

$P = \left(-\frac{3}{5}, -\frac{4}{5}\right)$ $x^2 + y^2 = 1$

Figure 4.21

Name	Abbreviation	Name	Abbreviation
sine	sin	cosecant	csc
cosine	cos	secant	sec
tangent	tan	cotangent	cot

Definitions of the Trigonometric Functions in Terms of a Unit Circle

If t is a real number and $P = (x, y)$ is a point on the unit circle that corresponds to t, then

$$\sin t = y \qquad\qquad \csc t = \frac{1}{y}, y \neq 0$$

$$\cos t = x \qquad\qquad \sec t = \frac{1}{x}, x \neq 0$$

$$\tan t = \frac{y}{x}, x \neq 0 \quad \cot t = \frac{x}{y}, y \neq 0.$$

Because this definition expresses function values in terms of coordinates of a point on a unit circle, the trigonometric functions are sometimes called the **circular functions**. Observe that the function values in the second column in the box are the reciprocals of the corresponding function values in the first column.

EXAMPLE 1   Finding Values of the Trigonometric Functions

In Figure 4.22, t is a real number equal to the length of the intercepted arc of an angle that measures t radians and $P = \left(-\frac{1}{2}, \frac{\sqrt{3}}{2}\right)$ is the point on the unit circle that corresponds to t. Use the figure to find the values of the trigonometric functions at t.

Solution The point P on the unit circle that corresponds to t has coordinates $\left(-\frac{1}{2}, \frac{\sqrt{3}}{2}\right)$. We use $x = -\frac{1}{2}$ and $y = \frac{\sqrt{3}}{2}$ to find the values of the trigonometric functions. Because radical expressions are usually written without radicals in the denominators, we simplify by rationalizing denominators where appropriate.

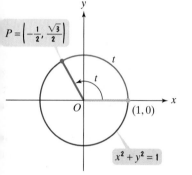

Figure 4.22

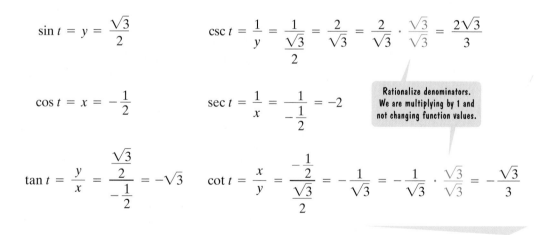

$$\sin t = y = \frac{\sqrt{3}}{2} \qquad\qquad \csc t = \frac{1}{y} = \frac{1}{\frac{\sqrt{3}}{2}} = \frac{2}{\sqrt{3}} = \frac{2}{\sqrt{3}} \cdot \frac{\sqrt{3}}{\sqrt{3}} = \frac{2\sqrt{3}}{3}$$

$$\cos t = x = -\frac{1}{2} \qquad\qquad \sec t = \frac{1}{x} = \frac{1}{-\frac{1}{2}} = -2$$

Rationalize denominators. We are multiplying by 1 and not changing function values.

$$\tan t = \frac{y}{x} = \frac{\frac{\sqrt{3}}{2}}{-\frac{1}{2}} = -\sqrt{3} \quad \cot t = \frac{x}{y} = \frac{-\frac{1}{2}}{\frac{\sqrt{3}}{2}} = -\frac{1}{\sqrt{3}} = -\frac{1}{\sqrt{3}} \cdot \frac{\sqrt{3}}{\sqrt{3}} = -\frac{\sqrt{3}}{3}$$

 Check Point 1 Use the figure on the right to find the values of the trigonometric functions at t.

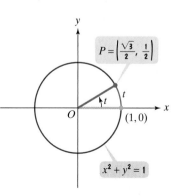

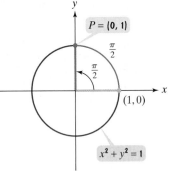

$P = (0, 1)$

$\frac{\pi}{2}$

$\frac{\pi}{2}$

$(1, 0)$

$x^2 + y^2 = 1$

Figure 4.23

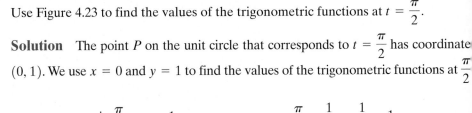

EXAMPLE 2   Finding Values of the Trigonometric Functions

Use Figure 4.23 to find the values of the trigonometric functions at $t = \dfrac{\pi}{2}$.

Solution   The point P on the unit circle that corresponds to $t = \dfrac{\pi}{2}$ has coordinate $(0, 1)$. We use $x = 0$ and $y = 1$ to find the values of the trigonometric functions at $\dfrac{\pi}{2}$

$$\sin \frac{\pi}{2} = y = 1 \qquad\qquad \csc \frac{\pi}{2} = \frac{1}{y} = \frac{1}{1} = 1$$

$$\cos \frac{\pi}{2} = x = 0 \qquad \begin{array}{c}\sec \frac{\pi}{2} \text{ and} \\ \tan \frac{\pi}{2} \text{ are} \\ \text{undefined.}\end{array} \qquad \sec \frac{\pi}{2} = \frac{1}{x} = \frac{1}{0}$$

$$\tan \frac{\pi}{2} = \frac{y}{x} = \frac{1}{0} \qquad\qquad \cot \frac{\pi}{2} = \frac{x}{y} = \frac{0}{1} = 0$$

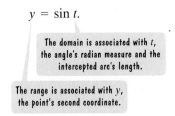 **Check Point 2** Use the figure on the right to find the values of the trigonometric functions at $t = \pi$.

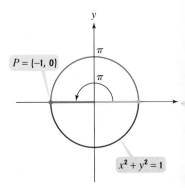

π

$P = (-1, 0)$

π

$x^2 + y^2 = 1$

 Recognize the domain and range of sine and cosine functions.

Domain and Range of Sine and Cosine Functions

The domain and range of each trigonometric function can be found from the un circle definition. At this point, let's look only at the sine and cosine functions,

$$\sin t = y \quad \text{and} \quad \cos t = x.$$

Figure 4.24 shows the sine function at t as the y-coordinate of a point along th unit circle:

$y = \sin t.$

The domain is associated with t, the angle's radian measure and the intercepted arc's length.

The range is associated with y, the point's second coordinate.

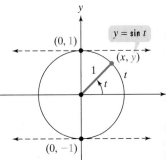

$y = \sin t$

$(0, 1)$

(x, y)

1

t

$(0, -1)$

Figure 4.24

Because t can be any real number, the domain of the sine function is $(-\infty, \infty)$, the set of all real numbers. The radius of the unit circle is 1 and the dashed horizontal lines in Figure 4.24 show that y cannot be less than -1 or great than 1. Thus, the range of the sine function is $[-1, 1]$, the set of all real numbers from -1 to 1, inclusive.

Figure 4.25 shows the cosine function at t as the x-coordinate of a point along the unit circle:

$$x = \cos t.$$

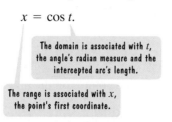

The domain is associated with t, the angle's radian measure and the intercepted arc's length.

The range is associated with x, the point's first coordinate.

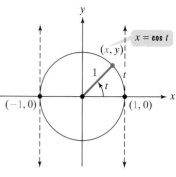

Figure 4.25

Because t can be any real number, the domain of the cosine function is $(-\infty, \infty)$. The radius of the unit circle is 1 and the dashed vertical lines in Figure 4.25 show that x cannot be less than -1 or greater than 1. Thus, the range of the cosine function is $[-1, 1]$.

The Domain and Range of the Sine and Cosine Functions

The domain of the sine function and the cosine function is $(-\infty, \infty)$, the set of all real numbers. The range of these functions is $[-1, 1]$, the set of all real numbers from -1 to 1, inclusive.

---

3 Find exact values of the trigonometric functions at $\dfrac{\pi}{4}$.

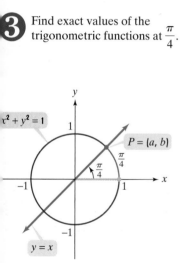

Figure 4.26

Exact Values of Trigonometric Functions at $t = \dfrac{\pi}{4}$

Trigonometric functions at $t = \dfrac{\pi}{4}$ occur frequently. How do we use the unit circle to find values of the trigonometric functions at $t = \dfrac{\pi}{4}$? Look at Figure 4.26. We must find the coordinates of point $P = (a, b)$ on the unit circle that correspond to $t = \dfrac{\pi}{4}$. Can you see that P lies on the line $y = x$? Thus, point P has equal x- and y-coordinates: $a = b$. We find these coordinates as follows:

$x^2 + y^2 = 1$	This is the equation of the unit circle.
$a^2 + b^2 = 1$	Point $P = (a, b)$ lies on the unit circle. Thus, its coordinates satisfy the circle's equation.
$a^2 + a^2 = 1$	Because $a = b$, substitute a for b in the previous equation.
$2a^2 = 1$	Add like terms.
$a^2 = \frac{1}{2}$	Divide both sides of the equation by 2.
$a = \sqrt{\frac{1}{2}}$	Because $a > 0$, take the positive square root of both sides.

We see that $a = \sqrt{\dfrac{1}{2}} = \dfrac{1}{\sqrt{2}}$. Because $a = b$, we also have $b = \dfrac{1}{\sqrt{2}}$. Thus, if $t = \dfrac{\pi}{4}$, point $P = \left(\dfrac{1}{\sqrt{2}}, \dfrac{1}{\sqrt{2}} \right)$ is the point on the unit circle that corresponds to t. Let's rationalize the denominator on each coordinate:

$$\frac{1}{\sqrt{2}} = \frac{1}{\sqrt{2}} \cdot \frac{\sqrt{2}}{\sqrt{2}} = \frac{\sqrt{2}}{2}.$$

We are multiplying by 1 and not changing the value of $\dfrac{1}{\sqrt{2}}$.

We use $\left(\dfrac{\sqrt{2}}{2}, \dfrac{\sqrt{2}}{2} \right)$ to find the values of the trigonometric functions at $t = \dfrac{\pi}{4}$.

EXAMPLE 3 **Finding Values of the Trigonometric Functions**

$$\text{at } t = \frac{\pi}{4}$$

Find $\sin \frac{\pi}{4}$, $\cos \frac{\pi}{4}$, and $\tan \frac{\pi}{4}$.

Solution The point P on the unit circle that corresponds to $t = \frac{\pi}{4}$ has coordinate $\left(\frac{\sqrt{2}}{2}, \frac{\sqrt{2}}{2} \right)$. We use $x = \frac{\sqrt{2}}{2}$ and $y = \frac{\sqrt{2}}{2}$ to find the values of the three trigonometric functions at $\frac{\pi}{4}$.

$$\sin \frac{\pi}{4} = y = \frac{\sqrt{2}}{2} \qquad \cos \frac{\pi}{4} = x = \frac{\sqrt{2}}{2} \qquad \tan \frac{\pi}{4} = \frac{y}{x} = \frac{\frac{\sqrt{2}}{2}}{\frac{\sqrt{2}}{2}} = 1$$

Check Point 3 Find $\csc \frac{\pi}{4}$, $\sec \frac{\pi}{4}$, and $\cot \frac{\pi}{4}$.

Because you will often see the trigonometric functions at $\frac{\pi}{4}$, it is a good idea to memorize the values shown in the following box. In the next section, you will learn to use a right triangle to obtain these values.

Trigonometric Functions at $\frac{\pi}{4}$

$$\sin \frac{\pi}{4} = \frac{\sqrt{2}}{2} \qquad\qquad \csc \frac{\pi}{4} = \sqrt{2}$$

$$\cos \frac{\pi}{4} = \frac{\sqrt{2}}{2} \qquad\qquad \sec \frac{\pi}{4} = \sqrt{2}$$

$$\tan \frac{\pi}{4} = 1 \qquad\qquad \cot \frac{\pi}{4} = 1$$

 Use even and odd trigonometric functions.

Even and Odd Trigonometric Functions

We have seen that a function is even if $f(-t) = f(t)$ and odd if $f(-t) = -f(t)$. We can use Figure 4.27 to show that the cosine function is an even function and the sine function is an odd function. By definition, the coordinates of the points P and Q in Figure 4.27 are as follows:

$$P: \ (\cos t, \sin t)$$
$$Q: \ (\cos(-t), \sin(-t)).$$

In Figure 4.27, the x-coordinates of P and Q are the same. Thus,

$$\cos(-t) = \cos t.$$

This shows that the cosine function is an even function. By contrast, the y-coordinates of P and Q are negatives of each other. Thus,

$$\sin(-t) = -\sin t.$$

This shows that the sine function is an odd function.

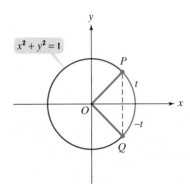

Figure 4.27

This argument is valid regardless of the length of t. Thus, the arc may terminate in any of the four quadrants or on any axis. Using the unit circle definition of the trigonometric functions, we obtain the following results:

Even and Odd Trigonometric Functions

The cosine and secant functions are *even*.

$$\cos(-t) = \cos t \qquad\qquad \sec(-t) = \sec t$$

The sine, cosecant, tangent, and cotangent functions are *odd*.

$$\sin(-t) = -\sin t \qquad\qquad \csc(-t) = -\csc t$$
$$\tan(-t) = -\tan t \qquad\qquad \cot(-t) = -\cot t$$

EXAMPLE 4   Using Even and Odd Functions to Find Values of Trigonometric Functions

Find the value of each trigonometric function:

a. $\cos\left(-\dfrac{\pi}{4}\right)$ **b.** $\tan\left(-\dfrac{\pi}{4}\right)$.

Solution

a. $\cos\left(-\dfrac{\pi}{4}\right) = \cos\dfrac{\pi}{4} = \dfrac{\sqrt{2}}{2}$ **b.** $\tan\left(-\dfrac{\pi}{4}\right) = -\tan\dfrac{\pi}{4} = -1$

Check Point 4 Find the value of each trigonometric function:

a. $\sec\left(-\dfrac{\pi}{4}\right)$ **b.** $\sin\left(-\dfrac{\pi}{4}\right)$.

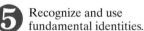
Recognize and use fundamental identities.

Fundamental Identities

Many relationships exist among the six trigonometric functions. These relationships are described using **trigonometric identities**. Trigonometric identities are equations that are true for all real numbers for which the trigonometric expressions in the equations are defined. For example, the definitions of the cosine and secant functions are given by

$$\cos t = x \quad\text{and}\quad \sec t = \frac{1}{x}, x \neq 0.$$

Substituting $\cos t$ for x in the equation on the right, we see that

$$\sec t = \frac{1}{\cos t}, \cos t \neq 0.$$

This identity is one of six **reciprocal identities**.

Reciprocal Identities

$$\sin t = \frac{1}{\csc t} \qquad\qquad \csc t = \frac{1}{\sin t}$$
$$\cos t = \frac{1}{\sec t} \qquad\qquad \sec t = \frac{1}{\cos t}$$
$$\tan t = \frac{1}{\cot t} \qquad\qquad \cot t = \frac{1}{\tan t}$$

Two other relationships that follow from the definitions of the trigonometric functions are called the **quotient identities**.

Quotient Identities

$$\tan t = \frac{\sin t}{\cos t} \qquad\qquad \cot t = \frac{\cos t}{\sin t}$$

If $\sin t$ and $\cos t$ are known, a quotient identity and three reciprocal identities make it possible to find the value of each of the four remaining trigonometric functions.

EXAMPLE 5   Using Quotient and Reciprocal Identities

Given $\sin t = \dfrac{2}{5}$ and $\cos t = \dfrac{\sqrt{21}}{5}$, find the value of each of the four remaining trigonometric functions.

Solution   We can find $\tan t$ by using the quotient identity that describes $\tan t$ as the quotient of $\sin t$ and $\cos t$.

$$\tan t = \frac{\sin t}{\cos t} = \frac{\frac{2}{5}}{\frac{\sqrt{21}}{5}} = \frac{2}{5} \cdot \frac{5}{\sqrt{21}} = \frac{2}{\sqrt{21}} = \frac{2}{\sqrt{21}} \cdot \frac{\sqrt{21}}{\sqrt{21}} = \frac{2\sqrt{21}}{21}$$

Rationalize the denominator.

We use the reciprocal identities to find the value of each of the remaining three functions.

$$\csc t = \frac{1}{\sin t} = \frac{1}{\frac{2}{5}} = \frac{5}{2}$$

$$\sec t = \frac{1}{\cos t} = \frac{1}{\frac{\sqrt{21}}{5}} = \frac{5}{\sqrt{21}} = \frac{5}{\sqrt{21}} \cdot \frac{\sqrt{21}}{\sqrt{21}} = \frac{5\sqrt{21}}{21}$$

Rationalize the denominator.

$$\cot t = \frac{1}{\tan t} = \frac{1}{\frac{2}{\sqrt{21}}} = \frac{\sqrt{21}}{2}$$

We found $\tan t = \dfrac{2}{\sqrt{21}}$. We could use $\tan t = \dfrac{2\sqrt{21}}{21}$, but then we would have to rationalize the denominator.

Check Point 5   Given $\sin t = \dfrac{2}{3}$ and $\cos t = \dfrac{\sqrt{5}}{3}$, find the value of each of the four remaining trigonometric functions.

Other relationships among trigonometric functions follow from the equation of the unit circle

$$x^2 + y^2 = 1.$$

Because $\cos t = x$ and $\sin t = y$, we see that

$$(\cos t)^2 + (\sin t)^2 = 1.$$

We will eliminate the parentheses in this identity by writing $\cos^2 t$ instead of $(\cos t)^2$ and $\sin^2 t$ instead of $(\sin t)^2$. With this notation, we can write the identity as

$$\cos^2 t + \sin^2 t = 1$$

or

$$\sin^2 t + \cos^2 t = 1. \qquad \text{The identity usually appears in this form.}$$

Two additional identities can be obtained from $x^2 + y^2 = 1$ by dividing both sides by x^2 and y^2, respectively. The three identities are called the **Pythagorean identities**.

Pythagorean Identities

$$\sin^2 t + \cos^2 t = 1 \qquad 1 + \tan^2 t = \sec^2 t \qquad 1 + \cot^2 t = \csc^2 t$$

EXAMPLE 6   Using a Pythagorean Identity

Given that $\sin t = \dfrac{3}{5}$ and $0 \leq t < \dfrac{\pi}{2}$, find the value of $\cos t$ using a trigonometric identity.

Solution   We can find the value of $\cos t$ by using the Pythagorean identity

$$\sin^2 t + \cos^2 t = 1.$$

$$\left(\frac{3}{5}\right)^2 + \cos^2 t = 1 \qquad \text{We are given that } \sin t = \frac{3}{5}.$$

$$\frac{9}{25} + \cos^2 t = 1 \qquad \text{Square } \frac{3}{5}: \left(\frac{3}{5}\right)^2 = \frac{3^2}{5^2} = \frac{9}{25}.$$

$$\cos^2 t = 1 - \frac{9}{25} \qquad \text{Subtract } \frac{9}{25} \text{ from both sides.}$$

$$\cos^2 t = \frac{16}{25} \qquad \text{Simplify: } 1 - \frac{9}{25} = \frac{25}{25} - \frac{9}{25} = \frac{16}{25}.$$

$$\cos t = \sqrt{\frac{16}{25}} = \frac{4}{5} \qquad \text{Because } 0 \leq t < \frac{\pi}{2}, \cos t, \text{ the x-coordinate of}$$

a point on the unit circle, is positive.

Thus, $\cos t = \dfrac{4}{5}$.

Check Point 6   Given that $\sin t = \dfrac{1}{2}$ and $0 \leq t < \dfrac{\pi}{2}$, find the value of $\cos t$ using a trigonometric identity.

6 Use periodic properties.

Periodic Functions

Certain patterns in nature repeat again and again. For example, the ocean level at a beach varies from low tide to high tide and then back to low tide approximately every 12 hours. If low tide occurs at noon, then high tide will be around 6 P.M. and low tide will occur again around midnight, and so on infinitely. If $f(t)$ represents the ocean level at the beach at any time t, then the level is the same 12 hours later. Thus,

$$f(t + 12) = f(t).$$

The word *periodic* means that this tidal behavior repeats infinitely. The *period*, 12 hours, is the time it takes to complete one full cycle.

Definition of a Periodic Function

A function f is **periodic** if there exists a positive number p such that

$$f(t + p) = f(t)$$

for all t in the domain of f. The smallest positive number p for which f is periodic is called the **period** of f.

The trigonometric functions are used to model periodic phenomena. Why? If we begin at any point P on the unit circle and travel a distance of 2π units along the perimeter, we will return to the same point P. Because the trigonometric

functions are defined in terms of the coordinates of that point P, we obtain th
following results:

Periodic Properties of the Sine and Cosine Functions

$$\sin(t + 2\pi) = \sin t \quad \text{and} \quad \cos(t + 2\pi) = \cos t$$

The sine and cosine functions are periodic functions and have period 2π.

Like the sine and cosine functions, the secant and cosecant functions hav
period 2π. However, the tangent and cotangent functions have a smaller period
Figure 4.28 shows that if we begin at any point $P(x, y)$ on the unit circle and trave
a distance of π units along the perimeter, we arrive at the point $Q(-x, -y)$. Th
tangent function, defined in terms of the coordinates of a point, is the same at (x, y)
and $(-x, -y)$.

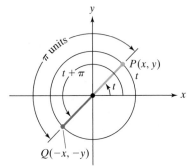

Figure 4.28 tan at P = tan at Q

$$\underset{\substack{\text{Tangent function} \\ \text{at } (x, y)}}{} \frac{y}{x} = \frac{-y}{-x} \underset{\substack{\text{Tangent function} \\ \pi \text{ radians later}}}{}$$

We see that $\tan(t + \pi) = \tan t$. The same observations apply to the cotangen
function.

Periodic Properties of the Tangent and Cotangent Functions

$$\tan(t + \pi) = \tan t \quad \text{and} \quad \cot(t + \pi) = \cot t$$

The tangent and cotangent functions are periodic functions and have period π.

EXAMPLE 7 Using Periodic Properties

Find the value of each trigonometric function:

a. $\sin \dfrac{9\pi}{4}$ **b.** $\tan\left(-\dfrac{5\pi}{4}\right)$.

Solution

a. $\sin \dfrac{9\pi}{4} = \sin\left(\dfrac{\pi}{4} + 2\pi\right) = \sin \dfrac{\pi}{4} = \dfrac{\sqrt{2}}{2}$

        $\sin(t + 2\pi) = \sin t$

b. $\tan\left(-\dfrac{5\pi}{4}\right) = -\tan \dfrac{5\pi}{4} = -\tan\left(\dfrac{\pi}{4} + \pi\right) = -\tan \dfrac{\pi}{4} = -1$

     The tangent function is $\tan(t + \pi) = \tan t$
     odd: $\tan(-t) = -\tan t$.

Check Point 7   Find the value of each trigonometric function:

a. $\cot \dfrac{5\pi}{4}$ **b.** $\cos\left(-\dfrac{9\pi}{4}\right)$.

Why do the trigonometric functions model phenomena that repeat *indefinitely*
By starting at point P on the unit circle and traveling a distance of 2π units, 4π unit
6π units, and so on, we return to the starting point P. Because the trigonometri
functions are defined in terms of the coordinates of that point P, if we add (or sub
tract) multiples of 2π to t, the values of the trigonometric functions of t do no

change. Furthermore, the values for the tangent and cotangent functions of t do not change if we add (or subtract) multiples of π to t.

> ### Repetitive Behavior of the Sine, Cosine, and Tangent Functions
> For any integer n and real number t,
>
> $$\sin(t + 2\pi n) = \sin t, \quad \cos(t + 2\pi n) = \cos t, \quad \text{and} \quad \tan(t + \pi n) = \tan t.$$

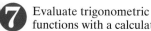

Evaluate trigonometric functions with a calculator.

Using a Calculator to Evaluate Trigonometric Functions

We used a unit circle to find values of the trigonometric functions at $\dfrac{\pi}{4}$. These are exact values. We can find approximate values of the trigonometric functions using a calculator.

The first step in using a calculator to evaluate trigonometric functions is to set the calculator to the correct *mode*, degrees or radians. The domains of the trigonometric functions in the unit circle are sets of real numbers. Therefore, we use the radian mode.

Most calculators have keys marked $\boxed{\text{SIN}}$, $\boxed{\text{COS}}$, and $\boxed{\text{TAN}}$. For example, to find the value of sin 1.2, set the calculator to the radian mode and enter 1.2 $\boxed{\text{SIN}}$ on most scientific calculators and $\boxed{\text{SIN}}$ 1.2 $\boxed{\text{ENTER}}$ on most graphing calculators. Consult the manual for your calculator.

To evaluate the cosecant, secant, and cotangent functions, use the key for the respective reciprocal function, $\boxed{\text{SIN}}$, $\boxed{\text{COS}}$, or $\boxed{\text{TAN}}$, and then use the reciprocal key. The reciprocal key is $\boxed{1/x}$ on many scientific calculators and $\boxed{x^{-1}}$ on many graphing calculators. For example, we can evaluate $\sec\dfrac{\pi}{12}$ using the following reciprocal relationship:

$$\sec\frac{\pi}{12} = \frac{1}{\cos\dfrac{\pi}{12}}.$$

Using the radian mode, enter one of the following keystroke sequences:

Many Scientific Calculators

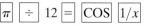

Many Graphing Calculators

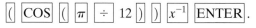

Rounding the display to four decimal places, we obtain $\sec\dfrac{\pi}{12} \approx 1.0353$.

EXAMPLE 8   Evaluating Trigonometric Functions with a Calculator

Use a calculator to find the value to four decimal places:

   a. $\cos\dfrac{\pi}{4}$ **b.** $\cot 1.2$.

Solution

Scientific Calculator Solution

Function	Mode	Keystrokes	Display, rounded to four decimal places
a. $\cos\dfrac{\pi}{4}$	Radian	$\boxed{\pi}$ $\boxed{\div}$ $\boxed{4}$ $\boxed{=}$ $\boxed{\text{COS}}$	0.7071
b. $\cot 1.2$	Radian	1.2 $\boxed{\text{TAN}}$ $\boxed{1/x}$	0.3888

Graphing Calculator Solution

Function	Mode	Keystrokes	Display, rounded to four decimal places
a. $\cos\dfrac{\pi}{4}$	Radian	COS (π ÷ 4) ENTER	0.7071
b. $\cot 1.2$	Radian	(TAN 1.2) x^{-1} ENTER	0.3888

Check Point 8 Use a calculator to find the value to four decimal places:

a. $\sin\dfrac{\pi}{4}$ **b.** $\csc 1.5$.

EXERCISE SET 4.2

Practice Exercises

In Exercises 1–4, a point $P(x, y)$ is shown on the unit circle corresponding to a real number t. Find the values of the trigonometric functions at t.

1.

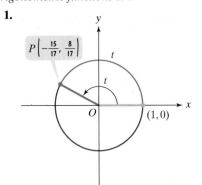

$P\left(-\dfrac{15}{17}, \dfrac{8}{17}\right)$

2.

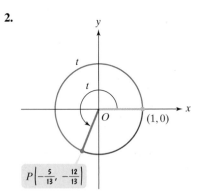

$P\left(-\dfrac{5}{13}, -\dfrac{12}{13}\right)$

3.

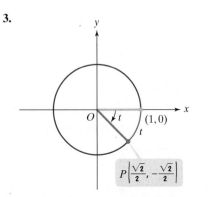

$P\left(\dfrac{\sqrt{2}}{2}, -\dfrac{\sqrt{2}}{2}\right)$

4.

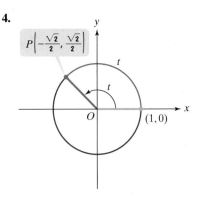

$P\left(-\dfrac{\sqrt{2}}{2}, \dfrac{\sqrt{2}}{2}\right)$

In Exercises 5–18, the unit circle has been divided into twelve equal arcs, corresponding to t-values of

$$0, \frac{\pi}{6}, \frac{\pi}{3}, \frac{\pi}{2}, \frac{2\pi}{3}, \frac{5\pi}{6}, \pi, \frac{7\pi}{6}, \frac{4\pi}{3}, \frac{3\pi}{2}, \frac{5\pi}{3}, \frac{11\pi}{6}, \text{ and } 2\pi.$$

Use the (x, y) coordinates in the figure to find the value of each trigonometric function at the indicated real number, t, or state that the expression is undefined.

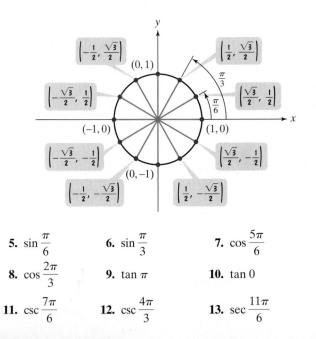

5. $\sin\dfrac{\pi}{6}$ **6.** $\sin\dfrac{\pi}{3}$ **7.** $\cos\dfrac{5\pi}{6}$

8. $\cos\dfrac{2\pi}{3}$ **9.** $\tan\pi$ **10.** $\tan 0$

11. $\csc\dfrac{7\pi}{6}$ **12.** $\csc\dfrac{4\pi}{3}$ **13.** $\sec\dfrac{11\pi}{6}$

14. $\sec \dfrac{5\pi}{3}$ **15.** $\sin \dfrac{3\pi}{2}$ **16.** $\cos \dfrac{3\pi}{2}$

17. $\sec \dfrac{3\pi}{2}$ **18.** $\tan \dfrac{3\pi}{2}$

In Exercises 19–24,

  a. *Use the unit circle shown for Exercises 5–18 to find the value of the trigonometric function.*

  b. *Use even and odd properties of trigonometric functions and your answer from part (a) to find the value of the same trigonometric function at the indicated real number.*

19. a. $\cos \dfrac{\pi}{6}$ **20. a.** $\cos \dfrac{\pi}{3}$

  b. $\cos\left(-\dfrac{\pi}{6}\right)$ **b.** $\cos\left(-\dfrac{\pi}{3}\right)$

21. a. $\sin \dfrac{5\pi}{6}$ **22. a.** $\sin \dfrac{2\pi}{3}$

  b. $\sin\left(-\dfrac{5\pi}{6}\right)$ **b.** $\sin\left(-\dfrac{2\pi}{3}\right)$

23. a. $\tan \dfrac{5\pi}{3}$ **24. a.** $\tan \dfrac{11\pi}{6}$

  b. $\tan\left(-\dfrac{5\pi}{3}\right)$ **b.** $\tan\left(-\dfrac{11\pi}{6}\right)$

In Exercises 25–28, $\sin t$ and $\cos t$ are given. Use identities to find $\tan t$, $\csc t$, $\sec t$, and $\cot t$. Where necessary, rationalize denominators.

25. $\sin t = \dfrac{8}{17}, \cos t = \dfrac{15}{17}$ **26.** $\sin t = \dfrac{3}{5}, \cos t = \dfrac{4}{5}$

27. $\sin t = \dfrac{1}{3}, \cos t = \dfrac{2\sqrt{2}}{3}$ **28.** $\sin t = \dfrac{2}{3}, \cos t = \dfrac{\sqrt{5}}{3}$

In Exercises 29–32, $0 \le t < \dfrac{\pi}{2}$ and $\sin t$ is given. Use the Pythagorean identity $\sin^2 t + \cos^2 t = 1$ to find $\cos t$.

29. $\sin t = \dfrac{6}{7}$ **30.** $\sin t = \dfrac{7}{8}$

31. $\sin t = \dfrac{\sqrt{39}}{8}$ **32.** $\sin t = \dfrac{\sqrt{21}}{5}$

In Exercises 33–38, use an identity to find the value of each expression. Do not use a calculator.

33. $\sin 1.7 \csc 1.7$ **34.** $\cos 2.3 \sec 2.3$

35. $\sin^2 \dfrac{\pi}{6} + \cos^2 \dfrac{\pi}{6}$ **36.** $\sin^2 \dfrac{\pi}{3} + \cos^2 \dfrac{\pi}{3}$

37. $\sec^2 \dfrac{\pi}{3} - \tan^2 \dfrac{\pi}{3}$ **38.** $\csc^2 \dfrac{\pi}{6} - \cot^2 \dfrac{\pi}{6}$

In Exercises 39–52, find the exact value of each trigonometric function. Do not use a calculator.

39. $\cos \dfrac{9\pi}{4}$ **40.** $\csc \dfrac{9\pi}{4}$

41. $\sin\left(-\dfrac{9\pi}{4}\right)$ **42.** $\sec\left(-\dfrac{9\pi}{4}\right)$

43. $\tan \dfrac{5\pi}{4}$ **44.** $\cot \dfrac{5\pi}{4}$

45. $\cot\left(-\dfrac{5\pi}{4}\right)$ **46.** $\tan\left(-\dfrac{9\pi}{4}\right)$

47. $-\tan\left(\dfrac{\pi}{4} + 15\pi\right)$ **48.** $-\cot\left(\dfrac{\pi}{4} + 17\pi\right)$

49. $\sin\left(-\dfrac{\pi}{4} - 1000\pi\right)$ **50.** $\sin\left(-\dfrac{\pi}{4} - 2000\pi\right)$

51. $\cos\left(-\dfrac{\pi}{4} - 1000\pi\right)$ **52.** $\cos\left(-\dfrac{\pi}{4} - 2000\pi\right)$

In Exercises 53–60, the unit circle has been divided into eight equal arcs, corresponding to t-values of

$$0, \dfrac{\pi}{4}, \dfrac{\pi}{2}, \dfrac{3\pi}{4}, \pi, \dfrac{5\pi}{4}, \dfrac{3\pi}{2}, \dfrac{7\pi}{4}, \text{ and } 2\pi.$$

  a. *Use the (x, y) coordinates in the figure to find the value of the trigonometric function.*

  b. *Use periodic properties and your answer from part (a) to find the value of the same trigonometric function at the indicated real number.*

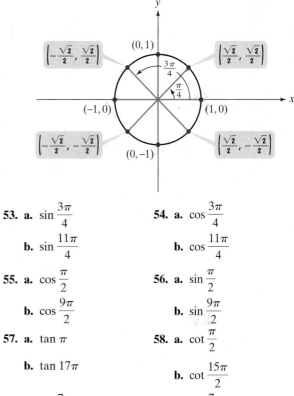

53. a. $\sin \dfrac{3\pi}{4}$ **54. a.** $\cos \dfrac{3\pi}{4}$

  b. $\sin \dfrac{11\pi}{4}$ **b.** $\cos \dfrac{11\pi}{4}$

55. a. $\cos \dfrac{\pi}{2}$ **56. a.** $\sin \dfrac{\pi}{2}$

  b. $\cos \dfrac{9\pi}{2}$ **b.** $\sin \dfrac{9\pi}{2}$

57. a. $\tan \pi$ **58. a.** $\cot \dfrac{\pi}{2}$

  b. $\tan 17\pi$ **b.** $\cot \dfrac{15\pi}{2}$

59. a. $\sin \dfrac{7\pi}{4}$ **60. a.** $\cos \dfrac{7\pi}{4}$

  b. $\sin \dfrac{47\pi}{4}$ **b.** $\cos \dfrac{47\pi}{4}$

In Exercises 61–70, use a calculator to find the value of the trigonometric function to four decimal places.

61. $\sin 0.8$ **62.** $\cos 0.6$

63. $\tan 3.4$ **64.** $\tan 3.7$

65. $\csc 1$ **66.** $\sec 1$

67. $\cos \dfrac{\pi}{10}$ **68.** $\sin \dfrac{3\pi}{10}$

69. $\cot \dfrac{\pi}{12}$ **70.** $\cot \dfrac{\pi}{18}$

Practice Plus

In Exercises 71–80, let

$$\sin t = a, \ \cos t = b, \ \text{and} \ \tan t = c.$$

Write each expression in terms of a, b, and c.

71. $\sin(-t) - \sin t$

72. $\tan(-t) - \tan t$

73. $4\cos(-t) - \cos t$

74. $3\cos(-t) - \cos t$

75. $\sin(t + 2\pi) - \cos(t + 4\pi) + \tan(t + \pi)$

76. $\sin(t + 2\pi) + \cos(t + 4\pi) - \tan(t + \pi)$

77. $\sin(-t - 2\pi) - \cos(-t - 4\pi) - \tan(-t - \pi)$

78. $\sin(-t - 2\pi) + \cos(-t - 4\pi) - \tan(-t - \pi)$

79. $\cos t + \cos(t + 1000\pi) - \tan t - \tan(t + 999\pi) - \sin t + 4\sin(t - 1000\pi)$

80. $-\cos t + 7\cos(t + 1000\pi) + \tan t + \tan(t + 999\pi) + \sin t + \sin(t - 1000\pi)$

Application Exercises

81. The number of hours of daylight, H, on day t of any given year (on January 1, $t = 1$) in Fairbanks, Alaska, can be modeled by the function

$$H(t) = 12 + 8.3\sin\left[\frac{2\pi}{365}(t - 80)\right].$$

a. March 21, the 80th day of the year, is the spring equinox. Find the number of hours of daylight in Fairbanks on this day.

b. June 21, the 172nd day of the year, is the summer solstice, the day with the maximum number of hours of daylight. To the nearest tenth of an hour, find the number of hours of daylight in Fairbanks on this day.

c. December 21, the 355th day of the year, is the winter solstice, the day with the minimum number of hours of daylight. Find, to the nearest tenth of an hour, the number of hours of daylight in Fairbanks on this day.

82. The number of hours of daylight, H, on day t of any given year (on January 1, $t = 1$) in San Diego, California, can be modeled by the function

$$H(t) = 12 + 2.4\sin\left[\frac{2\pi}{365}(t - 80)\right].$$

a. March 21, the 80th day of the year, is the spring equinox. Find the number of hours of daylight in San Diego on this day.

b. June 21, the 172nd day of the year, is the summer solstice, the day with the maximum number of hours of daylight. Find, to the nearest tenth of an hour, the number of hours of daylight in San Diego on this day.

c. December 21, the 355th day of the year, is the winter solstice, the day with the minimum number of hours of daylight. To the nearest tenth of an hour, find the number of hours of daylight in San Diego on this day.

83. People who believe in biorhythms claim that there are three cycles that rule our behavior—the physical, emotional, and mental. Each is a sine function of a certain period. The function for our emotional fluctuations is

$$E = \sin\frac{\pi}{14}t,$$

where t is measured in days starting at birth. Emotional fluctuations, E, are measured from -1 to 1, inclusive, with 1 representing peak emotional well-being, -1 representing the low for emotional well-being, and 0 representing feelin neither emotionally high nor low.

a. Find E corresponding to $t = 7, 14, 21, 28,$ and 35. Describ what you observe.

b. What is the period of the emotional cycle?

84. The height of the water, H, in feet, at a boat dock t hours afte 6 A.M. is given by

$$H = 10 + 4\sin\frac{\pi}{6}t.$$

a. Find the height of the water at the dock at 6 A.M., 9 A.M noon, 6 P.M., midnight, and 3 A.M.

b. When is low tide and when is high tide?

c. What is the period of this function and what does thi mean about the tides?

Writing in Mathematics

85. Why are the trigonometric functions sometimes calle circular functions?

86. Define the sine of t.

87. Given a point on the unit circle that corresponds to t, explai how to find $\tan t$.

88. What is the range of the sine function? Use the unit circle t explain where this range comes from.

89. Explain how to use the unit circle to find values of th trigonometric functions at $\frac{\pi}{4}$.

90. What do we mean by even trigonometric functions? Which o the six functions fall into this category?

91. Use words (not an equation) to describe one of th reciprocal identities.

92. Use words (not an equation) to describe one of the quotien identities.

93. Use words (not an equation) to describe one of th Pythagorean identities

94. What is a periodic function? Why are the sine and cosin functions periodic?

95. Explain how you can use the function for emotiona fluctuations in Exercise 83 to determine good days for havin dinner with your moody boss.

96. Describe a phenomenon that repeats infinitely. What is it period?

Critical Thinking Exercises

97. If $\pi < t < \frac{3\pi}{2}$, which of the following is true?

a. $\sin t > 0$ and $\tan t > 0$.

b. $\sin t < 0$ and $\tan t < 0$.

c. $\tan t > 0$ and $\cot t > 0$.

d. $\tan t < 0$ and $\cot t < 0$.

98. If $f(x) = \sin x$ and $f(a) = \frac{1}{4}$, find the value of

$$f(a) + f(a + 2\pi) + f(a + 4\pi) + f(a + 6\pi).$$

99. If $f(x) = \sin x$ and $f(a) = \frac{1}{4}$, find the value of $f(a) + 2f(-a)$.

100. The seats of a Ferris wheel are 40 feet from the wheel's cen ter. When you get on the ride, your seat is 5 feet above th ground. How far above the ground are you after rotatin through an angle of $\frac{17\pi}{4}$ radians? Round to the nearest foo

Right Triangle Trigonometry

Objectives

❶ Use right triangles to evaluate trigonometric functions.

❷ Find function values for $30° \left(\dfrac{\pi}{6}\right)$, $45° \left(\dfrac{\pi}{4}\right)$, and $60° \left(\dfrac{\pi}{3}\right)$.

❸ Use equal cofunctions of complements.

❹ Use right triangle trigonometry to solve applied problems.

In the last century, Ang Rita Sherpa climbed Mount Everest ten times, all without the use of bottled oxygen.

Mountain climbers have forever been fascinated by reaching the top of Mount Everest, sometimes with tragic results. The mountain, on Asia's Tibet-Nepal border, is Earth's highest, peaking at an incredible 29,035 feet. The heights of mountains can be found using trigonometric functions. Remember that the word *trigonometry* means *measurement of triangles*. Trigonometry is used in navigation, building, and engineering. For centuries, Muslims used trigonometry and the stars to navigate across the Arabian desert to Mecca, the birthplace of the prophet Muhammad, the founder of Islam. The ancient Greeks used trigonometry to record the locations of thousands of stars and worked out the motion of the Moon relative to Earth. Today, trigonometry is used to study the structure of DNA, the master molecule that determines how we grow from a single cell to a complex, fully developed adult.

❶ Use right triangles to evaluate trigonometric functions.

Right Triangle Definitions of Trigonometric Functions

We have seen that in a unit circle, the radian measure of a central angle is equal to the measure of the intercepted arc. Thus, the value of a trigonometric function at the real number t is its value at an angle of t radians

Figure 4.29(a) shows a central angle that measures $\dfrac{\pi}{3}$ radians and an intercepted arc of length $\dfrac{\pi}{3}$. Interpret $\dfrac{\pi}{3}$ as the measure of the central angle. In Figure 4.29(b), we construct a right triangle by dropping a line segment from point P perpendicular to the x-axis.

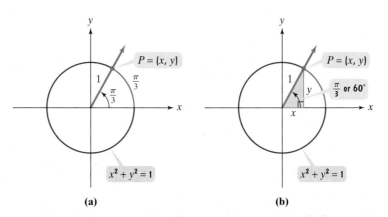

Figure 4.29 Interpreting trigonometric functions using a unit circle and a right triangle

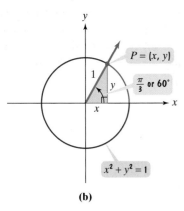

(b)

Figure 4.29(b) (repeated)

Now we can think of $\frac{\pi}{3}$, or 60°, as the measure of an acute angle in the right triangle in Figure 4.29(b). Because sin t is the second coordinate of point P and cos is the first coordinate of point P, we see that

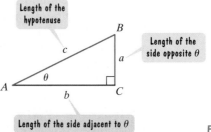

$$\sin\frac{\pi}{3} = \sin 60° = y = \frac{y}{1}$$

> This is the length of the side opposite the 60° angle in the right triangle.
> This is the length of the hypotenuse in the right triangle.

$$\cos\frac{\pi}{3} = \cos 60° = x = \frac{x}{1}.$$

> This is the length of the side adjacent to the 60° angle in the right triangle.
> This is the length of the hypotenuse in the right triangle.

In solving certain kinds of problems, it is helpful to interpret trigonometric functions in right triangles, where angles are limited to acute angles. Figure 4.30 shows a right triangle with one of its acute angles labeled θ. The side opposite the right angle, the hypotenuse, has length c. The other sides of the triangle are described by their position relative to the acute angle θ. One side is opposite θ. The length of this side is a. One side is adjacent to θ. The length of this side is b.

Length of the hypotenuse

Length of the side opposite θ

Length of the side adjacent to θ

Figure 4.30

Right Triangle Definitions of Trigonometric Functions

See Figure 4.30. The six **trigonometric functions of the acute angle θ** are defined as follows:

$$\sin\theta = \frac{\text{length of side opposite angle }\theta}{\text{length of hypotenuse}} = \frac{a}{c}$$

$$\csc\theta = \frac{\text{length of hypotenuse}}{\text{length of side opposite angle }\theta} = \frac{c}{a}$$

$$\cos\theta = \frac{\text{length of side adjacent to angle }\theta}{\text{length of hypotenuse}} = \frac{b}{c}$$

$$\sec\theta = \frac{\text{length of hypotenuse}}{\text{length of side adjacent to angle }\theta} = \frac{c}{b}$$

$$\tan\theta = \frac{\text{length of side opposite angle }\theta}{\text{length of side adjacent to angle }\theta} = \frac{a}{b}$$

$$\cot\theta = \frac{\text{length of side adjacent to angle }\theta}{\text{length of side opposite angle }\theta} = \frac{b}{a}$$

Each of the trigonometric functions of the acute angle θ is positive. Observe that the ratios in the second column in the box are the reciprocals of the corresponding ratios in the first column.

Study Tip

The word

SOHCAHTOA (pronounced: so-cah-tow-ah)

is a way to remember the right triangle definitions of the three basic trigonometric functions, sine, cosine, and tangent.

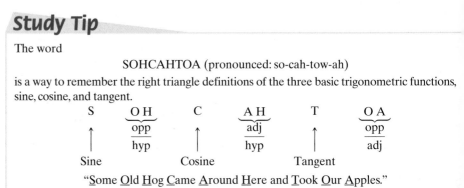

"Some Old Hog Came Around Here and Took Our Apples."

Figure 4.31 shows four right triangles of varying sizes. In each of the triangles, θ is the same acute angle, measuring approximately 56.3°. All four of these similar triangles have the same shape and the lengths of corresponding sides are in the same ratio. In each triangle, the tangent function has the same value for the angle θ: $\tan \theta = \frac{3}{2}$.

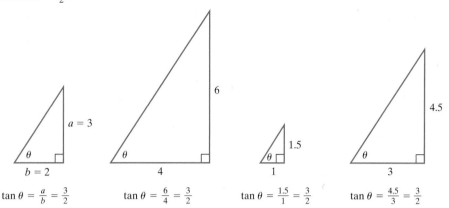

$$\tan \theta = \frac{a}{b} = \frac{3}{2} \qquad \tan \theta = \frac{6}{4} = \frac{3}{2} \qquad \tan \theta = \frac{1.5}{1} = \frac{3}{2} \qquad \tan \theta = \frac{4.5}{3} = \frac{3}{2}$$

Figure 4.31 A particular acute angle always gives the same ratio of opposite to adjacent sides.

In general, **the trigonometric function values of θ depend only on the size of angle θ and not on the size of the triangle.**

EXAMPLE 1   Evaluating Trigonometric Functions

Find the value of each of the six trigonometric functions of θ in Figure 4.32.

Figure 4.32

Solution We need to find the values of the six trigonometric functions of θ. However, we must know the lengths of all three sides of the triangle (a, b, and c) to evaluate all six functions. The values of a and b are given. We can use the Pythagorean Theorem, $c^2 = a^2 + b^2$, to find c.

$$\boxed{a = 5} \quad \boxed{b = 12}$$

$$c^2 = a^2 + b^2 = 5^2 + 12^2 = 25 + 144 = 169$$

$$c = \sqrt{169} = 13$$

Now that we know the lengths of the three sides of the triangle, we apply the definitions of the six trigonometric functions of θ. Referring to these lengths as opposite, adjacent, and hypotenuse, we have

$$\sin \theta = \frac{\text{opposite}}{\text{hypotenuse}} = \frac{5}{13} \qquad \csc \theta = \frac{\text{hypotenuse}}{\text{opposite}} = \frac{13}{5}$$

$$\cos \theta = \frac{\text{adjacent}}{\text{hypotenuse}} = \frac{12}{13} \qquad \sec \theta = \frac{\text{hypotenuse}}{\text{adjacent}} = \frac{13}{12}$$

$$\tan \theta = \frac{\text{opposite}}{\text{adjacent}} = \frac{5}{12} \qquad \cot \theta = \frac{\text{adjacent}}{\text{opposite}} = \frac{12}{5}.$$

Study Tip

The function values in the second column are reciprocals of those in the first column. You can obtain these values by exchanging the numerator and denominator of the corresponding ratios in the first column.

Check Point 1 Find the value of each of the six trigonometric functions of θ in the figure.

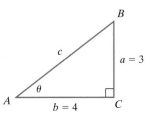

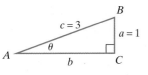

Figure 4.33

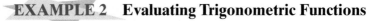

EXAMPLE 2   Evaluating Trigonometric Functions

Find the value of each of the six trigonometric functions of θ in Figure 4.33.

Solution   We begin by finding b.

$$a^2 + b^2 = c^2 \qquad \text{Use the Pythagorean Theorem.}$$

$$1^2 + b^2 = 3^2 \qquad \text{Figure 4.33 shows that } a = 1 \text{ and } c = 3.$$

$$1 + b^2 = 9 \qquad 1^2 = 1 \text{ and } 3^2 = 9.$$

$$b^2 = 8 \qquad \text{Subtract 1 from both sides.}$$

$$b = \sqrt{8} = 2\sqrt{2} \qquad \text{Take the principal square root and simplify:}$$
$$\sqrt{8} = \sqrt{4 \cdot 2} = \sqrt{4}\sqrt{2} = 2\sqrt{2}.$$

Now that we know the lengths of the three sides of the triangle, we apply the definitions of the six trigonometric functions of θ.

$$\sin \theta = \frac{\text{opposite}}{\text{hypotenuse}} = \frac{1}{3} \qquad\qquad \csc \theta = \frac{\text{hypotenuse}}{\text{opposite}} = \frac{3}{1} = 3$$

$$\cos \theta = \frac{\text{adjacent}}{\text{hypotenuse}} = \frac{2\sqrt{2}}{3} \qquad\qquad \sec \theta = \frac{\text{hypotenuse}}{\text{adjacent}} = \frac{3}{2\sqrt{2}}$$

$$\tan \theta = \frac{\text{opposite}}{\text{adjacent}} = \frac{1}{2\sqrt{2}} \qquad\qquad \cot \theta = \frac{\text{adjacent}}{\text{opposite}} = \frac{2\sqrt{2}}{1} = 2\sqrt{2}$$

We can simplify the values of $\tan \theta$ and $\sec \theta$ by rationalizing the denominators:

$$\tan \theta = \frac{1}{2\sqrt{2}} = \frac{1}{2\sqrt{2}} \cdot \frac{\sqrt{2}}{\sqrt{2}} = \frac{\sqrt{2}}{2 \cdot 2} = \frac{\sqrt{2}}{4} \qquad\qquad \sec \theta = \frac{3}{2\sqrt{2}} = \frac{3}{2\sqrt{2}} \cdot \frac{\sqrt{2}}{\sqrt{2}} = \frac{3\sqrt{2}}{2 \cdot 2} = \frac{3\sqrt{2}}{4}.$$

> We are multiplying by 1 and not changing the value of $\dfrac{1}{2\sqrt{2}}$.

> We are multiplying by 1 and not changing the value of $\dfrac{3}{2\sqrt{2}}$.

Check Point 2   Find the value of each of the six trigonometric functions of θ in the figure. Express each value in simplified form.

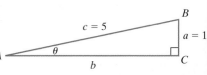

② Find function values for $30° \left(\dfrac{\pi}{6} \right)$, $45° \left(\dfrac{\pi}{4} \right)$, and $60° \left(\dfrac{\pi}{3} \right)$.

Function Values for Some Special Angles

In Section 4.2, we used the unit circle to find values of the trigonometric functions at $\dfrac{\pi}{4}$.

How can we find the values of the trigonometric functions at $\dfrac{\pi}{4}$, or 45°, using a right triangle? We construct a right triangle with a 45° angle, as shown in Figure 4.34 at the top of the next page. The triangle actually has two 45° angles. Thus, the triangle is isosceles—that is, it has two sides of the same length. Assume that each leg of the triangle has a length equal to 1. We can find the length of the hypotenuse using the Pythagorean Theorem.

$$(\text{length of hypotenuse})^2 = 1^2 + 1^2 = 2$$

$$\text{length of hypotenuse} = \sqrt{2}$$

With Figure 4.34, we can determine the trigonometric function values for 45°.

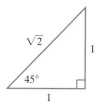

Figure 4.34 An isosceles right triangle

EXAMPLE 3 Evaluating Trigonometric Functions of 45°

Use Figure 4.34 to find sin 45°, cos 45°, and tan 45°.

Solution We apply the definitions of these three trigonometric functions. Where appropriate, we simplify by rationalizing denominators.

$$\sin 45° = \frac{\text{length of side opposite } 45°}{\text{length of hypotenuse}} = \frac{1}{\sqrt{2}} = \frac{1}{\sqrt{2}} \cdot \frac{\sqrt{2}}{\sqrt{2}} = \frac{\sqrt{2}}{2}$$

Rationalize denominators

$$\cos 45° = \frac{\text{length of side adjacent to } 45°}{\text{length of hypotenuse}} = \frac{1}{\sqrt{2}} = \frac{1}{\sqrt{2}} \cdot \frac{\sqrt{2}}{\sqrt{2}} = \frac{\sqrt{2}}{2}$$

$$\tan 45° = \frac{\text{length of side opposite } 45°}{\text{length of side adjacent to } 45°} = \frac{1}{1} = 1$$

Check Point 3 Use Figure 4.34 to find csc 45°, sec 45°, and cot 45°.

When you worked Check Point 3, did you actually use Figure 4.34 or did you use reciprocals to find the values?

$$\csc 45° = \sqrt{2} \qquad \sec 45° = \sqrt{2} \qquad \cot 45° = 1$$

Take the reciprocal of sin 45° = $\frac{1}{\sqrt{2}}$.   Take the reciprocal of cos 45° = $\frac{1}{\sqrt{2}}$.   Take the reciprocal of tan 45° = $\frac{1}{1}$.

Notice that if you use reciprocals, you should take the reciprocal of a function value before the denominator is rationalized. In this way, the reciprocal value will not contain a radical in the denominator.

Two other angles that occur frequently in trigonometry are 30°, or $\frac{\pi}{6}$ radian, and 60°, or $\frac{\pi}{3}$ radian, angles. We can find the values of the trigonometric functions of 30° and 60° by using a right triangle. To form this right triangle, draw an equilateral triangle—that is a triangle with all sides the same length. Assume that each side has a length equal to 2. Now take half of the equilateral triangle. We obtain the right triangle in Figure 4.35. This right triangle has a hypotenuse of length 2 and a leg of length 1. The other leg has length a, which can be found using the Pythagorean Theorem.

$$a^2 + 1^2 = 2^2$$
$$a^2 + 1 = 4$$
$$a^2 = 3$$
$$a = \sqrt{3}$$

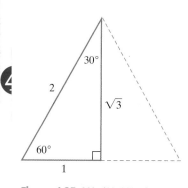

Figure 4.35 30°–60°–90° triangle

With the right triangle in Figure 4.35, we can determine the trigonometric functions for 30° and 60°.

EXAMPLE 4 Evaluating Trigonometric Functions of 30° and 60°

Use Figure 4.35 to find sin 60°, cos 60°, sin 30°, and cos 30°.

Solution We begin with 60°. Use the angle on the lower left in Figure 4.35.

$$\sin 60° = \frac{\text{length of side opposite } 60°}{\text{length of hypotenuse}} = \frac{\sqrt{3}}{2}$$

$$\cos 60° = \frac{\text{length of side adjacent to } 60°}{\text{length of hypotenuse}} = \frac{1}{2}$$

The Mountain Man

In the 1930s, a *National Geographic* team headed by Brad Washburn used trigonometry to create a map of the 5000-square-mile region of the Yukon, near the Canadian border. The team started with aerial photography. By drawing a network of angles on the photographs, the approximate locations of the major mountains and their rough heights were determined. The expedition then spent three months on foot to find the exact heights. Team members established two base points a known distance apart, one directly under the mountain's peak. By measuring the angle of elevation from one of the base points to the peak, the tangent function was used to determine the peak's height. The Yukon expedition was a major advance in the way maps are made.

EXERCISE SET 4.3

Practice Exercises

In Exercises 1–8, use the Pythagorean Theorem to find the length of the missing side of each right triangle. Then find the value of each of the six trigonometric functions of θ.

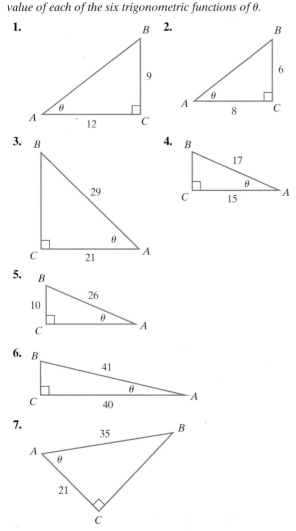

8.

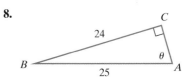

In Exercises 9–20, use the given triangles to evaluate each expression. If necessary, express the value without a square root in the denominator by rationalizing the denominator.

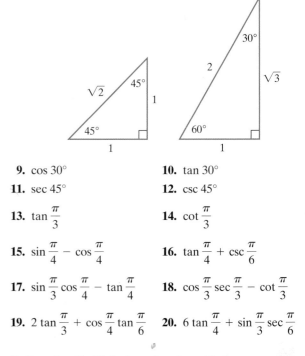

9. $\cos 30°$　　　　　　　**10.** $\tan 30°$

11. $\sec 45°$　　　　　　　**12.** $\csc 45°$

13. $\tan \dfrac{\pi}{3}$　　　　　　　**14.** $\cot \dfrac{\pi}{3}$

15. $\sin \dfrac{\pi}{4} - \cos \dfrac{\pi}{4}$　　　**16.** $\tan \dfrac{\pi}{4} + \csc \dfrac{\pi}{6}$

17. $\sin \dfrac{\pi}{3} \cos \dfrac{\pi}{4} - \tan \dfrac{\pi}{4}$　　**18.** $\cos \dfrac{\pi}{3} \sec \dfrac{\pi}{3} - \cot \dfrac{\pi}{3}$

19. $2 \tan \dfrac{\pi}{3} + \cos \dfrac{\pi}{4} \tan \dfrac{\pi}{6}$　　**20.** $6 \tan \dfrac{\pi}{4} + \sin \dfrac{\pi}{3} \sec \dfrac{\pi}{6}$

In Exercises 21–28, find a cofunction with the same value as the given expression.

21. $\sin 7°$　　　　　　　**22.** $\sin 19°$

23. $\csc 25°$　　　　　　　**24.** $\csc 35°$

25. $\tan \dfrac{\pi}{9}$　　　　　　　**26.** $\tan \dfrac{\pi}{7}$

27. $\cos \dfrac{2\pi}{5}$

28. $\cos \dfrac{3\pi}{8}$

In Exercises 29–34, find the measure of the side of the right triangle whose length is designated by a lowercase letter. Round answers to the nearest whole number.

29.

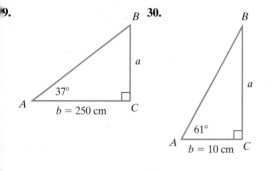

30.

31.

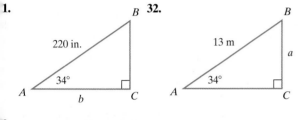

32.

33.

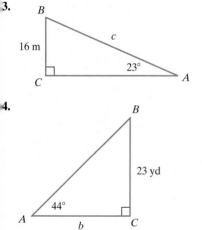

34.

In Exercises 35–38, use a calculator to find the value of the acute angle θ to the nearest degree.

35. $\sin \theta = 0.2974$

36. $\cos \theta = 0.8771$

37. $\tan \theta = 4.6252$

38. $\tan \theta = 26.0307$

In Exercises 39–42, use a calculator to find the value of the acute angle θ in radians, rounded to three decimal places.

39. $\cos \theta = 0.4112$

40. $\sin \theta = 0.9499$

41. $\tan \theta = 0.4169$

42. $\tan \theta = 0.5117$

Practice Plus

In Exercises 43–48, find the exact value of each expression. Do not use a calculator.

43. $\dfrac{\tan \dfrac{\pi}{3}}{2} - \dfrac{1}{\sec \dfrac{\pi}{6}}$

44. $\dfrac{1}{\cot \dfrac{\pi}{4}} - \dfrac{2}{\csc \dfrac{\pi}{6}}$

45. $1 + \sin^2 40° + \sin^2 50°$

46. $1 - \tan^2 10° + \csc^2 80°$

47. $\csc 37° \sec 53° - \tan 53° \cot 37°$

48. $\cos 12° \sin 78° + \cos 78° \sin 12°$

In Exercises 49–50, express each exact value as a single fraction. Do not use a calculator.

49. If $f(\theta) = 2 \cos \theta - \cos 2\theta$, find $f\left(\dfrac{\pi}{6}\right)$.

50. If $f(\theta) = 2 \sin \theta - \sin \dfrac{\theta}{2}$, find $f\left(\dfrac{\pi}{3}\right)$.

51. If θ is an acute angle and $\cot \theta = \dfrac{1}{4}$, find $\tan\left(\dfrac{\pi}{2} - \theta\right)$.

52. If θ is an acute angle and $\cos \theta = \dfrac{1}{3}$, find $\csc\left(\dfrac{\pi}{2} - \theta\right)$.

⭐ Application Exercises

53. To find the distance across a lake, a surveyor took the measurements shown in the figure. Use these measurements to determine how far it is across the lake. Round to the nearest yard.

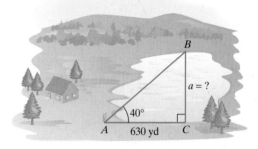

54. At a certain time of day, the angle of elevation of the sun is 40°. To the nearest foot, find the height of a tree whose shadow is 35 feet long.

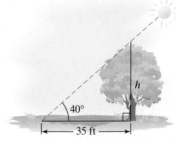

55. A tower that is 125 feet tall casts a shadow 172 feet long. Find the angle of elevation of the sun to the nearest degree.

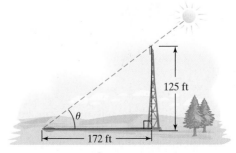

56. The Washington Monument is 555 feet high. If you stand one quarter of a mile, or 1320 feet, from the base of the monument and look to the top, find the angle of elevation to the nearest degree.

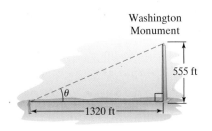

Washington
Monument

555 ft

θ

1320 ft

57. A plane rises from take-off and flies at an angle of 10° with the horizontal runway. When it has gained 500 feet, find the distance, to the nearest foot, the plane has flown.

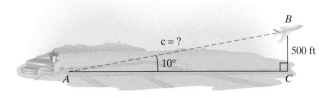

B

c = ?

500 ft

10°

A   C

58. A road is inclined at an angle of 5°. After driving 5000 feet along this road, find the driver's increase in altitude. Round to the nearest foot.

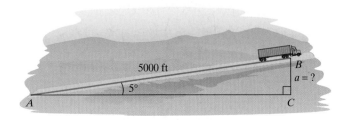

5000 ft

B
a = ?

5°

A   C

59. A telephone pole is 60 feet tall. A guy wire 75 feet long is attached from the ground to the top of the pole. Find the angle between the wire and the pole to the nearest degree.

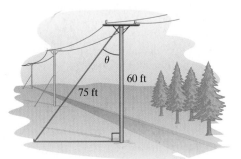

θ

60 ft

75 ft

60. A telephone pole is 55 feet tall. A guy wire 80 feet long is attached from the ground to the top of the pole. Find the angle between the wire and the pole to the nearest degree.

Writing in Mathematics

61. If you are given the lengths of the sides of a right triangle, describe how to find the sine of either acute angle.

62. Describe one similarity and one difference between the definitions of sin θ and cos θ, where θ is an acute angle of a right triangle.

63. Describe the triangle used to find the trigonometric functions of 45°.

64. Describe the triangle used to find the trigonometric functions of 30° and 60°.

65. Describe a relationship among trigonometric functions that is based on angles that are complements.

66. Describe what is meant by an angle of elevation and an angle of depression.

67. Stonehenge, the famous "stone circle" in England, was built between 2750 B.C. and 1300 B.C. using solid stone blocks weighing over 99,000 pounds each. It required 550 people to pull a single stone up a ramp inclined at a 9° angle. Describe how right triangle trigonometry can be used to determine the distance the 550 workers had to drag a stone in order to raise it to a height of 30 feet.

Technology Exercises

68. Use a calculator in the radian mode to fill in the values in the following table. Then draw a conclusion about $\dfrac{\sin \theta}{\theta}$ as θ approaches 0.

θ	0.4	0.3	0.2	0.1	0.01	0.001	0.0001	0.00001
sin θ								
$\dfrac{\sin \theta}{\theta}$								

69. Use a calculator in the radian mode to fill in the values in the following table. Then draw a conclusion about $\dfrac{\cos \theta - 1}{\theta}$ as θ approaches 0.

θ	0.4	0.3	0.2	0.1	0.01	0.001	0.0001	0.00001
cos θ								
$\dfrac{\cos \theta - 1}{\theta}$								

Critical Thinking Exercises

70. Which one of the following is true?

   a. $\dfrac{\tan 45°}{\tan 15°} = \tan 3°$

   b. $\tan^2 15° - \sec^2 15° = -1$

   c. $\sin 45° + \cos 45° = 1$

   d. $\tan^2 5° = \tan 25°$

71. Explain why the sine or cosine of an acute angle cannot be greater than or equal to 1.

72. Describe what happens to the tangent of an acute angle as the angle gets close to 90°. What happens at 90°?

73. From the top of a 250-foot lighthouse, a plane is sighted over-head and a ship is observed directly below the plane. The angle of elevation of the plane is 22° and the angle of depression of the ship is 35°. Find **a.** the distance of the ship from the lighthouse; **b.** the plane's height above the water. Round to the nearest foot.

SECTION 4.4   Trigonometric Functions of Any Angle

Objectives

1 Use the definitions of trigonometric functions of any angle.

2 Use the signs of the trigonometric functions.

3 Find reference angles.

4 Use reference angles to evaluate trigonometric functions.

Cycles govern many aspects of life—heartbeats, sleep patterns, seasons, and tides all follow regular, predictable cycles. Because of their periodic nature, trigonometric functions are used to model phenomena that occur in cycles. It is helpful to apply these models regardless of whether we think of the domains of trigonometric functions as sets of real numbers or sets of angles. In order to understand and use models for cyclic phenomena from an angle perspective, we need to move beyond right triangles.

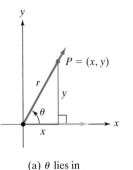

① Use the definitions of trigonometric functions of any angle.

Trigonometric Functions of Any Angle

In the last section, we evaluated trigonometric functions of acute angles, such as that shown in Figure 4.41(a). Note that this angle is in standard position. The point $P = (x, y)$ is a point r units from the origin on the terminal side of θ. A right triangle is formed by drawing a line segment from $P = (x, y)$ perpendicular to the x-axis. Note that y is the length of the side opposite θ and x is the length of the side adjacent to θ.

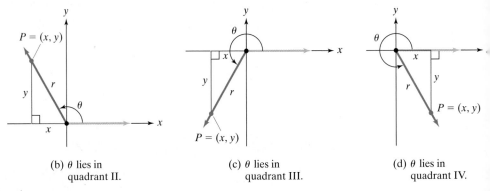

(a) θ lies in quadrant I.

(b) θ lies in quadrant II.

(c) θ lies in quadrant III.

(d) θ lies in quadrant IV.

Figure 4.41

Figures 4.41(b), (c), and (d) show angles in standard position, but they are not acute. We can extend our definitions of the six trigonometric functions to include such angles, as well as quadrantal angles. (Recall that a quadrantal angle has its terminal side on the x-axis or y-axis; such angles are *not* shown in Figure 4.41.) The point $P = (x, y)$ may be any point on the terminal side of the angle θ other than the origin, $(0, 0)$.

Definitions of Trigonometric Functions of Any Angle

Let θ be any angle in standard position and let $P = (x, y)$ be a point on the terminal side of θ. If $r = \sqrt{x^2 + y^2}$ is the distance from $(0, 0)$ to (x, y), as shown in Figure 4.41, the **six trigonometric functions of θ** are defined by the following ratios:

$$\sin \theta = \frac{y}{r} \qquad\qquad \csc \theta = \frac{r}{y}, y \neq 0$$

$$\cos \theta = \frac{x}{r} \qquad\qquad \sec \theta = \frac{r}{x}, x \neq 0$$

$$\tan \theta = \frac{y}{x}, x \neq 0 \qquad\qquad \cot \theta = \frac{x}{y}, y \neq 0.$$

The ratios in the second column are the reciprocals of the corresponding ratios in the first column.

Because the point $P = (x, y)$ is any point on the terminal side of θ other than the origin, $(0, 0)$, $r = \sqrt{x^2 + y^2}$ cannot be zero. Examine the six trigonometric functions defined above. Note that the denominator of the sine and cosine functions is r. Because $r \neq 0$, the sine and cosine functions are defined for any real value of the angle θ. This is not true for the other four trigonometric functions. Note that the denominator of the tangent and secant functions is x: $\tan \theta = \frac{y}{x}$ and $\sec \theta = \frac{r}{x}$. These functions are not defined if $x = 0$. If the point $P = (x, y)$ is on the y-axis, then $x = 0$. Thus, the tangent and secant functions are undefined for all quadrantal angles with terminal sides on the positive or negative y-axis. Likewise, if $P = (x, y)$ is on the x-axis, then $y = 0$, and the cotangent and cosecant functions are undefined: $\cot \theta = \frac{x}{y}$ and $\csc \theta = \frac{r}{y}$. The cotangent and cosecant functions are undefined for all quadrantal angles with terminal sides on the positive or negative x-axis.

EXAMPLE 1   Evaluating Trigonometric Functions

Let $P = (-3, -5)$ be a point on the terminal side of θ. Find each of the six trigonometric functions of θ.

Solution   The situation is shown in Figure 4.42. We need values for x, y, and r to evaluate all six trigonometric functions. We are given the values of x and y. Because $P = (-3, -5)$ is a point on the terminal side of θ, $x = -3$ and $y = -5$. Furthermore,

$$r = \sqrt{x^2 + y^2} = \sqrt{(-3)^2 + (-5)^2} = \sqrt{9 + 25} = \sqrt{34}.$$

Now that we know x, y, and r, we can find the six trigonometric functions of θ. Where appropriate, we will rationalize denominators.

$$\sin \theta = \frac{y}{r} = \frac{-5}{\sqrt{34}} = -\frac{5}{\sqrt{34}} \cdot \frac{\sqrt{34}}{\sqrt{34}} = -\frac{5\sqrt{34}}{34} \qquad \csc \theta = \frac{r}{y} = \frac{\sqrt{34}}{-5} = -\frac{\sqrt{34}}{5}$$

$$\cos \theta = \frac{x}{r} = \frac{-3}{\sqrt{34}} = -\frac{3}{\sqrt{34}} \cdot \frac{\sqrt{34}}{\sqrt{34}} = -\frac{3\sqrt{34}}{34} \qquad \sec \theta = \frac{r}{x} = \frac{\sqrt{34}}{-3} = -\frac{\sqrt{34}}{3}$$

$$\tan \theta = \frac{y}{x} = \frac{-5}{-3} = \frac{5}{3} \qquad\qquad\qquad \cot \theta = \frac{x}{y} = \frac{-3}{-5} = \frac{3}{5}$$

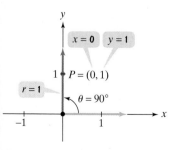

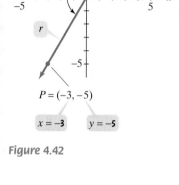

Figure 4.42

Check Point 1   Let $P = (1, -3)$ be a point on the terminal side of θ. Find each of the six trigonometric functions of θ.

How do we find the values of the trigonometric functions for a quadrantal angle? First, draw the angle in standard position. Second, choose a point P on the angle's terminal side. The trigonometric function values of θ depend only on the size of θ and not on the distance of point P from the origin. Thus, we will choose a point that is 1 unit from the origin. Finally, apply the definitions of the appropriate trigonometric functions.

EXAMPLE 2   Trigonometric Functions of Quadrantal Angles

Evaluate, if possible, the sine function and the tangent function at the following four quadrantal angles:

a. $\theta = 0° = 0$ **b.** $\theta = 90° = \dfrac{\pi}{2}$ **c.** $\theta = 180° = \pi$ **d.** $\theta = 270° = \dfrac{3\pi}{2}$.

Solution

a. If $\theta = 0° = 0$ radians, then the terminal side of the angle is on the positive x-axis. Let us select the point $P = (1, 0)$ with $x = 1$ and $y = 0$. This point is 1 unit from the origin, so $r = 1$. Figure 4.43 shows values of x, y, and r corresponding to $\theta = 0°$ or 0 radians. Now that we know x, y, and r, we can apply the definitions of the sine and tangent functions.

$$\sin 0° = \sin 0 = \frac{y}{r} = \frac{0}{1} = 0$$

$$\tan 0° = \tan 0 = \frac{y}{x} = \frac{0}{1} = 0$$

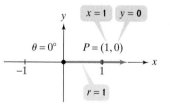

Figure 4.43

b. If $\theta = 90° = \dfrac{\pi}{2}$ radians, then the terminal side of the angle is on the positive y-axis. Let us select the point $P = (0, 1)$ with $x = 0$ and $y = 1$. This point is 1 unit from the origin, so $r = 1$. Figure 4.44 shows values of x, y, and r corresponding to $\theta = 90°$ or $\dfrac{\pi}{2}$. Now that we know x, y, and r, we can apply the definitions of the sine and tangent functions.

$$\sin 90° = \sin \frac{\pi}{2} = \frac{y}{r} = \frac{1}{1} = 1$$

$$\tan 90° = \tan \frac{\pi}{2} = \frac{y}{x} = \frac{1}{0}$$

Because division by 0 is undefined, $\tan 90°$ is undefined.

Figure 4.44

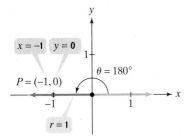

Figure 4.45

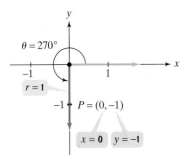

Figure 4.46

Discovery

Try finding tan 90° and tan 270° with your calculator. Describe what occurs.

c. If $\theta = 180° = \pi$ radians, then the terminal side of the angle is on the negative x-axis. Let us select the point $P = (-1, 0)$ with $x = -1$ and $y = 0$. This point is 1 unit from the origin, so $r = 1$. Figure 4.45 shows values of x, y, and r corresponding to $\theta = 180°$ or π. Now that we know x, y, and r, we can apply the definitions of the sine and tangent functions.

$$\sin 180° = \sin \pi = \frac{y}{r} = \frac{0}{1} = 0$$

$$\tan 180° = \tan \pi = \frac{y}{x} = \frac{0}{-1} = 0$$

d. If $\theta = 270° = \dfrac{3\pi}{2}$ radians, then the terminal side of the angle is on the negative y-axis. Let us select the point $P = (0, -1)$ with $x = 0$ and $y = -1$. This point is 1 unit from the origin, so $r = 1$. Figure 4.46 shows values of x, y, and r corresponding to $\theta = 270°$ or $\dfrac{3\pi}{2}$. Now that we know x, y, and r, we can apply the definitions of the sine and tangent functions.

$$\sin 270° = \sin \frac{3\pi}{2} = \frac{y}{r} = \frac{-1}{1} = -1$$

$$\tan 270° = \tan \frac{3\pi}{2} = \frac{y}{x} = \frac{-1}{0}$$

Because division by 0 is undefined, tan 270° is undefined.

Check Point 2 Evaluate, if possible, the cosine function and the cosecant function at the following four quadrantal angles:

a. $\theta = 0° = 0$ **b.** $\theta = 90° = \dfrac{\pi}{2}$

c. $\theta = 180° = \pi$ **d.** $\theta = 270° = \dfrac{3\pi}{2}$.

The Signs of the Trigonometric Functions

In Example 2, we evaluated trigonometric functions of quadrantal angles. However, we will now return to the trigonometric functions of nonquadrantal angles. **If θ is not a quadrantal angle, the sign of a trigonometric function depends on the quadrant in which θ lies.** In all four quadrants, r is positive. However, x and y can be positive or negative. For example, if θ lies in quadrant II, x is negative and y is positive. Thus, the only positive ratios in this quadrant are $\dfrac{y}{r}$ and its reciprocal, $\dfrac{r}{y}$. These ratios are the function values for the sine and cosecant, respectively. In short, if θ lies in quadrant II, $\sin \theta$ and $\csc \theta$ are positive. The other four trigonometric functions are negative.

Figure 4.47 summarizes the signs of the trigonometric functions. If θ lies in quadrant I, all six functions are positive. If θ lies in quadrant II, only $\sin \theta$ and $\csc \theta$ are positive. If θ lies in quadrant III, only $\tan \theta$ and $\cot \theta$ are positive. Finally, if θ lies in quadrant IV, only $\cos \theta$ and $\sec \theta$ are positive. Observe that the positive functions in each quadrant occur in reciprocal pairs.

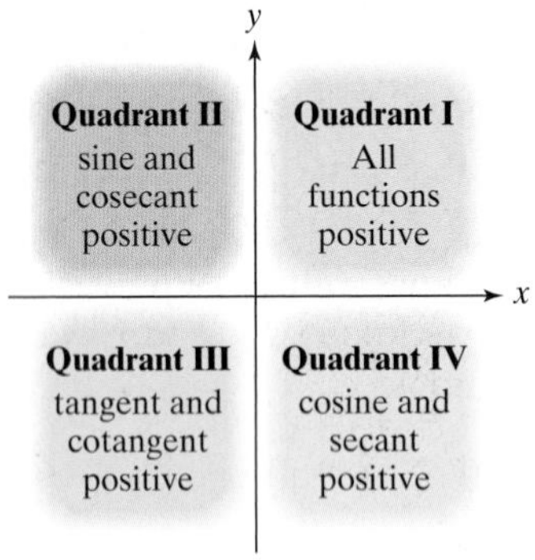

② Use the signs of the trigonometric functions.

Quadrant II	Quadrant I
sine and cosecant positive	All functions positive
Quadrant III	**Quadrant IV**
tangent and cotangent positive	cosine and secant positive

Figure 4.47 The signs of the trigonometric functions

Study Tip

Your author's high school trig teacher showed him this sentence to remember the signs of the trig functions:

All   **Students**   **Take**   **Calculus.**

| All trig functions are positive in **QI.** | Sine and its reciprocal, cosecant, are positive in **QII.** | Tangent and its reciprocal, cotangent, are positive in **QIII.** | Cosine and its reciprocal, secant, are positive in **QIV.** |

The sentence isn't true anymore, so you may prefer these memory devices:

All Snakes Tease Chickens.

A Smart Trig Class.

EXAMPLE 3   Finding the Quadrant in Which an Angle Lies

If $\tan \theta < 0$ and $\cos \theta > 0$, name the quadrant in which angle θ lies.

Solution   When $\tan \theta < 0$, θ lies in quadrant II or IV. When $\cos \theta > 0$, θ lies in quadrant I or IV. When both conditions are met ($\tan \theta < 0$ and $\cos \theta > 0$), θ must lie in quadrant IV.

Check Point 3   If $\sin \theta < 0$ and $\cos \theta < 0$, name the quadrant in which angle θ lies.

EXAMPLE 4   Evaluating Trigonometric Functions

Given $\tan \theta = -\frac{2}{3}$ and $\cos \theta > 0$, find $\cos \theta$ and $\csc \theta$.

Solution   Because the tangent is negative and the cosine is positive, θ lies in quadrant IV. This will help us to determine whether the negative sign in $\tan \theta = -\frac{2}{3}$ should be associated with the numerator or the denominator. Keep in mind that in quadrant IV, x is positive and y is negative. Thus,

> In quadrant IV, y is negative.

$$\tan \theta = -\frac{2}{3} = \frac{y}{x} = \frac{-2}{3}.$$

(See Figure 4.48.) Thus, $x = 3$ and $y = -2$. Furthermore,

$$r = \sqrt{x^2 + y^2} = \sqrt{3^2 + (-2)^2} = \sqrt{9 + 4} = \sqrt{13}.$$

Now that we know x, y, and r, we can find $\cos \theta$ and $\csc \theta$.

$$\cos \theta = \frac{x}{r} = \frac{3}{\sqrt{13}} = \frac{3}{\sqrt{13}} \cdot \frac{\sqrt{13}}{\sqrt{13}} = \frac{3\sqrt{13}}{13} \qquad \csc \theta = \frac{r}{y} = \frac{\sqrt{13}}{-2} = -\frac{\sqrt{13}}{2}$$

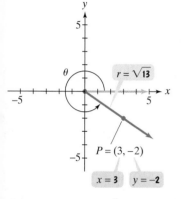

Figure 4.48 $\tan \theta = -\frac{2}{3}$ and $\cos \theta > 0$

Check Point 4   Given $\tan \theta = -\frac{1}{3}$ and $\cos \theta < 0$, find $\sin \theta$ and $\sec \theta$.

In Example 4, we used the quadrant in which θ lies to determine whether a negative sign should be associated with the numerator or the denominator. Here's a situation, similar to Example 4, where negative signs should be associated with *both* the numerator and the denominator:

$$\tan \theta = \frac{3}{5} \quad \text{and} \quad \cos \theta < 0.$$

Because the tangent is positive and the cosine is negative, θ lies in quadrant III. In quadrant III, x is negative and y is negative. Thus,

$$\tan \theta = \frac{3}{5} = \frac{y}{x} = \frac{-3}{-5}. \qquad \boxed{\text{We see that } x = -5 \text{ and } y = -3.}$$

③ Find reference angles.

Reference Angles

We will often evaluate trigonometric functions of positive angles greater than 90° and all negative angles by making use of a positive acute angle. This positive acute angle is called a *reference angle*.

> ### Definition of a Reference Angle
>
> Let θ be a nonacute angle in standard position that lies in a quadrant. Its **reference angle** is the positive acute angle θ' formed by the terminal side of θ and the x-axis.

Figure 4.49 shows the reference angle for θ lying in quadrants II, III, and IV. Notice that the formula used to find θ', the reference angle, varies according to the quadrant in which θ lies. You may find it easier to find the reference angle for a given angle by making a figure that shows the angle in standard position. The acute angle formed by the terminal side of this angle and the x-axis is the reference angle.

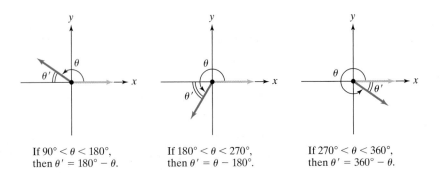

Figure 4.49 Reference angles, θ', for positive angles, θ, in quadrants II, III, and IV

If $90° < \theta < 180°$, then $\theta' = 180° - \theta$.

If $180° < \theta < 270°$, then $\theta' = \theta - 180°$.

If $270° < \theta < 360°$, then $\theta' = 360° - \theta$.

EXAMPLE 5 Finding Reference Angles

Find the reference angle, θ', for each of the following angles:

a. $\theta = 345°$ **b.** $\theta = \dfrac{5\pi}{6}$ **c.** $\theta = -135°$ **d.** $\theta = 2.5$.

Solution

a. A 345° angle in standard position is shown in Figure 4.50. Because 345° lies in quadrant IV, the reference angle is

$$\theta' = 360° - 345° = 15°.$$

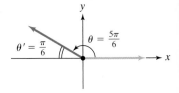

Figure 4.50

b. Because $\dfrac{5\pi}{6}$ lies between $\dfrac{\pi}{2} = \dfrac{3\pi}{6}$ and

$\pi = \dfrac{6\pi}{6}, \theta = \dfrac{5\pi}{6}$ lies in quadrant II. The angle is shown in Figure 4.51. The reference angle is

$$\theta' = \pi - \frac{5\pi}{6} = \frac{6\pi}{6} - \frac{5\pi}{6} = \frac{\pi}{6}.$$

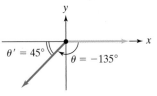

Figure 4.51

Discovery

Solve part (c) by first finding a positive coterminal angle for $-135°$ less than 360°. Use the positive coterminal angle to find the reference angle.

c. A $-135°$ angle in standard position is shown in Figure 4.52. The figure indicates that the positive acute angle formed by the terminal side of θ and the x-axis is 45°. The reference angle is

$$\theta' = 45°.$$

Figure 4.52

d. The angle $\theta = 2.5$ lies between $\dfrac{\pi}{2} \approx 1.57$ and $\pi \approx 3.14$. This means that $\theta = 2.5$ is in quadrant II, shown in Figure 4.53. The reference angle is

$$\theta' = \pi - 2.5 \approx 0.64.$$

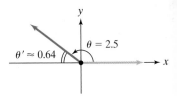

Figure 4.53

Check Point 5 Find the reference angle, θ', for each of the following angles:

a. $\theta = 210°$ **b.** $\theta = \dfrac{7\pi}{4}$ **c.** $\theta = -240°$ **d.** $\theta = 3.6$.

Finding reference angles for angles that are greater than $360°$ (2π) or less than $-360°$ (-2π) involves using coterminal angles. We have seen that coterminal angles have the same initial and terminal sides. Recall that coterminal angles can be obtained by increasing or decreasing an angle's measure by an integer multiple of $360°$ or 2π.

> **Finding Reference Angles for Angles Greater Than $360°$ (2π) or Less Than $-360°$ (-2π)**
>
> **1.** Find a positive angle α less than $360°$ or 2π that is coterminal with the given angle.
> **2.** Draw α in standard position.
> **3.** Use the drawing to find the reference angle for the given angle. The positive acute angle formed by the terminal side of α and the x-axis is the reference angle.

EXAMPLE 6 Finding Reference Angles

Find the reference angle for each of the following angles:

a. $\theta = 580°$ **b.** $\theta = \dfrac{8\pi}{3}$ **c.** $\theta = -\dfrac{13\pi}{6}$.

Solution

a. For a $580°$ angle, subtract $360°$ to find a positive coterminal angle less than $360°$.

$$580° - 360° = 220°$$

Figure 4.54 shows $\alpha = 220°$ in standard position. Because $220°$ lies in quadrant III, the reference angle is

$$\alpha' = 220° - 180° = 40°.$$

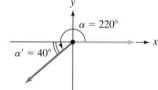

Figure 4.54

b. For an $\dfrac{8\pi}{3}$, or $2\dfrac{2}{3}\pi$ angle, subtract 2π to find a positive coterminal angle less than 2π.

$$\frac{8\pi}{3} - 2\pi = \frac{8\pi}{3} - \frac{6\pi}{3} = \frac{2\pi}{3}$$

Figure 4.55 shows $\alpha = \dfrac{2\pi}{3}$ in standard position.

Because $\dfrac{2\pi}{3}$ lies in quadrant II, the reference angle is

$$\alpha' = \pi - \frac{2\pi}{3} = \frac{3\pi}{3} - \frac{2\pi}{3} = \frac{\pi}{3}.$$

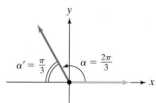

Figure 4.55

c. For a $-\dfrac{13\pi}{6}$, or $-2\dfrac{1}{6}\pi$ angle, add 4π to find a positive coterminal angle less than 2π.

$$-\frac{13\pi}{6} + 4\pi = -\frac{13\pi}{6} + \frac{24\pi}{6} = \frac{11\pi}{6}$$

Figure 4.56 shows $\alpha = \dfrac{11\pi}{6}$ in standard position.

Because $\dfrac{11\pi}{6}$ lies in quadrant IV, the reference angle is

$$\alpha' = 2\pi - \frac{11\pi}{6} = \frac{12\pi}{6} - \frac{11\pi}{6} = \frac{\pi}{6}.$$

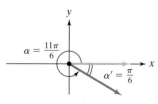

Figure 4.56

Discovery

Solve part (c) using the coterminal angle formed by adding 2π, rather than 4π, to the given angle.

Check
Point **6**   Find the reference angle for each of the following angles:

$$\textbf{a.} \;\; \theta = 665° \qquad \textbf{b.} \;\; \theta = \frac{15\pi}{4} \qquad \textbf{c.} \;\; \theta = -\frac{11\pi}{3}.$$

④ Use reference angles to evaluate trigonometric functions.

Evaluating Trigonometric Functions Using Reference Angles

The way that reference angles are defined makes them useful in evaluating trigonometric functions.

> ### Using Reference Angles to Evaluate Trigonometric Functions
> The values of the trigonometric functions of a given angle, θ, are the same as the values of the trigonometric functions of the reference angle, θ', except possibly for the sign. A function value of the acute reference angle, θ', is always positive. However, the same function value for θ may be positive or negative.

For example, we can use a reference angle, θ', to obtain an exact value for tan 120°. The reference angle for $\theta = 120°$ is $\theta' = 180° - 120° = 60°$. We know the exact value of the tangent function of the reference angle: $\tan 60° = \sqrt{3}$. We also know that the value of a trigonometric function of a given angle, θ, is the same as that of its reference angle, θ', except possibly for the sign. Thus, we can conclude that tan 120° equals $-\sqrt{3}$ or $\sqrt{3}$.

What sign should we attach to $\sqrt{3}$? A 120° angle lies in quadrant II, where only the sine and cosecant are positive. Thus, the tangent function is negative for a 120° angle. Therefore,

> Prefix by a negative sign to show tangent is negative in quadrant II.

$$\tan 120° = -\tan 60° = -\sqrt{3}.$$

> The reference angle for 120° is 60°.

In the previous section, we used two right triangles to find exact trigonometric values of 30°, 45°, and 60°. Using a procedure similar to finding tan 120°, we can now find the exact function values of all angles for which 30°, 45°, or 60° are reference angles.

> ### A Procedure for Using Reference Angles to Evaluate Trigonometric Functions
> The value of a trigonometric function of any angle θ is found as follows:
>
> 1. Find the associated reference angle, θ', and the function value for θ'.
> 2. Use the quadrant in which θ lies to prefix the appropriate sign to the function value in step 1.

Discovery

Draw the two right triangles involving 30°, 45°, and 60°. Indicate the length of each side. Use these lengths to verify the function values for the reference angles in the solution to Example 7.

EXAMPLE 7   Using Reference Angles to Evaluate Trigonometric Functions

Use reference angles to find the exact value of each of the following trigonometric functions:

$$\textbf{a.} \;\; \sin 135° \qquad \textbf{b.} \;\; \cos \frac{4\pi}{3} \qquad \textbf{c.} \;\; \cot\left(-\frac{\pi}{3}\right).$$

Solution

a. We use our two-step procedure to find sin 135°.

Step 1 Find the reference angle, θ', and sin θ'. Figure 4.57 shows 135° lies in quadrant II. The reference angle is

$$\theta' = 180° - 135° = 45°.$$

The function value for the reference angle is $\sin 45° = \dfrac{\sqrt{2}}{2}$.

Step 2 Use the quadrant in which θ lies to prefix the appropriate sign to the function value in step 1. The angle $\theta = 135°$ lies in quadrant II. Because the sine is positive in quadrant II, we put a $+$ sign before the function value of the reference angle. Thus,

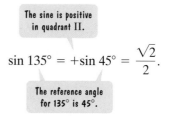

The sine is positive
in quadrant II.

$$\sin 135° = +\sin 45° = \dfrac{\sqrt{2}}{2}.$$

The reference angle
for 135° is 45°.

b. We use our two-step procedure to find $\cos \dfrac{4\pi}{3}$.

Step 1 Find the reference angle, θ', and cos θ'. Figure 4.58 shows that $\theta = \dfrac{4\pi}{3}$ lies in quadrant III. The reference angle is

$$\theta' = \dfrac{4\pi}{3} - \pi = \dfrac{4\pi}{3} - \dfrac{3\pi}{3} = \dfrac{\pi}{3}.$$

The function value for the reference angle is

$$\cos \dfrac{\pi}{3} = \dfrac{1}{2}.$$

Step 2 Use the quadrant in which θ lies to prefix the appropriate sign to the function value in step 1. The angle $\theta = \dfrac{4\pi}{3}$ lies in quadrant III. Because only the tangent and cotangent are positive in quadrant III, the cosine is negative in this quadrant. We put a $-$ sign before the function value of the reference angle. Thus,

The cosine is negative
in quadrant III.

$$\cos \dfrac{4\pi}{3} = -\cos \dfrac{\pi}{3} = -\dfrac{1}{2}.$$

The reference angle
for $\frac{4\pi}{3}$ is $\frac{\pi}{3}$.

c. We use our two-step procedure to find $\cot\left(-\dfrac{\pi}{3}\right)$.

Step 1 Find the reference angle, θ', and cot θ'. Figure 4.59 shows that $\theta = -\dfrac{\pi}{3}$ lies in quadrant IV. The reference angle is $\theta' = \dfrac{\pi}{3}$. The function value for the reference angle is $\cot \dfrac{\pi}{3} = \dfrac{\sqrt{3}}{3}$.

Step 2 Use the quadrant in which θ lies to prefix the appropriate sign to the function value in step 1. The angle $\theta = -\dfrac{\pi}{3}$ lies in quadrant IV. Because only

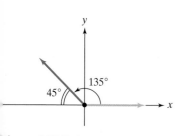

Figure 4.57 Reference angle for 135°

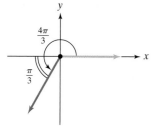

Figure 4.58 Reference angle for $\dfrac{4\pi}{3}$

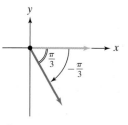

Figure 4.59 Reference angle for $-\dfrac{\pi}{3}$

the cosine and secant are positive in quadrant IV, the cotangent is negative i this quadrant. We put a − sign before the function value of the referenc angle. Thus,

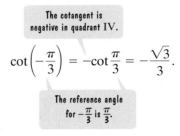

$$\cot\left(-\frac{\pi}{3}\right) = -\cot\frac{\pi}{3} = -\frac{\sqrt{3}}{3}.$$

The reference angle for $-\frac{\pi}{3}$ is $\frac{\pi}{3}$.

Check Point 7 Use reference angles to find the exact value of the following trigonometri functions:

a. $\sin 300°$ **b.** $\tan\frac{5\pi}{4}$ **c.** $\sec\left(-\frac{\pi}{6}\right).$

In our final example, we use positive coterminal angles less than 2π to find th reference angles.

EXAMPLE 8   Using Reference Angles to Evaluate Trigonometric Functions

Use reference angles to find the exact value of each of the following trigonometr functions:

a. $\tan\frac{14\pi}{3}$ **b.** $\sec\left(-\frac{17\pi}{4}\right).$

Solution

a. We use our two-step procedure to find $\tan\frac{14\pi}{3}$.

Step 1   Find the reference angle, θ', and $\tan\theta'$. Because the given angle, $\frac{14}{3}$ or $4\frac{2}{3}\pi$, exceeds 2π, subtract 4π to find a positive coterminal angle less than 2

$$\theta = \frac{14\pi}{3} - 4\pi = \frac{14\pi}{3} - \frac{12\pi}{3} = \frac{2\pi}{3}$$

Figure 4.60 shows $\theta = \frac{2\pi}{3}$ in standard position. The angle lies in quadrant The reference angle is

$$\theta' = \pi - \frac{2\pi}{3} = \frac{3\pi}{3} - \frac{2\pi}{3} = \frac{\pi}{3}.$$

The function value for the reference angle is $\tan\frac{\pi}{3} = \sqrt{3}.$

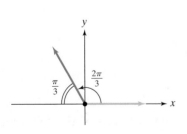

Figure 4.60 Reference angle for $\frac{2\pi}{3}$

Step 2   Use the quadrant in which θ lies to prefix the appropriate sign to th function value in step 1. The coterminal angle $\theta = \frac{2\pi}{3}$ lies in quadrant

Because the tangent is negative in quadrant II, we put a $-$ sign before the function value of the reference angle. Thus,

> The tangent is negative in quadrant II.

$$\tan \frac{14\pi}{3} = \tan \frac{2\pi}{3} = -\tan \frac{\pi}{3} = -\sqrt{3}.$$

> The reference angle for $\frac{2\pi}{3}$ is $\frac{\pi}{3}$.

b. We use our two-step procedure to find $\sec\left(-\dfrac{17\pi}{4}\right)$.

Step 1   Find the reference angle, θ', and $\sec\theta'$. Because the given angle, $-\dfrac{17\pi}{4}$ or $-4\dfrac{1}{4}\pi$, is less than -2π, add 6π (three multiples of 2π) to find a positive coterminal angle less than 2π.

$$\theta = -\frac{17\pi}{4} + 6\pi = -\frac{17\pi}{4} + \frac{24\pi}{4} = \frac{7\pi}{4}$$

Figure 4.61 shows $\theta = \dfrac{7\pi}{4}$ in standard position. The angle lies in quadrant IV. The reference angle is

$$\theta' = 2\pi - \frac{7\pi}{4} = \frac{8\pi}{4} - \frac{7\pi}{4} = \frac{\pi}{4}.$$

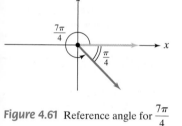

Figure 4.61 Reference angle for $\dfrac{7\pi}{4}$

The function value for the reference angle is $\sec \dfrac{\pi}{4} = \sqrt{2}$.

Step 2   Use the quadrant in which θ lies to prefix the appropriate sign to the function value in step 1. The coterminal angle $\theta = \dfrac{7\pi}{4}$ lies in quadrant IV. Because the secant is positive in quadrant IV, we put a $+$ sign before the function value of the reference angle. Thus,

> The secant is positive in quadrant IV.

$$\sec\left(-\frac{17\pi}{4}\right) = \sec \frac{7\pi}{4} = + \sec \frac{\pi}{4} = \sqrt{2}.$$

> The reference angle for $\frac{7\pi}{4}$ is $\frac{\pi}{4}$.

Check Point 8   Use reference angles to find the exact value of each of the following trigonometric functions:

a. $\cos \dfrac{17\pi}{6}$ **b.** $\sin\left(-\dfrac{22\pi}{3}\right)$.

Study Tip

Evaluating trigonometric functions like those in Example 8 and Check Point 8 involves using a number of concepts, including finding coterminal angles and reference angles, locating special angles, determining the signs of trigonometric functions in specific quadrants, and finding the trigonometric functions of special angles $\left(30° = \dfrac{\pi}{6}, 45° = \dfrac{\pi}{4}, \text{and } 60° = \dfrac{\pi}{3}\right)$. To be successful in trigonometry, it is often necessary to connect concepts. Here's an early reference sheet showing some of the concepts you should have at your fingertips (or memorized).

Degree and Radian Measures of Special and Quadrantal Angles

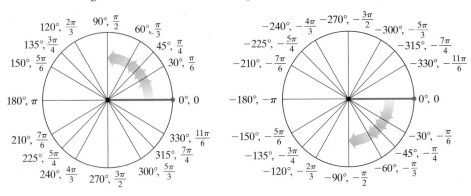

Special Right Triangles and Trigonometric Functions of Special Angles

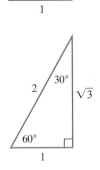

θ	$30° = \dfrac{\pi}{6}$	$45° = \dfrac{\pi}{4}$	$60° = \dfrac{\pi}{3}$
$\sin \theta$	$\dfrac{1}{2}$	$\dfrac{\sqrt{2}}{2}$	$\dfrac{\sqrt{3}}{2}$
$\cos \theta$	$\dfrac{\sqrt{3}}{2}$	$\dfrac{\sqrt{2}}{2}$	$\dfrac{1}{2}$
$\tan \theta$	$\dfrac{\sqrt{3}}{3}$	1	$\sqrt{3}$

Signs of the Trigonometric Functions

Quadrant II	Quadrant I
sine and cosecant positive	All functions positive
Quadrant III	Quadrant IV
tangent and cotangent positive	cosine and secant positive

Trigonometric Functions of Quadrantal Angles

θ	$0° = 0$	$90° = \dfrac{\pi}{2}$	$180° = \pi$	$270° = \dfrac{3\pi}{2}$
$\sin \theta$	0	1	0	-1
$\cos \theta$	1	0	-1	0
$\tan \theta$	0	undefined	0	undefined

Using Reference Angles to Evaluate Trigonometric Functions

$$\sin \theta = \boxed{} \sin \theta'$$
$$\cos \theta = \boxed{} \cos \theta'$$
$$\tan \theta = \boxed{} \tan \theta'$$

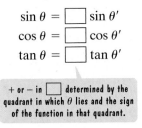

+ or − in ☐ determined by the quadrant in which θ lies and the sign of the function in that quadrant.

EXERCISE SET 4.4

Practice Exercises

In Exercises 1–8, a point on the terminal side of angle θ is given. Find the exact value of each of the six trigonometric functions of θ.

1. $(-4, 3)$ **2.** $(-12, 5)$ **3.** $(2, 3)$

4. $(3, 7)$ **5.** $(3, -3)$ **6.** $(5, -5)$

7. $(-2, -5)$ **8.** $(-1, -3)$

In Exercises 9–16, evaluate the trigonometric function at the quadrantal angle, or state that the expression is undefined.

9. $\cos \pi$ **10.** $\tan \pi$ **11.** $\sec \pi$

12. $\csc \pi$ **13.** $\tan \dfrac{3\pi}{2}$ **14.** $\cos \dfrac{3\pi}{2}$

15. $\cot \dfrac{\pi}{2}$ **16.** $\tan \dfrac{\pi}{2}$

In Exercises 17–22, let θ be an angle in standard position. Name the quadrant in which θ lies.

17. $\sin \theta > 0, \quad \cos \theta > 0$ **18.** $\sin \theta < 0, \quad \cos \theta > 0$

19. $\sin \theta < 0, \quad \cos \theta < 0$ **20.** $\tan \theta < 0, \quad \sin \theta < 0$

21. $\tan \theta < 0, \quad \cos \theta < 0$ **22.** $\cot \theta > 0, \quad \sec \theta < 0$

In Exercises 23–34, find the exact value of each of the remaining trigonometric functions of θ.

23. $\cos \theta = -\frac{3}{5}, \quad \theta$ in quadrant III

24. $\sin \theta = -\frac{12}{13}, \quad \theta$ in quadrant III

25. $\sin \theta = \frac{5}{13}, \quad \theta$ in quadrant II

26. $\cos \theta = \frac{4}{5}, \quad \theta$ in quadrant IV

27. $\cos \theta = \frac{8}{17}, \quad 270° < \theta < 360°$

28. $\cos \theta = \frac{1}{3}, \quad 270° < \theta < 360°$

29. $\tan \theta = -\frac{2}{3}, \quad \sin \theta > 0$ **30.** $\tan \theta = -\frac{1}{3}, \quad \sin \theta > 0$

31. $\tan \theta = \frac{4}{3}, \quad \cos \theta < 0$ **32.** $\tan \theta = \frac{5}{12}, \quad \cos \theta < 0$

33. $\sec \theta = -3, \quad \tan \theta > 0$ **34.** $\csc \theta = -4, \quad \tan \theta > 0$

In Exercises 35–60, find the reference angle for each angle.

35. $160°$ **36.** $170°$ **37.** $205°$

38. $210°$ **39.** $355°$ **40.** $351°$

41. $\dfrac{7\pi}{4}$ **42.** $\dfrac{5\pi}{4}$ **43.** $\dfrac{5\pi}{6}$

44. $\dfrac{5\pi}{7}$ **45.** $-150°$ **46.** $-250°$

47. $-335°$ **48.** $-359°$ **49.** 4.7

50. 5.5 **51.** $565°$ **52.** $553°$

53. $\dfrac{17\pi}{6}$ **54.** $\dfrac{11\pi}{4}$ **55.** $\dfrac{23\pi}{4}$

56. $\dfrac{17\pi}{3}$ **57.** $-\dfrac{11\pi}{4}$ **58.** $-\dfrac{17\pi}{6}$

59. $-\dfrac{25\pi}{6}$ **60.** $-\dfrac{13\pi}{3}$

In Exercises 61–86, use reference angles to find the exact value of each expression. Do not use a calculator.

61. $\cos 225°$ **62.** $\sin 300°$ **63.** $\tan 210°$

64. $\sec 240°$ **65.** $\tan 420°$ **66.** $\tan 405°$

67. $\sin \dfrac{2\pi}{3}$ **68.** $\cos \dfrac{3\pi}{4}$ **69.** $\csc \dfrac{7\pi}{6}$

70. $\cot \dfrac{7\pi}{4}$ **71.** $\tan \dfrac{9\pi}{4}$ **72.** $\tan \dfrac{9\pi}{2}$

73. $\sin(-240°)$ **74.** $\sin(-225°)$ **75.** $\tan\left(-\dfrac{\pi}{4}\right)$

76. $\tan\left(-\dfrac{\pi}{6}\right)$ **77.** $\sec 495°$ **78.** $\sec 510°$

79. $\cot \dfrac{19\pi}{6}$ **80.** $\cot \dfrac{13\pi}{3}$ **81.** $\cos \dfrac{23\pi}{4}$

82. $\cos \dfrac{35\pi}{6}$ **83.** $\tan\left(-\dfrac{17\pi}{6}\right)$ **84.** $\tan\left(-\dfrac{11\pi}{4}\right)$

85. $\sin\left(-\dfrac{17\pi}{3}\right)$ **86.** $\sin\left(-\dfrac{35\pi}{6}\right)$

Practice Plus

In Exercises 87–92, find the exact value of each expression. Write the answer as a single fraction. Do not use a calculator.

87. $\sin \dfrac{\pi}{3} \cos \pi - \cos \dfrac{\pi}{3} \sin \dfrac{3\pi}{2}$

88. $\sin \dfrac{\pi}{4} \cos 0 - \sin \dfrac{\pi}{6} \cos \pi$

89. $\sin \dfrac{11\pi}{4} \cos \dfrac{5\pi}{6} + \cos \dfrac{11\pi}{4} \sin \dfrac{5\pi}{6}$

90. $\sin \dfrac{17\pi}{3} \cos \dfrac{5\pi}{4} + \cos \dfrac{17\pi}{3} \sin \dfrac{5\pi}{4}$

91. $\sin \dfrac{3\pi}{2} \tan\left(-\dfrac{15\pi}{4}\right) - \cos\left(-\dfrac{5\pi}{3}\right)$

92. $\sin \dfrac{3\pi}{2} \tan\left(-\dfrac{8\pi}{3}\right) + \cos\left(-\dfrac{5\pi}{6}\right)$

In Exercises 93–98, let

$$f(x) = \sin x, \, g(x) = \cos x, \text{ and } h(x) = 2x.$$

Find the exact value of each expression. Do not use a calculator.

93. $f\left(\dfrac{4\pi}{3} + \dfrac{\pi}{6}\right) + f\left(\dfrac{4\pi}{3}\right) + f\left(\dfrac{\pi}{6}\right)$

94. $g\left(\dfrac{5\pi}{6} + \dfrac{\pi}{6}\right) + g\left(\dfrac{5\pi}{6}\right) + g\left(\dfrac{\pi}{6}\right)$

95. $(h \circ g)\left(\dfrac{17\pi}{3}\right)$ **96.** $(h \circ f)\left(\dfrac{11\pi}{4}\right)$

97. the average rate of change of f from $x_1 = \dfrac{5\pi}{4}$ to $x_2 = \dfrac{3\pi}{2}$

98. the average rate of change of g from $x_1 = \dfrac{3\pi}{4}$ to $x_2 = \pi$

In Exercises 99–104, find two values of θ, 0 ≤ θ < 2π, that satisfy each equation.

99. $\sin\theta = \dfrac{\sqrt{2}}{2}$

100. $\cos\theta = \dfrac{1}{2}$

101. $\sin\theta = -\dfrac{\sqrt{2}}{2}$

102. $\cos\theta = -\dfrac{1}{2}$

103. $\tan\theta = -\sqrt{3}$

104. $\tan\theta = -\dfrac{\sqrt{3}}{3}$

Writing in Mathematics

105. If you are given a point on the terminal side of angle θ, explain how to find sin θ.

106. Explain why tan 90° is undefined.

107. If cos θ > 0 and tan θ < 0, explain how to find the quadrant in which θ lies.

108. What is a reference angle? Give an example with your description.

109. Explain how reference angles are used to evaluate trigonometric functions. Give an example with your description.

CHAPTER 4
MID-CHAPTER CHECK POINT

What You Know: We learned to use radians to measure angles: One radian (approximately 57°) is the measure of the central angle that intercepts an arc equal in length to the radius of the circle. Using 180° = π radians, we converted degrees to radians (multiply by $\dfrac{\pi}{180°}$) and radians to degrees (multiply by $\dfrac{180°}{\pi}$). We defined the six trigonometric functions using coordinates of points along the unit circle, right triangles, and angles in standard position. Evaluating trigonometric functions using reference angles involved connecting a number of concepts, including finding coterminal and reference angles, locating special angles, determining the signs of the trigonometric functions in specific quadrants, and finding the function values at special angles. Use the important Study Tip on page 498 as a reference sheet to help connect these concepts.

In Exercises 1–2, convert each angle in degrees to radians. Express your answer as a multiple of π.

1. 10°

2. −105°

In Exercises 3–4, convert each angle in radians to degrees.

3. $\dfrac{5\pi}{12}$

4. $-\dfrac{13\pi}{20}$

In Exercises 5–7,

a. *Find a positive angle less than 360° or 2π that is coterminal with the given angle.*

b. *Draw the given angle in standard position.*

c. *Find the reference angle for the given angle.*

5. $\dfrac{11\pi}{3}$

6. $-\dfrac{19\pi}{4}$

7. 510°

8. Use the point shown on the unit circle to find each of the six trigonometric functions at t.

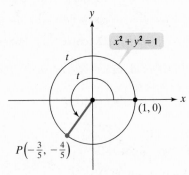

9. Use the triangle to find each of the six trigonometric functions of θ.

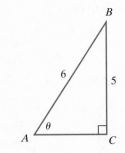

10. Use the point on the terminal side of θ to find each of the six trigonometric functions of θ.

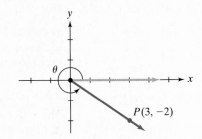

In Exercises 11–12, find the exact value of the remaining trigonometric functions of θ.

11. $\tan \theta = -\dfrac{3}{4}, \cos \theta < 0$ **12.** $\cos \theta = \dfrac{3}{7}, \sin \theta < 0$

In Exercises 13–14, find the measure of the side of the right triangle whose length is designated by a lowercase letter. Round the answer to the nearest whole number.

13.

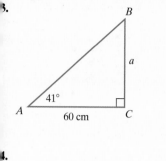

14.

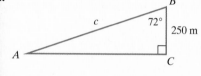

15. If $\cos \theta = \dfrac{1}{6}$ and θ is acute, find $\cot\left(\dfrac{\pi}{2} - \theta\right)$.

In Exercises 16–26, find the exact value of each expression. Do not use a calculator.

16. $\tan 30°$

17. $\cot 120°$

18. $\cos 240°$

19. $\sec \dfrac{11\pi}{6}$

20. $\sin^2 \dfrac{\pi}{7} + \cos^2 \dfrac{\pi}{7}$

21. $\sin\left(-\dfrac{2\pi}{3}\right)$

22. $\csc\left(\dfrac{22\pi}{3}\right)$

23. $\cos 495°$

24. $\tan\left(-\dfrac{17\pi}{6}\right)$

25. $\sin^2 \dfrac{\pi}{2} - \cos \pi$

26. $\cos\left(\dfrac{5\pi}{6} + 2\pi n\right) + \tan\left(\dfrac{5\pi}{6} + n\pi\right)$, n is an integer.

27. A circle has a radius of 40 centimeters. Find the length of the arc intercepted by a central angle of 36°. Express the answer in terms of π. Then round to two decimal places.

28. A merry-go-round makes 8 revolutions per minute. Find the linear speed, in feet per minute, of a horse 10 feet from the center. Express the answer in terms of π. Then round to one decimal place.

29. A plane takes off at an angle of 6°. After traveling for one mile, or 5280 feet, along this flight path, find the plane's height, to the nearest tenth of a foot, above the ground.

30. A tree that is 50 feet tall casts a shadow that is 60 feet long. Find the angle of elevation, to the nearest degree, of the sun.

SECTION 4.5   *Graphs of Sine and Cosine Functions*

Objectives

- Understand the graph of $y = \sin x$.
- Graph variations of $y = \sin x$.
- Understand the graph of $y = \cos x$.
- Graph variations of $y = \cos x$.
- Use vertical shifts of sine and cosine curves.
- Model periodic behavior.

Take a deep breath and relax. Many relaxation exercises involve slowing down our breathing. Some people suggest that the way we breathe affects every part of our lives. Did you know that graphs of trigonometric functions can be used to analyze the breathing cycle, which is our closest link to both life and death?

In this section, we use graphs of sine and cosine functions to visualize their properties. We use the traditional symbol x, rather than θ or t, to represent the independent variable. We use the symbol y for the dependent variable, or the function's value at x. Thus, we will be graphing $y = \sin x$ and $y = \cos x$ in rectangular coordinates. In all graphs of trigonometric functions, the independent variable, x, is measured in radians.

 Understand the graph of $y = \sin x$.

The Graph of $y = \sin x$

The trigonometric functions can be graphed in a rectangular coordinate system by plotting points whose coordinates satisfy the function. Thus, we graph $y = \sin x$ by listing some points on the graph. Because the period of the sine function is 2π,

we will graph the function on the interval $[0, 2\pi]$. The rest of the graph is made u
of repetitions of this portion.

Table 4.3 lists some values of (x, y) on the graph of $y = \sin x, 0 \le x \le 2\pi$.

Table 4.3 Values of (x, y) on the graph of $y = \sin x$

x	0	$\dfrac{\pi}{6}$	$\dfrac{\pi}{3}$	$\dfrac{\pi}{2}$	$\dfrac{2\pi}{3}$	$\dfrac{5\pi}{6}$	π	$\dfrac{7\pi}{6}$	$\dfrac{4\pi}{3}$	$\dfrac{3\pi}{2}$	$\dfrac{5\pi}{3}$	$\dfrac{11\pi}{6}$	2π
$y = \sin x$	0	$\dfrac{1}{2}$	$\dfrac{\sqrt{3}}{2}$	1	$\dfrac{\sqrt{3}}{2}$	$\dfrac{1}{2}$	0	$-\dfrac{1}{2}$	$-\dfrac{\sqrt{3}}{2}$	-1	$-\dfrac{\sqrt{3}}{2}$	$-\dfrac{1}{2}$	0

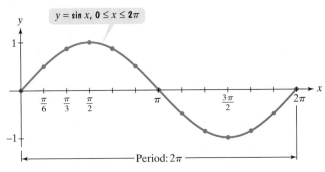

As x increases from 0 to $\dfrac{\pi}{2}$, y increases from 0 to 1.

As x increases from $\dfrac{\pi}{2}$ to π, y decreases from 1 to 0.

As x increases from π to $\dfrac{3\pi}{2}$, y decreases from 0 to −1.

As x increases from $\dfrac{3\pi}{2}$ to 2π, y increases from −1 to 0.

In plotting the points obtained in Table 4.3, we will use the approximatio
$\dfrac{\sqrt{3}}{2} \approx 0.87$. Rather than approximating π, we will mark off units on the x-axis i
terms of π. If we connect these points with a smooth curve, we obtain the grap
shown in Figure 4.62. The figure shows one period of the graph of $y = \sin x$.

$y = \sin x, 0 \le x \le 2\pi$

Figure 4.62 One period of the graph
of $y = \sin x$

We can obtain a more complete graph of $y = \sin x$ by continuing the portic
shown in Figure 4.62 to the left and to the right. The graph of the sine functio
called a **sine curve**, is shown in Figure 4.63. Any part of the graph that correspond
to one period (2π) is one cycle of the graph of $y = \sin x$.

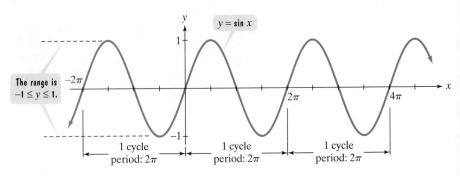

The range is $-1 \le y \le 1$.

$y = \sin x$

1 cycle period: 2π

1 cycle period: 2π

1 cycle period: 2π

Figure 4.63 The graph of $y = \sin x$

The graph of $y = \sin x$ allows us to visualize some of the properties of the si
function.

- The domain is $(-\infty, \infty)$, the set of all real numbers. The graph exten
indefinitely to the left and to the right with no gaps or holes.
- The range is $[-1, 1]$, the set of all real numbers between −1 and 1, inclusiv
The graph never rises above 1 or falls below −1.
- The period is 2π. The graph's pattern repeats in every interval of length 2
- The function is an odd function: $\sin(-x) = -\sin x$. This can be seen
observing that the graph is symmetric with respect to the origin.

Graphing Variations of $y = \sin x$

To graph variations of $y = \sin x$ by hand, it is helpful to find x-intercepts, maximum points, and minimum points. One complete cycle of the sine curve includes three x-intercepts, one maximum point, and one minimum point. The graph of $y = \sin x$ has x-intercepts at the beginning, middle, and end of its full period, shown in Figure 4.64. The curve reaches its maximum point $\frac{1}{4}$ of the way through the period. It reaches its minimum point $\frac{3}{4}$ of the

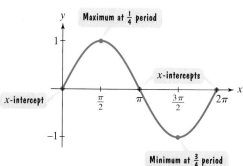

Figure 4.64 Key points in graphing the sine function

way through the period. Thus, key points in graphing sine functions are obtained by dividing the period into four equal parts. The x-coordinates of the five key points are as follows:

$$x_1 = \text{value of } x \text{ where the cycle begins}$$

$$x_2 = x_1 + \frac{\text{period}}{4}$$

$$x_3 = x_2 + \frac{\text{period}}{4}$$

$$x_4 = x_3 + \frac{\text{period}}{4}$$

$$x_5 = x_4 + \frac{\text{period}}{4}.$$

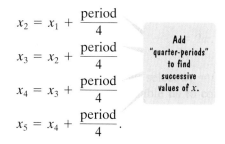

Add "quarter-periods" to find successive values of x.

The y-coordinates of the five key points are obtained by evaluating the given function at each of these values of x.

The graph of $y = \sin x$ forms the basis for graphing functions of the form

$$y = A \sin x.$$

For example, consider $y = 2 \sin x$, in which $A = 2$. We can obtain the graph of $y = 2 \sin x$ from that of $y = \sin x$ if we multiply each y-coordinate on the graph of $y = \sin x$ by 2. Figure 4.65 shows the graphs. The basic sine curve is *stretched* and ranges between -2 and 2, rather than between -1 and 1. However, both $y = \sin x$ and $y = 2 \sin x$ have a period of 2π.

In general, the graph of $y = A \sin x$ ranges between $-|A|$ and $|A|$. Thus, the range of the function is $-|A| \leq y \leq |A|$. If $|A| > 1$, the basic sine curve is *stretched*, as in Figure 4.65. If $|A| < 1$, the basic sine curve is *shrunk*. We call $|A|$ the **amplitude** of $y = A \sin x$. The maximum value of y on the graph of $y = A \sin x$ is $|A|$, the amplitude.

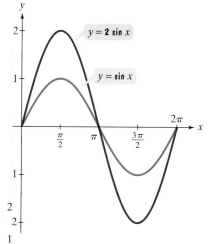

Figure 4.65 Comparing the graphs of $y = \sin x$ and $y = 2 \sin x$

Graphing Variations of $y = \sin x$

1. Identify the amplitude and the period.
2. Find the values of x for the five key points—the three x-intercepts, the maximum point, and the minimum point. Start with the value of x where the cycle begins and add quarter-periods—that is, $\dfrac{\text{period}}{4}$—to find successive values of x.
3. Find the values of y for the five key points by evaluating the function at each value of x from step 2.
4. Connect the five key points with a smooth curve and graph one complete cycle of the given function.
5. Extend the graph in step 4 to the left or right as desired.

② Graph variations of $y = \sin x$.

Now let us examine the graphs of functions of the form $y = A \sin(Bx - C)$ where $B > 0$. How do such graphs compare to those of functions of the form $y = A \sin Bx$? In both cases, the amplitude is $|A|$ and the period is $\dfrac{2\pi}{B}$. On complete cycle occurs if $Bx - C$ increases from 0 to 2π. This means that we can fin an interval containing one cycle by solving the following inequality:

$$0 \le Bx - C \le 2\pi. \qquad \text{$y = A \sin(Bx - C)$ completes one cycle as $Bx - C$ increases from 0 to 2π.}$$

$$C \le Bx \le C + 2\pi \qquad \text{Add C to all three parts.}$$

$$\frac{C}{B} \le x \le \frac{C}{B} + \frac{2\pi}{B} \qquad \text{Divide by B, where $B > 0$, and solve for x.}$$

> This is the x-coordinate on the left where the cycle begins.

> This is the x-coordinate on the right where the cycle ends. $\frac{2\pi}{B}$ is the period.

The voice balloon on the left indicates that the graph of $y = A \sin(Bx - C)$ is th graph of $y = A \sin Bx$ shifted horizontally by $\dfrac{C}{B}$. Thus, the number $\dfrac{C}{B}$ is the **pha: shift** associated with the graph.

The Graph of $y = A \sin(Bx - C)$

The graph of $y = A \sin(Bx - C)$ is obtained by horizontally shifting the graph of $y = A \sin Bx$ so that the starting point of the cycle is shifted from $x = 0$ to $x = \dfrac{C}{B}$. If $\dfrac{C}{B} > 0$, the shift is to the right. If $\dfrac{C}{B} < 0$, the shift is to the left. The number $\dfrac{C}{B}$ is called the **phase shift**.

$$\text{amplitude} = |A|$$

$$\text{period} = \frac{2\pi}{B}$$

$y = A \sin(Bx - C)$

Amplitude: $|A|$

Starting point: $x = \dfrac{C}{B}$

Period: $\dfrac{2\pi}{B}$

EXAMPLE 4   Graphing a Function of the Form $y = A \sin(Bx - C)$

Determine the amplitude, period, and phase shift of $y = 4 \sin\left(2x - \dfrac{2\pi}{3}\right)$. The graph one period of the function.

Solution

Step 1   Identify the amplitude, the period, and the phase shift. We must fir identify values for A, B, and C.

> The equation is of the form $y = A \sin(Bx - C)$.

$$y = 4 \sin\left(2x - \frac{2\pi}{3}\right)$$

Using the voice balloon, we see that $A = 4$, $B = 2$, and $C = \dfrac{2\pi}{3}$.

$$\text{amplitude:} \quad |A| = |4| = 4 \qquad \text{The maximum y is 4 and the minimum is -4.}$$

$$\text{period:} \quad \frac{2\pi}{B} = \frac{2\pi}{2} = \pi \qquad \text{Each cycle is of length π.}$$

$$\text{phase shift:} \quad \frac{C}{B} = \frac{\frac{2\pi}{3}}{2} = \frac{2\pi}{3} \cdot \frac{1}{2} = \frac{\pi}{3} \qquad \text{A cycle starts at $x = \frac{\pi}{3}$.}$$

Step 2   Find the x-values for the five key points. Begin by dividing the period, π, by 4.

$$\frac{\text{period}}{4} = \frac{\pi}{4}$$

Start with the value of x where the cycle begins: $x_1 = \dfrac{\pi}{3}$. Adding quarter-periods, $\dfrac{\pi}{4}$, the five x-values for the key points are

$$x_1 = \frac{\pi}{3}, \quad x_2 = \frac{\pi}{3} + \frac{\pi}{4} = \frac{4\pi}{12} + \frac{3\pi}{12} = \frac{7\pi}{12},$$

$$x_3 = \frac{7\pi}{12} + \frac{\pi}{4} = \frac{7\pi}{12} + \frac{3\pi}{12} = \frac{10\pi}{12} = \frac{5\pi}{6},$$

$$x_4 = \frac{5\pi}{6} + \frac{\pi}{4} = \frac{10\pi}{12} + \frac{3\pi}{12} = \frac{13\pi}{12},$$

$$x_5 = \frac{13\pi}{12} + \frac{\pi}{4} = \frac{13\pi}{12} + \frac{3\pi}{12} = \frac{16\pi}{12} = \frac{4\pi}{3}.$$

Study Tip

You can speed up the additions on the right by first writing the starting point, $\frac{\pi}{3}$, and the quarter-period, $\frac{\pi}{4}$, with a common denominator, 12.

starting point
$$= \frac{\pi}{3} = \frac{4\pi}{12}$$
quarter-period
$$= \frac{\pi}{4} = \frac{3\pi}{12}$$

Study Tip

You can check your computations for the x-values for the five key points. The difference between x_5 and x_1, or $x_5 - x_1$, should equal the period.

$$x_5 - x_1 = \frac{4\pi}{3} - \frac{\pi}{3} = \frac{3\pi}{3} = \pi$$

Because the period is π, this verifies that our five x-values are correct.

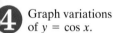 Graph variations of $y = \cos x$.

Step 3   Find the values of y for the five key points. We evaluate the function at each value of x from step 2.

Study Tip

If $B < 0$ in $y = A \cos$
$\cos(-\theta) = \cos\theta$ to rew
equation before obtaini
graph

Value of x	Value of y: $y = 4\sin\left(2x - \dfrac{2\pi}{3}\right)$	Coordinates of key point
$\dfrac{\pi}{3}$	$y = 4\sin\left(2 \cdot \dfrac{\pi}{3} - \dfrac{2\pi}{3}\right)$ $= 4\sin 0 = 4 \cdot 0 = 0$	$\left(\dfrac{\pi}{3}, 0\right)$
$\dfrac{7\pi}{12}$	$y = 4\sin\left(2 \cdot \dfrac{7\pi}{12} - \dfrac{2\pi}{3}\right)$ $= 4\sin\left(\dfrac{7\pi}{6} - \dfrac{2\pi}{3}\right)$ $= 4\sin\dfrac{3\pi}{6} = 4\sin\dfrac{\pi}{2} = 4 \cdot 1 = 4$	$\left(\dfrac{7\pi}{12}, 4\right)$ maximum point
$\dfrac{5\pi}{6}$	$y = 4\sin\left(2 \cdot \dfrac{5\pi}{6} - \dfrac{2\pi}{3}\right)$ $= 4\sin\left(\dfrac{5\pi}{3} - \dfrac{2\pi}{3}\right)$ $= 4\sin\dfrac{3\pi}{3} = 4\sin\pi = 4 \cdot 0 = 0$	$\left(\dfrac{5\pi}{6}, 0\right)$
$\dfrac{13\pi}{12}$	$y = 4\sin\left(2 \cdot \dfrac{13\pi}{12} - \dfrac{2\pi}{3}\right)$ $= 4\sin\left(\dfrac{13\pi}{6} - \dfrac{4\pi}{6}\right)$ $= 4\sin\dfrac{9\pi}{6} = 4\sin\dfrac{3\pi}{2} = 4(-1) = -4$	$\left(\dfrac{13\pi}{12}, -4\right)$ minimum point
$\dfrac{4\pi}{3}$	$y = 4\sin\left(2 \cdot \dfrac{4\pi}{3} - \dfrac{2\pi}{3}\right)$ $= 4\sin\dfrac{6\pi}{3} = 4\sin 2\pi = 4 \cdot 0 = 0$	$\left(\dfrac{4\pi}{3}, 0\right)$

Step 2   Find the x-values for the five key points. Begin by dividing the period, 4, by ◂

$$\frac{\text{period}}{4} = \frac{4}{4} = 1$$

Start with the value of x where the cycle begins: $x_1 = 0$. Adding quarter-periods, ▮ the five x-values for the key points are

$$x_1 = 0, \quad x_2 = 0 + 1 = 1, \quad x_3 = 1 + 1 = 2, \quad x_4 = 2 + 1 = 3, \quad x_5 = 3 + 1 = 4$$

Step 3   Find the values of y for the five key points. We evaluate the function ▮ each value of x from step 2.

Value of x	Value of y: $y = -3\cos\frac{\pi}{2}x$	Coordinates of key point	
0	$y = -3\cos\left(\frac{\pi}{2}\cdot 0\right)$ $= -3\cos 0 = -3\cdot 1 = -3$	$(0, -3)$	minimum point
1	$y = -3\cos\left(\frac{\pi}{2}\cdot 1\right)$ $= -3\cos\frac{\pi}{2} = -3\cdot 0 = 0$	$(1, 0)$	
2	$y = -3\cos\left(\frac{\pi}{2}\cdot 2\right)$ $= -3\cos\pi = -3(-1) = 3$	$(2, 3)$	maximum point
3	$y = -3\cos\left(\frac{\pi}{2}\cdot 3\right)$ $= -3\cos\frac{3\pi}{2} = -3\cdot 0 = 0$	$(3, 0)$	
4	$y = -3\cos\left(\frac{\pi}{2}\cdot 4\right)$ $= -3\cos 2\pi = -3\cdot 1 = -3$	$(4, -3)$	minimum point

In the interval $[0, 4]$, there are x-intercepts at 1 and 3. The minimum and maximum points are indicated by the voice balloons.

Step 4   Connect the five key points with a smooth curve and graph one complete cycle of the given function. The five key points for $y = -3\cos\frac{\pi}{2}x$ are shown in Figure 4.73. By connecting the points with a smooth curve, the blue portion shows one complete cycle of $y = -3\cos\frac{\pi}{2}x$ from 0 to 4.

Step 5   Extend the graph in step 4 to the left or right as desired. The blue portion of the graph in Figure 4.73 is for x from 0 to 4. In order to graph for $-4 \le x \le 4$, we continue this portion and extend the graph another full period to the left. This extension is shown in black in Figure 4.73.

Figure 4.73

③ Understan◂
of $y = \cos$

Technology

The graph of $y = -3\cos\frac{\pi}{2}x$ in a $[-4, 4, 1]$ by $[-4, 4, 1]$ viewing rectangle verifies our hand-drawn graph in Figure 4.73.

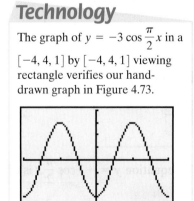

The range is $-1 \le y \le 1$.

$-\frac{3\pi}{2}$

Check Point 5   Determine the amplitude and period of $y = -4\cos\pi x$. Then graph th◂ function for $-2 \le x \le 2$.

Finally, let us examine the graphs of functions of the form $y = A\cos(Bx - C)$. Graphs of these functions shift the graph of $y = A\cos Bx$ horizontally by $\frac{C}{B}$.

> **The Graph of $y = A\cos(Bx - C)$**
>
> The graph of $y = A\cos(Bx - C)$ is obtained by horizontally shifting the graph of $y = A\cos Bx$ so that the starting point of the cycle is shifted from $x = 0$ to $x = \frac{C}{B}$. If $\frac{C}{B} > 0$, the shift is to the right. If $\frac{C}{B} < 0$, the shift is to the left.
>
> The number $\frac{C}{B}$ is called the **phase shift**.
>
> $$\text{amplitude} = |A|$$
>
> $$\text{period} = \frac{2\pi}{B}.$$
>
> $y = A\cos(Bx - C)$
> Amplitude: $|A|$
> Starting point: $x = \frac{C}{B}$
> Period: $\frac{2\pi}{B}$

EXAMPLE 6   Graphing a Function of the Form $y = A\cos(Bx - C)$

Determine the amplitude, period, and phase shift of $y = \frac{1}{2}\cos(4x + \pi)$. Then graph one period of the function.

Solution

Step 1   Identify the amplitude, the period, and the phase shift. We must first identify values for A, B, and C. To do this, we need to express the equation in the form $y = A\cos(Bx - C)$. Thus, we write $y = \frac{1}{2}\cos(4x + \pi)$ as $y = \frac{1}{2}\cos[4x - (-\pi)]$. Now we can identify values for A, B, and C.

The equation is of the form $y = A\cos(Bx - C)$.

$$y = \frac{1}{2}\cos[4x - (-\pi)]$$

Using the voice balloon, we see that $A = \frac{1}{2}$, $B = 4$, and $C = -\pi$.

amplitude: $|A| = \left|\frac{1}{2}\right| = \frac{1}{2}$ The maximum y is $\frac{1}{2}$ and the minimum is $-\frac{1}{2}$.

period: $\frac{2\pi}{B} = \frac{2\pi}{4} = \frac{\pi}{2}$ Each cycle is of length $\frac{\pi}{2}$.

phase shift: $\frac{C}{B} = -\frac{\pi}{4}$ A cycle starts at $x = -\frac{\pi}{4}$.

Step 2   Find the x-values for the five key points. Begin by dividing the period, $\frac{\pi}{2}$, by 4.

$$\frac{\text{period}}{4} = \frac{\frac{\pi}{2}}{4} = \frac{\pi}{8}$$

Start with the value of x where the cycle begins: $x_1 = -\dfrac{\pi}{4}$. Adding quarter-periods $\dfrac{\pi}{8}$, the five x-values for the key points are

$$x_1 = -\frac{\pi}{4}, \quad x_2 = -\frac{\pi}{4} + \frac{\pi}{8} = -\frac{2\pi}{8} + \frac{\pi}{8} = -\frac{\pi}{8}, \quad x_3 = -\frac{\pi}{8} + \frac{\pi}{8} = 0,$$

$$x_4 = 0 + \frac{\pi}{8} = \frac{\pi}{8}, \quad x_5 = \frac{\pi}{8} + \frac{\pi}{8} = \frac{2\pi}{8} = \frac{\pi}{4}.$$

Step 3 Find the values of y for the five key points. Take a few minutes and use your calculator to evaluate the function at each value of x from step 2. Show that the key points are

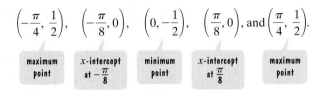

$$\left(-\frac{\pi}{4}, \frac{1}{2}\right), \quad \left(-\frac{\pi}{8}, 0\right), \quad \left(0, -\frac{1}{2}\right), \quad \left(\frac{\pi}{8}, 0\right), \text{ and } \left(\frac{\pi}{4}, \frac{1}{2}\right).$$

| maximum point | x-intercept at $-\frac{\pi}{8}$ | minimum point | x-intercept at $\frac{\pi}{8}$ | maximum point |

Technology

The graph of
$$y = \frac{1}{2}\cos(4x + \pi)$$
in a $\left[-\dfrac{\pi}{4}, \dfrac{\pi}{4}, \dfrac{\pi}{8}\right]$ by $[-1, 1, 1]$ viewing rectangle verifies our hand-drawn graph in Figure 4.74.

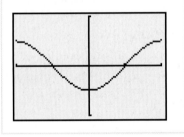

Step 4 Connect the five key points with a smooth curve and graph one complete cycle of the given function. The key points and the graph of $y = \frac{1}{2}\cos(4x + \pi)$ are shown in Figure 4.74.

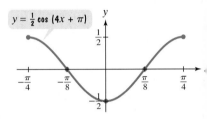

Figure 4.74

Check Point 6 Determine the amplitude, period, and phase shift of $y = \frac{3}{2}\cos(2x + \pi)$. Then graph one period of the function.

⑤ Use vertical shifts of sine and cosine curves.

Vertical Shifts of Sinusoidal Graphs

We now look at sinusoidal graphs of

$$y = A\sin(Bx - C) + D \quad \text{and} \quad y = A\cos(Bx - C) + D.$$

The constant D causes vertical shifts in the graphs of $y = A\sin(Bx - C)$ and $y = A\cos(Bx - C)$. If D is positive, the shift is D units upward. If D is negative, the shift is D units downward. These vertical shifts result in sinusoidal graphs oscillating about the horizontal line $y = D$ rather than about the x-axis. Thus, the maximum y is $D + |A|$ and the minimum y is $D - |A|$.

EXAMPLE 7 A Vertical Shift

Graph one period of the function $y = \frac{1}{2}\cos x - 1$.

Solution The graph of $y = \frac{1}{2}\cos x - 1$ is the graph of $y = \frac{1}{2}\cos x$ shifted one unit downward. The period of $y = \frac{1}{2}\cos x$ is 2π, which is also the period for the vertically shifted graph. The key points on the interval $[0, 2\pi]$ for $y = \frac{1}{2}\cos x - 1$ are found by first determining their x-coordinates. The quarter-period is $\dfrac{2\pi}{4}$, or $\dfrac{\pi}{2}$. The cycle begins at $x = 0$. As always, we add quarter-periods to generate x-values for each of the key points. The five x-values are

$$x_1 = 0, \quad x_2 = 0 + \frac{\pi}{2} = \frac{\pi}{2}, \quad x_3 = \frac{\pi}{2} + \frac{\pi}{2} = \pi,$$

$$x_4 = \pi + \frac{\pi}{2} = \frac{3\pi}{2}, \quad x_5 = \frac{3\pi}{2} + \frac{\pi}{2} = 2\pi.$$

The values of y for the five key points and their coordinates are determined as follows.

Value of x	Value of y: $y = \dfrac{1}{2}\cos x - 1$	Coordinates of key point
0	$y = \dfrac{1}{2}\cos 0 - 1$ $= \dfrac{1}{2} \cdot 1 - 1 = -\dfrac{1}{2}$	$\left(0, -\dfrac{1}{2}\right)$
$\dfrac{\pi}{2}$	$y = \dfrac{1}{2}\cos \dfrac{\pi}{2} - 1$ $= \dfrac{1}{2} \cdot 0 - 1 = -1$	$\left(\dfrac{\pi}{2}, -1\right)$
π	$y = \dfrac{1}{2}\cos \pi - 1$ $= \dfrac{1}{2}(-1) - 1 = -\dfrac{3}{2}$	$\left(\pi, -\dfrac{3}{2}\right)$
$\dfrac{3\pi}{2}$	$y = \dfrac{1}{2}\cos \dfrac{3\pi}{2} - 1$ $= \dfrac{1}{2} \cdot 0 - 1 = -1$	$\left(\dfrac{3\pi}{2}, -1\right)$
2π	$y = \dfrac{1}{2}\cos 2\pi - 1$ $= \dfrac{1}{2} \cdot 1 - 1 = -\dfrac{1}{2}$	$\left(2\pi, -\dfrac{1}{2}\right)$

The five key points for $y = \frac{1}{2}\cos x - 1$ are shown in Figure 4.75. By connecting the points with a smooth curve, we obtain one period of the graph.

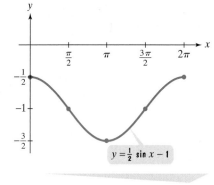

$y = \frac{1}{2}\sin x - 1$

Figure 4.75

Check Point 7 Graph one period of the function $y = 2\cos x + 1$.

Modeling Periodic Behavior

❻ Model periodic behavior.

Our breathing consists of alternating periods of inhaling and exhaling. Each complete pumping cycle of the human heart can be described using a sine function. Our brain waves during deep sleep are sinusoidal. Viewed in this way, trigonometry becomes an intimate experience.

Some graphing utilities have a SINe REGression feature. This feature gives the sine function in the form $y = A \sin(Bx + C) + D$ of best fit for wavelike data. At least four data points must be used. However, it is not always necessary to use technology. In our next example, we use our understanding of sinusoidal graphs model the process of breathing.

EXAMPLE 8 A Trigonometric Breath of Life

The graph in Figure 4.76 shows one complete normal breathing cycle. The cycle consists of inhaling and exhaling. It takes place every 5 seconds. Velocity of air flow is positive when we inhale and negative when we exhale. It is measured in liters per second. If y represents velocity of air flow after x seconds, find a function of the form $y = A \sin Bx$ that models air flow in a normal breathing cycle.

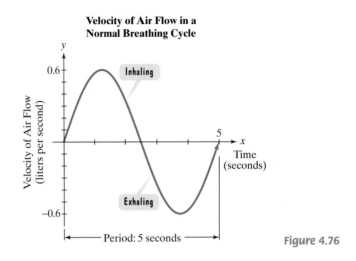

Velocity of Air Flow in a Normal Breathing Cycle

Figure 4.76

Solution We need to determine values for A and B in the equation $y = A \sin Bx$. A, the amplitude, is the maximum value of y. Figure 4.76 shows that this maximum value is 0.6. Thus, $A = 0.6$.

The value of B in $y = A \sin Bx$ can be found using the formula for the period, period $= \dfrac{2\pi}{B}$. The period of our breathing cycle is 5 seconds. Thus,

$$5 = \frac{2\pi}{B} \qquad \text{Our goal is to solve this equation for B.}$$

$$5B = 2\pi \qquad \text{Multiply both sides of the equation by B.}$$

$$B = \frac{2\pi}{5}. \qquad \text{Divide both sides of the equation by 5.}$$

We see that $A = 0.6$ and $B = \dfrac{2\pi}{5}$. Substitute these values into $y = A \sin Bx$. The breathing cycle is modeled by

$$y = 0.6 \sin \frac{2\pi}{5} x.$$

Check Point 8 Find an equation of the form $y = A \sin Bx$ that produces the graph shown in the figure on the right.

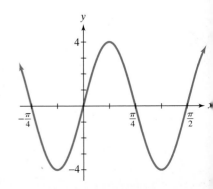

EXAMPLE 9   Modeling a Tidal Cycle

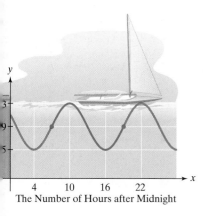

y

3-

9-

5-

4   10   16   22

The Number of Hours after Midnight

re 4.77

Figure 4.77 shows that the depth of water at a boat dock varies with the tides. The depth is 5 feet at low tide and 13 feet at high tide. On a certain day, low tide occurs at 4 A.M. and high tide at 10 A.M. If y represents the depth of the water, in feet, x hours after midnight, use a sine function of the form $y = A \sin(Bx - C) + D$ to model the water's depth.

Solution   We need to determine values for A, B, C, and D in the equation $y = A \sin(Bx - C) + D$. We can find these values using Figure 4.77. We begin with D.

To find D, we use the vertical shift. Because the water's depth ranges from a minimum of 5 feet to a maximum of 13 feet, the curve oscillates about the middle value, 9 feet. Thus, $D = 9$, which is the vertical shift.

At maximum depth, the water is 4 feet above 9 feet. Thus, A, the amplitude, is 4: $A = 4$.

To find B, we use the period. The blue portion of the graph shows that one complete tidal cycle occurs in $19 - 7$, or 12 hours. The period is 12. Thus,

$$12 = \frac{2\pi}{B} \qquad \text{\textit{Our goal is to solve this equation for B.}}$$

$$12B = 2\pi \qquad \text{\textit{Multiply both sides by B.}}$$

$$B = \frac{2\pi}{12} = \frac{\pi}{6}. \qquad \text{\textit{Divide both sides by 12.}}$$

To find C, we use the phase shift. The blue portion of the graph shows that the starting point of the cycle is shifted from 0 to 7. The phase shift, $\dfrac{C}{B}$, is 7.

$$7 = \frac{C}{B} \qquad \text{\textit{The phase shift of } y = A \sin(Bx - C) \text{ is } \frac{C}{B}.}$$

$$7 = \frac{C}{\dfrac{\pi}{6}} \qquad \text{\textit{From above, we have } B = \frac{\pi}{6}.}$$

$$\frac{7\pi}{6} = C \qquad \text{\textit{Multiply both sides of the equation by } \frac{\pi}{6}.}$$

We see that $A = 4$, $B = \dfrac{\pi}{6}$, $C = \dfrac{7\pi}{6}$, and $D = 9$. Substitute these values into $y = A \sin(Bx - C) + D$. The water's depth, in feet, x hours after midnight is modeled by

$$y = 4 \sin\left(\frac{\pi}{6}x - \frac{7\pi}{6}\right) + 9.$$

Technology

We can use a graphing utility to verify that the model in Example 9

$$y = 4 \sin\left(\frac{\pi}{6}x - \frac{7\pi}{6}\right) + 9$$

is correct. The graph of the function is shown in a $[0, 28, 4]$ by $[0, 15, 5]$ viewing rectangle.

Low tide: 5 feet at 4 A.M.

High tide: 13 feet at 10 A.M.

High tide

Low tide

15

10

5

0   4   8   12   16   20   24   28

Check Point 9   A region that is 30° north of the Equator averages a minimum of 10 hours of daylight in December. Hours of daylight are at a maximum of 14 hours in June. Let x represent the month of the year, with 1 for January, 2 for February, 3 for March, and 12 for December. If y represents the number of hours of daylight in month x, use a sine function of the form $y = A \sin(Bx - C) + D$ to model the hours of daylight.

EXERCISE SET 4.5

Practice Exercises

In Exercises 1–6, determine the amplitude of each function. Then graph the function and $y = \sin x$ in the same rectangular coordinate system for $0 \le x \le 2\pi$.

1. $y = 4 \sin x$

2. $y = 5 \sin x$

3. $y = \frac{1}{3} \sin x$

4. $y = \frac{1}{4} \sin x$

5. $y = -3 \sin x$

6. $y = -4 \sin x$

In Exercises 7–16, determine the amplitude and period of each function. Then graph one period of the function.

7. $y = \sin 2x$

8. $y = \sin 4x$

9. $y = 3 \sin \frac{1}{2} x$

10. $y = 2 \sin \frac{1}{4} x$

11. $y = 4 \sin \pi x$

12. $y = 3 \sin 2\pi x$

13. $y = -3 \sin 2\pi x$

14. $y = -2 \sin \pi x$

15. $y = -\sin \frac{2}{3} x$

16. $y = -\sin \frac{4}{3} x$

In Exercises 17–30, determine the amplitude, period, and phase shift of each function. Then graph one period of the function.

17. $y = \sin(x - \pi)$

18. $y = \sin\left(x - \frac{\pi}{2}\right)$

19. $y = \sin(2x - \pi)$

20. $y = \sin\left(2x - \frac{\pi}{2}\right)$

21. $y = 3 \sin(2x - \pi)$

22. $y = 3 \sin\left(2x - \frac{\pi}{2}\right)$

23. $y = \frac{1}{2} \sin\left(x + \frac{\pi}{2}\right)$

24. $y = \frac{1}{2} \sin(x + \pi)$

25. $y = -2 \sin\left(2x + \frac{\pi}{2}\right)$

26. $y = -3 \sin\left(2x + \frac{\pi}{2}\right)$

27. $y = 3 \sin(\pi x + 2)$

28. $y = 3 \sin(2\pi x + 4)$

29. $y = -2 \sin(2\pi x + 4\pi)$

30. $y = -3 \sin(2\pi x + 4\pi)$

In Exercises 31–34, determine the amplitude of each function. Then graph the function and $y = \cos x$ in the same rectangular coordinate system for $0 \le x \le 2\pi$.

31. $y = 2 \cos x$

32. $y = 3 \cos x$

33. $y = -2 \cos x$

34. $y = -3 \cos x$

In Exercises 35–42, determine the amplitude and period of each function. Then graph one period of the function.

35. $y = \cos 2x$

36. $y = \cos 4x$

37. $y = 4 \cos 2\pi x$

38. $y = 5 \cos 2\pi x$

39. $y = -4 \cos \frac{1}{2} x$

40. $y = -3 \cos \frac{1}{3} x$

41. $y = -\frac{1}{2} \cos \frac{\pi}{3} x$

42. $y = -\frac{1}{2} \cos \frac{\pi}{4} x$

In Exercises 43–52, determine the amplitude, period, and phase shift of each function. Then graph one period of the function.

43. $y = \cos\left(x - \frac{\pi}{2}\right)$

44. $y = \cos\left(x + \frac{\pi}{2}\right)$

45. $y = 3 \cos(2x - \pi)$

46. $y = 4 \cos(2x - \pi)$

47. $y = \frac{1}{2} \cos\left(3x + \frac{\pi}{2}\right)$

48. $y = \frac{1}{2} \cos(2x + \pi)$

49. $y = -3 \cos\left(2x - \frac{\pi}{2}\right)$

50. $y = -4 \cos\left(2x - \frac{\pi}{2}\right)$

51. $y = 2 \cos(2\pi x + 8\pi)$

52. $y = 3 \cos(2\pi x + 4\pi)$

In Exercises 53–60, use a vertical shift to graph one period of the function.

53. $y = \sin x + 2$

54. $y = \sin x - 2$

55. $y = \cos x - 3$

56. $y = \cos x + 3$

57. $y = 2 \sin \frac{1}{2} x + 1$

58. $y = 2 \cos \frac{1}{2} x + 1$

59. $y = -3 \cos 2\pi x + 2$

60. $y = -3 \sin 2\pi x + 2$

Practice Plus

In Exercises 61–66, find an equation for each graph.

61.

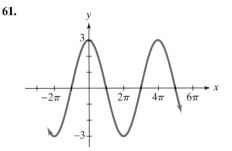

62.

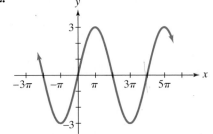

63.

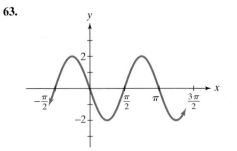

64.

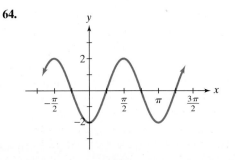

5.

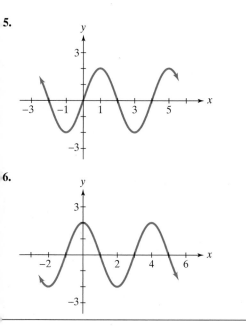

6.

In Exercises 67–70, graph one period of each function.

67. $y = \left| 2 \cos \dfrac{x}{2} \right|$

68. $y = \left| 3 \cos \dfrac{2x}{3} \right|$

69. $y = -\left| 3 \sin \pi x \right|$

70. $y = -\left| 2 \sin \dfrac{\pi x}{2} \right|$

In Exercises 71–74, graph f, g, and h in the same rectangular coordinate system for $0 \le x \le 2\pi$. Obtain the graph of h by adding or subtracting the corresponding y-coordinates on the graphs of f and g.

71. $f(x) = -2 \sin x,\ g(x) = \sin 2x,\ h(x) = (f + g)(x)$

72. $f(x) = 2 \cos x,\ g(x) = \cos 2x,\ h(x) = (f + g)(x)$

73. $f(x) = \sin x,\ g(x) = \cos 2x,\ h(x) = (f - g)(x)$

74. $f(x) = \cos x,\ g(x) = \sin 2x,\ h(x) = (f - g)(x)$

Application Exercises

In the theory of biorhythms, sine functions are used to measure a person's potential. You can obtain your biorhythm chart online by simply entering your date of birth, the date you want your biorhythm chart to begin, and the number of months you wish to be included in the plot. Shown below is your author's chart, beginning January 25, 2006, when he was 22,188 days old. We all have cycles with the same amplitudes and periods as those shown here. Each of our three basic cycles begins at birth. Use the biorhythm chart shown to solve Exercises 75–82. The longer tick marks correspond to the dates shown.

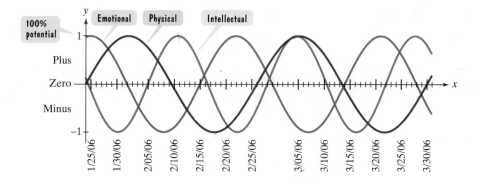

5. What is the period of the physical cycle?

6. What is the period of the emotional cycle?

7. What is the period of the intellectual cycle?

8. For the period shown, what is the worst day in February for your author to run in a marathon?

9. For the period shown, what is the best day in March for your author to meet an online friend for the first time?

0. For the period shown, what is the best day in February for your author to begin writing this trigonometry chapter?

1. If you extend these sinusoidal graphs to the end of the year, is there a day when your author should not even bother getting out of bed?

2. If you extend these sinusoidal graphs to the end of the year, are there any days where your author is at near-peak physical, emotional, and intellectual potential?

3. Rounded to the nearest hour, Los Angeles averages 14 hours of daylight in June, 10 hours in December, and 12 hours in

March and September. Let x represent the number of months after June and let y represent the number of hours of daylight in month x. Make a graph that displays the information from June of one year to June of the following year.

84. A clock with an hour hand that is 15 inches long is hanging on a wall. At noon, the distance between the tip of the hour hand and the ceiling is 23 inches. At 3 P.M., the distance is 38 inches; at 6 P.M., 53 inches; at 9 P.M., 38 inches; and at midnight the distance is again 23 inches. If y represents the distance between the tip of the hour hand and the ceiling x hours after noon, make a graph that displays the information for $0 \le x \le 24$.

85. The number of hours of daylight in Boston is given by

$$y = 3 \sin \dfrac{2\pi}{365}(x - 79) + 12,$$

where x is the number of days after January 1.

a. What is the amplitude of this function?

b. What is the period of this function?

c. How many hours of daylight are there on the longest day of the year?

d. How many hours of daylight are there on the shortest day of the year?

e. Graph the function for one period, starting on January 1.

86. The average monthly temperature, y, in degrees Fahrenheit, for Juneau, Alaska, can be modeled by $y = 16 \sin\left(\dfrac{\pi}{6}x - \dfrac{2\pi}{3}\right) + 40$, where x is the month of the year (January $= 1$, February $= 2, \ldots$ December $= 12$). Graph the function for $1 \le x \le 12$. What is the highest average monthly temperature? In which month does this occur?

87. The figure shows the depth of water at the end of a boat dock. The depth is 6 feet at low tide and 12 feet at high tide. On a certain day, low tide occurs at 6 A.M. and high tide at noon. If y represents the depth of the water x hours after midnight, use a cosine function of the form $y = A \cos Bx + D$ to model the water's depth.

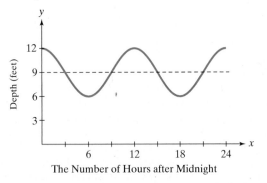

The Number of Hours after Midnight

88. The figure shows the depth of water at the end of a boat dock. The depth is 5 feet at high tide and 3 feet at low tide. On a certain day, high tide occurs at noon and low tide at 6 P.M. If y represents the depth of the water x hours after noon, use a cosine function of the form $y = A \cos Bx + D$ to model the water's depth.

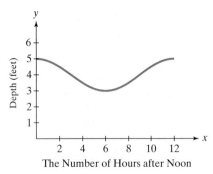

The Number of Hours after Noon

Writing in Mathematics

89. Without drawing a graph, describe the behavior of the basic sine curve.

90. What is the amplitude of the sine function? What does this tell you about the graph?

91. If you are given the equation of a sine function, how do you determine the period?

92. What does a phase shift indicate about the graph of a sine function? How do you determine the phase shift from the function's equation?

93. Describe a general procedure for obtaining the graph of $y = A \sin(Bx - C)$.

94. Without drawing a graph, describe the behavior of the basic cosine curve.

95. Describe a relationship between the graphs of $y = \sin x$ and $y = \cos x$.

96. Describe the relationship between the graphs of $y = A \cos(Bx - C)$ and $y = A \cos(Bx - C) + D$.

97. Biorhythm cycles provide interesting applications of sinusoidal graphs. But do you believe in the validity of biorhythms? Write a few sentences explaining why or why not.

Technology Exercises

98. Use a graphing utility to verify any five of the sine curves that you drew by hand in Exercises 7–30. The amplitude, period, and phase shift should help you to determine appropriate range settings.

99. Use a graphing utility to verify any five of the cosine curves that you drew by hand in Exercises 35–52.

100. Use a graphing utility to verify any two of the sinusoidal curves with vertical shifts that you drew in Exercises 53–60.

In Exercises 101–104, use a graphing utility to graph two periods of the function.

101. $y = 3 \sin(2x + \pi)$ **102.** $y = -2 \cos\left(2\pi x - \dfrac{\pi}{2}\right)$

103. $y = 0.2 \sin\left(\dfrac{\pi}{10}x + \pi\right)$ **104.** $y = 3 \sin(2x - \pi) + 5$

105. Use a graphing utility to graph $y = \sin x$ and $y = x - \dfrac{x^3}{6} + \dfrac{x^5}{120}$ in a $\left[-\pi, \pi, \dfrac{\pi}{2}\right]$ by $[-2, 2, 1]$ viewing rectangle. How do the graphs compare?

106. Use a graphing utility to graph $y = \cos x$ and $y = 1 - \dfrac{x^2}{2} + \dfrac{x^4}{24}$ in a $\left[-\pi, \pi, \dfrac{\pi}{2}\right]$ by $[-2, 2, 1]$ viewing rectangle. How do the graphs compare?

107. Use a graphing utility to graph
$$y = \sin x + \frac{\sin 2x}{2} + \frac{\sin 3x}{3} + \frac{\sin 4x}{4}$$
in a $\left[-2\pi, 2\pi, \dfrac{\pi}{2}\right]$ by $[-2, 2, 1]$ viewing rectangle. How do these waves compare to the smooth rolling waves of the basic sine curve?

108. Use a graphing utility to graph
$$y = \sin x - \frac{\sin 3x}{9} + \frac{\sin 5x}{25}$$
in a $\left[-2\pi, 2\pi, \dfrac{\pi}{2}\right]$ by $[-2, 2, 1]$ viewing rectangle. How do these waves compare to the smooth rolling waves of the basic sine curve?

109. The data show the average monthly temperatures for Washington, D.C.

   a. Use your graphing utility to draw a scatter plot of the data from $x = 1$ through $x = 12$.

   b. Use the SINe REGression feature to find the sinusoidal function of the form $y = A \sin(Bx + C) + D$ that best fits the data.

   c. Use your graphing utility to draw the sinusoidal function of best fit on the scatter plot.

x Month		Average Monthly Temperature, °F
1	(January)	34.6
2	(February)	37.5
3	(March)	47.2
4	(April)	56.5
5	(May)	66.4
6	(June)	75.6
7	(July)	80.0
8	(August)	78.5
9	(September)	71.3
10	(October)	59.7
11	(November)	49.8
12	(December)	39.4

Source: U.S. National Oceanic and Atmospheric Administration

110. Repeat Exercise 109 for data of your choice. The data can involve the average monthly temperatures for the region where you live or any data whose scatter plot takes the form of a sinusoidal function.

Critical Thinking Exercises

111. Determine the range of each of the following functions. Then give a viewing rectangle, or window, that shows two periods of the function's graph.

   a. $f(x) = 3 \sin\left(x + \dfrac{\pi}{6}\right) - 2$

   b. $g(x) = \sin 3\left(x + \dfrac{\pi}{6}\right) - 2$

112. Write the equation for a cosine function with amplitude π, period 1, and phase shift -2.

In Chapter 5, we will prove the following identities:

$$\sin^2 x = \frac{1}{2} - \frac{1}{2}\cos 2x$$

$$\cos^2 x = \frac{1}{2} + \frac{1}{2}\cos 2x.$$

Use these identities to solve Exercises 113–114.

113. Use the identity for $\sin^2 x$ to graph one period of $y = \sin^2 x$.

114. Use the identity for $\cos^2 x$ to graph one period of $y = \cos^2 x$.

Group Exercise

115. This exercise is intended to provide some fun with biorhythms, regardless of whether you believe they have any validity. We will use each member's chart to determine biorhythmic compatibility. Before meeting, each group member should go online and obtain his or her biorhythm chart. The date of the group meeting is the date on which your chart should begin. Include 12 months in the plot. At the meeting, compare differences and similarities among the intellectual sinusoidal curves. Using these comparisons, each person should find the one other person with whom he or she would be most intellectually compatible.

SECTION 4.6 *Graphs of Other Trigonometric Functions*

Objectives

❶ Understand the graph of $y = \tan x$.

❷ Graph variations of $y = \tan x$.

❸ Understand the graph of $y = \cot x$.

❹ Graph variations of $y = \cot x$.

❺ Understand the graphs of $y = \csc x$ and $y = \sec x$.

❻ Graph variations of $y = \csc x$ and $y = \sec x$.

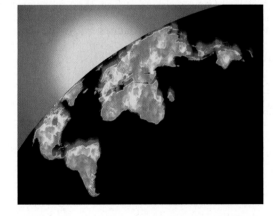

Recent advances in our understanding of climate have changed global warming from a subject for a disaster movie (the Statue of Liberty up to its chin in water) to a serious scientific and policy issue. Global warming is related to the burning of fossil fuels, which adds carbon dioxide to the atmosphere. In the 21st century, we will see whether our use of fossil fuels will add enough carbon dioxide to the atmosphere to change it (and our climate) in significant ways. In this section's exercise set, you will see how trigonometric graphs reveal interesting patterns in carbon dioxide concentration from 1990 through 2005. In the section itself trigonometric graphs will reveal patterns involving the tangent, cotangent, secant and cosecant functions.

① Understand the graph of $y = \tan x$.

The Graph of $y = \tan x$

The properties of the tangent function discussed in Section 4.2 will help us determine its graph. Because the tangent function has properties that are different from sinusoidal functions, its graph differs significantly from those of sine and cosine. Properties of the tangent function include the following:

- The period is π. It is only necessary to graph $y = \tan x$ over an interval of length π. The remainder of the graph consists of repetitions of that graph at intervals of π.

- The tangent function is an odd function: $\tan(-x) = -\tan x$. The graph is symmetric with respect to the origin.

- The tangent function is undefined at $\frac{\pi}{2}$. The graph of $y = \tan x$ has a vertical asymptote at $x = \frac{\pi}{2}$.

We obtain the graph of $y = \tan x$ using some points on the graph and origin symmetry. Table 4.5 lists some values of (x, y) on the graph of $y = \tan x$ on the interval $\left[0, \frac{\pi}{2}\right)$.

Table 4.5 Values of (x, y) on the graph of $y = \tan x$

x	0	$\frac{\pi}{6}$	$\frac{\pi}{4}$	$\frac{\pi}{3}$	$\frac{5\pi}{12}$ (75°)	$\frac{17\pi}{36}$ (85°)	$\frac{89\pi}{180}$ (89°)	1.57	$\frac{\pi}{2}$
$y = \tan x$	0	$\frac{\sqrt{3}}{3} \approx 0.6$	1	$\sqrt{3} \approx 1.7$	3.7	11.4	57.3	1255.8	undefined

As x increases from 0 to $\frac{\pi}{2}$, y increases slowly at first, then more and more rapidly.

The graph in Figure 4.78(a) is based on our observation that as x increases from 0 to $\frac{\pi}{2}$, y increases slowly at first, then more and more rapidly. Notice that y increases without bound as x approaches $\frac{\pi}{2}$. As the figure shows, the graph of $y = \tan x$ has a vertical asymptote at $x = \frac{\pi}{2}$.

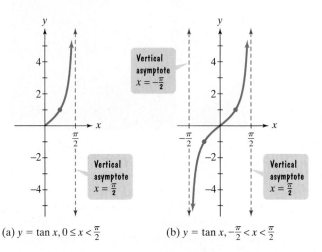

Figure 4.78 Graphing the tangent function

(a) $y = \tan x, 0 \le x < \frac{\pi}{2}$ (b) $y = \tan x, -\frac{\pi}{2} < x < \frac{\pi}{2}$

The graph of $y = \tan x$ can be completed on the interval $\left(-\dfrac{\pi}{2}, \dfrac{\pi}{2}\right)$ by using origin symmetry. Figure 4.78(b) shows the result of reflecting the graph in Figure 4.78(a) about the origin. The graph of $y = \tan x$ has another vertical asymptote at $x = -\dfrac{\pi}{2}$. Notice that y decreases without bound as x approaches $-\dfrac{\pi}{2}$.

Because the period of the tangent function is π, the graph in Figure 4.78(b) shows one complete period of $y = \tan x$. We obtain the complete graph of $y = \tan x$ by repeating the graph in Figure 4.78(b) to the left and right over intervals of π. The resulting graph and its main characteristics are shown in the following box:

The Tangent Curve: The Graph of $y = \tan x$ and Its Characteristics

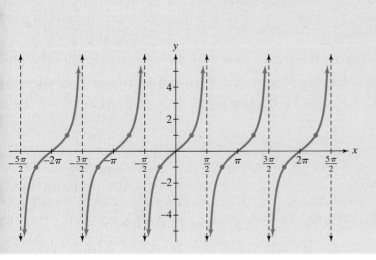

Characteristics

- **Period:** π
- **Domain:** All real numbers except odd multiples of $\dfrac{\pi}{2}$
- **Range:** All real numbers
- **Vertical asymptotes** at odd multiples of $\dfrac{\pi}{2}$
- **An x-intercept** occurs midway between each pair of consecutive asymptotes.
- **Odd function** with origin symmetry
- Points on the graph $\dfrac{1}{4}$ and $\dfrac{3}{4}$ of the way between consecutive asymptotes have y-coordinates of -1 and 1, respectively.

② Graph variations of $y = \tan x$.

Graphing Variations of $y = \tan x$

We use the characteristics of the tangent curve to graph tangent functions of the form $y = A \tan(Bx - C)$.

Graphing $y = A \tan(Bx - C)$

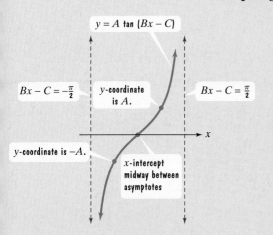

1. Find two consecutive asymptotes by finding an interval containing one period:
$$-\frac{\pi}{2} < Bx - C < \frac{\pi}{2}.$$
A pair of consecutive asymptotes occur at
$$Bx - C = -\frac{\pi}{2} \text{ and } Bx - C = \frac{\pi}{2}.$$

2. Identify an x-intercept, midway between the consecutive asymptotes.

3. Find the points on the graph $\dfrac{1}{4}$ and $\dfrac{3}{4}$ of the way between the consecutive asymptotes. These points have y-coordinates of $-A$ and A, respectively.

4. Use steps 1–3 to graph one full period of the function. Add additional cycles to the left or right as needed.

EXAMPLE 1 Graphing a Tangent Function

Graph $y = 2 \tan \dfrac{x}{2}$ for $-\pi < x < 3\pi$.

Solution Refer to Figure 4.79 as you read each step.

Step 1 Find two consecutive asymptotes. We do this by finding an interval containing one period.

$$-\frac{\pi}{2} < \frac{x}{2} < \frac{\pi}{2}$$
  Set up the inequality $-\dfrac{\pi}{2} <$ variable expression in tangent $< \dfrac{\pi}{2}$.

$$-\pi < x < \pi$$ Multiply all parts by 2 and solve for x.

An interval containing one period is $(-\pi, \pi)$. Thus, two consecutive asymptotes occur at $x = -\pi$ and $x = \pi$.

Step 2 Identify an x-intercept, midway between the consecutive asymptotes. Midway between $x = -\pi$ and $x = \pi$ is $x = 0$. An x-intercept is 0 and the graph passes through $(0, 0)$.

Step 3 Find points on the graph $\dfrac{1}{4}$ and $\dfrac{3}{4}$ of the way between the consecutive asymptotes. These points have y-coordinates of $-A$ and A. Because A, the coefficient of the tangent in $y = 2 \tan \dfrac{x}{2}$ is 2, these points have y-coordinates of -2 and 2. The graph passes through $\left(-\dfrac{\pi}{2}, -2\right)$ and $\left(\dfrac{\pi}{2}, 2\right)$.

Step 4 Use steps 1–3 to graph one full period of the function. We use the two consecutive asymptotes, $x = -\pi$ and $x = \pi$, an x-intercept of 0, and points midway between the x-intercept and asymptotes with y-coordinates of -2 and 2. We graph one period of $y = 2 \tan \dfrac{x}{2}$ from $-\pi$ to π. In order to graph for $-\pi < x < 3\pi$, we continue the pattern and extend the graph another full period to the right. The graph is shown in Figure 4.79.

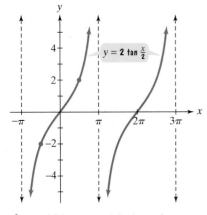

Figure 4.79 The graph is shown for two full periods.

Check Point 1 Graph $y = 3 \tan 2x$ for $-\dfrac{\pi}{4} < x < \dfrac{3\pi}{4}$.

EXAMPLE 2 Graphing a Tangent Function

Graph two full periods of $y = \tan\left(x + \dfrac{\pi}{4}\right)$.

Solution The graph of $y = \tan\left(x + \dfrac{\pi}{4}\right)$ is the graph of $y = \tan x$ shifted horizontally to the left $\dfrac{\pi}{4}$ units. Refer to Figure 4.80 as you read each step.

Step 1 Find two consecutive asymptotes. We do this by finding an interval containing one period.

$$-\frac{\pi}{2} < x + \frac{\pi}{4} < \frac{\pi}{2}$$ Set up the inequality $-\dfrac{\pi}{2} <$ variable expression in tangent $< \dfrac{\pi}{2}$.

$$-\frac{\pi}{2} - \frac{\pi}{4} < x < \frac{\pi}{2} - \frac{\pi}{4}$$ Subtract $\dfrac{\pi}{4}$ from all parts and solve for x.

$$-\frac{3\pi}{4} < x < \frac{\pi}{4}$$
  Simplify: $-\dfrac{\pi}{2} - \dfrac{\pi}{4} = -\dfrac{2\pi}{4} - \dfrac{\pi}{4} = -\dfrac{3\pi}{4}$
  and $\dfrac{\pi}{2} - \dfrac{\pi}{4} = \dfrac{2\pi}{4} - \dfrac{\pi}{4} = \dfrac{\pi}{4}$.

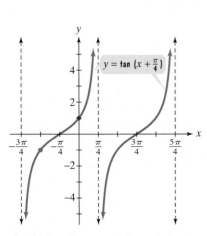

Figure 4.80 The graph is shown for two full periods.

An interval containing one period is $\left(-\dfrac{3\pi}{4}, \dfrac{\pi}{4}\right)$. Thus, two consecutive asymptotes occur at $x = -\dfrac{3\pi}{4}$ and $x = \dfrac{\pi}{4}$.

Step 2   Identify an x-intercept, midway between the consecutive asymptotes.

$$x\text{-intercept} = \dfrac{-\dfrac{3\pi}{4} + \dfrac{\pi}{4}}{2} = \dfrac{-\dfrac{2\pi}{4}}{2} = -\dfrac{2\pi}{8} = -\dfrac{\pi}{4}$$

An x-intercept is $-\dfrac{\pi}{4}$ and the graph passes through $\left(-\dfrac{\pi}{4}, 0 \right)$.

Step 3   Find points on the graph $\dfrac{1}{4}$ and $\dfrac{3}{4}$ of the way between the consecutive asymptotes. These points have y-coordinates of $-A$ and A. Because A, the coefficient of the tangent in $y = \tan\left(x + \dfrac{\pi}{4} \right)$ is 1, these points have y-coordinates of -1 and 1. They are shown as blue dots in Figure 4.80.

Step 4   Use steps 1–3 to graph one full period of the function. We use the two consecutive asymptotes, $x = -\dfrac{3\pi}{4}$ and $x = \dfrac{\pi}{4}$, to graph one full period of $y = \tan\left(x + \dfrac{\pi}{4} \right)$ from $-\dfrac{3\pi}{4}$ to $\dfrac{\pi}{4}$. We graph two full periods by continuing the pattern and extending the graph another full period to the right. The graph is shown in Figure 4.80.

Check Point 2   Graph two full periods of $y = \tan\left(x - \dfrac{\pi}{2} \right)$.

---

❸ Understand the graph of $y = \cot x$.

The Graph of $y = \cot x$

Like the tangent function, the cotangent function, $y = \cot x$, has a period of π. The graph and its main characteristics are shown in the following box:

The Cotangent Curve: The Graph of $y = \cot x$ and Its Characteristics

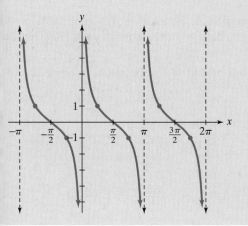

Characteristics

- **Period:** π
- **Domain:** All real numbers except integral multiples of π
- **Range:** All real numbers
- **Vertical asymptotes** at integral multiples of π
- **An x-intercept** occurs midway between each pair of consecutive asymptotes.
- **Odd function** with origin symmetry
- Points on the graph $\dfrac{1}{4}$ and $\dfrac{3}{4}$ of the way between consecutive asymptotes have y-coordinates of 1 and -1, respectively.

---

❹ Graph variations of $y = \cot x$.

Graphing Variations of $y = \cot x$

We use the characteristics of the cotangent curve to graph cotangent functions of the form $y = A \cot(Bx - C)$.

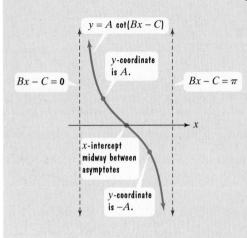

Graphing $y = A \cot(Bx - C)$

1. Find two consecutive asymptotes by finding an interval containing one full period:
$$0 < Bx - C < \pi.$$
A pair of consecutive asymptotes occur at
$$Bx - C = 0 \text{ and } Bx - C = \pi.$$

2. Identify an x-intercept, midway between the consecutive asymptotes.

3. Find the points on the graph $\frac{1}{4}$ and $\frac{3}{4}$ of the way between the consecutive asymptotes. These points have y-coordinates of A and $-A$, respectively.

4. Use steps 1–3 to graph one full period of the function. Add additional cycles to the left or right as needed.

EXAMPLE 3   Graphing a Cotangent Function

Graph $y = 3 \cot 2x$.

Solution   Refer to Figure 4.81 as you read each step.

Step 1   Find two consecutive asymptotes. We do this by finding an interval containing one period.

$$0 < 2x < \pi \qquad \text{Set up the inequality } 0 < \text{variable expression in cotangent} < \pi.$$

$$0 < x < \frac{\pi}{2} \qquad \text{Divide all parts by 2 and solve for } x.$$

An interval containing one period is $\left(0, \dfrac{\pi}{2}\right)$. Thus, two consecutive asymptotes occur at $x = 0$ and $x = \dfrac{\pi}{2}$.

Step 2   Identify an x-intercept, midway between the consecutive asymptotes. Midway between $x = 0$ and $x = \dfrac{\pi}{2}$ is $x = \dfrac{\pi}{4}$. An x-intercept is $\dfrac{\pi}{4}$ and the graph passes through $\left(\dfrac{\pi}{4}, 0\right)$.

Step 3   Find points on the graph $\frac{1}{4}$ and $\frac{3}{4}$ of the way between consecutive asymptotes. These points have y-coordinates of A and $-A$. Because A, the coefficient of the cotangent in $y = 3 \cot 2x$ is 3, these points have y-coordinates of 3 and -3. They are shown as blue dots in Figure 4.81.

Step 4   Use steps 1–3 to graph one full period of the function. We use the two consecutive asymptotes, $x = 0$ and $x = \dfrac{\pi}{2}$, to graph one full period of $y = 3 \cot 2x$. This curve is repeated to the left and right, as shown in Figure 4.81.

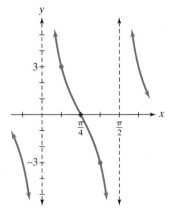

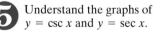
Figure 4.81 The graph of $y = 3 \cot 2x$

Check Point 3   Graph $y = \dfrac{1}{2} \cot \dfrac{\pi}{2} x$.

⑤ Understand the graphs of $y = \csc x$ and $y = \sec x$.

The Graphs of $y = \csc x$ and $y = \sec x$

We obtain the graphs of the cosecant and secant curves by using the reciprocal identities

$$\csc x = \frac{1}{\sin x} \quad \text{and} \quad \sec x = \frac{1}{\cos x}.$$

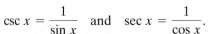

The identity $\csc x = \dfrac{1}{\sin x}$ tells us that the value of the cosecant function $y = \csc x$ at a given value of x equals the reciprocal of the corresponding value of the sine function, provided that the value of the sine function is not 0. If the value of $\sin x$ is 0, then at each of these values of x, the cosecant function is not defined. A vertical asymptote is associated with each of these values on the graph of $y = \csc x$.

We obtain the graph of $y = \csc x$ by taking reciprocals of the y-values in the graph of $y = \sin x$. Vertical asymptotes of $y = \csc x$ occur at the x-intercepts of $y = \sin x$. Likewise, we obtain the graph of $y = \sec x$ by taking the reciprocal of $y = \cos x$. Vertical asymptotes of $y = \sec x$ occur at the x-intercepts of $y = \cos x$. The graphs of $y = \csc x$ and $y = \sec x$ and their key characteristics are shown in the following boxes. We have used dashed red lines to graph $y = \sin x$ and $y = \cos x$ first, drawing vertical asymptotes through the x-intercepts.

The Cosecant Curve: The Graph of $y = \csc x$ and Its Characteristics

Characteristics

- **Period:** 2π

- **Domain:** All real numbers except integral multiples of π

- **Range:** All real numbers y such that $y \le -1$ or $y \ge 1: (-\infty, -1] \cup [1, \infty)$

- **Vertical asymptotes** at integral multiples of π

- **Odd function,** $\csc(-x) = -\csc x$, with origin symmetry

The Secant Curve: The Graph of $y = \sec x$ and Its Characteristics

Characteristics

- **Period:** 2π

- **Domain:** All real numbers except odd multiples of $\dfrac{\pi}{2}$

- **Range:** All real numbers y such that $y \le -1$ or $y \ge 1: (-\infty, -1] \cup [1, \infty)$

- **Vertical asymptotes** at odd multiples of $\dfrac{\pi}{2}$

- **Even function,** $\sec(-x) = \sec x$, with y-axis symmetry

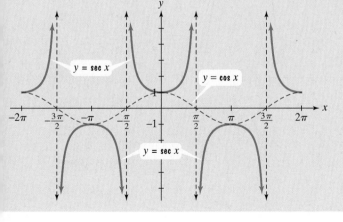

⑥ Graph variations of $y = \csc x$ and $y = \sec x$.

Graphing Variations of $y = \csc x$ and $y = \sec x$

We use graphs of functions involving the corresponding reciprocal functions to obtain graphs of cosecant and secant functions. To graph a cosecant or secant curve, begin by graphing the function where cosecant or secant is replaced by its reciprocal function. For example, to graph $y = 2 \csc 2x$, we use the graph of $y = 2 \sin 2x$. Likewise, to graph $y = -3 \sec \dfrac{x}{2}$, we use the graph of $y = -3 \cos \dfrac{x}{2}$.

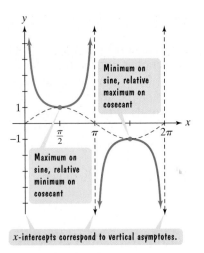

Minimum on sine, relative maximum on cosecant

Maximum on sine, relative minimum on cosecant

x-intercepts correspond to vertical asymptotes.

Figure 4.82

Figure 4.82 illustrates how we use a sine curve to obtain a cosecant curve. Notice that

- *x*-intercepts on the red sine curve correspond to vertical asymptotes of the blue cosecant curve.
- A maximum point on the red sine curve corresponds to a minimum point on a continuous portion of the blue cosecant curve.
- A minimum point on the red sine curve corresponds to a maximum point on a continuous portion of the blue cosecant curve.

EXAMPLE 4   Using a Sine Curve to Obtain a Cosecant Curve

Use the graph of $y = 2 \sin 2x$ in Figure 4.83 to obtain the graph of $y = 2 \csc 2x$.

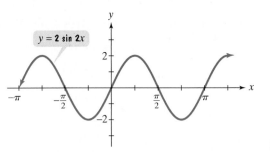

$y = 2 \sin 2x$

Figure 4.83

Solution   We begin our work in Figure 4.84 by showing the given graph, the graph of $y = 2 \sin 2x$, using dashed red lines. The *x*-intercepts of $y = 2 \sin 2x$ correspond to the vertical asymptotes of $y = 2 \csc 2x$. Thus, we draw vertical asymptotes through the *x*-intercepts, shown in Figure 4.84. Using the asymptotes as guides, we sketch the graph of $y = 2 \csc 2x$ in Figure 4.84.

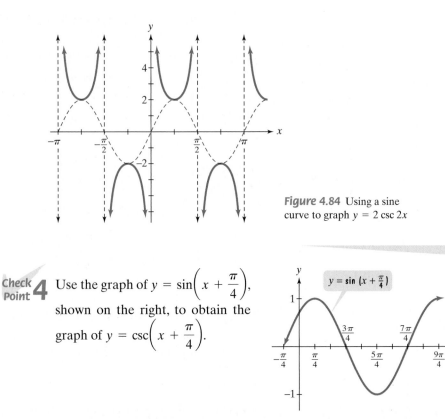

Figure 4.84 Using a sine curve to graph $y = 2 \csc 2x$

Check Point 4   Use the graph of $y = \sin\left(x + \dfrac{\pi}{4}\right)$, shown on the right, to obtain the graph of $y = \csc\left(x + \dfrac{\pi}{4}\right)$.

$y = \sin\left(x + \frac{\pi}{4}\right)$

We use a cosine curve to obtain a secant curve in exactly the same way we used a sine curve to obtain a cosecant curve. Thus,

- *x*-intercepts on the cosine curve correspond to vertical asymptotes on the secant curve.
- A maximum point on the cosine curve corresponds to a minimum point on a continuous portion of the secant curve.
- A minimum point on the cosine curve corresponds to a maximum point on a continuous portion of the secant curve.

EXAMPLE 5   Graphing a Secant Function

Graph $y = -3 \sec \dfrac{x}{2}$ for $-\pi < x < 5\pi$.

Solution   We begin by graphing the function $y = -3 \cos \dfrac{x}{2}$, where secant has been replaced by cosine, its reciprocal function. This equation is of the form $y = A \cos Bx$ with $A = -3$ and $B = \frac{1}{2}$.

$$\text{amplitude:} \quad |A| = |-3| = 3$$

> The maximum *y* is **3** and the minimum is **−3**.

$$\text{period:} \quad \frac{2\pi}{B} = \frac{2\pi}{\frac{1}{2}} = 4\pi$$

> Each cycle, including asymptotes, is of length 4π.

We use quarter-periods, $\dfrac{4\pi}{4}$, or π, to find the *x*-values for the five key points. Starting with $x = 0$, the *x*-values are $0, \pi, 2\pi, 3\pi$, and 4π. Evaluating the function at each of these values of *x*, the key points are

$$(0, -3), (\pi, 0), (2\pi, 3), (3\pi, 0), \text{ and } (4\pi, -3).$$

We use these key points to graph $y = -3 \cos \dfrac{x}{2}$ from 0 to 4π, shown using a dashed red line in Figure 4.85. In order to graph for $-\pi \le x \le 5\pi$, extend the dashed red graph π units to the left and π units to the right. Now use this dashed red graph to obtain the graph of the corresponding secant function, its reciprocal function. Draw vertical asymptotes through the *x*-intercepts. Using these asymptotes as guides, the graph of $y = -3 \sec \dfrac{x}{2}$ is shown in blue in Figure 4.85.

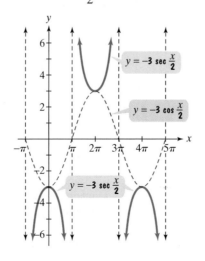

Figure 4.85 Using a cosine curve to graph $y = -3 \sec \dfrac{x}{2}$

Check Point 5 Graph $y = 2 \sec 2x$ for $-\dfrac{3\pi}{4} < x < \dfrac{3\pi}{4}$.

The Six Curves of Trigonometry

Table 4.6 summarizes the graphs of the six trigonometric functions. Below each of the graphs is a description of the domain, range, and period of the function.

Table 4.6   Graphs of the Six Trigonometric Functions

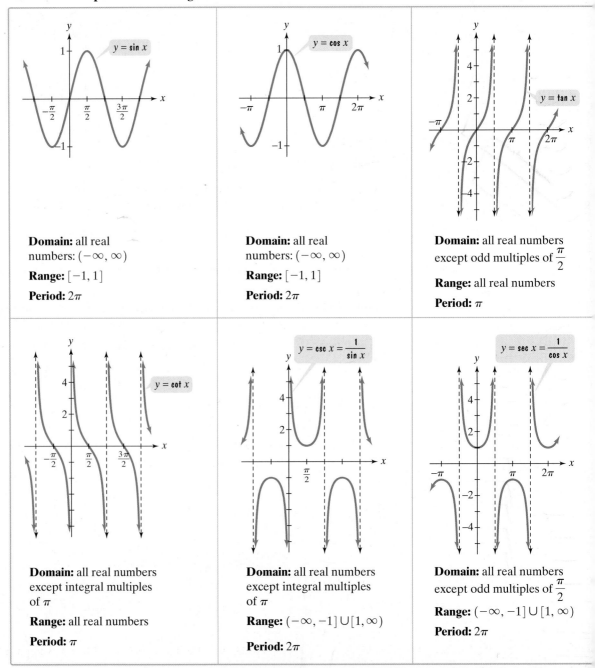

Domain: all real numbers: $(-\infty, \infty)$

Range: $[-1, 1]$

Period: 2π

Domain: all real numbers: $(-\infty, \infty)$

Range: $[-1, 1]$

Period: 2π

Domain: all real numbers except odd multiples of $\dfrac{\pi}{2}$

Range: all real numbers

Period: π

Domain: all real numbers except integral multiples of π

Range: all real numbers

Period: π

Domain: all real numbers except integral multiples of π

Range: $(-\infty, -1] \cup [1, \infty)$

Period: 2π

Domain: all real numbers except odd multiples of $\dfrac{\pi}{2}$

Range: $(-\infty, -1] \cup [1, \infty)$

Period: 2π

EXERCISE SET 4.6

Practice Exercises

In Exercises 1–4, the graph of a tangent function is given.
Select the equation for each graph from the following options:

$$y = \tan\left(x + \frac{\pi}{2}\right), \quad y = \tan(x + \pi), \quad y = -\tan x, \quad y = -\tan\left(x - \frac{\pi}{2}\right).$$

1.   **2.**   **3.**   **4.**

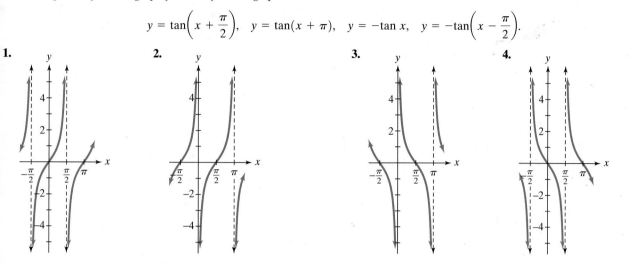

In Exercises 5–12, graph two periods of the given tangent function.

5. $y = 3 \tan \dfrac{x}{4}$ **6.** $y = 2 \tan \dfrac{x}{4}$ **7.** $y = \dfrac{1}{2} \tan 2x$ **8.** $y = 2 \tan 2x$

9. $y = -2 \tan \dfrac{1}{2}x$ **10.** $y = -3 \tan \dfrac{1}{2}x$ **11.** $y = \tan(x - \pi)$ **12.** $y = \tan\left(x - \dfrac{\pi}{4}\right)$

In Exercises 13–16, the graph of a cotangent function is given. Select the equation for each graph from the following options:

$$y = \cot\left(x + \frac{\pi}{2}\right), \quad y = \cot(x + \pi), \quad y = -\cot x, \quad y = -\cot\left(x - \frac{\pi}{2}\right).$$

13.   **14.**   **15.**   **16.**

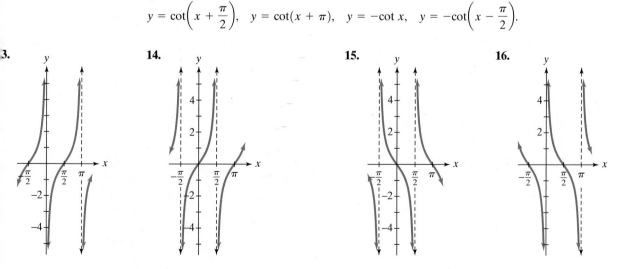

In Exercises 17–24, graph two periods of the given cotangent function.

17. $y = 2 \cot x$ **18.** $y = \dfrac{1}{2} \cot x$ **19.** $y = \dfrac{1}{2} \cot 2x$ **20.** $y = 2 \cot 2x$

21. $y = -3 \cot \dfrac{\pi}{2} x$ **22.** $y = -2 \cot \dfrac{\pi}{4} x$ **23.** $y = 3 \cot\left(x + \dfrac{\pi}{2}\right)$ **24.** $y = 3 \cot\left(x + \dfrac{\pi}{4}\right)$

In Exercises 25–28, use each graph to obtain the graph of the corresponding reciprocal function, cosecant or secant. Give the equation of the function for the graph that you obtain.

25.

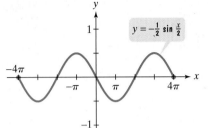

$y = -\frac{1}{2}\sin\frac{x}{2}$

26.

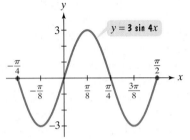

$y = 3\sin 4x$

27.

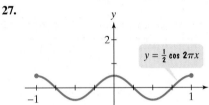

$y = \frac{1}{2}\cos 2\pi x$

28.

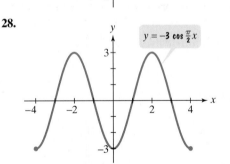

$y = -3\cos\frac{\pi}{2}x$

In Exercises 29–44, graph two periods of the given cosecant or secant function.

29. $y = 3\csc x$ **30.** $y = 2\csc x$

31. $y = \frac{1}{2}\csc\frac{x}{2}$ **32.** $y = \frac{3}{2}\csc\frac{x}{4}$

33. $y = 2\sec x$ **34.** $y = 3\sec x$

35. $y = \sec\frac{x}{3}$ **36.** $y = \sec\frac{x}{2}$

37. $y = -2\csc\pi x$ **38.** $y = -\frac{1}{2}\csc\pi x$

39. $y = -\frac{1}{2}\sec\pi x$ **40.** $y = -\frac{3}{2}\sec\pi x$

41. $y = \csc(x - \pi)$ **42.** $y = \csc\left(x - \frac{\pi}{2}\right)$

43. $y = 2\sec(x + \pi)$ **44.** $y = 2\sec\left(x + \frac{\pi}{2}\right)$

Practice Plus

In Exercises 45–52, graph two periods of each function.

45. $y = 2\tan\left(x - \frac{\pi}{6}\right) + 1$ **46.** $y = 2\cot\left(x + \frac{\pi}{6}\right) - 1$

47. $y = \sec\left(2x + \frac{\pi}{2}\right) - 1$ **48.** $y = \csc\left(2x - \frac{\pi}{2}\right) + 1$

49. $y = \csc|x|$ **50.** $y = \sec|x|$

51. $y = |\cot\frac{1}{2}x|$ **52.** $y = |\tan\frac{1}{2}x|$

In Exercises 53–54, let $f(x) = 2\sec x$, $g(x) = -2\tan x$, and $h(x) = 2x - \frac{\pi}{2}$.

53. Graph two periods of
$$y = (f \circ h)(x).$$

54. Graph two periods of
$$y = (g \circ h)(x).$$

In Exercises 55–58, use a graph to solve each equation for $-2\pi \le x \le 2\pi$.

55. $\tan x = -1$ **56.** $\cot x = -1$

57. $\csc x = 1$ **58.** $\sec x = 1$

Application Exercises

59. An ambulance with a rotating beam of light is parked 12 feet from a building. The function
$$d = 12\tan 2\pi t$$
describes the distance, d, in feet, of the rotating beam of light from point C after t seconds.

a. Graph the function on the interval $[0, 2]$.

b. For what values of t in $[0, 2]$ is the function undefined? What does this mean in terms of the rotating beam of light in the figure shown?

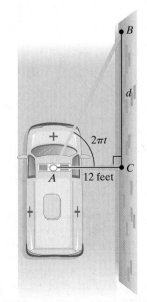

50. The angle of elevation from the top of a house to a jet flying 2 miles above the house is x radians. If d represents the horizontal distance, in miles, of the jet from the house, express d in terms of a trigonometric function of x. Then graph the function for $0 < x < \pi$.

51. Your best friend is marching with a band and has asked you to film him. The figure below shows that you have set yourself up 10 feet from the street where your friend will be passing from left to right. If d represents your distance, in feet, from your friend and x is the radian measure of the angle shown, express d in terms of a trigonometric function of x. Then graph the function for $-\frac{\pi}{2} < x < \frac{\pi}{2}$. Negative angles indicate that your marching buddy is on your left.

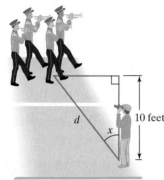

In Exercises 62–64, sketch a reasonable graph that models the given situation.

62. The number of hours of daylight per day in your hometown over a two-year period

63. The motion of a diving board vibrating 10 inches in each direction per second just after someone has dived off

64. The distance of a rotating beam of light from a point on a wall (See the figure for Exercise 59.)

Writing in Mathematics

55. Without drawing a graph, describe the behavior of the basic tangent curve.

56. If you are given the equation of a tangent function, how do you find a pair of consecutive asymptotes?

57. If you are given the equation of a tangent function, how do you identify an x-intercept?

58. Without drawing a graph, describe the behavior of the basic cotangent curve.

59. If you are given the equation of a cotangent function, how do you find a pair of consecutive asymptotes?

70. Explain how to determine the range of $y = \csc x$ from the graph. What is the range?

71. Explain how to use a sine curve to obtain a cosecant curve. Why can the same procedure be used to obtain a secant curve from a cosine curve?

72. Scientists record brain activity by attaching electrodes to the scalp and then connecting these electrodes to a machine. The brain activity recorded with this machine is shown in the three graphs at the top of the next column. Which trigonometric functions would be most appropriate for describing the oscillations in brain activity? Describe similarities and differences among these functions when modeling brain

activity when awake, during dreaming sleep, and during non-dreaming sleep.

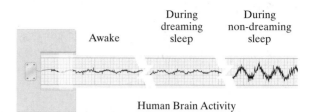

Human Brain Activity

Technology Exercises

In working Exercises 73–76, describe what happens at the asymptotes on the graphing utility. Compare the graphs in the connected and dot modes.

73. Use a graphing utility to verify any two of the tangent curves that you drew by hand in Exercises 5–12.

74. Use a graphing utility to verify any two of the cotangent curves that you drew by hand in Exercises 17–24.

75. Use a graphing utility to verify any two of the cosecant curves that you drew by hand in Exercises 29–44.

76. Use a graphing utility to verify any two of the secant curves that you drew by hand in Exercises 29–44.

In Exercises 77–82, use a graphing utility to graph each function. Use a range setting so that the graph is shown for at least two periods.

77. $y = \tan \frac{x}{4}$ **78.** $y = \tan 4x$

79. $y = \cot 2x$ **80.** $y = \cot \frac{x}{2}$

81. $y = \frac{1}{2} \tan \pi x$ **82.** $y = \frac{1}{2} \tan(\pi x + 1)$

In Exercises 83–86, use a graphing utility to graph each pair of functions in the same viewing rectangle. Use a range setting so that the graphs are shown for at least two periods.

83. $y = 0.8 \sin \frac{x}{2}$ and $y = 0.8 \csc \frac{x}{2}$

84. $y = -2.5 \sin \frac{\pi}{3} x$ and $y = -2.5 \csc \frac{\pi}{3} x$

85. $y = 4 \cos\left(2x - \frac{\pi}{6}\right)$ and $y = 4 \sec\left(2x - \frac{\pi}{6}\right)$

86. $y = -3.5 \cos\left(\pi x - \frac{\pi}{6}\right)$ and $y = -3.5 \sec\left(\pi x - \frac{\pi}{6}\right)$

87. Carbon dioxide particles in our atmosphere trap heat and raise the planet's temperature. The resultant gradually increasing temperature is called the greenhouse effect. Carbon dioxide accounts for about half of global warming. The function

$$y = 2.5 \sin 2\pi x + 0.0216x^2 + 0.654x + 316$$

models carbon dioxide concentration, y, in parts per million, where $x = 0$ represents January 1960; $x = \frac{1}{12}$, February 1960; $x = \frac{2}{12}$, March 1960; $\ldots, x = 1$, January 1961; $x = \frac{13}{12}$, February 1961; and so on. Use a graphing utility to graph the function in a $[30, 45, 5]$ by $[310, 420, 5]$ viewing rectangle. Describe what the graph reveals about carbon dioxide concentration from 1990 through 2005.

88. Graph $y = \sin \frac{1}{x}$ in a $[-0.2, 0.2, 0.01]$ by $[-1.2, 1.2, 0.01]$ viewing rectangle. What is happening as x approaches 0 from the left or the right? Explain this behavior.

Critical Thinking Exercises

In Exercises 89–90, write an equation for each blue graph.

89.

90.

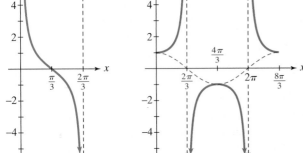

In Exercises 91–92, write the equation for a cosecant function satisfying the given conditions.

91. period: 3π; range: $(-\infty, -2] \cup [2, \infty)$

92. period: 2; range: $(-\infty, -\pi] \cup [\pi, \infty)$

93. Determine the range of the following functions. Then give a viewing rectangle, or window, that shows two periods of the function's graph.

a. $f(x) = \sec\left(3x + \dfrac{\pi}{2}\right)$ **b.** $g(x) = 3\sec\pi\left(x + \dfrac{1}{2}\right)$

94. For $x > 0$, what effect does 2^{-x} in $y = 2^{-x}\sin x$ have on the graph of $y = \sin x$? What kind of behavior can be modeled by a function such as $y = 2^{-x}\sin x$?

SECTION 4.7   *Inverse Trigonometric Functions*

Objectives

❶ Understand and use the inverse sine function.

❷ Understand and use the inverse cosine function.

❸ Understand and use the inverse tangent function.

❹ Use a calculator to evaluate inverse trigonometric functions.

❺ Find exact values of composite functions with inverse trigonometric functions.

In 2005, director George Lucas pulled out all the stops and completed the epic *Star Wars* odyssey with *Revenge of the Sith*. The movie is being shown at a local theater where you can experience the stunning force of its 2151 visual-effect shots (*Source: Time*) on a large screen. Where in the theater should you sit to maximize the visual impact of the director's fantastic galactic visions? In this section's exercise set, you will see how an inverse trigonometric function can enhance your movie-going experiences.

Study Tip

Here are some helpful things to remember from our earlier discussion of inverse functions.

- If no horizontal line intersects the graph of a function more than once, the function is one-to-one and has an inverse function.
- If the point (a, b) is on the graph of f, then the point (b, a) is on the graph of the inverse function, denoted f^{-1}. The graph of f^{-1} is a reflection of the graph of f about the line $y = x$.

① Understand and use the inverse sine function.

The Inverse Sine Function

Figure 4.86 shows the graph of $y = \sin x$. Can you see that every horizontal line that can be drawn between -1 and 1 intersects the graph infinitely many times? Thus, the sine function is not one-to-one and has no inverse function.

In Figure 4.87, we have taken a portion of the sine curve, restricting the domain of the sine function to $-\dfrac{\pi}{2} \le x \le \dfrac{\pi}{2}$. With this restricted domain, every horizontal line that can be drawn between -1 and 1 intersects the graph exactly once. Thus, the restricted function passes the horizontal line test and is one-to-one.

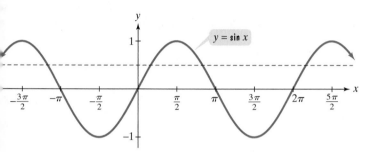

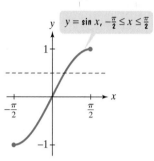

Figure 4.86 The horizontal line test shows that the sine function is not one-to-one and has no inverse function.

Figure 4.87 The restricted sine function passes the horizontal line test. It is one-to-one and has an inverse function.

On the restricted domain $-\dfrac{\pi}{2} \le x \le \dfrac{\pi}{2}$, $y = \sin x$ has an inverse function.

The inverse of the restricted sine function is called the **inverse sine function**. Two notations are commonly used to denote the inverse sine function:

$$y = \sin^{-1} x \quad \text{or} \quad y = \arcsin x.$$

In this book, we will use $y = \sin^{-1} x$. This notation has the same symbol as the inverse function notation $f^{-1}(x)$.

The Inverse Sine Function

The **inverse sine function**, denoted by $\sin^{-1}$, is the inverse of the restricted sine function $y = \sin x$, $-\dfrac{\pi}{2} \le x \le \dfrac{\pi}{2}$. Thus,

$$y = \sin^{-1} x \quad \text{means} \quad \sin y = x,$$

where $-\dfrac{\pi}{2} \le y \le \dfrac{\pi}{2}$ and $-1 \le x \le 1$. We read $y = \sin^{-1} x$ as "y equals the inverse sine at x."

Study Tip

The notation $y = \sin^{-1} x$ does not mean $y = \dfrac{1}{\sin x}$. The notation $y = \dfrac{1}{\sin x}$, or the reciprocal of the sine function, is written $y = (\sin x)^{-1}$ and means $y = \csc x$.

Inverse sine function   Reciprocal of sine function

$$y = \sin^{-1} x \qquad y = (\sin x)^{-1} = \dfrac{1}{\sin x} = \csc x$$

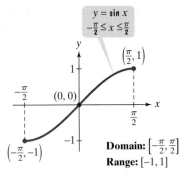
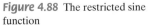

Domain: $\left[-\frac{\pi}{2}, \frac{\pi}{2}\right]$
Range: $[-1, 1]$

Figure 4.88 The restricted sine function

One way to graph $y = \sin^{-1} x$ is to take points on the graph of the restricted sine function and reverse the order of the coordinates. For example, Figure 4.88 shows that $\left(-\frac{\pi}{2}, -1\right)$, $(0, 0)$, and $\left(\frac{\pi}{2}, 1\right)$ are on the graph of the restricted sine function. Reversing the order of the coordinates gives $\left(-1, -\frac{\pi}{2}\right)$, $(0, 0)$, and $\left(1, \frac{\pi}{2}\right)$. We now use these three points to sketch the inverse sine function. The graph of $y = \sin^{-1} x$ is shown in Figure 4.89.

Another way to obtain the graph of $y = \sin^{-1} x$ is to reflect the graph of the restricted sine function about the line $y = x$, shown in Figure 4.90. The red graph is the restricted sine function and the blue graph is the graph of $y = \sin^{-1} x$.

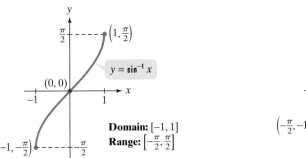

Domain: $[-1, 1]$
Range: $\left[-\frac{\pi}{2}, \frac{\pi}{2}\right]$

Figure 4.89 The graph of the inverse sine function

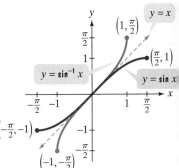

Figure 4.90 Using a reflection to obtain the graph of the inverse sine function

Exact values of $\sin^{-1} x$ can be found by thinking of $\mathbf{\sin^{-1} x}$ **as the angle in the interval** $\left[-\frac{\pi}{2}, \frac{\pi}{2}\right]$ **whose sine is** x. For example, we can use the two points on the blue graph of the inverse sine function in Figure 4.90 and write

$$\sin^{-1}(-1) = -\frac{\pi}{2} \quad \text{and} \quad \sin^{-1} 1 = \frac{\pi}{2}.$$

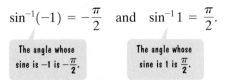

Because we are thinking of $\sin^{-1} x$ in terms of an angle, we will represent such an angle by θ.

Finding Exact Values of $\sin^{-1} x$

1. Let $\theta = \sin^{-1} x$.

2. Rewrite $\theta = \sin^{-1} x$ as $\sin \theta = x$, where $-\frac{\pi}{2} \leq \theta \leq \frac{\pi}{2}$.

3. Use the exact values in Table 4.7 to find the value of θ in $\left[-\frac{\pi}{2}, \frac{\pi}{2}\right]$ that satisfies $\sin \theta = x$.

Table 4.7 **Exact Values for** $\sin \theta$, $-\frac{\pi}{2} \leq \theta \leq \frac{\pi}{2}$

θ	$-\frac{\pi}{2}$	$-\frac{\pi}{3}$	$-\frac{\pi}{4}$	$-\frac{\pi}{6}$	0	$\frac{\pi}{6}$	$\frac{\pi}{4}$	$\frac{\pi}{3}$	$\frac{\pi}{2}$
$\sin \theta$	-1	$-\frac{\sqrt{3}}{2}$	$-\frac{\sqrt{2}}{2}$	$-\frac{1}{2}$	0	$\frac{1}{2}$	$\frac{\sqrt{2}}{2}$	$\frac{\sqrt{3}}{2}$	1

EXAMPLE 1   Finding the Exact Value of an Inverse Sine Function

Find the exact value of $\sin^{-1}\dfrac{\sqrt{2}}{2}$.

Solution

Step 1   Let $\theta = \sin^{-1}x$. Thus,

$$\theta = \sin^{-1}\frac{\sqrt{2}}{2}.$$

We must find the angle θ, $-\dfrac{\pi}{2} \leq \theta \leq \dfrac{\pi}{2}$, whose sine equals $\dfrac{\sqrt{2}}{2}$.

Step 2   Rewrite $\theta = \sin^{-1}x$ as $\sin \theta = x$, where $-\dfrac{\pi}{2} \leq \theta \leq \dfrac{\pi}{2}$. Using the

definition of the inverse sine function, we rewrite $\theta = \sin^{-1}\dfrac{\sqrt{2}}{2}$ as

$$\sin \theta = \frac{\sqrt{2}}{2}, \text{ where } -\frac{\pi}{2} \leq \theta \leq \frac{\pi}{2}.$$

Step 3   Use the exact values in Table 4.7 to find the value of θ in $\left[-\dfrac{\pi}{2}, \dfrac{\pi}{2}\right]$ that

satisfies $\sin \theta = x$. Table 4.7 on the previous page shows that the only angle in the

interval $\left[-\dfrac{\pi}{2}, \dfrac{\pi}{2}\right]$ that satisfies $\sin \theta = \dfrac{\sqrt{2}}{2}$ is $\dfrac{\pi}{4}$. Thus, $\theta = \dfrac{\pi}{4}$. Because θ, in step 1,

represents $\sin^{-1}\dfrac{\sqrt{2}}{2}$, we conclude that

$$\sin^{-1}\frac{\sqrt{2}}{2} = \frac{\pi}{4}. \qquad \textit{The angle in } \left[-\frac{\pi}{2}, \frac{\pi}{2}\right] \textit{ whose sine is } \frac{\sqrt{2}}{2} \textit{ is } \frac{\pi}{4}.$$

Check Point 1   Find the exact value of $\sin^{-1}\dfrac{\sqrt{3}}{2}$.

EXAMPLE 2   Finding the Exact Value of an Inverse Sine Function

Find the exact value of $\sin^{-1}\left(-\dfrac{1}{2}\right)$.

Solution

Step 1   Let $\theta = \sin^{-1}x$. Thus,

$$\theta = \sin^{-1}\left(-\frac{1}{2}\right).$$

We must find the angle θ, $-\dfrac{\pi}{2} \leq \theta \leq \dfrac{\pi}{2}$, whose sine equals $-\dfrac{1}{2}$.

Step 2   Rewrite $\theta = \sin^{-1}x$ as $\sin \theta = x$, where $-\dfrac{\pi}{2} \leq \theta \leq \dfrac{\pi}{2}$. We rewrite

$\theta = \sin^{-1}\left(-\dfrac{1}{2}\right)$ and obtain

$$\sin \theta = -\frac{1}{2}, \text{ where } -\frac{\pi}{2} \leq \theta \leq \frac{\pi}{2}.$$

Step 3   Use the exact values in Table 4.7 to find the value of θ in $\left[-\frac{\pi}{2}, \frac{\pi}{2}\right]$ tha

satisfies $\sin \theta = x$. Table 4.7 on page 536 shows that the only angle in the interva $\left[-\frac{\pi}{2}, \frac{\pi}{2}\right]$ that satisfies $\sin \theta = -\frac{1}{2}$ is $-\frac{\pi}{6}$. Thus,

$$\sin^{-1}\left(-\frac{1}{2}\right) = -\frac{\pi}{6}$$

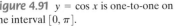 **Check Point 2**   Find the exact value of $\sin^{-1}\left(-\frac{\sqrt{2}}{2}\right)$.

Some inverse sine expressions cannot be evaluated. Because the domain of th inverse sine function is $[-1, 1]$, it is only possible to evaluate $\sin^{-1} x$ for values of in this domain. Thus, $\sin^{-1} 3$ cannot be evaluated. There is no angle whose sine is ?

② Understand and use the inverse cosine function.

The Inverse Cosine Function

Figure 4.91 shows how we restrict the domain of the cosine function so that becomes one-to-one and has an inverse function. Restrict the domain to the interva $[0, \pi]$, shown by the dark blue graph. Over this interval, the restricted cosin function passes the horizontal line test and has an inverse function.

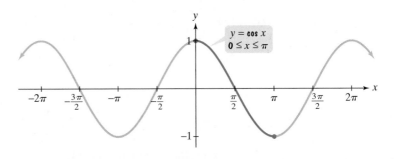

Figure 4.91 $y = \cos x$ is one-to-one on the interval $[0, \pi]$.

The Inverse Cosine Function
The **inverse cosine function**, denoted by $\cos^{-1}$, is the inverse of the restricted cosine function $y = \cos x, 0 \le x \le \pi$. Thus,

$$y = \cos^{-1} x \quad \text{means} \quad \cos y = x,$$

where $0 \le y \le \pi$ and $-1 \le x \le 1$.

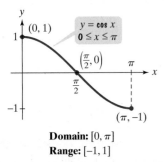

Figure 4.92 The restricted cosine function

One way to graph $y = \cos^{-1} x$ is to take points on the graph of the restricted cosine function and reverse the order of the coordinates. For example, Figure 4.92 shows that $(0, 1)$, $\left(\frac{\pi}{2}, 0\right)$, and $(\pi, -1)$ are on the graph of the restricted cosine function. Reversing the order of the coordinates gives $(1, 0)$, $\left(0, \frac{\pi}{2}\right)$, and $(-1, \pi)$.

We now use these three points to sketch the inverse cosine function. The graph of $y = \cos^{-1} x$ is shown in Figure 4.93. You can also obtain this graph by reflecting the graph of the restricted cosine function about the line $y = x$.

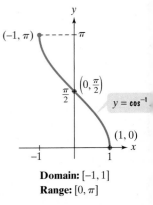

Figure 4.93 The graph of the inverse cosine function

Exact values of $\cos^{-1} x$ can be found by thinking of $\cos^{-1} x$ **as the angle in the interval $[0, \pi]$ whose cosine is x.**

> ### Finding Exact Values of $\cos^{-1} x$
>
> **1.** Let $\theta = \cos^{-1} x$.
> **2.** Rewrite $\theta = \cos^{-1} x$ as $\cos \theta = x$, where $0 \le \theta \le \pi$.
> **3.** Use the exact values in Table 4.8 to find the value of θ in $[0, \pi]$ that satisfies $\cos \theta = x$.

Table 4.8   Exact Values for $\cos \theta$, $0 \le \theta \le \pi$

θ	0	$\dfrac{\pi}{6}$	$\dfrac{\pi}{4}$	$\dfrac{\pi}{3}$	$\dfrac{\pi}{2}$	$\dfrac{2\pi}{3}$	$\dfrac{3\pi}{4}$	$\dfrac{5\pi}{6}$	π
$\cos \theta$	1	$\dfrac{\sqrt{3}}{2}$	$\dfrac{\sqrt{2}}{2}$	$\dfrac{1}{2}$	0	$-\dfrac{1}{2}$	$-\dfrac{\sqrt{2}}{2}$	$-\dfrac{\sqrt{3}}{2}$	-1

EXAMPLE 3   Finding the Exact Value of an Inverse Cosine Function

Find the exact value of $\cos^{-1}\!\left(-\dfrac{\sqrt{3}}{2}\right)$.

Solution

Step 1   Let $\theta = \cos^{-1} x$. Thus,

$$\theta = \cos^{-1}\!\left(-\frac{\sqrt{3}}{2}\right).$$

We must find the angle θ, $0 \le \theta \le \pi$, whose cosine equals $-\dfrac{\sqrt{3}}{2}$.

Step 2   Rewrite $\theta = \cos^{-1} x$ as $\cos \theta = x$, where $0 \le \theta \le \pi$. We obtain

$$\cos \theta = -\frac{\sqrt{3}}{2}, \text{ where } 0 \le \theta \le \pi.$$

Step 3   Use the exact values in Table 4.8 to find the value of θ in $[0, \pi]$ that satisfies $\cos \theta = x$. The table shows that the only angle in the interval $[0, \pi]$ that satisfies $\cos \theta = -\dfrac{\sqrt{3}}{2}$ is $\dfrac{5\pi}{6}$. Thus, $\theta = \dfrac{5\pi}{6}$ and

$$\cos^{-1}\!\left(-\frac{\sqrt{3}}{2}\right) = \frac{5\pi}{6}.$$ The angle in $[0, \pi]$ whose cosine is $-\dfrac{\sqrt{3}}{2}$ is $\dfrac{5\pi}{6}$.

Check Point 3   Find the exact value of $\cos^{-1}\!\left(-\dfrac{1}{2}\right)$.

    Understand and use the inverse tangent function.

The Inverse Tangent Function

Figure 4.94 at the top of the next page shows how we restrict the domain of the tangent function so that it becomes one-to-one and has an inverse function. Restrict the domain to the interval $\left(-\dfrac{\pi}{2}, \dfrac{\pi}{2}\right)$, shown by the solid blue graph. Over this interval, the restricted tangent function passes the horizontal line test and has an inverse function.

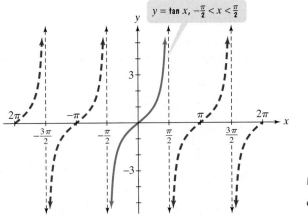

Figure 4.94 $y = \tan x$ is one-to-one on the interval $\left(-\dfrac{\pi}{2}, \dfrac{\pi}{2}\right)$.

The Inverse Tangent Function

The **inverse tangent function**, denoted by $\tan^{-1}$, is the inverse of the restricted tangent function $y = \tan x$, $-\dfrac{\pi}{2} < x < \dfrac{\pi}{2}$. Thus,

$$y = \tan^{-1} x \quad \text{means} \quad \tan y = x,$$

where $-\dfrac{\pi}{2} < y < \dfrac{\pi}{2}$ and $-\infty < x < \infty$.

We graph $y = \tan^{-1} x$ by taking points on the graph of the restricted function and reversing the order of the coordinates. Figure 4.95 shows that $\left(-\dfrac{\pi}{4}, -1\right)$, $(0, 0)$, and $\left(\dfrac{\pi}{4}, 1\right)$ are on the graph of the restricted tangent function. Reversing the order gives $\left(-1, -\dfrac{\pi}{4}\right)$, $(0, 0)$, and $\left(1, \dfrac{\pi}{4}\right)$. We now use these three points to graph the inverse tangent function. The graph of $y = \tan^{-1} x$ is shown in Figure 4.96. Notice that the vertical asymptotes become horizontal asymptotes for the graph of the inverse function.

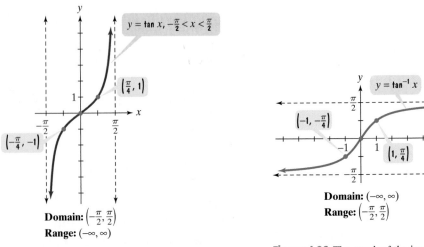

Domain: $\left(-\dfrac{\pi}{2}, \dfrac{\pi}{2}\right)$
Range: $(-\infty, \infty)$

Figure 4.95 The restricted tangent function

Domain: $(-\infty, \infty)$
Range: $\left(-\dfrac{\pi}{2}, \dfrac{\pi}{2}\right)$

Figure 4.96 The graph of the inverse tangent function

Exact values of $\tan^{-1} x$ can be found by thinking of **$\tan^{-1} x$ as the angle in the interval** $\left(-\dfrac{\pi}{2}, \dfrac{\pi}{2}\right)$ **whose tangent is x.**

Finding Exact Values of $\tan^{-1} x$

1. Let $\theta = \tan^{-1} x$.
2. Rewrite $\theta = \tan^{-1} x$ as $\tan \theta = x$, where $-\dfrac{\pi}{2} < \theta < \dfrac{\pi}{2}$.
3. Use the exact values in Table 4.9 to find the value of θ in $\left(-\dfrac{\pi}{2}, \dfrac{\pi}{2}\right)$ that satisfies $\tan \theta = x$.

Table 4.9   Exact Values for $\tan \theta$, $-\dfrac{\pi}{2} < \theta < \dfrac{\pi}{2}$

θ	$-\dfrac{\pi}{2}$	$-\dfrac{\pi}{3}$	$-\dfrac{\pi}{4}$	$-\dfrac{\pi}{6}$	0	$\dfrac{\pi}{6}$	$\dfrac{\pi}{4}$	$\dfrac{\pi}{3}$	$\dfrac{\pi}{2}$
$\tan \theta$	undef.	$-\sqrt{3}$	-1	$-\dfrac{\sqrt{3}}{3}$	0	$\dfrac{\sqrt{3}}{3}$	1	$\sqrt{3}$	undef.

EXAMPLE 4   Finding the Exact Value of an Inverse Tangent Function

Find the exact value of $\tan^{-1}\sqrt{3}$.

Solution

Step 1   Let $\theta = \tan^{-1} x$. Thus,
$$\theta = \tan^{-1}\sqrt{3}.$$
We must find the angle θ, $-\dfrac{\pi}{2} < \theta < \dfrac{\pi}{2}$, whose tangent equals $\sqrt{3}$.

Step 2   Rewrite $\theta = \tan^{-1} x$ as $\tan \theta = x$, where $-\dfrac{\pi}{2} < \theta < \dfrac{\pi}{2}$. We obtain
$$\tan \theta = \sqrt{3}, \text{ where } -\dfrac{\pi}{2} < \theta < \dfrac{\pi}{2}.$$

Step 3   Use the exact values in Table 4.9 to find the value of θ in $\left(-\dfrac{\pi}{2}, \dfrac{\pi}{2}\right)$ that satisfies $\tan \theta = x$. The table shows that the only angle in the interval $\left(-\dfrac{\pi}{2}, \dfrac{\pi}{2}\right)$ that satisfies $\tan \theta = \sqrt{3}$ is $\dfrac{\pi}{3}$. Thus, $\theta = \dfrac{\pi}{3}$ and
$$\tan^{-1}\sqrt{3} = \dfrac{\pi}{3}. \qquad \textit{The angle in } \left(-\dfrac{\pi}{2}, \dfrac{\pi}{2}\right) \textit{ whose tangent is } \sqrt{3} \textit{ is } \dfrac{\pi}{3}.$$

Study Tip

Do not confuse the domains of the restricted trigonometric functions with the intervals on which the nonrestricted functions complete one cycle.

Trigonometric Function	Domain of Restricted Function	Interval on Which Nonrestricted Function's Graph Completes One Period	
$y = \sin x$	$\left[-\dfrac{\pi}{2}, \dfrac{\pi}{2}\right]$	$[0, 2\pi]$	Period: 2π
$y = \cos x$	$[0, \pi]$	$[0, 2\pi]$	Period: 2π
$y = \tan x$	$\left(-\dfrac{\pi}{2}, \dfrac{\pi}{2}\right)$	$\left(-\dfrac{\pi}{2}, \dfrac{\pi}{2}\right)$	Period: π

These domain restrictions are the range for $y = \sin^{-1} x$, $y = \cos^{-1} x$, and $y = \tan^{-1} x$, respectively.

Check Point 4 Find the exact value of $\tan^{-1}(-1)$.

Table 4.10 summarizes the graphs of the three basic inverse trigonometric functions. Below each of the graphs is a description of the function's domain and range.

Table 4.10   Graphs of the Three Basic Inverse Trigonometric Functions

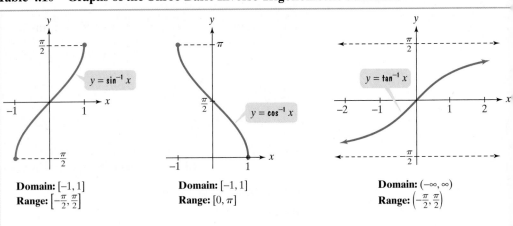

Domain: $[-1, 1]$	Domain: $[-1, 1]$	Domain: $(-\infty, \infty)$
Range: $\left[-\frac{\pi}{2}, \frac{\pi}{2}\right]$	Range: $[0, \pi]$	Range: $\left(-\frac{\pi}{2}, \frac{\pi}{2}\right)$

4 Use a calculator to evaluate inverse trigonometric functions.

Using a Calculator to Evaluate Inverse Trigonometric Functions

Calculators give approximate values of inverse trigonometric functions. Use the secondary keys marked $\boxed{\text{SIN}^{-1}}$, $\boxed{\text{COS}^{-1}}$, and $\boxed{\text{TAN}^{-1}}$. These keys are not buttons that you actually press. They are the secondary functions for the buttons labeled $\boxed{\text{SIN}}$, $\boxed{\text{COS}}$, and $\boxed{\text{TAN}}$, respectively. Consult your manual for the location of this feature.

EXAMPLE 5   Calculators and Inverse Trigonometric Functions

Use a calculator to find the value to four decimal places of each function:

a. $\sin^{-1}\dfrac{1}{4}$ **b.** $\tan^{-1}(-9.65)$.

Solution

Scientific Calculator Solution

Function	Mode	Keystrokes	Display, rounded to four places
a. $\sin^{-1}\dfrac{1}{4}$	Radian	1 $\boxed{\div}$ 4 $\boxed{=}$ $\boxed{\text{2nd}}$ $\boxed{\text{SIN}}$	0.2527
b. $\tan^{-1}(-9.65)$	Radian	9.65 $\boxed{+/-}$ $\boxed{\text{2nd}}$ $\boxed{\text{TAN}}$	-1.4675

Graphing Calculator Solution

Function	Mode	Keystrokes	Display, rounded to four places
a. $\sin^{-1}\dfrac{1}{4}$	Radian	$\boxed{\text{2nd}}$ $\boxed{\text{SIN}}$ $\boxed{(}$ 1 $\boxed{\div}$ 4 $\boxed{)}$ $\boxed{\text{ENTER}}$	0.2527
b. $\tan^{-1}(-9.65)$	Radian	$\boxed{\text{2nd}}$ $\boxed{\text{TAN}}$ $\boxed{(-)}$ 9.65 $\boxed{\text{ENTER}}$	-1.4675

Check Point 5 Use a calculator to find the value to four decimal places of each function:
a. $\cos^{-1}\dfrac{1}{3}$ **b.** $\tan^{-1}(-35.85)$.

What happens if you attempt to evaluate an inverse trigomometric function at a value that is not in its domain? In real number mode, most calculators will display an error message. For example, an error message can result if you attempt to approximate $\cos^{-1} 3$. There is no angle whose cosine is 3. The domain of the inverse cosine function is $[-1, 1]$ and 3 does not belong to this domain.

⑤ Find exact values of composite functions with inverse trigonometric functions.

Composition of Functions Involving Inverse Trigonometric Functions

In our earlier discussion of functions and their inverses, we saw that

$$f(f^{-1}(x)) = x \quad \text{and} \quad f^{-1}(f(x)) = x.$$

x must be in the domain of f^{-1}.

x must be in the domain of *f*.

We apply these properties to the sine, cosine, tangent, and their inverse functions to obtain the following properties:

Inverse Properties

The Sine Function and Its Inverse

$$\sin(\sin^{-1} x) = x \qquad \text{for every } x \text{ in the interval } [-1, 1]$$

$$\sin^{-1}(\sin x) = x \qquad \text{for every } x \text{ in the interval } \left[-\frac{\pi}{2}, \frac{\pi}{2}\right]$$

The Cosine Function and Its Inverse

$$\cos(\cos^{-1} x) = x \qquad \text{for every } x \text{ in the interval } [-1, 1]$$

$$\cos^{-1}(\cos x) = x \qquad \text{for every } x \text{ in the interval } [0, \pi]$$

The Tangent Function and Its Inverse

$$\tan(\tan^{-1} x) = x \qquad \text{for every real number } x$$

$$\tan^{-1}(\tan x) = x \qquad \text{for every } x \text{ in the interval } \left(-\frac{\pi}{2}, \frac{\pi}{2}\right)$$

The restrictions on *x* in the inverse properties are a bit tricky. For example,

$$\sin^{-1}\left(\sin \frac{\pi}{4}\right) = \frac{\pi}{4}.$$

$\sin^{-1}(\sin x) = x$ for *x* in $\left[-\frac{\pi}{2}, \frac{\pi}{2}\right]$.
Observe that $\frac{\pi}{4}$ is in this interval.

Can we use $\sin^{-1}(\sin x) = x$ to find the exact value of $\sin^{-1}\left(\sin \frac{5\pi}{4}\right)$? Is $\frac{5\pi}{4}$ in the interval $\left[-\frac{\pi}{2}, \frac{\pi}{2}\right]$? No. Thus, to evaluate $\sin^{-1}\left(\sin \frac{5\pi}{4}\right)$, we must first find $\sin \frac{5\pi}{4}$.

$\frac{5\pi}{4}$ is in quadrant III, where the sine is negative.

$$\sin \frac{5\pi}{4} = -\sin \frac{\pi}{4} = -\frac{\sqrt{2}}{2}$$

The reference angle for $\frac{5\pi}{4}$ is $\frac{\pi}{4}$.

We evaluate $\sin^{-1}\left(\sin \frac{5\pi}{4}\right)$ as follows:

$$\sin^{-1}\left(\sin \frac{5\pi}{4}\right) = \sin^{-1}\left(-\frac{\sqrt{2}}{2}\right) = -\frac{\pi}{4} \qquad \textit{If necessary, see Table 4.7 on page 536.}$$

To determine how to evaluate the composition of functions involving inverse trigonometric functions, first examine the value of x. You can use the inverse properties in the box shown on the previous page only if x is in the specified interval.

EXAMPLE 6 Evaluating Compositions of Functions and Their Inverses

Find the exact value, if possible:

a. $\cos(\cos^{-1} 0.6)$ **b.** $\sin^{-1}\left(\sin \dfrac{3\pi}{2}\right)$ **c.** $\cos(\cos^{-1} 1.5)$.

Solution

a. The inverse property $\cos(\cos^{-1} x) = x$ applies for every x in $[-1, 1]$. To evaluate $\cos(\cos^{-1} 0.6)$, observe that $x = 0.6$. This value of x lies in $[-1, 1]$, which is the domain of the inverse cosine function. This means that we can use the inverse property $\cos(\cos^{-1} x) = x$. Thus,

$$\cos(\cos^{-1} 0.6) = 0.6.$$

b. The inverse property $\sin^{-1}(\sin x) = x$ applies for every x in $\left[-\dfrac{\pi}{2}, \dfrac{\pi}{2}\right]$. To evaluate $\sin^{-1}\left(\sin \dfrac{3\pi}{2}\right)$, observe that $x = \dfrac{3\pi}{2}$. This value of x does not lie in $\left[-\dfrac{\pi}{2}, \dfrac{\pi}{2}\right]$. To evaluate this expression, we first find $\sin \dfrac{3\pi}{2}$.

$$\sin^{-1}\left(\sin \dfrac{3\pi}{2}\right) = \sin^{-1}(-1) = -\dfrac{\pi}{2} \quad \text{The angle in } \left[-\dfrac{\pi}{2}, \dfrac{\pi}{2}\right] \text{ whose sine is } -1 \text{ is } -\dfrac{\pi}{2}$$

c. The inverse property $\cos(\cos^{-1} x) = x$ applies for every x in $[-1, 1]$. To attempt to evaluate $\cos(\cos^{-1} 1.5)$, observe that $x = 1.5$. This value of x does not lie in $[-1, 1]$, which is the domain of the inverse cosine function. Thus, the expression $\cos(\cos^{-1} 1.5)$ is not defined because $\cos^{-1} 1.5$ is not defined.

Check Point 6 Find the exact value, if possible:

a. $\cos(\cos^{-1} 0.7)$ **b.** $\sin^{-1}(\sin \pi)$ **c.** $\cos[\cos^{-1}(-1.2)]$.

We can use points on terminal sides of angles in standard position to find exact values of expressions involving the composition of a function and a different inverse function. Here are two examples:

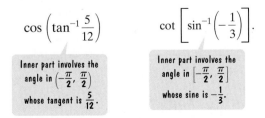

$$\cos\left(\tan^{-1} \dfrac{5}{12}\right) \qquad \cot\left[\sin^{-1}\left(-\dfrac{1}{3}\right)\right].$$

Inner part involves the angle in $\left(-\dfrac{\pi}{2}, \dfrac{\pi}{2}\right)$ whose tangent is $\dfrac{5}{12}$.

Inner part involves the angle in $\left[-\dfrac{\pi}{2}, \dfrac{\pi}{2}\right]$ whose sine is $-\dfrac{1}{3}$.

The inner part of each expression involves an angle. To evaluate such expressions, we represent such angles by θ. Then we use a sketch that illustrates our representation. Examples 7 and 8 show how to carry out such evaluations.

EXAMPLE 7   Evaluating a Composite Trigonometric Expression

Find the exact value of $\cos\left(\tan^{-1}\dfrac{5}{12}\right)$.

Solution   We let θ represent the angle in $\left(-\dfrac{\pi}{2},\dfrac{\pi}{2}\right)$ whose tangent is $\dfrac{5}{12}$. Thus,

$$\theta = \tan^{-1}\dfrac{5}{12}.$$

Using the definition of the inverse tangent function, we can rewrite this as

$$\tan\theta = \frac{5}{12}, \quad \text{where} \quad -\frac{\pi}{2} < \theta < \frac{\pi}{2}.$$

Because $\tan\theta$ is positive, θ must be an angle in $\left(0,\dfrac{\pi}{2}\right)$. Thus, θ is a first-quadrant angle. Figure 4.97 shows a right triangle in quadrant I with

$$\tan\theta = \frac{5}{12}. \qquad \boxed{\text{Side opposite } \theta, \text{ or } y} \\ \boxed{\text{Side adjacent to } \theta, \text{ or } x}$$

The hypotenuse of the triangle, r, or the distance from the origin to $(12, 5)$, is found using $r = \sqrt{x^2 + y^2}$.

$$r = \sqrt{x^2 + y^2} = \sqrt{12^2 + 5^2} = \sqrt{144 + 25} = \sqrt{169} = 13$$

We use the values for x and r to find the exact value of $\cos\left(\tan^{-1}\dfrac{5}{12}\right)$.

$$\cos\left(\tan^{-1}\frac{5}{12}\right) = \cos\theta = \frac{\text{side adjacent to } \theta, \text{ or } x}{\text{hypotenuse, or } r} = \frac{12}{13}$$

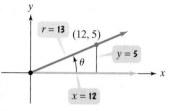

Figure 4.97 Representing $\tan\theta = \frac{5}{12}$

Check Point 7   Find the exact value of $\sin\left(\tan^{-1}\dfrac{3}{4}\right)$.

EXAMPLE 8   Evaluating a Composite Trigonometric Expression

Find the exact value of $\cot\left[\sin^{-1}\left(-\dfrac{1}{3}\right)\right]$.

Solution   We let θ represent the angle in $\left[-\dfrac{\pi}{2},\dfrac{\pi}{2}\right]$ whose sine is $-\dfrac{1}{3}$. Thus,

$$\theta = \sin^{-1}\left(-\frac{1}{3}\right) \quad \text{and} \quad \sin\theta = -\frac{1}{3}, \quad \text{where} \quad -\frac{\pi}{2} \le \theta \le \frac{\pi}{2}.$$

Because $\sin\theta$ is negative in $\sin\theta = -\dfrac{1}{3}$, θ must be an angle in $\left[-\dfrac{\pi}{2}, 0\right)$. Thus, θ is a negative angle that lies in quadrant IV. Figure 4.98 shows angle θ in quadrant IV with

$$\boxed{\text{In quadrant IV, } y \text{ is negative.}}$$

$$\sin\theta = -\frac{1}{3} = \frac{y}{r} = \frac{-1}{3}.$$

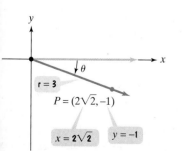

Figure 4.98 Representing $\sin\theta = -\frac{1}{3}$

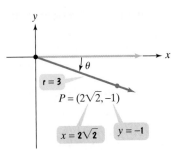

y

$r = 3$

$P = (2\sqrt{2}, -1)$

$x = 2\sqrt{2}$ $y = -1$

Figure 4.98 (repeated)

Thus, $y = -1$ and $r = 3$. The value of x can be found using $r = \sqrt{x^2 + y^2}$ o
$x^2 + y^2 = r^2$.

$$x^2 + (-1)^2 = 3^2 \qquad \text{Use } x^2 + y^2 = r^2 \text{ with } y = -1 \text{ and } r = 3$$
$$x^2 + 1 = 9 \qquad \text{Square } -1 \text{ and square } 3.$$
$$x^2 = 8 \qquad \text{Subtract 1 from both sides.}$$
$$x = \sqrt{8} = \sqrt{4 \cdot 2} = 2\sqrt{2} \qquad \text{Use the square root property. Remember that x is positive in quadrant IV.}$$

We use values for x and y to find the exact value of $\cot\left[\sin^{-1}\left(-\frac{1}{3}\right)\right]$.

$$\cot\left[\sin^{-1}\left(-\frac{1}{3}\right)\right] = \cot\theta = \frac{x}{y} = \frac{2\sqrt{2}}{-1} = -2\sqrt{2}$$

Check Point 8 Find the exact value of $\cos\left[\sin^{-1}\left(-\frac{1}{2}\right)\right]$.

Some composite functions with inverse trigonometric functions can b
simplified to algebraic expressions. To simplify such an expression, we represent th
inverse trigonometric function in the expression by θ. Then we use a right triangle

EXAMPLE 9 Simplifying an Expression Involving $\sin^{-1} x$

If $0 < x \le 1$, write $\cos(\sin^{-1} x)$ as an algebraic expression in x.

Solution We let θ represent the angle in $\left[-\frac{\pi}{2}, \frac{\pi}{2}\right]$ whose sine is x. Thus,

$$\theta = \sin^{-1} x \quad \text{and} \quad \sin\theta = x, \quad \text{where} \quad -\frac{\pi}{2} \le \theta \le \frac{\pi}{2}.$$

Because $0 < x \le 1$, $\sin\theta$ is positive. Thus, θ is a first-quadrant angle and can be rep
resented as an acute angle of a right triangle. Figure 4.99 shows a right triangle wit

$$\sin\theta = x = \frac{x}{1}. \qquad \text{Side opposite } \theta \quad \text{Hypotenuse}$$

The third side, a in Figure 4.99, can be found using the Pythagorean Theorem.

$$a^2 + x^2 = 1^2 \qquad \text{Apply the Pythagorean Theorem to the right triang in Figure 4.99.}$$
$$a^2 = 1 - x^2 \qquad \text{Subtract } x^2 \text{ from both sides.}$$
$$a = \sqrt{1 - x^2} \qquad \text{Use the square root property and solve for a. Remember that side a is positive}$$

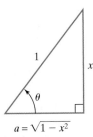

Figure 4.99
Representing
$\sin\theta = x$

$a = \sqrt{1-x^2}$

We use the right triangle in Figure 4.99 to write $\cos(\sin^{-1} x)$ as an algebra
expression.

$$\cos(\sin^{-1} x) = \cos\theta = \frac{\text{side adjacent to } \theta}{\text{hypotenuse}} = \frac{\sqrt{1-x^2}}{1} = \sqrt{1-x^2}$$

Check Point 9 If $x > 0$, write $\sec(\tan^{-1} x)$ as an algebraic expression in x.

The inverse secant function, $y = \sec^{-1} x$, is used in calculus. However, invers
cotangent and inverse cosecant functions are rarely used. Two of these remainir
inverse trigonometric functions are briefly developed in the exercise set th
follows.

EXERCISE SET 4.7

Practice Exercises

In Exercises 1–18, find the exact value of each expression.

1. $\sin^{-1}\dfrac{1}{2}$ **2.** $\sin^{-1} 0$

3. $\sin^{-1}\dfrac{\sqrt{2}}{2}$ **4.** $\sin^{-1}\dfrac{\sqrt{3}}{2}$

5. $\sin^{-1}\left(-\dfrac{1}{2}\right)$ **6.** $\sin^{-1}\left(-\dfrac{\sqrt{3}}{2}\right)$

7. $\cos^{-1}\dfrac{\sqrt{3}}{2}$ **8.** $\cos^{-1}\dfrac{\sqrt{2}}{2}$

9. $\cos^{-1}\left(-\dfrac{\sqrt{2}}{2}\right)$ **10.** $\cos^{-1}\left(-\dfrac{\sqrt{3}}{2}\right)$

11. $\cos^{-1} 0$ **12.** $\cos^{-1} 1$

13. $\tan^{-1}\dfrac{\sqrt{3}}{3}$ **14.** $\tan^{-1} 1$

15. $\tan^{-1} 0$ **16.** $\tan^{-1}(-1)$

17. $\tan^{-1}\left(-\sqrt{3}\right)$ **18.** $\tan^{-1}\left(-\dfrac{\sqrt{3}}{3}\right)$

In Exercises 19–30, use a calculator to find the value of each expression rounded to two decimal places.

19. $\sin^{-1} 0.3$ **20.** $\sin^{-1} 0.47$

21. $\sin^{-1}(-0.32)$ **22.** $\sin^{-1}(-0.625)$

23. $\cos^{-1}\dfrac{3}{8}$ **24.** $\cos^{-1}\dfrac{4}{9}$

25. $\cos^{-1}\dfrac{\sqrt{5}}{7}$ **26.** $\cos^{-1}\dfrac{\sqrt{7}}{10}$

27. $\tan^{-1}(-20)$ **28.** $\tan^{-1}(-30)$

29. $\tan^{-1}\left(-\sqrt{473}\right)$ **30.** $\tan^{-1}\left(-\sqrt{5061}\right)$

In Exercises 31–46, find the exact value of each expression, if possible. Do not use a calculator.

31. $\sin(\sin^{-1} 0.9)$ **32.** $\cos(\cos^{-1} 0.57)$

33. $\sin^{-1}\left(\sin\dfrac{\pi}{3}\right)$ **34.** $\cos^{-1}\left(\cos\dfrac{2\pi}{3}\right)$

35. $\sin^{-1}\left(\sin\dfrac{5\pi}{6}\right)$ **36.** $\cos^{-1}\left(\cos\dfrac{4\pi}{3}\right)$

37. $\tan(\tan^{-1} 125)$ **38.** $\tan(\tan^{-1} 380)$

39. $\tan^{-1}\left[\tan\left(-\dfrac{\pi}{6}\right)\right]$ **40.** $\tan^{-1}\left[\tan\left(-\dfrac{\pi}{3}\right)\right]$

41. $\tan^{-1}\left(\tan\dfrac{2\pi}{3}\right)$ **42.** $\tan^{-1}\left(\tan\dfrac{3\pi}{4}\right)$

43. $\sin^{-1}(\sin \pi)$ **44.** $\cos^{-1}(\cos 2\pi)$

45. $\sin(\sin^{-1} \pi)$ **46.** $\cos(\cos^{-1} 3\pi)$

In Exercises 47–62, use a sketch to find the exact value of each expression.

47. $\cos\left(\sin^{-1}\dfrac{4}{5}\right)$ **48.** $\sin\left(\tan^{-1}\dfrac{7}{24}\right)$

49. $\tan\left(\cos^{-1}\dfrac{5}{13}\right)$ **50.** $\cot\left(\sin^{-1}\dfrac{5}{13}\right)$

51. $\tan\left[\sin^{-1}\left(-\dfrac{3}{5}\right)\right]$ **52.** $\cos\left[\sin^{-1}\left(-\dfrac{4}{5}\right)\right]$

53. $\sin\left(\cos^{-1}\dfrac{\sqrt{2}}{2}\right)$ **54.** $\cos\left(\sin^{-1}\dfrac{1}{2}\right)$

55. $\sec\left[\sin^{-1}\left(-\dfrac{1}{4}\right)\right]$ **56.** $\sec\left[\sin^{-1}\left(-\dfrac{1}{2}\right)\right]$

57. $\tan\left[\cos^{-1}\left(-\dfrac{1}{3}\right)\right]$ **58.** $\tan\left[\cos^{-1}\left(-\dfrac{1}{4}\right)\right]$

59. $\csc\left[\cos^{-1}\left(-\dfrac{\sqrt{3}}{2}\right)\right]$ **60.** $\sec\left[\sin^{-1}\left(-\dfrac{\sqrt{2}}{2}\right)\right]$

61. $\cos\left[\tan^{-1}\left(-\dfrac{2}{3}\right)\right]$ **62.** $\sin\left[\tan^{-1}\left(-\dfrac{3}{4}\right)\right]$

In Exercises 63–72, use a right triangle to write each expression as an algebraic expression. Assume that x is positive and that the given inverse trigonometric function is defined for the expression in x.

63. $\tan(\cos^{-1} x)$ **64.** $\sin(\tan^{-1} x)$

65. $\cos(\sin^{-1} 2x)$ **66.** $\sin(\cos^{-1} 2x)$

67. $\cos\left(\sin^{-1}\dfrac{1}{x}\right)$ **68.** $\sec\left(\cos^{-1}\dfrac{1}{x}\right)$

69. $\cot\left(\tan^{-1}\dfrac{x}{\sqrt{3}}\right)$ **70.** $\cot\left(\tan^{-1}\dfrac{x}{\sqrt{2}}\right)$

71. $\sec\left(\sin^{-1}\dfrac{x}{\sqrt{x^2+4}}\right)$ **72.** $\cot\left(\sin^{-1}\dfrac{\sqrt{x^2-9}}{x}\right)$

73. a. Graph the restricted secant function, $y = \sec x$, by restricting x to the intervals $\left[0, \dfrac{\pi}{2}\right)$ and $\left(\dfrac{\pi}{2}, \pi\right]$.

b. Use the horizontal line test to explain why the restricted secant function has an inverse function.

c. Use the graph of the restricted secant function to graph $y = \sec^{-1} x$.

74. a. Graph the restricted cotangent function, $y = \cot x$, by restricting x to the interval $(0, \pi)$.

b. Use the horizontal line test to explain why the restricted cotangent function has an inverse function.

c. Use the graph of the restricted cotangent function to graph $y = \cot^{-1} x$.

Practice Plus

The graphs of $y = \sin^{-1} x$, $y = \cos^{-1} x$, and $y = \tan^{-1} x$ are shown in Table 4.10 on page 542. In Exercises 75–84, use transformations (vertical shifts, horizontal shifts, reflections, stretching, or shrinking) of these graphs to graph each function. Then use interval notation to give the function's domain and range.

75. $f(x) = \sin^{-1} x + \dfrac{\pi}{2}$ **76.** $f(x) = \cos^{-1} x + \dfrac{\pi}{2}$

77. $g(x) = \cos^{-1}(x + 1)$ **78.** $g(x) = \sin^{-1}(x + 1)$

79. $h(x) = -2\tan^{-1} x$ **80.** $h(x) = -3\tan^{-1} x$

81. $f(x) = \sin^{-1}(x-2) - \dfrac{\pi}{2}$ **82.** $f(x) = \cos^{-1}(x-2) - \dfrac{\pi}{2}$

83. $g(x) = \cos^{-1}\dfrac{x}{2}$ **84.** $g(x) = \sin^{-1}\dfrac{x}{2}$

In Exercises 85–92, determine the domain and the range of each function.

85. $f(x) = \sin(\sin^{-1} x)$ **86.** $f(x) = \cos(\cos^{-1} x)$

87. $f(x) = \cos^{-1}(\cos x)$ **88.** $f(x) = \sin^{-1}(\sin x)$

89. $f(x) = \sin^{-1}(\cos x)$ **90.** $f(x) = \cos^{-1}(\sin x)$

91. $f(x) = \sin^{-1} x + \cos^{-1} x$ **92.** $f(x) = \cos^{-1} x - \sin^{-1} x$

Application Exercises

93. Your neighborhood movie theater has a 25-foot-high screen located 8 feet above your eye level. If you sit too close to the screen, your viewing angle is too small, resulting in a distorted picture. By contrast, if you sit too far back, the image is quite small, diminishing the movie's visual impact. If you sit x feet back from the screen, your viewing angle, θ, is given by

$$\theta = \tan^{-1}\frac{33}{x} - \tan^{-1}\frac{8}{x}.$$

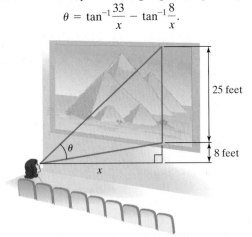

25 feet

8 feet

Find the viewing angle, in radians, at distances of 5 feet, 10 feet, 15 feet, 20 feet, and 25 feet.

94. The function $\theta = \tan^{-1}\dfrac{33}{x} - \tan^{-1}\dfrac{8}{x}$, described in Exercise 93, is graphed below in a $[0, 50, 10]$ by $[0, 1, 0.1]$ viewing rectangle. Use the graph to describe what happens to your viewing angle as you move farther back from the screen. How far back from the screen, to the nearest foot, should you sit to maximize your viewing angle? Verify this observation by finding the viewing angle one foot closer to the screen and one foot farther from the screen for this ideal viewing distance.

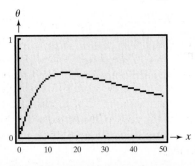

The formula

$$\theta = 2\tan^{-1}\frac{21.634}{x}$$

gives the viewing angle, θ, in radians, for a camera whose lens is x millimeters wide. Use this formula to solve Exercises 95–96.

95. Find the viewing angle, in radians and in degrees (to the nearest tenth of a degree), of a 28-millimeter lens.

96. Find the viewing angle, in radians and in degrees (to the nearest tenth of a degree), of a 300-millimeter telephoto lens.

For years, mathematicians were challenged by the following problem: What is the area of a region under a curve between two values of x? The problem was solved in the seventeenth century with the development of integral calculus. Using calculus, the area of the region under $y = \dfrac{1}{x^2 + 1}$, above the x-axis, and between $x = a$ and $x = b$ is $\tan^{-1} b - \tan^{-1} a$. Use this result, shown in the figure, to find the area of the region under $y = \dfrac{1}{x^2 + 1}$, above the x-axis, and between the values of a and b given in Exercises 97–98.

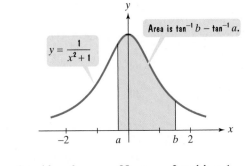

$y = \dfrac{1}{x^2 + 1}$

Area is $\tan^{-1} b - \tan^{-1} a$.

97. $a = 0$ and $b = 2$ **98.** $a = -2$ and $b = 1$

Writing in Mathematics

99. Explain why, without restrictions, no trigonometric function has an inverse function.

100. Describe the restriction on the sine function so that it has an inverse function.

101. How can the graph of $y = \sin^{-1} x$ be obtained from the graph of the restricted sine function?

102. Without drawing a graph, describe the behavior of the graph of $y = \sin^{-1} x$. Mention the function's domain and range in your description.

103. Describe the restriction on the cosine function so that it has an inverse function.

04. Without drawing a graph, describe the behavior of the graph of $y = \cos^{-1} x$. Mention the function's domain and range in your description.

05. Describe the restriction on the tangent function so that it has an inverse function.

06. Without drawing a graph, describe the behavior of the graph of $y = \tan^{-1} x$. Mention the function's domain and range in your description.

07. If $\sin^{-1}\left(\sin\dfrac{\pi}{3}\right) = \dfrac{\pi}{3}$, is $\sin^{-1}\left(\sin\dfrac{5\pi}{6}\right) = \dfrac{5\pi}{6}$? Explain your answer.

08. Explain how a right triangle can be used to find the exact value of $\sec\left(\sin^{-1}\frac{4}{5}\right)$.

09. Find the height of the screen and the number of feet that it is located above eye level in your favorite movie theater. Modify the formula given in Exercise 93 so that it applies to your theater. Then describe where in the theater you should sit so that a movie creates the greatest visual impact.

Technology Exercises

In Exercises 110–113, graph each pair of functions in the ame viewing rectangle. Use your knowledge of the domain and nge for the inverse trigonometric functions to select an appro- riate viewing rectangle. How is the graph of the second equation each exercise related to the graph of the first equation?

10. $y = \sin^{-1} x$ and $y = \sin^{-1} x + 2$

111. $y = \cos^{-1} x$ and $y = \cos^{-1}(x - 1)$

112. $y = \tan^{-1} x$ and $y = -2 \tan^{-1} x$

113. $y = \sin^{-1} x$ and $y = \sin^{-1}(x + 2) + 1$

114. Graph $y = \tan^{-1} x$ and its two horizontal asymptotes in a $[-3, 3, 1]$ by $\left[-\pi, \pi, \dfrac{\pi}{2}\right]$ viewing rectangle. Then change the range setting to $[-50, 50, 5]$ by $\left[-\pi, \pi, \dfrac{\pi}{2}\right]$. What do you observe?

115. Graph $y = \sin^{-1} x + \cos^{-1} x$ in a $[-2, 2, 1]$ by $[0, 3, 1]$ viewing rectangle. What appears to be true about the sum of the inverse sine and inverse cosine for values between -1 and 1, inclusive?

Critical Thinking Exercises

116. Solve $y = 2 \sin^{-1}(x - 5)$ for x in terms of y.

117. Solve for x: $2 \sin^{-1} x = \dfrac{\pi}{4}$.

118. Prove that if $x > 0$, $\tan^{-1} x + \tan^{-1}\dfrac{1}{x} = \dfrac{\pi}{2}$.

119. Derive the formula for θ, your viewing angle at the movie theater, in Exercise 93. *Hint* : Use the figure shown and represent the acute angle on the left in the smaller right triangle by α. Find expressions for $\tan \alpha$ and $\tan (\alpha + \theta)$.

ECTION 4.8 *Applications of Trigonometric Functions*

bjectives

Solve a right triangle.

Solve problems involving bearings.

Model simple harmonic motion.

In the late 1960s, popular musicians were searching for new sounds. Film composers were looking for ways to create unique sounds as well. From these efforts, synthesizers that electronically reproduce musical sounds were born. From providing the backbone of today's most popular music to providing the strange sounds for the most experimental music, synthesizers are at the forefront of music technology.

   If we did not understand the periodic nature of sinusoidal functions, the synthesizers used in almost all forms of music would not exist. In this section, we look at applications of trigonometric functions in solving right triangles and in modeling periodic phenomena such as sound.

① Solve a right triangle.

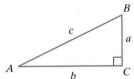

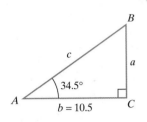

Figure 4.100 Labeling right triangles

Figure 4.101 Find B, a, and c.

Solving Right Triangles

Solving a right triangle means finding the missing lengths of its sides and the measurements of its angles. We will label right triangles so that side a is opposite angle A, side b is opposite angle B, and side c, the hypotenuse, is opposite right angle C. Figure 4.100 illustrates this labeling.

When solving a right triangle, we will use the sine, cosine, and tangent functions, rather than their reciprocals. Example 1 shows how to solve a right triangle when we know the length of a side and the measure of an acute angle.

EXAMPLE 1 Solving a Right Triangle

Solve the right triangle shown in Figure 4.101, rounding lengths to two decimal places.

Solution We begin by finding the measure of angle B. We do not need a trigonometric function to do so. Because $C = 90°$ and the sum of a triangle's angles is $180°$, we see that $A + B = 90°$. Thus,

$$B = 90° − A = 90° − 34.5° = 55.5°.$$

Now we need to find a. Because we have a known angle, an unknown opposite side, and a known adjacent side, we use the tangent function.

$$\tan 34.5° = \frac{a}{10.5}$$

         Side opposite the 34.5° angle

         Side adjacent to the 34.5° angle

Now we multiply both sides of this equation by 10.5 and solve for a.

$$a = 10.5 \tan 34.5° ≈ 7.22$$

Finally, we need to find c. Because we have a known angle, a known adjacent side, and an unknown hypotenuse, we use the cosine function.

$$\cos 34.5° = \frac{10.5}{c}$$

         Side adjacent to the 34.5° angle

         Hypotenuse

Now we multiply both sides of this equation by c and then solve for c.

$$c \cos 34.5° = 10.5 \qquad \text{Multiply both sides by } c.$$

$$c = \frac{10.5}{\cos 34.5°} ≈ 12.74 \qquad \text{Divide both sides by } \cos 34.5° \text{ and solve for } c.$$

In summary, $B = 55.5°$, $a ≈ 7.22$, and $c ≈ 12.74$.

Discovery

There is often more than one correct way to solve a right triangle. In Example 1, find a using angle $B = 55.5°$. Find c using the Pythagorean Theorem.

Check Point 1 In Figure 4.100, let $A = 62.7°$ and $a = 8.4$. Solve the right triangle, rounding lengths to two decimal places.

Trigonometry was first developed to measure heights and distances that were inconvenient or impossible to measure directly. In solving application problems, begin by making a sketch involving a right triangle that illustrates the problem's conditions. Then put your knowledge of solving right triangles to work and find the required distance or height.

EXAMPLE 2   Finding a Side of a Right Triangle

From a point on level ground 125 feet from the base of a tower, the angle of elevation is 57.2°. Approximate the height of the tower to the nearest foot.

Solution   A sketch is shown in Figure 4.102, where a represents the height of the tower. In the right triangle, we have a known angle, an unknown opposite side, and a known adjacent side. Therefore, we use the tangent function.

$$\tan 57.2° = \frac{a}{125}$$

Side opposite the 57.2° angle

Side adjacent to the 57.2° angle

Now we multiply both sides of this equation by 125 and solve for a.

$$a = 125 \tan 57.2° \approx 194$$

The tower is approximately 194 feet high.

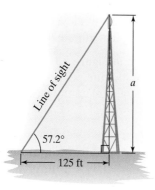

Figure 4.102 Determining height without using direct measurement

Check Point 2   From a point on level ground 80 feet from the base of the Eiffel Tower, the angle of elevation is 85.4°. Approximate the height of the Eiffel Tower to the nearest foot.

Example 3 illustrates how to find the measure of an acute angle of a right triangle if the lengths of two sides are known.

EXAMPLE 3   Finding an Angle of a Right Triangle

A kite flies at a height of 30 feet when 65 feet of string is out. If the string is in a straight line, find the angle that it makes with the ground. Round to the nearest tenth of a degree.

Solution   A sketch is shown in Figure 4.103, where A represents the angle the string makes with the ground. In the right triangle, we have an unknown angle, a known opposite side, and a known hypotenuse. Therefore, we use the sine function.

$$\sin A = \frac{30}{65}$$

Side opposite A

Hypotenuse

$$A = \sin^{-1}\frac{30}{65} \approx 27.5°$$

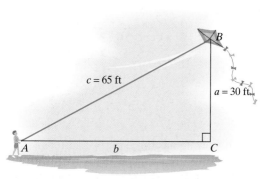

Figure 4.103 Flying a kite

The string makes an angle of approximately 27.5° with the ground.

Check Point 3   A guy wire is 13.8 yards long and is attached from the ground to a pole 6.7 yards above the ground. Find the angle, to the nearest tenth of a degree, that the wire makes with the ground.

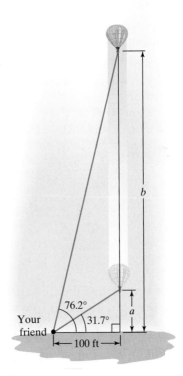

Figure 4.104 Ascending in a hot-air balloon

 Solve problems involving bearings.

Study Tip

The bearing from O to P can also be described using the phrase "the bearing of P from O."

Figure 4.105 An illustration of three bearings

EXAMPLE 4 Using Two Right Triangles to Solve a Problem

You are taking your first hot-air balloon ride. Your friend is standing on level ground 100 feet away from your point of launch, making a video of the terrified look on you rapidly ascending face. How rapidly? At one instant, the angle of elevation from th video camera to your face is 31.7°. One minute later, the angle of elevation is 76.2 How far did you travel, to the nearest tenth of a foot, during that minute?

Solution A sketch that illustrates the problem is shown in Figure 4.104. We nee to determine $b - a$, the distance traveled during the one-minute period. We find using the small right triangle. Because we have a known angle, an unknown opposit side, and a known adjacent side, we use the tangent function.

$$\tan 31.7° = \frac{a}{100}$$

Side opposite the 31.7° angle

Side adjacent to the 31.7° angle

$$a = 100 \tan 31.7° \approx 61.8$$

We find b using the tangent function in the large right triangle.

$$\tan 76.2° = \frac{b}{100}$$

Side opposite the 76.2° angle

Side adjacent to the 76.2° angle

$$b = 100 \tan 76.2° \approx 407.1$$

The balloon traveled $407.1 - 61.8$, or approximately 345.3 feet, during the minut

Check Point 4 You are standing on level ground 800 feet from Mt. Rushmore, looking the sculpture of Abraham Lincoln's face. The angle of elevation to the bo tom of the sculpture is 32° and the angle of elevation to the top is 35°. Fin the height of the sculpture of Lincoln's face to the nearest tenth of a foo

Trigonometry and Bearings

In navigation and surveying problems, the term *bearing* is used to specify th location of one point relative to another. The **bearing** from point O to point P the acute angle, measured in degrees, between ray OP and a north-south lin Figure 4.105 illustrates some examples of bearings. The north-south line and th east-west line intersect at right angles.

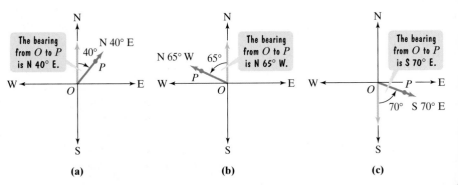

(a)   (b)   (c)

Each bearing has three parts: a letter (N or S), the measure of an acute angle, and letter (E or W). Here's how we write a bearing:

- If the acute angle is measured from the *north side* of the north-south line, the we write N first. [See Figure 4.105(a).] If the acute angle is measured from th *south side* of the north-south line, then we write S first. [See Figure 4.105(c]
- Second, we write the measure of the acute angle.
- If the acute angle is measured on the *east side* of the north-south line, then v write E last. [See Figure 4.105(a)]. If the acute angle is measured on the *west si* of the north-south line, then we write W last. [See Figure 4.105(b).]

EXAMPLE 5   Understanding Bearings

Use Figure 4.106 to find each of the following:

a. the bearing from O to B

b. the bearing from O to A.

Solution

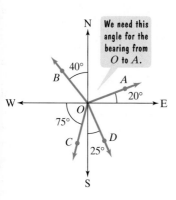

a. To find the bearing from O to B, we need the acute angle between the ray OB and the north-south line through O. The measurement of this angle is given to be 40°. Figure 4.106 shows that the angle is measured from the north side of the north-south line and lies west of the north-south line. Thus, the bearing from O to B is N 40° W.

b. To find the bearing from O to A, we need the acute angle between the ray OA and the north-south line through O. This angle is specified by the voice balloon in Figure 4.106. Because of the given 20° angle, this angle measures 90° − 20°, or 70°. This angle is measured from the north side of the north-south line. This angle is also east of the north-south line. Thus, the bearing from O to A is N 70° E.

Figure 4.106 Finding bearings

> **Check Point 5** Use Figure 4.106 to find each of the following:
>
> **a.** the bearing from O to D
>
> **b.** the bearing from O to C.

EXAMPLE 6   Finding the Bearing of a Boat

A boat leaves the entrance to a harbor and travels 25 miles on a bearing of N 42° E. Figure 4.107 shows that the captain then turns the boat 90°clockwise and travels 18 miles on a bearing of S 48° E. At that time:

a. How far is the boat, to the nearest tenth of a mile, from the harbor entrance?

b. What is the bearing, to the nearest tenth of a degree, of the boat from the harbor entrance?

Solution

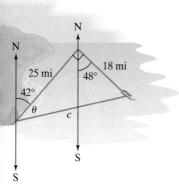

a. The boat's distance from the harbor entrance is represented by c in Figure 4.107. Because we know the length of two sides of the right triangle, we find c using the Pythagorean Theorem. We have

$$c^2 = a^2 + b^2 = 25^2 + 18^2 = 949$$
$$c = \sqrt{949} \approx 30.8.$$

The boat is approximately 30.8 miles from the harbor entrance.

Figure 4.107 Finding a boat's bearing from the harbor entrance

b. The bearing of the boat from the harbor entrance means the bearing from the entrance to the boat. Look at the north-south line passing through the harbor entrance on the left in Figure 4.107. The acute angle from this line to the ray on which the boat lies is $42° + \theta$. Because we are measuring the angle from the north side of the line and the boat is east of the harbor, its bearing from the harbor entrance is N$(42° + \theta)$E. To find θ, we use the right triangle shown in Figure 4.107 and the tangent function.

$$\tan \theta = \frac{\text{side opposite } \theta}{\text{side adjacent to } \theta} = \frac{18}{25}$$

$$\theta = \tan^{-1}\frac{18}{25}$$

Study Tip

When making a diagram showing bearings, draw a north-south line through each point at which a change in course occurs. The north side of the line lies above each point. The south side of the line lies below each point.

We can use a calculator in degree mode to find the value of θ: $\theta \approx 35.8°$. Thus, $42° + \theta = 42° + 35.8° = 77.8°$. The bearing of the boat from the harbor entrance is N 77.8° E.

Check Point 6 You leave the entrance to a system of hiking trails and hike 2.3 miles on bearing of S 31° W. Then the trail turns 90° clockwise and you hike 3. miles on a bearing of N 59° W. At that time:

a. How far are you, to the nearest tenth of a mile, from the entrance to the trail system?

b. What is your bearing, to the nearest tenth of a degree, from the entrance to the trail system?

 Model simple harmonic motion.

Simple Harmonic Motion

Because of their periodic nature, trigonometric functions are used to mode phenomena that occur again and again. This includes vibratory or oscillator motion, such as the motion of a vibrating guitar string, the swinging of a pendulum or the bobbing of an object attached to a spring. Trigonometric functions are also used to describe radio waves from your favorite FM station, television waves from your not-to-be-missed weekly sitcom, and sound waves from your most-prized CD

To see how trigonometric functions are used to model vibratory motion consider this: A ball is attached to a spring hung from the ceiling. You pull the ball down 4 inches and then release it. If we neglect the effects of friction and air resistance, the ball will continue bobbing up and down on the end of the spring These up-and-down oscillations are called **simple harmonic motion**.

To better understand this motion, we use a d-axis, where d represents distance This axis is shown in Figure 4.108. On this axis, the position of the ball before yo pull it down is $d = 0$. This rest position is called the **equilibrium position**. Now yo pull the ball down 4 inches to $d = -4$ and release it. Figure 4.109 shows a sequence of "photographs" taken at one-second time intervals illustrating the distance of the ball from its rest position, d.

The curve in Figure 4.109 shows how the ball's distance from its rest position changes over time. The curve is sinusoidal and the motion can be described using cosine or a sine function.

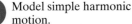
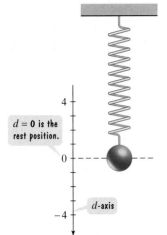

Figure 4.108 Using a d-axis to describe a ball's distance from its rest position

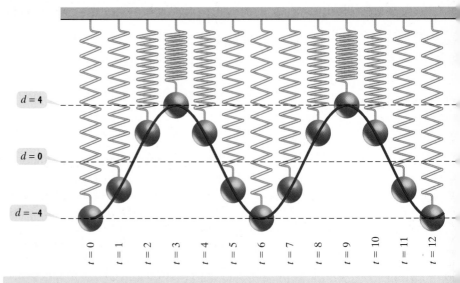

Figure 4.109 A sequence of "photographs" showing the bobbing ball's distance from the rest position, taken at one-second intervals

Simple Harmonic Motion

An object that moves on a coordinate axis is in **simple harmonic motion** if its distance from the origin, d, at time t is given by either

$$d = a \cos \omega t \quad \text{or} \quad d = a \sin \omega t.$$

The motion has **amplitude** $|a|$, the maximum displacement of the object from its rest position. The **period** of the motion is $\dfrac{2\pi}{\omega}$, where $\omega > 0$. The period gives the time it takes for the motion to go through one complete cycle.

Diminishing Motion with Increasing Time

Due to friction and other resistive forces, the motion of an oscillating object decreases over time. The function

$$d = 3e^{-0.1t} \cos 2t$$

models this type of motion. The graph of the function is shown in a $t = [0, 10, 1]$ by $d = [-3, 3, 1]$ viewing rectangle. Notice how the amplitude is decreasing with time as the moving object loses energy.

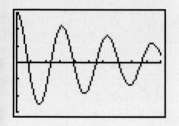

In describing simple harmonic motion, the equation with the cosine function, $d = a \cos \omega t$, is used if the object is at its greatest distance from rest position, the origin, at $t = 0$. By contrast, the equation with the sine function, $d = a \sin \omega t$, is used if the object is at its rest position, the origin, at $t = 0$.

EXAMPLE 7   Finding an Equation for an Object in Simple Harmonic Motion

A ball on a spring is pulled 4 inches below its rest position and then released. The period of the motion is 6 seconds. Write the equation for the ball's simple harmonic motion.

Solution   We need to write an equation that describes d, the distance of the ball from its rest position, after t seconds. (The motion is illustrated by the "photo" sequence in Figure 4.109 on page 554.) When the object is released ($t = 0$), the ball's distance from its rest position is 4 inches down. Because it is *down* 4 inches, d is negative: When $t = 0$, $d = -4$. Notice that the greatest distance from rest position occurs at $t = 0$. Thus, we will use the equation with the cosine function,

$$d = a \cos \omega t,$$

to model the ball's simple harmonic motion.

Now we determine values for a and ω. Recall that $|a|$ is the maximum displacement. Because the ball is initially below rest position, $a = -4$.

The value of ω in $d = a \cos \omega t$ can be found using the formula for the period.

$$\text{period} = \frac{2\pi}{\omega} = 6 \qquad \text{We are given that the period of the motion is 6 seconds.}$$

$$2\pi = 6\omega \qquad \text{Multiply both sides by } \omega.$$

$$\omega = \frac{2\pi}{6} = \frac{\pi}{3} \qquad \text{Divide both sides by 6 and solve for } \omega.$$

We see that $a = -4$ and $\omega = \frac{\pi}{3}$. Substitute these values into $d = a \cos \omega t$. The equation for the ball's simple harmonic motion is

$$d = -4 \cos \frac{\pi}{3} t.$$

Modeling Music

Sounds are caused by vibrating objects that result in variations in pressure in the surrounding air. Areas of high and low pressure moving through the air are modeled by the harmonic motion formulas. When these vibrations reach our eardrums, the eardrums' vibrations send signals to our brains which create the sensation of hearing.

French mathematician John Fourier (1768–1830) proved that all musical sounds—instrumental and vocal—could be modeled by sums involving sine functions. Modeling musical sounds with sinusoidal functions is used by synthesizers to electronically produce sounds unobtainable from ordinary musical instruments.

Check Point 7   A ball on a spring is pulled 6 inches below its rest position and then released. The period for the motion is 4 seconds. Write the equation for the ball's simple harmonic motion.

The period of the harmonic motion in Example 7 was 6 seconds. It takes 6 seconds for the moving object to complete one cycle. Thus, $\frac{1}{6}$ of a cycle is completed every second. We call $\frac{1}{6}$ the *frequency* of the moving object. **Frequency** describes the number of complete cycles per unit time and is the reciprocal of the period.

Frequency of an Object in Simple Harmonic Motion

An object in simple harmonic motion given by

$$d = a \cos \omega t \quad \text{or} \quad d = a \sin \omega t$$

has **frequency** f given by

$$f = \frac{\omega}{2\pi}, \omega > 0.$$

Equivalently,

$$f = \frac{1}{\text{period}}.$$

EXAMPLE 8 **Analyzing Simple Harmonic Motion**

Figure 4.110 shows a mass on a smooth table attached to a spring. The mass move in simple harmonic motion described by

$$d = 10 \cos \frac{\pi}{6} t,$$

with t measured in seconds and d in centimeters. Find **a.** the maximum displacemen **b.** the frequency, and **c.** the time required for one cycle.

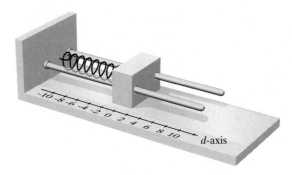

Figure 4.110 A mass attached to a spring, moving in simple harmonic motion

Solution   We begin by identifying values for a and ω.

$$d = 10 \cos \frac{\pi}{6} t$$

The form of this equation is
$d = a \cos \omega t$
with $a = 10$ and $\omega = \frac{\pi}{6}$.

a. The maximum displacement from the rest position is the amplitude. Becaus $a = 10$, the maximum displacement is 10 centimeters.

b. The frequency, f, is

$$f = \frac{\omega}{2\pi} = \frac{\frac{\pi}{6}}{2\pi} = \frac{\pi}{6} \cdot \frac{1}{2\pi} = \frac{1}{12}.$$

The frequency is $\frac{1}{12}$ cycle (or oscillation) per second.

c. The time required for one cycle is the period.

$$\text{period} = \frac{2\pi}{\omega} = \frac{2\pi}{\frac{\pi}{6}} = 2\pi \cdot \frac{6}{\pi} = 12$$

The time required for one cycle is 12 seconds. This value can also be obtaine by taking the reciprocal of the frequency in part (b).

 Check Point 8   An object moves in simple harmonic motion described by $d = 12 \cos \frac{\pi}{4}$

where t is measured in seconds and d in centimeters. Find **a.** the maximu displacement, **b.** the frequency, and **c.** the time required for one cycle.

7. ## Resisting Damage of Simple Harmonic Motion

8.

9.

Simple harmonic motion from an earthquake caused this highway in Oakland, California, to collapse. By studying the harmonic motion of the soil under the highway, engineers learn to build structures that can resist damage.

EXERCISE SET 4.8

Practice Exercises

In Exercises 1–12, solve the right triangle shown in the figure. Round lengths to two decimal places and express angles to the nearest tenth of a degree.

0.

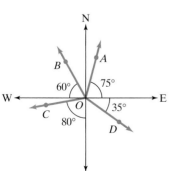

1. $A = 23.5°, b = 10$
2. $A = 41.5°, b = 20$
3. $A = 52.6°, c = 54$
4. $A = 54.8°, c = 80$

1.
5. $B = 16.8°, b = 30.5$
6. $B = 23.8°, b = 40.5$
7. $a = 30.4, c = 50.2$
8. $a = 11.2, c = 65.8$
9. $a = 10.8, b = 24.7$
10. $a = 15.3, b = 17.6$

52.
11. $b = 2, c = 7$
12. $b = 4, c = 9$

Use the figure shown to solve Exercises 13–16.

53.

54.

55.

13. Find the bearing from O to A.
14. Find the bearing from O to B.
56.
15. Find the bearing from O to C.
16. Find the bearing from O to D.

In Exercises 17–20, an object is attached to a coiled spring. In Exercises 17–18, the object is pulled down (negative direction from the rest position) and then released. In Exercises 19–20, the object is initially at its rest position. After that, it is pulled down and then released. Write an equation for the distance of the object from its rest position after t seconds.

	Distance from rest position at $t = 0$	Amplitude	Period
17.	6 centimeters	6 centimeters	4 seconds
18.	8 inches	8 inches	2 seconds
19.	0	3 inches	1.5 seconds
20.	0	5 centimeters	2.5 seconds

In Exercises 21–28, an object moves in simple harmonic motion described by the given equation, where t is measured in seconds and d in inches. In each exercise, find the following:
  a. *the maximum displacement*
  b. *the frequency*
  c. *the time required for one cycle.*

21. $d = 5 \cos \dfrac{\pi}{2}t$
22. $d = 10 \cos 2\pi t$

23. $d = -6 \cos 2\pi t$
24. $d = -8 \cos \dfrac{\pi}{2}t$

25. $d = \frac{1}{2} \sin 2t$
26. $d = \frac{1}{3} \sin 2t$

27. $d = -5 \sin \dfrac{2\pi}{3}t$
28. $d = -4 \sin \dfrac{3\pi}{2}t$

Practice Plus

In Exercises 29–36, find the length x to the nearest whole number.

29.

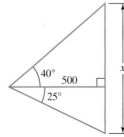

DEFINITIONS AND CONCEPTS	EXAMPLES

4.4 Trigonometric Functions of Any Angle

a. Definitions of the trigonometric functions of any angle are given in the box on page 488.

Ex. 1, p. 489;
Ex. 2, p. 489

b. Signs of the trigonometric functions: All functions are positive in quadrant I. If θ lies in quadrant II, $\sin\theta$ and $\csc\theta$ are positive. If θ lies in quadrant III, $\tan\theta$ and $\cot\theta$ are positive. If θ lies in quadrant IV, $\cos\theta$ and $\sec\theta$ are positive.

Ex. 3, p. 491;
Ex. 4, p. 491

c. If θ is a nonacute angle in standard position that lies in a quadrant, its reference angle is the positive acute angle θ' formed by the terminal side of θ and the x-axis. The reference angle for a given angle can be found by making a sketch that shows the angle in standard position. Figure 4.49 on page 492 shows reference angles for θ in quadrants II, III, and IV.

Ex. 5, p. 492;
Ex. 6, p. 493

d. The values of the trigonometric functions of a given angle are the same as the values of the functions of the reference angle, except possibly for the sign. A procedure for using reference angles to evaluate trigonometric functions is given in the lower box on page 494.

Ex. 7, p. 494;
Ex. 8, p. 496

4.5 and 4.6 Graphs of the Trigonometric Functions

a. Graphs of the six trigonometric functions, with a description of the domain, range, and period of each function, are given in Table 4.6 on page 530.

b. The graph of $y = A\sin(Bx - C)$ can be obtained using amplitude $= |A|$, period $= \dfrac{2\pi}{B}$, and phase shift $= \dfrac{C}{B}$. See the illustration in the box on page 508.

Ex. 1, p. 504;
Ex. 2, p. 505;
Ex. 3, p. 506;
Ex. 4, p. 508

c. The graph of $y = A\cos(Bx - C)$ can be obtained using amplitude $= |A|$, period $= \dfrac{2\pi}{B}$, and phase shift $= \dfrac{C}{B}$. See the illustration in the box on page 513.

Ex. 5, p. 511;
Ex. 6, p. 513

d. The constant D in $y = A\sin(Bx - C) + D$ and $y = A\cos(Bx - C) + D$ causes vertical shifts in the graphs in the preceding items (b) and (c). If $D > 0$, the shift is D units upward and if $D < 0$, the shift is D units downward. Oscillation is about the horizontal line $y = D$.

Ex. 7, p. 514

e. The graph of $y = A\tan(Bx - C)$ is obtained using the procedure in the box on page 523. Consecutive asymptotes $\left(\text{solve } -\dfrac{\pi}{2} < Bx - C < \dfrac{\pi}{2}; \text{ consecutive asymptotes occur at } Bx - C = -\dfrac{\pi}{2} \text{ and } Bx - C = \dfrac{\pi}{2} \right)$ and an x-intercept midway between them play a key role in the graphing process.

Ex. 1, p. 524;
Ex. 2, p. 524

f. The graph of $y = A\cot(Bx - C)$ is obtained using the procedure in the box on page 526. Consecutive asymptotes (solve $0 < Bx - C < \pi$; consecutive asymptotes occur at $Bx - C = 0$ and $Bx - C = \pi$) and an x-intercept midway between them play a key role in the graphing process.

Ex. 3, p. 526

g. To graph a cosecant curve, begin by graphing the corresponding sine curve. Draw vertical asymptotes through x-intercepts, using asymptotes as guides to sketch the graph. To graph a secant curve, first graph the corresponding cosine curve and use the same procedure.

Ex. 4, p. 528;
Ex. 5, p. 529

4.7 Inverse Trigonometric Functions

a. On the restricted domain $-\dfrac{\pi}{2} \le x \le \dfrac{\pi}{2}$, $y = \sin x$ has an inverse function, defined in the box on page 535. Think of $\sin^{-1} x$ as the angle in $\left[-\dfrac{\pi}{2}, \dfrac{\pi}{2} \right]$ whose sine is x. A procedure for finding exact values of $\sin^{-1} x$ is given in the box on page 536.

Ex. 1, p. 537;
Ex. 2, p. 537

41. If $\sin \theta = \dfrac{1}{4}$ and θ is acute, find $\tan\left(\dfrac{\pi}{2} - \theta\right)$.

42. A hiker climbs for a half mile up a slope whose inclination is 17°. How many feet of altitude, to the nearest foot, does the hiker gain?

43. To find the distance across a lake, a surveyor took the measurements in the figure shown. What is the distance across the lake? Round to the nearest meter.

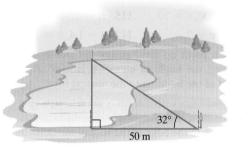

44. When a six-foot pole casts a four-foot shadow, what is the angle of elevation of the sun? Round to the nearest whole degree.

4.4

In Exercises 45–46, a point on the terminal side of angle θ is given. Find the exact value of each of the six trigonometric functions of θ, or state that the function is undefined.

45. $(-1, -5)$ **46.** $(0, -1)$

In Exercises 47–48, let θ be an angle in standard position. Name the quadrant in which θ lies.

47. $\tan \theta > 0$ and $\sec \theta > 0$

48. $\tan \theta > 0$ and $\cos \theta < 0$

In Exercises 49–51, find the exact value of each of the remaining trigonometric functions of θ.

49. $\cos \theta = \frac{2}{5}, \sin \theta < 0$ **50.** $\tan \theta = -\frac{1}{3}, \sin \theta > 0$

51. $\cot \theta = 3, \cos \theta < 0$

In Exercises 52–56, find the reference angle for each angle.

52. 265°   **53.** $\dfrac{5\pi}{8}$ **54.** $-410°$

55. $\dfrac{17\pi}{6}$ **56.** $-\dfrac{11\pi}{3}$

In Exercises 57–67, find the exact value of each expression. Do not use a calculator.

57. $\sin 240°$ **58.** $\tan 120°$ **59.** $\sec \dfrac{7\pi}{4}$

DEFINITIONS AND CONCEPTS   EXAMPLES

b. On the restricted domain $0 \le x \le \pi$, $y = \cos x$ has an inverse function, defined in the box on page 538. Think of $\cos^{-1} x$ as the angle in $[0, \pi]$ whose cosine is x. A procedure for finding exact values of $\cos^{-1} x$ is given in the box on page 539.   Ex. 3, p. 539

c. On the restricted domain $-\dfrac{\pi}{2} < x < \dfrac{\pi}{2}$, $y = \tan x$ has an inverse function, defined in the box on page 540.   Ex. 4, p. 541

Think of $\tan^{-1} x$ as the angle in $\left(-\dfrac{\pi}{2}, \dfrac{\pi}{2}\right)$ whose tangent is x. A procedure for finding exact values of $\tan^{-1} x$ is given in the box on page 541.

d. Graphs of the three basic inverse trigonometric functions, with a description of the domain and range of each function, are given in Table 4.10 on page 542.

e. Inverse properties are given in the box on page 543. Points on terminal sides of angles in standard position and right triangles are used to find exact values of the composition of a function and a different inverse function.   Ex. 6, p. 544;
Ex. 7, p. 545;
Ex. 8, p. 545;
Ex. 9, p. 546

4.8 Applications of Trigonometric Functions

a. Solving a right triangle means finding the missing lengths of its sides and the measurements of its angles. The Pythagorean Theorem, two acute angles whose sum is 90°, and appropriate trigonometric functions are used in this process.   Ex. 1, p. 550;
Ex. 2, p. 551;
Ex. 3, p. 551;
Ex. 4, p. 552

b. The bearing from point O to point P is the acute angle between ray OP and a north-south line, shown in Figure 4.105 on page 552.   Ex. 5, p. 553;
Ex. 6, p. 553

c. Simple harmonic motion, described in the box on page 554, is modeled by $d = a \cos \omega t$ or $d = a \sin \omega t$, with amplitude $= |a|$, period $= \dfrac{2\pi}{\omega}$, and frequency $= \dfrac{\omega}{2\pi} = \dfrac{1}{\text{period}}$.   Ex. 7, p. 555;
Ex. 8, p. 556

Study Tip

Much of the essential information in this chapter can be found in three places:

- Study Tip on page 498, showing special angles and how to obtain exact values of trigonometric functions at these angles
- Table 4.6 on page 530, showing the graphs of the six trigonometric functions, with their domains, ranges, and periods
- Table 4.10 on page 542, showing graphs of the three basic inverse trigonometric functions, with their domains and ranges.

Make copies of these pages and mount them on cardstock. Use this reference sheet as you work the review exercises until you have all the information on the reference sheet memorized for the chapter test.

Review Exercises

4.1

1. Find the radian measure of the central angle of a circle of radius 6 centimeters that intercepts an arc of length 27 centimeters.

In Exercises 2–4, convert each angle in degrees to radians. Express your answer as a multiple of π.

2. 15°   **3.** 120°   **4.** 315°

In Exercises 5–7, convert each angle in radians to degrees.

5. $\dfrac{5\pi}{3}$ **6.** $\dfrac{7\pi}{5}$ **7.** $-\dfrac{5\pi}{6}$

In Exercises 8–12, draw each angle in standard position.

8. $\dfrac{5\pi}{6}$ **9.** $-\dfrac{2\pi}{3}$ **10.** $\dfrac{8\pi}{3}$

11. 190°   **12.** $-135°$

*In Exercises 13–17, find a positive angle less than 360° or 2π th[at]
is coterminal with the given angle.*

13. 400°   **14.** −445°   **15.** $\dfrac{13\pi}{4}$

16. $\dfrac{31\pi}{6}$ **17.** $-\dfrac{8\pi}{3}$

18. Find the length of the arc on a circle of radius 10 [...]
    intercepted by a 135° central angle. Express arc length [in]
    terms of π. Then round your answer to two decimal pla[ces.]

19. The angular speed of a propeller on a wind generator is [...]
    revolutions per minute. Express this angular speed in rad[ians]
    per minute.

20. The propeller of an airplane has a radius of 3 feet. [The]
    propeller is rotating at 2250 revolutions per minute. Find [the]
    linear speed, in feet per minute, of the tip of the propeller.

4.2

*In Exercises 21–22, a point $P(x, y)$ is shown on the unit circle
corresponding to a real number t. Find the values of the
trigonometric functions at t.*

21.

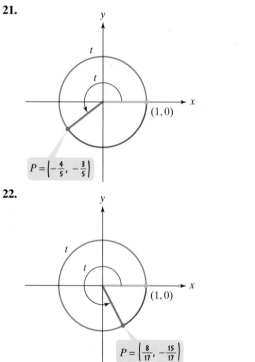

$P = \left(-\dfrac{4}{5}, -\dfrac{3}{5}\right)$

22.

$P = \left(\dfrac{8}{17}, -\dfrac{15}{17}\right)$

*In Exercises 23–26, use the figure shown to find the value of ea[ch]
trigonometric function at the indicated real number or state th[at]
the expression is undefined.*

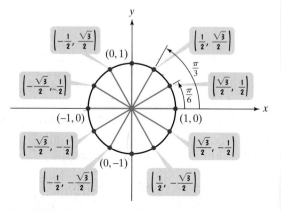

Cumulative Review Exercises (Chapters P–4)

Solve each equation or inequality in Exercises 1–6.

1. $x^2 = 18 + 3x$

2. $x^3 + 5x^2 - 4x - 20 = 0$

3. $\log_2 x + \log_2(x - 2) = 3$

4. $\sqrt{x - 3} + 5 = x$

5. $x^3 - 4x^2 + x + 6 = 0$

6. $|2x - 5| \le 11$

7. If $f(x) = \sqrt{x - 6}$, find $f^{-1}(x)$.

8. Divide $20x^3 - 6x^2 - 9x + 10$ by $5x + 2$.

9. Write as a single logarithm and evaluate: $\log 25 + \log 40$.

10. Convert $\dfrac{14\pi}{9}$ radians to degrees.

11. Find the maximum number of positive and negative real
    roots of the equation $3x^4 - 2x^3 + 5x^2 + x - 9 = 0$.

In Exercises 12–16, graph each equation.

12. $f(x) = \dfrac{x}{x^2 - 1}$

13. $(x - 2)^2 + y^2 = 1$

14. $y = (x - 1)(x + 2)^2$

15. $y = \sin\left(2x + \dfrac{\pi}{2}\right)$, from 0 to 2π

16. $y = 2 \tan 3x$; graph two complete cycles.

17. You invest in a new play. The cost includes an overhead o[f]
    $30,000, plus production costs of $2500 per performance. A [...]
    sold-out performance brings you $3125. How many sold-ou[t]
    performances must be played in order for you to break even[?]

18. Use the exponential growth model $A = A_0 e^{kt}$ to solve thi[s]
    exercise. In 1984, 25,000 cell phones were sold in the Unite[d]
    States and by 2003, 70,500,000 cell phones were sold.

    (*Source:* Consumer Electronics Association)

    a. Find the exponential function that models the data.

    b. By which year will 150,000,000 cell phones be sold in the
        United States?

19. The rate of heat lost through insulation varies inversely a[s]
    the thickness of the insulation. The rate of heat lost through [a]
    3.5-inch thickness of insulation is 2200 Btu per hour. What i[s]
    the rate of heat lost through a 5-inch thickness of the same
    insulation?

20. A tower is 200 feet tall. To the nearest degree, find the angle
    of elevation from a point 50 feet from the base of the towe[r]
    to the top of the tower.

CHAPTER 5

Analytic Trigonometry

equation is manipulated independe
the side containing the more compli
fundamental identities on the more c
it in a form identical to that of the o

No one method or technique c
ties can be verified by rewriting the
sines and cosines.

EXAMPLE 1 Changing to S

Verify the identity: $\sec x \cot x = $

Solution The left side of the equa
Thus, we work with the left side. Le
sines and cosines. Perhaps this strate
$\csc x$, the expression on the right.

$$\sec x \cot x = \frac{1}{\cos x} \cdot \frac{\cos}{\sin}$$

$$= \frac{1}{\cancel{\cos x}} \cdot \frac{\cancel{\text{co}}}{\sin}$$

$$= \frac{1}{\sin x}$$

$$= \csc x$$

By working with the left side and sin
we have verified the given identity.

Technology

You can use a graphing utility to provide evidence of an identity. Enter each s
Then use the TABLE feature or the graphs. The table should show that the f
values of x for which y_1, y_2, or both, are undefined. The graphs should appear
Let's check the identity in Example 1:

$$\sec x \cot x = \csc x.$$

$y_1 = \sec x \cot x$
Enter $\sec x$ as $\dfrac{1}{\cos x}$
and $\cot x$ as $\dfrac{1}{\tan x}$.

$y_2 = \csc x$
Enter $\csc x$ as $\dfrac{1}{\sin}$

Numeric Check
Display a table for y_1 and y_2. We started
our table at $-\pi$ and used $\Delta Tbl = \dfrac{\pi}{8}$.

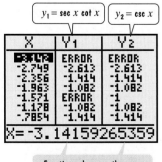

| $y_1 = \sec x \cot x$ | $y_2 = \csc x$ |

X	Y₁	Y₂
-3.142	ERROR	ERROR
-2.749	-2.613	-2.613
-2.356	-1.414	-1.414
-1.963	-1.082	-1.082
-1.571	ERROR	-1
-1.178	-1.082	-1.082
-.7854	-1.414	-1.414

X= -3.14159265359

Function values are the same
except for values of x for which
y_1, y_2, or both, are undefined.

THIS CHAPTER EMPHASIZES THE algebraic aspects of trigonometry. We derive important categories of identities involving trigonometric functions. These identities are used to simplify and analyze expressions that model phenomena as diverse as the distance achieved when throwing an object and musical sounds on a touch-tone phone. For example, we can find out critical information about an athlete's performance by using an identity to analyze an expression involving throwing distance. You will learn how to use trigonometric identities to better understand your periodic world.

YOU ENJOY WATCHING YOUR FRIEND participate in the shot put at college track and field events. After a few full turns in a circle, he throws ("puts") an 8-pound, 13-ounce shot from the shoulder. The range of his throwing distance continues to improve. Knowing that you are studying trigonometry, he asks if there is some way that a trigonometric expression might help achieve his best possible distance in the event.

This problem appears as Exercise 79 in Exercise Set 5.3.

SECTION 5.1 Verifying Trigonometric Identit...

Objective

❶ Use the fundamental trigonometric identities to verify identities.

Do you enjoy solving puzzles
solving skills that are important
for sheer pleasure releases o
well-being. Perhaps this is why

Thousands of relationshi
ing these relationships is like s
process. Thus, proving a trigo
your approach to problem sol
will become a better, more co
the feeling of satisfaction that

The Fundamental Identit

In Chapter 4, we used rigl
trigonometric functions. Alt
fundamental identities listed in
the expressions are defined.

Fundamental Trigonometr

Reciprocal Identities

$$\sin x = \frac{1}{\csc}$$

$$\csc x = \frac{1}{\sin}$$

Quotient Identities

$$\tan x$$

Pythagorean Identities

$$\sin^2 x + \cos^2 x = 1$$

Even-Odd Identities

$$\sin(-x) = -\sin x$$

$$\csc(-x) = -\csc x$$

Study Tip

Memorize the identities in the box. You may need to use variations of these fundamental identities. For example, instead of

$$\sin^2 x + \cos^2 x = 1,$$

you might want to use

$$\sin^2 x = 1 - \cos^2 x$$

or

$$\cos^2 x = 1 - \sin^2 x.$$

Therefore, it is important to know each relationship well so that mental algebraic manipulation is possible.

 Use the fundamental trigonometric identities to verify identities.

Using Fundamental Ident

The fundamental trigonometri
among trigonometric function:
identity can be simplified so t

Study Tip

When proving identities, be sure to write the variable associated with each trigonometric function. Do not get lazy and write

$$\sin \tan + \cos$$

for

$$\sin x \tan x + \cos x$$

because sin, tan, and cos are meaningless without specified variables.

Check Point 1 Verify the identity: $\csc x \tan x = \sec x$.

In verifying an identity, stay focused on your goal. When manipulating one side of the equation, continue to look at the other side to keep the desired form of the result in mind.

Study Tip

Verifying that an equation is an identity is different from solving an equation. You do not verify an identity by adding, subtracting, multiplying, or dividing each side by the same expression. If you do this, you have already assumed that the given statement is true. You do not know that it is true until after you have verified it.

EXAMPLE 2 Changing to Sines and Cosines to Verify an Identity

Verify the identity: $\sin x \tan x + \cos x = \sec x$.

Solution The left side is more complicated, so we start with it. Notice that the left side contains the sum of two terms, but the right side contains only one term. This means that somewhere during the verification process, the two terms on the left side must be added to form one term.

Let's begin by expressing the left side of the identity so that it contains only sines and cosines. Thus, we apply a quotient identity and replace $\tan x$ by $\frac{\sin x}{\cos x}$. Perhaps this strategy will enable us to transform the left side into $\sec x$, the expression on the right.

$$\sin x \tan x + \cos x = \sin x\left(\frac{\sin x}{\cos x}\right) + \cos x$$

Apply a quotient identity: $\tan x = \frac{\sin x}{\cos x}$.

$$= \frac{\sin^2 x}{\cos x} + \cos x.$$

Multiply.

$$= \frac{\sin^2 x}{\cos x} + \cos x \cdot \frac{\cos x}{\cos x}$$

The least common denominator is cos x. Write the second expression with a denominator of cos x.

$$= \frac{\sin^2 x}{\cos x} + \frac{\cos^2 x}{\cos x}$$

Multiply.

$$= \frac{\sin^2 x + \cos^2 x}{\cos x}$$

Add numerators and place this sum over the least common denominator.

$$= \frac{1}{\cos x}$$

Apply a Pythagorean identity: $\sin^2 x + \cos^2 x = 1$.

$$= \sec x$$

Apply a reciprocal identity: $\sec x = \frac{1}{\cos x}$.

By working with the left side and arriving at the right side, the identity is verified

Check Point 2 Verify the identity: $\cos x \cot x + \sin x = \csc x$.

Some identities are verified using factoring to simplify a trigonometric expression.

EXAMPLE 3   Using Factoring to Verify an Identity

Verify the identity: $\cos x - \cos x \sin^2 x = \cos^3 x$.

Solution   We start with the more complicated side, the left side. Factor out the greatest common factor, $\cos x$, from each of the two terms.

$$\cos x - \cos x \sin^2 x = \cos x(1 - \sin^2 x) \qquad \text{Factor cos x from the two terms.}$$

$$= \cos x \cdot \cos^2 x \qquad \text{Use a variation of sin}^2 \text{ x } + \text{ cos}^2 \text{ x } = 1.$$
$$\text{Solving for cos}^2 \text{ x, we obtain}$$
$$\text{cos}^2 \text{ x } = 1 - \text{ sin}^2 \text{ x.}$$

$$= \cos^3 x \qquad \text{Multiply.}$$

We worked with the left side and arrived at the right side. Thus, the identity is verified.

Check Point 3   Verify the identity: $\sin x - \sin x \cos^2 x = \sin^3 x$.

There is often more than one technique that can be used to verify an identity.

EXAMPLE 4   Using Two Techniques to Verify an Identity

Verify the identity: $\dfrac{1 + \sin \theta}{\cos \theta} = \sec \theta + \tan \theta$.

Solution

Method 1.   Separating a Single-Term Quotient into Two Terms
Let's separate the quotient on the left side into two terms using

$$\frac{a + b}{c} = \frac{a}{c} + \frac{b}{c}.$$

Perhaps this strategy will enable us to transform the left side into $\sec \theta + \tan \theta$, the sum on the right.

$$\frac{1 + \sin \theta}{\cos \theta} = \frac{1}{\cos \theta} + \frac{\sin \theta}{\cos \theta} \qquad \text{Divide each term in the numerator by cos } \theta.$$

$$= \sec \theta + \tan \theta \qquad \text{Apply a reciprocal identity and a quotient identity:}$$
$$\sec \theta = \frac{1}{\cos \theta} \text{ and } \tan \theta = \frac{\sin \theta}{\cos \theta}.$$

We worked with the left side and arrived at the right side. Thus, the identity is verified.

Method 2.   Changing to Sines and Cosines
Let's work with the right side of the identity and express it so that it contains only sines and cosines.

$$\sec \theta + \tan \theta = \frac{1}{\cos \theta} + \frac{\sin \theta}{\cos \theta} \qquad \text{Apply a reciprocal identity and a quotient identity:}$$
$$\sec \theta = \frac{1}{\cos \theta} \text{ and } \tan \theta = \frac{\sin \theta}{\cos \theta}.$$

$$= \frac{1 + \sin \theta}{\cos \theta} \qquad \text{Add numerators. Put this sum over the common}$$
$$\text{denominator.}$$

We worked with the right side and arrived at the left side. Thus, the identity is verified.

Check Point 4   Verify the identity: $\dfrac{1 + \cos \theta}{\sin \theta} = \csc \theta + \cot \theta$.

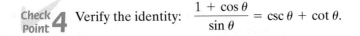

How do we verify identities in which sums or differences of fractions with trigonometric functions appear on one side? Use the least common denominator and combine the fractions. This technique is especially useful when the other side of the identity contains only one term.

EXAMPLE 5 Combining Fractional Expressions to Verify an Identity

Verify the identity: $\dfrac{\cos x}{1 + \sin x} + \dfrac{1 + \sin x}{\cos x} = 2 \sec x.$

Solution   We start with the more complicated side, the left side. The least common denominator of the fractions is $(1 + \sin x)(\cos x)$. We express each fraction in terms of this least common denominator by multiplying the numerator and denominator by the extra factor needed to form $(1 + \sin x)(\cos x)$.

$\dfrac{\cos x}{1 + \sin x} + \dfrac{1 + \sin x}{\cos x}$
  The least common denominator is $(1 + \sin x)(\cos x)$.

$= \dfrac{\cos x (\cos x)}{(1 + \sin x)(\cos x)} + \dfrac{(1 + \sin x)(1 + \sin x)}{(1 + \sin x)(\cos x)}$
  Rewrite each fraction with the least common denominator.

$= \dfrac{\cos^2 x}{(1 + \sin x)(\cos x)} + \dfrac{1 + 2 \sin x + \sin^2 x}{(1 + \sin x)(\cos x)}$
  Use the FOIL method to multiply $(1 + \sin x)(1 + \sin x)$.

$= \dfrac{\cos^2 x + 1 + 2 \sin x + \sin^2 x}{(1 + \sin x)(\cos x)}$
  Add numerators. Put this sum over the least common denominator.

$= \dfrac{(\sin^2 x + \cos^2 x) + 1 + 2 \sin x}{(1 + \sin x)(\cos x)}$
  Regroup terms to apply a Pythagorean identity.

$= \dfrac{1 + 1 + 2 \sin x}{(1 + \sin x)(\cos x)}$
  Apply a Pythagorean identity: $\sin^2 x + \cos^2 x = 1$.

$= \dfrac{2 + 2 \sin x}{(1 + \sin x)(\cos x)}$
  Add constant terms in the numerator: $1 + 1 = 2$.

$= \dfrac{2 \cancel{(1 + \sin x)}}{\cancel{(1 + \sin x)}(\cos x)}$
  Factor and simplify.

$= \dfrac{2}{\cos x}$

$= 2 \sec x$
  Apply a reciprocal identity: $\sec x = \dfrac{1}{\cos x}$.

We worked with the left side and arrived at the right side. Thus, the identity is verified.

Study Tip

Some students have difficulty verifying identities due to problems working with fractions. If this applies to you, review the section on rational expressions in Chapter P.

Check Point 5   Verify the identity: $\dfrac{\sin x}{1 + \cos x} + \dfrac{1 + \cos x}{\sin x} = 2 \csc x.$

Some identities are verified using a technique that may remind you of rationalizing a denominator.

EXAMPLE 6 Multiplying the Numerator and Denominator by the Same Factor to Verify an Identity

Verify the identity: $\dfrac{\sin x}{1 + \cos x} = \dfrac{1 - \cos x}{\sin x}.$

Solution   The suggestions given in the previous examples do not apply here. Everything is already expressed in terms of sines and cosines. Furthermore, there are no fractions to combine and neither side looks more complicated than the other. Let's solve the puzzle by working with the left side and making it look like the expression on the right. The expression on the right contains $1 - \cos x$ in the numerator. This suggests multiplying the numerator and denominator of the left side by $1 - \cos x$. By doing this, we obtain a factor of $1 - \cos x$ in the numerator, as in the numerator on the right.

$$\frac{\sin x}{1 + \cos x} = \frac{\sin x}{1 + \cos x} \cdot \frac{1 - \cos x}{1 - \cos x} \qquad \text{Multiply numerator and denominator by } 1 - \cos x.$$

$$= \frac{\sin x(1 - \cos x)}{1 - \cos^2 x} \qquad \text{Multiply. Use } (A + B)(A - B) = A^2 - B^2, \text{ with } A = 1 \text{ and } B = \cos x, \text{ to multiply denominators.}$$

$$= \frac{\sin x(1 - \cos x)}{\sin^2 x} \qquad \text{Use a variation of } \sin^2 x + \cos^2 x = 1. \text{ Solving for } \sin^2 x, \text{ we obtain } \sin^2 x = 1 - \cos^2 x.$$

$$= \frac{1 - \cos x}{\sin x} \qquad \text{Simplify: } \frac{\sin x}{\sin^2 x} = \frac{\cancel{\sin x}}{\cancel{\sin x} \cdot \sin x} = \frac{1}{\sin x}.$$

We worked with the left side and arrived at the right side. Thus, the identity is verified.

> **Discovery**
>
> Verify the identity in Example 6 by making the right side look like the left side. Start with the expression on the right. Multiply the numerator and denominator by $1 + \cos x$.

Check Point 6   Verify the identity: $\dfrac{\cos x}{1 + \sin x} = \dfrac{1 - \sin x}{\cos x}.$

EXAMPLE 7   Changing to Sines and Cosines to Verify an Identity

Verify the identity: $\dfrac{\tan x - \sin(-x)}{1 + \cos x} = \tan x.$

Solution   We begin with the left side. Our goal is to obtain $\tan x$, the expression on the right.

$$\frac{\tan x - \sin(-x)}{1 + \cos x} = \frac{\tan x - (-\sin x)}{1 + \cos x} \qquad \text{The sine function is odd: } \sin(-x) = -\sin x.$$

$$= \frac{\tan x + \sin x}{1 + \cos x} \qquad \text{Simplify.}$$

$$= \frac{\dfrac{\sin x}{\cos x} + \sin x}{1 + \cos x} \qquad \text{Apply a quotient identity: } \tan x = \dfrac{\sin x}{\cos x}.$$

$$= \frac{\dfrac{\sin x}{\cos x} + \dfrac{\sin x \cos x}{\cos x}}{1 + \cos x} \qquad \text{Express the terms in the numerator with the least common denominator, } \cos x.$$

$$= \frac{\dfrac{\sin x + \sin x \cos x}{\cos x}}{1 + \cos x} \qquad \text{Add in the numerator.}$$

$$= \frac{\sin x + \sin x \cos x}{\cos x} \div \frac{1 + \cos x}{1} \qquad \text{Rewrite the main fraction bar as } \div.$$

$$= \frac{\sin x + \sin x \cos x}{\cos x} \cdot \frac{1}{1 + \cos x} \qquad \text{Invert the divisor and multiply.}$$

$$= \frac{\sin x \cancel{(1 + \cos x)}}{\cos x} \cdot \frac{1}{\cancel{1 + \cos x}} \qquad \text{Factor and simplify.}$$

$$= \frac{\sin x}{\cos x} \qquad \text{Multiply the remaining factors in the numerator and in the denominator.}$$

$$= \tan x \qquad \text{Apply a quotient identity.}$$

> **Discovery**
>
> Try simplifying
>
> $$\frac{\dfrac{\sin x}{\cos x} + \sin x}{1 + \cos x}$$
>
> by multiplying the two terms in the numerator and the two terms in the denominator by $\cos x$. This method for simplifying the complex fraction involves multiplying the numerator and the denominator by the least common denominator of all fractions in the expression. Do you prefer this simplification procedure over the method used on the right?

The left side simplifies to $\tan x$, the right side. Thus, the identity is verified.

Check Point 7 Verify the identity: $\dfrac{\sec x + \csc(-x)}{\sec x \csc x} = \sin x - \cos x.$

Is every identity verified by working with only one side? No. You can some times work with each side separately and show that both sides are equal to the same trigonometric expression. This is illustrated in Example 8.

EXAMPLE 8 Working with Both Sides Separately to Verify an Identity

Verify the identity: $\dfrac{1}{1 + \cos\theta} + \dfrac{1}{1 - \cos\theta} = 2 + 2\cot^2\theta.$

Solution   We begin by working with the left side.

$\dfrac{1}{1 + \cos\theta} + \dfrac{1}{1 - \cos\theta}$	The least common denominator is $(1 + \cos\theta)(1 - \cos\theta).$
$= \dfrac{1(1 - \cos\theta)}{(1 + \cos\theta)(1 - \cos\theta)} + \dfrac{1(1 + \cos\theta)}{(1 + \cos\theta)(1 - \cos\theta)}$	Rewrite each fraction with the least common denominator.
$= \dfrac{1 - \cos\theta + 1 + \cos\theta}{(1 + \cos\theta)(1 - \cos\theta)}$	Add numerators. Put this sum over the least common denominator.
$= \dfrac{2}{(1 + \cos\theta)(1 - \cos\theta)}$	Simplify the numerator: $-\cos\theta + \cos\theta = 0$ and $1 + 1 = 2.$
$= \dfrac{2}{1 - \cos^2\theta}$	Multiply the factors in the denominator.

Now we work with the right side. Our goal is to transform this side into the simplified form attained for the left side, $\dfrac{2}{1 - \cos^2\theta}$.

$2 + 2\cot^2\theta = 2 + 2\left(\dfrac{\cos^2\theta}{\sin^2\theta}\right)$	Use a quotient identity: $\cot\theta = \dfrac{\cos\theta}{\sin\theta}.$
$= \dfrac{2\sin^2\theta}{\sin^2\theta} + \dfrac{2\cos^2\theta}{\sin^2\theta}$	Rewrite each fraction with the least common denominator, $\sin^2\theta.$
$= \dfrac{2\sin^2\theta + 2\cos^2\theta}{\sin^2\theta}$	Add numerators. Put this sum over the least common denominator.
$= \dfrac{2(\sin^2\theta + \cos^2\theta)}{\sin^2\theta}$	Factor out the greatest common factor, 2.
$= \dfrac{2}{\sin^2\theta}$	Apply a Pythagorean identity: $\sin^2\theta + \cos^2\theta = 1.$
$= \dfrac{2}{1 - \cos^2\theta}$	Use a variation of $\sin^2\theta + \cos^2\theta = 1$ and solve for $\sin^2\theta$: $\sin^2\theta = 1 - \cos^2\theta.$

The identity is verified because both sides are equal to $\dfrac{2}{1 - \cos^2\theta}$.

Check Point 8 Verify the identity: $\dfrac{1}{1 + \sin\theta} + \dfrac{1}{1 - \sin\theta} = 2 + 2\tan^2\theta.$

Guidelines for Verifying Trigonometric Identities

There is often more than one correct way to solve a puzzle, although one method may be shorter and more efficient than another. The same is true for verifying an identity. For example, how would you verify

$$\frac{\csc^2 x - 1}{\csc^2 x} = \cos^2 x?$$

One approach is to use a Pythagorean identity, $1 + \cot^2 x = \csc^2 x$, on the left side. Then change the resulting expression to sines and cosines.

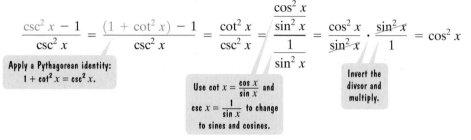

A more efficient strategy for verifying this identity may not be apparent at first glance. Work with the left side and divide each term in the numerator by the denominator, $\csc^2 x$.

$$\frac{\csc^2 x - 1}{\csc^2 x} = \frac{\csc^2 x}{\csc^2 x} - \frac{1}{\csc^2 x} = 1 - \sin^2 x = \cos^2 x$$

> **Apply a reciprocal identity:** $\sin x = \frac{1}{\csc x}$.

> **Use** $\sin^2 x + \cos^2 x = 1$ and solve for $\cos^2 x$.

With this strategy, we again obtain $\cos^2 x$, the expression on the right side, and it takes fewer steps than the first approach.

An even longer strategy, but one that works, is to replace each of the two occurrences of $\csc^2 x$ on the left side by $\dfrac{1}{\sin^2 x}$. This may be the approach that you first consider, particularly if you become accustomed to rewriting the more complicated side in terms of sines and cosines. The selection of an appropriate fundamental identity to solve the puzzle most efficiently is learned through lots of practice.

The more identities you prove, the more confident and efficient you will become. Although practice is the only way to learn how to verify identities, there are some guidelines developed throughout the section that should help you get started.

Guidelines for Verifying Trigonometric Identities

- Work with each side of the equation independently of the other side. Start with the more complicated side and transform it in a step-by-step fashion until it looks exactly like the other side.
- Analyze the identity and look for opportunities to apply the fundamental identities.
- Try using one or more of the following techniques:
  1. Rewrite the more complicated side in terms of sines and cosines.
  2. Factor out the greatest common factor.
  3. Separate a single-term quotient into two terms:
  $$\frac{a + b}{c} = \frac{a}{c} + \frac{b}{c} \quad \text{and} \quad \frac{a - b}{c} = \frac{a}{c} - \frac{b}{c}.$$
  4. Combine fractional expressions using the least common denominator.
  5. Multiply the numerator and the denominator by a binomial factor that appears on the other side of the identity.
- Don't be afraid to stop and start over again if you are not getting anywhere. Creative puzzle solvers know that strategies leading to dead ends often provide good problem-solving ideas.

EXERCISE SET 5.1

Practice Exercises

In Exercises 1–60, verify each identity.

1. $\sin x \sec x = \tan x$
2. $\cos x \csc x = \cot x$
3. $\tan(-x) \cos x = -\sin x$
4. $\cot(-x) \sin x = -\cos x$
5. $\tan x \csc x \cos x = 1$
6. $\cot x \sec x \sin x = 1$
7. $\sec x - \sec x \sin^2 x = \cos x$
8. $\csc x - \csc x \cos^2 x = \sin x$
9. $\cos^2 x - \sin^2 x = 1 - 2 \sin^2 x$
10. $\cos^2 x - \sin^2 x = 2 \cos^2 x - 1$
11. $\csc \theta - \sin \theta = \cot \theta \cos \theta$
12. $\tan \theta + \cot \theta = \sec \theta \csc \theta$
13. $\dfrac{\tan \theta \cot \theta}{\csc \theta} = \sin \theta$
14. $\dfrac{\cos \theta \sec \theta}{\cot \theta} = \tan \theta$
15. $\sin^2 \theta (1 + \cot^2 \theta) = 1$
16. $\cos^2 \theta (1 + \tan^2 \theta) = 1$
17. $\sin t \tan t = \dfrac{1 - \cos^2 t}{\cos t}$
18. $\cos t \cot t = \dfrac{1 - \sin^2 t}{\sin t}$
19. $\dfrac{\csc^2 t}{\cot t} = \csc t \sec t$
20. $\dfrac{\sec^2 t}{\tan t} = \sec t \csc t$
21. $\dfrac{\tan^2 t}{\sec t} = \sec t - \cos t$
22. $\dfrac{\cot^2 t}{\csc t} = \csc t - \sin t$
23. $\dfrac{1 - \cos \theta}{\sin \theta} = \csc \theta - \cot \theta$
24. $\dfrac{1 - \sin \theta}{\cos \theta} = \sec \theta - \tan \theta$
25. $\dfrac{\sin t}{\csc t} + \dfrac{\cos t}{\sec t} = 1$
26. $\dfrac{\sin t}{\tan t} + \dfrac{\cos t}{\cot t} = \sin t + \cos t$
27. $\tan t + \dfrac{\cos t}{1 + \sin t} = \sec t$
28. $\cot t + \dfrac{\sin t}{1 + \cos t} = \csc t$
29. $1 - \dfrac{\sin^2 x}{1 + \cos x} = \cos x$
30. $1 - \dfrac{\cos^2 x}{1 + \sin x} = \sin x$
31. $\dfrac{\cos x}{1 - \sin x} + \dfrac{1 - \sin x}{\cos x} = 2 \sec x$
32. $\dfrac{\sin x}{\cos x + 1} + \dfrac{\cos x - 1}{\sin x} = 0$
33. $\sec^2 x \csc^2 x = \sec^2 x + \csc^2 x$
34. $\csc^2 x \sec x = \sec x + \csc x \cot x$
35. $\dfrac{\sec x - \csc x}{\sec x + \csc x} = \dfrac{\tan x - 1}{\tan x + 1}$
36. $\dfrac{\csc x - \sec x}{\csc x + \sec x} = \dfrac{\cot x - 1}{\cot x + 1}$
37. $\dfrac{\sin^2 x - \cos^2 x}{\sin x + \cos x} = \sin x - \cos x$
38. $\dfrac{\tan^2 x - \cot^2 x}{\tan x + \cot x} = \tan x - \cot x$
39. $\tan^2 2x + \sin^2 2x + \cos^2 2x = \sec^2 2x$
40. $\cot^2 2x + \cos^2 2x + \sin^2 2x = \csc^2 2x$
41. $\dfrac{\tan 2\theta + \cot 2\theta}{\csc 2\theta} = \sec 2\theta$
42. $\dfrac{\tan 2\theta + \cot 2\theta}{\sec 2\theta} = \csc 2\theta$
43. $\dfrac{\tan x + \tan y}{1 - \tan x \tan y} = \dfrac{\sin x \cos y + \cos x \sin y}{\cos x \cos y - \sin x \sin y}$
44. $\dfrac{\cot x + \cot y}{1 - \cot x \cot y} = \dfrac{\cos x \sin y + \sin x \cos y}{\sin x \sin y - \cos x \cos y}$
45. $(\sec x - \tan x)^2 = \dfrac{1 - \sin x}{1 + \sin x}$
46. $(\csc x - \cot x)^2 = \dfrac{1 - \cos x}{1 + \cos x}$
47. $\dfrac{\sec t + 1}{\tan t} = \dfrac{\tan t}{\sec t - 1}$
48. $\dfrac{\csc t - 1}{\cot t} = \dfrac{\cot t}{\csc t + 1}$
49. $\dfrac{1 + \cos t}{1 - \cos t} = (\csc t + \cot t)^2$
50. $\dfrac{\cos^2 t + 4 \cos t + 4}{\cos t + 2} = \dfrac{2 \sec t + 1}{\sec t}$
51. $\cos^4 t - \sin^4 t = 1 - 2 \sin^2 t$
52. $\sin^4 t - \cos^4 t = 1 - 2 \cos^2 t$
53. $\dfrac{\sin \theta - \cos \theta}{\sin \theta} + \dfrac{\cos \theta - \sin \theta}{\cos \theta} = 2 - \sec \theta \csc \theta$
54. $\dfrac{\sin \theta}{1 - \cot \theta} - \dfrac{\cos \theta}{\tan \theta - 1} = \sin \theta + \cos \theta$
55. $(\tan^2 \theta + 1)(\cos^2 \theta + 1) = \tan^2 \theta + 2$
56. $(\cot^2 \theta + 1)(\sin^2 \theta + 1) = \cot^2 \theta + 2$
57. $(\cos \theta - \sin \theta)^2 + (\cos \theta + \sin \theta)^2 = 2$
58. $(3 \cos \theta - 4 \sin \theta)^2 + (4 \cos \theta + 3 \sin \theta)^2 = 25$
59. $\dfrac{\cos^2 x - \sin^2 x}{1 - \tan^2 x} = \cos^2 x$
60. $\dfrac{\sin x + \cos x}{\sin x} - \dfrac{\cos x - \sin x}{\cos x} = \sec x \csc x$

Practice Plus

In Exercises 61–66, half of an identity and the graph of this half are given. Use the graph to make a conjecture as to what the right side of the identity should be. Then prove your conjecture.

61. $\dfrac{(\sec x + \tan x)(\sec x - \tan x)}{\sec x} = ?$

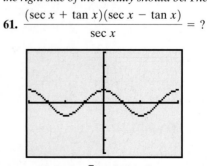

$[-2\pi, 2\pi, \frac{\pi}{2}]$ by $[-4, 4, 1]$

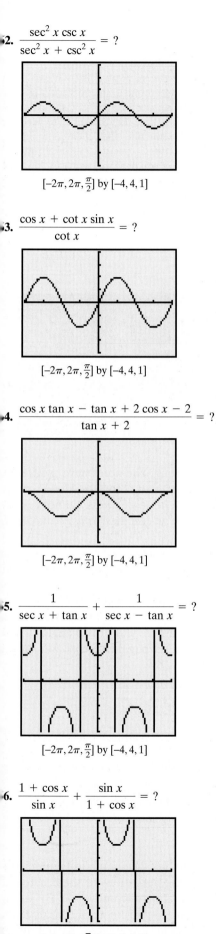

62. $\dfrac{\sec^2 x \csc x}{\sec^2 x + \csc^2 x} = ?$

$[-2\pi, 2\pi, \frac{\pi}{2}]$ by $[-4, 4, 1]$

63. $\dfrac{\cos x + \cot x \sin x}{\cot x} = ?$

$[-2\pi, 2\pi, \frac{\pi}{2}]$ by $[-4, 4, 1]$

64. $\dfrac{\cos x \tan x - \tan x + 2 \cos x - 2}{\tan x + 2} = ?$

$[-2\pi, 2\pi, \frac{\pi}{2}]$ by $[-4, 4, 1]$

65. $\dfrac{1}{\sec x + \tan x} + \dfrac{1}{\sec x - \tan x} = ?$

$[-2\pi, 2\pi, \frac{\pi}{2}]$ by $[-4, 4, 1]$

66. $\dfrac{1 + \cos x}{\sin x} + \dfrac{\sin x}{1 + \cos x} = ?$

$[-2\pi, 2\pi, \frac{\pi}{2}]$ by $[-4, 4, 1]$

In Exercises 67–74, rewrite each expression in terms of the given function or functions.

67. $\dfrac{\tan x + \cot x}{\csc x}; \cos x$ **68.** $\dfrac{\sec x + \csc x}{1 + \tan x}; \sin x$

69. $\dfrac{\cos x}{1 + \sin x} + \tan x; \cos x$ **70.** $\dfrac{1}{\sin x \cos x} - \cot x; \cot x$

71. $\dfrac{1}{1 - \cos x} - \dfrac{\cos x}{1 + \cos x}; \csc x$

72. $(\sec x + \csc x)(\sin x + \cos x) - 2 - \cot x; \tan x$

73. $\dfrac{1}{\csc x - \sin x}; \sec x \text{ and } \tan x$

74. $\dfrac{1 - \sin x}{1 + \sin x} - \dfrac{1 + \sin x}{1 - \sin x}; \sec x \text{ and } \tan x$

Writing in Mathematics

75. Explain how to verify an identity.

76. Describe two strategies that can be used to verify identities.

77. Describe how you feel when you successfully verify a difficult identity. What other activities do you engage in that evoke the same feelings?

78. A 10-point question on a quiz asks students to verify the identity

$$\frac{\sin^2 x - \cos^2 x}{\sin x + \cos x} = \sin x - \cos x.$$

One student begins with the left side and obtains the right side as follows:

$$\frac{\sin^2 x - \cos^2 x}{\sin x + \cos x} = \frac{\sin^2 x}{\sin x} - \frac{\cos^2 x}{\cos x} = \sin x - \cos x.$$

How many points (out of 10) would you give this student? Explain your answer.

Technology Exercises

In Exercises 79–87, graph each side of the equation in the same viewing rectangle. If the graphs appear to coincide, verify that the equation is an identity. If the graphs do not appear to coincide, this indicates the equation is not an identity. In these exercises, find a value of x for which both sides are defined but not equal.

79. $\tan x = \sec x(\sin x - \cos x) + 1$

80. $\sin x = -\cos x \tan(-x)$

81. $\sin\left(x + \dfrac{\pi}{4}\right) = \sin x + \sin\dfrac{\pi}{4}$

82. $\cos\left(x + \dfrac{\pi}{4}\right) = \cos x + \cos\dfrac{\pi}{4}$

83. $\cos(x + \pi) = \cos x$

84. $\sin(x + \pi) = \sin x$

85. $\dfrac{\sin x}{1 - \cos^2 x} = \csc x$

86. $\sin x - \sin x \cos^2 x = \sin^3 x$

87. $\sqrt{\sin^2 x + \cos^2 x} = \sin x + \cos x$

Critical Thinking Exercises

In Exercises 88–91, verify each identity.

88. $\dfrac{\sin^3 x - \cos^3 x}{\sin x - \cos x} = 1 + \sin x \cos x$

89. $\dfrac{\sin x - \cos x + 1}{\sin x + \cos x - 1} = \dfrac{\sin x + 1}{\cos x}$

90. $\ln|\sec x| = -\ln|\cos x|$ **91.** $\ln e^{\tan^2 x - \sec^2 x} = -1$

92. Use one of the fundamental identities in the box on page 570 to create an original identity.

Group Exercise

93. Group members are to write a helpful list of items for a pamphlet called "The Underground Guide to Verifying Identities." The pamphlet will be used primarily by students who sit, stare, and freak out every time they are asked to verify an identity. List easy ways to remember the fundamental identities. What helpful guidelines can you offer from the perspective of a student that you probably won't find in math books? If you have your own strategies that work particularly well, include them in the pamphlet.

SECTION 5.2 *Sum and Difference Formulas*

Objectives

❶ Use the formula for the cosine of the difference of two angles.

❷ Use sum and difference formulas for cosines and sines.

❸ Use sum and difference formulas for tangents.

Listen to the same note played on a piano and a violin. The notes have a different quality or "tone." Tone depends on the way an instrument vibrates. However, the less than 1% of the population with amusia, or true tone deafness, cannot tell the two sounds apart. Even simple, familiar tunes such as *Happy Birthday* and *Jingle Bells* are mystifying to amusics.

When a note is played, it vibrates at a specific fundamental frequency and has a particular amplitude. Amusics cannot tell the difference between two sounds from tuning forks modeled by $p = 3 \sin 2t$ and $p = 2\sin(2t + \pi)$, respectively. However, they can recognize the difference between the two equations. Notice that the second equation contains the sine of the sum of two angles. In this section, we will be developing identities involving the sums or differences of two angles. These formulas are called the **sum and difference formulas**. We begin with $\cos(\alpha - \beta)$, the cosine of the difference of two angles.

The Cosine of the Difference of Two Angles

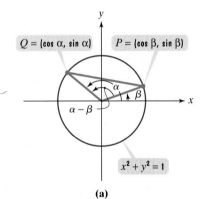

(a)

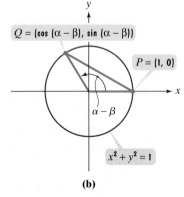

(b)

Figure 5.1 Using the unit circle and QP to develop a formula for $\cos(\alpha - \beta)$

> ### The Cosine of the Difference of Two Angles
>
> $$\cos(\alpha - \beta) = \cos \alpha \cos \beta + \sin \alpha \sin \beta$$
>
> The cosine of the difference of two angles equals the cosine of the first angle times the cosine of the second angle plus the sine of the first angle times the sine of the second angle.

We use Figure 5.1 to prove the identity in the box. The graph in Figure 5.1(a) shows a unit circle, $x^2 + y^2 = 1$. The figure uses the definitions of the cosine and sine functions as the x- and y-coordinates of points along the unit circle. For example, point P corresponds to angle β. By definition, the x-coordinate of P is $\cos \beta$ and the y-coordinate is $\sin \beta$. Similarly, point Q corresponds to angle α. By definition, the x-coordinate of Q is $\cos \alpha$ and the y-coordinate is $\sin \alpha$.

Note that if we draw a line segment between points P and Q, a triangle is formed. Angle $\alpha - \beta$ is one of the angles of this triangle. What happens if we rotate this triangle so that point P falls on the x-axis at $(1, 0)$? The result is shown in Figure 5.1(b). This rotation changes the coordinates of points P and Q. However, it has no effect on the length of line segment PQ.

We can use the distance formula, $d = \sqrt{(x_2 - x_1)^2 + (y_2 - y_1)^2}$, to find an expression for PQ in Figure 5.1(a) and in Figure 5.1(b). By equating the two expressions for PQ, we will obtain the identity for the cosine of the difference of two angles, $\alpha - \beta$. We first apply the distance formula in Figure 5.1(a).

$$PQ = \sqrt{(\cos \alpha - \cos \beta)^2 + (\sin \alpha - \sin \beta)^2}$$

Apply the distance formula, $d = \sqrt{(x_2 - x_1)^2 + (y_2 - y_1)^2}$, to find the distance between $(\cos \beta, \sin \beta)$ and $(\cos \alpha, \sin \alpha)$.

$$= \sqrt{\cos^2 \alpha - 2 \cos \alpha \cos \beta + \cos^2 \beta + \sin^2 \alpha - 2 \sin \alpha \sin \beta + \sin^2 \beta}$$

Square each expression using $(A - B)^2 = A^2 - 2AB + B^2$.

$$= \sqrt{(\sin^2 \alpha + \cos^2 \alpha) + (\sin^2 \beta + \cos^2 \beta) - 2 \cos \alpha \cos \beta - 2 \sin \alpha \sin \beta}$$

Regroup terms to apply a Pythagorean identity.

$$= \sqrt{1 + 1 - 2 \cos \alpha \cos \beta - 2 \sin \alpha \sin \beta}$$

Because $\sin^2 x + \cos^2 x = 1$, each expression in parentheses equals 1.

$$= \sqrt{2 - 2 \cos \alpha \cos \beta - 2 \sin \alpha \sin \beta}$$

Simplify.

Next, we apply the distance formula in Figure 5.1(b) to obtain a second expression for PQ. We let $(x_1, y_1) = (1, 0)$ and $(x_2, y_2) = (\cos(\alpha - \beta), \sin(\alpha - \beta))$.

$$PQ = \sqrt{[\cos(\alpha - \beta) - 1]^2 + [\sin(\alpha - \beta) - 0]^2}$$

Apply the distance formula to find the distance between $(1, 0)$ and $(\cos(\alpha - \beta), \sin(\alpha - \beta))$.

$$= \sqrt{\cos^2(\alpha - \beta) - 2 \cos(\alpha - \beta) + 1 + \sin^2(\alpha - \beta)}$$

Square each expression.

$$= \sqrt{\cos^2(\alpha - \beta) - 2 \cos(\alpha - \beta) + 1 + \sin^2(\alpha - \beta)}$$

Using a Pythagorean identity, $\sin^2(\alpha - \beta) + \cos^2(\alpha - \beta) = 1$.

$$= \sqrt{1 - 2 \cos(\alpha - \beta) + 1}$$

Use a Pythagorean identity.

$$= \sqrt{2 - 2 \cos(\alpha - \beta)}$$

Simplify.

Now we equate the two expressions for PQ.

$$\sqrt{2 - 2 \cos(\alpha - \beta)} = \sqrt{2 - 2 \cos \alpha \cos \beta - 2 \sin \alpha \sin \beta}$$

The rotation does not change the length of PQ.

$$2 - 2 \cos(\alpha - \beta) = 2 - 2 \cos \alpha \cos \beta - 2 \sin \alpha \sin \beta$$

Square both sides to eliminate radicals.

$$-2 \cos(\alpha - \beta) = -2 \cos \alpha \cos \beta - 2 \sin \alpha \sin \beta$$

Subtract 2 from both sides of the equation.

$$\cos(\alpha - \beta) = \cos \alpha \cos \beta + \sin \alpha \sin \beta$$

Divide both sides of the equation by -2.

This proves the identity for the cosine of the difference of two angles.

Now that we see where the identity for the cosine of the difference of two angles comes from, let's look at some applications of this result.

① Use the formula for the cosine of the difference of two angles.

EXAMPLE 1   Using the Difference Formula for Cosines to Find Exact Values

Find the exact value of cos 15°.

Solution   We know exact values for trigonometric functions of 60° and 45°. Thus, we write 15° as 60° − 45° and use the difference formula for cosines.

$$\cos 15° = \cos(60° − 45°)$$
$$= \cos 60° \cos 45° + \sin 60° \sin 45° \qquad \cos(\alpha − \beta) = \cos\alpha\cos\beta + \sin\alpha\sin\beta$$
$$= \frac{1}{2} \cdot \frac{\sqrt{2}}{2} + \frac{\sqrt{3}}{2} \cdot \frac{\sqrt{2}}{2} \qquad \text{Substitute exact values from memory or use special right triangles.}$$
$$= \frac{\sqrt{2}}{4} + \frac{\sqrt{6}}{4} \qquad \text{Multiply.}$$
$$= \frac{\sqrt{2} + \sqrt{6}}{4} \qquad \text{Add.}$$

Check Point 1   We know that $\cos 30° = \dfrac{\sqrt{3}}{2}$. Obtain this exact value using $\cos 30° = \cos(90° − 60°)$ and the difference formula for cosines.

EXAMPLE 2   Using the Difference Formula for Cosines to Find Exact Values

Find the exact value of cos 80° cos 20° + sin 80° sin 20°.

Solution   The given expression is the right side of the formula for $\cos(\alpha − \beta)$ with $\alpha = 80°$ and $\beta = 20°$.

$$\cos(\alpha − \beta) = \cos\alpha\cos\beta + \sin\alpha\sin\beta$$

$$\cos 80° \cos 20° + \sin 80° \sin 20° = \cos(80° − 20°) = \cos 60° = \frac{1}{2}$$

Check Point 2   Find the exact value of
$$\cos 70° \cos 40° + \sin 70° \sin 40°.$$

EXAMPLE 3   Verifying an Identity

Verify the identity: $\dfrac{\cos(\alpha − \beta)}{\sin\alpha\cos\beta} = \cot\alpha + \tan\beta.$

Solution   We work with the left side.

$$\frac{\cos(\alpha − \beta)}{\sin\alpha\cos\beta} = \frac{\cos\alpha\cos\beta + \sin\alpha\sin\beta}{\sin\alpha\cos\beta} \qquad \text{Use the formula for } \cos(\alpha − \beta).$$

$$= \frac{\cos\alpha\cos\beta}{\sin\alpha\cos\beta} + \frac{\sin\alpha\sin\beta}{\sin\alpha\cos\beta} \qquad \text{Divide each term in the numerator by } \sin\alpha\cos\beta.$$

$$= \frac{\cos\alpha}{\sin\alpha} \cdot \frac{\cos\beta}{\cos\beta} + \frac{\sin\alpha}{\sin\alpha} \cdot \frac{\sin\beta}{\cos\beta} \qquad \text{This step can be done mentally. We wanted you to see the substitutions that follow.}$$

$$= \cot\alpha \cdot 1 + 1 \cdot \tan\beta \qquad \text{Use quotient identities.}$$

$$= \cot\alpha + \tan\beta \qquad \text{Simplify.}$$

Sound Quality and Amusia

People with true tone deafness cannot hear the difference among tones produced by a tuning fork, a flute, an oboe, and a violin. They cannot dance or tell the difference between harmony and dissonance. People with amusia appear to have been born without the wiring necessary to process music. Intriguingly, they show no overt signs of brain damage and their brain scans appear normal. Thus, they can visually recognize the difference among sound waves that produce varying sound qualities.

Varying Sound Qualities

• Tuning fork: Sound waves are rounded and regular, giving a pure and gentle tone.

• Flute: Sound waves are smooth and give a fluid tone.

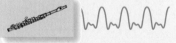

• Oboe: Rapid wave changes give a richer tone.

• Violin: Jagged waves give a brighter harsher tone.

We worked with the left side and arrived at the right side. Thus, the identity is verified.

Technology

The graphs of

$$y = \cos\left(\frac{\pi}{2} - x\right)$$

and

$$y = \sin x$$

are shown in the same viewing rectangle. The graphs are the same. The displayed math on the right shows the equivalence algebraically.

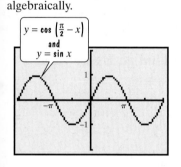

Check Point 3 Verify the identity: $\dfrac{\cos(\alpha - \beta)}{\cos \alpha \cos \beta} = 1 + \tan \alpha \tan \beta.$

The difference formula for cosines is used to establish other identities. For example, in our work with right triangles, we noted that cofunctions of complements are equal. Thus, because $\dfrac{\pi}{2} - \theta$ and θ are complements,

$$\cos\left(\frac{\pi}{2} - \theta\right) = \sin \theta.$$

We can use the formula for $\cos(\alpha - \beta)$ to prove this cofunction identity for all angles.

> Apply $\cos(\alpha - \beta)$ with $\alpha = \frac{\pi}{2}$ and $\theta = \beta$.
> $\cos(\alpha - \beta) = \cos \alpha \cos \beta + \sin \alpha \sin \beta$

$$\cos\left(\frac{\pi}{2} - \theta\right) = \cos\frac{\pi}{2} \cos \theta + \sin\frac{\pi}{2} \sin \theta$$

$$= 0 \cdot \cos \theta + 1 \cdot \sin \theta$$

$$= \sin \theta$$

Use sum and difference formulas for cosines and sines.

Sum and Difference Formulas for Cosines and Sines

Our formula for $\cos(\alpha - \beta)$ can be used to verify an identity for a sum involving cosines, as well as identities for a sum and a difference for sines.

Sum and Difference Formulas for Cosines and Sines

1. $\cos(\alpha + \beta) = \cos \alpha \cos \beta - \sin \alpha \sin \beta$

2. $\cos(\alpha - \beta) = \cos \alpha \cos \beta + \sin \alpha \sin \beta$

3. $\sin(\alpha + \beta) = \sin \alpha \cos \beta + \cos \alpha \sin \beta$

4. $\sin(\alpha - \beta) = \sin \alpha \cos \beta - \cos \alpha \sin \beta$

Up to now, we have concentrated on the second formula in the box. The first identity gives a formula for the cosine of the sum of two angles. It is proved as follows:

$$\cos(\alpha + \beta) = \cos[\alpha - (-\beta)]$$ Express addition as subtraction of an additive inverse.

$$= \cos \alpha \cos(-\beta) + \sin \alpha \sin(-\beta)$$ Use the difference formula for cosines.

$$= \cos \alpha \cos \beta + \sin \alpha(-\sin \beta)$$ Cosine is even: $\cos(-\beta) = \cos \beta$. Sine is odd: $\sin(-\beta) = -\sin \beta$.

$$= \cos \alpha \cos \beta - \sin \alpha \sin \beta.$$ Simplify.

Thus, the cosine of the sum of two angles equals the cosine of the first angle times the cosine of the second angle minus the sine of the first angle times the sine of the second angle.

The third identity in the box gives a formula for $\sin(\alpha + \beta)$, the sine of the sum of two angles. It is proved as follows:

$$\sin(\alpha + \beta) = \cos\left[\frac{\pi}{2} - (\alpha + \beta)\right]$$

Use a cofunction identity: $\sin \theta = \cos\left(\frac{\pi}{2} - \theta\right)$.

$$= \cos\left[\left(\frac{\pi}{2} - \alpha\right) - \beta\right]$$

Regroup.

$$= \cos\left(\frac{\pi}{2} - \alpha\right)\cos \beta + \sin\left(\frac{\pi}{2} - \alpha\right)\sin \beta$$

Use the difference formula for cosines.

$$= \sin \alpha \cos \beta + \cos \alpha \sin \beta.$$

Use cofunction identities.

Thus, the sine of the sum of two angles equals the sine of the first angle times the cosine of the second angle plus the cosine of the first angle times the sine of the second angle.

The final identity in the box, $\sin(\alpha - \beta) = \sin \alpha \cos \beta - \cos \alpha \sin \beta$, gives a formula for $\sin(\alpha - \beta)$, the sine of the difference of two angles. It is proved by writing $\sin(\alpha - \beta)$ as $\sin[\alpha + (-\beta)]$ and then using the formula for the sine of a sum.

EXAMPLE 4 Using the Sine of a Sum to Find an Exact Value

Find the exact value of $\sin\dfrac{7\pi}{12}$ using the fact that $\dfrac{7\pi}{12} = \dfrac{\pi}{3} + \dfrac{\pi}{4}$.

Solution We apply the formula for the sine of a sum.

$$\sin\frac{7\pi}{12} = \sin\left(\frac{\pi}{3} + \frac{\pi}{4}\right)$$

$$= \sin\frac{\pi}{3}\cos\frac{\pi}{4} + \cos\frac{\pi}{3}\sin\frac{\pi}{4} \qquad \sin(\alpha + \beta) = \sin \alpha \cos \beta + \cos \alpha \sin \beta$$

$$= \frac{\sqrt{3}}{2}\cdot\frac{\sqrt{2}}{2} + \frac{1}{2}\cdot\frac{\sqrt{2}}{2} \qquad \text{Substitute exact values.}$$

$$= \frac{\sqrt{6} + \sqrt{2}}{4} \qquad \text{Simplify.}$$

Check Point 4 Find the exact value of $\sin\dfrac{5\pi}{12}$ using the fact that

$$\frac{5\pi}{12} = \frac{\pi}{6} + \frac{\pi}{4}.$$

EXAMPLE 5 Finding Exact Values

Suppose that $\sin \alpha = \frac{12}{13}$ for a quadrant II angle α and $\sin \beta = \frac{3}{5}$ for a quadrant angle β. Find the exact value of each of the following:

a. $\cos \alpha$ **b.** $\cos \beta$ **c.** $\cos(\alpha + \beta)$ **d.** $\sin(\alpha + \beta)$.

Solution

a. We find $\cos \alpha$ using a sketch that illustrates

$$\sin \alpha = \frac{12}{13} = \frac{y}{r}.$$

Figure 5.2 shows a quadrant II angle α with $\sin \alpha = \frac{12}{13}$. We find x using $x^2 + y^2 = r^2$. Because α lies in quadrant II, x is negative.

$$x^2 + 12^2 = 13^2 \qquad\qquad x^2 + y^2 = r^2$$

$$x^2 + 144 = 169 \qquad\qquad \text{Square 12 and 13, respectively.}$$

$$x^2 = 25 \qquad\qquad\qquad \text{Subtract 144 from both sides.}$$

$$x = -\sqrt{25} = -5 \qquad\quad \text{If } x^2 = 25, \text{ then } x = \pm\sqrt{25} = \pm 5.$$

Choose $x = -\sqrt{25}$ because in quadrant II, x is negative.

Figure 5.2 $\sin \alpha = \frac{12}{13}$: α lies in quadrant II.

Thus,

$$\cos \alpha = \frac{x}{r} = \frac{-5}{13} = -\frac{5}{13}.$$

b. We find $\cos \beta$ using a sketch that illustrates

$$\sin \beta = \frac{3}{5} = \frac{y}{r}.$$

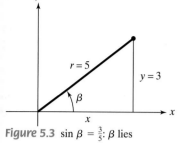

Figure 5.3 $\sin \beta = \frac{3}{5}$: β lies in quadrant I.

Figure 5.3 shows a quadrant I angle β with $\sin \beta = \frac{3}{5}$. We find x using $x^2 + y^2 = r^2$.

$$x^2 + 3^2 = 5^2 \qquad\qquad x^2 + y^2 = r^2$$
$$x^2 + 9 = 25 \qquad\qquad \text{Square 3 and 5, respectively.}$$
$$x^2 = 16 \qquad\qquad \text{Subtract 9 from both sides.}$$
$$x = \sqrt{16} = 4 \qquad \text{If } x^2 = 16, \text{ then } x = \pm\sqrt{16} = \pm 4. \text{ Choose } x = \sqrt{16} \text{ because in quadrant I, x is positive.}$$

Thus,

$$\cos \beta = \frac{x}{r} = \frac{4}{5}.$$

We use the given values and the exact values that we determined to find exact values for $\cos(\alpha + \beta)$ and $\sin(\alpha + \beta)$.

These values are given.	These are the values we found.

$$\sin \alpha = \frac{12}{13}, \ \sin \beta = \frac{3}{5} \qquad \cos \alpha = -\frac{5}{13}, \ \cos \beta = \frac{4}{5}$$

c. We use the formula for the cosine of a sum.

$$\cos(\alpha + \beta) = \cos \alpha \cos \beta - \sin \alpha \sin \beta$$
$$= \left(-\frac{5}{13}\right)\left(\frac{4}{5}\right) - \frac{12}{13}\left(\frac{3}{5}\right) = -\frac{56}{65}$$

d. We use the formula for the sine of a sum.

$$\sin(\alpha + \beta) = \sin \alpha \cos \beta + \cos \alpha \sin \beta$$
$$= \frac{12}{13}\cdot\frac{4}{5} + \left(-\frac{5}{13}\right)\cdot\frac{3}{5} = \frac{33}{65}$$

Check Point 5 Suppose that $\sin \alpha = \frac{4}{5}$ for a quadrant II angle α and $\sin \beta = \frac{1}{2}$ for a quadrant I angle β. Find the exact value of each of the following:

   a. $\cos \alpha$ **b.** $\cos \beta$ **c.** $\cos(\alpha + \beta)$ **d.** $\sin(\alpha + \beta)$.

EXAMPLE 6   Verifying Observations on a Graphing Utility

Figure 5.4 shows the graph of $y = \sin\left(x - \frac{3\pi}{2}\right)$ in a $\left[0, 2\pi, \frac{\pi}{2}\right]$ by $[-2, 2, 1]$ viewing rectangle.

   a. Describe the graph using another equation.
   b. Verify that the two equations are equivalent.

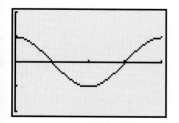

Figure 5.4 The graph of $y = \sin\left(x - \frac{3\pi}{2}\right)$ in a $\left[0, 2\pi, \frac{\pi}{2}\right]$ by $[-2, 2, 1]$ viewing rectangle

Solution

   a. The graph appears to be the cosine curve $y = \cos x$. It cycles through maximum, intercept, minimum, intercept, and back to maximum. Thus, $y = \cos x$ also describes the graph.

b. We must show that

$$\sin\left(x - \frac{3\pi}{2}\right) = \cos x.$$

We apply the formula for the sine of a difference on the left side.

$$\sin\left(x - \frac{3\pi}{2}\right) = \sin x \cos\frac{3\pi}{2} - \cos x \sin\frac{3\pi}{2} \qquad \begin{array}{l} sin(\alpha - \beta) = \\ sin\,\alpha\,cos\,\beta\,-\,cos\,\alpha\,sin\,\beta \end{array}$$

$$= \sin x \cdot 0 - \cos x(-1) \qquad cos\frac{3\pi}{2} = 0 \text{ and } sin\frac{3\pi}{2} = -1$$

$$= \cos x \qquad\qquad Simplify.$$

This verifies our observation that $y = \sin\left(x - \frac{3\pi}{2}\right)$ and $y = \cos x$ describe the same graph.

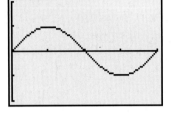

Figure 5.5

Check Point 6 Figure 5.5 shows the graph of $y = \cos\left(x + \frac{3\pi}{2}\right)$ in a $\left[0, 2\pi, \frac{\pi}{2}\right]$ by $[-2, 2, 1]$ viewing rectangle.

a. Describe the graph using another equation.

b. Verify that the two equations are equivalent.

3 Use sum and difference formulas for tangents.

Sum and Difference Formulas for Tangents

By writing $\tan(\alpha + \beta)$ as the quotient of $\sin(\alpha + \beta)$ and $\cos(\alpha + \beta)$, we can develop a formula for the tangent of a sum. Writing subtraction as addition of an inverse leads to a formula for the tangent of a difference.

Discovery

Derive the sum and difference formulas for tangents by working Exercises 55 and 56 in Exercise Set 5.2.

Sum and Difference Formulas for Tangents

$$\tan(\alpha + \beta) = \frac{\tan \alpha + \tan \beta}{1 - \tan \alpha \tan \beta}$$

The tangent of the sum of two angles equals the tangent of the first angle plus the tangent of the second angle divided by 1 minus their product.

$$\tan(\alpha - \beta) = \frac{\tan \alpha - \tan \beta}{1 + \tan \alpha \tan \beta}$$

The tangent of the difference of two angles equals the tangent of the first angle minus the tangent of the second angle divided by 1 plus their product.

EXAMPLE 7 Verifying an Identity

Verify the identity: $\tan\left(x - \frac{\pi}{4}\right) = \frac{\tan x - 1}{\tan x + 1}$.

Solution We work with the left side.

$$\tan\left(x - \frac{\pi}{4}\right) = \frac{\tan x - \tan\dfrac{\pi}{4}}{1 + \tan x \tan\dfrac{\pi}{4}} \qquad tan(\alpha - \beta) = \frac{tan\,\alpha\,-\,tan\,\beta}{1\,+\,tan\,\alpha\,tan\,\beta}$$

$$= \frac{\tan x - 1}{1 + \tan x \cdot 1} \qquad \tan\frac{\pi}{4} = 1$$

$$= \frac{\tan x - 1}{\tan x + 1}$$

Check Point 7 Verify the identity: $\tan(x + \pi) = \tan x$.

EXERCISE SET 5.2

Practice Exercises

Use the formula for the cosine of the difference of two angles to solve Exercises 1–12.

In Exercises 1–4, find the exact value of each expression.

1. $\cos(45° - 30°)$

2. $\cos(120° - 45°)$

3. $\cos\left(\dfrac{3\pi}{4} - \dfrac{\pi}{6}\right)$

4. $\cos\left(\dfrac{2\pi}{3} - \dfrac{\pi}{6}\right)$

In Exercises 5–8, each expression is the right side of the formula for $\cos(\alpha - \beta)$ with particular values for α and β.

  a. *Identify α and β in each expression.*

  b. *Write the expression as the cosine of an angle.*

  c. *Find the exact value of the expression.*

5. $\cos 50° \cos 20° + \sin 50° \sin 20°$

6. $\cos 50° \cos 5° + \sin 50° \sin 5°$

7. $\cos\dfrac{5\pi}{12}\cos\dfrac{\pi}{12} + \sin\dfrac{5\pi}{12}\sin\dfrac{\pi}{12}$ 8. $\cos\dfrac{5\pi}{18}\cos\dfrac{\pi}{9} + \sin\dfrac{5\pi}{18}\sin\dfrac{\pi}{9}$

In Exercises 9–12, verify each identity.

9. $\dfrac{\cos(\alpha - \beta)}{\cos \alpha \sin \beta} = \tan \alpha + \cot \beta$

10. $\dfrac{\cos(\alpha - \beta)}{\sin \alpha \sin \beta} = \cot \alpha \cot \beta + 1$

11. $\cos\left(x - \dfrac{\pi}{4}\right) = \dfrac{\sqrt{2}}{2}(\cos x + \sin x)$

12. $\cos\left(x - \dfrac{5\pi}{4}\right) = -\dfrac{\sqrt{2}}{2}(\cos x + \sin x)$

Use one or more of the six sum and difference identities to solve Exercises 13–54.

In Exercises 13–24, find the exact value of each expression.

13. $\sin(45° - 30°)$

14. $\sin(60° - 45°)$

15. $\sin 105°$

16. $\sin 75°$

17. $\cos(135° + 30°)$

18. $\cos(240° + 45°)$

19. $\cos 75°$

20. $\cos 105°$

21. $\tan\left(\dfrac{\pi}{6} + \dfrac{\pi}{4}\right)$

22. $\tan\left(\dfrac{\pi}{3} + \dfrac{\pi}{4}\right)$

23. $\tan\left(\dfrac{4\pi}{3} - \dfrac{\pi}{4}\right)$

24. $\tan\left(\dfrac{5\pi}{3} - \dfrac{\pi}{4}\right)$

In Exercises 25–32, write each expression as the sine, cosine, or tangent of an angle. Then find the exact value of the expression.

25. $\sin 25° \cos 5° + \cos 25° \sin 5°$

26. $\sin 40° \cos 20° + \cos 40° \sin 20°$

27. $\dfrac{\tan 10° + \tan 35°}{1 - \tan 10° \tan 35°}$

28. $\dfrac{\tan 50° - \tan 20°}{1 + \tan 50° \tan 20°}$

29. $\sin\dfrac{5\pi}{12}\cos\dfrac{\pi}{4} - \cos\dfrac{5\pi}{12}\sin\dfrac{\pi}{4}$

30. $\sin\dfrac{7\pi}{12}\cos\dfrac{\pi}{12} - \cos\dfrac{7\pi}{12}\sin\dfrac{\pi}{12}$

31. $\dfrac{\tan\dfrac{\pi}{5} - \tan\dfrac{\pi}{30}}{1 + \tan\dfrac{\pi}{5}\tan\dfrac{\pi}{30}}$

32. $\dfrac{\tan\dfrac{\pi}{5} + \tan\dfrac{4\pi}{5}}{1 - \tan\dfrac{\pi}{5}\tan\dfrac{4\pi}{5}}$

In Exercises 33–54, verify each identity.

33. $\sin\left(x + \dfrac{\pi}{2}\right) = \cos x$

34. $\sin\left(x + \dfrac{3\pi}{2}\right) = -\cos x$

35. $\cos\left(x - \dfrac{\pi}{2}\right) = \sin x$

36. $\cos(\pi - x) = -\cos x$

37. $\tan(2\pi - x) = -\tan x$

38. $\tan(\pi - x) = -\tan x$

39. $\sin(\alpha + \beta) + \sin(\alpha - \beta) = 2 \sin \alpha \cos \beta$

40. $\cos(\alpha + \beta) + \cos(\alpha - \beta) = 2 \cos \alpha \cos \beta$

41. $\dfrac{\sin(\alpha - \beta)}{\cos \alpha \cos \beta} = \tan \alpha - \tan \beta$

42. $\dfrac{\sin(\alpha + \beta)}{\cos \alpha \cos \beta} = \tan \alpha + \tan \beta$

43. $\tan\left(\theta + \dfrac{\pi}{4}\right) = \dfrac{\cos \theta + \sin \theta}{\cos \theta - \sin \theta}$

44. $\tan\left(\dfrac{\pi}{4} - \theta\right) = \dfrac{\cos \theta - \sin \theta}{\cos \theta + \sin \theta}$

45. $\cos(\alpha + \beta) \cos(\alpha - \beta) = \cos^2 \beta - \sin^2 \alpha$

46. $\sin(\alpha + \beta) \sin(\alpha - \beta) = \cos^2 \beta - \cos^2 \alpha$

47. $\dfrac{\sin(\alpha + \beta)}{\sin(\alpha - \beta)} = \dfrac{\tan \alpha + \tan \beta}{\tan \alpha - \tan \beta}$

48. $\dfrac{\cos(\alpha + \beta)}{\cos(\alpha - \beta)} = \dfrac{1 - \tan \alpha \tan \beta}{1 + \tan \alpha \tan \beta}$

49. $\dfrac{\cos(x + h) - \cos x}{h} = \cos x\dfrac{\cos h - 1}{h} - \sin x\dfrac{\sin h}{h}$

50. $\dfrac{\sin(x + h) - \sin x}{h} = \cos x\dfrac{\sin h}{h} + \sin x\dfrac{\cos h - 1}{h}$

51. $\sin 2\alpha = 2 \sin \alpha \cos \alpha$

  Hint: Write $\sin 2\alpha$ as $\sin(\alpha + \alpha)$.

52. $\cos 2\alpha = \cos^2 \alpha - \sin^2 \alpha$

  Hint: Write $\cos 2\alpha$ as $\cos(\alpha + \alpha)$.

53. $\tan 2\alpha = \dfrac{2 \tan \alpha}{1 - \tan^2 \alpha}$

  Hint: Write $\tan 2\alpha$ as $\tan(\alpha + \alpha)$.

54. $\tan\left(\dfrac{\pi}{4} + \alpha\right) - \tan\left(\dfrac{\pi}{4} - \alpha\right) = 2 \tan 2\alpha$

  Hint: Use the result in Exercise 53.

55. Derive the identity for $\tan(\alpha + \beta)$ using

$$\tan(\alpha + \beta) = \dfrac{\sin(\alpha + \beta)}{\cos(\alpha + \beta)}.$$

After applying the formulas for sums of sines and cosines, divide the numerator and denominator by $\cos \alpha \cos \beta$.

56. Derive the identity for $\tan(\alpha - \beta)$ using

$$\tan(\alpha - \beta) = \tan[\alpha + (-\beta)].$$

After applying the formula for the tangent of the sum of two angles, use the fact that the tangent is an odd function.

In Exercises 57–64, find the exact value of the following under the given conditions:

    a. $\cos(\alpha + \beta)$ **b.** $\sin(\alpha + \beta)$ **c.** $\tan(\alpha + \beta)$

57. $\sin \alpha = \frac{3}{5}, \alpha$ lies in quadrant I, and $\sin \beta = \frac{5}{13}, \beta$ lies in quadrant II.

58. $\sin \alpha = \frac{4}{5}, \alpha$ lies in quadrant I, and $\sin \beta = \frac{7}{25}, \beta$ lies in quadrant II.

59. $\tan \alpha = -\frac{3}{4}, \alpha$ lies in quadrant II, and $\cos \beta = \frac{1}{3}, \beta$ lies in quadrant I.

60. $\tan \alpha = -\frac{4}{3}, \alpha$ lies in quadrant II, and $\cos \beta = \frac{2}{3}, \beta$ lies in quadrant I.

61. $\cos \alpha = \frac{8}{17}, \alpha$ lies in quadrant IV, and $\sin \beta = -\frac{1}{2}, \beta$ lies in quadrant III.

62. $\cos \alpha = \frac{1}{2}, \alpha$ lies in quadrant IV, and $\sin \beta = -\frac{1}{3}, \beta$ lies in quadrant III.

63. $\tan \alpha = \frac{3}{4}, \pi < \alpha < \frac{3\pi}{2}$, and $\cos \beta = \frac{1}{4}, \frac{3\pi}{2} < \beta < 2\pi$.

64. $\sin \alpha = \frac{5}{6}, \frac{\pi}{2} < \alpha < \pi$, and $\tan \beta = \frac{3}{7}, \pi < \beta < \frac{3\pi}{2}$.

In Exercises 65–68, the graph with the given equation is shown in a $\left[0, 2\pi, \frac{\pi}{2}\right]$ by $[-2, 2, 1]$ viewing rectangle.

    a. *Describe the graph using another equation.*
    b. *Verify that the two equations are equivalent.*

65. $y = \sin(\pi - x)$

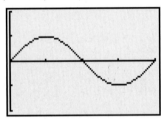

66. $y = \cos(x - 2\pi)$

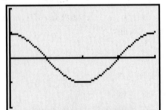

67. $y = \sin\left(x + \frac{\pi}{2}\right) + \sin\left(\frac{\pi}{2} - x\right)$

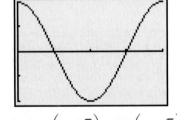

68. $y = \cos\left(x - \frac{\pi}{2}\right) - \cos\left(x + \frac{\pi}{2}\right)$

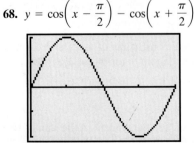

In Exercises 69–74, rewrite each expression as a simplified expression containing one term.

69. $\cos(\alpha + \beta) \cos \beta + \sin(\alpha + \beta) \sin \beta$

70. $\sin(\alpha - \beta) \cos \beta + \cos(\alpha - \beta) \sin \beta$

71. $\dfrac{\sin(\alpha + \beta) - \sin(\alpha - \beta)}{\cos(\alpha + \beta) + \cos(\alpha - \beta)}$ **72.** $\dfrac{\cos(\alpha - \beta) + \cos(\alpha + \beta)}{-\sin(\alpha - \beta) + \sin(\alpha + \beta)}$

73. $\cos\left(\frac{\pi}{6} + \alpha\right) \cos\left(\frac{\pi}{6} - \alpha\right) - \sin\left(\frac{\pi}{6} + \alpha\right) \sin\left(\frac{\pi}{6} - \alpha\right)$

    (Do not use four different identities to solve this exercise.)

74. $\sin\left(\frac{\pi}{3} - \alpha\right) \cos\left(\frac{\pi}{3} + \alpha\right) + \cos\left(\frac{\pi}{3} - \alpha\right) \sin\left(\frac{\pi}{3} + \alpha\right)$

    (Do not use four different identities to solve this exercise.)

In Exercises 75–78, half of an identity and the graph of this half are given. Use the graph to make a conjecture as to what the right side of the identity should be. Then prove your conjecture.

75. $\cos 2x \cos 5x + \sin 2x \sin 5x = ?$

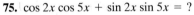

$[-2\pi, 2\pi, \frac{\pi}{2}]$ by $[-2, 2, 1]$

76. $\sin 5x \cos 2x - \cos 5x \sin 2x = ?$

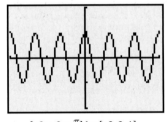

$[-2\pi, 2\pi, \frac{\pi}{2}]$ by $[-2, 2, 1]$

77. $\sin\dfrac{5x}{2} \cos 2x - \cos\dfrac{5x}{2} \sin 2x = ?$

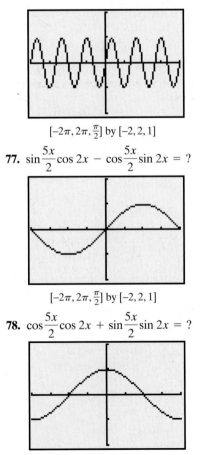

$[-2\pi, 2\pi, \frac{\pi}{2}]$ by $[-2, 2, 1]$

78. $\cos\dfrac{5x}{2} \cos 2x + \sin\dfrac{5x}{2} \sin 2x = ?$

$[-2\pi, 2\pi, \frac{\pi}{2}]$ by $[-2, 2, 1]$

Application Exercises

79. A ball attached to a spring is raised 2 feet and released with an initial vertical velocity of 3 feet per second. The distance of the ball from its rest position after t seconds is given by $d = 2 \cos t + 3 \sin t$. Show that

$$2 \cos t + 3 \sin t = \sqrt{13} \cos(t - \theta),$$

where θ lies in quadrant I and $\tan \theta = \frac{3}{2}$. Use the identity to find the amplitude and the period of the ball's motion.

80. A tuning fork is held a certain distance from your ears and struck. Your eardrums' vibrations after t seconds are given by $p = 3 \sin 2t$. When a second tuning fork is struck, the formula $p = 2 \sin(2t + \pi)$ describes the effects of the sound on the eardrums' vibrations. The total vibrations are given by $p = 3 \sin 2t + 2 \sin(2t + \pi)$.

a. Simplify p to a single term containing the sine.

b. If the amplitude of p is zero, no sound is heard. Based on your equation in part (a), does this occur with the two tuning forks in this exercise? Explain your answer.

Writing in Mathematics

In Exercises 81–87, use words to describe the formula for each of the following:

81. the cosine of the difference of two angles.

82. the cosine of the sum of two angles.

83. the sine of the sum of two angles.

84. the sine of the difference of two angles.

85. the tangent of the difference of two angles.

86. the tangent of the sum of two angles.

87. The distance formula and the definitions for cosine and sine are used to prove the formula for the cosine of the difference of two angles. This formula logically leads the way to the other sum and difference identities. Using this development of ideas and formulas, describe a characteristic of mathematical logic.

Technology Exercises

In Exercises 88–93, graph each side of the equation in the same viewing rectangle. If the graphs appear to coincide, verify that the equation is an identity. If the graphs do not appear to coincide, this indicates that the equation is not an identity. In these exercises, find a value of x for which both sides are defined but not equal.

88. $\cos\left(\dfrac{3\pi}{2} - x\right) = -\sin x$

89. $\tan(\pi - x) = -\tan x$

90. $\sin\left(x + \dfrac{\pi}{2}\right) = \sin x + \sin\dfrac{\pi}{2}$

91. $\cos\left(x + \dfrac{\pi}{2}\right) = \cos x + \cos\dfrac{\pi}{2}$

92. $\cos 1.2x \cos 0.8x - \sin 1.2x \sin 0.8x = \cos 2x$

93. $\sin 1.2x \cos 0.8x + \cos 1.2x \sin 0.8x = \sin 2x$

Critical Thinking Exercises

94. Verify the identity:

$$\dfrac{\sin(x - y)}{\cos x \cos y} + \dfrac{\sin(y - z)}{\cos y \cos z} + \dfrac{\sin(z - x)}{\cos z \cos x} = 0.$$

In Exercises 95–98, find the exact value of each expression. Do not use a calculator.

95. $\sin\left(\cos^{-1}\dfrac{1}{2} + \sin^{-1}\dfrac{3}{5}\right)$

96. $\sin\left[\sin^{-1}\dfrac{3}{5} - \cos^{-1}\left(-\dfrac{4}{5}\right)\right]$

97. $\cos\left(\tan^{-1}\dfrac{4}{3} + \cos^{-1}\dfrac{5}{13}\right)$

98. $\cos\left[\cos^{-1}\left(-\dfrac{\sqrt{3}}{2}\right) - \sin^{-1}\left(-\dfrac{1}{2}\right)\right]$

In Exercises 99–101, write each trigonometric expression as an algebraic expression (that is, without any trigonometric functions). Assume that x and y are positive and in the domain of the given inverse trigonometric function.

99. $\cos(\sin^{-1} x - \cos^{-1} y)$

100. $\sin(\tan^{-1} x - \sin^{-1} y)$

101. $\tan(\sin^{-1} x + \cos^{-1} y)$

Group Exercise

102. Remembering the six sum and difference identities can be difficult. Did you have problems with some exercises because the identity you were using in your head turned out to be an incorrect formula? Are there easy ways to remember the six new identities presented in this section? Group members should address this question, considering one identity at a time. For each formula, list ways to make it easier to remember.

Objectives

❶ Use the double-angle formulas.

❷ Use the power-reducing formulas.

❸ Use the half-angle formulas.

We have a long history of throwing things. Prior to 400 B.C., the Greeks competed i games that included discus throwing. In the seventeenth century, English soldier organized cannonball-throwing competitions. In 1827, a Yale University studen disappointed over failing an exam, took out his frustrations at the passing of collection plate in chapel. Seizing the monetary tray, he flung it in the direction of large open space on campus. Yale students see this act of frustration as the origin c the Frisbee.

In this section, we develop other important classes of identities, called th double-angle, power-reducing, and half-angle formulas. We will see how one of thes formulas can be used by athletes to increase throwing distance.

 Use the double-angle formulas.

Double-Angle Formulas

A number of basic identities follow from the sum formulas for sine, cosine, an tangent. The first category of identities involves **double-angle formulas**.

> ### Double-Angle Formulas
>
> $$\sin 2\theta = 2 \sin \theta \cos \theta$$
> $$\cos 2\theta = \cos^2 \theta - \sin^2 \theta$$
> $$\tan 2\theta = \frac{2 \tan \theta}{1 - \tan^2 \theta}$$

To prove each of these formulas, we replace α and β by θ in the sum formula for $\sin(\alpha + \beta)$, $\cos(\alpha + \beta)$, and $\tan(\alpha + \beta)$.

- $\sin 2\theta = \sin (\theta + \theta) = \sin \theta \cos \theta + \cos \theta \sin \theta = 2 \sin \theta \cos \theta$

  We use
  $\sin (\alpha + \beta) = \sin \alpha \cos \beta + \cos \alpha \sin \beta.$

- $\cos 2\theta = \cos (\theta + \theta) = \cos \theta \cos \theta - \sin \theta \sin \theta = \cos^2 \theta - \sin^2 \theta$

  We use
  $\cos (\alpha + \beta) = \cos \alpha \cos \beta - \sin \alpha \sin \beta.$

- $\tan 2\theta = \tan (\theta + \theta) = \dfrac{\tan \theta + \tan \theta}{1 - \tan \theta \tan \theta} = \dfrac{2 \tan \theta}{1 - \tan^2 \theta}$

  We use
  $\tan (\alpha + \beta) = \dfrac{\tan \alpha + \tan \beta}{1 - \tan \alpha \tan \beta}.$

Study Tip

The 2 that appears in each of the double-angle expressions cannot be pulled to the front and written as a coefficient.

Incorrect!

$$\sin 2\theta = 2 \sin \theta$$
$$\cos 2\theta = 2 \cos \theta$$
$$\tan 2\theta = 2 \tan \theta$$

The figure shows that the graphs of

$$y = \sin 2x$$

and

$$y = 2 \sin x$$

do not coincide: $\sin 2x \neq 2 \sin x$.

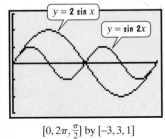

$[0, 2\pi, \frac{\pi}{2}]$ by $[-3, 3, 1]$

EXAMPLE 1   Using Double-Angle Formulas to Find Exact Values

If $\sin \theta = \frac{5}{13}$ and θ lies in quadrant II, find the exact value of each of the following:

    a. $\sin 2\theta$ **b.** $\cos 2\theta$ **c.** $\tan 2\theta$.

Solution   We begin with a sketch that illustrates

$$\sin \theta = \frac{5}{13} = \frac{y}{r}.$$

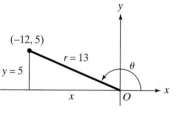

Figure 5.6 $\sin \theta = \frac{5}{13}$ and θ lies in quadrant II.

Figure 5.6 shows a quadrant II angle θ for which $\sin \theta = \frac{5}{13}$. We find x using $x^2 + y^2 = r^2$. Because θ lies in quadrant II, x is negative.

$$x^2 + 5^2 = 13^2 \qquad\qquad\qquad x^2 + y^2 = r^2$$
$$x^2 + 25 = 169 \qquad\qquad\qquad \text{Square 5 and 13, respectively.}$$
$$x^2 = 144 \qquad\qquad\qquad\quad \text{Subtract 25 from both sides.}$$
$$x = -\sqrt{144} = -12 \qquad \text{If } x^2 = 144, \text{ then } x = \pm\sqrt{144} = \pm 12.$$
$$\qquad\qquad\qquad\qquad\qquad\qquad \text{Choose } x = -\sqrt{144} \text{ because in}$$
$$\qquad\qquad\qquad\qquad\qquad\qquad \text{quadrant II, } x \text{ is negative.}$$

Now we can use values for x, y, and r to find the required values. We will use $\cos \theta = \frac{x}{r} = -\frac{12}{13}$ and $\tan \theta = \frac{y}{x} = -\frac{5}{12}$. We were given $\sin \theta = \frac{5}{13}$.

a. $\sin 2\theta = 2 \sin \theta \cos \theta = 2\left(\frac{5}{13}\right)\left(-\frac{12}{13}\right) = -\frac{120}{169}$

b. $\cos 2\theta = \cos^2 \theta - \sin^2 \theta = \left(-\frac{12}{13}\right)^2 - \left(\frac{5}{13}\right)^2 = \frac{144}{169} - \frac{25}{169} = \frac{119}{169}$

c. $\tan 2\theta = \dfrac{2 \tan \theta}{1 - \tan^2 \theta} = \dfrac{2\left(-\frac{5}{12}\right)}{1 - \left(-\frac{5}{12}\right)^2} = \dfrac{-\frac{5}{6}}{1 - \frac{25}{144}} = \dfrac{-\frac{5}{6}}{\frac{119}{144}} = \left(-\frac{5}{6}\right)\left(\frac{144}{119}\right) = -\frac{120}{119}$

Check Point 1   If $\sin \theta = \frac{4}{5}$ and θ lies in quadrant II, find the exact value of each of the following:

    a. $\sin 2\theta$ **b.** $\cos 2\theta$ **c.** $\tan 2\theta$.

EXAMPLE 2   Using the Double-Angle Formula for Tangent to Find an Exact Value

Find the exact value of $\dfrac{2 \tan 15°}{1 - \tan^2 15°}$.

Solution   The given expression is the right side of the formula for $\tan 2\theta$ with $\theta = 15°$.

$$\tan 2\theta = \frac{2 \tan \theta}{1 - \tan^2 \theta}$$

$$\frac{2 \tan 15°}{1 - \tan^2 15°} = \tan(2 \cdot 15°) = \tan 30° = \frac{\sqrt{3}}{3}$$

Check Point 2   Find the exact value of $\cos^2 15° - \sin^2 15°$.

There are three forms of the double-angle formula for $\cos 2\theta$. The form we have seen involves both the cosine and the sine:

$$\cos 2\theta = \cos^2 \theta - \sin^2 \theta.$$

There are situations where it is more efficient to express $\cos 2\theta$ in terms of just on trigonometric function. Using the Pythagorean identity $\sin^2 \theta + \cos^2 \theta = 1$, we ca write $\cos 2\theta = \cos^2 \theta - \sin^2 \theta$ in terms of the cosine only. We substitute $1 - \cos^2$ for $\sin^2 \theta$.

$$\cos 2\theta = \cos^2 \theta - \sin^2 \theta = \cos^2 \theta - (1 - \cos^2 \theta)$$
$$= \cos^2 \theta - 1 + \cos^2 \theta = 2\cos^2 \theta - 1$$

We can also use a Pythagorean identity to write $\cos 2\theta$ in terms of sine only We substitute $1 - \sin^2 \theta$ for $\cos^2 \theta$.

$$\cos 2\theta = \cos^2 \theta - \sin^2 \theta = 1 - \sin^2 \theta - \sin^2 \theta = 1 - 2\sin^2 \theta$$

Three Forms of the Double-Angle Formula for $\cos 2\theta$

$$\cos 2\theta = \cos^2 \theta - \sin^2 \theta$$
$$\cos 2\theta = 2\cos^2 \theta - 1$$
$$\cos 2\theta = 1 - 2\sin^2 \theta$$

EXAMPLE 3   Verifying an Identity

Verify the identity: $\cos 3\theta = 4\cos^3 \theta - 3\cos \theta$.

Solution   We begin by working with the left side. In order to obtain an expressio for $\cos 3\theta$, we use the sum formula and write 3θ as $2\theta + \theta$.

$\cos 3\theta = \cos(2\theta + \theta)$ — Write 3θ as $2\theta + \theta$.

$\qquad = \cos 2\theta \cos \theta - \sin 2\theta \sin \theta$ — $\cos(\alpha + \beta)$
$\qquad\qquad\qquad\qquad\qquad\qquad\qquad = \cos\alpha\cos\beta - \sin\alpha\sin\beta$

$\qquad \underbrace{2\cos^2\theta - 1}\qquad\underbrace{2\sin\theta\cos\theta}$

$\qquad = (2\cos^2\theta - 1)\cos\theta - 2\sin\theta\cos\theta\sin\theta$ — Substitute double-angle formulas. Because the right side of the given equation involves cosines only, use this form for $\cos 2\theta$.

$\qquad = 2\cos^3\theta - \cos\theta - 2\sin^2\theta\cos\theta$ — Multiply.

$\qquad\qquad\qquad\qquad\underbrace{1 - \cos^2\theta}$

$\qquad = 2\cos^3\theta - \cos\theta - 2(1 - \cos^2\theta)\cos\theta$ — To get cosines only, use $\sin^2\theta + \cos^2\theta = 1$ and substitute $1 - \cos^2\theta$ for $\sin^2\theta$.

$\qquad = 2\cos^3\theta - \cos\theta - 2\cos\theta + 2\cos^3\theta$ — Multiply.

$\qquad = 4\cos^3\theta - 3\cos\theta$ — Simplify:
$\qquad\qquad\qquad\qquad\qquad 2\cos^3\theta + 2\cos^3\theta = 4\cos^3\theta$ and
$\qquad\qquad\qquad\qquad\qquad -\cos\theta - 2\cos\theta = -3\cos\theta.$

By working with the left side and expressing it in a form identical to the right sid we have verified the identity.

Check Point 3   Verify the identity: $\sin 3\theta = 3\sin\theta - 4\sin^3\theta$.

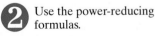

Power-Reducing Formulas

The double-angle formulas are used to derive the **power-reducing formulas**:

Power-Reducing Formulas

$$\sin^2 \theta = \frac{1 - \cos 2\theta}{2} \qquad \cos^2 \theta = \frac{1 + \cos 2\theta}{2} \qquad \tan^2 \theta = \frac{1 - \cos 2\theta}{1 + \cos 2\theta}$$

We can prove the first two formulas in the box by working with two forms of the double-angle formula for $\cos 2\theta$.

This is the form with sine only.   This is the form with cosine only.

$$\cos 2\theta = 1 - 2\sin^2\theta \qquad \cos 2\theta = 2\cos^2\theta - 1$$

Solve the formula on the left for $\sin^2\theta$. Solve the formula on the right for $\cos^2\theta$.

$$2\sin^2\theta = 1 - \cos 2\theta \qquad 2\cos^2\theta = 1 + \cos 2\theta$$

$$\sin^2\theta = \frac{1 - \cos 2\theta}{2} \qquad \cos^2\theta = \frac{1 + \cos 2\theta}{2}$$

Divide both sides of each equation by 2.

These are the first two formulas in the box. The third formula in the box is proved by writing the tangent as the quotient of the sine and the cosine.

$$\tan^2\theta = \frac{\sin^2\theta}{\cos^2\theta} = \frac{\dfrac{1 - \cos 2\theta}{2}}{\dfrac{1 + \cos 2\theta}{2}} = \frac{1 - \cos 2\theta}{2} \cdot \frac{\frac{1}{2}}{1 + \cos 2\theta} = \frac{1 - \cos 2\theta}{1 + \cos 2\theta}$$

Power-reducing formulas are quite useful in calculus. By reducing the power of trigonometric functions, calculus can better explore the relationship between a function and how it is changing at every single instant in time.

EXAMPLE 4   Reducing the Power of a Trigonometric Function

Write an equivalent expression for $\cos^4 x$ that does not contain powers of trigonometric functions greater than 1.

Solution   We will apply the formula for $\cos^2\theta$ twice.

$$\cos^4 x = (\cos^2 x)^2$$

$$= \left(\frac{1 + \cos 2x}{2}\right)^2$$

Use $\cos^2\theta = \dfrac{1 + \cos 2\theta}{2}$ with $\theta = x$.

$$= \frac{1 + 2\cos 2x + \cos^2 2x}{4}$$

Square the numerator: $(A+B)^2 = A^2 + 2AB + B^2$. Square the denominator.

$$= \frac{1}{4} + \frac{1}{2}\cos 2x + \frac{1}{4}\cos^2 2x$$

Divide each term in the numerator by 4.

We can reduce the power of $\cos^2 2x$ using $\cos^2\theta = \dfrac{1 + \cos 2\theta}{2}$ with $\theta = 2x$.

$$= \frac{1}{4} + \frac{1}{2}\cos 2x + \frac{1}{4}\left[\frac{1 + \cos 2(2x)}{2}\right]$$

Use the power-reducing formula for $\cos^2\theta$ with $\theta = 2x$.

$$= \frac{1}{4} + \frac{1}{2}\cos 2x + \frac{1}{8}(1 + \cos 4x)$$

Multiply.

$$= \frac{1}{4} + \frac{1}{2}\cos 2x + \frac{1}{8} + \frac{1}{8}\cos 4x$$

Distribute $\frac{1}{8}$ throughout parentheses.

$$= \frac{3}{8} + \frac{1}{2}\cos 2x + \frac{1}{8}\cos 4x$$

Simplify: $\frac{1}{4} + \frac{1}{8} = \frac{2}{8} + \frac{1}{8} = \frac{3}{8}$.

Thus, $\cos^4 x = \frac{3}{8} + \frac{1}{2}\cos 2x + \frac{1}{8}\cos 4x$. The expression for $\cos^4 x$ does not contain powers of trigonometric functions greater than 1.

Write an equivalent expression for $\sin^4 x$ that does not contain powers o trigonometric functions greater than 1.

 Use the half-angle formulas.

Study Tip

The $\frac{1}{2}$ that appears in each of the half-angle formulas cannot be pulled to the front and written as a coefficient.

Incorrect!

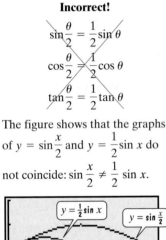

The figure shows that the graphs of $y = \sin \frac{x}{2}$ and $y = \frac{1}{2} \sin x$ do not coincide: $\sin \frac{x}{2} \neq \frac{1}{2} \sin x$.

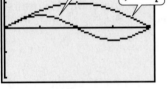

$[0, 2\pi, \frac{\pi}{2}]$ by $[-2, 2, 1]$

Half-Angle Formulas

Useful equivalent forms of the power-reducing formulas can be obtained by replacing θ with $\frac{\alpha}{2}$. Then solve for the trigonometric function on the left sides of the equations The resulting identities are called the **half-angle formulas:**

Half-Angle Formulas

$$\sin \frac{\alpha}{2} = \pm \sqrt{\frac{1 - \cos \alpha}{2}}$$

$$\cos \frac{\alpha}{2} = \pm \sqrt{\frac{1 + \cos \alpha}{2}}$$

$$\tan \frac{\alpha}{2} = \pm \sqrt{\frac{1 - \cos \alpha}{1 + \cos \alpha}}$$

The $\pm$ symbol in each formula does not mean that there are two possible values for each function. Instead, the $\pm$ indicates that you must determine the sign of the trigonometric function, $+$ or $-$, based on the quadrant in which the half-angle $\frac{\alpha}{2}$ lies.

If we know the exact value for the cosine of an angle, we can use the half-angl formulas to find exact values of sine, cosine, and tangent for half of that angle For example, we know that $\cos 225° = -\frac{\sqrt{2}}{2}$. In the next example, we find th exact value of the cosine of half of $225°$, or $\cos 112.5°$.

EXAMPLE 5 Using a Half-Angle Formula to Find an Exact Value

Find the exact value of $\cos 112.5°$.

Solution Because $112.5° = \frac{225°}{2}$, we use the half-angle formula for $\cos \frac{\alpha}{2}$ wit $\alpha = 225°$. What sign should we use when we apply the formula? Because $112.5°$ lie in quadrant II, where only the sine and cosecant are positive, $\cos 112.5° < 0$. Thu we use the $-$ sign in the half-angle formula.

$$\cos 112.5° = \cos \frac{225°}{2}$$

$$= -\sqrt{\frac{1 + \cos 225°}{2}} \qquad \text{Use } \cos \frac{\alpha}{2} = -\sqrt{\frac{1 + \cos \alpha}{2}} \text{ with } \alpha = 225°.$$

$$= -\sqrt{\frac{1 + \left(-\frac{\sqrt{2}}{2}\right)}{2}} \qquad \cos 225° = -\frac{\sqrt{2}}{2}$$

$$= -\sqrt{\frac{2 - \sqrt{2}}{4}} \qquad \text{Multiply the radicand by } \frac{2}{2}:$$
$$\frac{1 + \left(-\frac{\sqrt{2}}{2}\right)}{2} \cdot \frac{2}{2} = \frac{2 - \sqrt{2}}{4}.$$

$$= -\frac{\sqrt{2 - \sqrt{2}}}{2} \qquad \text{Simplify: } \sqrt{4} = 2.$$

Discovery

Use your calculator to find approximations for

$$-\frac{\sqrt{2 - \sqrt{2}}}{2}$$

and $\cos 112.5°$. What do you observe?

Study Tip

Keep in mind as you work with the half-angle formulas that the sign *outside* the radical is determined by the half angle $\frac{\alpha}{2}$. By contrast, the sign of $\cos \alpha$, which appears *under* the radical, is determined by the full angle α.

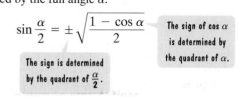

$$\sin \frac{\alpha}{2} = \pm \sqrt{\frac{1 - \cos \alpha}{2}}$$

The sign of $\cos \alpha$ is determined by the quadrant of α.

The sign is determined by the quadrant of $\frac{\alpha}{2}$.

Check Point 5   Use $\cos 210° = -\dfrac{\sqrt{3}}{2}$ to find the exact value of $\cos 105°$.

There are alternate formulas for $\tan \dfrac{\alpha}{2}$ that do not require us to determine what sign to use when applying the formula. These formulas are logically connected to the identities in Example 6 and Check Point 6.

EXAMPLE 6   Verifying an Identity

Verify the identity: $\tan \theta = \dfrac{1 - \cos 2\theta}{\sin 2\theta}$.

Solution   We work with the right side.

$$\frac{1 - \cos 2\theta}{\sin 2\theta} = \frac{1 - (1 - 2\sin^2 \theta)}{2 \sin \theta \cos \theta}$$

The form $\cos 2\theta = 1 - 2\sin^2 \theta$ is used because it produces only one term in the numerator. Use the double-angle formula for sine in the denominator.

$$= \frac{2 \sin^2 \theta}{2 \sin \theta \cos \theta}$$

Simplify the numerator.

$$= \frac{\sin \theta}{\cos \theta}$$

Divide the numerator and denominator by $2 \sin \theta$.

$$= \tan \theta$$

Use a quotient identity: $\tan \theta = \dfrac{\sin \theta}{\cos \theta}$.

The right side simplifies to $\tan \theta$, the expression on the left side. Thus, the identity is verified.

Check Point 6   Verify the identity: $\tan \theta = \dfrac{\sin 2\theta}{1 + \cos 2\theta}$.

Half-angle formulas for $\tan \dfrac{\alpha}{2}$ can be obtained using the identities in Example 6 and Check Point 6:

$$\tan \theta = \frac{1 - \cos 2\theta}{\sin 2\theta} \quad \text{and} \quad \tan \theta = \frac{\sin 2\theta}{1 + \cos 2\theta}.$$

Do you see how to do this? Replace each occurrence of θ with $\dfrac{\alpha}{2}$. This results in the following identities:

Half-Angle Formulas for Tangent

$$\tan \frac{\alpha}{2} = \frac{1 - \cos \alpha}{\sin \alpha}$$

$$\tan \frac{\alpha}{2} = \frac{\sin \alpha}{1 + \cos \alpha}$$

EXAMPLE 7 Verifying an Identity

Verify the identity: $\tan\dfrac{\alpha}{2} = \csc\alpha - \cot\alpha$.

Solution We begin with the right side.

$$\csc\alpha - \cot\alpha = \frac{1}{\sin\alpha} - \frac{\cos\alpha}{\sin\alpha} = \frac{1 - \cos\alpha}{\sin\alpha} = \tan\frac{\alpha}{2}$$

Express functions in terms of sines and cosines.

This is the first of the two half-angle formulas in the preceding box.

We worked with the right side and arrived at the left side. Thus, the identity is verified.

Check Point 7 Verify the identity: $\tan\dfrac{\alpha}{2} = \dfrac{\sec\alpha}{\sec\alpha\csc\alpha + \csc\alpha}$.

We conclude with a summary of the principal trigonometric identities developed in this section and the previous section. The fundamental identities can be found in the box on page 570.

Principal Trigonometric Identities

Sum and Difference Formulas

$$\sin(\alpha+\beta) = \sin\alpha\cos\beta + \cos\alpha\sin\beta \qquad \sin(\alpha-\beta) = \sin\alpha\cos\beta - \cos\alpha\sin\beta$$

$$\cos(\alpha+\beta) = \cos\alpha\cos\beta - \sin\alpha\sin\beta \qquad \cos(\alpha-\beta) = \cos\alpha\cos\beta + \sin\alpha\sin\beta$$

$$\tan(\alpha+\beta) = \frac{\tan\alpha + \tan\beta}{1 - \tan\alpha\tan\beta} \qquad \tan(\alpha-\beta) = \frac{\tan\alpha - \tan\beta}{1 + \tan\alpha\tan\beta}$$

Double-Angle Formulas

$$\sin 2\theta = 2\sin\theta\cos\theta$$

$$\cos 2\theta = \cos^2\theta - \sin^2\theta = 2\cos^2\theta - 1 = 1 - 2\sin^2\theta$$

$$\tan 2\theta = \frac{2\tan\theta}{1 - \tan^2\theta}$$

Power-Reducing Formulas

$$\sin^2\theta = \frac{1 - \cos 2\theta}{2} \qquad \cos^2\theta = \frac{1 + \cos 2\theta}{2} \qquad \tan^2\theta = \frac{1 - \cos 2\theta}{1 + \cos 2\theta}$$

Half-Angle Formulas

$$\sin\frac{\alpha}{2} = \pm\sqrt{\frac{1 - \cos\alpha}{2}} \qquad\qquad \cos\frac{\alpha}{2} = \pm\sqrt{\frac{1 + \cos\alpha}{2}}$$

$$\tan\frac{\alpha}{2} = \pm\sqrt{\frac{1 - \cos\alpha}{1 + \cos\alpha}} = \frac{1 - \cos\alpha}{\sin\alpha} = \frac{\sin\alpha}{1 + \cos\alpha}$$

Study Tip

To help remember the correct sign in the numerator in the first two power-reducing formulas and the first two half-angle formulas, remember *sinus-minus*–the sine is minus.

EXERCISE SET 5.3

Practice Exercises

In Exercises 1–6, use the figures to find the exact value of each trigonometric function.

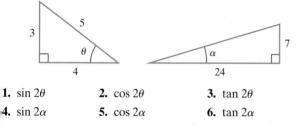

1. $\sin 2\theta$ **2.** $\cos 2\theta$ **3.** $\tan 2\theta$
4. $\sin 2\alpha$ **5.** $\cos 2\alpha$ **6.** $\tan 2\alpha$

In Exercises 7–14, use the given information to find the exact value of each of the following:

   a. $\sin 2\theta$ **b.** $\cos 2\theta$ **c.** $\tan 2\theta.$

7. $\sin \theta = \frac{15}{17}$, θ lies in quadrant II.

8. $\sin \theta = \frac{12}{13}$, θ lies in quadrant II.

9. $\cos \theta = \frac{24}{25}$, θ lies in quadrant IV.

10. $\cos \theta = \frac{40}{41}$, θ lies in quadrant IV.

11. $\cot \theta = 2$, θ lies in quadrant III.

12. $\cot \theta = 3$, θ lies in quadrant III.

13. $\sin \theta = -\frac{9}{41}$, θ lies in quadrant III.

14. $\sin \theta = -\frac{2}{3}$, θ lies in quadrant III.

In Exercises 15–22, write each expression as the sine, cosine, or tangent of a double angle. Then find the exact value of the expression.

15. $2 \sin 15° \cos 15°$ **16.** $2 \sin 22.5° \cos 22.5°$

17. $\cos^2 75° - \sin^2 75°$ **18.** $\cos^2 105° - \sin^2 105°$

19. $2 \cos^2 \frac{\pi}{8} - 1$ **20.** $1 - 2 \sin^2 \frac{\pi}{12}$

21. $\dfrac{2 \tan \frac{\pi}{12}}{1 - \tan^2 \frac{\pi}{12}}$ **22.** $\dfrac{2 \tan \frac{\pi}{8}}{1 - \tan^2 \frac{\pi}{8}}$

In Exercises 23–34, verify each identity.

23. $\sin 2\theta = \dfrac{2 \tan \theta}{1 + \tan^2 \theta}$ **24.** $\sin 2\theta = \dfrac{2 \cot \theta}{1 + \cot^2 \theta}$

25. $(\sin \theta + \cos \theta)^2 = 1 + \sin 2\theta$

26. $(\sin \theta - \cos \theta)^2 = 1 - \sin 2\theta$

27. $\sin^2 x + \cos 2x = \cos^2 x$

28. $1 - \tan^2 x = \dfrac{\cos 2x}{\cos^2 x}$ **29.** $\cot x = \dfrac{\sin 2x}{1 - \cos 2x}$

30. $\cot x = \dfrac{1 + \cos 2x}{\sin 2x}$

31. $\sin 2t - \tan t = \tan t \cos 2t$

32. $\sin 2t - \cot t = -\cot t \cos 2t$

33. $\sin 4t = 4 \sin t \cos^3 t - 4 \sin^3 t \cos t$

34. $\cos 4t = 8 \cos^4 t - 8 \cos^2 t + 1$

In Exercises 35–38, use the power-reducing formulas to rewrite each expression as an equivalent expression that does not contain powers of trigonometric functions greater than 1.

35. $6 \sin^4 x$ **36.** $10 \cos^4 x$
37. $\sin^2 x \cos^2 x$ **38.** $8 \sin^2 x \cos^2 x$

In Exercises 39–46, use a half-angle formula to find the exact value of each expression.

39. $\sin 15°$ **40.** $\cos 22.5°$ **41.** $\cos 157.5°$

42. $\sin 105°$ **43.** $\tan 75°$ **44.** $\tan 112.5°$

45. $\tan \dfrac{7\pi}{8}$ **46.** $\tan \dfrac{3\pi}{8}$

In Exercises 47–54, use the figures to find the exact value of each trigonometric function.

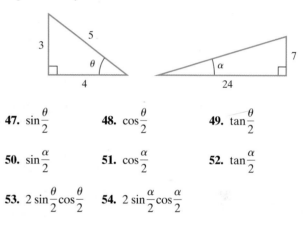

47. $\sin \dfrac{\theta}{2}$ **48.** $\cos \dfrac{\theta}{2}$ **49.** $\tan \dfrac{\theta}{2}$

50. $\sin \dfrac{\alpha}{2}$ **51.** $\cos \dfrac{\alpha}{2}$ **52.** $\tan \dfrac{\alpha}{2}$

53. $2 \sin \dfrac{\theta}{2} \cos \dfrac{\theta}{2}$ **54.** $2 \sin \dfrac{\alpha}{2} \cos \dfrac{\alpha}{2}$

In Exercises 55–58, use the given information to find the exact value of each of the following:

   a. $\sin \dfrac{\alpha}{2}$ **b.** $\cos \dfrac{\alpha}{2}$ **c.** $\tan \dfrac{\alpha}{2}.$

55. $\tan \alpha = \frac{4}{3}$, α lies in quadrant III.

56. $\tan \alpha = \frac{8}{15}$, α lies in quadrant III.

57. $\sec \alpha = -\frac{13}{5}$, α lies in quadrant II.

58. $\sec \alpha = -3$, α lies in quadrant II.

In Exercises 59–68, verify each identity.

59. $\sin^2 \dfrac{\theta}{2} = \dfrac{\sec \theta - 1}{2 \sec \theta}$ **60.** $\sin^2 \dfrac{\theta}{2} = \dfrac{\csc \theta - \cot \theta}{2 \csc \theta}$

61. $\cos^2 \dfrac{\theta}{2} = \dfrac{\sin \theta + \tan \theta}{2 \tan \theta}$ **62.** $\cos^2 \dfrac{\theta}{2} = \dfrac{\sec \theta + 1}{2 \sec \theta}$

63. $\tan \dfrac{\alpha}{2} = \dfrac{\tan \alpha}{\sec \alpha + 1}$ **64.** $2 \tan \dfrac{\alpha}{2} = \dfrac{\sin^2 \alpha + 1 - \cos^2 \alpha}{\sin \alpha (1 + \cos \alpha)}$

65. $\cot \dfrac{x}{2} = \dfrac{\sin x}{1 - \cos x}$ **66.** $\cot \dfrac{x}{2} = \dfrac{1 + \cos x}{\sin x}$

67. $\tan \dfrac{x}{2} + \cot \dfrac{x}{2} = 2 \csc x$ **68.** $\tan \dfrac{x}{2} - \cot \dfrac{x}{2} = -2 \cot x$

Practice Plus

In Exercises 69–78, half of an identity and the graph of this half are given. Use the graph to make a conjecture as to what the right side of the identity should be. Then prove your conjecture.

69. $\dfrac{\cot x - \tan x}{\cot x + \tan x} = ?$

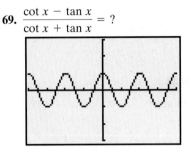

$[-2\pi, 2\pi, \frac{\pi}{2}]$ by $[-3, 3, 1]$

70. $\dfrac{2(\tan x - \cot x)}{\tan^2 x - \cot^2 x} = ?$

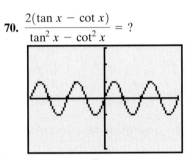

$[-2\pi, 2\pi, \frac{\pi}{2}]$ by $[-3, 3, 1]$

71. $\left(\sin\dfrac{x}{2} + \cos\dfrac{x}{2}\right)^2 = ?$

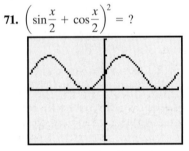

$[-2\pi, 2\pi, \frac{\pi}{2}]$ by $[-3, 3, 1]$

72. $\sin^2\dfrac{x}{2} - \cos^2\dfrac{x}{2} = ?$

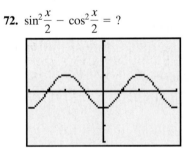

$[-2\pi, 2\pi, \frac{\pi}{2}]$ by $[-3, 3, 1]$

73. $\dfrac{\sin 2x}{\sin x} - \dfrac{\cos 2x}{\cos x} = ?$

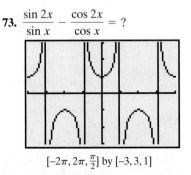

$[-2\pi, 2\pi, \frac{\pi}{2}]$ by $[-3, 3, 1]$

74. $\sin 2x \sec x = ?$

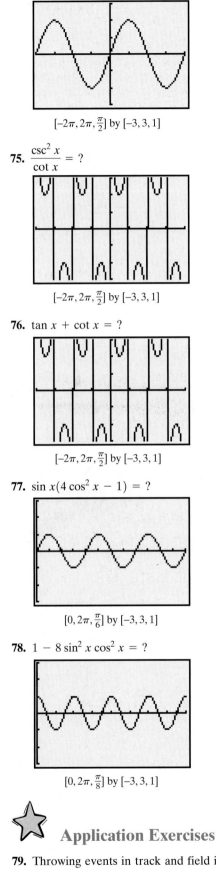

$[-2\pi, 2\pi, \frac{\pi}{2}]$ by $[-3, 3, 1]$

75. $\dfrac{\csc^2 x}{\cot x} = ?$

$[-2\pi, 2\pi, \frac{\pi}{2}]$ by $[-3, 3, 1]$

76. $\tan x + \cot x = ?$

$[-2\pi, 2\pi, \frac{\pi}{2}]$ by $[-3, 3, 1]$

77. $\sin x (4\cos^2 x - 1) = ?$

$[0, 2\pi, \frac{\pi}{6}]$ by $[-3, 3, 1]$

78. $1 - 8\sin^2 x \cos^2 x = ?$

$[0, 2\pi, \frac{\pi}{8}]$ by $[-3, 3, 1]$

Application Exercises

79. Throwing events in track and field include the shot put, the discus throw, the hammer throw, and the javelin throw. The distance that the athlete can achieve depends on the initial speed of the object thrown and the angle above the horizontal at which the object leaves the hand. This angle is represented by θ in the figure shown. The distance, d, in feet, that the athlete throws is modeled by the formula

$$d = \frac{v_0^2}{16}\sin\theta\cos\theta,$$

in which v_0 is the initial speed of the object thrown, in feet per second, and θ is the angle, in degrees, at which the object leaves the hand.

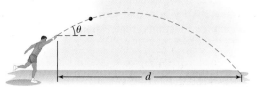

a. Use an identity to express the formula so that it contains the sine function only.

b. Use your formula from part (a) to find the angle, θ, that produces the maximum distance, d, for a given initial speed, v_0.

se this information to solve Exercises 80–81.

he speed of a supersonic aircraft is usually represented by a ʾach number, named after Austrian physicist Ernst Mach 838–1916). A Mach number is the speed of the aircraft, in miles ʾr hour, divided by the speed of sound, approximately 740 miles ʾr hour. Thus, a plane flying at twice the speed of sound has a ʾeed, M, of Mach 2.

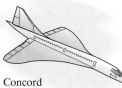

Concord
Mach 2.03

SR-71 Blackbird
Mach 3.3

ʾ an aircraft has a speed greater than Mach 1, a sonic boom is ʾard, created by sound waves that form a cone with a vertex ʾgle θ, shown in the figure.

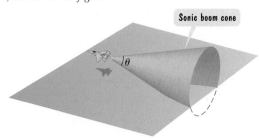

Sonic boom cone

ʾe relationship between the cone's vertex angle, θ, and the Mach ʾeed, M, of an aircraft that is flying faster than the speed of ʾund is given by

$$\sin\frac{\theta}{2} = \frac{1}{M}.$$

ʾ. If $\theta = \frac{\pi}{6}$, determine the Mach speed, M, of the aircraft. Express the speed as an exact value and as a decimal to the nearest tenth.

81. If $\theta = \frac{\pi}{4}$, determine the Mach speed, M, of the aircraft. Express the speed as an exact value and as a decimal to the nearest tenth.

Writing in Mathematics

In Exercises 82–89, use words to describe the formula for:

82. the sine of double an angle.

83. the cosine of double an angle. (Describe one of the three formulas.)

84. the tangent of double an angle.

85. the power-reducing formula for the sine squared of an angle.

86. the power-reducing formula for the cosine squared of an angle.

87. the sine of half an angle.

88. the cosine of half an angle.

89. the tangent of half an angle. (Describe one of the two formulas that does not involve a square root.)

90. Explain how the double-angle formulas are derived.

91. How can there be three forms of the double-angle formula for $\cos 2\theta$?

92. Without showing algebraic details, describe in words how to reduce the power of $\cos^4 x$.

93. Describe one or more of the techniques you use to help remember the identities in the box on page 596.

94. Your friend is about to compete as a shot-putter in a college field event. Using Exercise 79(b), write a short description to your friend on how to achieve the best distance possible in the throwing event.

Technology Exercises

In Exercises 95–98, graph each side of the equation in the same viewing rectangle. If the graphs appear to coincide, verify that the equation is an identity. If the graphs do not appear to coincide, find a value of x for which both sides are defined but not equal.

95. $3 - 6\sin^2 x = 3\cos 2x$ **96.** $4\cos^2\frac{x}{2} = 2 + 2\cos x$

97. $\sin\frac{x}{2} = \frac{1}{2}\sin x$ **98.** $\cos\frac{x}{2} = \frac{1}{2}\cos x$

*In Exercises 99–101, graph each equation in a $\left[-2\pi, 2\pi, \frac{\pi}{2}\right]$ by $[-3, 3, 1]$ viewing rectangle. Then **a.** Describe the graph using another equation, and **b.** Verify that the two equations are equivalent.*

99. $y = \dfrac{1 - 2\cos 2x}{2\sin x - 1}$ **100.** $y = \dfrac{2\tan\dfrac{x}{2}}{1 + \tan^2\dfrac{x}{2}}$

101. $y = \csc x - \cot x$

Critical Thinking Exercises

102. Verify the identity:

$$\sin^3 x + \cos^3 x = (\sin x + \cos x)\left(1 - \frac{\sin 2x}{2}\right).$$

In Exercises 103–106, find the exact value of each expression. Do not use a calculator.

103. $\sin\left(2\sin^{-1}\dfrac{\sqrt{3}}{2}\right)$ **104.** $\cos\left[2\tan^{-1}\left(-\dfrac{4}{3}\right)\right]$

105. $\cos^2\left(\dfrac{1}{2}\sin^{-1}\dfrac{3}{5}\right)$ **106.** $\sin^2\left(\dfrac{1}{2}\cos^{-1}\dfrac{3}{5}\right)$

107. Use a right triangle to write $\sin\left(2\sin^{-1}x\right)$ as an algebraic expression. Assume that x is positive and in the domain of the given inverse trigonometric function.

108. Use the power-reducing formulas to rewrite $\sin^6 x$ as an equivalent expression that does not contain powers of trigonometric functions greater than 1.

CHAPTER 5
MID-CHAPTER CHECK POINT

What You Know: Verifying an identity means showing that the expressions on each side are identical. Like solving puzzles, the process can be intriguing because there are sometimes several "best" ways to proceed. We presented some guidelines to help you get started (see page 577). We used fundamental trigonometric identities (see page 570), as well as sum and difference formulas, double-angle formulas, power-reducing formulas, and half-angle formulas (see page 596) to verify identities. We also used these formulas to find exact values of trigonometric functions.

Study Tip

Make copies of the boxes on pages 570 and 596 that contain the essential trigonometric identities. Mount these boxes on cardstock and add this reference sheet to the one you prepared for Chapter 4. (If you didn't prepare a reference sheet for Chapter 4, it's not too late: See the study tip on page 563.)

In Exercises 1–18, verify each identity.

1. $\cos x(\tan x + \cot x) = \csc x$

2. $\dfrac{\sin(x + \pi)}{\cos\left(x + \dfrac{3\pi}{2}\right)} = \tan^2 x - \sec^2 x$

3. $(\sin\theta + \cos\theta)^2 + (\sin\theta - \cos\theta)^2 = 2$

4. $\dfrac{\sin t - 1}{\cos t} = \dfrac{\cos t - \cot t}{\cos t \cot t}$

5. $\dfrac{1 - \cos 2x}{\sin 2x} = \tan x$

6. $\sin\theta\cos\theta + \cos^2\theta = \dfrac{\cos\theta(1 + \cot\theta)}{\csc\theta}$

7. $\dfrac{\sin x}{\tan x} + \dfrac{\cos x}{\cot x} = \sin x + \cos x$

8. $\sin^2\dfrac{t}{2} = \dfrac{\tan t - \sin t}{2\tan t}$

9. $\sin\alpha\cos\beta = \dfrac{1}{2}[\sin(\alpha + \beta) + \sin(\alpha - \beta)]$

10. $\dfrac{1 + \csc x}{\sec x} - \cot x = \cos x$

11. $\dfrac{\cot x - 1}{\cot x + 1} = \dfrac{1 - \tan x}{1 + \tan x}$

12. $2\sin^3\theta\cos\theta + 2\sin\theta\cos^3\theta = \sin 2\theta$

13. $\dfrac{\sin t + \cos t}{\sec t + \csc t} = \dfrac{\sin t}{\sec t}$

14. $\sec 2x = \dfrac{\sec^2 x}{2 - \sec^2 x}$

15. $\tan(\alpha + \beta)\tan(\alpha - \beta) = \dfrac{\tan^2\alpha - \tan^2\beta}{1 - \tan^2\alpha\tan^2\beta}$

16. $\csc\theta + \cot\theta = \dfrac{\sin\theta}{1 - \cos\theta}$

17. $\dfrac{1}{\csc 2x} = \dfrac{2\tan x}{1 + \tan^2 x}$ **18.** $\dfrac{\sec t - 1}{t\sec t} = \dfrac{1 - \cos t}{t}$

Use the following conditions to solve Exercises 19–22:

$$\sin\alpha = \frac{3}{5}, \qquad \frac{\pi}{2} < \alpha < \pi$$

$$\cos\beta = -\frac{12}{13}, \quad \pi < \beta < \frac{3\pi}{2}.$$

Find the exact value of each of the following.

19. $\cos(\alpha - \beta)$ **20.** $\tan(\alpha + \beta)$

21. $\sin 2\alpha$ **22.** $\cos\dfrac{\beta}{2}$

In Exercises 23–26, find the exact value of each expression. Do not use a calculator.

23. $\sin\left(\dfrac{3\pi}{4} + \dfrac{5\pi}{6}\right)$ **24.** $\cos^2 15° - \sin^2 15°$

25. $\cos\dfrac{5\pi}{12}\cos\dfrac{\pi}{12} + \sin\dfrac{5\pi}{12}\sin\dfrac{\pi}{12}$

26. $\tan 22.5°$

SECTION 5.4   *Product-to-Sum and Sum-to-Product Formulas*

Objectives

1 Use the product-to-sum formulas.

2 Use the sum-to-product formulas.

James K. Polk
Born November 2, 1795

Warren G. Harding
Born November 2, 1865

Of the 43 U.S. presidents, two share a birthday (same month and day). The probability of two or more people in a group sharing a birthday rises sharply as the group's size increases. Above 50 people, the probability approaches certainty. (You can verify the mathematics of this surprising result by studying Sections 10.6 and 10.7, and working Exercise 69 in Exercise Set 10.7.) So, come November 2, we salute Presidents Polk and Harding with

$$112, 163\text{-}, 112, 196\text{-}, 110, 8521\text{-}, 008, 121\text{-}.$$

Were you aware that each button on your touch-tone phone produces a unique sound? If we treat the commas as pauses and the hyphens as held notes, this sequence of numbers is *Happy Birthday* on a touch-tone phone.

Although *Happy Birthday* isn't Mozart or Sondheim, it is sinusoidal. Each of its touch-tone musical sounds can be described by the sum of two sine functions or the product of sines and cosines. In this section, we develop identities that enable us to use both descriptions. They are called the product-to-sum and sum-to-product formulas.

1 Use the product-to-sum formulas.

The Product-to-Sum Formulas

How do we write the products of sines and/or cosines as sums or differences? We use the following identities, which are called **product-to-sum formulas:**

> **Product-to-Sum Formulas**
>
> $$\sin \alpha \sin \beta = \tfrac{1}{2}[\cos(\alpha - \beta) - \cos(\alpha + \beta)]$$
> $$\cos \alpha \cos \beta = \tfrac{1}{2}[\cos(\alpha - \beta) + \cos(\alpha + \beta)]$$
> $$\sin \alpha \cos \beta = \tfrac{1}{2}[\sin(\alpha + \beta) + \sin(\alpha - \beta)]$$
> $$\cos \alpha \sin \beta = \tfrac{1}{2}[\sin(\alpha + \beta) - \sin(\alpha - \beta)]$$

Although these formulas are difficult to remember, they are fairly easy to derive. For example, let's derive the first identity in the box,

$$\sin \alpha \sin \beta = \tfrac{1}{2}[\cos(\alpha - \beta) - \cos(\alpha + \beta)].$$

We begin with the difference and sum formulas for the cosine, and subtract the second identity from the first:

$$
\begin{aligned}
\cos(\alpha - \beta) &= \cos\alpha\cos\beta + \sin\alpha\sin\beta \\
-[\cos(\alpha + \beta) &= \cos\alpha\cos\beta - \sin\alpha\sin\beta] \\
\hline
\cos(\alpha - \beta) - \cos(\alpha + \beta) &= \quad\quad 0 \quad\quad + 2\sin\alpha\sin\beta.
\end{aligned}
$$

| Subtract terms on the left side. | Subtract terms on the right side: $\cos\alpha\cos\beta - \cos\alpha\cos\beta = 0$. | Subtract terms on the right side: $\sin\alpha\sin\beta - (-\sin\alpha\sin\beta) = 2\sin\alpha\sin\beta$. |

Now we use this result to derive the product-to-sum formula for $\sin\alpha\sin\beta$.

$$2\sin\alpha\sin\beta = \cos(\alpha - \beta) - \cos(\alpha + \beta)$$

Reverse the sides in the preceding equation.

$$\sin\alpha\sin\beta = \tfrac{1}{2}[\cos(\alpha - \beta) - \cos(\alpha + \beta)]$$

Multiply each side by $\tfrac{1}{2}$.

This last equation is the desired formula. Likewise, we can derive the product-to-sum formula for cosine, $\cos\alpha\cos\beta = \tfrac{1}{2}[\cos(\alpha - \beta) + \cos(\alpha + \beta)]$. As we did for the previous derivation, begin with the difference and sum formulas for cosine. However, we *add* the formulas rather than subtracting them. Reversing both sides of this result and multiplying each side by $\tfrac{1}{2}$ produces the formula for $\cos\alpha\cos\beta$. The last two product-to-sum formulas, $\sin\alpha\cos\beta = \tfrac{1}{2}[\sin(\alpha + \beta) + \sin(\alpha - \beta)]$ and $\cos\alpha\cos\beta = \tfrac{1}{2}[\sin(\alpha + \beta) - \sin(\alpha - \beta)]$, are derived using the sum and difference formulas for sine in a similar manner.

Technology

The graphs of
$$y = \sin 8x \sin 3x$$
and
$$y = \tfrac{1}{2}(\cos 5x - \cos 11x)$$
are shown in a $\left[-2\pi, 2\pi, \dfrac{\pi}{2}\right]$ by $[-1, 1, 1]$ viewing rectangle. The graphs coincide. This supports our algebraic work in Example 1(a).

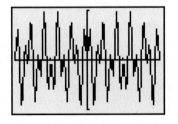

EXAMPLE 1   Using the Product-to-Sum Formulas

Express each of the following products as a sum or difference:

a. $\sin 8x \sin 3x$ **b.** $\sin 4x \cos x$.

Solution   The product-to-sum formula that we are using is shown in each of the voice balloons.

a.

$$\sin\alpha\sin\beta = \tfrac{1}{2}[\cos(\alpha - \beta) - \cos(\alpha + \beta)]$$

$$\sin 8x \sin 3x = \frac{1}{2}[\cos(8x - 3x) - \cos(8x + 3x)] = \frac{1}{2}(\cos 5x - \cos 11x)$$

b.

$$\sin\alpha\cos\beta = \tfrac{1}{2}[\sin(\alpha + \beta) + \sin(\alpha - \beta)]$$

$$\sin 4x \cos x = \frac{1}{2}[\sin(4x + x) + \sin(4x - x)] = \frac{1}{2}(\sin 5x + \sin 3x)$$

Check Point 1   Express each of the following products as a sum or difference:

a. $\sin 5x \sin 2x$ **b.** $\cos 7x \cos x$.

The Sum-to-Product Formulas

2 Use the sum-to-product formulas.

How do we write the sum or difference of sines and/or cosines as products? We use the following identities, which are called the **sum-to-product formulas:**

Sum-to-Product Formulas

$$\sin \alpha + \sin \beta = 2 \sin\frac{\alpha + \beta}{2}\cos\frac{\alpha - \beta}{2}$$

$$\sin \alpha - \sin \beta = 2 \sin\frac{\alpha - \beta}{2}\cos\frac{\alpha + \beta}{2}$$

$$\cos \alpha + \cos \beta = 2 \cos\frac{\alpha + \beta}{2}\cos\frac{\alpha - \beta}{2}$$

$$\cos \alpha - \cos \beta = -2 \sin\frac{\alpha + \beta}{2}\sin\frac{\alpha - \beta}{2}$$

We verify these formulas using the product-to-sum formulas. Let's verify the first sum-to-product formula

$$\sin \alpha + \sin \beta = 2 \sin\frac{\alpha + \beta}{2}\cos\frac{\alpha - \beta}{2}.$$

We start with the right side of the formula, the side with the product. We can apply the product-to-sum formula for $\sin \alpha \cos \beta$ to this expression. By doing so, we obtain the left side of the formula, $\sin \alpha + \sin \beta$. Here's how:

$$\sin \alpha \cos \beta = \tfrac{1}{2}[\sin(\alpha + \beta) + \sin(\alpha - \beta)]$$

$$2\,\sin\frac{\alpha + \beta}{2}\cos\frac{\alpha - \beta}{2} = 2 \cdot \frac{1}{2}\left[\sin\left(\frac{\alpha + \beta}{2} + \frac{\alpha - \beta}{2}\right) + \sin\left(\frac{\alpha + \beta}{2} - \frac{\alpha - \beta}{2}\right)\right]$$

$$= \sin\left(\frac{\alpha + \beta + \alpha - \beta}{2}\right) + \sin\left(\frac{\alpha + \beta - \alpha + \beta}{2}\right)$$

$$= \sin\frac{2\alpha}{2} + \sin\frac{2\beta}{2} = \sin \alpha + \sin \beta.$$

The three other sum-to-product formulas in the box are verified in a similar manner. Start with the right side and obtain the left side using an appropriate product-to-sum formula.

EXAMPLE 2 Using the Sum-to-Product Formulas

Express each sum or difference as a product:

   a. $\sin 9x + \sin 5x$ **b.** $\cos 4x - \cos 3x.$

Solution The sum-to-product formula that we are using is shown in each of the voice balloons.

a.

$$\sin \alpha + \sin \beta = 2 \sin\frac{\alpha + \beta}{2}\cos\frac{\alpha - \beta}{2}$$

$$\sin 9x + \sin 5x = 2 \sin\frac{9x + 5x}{2}\cos\frac{9x - 5x}{2}$$

$$= 2 \sin\frac{14x}{2}\cos\frac{4x}{2}$$

$$= 2 \sin 7x \cos 2x$$

b.

$$\cos \alpha - \cos \beta = -2 \sin \frac{\alpha + \beta}{2} \sin \frac{\alpha - \beta}{2}$$

$$\cos 4x - \cos 3x = -2 \sin \frac{4x + 3x}{2} \sin \frac{4x - 3x}{2}$$

$$= -2 \sin \frac{7x}{2} \sin \frac{x}{2}$$

Check Point 2 Express each sum as a product:

    a. $\sin 7x + \sin 3x$ **b.** $\cos 3x + \cos 2x$.

Some identities contain a fraction on one side with sums and differences of sines and/or cosines. Applying the sum-to-product formulas in the numerator and the denominator is often helpful in verifying these identities.

EXAMPLE 3   Using Sum-to-Product Formulas to Verify an Identity

Verify the identity: $\dfrac{\cos 3x - \cos 5x}{\sin 3x + \sin 5x} = \tan x$.

Solution   Because the left side is more complicated, we will work with it. We use sum-to-product formulas for the numerator and the denominator of the fraction on this side.

$$\frac{\cos 3x - \cos 5x}{\sin 3x + \sin 5x}$$

$$\cos \alpha - \cos \beta = -2 \sin \frac{\alpha + \beta}{2} \sin \frac{\alpha - \beta}{2}$$

$$= \frac{-2 \sin \dfrac{3x + 5x}{2} \sin \dfrac{3x - 5x}{2}}{\sin 3x + \sin 5x}$$

$$\sin \alpha + \sin \beta = 2 \sin \frac{\alpha + \beta}{2} \cos \frac{\alpha - \beta}{2}$$

$$= \frac{-2 \sin \dfrac{3x + 5x}{2} \sin \dfrac{3x - 5x}{2}}{2 \sin \dfrac{3x + 5x}{2} \cos \dfrac{3x - 5x}{2}}$$

$$= \frac{-2 \sin \dfrac{8x}{2} \sin \left(\dfrac{-2x}{2} \right)}{2 \sin \dfrac{8x}{2} \cos \left(\dfrac{-2x}{2} \right)}$$
    Perform the indicated additions and subtractions.

$$= \frac{-2 \sin 4x \sin(-x)}{2 \sin 4x \cos(-x)}$$
    Simplify.

$$= \frac{-(-\sin x)}{\cos x}$$
    The sine function is odd: $\sin(-x) = -\sin x$. The cosine function is even: $\cos(-x) = \cos x$.

$$= \frac{\sin x}{\cos x}$$
    Simplify.

$$= \tan x$$
    Apply a quotient identity: $\tan x = \dfrac{\sin x}{\cos x}$.

We worked with the left side and arrived at the right side. Thus, the identity is verified.

Check Point 3 Verify the identity: $\dfrac{\cos 3x - \cos x}{\sin 3x + \sin x} = -\tan x$.

Sinusoidal Sounds

Music is all around us. A mere snippet of a song from the past can trigger vivid memories, inducing emotions ranging from unabashed joy to deep sorrow. Trigonometric functions can explain how sound travels from its source and describe its pitch, loudness, and quality. Still unexplained is the remarkable influence music has on the brain, including the deepest question of all: Why do we appreciate music?

When a note is played, it disturbs nearby air molecules, creating regions of higher-than-normal pressure and regions of lower-than-normal pressure. If we graph pressure, y, versus time, t, we get a sine wave that represents the note. The frequency of the sine wave is the number of high-low disturbances, or vibrations, per second. The greater the frequency, the higher the pitch; the lesser the frequency, the lower the pitch.

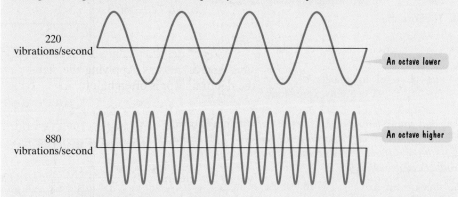

220 vibrations/second — An octave lower

880 vibrations/second — An octave higher

The amplitude of a note's sine wave is related to its loudness. The amplitude for the two sine waves shown above is the same. Thus, the notes have the same loudness, although they differ in pitch. The greater the amplitude, the louder the sound; the lesser the amplitude, the softer the sound. The amplitude and frequency are characteristic of every note—and thus of its graph—until the note dissipates.

EXERCISE SET 5.4

Practice Exercises

Because you may not be required to memorize the identities in this section, it's often tempting to pay no attention to them at all! Exercises 1–4 are provided to familiarize you with what these identities do. Fill in each blank using the word sum, difference, product, *or* quotient.

1. The formula

$$\sin \alpha \sin \beta = \frac{1}{2}[\cos(\alpha - \beta) - \cos(\alpha + \beta)]$$

can be used to change a _____ of two sines into the _____ of two cosine expressions.

2. The formula

$$\cos \alpha \cos \beta = \frac{1}{2}[\cos(\alpha - \beta) + \cos(\alpha + \beta)]$$

can be used to change a _____ of two cosines into the _____ of two cosine expressions.

3. The formula

$$\sin \alpha \cos \beta = \frac{1}{2}[\sin(\alpha + \beta) + \sin(\alpha - \beta)]$$

can be used to change a _____ of a sine and a cosine into the _____ of two sine expressions.

4. The formula

$$\cos \alpha \sin \beta = \frac{1}{2}[\sin(\alpha + \beta) - \sin(\alpha - \beta)]$$

can be used to change a _____ of a cosine and a sine into the _____ of two sine expressions.

Now that you've familiarized yourself with the formulas, in Exercises 5–12, use the appropriate formula to express each product as a sum or difference.

5. $\sin 6x \sin 2x$

6. $\sin 8x \sin 4x$

7. $\cos 7x \cos 3x$

8. $\cos 9x \cos 2x$

9. $\sin x \cos 2x$

10. $\sin 2x \cos 3x$

11. $\cos \dfrac{3x}{2} \sin \dfrac{x}{2}$

12. $\cos \dfrac{5x}{2} \sin \dfrac{x}{2}$

Exercises 13–16 are provided to familiarize you with the second set of identities presented in this section. Fill in each blank using the word sum, difference, product, *or* quotient.

13. The formula

$$\sin \alpha + \sin \beta = 2 \sin \frac{\alpha + \beta}{2} \cos \frac{\alpha - \beta}{2}$$

can be used to change a _____ of two sines into the _____ of a sine and a cosine expression.

14. The formula

$$\sin \alpha - \sin \beta = 2 \sin \frac{\alpha - \beta}{2} \cos \frac{\alpha + \beta}{2}$$

can be used to change a _____ of two sines into the _____ of a sine and a cosine expression.

15. The formula

$$\cos \alpha + \cos \beta = 2 \cos \frac{\alpha + \beta}{2} \cos \frac{\alpha - \beta}{2}$$

can be used to change a _____ of two cosines into the _____ of two cosine expressions.

16. The formula

$$\cos \alpha - \cos \beta = -2 \sin \frac{\alpha + \beta}{2} \sin \frac{\alpha - \beta}{2}$$

can be used to change a _____ of two cosines into the _____ of two sine expressions.

Now that you've familiarized yourself with the second set of formulas presented in this section, in Exercises 17–30, express each sum or difference as a product. If possible, find this product's exact value.

17. $\sin 6x + \sin 2x$

18. $\sin 8x + \sin 2x$

19. $\sin 7x - \sin 3x$

20. $\sin 11x - \sin 5x$

21. $\cos 4x + \cos 2x$

22. $\cos 9x - \cos 7x$

23. $\sin x + \sin 2x$

24. $\sin x - \sin 2x$

25. $\cos \dfrac{3x}{2} + \cos \dfrac{x}{2}$

26. $\sin \dfrac{3x}{2} + \sin \dfrac{x}{2}$

27. $\sin 75° + \sin 15°$

28. $\cos 75° - \cos 15°$

29. $\sin \dfrac{\pi}{12} - \sin \dfrac{5\pi}{12}$

30. $\cos \dfrac{\pi}{12} - \cos \dfrac{5\pi}{12}$

In Exercises 31–38, verify each identity.

31. $\dfrac{\sin 3x - \sin x}{\cos 3x - \cos x} = -\cot 2x$

32. $\dfrac{\sin x + \sin 3x}{\cos x + \cos 3x} = \tan 2x$

33. $\dfrac{\sin 2x + \sin 4x}{\cos 2x + \cos 4x} = \tan 3x$

34. $\dfrac{\cos 4x - \cos 2x}{\sin 2x - \sin 4x} = \tan 3x$

35. $\dfrac{\sin x - \sin y}{\sin x + \sin y} = \tan \dfrac{x - y}{2} \cot \dfrac{x + y}{2}$

36. $\dfrac{\sin x + \sin y}{\sin x - \sin y} = \tan \dfrac{x + y}{2} \cot \dfrac{x - y}{2}$

37. $\dfrac{\sin x + \sin y}{\cos x + \cos y} = \tan \dfrac{x + y}{2}$

38. $\dfrac{\sin x - \sin y}{\cos x - \cos y} = -\cot \dfrac{x + y}{2}$

Practice Plus

In Exercises 39–44, the graph with the given equation is shown in a $\left[0, 2\pi, \dfrac{\pi}{2}\right]$ by $[-2, 2, 1]$ viewing rectangle.

a. *Describe the graph using another equation.*

b. *Verify that the two equations are equivalent.*

39. $y = \dfrac{\sin x + \sin 3x}{2 \sin 2x}$

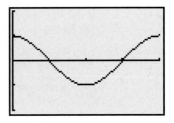

40. $y = \dfrac{\cos x - \cos 3x}{\sin x + \sin 3x}$

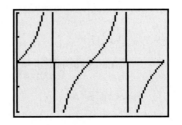

41. $y = \dfrac{\cos x - \cos 5x}{\sin x + \sin 5x}$

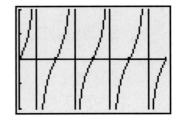

42. $y = \dfrac{\cos 5x - \cos 3x}{\sin 5x + \sin 3x}$

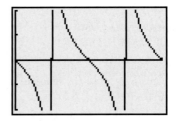

43. $y = \dfrac{\sin x - \sin 3x}{\cos x - \cos 3x}$

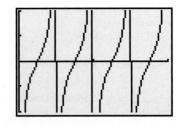

4. $y = \dfrac{\sin 2x + \sin 6x}{\cos 6x - \cos 2x}$

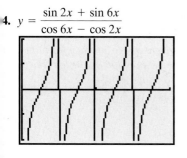

Writing in Mathematics

In Exercises 47–50, use words to describe the given formula.

47. $\sin \alpha \sin \beta = \frac{1}{2}[\cos(\alpha - \beta) - \cos(\alpha + \beta)]$

48. $\cos \alpha \cos \beta = \frac{1}{2}[\cos(\alpha - \beta) + \cos(\alpha + \beta)]$

49. $\sin \alpha + \sin \beta = 2 \sin\dfrac{\alpha + \beta}{2}\cos\dfrac{\alpha - \beta}{2}$

50. $\cos \alpha + \cos \beta = 2 \cos\dfrac{\alpha + \beta}{2}\cos\dfrac{\alpha - \beta}{2}$

51. Describe identities that can be verified using the sum-to-product formulas.

52. Why do the sounds produced by touching each button on a touch-tone phone have the same loudness? Answer the question using the equation described for Exercises 45 and 46, $y = \sin 2\pi lt + \sin 2\pi ht$, and determine the maximum value of y for each sound.

Application Exercises

Use this information to solve Exercises 45–46. The sound produced by touching each button on a touch-tone phone is described by

$$y = \sin 2\pi lt + \sin 2\pi ht,$$

where l and h are the low and high frequencies in the figure shown. For example, what sound is produced by touching 5? The low frequency is $l = 770$ cycles per second and the high frequency is $h = 1336$ cycles per second. The sound produced by touching 5 is described by

$$y = \sin 2\pi(770)t + \sin 2\pi(1336)t.$$

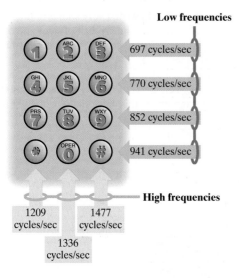

45. The touch-tone phone sequence for that most naive of melodies is given as follows:

Mary Had A Little Lamb
3212333,222,399,3212333322321.

a. Many numbers do not appear in this sequence, including 7. If you accidently touch 7 for one of the notes, describe this sound as the sum of sines.

b. Describe this accidental sound as a product of sines and cosines.

46. The touch-tone phone sequence for *Jingle Bells* is given as follows:

Jingle Bells
33,333,39123,666-663333322329,333,333,39123,666-6633,399621.

a. The first six notes of the song are produced by repeatedly touching 3. Describe this repeated sound as the sum of sines.

b. Describe the repeated sound as a product of sines and cosines.

Technology Exercises

In Exercises 53–56, graph each side of the equation in the same viewing rectangle. If the graphs appear to coincide, verify that the equation is an identity. If the graphs do not appear to coincide, find a value of x for which both sides are defined but not equal.

53. $\sin x + \sin 2x = \sin 3x$ **54.** $\cos x + \cos 2x = \cos 3x$

55. $\sin x + \sin 3x = 2 \sin 2x \cos x$

56. $\cos x + \cos 3x = 2 \cos 2x \cos x$

57. In Exercise 45(a), you wrote an equation for the sound produced by touching 7 on a touch-tone phone. Graph the equation in a $[0, 0.01, 0.001]$ by $[-2, 2, 1]$ viewing rectangle.

58. In Exercise 46(a), you wrote an equation for the sound produced by touching 3 on a touch-tone phone. Graph the equation in a $[0, 0.01, 0.001]$ by $[-2, 2, 1]$ viewing rectangle.

59. In this section, we saw how sums could be expressed as products. Sums of trigonometric functions can also be used to describe functions that are not trigonometric. French mathematician Jean Fourier (1768–1830) showed that *any function* can be described by a series of trigonometric functions. For example, the basic linear function $f(x) = x$ can also be represented by

$$f(x) = 2\left(\frac{\sin x}{1} - \frac{\sin 2x}{2} + \frac{\sin 3x}{3} - \frac{\sin 4x}{4} + \cdots\right).$$

a. Graph

$$y = 2\left(\frac{\sin x}{1}\right),$$

$$y = 2\left(\frac{\sin x}{1} - \frac{\sin 2x}{2}\right),$$

$$y = 2\left(\frac{\sin x}{1} - \frac{\sin 2x}{2} + \frac{\sin 3x}{3}\right)$$

and

$$y = 2\left(\frac{\sin x}{1} - \frac{\sin 2x}{2} + \frac{\sin 3x}{3} - \frac{\sin 4x}{4}\right)$$

in a $\left[-\pi, \pi, \dfrac{\pi}{2}\right]$ by $[-3, 3, 1]$ viewing rectangle. What patterns do you observe?

b. Graph

$$y = 2\left(\frac{\sin x}{1} - \frac{\sin 2x}{2} + \frac{\sin 3x}{3} - \frac{\sin 4x}{4} + \frac{\sin 5x}{5} - \frac{\sin 6x}{6}\right.$$

$$\left. + \frac{\sin 7x}{7} - \frac{\sin 8x}{8} + \frac{\sin 9x}{9} - \frac{\sin 10x}{10}\right)$$

in a $\left[-\pi, \pi, \frac{\pi}{2}\right]$ by $[-3, 3, 1]$ viewing rectangle. Is a portion of the graph beginning to look like the graph of $f(x) = x$? Obtain a better approximation for the line by graphing functions that contain more and more terms involving sines of multiple angles.

c. Use

$$x = 2\left(\frac{\sin x}{1} - \frac{\sin 2x}{2} + \frac{\sin 3x}{3} - \frac{\sin 4x}{4} + \cdots\right)$$

and substitute $\frac{\pi}{2}$ for x to obtain a formula for $\frac{\pi}{2}$. Show at least four nonzero terms. Then multiply both sides of your formula by 2 to write a nonending series of subtractions and additions that approaches π. Use this series to obtain an approximation for π that is more accurate than the one given by your graphing utility.

Critical Thinking Exercises

Use the identities for $\sin(\alpha + \beta)$ and $\sin(\alpha - \beta)$ to solve Exercises 60–61.

60. Add the left and right sides of the identities and derive the product-to-sum formula for $\sin \alpha \cos \beta$.

61. Subtract the left and right sides of the identities and derive the product-to-sum formula for $\cos \alpha \sin \beta$.

In Exercises 62–63, verify the given sum-to-product formula. Start with the right side and obtain the expression on the left side by using an appropriate product-to-sum formula.

62. $\sin \alpha - \sin \beta = 2 \sin\dfrac{\alpha - \beta}{2}\cos\dfrac{\alpha + \beta}{2}$

63. $\cos \alpha + \cos \beta = 2 \cos\dfrac{\alpha + \beta}{2}\cos\dfrac{\alpha - \beta}{2}$

In Exercises 64–65, verify each identity.

64. $\dfrac{\sin 2x + (\sin 3x + \sin x)}{\cos 2x + (\cos 3x + \cos x)} = \tan 2x$

65. $4 \cos x \cos 2x \sin 3x = \sin 2x + \sin 4x + \sin 6x$

Group Exercise

66. This activity should result in an unusual group display entitled "*Frere Jacques*, a New Perspective." Here is the touch-tone phone sequence:

Frere Jacques
4564,4564,69#,69#,#*#964,#*#964,414,414.

Group members should write every sound in the sequence as both the sum of sines and the product of sines and cosines. Use the sum of sines form and a graphing utility with a $[0, 0.01, 0.001]$ by $[-2, 2, 1]$ viewing rectangle to obtain a graph for every sound. Download these graphs. Use the graphs and equations to create your display in such a way that adults find the trigonometry of this naive melody interesting.

SECTION 5.5 *Trigonometric Equations*

Objectives

1 Find all solutions of a trigonometric equation.

2 Solve equations with multiple angles.

3 Solve trigonometric equations quadratic in form.

4 Use factoring to separate different functions in trigonometric equations.

5 Use identities to solve trigonometric equations.

6 Use a calculator to solve trigonometric equations.

Exponential functions display the manic energies of uncontrolled growth. By contrast, trigonometric functions repeat their behavior. Do they embody in their regularity some basic rhythm of the universe? The cycles of periodic phenomena provide events that we can comfortably count on. When will the moon look just as it does at this moment? When can I count on 13.5 hours of daylight? When will my breathing be exactly as it is right now? Models with trigonometric functions embrace the periodic rhythms of our world. Equations containing trigonometric functions are used to answer questions about these models.

① Find all solutions of a
trigonometric equation.

Trigonometric Equations and Their Solutions

A **trigonometric equation** is an equation that contains a trigonometric expression with a variable, such as sin x. We have seen that some trigonometric equations are identities, such as $\sin^2 x + \cos^2 x = 1$. These equations are true for every value of the variable for which the expressions are defined. In this section, we consider trigonometric equations that are true for only some values of the variable. The values that satisfy such an equation are its **solutions**. (There are trigonometric equations that have no solution.)

An example of a trigonometric equation is

$$\sin x = \tfrac{1}{2}.$$

A solution of this equation is $\frac{\pi}{6}$ because $\sin\frac{\pi}{6} = \frac{1}{2}$. By contrast, π is not a solution because $\sin \pi = 0 \neq \frac{1}{2}$.

Is $\frac{\pi}{6}$ the only solution of $\sin x = \frac{1}{2}$? The answer is no. Because of the periodic nature of the sine function, there are infinitely many values of x for which $\sin x = \frac{1}{2}$. Figure 5.7 shows five of the solutions, including $\frac{\pi}{6}$, for $-\frac{3\pi}{2} \leq x \leq \frac{7\pi}{2}$. Notice that the x-coordinates of the points where the graph of $y = \sin x$ intersects the line $y = \frac{1}{2}$ are the solutions of the equation $\sin x = \frac{1}{2}$.

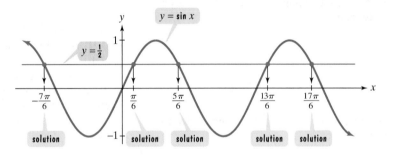

Figure 5.7 The equation $\sin x = \frac{1}{2}$ has five solutions when x is restricted to the interval $\left[-\frac{3\pi}{2}, \frac{7\pi}{2}\right]$.

How do we represent all solutions of $\sin x = \frac{1}{2}$? Because the period of the sine function is 2π, first find all solutions in $[0, 2\pi)$. The solutions are

$$x = \frac{\pi}{6} \quad \text{and} \quad x = \pi - \frac{\pi}{6} = \frac{5\pi}{6}.$$

> The sine is positive in quadrants I and II.

Any multiple of 2π can be added to these values and the sine is still $\frac{1}{2}$. Thus, all solutions of $\sin x = \frac{1}{2}$ are given by

$$x = \frac{\pi}{6} + 2n\pi \quad \text{or} \quad x = \frac{5\pi}{6} + 2n\pi$$

where n is any integer. By choosing any two integers, such as $n = 0$ and $n = 1$, we can find some solutions of $\sin x = \frac{1}{2}$. Thus, four of the solutions are determined as follows:

> Let $n = 0$.   Let $n = 1$.

$$x = \frac{\pi}{6} + 2\cdot 0\pi \qquad x = \frac{5\pi}{6} + 2\cdot 0\pi \qquad x = \frac{\pi}{6} + 2\cdot 1\pi \qquad x = \frac{5\pi}{6} + 2\cdot 1\pi$$

$$= \frac{\pi}{6} \qquad\qquad = \frac{5\pi}{6} \qquad\qquad = \frac{\pi}{6} + 2\pi \qquad\qquad x = \frac{5\pi}{6} + 2\pi$$

$$\qquad\qquad\qquad\qquad\qquad\qquad\qquad = \frac{\pi}{6} + \frac{12\pi}{6} = \frac{13\pi}{6} \qquad = \frac{5\pi}{6} + \frac{12\pi}{6} = \frac{17\pi}{6}.$$

These four solutions are shown among the five solutions in Figure 5.7.

Urban Canyons

A city's tall buildings and narrow streets reduce the amount of sunlight. If h is the average height of the buildings and w is the width of the street, the angle of elevation from the street to the top of the buildings is given by the trigonometric equation

$$\tan \theta = \frac{h}{w}.$$

A value of $\theta = 63°$ can result in an 85% loss of illumination.

Equations Involving a Single Trigonometric Function

To solve an equation containing a single trigonometric function:

- Isolate the function on one side of the equation.
- Solve for the variable.

EXAMPLE 1   Finding All Solutions of a Trigonometric Equation

Solve the equation: $3 \sin x - 2 = 5 \sin x - 1$.

Solution   The equation contains a single trigonometric function, $\sin x$.

Step 1   Isolate the function on one side of the equation. We can solve for $\sin x$ by collecting terms with $\sin x$ on the left side and constant terms on the right side.

$3 \sin x - 2 = 5 \sin x - 1$	This is the given equation.
$3 \sin x - 5 \sin x - 2 = 5 \sin x - 5 \sin x - 1$	Subtract 5 sin x from both sides.
$-2 \sin x - 2 = -1$	Simplify.
$-2 \sin x = 1$	Add 2 to both sides.
$\sin x = -\dfrac{1}{2}$	Divide both sides by -2 and solve for sin

Step 2   Solve for the variable. We must solve for x in $\sin x = -\dfrac{1}{2}$. Because $\sin \dfrac{\pi}{6} = \dfrac{1}{2}$, the solutions of $\sin x = -\dfrac{1}{2}$ in $[0, 2\pi)$ are

$$x = \pi + \frac{\pi}{6} = \frac{6\pi}{6} + \frac{\pi}{6} = \frac{7\pi}{6} \qquad x = 2\pi - \frac{\pi}{6} = \frac{12\pi}{6} - \frac{\pi}{6} = \frac{11\pi}{6}.$$

The sine is negative in quadrant III.

The sine is negative in quadrant IV.

Because the period of the sine function is 2π, the solutions of the equation are given by

$$x = \frac{7\pi}{6} + 2n\pi \quad \text{and} \quad x = \frac{11\pi}{6} + 2n\pi,$$

where n is any integer.

Check Point 1   Solve the equation: $5 \sin x = 3 \sin x + \sqrt{3}$.

Now we will concentrate on finding solutions of trigonometric equations for $0 \le x < 2\pi$. You can use a graphing utility to check the solutions of these equations. Graph the left side and graph the right side. The solutions are the x-coordinates of the points where the graphs intersect.

② Solve equations with multiple angles.

Equations Involving Multiple Angles

Here are examples of two equations that include multiple angles:

$$\tan 3x = 1 \qquad \sin \frac{x}{2} = \frac{\sqrt{3}}{2}.$$

The angle is a multiple of 3.

The angle is a multiple of $\frac{1}{2}$.

We will solve each equation for $0 \le x < 2\pi$. The period of the function plays an important role in ensuring that we do not leave out any solutions.

EXAMPLE 2   Solving an Equation with a Multiple Angle

Solve the equation: $\tan 3x = 1, \quad 0 \le x < 2\pi$.

Solution   The period of the tangent function is π. In the interval $[0, \pi)$, the only value for which the tangent function is 1 is $\dfrac{\pi}{4}$. This means that $3x = \dfrac{\pi}{4}$. Because the period is π, all the solutions to $\tan 3x = 1$ are given by

$$3x = \frac{\pi}{4} + n\pi \qquad \text{\footnotesize \textit{n} is any integer.}$$

$$x = \frac{\pi}{12} + \frac{n\pi}{3} \qquad \text{\footnotesize Divide both sides by 3 and solve for x.}$$

In the interval $[0, 2\pi)$, we obtain the solutions of $\tan 3x = 1$ as follows:

Let $n = 0$.

$x = \dfrac{\pi}{12} + \dfrac{0\pi}{3}$

$\quad = \dfrac{\pi}{12}$

Let $n = 1$.

$x = \dfrac{\pi}{12} + \dfrac{1\pi}{3}$

$\quad = \dfrac{\pi}{12} + \dfrac{4\pi}{12} = \dfrac{5\pi}{12}$

Let $n = 2$.

$x = \dfrac{\pi}{12} + \dfrac{2\pi}{3}$

$\quad = \dfrac{\pi}{12} + \dfrac{8\pi}{12} = \dfrac{9\pi}{12} = \dfrac{3\pi}{4}$

Let $n = 3$.

$x = \dfrac{\pi}{12} + \dfrac{3\pi}{3}$

$\quad = \dfrac{\pi}{12} + \dfrac{12\pi}{12} = \dfrac{13\pi}{12}$

Let $n = 4$.

$x = \dfrac{\pi}{12} + \dfrac{4\pi}{3}$

$\quad = \dfrac{\pi}{12} + \dfrac{16\pi}{12} = \dfrac{17\pi}{12}$

Let $n = 5$.

$x = \dfrac{\pi}{12} + \dfrac{5\pi}{3}$

$\quad = \dfrac{\pi}{12} + \dfrac{20\pi}{12} = \dfrac{21\pi}{12} = \dfrac{7\pi}{4}.$

If you let $n = 6$, you will obtain $x = \dfrac{25\pi}{12}$. This value exceeds 2π. In the interval $[0, 2\pi)$, the solutions of $\tan 3x = 1$ are $\dfrac{\pi}{12}, \dfrac{5\pi}{12}, \dfrac{3\pi}{4}, \dfrac{13\pi}{12}, \dfrac{17\pi}{12}$, and $\dfrac{7\pi}{4}$. These solutions are illustrated by the six intersection points in the technology box.

Check Point 2   Solve the equation: $\tan 2x = \sqrt{3}, 0 \le x < 2\pi$.

EXAMPLE 3   Solving an Equation with a Multiple Angle

Solve the equation: $\sin\dfrac{x}{2} = \dfrac{\sqrt{3}}{2}, 0 \le x < 2\pi$.

Solution   The period of the sine function is 2π. In the interval $[0, 2\pi)$, there are two values at which the sine function is $\dfrac{\sqrt{3}}{2}$. One of these values is $\dfrac{\pi}{3}$. The sine is positive in quadrant II; thus, the other value is $\pi - \dfrac{\pi}{3}$, or $\dfrac{2\pi}{3}$. This means that $\dfrac{x}{2} = \dfrac{\pi}{3}$ or $\dfrac{x}{2} = \dfrac{2\pi}{3}$. Because the period is 2π, all the solutions of $\sin\dfrac{x}{2} = \dfrac{\sqrt{3}}{2}$ are given by

$$\frac{x}{2} = \frac{\pi}{3} + 2n\pi \qquad \text{or} \qquad \frac{x}{2} = \frac{2\pi}{3} + 2n\pi. \quad \text{\footnotesize \textit{n} is any integer.}$$

$$x = \frac{2\pi}{3} + 4n\pi \qquad\qquad x = \frac{4\pi}{3} + 4n\pi. \quad \text{\footnotesize Multiply both sides by 2 and solve for x.}$$

Technology

Shown below are the graphs of
$$y = \tan 3x$$
and
$$y = 1$$
in a $\left[0, 2\pi, \dfrac{\pi}{2}\right]$ by $[-3, 3, 1]$ viewing rectangle. The solutions of
$$\tan 3x = 1$$
in $[0, 2\pi)$ are shown by the x-coordinates of the six intersection points.

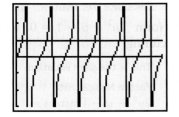

Shown below are the graphs of
$$y = \sin x \cos x$$
and
$$y = \tfrac{1}{2}$$
in a $\left[0, 2\pi, \dfrac{\pi}{2}\right]$ by $[-1, 1, 1]$
viewing rectangle.

The solutions of
$$\sin x \cos x = \tfrac{1}{2}$$
are shown by the x-coordinates
of the two intersection points.

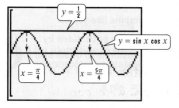

Notice that we have an equation, $\sin 2x = 1$, with $2x$, a multiple angle. The period of the sine function is 2π. In the interval $[0, 2\pi)$, the only value for which the sine function is 1 is $\dfrac{\pi}{2}$. This means that $2x = \dfrac{\pi}{2}$. Because the period is 2π, all the solutions of $\sin 2x = 1$ are given by

$$2x = \frac{\pi}{2} + 2n\pi \qquad \text{\textit{n} is any integer.}$$

$$x = \frac{\pi}{4} + n\pi \qquad \text{Divide both sides by 2 and solve for x.}$$

The solutions of $\sin x \cos x = \dfrac{1}{2}$ in the interval $[0, 2\pi)$ are obtained by letting $n = 0$ and $n = 1$. The solutions are $\dfrac{\pi}{4}$ and $\dfrac{5\pi}{4}$.

Check Point 9 Solve the equation: $\sin x \cos x = -\dfrac{1}{2}$, $0 \le x < 2\pi$.

Let's look at another equation that contains two different functions, $\sin x - \cos x = 1$. Can you think of an identity that can be used to produce only one function? Perhaps $\sin^2 x + \cos^2 x = 1$ might be helpful. The next example shows how we can use this identity after squaring both sides of the given equation. Remember that if we raise both sides of an equation to an even power, we have the possibility of introducing extraneous solutions. Thus, we must check each proposed solution in the given equation. Alternatively, we can use a graphing utility to verify actual solutions.

A graphing utility can be used instead of the algebraic check on the next page. Shown are the graphs of
$$y = \sin x - \cos x$$
and
$$y = 1$$
in a $\left[0, 2\pi, \dfrac{\pi}{2}\right]$ by $[-2, 2, 1]$
viewing rectangle. The actual solutions of
$$\sin x - \cos x = 1$$
are shown by the x-coordinates of the two intersection points, $\dfrac{\pi}{2}$ and π.

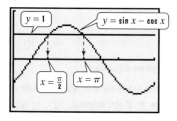

EXAMPLE 10   Using an Identity to Solve a Trigonometric Equation

Solve the equation: $\sin x - \cos x = 1$, $0 \le x < 2\pi$.

Solution We square both sides of the equation in anticipation of using $\sin^2 x + \cos^2 x = 1$.

$$\sin x - \cos x = 1 \qquad \text{This is the given equation.}$$

$$(\sin x - \cos x)^2 = 1^2 \qquad \text{Square both sides.}$$

$$\sin^2 x - 2 \sin x \cos x + \cos^2 x = 1 \qquad \begin{array}{l}\text{Square the left side using}\\ (A - B)^2 = A^2 - 2AB + B^2.\end{array}$$

$$\sin^2 x + \cos^2 x - 2 \sin x \cos x = 1 \qquad \text{Rearrange terms.}$$

$$1 - 2 \sin x \cos x = 1 \qquad \begin{array}{l}\text{Apply a Pythagorean identity:}\\ \sin^2 x + \cos^2 x = 1.\end{array}$$

$$-2 \sin x \cos x = 0 \qquad \text{Subtract 1 from both sides of the equation.}$$

$$\sin x \cos x = 0 \qquad \text{Divide both sides of the equation by } -2.$$

$$\sin x = 0 \quad \text{or} \quad \cos x = 0 \qquad \text{Set each factor equal to 0.}$$

$$x = 0 \quad x = \pi \qquad x = \frac{\pi}{2} \quad x = \frac{3\pi}{2} \qquad \text{Solve for x in } [0, 2\pi).$$

We check these proposed solutions to see if any are extraneous.

Check 0:

$\sin x - \cos x = 1$

$\sin 0 - \cos 0 \overset{?}{=} 1$

$0 - 1 \overset{?}{=} 1$

$-1 = 1,$ false

0 is extraneous.

Check $\dfrac{\pi}{2}$:

$\sin x - \cos x = 1$

$\sin \dfrac{\pi}{2} - \cos \dfrac{\pi}{2} \overset{?}{=} 1$

$1 - 0 \overset{?}{=} 1$

$1 = 1,$ true

Check π:

$\sin x - \cos x = 1$

$\sin \pi - \cos \pi \overset{?}{=} 1$

$0 - (-1) \overset{?}{=} 1$

$1 = 1,$ true

Check $\dfrac{3\pi}{2}$:

$\sin x - \cos x = 1$

$\sin \dfrac{3\pi}{2} - \cos \dfrac{3\pi}{2} \overset{?}{=} 1$

$-1 - 0 \overset{?}{=} 1$

$-1 = 1,$ false

$\dfrac{3\pi}{2}$ is extraneous.

The actual solutions in the interval $[0, 2\pi)$ are $\dfrac{\pi}{2}$ and π.

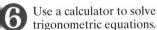 **Check Point 10** Solve the equation: $\cos x - \sin x = -1, \quad 0 \le x < 2\pi$.

 Use a calculator to solve trigonometric equations.

Using a Calculator to Solve Trigonometric Equations

In all our previous examples, the equations had solutions that were found by knowing the exact values of trigonometric functions of special angles, such as $\dfrac{\pi}{6}, \dfrac{\pi}{4},$ and $\dfrac{\pi}{3}$. However, not all trigonometric equations involve these special angles. For those that do not, we will use the secondary keys marked $\boxed{\text{SIN}^{-1}}$, $\boxed{\text{COS}^{-1}}$, and $\boxed{\text{TAN}^{-1}}$ on a calculator. Recall that on most calculators, the inverse trigonometric function keys are the secondary functions for the buttons labeled $\boxed{\text{SIN}}$, $\boxed{\text{COS}}$, and $\boxed{\text{TAN}}$, respectively.

EXAMPLE 11   Solving Trigonometric Equations with a Calculator

Solve each equation, correct to four decimal places, for $0 \le x < 2\pi$:

a. $\tan x = 12.8044$ **b.** $\cos x = -0.4317$.

Solution   We begin by using a calculator to find $\theta, 0 \le \theta < \dfrac{\pi}{2}$ satisfying the following equations:

$$\tan \theta = 12.8044 \qquad \cos \theta = 0.4317.$$

These numbers are the absolute values of the given range values.

Once θ is determined, we use our knowledge of the signs of the trigonometric functions to find x in $[0, 2\pi)$ satisfying $\tan x = 12.8044$ and $\cos x = -0.4317$.

a. $\tan x = 12.8044$ This is the given equation.

$\tan \theta = 12.8044$ Use a calculator to solve this equation for $\theta, 0 \le \theta < \dfrac{\pi}{2}$.

$\theta = \tan^{-1}(12.8044) \approx 1.4929$ 12.8044 $\boxed{\text{2nd}}$ $\boxed{\text{TAN}}$ or $\boxed{\text{2nd}}$ $\boxed{\text{TAN}}$ 12.8044 $\boxed{\text{ENTER}}$

$\tan x = 12.8044$ Return to the given equation. Because the tangent is positive, x lies in quadrant I or III.

$x \approx 1.4929 \qquad x \approx \pi + 1.4929 \approx 4.6345$ Solve for x, $0 \le x < 2\pi$.

The tangent is positive in quadrant I.   The tangent is positive in quadrant III.

Correct to four decimal places, the solutions of $\tan x = 12.8044$ in the interval $[0, 2\pi)$ are 1.4929 and 4.6345.

b. $\cos x = -0.4317$ This is the given equation.

$\cos \theta = 0.4317$ Use a calculator to solve this equation for $\theta, 0 \le \theta < \dfrac{\pi}{2}$.

$\theta = \cos^{-1}(0.4317) \approx 1.1244$ .4317 2nd COS or 2nd COS .4317 ENTER

$\cos x = -0.4317$ Return to the given equation. Because the cosine is negative, x lies in quadrant II or III.

$x \approx \pi - 1.1244 \approx 2.0172$ $x \approx \pi + 1.1244 \approx 4.2660$ Solve for x, $0 \le x < 2\pi$

> The cosine is negative in quadrant II.   The cosine is negative in quadrant III.

Correct to four decimal places, the solutions of $\cos x = -0.4317$ in the interval $[0, 2\pi)$ are 2.0172 and 4.2660.

Check Point 11 Solve each equation, correct to four decimal places, for $0 \le x < 2\pi$:
    a. $\tan x = 3.1044$ **b.** $\sin x = -0.2315$.

EXAMPLE 12   Solving a Trigonometric Equation Using the Quadratic Formula and a Calculator

Solve the equation, correct to four decimal places, for $0 \le x < 2\pi$:

$$\sin^2 x - \sin x - 1 = 0.$$

Solution The given equation is in quadratic form $u^2 - u - 1 = 0$ with $u = \sin x$. We use the quadratic formula to solve for $\sin x$ because $u^2 - u - 1$ cannot be factored. Begin by identifying the values for $a, b,$ and c.

$$\sin^2 x - \sin x - 1 = 0$$

> $a = 1$ $b = -1$ $c = -1$

Substituting these values into the quadratic formula and simplifying gives the values for $\sin x$. Once we obtain these values, we will solve for x.

$$\sin x = \frac{-b \pm \sqrt{b^2 - 4ac}}{2a} = \frac{-(-1) \pm \sqrt{(-1)^2 - 4(1)(-1)}}{2(1)} = \frac{1 \pm \sqrt{1 - (-4)}}{2} = \frac{1 \pm \sqrt{5}}{2}$$

$$\sin x = \frac{1 + \sqrt{5}}{2} \approx 1.6180$$ or $$\sin x = \frac{1 - \sqrt{5}}{2} \approx -0.6180$$

> This equation has no solution because sin x cannot be greater than 1.

> The sine is negative in quadrants III and IV. Use a calculator to solve $\sin \theta = 0.6180$, $0 \le \theta < \dfrac{\pi}{2}$.

Using a calculator to solve $\sin \theta = 0.6180$, we have

$$\theta = \sin^{-1}(0.6180) \approx 0.6662.$$

We use 0.6662 to solve $\sin x = -0.6180$, $0 \le x < 2\pi$.

$$x \approx \pi + 0.6662 \approx 3.8078 \qquad x \approx 2\pi - 0.6662 \approx 5.6170$$

> The sine is negative in quadrant III.   The sine is negative in quadrant IV.

Correct to four decimal places, the solutions of $\sin^2 x - \sin x - 1 = 0$ in the interval $[0, 2\pi)$ are 3.8078 and 5.6170.

Check Point **12** Solve the equation, correct to four decimal places, for $0 \le x < 2\pi$:
$$\cos^2 x + 5 \cos x + 3 = 0.$$

EXERCISE SET 5.5

Practice Exercises

In Exercises 1–10, use substitution to determine whether the given x-value is a solution of the equation.

1. $\cos x = \dfrac{\sqrt{2}}{2}, \quad x = \dfrac{\pi}{4}$ **2.** $\tan x = \sqrt{3}, \quad x = \dfrac{\pi}{3}$

3. $\sin x = \dfrac{\sqrt{3}}{2}, \quad x = \dfrac{\pi}{6}$ **4.** $\sin x = \dfrac{\sqrt{2}}{2}, \quad x = \dfrac{\pi}{3}$

5. $\cos x = -\dfrac{1}{2}, \quad x = \dfrac{2\pi}{3}$ **6.** $\cos x = -\dfrac{1}{2}, \quad x = \dfrac{4\pi}{3}$

7. $\tan 2x = -\dfrac{\sqrt{3}}{3}, \quad x = \dfrac{5\pi}{12}$

8. $\cos \dfrac{2x}{3} = -\dfrac{1}{2}, \quad x = \pi$

9. $\cos x = \sin 2x, \quad x = \dfrac{\pi}{3}$

10. $\cos x + 2 = \sqrt{3} \sin x, \quad x = \dfrac{\pi}{6}$

In Exercises 11–24, find all solutions of each equation.

11. $\sin x = \dfrac{\sqrt{3}}{2}$ **12.** $\cos x = \dfrac{\sqrt{3}}{2}$

13. $\tan x = 1$ **14.** $\tan x = \sqrt{3}$

15. $\cos x = -\dfrac{1}{2}$ **16.** $\sin x = -\dfrac{\sqrt{2}}{2}$

17. $\tan x = 0$ **18.** $\sin x = 0$

19. $2 \cos x + \sqrt{3} = 0$ **20.** $2 \sin x + \sqrt{3} = 0$

21. $4 \sin \theta - 1 = 2 \sin \theta$ **22.** $5 \sin \theta + 1 = 3 \sin \theta$

23. $3 \sin \theta + 5 = -2 \sin \theta$ **24.** $7 \cos \theta + 9 = -2 \cos \theta$

Exercises 25–38 involve equations with multiple angles. Solve each equation on the interval $[0, 2\pi)$.

25. $\sin 2x = \dfrac{\sqrt{3}}{2}$ **26.** $\cos 2x = \dfrac{\sqrt{2}}{2}$

27. $\cos 4x = -\dfrac{\sqrt{3}}{2}$ **28.** $\sin 4x = -\dfrac{\sqrt{2}}{2}$

29. $\tan 3x = \dfrac{\sqrt{3}}{3}$ **30.** $\tan 3x = \sqrt{3}$

31. $\tan \dfrac{x}{2} = \sqrt{3}$ **32.** $\tan \dfrac{x}{2} = \dfrac{\sqrt{3}}{3}$

33. $\sin \dfrac{2\theta}{3} = -1$ **34.** $\cos \dfrac{2\theta}{3} = -1$

35. $\sec \dfrac{3\theta}{2} = -2$ **36.** $\cot \dfrac{3\theta}{2} = -\sqrt{3}$

37. $\sin\left(2x + \dfrac{\pi}{6}\right) = \dfrac{1}{2}$ **38.** $\sin\left(2x - \dfrac{\pi}{4}\right) = \dfrac{\sqrt{2}}{2}$

Exercises 39–52 involve trigonometric equations quadratic in form. Solve each equation on the interval $[0, 2\pi)$.

39. $2 \sin^2 x - \sin x - 1 = 0$

40. $2 \sin^2 x + \sin x - 1 = 0$

41. $2 \cos^2 x + 3 \cos x + 1 = 0$

42. $\cos^2 x + 2 \cos x - 3 = 0$

43. $2 \sin^2 x = \sin x + 3$ **44.** $2 \sin^2 x = 4 \sin x + 6$

45. $\sin^2 \theta - 1 = 0$ **46.** $\cos^2 \theta - 1 = 0$

47. $4 \cos^2 x - 1 = 0$ **48.** $4 \sin^2 x - 3 = 0$

49. $9 \tan^2 x - 3 = 0$ **50.** $3 \tan^2 x - 9 = 0$

51. $\sec^2 x - 2 = 0$ **52.** $4 \sec^2 x - 2 = 0$

In Exercises 53–62, solve each equation on the interval $[0, 2\pi)$.

53. $(\tan x - 1)(\cos x + 1) = 0$

54. $(\tan x + 1)(\sin x - 1) = 0$

55. $\left(2 \cos x + \sqrt{3}\right)(2 \sin x + 1) = 0$

56. $\left(2 \cos x - \sqrt{3}\right)(2 \sin x - 1) = 0$

57. $\cot x(\tan x - 1) = 0$ **58.** $\cot x(\tan x + 1) = 0$

59. $\sin x + 2 \sin x \cos x = 0$ **60.** $\cos x - 2 \sin x \cos x = 0$

61. $\tan^2 x \cos x = \tan^2 x$ **62.** $\cot^2 x \sin x = \cot^2 x$

In Exercises 63–84, use an identity to solve each equation on the interval $[0, 2\pi)$.

63. $2 \cos^2 x + \sin x - 1 = 0$ **64.** $2 \cos^2 x - \sin x - 1 = 0$

65. $\sin^2 x - 2 \cos x - 2 = 0$ **66.** $4 \sin^2 x + 4 \cos x - 5 = 0$

67. $4 \cos^2 x = 5 - 4 \sin x$ **68.** $3 \cos^2 x = \sin^2 x$

69. $\sin 2x = \cos x$ **70.** $\sin 2x = \sin x$

71. $\cos 2x = \cos x$ **72.** $\cos 2x = \sin x$

73. $\cos 2x + 5 \cos x + 3 = 0$ **74.** $\cos 2x + \cos x + 1 = 0$

75. $\sin x \cos x = \dfrac{\sqrt{2}}{4}$ **76.** $\sin x \cos x = \dfrac{\sqrt{3}}{4}$

77. $\sin x + \cos x = 1$ **78.** $\sin x + \cos x = -1$

79. $\sin\left(x + \dfrac{\pi}{4}\right) + \sin\left(x - \dfrac{\pi}{4}\right) = 1$

80. $\sin\left(x + \dfrac{\pi}{3}\right) + \sin\left(x - \dfrac{\pi}{3}\right) = 1$

81. $\sin 2x \cos x + \cos 2x \sin x = \dfrac{\sqrt{2}}{2}$

82. $\sin 3x \cos 2x + \cos 3x \sin 2x = 1$

83. $\tan x + \sec x = 1$ **84.** $\tan x - \sec x = 1$

In Exercises 85–96, use a calculator to solve each equation, correct to four decimal places, on the interval $[0, 2\pi)$.

85. $\sin x = 0.8246$ **86.** $\sin x = 0.7392$

87. $\cos x = -\dfrac{2}{5}$

88. $\cos x = -\dfrac{4}{7}$

89. $\tan x = -3$

90. $\tan x = -5$

91. $\cos^2 x - \cos x - 1 = 0$

92. $3 \cos^2 x - 8 \cos x - 3 = 0$

93. $4 \tan^2 x - 8 \tan x + 3 = 0$

94. $\tan^2 x - 3 \tan x + 1 = 0$

95. $7 \sin^2 x - 1 = 0$

96. $5 \sin^2 x - 1 = 0$

In Exercises 97–116, use the most appropriate method to solve each equation on the interval $[0, 2\pi)$. Use exact values where possible or give approximate solutions correct to four decimal places.

97. $2 \cos 2x + 1 = 0$

98. $2 \sin 3x + \sqrt{3} = 0$

99. $\sin 2x + \sin x = 0$

100. $\sin 2x + \cos x = 0$

101. $3 \cos x - 6\sqrt{3} = \cos x - 5\sqrt{3}$

102. $\cos x - 5 = 3 \cos x + 6$

103. $\tan x = -4.7143$

104. $\tan x = -6.2154$

105. $2 \sin^2 x = 3 - \sin x$

106. $2 \sin^2 x = 2 - 3 \sin x$

107. $\cos x \csc x = 2 \cos x$

108. $\tan x \sec x = 2 \tan x$

109. $5 \cot^2 x - 15 = 0$

110. $5 \sec^2 x - 10 = 0$

111. $\cos^2 x + 2 \cos x - 2 = 0$

112. $\cos^2 x + 5 \cos x - 1 = 0$

113. $5 \sin x = 2 \cos^2 x - 4$ **114.** $7 \cos x = 4 - 2 \sin^2 x$

115. $2 \tan^2 x + 5 \tan x + 3 = 0$

116. $3 \tan^2 x - \tan x - 2 = 0$

Practice Plus

In Exercises 117–120, graph f and g in the same rectangular coordinate system for $0 \leq x \leq 2\pi$. Then solve a trigonometric equation to determine points of intersection and identify these points on your graphs.

117. $f(x) = 3 \cos x,\ g(x) = \cos x - 1$

118. $f(x) = 3 \sin x,\ g(x) = \sin x - 1$

119. $f(x) = \cos 2x,\ g(x) = -2 \sin x$

120. $f(x) = \cos 2x,\ g(x) = 1 - \sin x$

In Exercises 121–126, solve each equation on the interval $[0, 2\pi)$.

121. $|\cos x| = \dfrac{\sqrt{3}}{2}$

122. $|\sin x| = \dfrac{1}{2}$

123. $10 \cos^2 x + 3 \sin x - 9 = 0$

124. $3 \cos^2 x - \sin x = \cos^2 x$

125. $2 \cos^3 x + \cos^2 x - 2 \cos x - 1 = 0$ (*Hint:* Use factoring by grouping.)

126. $2 \sin^3 x - \sin^2 x - 2 \sin x + 1 = 0$ (*Hint:* Use factoring by grouping.)

In Exercises 127–128, find the x-intercepts, correct to four decimal places, of the graph of each function. Then use the x-intercepts to match the function with its graph. The graphs are labeled (a) and (b).

127. $f(x) = \tan^2 x - 3 \tan x + 1$

128. $g(x) = 4 \tan^2 x - 8 \tan x + 3$

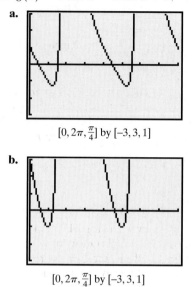

a.

$[0, 2\pi, \frac{\pi}{4}]$ by $[-3, 3, 1]$

b.

$[0, 2\pi, \frac{\pi}{4}]$ by $[-3, 3, 1]$

⭐ Application Exercises

Use this information to solve Exercises 129–130. Our cycle of normal breathing takes place every 5 seconds. Velocity of air flow, y, measured in liters per second, after x seconds is modeled by

$$y = 0.6 \sin \frac{2\pi}{5} x.$$

Velocity of air flow is positive when we inhale and negative when we exhale.

129. Within each breathing cycle, when are we inhaling at a rate of 0.3 liter per second? Round to the nearest tenth of a second.

130. Within each breathing cycle, when are we exhaling at a rate of 0.3 liter per second? Round to the nearest tenth of a second.

Use this information to solve Exercises 131–132. The number of hours of daylight in Boston is given by

$$y = 3 \sin \left[\frac{2\pi}{365} (x - 79) \right] + 12,$$

where x is the number of days after January 1.

131. Within a year, when does Boston have 10.5 hours of daylight? Give your answer in days after January 1 and round to the nearest day.

132. Within a year, when does Boston have 13.5 hours of daylight? Give your answer in days after January 1 and round to the nearest day.

Use this information to solve Exercises 133–134. A ball on a spring is pulled 4 inches below its rest position and then released. After t seconds, the ball's distance, d, in inches from its rest position is given by

$$d = -4 \cos \frac{\pi}{3} t.$$

133. Find all values of t for which the ball is 2 inches above its rest position.

134. Find all values of t for which the ball is 2 inches below its rest position.

Use this information to solve Exercises 135–136. When throwing an object, the distance achieved depends on its initial velocity, v_0, and the angle above the horizontal at which the object is thrown, θ. The distance, d, in feet, that describes the range covered is given by

$$d = \frac{v_0^2}{16} \sin \theta \cos \theta,$$

where v_0 is measured in feet per second.

135. You and your friend are throwing a baseball back and forth. If you throw the ball with an initial velocity of $v_0 = 90$ feet per second, at what angle of elevation, θ, to the nearest degree, should you direct your throw so that it can be easily caught by your friend located 170 feet away?

136. In Exercise 135, you increase the distance between you and your friend to 200 feet. With this increase, at what angle of elevation, θ, to the nearest degree, should you direct your throw?

Writing in Mathematics

137. What are the solutions of a trigonometric equation?

138. Describe the difference between verifying a trigonometric identity and solving a trigonometric equation.

139. Without actually solving the equation, describe how to solve

$$3 \tan x - 2 = 5 \tan x - 1.$$

140. In the interval $[0, 2\pi)$, the solutions of $\sin x = \cos 2x$ are $\frac{\pi}{6}, \frac{5\pi}{6}$, and $\frac{3\pi}{2}$. Explain how to use graphs generated by a graphing utility to check these solutions.

141. Suppose you are solving equations in the interval $[0, 2\pi)$. Without actually solving equations, what is the difference between the number of solutions of $\sin x = \frac{1}{2}$ and $\sin 2x = \frac{1}{2}$? How do you account for this difference?

In Exercises 142–143, describe a general strategy for solving each equation. Do not solve the equation.

142. $2 \sin^2 x + 5 \sin x + 3 = 0$

143. $\sin 2x = \sin x$

144. Describe a natural periodic phenomenon. Give an example of a question that can be answered by a trigonometric equation in the study of this phenomenon.

145. Some people experience depression with loss of sunlight. Use the essay on page 610 to determine whether such a person should live on a city street that is 80 feet wide with buildings whose heights average 400 feet. Explain your answer and include θ, to the nearest degree, in your argument.

Technology Exercises

146. Use a graphing utility to verify the solutions of any five equations that you solved in Exercises 63–84.

In Exercises 147–151, use a graphing utility to approximate the solutions of each equation in the interval $[0, 2\pi)$. Round to the nearest hundredth of a radian.

147. $15 \cos^2 x + 7 \cos x - 2 = 0$

148. $\cos x = x$

149. $2 \sin^2 x = 1 - 2 \sin x$

150. $\sin 2x = 2 - x^2$

151. $\sin x + \sin 2x + \sin 3x = 0$

Critical Thinking Exercises

152. Which one of the following is true?
   a. The equation $(\sin x - 3)(\cos x + 2) = 0$ has no solution.

   b. The equation $\tan x = \frac{\pi}{2}$ has no solution.

   c. A trigonometric equation with an infinite number of solutions is an identity.

   d. The equations $\sin 2x = 1$ and $\sin 2x = \frac{1}{2}$ have the same number of solutions on the interval $[0, 2\pi)$.

In Exercises 153–155, solve each equation on the interval $[0, 2\pi)$. Do not use a calculator.

153. $2 \cos x - 1 + 3 \sec x = 0$

154. $\sin 3x + \sin x + \cos x = 0$

155. $\sin x + 2 \sin \frac{x}{2} = \cos \frac{x}{2} + 1$

Chapter 5
Summary, Review, and Test

Summary

DEFINITIONS AND CONCEPTS	EXAMPLES
5.1 Verifying Trigonometric Identities	
a. Identities are trigonometric equations that are true for all values of the variable for which the expressions are defined.	
b. Fundamental trigonometric identities are given in the box on page 570.	
c. Guidelines for verifying trigonometric identities are given in the box on page 577.	Ex. 1, p. 571; Ex. 2, p. 572; Ex. 3, p. 573; Ex. 4, p. 573; Ex. 5, p. 574; Ex. 6, p. 574; Ex. 7, p. 575; Ex. 8, p. 576
5.2 Sum and Difference Formulas	
a. Sum and difference formulas are given in the box on page 583 and the box on page 586.	
b. Sum and difference formulas can be used to find exact values of trigonometric functions.	Ex. 1, p. 582; Ex. 2, p. 582; Ex. 4, p. 584; Ex. 5, p. 584
c. Sum and difference formulas can be used to verify trigonometric identities.	Ex. 3, p. 582; Ex. 6, p. 585; Ex. 7, p. 586
5.3 Double-Angle, Power-Reducing, and Half-Angle Formulas	
a. Double-angle, power-reducing, and half-angle formulas are given in the box on page 596.	
b. Double-angle and half-angle formulas can be used to find exact values of trigonometric functions.	Ex. 1, p. 591; Ex. 2, p. 591; Ex. 5, p. 594
c. Double-angle and half-angle formulas can be used to verify trigonometric identities.	Ex. 3, p. 592; Ex. 6, p. 595; Ex. 7, p. 596
d. Power-reducing formulas can be used to reduce the powers of trigonometric functions.	Ex. 4, p. 593
5.4 Product-to-Sum and Sum-to-Product Formulas	
a. The product-to-sum formulas are given in the box on page 601.	Ex. 1, p. 602
b. The sum-to-product formulas are given in the box on page 603. These formulas are useful to verify identities with fractions that contain sums and differences of sines and/or cosines.	Ex. 2, p. 603; Ex. 3, p. 604
5.5 Trigonometric Equations	
a. The values that satisfy a trigonometric equation are its solutions.	
b. To solve an equation containing a single trigonometric function, isolate the function on one side and solve for the variable.	Ex. 1, p. 610
c. When solving equations involving multiple angles, the period plays an important role in ensuring that we do not leave out any solutions.	Ex. 2, p. 611; Ex. 3, p. 611

DEFINITIONS AND CONCEPTS	**EXAMPLES**
d. Trigonometric equations quadratic in form can be expressed as $au^2 + bu + c = 0$, where u is a trigonometric function and $a \neq 0$. Such equations can be solved by factoring, the square root property, or the quadratic formula.	Ex. 4, p. 612; Ex. 5, p. 612; Ex. 12, p. 618
e. Factoring can be used to separate two different trigonometric functions in an equation.	Ex. 6, p. 613
f. Identities are used to solve some trigonometric equations.	Ex. 7, p. 614; Ex. 8, p. 615; Ex. 9, p. 615; Ex. 10, p. 616
g. Some trigonometric equations have solutions that cannot be determined by knowing the exact values of trigonometric functions of special angles. Such equations are solved using a calculator's inverse trigonometric function feature.	Ex. 11, p. 617; Ex. 12, p. 618

Review Exercises

5.1

In Exercises 1–13, verify each identity.

1. $\sec x - \cos x = \tan x \sin x$

2. $\cos x + \sin x \tan x = \sec x$

3. $\sin^2 \theta (1 + \cot^2 \theta) = 1$

4. $(\sec \theta - 1)(\sec \theta + 1) = \tan^2 \theta$

5. $\dfrac{1 - \tan x}{\sin x} = \csc x - \sec x$

6. $\dfrac{1}{\sin t - 1} + \dfrac{1}{\sin t + 1} = -2 \tan t \sec t$

7. $\dfrac{1 + \sin t}{\cos^2 t} = \tan^2 t + 1 + \tan t \sec t$

8. $\dfrac{\cos x}{1 - \sin x} = \dfrac{1 + \sin x}{\cos x}$

9. $1 - \dfrac{\sin^2 x}{1 + \cos x} = \cos x$

10. $(\tan \theta + \cot \theta)^2 = \sec^2 \theta + \csc^2 \theta$

11. $\dfrac{1}{\sin \theta + \cos \theta} + \dfrac{1}{\sin \theta - \cos \theta} = \dfrac{2 \sin \theta}{\sin^4 \theta - \cos^4 \theta}$

12. $\dfrac{\cos t}{\cot t - 5 \cos t} = \dfrac{1}{\csc t - 5}$

13. $\dfrac{1 - \cos t}{1 + \cos t} = (\csc t - \cot t)^2$

5.2 and 5.3

In Exercises 14–19, use a sum or difference formula to find the exact value of each expression.

14. $\cos(45° + 30°)$

15. $\sin 195°$

16. $\tan\left(\dfrac{4\pi}{3} - \dfrac{\pi}{4}\right)$

17. $\tan \dfrac{5\pi}{12}$

18. $\cos 65° \cos 5° + \sin 65° \sin 5°$

19. $\sin 80° \cos 50° - \cos 80° \sin 50°$

In Exercises 20–31, verify each identity.

20. $\sin\left(x + \dfrac{\pi}{6}\right) - \cos\left(x + \dfrac{\pi}{3}\right) = \sqrt{3} \sin x$

21. $\tan\left(x + \dfrac{3\pi}{4}\right) = \dfrac{\tan x - 1}{1 + \tan x}$

22. $\sec(\alpha + \beta) = \dfrac{\sec \alpha \sec \beta}{1 - \tan \alpha \tan \beta}$

23. $\dfrac{\cos(\alpha - \beta)}{\cos \alpha \cos \beta} = 1 + \tan \alpha \tan \beta$

24. $\cos^4 t - \sin^4 t = \cos 2t$

25. $\sin t - \cos 2t = (2 \sin t - 1)(\sin t + 1)$

26. $\dfrac{\sin 2\theta - \sin \theta}{\cos 2\theta + \cos \theta} = \dfrac{1 - \cos \theta}{\sin \theta}$

27. $\dfrac{\sin 2\theta}{1 - \sin^2 \theta} = 2 \tan \theta$

28. $\tan 2t = 2 \sin t \cos t \sec 2t$

29. $\cos 4t = 1 - 8 \sin^2 t \cos^2 t$

30. $\tan \dfrac{x}{2}(1 + \cos x) = \sin x$

31. $\tan \dfrac{x}{2} = \dfrac{\sec x - 1}{\tan x}$

In Exercises 32–34, the graph with the given equation is shown in a $\left[0, 2\pi, \frac{\pi}{2}\right]$ by $[-2, 2, 1]$ viewing rectangle.

   a. *Describe the graph using another equation.*

   b. *Verify that the two equations are equivalent.*

32. $y = \sin\left(x - \dfrac{3\pi}{2}\right)$

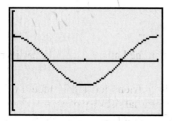

33. $y = \cos\left(x + \dfrac{\pi}{2}\right)$

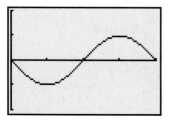

34. $y = \dfrac{\tan x - 1}{1 - \cot x}$

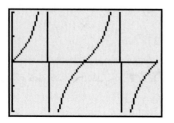

In Exercises 35–38, find the exact value of the following under the given conditions:

   a. $\sin(\alpha + \beta)$

   b. $\cos(\alpha - \beta)$

   c. $\tan(\alpha + \beta)$

   d. $\sin 2\alpha$

   e. $\cos\dfrac{\beta}{2}$

35. $\sin \alpha = \frac{3}{5}$, α lies in quadrant I, and $\sin \beta = \frac{12}{13}$, β lies in quadrant II.

36. $\tan \alpha = \frac{4}{3}$, α lies in quadrant III, and $\tan \beta = \frac{5}{12}$, β lies in quadrant I.

37. $\tan \alpha = -3$, α lies in quadrant II, and $\cot \beta = -3$, β lies in quadrant IV.

38. $\sin \alpha = -\frac{1}{3}$, α lies in quadrant III, and $\cos \beta = -\frac{1}{3}$, β lies in quadrant III.

In Exercises 39–42, use double- and half-angle formulas to find the exact value of each expression.

39. $\cos^2 15° - \sin^2 15°$

40. $\dfrac{2 \tan \dfrac{5\pi}{12}}{1 - \tan^2 \dfrac{5\pi}{12}}$

41. $\sin 22.5°$

42. $\tan \dfrac{\pi}{12}$

5.4

In Exercises 43–44, express each product as a sum or difference.

43. $\sin 6x \sin 4x$

44. $\sin 7x \cos 3x$

In Exercises 45–46, express each sum or difference as a product. If possible, find this product's exact value.

45. $\sin 2x - \sin 4x$

46. $\cos 75° + \cos 15°$

In Exercises 47–48, verify each identity.

47. $\dfrac{\cos 3x + \cos 5x}{\cos 3x - \cos 5x} = \cot x \cot 4x$

48. $\dfrac{\sin 2x + \sin 6x}{\sin 2x - \sin 6x} = -\tan 4x \cot 2x$

49. The graph with the given equation is shown in a $\left[0, 2\pi, \dfrac{\pi}{2}\right]$ by $[-2, 2, 1]$ viewing rectangle.

$$y = \dfrac{\cos 3x + \cos x}{\sin 3x - \sin x}$$

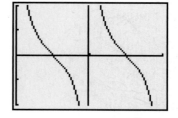

   a. Describe the graph using another equation.

   b. Verify that the two equations are equivalent.

5.5

In Exercises 50–53, find all solutions of each equation.

50. $\cos x = -\dfrac{1}{2}$

51. $\sin x = \dfrac{\sqrt{2}}{2}$

62. $2 \sin x + 1 = 0$

63. $\sqrt{3} \tan x - 1 = 0$

In Exercises 54–67, solve each equation on the interval $[0, 2\pi)$. Use exact values where possible or give approximate solutions correct to four decimal places.

54. $\cos 2x = -1$

55. $\sin 3x = 1$

56. $\tan \dfrac{x}{2} = -1$

57. $\tan x = 2 \cos x \tan x$

58. $\cos^2 x - 2 \cos x = 3$

59. $2 \cos^2 x - \sin x = 1$

60. $4 \sin^2 x = 1$

61. $\cos 2x - \sin x = 1$

62. $\sin 2x = \sqrt{3} \sin x$

63. $\sin x = \tan x$

64. $\sin x = -0.6031$

65. $5 \cos^2 x - 3 = 0$

66. $\sec^2 x = 4 \tan x - 2$

67. $2 \sin^2 x + \sin x - 2 = 0$

68. A ball on a spring is pulled 6 inches below its rest position and then released. After t seconds, the ball's distance, d, in inches from its rest position is given by

$$d = -6 \cos \frac{\pi}{2} t.$$

Find all values of t for which the ball is 3 inches below its rest position.

69. You are playing catch with a friend located 100 feet away. If you throw the ball with an initial velocity of $v_0 = 90$ feet per second, at what angle of elevation, θ, to the nearest degree should you direct your throw so that it can be caught easily? Use the formula

$$d = \frac{v_0^2}{16} \sin \theta \cos \theta.$$

Chapter 5 Test

Use the following conditions to solve Exercises 1–4:

$$\sin \alpha = \tfrac{4}{5}, \alpha \text{ lies in quadrant II.}$$
$$\cos \beta = \tfrac{5}{13}, \beta \text{ lies in quadrant I.}$$

Find the exact value of each of the following.

1. $\cos(\alpha + \beta)$

2. $\tan(\alpha - \beta)$

3. $\sin 2\alpha$

4. $\cos \dfrac{\beta}{2}$

5. Use $105° = 135° - 30°$ to find the exact value of $\sin 105°$.

In Exercises 6–11, verify each identity.

6. $\cos x \csc x = \cot x$

7. $\dfrac{\sec x}{\cot x + \tan x} = \sin x$

8. $1 - \dfrac{\cos^2 x}{1 + \sin x} = \sin x$

9. $\cos\left(\theta + \dfrac{\pi}{2}\right) = -\sin \theta$

10. $\dfrac{\sin(\alpha - \beta)}{\sin \alpha \cos \beta} = 1 - \cot \alpha \tan \beta$

11. $\sin t \cos t (\tan t + \cot t) = 1$

In Exercises 12–18, solve each equation on the interval $[0, 2\pi)$. Use exact values where possible or give approximate solutions correct to four decimal places.

12. $\sin 3x = -\tfrac{1}{2}$

13. $\sin 2x + \cos x = 0$

14. $2 \cos^2 x - 3 \cos x + 1 = 0$

15. $2 \sin^2 x + \cos x = 1$

16. $\cos x = -0.8092$

17. $\tan x \sec x = 3 \tan x$

18. $\tan^2 x - 3 \tan x - 2 = 0$

Cumulative Review Exercises (Chapters P–5)

Solve each equation or inequality in Exercises 1–5.

1. $x^3 + x^2 - x + 15 = 0$

2. $11^{x-1} = 125$

3. $x^2 + 2x - 8 > 0$

4. $\cos 2x + 3 = 5 \cos x, \quad 0 \le x < 2\pi$

5. $\tan x + \sec^2 x = 3, \quad 0 \le x < 2\pi$

In Exercises 6–11, graph each equation.

6. $y = \sqrt{x + 2} - 1$; Use transformations of the graph of $y = \sqrt{x}$.

7. $(x - 1)^2 + (y + 2)^2 = 9$

8. $y + 2 = \frac{1}{3}(x - 1)$

9. $y = 3 \cos 2x, \quad -2\pi \le x \le 2\pi$

10. $y = 2 \sin \frac{x}{2} + 1, \quad -2\pi \le x \le 2\pi$

11. $f(x) = (x - 1)^2(x - 3)$

12. If $f(x) = x^2 + 3x - 1$, find $\dfrac{f(a + h) - f(a)}{h}$.

13. Find the exact value of $\sin 225°$.

14. Verify the identity: $\sec^4 x - \sec^2 x = \tan^4 x + \tan^2 x$.

15. Convert $320°$ to radians.

16. How long would it take for any amount of money compounded continuously at 5.75% per year, to triple Round to the nearest tenth of a year.

17. If $f(x) = \dfrac{2x + 1}{x - 3}$, find $f^{-1}(x)$.

18. If C is a right angle in triangle ABC with $A = 23°$ and $a = 12$, solve the triangle.

19. A formula for calculating an infant's dosage for medication is

$$\text{Infant's dose} = \frac{\text{age of infant in months}}{150} \times \text{adult dose}.$$

If a 12-month-old infant is to receive 8.5 mg of medication find the equivalent adult dose to the nearest milligram.

20. From a point on the ground 12 feet from the base of a flagpole the angle of elevation to the top of the pole is $53°$. Approximate the height of the flagpole to the nearest tenth of a foot

Additional Topics in Trigonometry

THESE DAYS, COMPUTERS AND trigonometric functions are everywhere. Trigonometry plays a critical role in analyzing the forces that surround your every move. Using trigonometry to understand how forces are measured is one of the topics in this chapter that focuses on additional applications of trigonometry.

YOU ENJOY RUNNING, ALTHOUGH LATELY you experience discomfort at various points of impact. Your doctor suggests a computer analysis. By attaching sensors to your running shoes as you jog along a treadmill, the computer provides a printout of the magnitude and direction of the forces as your feet hit the ground. Based on this analysis, customized orthotics can be made to fit inside your shoes to minimize the impact.

From running to standing still, our bodies are surrounded by forces, illustrated in the Section 6.6 opener and analyzed in the boxed essay on page 682.

SECTION 6.1 *The Law of Sines*

Objectives

❶ Use the Law of Sines to solve oblique triangles.

❷ Use the Law of Sines to solve, if possible, the triangle or triangles in the ambiguous case.

❸ Find the area of an oblique triangle using the sine function.

❹ Solve applied problems using the Law of Sines.

Point Reyes National Seashore, 40 miles north of San Francisco, consists of 75,000 acres with miles of pristine surf-pummeled beaches, forested ridges, and bay flanked by white cliffs. A few people, inspired by nature in the raw, live on private property adjoining the National Seashore. In 1995, a fire in the park burned 12,350 acres and destroyed 45 homes.

Fire is a necessary part of the life cycle in many wilderness areas. It is also an ongoing threat to those who choose to live surrounded by nature's unspoiled beauty. In this section, we see how trigonometry can be used to locate small wilderness fires before they become raging infernos. To do this, we begin by considering triangles other than right triangles.

The Law of Sines and Its Derivation

An **oblique triangle** is a triangle that does not contain a right angle. Figure 6.1 shows that an oblique triangle has either three acute angles or two acute angles and one obtuse angle. Notice that the angles are labeled A, B, and C. The sides opposite each angle are labeled as a, b, and c, respectively.

> **Study Tip**
>
> Up until now, our work with triangles has involved right triangles. **Do not apply relationships that are valid for right triangles to oblique triangles**. Avoid the error of using the Pythagorean Theorem, $a^2 + b^2 = c^2$, to find a missing side of an oblique triangle. This relationship among the three sides applies only to right triangles.

> **Study Tip**
>
> Up until now, our work with triangles has involved right triangles. **Do not apply relationships that are valid for right triangles to oblique triangles**. Avoid the error of using the Pythagorean Theorem, $a^2 + b^2 = c^2$, to find a missing side of an oblique triangle. This relationship among the three sides applies only to right triangles.

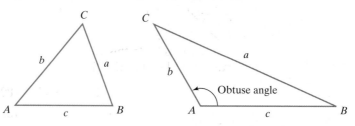

Figure 6.1 Oblique triangles

The relationships among the sides and angles of right triangles defined by the trigonometric functions are not valid for oblique triangles. Thus, we must observe and develop new relationships in order to work with oblique triangles.

Many relationships exist among the sides and angles in oblique triangles. One such relationship is called the **Law of Sines**.

> **Study Tip**
>
> The Law of Sines can be expressed with the sines in the numerator:
>
> $$\frac{\sin A}{a} = \frac{\sin B}{b} = \frac{\sin C}{c}.$$

The Law of Sines

If A, B, and C are the measures of the angles of a triangle, and a, b, and c are the lengths of the sides opposite these angles, then

$$\frac{a}{\sin A} = \frac{b}{\sin B} = \frac{c}{\sin C}.$$

The ratio of the length of the side of any triangle to the sine of the angle opposite that side is the same for all three sides of the triangle.

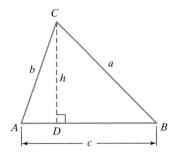

Figure 6.2 Drawing an altitude to prove the Law of Sines

To prove the Law of Sines, we draw an altitude of length h from one of the vertices of the triangle. In Figure 6.2, the altitude is drawn from vertex C. Two smaller triangles are formed, triangles ACD and BCD. Note that both are right triangles. Thus, we can use the definition of the sine of an angle of a right triangle.

$$\sin B = \frac{h}{a} \qquad \sin A = \frac{h}{b} \qquad \sin \theta = \frac{opposite}{hypotenuse}$$

$$h = a \sin B \qquad h = b \sin A \qquad \text{Solve each equation for } h.$$

Because we have found two expressions for h, we can set these expressions equal to each other.

$$a \sin B = b \sin A \qquad \text{Equate the expressions for } h.$$

$$\frac{a \sin B}{\sin A \sin B} = \frac{b \sin A}{\sin A \sin B} \qquad \text{Divide both sides by } \sin A \sin B.$$

$$\frac{a}{\sin A} = \frac{b}{\sin B} \qquad \text{Simplify.}$$

This proves part of the Law of Sines. If we use the same process and draw an altitude of length h from vertex A, we obtain the following result:

$$\frac{b}{\sin B} = \frac{c}{\sin C}.$$

When this equation is combined with the previous equation, we obtain the Law of Sines. Because the sine of an angle is equal to the sine of 180° minus that angle, the Law of Sines is derived in a similar manner if the oblique triangle contains an obtuse angle.

① Use the Law of Sines to solve oblique triangles.

Solving Oblique Triangles

Solving an oblique triangle means finding the lengths of its sides and the measurements of its angles. The Law of Sines can be used to solve a triangle in which one side and two angles are known. The three known measurements can be abbreviated using SAA (a side and two angles are known) or ASA (two angles and the side between them are known).

EXAMPLE 1   Solving an SAA Triangle Using the Law of Sines

Solve the triangle shown in Figure 6.3 with $A = 46°$, $C = 63°$, and $c = 56$ inches. Round lengths of sides to the nearest tenth.

Solution   We begin by finding B, the third angle of the triangle. We do not need the Law of Sines to do this. Instead, we use the fact that the sum of the measures of the interior angles of a triangle is 180°.

$$A + B + C = 180°$$

$$46° + B + 63° = 180° \qquad \text{Substitute the given values:}$$
$$\qquad\qquad\qquad\qquad A = 46° \text{ and } C = 63°.$$

$$109° + B = 180° \qquad \text{Add.}$$

$$B = 71° \qquad \text{Subtract 109° from both sides.}$$

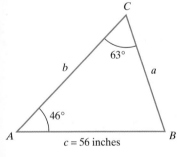

Figure 6.3 Solving an oblique SAA triangle

When we use the Law of Sines, we must be given one of the three ratios. In this example, we are given c and C: $c = 56$ and $C = 63°$. Thus, we use the ratio $\dfrac{c}{\sin C}$, or $\dfrac{56}{\sin 63°}$, to find the other two sides. Use the Law of Sines to find a.

$$\frac{a}{\sin A} = \frac{c}{\sin C} \qquad \text{The ratio of any side to the sine of its opposite angle equals the ratio of any other side to the sine of its opposite angle.}$$

$$\frac{a}{\sin 46°} = \frac{56}{\sin 63°} \qquad A = 46°, c = 56, \text{ and } C = 63°.$$

$$a = \frac{56 \sin 46°}{\sin 63°} \qquad \text{Multiply both sides by } \sin 46° \text{ and solve for } a.$$

$$a \approx 45.2 \text{ inches} \qquad \text{Use a calculator.}$$

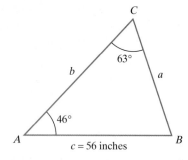

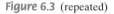

Figure 6.3 (repeated)

Use the Law of Sines again, this time to find b.

$$\frac{b}{\sin B} = \frac{c}{\sin C}$$ We use the given ratio, $\frac{c}{\sin C}$, to find b.

$$\frac{b}{\sin 71°} = \frac{56}{\sin 63°}$$ We found that $B = 71°$. We are given $c = 56$ and $C = 63°$.

$$b = \frac{56 \sin 71°}{\sin 63°}$$ Multiply both sides by $\sin 71°$ and solve for b.

$$b \approx 59.4 \text{ inches}$$ Use a calculator.

The solution is $B = 71°$, $a \approx 45.2$ inches, and $b \approx 59.4$ inches.

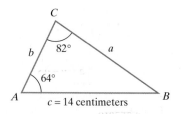

Figure 6.4

Check Point 1 Solve the triangle shown in Figure 6.4 with $A = 64°$, $C = 82°$, and $c = 1$ centimeters. Round as in Example 1.

EXAMPLE 2 Solving an ASA Triangle Using the Law of Sines

Solve triangle ABC if $A = 50°$, $C = 33.5°$, and $b = 76$. Round measures to the nearest tenth.

Solution We begin by drawing a picture of triangle ABC and labeling it with the given information. Figure 6.5 shows the triangle that we must solve. We begin by finding B.

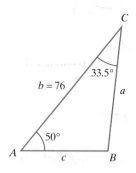

Figure 6.5 Solving an ASA triangle

$$A + B + C = 180°$$ The sum of the measures of a triangle's interior angles is 180°.

$$50° + B + 33.5° = 180°$$ $A = 50°$ and $C = 33.5°$.

$$83.5° + B = 180°$$ Add.

$$B = 96.5°$$ Subtract 83.5° from both sides.

Keep in mind that we must be given one of the three ratios to apply the Law of Sines. In this example, we are given that $b = 76$ and we found that $B = 96.5°$. Thus we use the ratio $\frac{b}{\sin B}$, or $\frac{76}{\sin 96.5°}$, to find the other two sides. Use the Law of Sines to find a and c.

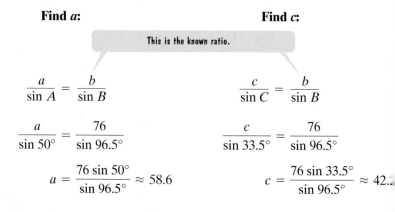

Find a:

This is the known ratio.

$$\frac{a}{\sin A} = \frac{b}{\sin B}$$

$$\frac{a}{\sin 50°} = \frac{76}{\sin 96.5°}$$

$$a = \frac{76 \sin 50°}{\sin 96.5°} \approx 58.6$$

Find c:

$$\frac{c}{\sin C} = \frac{b}{\sin B}$$

$$\frac{c}{\sin 33.5°} = \frac{76}{\sin 96.5°}$$

$$c = \frac{76 \sin 33.5°}{\sin 96.5°} \approx 42.2$$

The solution is $B = 96.5°$, $a \approx 58.6$, and $c \approx 42.2$.

Check Point 2 Solve triangle ABC if $A = 40°$, $C = 22.5°$, and $b = 12$. Round as in Example 2.

Use the Law of Sines to solve, if possible, the triangle or triangles in the ambiguous case.

The Ambiguous Case (SSA)

If we are given two sides and an angle opposite one of them (SSA), does this determine a unique triangle? Can we solve this case using the Law of Sines? Such a case is called the **ambiguous case** because the given information may result in one triangle, two triangles, or no triangle at all. For example, in Figure 6.6, we are given a, b, and A. Because a is shorter than h, it is not long enough to form a triangle. The number of possible triangles, if any, that can be formed in the SSA case depends on h, the length of the altitude, where $h = b \sin A$.

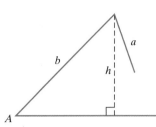

Figure 6.6 Given SSA, no triangle may result.

The Ambiguous Case (SSA)

Consider a triangle in which a, b, and A are given. This information may result in

One Triangle	One Right Triangle	No Triangle	Two Triangles

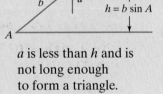

One Triangle

$h = b \sin A$

a is greater than h and a is greater than b. One triangle is formed.

One Right Triangle

$h = b \sin A$

$a = h$ and is just the right length to form a right triangle.

No Triangle

$h = b \sin A$

a is less than h and is not long enough to form a triangle.

Two Triangles

$h = b \sin A$

a is greater than h and a is less than b. Two distinct triangles are formed.

In an SSA situation, it is not necessary to draw an accurate sketch like those shown in the box. The Law of Sines determines the number of triangles, if any, and gives the solution for each triangle.

EXAMPLE 3   Solving an SSA Triangle Using the Law of Sines (One Solution)

Solve triangle ABC if $A = 43°$, $a = 81$, and $b = 62$. Round lengths of sides to the nearest tenth and angle measures to the nearest degree.

Solution   We begin with the sketch in Figure 6.7. The known ratio is $\dfrac{a}{\sin A}$, or $\dfrac{81}{\sin 43°}$. Because side b is given, we use the Law of Sines to find angle B.

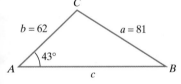

Figure 6.7 Solving an SSA triangle; the ambiguous case

$$\frac{a}{\sin A} = \frac{b}{\sin B} \qquad \text{Apply the Law of Sines.}$$

$$\frac{81}{\sin 43°} = \frac{62}{\sin B} \qquad a = 81, b = 62, \text{ and } A = 43°.$$

$$81 \sin B = 62 \sin 43° \qquad \text{Cross multiply: If } \frac{a}{b} = \frac{c}{d}, \text{ then } ad = bc.$$

$$\sin B = \frac{62 \sin 43°}{81} \qquad \text{Divide both sides by 81 and solve for sin B.}$$

$$\sin B \approx 0.5220 \qquad \text{Use a calculator.}$$

There are two angles B between $0°$ and $180°$ for which $\sin B \approx 0.5220$.

$$B_1 \approx 31° \qquad B_2 \approx 180° - 31° = 149°$$

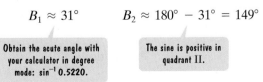

Obtain the acute angle with your calculator in degree mode: $\sin^{-1} 0.5220$.

The sine is positive in quadrant II.

Look at Figure 6.7. Given that $A = 43°$, can you see that $B_2 \approx 149°$ is impossible? By adding $149°$ to the given angle, $43°$, we exceed a $180°$ sum:

$$43° + 149° = 192°.$$

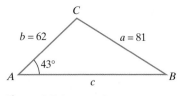

Figure 6.7 (repeated)

Thus, the only possibility is that $B_1 \approx 31°$. We find C using this approximation for B and the measure that was given for A: $A = 43°$.

$$C = 180° - B_1 - A \approx 180° - 31° - 43° = 106°$$

Side c that lies opposite this $106°$ angle can now be found using the Law of Sines.

$$\frac{c}{\sin C} = \frac{a}{\sin A} \qquad \text{Apply the Law of Sines.}$$

$$\frac{c}{\sin 106°} = \frac{81}{\sin 43°} \qquad a = 81, C \approx 106°, \text{ and } A = 43°.$$

$$c = \frac{81 \sin 106°}{\sin 43°} \approx 114.2 \qquad \text{Multiply both sides by } \sin 106° \text{ and solve for } c.$$

There is one triangle and the solution is B_1(or B) $\approx 31°, C \approx 106°$, and $c \approx 114.2$

Check Point 3 Solve triangle ABC if $A = 57°, a = 33$, and $b = 26$. Round as in Example 3

EXAMPLE 4 Solving an SSA Triangle Using the Law of Sines (No Solution)

Solve triangle ABC if $A = 75°, a = 51$, and $b = 71$.

Solution The known ratio is $\dfrac{a}{\sin A}$, or $\dfrac{51}{\sin 75°}$. Because side b is given, we use the Law of Sines to find angle B.

$$\frac{a}{\sin A} = \frac{b}{\sin B} \qquad \text{Use the Law of Sines.}$$

$$\frac{51}{\sin 75°} = \frac{71}{\sin B} \qquad \text{Substitute the given values.}$$

$$51 \sin B = 71 \sin 75° \qquad \text{Cross multiply: If } \frac{a}{b} = \frac{c}{d}, \text{ then } ad = bc.$$

$$\sin B = \frac{71 \sin 75°}{51} \approx 1.34 \qquad \text{Divide by 51 and solve for } \sin B.$$

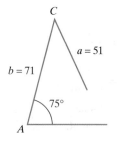

Figure 6.8 a is not long enough to form a triangle.

Because the sine can never exceed 1, there is no angle B for which $\sin B \approx 1.34$ There is no triangle with the given measurements, as illustrated in Figure 6.8.

Check Point 4 Solve triangle ABC if $A = 50°, a = 10$, and $b = 20$.

EXAMPLE 5 Solving an SSA Triangle Using the Law of Sines (Two Solutions)

Solve triangle ABC if $A = 40°, a = 54$, and $b = 62$. Round lengths of sides to the nearest tenth and angle measures to the nearest degree.

Solution The known ratio is $\dfrac{a}{\sin A}$, or $\dfrac{54}{\sin 40°}$. We use the Law of Sines to find angle B.

$$\frac{a}{\sin A} = \frac{b}{\sin B} \qquad \text{Use the Law of Sines.}$$

$$\frac{54}{\sin 40°} = \frac{62}{\sin B} \qquad \text{Substitute the given values.}$$

$$54 \sin B = 62 \sin 40° \qquad \text{Cross multiply: If } \frac{a}{b} = \frac{c}{d}, \text{ then } ad = bc.$$

$$\sin B = \frac{62 \sin 40°}{54} \approx 0.7380 \qquad \text{Divide by 54 and solve for } \sin B.$$

There are two angles B between $0°$ and $180°$ for which $\sin B \approx 0.7380$.

$$B_1 \approx 48° \qquad\qquad B_2 \approx 180° - 48° = 132°$$

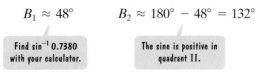

If you add either angle to the given angle, $40°$, the sum does not exceed $180°$. Thus, there are two triangles with the given conditions, shown in Figure 6.9(a). The triangles, AB_1C_1 and AB_2C_2, are shown separately in Figure 6.9(b) and 6.9(c).

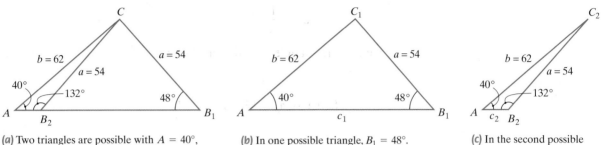

(a) Two triangles are possible with $A = 40°$, $a = 54$, and $b = 62$.

(b) In one possible triangle, $B_1 = 48°$.

(c) In the second possible triangle, $B_2 = 132°$.

Figure 6.9

Study Tip

The two triangles shown in Figure 6.9 are helpful in organizing the solutions. However, if you keep track of the two triangles, one with the given information and $B_1 = 48°$, and the other with the given information and $B_2 = 132°$, you do not have to draw the figure to solve the triangles.

We find angles C_1 and C_2 using a $180°$ angle sum in each of the two triangles.

$$
\begin{aligned}
C_1 &= 180° - A - B_1 & C_2 &= 180° - A - B_2 \\
&\approx 180° - 40° - 48° & &\approx 180° - 40° - 132° \\
&= 92° & &= 8°
\end{aligned}
$$

We use the Law of Sines to find c_1 and c_2.

$$
\begin{aligned}
\frac{c_1}{\sin C_1} &= \frac{a}{\sin A} & \frac{c_2}{\sin C_2} &= \frac{a}{\sin A} \\
\frac{c_1}{\sin 92°} &= \frac{54}{\sin 40°} & \frac{c_2}{\sin 8°} &= \frac{54}{\sin 40°} \\
c_1 &= \frac{54 \sin 92°}{\sin 40°} \approx 84.0 & c_2 &= \frac{54 \sin 8°}{\sin 40°} \approx 11.7
\end{aligned}
$$

There are two triangles. In one triangle, the solution is $B_1 \approx 48°$, $C_1 \approx 92°$, and $c_1 \approx 84.0$. In the other triangle, $B_2 \approx 132°$, $C_2 \approx 8°$, and $c_2 \approx 11.7$.

 Check Point 5 Solve triangle ABC if $A = 35°$, $a = 12$, and $b = 16$. Round as in Example 5.

The Area of an Oblique Triangle

 Find the area of an oblique triangle using the sine function.

A formula for the area of an oblique triangle can be obtained using the procedure for proving the Law of Sines. We draw an altitude of length h from one of the vertices of the triangle, as shown in Figure 6.10. We apply the definition of the sine of angle A, $\dfrac{\text{opposite}}{\text{hypotenuse}}$, in right triangle ACD:

$$\sin A = \frac{h}{b}, \quad \text{so} \quad h = b \sin A.$$

The area of a triangle is $\frac{1}{2}$ the product of any side and the altitude drawn to that side. Using the altitude h in Figure 6.10, we have

$$\text{Area} = \frac{1}{2}ch = \frac{1}{2}cb \sin A.$$

Use the result from above: $h = b \sin A$.

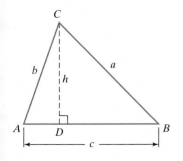

Figure 6.10

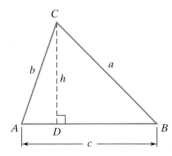

Figure 6.10 (repeated)

This result, Area $= \frac{1}{2}cb \sin A$, or $\frac{1}{2}bc \sin A$, indicates that the area of the triangle i one-half the product of b and c times the sine of their included angle. If we draw altitudes from the other two vertices, we can use any two sides to compute the area

Area of An Oblique Triangle

The area of a triangle equals one-half the product of the lengths of two sides times the sine of their included angle. In Figure 6.10, this wording can be expressed by the formulas

$$\text{Area} = \tfrac{1}{2}bc \sin A = \tfrac{1}{2}ab \sin C = \tfrac{1}{2}ac \sin B.$$

EXAMPLE 6   Finding the Area of an Oblique Triangle

Find the area of a triangle having two sides of lengths 24 meters and 10 meters and an included angle of 62°. Round to the nearest square meter.

Solution   The triangle is shown in Figure 6.11. Its area is half the product of the lengths of the two sides times the sine of the included angle.

$$\text{Area} = \tfrac{1}{2}(24)(10)(\sin 62°) \approx 106$$

The area of the triangle is approximately 106 square meters.

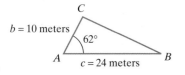

Figure 6.11 Finding the area of an SAS triangle

Check Point 6   Find the area of a triangle having two sides of lengths 8 meters and 12 meters and an included angle of 135°. Round to the nearest square meter

 Solve applied problems using the Law of Sines.

Applications of the Law of Sines

We have seen how the trigonometry of right triangles can be used to solve many different kinds of applied problems. The Law of Sines enables us to work with triangles that are not right triangles. As a result, this law can be used to solve problems involving surveying, engineering, astronomy, navigation, and the environment Example 7 illustrates the use of the Law of Sines in detecting potentially devastating fires

EXAMPLE 7   An Application of the Law of Sines

Two fire-lookout stations are 20 miles apart, with station B directly east of station A Both stations spot a fire on a mountain to the north. The bearing from station A to the fire is N50°E (50° east of north). The bearing from station B to the fire is N36°W (36° west of north). How far, to the nearest tenth of a mile, is the fire from station A?

Solution   Figure 6.12 shows the information given in the problem. The distance from station A to the fire is represented by b. Notice that the angles describing the bearing from each station to the fire, 50° and 36°, are not interior angles of triangle ABC. Using a north-south line, the interior angles are found as follows:

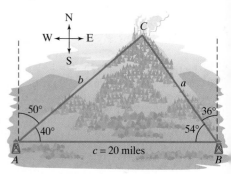

Figure 6.12

$$A = 90° - 50° = 40° \qquad B = 90° - 36° = 54°.$$

To find b using the Law of Sines, we need a known side and an angle opposite that side. Because $c = 20$ miles, we find angle C using a 180° angle sum in the triangle Thus,

$$C = 180° - A - B = 180° - 40° - 54° = 86°.$$

The ratio $\dfrac{c}{\sin C}$, or $\dfrac{20}{\sin 86°}$, is now known. We use this ratio and the Law of Sines to find b.

$$\frac{b}{\sin B} = \frac{c}{\sin C}$$ Use the Law of Sines.

$$\frac{b}{\sin 54°} = \frac{20}{\sin 86°}$$ $c = 20, B = 54°, $ and $C = 86°.$

$$b = \frac{20 \sin 54°}{\sin 86°} \approx 16.2$$ Multiply both sides by $\sin 54°$ and solve for b.

The fire is approximately 16.2 miles from station A.

Check Point 7 Two fire-lookout stations are 13 miles apart, with station B directly east of station A. Both stations spot a fire. The bearing of the fire from station A is N35°E and the bearing of the fire from station B is N49°W. How far, to the nearest tenth of a mile, is the fire from station B?

EXERCISE SET 6.1

Practice Exercises

In Exercises 1–8, solve each triangle. Round lengths of sides to the nearest tenth and angle measures to the nearest degree.

1.

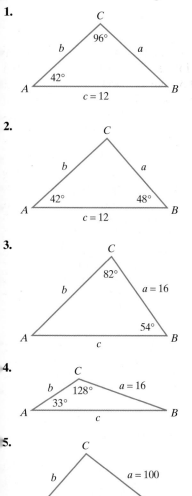

2.

3.

4.

5.

6.

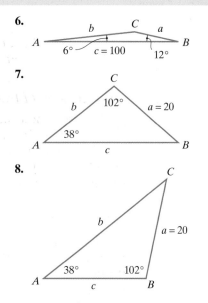

7.

8.

In Exercises 9–16, solve each triangle. Round lengths to the nearest tenth and angle measures to the nearest degree.

9. $A = 44°, B = 25°, a = 12$

10. $A = 56°, C = 24°, a = 22$

11. $B = 85°, C = 15°, b = 40$

12. $A = 85°, B = 35°, c = 30$

13. $A = 115°, C = 35°, c = 200$

14. $B = 5°, C = 125°, b = 200$

15. $A = 65°, B = 65°, c = 6$

16. $B = 80°, C = 10°, a = 8$

In Exercises 17–32, two sides and an angle (SSA) of a triangle are given. Determine whether the given measurements produce one triangle, two triangles, or no triangle at all. Solve each triangle that results. Round to the nearest tenth and the nearest degree for sides and angles, respectively.

17. $a = 20, b = 15, A = 40°$ **18.** $a = 30, b = 20, A = 50°$

19. $a = 10, c = 8.9, A = 63°$ **20.** $a = 57.5, c = 49.8, A = 136°$

21. $a = 42.1, c = 37, A = 112°$

22. $a = 6.1, b = 4, A = 162°$

23. $a = 10, b = 40, A = 30°$

24. $a = 10, b = 30, A = 150°$

25. $a = 16, b = 18, A = 60°$

26. $a = 30, b = 40, A = 20°$

27. $a = 12, b = 16.1, A = 37°$

28. $a = 7, b = 28, A = 12°$

29. $a = 22, c = 24.1, A = 58°$

30. $a = 95, c = 125, A = 49°$

31. $a = 9.3, b = 41, A = 18°$

32. $a = 1.4, b = 2.9, A = 142°$

In Exercises 33–38, find the area of the triangle having the given measurements. Round to the nearest square unit.

33. $A = 48°, b = 20$ feet, $c = 40$ feet

34. $A = 22°, b = 20$ feet, $c = 50$ feet

35. $B = 36°, a = 3$ yards, $c = 6$ yards

36. $B = 125°, a = 8$ yards, $c = 5$ yards

37. $C = 124°, a = 4$ meters, $b = 6$ meters

38. $C = 102°, a = 16$ meters, $b = 20$ meters

Practice Plus

In Exercises 39–40, find h to the nearest tenth.

39.

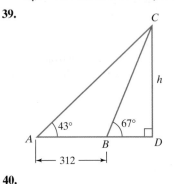

40.

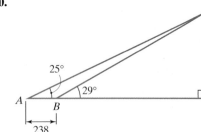

In Exercises 41–42, find a to the nearest tenth.

41.

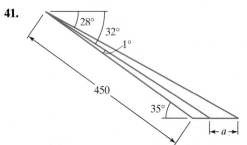

42.

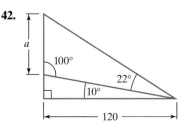

In Exercises 43–44, use the given measurements to solve the following triangle. Round lengths of sides to the nearest tenth and angle measures to the nearest degree.

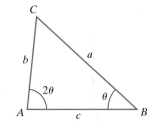

43. $a = 300, b = 200$ **44.** $a = 400, b = 300$

In Exercises 45–46, find the area of the triangle with the given vertices. Round to the nearest square unit.

45. $(-3, -2), (2, -2), (1, 2)$ **46.** $(-2, -3), (-2, 2), (2, 1)$

Application Exercises

47. Two fire-lookout stations are 10 miles apart, with station B directly east of station A. Both stations spot a fire. The bearing of the fire from station A is N25°E and the bearing of the fire from station B is N56°W. How far, to the nearest tenth of a mile, is the fire from each lookout station?

48. The Federal Communications Commission is attempting to locate an illegal radio station. It sets up two monitoring stations, A and B, with station B 40 miles east of station A. Station A measures the illegal signal from the radio station as coming from a direction of 48° east of north. Station B measures the signal as coming from a point 34° west of north. How far is the illegal radio station from monitoring stations A and B? Round to the nearest tenth of a mile.

49. The figure shows a 1200-yard-long sand beach and an oil platform in the ocean. The angle made with the platform from one end of the beach is 85° and from the other end is 76°. Find the distance of the oil platform, to the nearest tenth of a yard, from each end of the beach.

0. A surveyor needs to determine the distance between two points that lie on opposite banks of a river. The figure shows that 300 yards are measured along one bank. The angles from each end of this line segment to a point on the opposite bank are 62° and 53°. Find the distance between *A* and *B* to the nearest tenth of a yard.

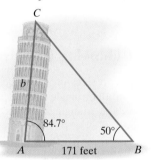

1. Closed to tourists since 1990, the Leaning Tower of Pisa in Italy leans at an angle of about 84.7°. The figure shows that 171 feet from the base of the tower, the angle of elevation to the top is 50°. Find the distance, to the nearest tenth of a foot, from the base to the top of the tower.

2. A pine tree growing on a hillside makes a 75° angle with the hill. From a point 80 feet up the hill, the angle of elevation to the top of the tree is 62° and the angle of depression to the bottom is 23°. Find, to the nearest tenth of a foot, the height of the tree.

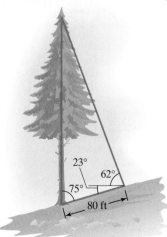

3. The figure shows a shot-put ring. The shot is tossed from *A* and lands at *B*. Using modern electronic equipment, the distance of the toss can be measured without the use of measuring tapes. When the shot lands at *B*, an electronic transmitter

placed at *B* sends a signal to a device in the official's booth above the track. The device determines the angles at *B* and *C*. At a track meet, the distance from the official's booth to the shot-put ring is 562 feet. If *B* = 85.3° and *C* = 5.7°, determine the length of the toss to the nearest tenth of a foot.

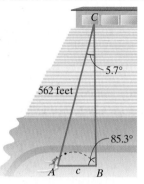

54. A pier forms an 85° angle with a straight shore. At a distance of 100 feet from the pier, the line of sight to the tip forms a 37° angle. Find the length of the pier to the nearest tenth of a foot.

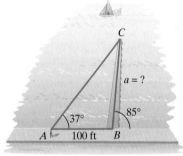

55. When the angle of elevation of the sun is 62°, a telephone pole that is tilted at an angle of 8° directly away from the sun casts a shadow 20 feet long. Determine the length of the pole to the nearest tenth of a foot.

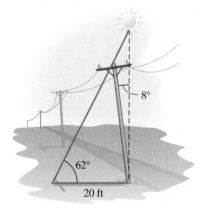

56. A leaning wall is inclined 6° from the vertical. At a distance of 40 feet from the wall, the angle of elevation to the top is 22°. Find the height of the wall to the nearest tenth of a foot.

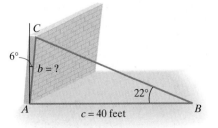

57. Redwood trees in California's Redwood National Park are hundreds of feet tall. The height of one of these trees is represented by h in the figure shown.

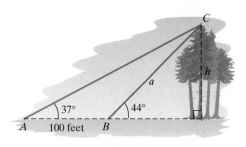

a. Use the measurements shown to find a, to the nearest tenth of a foot, in oblique triangle ABC.

b. Use the right triangle shown to find the height, to the nearest tenth of a foot, of a typical redwood tree in the park.

58. The figure shows a cable car that carries passengers from A to C. Point A is 1.6 miles from the base of the mountain. The angles of elevation from A and B to the mountain's peak are $22°$ and $66°$, respectively.

a. Determine, to the nearest tenth of a foot, the distance covered by the cable car.

b. Find a, to the nearest tenth of a foot, in oblique triangle ABC.

c. Use the right triangle to find the height of the mountain to the nearest tenth of a foot.

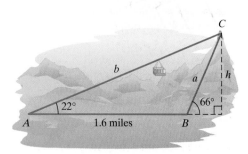

59. Lighthouse B is 7 miles west of lighthouse A. A boat leaves A and sails 5 miles. At this time, it is sighted from B. If the bearing of the boat from B is N62°E, how far from B is the boat? Round to the nearest tenth of a mile.

60. After a wind storm, you notice that your 16-foot flagpole may be leaning, but you are not sure. From a point on the ground 15 feet from the base of the flagpole, you find that the angle of elevation to the top is 48°. Is the flagpole leaning? If so, find the acute angle, to the nearest degree, that the flagpole makes with the ground.

Writing in Mathematics

61. What is an oblique triangle?

62. Without using symbols, state the Law of Sines in your ow[n] words.

63. Briefly describe how the Law of Sines is proved.

64. What does it mean to solve an oblique triangle?

65. What do the abbreviations SAA and ASA mean?

66. Why is SSA called the ambiguous case?

67. How is the sine function used to find the area of an obliqu[e] triangle?

68. Write an original problem that can be solved using the La[w] of Sines. Then solve the problem.

69. Use Exercise 53 to describe how the Law of Sines is used fo[r] throwing events at track and field meets. Why aren't tap[e] measures used to determine tossing distance?

70. You are cruising in your boat parallel to the coast, looking [at] a lighthouse. Explain how you can use your boat's speed an[d] a device for measuring angles to determine the distance [at] any instant from your boat to the lighthouse.

Critical Thinking Exercises

71. If you are given two sides of a triangle and their include[d] angle, you can find the triangle's area. Can the Law of Sine[s] be used to solve the triangle with this given information[?] Explain your answer.

72. Two buildings of equal height are 800 feet apart. An observe[r] on the street between the buildings measures the angles [of] elevation to the tops of the buildings as 27° and 41°, respe[c]tively. How high, to the nearest foot, are the buildings?

73. The figure shows the design for the top of the wing of a j[et] fighter. The fuselage is 5 feet wide. Find the wing span CC' t[o] the nearest tenth of a foot.

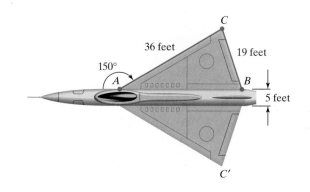

SECTION 6.2 The Law of Cosines

Objectives

❶ Use the Law of Cosines to solve oblique triangles.

❷ Solve applied problems using the Law of Cosines.

❸ Use Heron's formula to find the area of a triangle.

Paleontologists use trigonometry to study the movements made by dinosaurs millions of years ago. Figure 6.13, based on data collected at Dinosaur Valley State Park in Glen Rose, Texas, shows footprints made by a two-footed carnivorous (meat-eating) dinosaur and the hindfeet of a herbivorous (plant-eating) dinosaur.

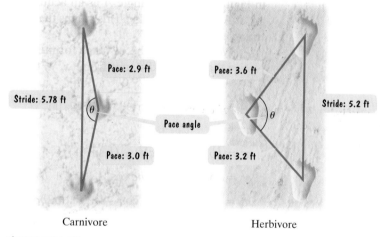

Carnivore   Herbivore

Figure 6.13 Dinosaur Footprints

Source: Glen J. Kuban, *An Overview of Dinosaur Tracking*

For each dinosaur, the figure indicates the *pace* and the *stride*. The pace is the distance from the left footprint to the right footprint, and vice versa. The stride is the distance from the left footprint to the next left footprint or from the right footprint to the next right footprint. Also shown in Figure 6.13 is the pace angle, designated by θ. Notice that neither dinosaur moves with a pace angle of 180°, meaning that the footprints are directly in line. The footprints show a "zig-zig" pattern that is numerically described by the pace angle. A dinosaur that is an efficient walker has a pace angle close to 180°, minimizing zig-zag motion and maximizing forward motion.

How can we determine the pace angles for the carnivore and the herbivore in Figure 6.13? Problems such as this, in which we know the measures of three sides of a triangle and we need to find the measurement of a missing angle, cannot be solved by the Law of Sines. To numerically describe which dinosaur in Figure 6.13 made more forward progress with each step, we turn to the Law of Cosines.

The Law of Cosines and Its Derivation

We now look at another relationship that exists among the sides and angles in an oblique triangle. **The Law of Cosines** is used to solve triangles in which two sides and the included angle (SAS) are known, or those in which three sides (SSS) are known.

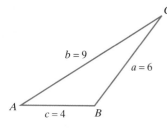

Figure 6.17 Solving an SSS triangle

EXAMPLE 2 Solving an SSS Triangle

Solve triangle ABC if $a = 6, b = 9,$ and $c = 4$. Round angle measures to the nearest degree.

Solution We are given three sides. Therefore, we apply the three-step procedure for solving an SSS triangle. The triangle is shown in Figure 6.17.

Step 1   Use the Law of Cosines to find the angle opposite the longest side. The longest side is $b = 9$. Thus, we will find angle B.

$$b^2 = a^2 + c^2 - 2ac \cos B \qquad \text{Apply the Law of Cosines to find B.}$$

$$2ac \cos B = a^2 + c^2 - b^2 \qquad \text{Solve for cos B.}$$

$$\cos B = \frac{a^2 + c^2 - b^2}{2ac}$$

$$\cos B = \frac{6^2 + 4^2 - 9^2}{2 \cdot 6 \cdot 4} = -\frac{29}{48} \qquad a = 6, b = 9, \text{ and } c = 4.$$

Using a calculator, $\cos^{-1}\left(\frac{29}{48}\right) \approx 53°$. Because $\cos B$ is negative, B is an obtuse angle. Thus,

$$B \approx 180° - 53° = 127°. \qquad \text{Because the domain of } y = \cos^{-1} x \text{ is } [0, \pi], \text{ you can use a calculator to find } \cos^{-1}\left(-\frac{29}{48}\right) \approx 127°.$$

Step 2   Use the Law of Sines to find either of the two remaining acute angles. We will find angle A.

$$\frac{a}{\sin A} = \frac{b}{\sin B} \qquad \text{Apply the Law of Sines.}$$

$$\frac{6}{\sin A} = \frac{9}{\sin 127°} \qquad \begin{array}{l}\text{We are given } a = 6 \text{ and } b = 9. \text{ We found} \\ \text{that } B \approx 127°.\end{array}$$

$$9 \sin A = 6 \sin 127° \qquad \text{Cross multiply.}$$

$$\sin A = \frac{6 \sin 127°}{9} \approx 0.5324 \qquad \text{Divide by 9 and solve for sin A.}$$

$$A \approx 32° \qquad \text{Find } \sin^{-1} 0.5324 \text{ using a calculator.}$$

Step 3   Find the third angle. Subtract the measures of the angles found in steps and 2 from 180°.

$$C = 180° - B - A \approx 180° - 127° - 32° = 21°$$

The solution is $B \approx 127°$, $A \approx 32°$, and $C \approx 21°$.

Check Point 2 Solve triangle ABC if $a = 8, b = 10,$ and $c = 5$. Round angle measures to the nearest degree.

Applications of the Law of Cosines

 Solve applied problems using the Law of Cosines.

Applied problems involving SAS and SSS triangles can be solved using the Law of Cosines.

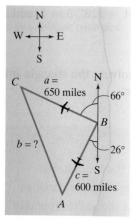

Figure 6.18

EXAMPLE 3 An Application of the Law of Cosines

Two airplanes leave an airport at the same time on different runways. One flies on a bearing of N66°W at 325 miles per hour. The other airplane flies on a bearing of S26°W at 300 miles per hour. How far apart will the airplanes be after two hours?

Solution After two hours, the plane flying at 325 miles per hour travels 325 · miles, or 650 miles. Similarly, the plane flying at 300 miles per hour travels 600 mile. The situation is illustrated in Figure 6.18.

   Let $b =$ the distance between the planes after two hours. We can use a north-south line to find angle B in triangle ABC. Thus,

$$B = 180° - 66° - 26° = 88°.$$

We now have $a = 650$, $c = 600$, and $B = 88°$. We use the Law of Cosines to find b in this SAS situation.

$$b^2 = a^2 + c^2 - 2ac \cos B \qquad \text{Apply the Law of Cosines.}$$

$$b^2 = 650^2 + 600^2 - 2(650)(600) \cos 88° \qquad \text{Substitute: } a = 650, c = 600, \text{ and}$$
$$B = 88°.$$

$$\approx 755{,}278 \qquad \text{Use a calculator.}$$

$$b \approx \sqrt{755{,}278} \approx 869 \qquad \text{Take the square root and solve for } b.$$

After two hours, the planes are approximately 869 miles apart.

Check Point 3 Two airplanes leave an airport at the same time on different runways. One flies directly north at 400 miles per hour. The other airplane flies on a bearing of N75°E at 350 miles per hour. How far apart will the airplanes be after two hours?

3 Use Heron's formula to find the area of a triangle.

Heron's Formula

Approximately 2000 years ago, the Greek mathematician Heron of Alexandria derived a formula for the area of a triangle in terms of the lengths of its sides. A more modern derivation uses the Law of Cosines and can be found in the appendix.

> **Heron's Formula for the Area of a Triangle**
>
> The area of a triangle with sides a, b, and c is
>
> $$\text{Area} = \sqrt{s(s - a)(s - b)(s - c)},$$
>
> where s is one-half its perimeter: $s = \frac{1}{2}(a + b + c)$.

EXAMPLE 4   Using Heron's Formula

Find the area of the triangle with $a = 12$ yards, $b = 16$ yards, and $c = 24$ yards. Round to the nearest square yard.

Solution   Begin by calculating one-half the perimeter:

$$s = \frac{1}{2}(a + b + c) = \frac{1}{2}(12 + 16 + 24) = 26.$$

Use Heron's formula to find the area:

$$\text{Area} = \sqrt{s(s - a)(s - b)(s - c)}$$
$$= \sqrt{26(26 - 12)(26 - 16)(26 - 24)}$$
$$= \sqrt{7280} \approx 85.$$

The area of the triangle is approximately 85 square yards.

Check Point 4 Find the area of the triangle with $a = 6$ meters, $b = 16$ meters, and $c = 18$ meters. Round to the nearest square meter.

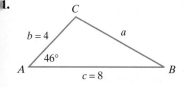

EXERCISE SET 6.2

Practice Exercises

In Exercises 1–8, solve each triangle. Round lengths of sides to the nearest tenth and angle measures to the nearest degree.

1.

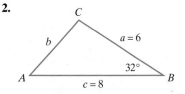

C, b = 4, a, 46°, A, c = 8, B

2.

C, b, a = 6, 32°, A, c = 8, B

3.

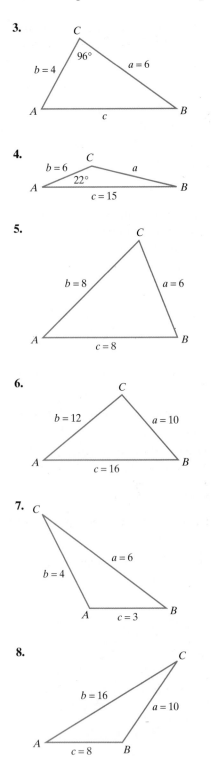

4.

5.

6.

7.

8.

In Exercises 9–24, solve each triangle. Round lengths to the nearest tenth and angle measures to the nearest degree.

9. $a = 5, b = 7, C = 42°$ **10.** $a = 10, b = 3, C = 15°$

11. $b = 5, c = 3, A = 102°$ **12.** $b = 4, c = 1, A = 100°$

13. $a = 6, c = 5, B = 50°$ **14.** $a = 4, c = 7, B = 55°$

15. $a = 5, c = 2, B = 90°$ **16.** $a = 7, c = 3, B = 90°$

17. $a = 5, b = 7, c = 10$ **18.** $a = 4, b = 6, c = 9$

19. $a = 3, b = 9, c = 8$

20. $a = 4, b = 7, c = 6$

21. $a = 3, b = 3, c = 3$

22. $a = 5, b = 5, c = 5$

23. $a = 63, b = 22, c = 50$

24. $a = 66, b = 25, c = 45$

In Exercises 25–30, use Heron's formula to find the area of each triangle. Round to the nearest square unit.

25. $a = 4$ feet, $b = 4$ feet, $c = 2$ feet

26. $a = 5$ feet, $b = 5$ feet, $c = 4$ feet

27. $a = 14$ meters, $b = 12$ meters, $c = 4$ meters

28. $a = 16$ meters, $b = 10$ meters, $c = 8$ meters

29. $a = 11$ yards, $b = 9$ yards, $c = 7$ yards

30. $a = 13$ yards, $b = 9$ yards, $c = 5$ yards

Practice Plus

In Exercises 31–32, solve each triangle. Round lengths of sides to the nearest tenth and angle measures to the nearest degree.

31.

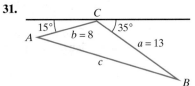

32.

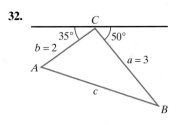

In Exercises 33–34, the three circles are arranged so that they touch each other, as shown in the figure. Use the given radii for the circles with centers A, B, and C, respectively, to solve triangle ABC. Round angle measures to the nearest degree.

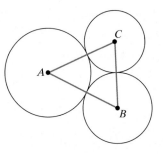

33. $5.0, 4.0, 3.5$ **34.** $7.5, 4.3, 3.0$

n Exercises 35–36, the three given points are the vertices of a
iangle. Solve each triangle, rounding lengths of sides to the
earest tenth and angle measures to the nearest degree.

5. $A(0, 0), B(-3, 4), C(3, -1)$

6. $A(0, 0), B(4, -3), C(1, -5)$

⭐ Application Exercises

7. Use Figure 6.13 on page 639 to find the pace angle, to the nearest degree, for the carnivore. Does the angle indicate that this dinosaur was an efficient walker? Describe your answer.

8. Use Figure 6.13 on page 639 to find the pace angle, to the nearest degree, for the herbivore. Does the angle indicate that this dinosaur was an efficient walker? Describe your answer.

9. Two ships leave a harbor at the same time. One ship travels on a bearing of S12°W at 14 miles per hour. The other ship travels on a bearing of N75°E at 10 miles per hour. How far apart will the ships be after three hours? Round to the nearest tenth of a mile.

0. A plane leaves airport A and travels 580 miles to airport B on a bearing of N34°E. The plane later leaves airport B and travels to airport C 400 miles away on a bearing of S74°E. Find the distance from airport A to airport C to the nearest tenth of a mile.

1. Find the distance across the lake from A to C, to the nearest yard, using the measurements shown in the figure.

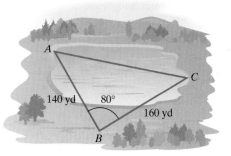

2. To find the distance across a protected cove at a lake, a surveyor makes the measurements shown in the figure. Use these measurements to find the distance from A to B to the nearest yard.

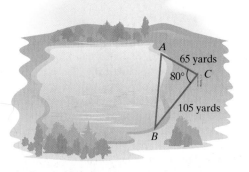

The diagram shows three islands in Florida Bay. You rent a boat and plan to visit each of these remote islands. Use the diagram to solve Exercises 43–44.

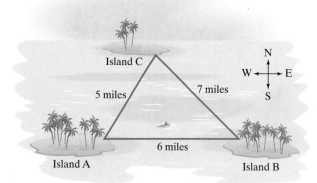

43. If you are on island A, on what bearing should you navigate to go to island C?

44. If you are on island B, on what bearing should you navigate to go to island C?

45. You are on a fishing boat that leaves its pier and heads east. After traveling for 25 miles, there is a report warning of rough seas directly south. The captain turns the boat and follows a bearing of S40°W for 13.5 miles.

   a. At this time, how far are you from the boat's pier? Round to the nearest tenth of a mile.

   b. What bearing could the boat have originally taken to arrive at this spot?

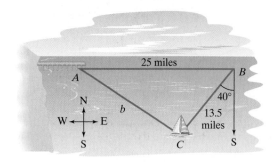

46. You are on a fishing boat that leaves its pier and heads east. After traveling for 30 miles, there is a report warning of rough seas directly south. The captain turns the boat and follows a bearing of S45°W for 12 miles.

   a. At this time, how far are you from the boat's pier? Round to the nearest tenth of a mile.

   b. What bearing could the boat have originally taken to arrive at this spot?

47. The figure shows a 400-foot tower on the side of a hill that forms a 7° angle with the horizontal. Find the length of each of the two guy wires that are anchored 80 feet uphill and downhill from the tower's base and extend to the top of the tower. Round to the nearest tenth of a foot.

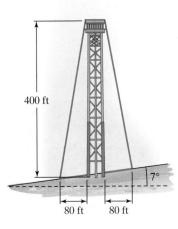

400 ft

7°

80 ft   80 ft

48. The figure shows a 200-foot tower on the side of a hill that forms a 5° angle with the horizontal. Find the length of each of the two guy wires that are anchored 150 feet uphill and downhill from the tower's base and extend to the top of the tower. Round to the nearest tenth of a foot.

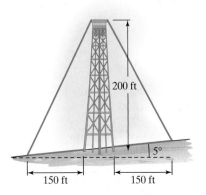

200 ft

5°

150 ft   150 ft

49. A Major League baseball diamond has four bases forming a square whose sides measure 90 feet each. The pitcher's mound is 60.5 feet from home plate on a line joining home plate and second base. Find the distance from the pitcher's mound to first base. Round to the nearest tenth of a foot.

50. A Little League baseball diamond has four bases forming a square whose sides measure 60 feet each. The pitcher's mound is 46 feet from home plate on a line joining home plate and second base. Find the distance from the pitcher's mound to third base. Round to the nearest tenth of a foot.

51. A piece of commercial real estate is priced at $3.50 per square foot. Find the cost, to the nearest dollar, of a triangular lot measuring 240 feet by 300 feet by 420 feet.

52. A piece of commercial real estate is priced at $4.50 per square foot. Find the cost, to the nearest dollar, of a triangular lot measuring 320 feet by 510 feet by 410 feet.

Writing in Mathematics

53. Without using symbols, state the Law of Cosines in your ow words.

54. Why can't the Law of Sines be used in the first step to solv an SAS triangle?

55. Describe a strategy for solving an SAS triangle.

56. Describe a strategy for solving an SSS triangle.

57. Under what conditions would you use Heron's formula t find the area of a triangle?

58. Describe an applied problem that can be solved using th Law of Cosines, but not the Law of Sines.

59. The pitcher on a Little League team is studying angles i geometry and has a question. "Coach, suppose I'm on th pitcher's mound facing home plate. I catch a fly ball hit in m direction. If I turn to face first base and throw the ba through how many degrees should I turn for a direct throw? Use the information given in Exercise 50 and write an answe to the pitcher's question. Without getting too technica describe to the pitcher how you obtained this angle.

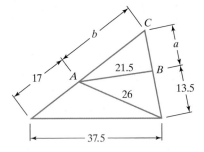

Critical Thinking Exercises

60. The lengths of the diagonals of a parallelogram are 20 inche and 30 inches. The diagonals intersect at an angle of 35°. Fin the lengths of the parallelogram's sides. (*Hint*: Diagonals of parallelogram bisect one another.)

61. Use the figure to solve triangle *ABC*. Round lengths of side to the nearest tenth and angle measures to the neare degree.

17   b   *A*   21.5   *B*   *C*   *a*   13.5   26   37.5

62. The minute hand and the hour hand of a clock have lengt *m* inches and *h* inches, respectively. Determine the distanc between the tips of the hands at 10:00 in terms of *m* and

Group Exercise

63. The group should design five original problems that can b solved using the Laws of Sines and Cosines. At least tw problems should be solved using the Law of Sines, o should be the ambiguous case, and at least two problem should be solved using the Law of Cosines. At least one prob lem should be an application problem using the Law of Sin and at least one problem should involve an application usi the Law of Cosines. The group should turn in both th problems and their solutions.

SECTION 6.3 *Polar Coordinates*

Objectives

① Plot points in the polar coordinate system.

② Find multiple sets of polar coordinates for a given point.

③ Convert a point from polar to rectangular coordinates.

④ Convert a point from rectangular to polar coordinates.

⑤ Convert an equation from rectangular to polar coordinates.

⑥ Convert an equation from polar to rectangular coordinates.

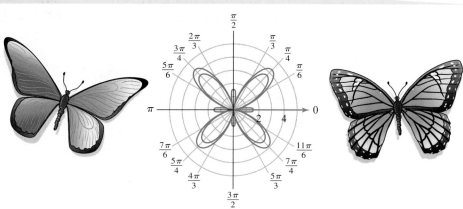

Butterflies are among the most celebrated of all insects. It's hard not to notice their beautiful colors and graceful flight. Their symmetry can be explored with trigonometric functions and a system for plotting points called the *polar coordinate system*. In many cases, polar coordinates are simpler and easier to use than rectangular coordinates.

Plotting Points in the Polar Coordinate System

① Plot points in the polar coordinate system.

The foundation of the polar coordinate system is a horizontal ray that extends to the right. The ray is called the **polar axis** and is shown in Figure 6.19. The endpoint of the ray is called the **pole**.

A point P in the polar coordinate system is represented by an ordered pair of numbers (r, θ). Figure 6.20 shows $P = (r, \theta)$ in the polar coordinate system.

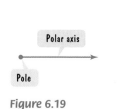

Figure 6.19

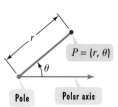

Figure 6.20 Representing a point in the polar coordinate system

- r is a directed distance from the pole to P. (We shall see that r can be positive, negative, or zero.)
- θ is an angle from the polar axis to the line segment from the pole to P. This angle can be measured in degrees or radians. Positive angles are measured counterclockwise from the polar axis. Negative angles are measured clockwise from the polar axis.

We refer to the ordered pair (r, θ) as the **polar coordinates** of P.

Let's look at a specific example. Suppose that the polar coordinates of a point P are $\left(3, \dfrac{\pi}{4}\right)$. Because θ is positive, we locate this point by drawing $\theta = \dfrac{\pi}{4}$ counterclockwise from the polar axis. Then we count out a distance of three units along the terminal side of the angle to reach the point P. Figure 6.21 shows that $(r, \theta) = \left(3, \dfrac{\pi}{4}\right)$ lies three units from the pole on the terminal side of the angle $\theta = \dfrac{\pi}{4}$.

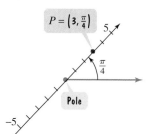

Figure 6.21 Locating a point in polar coordinates

Chapter 6 • Additional Topics in Trigonometry

The sign of r is important in locating $P = (r, \theta)$ in polar coordinates.

The Sign of r and a Point's Location in Polar Coordinates

The point $P = (r, \theta)$ is located $|r|$ units from the pole. If $r > 0$, the point lies on the terminal side of θ. If $r < 0$, the point lies along the ray opposite the terminal side of θ. If $r = 0$, the point lies at the pole, regardless of the value of θ.

EXAMPLE 1 Plotting Points in a Polar Coordinate System

Plot the points with the following polar coordinates:

a. $(2, 135°)$ **b.** $\left(-3, \dfrac{3\pi}{2}\right)$ **c.** $\left(-1, -\dfrac{\pi}{4}\right)$.

Solution

Study Tip

Wondering where the concentric circles in Figure 6.22 came from and why we've shown them? The circles are drawn to help plot each point at the appropriate distance from the pole.

a. To plot the point $(r, \theta) = (2, 135°)$, begin with the $135°$ angle. Because $135°$ i
a positive angle, draw $\theta = 135°$ counterclockwise from the polar axis. Now
consider $r = 2$. Because $r > 0$, plot the point by going out two units on th
terminal side of θ. Figure 6.22(a) shows the point.

b. To plot the point $(r, \theta) = \left(-3, \dfrac{3\pi}{2}\right)$, begin with the $\dfrac{3\pi}{2}$ angle. Because $\dfrac{3\pi}{2}$ is
positive angle, we draw $\theta = \dfrac{3\pi}{2}$ counterclockwise from the polar axis. Now
consider $r = -3$. Because $r < 0$, plot the point by going out three units alon
the ray *opposite* the terminal side of θ. Figure 6.22(b) shows the point.

c. To plot the point $(r, \theta) = \left(-1, -\dfrac{\pi}{4}\right)$, begin with the $-\dfrac{\pi}{4}$ angle. Because $-\dfrac{\pi}{4}$
is a negative angle, draw $\theta = -\dfrac{\pi}{4}$ clockwise from the polar axis. Now conside
$r = -1$. Because $r < 0$, plot the point by going out one unit along the ra
opposite the terminal side of θ. Figure 6.22(c) shows the point.

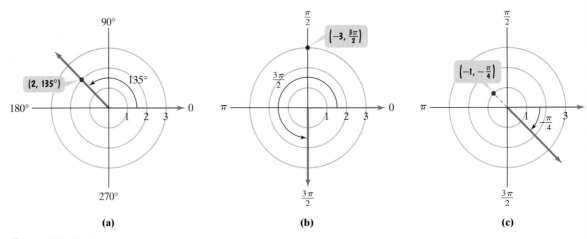

(a)   (b)   (c)

Figure 6.22 Plotting points

Check Point 1 Plot the points with the following polar coordinates:

a. $(3, 315°)$ **b.** $(-2, \pi)$ **c.** $\left(-1, -\dfrac{\pi}{2}\right)$.

② Find multiple sets of polar coordinates for a given point.

Discovery

Illustrate the statements in the voice balloons by plotting the points with the following polar coordinates:

a. $\left(1, \dfrac{\pi}{2}\right)$ and $\left(1, \dfrac{5\pi}{2}\right)$

b. $\left(3, \dfrac{\pi}{4}\right)$ and $\left(-3, \dfrac{5\pi}{4}\right)$.

Multiple Representations of Points in the Polar Coordinate System

In rectangular coordinates, each point (x, y) has exactly one representation. By contrast, any point in polar coordinates can be represented in infinitely many ways. For example,

$$(r, \theta) = (r, \theta + 2\pi) \quad \text{and} \quad (r, \theta) = (-r, \theta + \pi).$$

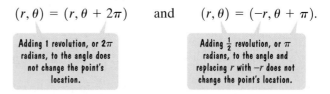

Adding 1 revolution, or 2π radians, to the angle does not change the point's location.

Adding $\frac{1}{2}$ revolution, or π radians, to the angle and replacing r with $-r$ does not change the point's location.

Thus, to find two other representations for the point (r, θ),

- Add 2π to the angle and do not change r.
- Add π to the angle and replace r with $-r$.

Continually adding or subtracting 2π in either of these representations does not change the point's location.

> **Multiple Representations of Points**
>
> If n is any integer, the point (r, θ) can be represented as
> $$(r, \theta) = (r, \theta + 2n\pi) \quad \text{or} \quad (r, \theta) = (-r, \theta + \pi + 2n\pi).$$

EXAMPLE 2 Finding Other Polar Coordinates for a Given Point

The point $\left(2, \dfrac{\pi}{3}\right)$ is plotted in Figure 6.23. Find another representation of this point in which

a. r is positive and $2\pi < \theta < 4\pi$.

b. r is negative and $0 < \theta < 2\pi$.

c. r is positive and $-2\pi < \theta < 0$.

Solution

a. We want $r > 0$ and $2\pi < \theta < 4\pi$. Using $\left(2, \dfrac{\pi}{3}\right)$, add 2π to the angle and do not change r.

$$\left(2, \frac{\pi}{3}\right) = \left(2, \frac{\pi}{3} + 2\pi\right) = \left(2, \frac{\pi}{3} + \frac{6\pi}{3}\right) = \left(2, \frac{7\pi}{3}\right)$$

b. We want $r < 0$ and $0 < \theta < 2\pi$. Using $\left(2, \dfrac{\pi}{3}\right)$, add π to the angle and replace r with $-r$.

$$\left(2, \frac{\pi}{3}\right) = \left(-2, \frac{\pi}{3} + \pi\right) = \left(-2, \frac{\pi}{3} + \frac{3\pi}{3}\right) = \left(-2, \frac{4\pi}{3}\right)$$

c. We want $r > 0$ and $-2\pi < \theta < 0$. Using $\left(2, \dfrac{\pi}{3}\right)$, subtract 2π from the angle and do not change r.

$$\left(2, \frac{\pi}{3}\right) = \left(2, \frac{\pi}{3} - 2\pi\right) = \left(2, \frac{\pi}{3} - \frac{6\pi}{3}\right) = \left(2, -\frac{5\pi}{3}\right)$$

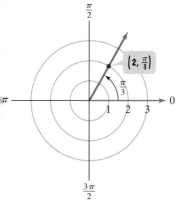

Figure 6.23 Finding other representations of a given point

Check Point 2 Find another representation of $\left(5, \dfrac{\pi}{4}\right)$ in which

a. r is positive and $2\pi < \theta < 4\pi$.

b. r is negative and $0 < \theta < 2\pi$.

c. r is positive and $-2\pi < \theta < 0$.

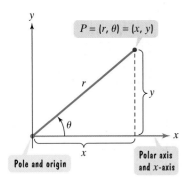

Figure 6.24 Polar and rectangular coordinate systems

Relations between Polar and Rectangular Coordinates

We now consider both polar and rectangular coordinates simultaneously. Figure 6.24 shows the two coordinate systems. The polar axis coincides with the positive x-axis and the pole coincides with the origin. A point P, other than the origin, has rectangular coordinates (x, y) and polar coordinates (r, θ), as indicated in the figure. We wish to find equations relating the two sets of coordinates. From the figure, we see that

$$x^2 + y^2 = r^2$$

$$\sin \theta = \frac{y}{r} \qquad \cos \theta = \frac{x}{r} \qquad \tan \theta = \frac{y}{x}.$$

These relationships hold when P is in any quadrant and when $r > 0$ or $r < 0$.

Relations between Polar and Rectangular Coordinates

$$x = r \cos \theta$$
$$y = r \sin \theta$$
$$x^2 + y^2 = r^2$$
$$\tan \theta = \frac{y}{x}$$

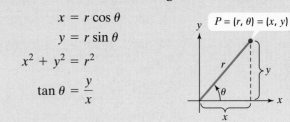

③ Convert a point from polar to rectangular coordinates.

Point Conversion from Polar to Rectangular Coordinates

To convert a point from polar coordinates (r, θ) to rectangular coordinates (x, y) use the formulas $x = r \cos \theta$ and $y = r \sin \theta$.

EXAMPLE 3 Polar-to-Rectangular Point Conversion

Find the rectangular coordinates of the points with the following polar coordinates:

a. $\left(2, \frac{3\pi}{2}\right)$ **b.** $\left(-8, \frac{\pi}{3}\right)$.

Solution We find (x, y) by substituting the given values for r and θ into $x = r \cos \theta$ and $y = r \sin \theta$.

a. We begin with the rectangular coordinates of the point $(r, \theta) = \left(2, \frac{3\pi}{2}\right)$.

$$x = r \cos \theta = 2 \cos \frac{3\pi}{2} = 2 \cdot 0 = 0$$

$$y = r \sin \theta = 2 \sin \frac{3\pi}{2} = 2(-1) = -2$$

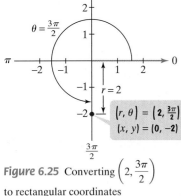

Figure 6.25 Converting $\left(2, \frac{3\pi}{2}\right)$ to rectangular coordinates

The rectangular coordinates of $\left(2, \frac{3\pi}{2}\right)$ are $(0, -2)$. See Figure 6.25.

b. We now find the rectangular coordinates of the point $(r, \theta) = \left(-8, \frac{\pi}{3}\right)$.

$$x = r \cos \theta = -8 \cos \frac{\pi}{3} = -8\left(\frac{1}{2}\right) = -4$$

$$y = r \sin \theta = -8 \sin \frac{\pi}{3} = -8\left(\frac{\sqrt{3}}{2}\right) = -4\sqrt{3}$$

The rectangular coordinates of $\left(-8, \frac{\pi}{3}\right)$ are $\left(-4, -4\sqrt{3}\right)$.

Technology

Some graphing utilities can convert a point from polar coordinates to rectangular coordinates. Consult your manual. The screen on the right verifies the polar-rectangular conversion in Example 3(a). It shows that the rectangular coordinates of

$(r, \theta) = \left(2, \dfrac{3\pi}{2}\right)$ are $(0, -2)$. Notice that the x- and y-coordinates are displayed separately.

```
P►Rx(2,3π/2)
                    0
P►Ry(2,3π/2)
                   -2
```

Check Point 3   Find the rectangular coordinates of the points with the following polar coordinates:

    a. $(3, \pi)$ **b.** $\left(-10, \dfrac{\pi}{6}\right)$.

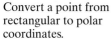
Convert a point from rectangular to polar coordinates.

Point Conversion from Rectangular to Polar Coordinates

Conversion from rectangular coordinates (x, y) to polar coordinates (r, θ) is a bit more complicated. Keep in mind that there are infinitely many representations for a point in polar coordinates. If the point (x, y) lies in one of the four quadrants, we will use a representation in which

- r is positive, and
- θ is the smallest positive angle with the terminal side passing through (x, y).

These conventions provide the following procedure:

> **Converting a Point from Rectangular to Polar Coordinates**
> **($r > 0$ and $0 \le \theta < 2\pi$)**
>
> **1.** Plot the point (x, y).
> **2.** Find r by computing the distance from the origin to (x, y): $r = \sqrt{x^2 + y^2}$.
> **3.** Find θ using $\tan \theta = \dfrac{y}{x}$ with the terminal side θ passing through (x, y).

EXAMPLE 4   Rectangular-to-Polar Point Conversion

Find polar coordinates of the point whose rectangular coordinates are $\left(-1, \sqrt{3}\right)$.

Solution   We begin with $(x, y) = \left(-1, \sqrt{3}\right)$ and use our three-step procedure to find a set of polar coordinates (r, θ).

Step 1   Plot the point (x, y). The point $\left(-1, \sqrt{3}\right)$ is plotted in quadrant II in Figure 6.26.

Step 2   Find r by computing the distance from the origin to (x, y).
$$r = \sqrt{x^2 + y^2} = \sqrt{(-1)^2 + \left(\sqrt{3}\right)^2} = \sqrt{1 + 3} = \sqrt{4} = 2$$

Step 3   Find θ using $\tan \theta = \dfrac{y}{x}$ with the terminal side of θ passing through (x, y).

$$\tan \theta = \frac{y}{x} = \frac{\sqrt{3}}{-1} = -\sqrt{3}$$

We know that $\tan \dfrac{\pi}{3} = \sqrt{3}$. Because θ lies in quadrant II,

$$\theta = \pi - \frac{\pi}{3} = \frac{3\pi}{3} - \frac{\pi}{3} = \frac{2\pi}{3}.$$

One representation of $\left(-1, \sqrt{3}\right)$ in polar coordinates is $(r, \theta) = \left(2, \dfrac{2\pi}{3}\right)$.

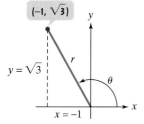

Figure 6.26 Converting $\left(-1, \sqrt{3}\right)$ to polar coordinates

Technology

The screen shows the rectangular-polar conversion for $(-1, \sqrt{3})$ on a graphing utility. In Example 4, we showed that $(x, y) = (-1, \sqrt{3})$ can be represented in polar coordinates as $(r, \theta) = \left(2, \dfrac{2\pi}{3}\right)$.

Using $\dfrac{2\pi}{3} \approx 2.09439510239$ verifies that our conversion is correct. Notice that the r- and (approximate) θ-coordinates are displayed separately.

```
R▸Pr(-1,√(3))
                    2
R▸Pθ(-1,√(3))
          2.094395102
```

Check Point 4 Find polar coordinates of the point whose rectangular coordinates ar $\left(1, -\sqrt{3}\right)$.

If a point (x, y) lies on a positive or negative axis, we use a representation i which

- r is positive, and
- θ is the smallest quadrantal angle that lies on the same positive or negativ axis as (x, y).

In these cases, you can find r and θ by plotting (x, y) and inspecting the figure. Let see how this is done.

EXAMPLE 5   Rectangular-to-Polar Point Conversion

Find polar coordinates of the point whose rectangular coordinates are $(-2, 0)$.

Solution   We begin with $(x, y) = (-2, 0)$ and find a set of polar coordinate (r, θ).

Step 1   Plot the point (x, y). The point $(-2, 0)$ is plotted in Figure 6.27.

Step 2   Find r, the distance from the origin to (x, y). Can you tell by looking a Figure 6.27 that this distance is 2?
$$r = \sqrt{x^2 + y^2} = \sqrt{(-2)^2 + 0^2} = \sqrt{4} = 2$$

Step 3   Find θ with θ lying on the same positive or negative axis as (x, y). Th point $(-2, 0)$ is on the negative x-axis. Thus, θ lies on the negative x-axis and $\theta = \pi$ One representation of $(-2, 0)$ in polar coordinates is $(2, \pi)$.

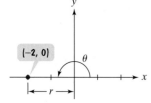

Figure 6.27 Converting $(-2, 0)$ to polar coordinates

Check Point 5 Find polar coordinates of the point whose rectangular coordinate are $(0, -4)$.

Convert an equation from rectangular to polar coordinates.

Equation Conversion from Rectangular to Polar Coordinates

A **polar equation** is an equation whose variables are r and θ. Two examples of pola equations are

$$r = \frac{5}{\cos \theta + \sin \theta} \quad \text{and} \quad r = 3 \csc \theta.$$

To convert a rectangular equation in x and y to a polar equation in r and θ, replace with $r \cos \theta$ and y with $r \sin \theta$.

EXAMPLE 6   Converting Equations from Rectangular to Polar Coordinates

Convert each rectangular equation to a polar equation that expresses r in terms of θ

a. $x + y = 5$ **b.** $(x - 1)^2 + y^2 = 1$.

Solution   Our goal is to obtain equations in which the variables are r and θ, rather than x and y. We use $x = r\cos\theta$ and $y = r\sin\theta$. We then solve the equations for r, obtaining equivalent equations that give r in terms of θ.

a.

$$x + y = 5$$

This is the given equation in rectangular coordinates. The graph is a line passing through $(5, 0)$ and $(0, 5)$.

$$r\cos\theta + r\sin\theta = 5$$

Replace x with $r\cos\theta$ and y with $r\sin\theta$.

$$r(\cos\theta + \sin\theta) = 5$$

Factor out r.

$$r = \frac{5}{\cos\theta + \sin\theta}$$

Divide both sides of the equation by $\cos\theta + \sin\theta$ and solve for r.

Thus, the polar equation for $x + y = 5$ is $r = \dfrac{5}{\cos\theta + \sin\theta}$.

b.

$$(x - 1)^2 + y^2 = 1$$

This is the given equation in rectangular coordinates. The graph is a circle with radius 1 and center at $(h, k) = (1, 0)$.

> The standard form of a circle's equation is $(x - h)^2 + (y - k)^2 = r^2$, with radius r and center at (h, k).

$$(r\cos\theta - 1)^2 + (r\sin\theta)^2 = 1$$

Replace x with $r\cos\theta$ and y with $r\sin\theta$.

$$r^2\cos^2\theta - 2r\cos\theta + 1 + r^2\sin^2\theta = 1$$

Use $(A - B)^2 = A^2 - 2AB + B^2$ to square $r\cos\theta - 1$.

$$r^2\cos^2\theta + r^2\sin^2\theta - 2r\cos\theta = 0$$

Subtract 1 from both sides and rearrange terms.

$$r^2 - 2r\cos\theta = 0$$

Simplify: $r^2\cos^2\theta + r^2\sin^2\theta = r^2(\cos^2\theta + \sin^2\theta) = r^2 \cdot 1 = r^2$.

$$r(r - 2\cos\theta) = 0$$

Factor out r.

$$r = 0 \quad \text{or} \quad r - 2\cos\theta = 0$$

Set each factor equal to 0.

$$r = 2\cos\theta$$

Solve for r.

The graph of $r = 0$ is a single point, the pole. Because the pole also satisfies the equation $r = 2\cos\theta$ (for $\theta = \frac{\pi}{2}, r = 0$), it is not necessary to include the equation $r = 0$. Thus, the polar equation for $(x - 1)^2 + y^2 = 1$ is $r = 2\cos\theta$.

Check Point 6   Convert each rectangular equation to a polar equation that expresses r in terms of θ:

a. $3x - y = 6$ **b.** $x^2 + (y + 1)^2 = 1$.

⑥ Convert an equation from polar to rectangular coordinates.

Equation Conversion from Polar to Rectangular Coordinates

When we convert an equation from polar to rectangular coordinates, our goal is to obtain an equation in which the variables are x and y, rather than r and θ. We use one or more of the following equations:

$$r^2 = x^2 + y^2 \qquad r\cos\theta = x \qquad r\sin\theta = y \qquad \tan\theta = \frac{y}{x}.$$

To use these equations, it is sometimes necessary to do something to the given polar equation. This could include squaring both sides, using an identity, taking the tangent of both sides, or multiplying both sides by r.

EXAMPLE 7 Converting Equations from Polar to Rectangular Form

Convert each polar equation to a rectangular equation in x and y:

a. $r = 5$ **b.** $\theta = \dfrac{\pi}{4}$ **c.** $r = 3 \csc \theta$ **d.** $r = -6 \cos \theta$.

Solution In each case, let's express the rectangular equation in a form tha~~t~~ enables us to recognize its graph.

a. We use $r^2 = x^2 + y^2$ to convert the polar equation $r = 5$ to a rectangula~~r~~ equation.

$r = 5$	This is the given polar equation.
$r^2 = 25$	Square both sides.
$x^2 + y^2 = 25$	Use $r^2 = x^2 + y^2$ on the left side.

The rectangular equation for $r = 5$ is $x^2 + y^2 = 25$. The graph is a circle wit~~h~~ center at $(0, 0)$ and radius 5.

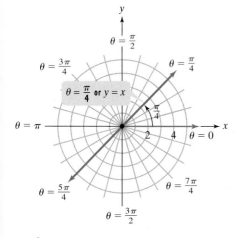

b. We use $\tan \theta = \dfrac{y}{x}$ to convert the polar equation $\theta = \dfrac{\pi}{4}$ to a rectangula~~r~~ equation in x and y.

$\theta = \dfrac{\pi}{4}$	This is the given polar equation.
$\tan \theta = \tan \dfrac{\pi}{4}$	Take the tangent of both sides.
$\tan \theta = 1$	$\tan \dfrac{\pi}{4} = 1$
$\dfrac{y}{x} = 1$	Use $\tan \theta = \dfrac{y}{x}$ on the left side.
$y = x$	Multiply both sides by x.

The rectangular equation for $\theta = \dfrac{\pi}{4}$ is $y = x$. The graph is a line that bisect~~s~~ quadrants I and III. Figure 6.28 shows the line drawn in a polar coordinate system.

Figure 6.28

c. We use $r \sin \theta = y$ to convert the polar equation $r = 3 \csc \theta$ to a rectangula~~r~~ equation. To do this, we express the cosecant in terms of the sine.

$r = 3 \csc \theta$	This is the given polar equation.
$r = \dfrac{3}{\sin \theta}$	$\csc \theta = \dfrac{1}{\sin \theta}$
$r \sin \theta = 3$	Multiply both sides by sin θ.
$y = 3$	Use $r \sin \theta = y$ on the left side.

The rectangular equation for $r = 3 \csc \theta$ is $y = 3$. The graph is a horizonta~~l~~ line 3 units above the x-axis. Figure 6.29 shows the line drawn in a polar coor~~-~~ dinate system.

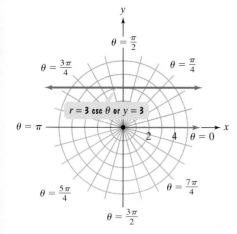

Figure 6.29

d. To convert $r = -6 \cos \theta$ to rectangular coordinates, we multiply both sides b~~y~~ r. Then we use $r^2 = x^2 + y^2$ on the left side and $r \cos \theta = x$ on the right side.

$r = -6 \cos \theta$	This is the given polar equation.
$r^2 = -6r \cos \theta$	Multiply both sides by r.
$x^2 + y^2 = -6x$	Convert to rectangular coordinates: $r^2 = x^2 + y^2$ and $r \cos \theta = x$.
$x^2 + 6x + y^2 = 0$	Add 6x to both sides.
$x^2 + 6x + 9 + y^2 = 9$	Complete the square on x: $\frac{1}{2} \cdot 6 = 3$ and $3^2 = 9$.
$(x + 3)^2 + y^2 = 9$	Factor.

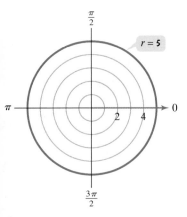

The rectangular equation for $r = -6 \cos \theta$ is $(x + 3)^2 + y^2 = 9$.
This last equation is the standard form of the equation of a circle, $(x - h)^2 + (y - k)^2 = r^2$, with radius r and center at (h, k). Thus, the graph of $(x + 3)^2 + y^2 = 9$ is a circle with center at $(-3, 0)$ and radius 3.

Converting a polar equation to a rectangular equation may be a useful way to develop or check a graph. For example, the graph of the polar equation $r = 5$ consists of all points that are five units from the pole. Thus, the graph is a circle centered at the pole with radius 5. The rectangular equation for $r = 5$, namely $x^2 + y^2 = 25$, has precisely the same graph (see Figure 6.30). We will discuss graphs of polar equations in the next section.

Figure 6.30 The equations $r = 5$ and $x^2 + y^2 = 25$ have the same graph.

Check Point 7 Convert each polar equation to a rectangular equation in x and y:

a. $r = 4$ **b.** $\theta = \dfrac{3\pi}{4}$ **c.** $r = -2 \sec \theta$ **d.** $r = 10 \sin \theta$.

EXERCISE SET 6.3

Practice Exercises

In Exercises 1–10, indicate if the point with the given polar coordinates is represented by A, B, C, or D on the graph.

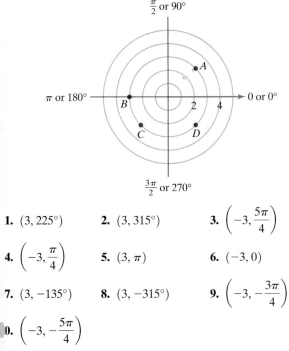

1. $(3, 225°)$ **2.** $(3, 315°)$ **3.** $\left(-3, \dfrac{5\pi}{4}\right)$

4. $\left(-3, \dfrac{\pi}{4}\right)$ **5.** $(3, \pi)$ **6.** $(-3, 0)$

7. $(3, -135°)$ **8.** $(3, -315°)$ **9.** $\left(-3, -\dfrac{3\pi}{4}\right)$

10. $\left(-3, -\dfrac{5\pi}{4}\right)$

In Exercises 11–20, use a polar coordinate system like the one shown for Exercises 1–10 to plot each point with the given polar coordinates.

11. $(2, 45°)$ **12.** $(1, 45°)$ **13.** $(3, 90°)$

14. $(2, 270°)$ **15.** $\left(3, \dfrac{4\pi}{3}\right)$ **16.** $\left(3, \dfrac{7\pi}{6}\right)$

17. $(-1, \pi)$ **18.** $\left(-1, \dfrac{3\pi}{2}\right)$ **19.** $\left(-2, -\dfrac{\pi}{2}\right)$

20. $(-3, -\pi)$

In Exercises 21–26, use a polar coordinate system like the one shown for Exercises 1–10 to plot each point with the given polar coordinates. Then find another representation (r, θ) of this point in which

a. $r > 0$, $2\pi < \theta < 4\pi$.

b. $r < 0$, $0 < \theta < 2\pi$.

c. $r > 0$, $-2\pi < \theta < 0$.

21. $\left(5, \dfrac{\pi}{6}\right)$ **22.** $\left(8, \dfrac{\pi}{6}\right)$ **23.** $\left(10, \dfrac{3\pi}{4}\right)$

24. $\left(12, \dfrac{2\pi}{3}\right)$ **25.** $\left(4, \dfrac{\pi}{2}\right)$ **26.** $\left(6, \dfrac{\pi}{2}\right)$

In Exercises 27–32, select the representations that do not change the location of the given point.

27. $(7, 140°)$

 a. $(-7, 320°)$ **b.** $(-7, -40°)$
 c. $(-7, 220°)$ **d.** $(7, -220°)$

28. $(4, 120°)$

 a. $(-4, 300°)$ **b.** $(-4, -240°)$
 c. $(4, -240°)$ **d.** $(4, 480°)$

29. $\left(2, -\dfrac{3\pi}{4}\right)$

 a. $\left(2, -\dfrac{7\pi}{4}\right)$ **b.** $\left(2, \dfrac{5\pi}{4}\right)$

 c. $\left(-2, -\dfrac{\pi}{4}\right)$ **d.** $\left(-2, -\dfrac{7\pi}{4}\right)$

30. $\left(-2, \dfrac{7\pi}{6}\right)$

 a. $\left(-2, -\dfrac{5\pi}{6}\right)$ **b.** $\left(-2, -\dfrac{\pi}{6}\right)$

 c. $\left(2, -\dfrac{\pi}{6}\right)$ **d.** $\left(2, \dfrac{\pi}{6}\right)$

31. $\left(-5, -\dfrac{\pi}{4}\right)$

   a. $\left(-5, \dfrac{7\pi}{4}\right)$ **b.** $\left(5, -\dfrac{5\pi}{4}\right)$

   c. $\left(-5, \dfrac{11\pi}{4}\right)$ **d.** $\left(5, \dfrac{\pi}{4}\right)$

32. $(-6, 3\pi)$

   a. $(6, 2\pi)$ **b.** $(6, -\pi)$

   c. $(-6, \pi)$ **d.** $(-6, -2\pi)$

In Exercises 33–40, polar coordinates of a point are given. Find the rectangular coordinates of each point.

33. $(4, 90°)$ **34.** $(6, 180°)$ **35.** $\left(2, \dfrac{\pi}{3}\right)$

36. $\left(2, \dfrac{\pi}{6}\right)$ **37.** $\left(-4, \dfrac{\pi}{2}\right)$ **38.** $\left(-6, \dfrac{3\pi}{2}\right)$

39. $(7.4, 2.5)$ **40.** $(8.3, 4.6)$

In Exercises 41–48, the rectangular coordinates of a point are given. Find polar coordinates of each point.

41. $(-2, 2)$ **42.** $(2, -2)$

43. $\left(2, -2\sqrt{3}\right)$ **44.** $\left(-2\sqrt{3}, 2\right)$

45. $\left(-\sqrt{3}, -1\right)$ **46.** $\left(-1, -\sqrt{3}\right)$

47. $(5, 0)$ **48.** $(0, -6)$

In Exercises 49–58, convert each rectangular equation to a polar equation that expresses r in terms of θ.

49. $3x + y = 7$ **50.** $x + 5y = 8$

51. $x = 7$ **52.** $y = 3$

53. $x^2 + y^2 = 9$ **54.** $x^2 + y^2 = 16$

55. $(x - 2)^2 + y^2 = 4$ **56.** $x^2 + (y + 3)^2 = 9$

57. $y^2 = 6x$ **58.** $x^2 = 6y$

In Exercises 59–74, convert each polar equation to a rectangular equation. Then use a rectangular coordinate system to graph the rectangular equation.

59. $r = 8$ **60.** $r = 10$

61. $\theta = \dfrac{\pi}{2}$ **62.** $\theta = \dfrac{\pi}{3}$

63. $r \sin\theta = 3$ **64.** $r \cos\theta = 7$

65. $r = 4 \csc\theta$ **66.** $r = 6 \sec\theta$

67. $r = \sin\theta$ **68.** $r = \cos\theta$

69. $r = 12 \cos\theta$ **70.** $r = -4 \sin\theta$

71. $r = 6 \cos\theta + 4 \sin\theta$ **72.** $r = 8 \cos\theta + 2 \sin\theta$

73. $r^2 \sin 2\theta = 2$ **74.** $r^2 \sin 2\theta = 4$

Practice Plus

In Exercises 75–78, show that each statement is true by converting the given polar equation to a rectangular equation.

75. Show that the graph of $r = a \sec\theta$ is a vertical line a units to the right of the y-axis if $a > 0$ and $|a|$ units to the left of the y-axis if $a < 0$.

76. Show that the graph of $r = a \csc\theta$ is a horizontal line a units above the x-axis if $a > 0$ and $|a|$ units below the x-axis if $a < 0$.

77. Show that the graph of $r = a \sin\theta$ is a circle with center at $\left(0, \dfrac{a}{2}\right)$ and radius $\dfrac{a}{2}$.

78. Show that the graph of $r = a \cos\theta$ is a circle with center at $\left(\dfrac{a}{2}, 0\right)$ and radius $\dfrac{a}{2}$.

In Exercises 79–80, convert each polar equation to a rectangular equation. Then determine the graph's slope and y-intercept.

79. $r \sin\left(\theta - \dfrac{\pi}{4}\right) = 2$ **80.** $r \cos\left(\theta + \dfrac{\pi}{6}\right) = 8$

In Exercises 81–82, find the rectangular coordinates of each pair of points. Then find the distance, in simplified radical form, between the points.

81. $\left(2, \dfrac{2\pi}{3}\right)$ and $\left(4, \dfrac{\pi}{6}\right)$ **82.** $(6, \pi)$ and $\left(5, \dfrac{7\pi}{4}\right)$

Application Exercises

Use the figure of the merry-go-round to solve Exercises 83–84. There are four circles of horses. Each circle is three feet from the next circle. The radius of the inner circle is 6 feet.

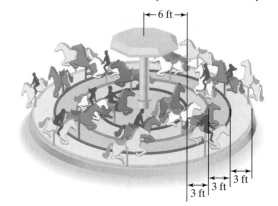

83. If a horse in the outer circle is $\frac{2}{3}$ of the way around the merry-go-round, give its polar coordinates.

84. If a horse in the inner circle is $\frac{5}{6}$ of the way around the merry-go-round, give its polar coordinates.

The wind is blowing at 10 knots. Sailboat racers look for a sailing angle to the 10-knot wind that produces maximum sailing speed. In this application, (r, θ) describes the sailing speed, r, in knots, at an angle θ to the 10-knot wind. Use this information to solve Exercises 85–87.

85. Interpret the polar coordinates: $(6.3, 50°)$.

86. Interpret the polar coordinates: $(7.4, 85°)$.

87. Four points in this 10-knot-wind situation are $(6.3, 50°)$, $(7.4, 85°)$, $(7.5, 105°)$, and $(7.3, 135°)$. Based on these points, which sailing angle to the 10-knot wind would you recommend to a serious sailboat racer? What sailing speed is achieved at this angle?

Writing in Mathematics

88. Explain how to plot (r, θ) if $r > 0$ and $\theta > 0$.

89. Explain how to plot (r, θ) if $r < 0$ and $\theta > 0$.

90. If you are given polar coordinates of a point, explain how to find two additional sets of polar coordinates for the point.

91. Explain how to convert a point from polar to rectangular coordinates. Provide an example with your explanation.

92. Explain how to convert a point from rectangular to polar coordinates. Provide an example with your explanation.

3. Explain how to convert from a rectangular equation to a polar equation.

4. In converting $r = 5$ from a polar equation to a rectangular equation, describe what should be done to both sides of the equation and why this should be done.

5. In converting $r = \sin \theta$ from a polar equation to a rectangular equation, describe what should be done to both sides of the equation and why this should be done.

6. Suppose that (r, θ) describes the sailing speed, r, in knots, at an angle θ to a wind blowing at 20 knots. You have a list of all ordered pairs (r, θ) for integral angles from $\theta = 0°$ to $\theta = 180°$. Describe a way to present this information so that a serious sailboat racer can visualize sailing speeds at different sailing angles to the wind.

Technology Exercises

In Exercises 97–99, polar coordinates of a point are given. Use a graphing utility to find the rectangular coordinates of each point to three decimal places.

97. $\left(4, \dfrac{2\pi}{3}\right)$ **98.** $(5.2, 1.7)$ **99.** $(-4, 1.088)$

In Exercises 100–102, the rectangular coordinates of a point are given. Use a graphing utility to find polar coordinates of each point to three decimal places.

100. $(-5, 2)$ **101.** $\left(\sqrt{5}, 2\right)$

102. $(-4.308, -7.529)$

Critical Thinking Exercises

103. Prove that the distance, d, between two points with polar coordinates (r_1, θ_1) and (r_2, θ_2) is

$$d = \sqrt{r_1^2 + r_2^2 - 2r_1 r_2 \cos(\theta_2 - \theta_1)}.$$

104. Use the formula in Exercise 103 to find the distance between $\left(2, \dfrac{5\pi}{6}\right)$ and $\left(4, \dfrac{\pi}{6}\right)$. Express the answer in simplified radical form.

SECTION 6.4   *Graphs of Polar Equations*

Objectives

1 Use point plotting to graph polar equations.

2 Use symmetry to graph polar equations.

The America's Cup is the supreme event in ocean sailing. Competition is fierce and the costs are huge. Competitors look to mathematics to provide the critical innovation that can make the difference between winning and losing. In this section's exercise set, you will see how graphs of polar equations play a role in sailing faster using mathematics.

Using Polar Grids to Graph Polar Equations

Recall that a **polar equation** is an equation whose variables are r and θ. The **graph of a polar equation** is the set of all points whose polar coordinates satisfy the equation. We use **polar grids** like the one shown in Figure 6.31 to graph polar equations. The grid consists of circles with centers at the pole. This polar grid shows five such circles. A polar grid also shows lines passing through the pole. In this grid, each line represents an angle for which we know the exact values of the trigonometric functions.

Many polar coordinate grids show more circles and more lines through the pole than in Figure 6.31. See if your campus bookstore has paper with polar grids and use the polar graph paper throughout this section.

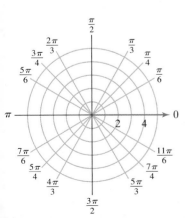

Figure 6.31 A polar coordinate grid

Use point plotting to graph polar equations.

Graphing a Polar Equation by Point Plotting

One method for graphing a polar equation such as $r = 4 \cos \theta$ is the **point-plotting method**. First, we make a table of values that satisfy the equation. Next, we plot these ordered pairs as points in the polar coordinate system. Finally, we connect the points with a smooth curve. This often gives us a picture of all ordered pairs (r, θ) that satisfy the equation.

EXAMPLE 1   Graphing an Equation Using the Point-Plotting Method

Graph the polar equation $r = 4 \cos \theta$ with θ in radians.

Solution   We construct a partial table of coordinates using multiples of $\frac{\pi}{6}$. Then we plot the points and join them with a smooth curve, as shown in Figure 6.32.

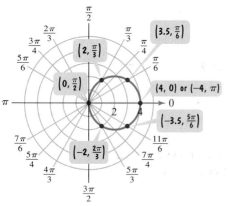

Figure 6.32 The graph of $r = 4 \cos \theta$

θ	$r = 4 \cos \theta$	(r, θ)
0	$4 \cos 0 = 4 \cdot 1 = 4$	$(4, 0)$
$\frac{\pi}{6}$	$4 \cos \frac{\pi}{6} = 4 \cdot \frac{\sqrt{3}}{2} = 2\sqrt{3} \approx 3.5$	$\left(3.5, \frac{\pi}{6}\right)$
$\frac{\pi}{3}$	$4 \cos \frac{\pi}{3} = 4 \cdot \frac{1}{2} = 2$	$\left(2, \frac{\pi}{3}\right)$
$\frac{\pi}{2}$	$4 \cos \frac{\pi}{2} = 4 \cdot 0 = 0$	$\left(0, \frac{\pi}{2}\right)$
$\frac{2\pi}{3}$	$4 \cos \frac{2\pi}{3} = 4\left(-\frac{1}{2}\right) = -2$	$\left(-2, \frac{2\pi}{3}\right)$
$\frac{5\pi}{6}$	$4 \cos \frac{5\pi}{6} = 4\left(-\frac{\sqrt{3}}{2}\right) = -2\sqrt{3} \approx -3.5$	$\left(-3.5, \frac{5\pi}{6}\right)$
π	$4 \cos \pi = 4(-1) = -4$	$(-4, \pi)$
Values of r repeat.		

Technology

A graphing utility can be used to obtain the graph of a polar equation. Use the polar mode with angle measure in radians. You must enter the minimum and maximum values for θ and an increment setting for θ, called θ step. θ step determines the number of points that the graphing utility will plot. Make θ step relatively small so that a significant number of points are plotted.

Shown is the graph of $r = 4 \cos \theta$ in a $[-7.5, 7.5, 1]$ by $[-5, 5, 1]$ viewing rectangle with

$$\theta \text{ min} = 0$$
$$\theta \text{ max} = 2\pi$$
$$\theta \text{ step} = \frac{\pi}{48}.$$

A square setting was used.

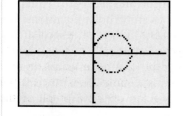

The graph of $r = 4 \cos \theta$ in Figure 6.32 looks like a circle of radius 2 whose center is at the point $(x, y) = (2, 0)$. We can verify this observation by changing the polar equation to a rectangular equation.

$r = 4 \cos \theta$	This is the given polar equation.
$r^2 = 4r \cos \theta$	Multiply both sides by r.
$x^2 + y^2 = 4x$	Convert to rectangular coordinates: $r^2 = x^2 + y^2$ and $r \cos \theta = x$.
$x^2 - 4x + y^2 = 0$	Subtract 4x from both sides.
$x^2 - 4x + 4 + y^2 = 4$	Complete the square on x: $\frac{1}{2}(-4) = -2$ and $(-2)^2 = 4$. Add 4 to both sides.
$(x - 2)^2 + y^2 = 2^2$	Factor.

This last equation is the standard form of the equation of a circle $(x - h)^2 + (y - k)^2 = r^2$, with radius r and center at (h, k). Thus, the radius is 2 and the center is at $(h, k) = (2, 0)$.

In general, circles have simpler equations in polar form than in rectangular form.

Circles in Polar Coordinates

The graphs of

$$r = a \cos \theta \quad \text{and} \quad r = a \sin \theta$$

are circles.

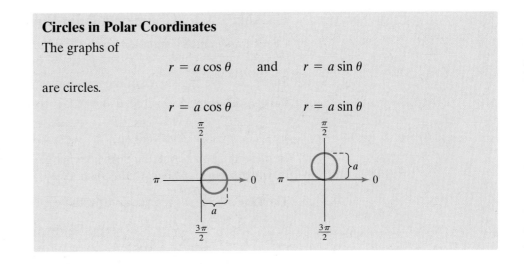

Check Point 1 Graph the equation $r = 4 \sin \theta$ with θ in radians. Use multiples of $\dfrac{\pi}{6}$ from 0 to π to generate coordinates for points (r, θ).

 Use symmetry to graph polar equations.

Graphing a Polar Equation Using Symmetry

If the graph of a polar equation exhibits symmetry, you may be able to graph it more quickly. Three types of symmetry can be helpful.

Tests for Symmetry in Polar Coordinates

Symmetry with Respect to the Polar Axis (x-Axis)

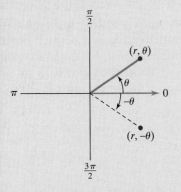

Replace θ with $-\theta$. If an equivalent equation results, the graph is symmetric with respect to the polar axis.

Symmetry with Respect to the Line $\theta = \dfrac{\pi}{2}$ (y-Axis)

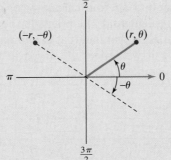

Replace (r, θ) with $(-r, -\theta)$. If an equivalent equation results, the graph is symmetric with respect to $\theta = \dfrac{\pi}{2}$.

Symmetry with Respect to the Pole (Origin)

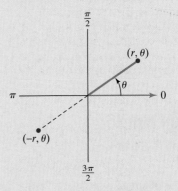

Replace r with $-r$. If an equivalent equation results, the graph is symmetric with respect to the pole.

If a polar equation passes a symmetry test, then its graph exhibits that symmetry. By contrast, if a polar equation fails a symmetry test, then its graph *may or may not* have that kind of symmetry. Thus, the graph of a polar equation may have a symmetry even if it fails a test for that particular symmetry. Nevertheless, the symmetry tests are useful. If we detect symmetry, we can obtain a graph of the equation by plotting fewer points.

EXAMPLE 2 Graphing a Polar Equation Using Symmetry

Check for symmetry and then graph the polar equation:
$$r = 1 - \cos\theta.$$

Solution We apply each of the tests for symmetry.

Polar Axis: Replace θ with $-\theta$ in $r = 1 - \cos\theta$:

$$r = 1 - \cos(-\theta) \qquad \text{Replace } \theta \text{ with } -\theta \text{ in } r = 1 - \cos\theta.$$
$$r = 1 - \cos\theta \qquad \text{The cosine function is even: } \cos(-\theta) = \cos\theta.$$

Because the polar equation does not change when θ is replaced with $-\theta$, the graph is symmetric with respect to the polar axis.

The Line $\theta = \dfrac{\pi}{2}$: Replace (r, θ) with $(-r, -\theta)$ in $r = 1 - \cos\theta$:

$$-r = 1 - \cos(-\theta) \qquad \text{Replace } r \text{ with } -r \text{ and } \theta \text{ with } -\theta \text{ in } r = 1 - \cos\theta.$$
$$-r = 1 - \cos\theta \qquad \cos(-\theta) = \cos\theta.$$
$$r = \cos\theta - 1 \qquad \text{Multiply both sides by } -1.$$

Because the polar equation $r = 1 - \cos\theta$ changes to $r = \cos\theta - 1$ when (r, θ) is replaced with $(-r, -\theta)$, the equation fails this symmetry test. The graph may or may not be symmetric with respect to the line $\theta = \dfrac{\pi}{2}$.

The Pole: Replace r with $-r$ in $r = 1 - \cos\theta$:

$$-r = 1 - \cos\theta \qquad \text{Replace } r \text{ with } -r \text{ in } r = 1 - \cos\theta.$$
$$r = \cos\theta - 1 \qquad \text{Multiply both sides by } -1.$$

Because the polar equation $r = 1 - \cos\theta$ changes to $r = \cos\theta - 1$ when r is replaced with $-r$, the equation fails this symmetry test. The graph may or may not be symmetric with respect to the pole.

Now we are ready to graph $r = 1 - \cos\theta$. Because the period of the cosine function is 2π, we need not consider values of θ beyond 2π. Recall that we discovered the graph of the equation $r = 1 - \cos\theta$ has symmetry with respect to the polar axis. Because the graph has this symmetry, we can obtain a complete graph by plotting fewer points. Let's start by finding the values of r for values of θ from 0 to π.

The values for r and θ are in the table above Figure 6.33(a). These values can be obtained using your calculator or possibly with the TABLE feature on some graphing calculators. The points in the table are plotted in Figure 6.33(a). Examine the graph. Keep in mind that the graph must be symmetric with respect to the polar axis. Thus, if we reflect the graph in Figure 6.33(a) about the polar axis, we will obtain a complete graph of $r = 1 - \cos\theta$. This graph is shown in Figure 6.33(b).

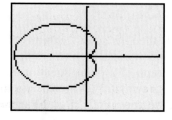
θ	0	$\dfrac{\pi}{6}$	$\dfrac{\pi}{3}$	$\dfrac{\pi}{2}$	$\dfrac{2\pi}{3}$	$\dfrac{5\pi}{6}$	π
r	0	0.13	0.5	1	1.5	1.87	2

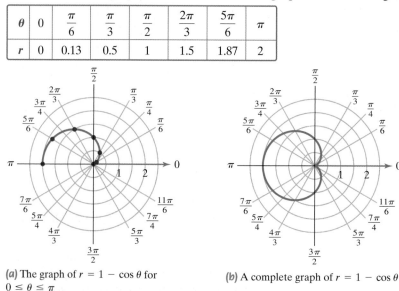

(a) The graph of $r = 1 - \cos\theta$ for $0 \le \theta \le \pi$

(b) A complete graph of $r = 1 - \cos\theta$

Figure 6.33 Graphing $r = 1 - \cos\theta$

Check Point **2** Check for symmetry and then graph the polar equation:

$$r = 1 + \cos \theta.$$

EXAMPLE 3   Graphing a Polar Equation

Graph the polar equation: $r = 1 + 2 \sin \theta$.

Solution   We first check for symmetry.

$$r = 1 + 2 \sin \theta$$

Polar Axis	**The Line $\theta = \dfrac{\pi}{2}$**	**The Pole**
Replace θ with $-\theta$.	Replace (r, θ) with $(-r, -\theta)$.	Replace r with $-r$.
$r = 1 + 2 \sin (-\theta)$	$-r = 1 + 2 \sin (-\theta)$	$-r = 1 + 2 \sin \theta$
$r = 1 + 2 (-\sin \theta)$	$-r = 1 - 2 \sin \theta$	$r = -1 - 2 \sin \theta$
$r = 1 - 2 \sin \theta$	$-r = -1 + 2 \sin \theta$	

None of these equations are equivalent to $r = 1 + 2 \sin \theta$. Thus, the graph may or may not have each of these kinds of symmetry.

Now we are ready to graph $r = 1 + 2 \sin \theta$. Because the period of the sine function is 2π, we need not consider values of θ beyond 2π. We identify points on the graph of $r = 1 + 2 \sin \theta$ by assigning values to θ and calculating the corresponding values of r. The values for r and θ are in the tables above Figure 6.34(a), Figure 6.34(b), and Figure 6.34(c). The complete graph of $r = 1 + 2 \sin \theta$ is shown in Figure 6.34(c). The inner loop indicates that the graph passes through the pole twice.

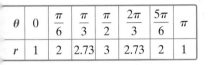

θ	0	$\dfrac{\pi}{6}$	$\dfrac{\pi}{3}$	$\dfrac{\pi}{2}$	$\dfrac{2\pi}{3}$	$\dfrac{5\pi}{6}$	π
r	1	2	2.73	3	2.73	2	1

θ	$\dfrac{7\pi}{6}$	$\dfrac{4\pi}{3}$	$\dfrac{3\pi}{2}$
r	0	-0.73	-1

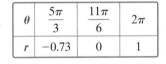

θ	$\dfrac{5\pi}{3}$	$\dfrac{11\pi}{6}$	2π
r	-0.73	0	1

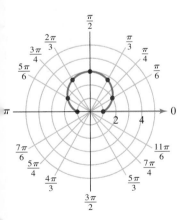

(a) The graph of $r = 1 + 2 \sin \theta$ for $0 \le \theta \le \pi$

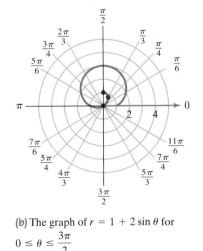

(b) The graph of $r = 1 + 2 \sin \theta$ for $0 \le \theta \le \dfrac{3\pi}{2}$

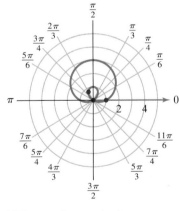

(c) The complete graph of $r = 1 + 2 \sin \theta$ for $0 \le \theta \le 2\pi$

Figure 6.34 Graphing $r = 1 + 2 \sin \theta$

Although the polar equation $r = 1 + 2 \sin \theta$ failed the test for symmetry with respect to the line $\theta = \frac{\pi}{2}$ (the y-axis), its graph in Figure 6.34(c) reveals this kind of symmetry.

We're not quite sure if the polar graph in Figure 6.34(c) looks like a snail. However, the graph is called a *limaçon*, which is a French word for snail. Limaçons come with and without inner loops.

Limaçons

The graphs of

$$r = a + b \sin \theta, \quad r = a - b \sin \theta,$$
$$r = a + b \cos \theta, \quad r = a - b \cos \theta, \quad a > 0, b > 0$$

are called **limaçons**. The ratio $\dfrac{a}{b}$ determines a limaçon's shape.

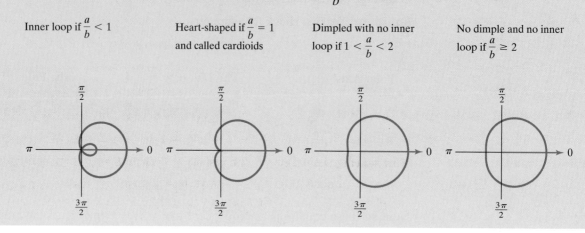

Inner loop if $\dfrac{a}{b} < 1$ Heart-shaped if $\dfrac{a}{b} = 1$ and called cardioids   Dimpled with no inner loop if $1 < \dfrac{a}{b} < 2$ No dimple and no inner loop if $\dfrac{a}{b} \geq 2$

> **Check Point 3** Graph the polar equation: $r = 1 - 2 \sin \theta$.

EXAMPLE 4   Graphing a Polar Equation

Graph the polar equation: $r = 4 \sin 2\theta$.

Solution We first check for symmetry.

$$r = 4 \sin 2\theta$$

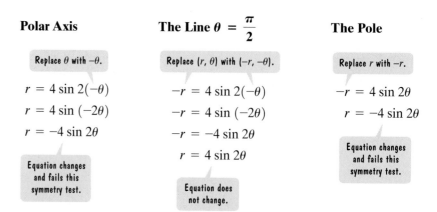

Polar Axis

Replace θ with $-\theta$.

$r = 4 \sin 2(-\theta)$

$r = 4 \sin (-2\theta)$

$r = -4 \sin 2\theta$

Equation changes and fails this symmetry test.

The Line $\theta = \dfrac{\pi}{2}$

Replace (r, θ) with $(-r, -\theta)$.

$-r = 4 \sin 2(-\theta)$

$-r = 4 \sin (-2\theta)$

$-r = -4 \sin 2\theta$

$r = 4 \sin 2\theta$

Equation does not change.

The Pole

Replace r with $-r$.

$-r = 4 \sin 2\theta$

$r = -4 \sin 2\theta$

Equation changes and fails this symmetry test.

Thus, we can be sure that the graph is symmetric with respect to $\theta = \dfrac{\pi}{2}$. The graph may or may not be symmetric with respect to the polar axis or the pole.

Now we are ready to graph $r = 4 \sin 2\theta$. In Figure 6.35, we plot points on the graph of $r = 4 \sin 2\theta$ using values of θ from 0 to π and the corresponding values of r. These coordinates are shown in the tables at the left of the graph.

θ	0	$\dfrac{\pi}{6}$	$\dfrac{\pi}{4}$	$\dfrac{\pi}{3}$	$\dfrac{\pi}{2}$
r	0	3.46	4	3.46	0

θ	$\dfrac{2\pi}{3}$	$\dfrac{3\pi}{4}$	$\dfrac{5\pi}{6}$	π
r	-3.46	-4	-3.46	0

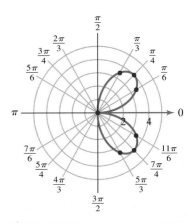

Figure 6.35 The graph of $r = 4\sin 2\theta$ for $0 \le \theta \le \pi$

Now we can use symmetry with respect to the line $\theta = \dfrac{\pi}{2}$ (the y-axis) to complete the graph. By reflecting the graph in Figure 6.35 about the y-axis, we obtain the complete graph of $r = 4\sin 2\theta$ from 0 to 2π. The graph is shown in Figure 6.36 .

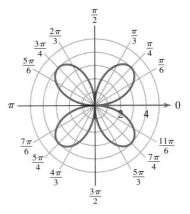

Figure 6.36 The graph of $r = 4\sin 2\theta$ for $0 \le \theta \le 2\pi$

Although the polar equation $r = 4\sin 2\theta$ failed the tests for symmetry with respect to the polar axis (the x-axis) and the pole (the origin), its graph in Figure 6.36 reveals all three types of symmetry.

The curve in Figure 6.36 is called a **rose with four petals**. We can use a trigonometric equation to confirm the four angles that give the location of the petal points. The petal points of $r = 4\sin 2\theta$ are located at values of θ for which $r = 4$ or $r = -4$.

$4\sin 2\theta = 4$	or $\qquad 4\sin 2\theta = -4$	Use $r = 4\sin 2\theta$ and set r equal to 4 or -4.
$\sin 2\theta = 1$	$\sin 2\theta = -1$	Divide both sides by 4.
$2\theta = \dfrac{\pi}{2} + 2n\pi$	$2\theta = \dfrac{3\pi}{2} + 2n\pi$	Solve for 2θ, where n is any integer.
$\theta = \dfrac{\pi}{4} + n\pi$	$\theta = \dfrac{3\pi}{4} + n\pi$	Divide both sides by 2 and solve for θ.

If $n = 0$, $\theta = \dfrac{\pi}{4}$.
If $n = 1$, $\theta = \dfrac{5\pi}{4}$.

If $n = 0$, $\theta = \dfrac{3\pi}{4}$.
If $n = 1$, $\theta = \dfrac{7\pi}{4}$.

Figure 6.36 confirms that four angles giving the locations of the petal points are $\dfrac{\pi}{4}, \dfrac{3\pi}{4}, \dfrac{5\pi}{4},$ and $\dfrac{7\pi}{4}$.

Technology

The graph of
$$r = 4 \sin 2\theta$$
was obtained using a $[-4, 4, 1]$ by $[-4, 4, 1]$ viewing rectangle and
$$\theta \min = 0, \quad \theta \max = 2\pi,$$
$$\theta \text{ step} = \frac{\pi}{48}.$$

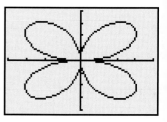

Rose Curves

The graphs of
$$r = a \sin n\theta \quad \text{and} \quad r = a \cos n\theta, \quad a \neq 0,$$
are called **rose curves**. If n is even, the rose has $2n$ petals. If n is odd, the rose has n petals.

$r = a \sin 2\theta$ Rose curve with 4 petals	$r = a \cos 3\theta$ Rose curve with 3 petals	$r = a \cos 4\theta$ Rose curve with 8 petals	$r = a \sin 5\theta$ Rose curve with 5 petals

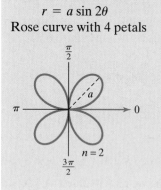

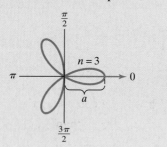

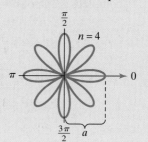

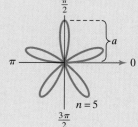

Check Point 4 Graph the polar equation: $r = 3 \cos 2\theta$.

EXAMPLE 5   Graphing a Polar Equation

Graph the polar equation: $r^2 = 4 \sin 2\theta$.

Solution   We first check for symmetry.
$$r^2 = 4 \sin 2\theta$$

Polar Axis	**The Line** $\theta = \dfrac{\pi}{2}$	**The Pole**
Replace θ with $-\theta$.	Replace (r, θ) with $(-r, -\theta)$.	Replace r with $-r$.
$r^2 = 4 \sin 2(-\theta)$	$(-r)^2 = 4 \sin 2(-\theta)$	$(-r)^2 = 4 \sin 2\theta$
$r^2 = 4 \sin (-2\theta)$	$r^2 = 4 \sin (-2\theta)$	$r^2 = 4 \sin 2\theta$
$r^2 = -4 \sin 2\theta$	$r^2 = -4 \sin 2\theta$	
Equation changes and fails this symmetry test.	Equation changes and fails this symmetry test.	Equation does not change.

Thus, we can be sure that the graph is symmetric with respect to the pole. The graph may or may not be symmetric with respect to the polar axis or the line $\theta = \dfrac{\pi}{2}$.

Now we are ready to graph $r^2 = 4 \sin 2\theta$. In Figure 6.37(a), we plot points on the graph by using values of θ from 0 to $\dfrac{\pi}{2}$ and the corresponding values of r. These coordinates are shown in the table above Figure 6.37(a). Notice that the points in Figure 6.37(a) are shown for $r \geq 0$. Because the graph is symmetric with respect to the pole, we can reflect the graph in Figure 6.37(a) about the pole and obtain the graph in Figure 6.37(b).

θ	0	$\dfrac{\pi}{6}$	$\dfrac{\pi}{4}$	$\dfrac{\pi}{3}$	$\dfrac{\pi}{2}$
r	0	± 1.9	± 2	± 1.9	0

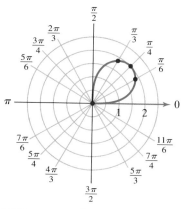

(a) The graph of $r^2 = 4 \sin 2\theta$ for $0 \leq \theta \leq \dfrac{\pi}{2}$ and $r \geq 0$

(b) Using symmetry with respect to the pole on the graph of $r^2 = 4 \sin 2\theta$

Figure 6.37 Graphing $r^2 = 4 \sin 2\theta$

Does Figure 6.37(b) show a complete graph of $r^2 = 4 \sin 2\theta$ or do we need to continue graphing for angles greater than $\dfrac{\pi}{2}$? If θ is in quadrant II, 2θ is in quadrant III or IV, where $\sin 2\theta$ is negative. Thus, $4 \sin 2\theta$ is negative. However, $r^2 = 4 \sin 2\theta$ and r^2 cannot be negative. The same observation applies to quadrant IV. This means that there are no points on the graph in quadrants II or IV. Thus, Figure 6.37(b) shows the complete graph of $r^2 = 4 \sin 2\theta$.

The curve in Figure 6.37(b) is shaped like a propeller and is called a *lemniscate*.

Lemniscates

The graphs of

$$r^2 = a^2 \sin 2\theta \quad \text{and} \quad r^2 = a^2 \cos 2\theta, \quad a \neq 0$$

are called **lemniscates**.

$r^2 = a^2 \sin 2\theta$ is symmetric with respect to the pole.

$r^2 = a^2 \cos 2\theta$ is symmetric with respect to the polar axis, $\theta = \dfrac{\pi}{2}$, and the pole.

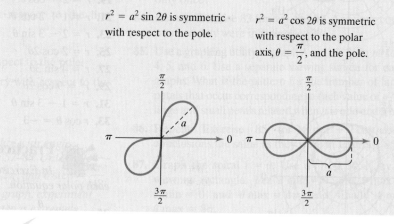

Check Point 5 Graph the polar equation: $r^2 = 4 \cos 2\theta$.

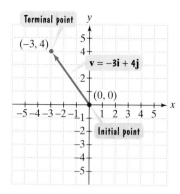

Figure 6.56 Sketching $\mathbf{v} = -3\mathbf{i} + 4\mathbf{j}$ in the rectangular coordinate system

EXAMPLE 2   Representing a Vector in Rectangular Coordinates and Finding Its Magnitude

Sketch the vector $\mathbf{v} = -3\mathbf{i} + 4\mathbf{j}$ and find its magnitude.

Solution   For the given vector $\mathbf{v} = -3\mathbf{i} + 4\mathbf{j}$, $a = -3$ and $b = 4$. The vector can be represented with its initial point at the origin, $(0, 0)$, as shown in Figure 6.56. The vector's terminal point is then $(a, b) = (-3, 4)$. We sketch the vector by drawing an arrow from $(0, 0)$ to $(-3, 4)$. We determine the magnitude of the vector by using the distance formula. Thus, the magnitude is

$$\|\mathbf{v}\| = \sqrt{a^2 + b^2} = \sqrt{(-3)^2 + 4^2} = \sqrt{9 + 16} = \sqrt{25} = 5.$$

Check Point 2   Sketch the vector $\mathbf{v} = 3\mathbf{i} - 3\mathbf{j}$ and find its magnitude.

The vector in Example 2 was represented with its initial point at the origin. A vector whose initial point is at the origin is called a **position vector**. Any vector in rectangular coordinates whose initial point is not at the origin can be shown to be equal to a position vector. As shown in the following box, this gives us a way to represent vectors between any two points.

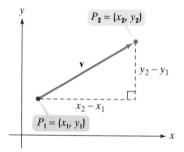

(a)

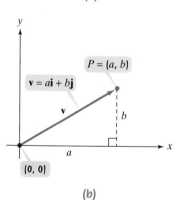

(b)

Figure 6.57

> **Representing Vectors in Rectangular Coordinates**
>
> Vector $\mathbf{v}$ with initial point $P_1 = (x_1, y_1)$ and terminal point $P_2 = (x_2, y_2)$ is equal to the position vector
>
> $$\mathbf{v} = (x_2 - x_1)\mathbf{i} + (y_2 - y_1)\mathbf{j}.$$

We can use congruent triangles, triangles with the same size and shape, to derive this formula. Begin with the right triangle in Figure 6.57(a). This triangle shows vector $\mathbf{v}$ from $P_1 = (x_1, y_1)$ to $P_2 = (x_2, y_2)$. In Figure 6.57(b), we move vector $\mathbf{v}$, without changing its magnitude or its direction, so that its initial point is at the origin. Using this position vector in Figure 6.57(b), we see that

$$\mathbf{v} = a\mathbf{i} + b\mathbf{j},$$

where a and b are the components of $\mathbf{v}$. The equal vectors and the right angles in the right triangles in Figures 6.57(a) and (b) result in congruent triangles. The corresponding sides of these congruent triangles are equal, so that $a = x_2 - x_1$ and $b = y_2 - y_1$. This means that $\mathbf{v}$ may be expressed as

$$\mathbf{v} = a\mathbf{i} + b\mathbf{j} = (x_2 - x_1)\mathbf{i} + (y_2 - y_1)\mathbf{j}.$$

> Horizontal component: x-coordinate of terminal point minus x-coordinate of initial point

> Vertical component: y-coordinate of terminal point minus y-coordinate of initial point

Thus, any vector between two points in rectangular coordinates can be expressed in terms of $\mathbf{i}$ and $\mathbf{j}$. In rectangular coordinates, the term *vector* refers to the position vector expressed in terms of $\mathbf{i}$ and $\mathbf{j}$ that is equal to it.

EXAMPLE 3   Representing a Vector in Rectangular Coordinates

Let $\mathbf{v}$ be the vector from initial point $P_1 = (3, -1)$ to terminal point $P_2 = (-2, 5)$. Write $\mathbf{v}$ in terms of $\mathbf{i}$ and $\mathbf{j}$.

Solution   We identify the values for the variables in the formula.

$$P_1 = (3, -1) \qquad P_2 = (-2, 5)$$

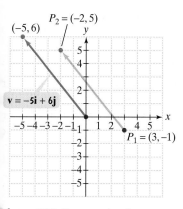

Figure 6.58 Representing the vector from $(3, -1)$ to $(-2, 5)$ as a position vector

Using these values, we write **v** in terms of **i** and **j** as follows:

$$\mathbf{v} = (x_2 - x_1)\mathbf{i} + (y_2 - y_1)\mathbf{j} = (-2 - 3)\mathbf{i} + [5 - (-1)]\mathbf{j} = -5\mathbf{i} + 6\mathbf{j}.$$

Figure 6.58 shows the vector from $P_1 = (3, -1)$ to $P_2 = (-2, 5)$ represented in terms of **i** and **j** and as a position vector.

Study Tip

When finding the distance from $P_1 = (x_1, y_1)$ to $P_2 = (x_2, y_2)$, the order in which the subtractions are performed makes no difference:

$$d = \sqrt{(x_2 - x_1)^2 + (y_2 - y_1)^2} \quad \text{or} \quad d = \sqrt{(x_1 - x_2)^2 + (y_1 - y_2)^2}.$$

When writing the vector from $P_1 = (x_1, y_1)$ to $P_2 = (x_2, y_2)$, P_2 must be the terminal point and the order in the subtractions is important:

$$\mathbf{v} = (x_2 - x_1)\mathbf{i} + (y_2 - y_1)\mathbf{j}.$$

(x_2, y_2), the terminal point, is used first in each subtraction.

Check Point 3 Let **v** be the vector from initial point $P_1 = (-1, 3)$ to terminal point $P_2 = (2, 7)$. Write **v** in terms of **i** and **j**.

④ Perform operations with vectors in terms of **i** and **j**.

Operations with Vectors in Terms of i and j

If vectors are expressed in terms of **i** and **j**, we can easily carry out operations such as vector addition, vector subtraction, and scalar multiplication. Recall the geometric definitions of these operations given earlier. Based on these ideas, we can add and subtract vectors using the following procedure:

Adding and Subtracting Vectors in Terms of i and j

If $\mathbf{v} = a_1\mathbf{i} + b_1\mathbf{j}$ and $\mathbf{w} = a_2\mathbf{i} + b_2\mathbf{j}$, then

$$\mathbf{v} + \mathbf{w} = (a_1 + a_2)\mathbf{i} + (b_1 + b_2)\mathbf{j}$$
$$\mathbf{v} - \mathbf{w} = (a_1 - a_2)\mathbf{i} + (b_1 - b_2)\mathbf{j}.$$

EXAMPLE 4   Adding and Subtracting Vectors

If $\mathbf{v} = 5\mathbf{i} + 4\mathbf{j}$ and $\mathbf{w} = 6\mathbf{i} - 9\mathbf{j}$, find each of the following vectors:

a. $\mathbf{v} + \mathbf{w}$ **b.** $\mathbf{v} - \mathbf{w}$.

Solution

a. $\mathbf{v} + \mathbf{w} = (5\mathbf{i} + 4\mathbf{j}) + (6\mathbf{i} - 9\mathbf{j})$ These are the given vectors.

$\qquad\qquad = (5 + 6)\mathbf{i} + [4 + (-9)]\mathbf{j}$ Add the horizontal components.
Add the vertical components.

$\qquad\qquad = 11\mathbf{i} - 5\mathbf{j}$ Simplify.

b. $\mathbf{v} - \mathbf{w} = (5\mathbf{i} + 4\mathbf{j}) - (6\mathbf{i} - 9\mathbf{j})$ These are the given vectors.

$\qquad\qquad = (5 - 6)\mathbf{i} + [4 - (-9)]\mathbf{j}$ Subtract the horizontal components.
Subtract the vertical components.

$\qquad\qquad = -\mathbf{i} + 13\mathbf{j}$ Simplify.

Check Point 4 If $\mathbf{v} = 7\mathbf{i} + 3\mathbf{j}$ and $\mathbf{w} = 4\mathbf{i} - 5\mathbf{j}$, find each of the following vectors:

a. $\mathbf{v} + \mathbf{w}$ **b.** $\mathbf{v} - \mathbf{w}$.

How do we perform scalar multiplication if vectors are expressed in terms of and **j**? We use the following procedure to multiply the vector **v** by the scalar k:

Scalar Multiplication with a Vector in Terms of i and j

If $\mathbf{v} = a\mathbf{i} + b\mathbf{j}$ and k is a real number, then the scalar multiplication of the vector **v** and the scalar k is

$$k\mathbf{v} = (ka)\mathbf{i} + (kb)\mathbf{j}.$$

EXAMPLE 5 Scalar Multiplication

If $\mathbf{v} = 5\mathbf{i} + 4\mathbf{j}$, find each of the following vectors:

    a. 6**v**   **b.** $-3\mathbf{v}$.

Solution

    a. $6\mathbf{v} = 6(5\mathbf{i} + 4\mathbf{j})$ *The scalar multiplication is expressed with the given vector.*

               $= (6 \cdot 5)\mathbf{i} + (6 \cdot 4)\mathbf{j}$ *Multiply each component by 6.*
               $= 30\mathbf{i} + 24\mathbf{j}$ *Simplify.*

    b. $-3\mathbf{v} = -3(5\mathbf{i} + 4\mathbf{j})$ *The scalar multiplication is expressed with the given vector.*

               $= (-3 \cdot 5)\mathbf{i} + (-3 \cdot 4)\mathbf{j}$ *Multiply each component by -3.*
               $= -15\mathbf{i} - 12\mathbf{j}$ *Simplify.*

Check Point 5 If $\mathbf{v} = 7\mathbf{i} + 10\mathbf{j}$, find each of the following vectors:

    a. 8**v**   **b.** $-5\mathbf{v}$.

EXAMPLE 6 Vector Operations

If $\mathbf{v} = 5\mathbf{i} + 4\mathbf{j}$ and $\mathbf{w} = 6\mathbf{i} - 9\mathbf{j}$, find $4\mathbf{v} - 2\mathbf{w}$.

Solution

$4\mathbf{v} - 2\mathbf{w} = 4(5\mathbf{i} + 4\mathbf{j}) - 2(6\mathbf{i} - 9\mathbf{j})$ *Operations are expressed with the given vectors.*
             $= 20\mathbf{i} + 16\mathbf{j} - 12\mathbf{i} + 18\mathbf{j}$ *Perform each scalar multiplication.*
             $= (20 - 12)\mathbf{i} + (16 + 18)\mathbf{j}$ *Add horizontal and vertical components to perform the vector addition.*
             $= 8\mathbf{i} + 34\mathbf{j}$ *Simplify.*

Check Point 6 If $\mathbf{v} = 7\mathbf{i} + 3\mathbf{j}$ and $\mathbf{w} = 4\mathbf{i} - 5\mathbf{j}$, find $6\mathbf{v} - 3\mathbf{w}$.

Properties involving vector operations resemble familiar properties of real numbers. For example, the order in which vectors are added makes no difference

$$\mathbf{u} + \mathbf{v} = \mathbf{v} + \mathbf{u}.$$

Does this remind you of the commutative property $a + b = b + a$?

Just as 0 plays an important role in the properties of real numbers, the **zero vector 0** plays exactly the same role in the properties of vectors.

The Zero Vector

The vector whose magnitude is 0 is called the **zero vector, 0.** The zero vector is assigned no direction. It can be expressed in terms of **i** and **j** using

$$\mathbf{0} = 0\mathbf{i} + 0\mathbf{j}.$$

Properties of vector addition and scalar multiplication are given as follows:

> ## Properties of Vector Addition and Scalar Multiplication
>
> If **u**, **v**, and **w** are vectors, and c and d are scalars, then the following properties are true.
>
> ### Vector Addition Properties
>
> 1. $\mathbf{u} + \mathbf{v} = \mathbf{v} + \mathbf{u}$ Commutative Property
> 2. $(\mathbf{u} + \mathbf{v}) + \mathbf{w} = \mathbf{u} + (\mathbf{v} + \mathbf{w})$ Associative Property
> 3. $\mathbf{u} + \mathbf{0} = \mathbf{0} + \mathbf{u} = \mathbf{u}$ Additive Identity
> 4. $\mathbf{u} + (-\mathbf{u}) = (-\mathbf{u}) + \mathbf{u} = \mathbf{0}$ Additive Inverse
>
> ### Scalar Multiplication Properties
>
> 1. $(cd)\mathbf{u} = c(d\mathbf{u})$ Associative Property
> 2. $c(\mathbf{u} + \mathbf{v}) = c\mathbf{u} + c\mathbf{v}$ Distributive Property
> 3. $(c + d)\mathbf{u} = c\mathbf{u} + d\mathbf{u}$ Distributive Property
> 4. $1\mathbf{u} = \mathbf{u}$ Multiplicative Identity
> 5. $0\mathbf{u} = \mathbf{0}$ Multiplication Property of Zero
> 6. $\|c\mathbf{v}\| = |c|\|\mathbf{v}\|$ Magnitude Property

5 Find the unit vector in the direction of **v**.

Unit Vectors

A **unit vector** is defined to be a vector whose magnitude is one. In many applications of vectors, it is helpful to find the unit vector that has the same direction as a given vector.

Discovery

To find out why the procedure in the box produces a unit vector, work Exercise 105 in Exercise Set 6.6.

> ## Finding the Unit Vector that Has the Same Direction as a Given Nonzero Vector v
>
> For any nonzero vector **v**, the vector
>
> $$\frac{\mathbf{v}}{\|\mathbf{v}\|}$$
>
> is the unit vector that has the same direction as **v**. To find this vector, divide **v** by its magnitude.

EXAMPLE 7   Finding a Unit Vector

Find the unit vector in the same direction as $\mathbf{v} = 5\mathbf{i} - 12\mathbf{j}$. Then verify that the vector has magnitude 1.

Solution   We find the unit vector in the same direction as **v** by dividing **v** by its magnitude. We first find the magnitude of **v**.

$$\|\mathbf{v}\| = \sqrt{a^2 + b^2} = \sqrt{5^2 + (-12)^2} = \sqrt{25 + 144} = \sqrt{169} = 13$$

The unit vector in the same direction as **v** is

$$\frac{\mathbf{v}}{\|\mathbf{v}\|} = \frac{5\mathbf{i} - 12\mathbf{j}}{13} = \frac{5}{13}\mathbf{i} - \frac{12}{13}\mathbf{j}.$$ This is the scalar multiplication of v and $\frac{1}{13}$.

Now we must verify that the magnitude of this vector is 1. Recall that the magnitude of $a\mathbf{i} + b\mathbf{j}$ is $\sqrt{a^2 + b^2}$. Thus, the magnitude of $\frac{5}{13}\mathbf{i} - \frac{12}{13}\mathbf{j}$ is

$$\sqrt{\left(\frac{5}{13}\right)^2 + \left(-\frac{12}{13}\right)^2} = \sqrt{\frac{25}{169} + \frac{144}{169}} = \sqrt{\frac{169}{169}} = \sqrt{1} = 1.$$

Check Point 7   Find the unit vector in the same direction as $\mathbf{v} = 4\mathbf{i} - 3\mathbf{j}$. Then verify that the vector has magnitude 1.

76. If a force of 30 pounds is needed to pull the box up the ramp, find the weight of the box.

In Exercises 77–78, round answers to the nearest pound.

77. a. Find the magnitude of the force required to keep a 3500-pound car from sliding down a hill inclined at $5.5°$ from the horizontal.

  b. Find the magnitude of the force of the car against the hill.

78. a. Find the magnitude of the force required to keep a 280-pound barrel from sliding down a ramp inclined at $12.5°$ from the horizontal.

  b. Find the magnitude of the force of the barrel against the ramp.

The forces $\mathbf{F}_1$, $\mathbf{F}_2$, $\mathbf{F}_3$, ..., $\mathbf{F}_n$ acting on an object are in **equilibrium** *if the resultant force is the zero vector:*

$$\mathbf{F}_1 + \mathbf{F}_2 + \mathbf{F}_3 + \cdots + \mathbf{F}_n = \mathbf{0}.$$

In Exercises 79–82, the given forces are acting on an object.
  a. *Find the resultant force.*
  b. *What additional force is required for the given forces to be in equilibrium?*

79. $\mathbf{F}_1 = 3\mathbf{i} - 5\mathbf{j}$, $\quad \mathbf{F}_2 = 6\mathbf{i} + 2\mathbf{j}$

80. $\mathbf{F}_1 = -2\mathbf{i} + 3\mathbf{j}$, $\quad \mathbf{F}_2 = \mathbf{i} - \mathbf{j}$, $\quad \mathbf{F}_3 = 5\mathbf{i} - 12\mathbf{j}$

81.

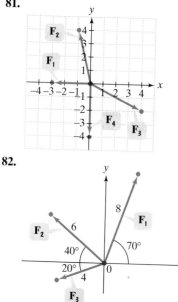

82.

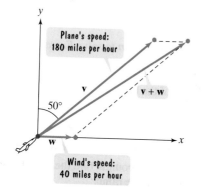

83. The figure shows a small plane flying at a speed of 180 miles per hour on a bearing of N50°E. The wind is blowing from west to east at 40 miles per hour. The figure indicates that $\mathbf{v}$ represents the velocity of the plane in still air and $\mathbf{w}$ represents the velocity of the wind.

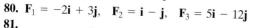

  a. Express $\mathbf{v}$ and $\mathbf{w}$ in terms of their magnitudes and direction angles.

  b. Find the resultant vector, $\mathbf{v} + \mathbf{w}$.

  c. The magnitude of $\mathbf{v} + \mathbf{w}$, called the **ground speed** of the plane, gives its speed relative to the ground. Approximate the ground speed to the nearest mile per hour.

  d. The direction angle of $\mathbf{v} + \mathbf{w}$ gives the plane's true course relative to the ground. Approximate the true course to the nearest tenth of a degree. What is the plane's true bearing?

84. Use the procedure outlined in Exercise 83 to solve this exercise. A plane is flying at a speed of 400 miles per hour on a bearing of N50°W. The wind is blowing at 30 miles per hour on a bearing of N25°E.

  a. Approximate the plane's ground speed to the nearest mile per hour.

  b. Approximate the plane's true course to the nearest tenth of a degree. What is its true bearing?

85. A plane is flying at a speed of 320 miles per hour on a bearing of N70°E. Its ground speed is 370 miles per hour and its true course is $30°$. Find the speed, to the nearest mile per hour, and the direction angle, to the nearest tenth of a degree, of the wind.

86. A plane is flying at a speed of 540 miles per hour on a bearing of S36°E. Its ground speed is 500 miles per hour and its true bearing is S44°E. Find the speed, to the nearest mile per hour, and the direction angle, to the nearest tenth of a degree, of the wind.

Writing in Mathematics

87. What is a directed line segment?

88. What are equal vectors?

89. If vector $\mathbf{v}$ is represented by an arrow, how is -3 represented?

90. If vectors $\mathbf{u}$ and $\mathbf{v}$ are represented by arrows, describe how the vector sum $\mathbf{u} + \mathbf{v}$ is represented.

91. What is the vector $\mathbf{i}$?

92. What is the vector $\mathbf{j}$?

93. What is a position vector? How is a position vector represented using $\mathbf{i}$ and $\mathbf{j}$?

94. If $\mathbf{v}$ is a vector between any two points in the rectangular coordinate system, explain how to write $\mathbf{v}$ in terms of $\mathbf{i}$ and

95. If two vectors are expressed in terms of $\mathbf{i}$ and $\mathbf{j}$, explain how to find their sum.

96. If two vectors are expressed in terms of $\mathbf{i}$ and $\mathbf{j}$, explain how to find their difference.

97. If a vector is expressed in terms of $\mathbf{i}$ and $\mathbf{j}$, explain how to find the scalar multiplication of the vector and a given scalar k.

98. What is the zero vector?

99. Describe one similarity between the zero vector and the number 0.

100. Explain how to find the unit vector in the direction of any given vector $\mathbf{v}$.

101. Explain how to write a vector in terms of its magnitude and direction.

102. You are on an airplane. The pilot announces the plane's speed over the intercom. Which speed do you think is being reported: the speed of the plane in still air or the speed after the effect of the wind has been accounted for? Explain your answer.

103. Use vectors to explain why it is difficult to hold a heavy stack of books perfectly still for a long period of time. As you become exhausted, what eventually happens? What does this mean in terms of the forces acting on the books?

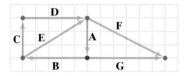

Critical Thinking Exercises

104. Use the figure shown to select a true statement.

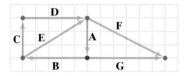

a. **A + B = E**

b. **D + A + B + C = 0**

c. **B − E = G − F**

d. $\|\mathbf{A}\| \neq \|\mathbf{C}\|$

105. Let $\mathbf{v} = a\mathbf{i} + b\mathbf{j}$. Show that $\dfrac{\mathbf{v}}{\|\mathbf{v}\|}$ is a unit vector in the direction of $\mathbf{v}$.

In Exercises 106–107, refer to the navigational compass shown in the figure. The compass is marked clockwise in degrees that start at north 0°.

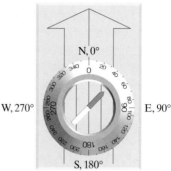

106. An airplane has an air speed of 240 miles per hour and a compass heading of 280°. A steady wind of 30 miles per hour is blowing in the direction of 265°. What is the plane's true speed relative to the ground? What is its compass heading relative to the ground?

107. Two tugboats are pulling on a large ship that has gone aground. One tug pulls with a force of 2500 pounds in a compass direction of 55°. The second tug pulls with a force of 2000 pounds in a compass direction of 95°. Find the magnitude and the compass direction of the resultant force.

108. You want to fly your small plane due north, but there is a 75 kilometer wind blowing from west to east.
    a. Find the direction angle for where you should head the plane if your speed relative to the ground is 310 kilometers per hour.
    b. If you increase your air speed, should the direction angle in part (a) increase or decrease? Explain your answer.

SECTION 6.7 *The Dot Product*

Objectives

❶ Find the dot product of two vectors.

❷ Find the angle between two vectors.

❸ Use the dot product to determine if two vectors are orthogonal.

❹ Find the projection of a vector onto another vector.

❺ Express a vector as the sum of two orthogonal vectors.

❻ Compute work.

Talk about hard work! I can see the weightlifter's muscles quivering from the exertion of holding the barbell in a stationary position above his head. Still, I'm not sure if he's doing as much work as I am, sitting at my desk with my brain quivering from studying trigonometric functions and their applications.

Would it surprise you to know that neither you nor the weightlifter are doing any work at all? The definition of work in physics and mathematics is not the same as what we mean by "work" in everyday use. To understand what is involved in real work, we turn to a new vector operation called the dot product.

 Find the dot product of two vectors.

The Dot Product of Two Vectors

The operations of vector addition and scalar multiplication result in vectors. By contrast, the *dot product* of two vectors results in a scalar (a real number), rather than a vector.

Definition of the Dot Product

If $\mathbf{v} = a_1\mathbf{i} + b_1\mathbf{j}$ and $\mathbf{w} = a_2\mathbf{i} + b_2\mathbf{j}$ are vectors, the **dot product $\mathbf{v} \cdot \mathbf{w}$** is defined as follows:

$$\mathbf{v} \cdot \mathbf{w} = a_1a_2 + b_1b_2.$$

The dot product of two vectors is the sum of the products of their horizontal components and their vertical components.

EXAMPLE 1   Finding Dot Products

If $\mathbf{v} = 5\mathbf{i} - 2\mathbf{j}$ and $\mathbf{w} = -3\mathbf{i} + 4\mathbf{j}$, find each of the following dot products:

a. $\mathbf{v} \cdot \mathbf{w}$ b. $\mathbf{w} \cdot \mathbf{v}$ c. $\mathbf{v} \cdot \mathbf{v}$.

Solution   To find each dot product, multiply the two horizontal components, and then multiply the two vertical components. Finally, add the two products.

a. $\mathbf{v} \cdot \mathbf{w} = 5(-3) + (-2)(4) = -15 - 8 = -23$

Multiply the horizontal components and multiply the vertical components of $\mathbf{v} = 5\mathbf{i} - 2\mathbf{j}$ and $\mathbf{w} = -3\mathbf{i} + 4\mathbf{j}$.

b. $\mathbf{w} \cdot \mathbf{v} = -3(5) + 4(-2) = -15 - 8 = -23$

Multiply the horizontal components and multiply the vertical components of $\mathbf{w} = -3\mathbf{i} + 4\mathbf{j}$ and $\mathbf{v} = 5\mathbf{i} - 2\mathbf{j}$.

c. $\mathbf{v} \cdot \mathbf{v} = 5(5) + (-2)(-2) = 25 + 4 = 29$

Multiply the horizontal components and multiply the vertical components of $\mathbf{v} = 5\mathbf{i} - 2\mathbf{j}$ and $\mathbf{v} = 5\mathbf{i} - 2\mathbf{j}$.

Check Point 1   If $\mathbf{v} = 7\mathbf{i} - 4\mathbf{j}$ and $\mathbf{w} = 2\mathbf{i} - \mathbf{j}$, find each of the following dot products:

a. $\mathbf{v} \cdot \mathbf{w}$ b. $\mathbf{w} \cdot \mathbf{v}$ c. $\mathbf{w} \cdot \mathbf{w}$.

In Example 1 and Check Point 1, did you notice that $\mathbf{v} \cdot \mathbf{w}$ and $\mathbf{w} \cdot \mathbf{v}$ produce the same scalar? The fact that $\mathbf{v} \cdot \mathbf{w} = \mathbf{w} \cdot \mathbf{v}$ follows from the definition of the dot product. Properties of the dot product are given in the following box. Proofs for some of these properties are given in the appendix.

Properties of the Dot Product

If **u**, **v**, and **w** are vectors, and c is a scalar, then

1. $\mathbf{u} \cdot \mathbf{v} = \mathbf{v} \cdot \mathbf{u}$
2. $\mathbf{u} \cdot (\mathbf{v} + \mathbf{w}) = \mathbf{u} \cdot \mathbf{v} + \mathbf{u} \cdot \mathbf{w}$
3. $\mathbf{0} \cdot \mathbf{v} = 0$
4. $\mathbf{v} \cdot \mathbf{v} = \|\mathbf{v}\|^2$
5. $(c\mathbf{u}) \cdot \mathbf{v} = c(\mathbf{u} \cdot \mathbf{v}) = \mathbf{u} \cdot (c\mathbf{v})$

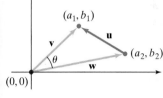

Figure 6.64

The Angle between Two Vectors

The Law of Cosines can be used to derive another formula for the dot product. This formula will give us a way to find the angle between two vectors.

Figure 6.64 shows vectors $\mathbf{v} = a_1\mathbf{i} + b_1\mathbf{j}$ and $\mathbf{w} = a_2\mathbf{i} + b_2\mathbf{j}$. By the definition of the dot product, we know that $\mathbf{v} \cdot \mathbf{w} = a_1 a_2 + b_1 b_2$. Our new formula for the dot product involves the angle between the vectors, shown as θ in the figure. Apply the Law of Cosines to the triangle shown in the figure.

$$\|\mathbf{u}\|^2 = \|\mathbf{v}\|^2 + \|\mathbf{w}\|^2 - 2\|\mathbf{v}\|\,\|\mathbf{w}\|\cos\theta \qquad \textsf{Use the Law of Cosines.}$$

$$\boxed{\begin{array}{l} \mathbf{u} = (a_1 - a_2)\mathbf{i} + (b_1 - b_2)\mathbf{j} \\ \|\mathbf{u}\| = \sqrt{(a_1 - a_2)^2 + (b_1 - b_2)^2} \end{array}} \quad \boxed{\begin{array}{l} \mathbf{v} = a_1\mathbf{i} + b_1\mathbf{j} \\ \|\mathbf{v}\| = \sqrt{a_1^{\,2} + b_1^{\,2}} \end{array}} \quad \boxed{\begin{array}{l} \mathbf{w} = a_2\mathbf{i} + b_2\mathbf{j} \\ \|\mathbf{w}\| = \sqrt{a_2^{\,2} + b_2^{\,2}} \end{array}}$$

$$(a_1 - a_2)^2 + (b_1 - b_2)^2 = (a_1^2 + b_1^2) + (a_2^2 + b_2^2) - 2\|\mathbf{v}\|\|\mathbf{w}\|\cos\theta$$

Substitute the squares of the magnitudes of vectors u, v, and w into the Law of Cosines.

$$a_1^2 - 2a_1 a_2 + a_2^2 + b_1^2 - 2b_1 b_2 + b_2^2 = a_1^2 + b_1^2 + a_2^2 + b_2^2 - 2\|\mathbf{v}\|\|\mathbf{w}\|\cos\theta$$

Square the binomials using $(A - B)^2 = A^2 - 2AB + B^2$.

$$-2a_1 a_2 - 2b_1 b_2 = -2\|\mathbf{v}\|\|\mathbf{w}\|\cos\theta$$

Subtract a_1^2, a_2^2, b_1^2, and b_2^2 from both sides of the equation.

$$a_1 a_2 + b_1 b_2 = \|\mathbf{v}\|\,\|\mathbf{w}\|\cos\theta$$

Divide both sides by -2.

$$\boxed{\begin{array}{c} \textsf{By definition,} \\ \mathbf{v} \cdot \mathbf{w} = a_1 a_2 + b_1 b_2. \end{array}}$$

$$\mathbf{v} \cdot \mathbf{w} = \|\mathbf{v}\|\|\mathbf{w}\|\cos\theta$$

Substitute v · w for the expression on the left side of the equation.

Alternative Formula for the Dot Product

If **v** and **w** are two nonzero vectors and θ is the smallest nonnegative angle between them, then

$$\mathbf{v} \cdot \mathbf{w} = \|\mathbf{v}\|\|\mathbf{w}\|\cos\theta.$$

2 Find the angle between two vectors.

Solving the formula in the box for $\cos\theta$ gives us a formula for finding the angle between two vectors:

Formula for the Angle between Two Vectors

If **v** and **w** are two nonzero vectors and θ is the smallest nonnegative angle between **v** and **w**, then

$$\cos\theta = \frac{\mathbf{v} \cdot \mathbf{w}}{\|\mathbf{v}\|\|\mathbf{w}\|} \quad \text{and} \quad \theta = \cos^{-1}\left(\frac{\mathbf{v} \cdot \mathbf{w}}{\|\mathbf{v}\|\|\mathbf{w}\|}\right).$$

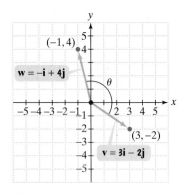

Figure 6.65 Finding the angle between two vectors

EXAMPLE 2 Finding the Angle between Two Vectors

Find the angle θ between the vectors $\mathbf{v} = 3\mathbf{i} - 2\mathbf{j}$ and $\mathbf{w} = -\mathbf{i} + 4\mathbf{j}$, shown in Figure 6.65. Round to the nearest tenth of a degree.

Solution Use the formula for the angle between two vectors.

$$\cos \theta = \frac{\mathbf{v} \cdot \mathbf{w}}{\|\mathbf{v}\|\|\mathbf{w}\|}$$
Begin with the formula for the cosine of the angle between two vectors.

$$= \frac{(3\mathbf{i} - 2\mathbf{j}) \cdot (-\mathbf{i} + 4\mathbf{j})}{\sqrt{3^2 + (-2)^2}\sqrt{(-1)^2 + 4^2}}$$
Substitute the given vectors in the numerator. Find the magnitude of each vector in the denominator.

$$= \frac{3(-1) + (-2)(4)}{\sqrt{13}\sqrt{17}}$$
Find the dot product in the numerator. Simplify in the denominator.

$$= -\frac{11}{\sqrt{221}}$$
Perform the indicated operations.

The angle θ between the vectors is

$$\theta = \cos^{-1}\left(-\frac{11}{\sqrt{221}}\right) \approx 137.7°.$$
Use a calculator.

Check Point 2 Find the angle between the vectors $\mathbf{v} = 4\mathbf{i} - 3\mathbf{j}$ and $\mathbf{w} = \mathbf{i} + 2\mathbf{j}$. Round to the nearest tenth of a degree.

 Use the dot product to determine if two vectors are orthogonal.

Parallel and Orthogonal Vectors

Two vectors are **parallel** when the angle θ between the vectors is $0°$ or $180°$. If $\theta = 0°$, the vectors point in the same direction. If $\theta = 180°$, the vectors point in opposite directions. Figure 6.66 shows parallel vectors.

$\theta = 0°$ and $\cos \theta = 1$. Vectors point in the same direction.

$\theta = 180°$ and $\cos \theta = -1$. Vectors point in opposite directions.

Figure 6.66 Parallel vectors

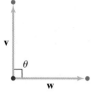

Figure 6.67 Orthogonal vectors: $\theta = 90°$ and $\cos \theta = 0$

Two vectors are **orthogonal** when the angle between the vectors is $90°$, shown in Figure 6.67. (The word "orthogonal," rather than "perpendicular," is used to describe vectors that meet at right angles.) We know that $\mathbf{v} \cdot \mathbf{w} = \|\mathbf{v}\|\|\mathbf{w}\| \cos \theta$. If $\mathbf{v}$ and $\mathbf{w}$ are orthogonal, then

$$\mathbf{v} \cdot \mathbf{w} = \|\mathbf{v}\|\|\mathbf{w}\| \cos 90° = \|\mathbf{v}\|\|\mathbf{w}\|(0) = 0.$$

Conversely, if $\mathbf{v}$ and $\mathbf{w}$ are vectors such that $\mathbf{v} \cdot \mathbf{w} = 0$, then $\|\mathbf{v}\| = 0$ or $\|\mathbf{w}\| = 0$ or $\cos \theta = 0$. If $\cos \theta = 0$, then $\theta = 90°$, so $\mathbf{v}$ and $\mathbf{w}$ are orthogonal.

This discussion is summarized as follows:

The Dot Product and Orthogonal Vectors

Two nonzero vectors $\mathbf{v}$ and $\mathbf{w}$ are orthogonal if and only if $\mathbf{v} \cdot \mathbf{w} = 0$.
Because $\mathbf{0} \cdot \mathbf{v} = 0$, the zero vector is orthogonal to every vector $\mathbf{v}$.

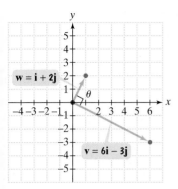

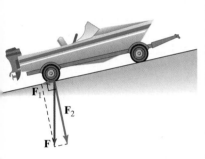

Figure 6.68 Orthogonal vectors

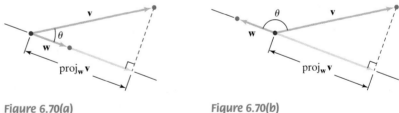

Figure 6.69

EXAMPLE 3   Determining Whether Vectors Are Orthogonal

Are the vectors $\mathbf{v} = 6\mathbf{i} - 3\mathbf{j}$ and $\mathbf{w} = \mathbf{i} + 2\mathbf{j}$ orthogonal?

Solution   The vectors are orthogonal if their dot product is 0. Begin by finding $\mathbf{v} \cdot \mathbf{w}$.

$$\mathbf{v} \cdot \mathbf{w} = (6\mathbf{i} - 3\mathbf{j}) \cdot (\mathbf{i} + 2\mathbf{j}) = 6(1) + (-3)(2) = 6 - 6 = 0$$

The dot product is 0. Thus, the given vectors are orthogonal. They are shown in Figure 6.68.

Check Point 3   Are the vectors $\mathbf{v} = 2\mathbf{i} + 3\mathbf{j}$ and $\mathbf{w} = 6\mathbf{i} - 4\mathbf{j}$ orthogonal?

Projection of a Vector Onto Another Vector

You know how to add two vectors to obtain a resultant vector. We now reverse this process by expressing a vector as the sum of two orthogonal vectors. By doing this, you can determine how much force is applied in a particular direction. For example, Figure 6.69 shows a boat on a tilted ramp. The force due to gravity, $\mathbf{F}$, is pulling straight down on the boat. Part of this force, $\mathbf{F}_1$, is pushing the boat down the ramp. Another part of this force, $\mathbf{F}_2$, is pressing the boat against the ramp, at a right angle to the incline. These two orthogonal vectors, $\mathbf{F}_1$ and $\mathbf{F}_2$, are called the **vector components** of $\mathbf{F}$. Notice that

$$\mathbf{F} = \mathbf{F}_1 + \mathbf{F}_2.$$

A method for finding $\mathbf{F}_1$ and $\mathbf{F}_2$ involves projecting a vector onto another vector.

Figure 6.70 shows two nonzero vectors, $\mathbf{v}$ and $\mathbf{w}$, with the same initial point. The angle between the vectors, θ, is acute in Figure 6.70(a) and obtuse in Figure 6.70(b). A third vector, called the **vector projection of v onto w**, is also shown in each figure, denoted by $\text{proj}_{\mathbf{w}}\mathbf{v}$.

Figure 6.70(a)   **Figure 6.70(b)**

How is the vector projection of $\mathbf{v}$ onto $\mathbf{w}$ formed? Draw the line segment from the terminal point of $\mathbf{v}$ that forms a right angle with a line through $\mathbf{w}$, shown in red. The projection of $\mathbf{v}$ onto $\mathbf{w}$ lies on a line through $\mathbf{w}$, and is parallel to vector $\mathbf{w}$. This vector begins at the common initial point of $\mathbf{v}$ and $\mathbf{w}$. It ends at the point where the dashed red line segment intersects the line through $\mathbf{w}$.

Our goal is to determine an expression for $\text{proj}_{\mathbf{w}}\mathbf{v}$. We begin with its magnitude. By the definition of the cosine function,

$$\cos \theta = \frac{\|\text{proj}_{\mathbf{w}}\mathbf{v}\|}{\|\mathbf{v}\|}.$$

> This is the magnitude of the vector projection of v onto w.

$\|\mathbf{v}\| \cos \theta = \|\text{proj}_{\mathbf{w}}\mathbf{v}\|$ Multiply both sides by $\|\mathbf{v}\|$.

$\|\text{proj}_{\mathbf{w}}\mathbf{v}\| = \|\mathbf{v}\| \cos \theta$ Reverse the two sides.

We can rewrite the right side of this equation and obtain another expression for the magnitude of the vector projection of $\mathbf{v}$ onto $\mathbf{w}$. To do so, use the alternate formula for the dot product, $\mathbf{v} \cdot \mathbf{w} = \|\mathbf{v}\|\|\mathbf{w}\| \cos \theta$.

Divide both sides of $\mathbf{v} \cdot \mathbf{w} = \|\mathbf{v}\|\|\mathbf{w}\| \cos \theta$ by $\|\mathbf{w}\|$:

$$\frac{\mathbf{v} \cdot \mathbf{w}}{\|\mathbf{w}\|} = \|\mathbf{v}\| \cos \theta.$$

The expression on the right side of this equation, $\|\mathbf{v}\| \cos \theta$, is the same expression that appears in the formula for $\|\text{proj}_{\mathbf{w}}\mathbf{v}\|$. Thus,

$$\|\text{proj}_{\mathbf{w}}\mathbf{v}\| = \|\mathbf{v}\| \cos \theta = \frac{\mathbf{v} \cdot \mathbf{w}}{\|\mathbf{w}\|}.$$

We use the formula for the magnitude of $\text{proj}_{\mathbf{w}}\mathbf{v}$ to find the vector itself. This is done by finding the scalar product of the magnitude and the unit vector in the direction of $\mathbf{w}$.

$$\text{proj}_{\mathbf{w}}\mathbf{v} = \left(\frac{\mathbf{v} \cdot \mathbf{w}}{\|\mathbf{w}\|}\right)\left(\frac{\mathbf{w}}{\|\mathbf{w}\|}\right) = \frac{\mathbf{v} \cdot \mathbf{w}}{\|\mathbf{w}\|^2}\mathbf{w}$$

This is the magnitude of the vector projection of v onto w.

This is the unit vector in the direction of w.

The Vector Projection of v Onto w

If $\mathbf{v}$ and $\mathbf{w}$ are two nonzero vectors, the vector projection of $\mathbf{v}$ onto $\mathbf{w}$ is

$$\text{proj}_{\mathbf{w}}\mathbf{v} = \frac{\mathbf{v} \cdot \mathbf{w}}{\|\mathbf{w}\|^2}\mathbf{w}.$$

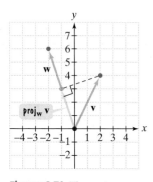

Figure 6.71 The vector projection of **v** onto **w**

EXAMPLE 4 **Finding the Vector Projection of One Vector Onto Another**

If $\mathbf{v} = 2\mathbf{i} + 4\mathbf{j}$ and $\mathbf{w} = -2\mathbf{i} + 6\mathbf{j}$, find the vector projection of $\mathbf{v}$ onto $\mathbf{w}$.

Solution The vector projection of $\mathbf{v}$ onto $\mathbf{w}$ is found using the formula for $\text{proj}_{\mathbf{w}}\mathbf{v}$.

$$\text{proj}_{\mathbf{w}}\mathbf{v} = \frac{\mathbf{v} \cdot \mathbf{w}}{\|\mathbf{w}\|^2}\mathbf{w} = \frac{(2\mathbf{i} + 4\mathbf{j}) \cdot (-2\mathbf{i} + 6\mathbf{j})}{\left(\sqrt{(-2)^2 + 6^2}\right)^2}\mathbf{w}$$

$$= \frac{2(-2) + 4(6)}{\left(\sqrt{40}\right)^2}\mathbf{w} = \frac{20}{40}\mathbf{w} = \tfrac{1}{2}(-2\mathbf{i} + 6\mathbf{j}) = -\mathbf{i} + 3\mathbf{j}$$

The three vectors, $\mathbf{v}$, $\mathbf{w}$, and $\text{proj}_{\mathbf{w}}\mathbf{v}$, are shown in Figure 6.71.

Check Point 4 If $\mathbf{v} = 2\mathbf{i} - 5\mathbf{j}$ and $\mathbf{w} = \mathbf{i} - \mathbf{j}$, find the vector projection of $\mathbf{v}$ onto $\mathbf{w}$.

 Express a vector as the sum of two orthogonal vectors.

We use the vector projection of $\mathbf{v}$ onto $\mathbf{w}$, $\text{proj}_{\mathbf{w}}\mathbf{v}$, to express $\mathbf{v}$ as the sum of two orthogonal vectors.

The Vector Components of v

Let $\mathbf{v}$ and $\mathbf{w}$ be two nonzero vectors. Vector $\mathbf{v}$ can be expressed as the sum of two orthogonal vectors, $\mathbf{v}_1$ and $\mathbf{v}_2$, where $\mathbf{v}_1$ is parallel to $\mathbf{w}$ and $\mathbf{v}_2$ is orthogonal to $\mathbf{w}$.

$$\mathbf{v}_1 = \text{proj}_{\mathbf{w}}\mathbf{v} = \frac{\mathbf{v} \cdot \mathbf{w}}{\|\mathbf{w}\|^2}\mathbf{w}, \quad \mathbf{v}_2 = \mathbf{v} - \mathbf{v}_1$$

Thus, $\mathbf{v} = \mathbf{v}_1 + \mathbf{v}_2$. The vectors $\mathbf{v}_1$ and $\mathbf{v}_2$ are called the **vector components** of $\mathbf{v}$. The process of expressing $\mathbf{v}$ as $\mathbf{v}_1 + \mathbf{v}_2$ is called the **decomposition** of $\mathbf{v}$ into $\mathbf{v}_1$ and $\mathbf{v}_2$.

EXAMPLE 5 Decomposing a Vector into Two Orthogonal Vectors

Let $\mathbf{v} = 2\mathbf{i} + 4\mathbf{j}$ and $\mathbf{w} = -2\mathbf{i} + 6\mathbf{j}$. Decompose $\mathbf{v}$ into two vectors, $\mathbf{v}_1$ and $\mathbf{v}_2$, where $\mathbf{v}_1$ is parallel to $\mathbf{w}$ and $\mathbf{v}_2$ is orthogonal to $\mathbf{w}$.

Solution These are the vectors we worked with in Example 4. We use the formulas in the preceding box.

$$\mathbf{v}_1 = \text{proj}_{\mathbf{w}}\mathbf{v} = -\mathbf{i} + 3\mathbf{j} \qquad \text{We obtained this vector in Example 4.}$$
$$\mathbf{v}_2 = \mathbf{v} - \mathbf{v}_1 = (2\mathbf{i} + 4\mathbf{j}) - (-\mathbf{i} + 3\mathbf{j}) = 3\mathbf{i} + \mathbf{j}$$

Check Point 5 Let $\mathbf{v} = 2\mathbf{i} - 5\mathbf{j}$ and $\mathbf{w} = \mathbf{i} - \mathbf{j}$. (These are the vectors from Check Point 4.) Decompose $\mathbf{v}$ into two vectors, $\mathbf{v}_1$ and $\mathbf{v}_2$, where $\mathbf{v}_1$ is parallel to $\mathbf{w}$ and $\mathbf{v}_2$ is orthogonal to $\mathbf{w}$.

⑥ Compute work.

Work: An Application of the Dot Product

The bad news: Your car just died. The good news: It died on a level road just 200 feet from a gas station. Exerting a constant force of 90 pounds, and not necessarily whistling as you work, you manage to push the car to the gas station.

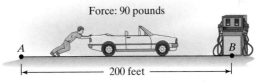

Force: 90 pounds

200 feet

Although you did not whistle, you certainly did work pushing the car 200 feet from point A to point B. How much work did you do? If a constant force $\mathbf{F}$ is applied to an object, moving it from point A to point B in the direction of the force, the work, W, done is

$$W = (\text{magnitude of force})(\text{distance from } A \text{ to } B).$$

You pushed with a force of 90 pounds for a distance of 200 feet. The work done by your force is

$$W = (90 \text{ pounds})(200 \text{ feet})$$

or 18,000 foot-pounds. Work is often measured in foot-pounds or in newton-meters.

The photo on the left shows an adult pulling a small child in a wagon. Work is being done. However, the situation is not quite the same as pushing your car. Pushing the car, the force you applied was along the line of motion. By contrast, the force of the adult pulling the wagon is not applied along the line of the wagon's motion. In this case, the dot product is used to determine the work done by the force.

Definition of Work
The work, W, done by a force $\mathbf{F}$ moving an object from A to B is
$$W = \mathbf{F} \cdot \overrightarrow{AB}.$$

When computing work, it is often easier to use the alternative formula for the dot product. Thus,

$$W = \mathbf{F} \cdot \overrightarrow{AB} = \|\mathbf{F}\| \, \|\overrightarrow{AB}\| \cos \theta.$$

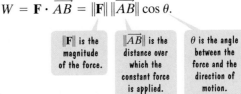

$\|\mathbf{F}\|$ is the magnitude of the force. $\|\overrightarrow{AB}\|$ is the distance over which the constant force is applied. θ is the angle between the force and the direction of motion.

It is correct to refer to W as either the work done or the work done by the force.

EXAMPLE 6 Computing Work

A child pulls a sled along level ground by exerting a force of 30 pounds on a rope that makes an angle of 35° with the horizontal. How much work is done pulling the sled 200 feet?

Solution The situation is illustrated in Figure 6.72. The work done is

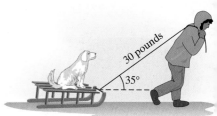

Figure 6.72 Computing work done pulling the sled 200 feet

$$W = \|\mathbf{F}\| \|\overrightarrow{AB}\| \cos \theta = (30)(200) \cos 35° \approx 4915.$$

| Magnitude of the force is 30 pounds. | Distance is 200 feet. | The angle between the force and the sled's motion is 35°. |

Thus, the work done is approximately 4915 foot-pounds.

Check Point 6 A child pulls a wagon along level ground by exerting a force of 20 pounds on a handle that makes an angle of 30° with the horizontal. How much work is done pulling the wagon 150 feet?

EXERCISE SET 6.7

 Practice Exercises

In Exercises 1–8, use the given vectors to find $\mathbf{v} \cdot \mathbf{w}$ *and* $\mathbf{v} \cdot \mathbf{v}$.

1. $\mathbf{v} = 3\mathbf{i} + \mathbf{j}, \quad \mathbf{w} = \mathbf{i} + 3\mathbf{j}$
2. $\mathbf{v} = 3\mathbf{i} + 3\mathbf{j}, \quad \mathbf{w} = \mathbf{i} + 4\mathbf{j}$
3. $\mathbf{v} = 5\mathbf{i} - 4\mathbf{j}, \quad \mathbf{w} = -2\mathbf{i} - \mathbf{j}$
4. $\mathbf{v} = 7\mathbf{i} - 2\mathbf{j}, \quad \mathbf{w} = -3\mathbf{i} - \mathbf{j}$
5. $\mathbf{v} = -6\mathbf{i} - 5\mathbf{j}, \quad \mathbf{w} = -10\mathbf{i} - 8\mathbf{j}$
6. $\mathbf{v} = -8\mathbf{i} - 3\mathbf{j}, \quad \mathbf{w} = -10\mathbf{i} - 5\mathbf{j}$
7. $\mathbf{v} = 5\mathbf{i}, \quad \mathbf{w} = \mathbf{j}$ 8. $\mathbf{v} = \mathbf{i}, \quad \mathbf{w} = -5\mathbf{j}$

In Exercises 9–16, let

$$\mathbf{u} = 2\mathbf{i} - \mathbf{j}, \quad \mathbf{v} = 3\mathbf{i} + \mathbf{j}, \quad \text{and} \quad \mathbf{w} = \mathbf{i} + 4\mathbf{j}.$$

Find each specified scalar.

9. $\mathbf{u} \cdot (\mathbf{v} + \mathbf{w})$ 10. $\mathbf{v} \cdot (\mathbf{u} + \mathbf{w})$
11. $\mathbf{u} \cdot \mathbf{v} + \mathbf{u} \cdot \mathbf{w}$ 12. $\mathbf{v} \cdot \mathbf{u} + \mathbf{v} \cdot \mathbf{w}$
13. $(4\mathbf{u}) \cdot \mathbf{v}$ 14. $(5\mathbf{v}) \cdot \mathbf{w}$
15. $4(\mathbf{u} \cdot \mathbf{v})$ 16. $5(\mathbf{v} \cdot \mathbf{w})$

In Exercises 17–22, find the angle between $\mathbf{v}$ *and* $\mathbf{w}$. *Round to the nearest tenth of a degree.*

17. $\mathbf{v} = 2\mathbf{i} - \mathbf{j}, \quad \mathbf{w} = 3\mathbf{i} + 4\mathbf{j}$
18. $\mathbf{v} = -2\mathbf{i} + 5\mathbf{j}, \quad \mathbf{w} = 3\mathbf{i} + 6\mathbf{j}$
19. $\mathbf{v} = -3\mathbf{i} + 2\mathbf{j}, \quad \mathbf{w} = 4\mathbf{i} - \mathbf{j}$
20. $\mathbf{v} = \mathbf{i} + 2\mathbf{j}, \quad \mathbf{w} = 4\mathbf{i} - 3\mathbf{j}$
21. $\mathbf{v} = 6\mathbf{i}, \quad \mathbf{w} = 5\mathbf{i} + 4\mathbf{j}$
22. $\mathbf{v} = 3\mathbf{j}, \quad \mathbf{w} = 4\mathbf{i} + 5\mathbf{j}$

In Exercises 23–32, use the dot product to determine whether $\mathbf{v}$ *and* $\mathbf{w}$ *are orthogonal.*

23. $\mathbf{v} = \mathbf{i} + \mathbf{j}, \quad \mathbf{w} = \mathbf{i} - \mathbf{j}$ 24. $\mathbf{v} = \mathbf{i} + \mathbf{j}, \quad \mathbf{w} = -\mathbf{i} + \mathbf{j}$
25. $\mathbf{v} = 2\mathbf{i} + 8\mathbf{j}, \quad \mathbf{w} = 4\mathbf{i} - \mathbf{j}$
26. $\mathbf{v} = 8\mathbf{i} - 4\mathbf{j}, \quad \mathbf{w} = -6\mathbf{i} - 12\mathbf{j}$
27. $\mathbf{v} = 2\mathbf{i} - 2\mathbf{j}, \quad \mathbf{w} = -\mathbf{i} + \mathbf{j}$
28. $\mathbf{v} = 5\mathbf{i} - 5\mathbf{j}, \quad \mathbf{w} = \mathbf{i} - \mathbf{j}$
29. $\mathbf{v} = 3\mathbf{i}, \quad \mathbf{w} = -4\mathbf{i}$ 30. $\mathbf{v} = 5\mathbf{i}, \quad \mathbf{w} = -6\mathbf{i}$
31. $\mathbf{v} = 3\mathbf{i}, \quad \mathbf{w} = -4\mathbf{j}$ 32. $\mathbf{v} = 5\mathbf{i}, \quad \mathbf{w} = -6\mathbf{j}$

In Exercise 33–38, find $\text{proj}_{\mathbf{w}}\mathbf{v}$. *Then decompose* $\mathbf{v}$ *into two vectors,* $\mathbf{v}_1$ *and* $\mathbf{v}_2$, *where* $\mathbf{v}_1$ *is parallel to* $\mathbf{w}$ *and* $\mathbf{v}_2$ *is orthogonal to* $\mathbf{w}$.

33. $\mathbf{v} = 3\mathbf{i} - 2\mathbf{j}, \quad \mathbf{w} = \mathbf{i} - \mathbf{j}$
34. $\mathbf{v} = 3\mathbf{i} - 2\mathbf{j}, \quad \mathbf{w} = 2\mathbf{i} + \mathbf{j}$
35. $\mathbf{v} = \mathbf{i} + 3\mathbf{j}, \quad \mathbf{w} = -2\mathbf{i} + 5\mathbf{j}$
36. $\mathbf{v} = 2\mathbf{i} + 4\mathbf{j}, \quad \mathbf{w} = -3\mathbf{i} + 6\mathbf{j}$
37. $\mathbf{v} = \mathbf{i} + 2\mathbf{j}, \quad \mathbf{w} = 3\mathbf{i} + 6\mathbf{j}$
38. $\mathbf{v} = 2\mathbf{i} + \mathbf{j}, \quad \mathbf{w} = 6\mathbf{i} + 3\mathbf{j}$

Practice Plus

In Exercises 39–42, let

$$\mathbf{u} = -\mathbf{i} + \mathbf{j}, \quad \mathbf{v} = 3\mathbf{i} - 2\mathbf{j}, \quad \text{and} \quad \mathbf{w} = -5\mathbf{j}.$$

Find each specified scalar or vector.

39. $5\mathbf{u} \cdot (3\mathbf{v} - 4\mathbf{w})$ 40. $4\mathbf{u} \cdot (5\mathbf{v} - 3\mathbf{w})$
41. $\text{proj}_{\mathbf{u}}(\mathbf{v} + \mathbf{w})$ 42. $\text{proj}_{\mathbf{u}}(\mathbf{v} - \mathbf{w})$

In Exercises 43–44, find the angle, in degrees, between **v** *and* **w**.

43. $\mathbf{v} = 2\cos\dfrac{4\pi}{3}\mathbf{i} + 2\sin\dfrac{4\pi}{3}\mathbf{j},\quad \mathbf{w} = 3\cos\dfrac{3\pi}{2}\mathbf{i} + 3\sin\dfrac{3\pi}{2}\mathbf{j}$

44. $\mathbf{v} = 3\cos\dfrac{5\pi}{3}\mathbf{i} + 3\sin\dfrac{5\pi}{3}\mathbf{j},\quad \mathbf{w} = 2\cos\pi\mathbf{i} + 2\sin\pi\mathbf{j}$

In Exercises 45–50, determine whether **v** *and* **w** *are parallel, orthogonal, or neither.*

45. $\mathbf{v} = 3\mathbf{i} - 5\mathbf{j},\quad \mathbf{w} = 6\mathbf{i} - 10\mathbf{j}$

46. $\mathbf{v} = -2\mathbf{i} + 3\mathbf{j},\quad \mathbf{w} = -6\mathbf{i} + 9\mathbf{j}$

47. $\mathbf{v} = 3\mathbf{i} - 5\mathbf{j},\quad \mathbf{w} = 6\mathbf{i} + 10\mathbf{j}$

48. $\mathbf{v} = -2\mathbf{i} + 3\mathbf{j},\quad \mathbf{w} = -6\mathbf{i} - 9\mathbf{j}$

49. $\mathbf{v} = 3\mathbf{i} - 5\mathbf{j},\quad \mathbf{w} = 6\mathbf{i} + \dfrac{18}{5}\mathbf{j}$

50. $\mathbf{v} = -2\mathbf{i} + 3\mathbf{j},\quad \mathbf{w} = -6\mathbf{i} - 4\mathbf{j}$

Application Exercises

51. The components of $\mathbf{v} = 240\mathbf{i} + 300\mathbf{j}$ represent the respective number of gallons of regular and premium gas sold at a station. The components of $\mathbf{w} = 2.90\mathbf{i} + 3.07\mathbf{j}$ represent the respective prices per gallon for each kind of gas. Find $\mathbf{v} \cdot \mathbf{w}$ and describe what the answer means in practical terms.

52. The components of $\mathbf{v} = 180\mathbf{i} + 450\mathbf{j}$ represent the respective number of one-day and three-day videos rented from a video store. The components of $\mathbf{w} = 3\mathbf{i} + 2\mathbf{j}$ represent the prices to rent the one-day and three-day videos, respectively. Find $\mathbf{v} \cdot \mathbf{w}$ and describe what the answer means in practical terms.

53. Find the work done in pushing a car along a level road from point A to point B, 80 feet from A, while exerting a constant force of 95 pounds. Round to the nearest foot-pound.

54. Find the work done when a crane lifts a 6000-pound boulder through a vertical distance of 12 feet. Round to the nearest foot-pound.

55. A wagon is pulled along level ground by exerting a force of 40 pounds on a handle that makes an angle of 32° with the horizontal. How much work is done pulling the wagon 100 feet? Round to the nearest foot-pound.

56. A wagon is pulled along level ground by exerting a force of 25 pounds on a handle that makes an angle of 38° with the horizontal. How much work is done pulling the wagon 100 feet? Round to the nearest foot-pound.

57. A force of 60 pounds on a rope is used to pull a box up a ramp inclined at 12° from the horizontal. The figure shows that the rope forms an angle of 38° with the horizontal. How much work is done pulling the box 20 feet along the ramp?

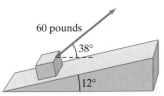

58. A force of 80 pounds on a rope is used to pull a box up a ramp inclined at 10° from the horizontal. The rope forms an angle of 33° with the horizontal. How much work is done pulling the box 25 feet along the ramp?

59. A force is given by the vector $\mathbf{F} = 3\mathbf{i} + 2\mathbf{j}$. The force moves an object along a straight line from the point $(4, 9)$ to the point $(10, 20)$. Find the work done if the distance is measured in feet and the force is measured in pounds.

60. A force is given by the vector $\mathbf{F} = 5\mathbf{i} + 7\mathbf{j}$. The force moves an object along a straight line from the point $(8, 11)$ to the point $(18, 20)$. Find the work done if the distance is measured in meters and the force is measured in newtons.

61. A force of 4 pounds acts in the direction of 50° to the horizontal. The force moves an object along a straight line from the point $(3, 7)$ to the point $(8, 10)$, with distance measured in feet. Find the work done by the force.

62. A force of 6 pounds acts in the direction of 40° to the horizontal. The force moves an object along a straight line from the point $(5, 9)$ to the point $(8, 20)$, with the distance measured in feet. Find the work done by the force.

63. Refer to Figure 6.69 on page 697. Suppose that the boat weighs 700 pounds and is on a ramp inclined at 30°. Represent the force due to gravity, $\mathbf{F}$, using

$$\mathbf{F} = -700\mathbf{j}.$$

 a. Write a unit vector along the ramp in the upward direction.

 b. Find the vector projection of $\mathbf{F}$ onto the unit vector from part (a).

 c. What is the magnitude of the vector projection in part (b)? What does this represent?

64. Refer to Figure 6.69 on page 697. Suppose that the boat weighs 650 pounds and is on a ramp inclined at 30°. Represent the force due to gravity, $\mathbf{F}$, using

$$\mathbf{F} = -650\mathbf{j}.$$

 a. Write a unit vector along the ramp in the upward direction.

 b. Find the vector projection of $\mathbf{F}$ onto the unit vector from part (a).

 c. What is the magnitude of the vector projection in part (b)? What does this represent?

Writing in Mathematics

65. Explain how to find the dot product of two vectors.

66. Using words and no symbols, describe how to find the dot product of two vectors with the alternative formula

$$\mathbf{v} \cdot \mathbf{w} = \|\mathbf{v}\|\|\mathbf{w}\| \cos \theta.$$

67. Describe how to find the angle between two vectors.

68. What are parallel vectors?

69. What are orthogonal vectors?

70. How do you determine if two vectors are orthogonal?

71. Draw two vectors, **v** and **w**, with the same initial point. Show the vector projection of **v** onto **w** in your diagram. Then describe how you identified this vector.

72. How do you determine the work done by a force $\mathbf{F}$ in moving an object from A to B when the direction of the force is not along the line of motion?

73. A weightlifter is holding a barbell perfectly still above his head, his body shaking from the effort. How much work is the weightlifter doing? Explain your answer.

74. Describe one way in which the everyday use of the word "work" is different from the definition of work given in this section.

In Exercises 75–77, use the vectors

$$\mathbf{u} = a_1\mathbf{i} + b_1\mathbf{j}, \quad \mathbf{v} = a_2\mathbf{i} + b_2\mathbf{j}, \quad \text{and} \quad \mathbf{w} = a_3\mathbf{i} + b_3\mathbf{j},$$

to prove the given property.

75. $\mathbf{u} \cdot \mathbf{v} = \mathbf{v} \cdot \mathbf{u}$

76. $(c\mathbf{u}) \cdot \mathbf{v} = c(\mathbf{u} \cdot \mathbf{v})$

77. $\mathbf{u} \cdot (\mathbf{v} + \mathbf{w}) = \mathbf{u} \cdot \mathbf{v} + \mathbf{u} \cdot \mathbf{w}$

78. If $\mathbf{v} = -2\mathbf{i} + 5\mathbf{j}$, find a vector orthogonal to $\mathbf{v}$.

79. Find a value of b so that $15\mathbf{i} - 3\mathbf{j}$ and $-4\mathbf{i} + b\mathbf{j}$ are orthogonal.

80. Prove that the projection of $\mathbf{v}$ onto $\mathbf{i}$ is $(\mathbf{v} \cdot \mathbf{i})\mathbf{i}$.

81. Find two vectors $\mathbf{v}$ and $\mathbf{w}$ such that the projection of $\mathbf{v}$ onto $\mathbf{w}$ is $\mathbf{v}$.

Group Exercise

82. Group members should research and present a report on unusual and interesting applications of vectors.

Chapter 6
Summary, Review, and Test

Summary

DEFINITIONS AND CONCEPTS	EXAMPLES

6.1 **and** 6.2 The Law of Sines; The Law of Cosines

a. The Law of Sines

$$\frac{a}{\sin A} = \frac{b}{\sin B} = \frac{c}{\sin C}$$

Ex. 1, p. 629;
Ex. 2, p. 630;
Ex. 3, p. 631;
Ex. 4, p. 632;
Ex. 5, p. 632

b. The Law of Sines is used to solve SAA, ASA, and SSA (the ambiguous case) triangles. The ambiguous case may result in no triangle, one triangle, or two triangles; see the box on page 631.

c. The area of a triangle equals one-half the product of the lengths of two sides times the sine of their included angle.

Ex. 6, p. 634

d. The Law of Cosines

$$a^2 = b^2 + c^2 - 2bc \cos A$$
$$b^2 = a^2 + c^2 - 2ac \cos B$$
$$c^2 = a^2 + b^2 - 2ab \cos C$$

e. The Law of Cosines is used to find the side opposite the given angle in an SAS triangle; see the box on page 640. The Law of Cosines is also used to find the angle opposite the longest side in an SSS triangle; see the box on page 641.

Ex. 1, p. 641;
Ex. 2, p. 642

f. Heron's Formula for the Area of a Triangle

The area of a triangle with sides $a, b,$ and c is $\sqrt{s(s-a)(s-b)(s-c)}$, where s is one-half its perimeter: $s = \frac{1}{2}(a + b + c)$.

Ex. 4, p. 643

6.3 **and** 6.4 Polar Coordinates; Graphs of Polar Equations

a. A point P in the polar coordinate system is represented by (r, θ), where r is the directed distance from the pole to the point and θ is the angle from the polar axis to line segment OP. The elements of the ordered pair (r, θ) are called the polar coordinates of P. See Figure 6.20 on page 647. When r in (r, θ) is negative, a point is located $|r|$ units along the ray opposite the terminal side of θ. Important information about the sign of r and the location of the point (r, θ) is found in the box on page 648.

Ex. 1, p. 648

b. Multiple Representations of Points

If n is any integer, $(r, \theta) = (r, \theta + 2n\pi)$ or $(r, \theta) = (-r, \theta + \pi + 2n\pi)$.

Ex. 2, p. 649

c. Relations between Polar and Rectangular Coordinates

$$x = r \cos \theta, \quad y = r \sin \theta, \quad x^2 + y^2 = r^2, \quad \tan \theta = \frac{y}{x}$$

DEFINITIONS AND CONCEPTS	**EXAMPLES**
d. To convert a point from polar coordinates (r, θ) to rectangular coordinates (x, y), use $x = r \cos \theta$ and $y = r \sin \theta$.	Ex. 3, p. 650
e. To convert a point from rectangular coordinates (x, y) to polar coordinates (r, θ), use the procedure in the box on page 651.	Ex. 4, p. 651; Ex. 5, p. 652
f. To convert a rectangular equation to a polar equation, replace x with $r \cos \theta$ and y with $r \sin \theta$.	Ex. 6, p. 652
g. To convert a polar equation to a rectangular equation, use one or more of $$r^2 = x^2 + y^2, \quad r \cos \theta = x, \quad r \sin \theta = y, \quad \text{and} \quad \tan \theta = \frac{y}{x}.$$ It is often necessary to do something to the given polar equation before using the preceding expressions.	Ex. 7, p. 654
h. A polar equation is an equation whose variables are r and θ. The graph of a polar equation is the set of all points whose polar coordinates satisfy the equation.	Ex. 1, p. 658
i. Polar equations can be graphed using point plotting and symmetry (see the box on page 659).	Ex. 2, p. 660
j. The graphs of $r = a \cos \theta$ and $r = a \sin \theta$ are circles. See the box on page 659. The graphs of $r = a \pm b \sin \theta$ and $r = a \pm b \cos \theta$ are called limaçons ($a > 0$ and $b > 0$), shown in the box on page 662. The graphs of $r = a \sin n\theta$ and $r = a \cos n\theta$, $a \neq 0$, are rose curves with $2n$ petals if n is even and n petals if n is odd. See the box on page 664. The graphs of $r^2 = a^2 \sin 2\theta$ and $r^2 = a^2 \cos 2\theta$, $a \neq 0$, are called lemniscates and are shown in the box on page 665.	Ex. 3, p. 661; Ex. 4, p. 662; Ex. 5, p. 664

6.5 Complex Numbers in Polar Form; DeMoivre's Theorem

a. The complex number $z = a + bi$ is represented as a point (a, b) in the complex plane, shown in Figure 6.38 on page 669.	Ex. 1, p. 670
b. The absolute value of $z = a + bi$ is $|z| = |a + bi| = \sqrt{a^2 + b^2}$.	Ex. 2, p. 670
c. The polar form of $z = a + bi$ is $z = r(\cos \theta + i \sin \theta)$, where $a = r \cos \theta$, $b = r \sin \theta$, $r = \sqrt{a^2 + b^2}$, and $\tan \theta = \frac{b}{a}$. We call r the modulus and θ the argument of z, with $0 \leq \theta < 2\pi$.	Ex. 3, p. 671; Ex. 4, p. 672
d. Multiplying Complex Numbers in Polar Form: Multiply moduli and add arguments. See the box on page 672.	Ex. 5, p. 673
e. Dividing Complex Numbers in Polar Form: Divide moduli and subtract arguments. See the box on page 673.	Ex. 6, p. 673
f. DeMoivre's Theorem is used to find powers of complex numbers in polar form. $$[r(\cos \theta + i \sin \theta)]^n = r^n(\cos n\theta + i \sin n\theta)$$	Ex. 7, p. 674; Ex. 8, p. 675
g. DeMoivre's Theorem can be used to find roots of complex numbers in polar form. The n distinct nth roots of $r(\cos \theta + i \sin \theta)$ are $$\sqrt[n]{r}\left[\cos\left(\frac{\theta + 2\pi k}{n}\right) + i \sin\left(\frac{\theta + 2\pi k}{n}\right)\right]$$ or $$\sqrt[n]{r}\left[\cos\left(\frac{\theta + 360°k}{n}\right) + i \sin\left(\frac{\theta + 360°k}{n}\right)\right],$$ where $k = 0, 1, 2, \ldots, n - 1$.	Ex. 9, p. 676; Ex. 10, p. 677

DEFINITIONS AND CONCEPTS	EXAMPLES

6.6 Vectors

a. A vector is a directed line segment.

b. Equal vectors have the same magnitude and the same direction.

Ex. 1, p. 681

c. The vector $k\mathbf{v}$, the scalar multiple of the vector $\mathbf{v}$ and the scalar k, has magnitude $|k|\|\mathbf{v}\|$. The direction of $k\mathbf{v}$ is the same as that of $\mathbf{v}$ if $k > 0$ and opposite $\mathbf{v}$ if $k < 0$.

Figure 6.52, p. 682

d. The sum $\mathbf{u} + \mathbf{v}$, called the resultant vector, can be expressed geometrically. Position $\mathbf{u}$ and $\mathbf{v}$ so that the terminal point of $\mathbf{u}$ coincides with the initial point of $\mathbf{v}$. The vector $\mathbf{u} + \mathbf{v}$ extends from the initial point of $\mathbf{u}$ to the terminal point of $\mathbf{v}$.

Figure 6.53, p. 682

e. The difference of two vectors, $\mathbf{u} - \mathbf{v}$, is defined as $\mathbf{u} + (-\mathbf{v})$.

Figure 6.54, p. 683

f. The vector $\mathbf{i}$ is the unit vector whose direction is along the positive x-axis. The vector $\mathbf{j}$ is the unit vector whose direction is along the positive y-axis.

g. Vector $\mathbf{v}$, from $(0,0)$ to (a,b), called a position vector, is represented as $\mathbf{v} = a\mathbf{i} + b\mathbf{j}$, where a is the horizontal component and b is the vertical component. The magnitude of $\mathbf{v}$ is given by $\|\mathbf{v}\| = \sqrt{a^2 + b^2}$.

Ex. 2, p. 684

h. Vector $\mathbf{v}$ from (x_1, y_1) to (x_2, y_2) is equal to the position vector $\mathbf{v} = (x_2 - x_1)\mathbf{i} + (y_2 - y_1)\mathbf{j}$. In rectangular coordinates, the term "vector" refers to the position vector in terms of $\mathbf{i}$ and $\mathbf{j}$ that is equal to it.

Ex. 3, p. 684

i. Operations with Vectors in Terms of $\mathbf{i}$ and $\mathbf{j}$
If $\mathbf{v} = a_1\mathbf{i} + b_1\mathbf{j}$ and $\mathbf{w} = a_2\mathbf{i} + b_2\mathbf{j}$, then
$$\bullet\ \mathbf{v} + \mathbf{w} = (a_1 + a_2)\mathbf{i} + (b_1 + b_2)\mathbf{j}$$
$$\bullet\ \mathbf{v} - \mathbf{w} = (a_1 - a_2)\mathbf{i} + (b_1 - b_2)\mathbf{j}$$
$$\bullet\ k\mathbf{v} = (ka_1)\mathbf{i} + (kb_1)\mathbf{j}$$

Ex. 4, p. 685; Ex. 5, p. 686; Ex. 6, p. 686

j. The zero vector $\mathbf{0}$ is the vector whose magnitude is 0 and is assigned no direction. Many properties of vector addition and scalar multiplication involve the zero vector. Some of these properties are listed in the box on page 687.

k. The vector $\dfrac{\mathbf{v}}{\|\mathbf{v}\|}$ is the unit vector that has the same direction as $\mathbf{v}$.

Ex. 7, p. 687

l. A vector with magnitude $\|\mathbf{v}\|$ and direction angle θ, the angle that $\mathbf{v}$ makes with the positive x-axis, can be expressed in terms of its magnitude and direction angle as
$$\mathbf{v} = \|\mathbf{v}\| \cos \theta \mathbf{i} + \|\mathbf{v}\| \sin \theta \mathbf{j}.$$

Ex. 8, p. 688; Ex. 9, p. 689

6.7 The Dot Product

a. Definition of the Dot Product
If $\mathbf{v} = a_1\mathbf{i} + b_1\mathbf{j}$ and $\mathbf{w} = a_2\mathbf{i} + b_2\mathbf{j}$, the dot product of $\mathbf{v}$ and $\mathbf{w}$ is defined by $\mathbf{v} \cdot \mathbf{w} = a_1a_2 + b_1b_2$.

Ex. 1, p. 694

b. Alternative Formula for the Dot Product: $\mathbf{v} \cdot \mathbf{w} = \|\mathbf{v}\|\|\mathbf{w}\| \cos \theta$, where θ is the smallest nonnegative angle between $\mathbf{v}$ and $\mathbf{w}$

c. Angle between Two Vectors
$$\cos \theta = \frac{\mathbf{v} \cdot \mathbf{w}}{\|\mathbf{v}\|\|\mathbf{w}\|} \quad \text{and} \quad \theta = \cos^{-1}\left(\frac{\mathbf{v} \cdot \mathbf{w}}{\|\mathbf{v}\|\|\mathbf{w}\|}\right)$$

Ex. 2, p. 696

d. Two vectors are orthogonal when the angle between them is $90°$. To show that two vectors are orthogonal, show that their dot product is zero.

Ex. 3, p. 697

e. The vector projection of $\mathbf{v}$ onto $\mathbf{w}$ is given by
$$\text{proj}_{\mathbf{w}}\mathbf{v} = \frac{\mathbf{v} \cdot \mathbf{w}}{\|\mathbf{w}\|^2}\mathbf{w}.$$

Ex. 4, p. 698

f. Expressing a vector as the sum of two orthogonal vectors, called the vector components, is shown in the box on page 698.

Ex. 5, p. 699

g. The work, W, done by a force $\mathbf{F}$ moving an object from A to B is $W = \mathbf{F} \cdot \overrightarrow{AB}$.
Thus, $W = \|\mathbf{F}\|\|\overrightarrow{AB}\| \cos \theta$, where θ is the angle between the force and the direction of motion.

Ex. 6, p. 700

Review Exercises

6.1 and 6.2

In Exercises 1–12, solve each triangle. Round lengths to the nearest tenth and angle measures to the nearest degree. If no triangle exists, state "no triangle." If two triangles exist, solve each triangle.

1. $A = 70°$, $B = 55°$, $a = 12$
2. $B = 107°$, $C = 30°$, $c = 126$
3. $B = 66°$, $a = 17$, $c = 12$
4. $a = 117$, $b = 66$, $c = 142$
5. $A = 35°$, $B = 25°$, $c = 68$
6. $A = 39°$, $a = 20$, $b = 26$
7. $C = 50°$, $a = 3$, $c = 1$
8. $A = 162°$, $b = 11.2$, $c = 48.2$
9. $a = 26.1$, $b = 40.2$, $c = 36.5$
10. $A = 40°$, $a = 6$, $b = 4$
11. $B = 37°$, $a = 12.4$, $b = 8.7$
12. $A = 23°$, $a = 54.3$, $b = 22.1$

In Exercises 13–16, find the area of the triangle having the given measurements. Round to the nearest square unit.

13. $C = 42°$, $a = 4$ feet, $b = 6$ feet
14. $A = 22°$, $b = 4$ feet, $c = 5$ feet
15. $a = 2$ meters, $b = 4$ meters, $c = 5$ meters
16. $a = 2$ meters, $b = 2$ meters, $c = 2$ meters

17. The A-frame cabin shown below is 35 feet wide. The roof of the cabin makes a 60° angle with the cabin's base. Find the length of one side of the roof from its ground level to the peak. Round to the nearest tenth of a foot.

18. Two cars leave a city at the same time and travel along straight highways that differ in direction by 80°. One car averages 60 miles per hour and the other averages 50 miles per hour. How far apart will the cars be after 30 minutes? Round to the nearest tenth of a mile.

19. Two airplanes leave an airport at the same time on different runways. One flies on a bearing of N66.5°W at 325 miles per hour. The other airplane flies on a bearing of S26.5°W at 300 miles per hour. How far apart will the airplanes be after two hours?

20. The figure shows three roads that intersect to bound a triangular piece of land. Find the lengths of the other two sides of the land to the nearest foot.

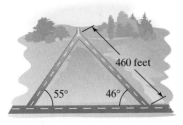

21. A commercial piece of real estate is priced at $5.25 per square foot. Find the cost, to the nearest dollar, of a triangular lot measuring 260 feet by 320 feet by 450 feet.

6.3 and 6.4

In Exercises 22–27, plot each point in polar coordinates and find its rectangular coordinates.

22. $(4, 60°)$
23. $(3, 150°)$
24. $\left(-4, \dfrac{4\pi}{3}\right)$
25. $\left(-2, \dfrac{5\pi}{4}\right)$
26. $\left(-4, -\dfrac{\pi}{2}\right)$
27. $\left(-2, -\dfrac{\pi}{4}\right)$

In Exercises 28–30, plot each point in polar coordinates. Then find another representation (r, θ) of this point in which:

    a. $r > 0$, $2\pi < \theta < 4\pi$.

    b. $r < 0$, $0 < \theta < 2\pi$.

    c. $r > 0$, $-2\pi < \theta < 0$.

28. $\left(3, \dfrac{\pi}{6}\right)$
29. $\left(2, \dfrac{2\pi}{3}\right)$
30. $\left(3.5, \dfrac{\pi}{2}\right)$

In Exercises 31–36, the rectangular coordinates of a point are given. Find polar coordinates of each point.

31. $(-4, 4)$
32. $(3, -3)$
33. $(5, 12)$
34. $(-3, 4)$
35. $(0, -5)$
36. $(1, 0)$

In Exercises 37–39, convert each rectangular equation to a polar equation that expresses r in terms of θ.

37. $2x + 3y = 8$
38. $x^2 + y^2 = 100$
39. $(x - 6)^2 + y^2 = 36$

In Exercises 40–46, convert each polar equation to a rectangular equation. Then use your knowledge of the rectangular equation to graph the polar equation in a polar coordinate system.

40. $r = 3$
41. $\theta = \dfrac{3\pi}{4}$
42. $r \cos \theta = -1$
43. $r = 5 \csc \theta$
44. $r = 3 \cos \theta$
45. $4r \cos \theta + r \sin \theta = 8$
46. $r^2 \sin 2\theta = -2$

In Exercises 47–49, test for symmetry with respect to

    a. *the polar axis.*
    b. *the line $\theta = \dfrac{\pi}{2}$.*

    c. *the pole.*

47. $r = 5 + 3 \cos \theta$
48. $r = 3 \sin \theta$
49. $r^2 = 9 \cos 2\theta$

In Exercises 50–56, graph each polar equation. Be sure to test for symmetry.

50. $r = 3 \cos \theta$
51. $r = 2 + 2 \sin \theta$
52. $r = \sin 2\theta$
53. $r = 2 + \cos \theta$
54. $r = 1 + 3 \sin \theta$
55. $r = 1 - 2 \cos \theta$
56. $r^2 = \cos 2\theta$

6.5

In Exercises 57–60, plot each complex number. Then write the complex number in polar form. You may express the argument in degrees or radians.

57. $1 - i$

58. $-2\sqrt{3} + 2i$

59. $-3 - 4i$

60. $-5i$

In Exercises 61–64, write each complex number in rectangular form. If necessary, round to the nearest tenth.

61. $8(\cos 60° + i \sin 60°)$

62. $4(\cos 210° + i \sin 210°)$

63. $6\left(\cos \dfrac{2\pi}{3} + i \sin \dfrac{2\pi}{3}\right)$

64. $0.6(\cos 100° + i \sin 100°)$

In Exercises 65–67, find the product of the complex numbers. Leave answers in polar form.

65. $z_1 = 3(\cos 40° + i \sin 40°)$
$z_2 = 5(\cos 70° + i \sin 70°)$

66. $z_1 = \cos 210° + i \sin 210°$
$z_2 = \cos 55° + i \sin 55°$

67. $z_1 = 4\left(\cos \dfrac{3\pi}{7} + i \sin \dfrac{3\pi}{7}\right)$
$z_2 = 10\left(\cos \dfrac{4\pi}{7} + i \sin \dfrac{4\pi}{7}\right)$

In Exercises 68–70, find the quotient $\dfrac{z_1}{z_2}$ of the complex numbers. Leave answers in polar form.

68. $z_1 = 10(\cos 10° + i \sin 10°)$
$z_2 = 5(\cos 5° + i \sin 5°)$

69. $z_1 = 5\left(\cos \dfrac{4\pi}{3} + i \sin \dfrac{4\pi}{3}\right)$
$z_2 = 10\left(\cos \dfrac{\pi}{3} + i \sin \dfrac{\pi}{3}\right)$

70. $z_1 = 2\left(\cos \dfrac{5\pi}{3} + i \sin \dfrac{5\pi}{3}\right)$
$z_2 = \cos \dfrac{\pi}{2} + i \sin \dfrac{\pi}{2}$

In Exercises 71–75, use DeMoivre's Theorem to find the indicated power of the complex number. Write answers in rectangular form.

71. $[2(\cos 20° + i \sin 20°)]^3$

72. $[4(\cos 50° + i \sin 50°)]^3$

73. $\left[\dfrac{1}{2}\left(\cos \dfrac{\pi}{14} + i \sin \dfrac{\pi}{14}\right)\right]^7$

74. $\left(1 - \sqrt{3}i\right)^2$

75. $(-2 - 2i)^5$

In Exercises 76–77, find all the complex roots. Write roots in polar form with θ in degrees.

76. The complex square roots of $49(\cos 50° + i \sin 50°)$

77. The complex cube roots of $125(\cos 165° + i \sin 165°)$

In Exercises 78–81, find all the complex roots. Write roots in rectangular form.

78. The complex fourth roots of $16\left(\cos \dfrac{2\pi}{3} + i \sin \dfrac{2\pi}{3}\right)$

79. The complex cube roots of $8i$

80. The complex cube roots of -1

81. The complex fifth roots of $-1 - i$

6.6

In Exercises 82–84, sketch each vector as a position vector and find its magnitude.

82. $\mathbf{v} = -3\mathbf{i} - 4\mathbf{j}$

83. $\mathbf{v} = 5\mathbf{i} - 2\mathbf{j}$

84. $\mathbf{v} = -3\mathbf{j}$

In Exercises 85–86, let $\mathbf{v}$ be the vector from initial point P_1 to terminal point P_2. Write $\mathbf{v}$ in terms of $\mathbf{i}$ and $\mathbf{j}$.

85. $P_1 = (2, -1)$, $P_2 = (5, -3)$

86. $P_1 = (-3, 0)$, $P_2 = (-2, -2)$

In Exercises 87–90, let
$$\mathbf{v} = \mathbf{i} - 5\mathbf{j} \quad \text{and} \quad \mathbf{w} = -2\mathbf{i} + 7\mathbf{j}.$$
Find each specified vector or scalar.

87. $\mathbf{v} + \mathbf{w}$

88. $\mathbf{w} - \mathbf{v}$

89. $6\mathbf{v} - 3\mathbf{w}$

90. $\|-2\mathbf{v}\|$

In Exercises 91–92, find the unit vector that has the same direction as the vector $\mathbf{v}$.

91. $\mathbf{v} = 8\mathbf{i} - 6\mathbf{j}$

92. $\mathbf{v} = -\mathbf{i} + 2\mathbf{j}$

93. The magnitude and direction angle of $\mathbf{v}$ are $\|\mathbf{v}\| = 12$ and $\theta = 60°$. Express $\mathbf{v}$ in terms of $\mathbf{i}$ and $\mathbf{j}$.

94. The magnitude and direction of two forces acting on an object are 100 pounds, N25°E, and 200 pounds, N80°E, respectively. Find the magnitude, to the nearest pound, and the direction angle, to the nearest tenth of a degree, of the resultant force.

95. Your boat is moving at a speed of 15 miles per hour at an angle of 25° upstream on a river flowing at 4 miles per hour. The situation is illustrated in the figure below.

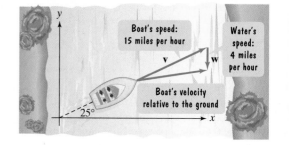

a. Find the vector representing your boat's velocity relative to the ground.

b. What is the speed of your boat, to the nearest mile per hour, relative to the ground?

c. What is the boat's direction angle, to the nearest tenth of a degree, relative to the ground?

6.7

96. If $\mathbf{u} = 5\mathbf{i} + 2\mathbf{j}$, $\mathbf{v} = \mathbf{i} - \mathbf{j}$, and $\mathbf{w} = 3\mathbf{i} - 7\mathbf{j}$, find $\mathbf{u} \cdot (\mathbf{v} + \mathbf{w})$.

In Exercises 97–99, find the dot product $\mathbf{v} \cdot \mathbf{w}$. Then find the angle between $\mathbf{v}$ and $\mathbf{w}$ to the nearest tenth of a degree.

97. $\mathbf{v} = 2\mathbf{i} + 3\mathbf{j}$, $\mathbf{w} = 7\mathbf{i} - 4\mathbf{j}$

98. $\mathbf{v} = 2\mathbf{i} + 4\mathbf{j}$, $\mathbf{w} = 6\mathbf{i} - 11\mathbf{j}$

99. $\mathbf{v} = 2\mathbf{i} + \mathbf{j}$, $\mathbf{w} = \mathbf{i} - \mathbf{j}$

In Exercises 100–101, use the dot product to determine whether **v** and **w** are orthogonal.

100. $\mathbf{v} = 12\mathbf{i} - 8\mathbf{j}, \quad \mathbf{w} = 2\mathbf{i} + 3\mathbf{j}$

101. $\mathbf{v} = \mathbf{i} + 3\mathbf{j}, \quad \mathbf{w} = -3\mathbf{i} - \mathbf{j}$

In Exercises 102–103, find $\text{proj}_\mathbf{w}\ \mathbf{v}$. Then decompose **v** into two vectors, $\mathbf{v}_1$ and $\mathbf{v}_2$, where $\mathbf{v}_1$ is parallel to **w** and $\mathbf{v}_2$ is orthogonal to **w**.

102. $\mathbf{v} = -2\mathbf{i} + 5\mathbf{j}, \quad \mathbf{w} = 5\mathbf{i} + 4\mathbf{j}$

103. $\mathbf{v} = -\mathbf{i} + 2\mathbf{j}, \quad \mathbf{w} = 3\mathbf{i} - \mathbf{j}$

104. A heavy crate is dragged 50 feet along a level floor. Find the work done if a force of 30 pounds at an angle of 42° is used.

105. Explain why the weightlifter does more work in raising 300 kilograms above her head than Atlas, who is supporting the entire world.

Chapter 6 Test

1. In oblique triangle ABC, $A = 34°$, $B = 68°$, and $a = 4.8$. Find b to the nearest tenth.

2. In oblique triangle ABC, $C = 68°$, $a = 5$, and $b = 6$. Find c to the nearest tenth.

3. In oblique triangle ABC, $a = 17$ inches, $b = 45$ inches, and $c = 32$ inches. Find the area of the triangle to the nearest square inch.

4. Plot $\left(4, \dfrac{5\pi}{4}\right)$ in the polar coordinate system. Then write two other ordered pairs (r, θ) that name this point.

5. If the rectangular coordinates of a point are $(1, -1)$, find polar coordinates of the point.

6. Convert $x^2 + (y + 8)^2 = 64$ to a polar equation that expresses r in terms of θ.

7. Convert to a rectangular equation and then graph: $r = -4 \sec \theta$.

In Exercises 8–9, graph each polar equation.

8. $r = 1 + \sin \theta$

9. $r = 1 + 3 \cos \theta$

10. Write $-\sqrt{3} + i$ in polar form.

In Exercises 11–13, perform the indicated operation. Leave answers in polar form.

11. $5(\cos 15° + i \sin 15°) \cdot 10(\cos 5° + i \sin 5°)$

12. $\dfrac{2\left(\cos \dfrac{\pi}{2} + i \sin \dfrac{\pi}{2}\right)}{4\left(\cos \dfrac{\pi}{3} + i \sin \dfrac{\pi}{3}\right)}$

13. $[2(\cos 10° + i \sin 10°)]^5$

14. Find the three cube roots of 27. Write roots in rectangular form.

15. If $P_1 = (-2, 3)$, $P_2 = (-1, 5)$, and **v** is the vector from P_1 to P_2,

   a. Write **v** in terms of **i** and **j**.

   b. Find $\|\mathbf{v}\|$.

In Exercises 16–19, let

$$\mathbf{v} = -5\mathbf{i} + 2\mathbf{j} \quad \text{and} \quad \mathbf{w} = 2\mathbf{i} - 4\mathbf{j}.$$

Find the specified vector, scalar, or angle.

16. $3\mathbf{v} - 4\mathbf{w}$

17. $\mathbf{v} \cdot \mathbf{w}$

18. the angle between **v** and **w**, to the nearest degree

19. $\text{proj}_\mathbf{w}\mathbf{v}$

20. A small fire is sighted from ranger stations A and B. Station B is 1.6 miles due east of station A. The bearing of the fire from station A is N40°E and the bearing of the fire from station B is N50°W. How far, to the nearest tenth of a mile, is the fire from station A?

21. The magnitude and direction of two forces acting on an object are 250 pounds, N60°E, and 150 pounds, S45°E. Find the magnitude, to the nearest pound, and the direction angle, to the nearest tenth of a degree, of the resultant force.

22. A child is pulling a wagon with a force of 40 pounds. How much work is done in moving the wagon 60 feet if the handle makes an angle of 35° with the ground? Round to the nearest foot-pound.

SECTION 7.1 *Systems of Linear Equations in Two Variables*

Objectives

❶ Decide whether an ordered pair is a solution of a linear system.

❷ Solve linear systems by substitution.

❸ Solve linear systems by addition.

❹ Identify systems that do not have exactly one ordered-pair solution.

❺ Solve problems using systems of linear equations.

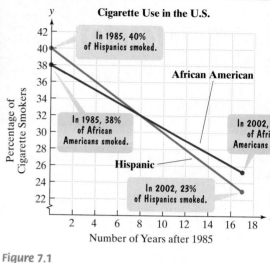

Figure 7.1

Source: Dept. of Health and Human Services

Although we still see celebrities smoking in movies, in music videos, and c television, there has been a remarkable decline in the percentage of cigaret smokers in the United States. The decline among African Americans and Hispanic illustrated in Figure 7.1, can be analyzed using a pair of linear models in tw variables.

In the first two sections of this chapter, you will learn to model your world wit two equations in two variables and three equations in three variables. The methoc you learn for solving these systems provide the foundation for solving comple problems involving thousands of equations containing thousands of variables. In th exercise set, you will apply these methods to analyze the linear decrease in cigarett use among whites, African Americans, and Hispanics.

 Decide whether an ordered pair is a solution of a linear system.

Systems of Linear Equations and Their Solutions

We have seen that all equations in the form $Ax + By = C$ are straight lines whe graphed. Two such equations are called a **system of linear equations** or a **linear systen** **A solution to a system of linear equations in two variables** is an ordered pair tha satisfies both equations in the system. For example, $(3, 4)$ satisfies the system

$$x + y = 7 \qquad \text{(3 + 4 is, indeed, 7.)}$$

$$x - y = -1. \qquad \text{(3 − 4 is, indeed, −1.)}$$

Thus, $(3, 4)$ satisfies both equations and is a solution of the system. The solution ca be described by saying that $x = 3$ and $y = 4$. The solution can also be describe using set notation. The solution set to the system is $\{(3, 4)\}$—that is, the se consisting of the ordered pair $(3, 4)$.

A system of linear equations can have exactly one solution, no solution, o infinitely many solutions. We begin with systems that have exactly one solution.

EXAMPLE 1 Determining Whether Ordered Pairs Are Solutions of a System

Consider the system:

$$x + 2y = 2$$

$$x - 2y = 6.$$

Determine if each ordered pair is a solution of the system:

    a. $(4, -1)$ **b.** $(-4, 3)$.

Solution

a. We begin by determining whether $(4, -1)$ is a solution. Because 4 is the x-coordinate and -1 is the y-coordinate of $(4, -1)$, we replace x with 4 and y with -1.

$$x + 2y = 2 \qquad\qquad x - 2y = 6$$
$$4 + 2(-1) \overset{?}{=} 2 \qquad\qquad 4 - 2(-1) \overset{?}{=} 6$$
$$4 + (-2) \overset{?}{=} 2 \qquad\qquad 4 - (-2) \overset{?}{=} 6$$
$$2 = 2, \quad \text{true} \qquad\qquad 4 + 2 \overset{?}{=} 6$$
$$6 = 6, \quad \text{true}$$

The pair $(4, -1)$ satisfies both equations: It makes each equation true. Thus, the ordered pair is a solution of the system.

b. To determine whether $(-4, 3)$ is a solution, we replace x with -4 and y with 3.

$$x + 2y = 2 \qquad\qquad x - 2y = 6$$
$$-4 + 2 \cdot 3 \overset{?}{=} 2 \qquad\qquad -4 - 2 \cdot 3 \overset{?}{=} 6$$
$$-4 + 6 \overset{?}{=} 2 \qquad\qquad -4 - 6 \overset{?}{=} 6$$
$$2 = 2, \quad \text{true} \qquad\qquad -10 = 6, \quad \text{false}$$

The pair $(-4, 3)$ fails to satisfy *both* equations: It does not make both equations true. Thus, the ordered pair is not a solution of the system.

Study Tip

When solving linear systems by graphing, neatly drawn graphs are essential for determining points of intersection.

- Use rectangular coordinate graph paper.
- Use a ruler or straightedge.
- Use a pencil with a sharp point.

The solution of a system of linear equations can sometimes be found by graphing both of the equations in the same rectangular coordinate system. For a system with one solution, the **coordinates of the point of intersection give the system's solution**. For example, the system in Example 1,

$$x + 2y = 2$$
$$x - 2y = 6$$

is graphed in Figure 7.2. The solution of the system, $(4, -1)$, corresponds to the point of intersection of the lines.

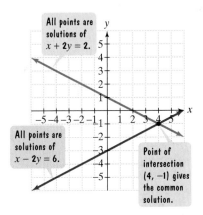

Figure 7.2 Visualizing a system's solution

 Consider the system:

$$2x - 3y = -4$$
$$2x + y = 4.$$

Determine if each ordered pair is a solution of the system:

    a. $(1, 2)$ **b.** $(7, 6)$.

 Solve linear systems by substitution.

Eliminating a Variable Using the Substitution Method

Finding the solution to a linear system by graphing equations may not be easy to do. For example, a solution of $\left(-\frac{2}{3}, \frac{157}{29}\right)$ would be difficult to "see" as an intersection point on a graph.

    Let's consider a method that does not depend on finding a system's solution visually: the substitution method. This method involves converting the system to one equation in one variable by an appropriate substitution.

Solving Linear Systems by Substitution

1. Solve either of the equations for one variable in terms of the other. (If one of the equations is already in this form, you can skip this step.)
2. Substitute the expression found in step 1 into the *other* equation. This will result in an equation in one variable.
3. Solve the equation containing one variable.
4. Back-substitute the value found in step 3 into one of the original equations. Simplify and find the value of the remaining variable.
5. Check the proposed solution in both of the system's given equations.

EXAMPLE 2 Solving a System by Substitution

Solve by the substitution method:

$$5x - 4y = 9$$
$$x - 2y = -3.$$

Solution

Step 1 Solve either of the equations for one variable in terms of the other. We begin by isolating one of the variables in either of the equations. By solving for x in the second equation, which has a coefficient of 1, we can avoid fractions.

$$x - 2y = -3 \qquad \text{This is the second equation in the given system.}$$
$$x = 2y - 3 \qquad \text{Solve for } x \text{ by adding } 2y \text{ to both sides.}$$

Step 2 Substitute the expression from step 1 into the other equation. We substitute $2y - 3$ for x in the first equation.

$$x = \boxed{2y - 3} \qquad\qquad 5\boxed{x} - 4y = 9$$

This gives us an equation in one variable, namely

$$5(2y - 3) - 4y = 9.$$

The variable x has been eliminated.

Step 3 Solve the resulting equation containing one variable.

$$5(2y - 3) - 4y = 9 \qquad \text{This is the equation containing one variable.}$$
$$10y - 15 - 4y = 9 \qquad \text{Apply the distributive property.}$$
$$6y - 15 = 9 \qquad \text{Combine like terms.}$$
$$6y = 24 \qquad \text{Add 15 to both sides.}$$
$$y = 4 \qquad \text{Divide both sides by 6.}$$

Step 4 Back-substitute the obtained value into one of the original equations. We back-substitute 4 for y into one of the original equations to find x. Let's use both equations to show that we obtain the same value for x in either case.

Using the first equation:	Using the second equation:
$5x - 4y = 9$	$x - 2y = -3$
$5x - 4(4) = 9$	$x - 2(4) = -3$
$5x - 16 = 9$	$x - 8 = -3$
$5x = 25$	$x = 5$
$x = 5$	

With $x = 5$ and $y = 4$, the proposed solution is $(5, 4)$.

Step 5 Check. Take a moment to show that $(5, 4)$ satisfies both given equations. The solution set is $\{(5, 4)\}$.

Technology

A graphing utility can be used to solve the system in Example 2. Solve each equation for y, graph the equations, and use the intersection feature. The utility displays the solution $(5, 4)$ as $x = 5$, $y = 4$.

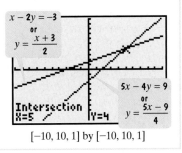

$x - 2y = -3$
or
$y = \dfrac{x + 3}{2}$

$5x - 4y = 9$
or
$y = \dfrac{5x - 9}{4}$

$[-10, 10, 1]$ by $[-10, 10, 1]$

Check Point 2   Solve by the substitution method:

$$3x + 2y = 4$$
$$2x + y = 1.$$

③ Solve linear systems by addition.

Eliminating a Variable Using the Addition Method

The substitution method is most useful if one of the given equations has an isolated variable. A second, and frequently the easiest, method for solving a linear system is the addition method. Like the substitution method, the addition method involves eliminating a variable and ultimately solving an equation containing only one variable. However, this time we eliminate a variable by adding the equations.

For example, consider the following system of linear equations:

$$3x - 4y = 11$$
$$-3x + 2y = -7.$$

When we add these two equations, the x-terms are eliminated. This occurs because the coefficients of the x-terms, 3 and -3, are opposites (additive inverses) of each other:

$$3x - 4y = 11$$
$$\underline{-3x + 2y = -7}$$

Add: $-2y = 4$ |The sum is an equation in one variable.|

$y = -2$ *Solve for y by dividing both sides by −2.*

Now we can back-substitute -2 for y into one of the original equations to find x. It does not matter which equation you use; you will obtain the same value for x in either case. If we use either equation, we can show that $x = 1$ and the solution $(1, -2)$ satisfies both equations in the system.

When we use the addition method, we want to obtain two equations whose sum is an equation containing only one variable. The key step is to **obtain, for one of the variables, coefficients that differ only in sign**. To do this, we may need to multiply one or both equations by some nonzero number so that the coefficients of one of the variables, x or y, become opposites. Then when the two equations are added, this variable is eliminated.

Study Tip

Although the addition method is also known as the elimination method, variables are eliminated when using both the substitution and addition methods. The name *addition method* specifically tells us that the elimination of a variable is accomplished by adding two equations.

Solving Linear Systems by Addition

1. If necessary, rewrite both equations in the form $Ax + By = C$.
2. If necessary, multiply either equation or both equations by appropriate nonzero numbers so that the sum of the x-coefficients or the sum of the y-coefficients is 0.
3. Add the equations in step 2. The sum is an equation in one variable.
4. Solve the equation in one variable.
5. Back-substitute the value obtained in step 4 into either of the given equations and solve for the other variable.
6. Check the solution in both of the original equations.

EXAMPLE 3   Solving a System by the Addition Method

Solve by the addition method:

$$3x + 2y = 48$$
$$9x - 8y = -24.$$

Solution

Step 1   Rewrite both equations in the form $Ax + By = C$. Both equations are already in this form. Variable terms appear on the left and constants appear on the right.

Step 2   If necessary, multiply either equation or both equations by appropriate numbers so that the sum of the x-coefficients or the sum of the y-coefficients is 0. We can eliminate x or y. Let's eliminate x. Consider the terms in x in each equation; that is, $3x$ and $9x$. To eliminate x, we can multiply each term of the first equation by -3 and then add the equations.

$$3x + 2y = 48 \quad \xrightarrow{\text{Multiply by } -3.} \quad -9x - 6y = -144$$
$$9x - 8y = -24 \quad \xrightarrow{\text{No change}} \quad 9x - 8y = -24$$

Step 3   Add the equations. $\qquad\qquad$ Add: $\quad -14y = -168$

Step 4   Solve the equation in one variable. We solve $-14y = -168$ by dividing both sides by -14.

$$\frac{-14y}{-14} = \frac{-168}{-14} \qquad \text{Divide both sides by } -14.$$
$$y = 12 \qquad \text{Simplify.}$$

Step 5   Back-substitute and find the value for the other variable. We can back-substitute 12 for y into either one of the given equations. We'll use the first one.

$$3x + 2y = 48 \qquad \text{This is the first equation in the given system.}$$
$$3x + 2(12) = 48 \qquad \text{Substitute 12 for y.}$$
$$3x + 24 = 48 \qquad \text{Multiply.}$$
$$3x = 24 \qquad \text{Subtract 24 from both sides.}$$
$$x = 8 \qquad \text{Divide both sides by 3.}$$

We found that $y = 12$ and $x = 8$. The proposed solution is $(8, 12)$.

Step 6   Check. Take a few minutes to show that $(8, 12)$ satisfies both of the original equations in the system. The solution set is $\{(8, 12)\}$.

 Solve by the addition method:

$$4x + 5y = 3$$
$$2x - 3y = 7.$$

Some linear systems have solutions that are not integers. If the value of one variable turns out to be a "messy" fraction, back-substitution might lead to cumbersome arithmetic. If this happens, you can return to the original system and use the addition method to find the value of the other variable.

EXAMPLE 4   Solving a System by the Addition Method

Solve by the addition method:

$$2x = 7y - 17$$
$$5y = 17 - 3x.$$

Solution

Step 1   Rewrite both equations in the form $Ax + By = C$. We first arrange the system so that variable terms appear on the left and constants appear on the right. We obtain

$$2x - 7y = -17 \qquad \text{Subtract 7y from both sides of the first equation.}$$
$$3x + 5y = 17. \qquad \text{Add 3x to both sides of the second equation.}$$

Step 2   If necessary, multiply either equation or both equations by appropriate numbers so that the sum of the x-coefficients or the sum of the y-coefficients is 0. We can eliminate x or y. Let's eliminate x by multiplying the first equation by 3 and the second equation by -2.

$$2x - 7y = -17 \quad \xrightarrow{\text{Multiply by 3.}} \quad 6x - 21y = -51$$
$$3x + 5y = 17 \quad \xrightarrow{\text{Multiply by }-2.} \quad \underline{-6x - 10y = -34}$$

Step 3   Add the equations. $\qquad\qquad\qquad$ Add: $\qquad -31y = -85$

Step 4   Solve the equation in one variable. We solve $-31y = -85$ by dividing both sides by -31.

$$\frac{-31y}{-31} = \frac{-85}{-31} \qquad \text{Divide both sides by } -31.$$

$$y = \frac{85}{31} \qquad \text{Simplify.}$$

Step 5   Back-substitute and find the value for the other variable. Back-substitution of $\frac{85}{31}$ for y into either of the given equations results in cumbersome arithmetic. Instead, let's use the addition method on the given system in the form $Ax + By = C$ to find the value for x. Thus, we eliminate y by multiplying the first equation by 5 and the second equation by 7.

$$2x - 7y = -17 \quad \xrightarrow{\text{Multiply by 5.}} \quad 10x - 35y = -85$$
$$3x + 5y = 17 \quad \xrightarrow{\text{Multiply by 7.}} \quad \underline{21x + 35y = 119}$$
$$\text{Add: } 31x = 34$$

$$x = \frac{34}{31} \qquad \text{Divide both sides by 31.}$$

We found that $y = \dfrac{85}{31}$ and $x = \dfrac{34}{31}$. The proposed solution is $\left(\dfrac{34}{31}, \dfrac{85}{31}\right)$.

Step 6   Check. For this system, a calculator is helpful in showing that $\left(\frac{34}{31}, \frac{85}{31}\right)$ satisfies both of the original equations in the system. The solution set is $\left\{\left(\frac{34}{31}, \frac{85}{31}\right)\right\}$.

Check Point 4   Solve by the addition method:

$$2x = 9 + 3y$$
$$4y = 8 - 3x.$$

Identify systems that do not have exactly one ordered-pair solution.

Linear Systems Having No Solution or Infinitely Many Solutions

We have seen that a system of linear equations in two variables represents a pair of lines. The lines either intersect at one point, are parallel, or are identical. Thus, there are three possibilities for the number of solutions to a system of two linear equations.

The Number of Solutions to a System of Two Linear Equations

The number of solutions to a system of two linear equations in two variables is given by one of the following. (See Figure 7.3.)

Number of Solutions	What This Means Graphically
Exactly one ordered pair solution	The two lines intersect at one point.
No solution	The two lines are parallel.
Infinitely many solutions	The two lines are identical.

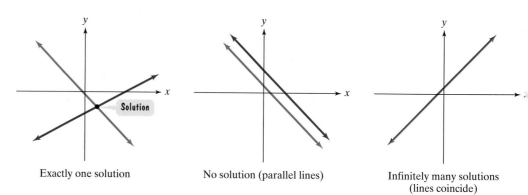

Exactly one solution   No solution (parallel lines)   Infinitely many solutions (lines coincide)

Figure 7.3 Possible graphs for a system of two linear equations in two variables

A linear system with no solution is called an **inconsistent system**. If you attempt to solve such a system by substitution or addition, you will eliminate both variables. A false statement, such as $0 = 12$, will be the result.

EXAMPLE 5   A System with No Solution

Solve the system:

$$4x + 6y = 12$$
$$6x + 9y = 12.$$

Solution   Because no variable is isolated, we will use the addition method. To obtain coefficients of x that differ only in sign, we multiply the first equation by 3 and multiply the second equation by -2.

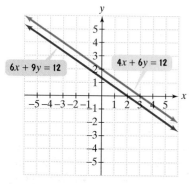

$$
\begin{array}{ll}
4x + 6y = 12 & \xrightarrow{\text{Multiply by 3.}} \\
6x + 9y = 12 & \xrightarrow{\text{Multiply by } -2.} \\
& \qquad\qquad \text{Add:}
\end{array}
\quad
\begin{array}{r}
12x + 18y = 36 \\
-12x - 18y = -24 \\
\hline
0 = 12
\end{array}
$$

There are no values of x and y for which $0 = 12$. No values of x and y satisfy $0x + 0y = 12$.

The false statement $0 = 12$ indicates that the system is inconsistent and has no solution. The solution set is the empty set, $\varnothing$.

Figure 7.4 The graph of an inconsistent system

The lines corresponding to the two equations in Example 5 are shown in Figure 7.4. The lines are parallel and have no point of intersection.

Discovery

Show that the graphs of $4x + 6y = 12$ and $6x + 9y = 12$ must be parallel lines by solving each equation for y. What is the slope and y-intercept for each line? What does this mean? If a linear system is inconsistent, what must be true about the slopes and y-intercepts for the system's graphs?

Check Point 5 Solve the system:

$$5x - 2y = 4$$
$$-10x + 4y = 7.$$

A linear system that has at least one solution is called a **consistent system**. Lines that intersect and lines that coincide both represent consistent systems. If the lines coincide, then the consistent system has infinitely many solutions, represented by every point on either line.

The equations in a linear system with infinitely many solutions are called **dependent**. If you attempt to solve such a system by substitution or addition, you will eliminate both variables. However, a true statement, such as $10 = 10$, will be the result.

EXAMPLE 6   A System with Infinitely Many Solutions

Solve the system:

$$y = 3x - 2$$
$$15x - 5y = 10.$$

Solution   Because the variable y is isolated in $y = 3x - 2$, the first equation, we can use the substitution method. We substitute the expression for y into the second equation.

$$y = \boxed{3x - 2} \qquad 15x - 5\boxed{y} = 10 \quad \text{Substitute } 3x - 2 \text{ for } y.$$

$$15x - 5(3x - 2) = 10 \quad \text{The substitution results in an equation in one variable.}$$

$$15x - 15x + 10 = 10 \quad \text{Apply the distributive property.}$$

> This statement is true for all values of x and y.

$$10 = 10 \quad \text{Simplify.}$$

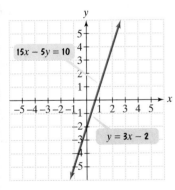

Figure 7.5 The graph of a system with infinitely many solutions

In our final step, both variables have been eliminated and the resulting statement, $10 = 10$, is true. This true statement indicates that the system has infinitely many solutions. The solution set consists of all points (x, y) lying on either of the coinciding lines, $y = 3x - 2$ or $15x - 5y = 10$, as shown in Figure 7.5.

We express the solution set for the system in one of two equivalent ways:

$$\{(x, y) \mid y = 3x - 2\} \qquad \text{or} \qquad \{(x, y) \mid 15x - 5y = 10\}.$$

> The set of all ordered pairs (x, y) such that $y = 3x - 2$

> The set of all ordered pairs (x, y) such that $15x - 5y = 10$

Study Tip

Although the system in Example 6 has infinitely many solutions, this does not mean that any ordered pair of numbers you can form will be a solution. The ordered pair (x, y) must satisfy one of the system's equations, $y = 3 - 2x$ or $15x - 5y = 10$, and there are infinitely many such ordered pairs. Because the graphs are coinciding lines, the ordered pairs that are solutions of one of the equations are also solutions of the other equation.

Check
Point 6 Solve the system:

$$x = 4y - 8$$
$$5x - 20y = -40.$$

5 Solve problems using systems of linear equations.

Applications

We begin with applications that involve two unknown quantities. We will let x and y represent these quantities. We then model the verbal conditions of the problem with a system of linear equations in x and y.

Strategy for Problem Solving Using Systems of Equations

Step 1 Read the problem carefully. Attempt to state the problem in your own words and state what the problem is looking for. Use variables to represent unknown quantities.

Step 2 Write a system of equations that models the problem's conditions.

Step 3 Solve the system and answer the problem's question.

Step 4 Check the proposed solution in the original wording of the problem.

Chemists and pharmacists often have to change the concentration of solutions and other mixtures. In these situations, the amount of a particular ingredient in the solution or mixture is expressed as a percentage of the total.

EXAMPLE 7 Solving a Mixture Problem

A chemist working on a flu vaccine needs to mix a 10% sodium-iodine solution with a 60% sodium-iodine solution to obtain 50 milliliters of a 30% sodium-iodine solution. How many milliliters of the 10% solution and of the 60% solution should be mixed?

Solution

Step 1 Use variables to represent unknown quantities.
Let x = the number of milliliters of the 10% solution to be used in the mixture.
Let y = the number of milliliters of the 60% solution to be used in the mixture.

Step 2 Write a system of equations that models the problem's conditions. The situation is illustrated in Figure 7.6.

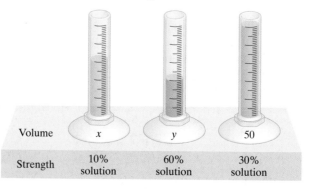

Figure 7.6

The chemist needs 50 milliliters of a 30% sodium-iodine solution. We form a table that shows the amount of sodium-iodine in each of the three solutions.

Solution	Number of milliliters	×	Percent of Sodium-Iodine	=	Amount of Sodium-Iodine
10% Solution	x		10% = 0.1		$0.1x$
60% Solution	y		60% = 0.6		$0.6y$
30% Mixture	50		30% = 0.3		$0.3(50) = 15$

The chemist needs to obtain a 50-milliliter mixture.

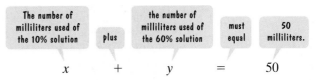

$$x \quad + \quad y \quad = \quad 50$$

The 50-milliliter mixture must be 30% sodium-iodine. The amount of sodium-iodine must be 30% of 50, or $(0.3)(50) = 15$ milliliters.

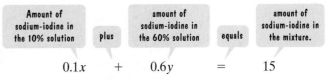

$$0.1x \quad + \quad 0.6y \quad = \quad 15$$

Step 3 Solve the system and answer the problem's question. The system

$$x + y = 50$$
$$0.1x + 0.6y = 15$$

can be solved by substitution or addition. Let's use substitution. Solving the first equation for y, we obtain $y = 50 - x$.

$$y = \boxed{50 - x} \qquad 0.1x + 0.6\boxed{y} = 15$$

We substitute $50 - x$ for y in the second equation, $0.1x + 0.6y = 15$. This gives us an equation in one variable.

$0.1x + 0.6(50 - x) = 15$	This equation contains one variable, x.
$0.1x + 30 - 0.6x = 15$	Apply the distributive property.
$-0.5x + 30 = 15$	Combine like terms.
$-0.5x = -15$	Subtract 30 from both sides.
$x = \dfrac{-15}{-0.5} = 30$	Divide both sides by -0.5.

Back-substituting 30 for x in either of the system's equations ($x + y = 50$ is easier to use) gives $y = 20$. Because x represents the number of milliliters of the 10% solution and y the number of milliliters of the 60% solution, the chemist should mix 30 milliliters of the 10% solution with 20 milliliters of the 60% solution.

Step 4 Check the proposed solution in the original wording of the problem. The problem states that the chemist needs 50 milliliters of a 30% sodium-iodine solution. The amount of sodium-iodine in this mixture is $0.3(50)$, or 15 milliliters. The amount of sodium-iodine in 30 milliliters of the 10% solution is $0.1(30)$, or 3 milliliters. The amount of sodium-iodine in 20 milliliters of the 60% solution is $0.6(20) = 12$ milliliters. The amount of sodium-iodine in the two solutions used in the mixture is 3 milliliters + 12 milliliters, or 15 milliliters, exactly as it should be.

Check Point 7 A chemist needs to mix an 18% acid solution with a 45% acid solution to obtain 12 liters of a 36% acid solution. How many liters of each of the acid solutions must be used?

We have seen that if an object moves at an average velocity v, the distance, s, covered in time t is given by the formula

$$s = vt \qquad \text{Distance equals velocity times time.}$$

Recall that objects that move in accordance with this formula are said to be in **uniform motion.** Wind and water current have the effect of increasing or decreasing a traveler's velocity.

Study Tip

It is not always necessary to use x and y to represent a problem's variables. Select letters that help you remember what the variables represent. For example, in Example 8, you may prefer using p and w rather than x and y:

p = plane's average velocity in still air

w = wind's average velocity.

EXAMPLE 8   Solving a Uniform Motion Problem

When a small airplane flies with the wind, it can travel 450 miles in 3 hours. When the same airplane flies in the opposite direction against the wind, it takes 5 hours to fly the same distance. Find the average velocity of the plane in still air and the average velocity of the wind.

Solution

Step 1   Use variables to represent unknown quantities.
Let x = the average velocity of the plane in still air.
Let y = the average velocity of the wind.

Step 2   Write a system of equations that models the problem's conditions. As it travels with the wind, the plane's average velocity is increased. The net average velocity is its average velocity in still air, x, plus the average velocity of the wind, y, given by the expression $x + y$. As it travels against the wind, the plane's average velocity is decreased. The net average velocity is its average velocity in still air, x, minus the average velocity of the wind, y, given by the expression $x - y$. Here is a chart that summarizes the problem's information and includes the increased and decreased velocities:

	Velocity	×	Time	=	Distance
Trip with the Wind	$x + y$		3		$3(x + y)$
Trip against the Wind	$x - y$		5		$5(x - y)$

The problem states that the distance in each direction is 450 miles. We use this information to write our system of equations.

The distance of the trip with the wind   is   450 miles.

$$3(x + y) = 450$$

The distance of the trip against the wind   is   450 miles.

$$5(x - y) = 450$$

Step 3   Solve the system and answer the problem's question. We can simplify the system by dividing both sides of the equations by 3 and 5, respectively.

$$3(x + y) = 450 \quad \xrightarrow{\text{Divide by 3.}} \quad x + y = 150$$
$$5(x - y) = 450 \quad \xrightarrow{\text{Divide by 5.}} \quad x - y = 90$$

Solve the system on the right by the addition method.

$$
\begin{aligned}
x + y &= 150 \\
x - y &= 90 \\
\hline
\text{Add:} \quad 2x &= 240 \\
x &= 120 \quad \text{Divide both sides by 2.}
\end{aligned}
$$

Back-substituting 120 for x in either of the system's equations gives $y = 30$. Because $x = 120$ and $y = 30$, the average velocity of the plane in still air is 120 miles per hour and the average velocity of the wind is 30 miles per hour.

Step 4   Check the proposed solution in the original wording of the problem. The problem states that the distance in each direction is 450 miles. The average velocity of the plane with the wind is $120 + 30 = 150$ miles per hour. In 3 hours, it travels $150 \cdot 3$, or 450 miles, which checks with the stated condition. Furthermore, the average velocity of the plane against the wind is $120 - 30 = 90$ miles per hour. In 5 hours, it travels $90 \cdot 5 = 450$ miles, which is the stated distance.

Check
Point **8**   With the current, a motorboat can travel 84 miles in 2 hours. Against the current, the same trip takes 3 hours. Find the average velocity of the boat in still water and the average velocity of the current.

Functions of Business: Break-Even Analysis

Suppose that a company produces and sells x units of a product. Its *revenue function* is the money generated by selling x units of the product. Its *cost function* is the cost of producing x units of the product.

Revenue and Cost Functions

A company produces and sells x units of a product.

Revenue Function

$$R(x) = (\text{price per unit sold})x$$

Cost Function

$$C(x) = \text{fixed cost} + (\text{cost per unit produced})x$$

The point of intersection of the graphs of the revenue and cost functions is called the **break-even point.** The x-coordinate of the point reveals the number of units that a company must produce and sell so that money coming in, the revenue, is equal to money going out, the cost. The y-coordinate of the break-even point gives the amount of money coming in and going out. Example 9 illustrates the use of the substitution method in determining a company's break-even point.

EXAMPLE 9   Finding a Break-Even Point

A company is planning to manufacture radically different wheelchairs. Fixed cost will be $500,000 and it will cost $400 to produce each wheelchair. Each wheelchair will be sold for $600.

  a. Write the cost function, C, of producing x wheelchairs.

  b. Write the revenue function, R, from the sale of x wheelchairs.

  c. Determine the break-even point. Describe what this means.

Solution

  a. The cost function is the sum of the fixed cost and variable cost.

Fixed cost of $500,000	plus	Variable cost: $400 for each chair produced

$$C(x) = 500,000 + 400x$$

  b. The revenue function is the money generated from the sale of x wheelchairs.

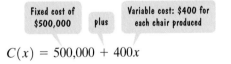

Revenue per chair, $600, times	the number of chairs sold

$$R(x) = 600x$$

  c. The break-even point occurs where the graphs of C and R intersect. Thus, we find this point by solving the system

$$\begin{aligned} C(x) &= 500,000 + 400x \\ R(x) &= 600x \end{aligned} \quad \text{or} \quad \begin{aligned} y &= 500,000 + 400x \\ y &= 600x. \end{aligned}$$

Using substitution, we can substitute $600x$ for y in the first equation:

$$600x = 500,000 + 400x$$ Substitute 600x for y in
                                y = 500,000 + 400x.

$$200x = 500,000$$ Subtract 400x from both sides.

$$x = 2500$$ Divide both sides by 200.

Back-substituting 2500 for x in either of the system's equations (or functions) $C(x) = 500,000 + 400x$ or $R(x) = 600x$, we obtain

$$R(2500) = 600(2500) = 1,500,000.$$

We used $R(x) = 600x$.

The break-even point is (2500, 1,500,000). This means that the company will break even if it produces and sells 2500 wheelchairs. At this level, the money coming in is equal to the money going out: $1,500,000.

Figure 7.7 shows the graphs of the revenue and cost functions for the wheelchair business. Similar graphs and models apply no matter how small or large a business venture may be.

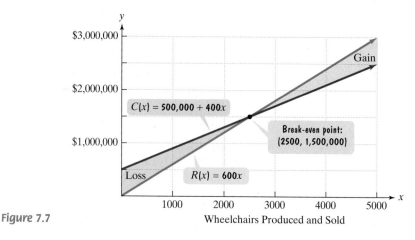

Figure 7.7   Wheelchairs Produced and Sold

The intersection point confirms that the company breaks even by producing and selling 2500 wheelchairs. Can you see what happens for $x < 2500$? The red cost graph lies above the blue revenue graph. The cost is greater than the revenue and the business is losing money. Thus, if they sell fewer than 2500 wheelchairs, the result is a *loss*. By contrast, look at what happens for $x > 2500$. The blue revenue graph lies above the red cost graph. The revenue is greater than the cost and the business is making money. Thus, if they sell more than 2500 wheelchairs, the result is a *gain*.

 Check Point 9   A company that manufactures running shoes has a fixed cost of $300,000. Additionally, it costs $30 to produce each pair of shoes. They are sold a $80 per pair.

a. Write the cost function, C, of producing x pairs of running shoes.

b. Write the revenue function, R, from the sale of x pairs of running shoes.

c. Determine the break-even point. Describe what this means.

What does every entrepreneur, from a kid selling lemonade to Donald Trump, want to do? Generate profit, of course. The *profit* made is the money taken in, or the revenue, minus the money spent, or the cost. This relationship between revenue and cost allows us to define the *profit function, $P(x)$*.

The Profit Function

The profit, $P(x)$, generated after producing and selling x units of a product is given by the **profit function**

$$P(x) = R(x) - C(x),$$

where R and C are the revenue and cost functions, respectively.

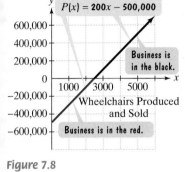

The profit function for the wheelchair business in Example 7 is

$$\begin{aligned} P(x) &= R(x) - C(x) \\ &= 600x - (500{,}000 + 400x) \\ &= 200x - 500{,}000. \end{aligned}$$

The graph of this profit function is shown in Figure 7.8. The red portion lies below the x-axis and shows a loss when fewer than 2500 wheelchairs are sold. The business is "in the red." The black portion lies above the x-axis and shows a gain when more than 2500 wheelchairs are sold. The wheelchair business is "in the black."

Figure 7.8

EXERCISE SET 7.1

Practice Exercises

In Exercises 1–4, determine whether the given ordered pair is a solution of the system.

1. $(2, 3)$
$x + 3y = 11$
$x - 5y = -13$

2. $(-3, 5)$
$9x + 7y = 8$
$8x - 9y = -69$

3. $(2, 5)$
$2x + 3y = 17$
$x + 4y = 16$

4. $(8, 5)$
$5x - 4y = 20$
$3y = 2x + 1$

In Exercises 5–18, solve each system by the substitution method.

5. $x + y = 4$
$y = 3x$

6. $x + y = 6$
$y = 2x$

7. $x + 3y = 8$
$y = 2x - 9$

8. $2x - 3y = -13$
$y = 2x + 7$

9. $x = 4y - 2$
$x = 6y + 8$

10. $x = 3y + 7$
$x = 2y - 1$

11. $5x + 2y = 0$
$x - 3y = 0$

12. $4x + 3y = 0$
$2x - y = 0$

13. $2x + 5y = -4$
$3x - y = 11$

14. $2x + 5y = 1$
$-x + 6y = 8$

15. $2x - 3y = 8 - 2x$
$3x + 4y = x + 3y + 14$

16. $3x - 4y = x - y + 4$
$2x + 6y = 5y - 4$

17. $y = \dfrac{1}{3}x + \dfrac{2}{3}$
$y = \dfrac{5}{7}x - 2$

18. $y = -\dfrac{1}{2}x + 2$
$y = \dfrac{3}{4}x + 7$

In Exercises 19–30, solve each system by the addition method.

19. $x + y = 1$
$x - y = 3$

20. $x + y = 6$
$x - y = -2$

21. $2x + 3y = 6$
$2x - 3y = 6$

22. $3x + 2y = 14$
$3x - 2y = 10$

23. $x + 2y = 2$
$-4x + 3y = 25$

24. $2x - 7y = 2$
$3x + y = -20$

25. $4x + 3y = 15$
$2x - 5y = 1$

26. $3x - 7y = 13$
$6x + 5y = 7$

27. $3x - 4y = 11$
$2x + 3y = -4$

28. $2x + 3y = -16$
$5x - 10y = 30$

29. $3x = 4y + 1$
$3y = 1 - 4x$

30. $5x = 6y + 40$
$2y = 8 - 3x$

In Exercises 31–42, solve by the method of your choice. Identify systems with no solution and systems with infinitely many solutions, using set notation to express their solution sets.

31. $x = 9 - 2y$
$x + 2y = 13$

32. $6x + 2y = 7$
$y = 2 - 3x$

33. $y = 3x - 5$
$21x - 35 = 7y$

34. $9x - 3y = 12$
$y = 3x - 4$

35. $3x - 2y = -5$
$4x + y = 8$

36. $2x + 5y = -4$
$3x - y = 11$

37. $x + 3y = 2$
$3x + 9y = 6$

38. $4x - 2y = 2$
$2x - y = 1$

39. $\dfrac{x}{4} - \dfrac{y}{4} = -1$
$x + 4y = -9$

40. $\dfrac{x}{6} - \dfrac{y}{2} = \dfrac{1}{3}$
$x + 2y = -3$

41. $2x = 3y + 4$
$4x = 3 - 5y$

42. $4x = 3y + 8$
$2x = -14 + 5y$

In Exercises 43–46, let x represent one number and let y represent the other number. Use the given conditions to write a system of equations. Solve the system and find the numbers.

43. The sum of two numbers is 7. If one number is subtracted from the other, their difference is −1. Find the numbers.

44. The sum of two numbers is 2. If one number is subtracted from the other, their difference is 8. Find the numbers.

45. Three times a first number decreased by a second number is 1. The first number increased by twice the second number is 12. Find the numbers.

46. The sum of three times a first number and twice a second number is 8. If the second number is subtracted from twice the first number, the result is 3. Find the numbers.

Practice Plus

In Exercises 47–48, solve each system by the method of your choice.

47. $\dfrac{x + 2}{2} - \dfrac{y + 4}{3} = 3$
$\dfrac{x + y}{5} = \dfrac{x - y}{2} - \dfrac{5}{2}$

48. $\dfrac{x - y}{3} = \dfrac{x + y}{2} - \dfrac{1}{2}$
$\dfrac{x + 2}{2} - 4 = \dfrac{y + 4}{3}$

In Exercises 49–50, solve each system for x and y, expressing either value in terms of a or b, if necessary. Assume that $a \neq 0$ and $b \neq 0$.

49. $5ax + 4y = 17$
$ax + 7y = 22$

50. $4ax + by = 3$
$6ax + 5by = 8$

51. For the linear function $f(x) = mx + b, f(-2) = 11$ and $f(3) = -9$. Find m and b.

52. For the linear function $f(x) = mx + b, f(-3) = 23$ and $f(2) = -7$. Find m and b.

Use the graphs of the linear functions to solve Exercises 53–54.

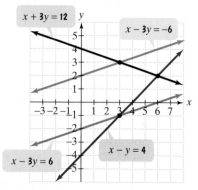

53. Write the linear system whose solution set is $\{(6, 2)\}$. Express each equation in the system in slope-intercept form.

54. Write the linear system whose solution set is $\emptyset$. Express each equation in the system in slope-intercept form.

Application Exercises

55. A wine company needs to blend a California wine with a 5% alcohol content and a French wine with a 9% alcohol content to obtain 200 gallons of wine with a 7% alcohol content. How may gallons of each kind of wine must be used?

56. A jeweler needs to mix an alloy with a 16% gold content and an alloy with a 28% gold content to obtain 32 ounces of a new alloy with a 25% gold content. How many ounces of each of the original alloys must be used?

57. For thousands of years, gold has been considered one of Earth's most precious metals. One hundred percent pure gold is 24-karat gold, which is too soft to be made into jewelry. In the United States, most gold jewelry is 14-karat gold, approximately 58% gold. If 18-karat gold is 75% gold and 12-karat gold is 50% gold, how much of each should be used to make a 14-karat gold bracelet weighing 300 grams?

58. In the "Peanuts" cartoon shown, solve the problem that is sending Peppermint Patty into an agitated state. How much cream and how much milk, to the nearest thousandth of a gallon, must be mixed together to obtain 50 gallons of cream that contains 12.5% butterfat?

© 1978 United Media/United Feature Syndicate, Inc.

59. The manager of a candystand at a large multiplex cinema has a popular candy that sells for $1.60 per pound. The manager notices a different candy worth $2.10 per pound that is not selling well. The manager decides to form a mixture of both types of candy to help clear the inventory of the more expensive type. How many pounds of each kind of candy should be used to create a 75-pound mixture selling for $1.90 per pound?

60. A grocer needs to mix raisins at $2.00 per pound with granola at $3.25 per pound to obtain 10 pounds of a mixture that costs $2.50 per pound. How many pounds of raisins and how many pound of granola must be used?

61. When a small plane flies with the wind, it can travel 800 miles in 5 hours. When the plane flies in the opposite direction, against the wind. it takes 8 hours to fly the same distance. Find the average velocity of the plane in still air and the average velocity of the wind.

62. When a plane flies with the wind, it can travel 4200 miles in 6 hours. When the plane flies in the opposite direction, against the wind, it takes 7 hours to fly the same distance. Find the average velocity of the plane in still air and the average velocity of the wind.

63. A boat's crew rowed 16 kilometers downstream, with the current, in 2 hours. The return trip upstream, against the current, covered the same distance, but took 4 hours. Find the crew's average rowing velocity in still water and the average velocity of the current.

64. A motorboat traveled 36 miles downstream, with the current, in 1.5 hours. The return trip upstream, against the current, covered the same distance, but took 2 hours. Find the boat's average velocity in still water and the average velocity of the current.

65. With the current, you can canoe 24 miles in 4 hours. Against the same current, you can canoe only $\frac{3}{4}$ of this distance in 6 hours. Find your average velocity in still water and the average velocity of the current.

66. With the current, you can row 24 miles in 3 hours. Against the same current, you can row only $\frac{3}{4}$ of this distance in 4 hours. Find your average rowing velocity in still water and the average velocity of the current.

The figure shows the graphs of the cost and revenue functions for a company that manufactures and sells small radios. Use the information in the figure to solve Exercises 67–72.

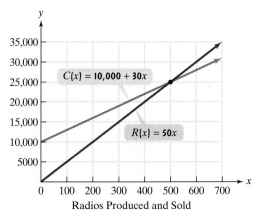

Radios Produced and Sold

67. How many radios must be produced and sold for the company to break even?

68. More than how many radios must be produced and sold for the company to have a profit?

69. Use the formulas shown in the voice balloons to find $R(200) - C(200)$. Describe what this means for the company.

70. Use the formulas shown in the voice balloons to find $R(300) - C(300)$. Describe what this means for the company.

71. a. Use the formulas shown in the voice balloons to write the company's profit function, P, from producing and selling x radios.

   b. Find the company's profit if 10,000 radios are produced and sold.

72. a. Use the formulas shown in the voice balloons to write the company's profit function, P, from producing and selling x radios.

   b. Find the company's profit if 20,000 radios are produced and sold.

Exercises 73–76 describe a number of business ventures. For each exercise,
   a. *Write the cost function, C.*
   b. *Write the revenue function, R.*
   c. *Determine the break-even point. Describe what this means.*

73. A company that manufactures small canoes has a fixed cost of $18,000. It costs $20 to produce each canoe. The selling price is $80 per canoe. (In solving this exercise, let x represent the number of canoes produced and sold.)

74. A company that manufactures bicycles has a fixed cost of $100,000. It costs $100 to produce each bicycle. The selling price is $300 per bike. (In solving this exercise, let x represent the number of bicycles produced and sold.)

75. You invest in a new play. The cost includes an overhead of $30,000, plus production costs of $2500 per performance. A sold-out performance brings in $3125. (In solving this exercise, let x represent the number of sold-out performances.)

76. You invested $30,000 and started a business writing greeting cards. Supplies cost 2¢ per card and you are selling each card for 50¢. (In solving this exercise, let x represent the number of cards produced and sold.)

An important application of systems of equations arises in connection with supply and demand. As the price of a product increases, the demand for that product decreases. However, at higher prices, suppliers are willing to produce greater quantities of the product. Exercises 77–78 involve supply and demand.

77. A chain of electronics stores sells hand-held color televisions. The weekly demand and supply models are given as follows:

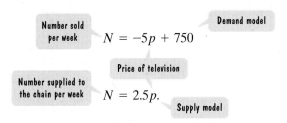

   a. How many hand-held color televisions can be sold and supplied at $120 per television?

   b. Find the price at which supply and demand are equal. At this price, how many televisions can be supplied and sold each week?

78. At a price of p dollars per ticket, the number of tickets to a rock concert that can be sold is given by the demand model $N = -25p + 7800$. At a price of p dollars per ticket, the number of tickets that the concert's promoters are willing to make available is given by the supply model $N = 5p + 6000$. (*Exercise continues on the next page.*)

a. How many tickets can be sold and supplied for $50 per ticket?

b. Find the ticket price at which supply and demand are equal. At this price, how many tickets will be supplied and sold?

79. The graphs shown below are based on 543 adults polled nationally by *Newsweek*.

Are You in Favor of the Death Penalty for a Person Convicted of Murder?

For

79%   76   80   77   71   66

Against

16%   18   16   13   22   28

Oct.   June   Sept.   May   Feb.   Feb.
1988   1991   1994   1995   1999   2000

Source: Newsweek

The function $13x + 12y = 992$ models the percent, y, in favor of the death penalty x years after 1988. The function $-x + y = 16$ models the percent, y, against the death penalty x years after 1988. If the trends shown by the graphs continue, in which year will the percentage of Americans in favor of the death penalty be the same as the percentage of Americans who oppose it? For that year, what percent will be for the death penalty and what percent will be against it?

80. One of the most dramatic developments in the work force has been the increase in the number of women, at approximately $\frac{1}{2}$% per year. By contrast, the percentage of men is decreasing by $\frac{1}{4}$% per year. The graphs shown below illustrate these changes.

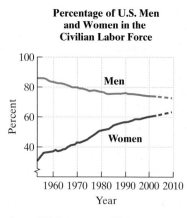

Percentage of U.S. Men and Women in the Civilian Labor Force

Men

Women

1960 1970 1980 1990 2000 2010
Year

Source: U.S. Department of Labor

The function $y = 0.52x + 35.7$ models the percentage, y, of U.S. women in the work force x years after 1955. The function $0.25x + y = 85.4$ models the percentage, y, of U.S. men in the work force x years after 1955. Use these models to determine when the percentage of women in the work force will be the same as the percentage of men in the

work force. Round to the nearest year. What percentage of women and what percentage of men will be in the work force at that time?

81. Although Social Security is a problem, some projections indicate that there's a much bigger time bomb ticking in the federal budget, and that's Medicare. In 2000, the cost of Social Security was 5.48% of the gross domestic product, increasing by 0.04% of the GDP per year. In 2000, the cost of Medicare was 1.84% of the gross domestic product, increasing by 0.17% of the GDP per year.

(*Source:* Congressional Budget Office)

a. Write a function that models the cost of Social Security a a percentage of the GDP x years after 2000.

b. Write a function that models the cost of Medicare as a percentage of the GDP x years after 2000.

c. In which year will the cost of Medicare and Social Security be the same? For that year, what will be the cost of each program as a percentage of the GDP? Which program will have the greater cost after that year?

82. The graph indicates that in 1984, there were 72 meals per person at take-out restaurants. For the period shown, this number increased by an average of 2.25 meals per person pe year. In 1984, there were 94 meals per person at on-premise dining facilities and this number decreased by an average c 0.55 meals per person per year.

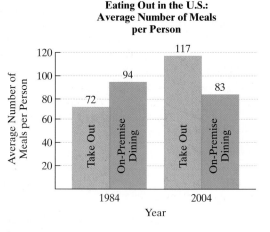

Eating Out in the U.S.: Average Number of Meals per Person

117

94

83

72

1984   2004
Year

Source: The NPD Group

a. Write a function that models the average numbe of meals per person at take-out restaurants x year after 1984.

b. Write a function that models the average number o meals per person at on-premise dining facilities x year after 1984.

c. In which year, to the nearest whole year, was the averag number of meals per person for take-out and on-premis restaurants the same? For that year, how many meals pe person, to the nearest whole number, were there for eac kind of restaurant? Which kind of restaurant had th greater number of meals per person after that year?

The bar graph shows the percentage of Americans who used cigarettes, by ethnicity, in 1985 and 2002. For each of the groups shown, cigarette use has been linearly decreasing. Use this information to solve Exercises 83–84.

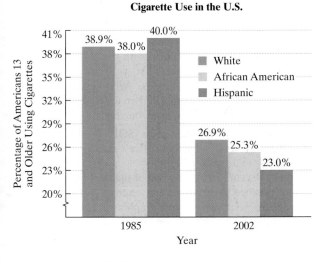

Cigarette Use in the U.S.

Source: Department of Health and Human Services

83. In this exercise, let x represent the number of years after 1985 and let y represent the percentage of Americans in one of the groups shown who used cigarettes.

   a. Use the data points $(0, 38)$ and $(17, 25.3)$ to find the slope-intercept equation of the line that models the percentage of African Americans who used cigarettes, y, x years after 1985. Round the value of m to two decimal places.

   b. Use the data points $(0, 40)$ and $(17, 23)$ to find the slope-intercept equation of the line that models the percentage of Hispanics who used cigarettes, y, x years after 1985.

   c. Use the models from parts (a) and (b) to find the year during which cigarette use was the same for African Americans and Hispanics. What percentage of each group used cigarettes during that year?

84. In this exercise, let x represent the number of years after 1985 and let y represent the percentage of Americans in one of the groups shown who used cigarettes.

   a. Use the data points $(0, 38.9)$ and $(17, 26.9)$ to find the slope-intercept equation of the line that models the percentage of whites who used cigarettes, y, x years after 1985. Round the value of m to two decimal places.

   b. Use the data points $(0, 40)$ and $(17, 23)$ to find the slope-intercept equation of the line that models the percentage of Hispanics who used cigarettes, y, x years after 1985.

   c. Use the models from parts (a) and (b) to find the year, to the nearest whole year, during which cigarette use was the same for whites and Hispanics. What percentage of each group, to the nearest percent, used cigarettes during that year?

Use a system of linear equations to solve Exercises 85–92.

The graph shows the calories in some favorite fast foods. Use the information in Exercises 85–86 to find the exact caloric content of the specified foods.

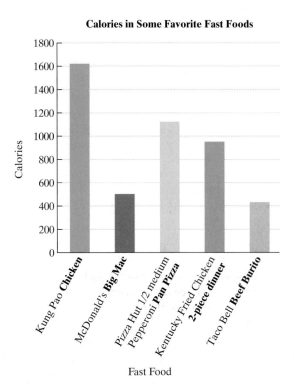

Calories in Some Favorite Fast Foods

Source: Center for Science in the Public Interest

85. One pan pizza and two beef burritos provide 1980 calories. Two pan pizzas and one beef burrito provide 2670 calories. Find the caloric content of each item.

86. One Kung Pao chicken and two Big Macs provide 2620 calories. Two Kung Pao chickens and one Big Mac provide 3740 calories. Find the caloric content of each item.

87. Cholesterol intake should be limited to 300 mg or less each day. One serving of scrambled eggs from McDonalds and one Double Beef Whopper from Burger King exceed this intake by 241 mg. Two servings of scrambled eggs and three Double Beef Whoppers provide 1257 mg of cholesterol. Determine the cholesterol content in each item.

88. Two medium eggs and three cups of ice cream contain 701 milligrams of cholesterol. One medium egg and one cup of ice cream exceed the suggested daily cholesterol intake of 300 milligrams by 25 milligrams. Determine the cholesterol content in each item.

89. A hotel has 200 rooms. Those with kitchen facilities rent for $100 per night and those without kitchen facilities rent for $80 per night. On a night when the hotel was completely occupied, revenues were $17,000. How many of each type of room does the hotel have?

90. A new restaurant is to contain two-seat tables and four-seat tables. Fire codes limit the restaurant's maximum occupancy to 56 customers. If the owners have hired enough servers to handle 17 tables of customers, how many of each kind of table should they purchase?

91. A rectangular lot whose perimeter is 360 feet is fenced along three sides. An expensive fencing along the lot's length costs $20 per foot and an inexpensive fencing along the two side widths costs only $8 per foot. The total cost of the fencing along the three sides comes to $3280. What are the lot's dimensions?

92. A rectangular lot whose perimeter is 320 feet is fenced along three sides. An expensive fencing along the lot's length costs $16 per foot and an inexpensive fencing along the two side widths costs only $5 per foot. The total cost of the fencing along the three sides comes to $2140. What are the lot's dimensions?

In Exercises 93–94, an isosceles triangle containing two angles with equal measure is shown. The degree measure of each triangle's three interior angles and an exterior angle is represented with variables. Find the measure of the three interior angles.

93.

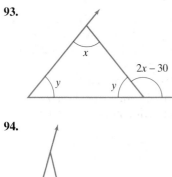

94.

Writing in Mathematics

95. What is a system of linear equations? Provide an example with your description.

96. What is the solution of a system of linear equations?

97. Explain how to solve a system of equations using the substitution method. Use $y = 3 - 3x$ and $3x + 4y = 6$ to illustrate your explanation.

98. Explain how to solve a system of equations using the addition method. Use $3x + 5y = -2$ and $2x + 3y = 0$ to illustrate your explanation.

99. When is it easier to use the addition method rather than the substitution method to solve a system of equations?

100. When using the addition or substitution method, how can you tell if a system of linear equations has infinitely many solutions? What is the relationship between the graphs of the two equations?

101. When using the addition or substitution method, how can you tell if a system of linear equations has no solution? What is the relationship between the graphs of the two equations?

102. Describe the break-even point for a business.

Technology Exercises

103. Verify your solutions to any five exercises in Exercises 5–4? by using a graphing utility to graph the two equations in the system in the same viewing rectangle. Then use the intersection feature to display the solution.

104. Some graphing utilities can give the solution to a linear system of equations. (Consult your manual for details.) This capability is usually accessed with the $\boxed{\text{SIMULT}}$ (simultaneous equations) feature. First, you will enter 2, for two equations in two variables. With each equation in $Ax + By = C$ form, you will then enter the coefficients for x and y and the constant term, one equation at a time. After entering all six numbers, press $\boxed{\text{SOLVE}}$. The solution will be displayed on the screen. (The x-value may be displayed as $x_1 =$ and the y-value as $x_2 =$.) Use this capability to verify the solution to any five of the exercises you solved in the practice exercises of this exercise set. Describe what happens when you use your graphing utility on a system with no solution or infinitely many solutions.

Critical Thinking Exercises

105. Write a system of equations having $\{(-2, 7)\}$ as a solution set. (More than one system is possible.)

106. Solve the system for x and y in terms of a_1, b_1, c_1, a_2, b_2, and c_2:

$$a_1 x + b_1 y = c_1$$
$$a_2 x + b_2 y = c_2.$$

107. Two identical twins can only be distinguished by the characteristic that one always tells the truth and the other always lies. One twin tells you of a lucky number pair: "When I multiply my first lucky number by 3 and my second lucky number by 6, the addition of the resulting numbers produces a sum of 12. When I add my first lucky number and twice my second lucky number, the sum is 5." Which twin is talking?

108. A marching band has 52 members and there are 24 in the pom-pom squad. They wish to form several hexagons and squares like those diagrammed below. Can it be done with no people left over?

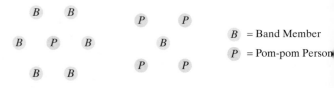

Group Exercise

109. The group should write four different word problems that can be solved using a system of linear equations in two variables. All of the problems should be on different topics. The group should turn in the four problems and their algebraic solutions.

Systems of Linear Equations in Three Variables

Objectives

1 Verify the solution of a system of linear equations in three variables.

2 Solve systems of linear equations in three variables.

3 Solve problems using systems in three variables.

All animals sleep, but the length of time they sleep varies widely: Cattle sleep for only a few minutes at a time. We humans seem to need more sleep than other animals, up to eight hours a day. Without enough sleep, we have difficulty concentrating, make mistakes in routine tasks, lose energy, and feel bad-tempered. There is a relationship between hours of sleep and death rate per year per 100,000 people. How many hours of sleep will put you in the group with the minimum death rate? In this section, we will answer this question by solving a system of linear equations with more than two variables.

 Verify the solution of a system of linear equations in three variables.

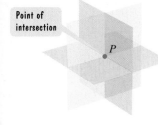

Point of intersection

Figure 7.9

Systems of Linear Equations in Three Variables and Their Solutions

An equation such as $x + 2y - 3z = 9$ is called a *linear equation in three variables*. In general, any equation of the form

$$Ax + By + Cz = D,$$

where A, B, C, and D are real numbers such that A, B, and C are not all 0, is a **linear equation in three variables: x, y, and z**. The graph of this linear equation in three variables is a plane in three-dimensional space.

   The process of solving a system of three linear equations in three variables is geometrically equivalent to finding the point of intersection (assuming that there is one) of three planes in space. (See Figure 7.9.) A **solution** of a system of linear equations in three variables is an ordered triple of real numbers that satisfies all equations of the system. The **solution set** of the system is the set of all its solutions.

EXAMPLE 1   Determining Whether an Ordered Triple Satisfies a System

Show that the ordered triple $(-1, 2, -2)$ is a solution of the system:

$$x + 2y - 3z = 9$$
$$2x - y + 2z = -8$$
$$-x + 3y - 4z = 15.$$

Solution   Because -1 is the x-coordinate, 2 is the y-coordinate, and -2 is the z-coordinate of $(-1, 2, -2)$, we replace x with -1, y with 2, and z with -2 in each of the three equations.

$$x + 2y - 3z = 9$$
$$-1 + 2(2) - 3(-2) \stackrel{?}{=} 9$$
$$-1 + 4 + 6 \stackrel{?}{=} 9$$
$$9 = 9, \text{ true}$$

$$2x - y + 2z = -8$$
$$2(-1) - 2 + 2(-2) \stackrel{?}{=} -8$$
$$-2 - 2 - 4 \stackrel{?}{=} -8$$
$$-8 = -8, \text{ true}$$

$$-x + 3y - 4z = 15$$
$$-(-1) + 3(2) - 4(-2) \stackrel{?}{=} 15$$
$$1 + 6 + 8 \stackrel{?}{=} 15$$
$$15 = 15, \text{ true}$$

The ordered triple $(-1, 2, -2)$ satisfies the three equations: It makes each equation true. Thus, the ordered triple is a solution of the system.

Check Point 1 Show that the ordered triple $(-1, -4, 5)$ is a solution of the system:

$$x - 2y + 3z = 22$$
$$2x - 3y - z = 5$$
$$3x + y - 5z = -32.$$

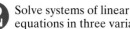 Solve systems of linear equations in three variables.

Solving Systems of Linear Equations in Three Variables by Eliminating Variables

The method for solving a system of linear equations in three variables is similar to that used on systems of linear equations in two variables. We use addition to eliminate any variable, reducing the system to two equations in two variables. Once we obtain a system of two equations in two variables, we use addition or substitution to eliminate a variable. The result is a single equation in one variable. We solve this equation to get the value of the remaining variable. Other variable values are found by back-substitution.

Study Tip

It does not matter which variable you eliminate, as long as you do it in two different pairs of equations.

Solving Linear Systems in Three Variables by Eliminating Variables

1. Reduce the system to two equations in two variables. This is usually accomplished by taking two different pairs of equations and using the addition method to eliminate the same variable from both pairs.
2. Solve the resulting system of two equations in two variables using addition or substitution. The result is an equation in one variable that gives the value of that variable.
3. Back-substitute the value of the variable found in step 2 into either of the equations in two variables to find the value of the second variable.
4. Use the values of the two variables from steps 2 and 3 to find the value of the third variable by back-substituting into one of the original equations.
5. Check the proposed solution in each of the original equations.

EXAMPLE 2   Solving a System in Three Variables

Solve the system:

$$5x - 2y - 4z = 3 \qquad \text{Equation 1}$$
$$3x + 3y + 2z = -3 \qquad \text{Equation 2}$$
$$-2x + 5y + 3z = 3. \qquad \text{Equation 3}$$

Solution   There are many ways to proceed. Because our initial goal is to reduce the system to two equations in two variables, **the central idea is to take two different pairs of equations and eliminate the same variable from both pairs.**

Step 1   Reduce the system to two equations in two variables. We choose any two equations and use the addition method to eliminate a variable. Let's eliminate z using Equations 1 and 2. We do so by multiplying Equation 2 by 2. Then we add equations.

(Equation 1) $5x - 2y - 4z = 3$ $\xrightarrow{\text{No change}}$ $5x - 2y - 4z = 3$

(Equation 2) $3x + 3y + 2z = -3$ $\xrightarrow{\text{Multiply by 2.}}$ $\underline{6x + 6y + 4z = -6}$

Add: $11x + 4y \qquad = -3$ Equation 4

Now we must eliminate the *same* variable using another pair of equations. We can eliminate z from Equations 2 and 3. First, we multiply Equation 2 by -3. Next, we multiply Equation 3 by 2. Finally, we add equations.

(Equation 2) $3x + 3y + 2z = -3$ Multiply by -3. $-9x - 9y - 6z = 9$

(Equation 3) $-2x + 5y + 3z = 3$ Multiply by 2. $\underline{-4x + 10y + 6z = 6}$

Add: $-13x + y = 15$ Equation 5

Equations 4 and 5 give us a system of two equations in two variables:

$$11x + 4y = -3 \qquad \text{Equation 4}$$
$$-13x + y = 15. \qquad \text{Equation 5}$$

Step 2   Solve the resulting system of two equations in two variables. We will use the addition method to solve Equations 4 and 5 for x and y. To do so, we multiply Equation 5 on both sides by -4 and add this to Equation 4.

(Equation 4) $11x + 4y = -3$ No change $11x + 4y = -3$

(Equation 5) $-13x + y = 15$ Multiply by -4. $\underline{52x - 4y = -60}$

Add: $63x = -63$

$x = -1$ Divide both sides by 63.

Step 3   Use back-substitution in one of the equations in two variables to find the value of the second variable. We back-substitute -1 for x in either Equation 4 or 5 to find the value of y.

$$-13x + y = 15 \qquad \text{Equation 5}$$
$$-13(-1) + y = 15 \qquad \text{Substitute } -1 \text{ for } x.$$
$$13 + y = 15 \qquad \text{Multiply.}$$
$$y = 2 \qquad \text{Subtract 13 from both sides.}$$

Step 4   Back-substitute the values found for two variables into one of the original equations to find the value of the third variable. We can now use any one of the original equations and back-substitute the values of x and y to find the value for z. We will use Equation 2.

$$3x + 3y + 2z = -3 \qquad \text{Equation 2}$$
$$3(-1) + 3(2) + 2z = -3 \qquad \text{Substitute } -1 \text{ for } x \text{ and } 2 \text{ for } y.$$
$$3 + 2z = -3 \qquad \text{Multiply and then add:}$$
$$\qquad\qquad\qquad\qquad 3(-1) + 3(2) = -3 + 6 = 3.$$
$$2z = -6 \qquad \text{Subtract 3 from both sides.}$$
$$z = -3 \qquad \text{Divide both sides by 2.}$$

With $x = -1$, $y = 2$, and $z = -3$, the proposed solution is the ordered triple $(-1, 2, -3)$.

Step 5   Check. Check the proposed solution, $(-1, 2, -3)$, by substituting the values for x, y, and z into each of the three original equations. These substitutions yield three true statements. Thus, the solution set is $\{(-1, 2, -3)\}$.

Check Point 2   Solve the system:

$$x + 4y - z = 20$$
$$3x + 2y + z = 8$$
$$2x - 3y + 2z = -16.$$

In some examples, one of the variables is already eliminated from a given equation. In this case, the same variable should be eliminated from the other two equations, thereby making it possible to omit one of the elimination steps. We illustrate this idea in Example 3.

EXAMPLE 3 Solving a System of Equations with a Missing Term

Solve the system:

$$\begin{array}{ll} x + + z = 8 & \text{Equation 1} \\ x + y + 2z = 17 & \text{Equation 2} \\ x + 2y + z = 16. & \text{Equation 3} \end{array}$$

Solution

Step 1 Reduce the system to two equations in two variables. Because Equation 1 contains only x and z, we could omit one of the elimination steps by eliminating y using Equations 2 and 3. This will give us two equations in x and z. To eliminate y using Equations 2 and 3, we multiply Equation 2 by -2 and add Equation 3.

$$\begin{array}{lll} \text{(Equation 2)} \quad x + y + 2z = 17 & \xrightarrow{\text{Multiply by } -2.} & -2x - 2y - 4z = -34 \\ \text{(Equation 3)} \quad x + 2y + z = 16 & \xrightarrow{\text{No change}} & \underline{x + 2y + z = 16} \\ & \text{Add:} & -x - 3z = -18 \quad \text{Equation 4} \end{array}$$

Equation 4 and the given Equation 1 provide us with a system of two equations in two variables:

$$\begin{array}{ll} x + z = 8 & \text{Equation 1} \\ -x - 3z = -18. & \text{Equation 4} \end{array}$$

Step 2 Solve the resulting system of two equations in two variables. We will solve Equations 1 and 4 for x and z.

$$\begin{array}{lll} & x + z = 8 & \text{Equation 1} \\ & \underline{-x - 3z = -18} & \text{Equation 4} \\ \text{Add:} & -2z = -10 & \\ & z = 5 & \text{Divide both sides by } -2. \end{array}$$

Step 3 Use back-substitution in one of the equations in two variables to find the value of the second variable. To find x, we back-substitute 5 for z in either Equation 1 or 4. We will use Equation 1.

$$\begin{array}{ll} x + z = 8 & \text{Equation 1} \\ x + 5 = 8 & \text{Substitute 5 for z.} \\ x = 3 & \text{Subtract 5 from both sides.} \end{array}$$

Step 4 Back-substitute the values found for two variables into one of the original equations to find the value of the third variable. To find y, we back-substitute 3 for x and 5 for z into Equation 2 or 3. We cannot use Equation 1 because y is missing in this equation. We will use Equation 2.

$$\begin{array}{ll} x + y + 2z = 17 & \text{Equation 2} \\ 3 + y + 2(5) = 17 & \text{Substitute 3 for x and 5 for z.} \\ y + 13 = 17 & \text{Multiply and add.} \\ y = 4 & \text{Subtract 13 from both sides.} \end{array}$$

We found that $z = 5$, $x = 3$, and $y = 4$. Thus, the proposed solution is the ordered triple $(3, 4, 5)$.

Step 5 Check. Substituting 3 for x, 4 for y, and 5 for z into each of the three original equations yields three true statements. Consequently, the solution set is $\{(3, 4, 5)\}$.

Check Point 3 Solve the system:

$$\begin{array}{l} 2y - z = 7 \\ x + 2y + z = 17 \\ 2x - 3y + 2z = -1. \end{array}$$

A system of linear equations in three variables represents three planes. The three planes may not always intersect at one point. The planes may have no common point of intersection and represent an inconsistent system with no solution. By contrast, the planes may coincide or intersect along a line. In these cases, the planes have infinitely many points in common and represent systems with infinitely many solutions. Systems of linear equations in three variables that are inconsistent or that contain dependent equations will be discussed in Chapter 8.

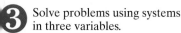 Solve problems using systems in three variables.

Applications

Systems of equations may allow us to find models for data without using a graphing utility. Three data points that do not lie on or near a line determine the graph of a quadratic function of the form $y = ax^2 + bx + c, a \neq 0$. Quadratic functions often model situations in which values of y are decreasing and then increasing, suggesting the cuplike shape of a parabola.

EXAMPLE 4   Modeling Data Relating Sleep and Death Rate

In a study relating sleep and death rate, the following data were obtained. Use the function $y = ax^2 + bx + c$ to model the data.

x (Average Number of Hours of Sleep)	y (Death Rate per Year per 100,000 Males)
4	1682
7	626
9	967

Solution   We need to find values for a, b, and c in $y = ax^2 + bx + c$. We can do so by solving a system of three linear equations in a, b, and c. We obtain the three equations by using the values of x and y from the data as follows:

$y = ax^2 + bx + c$ Use the quadratic function to model the data.

When $x = 4, y = 1682$: $1682 = a \cdot 4^2 + b \cdot 4 + c$ or $16a + 4b + c = 1682$

When $x = 7, y = 626$: $626 = a \cdot 7^2 + b \cdot 7 + c$ or $49a + 7b + c = 626$

When $x = 9, y = 967$: $967 = a \cdot 9^2 + b \cdot 9 + c$ or $81a + 9b + c = 967.$

The easiest way to solve this system is to eliminate c from two pairs of equations, obtaining two equations in a and b. Solving this system gives $a = 104.5$, $b = -1501.5$, and $c = 6016$. We now substitute the values for a, b, and c into $y = ax^2 + bx + c$. The function that models the given data is

$$y = 104.5x^2 - 1501.5x + 6016.$$

We can use the model that we obtained in Example 4 to find the death rate of males who average, say, 6 hours of sleep. First, write the model in function notation:

$$f(x) = 104.5x^2 - 1501.5x + 6016.$$

Substitute 6 for x:

$$f(6) = 104.5(6)^2 - 1501.5(6) + 6016 = 769.$$

According to the model, the death rate for males who average 6 hours of sleep is 769 deaths per 100,000 males.

Technology

The graph of
$y = 104.5x^2 - 1501.5x + 6016$
is displayed in a $[3, 12, 1]$ by $[500, 2000, 100]$ viewing rectangle. The minimum function feature shows that the lowest point on the graph, the vertex, is approximately $(7.2, 622.5)$. Men who average 7.2 hours of sleep are in the group with the lowest death rate, approximately 622.5 deaths per 100,000 males.

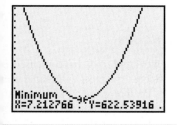

Minimum
X=7.212766   Y=622.53916

Check Point 4   Find the quadratic function $y = ax^2 + bx + c$ whose graph passes through the points $(1, 4)$, $(2, 1)$, and $(3, 4)$.

EXERCISE SET 7.2

Practice Exercises

In Exercises 1–4, determine if the given ordered triple is a solution of the system.

1. $(2, -1, 3)$

$$x + y + z = 4$$
$$x - 2y - z = 1$$
$$2x - y - 2z = -1$$

2. $(5, -3, -2)$

$$x + y + z = 0$$
$$x + 2y - 3z = 5$$
$$3x + 4y + 2z = -1$$

3. $(4, 1, 2)$

$$x - 2y = 2$$
$$2x + 3y = 11$$
$$y - 4z = -7$$

4. $(-1, 3, 2)$

$$x - 2z = -5$$
$$y - 3z = -3$$
$$2x - z = -4$$

Solve each system in Exercises 5–18.

5. $x + y + 2z = 11$
$x + y + 3z = 14$
$x + 2y - z = 5$

6. $2x + y - 2z = -1$
$3x - 3y - z = 5$
$x - 2y + 3z = 6$

7. $4x - y + 2z = 11$
$x + 2y - z = -1$
$2x + 2y - 3z = -1$

8. $x - y + 3z = 8$
$3x + y - 2z = -2$
$2x + 4y + z = 0$

9. $3x + 2y - 3z = -2$
$2x - 5y + 2z = -2$
$4x - 3y + 4z = 10$

10. $2x + 3y + 7z = 13$
$3x + 2y - 5z = -22$
$5x + 7y - 3z = -28$

11. $2x - 4y + 3z = 17$
$x + 2y - z = 0$
$4x - y - z = 6$

12. $x + z = 3$
$x + 2y - z = 1$
$2x - y + z = 3$

13. $2x + y = 2$
$x + y - z = 4$
$3x + 2y + z = 0$

14. $x + 3y + 5z = 20$
$y - 4z = -16$
$3x - 2y + 9z = 36$

15. $x + y = -4$
$y - z = 1$
$2x + y + 3z = -21$

16. $x + y = 4$
$x + z = 4$
$y + z = 4$

17. $3(2x + y) + 5z = -1$
$2(x - 3y + 4z) = -9$
$4(1 + x) = -3(z - 3y)$

18. $7z - 3 = 2(x - 3y)$
$5y + 3z - 7 = 4x$
$4 + 5z = 3(2x - y)$

In Exercises 19–22, find the quadratic function $y = ax^2 + bx + c$ whose graph passes through the given points.

19. $(-1, 6), (1, 4), (2, 9)$ **20.** $(-2, 7), (1, -2), (2, 3)$

21. $(-1, -4), (1, -2), (2, 5)$ **22.** $(1, 3), (3, -1), (4, 0)$

In Exercises 23–24, let x represent the first number, y the second number, and z the third number. Use the given conditions to write a system of equations. Solve the system and find the numbers.

23. The sum of three numbers is 16. The sum of twice the first number, 3 times the second number, and 4 times the third number is 46. The difference between 5 times the first number and the second number is 31. Find the three numbers.

24. The following is known about three numbers: Three times the first number plus the second number plus twice the third number is 5. If 3 times the second number is subtracted from the sum of the first number and 3 times the third number, the result is 2. If the third number is subtracted from 2 times the first number and 3 times the second number, the result is 1. Find the numbers.

Practice Plus

Solve each system in Exercises 25–26.

25. $\dfrac{x + 2}{6} - \dfrac{y + 4}{3} + \dfrac{z}{2} = 0$

$\dfrac{x + 1}{2} + \dfrac{y - 1}{2} - \dfrac{z}{4} = \dfrac{9}{2}$

$\dfrac{x - 5}{4} + \dfrac{y + 1}{3} + \dfrac{z - 2}{2} = \dfrac{19}{4}$

26. $\dfrac{x + 3}{2} - \dfrac{y - 1}{2} + \dfrac{z + 2}{4} = \dfrac{3}{2}$

$\dfrac{x - 5}{2} + \dfrac{y + 1}{3} - \dfrac{z}{4} = -\dfrac{25}{6}$

$\dfrac{x - 3}{4} - \dfrac{y + 1}{2} + \dfrac{z - 3}{2} = -\dfrac{5}{2}$

In Exercises 27–28, find the equation of the quadratic function $y = ax^2 + bx + c$ whose graph is shown. Select three points whose coordinates appear to be integers.

27.

28.

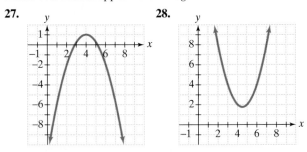

In Exercises 29–30, solve each system for (x, y, z) in terms of the nonzero constants a, b, and c.

29. $ax - by - 2cz = 21$
$ax + by + cz = 0$
$2ax - by + cz = 14$

30. $ax - by + 2cz = -4$
$ax + 3by - cz = 1$
$2ax + by + 3cz = 2$

Application Exercises

31. Although headlines about illegal steroids have focused on professional and Olympic athletes, the most vulnerable user may be high school students. The bar graph at the top of the next page shows the percentage of U.S. high school seniors who had taken steroids from 2000 through 2003.

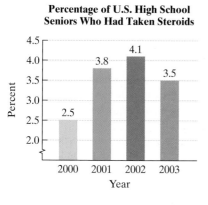

Percentage of U.S. High School Seniors Who Had Taken Steroids

Source: University of Michigan

a. Write the data for 2000, 2002, and 2003 as ordered pairs (x, y), where x is the number of years after 2000 and y is the percentage of seniors who had taken steroids in that year.

b. The three data points in part (a) can be modeled by the quadratic function $y = ax^2 + bx + c$. Substitute each ordered pair into this function, one ordered pair at a time, and write a system of linear equations in three variables that can be used to find values for a, b, and c.

c. Solve the system in part (b) and write a quadratic function that models the percentage of U.S. high school seniors who had taken steroids x years after 2000.

32. The bar graph shows the percentage of people in the United States living below the poverty level for selected years from 1990 through 2003.

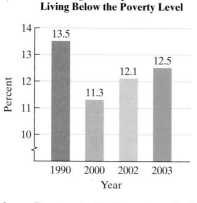

Percentage of People in the U.S. Living Below the Poverty Level

Source: Department of Health and Human Services

a. Write the data for 1990, 2002, and 2003 as ordered pairs (x, y), where x is the number of years after 1990 and y is the percentage of people living below the poverty level.

b. The three data points in part (a) can be modeled by the quadratic function $y = ax^2 + bx + c$. Substitute each ordered pair into this function, one ordered pair at a time, and write a system of linear equations in three variables that can be used to find values for a, b, and c. It is not necessary to solve the system.

33. You throw a ball straight up from a rooftop. The ball misses the rooftop on its way down and eventually strikes the ground. A mathematical model can be used to describe the relationship for the ball's height above the ground, y, after x seconds. Consider the following data:

x, seconds after the ball is thrown	y, ball's height, in feet, above the ground
1	224
3	176
4	104

a. Find the quadratic function $y = ax^2 + bx + c$ whose graph passes through the given points.

b. Use the function in part (a) to find the value for y when $x = 5$. Describe what this means.

34. A mathematical model can be used to describe the relationship between the number of feet a car travels once the brakes are applied, y, and the number of seconds the car is in motion after the brakes are applied, x. A research firm collects the following data:

x, seconds in motion after brakes are applied	y, feet car travels once the brakes are applied
1	46
2	84
3	114

a. Find the quadratic function $y = ax^2 + bx + c$ whose graph passes through the given points.

b. Use the function in part (a) to find the value for y when $x = 6$. Describe what this means.

Use a system of linear equations in three variables to solve Exercises 35–41.

35. In current U.S. dollars, John D. Rockefeller's 1913 fortune of $900 million would be worth about $189 billion. The bar graph shows that Rockefeller is the wealthiest among the world's five richest people of all time. The combined estimated wealth, in current billions of U.S. dollars, of Andrew Carnegie, Cornelius Vanderbilt, and Bill Gates is $244 billion. The difference between Carnegie's estimated wealth and Vanderbilt's is $4 billion. The difference between Vanderbilt's estimated wealth and Gates's is $48 billion. Find the estimated wealth, in current billions of U.S. dollars, of Carnegie, Vanderbilt, and Gates.

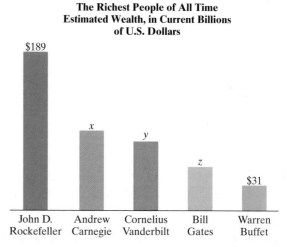

The Richest People of All Time Estimated Wealth, in Current Billions of U.S. Dollars

Source: Scholastic Book of World Records

36. The circle graph shows the percentage of Americans who drink caffeinated beverages on a daily basis and the number of cups consumed per day.

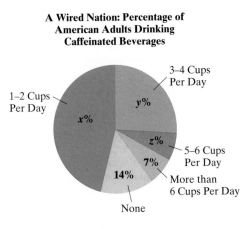

A Wired Nation: Percentage of American Adults Drinking Caffeinated Beverages

Source: Harris Interactive

72% of American adults drink from one to four cups of caffeinated beverages per day and 40% drink three or more cups per day. Find the percentage who drink from one to two cups, from three to four cups, and from five to six cups of caffeinated beverages per day.

37. At a college production of *Streetcar Named Desire*, 400 tickets were sold. The ticket prices were $8, $10, and $12, and the total income from ticket sales was $3700. How many tickets of each type were sold if the combined number of $8 and $10 tickets sold was 7 times the number of $12 tickets sold?

38. A certain brand of razor blades comes in packages of 6, 12, and 24 blades, costing $2, $3, and $4 per package, respectively. A store sold 12 packages containing a total of 162 razor blades and took in $35. How many packages of each type were sold?

39. A person invested $6700 for one year, part at 8%, part at 10%, and the remainder at 12%. The total annual income from these investments was $716. The amount of money invested at 12% was $300 more than the amount invested at 8% and 10% combined. Find the amount invested at each rate.

40. A person invested $17,000 for one year, part at 10%, part at 12%, and the remainder at 15%. The total annual income from these investments was $2110. The amount of money invested at 12% was $1000 less than the amount invested at 10% and 15% combined. Find the amount invested at each rate.

41. In the following triangle, the degree measures of the three interior angles and two of the exterior angles are represented with variables. Find the measure of each interior angle.

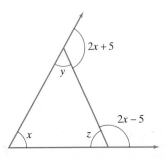

Writing in Mathematics

42. What is a system of linear equations in three variables?

43. How do you determine whether a given ordered triple is a solution of a system in three variables?

44. Describe in general terms how to solve a system in three variables.

45. AIDS is taking a deadly toll on southern Africa. Describe how to use the techniques that you learned in this section to obtain a model for African life span using projections with AIDS, shown by the red graph in the figure. Let x represent the number of years after 1985 and let y represent African life span in that year.

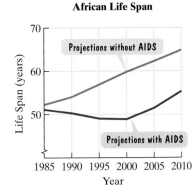

African Life Span

Source: United Nations

Technology Exercises

46. Does your graphing utility have a feature that allows you to solve linear systems by entering coefficients and constant terms? If so, use this feature to verify the solutions to any five exercises that you worked by hand from Exercises 5–16.

47. Verify your results in Exercises 19–22 by using a graphing utility to graph the resulting parabola. Trace along the curve and convince yourself that the three points given in the exercise lie on the parabola.

Critical Thinking Exercises

48. Describe how the system

$$x + y - z - 2w = -8$$
$$x - 2y + 3z + w = 18$$
$$2x + 2y + 2z - 2w = 10$$
$$2x + y - z + w = 3$$

could be solved. Is it likely that in the near future a graphing utility will be available to provide a geometric solution (using intersecting graphs) to this system? Explain.

49. A modernistic painting consists of triangles, rectangles, and pentagons, all drawn so as to not overlap or share sides. Within each rectangle are drawn 2 red roses and each pentagon contains 5 carnations. How many triangles, rectangles, and pentagons appear in the painting if the painting contains a total of 40 geometric figures, 153 sides of geometric figures, and 72 flowers?

Group Exercise

50. Group members should develop appropriate functions that model each of the projections shown in Exercise 45.

SECTION 7.3   *Partial Fractions*

Objectives

1 Decompose $\dfrac{P}{Q}$, where Q has only distinct linear factors.

2 Decompose $\dfrac{P}{Q}$, where Q has repeated linear factors.

3 Decompose $\dfrac{P}{Q}$, where Q has a nonrepeated prime quadratic factor.

4 Decompose $\dfrac{P}{Q}$, where Q has a prime, repeated quadratic factor.

The rising and setting of the sun suggest the obvious: Things change over time. Calculus is the study of rates of change, allowing the motion of the rising sun to be measured by "freezing the frame" at one instant in time. If you are given a function, calculus reveals its rate of change at any "frozen" instant. In this section, you will learn an algebraic technique used in calculus to find a function if its rate of change is known. The technique involves expressing a given function in terms of simpler functions.

The Idea behind Partial Fraction Decomposition

We know how to use common denominators to write a sum or difference of rational expressions as a single rational expression. For example,

$$\frac{3}{x-4} - \frac{2}{x+2} = \frac{3(x+2) - 2(x-4)}{(x-4)(x+2)}$$

$$= \frac{3x + 6 - 2x + 8}{(x-4)(x+2)} = \frac{x + 14}{(x-4)(x+2)}.$$

For solving the kind of calculus problem described in the section opener, we must reverse this process:

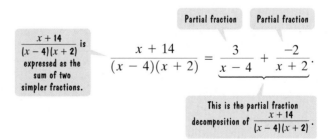

Each of the two fractions on the right is called a **partial fraction**. The sum of these fractions is called the **partial fraction decomposition** of the rational expression on the left-hand side.

   Partial fraction decompositions can be written for rational expressions of the form $\dfrac{P(x)}{Q(x)}$, where P and Q have no common factors and the highest power in the numerator is less than the highest power in the denominator. In this section, we will show you how to write the partial fraction decompositions for each of the following rational expressions:

$$\frac{9x^2 - 9x + 6}{(2x - 1)(x + 2)(x - 2)}$$

$P(x) = 9x^2 - 9x + 6$; highest power = **2**

$Q(x) = (2x - 1)(x + 2)(x - 2)$; multiplying factors, highest power = **3**.

$$\frac{5x^3 - 3x^2 + 7x - 3}{(x^2 + 1)^2}.$$

$P(x) = 5x^3 - 3x^2 + 7x - 3$; highest power = **3**

$Q(x) = (x^2 + 1)^2$; squaring the expression, highest power = **4**.

The partial fraction decomposition of a rational expression depends on the factors of the denominator. We consider four cases involving different kinds of factors in the denominator:

1. The denominator is a product of distinct linear factors.
2. The denominator is a product of linear factors, some of which are repeated.
3. The denominator has prime quadratic factors, none of which is repeated.
4. The denominator has a repeated prime quadratic factor.

① Decompose $\dfrac{P}{Q}$, where Q has only distinct linear factors.

The Partial Fraction Decomposition of a Rational Expression with Distinct Linear Factors in the Denominator

If the denominator of a rational expression has a linear factor of the form $ax + b$, then the partial fraction decomposition will contain a term of the form

$$\frac{A}{ax + b}.$$

Constant

Linear factor

Each distinct linear factor in the denominator produces a partial fraction of the form *constant over linear factor*. For example,

$$\frac{9x^2 - 9x + 6}{(2x - 1)(x + 2)(x - 2)} = \frac{A}{2x - 1} + \frac{B}{x + 2} + \frac{C}{x - 2}.$$

We write a constant over each linear factor in the denominator.

The Partial Fraction Decomposition of $\dfrac{P(x)}{Q(x)}$: $Q(x)$ Has Distinct Linear Factors

The form of the partial fraction decomposition for a rational expression with distinct linear factors in the denominator is

$$\frac{P(x)}{(a_1x + b_1)(a_2x + b_2)(a_3x + b_3)\cdots(a_nx + b_n)}$$

$$= \frac{A_1}{a_1x + b_1} + \frac{A_2}{a_2x + b_2} + \frac{A_3}{a_3x + b_3} + \cdots + \frac{A_n}{a_nx + b_n}.$$

EXAMPLE 1   Partial Fraction Decomposition with Distinct Linear Factors

Find the partial fraction decomposition of

$$\frac{x + 14}{(x - 4)(x + 2)}.$$

Solution   We begin by setting up the partial fraction decomposition with the unknown constants. Write a constant over each of the two distinct linear factors in the denominator.

$$\frac{x + 14}{(x - 4)(x + 2)} = \frac{A}{x - 4} + \frac{B}{x + 2}$$

Our goal is to find A and B. We do this by multiplying both sides of the equation by the least common denominator, $(x - 4)(x + 2)$.

$$(x - 4)(x + 2)\frac{x + 14}{(x - 4)(x + 2)} = (x - 4)(x + 2)\left(\frac{A}{x - 4} + \frac{B}{x + 2}\right)$$

We use the distributive property on the right side.

$$\cancel{(x - 4)}\,\cancel{(x + 2)}\frac{x + 14}{\cancel{(x - 4)}\,\cancel{(x + 2)}}$$

$$= \cancel{(x - 4)}(x + 2)\frac{A}{\cancel{(x - 4)}} + (x - 4)\cancel{(x + 2)}\frac{B}{\cancel{(x + 2)}}$$

Dividing out common factors in numerators and denominators, we obtain

$$x + 14 = A(x + 2) + B(x - 4).$$

To find values for A and B that make both sides equal, we'll express the sides in exactly the same form by writing the variable x-terms and then writing the constant terms. Apply the distributive property on the right side.

$$x + 14 = Ax + 2A + Bx - 4B \qquad \text{Distribute A and B over the parentheses.}$$

$$x + 14 = Ax + Bx + 2A - 4B \qquad \text{Rearrange terms.}$$

$$1x + 14 = (A + B)x + (2A - 4B) \qquad \text{Rewrite to identify the coefficient of } x$$
$$\text{and the constant term.}$$

As shown by the arrows, if two polynomials are equal, coefficients of like powers of x must be equal ($A + B = 1$) and their constant terms must be equal ($2A - 4B = 14$). Consequently, A and B satisfy the following two equations:

$$A + B = 1$$
$$2A - 4B = 14.$$

We can use the addition method to solve this linear system in two variables. By multiplying the first equation by -2 and adding equations, we obtain $A = 3$ and $B = -2$. Thus,

$$\frac{x + 14}{(x - 4)(x + 2)} = \frac{A}{x - 4} + \frac{B}{x + 2} = \frac{3}{x - 4} + \frac{-2}{x + 2}\left(\text{or } \frac{3}{x - 4} - \frac{2}{x + 2}\right).$$

Study Tip

You will encounter some examples in which the denominator of the given rational expression is not already factored. If necessary, begin by factoring the denominator. Then apply the six steps needed to obtain the partial fraction decomposition.

Steps in Partial Fraction Decomposition

1. Set up the partial fraction decomposition with the unknown constants A, B, C, etc., in the numerator of the decomposition.
2. Multiply both sides of the resulting equation by the least common denominator.
3. Simplify the right-hand side of the equation.
4. Write both sides in descending powers, equate coefficients of like powers of x, and equate constant terms.
5. Solve the resulting linear system for A, B, C, etc.
6. Substitute the values for A, B, C, etc., into the equation in step 1 and write the partial fraction decomposition.

Check Point 1 Find the partial fraction decomposition of $\dfrac{5x - 1}{(x - 3)(x + 4)}$.

② Decompose $\dfrac{P}{Q}$, where Q has repeated linear factors.

The Partial Fraction Decomposition of a Rational Expression with Linear Factors in the Denominator, Some of Which Are Repeated

Suppose that $(ax + b)^n$ is a factor of the denominator. This means that the linear factor $ax + b$ is repeated n times. When this occurs, the partial fraction decomposition will contain a sum of n fractions for this factor of the denominator.

The Partial Fraction Decomposition of $\dfrac{P(x)}{Q(x)}$: $Q(x)$ Has Repeated Linear Factors

The form of the partial fraction decomposition for a rational expression containing the linear factor $ax + b$ occuring n times as its denominator is

$$\frac{P(x)}{(ax + b)^n} = \frac{A_1}{ax + b} + \frac{A_2}{(ax + b)^2} + \frac{A_3}{(ax + b)^3} + \cdots + \frac{A_n}{(ax + b)^n}.$$

> Include one fraction with a constant numerator for each power of $ax + b$.

EXAMPLE 2 Partial Fraction Decomposition with Repeated Linear Factors

Find the partial fraction decomposition of $\dfrac{x - 18}{x(x - 3)^2}$.

Solution

Step 1 Set up the partial fraction decomposition with the unknown constants. Because the linear factor $x - 3$ occurs twice, we must include one fraction with constant numerator for each power of $x - 3$.

$$\frac{x - 18}{x(x - 3)^2} = \frac{A}{x} + \frac{B}{x - 3} + \frac{C}{(x - 3)^2}$$

Step 2 Multiply both sides of the resulting equation by the least common denominator. We clear fractions, multiplying both sides by $x(x - 3)^2$, the least common denominator.

$$x(x - 3)^2 \left[\frac{x - 18}{x(x - 3)^2} \right] = x(x - 3)^2 \left[\frac{A}{x} + \frac{B}{x - 3} + \frac{C}{(x - 3)^2} \right]$$

We use the distributive property on the right side.

$$\cancel{x}\cancel{(x-3)^2} \cdot \frac{x - 18}{\cancel{x}\cancel{(x-3)^2}} = \cancel{x}(x - 3)^2 \cdot \frac{A}{\cancel{x}} + x(x - 3)^2 \cdot \frac{B}{(x - 3)} + x\cancel{(x-3)^2} \cdot \frac{C}{\cancel{(x-3)}}$$

Dividing out common factors in numerators and denominators, we obtain

$$x - 18 = A(x - 3)^2 + Bx(x - 3) + Cx.$$

Step 3 Simplify the right side of the equation. Square $x - 3$. Then apply the distributive property.

$$x - 18 = A(x^2 - 6x + 9) + Bx(x - 3) + Cx \qquad \text{Square } x - 3 \text{ using } (A - B)^2 = A^2 - 2AB + B^2$$

$$x - 18 = Ax^2 - 6Ax + 9A + Bx^2 - 3Bx + Cx \qquad \text{Apply the distributive property.}$$

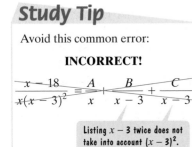

Step 4   Write both sides in descending powers, equate coefficients of like powers of x, and equate constant terms. The left side, $x - 18$, is in descending powers of x: $x - 18x^0$. We will write the right side in descending powers of x.

$$x - 18 = Ax^2 + Bx^2 - 6Ax - 3Bx + Cx + 9A \qquad \text{Rearrange terms on the right side.}$$

Express both sides in the same form.

$$0x^2 + 1x - 18 = (A + B)x^2 + (-6A - 3B + C)x + 9A \qquad \text{Rewrite to identify coefficients and the constant term.}$$

Equating coefficients of like powers of x and equating constant terms results in the following system of linear equations:

$$
\begin{aligned}
A + B &= 0 \\
-6A - 3B + C &= 1 \\
9A &= -18.
\end{aligned}
$$

Step 5   Solve the resulting system for A, B, and C. Dividing both sides of the last equation by 9, we obtain $A = -2$. Substituting -2 for A in the first equation, $A + B = 0$, gives $-2 + B = 0$, so $B = 2$. We find C by substituting -2 for A and 2 for B in the middle equation, $-6A - 3B + C = 1$. We obtain $C = -5$.

Step 6   Substitute the values of A, B, and C, and write the partial fraction decomposition. With $A = -2$, $B = 2$, and $C = -5$, the required partial fraction decomposition is

$$\frac{x - 18}{x(x - 3)^2} = \frac{A}{x} + \frac{B}{x - 3} + \frac{C}{(x - 3)^2} = -\frac{2}{x} + \frac{2}{x - 3} - \frac{5}{(x - 3)^2}.$$

Check Point 2   Find the partial fraction decomposition of $\dfrac{x + 2}{x(x - 1)^2}$.

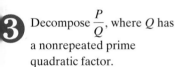

③ Decompose $\dfrac{P}{Q}$, where Q has a nonrepeated prime quadratic factor.

The Partial Fraction Decomposition of a Rational Expression with Prime, Nonrepeated Quadratic Factors in the Denominator

Our final two cases of partial fraction decomposition involve prime quadratic factors of the form $ax^2 + bx + c$. Based on our work with the discriminant, we know that $ax^2 + bx + c$ is prime and cannot be factored over the integers if $b^2 - 4ac < 0$ or if $b^2 - 4ac$ is not a perfect square.

The Partial Fraction Decomposition of $\dfrac{P(x)}{Q(x)}$: $Q(x)$ Has a Nonrepeated, Prime Quadratic Factor

If $ax^2 + bx + c$ is a prime quadratic factor of $Q(x)$, the partial fraction decomposition will contain a term of the form

$$\frac{Ax + B}{ax^2 + bx + c}.$$

Linear numerator

Quadratic factor

The voice balloons in the box show that each distinct prime quadratic factor in the denominator produces a partial fraction of the form *linear numerator over quadratic factor*. For example,

$$\frac{3x^2 + 17x + 14}{(x - 2)(x^2 + 2x + 4)} = \frac{A}{x - 2} + \frac{Bx + C}{x^2 + 2x + 4}.$$

We write a constant over the linear factor in the denominator.

We write a linear numerator over the prime quadratic factor in the denominator.

742   Chapter 7 • Systems of Equations and Inequalities

Our next example illustrates how a linear system in three variables is used to determine values for A, B, and C.

EXAMPLE 3   Partial Fraction Decomposition

Find the partial fraction decomposition of

$$\frac{3x^2 + 17x + 14}{(x - 2)(x^2 + 2x + 4)}.$$

Solution

Step 1   Set up the partial fraction decomposition with the unknown constants. We put a constant (A) over the linear factor and a linear expression $(Bx + C)$ over the prime quadratic factor.

$$\frac{3x^2 + 17x + 14}{(x - 2)(x^2 + 2x + 4)} = \frac{A}{x - 2} + \frac{Bx + C}{x^2 + 2x + 4}$$

Step 2   Multiply both sides of the resulting equation by the least common denominator. We clear fractions, multiplying both sides by $(x - 2)(x^2 + 2x + 4)$, the least common denominator.

$$(x - 2)(x^2 + 2x + 4)\left[\frac{3x^2 + 17x + 14}{(x - 2)(x^2 + 2x + 4)}\right] = (x - 2)(x^2 + 2x + 4)\left[\frac{A}{x - 2} + \frac{Bx + C}{x^2 + 2x + 4}\right]$$

We use the distributive property on the right side.

$$(x - 2)(x^2 + 2x + 4) \cdot \frac{3x^2 + 17x + 14}{(x - 2)(x^2 + 2x + 4)}$$

$$= (x - 2)(x^2 + 2x + 4) \cdot \frac{A}{x - 2} + (x - 2)(x^2 + 2x + 4) \cdot \frac{Bx + C}{x^2 + 2x + 4}$$

Dividing out common factors in numerators and denominators, we obtain

$$3x^2 + 17x + 14 = A(x^2 + 2x + 4) + (Bx + C)(x - 2).$$

Step 3   Simplify the right side of the equation. We multiply on the right side by distributing A over each term in parentheses and multiplying $(Bx + C)(x - 2)$ using the FOIL method.

$$3x^2 + 17x + 14 = Ax^2 + 2Ax + 4A + Bx^2 - 2Bx + Cx - 2C$$

Step 4   Write both sides in descending powers, equate coefficients of like powers of x, and equate constant terms. The left side, $3x^2 + 17x + 14$, is in descending powers of x. We write the right side in descending powers of x

$$3x^2 + 17x + 14 = Ax^2 + Bx^2 + 2Ax - 2Bx + Cx + 4A - 2C$$

and express both sides in the same form.

$$3x^2 + 17x + 14 = (A + B)x^2 + (2A - 2B + C)x + (4A - 2C)$$

Equating coefficients of like powers of x and equating constant terms results in the following system of linear equations:

$$A + B = 3$$
$$2A - 2B + C = 17$$
$$4A - 2C = 14.$$

Step 5   Solve the resulting system for A, B, and C. Because the first equation involves A and B, we can obtain another equation in A and B by eliminating C from the second and third equations. Multiply the second equation by 2 and add equations. Solving in this manner, we obtain $A = 5$, $B = -2$, and $C = 3$.

Step 6   Substitute the values of A, B, and C, and write the partial fraction decomposition. With $A = 5$, $B = -2$, and $C = 3$, the required partial fraction decomposition is

$$\frac{3x^2 + 17x + 14}{(x - 2)(x^2 + 2x + 4)} = \frac{A}{x - 2} + \frac{Bx + C}{x^2 + 2x + 4} = \frac{5}{x - 2} + \frac{-2x + 3}{x^2 + 2x + 4}.$$

Technology

You can use the $\boxed{\text{TABLE}}$ feature of a graphing utility to check a partial fraction decomposition. To check the result of Example 3, enter the given rational function and its partial fraction decomposition:

$$y_1 = \frac{3x^2 + 17x + 14}{(x - 2)(x^2 + 2x + 4)}$$

$$y_2 = \frac{5}{x - 2} + \frac{-2x + 3}{x^2 + 2x + 4}.$$

X	Y₁	Y₂
-3	.28571	.28571
-2	.5	.5
-1	0	0
0	-1.75	-1.75
1	-4.857	-4.857
2	ERROR	ERROR
3	4.8421	4.8421

X=-3

No matter how far up or down we scroll, $y_1 = y_2$, so the decomposition appears to be correct.

Check Point 3 Find the partial fraction decomposition of

$$\frac{8x^2 + 12x - 20}{(x + 3)(x^2 + x + 2)}.$$

④ Decompose $\dfrac{P}{Q}$, where Q has a prime, repeated quadratic factor.

The Partial Fraction Decomposition of a Rational Expression with a Prime, Repeated Quadratic Factor in the Denominator

Suppose that $(ax^2 + bx + c)^n$ is a factor of the denominator and that $ax^2 + bx + c$ cannot be factored further. This means that the quadratic factor $ax^2 + bx + c$ occurs n times. When this occurs, the partial fraction decomposition will contain a linear numerator for each power of $ax^2 + bx + c$.

The Partial Fraction Decomposition of $\dfrac{P(x)}{Q(x)}$: $Q(x)$ Has a Prime, Repeated Quadratic Factor

The form of the partial fraction decomposition for a rational expression containing the prime factor $ax^2 + bx + c$ occuring n times as its denominator is

$$\frac{P(x)}{(ax^2 + bx + c)^n} = \frac{A_1x + B_1}{ax^2 + bx + c} + \frac{A_2x + B_2}{(ax^2 + bx + c)^2} + \frac{A_3x + B_3}{(ax^2 + bx + c)^3} + \cdots + \frac{A_nx + B_n}{(ax^2 + bx + c)^n}.$$

Include one fraction with a linear numerator for each power of $ax^2 + bx + c$.

EXAMPLE 4 Partial Fraction Decomposition with a Repeated Quadratic Factor

Find the partial fraction decomposition of

$$\frac{5x^3 - 3x^2 + 7x - 3}{\left(x^2 + 1\right)^2}.$$

Solution

Step 1 Set up the partial fraction decomposition with the unknown constants. Because the quadratic factor $x^2 + 1$ occurs twice, we must include one fraction with a linear numerator for each power of $x^2 + 1$.

$$\frac{5x^3 - 3x^2 + 7x - 3}{\left(x^2 + 1\right)^2} = \frac{Ax + B}{x^2 + 1} + \frac{Cx + D}{\left(x^2 + 1\right)^2}$$

Step 2 Multiply both sides of the resulting equation by the least common denominator. We clear fractions, multiplying both sides by $\left(x^2 + 1\right)^2$, the least common denominator.

$$\left(x^2 + 1\right)^2 \left[\frac{5x^3 - 3x^2 + 7x - 3}{\left(x^2 + 1\right)^2}\right] = \left(x^2 + 1\right)^2 \left[\frac{Ax + B}{x^2 + 1} + \frac{Cx + D}{\left(x^2 + 1\right)^2}\right]$$

Now we multiply and simplify.

$$5x^3 - 3x^2 + 7x - 3 = (x^2 + 1)(Ax + B) + Cx + D$$

Step 3 Simplify the right side of the equation. We multiply $(x^2 + 1)(Ax + B)$ using the FOIL method.

$$5x^3 - 3x^2 + 7x - 3 = Ax^3 + Bx^2 + Ax + B + Cx + D$$

Step 4 Write both sides in descending powers, equate coefficients of like powers of x, and equate constant terms.

$$5x^3 - 3x^2 + 7x - 3 = Ax^3 + Bx^2 + Ax + Cx + B + D$$

$$5x^3 - 3x^2 + 7x - 3 = Ax^3 + Bx^2 + (A + C)x + (B + D)$$

Equating coefficients of like powers of x and equating constant terms results in the following system of linear equations:

$$\begin{aligned} A &= 5 \\ B &= -3 \\ A + C &= 7 \qquad \text{With } A = 5, \text{ we immediately obtain } C = 2. \\ B + D &= -3. \qquad \text{With } B = -3, \text{ we immediately obtain } D = 0. \end{aligned}$$

Step 5 Solve the resulting system for A, B, C, and D. Based on our observation in step 4, $A = 5$, $B = -3$, $C = 2$, and $D = 0$.

Step 6 Substitute the values of A, B, C, and D, and write the partial fraction decomposition.

$$\frac{5x^3 - 3x^2 + 7x - 3}{\left(x^2 + 1\right)^2} = \frac{Ax + B}{x^2 + 1} + \frac{Cx + D}{\left(x^2 + 1\right)^2} = \frac{5x - 3}{x^2 + 1} + \frac{2x}{\left(x^2 + 1\right)^2}$$

Check Point 4 Find the partial fraction decomposition of $\dfrac{2x^3 + x + 3}{\left(x^2 + 1\right)^2}$.

EXERCISE SET 7.3

 **Practice Exercises**

In Exercises 1–8, write the form of the partial fraction decomposition of the rational expression. It is not necessary to solve for the constants.

1. $\dfrac{11x - 10}{(x - 2)(x + 1)}$

2. $\dfrac{5x + 7}{(x - 1)(x + 3)}$

3. $\dfrac{6x^2 - 14x - 27}{(x + 2)(x - 3)^2}$

4. $\dfrac{3x + 16}{(x + 1)(x - 2)^2}$

5. $\dfrac{5x^2 - 6x + 7}{(x - 1)(x^2 + 1)}$

6. $\dfrac{5x^2 - 9x + 19}{(x - 4)(x^2 + 5)}$

7. $\dfrac{x^3 + x^2}{(x^2 + 4)^2}$

8. $\dfrac{7x^2 - 9x + 3}{(x^2 + 7)^2}$

In Exercises 9–42, write the partial fraction decomposition of each rational expression.

9. $\dfrac{x}{(x - 3)(x - 2)}$

10. $\dfrac{1}{x(x - 1)}$

11. $\dfrac{3x + 50}{(x - 9)(x + 2)}$

12. $\dfrac{5x - 1}{(x - 2)(x + 1)}$

13. $\dfrac{7x - 4}{x^2 - x - 12}$

14. $\dfrac{9x + 21}{x^2 + 2x - 15}$

15. $\dfrac{4}{2x^2 - 5x - 3}$

16. $\dfrac{x}{x^2 + 2x - 3}$

17. $\dfrac{4x^2 + 13x - 9}{x(x - 1)(x + 3)}$

18. $\dfrac{4x^2 - 5x - 15}{x(x + 1)(x - 5)}$

19. $\dfrac{4x^2 - 7x - 3}{x^3 - x}$

20. $\dfrac{2x^2 - 18x - 12}{x^3 - 4x}$

21. $\dfrac{6x - 11}{(x - 1)^2}$

22. $\dfrac{x}{(x + 1)^2}$

23. $\dfrac{x^2 - 6x + 3}{(x - 2)^3}$

24. $\dfrac{2x^2 + 8x + 3}{(x + 1)^3}$

25. $\dfrac{x^2 + 2x + 7}{x(x - 1)^2}$

26. $\dfrac{3x^2 + 49}{x(x + 7)^2}$

27. $\dfrac{x^2}{(x - 1)^2(x + 1)}$

28. $\dfrac{x^2}{(x - 1)^2(x + 1)^2}$

29. $\dfrac{5x^2 - 6x + 7}{(x - 1)(x^2 + 1)}$

30. $\dfrac{5x^2 - 9x + 19}{(x - 4)(x^2 + 5)}$

31. $\dfrac{5x^2 + 6x + 3}{(x + 1)(x^2 + 2x + 2)}$

32. $\dfrac{9x + 2}{(x - 2)(x^2 + 2x + 2)}$

33. $\dfrac{x + 4}{x^2(x^2 + 4)}$

34. $\dfrac{10x^2 + 2x}{(x - 1)^2(x^2 + 2)}$

35. $\dfrac{6x^2 - x + 1}{x^3 + x^2 + x + 1}$

36. $\dfrac{3x^2 - 2x + 8}{x^3 + 2x^2 + 4x + 8}$

37. $\dfrac{x^3 + x^2 + 2}{(x^2 + 2)^2}$

38. $\dfrac{x^2 + 2x + 3}{(x^2 + 4)^2}$

39. $\dfrac{x^3 - 4x^2 + 9x - 5}{(x^2 - 2x + 3)^2}$

40. $\dfrac{3x^3 - 6x^2 + 7x - 2}{(x^2 - 2x + 2)^2}$

41. $\dfrac{4x^2 + 3x + 14}{x^3 - 8}$

42. $\dfrac{3x - 5}{x^3 - 1}$

 **Practice Plus**

In Exercises 43–46, perform each long division and write the partial fraction decomposition of the remainder term.

43. $\dfrac{x^5 + 2}{x^2 - 1}$

44. $\dfrac{x^5}{x^2 - 4x + 4}$

45. $\dfrac{x^4 - x^2 + 2}{x^3 - x^2}$

46. $\dfrac{x^4 + 2x^3 - 4x^2 + x - 3}{x^2 - x - 2}$

In Exercises 47–50, write the partial fraction decomposition of each rational expression.

47. $\dfrac{1}{x^2 - c^2}$ $(c \neq 0)$

48. $\dfrac{ax + b}{x^2 - c^2}$ $(c \neq 0)$

49. $\dfrac{ax + b}{(x - c)^2}$ $(c \neq 0)$

50. $\dfrac{1}{x^2 - ax - bx + ab}$ $(a \neq b)$

 **Application Exercises**

51. Find the partial fraction decomposition for $\dfrac{1}{x(x + 1)}$ and use the result to find the following sum:
$$\frac{1}{1 \cdot 2} + \frac{1}{2 \cdot 3} + \frac{1}{3 \cdot 4} + \cdots + \frac{1}{99 \cdot 100}.$$

52. Find the partial fraction decomposition for $\dfrac{2}{x(x + 2)}$ and use the result to find the following sum:
$$\frac{2}{1 \cdot 3} + \frac{2}{3 \cdot 5} + \frac{2}{5 \cdot 7} + \cdots + \frac{2}{99 \cdot 101}.$$

Writing in Mathematics

53. Explain what is meant by the partial fraction decomposition of a rational expression.

54. Explain how to find the partial fraction decomposition of a rational expression with distinct linear factors in the denominator.

55. Explain how to find the partial fraction decomposition of a rational expression with a repeated linear factor in the denominator.

56. Explain how to find the partial fraction decomposition of a rational expression with a prime quadratic factor in the denominator.

57. Explain how to find the partial fraction decomposition of a rational expression with a repeated, prime quadratic factor in the denominator.

58. How can you verify your result for the partial fraction decomposition for a given rational expression without using a graphing utility?

Technology Exercise

59. Use the [TABLE] feature of a graphing utility to verify any three of the decompositions that you obtained in Exercises 9–42.

Critical Thinking Exercises

60. Use an extension of the Study Tip at the top of page 744 to describe how to set up the partial fraction decomposition of rational expression that contains powers of a prime cubic facto in the denominator. Give an example of such a decomposition

61. Find the partial fraction decomposition of

$$\frac{4x^2 + 5x - 9}{x^3 - 6x - 9}.$$

SECTION 7.4   Systems of Nonlinear Equations in Two Variables

Objectives

❶ Recognize systems of nonlinear equations in two variables.

❷ Solve nonlinear systems by substitution.

❸ Solve nonlinear systems by addition.

❹ Solve problems using systems of nonlinear equations.

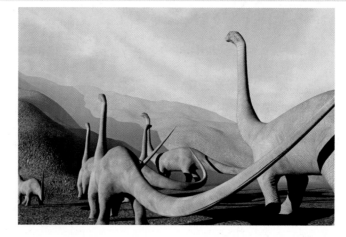

Scientists debate the probability that a "doomsday rock" will collide with Earth. It ha been estimated that an asteroid, a tiny planet that revolves around the sun, crashes int Earth about once every 250,000 years, and that such a collision would have disastrou results. In 1908, a small fragment struck Siberia, leveling thousands of acres of tree One theory about the extinction of dinosaurs 65 million years ago involves Earth collision with a large asteroid and the resulting drastic changes in Earth's climate.

Understanding the path of Earth and the path of a comet is essential t detecting threatening space debris. Orbits about the sun are not described by linea equations in the form $Ax + By = C$. The ability to solve systems that contain mor than just linear equations provides NASA scientists watching for troublesome as teroids with a way to locate possible collision points with Earth's orbit.

Recognize systems of nonlinear equations in two variables.

Systems of Nonlinear Equations and Their Solutions

A **system of** two **nonlinear equations** in two variables, also called a **nonlinear system** contains at least one equation that cannot be expressed in the form $Ax + By = C$ Here are two examples:

$$x^2 = 2y + 10$$
$$3x - y = 9$$

> Not in the form $Ax + By = C$. The term x^2 is not linear.

$$y = x^2 + 3$$
$$x^2 + y^2 = 9.$$

> Neither equation is in the form $Ax + By = C$. The terms x^2 and y^2 are not linear.

A **solution** of a nonlinear system in two variables is an ordered pair of rea numbers that satisfies both equations in the system. The **solution set** of the system i the set of all such ordered pairs. As with linear systems in two variables, the solutio of a nonlinear system (if there is one) corresponds to the intersection point(s) of th graphs of the equations in the system. Unlike linear systems, the graphs can b circles, parabolas, or anything other than two lines. We will solve nonlinear system using the substitution method and the addition method.

② Solve nonlinear systems by substitution.

Eliminating a Variable Using the Substitution Method

The substitution method involves converting a nonlinear system to one equation in one variable by an appropriate substitution. The steps in the solution process are exactly the same as those used to solve a linear system by substitution. However, when you obtain an equation in one variable, this equation may not be linear. In our first example, this equation is quadratic.

EXAMPLE 1   Solving a Nonlinear System by the Substitution Method

Solve by the substitution method:

$$x^2 = 2y + 10 \qquad \text{(The graph is a parabola.)}$$
$$3x - y = 9. \qquad \text{(The graph is a line.)}$$

Solution

Step 1   Solve one of the equations for one variable in terms of the other. We begin by isolating one of the variables raised to the first power in either of the equations. By solving for y in the second equation, which has a coefficient of -1, we can avoid fractions.

$$3x - y = 9 \qquad \text{This is the second equation in the given system.}$$
$$3x = y + 9 \qquad \text{Add } y \text{ to both sides.}$$
$$3x - 9 = y \qquad \text{Subtract 9 from both sides.}$$

Step 2   Substitute the expression from step 1 into the other equation. We substitute $3x - 9$ for y in the first equation.

$$y = \boxed{3x - 9} \qquad x^2 = 2\boxed{y} + 10$$

This gives us an equation in one variable, namely

$$x^2 = 2(3x - 9) + 10.$$

The variable y has been eliminated.

Step 3   Solve the resulting equation containing one variable.

$$x^2 = 2(3x - 9) + 10 \qquad \text{This is the equation containing one variable.}$$
$$x^2 = 6x - 18 + 10 \qquad \text{Use the distributive property.}$$
$$x^2 = 6x - 8 \qquad \text{Combine numerical terms on the right.}$$
$$x^2 - 6x + 8 = 0 \qquad \text{Move all terms to one side and set the quadratic equation equal to 0.}$$
$$(x - 4)(x - 2) = 0 \qquad \text{Factor.}$$
$$x - 4 = 0 \quad \text{or} \quad x - 2 = 0 \qquad \text{Set each factor equal to 0.}$$
$$x = 4 \qquad\qquad x = 2 \qquad \text{Solve for x.}$$

Step 4   Back-substitute the obtained values into the equation from step 1. Now that we have the x-coordinates of the solutions, we back-substitute 4 for x and 2 for x into the equation $y = 3x - 9$.

$$\text{If } x \text{ is 4,} \qquad y = 3(4) - 9 = 3, \qquad \text{so } (4, 3) \text{ is a solution.}$$
$$\text{If } x \text{ is 2,} \qquad y = 3(2) - 9 = -3, \quad \text{so } (2, -3) \text{ is a solution.}$$

Step 5   Check the proposed solutions in both of the system's given equations. We begin by checking $(4, 3)$. Replace x with 4 and y with 3.

$$x^2 = 2y + 10 \qquad\qquad 3x - y = 9 \qquad\qquad \text{These are the given equations.}$$
$$4^2 \overset{?}{=} 2(3) + 10 \qquad\qquad 3(4) - 3 \overset{?}{=} 9 \qquad\qquad \text{Let x = 4 and y = 3.}$$
$$16 \overset{?}{=} 6 + 10 \qquad\qquad 12 - 3 \overset{?}{=} 9 \qquad\qquad \text{Simplify.}$$
$$16 = 16, \quad \text{true} \qquad\qquad 9 = 9, \quad \text{true} \qquad\qquad \text{True statements result.}$$

The ordered pair $(4, 3)$ satisfies both equations. Thus, $(4, 3)$ is a solution of the system.

Step 3 Solve the system and answer the problem's question. We must solve the system

$$2x + 3y = 36 \qquad \text{Equation 1}$$
$$xy = 54. \qquad \text{Equation 2}$$

We will use substitution. Because Equation 1 has no coefficients of 1 or -1, we will work with Equation 2 and solve for y. Dividing both sides of $xy = 54$ by x, we obtain

$$y = \frac{54}{x}.$$

Now we substitute $\frac{54}{x}$ for y in Equation 1 and solve for x.

$$2x + 3y = 36 \qquad \text{This is Equation 1.}$$

$$2x + 3 \cdot \frac{54}{x} = 36 \qquad \text{Substitute } \frac{54}{x} \text{ for y.}$$

$$2x + \frac{162}{x} = 36 \qquad \text{Multiply.}$$

$$x\left(2x + \frac{162}{x}\right) = 36 \cdot x \qquad \text{Clear fractions by multiplying both sides by x.}$$

$$2x^2 + 162 = 36x \qquad \text{Use the distributive property on the left side.}$$

$$2x^2 - 36x + 162 = 0 \qquad \text{Subtract 36x from both sides and set the quadratic equation equal to 0.}$$

$$2(x^2 - 18x + 81) = 0 \qquad \text{Factor out 2.}$$

$$2(x - 9)^2 = 0 \qquad \text{Factor completely using } A^2 - 2AB + B^2 = (A - B)^2.$$

$$x - 9 = 0 \qquad \text{Set the repeated factor equal to zero.}$$

$$x = 9 \qquad \text{Solve for x.}$$

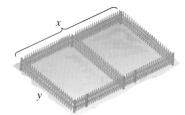

We back-substitute this value of x into $y = \frac{54}{x}$.

$$\text{If } x = 9, \quad y = \frac{54}{9} = 6.$$

Figure 7.14 (repeated)

This means that the dimensions of the enclosure in Figure 7.14 are 9 yards by 6 yards.

Step 4 Check the proposed solution in the original wording of the problem. Take a moment to check that a length of 9 yards and a width of 6 yards results in 36 yards of fencing and an area of 54 square yards.

Check Point 5 Find the length and width of a rectangle whose perimeter is 20 feet and whose area is 21 square feet.

EXERCISE SET 7.4

 Practice Exercises

In Exercises 1–18, solve each system by the substitution method.

1. $x + y = 2$
  $y = x^2 - 4$

2. $x - y = -1$
  $y = x^2 + 1$

3. $x + y = 2$
  $y = x^2 - 4x + 4$

4. $2x + y = -5$
  $y = x^2 + 6x + 7$

5. $y = x^2 - 4x - 10$
  $y = -x^2 - 2x + 14$

6. $y = x^2 + 4x + 5$
  $y = x^2 + 2x - 1$

7. $x^2 + y^2 = 25$
  $x - y = 1$

8. $x^2 + y^2 = 5$
  $3x - y = 5$

9. $xy = 6$
  $2x - y = 1$

10. $xy = -12$
  $x - 2y + 14 = 0$

1. $y^2 = x^2 - 9$
$\quad 2y = x - 3$

3. $xy = 3$
$\quad x^2 + y^2 = 10$

5. $x + y = 1$
$\quad x^2 + xy - y^2 = -5$

7. $x + y = 1$
$\quad (x - 1)^2 + (y + 2)^2 = 10$

12. $x^2 + y = 4$
$\quad 2x + y = 1$

14. $xy = 4$
$\quad x^2 + y^2 = 8$

16. $x + y = -3$
$\quad x^2 + 2y^2 = 12y + 18$

18. $2x + y = 4$
$\quad (x + 1)^2 + (y - 2)^2 = 4$

In Exercises 19–28, solve each system by the addition method.

19. $x^2 + y^2 = 13$
$\quad x^2 - y^2 = 5$

21. $x^2 - 4y^2 = -7$
$\quad 3x^2 + y^2 = 31$

23. $3x^2 + 4y^2 - 16 = 0$
$\quad 2x^2 - 3y^2 - 5 = 0$

25. $x^2 + y^2 = 25$
$\quad (x - 8)^2 + y^2 = 41$

27. $y^2 - x = 4$
$\quad x^2 + y^2 = 4$

20. $4x^2 - y^2 = 4$
$\quad 4x^2 + y^2 = 4$

22. $3x^2 - 2y^2 = -5$
$\quad 2x^2 - y^2 = -2$

24. $16x^2 - 4y^2 - 72 = 0$
$\quad x^2 - y^2 - 3 = 0$

26. $x^2 + y^2 = 5$
$\quad x^2 + (y - 8)^2 = 41$

28. $x^2 - 2y = 8$
$\quad x^2 + y^2 = 16$

In Exercises 29–42, solve each system by the method of your choice.

29. $3x^2 + 4y^2 = 16$
$\quad 2x^2 - 3y^2 = 5$

31. $2x^2 + y^2 = 18$
$\quad xy = 4$

33. $x^2 + 4y^2 = 20$
$\quad x + 2y = 6$

35. $x^3 + y = 0$
$\quad x^2 - y = 0$

37. $x^2 + (y - 2)^2 = 4$
$\quad x^2 - 2y = 0$

39. $y = (x + 3)^2$
$\quad x + 2y = -2$

41. $x^2 + y^2 + 3y = 22$
$\quad 2x + y = -1$

30. $x + y^2 = 4$
$\quad x^2 + y^2 = 16$

32. $x^2 + 4y^2 = 20$
$\quad xy = 4$

34. $3x^2 - 2y^2 = 1$
$\quad 4x - y = 3$

36. $x^3 + y = 0$
$\quad 2x^2 - y = 0$

38. $x^2 - y^2 - 4x + 6y - 4 = 0$
$\quad x^2 + y^2 - 4x - 6y + 12 = 0$

40. $(x - 1)^2 + (y + 1)^2 = 5$
$\quad 2x - y = 3$

42. $x - 3y = -5$
$\quad x^2 + y^2 - 25 = 0$

In Exercises 43–46, let x represent one number and let y represent the other number. Use the given conditions to write a system of nonlinear equations. Solve the system and find the numbers.

43. The sum of two numbers is 10 and their product is 24. Find the numbers.

44. The sum of two numbers is 20 and their product is 96. Find the numbers.

45. The difference between the squares of two numbers is 3. Twice the square of the first number increased by the square of the second number is 9. Find the numbers.

46. The difference between the squares of two numbers is 5. Twice the square of the second number subtracted from three times the square of the first number is 19. Find the numbers.

Practice Plus

In Exercises 47–52, solve each system by the method of your choice.

47. $2x^2 + xy = 6$
$\quad x^2 + 2xy = 0$

49. $-4x + y = 12$
$\quad y = x^3 + 3x^2$

51. $\dfrac{3}{x^2} + \dfrac{1}{y^2} = 7$
$\quad \dfrac{5}{x^2} - \dfrac{2}{y^2} = -3$

48. $4x^2 + xy = 30$
$\quad x^2 + 3xy = -9$

50. $-9x + y = 45$
$\quad y = x^3 + 5x^2$

52. $\dfrac{2}{x^2} + \dfrac{1}{y^2} = 11$
$\quad \dfrac{4}{x^2} - \dfrac{2}{y^2} = -14$

In Exercises 53–54, make a rough sketch in a rectangular coordinate system of the graphs representing the equations in each system.

53. The system, whose graphs are a line with positive slope and a parabola whose equation has a positive leading coefficient, has two solutions.

54. The system, whose graphs are a line with negative slope and a parabola whose equation has a negative leading coefficient, has one solution.

Application Exercises

55. A planet's orbit follows a path described by $16x^2 + 4y^2 = 64$. A comet follows the parabolic path $y = x^2 - 4$. Where might the comet intersect the orbiting planet?

56. A system for tracking ships indicates that a ship lies on a path described by $2y^2 - x^2 = 1$. The process is repeated and the ship is found to lie on a path described by $2x^2 - y^2 = 1$. If it is known that the ship is located in the first quadrant of the coordinate system, determine its exact location.

57. Find the length and width of a rectangle whose perimeter is 36 feet and whose area is 77 square feet.

58. Find the length and width of a rectangle whose perimeter is 40 feet and whose area is 96 square feet.

Use the formula for the area of a rectangle and the Pythagorean Theorem to solve Exercises 59–60.

59. A small television has a picture with a diagonal measure of 10 inches and a viewing area of 48 square inches. Find the length and width of the screen.

60. The area of a rug is 108 square feet and the length of its diagonal is 15 feet. Find the length and width of the rug.

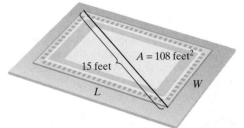

61. The figure shows a square floor plan with a smaller square area that will accommodate a combination fountain and pool. The floor with the fountain-pool area removed has an area of 21 square meters and a perimeter of 24 meters. Find the dimensions of the floor and the dimensions of the square that will accommodate the pool.

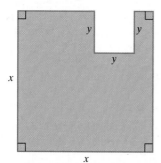

62. The area of the rectangular piece of cardboard shown on the left is 216 square inches. The cardboard is used to make an open box by cutting a 2-inch square from each corner and turning up the sides. If the box is to have a volume of 224 cubic inches, find the length and width of the cardboard that must be used.

63. The graphs show the number, in thousands, of bachelor's degrees in science and engineering received by U.S. men and women from 1991 through 2001.

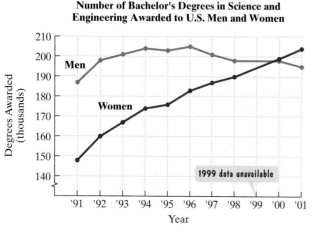

Source: National Science Foundation

The data can be modeled by quadratic and linear functions:

Men $y = -0.5x^2 + 5.8x + 185.8$
Women $5.4x - y + 146.5 = 0.$

In each function, x represents the number of years after 1990 and y represents the number, in thousands, of bachelor's degrees that were awarded. According to these functions, in which year, to the nearest whole year, did men and women receive the same number of bachelor's degrees in science and engineering? How well does this describe the information displayed by the graphs?

Writing in Mathematics

64. What is a system of nonlinear equations? Provide an example with your description.

65. Explain how to solve a nonlinear system using the substitution method. Use $x^2 + y^2 = 9$ and $2x - y = 3$ to illustrate your explanation.

66. Explain how to solve a nonlinear system using the addition method. Use $x^2 - y^2 = 5$ and $3x^2 - 2y^2 = 19$ to illustrate your explanation.

67. The daily demand and supply models for a carrot cake supplied by a bakery to a convenience store are given by the

demand model $N = 40 - 3p$ and the supply model $N = \dfrac{p^2}{10}$

in which p is the price of the cake and N is the number of cakes sold or supplied each day to the convenience store. Explain how to determine the price at which supply and demand are equal. Then describe how to find how many carrot cakes can be supplied and sold each day at this price.

Technology Exercises

68. Verify your solutions to any five exercises from Exercises 1–42 by using a graphing utility to graph the two equations in the system in the same viewing rectangle. Then use the intersection feature to verify the solutions.

69. Write a system of equations, one equation whose graph is a line and the other whose graph is a parabola, that has no ordered pairs that are real numbers in its solution set. Graph the equations using a graphing utility and verify that you are correct.

Critical Thinking Exercises

70. Which one of the following is true?
   a. A system of two equations in two variables whose graphs are a circle and a line can have four real ordered-pair solutions.
   b. A system of two equations in two variables whose graphs are a parabola and a circle can have four real ordered-pair solutions.
   c. A system of two equations in two variables whose graphs are two circles must have at least two real ordered-pair solutions.
   d. A system of two equations in two variables whose graphs are a parabola and a circle cannot have only one real ordered-pair solution.

71. The points of intersection of the graphs of $xy = 20$ and $x^2 + y^2 = 41$ are joined to form a rectangle. Find the area of the rectangle.

72. Find a and b in this figure.

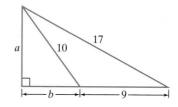

Solve the systems in Exercises 73–74.

73. $\log_y x = 3$
    $\log_y(4x) = 5$

74. $\log x^2 = y + 3$
    $\log x = y - 1$

CHAPTER 7
MID-CHAPTER CHECK POINT

What You Know: We learned to solve systems of equations. We solved linear and nonlinear systems in two variables by the substitution method and by the addition method. We solved linear systems in three variables by eliminating a variable, reducing the system to two equations in two variables. We saw that some linear systems, called inconsistent systems, have no solution, whereas other linear systems, called dependent systems, have infinitely many solutions. We applied systems to a variety of situations, including finding the break-even point for a business, finding a quadratic function from three points on its graph, and finding a rational function's partial fraction decomposition.

In Exercises 1–12, solve each system by the method of your choice.

1. $x = 3y - 7$
$\quad 4x + 3y = 2$

2. $3x + 4y = -5$
$\quad 2x - 3y = 8$

3. $\dfrac{2x}{3} + \dfrac{y}{5} = 6$
$\quad \dfrac{x}{6} - \dfrac{y}{2} = -4$

4. $y = 4x - 5$
$\quad 8x - 2y = 10$

5. $2x + 5y = 3$
$\quad 3x - 2y = 1$

6. $\dfrac{x}{12} - y = \dfrac{1}{4}$
$\quad 4x - 48y = 16$

7. $2x - y + 2z = -8$
$\quad x + 2y - 3z = 9$
$\quad 3x - y - 4z = 3$

8. $x - \quad\; 3z = -5$
$\quad 2x - y + 2z = 16$
$\quad 7x - 3y - 5z = 19$

9. $x^2 + y^2 = 9$
$\quad x + 2y - 3 = 0$

10. $3x^2 + 2y^2 = 14$
$\quad 2x^2 - y^2 = 7$

11. $y = x^2 - 6$
$\quad x^2 + y^2 = 8$

12. $x - 2y = 4$
$\quad 2y^2 + xy = 8$

In Exercises 13–16, write the partial fraction decomposition of each rational expression.

13. $\dfrac{x^2 - 6x + 3}{(x - 2)^3}$

14. $\dfrac{10x^2 + 9x - 7}{(x + 2)(x^2 - 1)}$

15. $\dfrac{x^2 + 4x - 23}{(x + 3)(x^2 + 4)}$

16. $\dfrac{x^3}{(x^2 + 4)^2}$

17. A company is planning to manufacture PDAs (personal digital assistants). The fixed cost will be $400,000 and it will cost $20 to produce each PDA. Each PDA will be sold for $100.
 a. Write the cost function, C, of producing x PDAs.
 b. Write the revenue function, R, from the sale of x PDAs.
 c. Write the profit function, P, from producing and selling x PDAs.
 d. Determine the break-even point. Describe what this means.

18. Roses sell for $3 each and carnations for $1.50 each. If a mixed bouquet of 20 flowers consisting of roses and carnations costs $39, how many of each type of flower is in the bouquet?

19. At the north campus of a small liberal arts college, 10% of the students are women. At the south campus, 50% of the students are women. The campuses are merged into one east campus. If 40% of the 1200 students at the east campus are women, how many students did the north and south campuses have before the merger?

20. With the current, you can row 9 miles in 2 hours. Against the current, your return trip takes 6 hours. Find your average rowing velocity in still water and the average velocity of the current.

21. Find the measure of each angle whose degree measure is represented with a variable.

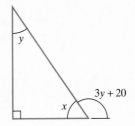

22. Find the quadratic function $y = ax^2 + bx + c$ whose graph passes through the points $(-1, 0)$, $(1, 4)$, and $(2, 3)$.

23. Find the length and width of a rectangle whose perimeter is 21 meters and whose area is 20 square meters.

SECTION 7.5   Systems of Inequalities

Objectives

❶ Graph a linear inequality in two variables.

❷ Graph a nonlinear inequality in two variables.

❸ Graph a system of inequalities.

❹ Solve applied problems involving systems of inequalities.

Had a good workout lately? If so, could you tell if you were overdoing it or not pushing yourself hard enough? In this section's exercise set, we will use systems of inequalities in two variables to help you establish a target zone for your workouts.

Linear Inequalities in Two Variables and Their Solutions

We have seen that equations in the form $Ax + By = C$ are straight lines when graphed. If we change the symbol $=$ to $>$, $<$, $\geq$, or $\leq$, we obtain a **linear inequality in two variables**. Some examples of linear inequalities in two variables are $x + y > 2$, $3x - 5y \leq 15$, and $2x - y < 4$.

A **solution of an inequality in two variables**, x and y, is an ordered pair of real numbers with the following property: When the x-coordinate is substituted for x and the y-coordinate is substituted for y in the inequality, we obtain a true statement. For example, $(3, 2)$ is a solution of the inequality $x + y > 1$. When 3 is substituted for x and 2 is substituted for y, we obtain the true statement $3 + 2 > 1$, or $5 > 1$. Because there are infinitely many pairs of numbers that have a sum greater than 1, the inequality $x + y > 1$ has infinitely many solutions. Each ordered-pair solution is said to **satisfy** the inequality. Thus, $(3, 2)$ satisfies the inequality $x + y > 1$.

The Graph of a Linear Inequality in Two Variables

 Graph a linear inequality in two variables.

We know that the graph of an equation in two variables is the set of all points whose coordinates satisfy the equation. Similarly, the **graph of an inequality in two variables** is the set of all points whose coordinates satisfy the inequality.

Let's use Figure 7.15 to get an idea of what the graph of a linear inequality in two variables looks like. Part of the figure shows the graph of the linear equation $x + y = 2$. The line divides the points in the rectangular coordinate system into three sets. First, there is the set of points along the line, satisfying $x + y = 2$. Next, there is the set of points in the green region above the line. Points in the green region satisfy the linear inequality $x + y > 2$. Finally, there is the set of points in the purple region below the line. Points in the purple region satisfy the linear inequality $x + y < 2$.

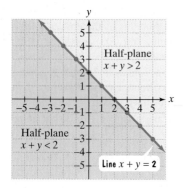

Figure 7.15

A **half-plane** is the set of all the points on one side of a line. In Figure 7.15, the green region is a half-plane. The purple region is also a half-plane. A half-plane is the graph of a linear inequality that involves $>$ or $<$. The graph of an inequality that involves $\geq$ or $\leq$ is a half-plane and a line. A solid line is used to show that a line is part of a graph. A dashed line is used to show that a line is not part of a graph.

Graphing a Linear Inequality in Two Variables

1. Replace the inequality symbol with an equal sign and graph the corresponding linear equation. Draw a solid line if the original inequality contains a $\leq$ or $\geq$ symbol. Draw a dashed line if the original inequality contains a $<$ or $>$ symbol.
2. Choose a test point from one of the half-planes. (Do not choose a point on the line.) Substitute the coordinates of the test point into the inequality.
3. If a true statement results, shade the half-plane containing this test point. If a false statement results, shade the half-plane not containing this test point.

EXAMPLE 1 Graphing a Linear Inequality in Two Variables

Graph: $2x - 3y \geq 6$.

Solution

Step 1 Replace the inequality symbol by $=$ and graph the linear equation. We need to graph $2x - 3y = 6$. We can use intercepts to graph this line.

We set $y = 0$ to find the x-intercept.	We set $x = 0$ to find the y-intercept.
$2x - 3y = 6$	$2x - 3y = 6$
$2x - 3 \cdot 0 = 6$	$2 \cdot 0 - 3y = 6$
$2x = 6$	$-3y = 6$
$x = 3$	$y = -2$

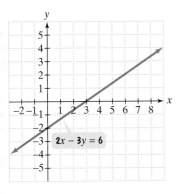

Figure 7.16 Preparing to graph $2x - 3y \geq 6$

The x-intercept is 3, so the line passes through $(3, 0)$. The y-intercept is -2, so the line passes through $(0, -2)$. Using the intercepts, the line is shown in Figure 7.16 as a solid line. This is because the inequality $2x - 3y \geq 6$ contains a $\geq$ symbol, in which equality is included.

Step 2   Choose a test point from one of the half-planes and not from the line. Substitute its coordinates into the inequality. The line $2x - 3y = 6$ divides the plane into three parts—the line itself and two half-planes. The points in one half-plane satisfy $2x - 3y > 6$. The points in the other half-plane satisfy $2x - 3y < 6$. We need to find which half-plane belongs to the solution of $2x - 3y \geq 6$. To do so, we test a point from either half-plane. The origin, $(0, 0)$, is the easiest point to test.

$$2x - 3y \geq 6 \qquad \text{\small This is the given inequality.}$$
$$2 \cdot 0 - 3 \cdot 0 \overset{?}{\geq} 6 \qquad \text{\small Test } (0, 0) \text{ by substituting 0 for } x \text{ and 0 for } y.$$
$$0 - 0 \overset{?}{\geq} 6 \qquad \text{\small Multiply.}$$
$$0 \geq 6 \qquad \text{\small This statement is false.}$$

Step 3   If a false statement results, shade the half-plane not containing the test point. Because 0 is not greater than or equal to 6, the test point, $(0, 0)$, is not part of the solution set. Thus, the half-plane below the solid line $2x - 3y = 6$ is part of the solution set. The solution set is the line and the half-plane that does not contain the point $(0, 0)$, indicated by shading this half-plane. The graph is shown using green shading and a blue line in Figure 7.17.

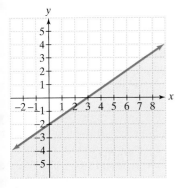

Figure 7.17 The graph of $2x - 3y \geq 6$

Check Point 1   Graph: $4x - 2y \geq 8$.

When graphing a linear inequality, test a point that lies in one of the half-planes and *not on the line dividing the half-planes*. The test point $(0, 0)$ is convenient because it is easy to calculate when 0 is substituted for each variable. However, if $(0, 0)$ lies on the dividing line and not in a half-plane, a different test point must be selected.

EXAMPLE 2   Graphing a Linear Inequality in Two Variables

Graph: $y > -\dfrac{2}{3}x$.

Solution

Step 1   Replace the inequality symbol by = and graph the linear equation. Because we are interested in graphing $y > -\frac{2}{3}x$, we begin by graphing $y = -\frac{2}{3}x$. We can use the slope and the y-intercept to graph this linear function.

$$y = -\frac{2}{3}x + 0$$

Slope $= \dfrac{-2}{3} = \dfrac{\text{rise}}{\text{run}}$ y-intercept $= 0$

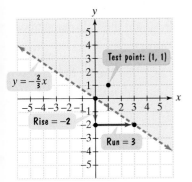

Figure 7.18 The graph of $y > -\frac{2}{3}x$

The y-intercept is 0, so the line passes through $(0, 0)$. Using the y-intercept and the slope, the line is shown in Figure 7.18 as a dashed line. This is because the inequality $y > -\frac{2}{3}x$ contains a $>$ symbol, in which equality is not included.

Step 2   Choose a test point from one of the half-planes and not from the line. Substitute its coordinates into the inequality. We cannot use $(0, 0)$ as a test point because it lies on the line and not in a half-plane. Let's use $(1, 1)$, which lies in the half-plane above the line.

$$y > -\frac{2}{3}x \qquad \text{\small This is the given inequality.}$$
$$1 \overset{?}{>} -\frac{2}{3} \cdot 1 \qquad \text{\small Test } (1, 1) \text{ by substituting 1 for } x \text{ and 1 for } y.$$
$$1 > -\frac{2}{3} \qquad \text{\small This statement is true.}$$

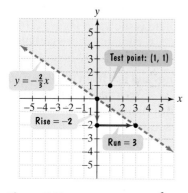

Figure 7.18 The graph of $y > -\frac{2}{3}x$ (repeated)

Step 3 **If a true statement results, shade the half-plane containing the test point.** Because 1 is greater than $-\frac{2}{3}$, the test point $(1, 1)$ is part of the solution set. All the points on the same side of the line $y = -\frac{2}{3}x$ as the point $(1, 1)$ are members of the solution set. The solution set is the half-plane that contains the point $(1, 1)$, indicated by shading this half-plane. The graph is shown using green shading and a dashed blue line in Figure 7.18.

Technology

Most graphing utilities can graph inequalities in two variables with the $\boxed{\text{SHADE}}$ feature. The procedure varies by model, so consult your manual. For most graphing utilities, you must first solve for y if it is not already isolated. The figure shows the graph of $y > -\frac{2}{3}x$. Most displays do not distinguish between dashed and solid boundary lines.

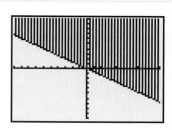

Check Point 2 Graph: $y > -\dfrac{3}{4}x$.

Graphing Linear Inequalities without Using Test Points

You can graph inequalities in the form $y > mx + b$ or $y < mx + b$ without using test points. The inequality symbol indicates which half-plane to shade.

- If $y > mx + b$, shade the half-plane above the line $y = mx + b$.
- If $y < mx + b$, shade the half-plane below the line $y = mx + b$.

It is also not necessary to use test points when graphing inequalities involving half-planes on one side of a vertical or a horizontal line.

For the Vertical Line $x = a$:

- If $x > a$, shade the half-plane to the right of $x = a$.
- If $x < a$, shade the half-plane to the left of $x = a$.

For the Horizontal Line $y = b$:

- If $y > b$, shade the half-plane above $y = b$.
- If $y < b$, shade the half-plane below $y = b$.

Study Tip

Continue using test points to graph inequalities in the form $Ax + By > C$ or $Ax + By < C$. The graph of $Ax + By > C$ can lie above or below the line of $Ax + By = C$, depending on the value of B. The same comment applies to the graph of $Ax + By < C$.

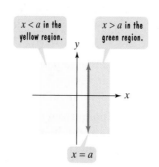

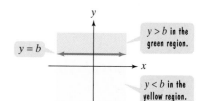

EXAMPLE 3 Graphing Inequalities without Using Test Points

Graph each inequality in a rectangular coordinate system:

a. $y \le -3$ **b.** $x > 2$.

Solution

a. $y \leq -3$

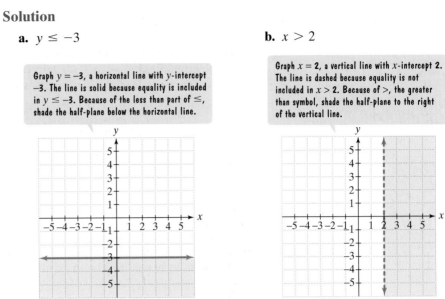

Graph $y = -3$, a horizontal line with y-intercept -3. The line is solid because equality is included in $y \leq -3$. Because of the less than part of $\leq$, shade the half-plane below the horizontal line.

b. $x > 2$

Graph $x = 2$, a vertical line with x-intercept 2. The line is dashed because equality is not included in $x > 2$. Because of $>$, the greater than symbol, shade the half-plane to the right of the vertical line.

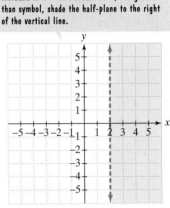

Check Point 3 Graph each inequality in a rectangular coordinate system:

   a. $y > 1$ **b.** $x \leq -2$.

  Graph a nonlinear inequality in two variables.

Graphing a Nonlinear Inequality in Two Variables

Example 4 illustrates that a nonlinear inequality in two variables is graphed in the same way that we graph a linear inequality.

EXAMPLE 4   Graphing a Nonlinear Inequality in Two Variables

Graph: $x^2 + y^2 \leq 9$.

Solution

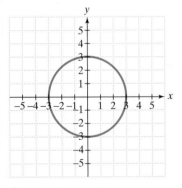

Figure 7.19 Preparing to graph $x^2 + y^2 \leq 9$

Step 1   Replace the inequality symbol with = and graph the nonlinear equation. We need to graph $x^2 + y^2 = 9$. The graph is a circle of radius 3 with its center at the origin. The graph is shown in Figure 7.19 as a solid circle because equality is included in the $\leq$ symbol.

Step 2   Choose a test point from one of the regions and not from the circle. Substitute its coordinates into the inequality. The circle divides the plane into three parts—the circle itself, the region inside the circle, and the region outside the circle. We need to determine whether the region inside or outside the circle is included in the solution. To do so, we will use the test point $(0, 0)$ from inside the circle.

$x^2 + y^2 \leq 9$	This is the given inequality.
$0^2 + 0^2 \overset{?}{\leq} 9$	Test $(0, 0)$ by substituting 0 for x and 0 for y.
$0 + 0 \overset{?}{\leq} 9$	Square 0: $0^2 = 0$.
$0 \leq 9$	Add. This statement is true.

Step 3   If a true statement results, shade the region containing the test point. The true statement tells us that all the points inside the circle satisfy $x^2 + y^2 \leq 9$. The graph is shown using green shading and a solid blue circle in Figure 7.20.

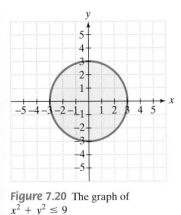

Figure 7.20 The graph of $x^2 + y^2 \leq 9$

Check Point 4 Graph: $x^2 + y^2 \geq 16$.

③ Graph a system of inequalities.

Systems of Inequalities in Two Variables

Just as two linear equations make up a system of linear equations, two or more linear inequalities make up a **system of linear inequalities**. Here is an example of a system of linear inequalities:

$$x - y < 1$$
$$2x + 3y \geq 12.$$

A **solution of a system of linear inequalities** in two variables is an ordered pair that satisfies each inequality in the system. The set of all such ordered pairs is the **solution set** of the system. Thus, to graph a system of inequalities in two variables, begin by graphing each individual inequality in the same rectangular coordinate system. Then find the region, if there is one, that is common to every graph in the system. This region of intersection gives a picture of the system's solution set.

EXAMPLE 5   Graphing a System of Linear Inequalities

Graph the solution set of the system:

$$x - y < 1$$
$$2x + 3y \geq 12.$$

Solution   Replacing each inequality symbol with an equal sign indicates that we need to graph $x - y = 1$ and $2x + 3y = 12$. We can use intercepts to graph these lines.

$x - y = 1$	$2x + 3y = 12$
x-intercept: $x - 0 = 1$	x-intercept: $2x + 3 \cdot 0 = 12$
$x = 1$	$2x = 12$
The line passes through $(1, 0)$.	$x = 6$
	The line passes through $(6, 0)$.
y-intercept: $0 - y = 1$	y-intercept: $2 \cdot 0 + 3y = 12$
$-y = 1$	$3y = 12$
$y = -1$	$y = 4$
The line passes through $(0, -1)$.	The line passes through $(0, 4)$.

Set $y = 0$ in each equation.

Set $x = 0$ in each equation.

Now we are ready to graph the solution set of the system of linear inequalities.

Graph $x - y < 1$. The blue line, $x - y = 1$, is dashed: Equality is not included in $x - y < 1$. Because $(0, 0)$ makes the inequality true $(0 - 0 < 1$, or $0 < 1$, is true), shade the half-plane containing $(0, 0)$ in yellow.

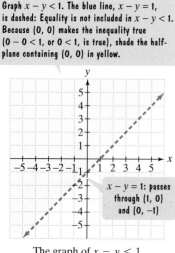

The graph of $x - y < 1$

Add the graph of $2x + 3y \geq 12$. The red line, $2x + 3y = 12$, is solid: Equality is included in $2x + 3y \geq 12$. Because $(0, 0)$ makes the inequality false $(2 \cdot 0 + 3 \cdot 0 \geq 12$, or $0 \geq 12$, is false), shade the half-plane not containing $(0, 0)$ using green vertical shading.

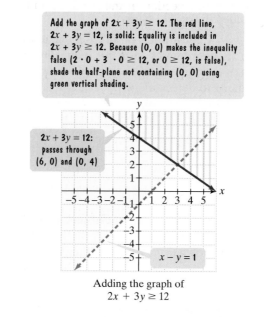

$2x + 3y = 12$: passes through $(6, 0)$ and $(0, 4)$

$x - y = 1$

Adding the graph of $2x + 3y \geq 12$

The solution set of the system is graphed as the intersection (the overlap) of the two half-planes. This is the region in which the yellow shading and the green vertical shading overlap.

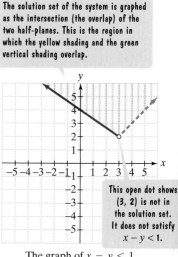

This open dot shows $(3, 2)$ is not in the solution set. It does not satisfy $x - y < 1$.

The graph of $x - y < 1$ and $2x + 3y \geq 12$

$x - y = 1$: passes through $(1, 0)$ and $(0, -1)$

Check Point 5 Graph the solution set of the system:

$$x - 3y < 6$$
$$2x + 3y \geq -6.$$

EXAMPLE 6   Graphing a System of Inequalities

Graph the solution set of the system:

$$y \geq x^2 - 4$$
$$x - y \geq 2.$$

Solution   We begin by graphing $y \geq x^2 - 4$. Because equality is included in $\geq$, we graph $y = x^2 - 4$ as a solid parabola. Because $(0, 0)$ makes the inequality $y \geq x^2 - 4$ true (we obtain $0 \geq -4$), we shade the interior portion of the parabola containing $(0, 0)$, shown in yellow in Figure 7.21.

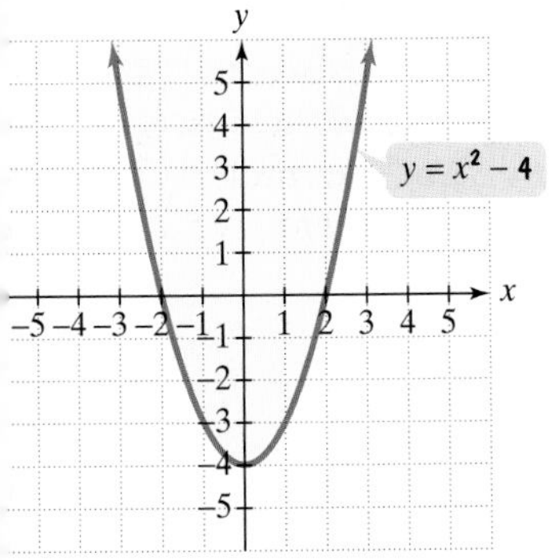

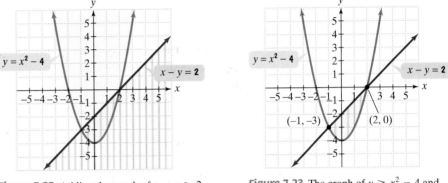

Figure 7.21 The graph of $y \geq x^2 - 4$

Figure 7.22 Adding the graph of $x - y \geq 2$

Figure 7.23 The graph of $y \geq x^2 - 4$ and $x - y \geq 2$

Now we graph $x - y \geq 2$ in the same rectangular coordinate system. First we graph the line $x - y = 2$ using its x-intercept, 2, and its y-intercept, -2. Because $(0, 0)$ makes the inequality $x - y \geq 2$ false (we obtain $0 \geq 2$), we shade the half-plane below the line. This is shown in Figure 7.22 using green vertical shading.

The solution of the system is shown in Figure 7.23 by the intersection (the overlap) of the solid yellow and green vertical shadings. The graph of the system's solution set consists of the region enclosed by the parabola and the line. To find the points of intersection of the parabola and the line, use the substitution method to solve the nonlinear system

$$y = x^2 - 4$$
$$x - y = 2.$$

Take a moment to show that the solutions are $(-1, -3)$ and $(2, 0)$, as shown in Figure 7.23.

Check Point 6 Graph the solution set of the system:

$$y \geq x^2 - 4$$
$$x + y \leq 2.$$

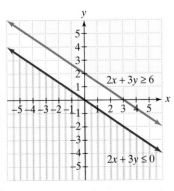

Figure 7.24 A system of inequalities with no solution

A system of inequalities has no solution if there are no points in the rectangular coordinate system that simultaneously satisfy each inequality in the system. For example, the system

$$2x + 3y \geq 6$$
$$2x + 3y \leq 0,$$

whose separate graphs are shown in Figure 7.24, has no overlapping region. Thus, the system has no solution. The solution set is $\emptyset$, the empty set.

EXAMPLE 7   Graphing a System of Inequalities

Graph the solution set of the system:

$$x - y < 2$$
$$-2 \leq x < 4$$
$$y < 3.$$

Solution   We begin by graphing $x - y < 2$, the first given inequality. The line
$x - y = 2$ has an x-intercept of 2 and a y-intercept of -2. The test point $(0, 0)$
makes the inequality $x - y < 2$ true, and its graph is shown in Figure 7.25.

Now, let's consider the second given inequality, $-2 \leq x < 4$. Replacing the
inequality symbols by $=$, we obtain $x = -2$ and $x = 4$, graphed as red vertical lines
in Figure 7.26. The line of $x = 4$ is not included. Using $(0, 0)$ as a test point and substituting the x-coordinate, 0, into $-2 \leq x < 4$, we obtain the true statement
$-2 \leq 0 < 4$. We therefore shade the region between the vertical lines. We must
intersect this region with the yellow region in Figure 7.25. The resulting region is
shown in yellow and green vertical shading in Figure 7.26.

Finally, let's consider the third given inequality, $y < 3$. Replacing the inequality
symbol by $=$, we obtain $y = 3$, which graphs as a horizontal line. Because of the less
than symbol in $y < 3$, the graph consists of the half-plane below the line $y = 3$. We
must intersect this half-plane with the region in Figure 7.26. The resulting region is
shown in yellow and green vertical shading in Figure 7.27. This region represents the
graph of the solution set of the given system.

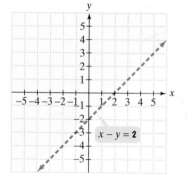

Figure 7.25 The graph of $x - y < 2$

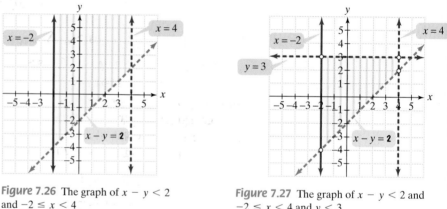

Figure 7.26 The graph of $x - y < 2$
and $-2 \leq x < 4$

Figure 7.27 The graph of $x - y < 2$ and
$-2 \leq x < 4$ and $y < 3$

In Figure 7.27, it may be difficult to tell where the graph of $x - y = 2$ intersects the
vertical line $x = 4$. Using the substitution method, it can be determined that this
intersection point is $(4, 2)$. Take a moment to verify that the four intersection points
in Figure 7.27 are, clockwise from upper left, $(-2, 3)$, $(4, 3)$, $(4, 2)$, and $(-2, -4)$.
These points are shown as open dots because none satisfies all three of the system's
inequalities.

 Check Point 7   Graph the solution set of the system:

$$x + y < 2$$
$$-2 \leq x < 1$$
$$y > -3.$$

❹ Solve applied problems
involving systems of
inequalities.

Applications

Temperature and precipitation affect whether or not trees and forests can grow. At
certain levels of precipitation and temperature, only grasslands and deserts will exist.
Figure 7.28 shows three kinds of regions—deserts, grasslands, and forests—that result
from various ranges of temperature, T, and precipitation, P.

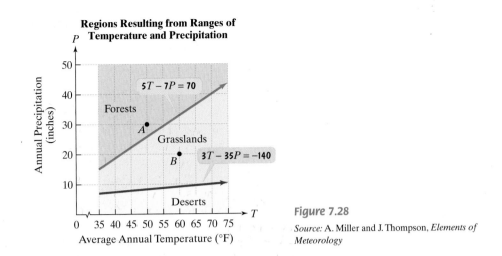

Figure 7.28

Source: A. Miller and J. Thompson, *Elements of Meteorology*

Systems of inequalities can be used to describe where forests, grasslands, and deserts occur. Because these regions occur when the average annual temperature, T, is 35°F or greater, each system contains the inequality $T \geq 35$.

Forests occur if	**Grasslands occur if**	**Deserts occur if**
$T \geq 35$	$T \geq 35$	$T \geq 35$
$5T - 7P < 70.$	$5T - 7P \geq 70$	$3T - 35P > -140.$
	$3T - 35P \leq -140.$	

EXAMPLE 8   Forests and Systems of Inequalities

Show that point A in Figure 7.28 is a solution of the system of inequalities that describes where forests occur.

Solution   Point A has coordinates $(50, 30)$. This means that if a region has an average annual temperature of 50°F and an average annual precipitation of 30 inches, a forest occurs. We can show that $(50, 30)$ satisfies the system of inequalities for forests by substituting 50 for T and 30 for P in each inequality in the system.

$$T \geq 35 \qquad\qquad 5T - 7P < 70$$
$$50 \geq 35, \quad \text{true} \qquad 5 \cdot 50 - 7 \cdot 30 \overset{?}{<} 70$$
$$250 - 210 \overset{?}{<} 70$$
$$40 < 70, \quad \text{true}$$

The coordinates $(50, 30)$ make each inequality true. Thus, $(50, 30)$ satisfies the system for forests.

Check Point 8   Show that point B in Figure 7.28 is a solution of the system of inequalities that describes where grasslands occur.

EXERCISE SET 7.5

Practice Exercises

In Exercises 1–26, graph each inequality.

1. $x + 2y \leq 8$

2. $3x - 6y \leq 12$

3. $x - 2y > 10$

4. $2x - y > 4$

5. $y \leq \dfrac{1}{3}x$

6. $y \leq \dfrac{1}{4}x$

7. $y > 2x - 1$

8. $y > 3x + 2$

9. $x \leq 1$

10. $x \leq -3$

11. $y > 1$

12. $y > -3$

13. $x^2 + y^2 \le 1$

14. $x^2 + y^2 \le 4$

15. $x^2 + y^2 > 25$

16. $x^2 + y^2 > 36$

17. $(x - 2)^2 + (y + 1)^2 < 9$

18. $(x + 2)^2 + (y - 1)^2 < 16$

19. $y < x^2 - 1$

20. $y < x^2 - 9$

21. $y \ge x^2 - 9$

22. $y \ge x^2 - 1$

23. $y > 2^x$

24. $y \le 3^x$

25. $y \ge \log_2(x + 1)$

26. $y \ge \log_3(x - 1)$

In Exercises 27–62, graph the solution set of each system of inequalities or indicate that the system has no solution.

27. $3x + 6y \le 6$
$2x + y \le 8$

28. $x - y \ge 4$
$x + y \le 6$

29. $2x - 5y \le 10$
$3x - 2y > 6$

30. $2x - y \le 4$
$3x + 2y > -6$

31. $y > 2x - 3$
$y < -x + 6$

32. $y < -2x + 4$
$y < x - 4$

33. $x + 2y \le 4$
$y \ge x - 3$

34. $x + y \le 4$
$y \ge 2x - 4$

35. $x \le 2$
$y \ge -1$

36. $x \le 3$
$y \le -1$

37. $-2 \le x < 5$

38. $-2 < y \le 5$

39. $x - y \le 1$
$x \ge 2$

40. $4x - 5y \ge -20$
$x \ge -3$

41. $x + y > 4$
$x + y < -1$

42. $x + y > 3$
$x + y < -2$

43. $x + y > 4$
$x + y > -1$

44. $x + y > 3$
$x + y > -2$

45. $y \ge x^2 - 1$
$x - y \ge -1$

46. $y \ge x^2 - 4$
$x - y \ge 2$

47. $x^2 + y^2 \le 16$
$x + y > 2$

48. $x^2 + y^2 \le 4$
$x + y > 1$

49. $x^2 + y^2 > 1$
$x^2 + y^2 < 4$

50. $x^2 + y^2 > 1$
$x^2 + y^2 < 9$

51. $(x - 1)^2 + (y + 1)^2 < 25$
$(x - 1)^2 + (y + 1)^2 \ge 16$

52. $(x + 1)^2 + (y - 1)^2 < 16$
$(x + 1)^2 + (y - 1)^2 \ge 4$

53. $x^2 + y^2 \le 1$
$y - x^2 > 0$

54. $x^2 + y^2 < 4$
$y - x^2 \ge 0$

55. $x^2 + y^2 < 16$
$y \ge 2^x$

56. $x^2 + y^2 \le 16$
$y < 2^x$

57. $x - y \le 2$
$x > -2$
$y \le 3$

58. $3x + y \le 6$
$x > -2$
$y \le 4$

59. $x \ge 0$
$y \ge 0$
$2x + 5y < 10$
$3x + 4y \le 12$

60. $x \ge 0$
$y \ge 0$
$2x + y < 4$
$2x - 3y \le 6$

61. $3x + y \le 6$
$2x - y \le -1$
$x > -2$
$y < 4$

62. $2x + y \le 6$
$x + y > 2$
$1 \le x \le 2$
$y < 3$

Practice Plus

In Exercises 63–64, write each sentence as an inequality in two variables. Then graph the inequality.

63. The y-variable is at least 4 more than the product of -2 and the x-variable.

64. The y-variable is at least 2 more than the product of -3 and the x-variable.

In Exercises 65–68, write the given sentences as a system of inequalities in two variables. Then graph the system.

65. The sum of the x-variable and the y-variable is at most 4. The y-variable added to the product of 3 and the x-variable does not exceed 6.

66. The sum of the x-variable and the y-variable is at most 3. The y-variable added to the product of 4 and the x-variable does not exceed 6.

67. The sum of the x-variable and the y-variable is no more than 2. The y-variable is no less than the difference between the square of the x-variable and 4.

68. The sum of the squares of the x-variable and the y-variable is no more than 25. The sum of twice the y-variable and the x-variable is no less than 5.

In Exercises 69–70, rewrite each inequality in the system without absolute value bars. Then graph the rewritten system in rectangular coordinates.

69. $|x| \le 2$
$|y| \le 3$

70. $|x| \le 1$
$|y| \le 2$

*The graphs of solution sets of systems of inequalities involve finding the intersection of the solution sets of two or more inequalities. By contrast, in Exercises 71–72, you will be graphing the **union** of the solution sets of two inequalities.*

71. Graph the union of $y > \frac{3}{2}x - 2$ and $y < 4$.

72. Graph the union of $x - y \ge -1$ and $5x - 2y \le 10$.

Without graphing, in Exercises 73–76, determine if each system has no solution or infinitely many solutions.

73. $3x + y < 9$
$3x + y > 9$

74. $6x - y \le 24$
$6x - y > 24$

75. $(x + 4)^2 + (y - 3)^2 \le 9$
$(x + 4)^2 + (y - 3)^2 \ge 9$

76. $(x - 4)^2 + (y + 3)^2 \le 24$
$(x - 4)^2 + (y + 3)^2 \ge 24$

Application Exercises

Maximum heart rate, H, in beats per minute is a function of age, a, modeled by the formula

$$H = 220 - a,$$

where $10 \leq a \leq 70$. The bar graph shows the target heart rate ranges as a percentage of the maximum heart rate for four types of exercise goals.

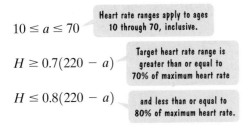

Target Heart Rate Ranges for Exercise Goals

Exercise Goal
- Boost performance as a competitive athlete
- Improve cardiovascular conditioning
- Lose weight
- Improve overall health and reduce risk of heart attack

Percentage of Maximum Heart Rate (40% 50% 60% 70% 80% 90% 100%)

Source: Vitality

In Exercises 77–80, systems of inequalities will be used to model three of the target heart rate ranges shown in the bar graph. We begin with the target heart rate range for cardiovascular conditioning, modeled by the following system of inequalities:

$10 \leq a \leq 70$ — Heart rate ranges apply to ages 10 through 70, inclusive.

$H \geq 0.7(220 - a)$ — Target heart rate range is greater than or equal to 70% of maximum heart rate

$H \leq 0.8(220 - a)$ — and less than or equal to 80% of maximum heart rate.

The graph of this system is shown in the figure. Use the graph to solve Exercises 77–78.

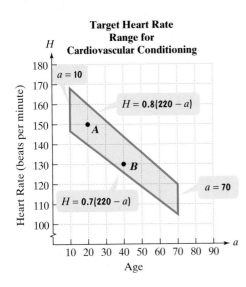

Target Heart Rate Range for Cardiovascular Conditioning

Heart Rate (beats per minute) vs. Age

$a = 10$
$H = 0.8(220 - a)$
$a = 70$
$H = 0.7(220 - a)$

77. a. What are the coordinates of point A and what does this mean in terms of age and heart rate?

b. Show that point A is a solution of the system of inequalities.

78. a. What are the coordinates of point B and what does this mean in terms of age and heart rate?

b. Show that point B is a solution of the system of inequalities.

79. Write a system of inequalities that models the target heart rate range for the goal of losing weight.

80. Write a system of inequalities that models the target heart rate range for improving overall health.

81. Many elevators have a capacity of 2000 pounds.

a. If a child averages 50 pounds and an adult 150 pounds, write an inequality that describes when x children and y adults will cause the elevator to be overloaded.

b. Graph the inequality. Because x and y must be positive, limit the graph to quadrant I only.

c. Select an ordered pair satisfying the inequality. What are its coordinates and what do they represent in this situation?

82. A patient is not allowed to have more than 330 milligrams of cholesterol per day from a diet of eggs and meat. Each egg provides 165 milligrams of cholesterol. Each ounce of meat provides 110 milligrams.

a. Write an inequality that describes the patient's dietary restrictions for x eggs and y ounces of meat.

b. Graph the inequality. Because x and y must be positive, limit the graph to quadrant I only.

c. Select an ordered pair satisfying the inequality. What are its coordinates and what do they represent in this situation?

83. On your next vacation, you will divide lodging between large resorts and small inns. Let x represent the number of nights spent in large resorts. Let y represent the number of nights spent in small inns.

a. Write a system of inequalities that models the following conditions:

You want to stay at least 5 nights. At least one night should be spent at a large resort. Large resorts average $200 per night and small inns average $100 per night. Your budget permits no more than $700 for lodging.

b. Graph the solution set of the system of inequalities in part (a).

c. Based on your graph in part (b), what is the greatest number of nights you could spend at a large resort and still stay within your budget?

84. A person with no more than $15,000 to invest plans to place the money in two investments. One investment is high risk, high yield; the other is low risk, low yield. At least $2000 is to be placed in the high-risk investment. Furthermore, the amount invested at low risk should be at least three times the amount invested at high risk. Find and graph a system of inequalities that describes all possibilities for placing the money in the high- and low-risk investments.

The graph of an inequality in two variables is a region in the rectangular coordinate system. Regions in coordinate systems have numerous applications. For example, the regions in the following two graphs indicate whether a person is obese, overweight, borderline overweight, normal weight, or underweight.

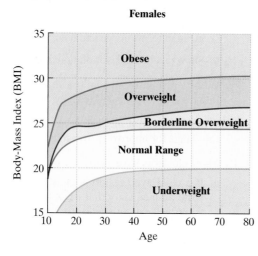

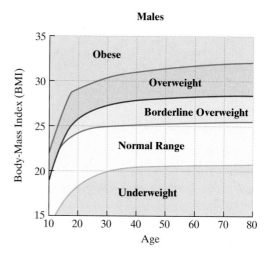

Source: Centers for Disease Control and Prevention

The horizontal axis shows a person's age. The vertical axis shows that person's body-mass index (BMI), computed using the following formula:

$$\text{BMI} = \frac{703W}{H^2}.$$

The variable W represents weight, in pounds. The variable H represents height, in inches. Use this information to solve Exercises 85–86.

85. A man is 20 years old, 72 inches (6 feet) tall, and weighs 200 pounds.

   a. Compute the man's BMI. Round to the nearest tenth.

   b. Use the man's age and his BMI to locate this information as a point in the coordinate system for males. Is this person obese, overweight, borderline overweight, normal weight, or underweight?

86. A woman is 25 years old, 66 inches (5 feet, 6 inches) tall, and weighs 105 pounds.

   a. Compute the woman's BMI. Round to the nearest tenth.

   b. Use the woman's age and her BMI to locate this information as a point in the coordinate system for females. Is this person obese, overweight, borderline overweight, normal weight, or underweight?

Writing in Mathematics

87. What is a linear inequality in two variables? Provide an example with your description.

88. How do you determine if an ordered pair is a solution of an inequality in two variables, x and y?

89. What is a half-plane?

90. What does a solid line mean in the graph of an inequality?

91. What does a dashed line mean in the graph of an inequality?

92. Compare the graphs of $3x - 2y > 6$ and $3x - 2y \le 6$. Discuss similarities and differences between the graphs.

93. What is a system of linear inequalities?

94. What is a solution of a system of linear inequalities?

95. Explain how to graph the solution set of a system of inequalities.

96. What does it mean if a system of linear inequalities has no solution?

Technology Exercises

Graphing utilities can be used to shade regions in the rectangular coordinate system, thereby graphing an inequality in two variables. Read the section of the user's manual for your graphing utility that describes how to shade a region. Then use your graphing utility to graph the inequalities in Exercises 97–102.

97. $y \le 4x + 4$

98. $y \ge \frac{2}{3}x - 2$

99. $y \ge x^2 - 4$

100. $y \ge \frac{1}{2}x^2 - 2$

101. $2x + y \le 6$

102. $3x - 2y \ge 6$

103. Does your graphing utility have any limitations in terms of graphing inequalities? If so, what are they?

104. Use a graphing utility with a $\boxed{\text{SHADE}}$ feature to verify any five of the graphs that you drew by hand in Exercises 1–26.

105. Use a graphing utility with a $\boxed{\text{SHADE}}$ feature to verify any five of the graphs that you drew by hand for the systems in Exercises 27–62.

Critical Thinking Exercises

In Exercises 106–109, write a system of inequalities for each graph.

106.

107.

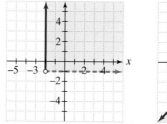

08.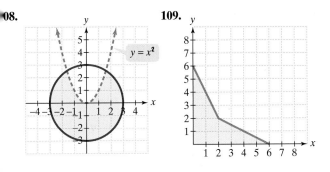

109.

110. Write a system of inequalities whose solution set includes every point in the rectangular coordinate system.

111. Sketch the graph of the solution set for the following system of inequalities:

$$y \geq nx + b \ (n < 0, b > 0)$$
$$y \leq mx + b \ (m > 0, b > 0).$$

SECTION 7.6 *Linear Programming*

Objectives

1 Write an objective function describing a quantity that must be maximized or minimized.

2 Use inequalities to describe limitations in a situation.

3 Use linear programming to solve problems.

West Berlin children at Tempelhof airport watch fleets of U.S. airplanes bringing in supplies to circumvent the Soviet blockade. The airlift began June 28, 1948 and continued for 15 months.

The Berlin Airlift (1948–1949) was an operation by the United States and Great Britain in response to military action by the former Soviet Union: Soviet troops closed all roads and rail lines between West Germany and Berlin, cutting off supply routes to the city. The Allies used a mathematical technique developed during World War II to maximize the amount of supplies transported. During the 15-month airlift, 278,228 flights provided basic necessities to blockaded Berlin, saving one of the world's great cities.

In this section, we will look at an important application of systems of linear inequalities. Such systems arise in **linear programming**, a method for solving problems in which a particular quantity that must be maximized or minimized is limited by other factors. Linear programming is one of the most widely used tools in management science. It helps businesses allocate resources to manufacture products in a way that will maximize profit. Linear programming accounts for more than 50% and perhaps as much as 90% of all computing time used for management decisions in business. The Allies used linear programming to save Berlin.

Objective Functions in Linear Programming

1 Write an objective function describing a quantity that must be maximized or minimized.

Many problems involve quantities that must be maximized or minimized. Businesses are interested in maximizing profit. An operation in which bottled water and medical kits are shipped to earthquake victims needs to maximize the number of victims helped by this shipment. An **objective function** is an algebraic expression in two or more variables describing a quantity that must be maximized or minimized.

EXAMPLE 1   Writing an Objective Function

Bottled water and medical supplies are to be shipped to victims of an earthquake by plane. Each container of bottled water will serve 10 people and each medical kit will aid 6 people. If *x* represents the number of bottles of water to be shipped and *y* represents the number of medical kits, write the objective function that describes the number of people who can be helped.

Solution Because each bottle of water serves 10 people and each medical kit aid
6 people, we have

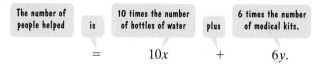

| The number of people helped | is | 10 times the number of bottles of water | plus | 6 times the number of medical kits. |

$$= \quad 10x \quad + \quad 6y.$$

Using z to represent the number of people helped, the objective function is

$$z = 10x + 6y.$$

Unlike the functions that we have seen so far, the objective function is an equation
in three variables. For a value of x and a value of y, there is one and only one value
of z. Thus, z is a function of x and y.

Check Point 1 A company manufactures bookshelves and desks for computers. Let x represent the number of bookshelves manufactured daily and y the number of desks manufactured daily. The company's profits are $25 per bookshelf and $55 per desk. Write the objective function that describes the company's total daily profit, z, from x bookshelves and y desks. (Check Points 2 through 4 are related to this situation, so keep track of your answers.)

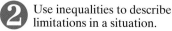

② Use inequalities to describe limitations in a situation.

Constraints in Linear Programming

Ideally, the number of earthquake victims helped in Example 1 should increase without restriction so that every victim receives water and medical kits. However, the planes that ship these supplies are subject to weight and volume restrictions. In linear programming problems, such restrictions are called **constraints**. Each constraint is expressed as a linear inequality. The list of constraints forms a system of linear inequalities.

EXAMPLE 2 Writing a Constraint

Each plane can carry no more than 80,000 pounds. The bottled water weighs 20 pounds per container and each medical kit weighs 10 pounds. Let x represent the number of bottles of water to be shipped and y the number of medical kits. Write an inequality that describes this constraint.

Solution Because each plane can carry no more than 80,000 pounds, we have

| The total weight of the water bottles | plus | the total weight of the medical kits | must be less than or equal to | 80,000 pounds. |

$$20x \quad + \quad 10y \quad \leq \quad 80,000.$$

└─ Each bottle weighs 20 pounds.   └─ Each kit weighs 10 pounds.

The plane's weight constraint is described by the inequality

$$20x + 10y \leq 80,000.$$

Check Point 2 To maintain high quality, the company in Check Point 1 should not manufacture more than a total of 80 bookshelves and desks per day. Write an inequality that describes this constraint.

In addition to a weight constraint on its cargo, each plane has a limited amount of space in which to carry supplies. Example 3 demonstrates how to express this constraint.

EXAMPLE 3   Writing a Constraint

Each plane can carry a total volume of supplies that does not exceed 6000 cubic feet. Each water bottle is 1 cubic foot and each medical kit also has a volume of 1 cubic foot. With x still representing the number of water bottles and y the number of medical kits, write an inequality that describes this second constraint.

Solution Because each plane can carry a volume of supplies that does not exceed 6000 cubic feet, we have

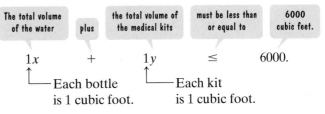

| The total volume of the water | plus | the total volume of the medical kits | must be less than or equal to | 6000 cubic feet. |

$$1x \quad + \quad 1y \quad \le \quad 6000.$$

└── Each bottle is 1 cubic foot.   └── Each kit is 1 cubic foot.

The plane's volume constraint is described by the inequality $x + y \le 6000$.

In summary, here's what we have described so far in this aid-to-earthquake-victims situation:

$$z = 10x + 6y$$ *This is the objective function describing the number of people helped with x bottles of water and y medical kits.*

$$20x + 10y \le 80,000$$ *These are the constraints based on each plane's weight and*
$$x + y \le 6000.$$ *volume limitations.*

Check Point 3 To meet customer demand, the company in Check Point 1 must manufacture between 30 and 80 bookshelves per day, inclusive. Furthermore, the company must manufacture at least 10 and no more than 30 desks per day. Write an inequality that describes each of these sentences. Then summarize what you have described about this company by writing the objective function for its profits and the three constraints.

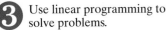

Use linear programming to solve problems.

Solving Problems with Linear Programming

The problem in the earthquake situation described previously is to maximize the number of victims who can be helped, subject to each plane's weight and volume constraints. The process of solving this problem is called *linear programming*, based on a theorem that was proven during World War II.

Solving a Linear Programming Problem

Let $z = ax + by$ be an objective function that depends on x and y. Furthermore, z is subject to a number of constraints on x and y. If a maximum or minimum value of z exists, it can be determined as follows:

1. Graph the system of inequalities representing the constraints.
2. Find the value of the objective function at each corner, or **vertex**, of the graphed region. The maximum and minimum of the objective function occur at one or more of the corner points.

EXAMPLE 4   Solving a Linear Programming Problem

Determine how many bottles of water and how many medical kits should be sent on each plane to maximize the number of earthquake victims who can be helped.

Solution We must maximize $z = 10x + 6y$ subject to the following constraints:

$$20x + 10y \le 80,000$$
$$x + y \le 6000.$$

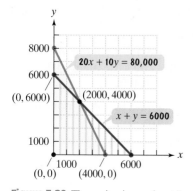

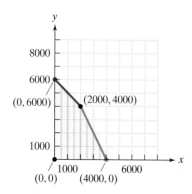

Figure 7.29 The region in quadrant I representing the constraints $20x + 10y \leq 80,000$ and $x + y \leq 6000$

Figure 7.30

Step 1 Graph the system of inequalities representing the constraints. Because (the number of bottles of water per plane) and y (the number of medical kits per plane) must be nonnegative, we need to graph the system of inequalities in quadrant I and its boundary only.

To graph the inequality $20x + 10y \leq 80,000$, we graph the equation $20x + 10y = 80,000$ as a solid blue line (Figure 7.29). Setting $y = 0$, the x-intercept is 4000 and setting $x = 0$, the y-intercept is 8000. Using $(0, 0)$ as a test point, the inequality is satisfied, so we shade below the blue line, as shown in yellow in Figure 7.29.

Now we graph $x + y \leq 6000$ by first graphing $x + y = 6000$ as a solid red line. Setting $y = 0$, the x-intercept is 6000. Setting $x = 0$, the y-intercept is 6000. Using $(0, 0)$ as a test point, the inequality is satisfied, so we shade below the red line as shown using green vertical shading in Figure 7.29.

We use the addition method to find where the lines $20x + 10y = 80,000$ and $x + y = 6000$ intersect.

$$
\begin{array}{lll}
20x + 10y = 80,000 & \xrightarrow{\text{No change}} & 20x + 10y = 80,000 \\
x + y = 6000 & \xrightarrow{\text{Multiply by }-10.} & -10x - 10y = -60,000 \\
& \text{Add:} & 10x = 20,000 \\
& & x = 2000
\end{array}
$$

Back-substituting 2000 for x in $x + y = 6000$, we find $y = 4000$, so the intersection point is $(2000, 4000)$.

The system of inequalities representing the constraints is shown by the region in which the yellow shading and the green vertical shading overlap in Figure 7.29. The graph of the system of inequalities is shown again in Figure 7.30. The red and blue line segments are included in the graph.

Step 2 Find the value of the objective function at each corner of the graphed region. The maximum and minimum of the objective function occur at one or more of the corner points. We must evaluate the objective function, $z = 10x + 6y$, at the four corners, or vertices, of the region in Figure 7.30.

Corner (x, y)	Objective Function $z = 10x + 6y$
$(0, 0)$	$z = 10(0) + 6(0) = 0$
$(4000, 0)$	$z = 10(4000) + 6(0) = 40,000$
$(2000, 4000)$	$z = 10(2000) + 6(4000) = 44,000$ ← maximum
$(0, 6000)$	$z = 10(0) + 6(6000) = 36,000$

Thus, the maximum value of z is 44,000, and this occurs when $x = 2000$ and $y = 4000$. In practical terms, this means that the maximum number of earthquake victims who can be helped with each plane shipment is 44,000. This can be accomplished by sending 2000 water bottles and 4000 medical kits per plane.

 Check Point 4 For the company in Check Points 1–3, how many bookshelves and how many desks should be manufactured per day to obtain maximum profit? What is the maximum daily profit?

EXAMPLE 5 Solving a Linear Programming Problem

Find the maximum value of the objective function

$$z = 2x + y$$

subject to the following constraints:

$$x \geq 0, y \geq 0$$
$$x + 2y \leq 5$$
$$x - y \leq 2.$$

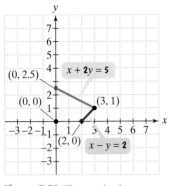

Figure 7.31 The graph of $x + 2y \le 5$ and $x - y \le 2$ in quadrant I

Solution   We begin by graphing the region in quadrant I $(x \ge 0, y \ge 0)$ formed by the constraints. The graph is shown in Figure 7.31.

Now we evaluate the objective function at the four vertices of this region.

Objective function: $z = 2x + y$

At $(0, 0)$: $z = 2 \cdot 0 + 0 = 0$

At $(2, 0)$: $z = 2 \cdot 2 + 0 = 4$

At $(3, 1)$: $z = 2 \cdot 3 + 1 = 7$ **Maximum value of z**

At $(0, 2.5)$: $z = 2 \cdot 0 + 2.5 = 2.5$

Thus, the maximum value of z is 7, and this occurs when $x = 3$ and $y = 1$.

We can see why the objective function in Example 5 has a maximum value that occurs at a vertex by solving the equation for y.

$$z = 2x + y$$ This is the objective function of Example 5.

$$y = -2x + z$$ Solve for y. Recall that the slope-intercept form of a line is $y = mx + b$.

Slope $= -2$ y-intercept $= z$

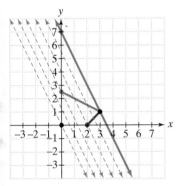

Figure 7.32 The line with slope -2 with the greatest y-intercept that intersects the shaded region passes through one of its vertices.

In this form, z represents the y-intercept of the objective function. The equation describes infinitely many parallel lines, each with slope -2. The process in linear programming involves finding the maximum z-value for all lines that intersect the region determined by the constraints. Of all the lines whose slope is -2, we're looking for the one with the greatest y-intercept that intersects the given region. As we see in Figure 7.32, such a line will pass through one (or possibly more) of the vertices of the region.

Check Point 5   Find the maximum value of the objective function $z = 3x + 5y$ subject to the constraints $x \ge 0, y \ge 0, x + y \ge 1, x + y \le 6$.

EXERCISE SET 7.6

Practice Exercises

In Exercises 1–4, find the value of the objective function at each corner of the graphed region. What is the maximum value of the objective function? What is the minimum value of the objective function?

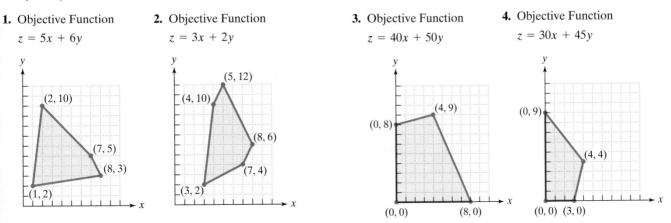

1. Objective Function
$z = 5x + 6y$

2. Objective Function
$z = 3x + 2y$

3. Objective Function
$z = 40x + 50y$

4. Objective Function
$z = 30x + 45y$

In Exercises 5–14, an objective function and a system of linear inequalities representing constraints are given.

a. Graph the system of inequalities representing the constraints.

b. Find the value of the objective function at each corner of the graphed region.

c. Use the values in part (b) to determine the maximum value of the objective function and the values of x and y for which the maximum occurs.

5. Objective Function $z = 3x + 2y$
   Constraints $x \geq 0, y \geq 0$
        $2x + y \leq 8$
        $x + y \geq 4$

6. Objective Function $z = 2x + 3y$
   Constraints $x \geq 0, y \geq 0$
        $2x + y \leq 8$
        $2x + 3y \leq 12$

7. Objective Function $z = 4x + y$
   Constraints $x \geq 0, y \geq 0$
        $2x + 3y \leq 12$
        $x + y \geq 3$

8. Objective Function $z = x + 6y$
   Constraints $x \geq 0, y \geq 0$
        $2x + y \leq 10$
        $x - 2y \geq -10$

9. Objective Function $z = 3x - 2y$
   Constraints $1 \leq x \leq 5$
        $y \geq 2$
        $x - y \geq -3$

10. Objective Function $z = 5x - 2y$
   Constraints $0 \leq x \leq 5$
        $0 \leq y \leq 3$
        $x + y \geq 2$

11. Objective Function $z = 4x + 2y$
   Constraints $x \geq 0, y \geq 0$
        $2x + 3y \leq 12$
        $3x + 2y \leq 12$
        $x + y \geq 2$

12. Objective Function $z = 2x + 4y$
   Constraints $x \geq 0, y \geq 0$
        $x + 3y \geq 6$
        $x + y \geq 3$
        $x + y \leq 9$

13. Objective Function $z = 10x + 12y$
   Constraints $x \geq 0, y \geq 0$
        $x + y \leq 7$
        $2x + y \leq 10$
        $2x + 3y \leq 18$

14. Objective Function $z = 5x + 6y$
   Constraints $x \geq 0, y \geq 0$
        $2x + y \geq 10$
        $x + 2y \geq 10$
        $x + y \leq 10$

Application Exercises

15. A television manufacturer makes console and wide-screen televisions. The profit per unit is $125 for the console televisions and $200 for the wide-screen televisions.

a. Let $x =$ the number of consoles manufactured in a month and let $y =$ the number of wide-screens manufactured in a month. Write the objective function that describes the total monthly profit.

b. The manufacturer is bound by the following constraints:
   • Equipment in the factory allows for making at most 450 console televisions in one month.
   • Equipment in the factory allows for making at most 200 wide-screen televisions in one month.
   • The cost to the manufacturer per unit is $600 for the console televisions and $900 for the wide-screen televisions. Total monthly costs cannot exceed $360,000.

Write a system of three inequalities that describes these constraints.

c. Graph the system of inequalities in part (b). Use only the first quadrant and its boundary, because x and y must both be nonnegative.

d. Evaluate the objective function for total monthly profit at each of the five vertices of the graphed region. [The vertices should occur at $(0, 0)$, $(0, 200)$, $(300, 200)$, $(450, 100)$, and $(450, 0)$.]

e. Complete the missing portions of this statement: The television manufacturer will make the greatest profit by manufacturing _____ console televisions each month and _____ wide-screen televisions each month. The maximum monthly profit is $ _____.

16. a. A student earns $10 per hour for tutoring and $7 per hour as a teacher's aid. Let $x =$ the number of hours each week spent tutoring and let $y =$ the number of hours each week spent as a teacher's aid. Write the objective function that describes total weekly earnings.

b. The student is bound by the following constraints:
   • To have enough time for studies, the student can work no more than 20 hours per week.
   • The tutoring center requires that each tutor spend at least three hours per week tutoring.
   • The tutoring center requires that each tutor spend no more than eight hours per week tutoring.

Write a system of three inequalities that describes these constraints.

c. Graph the system of inequalities in part (b). Use only the first quadrant and its boundary, because x and y are nonnegative.

d. Evaluate the objective function for total weekly earnings at each of the four vertices of the graphed region. [The vertices should occur at $(3, 0)$, $(8, 0)$, $(3, 17)$, and $(8, 12)$.]

e. Complete the missing portions of this statement: The student can earn the maximum amount per week by tutoring for _____ hours per week and working as a teacher's aid for _____ hours per week. The maximum amount that the student can earn each week is $ _____.

Use the two steps for solving a linear programming problem, given in the box on page 769, to solve the problems in Exercises 17–23.

17. A manufacturer produces two models of mountain bicycles. The times (in hours) required for assembling and painting each model are given in the following table:

	Model A	Model B
Assembling	5	4
Painting	2	3

The maximum total weekly hours available in the assembly department and the paint department are 200 hours and 108 hours, respectively. The profits per unit are $25 for model A and $15 for model B. How many of each type should be produced to maximize profit?

18. A large institution is preparing lunch menus containing foods A and B. The specifications for the two foods are given in the following table:

Food	Units of Fat per Ounce	Units of Carbohydrates per Ounce	Units of Protein per Ounce
A	1	2	1
B	1	1	1

Each lunch must provide at least 6 units of fat per serving, no more than 7 units of protein, and at least 10 units of carbohydrates. The institution can purchase food A for $0.12 per ounce and food B for $0.08 per ounce. How many ounces of each food should a serving contain to meet the dietary requirements at the least cost?

19. Food and clothing are shipped to victims of a natural disaster. Each carton of food will feed 12 people, while each carton of clothing will help 5 people. Each 20-cubic-foot box of food weighs 50 pounds and each 10-cubic-foot box of clothing weighs 20 pounds. The commercial carriers transporting food and clothing are bound by the following constraints:

- The total weight per carrier cannot exceed 19,000 pounds.
- The total volume must be less than 8000 cubic feet.

How many cartons of food and clothing should be sent with each plane shipment to maximize the number of people who can be helped?

20. On June 24, 1948, the former Soviet Union blocked all land and water routes through East Germany to Berlin. A gigantic airlift was organized using American and British planes to supply food, clothing, and other supplies to the more than 2 million people in West Berlin. The cargo capacity was 30,000 cubic feet for an American plane and 20,000 cubic feet for a British plane. To break the Soviet blockade, the Western Allies had to maximize cargo capacity, but were subject to the following restrictions:

- No more than 44 planes could be used.
- The larger American planes required 16 personnel per flight, double that of the requirement for the British planes. The total number of personnel available could not exceed 512.

- The cost of an American flight was $9000 and the cost of a British flight was $5000. Total weekly costs could not exceed $300,000.

Find the number of American and British planes that were used to maximize cargo capacity.

21. A theater is presenting a program on drinking and driving for students and their parents. The proceeds will be donated to a local alcohol information center. Admission is $2.00 for parents and $1.00 for students. However, the situation has two constraints: The theater can hold no more than 150 people and every two parents must bring at least one student. How many parents and students should attend to raise the maximum amount of money?

22. You are about to take a test that contains computation problems worth 6 points each and word problems worth 10 points each. You can do a computation problem in 2 minutes and a word problem in 4 minutes. You have 40 minutes to take the test and may answer no more than 12 problems. Assuming you answer all the problems attempted correctly, how many of each type of problem must you answer to maximize your score? What is the maximum score?

23. In 1978, a ruling by the Civil Aeronautics Board allowed Federal Express to purchase larger aircraft. Federal Express's options included 20 Boeing 727s that United Airlines was retiring and/or the French-built Dassault Fanjet Falcon 20. To aid in their decision, executives at Federal Express analyzed the following data:

	Boeing 727	Falcon 20
Direct Operating Cost	$1400 per hour	$500 per hour
Payload	42,000 pounds	6000 pounds

Federal Express was faced with the following constraints:

- Hourly operating cost was limited to $35,000.
- Total payload had to be at least 672,000 pounds.
- Only twenty 727s were available.

Given the constraints, how many of each kind of aircraft should Federal Express have purchased to maximize the number of aircraft?

Writing in Mathematics

24. What kinds of problems are solved using the linear programming method?

25. What is an objective function in a linear programming problem?

26. What is a constraint in a linear programming problem? How is a constraint represented?

27. In your own words, describe how to solve a linear programming problem.

28. Describe a situation in your life in which you would really like to maximize something, but you are limited by at least two constraints. Can linear programming be used in this situation? Explain your answer.

Critical Thinking Exercises

29. Suppose that you inherit $10,000. The will states how you must invest the money. Some (or all) of the money must be invested in stocks and bonds. The requirements are that at least $3000 be invested in bonds, with expected returns of $0.08 per dollar, and at least $2000 be invested in stocks, with expected returns of $0.12 per dollar. Because the stocks are medium risk, the final stipulation requires that the investment in bonds should never be less than the investment in stocks. How should the money be invested so as to maximize your expected returns?

30. Consider the objective function $z = Ax + By$ ($A > 0$ and $B > 0$) subject to the following constraints: $2x + 3y \le 9$, $x - y \le 2$, $x \ge 0$, and $y \ge 0$. Prove that the objective function will have the same maximum value at the vertices $(3, 1)$ and $(0, 3)$ if $A = \frac{2}{3}B$.

Group Exercises

31. Group members should choose a particular field of interest. Research how linear programming is used to solve problems in that field. If possible, investigate the solution of a specific practical problem. Present a report on your findings, including the contributions of George Dantzig, Narendra Karmarkar, and L. G. Khachion to linear programming.

32. Members of the group should interview a business executive who is in charge of deciding the product mix for a business. How are production policy decisions made? Are other methods used in conjunction with linear programming? What are these methods? What sort of academic background, particularly in mathematics, does this executive have? Present a group report addressing these questions, emphasizing the role of linear programming for the business.

Chapter 7
Summary, Review, and Test

Summary

DEFINITIONS AND CONCEPTS	EXAMPLES

7.1 Systems of Linear Equations in Two Variables

a. Two equations in the form $Ax + By = C$ are called a system of linear equations. A solution of the system is an ordered pair that satisfies both equations in the system.

Ex. 1, p. 710

b. Systems of linear equations in two variables can be solved by eliminating a variable, using the substitution method (see the box on page 712) or the addition method (see the box on page 713).

Ex. 2, p. 712;
Ex. 3, p. 714;
Ex. 4, p. 715

c. Some linear systems have no solution and are called inconsistent systems; others have infinitely many solutions. The equations in a linear system with infinitely many solutions are called dependent. For details, see the box on page 716.

Ex. 5, p. 716;
Ex. 6, p. 717

d. Functions of Business

Ex. 9, p. 721;
Figure 7.8, p. 723

Revenue Function

$$R(x) = (\text{price per unit sold})x$$

Cost Function

$$C(x) = \text{fixed cost} + (\text{cost per unit produced})x$$

Profit Function

$$P(x) = R(x) - C(x)$$

The point of intersection of the graphs of R and C is the break-even point. The x-coordinate of the point reveals the number of units that a company must produce and sell so that the money coming in, the revenue, is equal to the money going out, the cost. The y-coordinate gives the amount of money coming in and going out.

DEFINITIONS AND CONCEPTS	EXAMPLES

7.2 Systems of Linear Equations in Three Variables

a. Three equations in the form $Ax + By + Cz = D$ are called a system of linear equations in three variables. A solution of the system is an ordered triple that satisfies all three equations in the system. — Ex. 1, p. 729

b. A system of linear equations in three variables can be solved by eliminating variables. Use the addition method to eliminate any variable, reducing the system to two equations in two variables. Use substitution or the addition method to solve the resulting system in two variables. Details are found in the box on page 730. — Ex. 2, p. 730; Ex. 3, p. 732

c. Three points that do not lie on a line determine the graph of a quadratic function $y = ax^2 + bx + c$. Use the three given points to create a system of three equations. Solve the system to find a, b, and c. — Ex 4, p. 733

7.3 Partial Fractions

a. Partial fraction decomposition is used on rational expressions in which the numerator and denominator have no common factors and the highest power in the numerator is less than the highest power in the denominator. The steps in partial fraction decomposition are given in the box on page 739.

b. Include one partial fraction with a constant numerator for each distinct linear factor in the denominator. Include one partial fraction with a constant numerator for each power of a repeated linear factor in the denominator. — Ex. 1, p. 738; Ex. 2, p. 740

c. Include one partial fraction with a linear numerator for each distinct prime quadratic factor in the denominator. Include one partial fraction with a linear numerator for each power of a prime, repeated quadratic factor in the denominator. — Ex. 3, p. 742; Ex. 4, p. 744

7.4 Systems of Nonlinear Equations in Two Variables

a. A system of two nonlinear equations in two variables contains at least one equation that cannot be expressed as $Ax + By = C$.

b. Systems of nonlinear equations in two variables can be solved algebraically by eliminating all occurrences of one of the variables by the substitution or addition methods. — Ex. 1, p. 747; Ex. 2, p. 748; Ex. 3, p. 749; Ex. 4, p. 750

7.5 Systems of Inequalities

a. A linear inequality in two variables can be written in the form $Ax + By > C$, $Ax + By \geq C$, $Ax + By < C$, or $Ax + By \leq C$.

b. The procedure for graphing a linear inequality in two variables is given in the box on page 756. A nonlinear inequality in two variables is graphed using the same procedure. — Ex. 1, p. 756; Ex. 2, p. 757; Ex. 3, p. 758; Ex. 4, p. 759

c. To graph the solution set of a system of inequalities, graph each inequality in the system in the same rectangular coordinate system. Then find the region, if there is one, that is common to every graph in the system. — Ex. 5, p. 760; Ex. 6, p. 761; Ex. 7, p. 762

7.6 Linear Programming

a. An objective function is an algebraic expression in three variables describing a quantity that must be maximized or minimized. — Ex. 1, p. 767

b. Constraints are restrictions, expressed as linear inequalities. — Ex. 2, p. 768; Ex. 3, p. 769

c. Steps for solving a linear programming problem are given in the box on page 769. — Ex. 4, p. 769; Ex. 5, p. 770

Perform matrix row operations.

A matrix with 1s down the main diagonal and 0s below the 1s is said to be in **row-echelon form**. How do we produce a matrix in this form? We use **row operations** on the augmented matrix. These row operations are just like what you did when solving a linear system by the addition method. The difference is that we no longer write the variables, usually represented by x, y, and z.

Matrix Row Operations

The following row operations produce matrices that represent systems with the same solution set:

1. Two rows of a matrix may be interchanged. This is the same as interchanging two equations in a linear system.

2. The elements in any row may be multiplied by a nonzero number. This is the same as multiplying both sides of an equation by a nonzero number.

3. The elements in any row may be multiplied by a nonzero number, and these products may be added to the corresponding elements in any other row. This is the same as multiplying both sides of an equation by a nonzero number and then adding equations to eliminate a variable.

Two matrices are **row equivalent** if one can be obtained from the other by a sequence of row operations.

Each matrix row operation in the preceding box can be expressed symbolically as follows:

1. Interchange the elements in the ith and jth rows: $R_i \leftrightarrow R_j$.

2. Multiply each element in the ith row by k: kR_i.

3. Add k times the elements in row i to the corresponding elements in row j: $kR_i + R_j$.

EXAMPLE 2   Performing Matrix Row Operations

Use the matrix

$$\begin{bmatrix} 3 & 18 & -12 & | & 21 \\ 1 & 2 & -3 & | & 5 \\ -2 & -3 & 4 & | & -6 \end{bmatrix}$$

and perform each indicated row operation:

a. $R_1 \leftrightarrow R_2$ **b.** $\frac{1}{3}R_1$ **c.** $2R_2 + R_3$.

Solution

a. The notation $R_1 \leftrightarrow R_2$ means to interchange the elements in row 1 and row 2. This results in the row-equivalent matrix

$$\begin{bmatrix} 1 & 2 & -3 & | & 5 \\ 3 & 18 & -12 & | & 21 \\ -2 & -3 & 4 & | & -6 \end{bmatrix}.$$

This was row 2; now it's row 1.
This was row 1; now it's row 2.

b. The notation $\frac{1}{3}R_1$ means to multiply each element in row 1 by $\frac{1}{3}$. This results in the row-equivalent matrix

$$\begin{bmatrix} \frac{1}{3}(3) & \frac{1}{3}(18) & \frac{1}{3}(-12) & | & \frac{1}{3}(21) \\ 1 & 2 & -3 & | & 5 \\ -2 & -3 & 4 & | & -6 \end{bmatrix} = \begin{bmatrix} 1 & 6 & -4 & | & 7 \\ 1 & 2 & -3 & | & 5 \\ -2 & -3 & 4 & | & -6 \end{bmatrix}.$$

c. The notation $2R_2 + R_3$ means to add 2 times the elements in row 2 to the corresponding elements in row 3. Replace the elements in row 3 by these sums. First, we find 2 times the elements in row 2, namely, 1, 2, −3 and 5:

$$2(1) \text{ or } 2, \qquad 2(2) \text{ or } 4, \qquad 2(-3) \text{ or } -6, \qquad 2(5) \text{ or } 10.$$

Now we add these products to the corresponding elements in row 3. Although we use row 2 to find the products, row 2 does not change. It is the elements in row 3 that change, resulting in the row-equivalent matrix

Replace row 3 by the
sum of itself and
2 times row 2.

$$\begin{bmatrix} 3 & 18 & -12 & 21 \\ 1 & 2 & -3 & 5 \\ -2+2=0 & -3+4=1 & 4+(-6)=-2 & -6+10=4 \end{bmatrix} = \begin{bmatrix} 3 & 18 & -12 & 21 \\ 1 & 2 & -3 & 5 \\ 0 & 1 & -2 & 4 \end{bmatrix}.$$

Check Point 2 Use the matrix

$$\begin{bmatrix} 4 & 12 & -20 & 8 \\ 1 & 6 & -3 & 7 \\ -3 & -2 & 1 & -9 \end{bmatrix}$$

and perform each indicated row operation:

a. $R_1 \leftrightarrow R_2$ **b.** $\frac{1}{4}R_1$ **c.** $3R_2 + R_3$.

③ Use matrices and Gaussian elimination to solve systems.

The process that we use to solve linear systems using matrix row operations is called **Gaussian elimination**, after the German mathematician Carl Friedrich Gauss (1777–1855). Here are the steps used in Gaussian elimination:

Solving Linear Systems Using Gaussian Elimination

1. Write the augmented matrix for the system.
2. Use matrix row operations to simplify the matrix to a row-equivalent matrix in row-echelon form, with 1s down the main diagonal from upper left to lower right, and 0s below the 1s.

$$\begin{bmatrix} 1 & * & * & * \\ * & * & * & * \\ * & * & * & * \end{bmatrix} \rightarrow \begin{bmatrix} 1 & * & * & * \\ 0 & * & * & * \\ 0 & * & * & * \end{bmatrix} \rightarrow \begin{bmatrix} 1 & * & * & * \\ 0 & 1 & * & * \\ 0 & * & * & * \end{bmatrix} \rightarrow \begin{bmatrix} 1 & * & * & * \\ 0 & 1 & * & * \\ 0 & 0 & * & * \end{bmatrix} \rightarrow \begin{bmatrix} 1 & * & * & * \\ 0 & 1 & * & * \\ 0 & 0 & 1 & * \end{bmatrix}$$

Get 1 in the upper left-hand corner. | Use the 1 in the first column to get 0s below it. | Get 1 in the second row, second column position. | Use the 1 in the second column to get 0 below it. | Get 1 in the third row, third column position.

3. Write the system of linear equations corresponding to the matrix in step 2 and use back-substitution to find the system's solution.

EXAMPLE 3   Gaussian Elimination with Back-Substitution

Use matrices to solve the system:

$$\begin{aligned} 3x + y + 2z &= 31 \\ x + y + 2z &= 19 \\ x + 3y + 2z &= 25. \end{aligned}$$

Solution

Step 1 Write the augmented matrix for the system.

Linear System	**Augmented Matrix**
$\begin{aligned} 3x + y + 2z &= 31 \\ x + y + 2z &= 19 \\ x + 3y + 2z &= 25 \end{aligned}$	$\begin{bmatrix} 3 & 1 & 2 & 31 \\ 1 & 1 & 2 & 19 \\ 1 & 3 & 2 & 25 \end{bmatrix}$

Step 2   Use matrix row operations to simplify the matrix to row-echelon form
with 1s down the main diagonal from upper left to lower right, and 0s below the 1s.
Our first step in achieving this goal is to get 1 in the top position of the first column.

We want 1 in
this position.
$$\left[\begin{array}{ccc|c} 3 & 1 & 2 & 31 \\ 1 & 1 & 2 & 19 \\ 1 & 3 & 2 & 25 \end{array}\right]$$

To get 1 in this position, we interchange row 1 and row 2: $R_1 \leftrightarrow R_2$. (We could also
interchange row 1 and row 3 to attain our goal.)

$$\left[\begin{array}{ccc|c} 1 & 1 & 2 & 19 \\ 3 & 1 & 2 & 31 \\ 1 & 3 & 2 & 25 \end{array}\right]$$

This was row 2; now it's row 1.

This was row 1; now it's row 2.

Now we want to get 0s below the 1 in the first column.

We want 0 in
these positions.
$$\left[\begin{array}{ccc|c} 1 & 1 & 2 & 19 \\ 3 & 1 & 2 & 31 \\ 1 & 3 & 2 & 25 \end{array}\right]$$

To get a 0 where there is now a 3, multiply the top row of numbers by -3 and add
these products to the second row of numbers: $-3R_1 + R_2$. To get a 0 where there is
now a 1, multiply the top row of numbers by -1 and add these products to the third
row of numbers: $-1R_1 + R_3$. Although we are using row 1 to find the products, the
numbers in row 1 do not change.

Replace row 2 by
$-3R_1 + R_2$.

Replace row 3 by
$-1R_1 + R_3$.
$$\left[\begin{array}{ccc|c} 1 & 1 & 2 & 19 \\ -3(1)+3 & -3(1)+1 & -3(2)+2 & -3(19)+31 \\ -1(1)+1 & -1(1)+3 & -1(2)+2 & -1(19)+25 \end{array}\right] = \left[\begin{array}{ccc|c} 1 & 1 & 2 & 19 \\ 0 & -2 & -4 & -26 \\ 0 & 2 & 0 & 6 \end{array}\right]$$

We want 1 in this position.

We move on to the second column. To get 1 in the desired position, we multiply $-\frac{1}{2}$
by its reciprocal, $-\frac{1}{2}$. Therefore, we multiply all the numbers in the second row by
$-\frac{1}{2}$: $-\frac{1}{2}R_2$.

$-\frac{1}{2}R_2$
$$\left[\begin{array}{ccc|c} 1 & 1 & 2 & 19 \\ -\frac{1}{2}(0) & -\frac{1}{2}(-2) & -\frac{1}{2}(-4) & -\frac{1}{2}(-26) \\ 0 & 2 & 0 & 6 \end{array}\right] = \left[\begin{array}{ccc|c} 1 & 1 & 2 & 19 \\ 0 & 1 & 2 & 13 \\ 0 & 2 & 0 & 6 \end{array}\right].$$

We want 0 in this position.

We are not yet done with the second column. The voice balloon shows that we want
to get a 0 where there is now a 2. If we multiply the second row of numbers by -2
and add these products to the third row of numbers, we will get 0 in this position:
$-2R_2 + R_3$. Although we are using the numbers in row 2 to find the products, the
numbers in row 2 do not change.

Replace row 3 by
$-2R_2 + R_3$.
$$\left[\begin{array}{ccc|c} 1 & 1 & 2 & 19 \\ 0 & 1 & 2 & 13 \\ -2(0)+0 & -2(1)+2 & -2(2)+0 & -2(13)+6 \end{array}\right] = \left[\begin{array}{ccc|c} 1 & 1 & 2 & 19 \\ 0 & 1 & 2 & 13 \\ 0 & 0 & -4 & -20 \end{array}\right]$$

We want 1 in this position.

We move on to the third column. To get 1 in the desired position, we multiply -4 by
its reciprocal, $-\frac{1}{4}$. Therefore, we multiply all the numbers in the third row by
$-\frac{1}{4}$: $-\frac{1}{4}R_3$.

$-\frac{1}{4}R_3$
$$\left[\begin{array}{ccc|c} 1 & 1 & 2 & 19 \\ 0 & 1 & 2 & 13 \\ -\frac{1}{4}(0) & -\frac{1}{4}(0) & -\frac{1}{4}(-4) & -\frac{1}{4}(-20) \end{array}\right] = \left[\begin{array}{ccc|c} 1 & 1 & 2 & 19 \\ 0 & 1 & 2 & 13 \\ 0 & 0 & 1 & 5 \end{array}\right].$$

We now have the desired matrix in row-echelon form, with 1s down the main diagonal and 0s below the 1s.

Step 3   Write the system of linear equations corresponding to the matrix in step 2 and use back-substitution to find the system's solution. The system represented by the matrix in step 2 is

$$\begin{bmatrix} 1 & 1 & 2 & | & 19 \\ 0 & 1 & 2 & | & 13 \\ 0 & 0 & 1 & | & 5 \end{bmatrix} \rightarrow \begin{array}{l} 1x + 1y + 2z = 19 \\ 0x + 1y + 2z = 13 \\ 0x + 0y + 1z = 5 \end{array} \quad \text{or} \quad \begin{array}{rl} x + y + 2z = 19 & (1) \\ y + 2z = 13. & (2) \\ z = 5 & (3) \end{array}$$

We immediately see from equation (3) that the value for z is 5. To find y, we back-substitute 5 for z in the second equation.

$$y + 2z = 13 \qquad \text{Equation (2)}$$
$$y + 2(5) = 13 \qquad \text{Substitute 5 for z.}$$
$$y = 3 \qquad \text{Solve for y.}$$

Finally, back-substitute 3 for y and 5 for z in the first equation.

$$x + y + 2z = 19 \qquad \text{Equation (1)}$$
$$x + 3 + 2(5) = 19 \qquad \text{Substitute 3 for y and 5 for z.}$$
$$x + 13 = 19 \qquad \text{Multiply and add.}$$
$$x = 6 \qquad \text{Subtract 13 from both sides.}$$

The solution set of the original system is $\{(6, 3, 5)\}$. Check to see that the solution satisfies all three equations in the given system.

Check Point 3   Use matrices to solve the system:

$$\begin{array}{rl} 2x + y + 2z &= 18 \\ x - y + 2z &= 9 \\ x + 2y - z &= 6. \end{array}$$

Modern supercomputers are capable of solving systems with more than 600,000 variables. The augmented matrices for such systems are huge, but the solution using matrices is exactly like what we did in Example 3. Work with the augmented matrix, one column at a time. First, get 1 in the desired position. Then get 0s below the 1. Let's see how this works for a linear system involving four equations in four variables.

EXAMPLE 4   Gaussian Elimination with Back-Substitution

Use matrices to solve the system:

$$\begin{array}{rl} 2w + x + 3y - z &= 6 \\ w - x + 2y - 2z &= -1 \\ w - x - y + z &= -4 \\ -w + 2x - 2y - z &= -7. \end{array}$$

Solution

Step 1   Write the augmented matrix for the system.

Linear System	Augmented Matrix

$$\begin{array}{rl} 2w + x + 3y - z &= 6 \\ w - x + 2y - 2z &= -1 \\ w - x - y + z &= -4 \\ -w + 2x - 2y - z &= -7 \end{array} \qquad \begin{bmatrix} 2 & 1 & 3 & -1 & | & 6 \\ 1 & -1 & 2 & -2 & | & -1 \\ 1 & -1 & -1 & 1 & | & -4 \\ -1 & 2 & -2 & -1 & | & -7 \end{bmatrix}$$

Step 2 **Use matrix row operations to simplify the matrix to row-echelon form, wit▮ 1s down the diagonal from upper left to lower right, and 0s below the 1s.** Workin▮ one column at a time, we must obtain 1 in the diagonal position. Then we use this 1 t▮ get 0s below it. Thus, our first step in achieving this goal is to get 1 in the top positio▮ of the first column. To do this, we interchange row 1 and row 2: $R_1 \leftrightarrow R_2$.

$$\begin{bmatrix} 1 & -1 & 2 & -2 & | & -1 \\ 2 & 1 & 3 & -1 & | & 6 \\ 1 & -1 & -1 & 1 & | & -4 \\ -1 & 2 & -2 & -1 & | & -7 \end{bmatrix}$$

We want 0s in these positions.

This was row 2; now it's row 1.

This was row 1; now it's row 2.

Now we use the 1 at the top of the first column to get 0s below it.

Use the previous matrix and:
Replace row 2 by $-2R_1 + R_2$.
Replace row 3 by $-1R_1 + R_3$.
Replace row 4 by $1R_1 + R_4$.

$$\begin{bmatrix} 1 & -1 & 2 & -2 & | & -1 \\ 0 & 3 & -1 & 3 & | & 8 \\ 0 & 0 & -3 & 3 & | & -3 \\ 0 & 1 & 0 & -3 & | & -8 \end{bmatrix}$$

We want 1 in this position.

We move on to the second column. We can obtain 1 in the desired position b▮ multiplying the numbers in the second row by $\frac{1}{3}$, the reciprocal of 3.

$$\begin{bmatrix} 1 & -1 & 2 & -2 & | & -1 \\ \frac{1}{3}(0) & \frac{1}{3}(3) & \frac{1}{3}(-1) & \frac{1}{3}(3) & | & \frac{1}{3}(8) \\ 0 & 0 & -3 & 3 & | & -3 \\ 0 & 1 & 0 & -3 & | & -8 \end{bmatrix} = \begin{bmatrix} 1 & -1 & 2 & -2 & | & -1 \\ 0 & 1 & -\frac{1}{3} & 1 & | & \frac{8}{3} \\ 0 & 0 & -3 & 3 & | & -3 \\ 0 & 1 & 0 & -3 & | & -8 \end{bmatrix} \quad \frac{1}{3}R_2$$

We want 0s in these positions.
The top position already has a 0.

Now we use the 1 in the second row, second column position to get 0s below it.

Replace row 4 in the previous matrix by $-1R_2 + R_4$.

$$\begin{bmatrix} 1 & -1 & 2 & -2 & | & -1 \\ 0 & 1 & -\frac{1}{3} & 1 & | & \frac{8}{3} \\ 0 & 0 & -3 & 3 & | & -3 \\ 0 & 0 & \frac{1}{3} & -4 & | & -\frac{32}{3} \end{bmatrix}$$

We want 1 in this position.

We move on to the third column. We can obtain 1 in the desired position by mult▮ plying the numbers in the third row by $-\frac{1}{3}$, the reciprocal of -3.

$$\begin{bmatrix} 1 & -1 & 2 & -2 & | & -1 \\ 0 & 1 & -\frac{1}{3} & 1 & | & \frac{8}{3} \\ -\frac{1}{3}(0) & -\frac{1}{3}(0) & -\frac{1}{3}(-3) & -\frac{1}{3}(3) & | & -\frac{1}{3}(-3) \\ 0 & 0 & \frac{1}{3} & -4 & | & -\frac{32}{3} \end{bmatrix} = \begin{bmatrix} 1 & -1 & 2 & -2 & | & -1 \\ 0 & 1 & -\frac{1}{3} & 1 & | & \frac{8}{3} \\ 0 & 0 & 1 & -1 & | & 1 \\ 0 & 0 & \frac{1}{3} & -4 & | & -\frac{32}{3} \end{bmatrix} \quad -\frac{1}{3}R_3$$

We want 0 in this position.

Now we use the 1 in the third column to get 0 below it.

Replace row 4 in the previous matrix by $-\frac{1}{3}R_3 + R_4$.

$$\begin{bmatrix} 1 & -1 & 2 & -2 & | & -1 \\ 0 & 1 & -\frac{1}{3} & 1 & | & \frac{8}{3} \\ 0 & 0 & 1 & -1 & | & 1 \\ 0 & 0 & 0 & -\frac{11}{3} & | & -11 \end{bmatrix}$$

We want 1 in this position.

We move on to the fourth column. Because we want 1s down the diagonal from upper left to lower right, we want 1 where there is now $-\frac{11}{3}$. We can obtain 1 in this position by multiplying the numbers in the fourth row by $-\frac{3}{11}$.

$$\begin{bmatrix} 1 & -1 & 2 & -2 & | & -1 \\ 0 & 1 & -\frac{1}{3} & 1 & | & \frac{8}{3} \\ 0 & 0 & 1 & -1 & | & 1 \\ -\frac{3}{11}(0) & -\frac{3}{11}(0) & -\frac{3}{11}(0) & -\frac{3}{11}\left(-\frac{11}{3}\right) & | & -\frac{3}{11}(-11) \end{bmatrix}$$

$$= \begin{bmatrix} 1 & -1 & 2 & -2 & | & -1 \\ 0 & 1 & -\frac{1}{3} & 1 & | & \frac{8}{3} \\ 0 & 0 & 1 & -1 & | & 1 \\ 0 & 0 & 0 & 1 & | & 3 \end{bmatrix} \quad -\frac{3}{11}R_4$$

We now have the desired matrix in row-echelon form, with 1s down the main diagonal and 0s below the 1s. An equivalent row-echelon matrix can be obtained using a graphing utility and the $\boxed{\text{REF}}$ command on the augmented matrix.

Step 3 Write the system of linear equations corresponding to the matrix in step 2 and use back-substitution to find the system's solution. The system represented by the matrix in step 2 is

$$\begin{bmatrix} 1 & -1 & 2 & -2 & | & -1 \\ 0 & 1 & -\frac{1}{3} & 1 & | & \frac{8}{3} \\ 0 & 0 & 1 & -1 & | & 1 \\ 0 & 0 & 0 & 1 & | & 3 \end{bmatrix} \rightarrow \begin{array}{l} 1w - 1x + 2y - 2z = -1 \\ 0w + 1x - \frac{1}{3}y + 1z = \frac{8}{3} \\ 0w + 0x + 1y - 1z = 1 \\ 0w + 0x + 0y + 1z = 3 \end{array} \quad \text{or} \quad \begin{array}{r} w - x + 2y - 2z = -1 \\ x - \frac{1}{3}y + z = \frac{8}{3} \\ y - z = 1 \\ z = 3. \end{array}$$

We immediately see that the value for z is 3. We can now use back-substitution to find the values for y, x, and w.

These are the four equations from the last column.

$$z = 3 \quad\bigg|\quad \begin{array}{l} y - z = 1 \\ \\ y - 3 = 1 \\ \\ y = 4 \end{array} \quad\bigg|\quad \begin{array}{l} x - \frac{1}{3}y + z = \frac{8}{3} \\ \\ x - \frac{1}{3}(4) + 3 = \frac{8}{3} \\ \\ x + \frac{5}{3} = \frac{8}{3} \\ \\ x = 1 \end{array} \quad\bigg|\quad \begin{array}{l} w - x + 2y - 2z = -1 \\ \\ w - 1 + 2(4) - 2(3) = -1 \\ \\ w - 1 + 8 - 6 = -1 \\ \\ w + 1 = -1 \\ \\ w = -2 \end{array}$$

Let's agree to write the solution set for the system in the alphabetical order in which the variables for the given system appeared from left to right, namely (w, x, y, z). Thus, the solution set is $\{(-2, 1, 4, 3)\}$. We can verify this solution set by substituting the value for each variable into the original system of equations and obtaining four true statements.

Check Point **4**   Use matrices to solve the system:

$$\begin{array}{r} w - 3x - 2y + z = -3 \\ 2w - 7x - y + 2z = 1 \\ 3w - 7x - 3y + 3z = -5 \\ 5w + x + 4y - 2z = 18. \end{array}$$

4 Use matrices and Gauss-Jordan elimination to solve systems.

Gauss-Jordan Elimination

Using Gaussian elimination, we obtain a matrix in row-echelon form, with 1s down the main diagonal and 0s below the 1s. A second method, called **Gauss-Jordan elimination**, after Carl Friedrich Gauss and Wilhelm Jordan (1842–1899), continue the process until a matrix with 1s down the main diagonal and 0s in every position *above and below* each 1 is found. Such a matrix is said to be in **reduced row-echelon form**. For a system of linear equations in three variables, x, y, and z, we must get the augmented matrix into the form

$$\begin{bmatrix} 1 & 0 & 0 & | & a \\ 0 & 1 & 0 & | & b \\ 0 & 0 & 1 & | & c \end{bmatrix}.$$

Based on this matrix, we conclude that $x = a$, $y = b$, and $z = c$.

Solving Linear Systems Using Gauss-Jordan Elimination

1. Write the augmented matrix for the system.
2. Use matrix row operations to simplify the matrix to a row-equivalent matrix in reduced row-echelon form, with 1s down the main diagonal from upper left to lower right, and 0s above and below the 1s.
   a. Get 1 in the upper left-hand corner.
   b. Use the 1 in the first column to get 0s below it.
   c. Get 1 in the second row, second column.
   d. Use the 1 in the second column to make the remaining entries in the second column 0.
   e. Get 1 in the third row, third column.
   f. Use the 1 in the third column to make the remaining entries in the third column 0.
   g. Continue this procedure as far as possible.
3. Use the reduced row-echelon form of the matrix in step 2 to write the system's solution set. (Back-substitution is not necessary.)

Technology

Most graphing utilities can convert a matrix to reduced row-echelon form. Enter the system's augmented matrix and name it A. Then use the $\boxed{\text{RREF}}$ (reduced row-echelon form) command on matrix A.

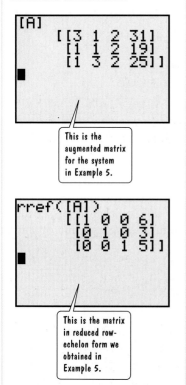

This is the augmented matrix for the system in Example 5.

This is the matrix in reduced row-echelon form we obtained in Example 5.

EXAMPLE 5   Using Gauss-Jordan Elimination

Use Gauss-Jordan elimination to solve the system:

$$\begin{aligned} 3x + y + 2z &= 31 \\ x + y + 2z &= 19 \\ x + 3y + 2z &= 25. \end{aligned}$$

Solution   In Example 3, we used Gaussian elimination to obtain the following matrix

We want 0s in these positions.

$$\begin{bmatrix} 1 & 1 & 2 & | & 19 \\ 0 & 1 & 2 & | & 13 \\ 0 & 0 & 1 & | & 5 \end{bmatrix}.$$

To use Gauss-Jordan elimination, we need 0s both below and above the 1s in the main diagonal. We use the 1 in the second row, second column to get a 0 above it.

Replace row 1 in the previous matrix by $-1R_2 + R_1$.

$$\begin{bmatrix} 1 & 0 & 0 & | & 6 \\ 0 & 1 & 2 & | & 13 \\ 0 & 0 & 1 & | & 5 \end{bmatrix}$$

We want 0s in these positions.

We use the 1 in the third column to get 0s above it.

$$\begin{bmatrix} 1 & 0 & 0 & | & 6 \\ 0 & 1 & 0 & | & 3 \\ 0 & 0 & 1 & | & 5 \end{bmatrix}$$

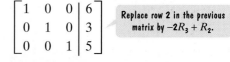

Replace row 2 in the previous matrix by $-2R_3 + R_2$.

This last matrix corresponds to

$$x = 6, \quad y = 3, \quad z = 5.$$

As we found in Example 3, the solution set is $\{(6, 3, 5)\}$.

Check Point 5 Solve the system in Check Point 3 using Gauss-Jordan elimination. Begin by working with the matrix that you obtained in Check Point 3.

EXERCISE SET 8.1

Practice Exercises

In Exercises 1–8, write the augmented matrix for each system of linear equations.

1. $2x + y + 2z = 2$
$3x - 5y - z = 4$
$x - 2y - 3z = -6$

2. $3x - 2y + 5z = 31$
$x + 3y - 3z = -12$
$-2x - 5y + 3z = 11$

3. $x - y + z = 8$
$y - 12z = -15$
$z = 1$

4. $x - 2y + 3z = 9$
$y + 3z = 5$
$z = 2$

5. $5x - 2y - 3z = 0$
$x + y = 5$
$2x - 3z = 4$

6. $x - 2y + z = 10$
$3x + y = 5$
$7x + 2z = 2$

7. $2w + 5x - 3y + z = 2$
$3x + y = 4$
$w - x + 5y = 9$
$5w - 5x - 2y = 1$

8. $4w + 7x - 8y + z = 3$
$5x + y = 5$
$w - x - y = 17$
$2w - 2x + 11y = 4$

In Exercises 9–12, write the system of linear equations represented by the augmented matrix. Use x, y, z, and, if necessary, w, x, y, and z, for the variables.

9. $\begin{bmatrix} 5 & 0 & 3 & | & -11 \\ 0 & 1 & -4 & | & 12 \\ 7 & 2 & 0 & | & 3 \end{bmatrix}$

10. $\begin{bmatrix} 7 & 0 & 4 & | & -13 \\ 0 & 1 & -5 & | & 11 \\ 2 & 7 & 0 & | & 6 \end{bmatrix}$

11. $\begin{bmatrix} 1 & 1 & 4 & 1 & | & 3 \\ -1 & 1 & -1 & 0 & | & 7 \\ 2 & 0 & 0 & 5 & | & 11 \\ 0 & 0 & 12 & 4 & | & 5 \end{bmatrix}$

12. $\begin{bmatrix} 4 & 1 & 5 & 1 & | & 6 \\ 1 & -1 & 0 & -1 & | & 8 \\ 3 & 0 & 0 & 7 & | & 4 \\ 0 & 0 & 11 & 5 & | & 3 \end{bmatrix}$

In Exercises 13–18, write the system of linear equations represented by the augmented matrix. Use x, y, z, and, if necessary, w, x, y, and z, for the variables. Once the system is written, use back-substitution to find its solution.

13. $\begin{bmatrix} 1 & 0 & -4 & | & 5 \\ 0 & 1 & -12 & | & 13 \\ 0 & 0 & 1 & | & -\frac{1}{2} \end{bmatrix}$

14. $\begin{bmatrix} 1 & 2 & 1 & | & 0 \\ 0 & 1 & 0 & | & -2 \\ 0 & 0 & 1 & | & 3 \end{bmatrix}$

15. $\begin{bmatrix} 1 & \frac{1}{2} & 1 & | & \frac{11}{2} \\ 0 & 1 & \frac{3}{2} & | & 7 \\ 0 & 0 & 1 & | & 4 \end{bmatrix}$

16. $\begin{bmatrix} 1 & 1 & 0 & | & 3 \\ 0 & 1 & \frac{3}{2} & | & -2 \\ 0 & 0 & 1 & | & 0 \end{bmatrix}$

17. $\begin{bmatrix} 1 & -1 & 1 & 1 & | & 3 \\ 0 & 1 & -2 & -1 & | & 0 \\ 0 & 0 & 1 & 6 & | & 17 \\ 0 & 0 & 0 & 1 & | & 3 \end{bmatrix}$

18. $\begin{bmatrix} 1 & 2 & -1 & 0 & | & 2 \\ 0 & 1 & 1 & -2 & | & -3 \\ 0 & 0 & 1 & -1 & | & -2 \\ 0 & 0 & 0 & 1 & | & 3 \end{bmatrix}$

In Exercises 19–24, perform each matrix row operation and write the new matrix.

19. $\begin{bmatrix} 2 & -6 & 4 & | & 10 \\ 1 & 5 & -5 & | & 0 \\ 3 & 0 & 4 & | & 7 \end{bmatrix}$ $\frac{1}{2}R_1$

20. $\begin{bmatrix} 3 & -12 & 6 & | & 9 \\ 1 & -4 & 4 & | & 0 \\ 2 & 0 & 7 & | & 4 \end{bmatrix}$ $\frac{1}{3}R_1$

21. $\begin{bmatrix} 1 & -3 & 2 & | & 0 \\ 3 & 1 & -1 & | & 7 \\ 2 & -2 & 1 & | & 3 \end{bmatrix}$ $-3R_1 + R_2$

22. $\begin{bmatrix} 1 & -1 & 5 & | & -6 \\ 3 & 3 & -1 & | & 10 \\ 1 & 3 & 2 & | & 5 \end{bmatrix}$ $-3R_1 + R_2$

23. $\begin{bmatrix} 1 & -1 & 1 & 1 & | & 3 \\ 0 & 1 & -2 & -1 & | & 0 \\ 2 & 0 & 3 & 4 & | & 11 \\ 5 & 1 & 2 & 4 & | & 6 \end{bmatrix}$ $\begin{matrix} -2R_1 + R_3 \\ -5R_1 + R_4 \end{matrix}$

24. $\begin{bmatrix} 1 & -5 & 2 & -2 & | & 4 \\ 0 & 1 & -3 & -1 & | & 0 \\ 3 & 0 & 2 & -1 & | & 6 \\ -4 & 1 & 4 & 2 & | & -3 \end{bmatrix}$ $\begin{matrix} -3R_1 + R_3 \\ 4R_1 + R_4 \end{matrix}$

In Exercises 25–26, a few steps in the process of simplifying the given matrix to row-echelon form, with 1s down the diagonal from upper left to lower right, and 0s below the 1s, are shown. Fill in the missing numbers in the steps that are shown.

25. $\begin{bmatrix} 1 & -1 & 1 & | & 8 \\ 2 & 3 & -1 & | & -2 \\ 3 & -2 & -9 & | & 9 \end{bmatrix} \rightarrow \begin{bmatrix} 1 & -1 & 1 & | & 8 \\ 0 & 5 & \square & | & \square \\ 0 & 1 & \square & | & \square \end{bmatrix}$

$\rightarrow \begin{bmatrix} 1 & -1 & 1 & | & 8 \\ 0 & 1 & \square & | & \square \\ 0 & 1 & \square & | & \square \end{bmatrix}$

26. $\begin{bmatrix} 1 & -2 & 3 & | & 4 \\ 2 & 1 & -4 & | & 3 \\ -3 & 4 & -1 & | & -2 \end{bmatrix} \rightarrow \begin{bmatrix} 1 & -2 & 3 & | & 4 \\ 0 & 5 & \square & | & \square \\ 0 & -2 & \square & | & \square \end{bmatrix}$

$\rightarrow \begin{bmatrix} 1 & -2 & 3 & | & 4 \\ 0 & 1 & \square & | & \square \\ 0 & -2 & \square & | & \square \end{bmatrix}$

In Exercises 27–44, solve each system of equations using matrices. Use Gaussian elimination with back-substitution or Gauss-Jordan elimination.

27.
$$x + y - z = -2$$
$$2x - y + z = 5$$
$$-x + 2y + 2z = 1$$

28.
$$x - 2y - z = 2$$
$$2x - y + z = 4$$
$$-x + y - 2z = -4$$

29.
$$x + 3y = 0$$
$$x + y + z = 1$$
$$3x - y - z = 11$$

30.
$$3y - z = -1$$
$$x + 5y - z = -4$$
$$-3x + 6y + 2z = 11$$

31.
$$2x - y - z = 4$$
$$x + y - 5z = -4$$
$$x - 2y = 4$$

32.
$$x - 3z = -2$$
$$2x + 2y + z = 4$$
$$3x + y - 2z = 5$$

33.
$$x + y + z = 4$$
$$x - y - z = 0$$
$$x - y + z = 2$$

34.
$$3x + y - z = 0$$
$$x + y + 2z = 6$$
$$2x + 2y + 3z = 10$$

35.
$$x + 2y = z - 1$$
$$x = 4 + y - z$$
$$x + y - 3z = -2$$

36.
$$2x + y = z + 1$$
$$2x = 1 + 3y - z$$
$$x + y + z = 4$$

37.
$$3a - b - 4c = 3$$
$$2a - b + 2c = -8$$
$$a + 2b - 3c = 9$$

38.
$$3a + b - c = 0$$
$$2a + 3b - 5c = 1$$
$$a - 2b + 3c = -4$$

39.
$$2x + 2y + 7z = -1$$
$$2x + y + 2z = 2$$
$$4x + 6y + z = 15$$

40.
$$3x + 2y + 3z = 3$$
$$4x - 5y + 7z = 1$$
$$2x + 3y - 2z = 6$$

41.
$$w + x + y + z = 4$$
$$2w + x - 2y - z = 0$$
$$w - 2x - y - 2z = -2$$
$$3w + 2x + y + 3z = 4$$

42.
$$w + x + y + z = 5$$
$$w + 2x - y - 2z = -1$$
$$w - 3x - 3y - z = -1$$
$$2w - x + 2y - z = -2$$

43.
$$3w - 4x + y + z = 9$$
$$w + x - y - z = 0$$
$$2w + x + 4y - 2z = 3$$
$$-w + 2x + y - 3z = 3$$

44.
$$2w + y - 3z = 8$$
$$w - x + 4z = -10$$
$$3w + 5x - y - z = 20$$
$$w + x - y - z = 6$$

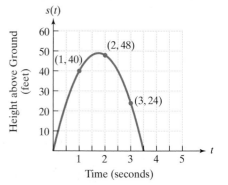

Practice Plus

45. Find the quadratic function $f(x) = ax^2 + bx + c$ for which $f(-2) = -4$, $f(1) = 2$, and $f(2) = 0$.

46. Find the quadratic function $f(x) = ax^2 + bx + c$ for which $f(-1) = 5$, $f(1) = 3$, and $f(2) = 5$.

47. Find the cubic function $f(x) = ax^3 + bx^2 + cx + d$ for which $f(-1) = 0$, $f(1) = 2$, $f(2) = 3$, and $f(3) = 12$.

48. Find the cubic function $f(x) = ax^3 + bx^2 + cx + d$ for which $f(-1) = 3$, $f(1) = 1$, $f(2) = 6$, and $f(3) = 7$.

49. Solve the system:

$$2 \ln w + \ln x + 3 \ln y - 2 \ln z = -6$$
$$4 \ln w + 3 \ln x + \ln y - \ln z = -2$$
$$\ln w + \ln x + \ln y + \ln z = -5$$
$$\ln w + \ln x - \ln y - \ln z = 5.$$

(*Hint:* Let $A = \ln w$, $B = \ln x$, $C = \ln y$, and $D = \ln z$. Solve the system for A, B, C, and D. Then use the logarithmic equations to find w, x, y, and z.)

50. Solve the system:

$$\ln w + \ln x + \ln y + \ln z = -1$$
$$-\ln w + 4 \ln x + \ln y - \ln z = 0$$
$$\ln w - 2 \ln x + \ln y - 2 \ln z = 11$$
$$-\ln w - 2 \ln x + \ln y + 2 \ln z = -3.$$

(*Hint:* Let $A = \ln w$, $B = \ln x$, $C = \ln y$, and $D = \ln$ Solve the system for A, B, C, and D. Then use the logarithm equations to find w, x, y, and z.)

Application Exercises

51. A ball is thrown straight upward. A position function

$$s(t) = \tfrac{1}{2}at^2 + v_0 t + s_0$$

can be used to describe the ball's height, $s(t)$, in feet, afte t seconds.

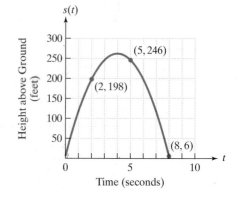

a. Use the points labeled in the graph to find the values of v_0, and s_0. Solve the system of linear equations involvin a, v_0, and s_0 using matrices.

b. Find and interpret $s(3.5)$. Identify your solution as a poi on the graph shown.

c. After how many seconds does the ball reach its maximu height? What is its maximum height?

52. A football is kicked straight upward. A position function

$$s(t) = \tfrac{1}{2}at^2 + v_0 t + s_0$$

can be used to describe the ball's height, $s(t)$, in feet, afte t seconds.

a. Use the points labeled in the graph to find the values of v_0, and s_0. Solve the system of linear equations involvin a, v_0, and s_0 using matrices.

b. Find and interpret $s(7)$. Identify your solution as a poi on the graph shown.

c. After how many seconds does the ball reach its maximu height? What is its maximum height?

Write a system of linear equations in three variables to solve Exercises 53–56. Then use matrices to solve the system.

Exercises 53–54 are based on a Time/CNN telephone poll that included never-married single women between the ages of 18 and 49 and never-married men between the ages of 18 and 49. The circle graphs show the results for one of the questions in the poll.

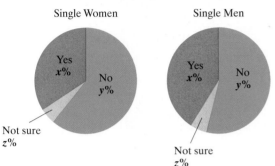

If You Couldn't Find the Perfect Mate, Would You Marry Someone Else?

Single Women — Yes $x\%$, No $y\%$, Not sure $z\%$

Single Men — Yes $x\%$, No $y\%$, Not sure $z\%$

53. For single women in the poll, the percentage who said no exceeded the combined percentages for those who said yes and those who said not sure by 22%. If the percentage who said yes is doubled, it is 7% more than the percentage who said no. Find the percentage of single women who responded yes, no, and not sure.

54. For single men in the poll, the percentage who said no exceeded the combined percentages for those who said yes and those who said not sure by 8%. If the percentage who said yes is doubled, it is 28% more than the percentage who said no. Find the percentage of single men who responded yes, no, and not sure.

55. Three foods have the following nutritional content per ounce.

	Calories	Protein (in grams)	Vitamin C (in milligrams)
Food A	40	5	30
Food B	200	2	10
Food C	400	4	300

If a meal consisting of the three foods allows exactly 660 calories, 25 grams of protein, and 425 milligrams of vitamin C, how many ounces of each kind of food should be used?

56. A furniture company produces three types of desks: a children's model, an office model, and a deluxe model. Each desk is manufactured in three stages: cutting, construction, and finishing. The time requirements for each model and manufacturing stage are given in the following table.

	Children's model	Office model	Deluxe model
Cutting	2 hr	3 hr	2 hr
Construction	2 hr	1 hr	3 hr
Finishing	1 hr	1 hr	2 hr

Each week the company has available a maximum of 100 hours for cutting, 100 hours for construction, and 65 hours for finishing. If all available time must be used, how many of each type of desk should be produced each week?

Writing in Mathematics

57. What is a matrix?

58. Describe what is meant by the augmented matrix of a system of linear equations.

59. In your own words, describe each of the three matrix row operations. Give an example with each of the operations.

60. Describe how to use row operations and matrices to solve a system of linear equations.

61. What is the difference between Gaussian elimination and Gauss-Jordan elimination?

Technology Exercises

62. Most graphing utilities can perform row operations on matrices. Consult the owner's manual for your graphing utility to learn proper keystrokes for performing these operations. Then duplicate the row operations of any three exercises that you solved from Exercises 19–24.

63. If your graphing utility has a [REF] (row-echelon form) command or a [RREF] (reduced row-echelon form) command, use this feature to verify your work with any five systems that you solved from Exercises 27–44.

64. Solve using a graphing utility's [REF] or [RREF] command:

$$\begin{aligned}
2x_1 - 2x_2 + 3x_3 - x_4 &= 12 \\
x_1 + 2x_2 - x_3 + 2x_4 - x_5 &= -7 \\
x_1 + x_3 + x_4 - 5x_5 &= 1 \\
-x_1 + x_2 - x_3 - 2x_4 - 3x_5 &= 0 \\
x_1 - x_2 - x_4 + x_5 &= 4.
\end{aligned}$$

Critical Thinking Exercises

65. Which one of the following is true?

a. A matrix row operation such as $-\frac{4}{5}R_1 + R_2$ is not permitted because of the negative fraction.

b. The augmented matrix for the system

$$\begin{aligned}
x - 3y &= 5 \\
y - 2z &= 7 \quad \text{is} \quad \begin{bmatrix} 1 & -3 & 5 \\ 1 & -2 & 7 \\ 2 & 1 & 4 \end{bmatrix}. \\
2x + z &= 4
\end{aligned}$$

c. In solving a linear system of three equations in three variables, we begin with the augmented matrix and use row operations to obtain a row-equivalent matrix with 0s down the diagonal from left to right and 1s below each 0.

d. None of the above is true.

66. The table shows the daily production level and profit for a business.

x (Number of Units Produced Daily)	30	50	100
y (Daily Profit)	$5900	$7500	$4500

Use the quadratic function $y = ax^2 + bx + c$ to determine the number of units that should be produced each day for maximum profit. What is the maximum daily profit?

SECTION 8.2   Inconsistent and Dependent Systems and Their Applications

Objectives

❶ Apply Gaussian elimination to systems without unique solutions.

❷ Apply Gaussian elimination to systems with more variables than equations.

❸ Solve problems involving systems without unique solutions.

Traffic jams getting you down? Powerful computers, able to solve systems wit hundreds of thousands of variables in a single bound, may promise a gridlock-fre future. The computer in your car could be linked to a central computer that manage traffic flow by controlling traffic lights, rerouting you away from traffic congestion issuing weather reports, and selecting the best route to your destination. New tech nologies could eventually drive your car at a steady 75 miles per hour alon automated highways as you comfortably nap. In this section, we look at the role c linear systems without unique solutions in a future free of traffic jams.

Linear systems can have one solution, no solution, or infinitely many solution We can use Gaussian elimination on systems with three or more variables to deter mine how many solutions such systems may have. In the case of systems with n solution or infinitely many solutions, it is impossible to rewrite the augmented matri in the desired form with 1s down the main diagonal from upper left to lower right, an 0s below the 1s. Let's see what this means by looking at a system that has no solutio

 Apply Gaussian elimination to systems without unique solutions.

EXAMPLE 1   A System with No Solution

Use Gaussian elimination to solve the system:

$$x - y - 2z = 2$$
$$2x - 3y + 6z = 5$$
$$3x - 4y + 4z = 12.$$

Solution

Step 1   Write the augmented matrix for the system.

Linear System	**Augmented Matrix**
$x - y - 2z = 2$ $2x - 3y + 6z = 5$ $3x - 4y + 4z = 12$	$\begin{bmatrix} 1 & -1 & -2 & 2 \\ 2 & -3 & 6 & 5 \\ 3 & -4 & 4 & 12 \end{bmatrix}$

Step 2   Attempt to simplify the matrix to row-echelon form, with 1s down th main diagonal and 0s below the 1s. Notice that the augmented matrix already has 1 in the top position of the first column. Now we want 0s below the 1. To get the firs 0, multiply row 1 by -2 and add these products to row 2. To get the second 0, multi ply row 1 by -3 and add these products to row 3. Performing these operations, w obtain the following matrix:

We want 1 in this position. $\begin{bmatrix} 1 & -1 & -2 & 2 \\ 0 & -1 & 10 & 1 \\ 0 & -1 & 10 & 6 \end{bmatrix}.$

Use the augmented matrix and:
Replace row 2 by $-2R_1 + R_2$.
Replace row 3 by $-3R_1 + R_3$.

Moving on to the second column, we obtain 1 in the desired position by multiplying row 2 by -1.

$$\begin{bmatrix} 1 & -1 & -2 & 2 \\ -1(0) & -1(-1) & -1(10) & -1(1) \\ 0 & -1 & 10 & 6 \end{bmatrix} = \begin{bmatrix} 1 & -1 & -2 & 2 \\ 0 & 1 & -10 & -1 \\ 0 & -1 & 10 & 6 \end{bmatrix} \quad -1R_2$$

We want 0 in this position.

Now we want a 0 below the 1 in column 2. To get the 0, multiply row 2 by 1 and add these products to row 3. (Equivalently, add row 2 to row 3.) We obtain the following matrix:

$$\begin{bmatrix} 1 & -1 & -2 & 2 \\ 0 & 1 & -10 & -1 \\ 0 & 0 & 0 & 5 \end{bmatrix}.$$

Replace row **3** in the previous matrix by $1R_2 + R_3$.

It is impossible to convert this last matrix to the desired form of 1s down the main diagonal. If we translate the last row back into equation form, we get

$$0x + 0y + 0z = 5,$$

There are no values of x, y, and z for which $0 = 5$.

which is false. Regardless of which values we select for x, y, and z, the last equation can never be a true statement. Consequently, the system has no solution. The solution set is $\varnothing$, the empty set.

Check Point 1   Use Gaussian elimination to solve the system:

$$x - 2y - z = -5$$
$$2x - 3y - z = 0$$
$$3x - 4y - z = 1.$$

Recall that the graph of a system of three linear equations in three variables consists of three planes. When these planes intersect in a single point, the system has precisely one ordered-triple solution. When the planes have no point in common, the system has no solution, like the one in Example 1. Figure 8.1 illustrates some of the geometric possibilities for these inconsistent systems.

Three planes are parallel with no common intersection point.

Two planes are parallel with no common intersection point.

Planes intersect two at a time. There is no intersection point common to all three planes.

Figure 8.1 Three planes may have no common point of intersection.

Now let's see what happens when we apply Gaussian elimination to a system with infinitely many solutions. Representing the solution set for these systems can be a bit tricky.

EXAMPLE 2 **A System with an Infinite Number of Solutions**

Use Gaussian elimination to solve the following system:

$$3x - 4y + 4z = 7$$
$$x - y - 2z = 2$$
$$2x - 3y + 6z = 5.$$

Solution As always, we start with the augmented matrix.

$$\begin{bmatrix} 3 & -4 & 4 & | & 7 \\ 1 & -1 & -2 & | & 2 \\ 2 & -3 & 6 & | & 5 \end{bmatrix} \xrightarrow[\substack{\text{Reverse rows} \\ \text{1 and 2.}}]{R_1 \leftrightarrow R_2} \begin{bmatrix} 1 & -1 & -2 & | & 2 \\ 3 & -4 & 4 & | & 7 \\ 2 & -3 & 6 & | & 5 \end{bmatrix} \xrightarrow[\substack{\text{Replace row 3} \\ \text{by } -2R_1 + R_3.}]{\substack{\text{Replace row 2} \\ \text{by } -3R_1 + R_2.}}$$

$$\begin{bmatrix} 1 & -1 & -2 & | & 2 \\ 0 & -1 & 10 & | & 1 \\ 0 & -1 & 10 & | & 1 \end{bmatrix} \xrightarrow[\substack{\text{Multiply row} \\ \text{2 by } -1.}]{-1R_2} \begin{bmatrix} 1 & -1 & -2 & | & 2 \\ 0 & 1 & -10 & | & -1 \\ 0 & -1 & 10 & | & 1 \end{bmatrix} \xrightarrow[\substack{\text{Replace row 3} \\ \text{by } 1R_2 + R_3.}]{}$$

$$\begin{bmatrix} 1 & -1 & -2 & | & 2 \\ 0 & 1 & -10 & | & -1 \\ 0 & 0 & 0 & | & 0 \end{bmatrix}$$

If we translate row 3 of the matrix into equation form, we obtain

$$0x + 0y + 0z = 0$$

or

$$0 = 0.$$

This equation results in a true statement regardless of which values we select for x, y, and z. Consequently, the equation $0x + 0y + 0z = 0$ is *dependent* on the other two equations in the system in the sense that it adds no new information about the variables. Thus, we can drop it from the system, which can now be expressed in the form

$$\begin{bmatrix} 1 & -1 & -2 & | & 2 \\ 0 & 1 & -10 & | & -1 \end{bmatrix}.$$ This is the last matrix from above with row 3 omitted.

The original system is equivalent to the system

$$x - y - 2z = 2$$ This is the system represented
$$y - 10z = -1.$$ by the above matrix.

Although neither of these equations gives a value for z, we can use them to express x and y in terms of z. From the last equation, we obtain

$$y = 10z - 1.$$ Add 10z to both sides and isolate y.

Back-substituting for y into the previous equation, we can find x in terms of z.

$$x - y - 2z = 2$$ This is the first equation obtained from the final matrix.
$$x - (10z - 1) - 2z = 2$$ Because y = 10z − 1, substitute 10z − 1 for y.
$$x - 10z + 1 - 2z = 2$$ Apply the distributive property.
$$x - 12z + 1 = 2$$ Combine like terms.
$$x = 12z + 1$$ Solve for x in terms of z.

We have now found two equations expressing x and y in terms of z:

$$x = 12z + 1$$
$$y = 10z - 1.$$

Because no value is determined for z, we can find a solution of the system by letting z equal any real number and then using these equations to obtain x and y. For example, if $z = 1$, then

$$x = 12z + 1 = 12(1) + 1 = 13 \text{ and}$$
$$y = 10z - 1 = 10(1) - 1 = 9.$$

Consequently, $(13, 9, 1)$ is a solution of the system. On the other hand, if we let $z = -1$, then

$$x = 12z + 1 = 12(-1) + 1 = -11 \text{ and}$$
$$y = 10z - 1 = 10(-1) - 1 = -11.$$

Thus, $(-11, -11, -1)$ is another solution of the system.

We see that for any arbitrary choice of z, every ordered triple of the form $(12z + 1, 10z - 1, z)$ is a solution of the system. The solution set of this system with dependent equations is

$$\{(12z + 1, 10z - 1, z)\}.$$

We have seen that when three planes have no point in common, the corresponding system has no solution. When the system has infinitely many solutions, like the one in Example 2, the three planes intersect in more than one point. Figure 8.2 illustrates geometric possibilities for systems with dependent equations.

The planes intersect
along a common line.

The planes coincide.

Figure 8.2 Three planes may intersect at infinitely many points.

 2 Use Gaussian elimination to solve the following system:

$$x - 2y - z = 5$$
$$2x - 5y + 3z = 6$$
$$x - 3y + 4z = 1.$$

 Apply Gaussian elimination to systems with more variables than equations.

Nonsquare Systems

Up to this point, we have encountered only *square* systems in which the number of equations is equal to the number of variables. In a **nonsquare system**, the number of variables differs from the number of equations. In Example 3, we have two equations and three variables.

EXAMPLE 3   A System with Fewer Equations Than Variables

Use Gaussian elimination to solve the system:

$$3x + 7y + 6z = 26$$
$$x + 2y + z = 8.$$

Solution   We begin with the augmented matrix.

$$\begin{bmatrix} 3 & 7 & 6 & | & 26 \\ 1 & 2 & 1 & | & 8 \end{bmatrix} \xrightarrow{R_1 \leftrightarrow R_2} \begin{bmatrix} 1 & 2 & 1 & | & 8 \\ 3 & 7 & 6 & | & 26 \end{bmatrix} \xrightarrow[\text{by } -3R_1 + R_2.]{\text{Replace row 2}} \begin{bmatrix} 1 & 2 & 1 & | & 8 \\ 0 & 1 & 3 & | & 2 \end{bmatrix}$$

Because we now have 1s down the diagonal that begins with the upper-left entry and a 0 below the leading 1, we translate the matrix back into equation form.

$$x + 2y + z = 8 \qquad \text{Equation 1}$$
$$y + 3z = 2 \qquad \text{Equation 2}$$

We can let z equal any real number and use back-substitution to express x and y in terms of z.

Equation 2	**Equation 1**
$y + 3z = 2$	$x + 2y + z = 8$
$y = -3z + 2$	$x + 2(-3z + 2) + z = 8$
	$x - 6z + 4 + z = 8$
	$x - 5z + 4 = 8$
	$x = 5z + 4$

For any arbitrary choice of z, every ordered triple of the form $(5z + 4, -3z + 2, z)$ is a solution of the system. We can express the system's solution set as

$$\{(5z + 4, -3z + 2, z)\}.$$

Discovery

Let $z = 1$ for the solution set
$$\{(5z + 4, -3z + 2, z)\}.$$
What solution do you obtain? Substitute these three values in the two original equations:
$$3x + 7y + 6z = 26$$
$$x + 2y + z = 8.$$
Show that each equation is satisfied. Repeat this process for two other values of z.

Check Point 3   Use Gaussian elimination to solve the system:

$$x + 2y + 3z = 70$$
$$x + y + z = 60.$$

③ Solve problems involving systems without unique solutions.

Applications

How will computers be programmed to control traffic flow and avoid congestion? They will be required to solve systems continually based on the following premise: If traffic is to keep moving, during any period of time the number of cars entering an intersection must equal the number of cars leaving that intersection. Let's see what this means by looking at the intersections of four one-way city streets.

EXAMPLE 4   Traffic Control

Figure 8.3 shows the intersections of four one-way streets. As you study the figure, notice that 300 cars per hour want to enter intersection I_1 from the north on 27th Avenue. Also, 200 cars per hour want to head east from intersection I_2 on Palm Drive. The letters w, x, y, and z stand for the number of cars passing between the intersections.

a. If the traffic is to keep moving, at each intersection the number of cars entering per hour must equal the number of cars leaving per hour. Use this idea to set up a linear system of equations involving w, x, y, and z.

b. Use Gaussian elimination to solve the system.

c. If construction on 27th Avenue limits z to 50 cars per hour, how many cars per hour must pass between the other intersections to keep traffic flowing?

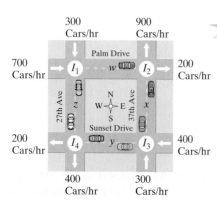

Figure 8.3 The intersections of four one-way streets

Solution

a. Set up the system by considering one intersection at a time, referring to Figure 8.3.

For Intersection I_1: Because $300 + 700 = 1000$ cars enter I_1 and $w + z$ cars leave the intersection, then $w + z = 1000$.

For Intersection I_2: Because $w + x$ cars enter the intersection and $200 + 900 = 1100$ cars leave I_2, then $w + x = 1100$.

For Intersection I_3: Figure 8.3 indicates that $300 + 400 = 700$ cars enter and $x + y$ leave, so $x + y = 700$.

For Intersection I_4: With $y + z$ cars entering and $200 + 400 = 600$ cars exiting, traffic will keep flowing if $y + z = 600$.

The system of equations that describes this situation is given by

$$w + z = 1000$$
$$w + x = 1100$$
$$x + y = 700$$
$$y + z = 600.$$

b. To solve this system using Gaussian elimination, we begin with the augmented matrix.

**System of Linear Equations
(showing missing variables
with 0 coefficients)**

$$1w + 0x + 0y + 1z = 1000$$
$$1w + 1x + 0y + 0z = 1100$$
$$0w + 1x + 1y + 0z = 700$$
$$0w + 0x + 1y + 1z = 600$$

Augmented Matrix

$$\begin{bmatrix} 1 & 0 & 0 & 1 & | & 1000 \\ 1 & 1 & 0 & 0 & | & 1100 \\ 0 & 1 & 1 & 0 & | & 700 \\ 0 & 0 & 1 & 1 & | & 600 \end{bmatrix}$$

We can now use row operations to obtain the following matrix:

$$\begin{bmatrix} 1 & 0 & 0 & 1 & | & 1000 \\ 0 & 1 & 0 & -1 & | & 100 \\ 0 & 0 & 1 & 1 & | & 600 \\ 0 & 0 & 0 & 0 & | & 0 \end{bmatrix}.$$

$w + z = 1000$

$x - z = 100$

$y + z = 600$

The last row of the matrix shows that the system in the voice balloons has dependent equations and infinitely many solutions. To write the solution set containing these infinitely many solutions, let z equal any real number. Use the three equations in the voice balloons to express $w, x,$ and y in terms of z:

$$w = 1000 - z, \quad x = 100 + z, \quad \text{and} \quad y = 600 - z.$$

With z arbitrary, the alphabetical ordered solution (w, x, y, z) enables us to express the system's solution set as

$$\{(1000 - z, 100 + z, 600 - z, z)\}.$$

c. We are given that construction limits z to 50 cars per hour. Because $z = 50$, we substitute 50 for z in the system's ordered solution:

$$(1000 - z, 100 + z, 600 - z, z) \qquad \textit{Use the system's solution.}$$
$$= (1000 - 50, 100 + 50, 600 - 50, 50) \qquad z = 50$$
$$= (950, 150, 550, 50).$$

Thus, $w = 950, x = 150,$ and $y = 550$. (See Figure 8.4.) With construction on 27th Avenue, this means that to keep traffic flowing, 950 cars per hour must be routed between I_1 and I_2, 150 per hour between I_3 and I_2, and 550 per hour between I_3 and I_4.

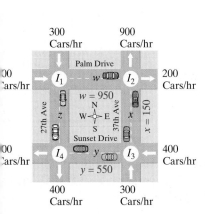

Figure 8.4 With z limited to 50 cars per hour, values for $w, x,$ and y are determined.

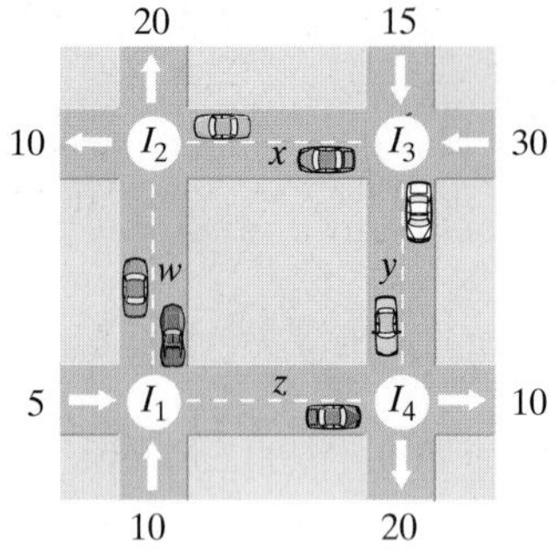

Figure 8.5

Check Point 4 Figure 8.5 shows a system of four one-way streets. The numbers in the figure denote the number of cars per minute that travel in the direction shown.

a. Use the requirement that the number of cars entering each of the intersections per minute must equal the number of cars leaving per minute to set up a system of equations in w, x, y, and z.

b. Use Gaussian elimination to solve the system.

c. If construction limits z to 10 cars per minute, how many cars per minute must pass between the other intersections to keep traffic flowing?

EXERCISE SET 8.2

Practice Exercises

In Exercises 1–24, use Gaussian elimination to find the complete solution to each system of equations, or show that none exists.

1. $\begin{aligned} 5x + 12y + z &= 10 \\ 2x + 5y + 2z &= -1 \\ x + 2y - 3z &= 5 \end{aligned}$

2. $\begin{aligned} 2x - 4y + z &= 3 \\ x - 3y + z &= 5 \\ 3x - 7y + 2z &= 12 \end{aligned}$

3. $\begin{aligned} 5x + 8y - 6z &= 14 \\ 3x + 4y - 2z &= 8 \\ x + 2y - 2z &= 3 \end{aligned}$

4. $\begin{aligned} 5x - 11y + 6z &= 12 \\ -x + 3y - 2z &= -4 \\ 3x - 5y + 2z &= 4 \end{aligned}$

5. $\begin{aligned} 3x + 4y + 2z &= 3 \\ 4x - 2y - 8z &= -4 \\ x + y - z &= 3 \end{aligned}$

6. $\begin{aligned} 2x - y - z &= 0 \\ x + 2y + z &= 3 \\ 3x + 4y + 2z &= 8 \end{aligned}$

7. $\begin{aligned} 8x + 5y + 11z &= 30 \\ -x - 4y + 2z &= 3 \\ 2x - y + 5z &= 12 \end{aligned}$

8. $\begin{aligned} x + y - 10z &= -4 \\ x - 7z &= -5 \\ 3x + 5y - 36z &= -10 \end{aligned}$

9. $\begin{aligned} w - 2x - y - 3z &= -9 \\ w + x - y &= 0 \\ 3w + 4x + z &= 6 \\ 2x - 2y + z &= 3 \end{aligned}$

10. $\begin{aligned} 2w + x - 2y - z &= 3 \\ w - 2x + y + z &= 4 \\ -w - 8x + 7y + 5z &= 13 \\ 3w + x - 2y + 2z &= 6 \end{aligned}$

11. $\begin{aligned} 2w + x - y &= 3 \\ w - 3x + 2y &= -4 \\ 3w + x - 3y + z &= 1 \\ w + 2x - 4y - z &= -2 \end{aligned}$

12. $\begin{aligned} 2w - x + 3y + z &= 0 \\ 3w + 2x + 4y - z &= 0 \\ 5w - 2x - 2y - z &= 0 \\ 2w + 3x - 7y - 5z &= 0 \end{aligned}$

13. $\begin{aligned} w - 3x + y - 4z &= 4 \\ -2w + x + 2y &= -2 \\ 3w - 2x + y - 6z &= 2 \\ -w + 3x + 2y - z &= -6 \end{aligned}$

14. $\begin{aligned} 3w + 2x - y + 2z &= -12 \\ 4w - x + y + 2z &= 1 \\ w + x + y + z &= -2 \\ -2w + 3x + 2y - 3z &= 10 \end{aligned}$

15. $\begin{aligned} 2x + y - z &= 2 \\ 3x + 3y - 2z &= 3 \end{aligned}$

16. $\begin{aligned} 3x + 2y - z &= 5 \\ x + 2y - z &= 1 \end{aligned}$

17. $\begin{aligned} x + 2y + 3z &= 5 \\ y - 5z &= 0 \end{aligned}$

18. $\begin{aligned} 3x - y + 4z &= 8 \\ y + 2z &= 1 \end{aligned}$

19. $\begin{aligned} x + y - 2z &= 2 \\ 3x - y - 6z &= -7 \end{aligned}$

20. $\begin{aligned} -2x - 5y + 10z &= 19 \\ x + 2y - 4z &= 12 \end{aligned}$

21. $\begin{aligned} w + x - y + z &= -2 \\ 2w - x + 2y - z &= 7 \\ -w + 2x + y + 2z &= -1 \end{aligned}$

22. $\begin{aligned} 2w - 3x + 4y + z &= 7 \\ w - x + 3y - 5z &= 10 \\ 3w + x - 2y - 2z &= 6 \end{aligned}$

23. $\begin{aligned} w + 2x + 3y - z &= 7 \\ 2x - 3y + z &= 4 \\ w - 4x + y &= 3 \end{aligned}$

24. $\begin{aligned} w - x + z &= 0 \\ w - 4x + y + 2z &= 0 \\ 3w - y + 2z &= 0 \end{aligned}$

Practice Plus

In Exercises 25–28, the first screen shows the augmented matrix, A, for a nonsquare linear system of three equations in four variables, w, x, y, and z. The second screen shows the reduced row-echelon form of matrix A. For each exercise,

a. *Write the system represented by A.*

b. *Use the reduced row-echelon form of A to find the system's complete solution.*

25.

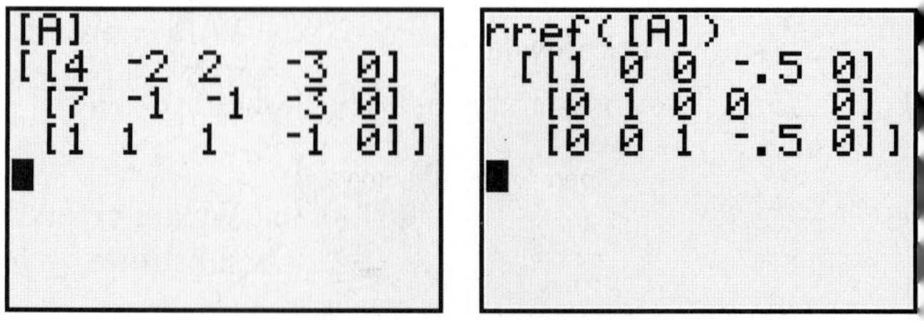

26.

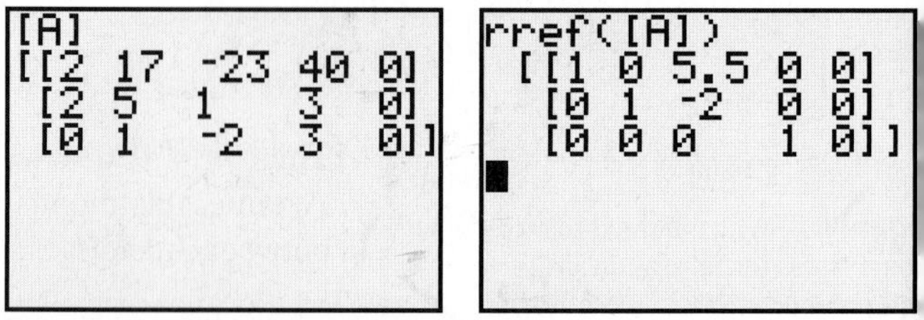

7.

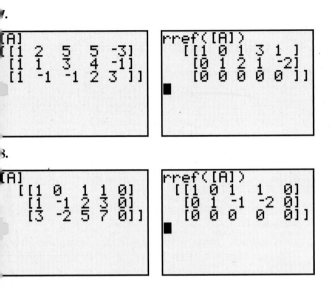

8.

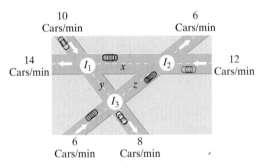

Application Exercises

The figure for Exercises 29–32 shows the intersection of three one-way streets. To keep traffic moving, the number of cars per minute entering an intersection must equal the number exiting that intersection. For intersection I_1, $x + 10$ cars enter and $y + 14$ cars exit per minute. Thus, $x + 10 = y + 14$.

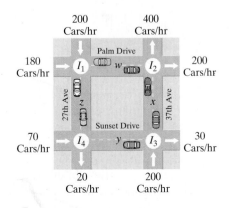

29. Write an equation for intersection I_2 that keeps traffic moving.

30. Write an equation for intersection I_3 that keeps traffic moving.

31. Use Gaussian elimination to solve the system formed by the equation given prior to Exercise 29 and the two equations that you obtained in Exercises 29–30.

32. Use your ordered solution obtained in Exercise 31 to solve this exercise. If construction limits z to 4 cars per minute, how many cars per minute must pass between the other intersections to keep traffic flowing?

33. The figure shows the intersection of four one-way streets.

a. Set up a system of equations that keeps traffic moving.

b. Use Gaussian elimination to solve the system.

c. If construction limits z to 50 cars per hour, how many cars per hour must pass between the other intersections to keep traffic moving?

34. The vitamin content per ounce for three foods is given in the following table.

	Milligrams per Ounce		
	Thiamin	**Riboflavin**	**Niacin**
Food A	3	7	1
Food B	1	5	3
Food C	3	8	2

a. Use matrices to show that no combination of these foods can provide exactly 14 mg of thiamin, 32 mg of riboflavin, and 9 mg of niacin.

b. Use matrices to describe in practical terms what happens if the riboflavin requirement is increased by 5 mg and the other requirements stay the same.

35. Three foods have the following nutritional content per ounce.

	Units per Ounce		
	Vitamin A	**Iron**	**Calcium**
Food 1	20	20	10
Food 2	30	10	10
Food 3	10	10	30

a. A diet must consist precisely of 220 units of vitamin A, 180 units of iron, and 340 units of calcium. However, the dietician runs out of Food 1. Use a matrix approach to show that under these conditions the dietary requirements cannot be met.

b. Now suppose that all three foods are available, but due to problems with vitamin A for pregnant women, a hospital dietician no longer wants to include this vitamin in the diet. Use matrices to give two possible ways to meet the iron and calcium requirements with the three foods.

36. A company that manufactures products A, B, and C does both manufacturing and testing. The hours needed to manufacture and test each product are shown in the table.

	Hours Needed Weekly to Manufacture	Hours Needed Weekly to Test
Product A	7	2
Product B	6	2
Product C	3	1

The company has exactly 67 hours per week available for manufacturing and 20 hours per week available for testing. Give two different combinations for the number of products that can be manufactured and tested weekly.

Writing in Mathematics

37. Describe what happens when Gaussian elimination is used to solve an inconsistent system.

38. Describe what happens when Gaussian elimination is used to solve a system with dependent equations.

39. In solving a system of dependent equations in three variables, one student simply said that there are infinitely many solutions. A second student expressed the solution set as $\{(4z + 3, 5z - 1, z)\}$. Which is the better form of expressing the solution set and why?

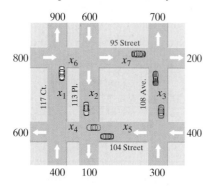

Technology Exercise

40. a. The figure shows the intersections of a number of one-way streets. The numbers given represent traffic flow at a peak period (from 4 P.M. to 5:30 P.M.). Use the figure to write a linear system of six equations in seven variables based on the idea that at each intersection the number of cars entering must equal the number of cars leaving.

   b. Use a graphing utility with a $\boxed{\text{REF}}$ or $\boxed{\text{RREF}}$ command to find the complete solution to the system.

Critical Thinking Exercise

41. Consider the linear system

$$
\begin{aligned}
x + 3y + z &= a^2 \\
2x + 5y + 2az &= 0 \\
x + y + a^2z &= -9.
\end{aligned}
$$

For which values of a will the system be inconsistent?

Group Exercise

42. Before beginning this exercise, the group needs to read an solve Exercise 40.

   a. A political group is planning a demonstration on 95t Street between 113th Place and 117th Court for 5 P.M Wednesday. The problem becomes one of minimizin traffic flow on 95th Street (between 113th and 117th without causing traffic tie-ups on other streets. One possible solution is to close off traffic on 95th Street betwee 113th and 117th (let $x_6 = 0$). What can group member conclude about x_7 under these conditions?

   b. Working with a matrix allows us to simplify the problem caused by the political demonstration, but it did no actually solve the problem. There are an infinite numbe of solutions; each value of x_7 we choose gives us a new picture. We also assumed x_6 was equal to 0; changing tha assumption would also lead to different solutions. Wit your group, design another solution to the traffic flow problem caused by the political demonstration.

SECTION 8.3 *Matrix Operations and Their Applications*

Objectives

❶ Use matrix notation.

❷ Understand what is meant by equal matrices.

❸ Add and subtract matrices.

❹ Perform scalar multiplication.

❺ Solve matrix equations.

❻ Multiply matrices.

❼ Describe applied situations with matrix operations.

Turn on your computer and read your e-mail or write a paper. When you need to do research, use the Internet to browse through art museums and photography exhibits. When you need a break, load a flight simulator program and fly through a photorealistic computer world. As different as these experiences may be, they al share one thing—you're looking at images based on matrices. Matrices have appli cations in numerous fields, including the new technology of digital photography in which pictures are represented by numbers rather than film. In this section, we turn our attention to matrix algebra and some of its applications.

① Use matrix notation.

Notations for Matrices

We have seen that an array of numbers, arranged in rows and columns and placed in brackets, is called a matrix. We can represent a matrix in two different ways.

- A capital letter, such as A, B, or C, can denote a matrix.
- A lowercase letter enclosed in brackets, such as that shown below, can denote a matrix.

$$A = [a_{ij}]$$

> Matrix A with elements a_{ij}

A general element in matrix A is denoted by a_{ij}. This refers to the element in the ith row and jth column. For example, a_{32} is the element of A located in the third row, second column.

A matrix of **order $m \times n$** has m rows and n columns. If $m = n$, a matrix has the same number of rows as columns and is called a **square matrix**.

EXAMPLE 1   Matrix Notation

Let

$$A = \begin{bmatrix} 3 & 2 & 0 \\ -4 & -5 & -\frac{1}{5} \end{bmatrix}.$$

a. What is the order of A?

b. If $A = [a_{ij}]$, identify a_{23} and a_{12}.

Solution

a. The matrix has 2 rows and 3 columns, so it is of order 2×3.

b. The element a_{23} is in the second row and third column. Thus, $a_{23} = -\frac{1}{5}$.

The element a_{12} is in the first row and second column. Consequently, $a_{12} = 2$.

Check Point 1 Let

$$A = \begin{bmatrix} 5 & -2 \\ -3 & \pi \\ 1 & 6 \end{bmatrix}.$$

a. What is the order of A?   **b.** Identify a_{12} and a_{31}.

② Understand what is meant by equal matrices.

Equality of Matrices

Two matrices are **equal** if and only if they have the same order and corresponding elements are equal.

> **Definition of Equality of Matrices**
>
> Two matrices A and B are **equal** if and only if they have the same order $m \times n$ and $a_{ij} = b_{ij}$ for $i = 1, 2, \ldots, m$ and $j = 1, 2, \ldots, n$.

For example, if $A = \begin{bmatrix} x & y+1 \\ z & 6 \end{bmatrix}$ and $B = \begin{bmatrix} 1 & 5 \\ 3 & 6 \end{bmatrix}$, then $A = B$ if and only if $x = 1$, $y + 1 = 5$ (so $y = 4$), and $z = 3$.

Add and subtract matrices.

Matrix Addition and Subtraction

Table 8.1 shows that matrices of the same order can be added or subtracted by simply adding or subtracting corresponding elements.

Table 8.1   Adding and Subtracting Matrices

Let $A = [a_{ij}]$ and $B = [b_{ij}]$ be matrices of order $m \times n$.		
Definition	**The Definition in Words**	**Example**
Matrix Addition $A + B = [a_{ij} + b_{ij}]$	Matrices of the same order are added by adding the elements in corresponding positions.	$\begin{bmatrix} 1 & -2 \\ 3 & 5 \end{bmatrix} + \begin{bmatrix} -1 & 6 \\ 0 & 4 \end{bmatrix}$ $= \begin{bmatrix} 1 + (-1) & -2 + 6 \\ 3 + 0 & 5 + 4 \end{bmatrix} = \begin{bmatrix} 0 & 4 \\ 3 & 9 \end{bmatrix}$
Matrix Subtraction $A - B = [a_{ij} - b_{ij}]$	Matrices of the same order are subtracted by subtracting the elements in corresponding positions.	$\begin{bmatrix} 1 & -2 \\ 3 & 5 \end{bmatrix} - \begin{bmatrix} -1 & 6 \\ 0 & 4 \end{bmatrix}$ $= \begin{bmatrix} 1 - (-1) & -2 - 6 \\ 3 - 0 & 5 - 4 \end{bmatrix} = \begin{bmatrix} 2 & -8 \\ 3 & 1 \end{bmatrix}$

The sum or difference of two matrices of different orders is undefined. For example, consider the matrices

$$A = \begin{bmatrix} 0 & 3 \\ 4 & 3 \end{bmatrix} \quad \text{and} \quad B = \begin{bmatrix} 1 & 9 \\ 4 & 5 \\ 2 & 3 \end{bmatrix}.$$

The order of A is 2×2; the order of B is 3×2. These matrices are of different orders and cannot be added or subtracted.

Technology

Graphing utilities can add and subtract matrices. Enter the matrices and name them $[A]$ and $[B]$. Then use a keystroke sequence similar to

$\boxed{[A]}\boxed{+}\boxed{[B]}\boxed{\text{ENTER}}$

$\boxed{[A]}\boxed{-}\boxed{[B]}\boxed{\text{ENTER}}$

Consult your manual and verify the results in Example 2.

EXAMPLE 2   Adding and Subtracting Matrices

Perform the indicated matrix operations:

a. $\begin{bmatrix} 0 & 5 & 3 \\ -2 & 6 & -8 \end{bmatrix} + \begin{bmatrix} -2 & 3 & 5 \\ 7 & -9 & 6 \end{bmatrix}$

b. $\begin{bmatrix} -6 & 7 \\ 2 & -3 \end{bmatrix} - \begin{bmatrix} -5 & 6 \\ 0 & -4 \end{bmatrix}.$

Solution

a.
$$\begin{bmatrix} 0 & 5 & 3 \\ -2 & 6 & -8 \end{bmatrix} + \begin{bmatrix} -2 & 3 & 5 \\ 7 & -9 & 6 \end{bmatrix}$$

$$= \begin{bmatrix} 0 + (-2) & 5 + 3 & 3 + 5 \\ -2 + 7 & 6 + (-9) & -8 + 6 \end{bmatrix}$$
Add the corresponding elements in the 2 × 3 matrices.

$$= \begin{bmatrix} -2 & 8 & 8 \\ 5 & -3 & -2 \end{bmatrix}$$
Simplify.

b.
$$\begin{bmatrix} -6 & 7 \\ 2 & -3 \end{bmatrix} - \begin{bmatrix} -5 & 6 \\ 0 & -4 \end{bmatrix}$$

$$= \begin{bmatrix} -6 - (-5) & 7 - 6 \\ 2 - 0 & -3 - (-4) \end{bmatrix}$$
Subtract the corresponding elements in the 2 × 2 matrices.

$$= \begin{bmatrix} -1 & 1 \\ 2 & 1 \end{bmatrix}$$
Simplify.

Check Point 2 Perform the indicated matrix operations:

a. $\begin{bmatrix} -4 & 3 \\ 7 & -6 \end{bmatrix} + \begin{bmatrix} 6 & -3 \\ 2 & -4 \end{bmatrix}$ b. $\begin{bmatrix} 5 & 4 \\ -3 & 7 \\ 0 & 1 \end{bmatrix} - \begin{bmatrix} -4 & 8 \\ 6 & 0 \\ -5 & 3 \end{bmatrix}$.

A matrix whose elements are all equal to 0 is called a **zero matrix**. If A is an $m \times n$ matrix and 0 is the $m \times n$ zero matrix, then $A + 0 = A$. For example,

$$\begin{bmatrix} -5 & 2 \\ 3 & 6 \end{bmatrix} + \begin{bmatrix} 0 & 0 \\ 0 & 0 \end{bmatrix} = \begin{bmatrix} -5 & 2 \\ 3 & 6 \end{bmatrix}.$$

The $m \times n$ zero matrix is called the **additive identity** for $m \times n$ matrices.

For any matrix A, the **additive inverse** of A, written $-A$, is the matrix of the same order of A such that every element of $-A$ is the opposite of the corresponding element of A. Because corresponding elements are added in matrix addition, $A + (-A)$ is a zero matrix. For example,

$$\begin{bmatrix} -5 & 2 \\ 3 & 6 \end{bmatrix} + \begin{bmatrix} 5 & -2 \\ -3 & -6 \end{bmatrix} = \begin{bmatrix} 0 & 0 \\ 0 & 0 \end{bmatrix}.$$

Properties of matrix addition are similar to properties for adding real numbers.

Properties of Matrix Addition

If A, B, and C are $m \times n$ matrices and 0 is the $m \times n$ zero matrix, then the following properties are true.

1. $A + B = B + A$ Commutative Property of Addition

2. $(A + B) + C = A + (B + C)$ Associative Property of Addition

3. $A + 0 = 0 + A = A$ Additive Identity Property

4. $A + (-A) = (-A) + A = 0$ Additive Inverse Property

④ Perform scalar multiplication.

Scalar Multiplication

A matrix of order 1×1, such as [6], contains only one entry. To distinguish this matrix from the number 6, we refer to 6 as a **scalar**. In general, in our work with matrices, we will refer to real numbers as scalars.

To multiply a matrix A by a scalar c, we multiply each entry in A by c. For example,

$$4\begin{bmatrix} 2 & 5 \\ -3 & 0 \end{bmatrix} = \begin{bmatrix} 4(2) & 4(5) \\ 4(-3) & 4(0) \end{bmatrix} = \begin{bmatrix} 8 & 20 \\ -12 & 0 \end{bmatrix}.$$

Scalar   Matrix

Definition of Scalar Multiplication

If $A = [a_{ij}]$ is a matrix of order $m \times n$ and c is a scalar, then the matrix cA is the $m \times n$ matrix given by

$$cA = [ca_{ij}].$$

This matrix is obtained by multiplying each element of A by the real number c. We call cA a **scalar multiple** of A.

EXAMPLE 3   Scalar Multiplication

If $A = \begin{bmatrix} -1 & 4 \\ 3 & 0 \end{bmatrix}$ and $B = \begin{bmatrix} 2 & -3 \\ 5 & -6 \end{bmatrix}$, find the following matrices:

a. $-5B$ **b.** $2A + 3B$.

Solution

a. $-5B = -5\begin{bmatrix} 2 & -3 \\ 5 & -6 \end{bmatrix} = \begin{bmatrix} -5(2) & -5(-3) \\ -5(5) & -5(-6) \end{bmatrix} = \begin{bmatrix} -10 & 15 \\ -25 & 30 \end{bmatrix}$

> Multiply each element by −5.

b. $2A + 3B = 2\begin{bmatrix} -1 & 4 \\ 3 & 0 \end{bmatrix} + 3\begin{bmatrix} 2 & -3 \\ 5 & -6 \end{bmatrix}$

$= \begin{bmatrix} 2(-1) & 2(4) \\ 2(3) & 2(0) \end{bmatrix} + \begin{bmatrix} 3(2) & 3(-3) \\ 3(5) & 3(-6) \end{bmatrix}$

> Multiply each element in A by 2.

> Multiply each element in B by 3.

$= \begin{bmatrix} -2 & 8 \\ 6 & 0 \end{bmatrix} + \begin{bmatrix} 6 & -9 \\ 15 & -18 \end{bmatrix} = \begin{bmatrix} -2 + 6 & 8 + (-9) \\ 6 + 15 & 0 + (-18) \end{bmatrix}$

> Perform the addition of these 2 x 2 matrices by adding corresponding elements.

$= \begin{bmatrix} 4 & -1 \\ 21 & -18 \end{bmatrix}$

Technology

You can verify the algebraic solution in Example 3(b) by first entering the matrices [A] and [B] into your graphing utility. The screen below shows the required computation.

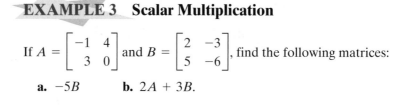

```
2[A]+3[B]
        [[4   -1 ]
        [21  -18]]
■
```

Check Point 3 If $A = \begin{bmatrix} -4 & 1 \\ 3 & 0 \end{bmatrix}$ and $B = \begin{bmatrix} -1 & -2 \\ 8 & 5 \end{bmatrix}$, find the following matrices:

a. $-6B$ **b.** $3A + 2B$.

Properties of scalar multiplication are similar to properties for multiplying real numbers.

Discovery

Verify each of the four properties listed in the box using

$A = \begin{bmatrix} 2 & -4 \\ -5 & 3 \end{bmatrix}$,

$B = \begin{bmatrix} 4 & 0 \\ 1 & -6 \end{bmatrix}$,

$c = 4$, and $d = 2$.

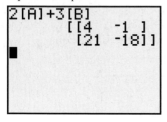

⑤ Solve matrix equations.

Properties of Scalar Multiplication

If A and B are $m \times n$ matrices, and c and d are scalars, then the following properties are true.

1. $(cd)A = c(dA)$ Associative Property of Scalar Multiplication

2. $1A = A$ Scalar Identity Property

3. $c(A + B) = cA + cB$ Distributive Property

4. $(c + d)A = cA + dA$ Distributive Property

Have you noticed the many similarities between addition of real numbers and matrix addition, subtraction of real numbers and matrix subtraction, and multiplication of real numbers and scalar multiplication? Example 4 shows how these similarities can be used to solve matrix equations involving matrix addition, matrix subtraction, and scalar multiplication.

EXAMPLE 4   Solving a Matrix Equation

Solve for X in the matrix equation

$$2X + A = B,$$

where $A = \begin{bmatrix} 1 & -5 \\ 0 & 2 \end{bmatrix}$ and $B = \begin{bmatrix} -6 & 5 \\ 9 & 1 \end{bmatrix}$.

Solution   We begin by solving the matrix equation for X.

$$2X + A = B \qquad \text{This is the given matrix equation.}$$
$$2X = B - A \qquad \text{Subtract matrix A from both sides.}$$

We multiply both sides by $\frac{1}{2}$ rather than divide both sides by 2. This is in anticipation of performing scalar multiplication.

$$X = \frac{1}{2}(B - A) \qquad \text{Multiply both sides by } \tfrac{1}{2} \text{ and solve for matrix X.}$$

Now we use the matrices A and B to find the matrix X.

$$X = \frac{1}{2}\left(\begin{bmatrix} -6 & 5 \\ 9 & 1 \end{bmatrix} - \begin{bmatrix} 1 & -5 \\ 0 & 2 \end{bmatrix} \right) \qquad \text{Substitute the matrices into } X = \tfrac{1}{2}(B - A).$$

$$= \frac{1}{2}\begin{bmatrix} -7 & 10 \\ 9 & -1 \end{bmatrix} \qquad \text{Subtract matrices by subtracting corresponding elements.}$$

$$= \begin{bmatrix} -\dfrac{7}{2} & 5 \\ \dfrac{9}{2} & -\dfrac{1}{2} \end{bmatrix} \qquad \text{Perform the scalar multiplication by multiplying each element by } \tfrac{1}{2}.$$

Take a few minutes to show that this matrix satisfies the given equation $2X + A = B$. Substitute the matrix for X and the given matrices for A and B into the equation. The matrices on each side of the equal sign, $2X + A$ and B, should be equal.

Check Point 4   Solve for X in the matrix equation $3X + A = B$, where

$$A = \begin{bmatrix} 2 & -8 \\ 0 & 4 \end{bmatrix} \quad \text{and} \quad B = \begin{bmatrix} -10 & 1 \\ -9 & 17 \end{bmatrix}.$$

⑥ Multiply matrices.

Matrix Multiplication

We do not multiply two matrices by multiplying the corresponding entries of matrices. Instead, we must think of matrix multiplication as *row-by-column multiplication*. To better understand how this works, let's begin with the definition of matrix multiplication for matrices of order 2×2.

Definition of Matrix Multiplication: 2×2 Matrices

Row 1 of A × Column 1 of B Row 1 of A × Column 2 of B

$$AB = \begin{bmatrix} a & b \\ c & d \end{bmatrix}\begin{bmatrix} e & f \\ g & h \end{bmatrix} = \begin{bmatrix} ae + bg & af + bh \\ ce + dg & cf + dh \end{bmatrix}$$

Row 2 of A × Column 1 of B Row 2 of A × Column 2 of B

Notice that we obtain the element in the ith row and jth column in AB by perform-ing computations with elements in the ith row of A and the jth column of B. For example, we obtain the element in the first row and first column of AB by performing computations with elements in the first row of A and the first column of B.

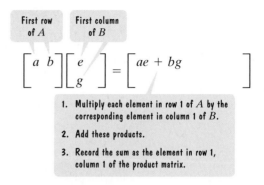

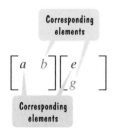

Figure 8.6 Finding corresponding elements when multiplying matrices

You may wonder how to find the corresponding elements in step 1 in the voice balloon. The element at the far left of row 1 corresponds to the element at the top of column 1. The second element from the left of row 1 corresponds to the second element from the top of column 1. This is illustrated in Figure 8.6.

Study Tip

Writing the location of each element in the product matrix AB may help you to remember how to multiply 2×2 matrices.

$$AB = \begin{bmatrix} p_{11} & p_{12} \\ p_{21} & p_{22} \end{bmatrix}$$

Row 1 (of A) × Column 1 (of B)   Row 1 × Column 2

Row 2 × Column 1   Row 2 × Column 2

EXAMPLE 5 Multiplying Matrices

Find AB, given

$$A = \begin{bmatrix} 2 & 3 \\ 4 & 7 \end{bmatrix} \quad \text{and} \quad B = \begin{bmatrix} 0 & 1 \\ 5 & 6 \end{bmatrix}.$$

Solution We will perform a row-by-column computation.

$$AB = \begin{bmatrix} 2 & 3 \\ 4 & 7 \end{bmatrix}\begin{bmatrix} 0 & 1 \\ 5 & 6 \end{bmatrix}$$

Row 1 of A × Column 1 of B Row 1 of A × Column 2 of B

$$= \begin{bmatrix} 2(0) + 3(5) & 2(1) + 3(6) \\ 4(0) + 7(5) & 4(1) + 7(6) \end{bmatrix} = \begin{bmatrix} 15 & 20 \\ 35 & 46 \end{bmatrix}$$

Row 2 of A × Column 1 of B Row 2 of A × Column 2 of B

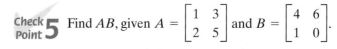

Check Point 5 Find AB, given $A = \begin{bmatrix} 1 & 3 \\ 2 & 5 \end{bmatrix}$ and $B = \begin{bmatrix} 4 & 6 \\ 1 & 0 \end{bmatrix}$.

We can generalize the process of Example 5 to multiplying an $m \times n$ matrix and an $n \times p$ matrix. **For the product of two matrices to be defined, the number of columns of the first matrix must equal the number of rows of the second matrix.**

First Matrix   **Second Matrix**
$m \times n$ $n \times p$

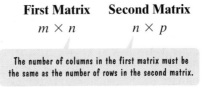

The number of columns in the first matrix must be the same as the number of rows in the second matrix.

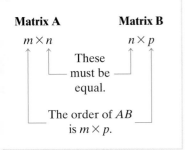
Definition of Matrix Multiplication

The **product** of an $m \times n$ matrix, A, and an $n \times p$ matrix, B, is an $m \times p$ matrix, AB, whose elements are found as follows. The element in the ith row and jth column of AB is found by multiplying each element in the ith row of A by the corresponding element in the jth column of B and adding the products.

To find a product AB, each row of A must have the same number of elements as each column of B. We obtain p_{ij}, the element in the ith row and jth column in AB, by performing computations with elements in the ith row of A and the jth column of B:

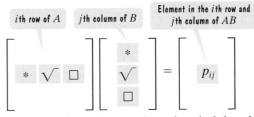

ith row of A jth column of B Element in the ith row and jth column of AB

When multiplying corresponding elements, keep in mind that the element at the far left of row i corresponds to the element at the top of column j. The element second from the left of row i corresponds to the element second from the top of column j. Likewise, the element third from the left of row i corresponds to the element third from the top of column j, and so on.

EXAMPLE 6   Multiplying Matrices

Matrices A and B are defined as follows:

$$A = \begin{bmatrix} 1 & 2 & 3 \end{bmatrix} \qquad B = \begin{bmatrix} 4 \\ 5 \\ 6 \end{bmatrix}.$$

Find each product: **a.** AB **b.** BA.

Solution

a. Matrix A is a 1×3 matrix and matrix B is a 3×1 matrix. Thus, the product is a 1×1 matrix.

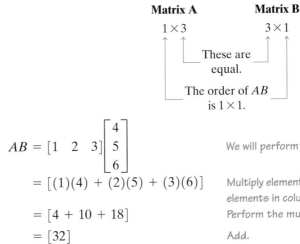

Matrix A   **Matrix B**
1×3 3×1

These are equal.

The order of AB is 1×1.

$$AB = \begin{bmatrix} 1 & 2 & 3 \end{bmatrix} \begin{bmatrix} 4 \\ 5 \\ 6 \end{bmatrix}$$ We will perform a row-by-column computation.

$$= \begin{bmatrix} (1)(4) + (2)(5) + (3)(6) \end{bmatrix}$$ Multiply elements in row 1 of A by corresponding elements in column 1 of B and add the products.

$$= \begin{bmatrix} 4 + 10 + 18 \end{bmatrix}$$ Perform the multiplications.

$$= \begin{bmatrix} 32 \end{bmatrix}$$ Add.

Technology

The screens illustrate the solution of Example 6 using a graphing utility.

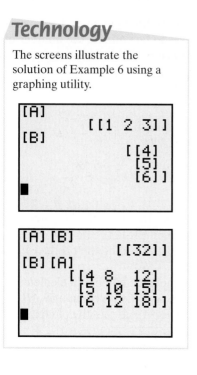

b. Matrix B is a 3×1 matrix and matrix A is a 1×3 matrix. Thus, the produc BA is a 3×3 matrix.

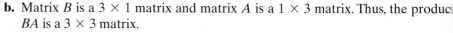

Matrix A   **Matrix B**
3×1 1×3

These are equal.

The order of AB is 3×3.

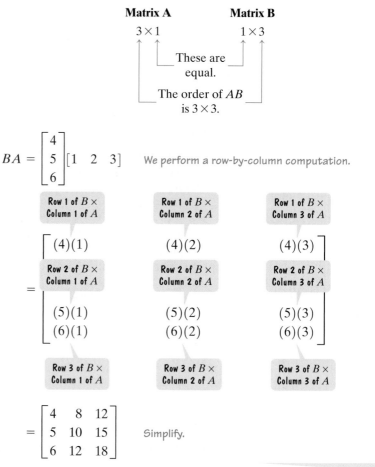

$$BA = \begin{bmatrix} 4 \\ 5 \\ 6 \end{bmatrix} \begin{bmatrix} 1 & 2 & 3 \end{bmatrix} \quad \text{We perform a row-by-column computation.}$$

$$= \begin{bmatrix} (4)(1) & (4)(2) & (4)(3) \\ (5)(1) & (5)(2) & (5)(3) \\ (6)(1) & (6)(2) & (6)(3) \end{bmatrix}$$

Row 1 of $B \times$ Column 1 of A | Row 1 of $B \times$ Column 2 of A | Row 1 of $B \times$ Column 3 of A
Row 2 of $B \times$ Column 1 of A | Row 2 of $B \times$ Column 2 of A | Row 2 of $B \times$ Column 3 of A
Row 3 of $B \times$ Column 1 of A | Row 3 of $B \times$ Column 2 of A | Row 3 of $B \times$ Column 3 of A

$$= \begin{bmatrix} 4 & 8 & 12 \\ 5 & 10 & 15 \\ 6 & 12 & 18 \end{bmatrix} \quad \text{Simplify.}$$

Arthur Cayley

Matrices were first studied intensively by the English mathematician Arthur Cayley (1821–1895). Before reaching the age of 25, he published 25 papers, setting a pattern of prolific creativity that lasted throughout his life. Cayley was a lawyer, painter, mountaineer, and Cambridge professor whose greatest invention was that of matrices and matrix theory. Cayley's matrix algebra, especially the noncommutativity of multiplication ($AB \neq BA$), opened up a new area of mathematics called abstract algebra.

In Example 6, did you notice that AB and BA are different matrices? For mos matrices A and B, $AB \neq BA$. Because **matrix multiplication is not commutative**, be careful about the order in which matrices appear when performing this operation

Check Point 6 If $A = \begin{bmatrix} 2 & 0 & 4 \end{bmatrix}$ and $B = \begin{bmatrix} 1 \\ 3 \\ 7 \end{bmatrix}$, find AB and BA.

EXAMPLE 7   Multiplying Matrices

Where possible, find each product:

a. $\begin{bmatrix} 4 & 2 \\ 1 & 3 \end{bmatrix} \begin{bmatrix} 1 & 2 & 3 & 4 \\ 0 & 2 & -1 & 6 \end{bmatrix}$ **b.** $\begin{bmatrix} 1 & 2 & 3 & 4 \\ 0 & 2 & -1 & 6 \end{bmatrix} \begin{bmatrix} 4 & 2 \\ 1 & 3 \end{bmatrix}$.

Solution

a. The first matrix is a 2×2 matrix and the second is a 2×4 matrix. The product will be a 2×4 matrix.

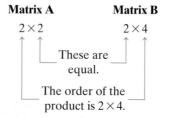

Matrix A   **Matrix B**
2×2 2×4

These are equal.

The order of the product is 2×4.

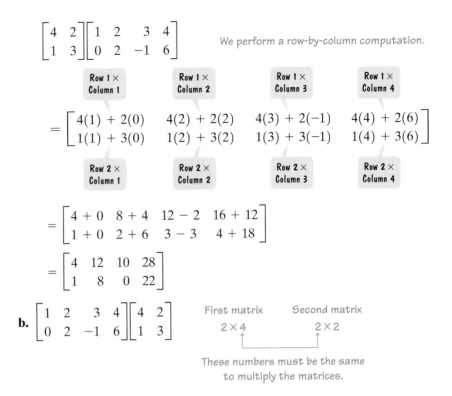

$$\begin{bmatrix} 4 & 2 \\ 1 & 3 \end{bmatrix}\begin{bmatrix} 1 & 2 & 3 & 4 \\ 0 & 2 & -1 & 6 \end{bmatrix}$$

We perform a row-by-column computation.

Row 1 × Column 1   Row 1 × Column 2   Row 1 × Column 3   Row 1 × Column 4

$$=\begin{bmatrix} 4(1)+2(0) & 4(2)+2(2) & 4(3)+2(-1) & 4(4)+2(6) \\ 1(1)+3(0) & 1(2)+3(2) & 1(3)+3(-1) & 1(4)+3(6) \end{bmatrix}$$

Row 2 × Column 1   Row 2 × Column 2   Row 2 × Column 3   Row 2 × Column 4

$$=\begin{bmatrix} 4+0 & 8+4 & 12-2 & 16+12 \\ 1+0 & 2+6 & 3-3 & 4+18 \end{bmatrix}$$

$$=\begin{bmatrix} 4 & 12 & 10 & 28 \\ 1 & 8 & 0 & 22 \end{bmatrix}$$

b. $\begin{bmatrix} 1 & 2 & 3 & 4 \\ 0 & 2 & -1 & 6 \end{bmatrix}\begin{bmatrix} 4 & 2 \\ 1 & 3 \end{bmatrix}$

First matrix   Second matrix

2×4 2×2

These numbers must be the same to multiply the matrices.

The number of columns in the first matrix does not equal the number of rows in the second matrix. Thus, the product of these two matrices is undefined.

Check Point 7 Where possible, find each product:

a. $\begin{bmatrix} 1 & 3 \\ 0 & 2 \end{bmatrix}\begin{bmatrix} 2 & 3 & -1 & 6 \\ 0 & 5 & 4 & 1 \end{bmatrix}$ **b.** $\begin{bmatrix} 2 & 3 & -1 & 6 \\ 0 & 5 & 4 & 1 \end{bmatrix}\begin{bmatrix} 1 & 3 \\ 0 & 2 \end{bmatrix}.$

Although matrix multiplication is not commutative, it does obey many of the properties of real numbers.

Properties of Matrix Multiplication

If A, B, and C are matrices and c is a scalar, then the following properties are true. (Assume the order of each matrix is such that all operations in these properties are defined.)

1. $(AB)C = A(BC)$ Associative Property of Matrix Multiplication

2. $A(B + C) = AB + AC$ Distributive Properties of Matrix
$(A + B)C = AC + BC$ Multiplication

3. $c(AB) = (cA)B$ Associative Property of Scalar Multiplication

 Describe applied situations with matrix operations.

Applications

All of the still images that you see on the Web have been created or manipulated on a computer in a digital format—made up of hundreds of thousands, or even millions, of tiny squares called **pixels**. Pixels are created by dividing an image into a grid. The computer can change the brightness of every square or pixel in this grid. A digital camera captures photos in this digital format. Also, you can scan pictures to convert them into digital format. Example 8 illustrates the role that matrices play in this new technology.

EXAMPLE 8   Matrices and Digital Photography

The letter L in Figure 8.7 is shown using 9 pixels in a 3 × 3 grid. The colors possible in the grid are shown in Figure 8.8. Each color is represented by a specific number 0, 1, 2, or 3.

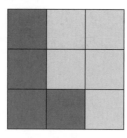

Figure 8.7 The letter L

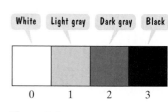

Figure 8.8 Color levels

a. Find a matrix that represents a digital photograph of this letter L.

b. Increase the contrast of the letter L by changing the dark gray to black and the light gray to white. Use matrix addition to accomplish this.

Solution

a. Look at the L and the background in Figure 8.7. Because the L is dark gray, color level 2, and the background is light gray, color level 1, a digital photograph of Figure 8.7 can be represented by the matrix

$$\begin{bmatrix} 2 & 1 & 1 \\ 2 & 1 & 1 \\ 2 & 2 & 1 \end{bmatrix}.$$

b. We can make the L black, color level 3, by increasing each 2 in the above matrix to 3. We can make the background white, color level 0, by decreasing each 1 in the above matrix to 0. This is accomplished using the following matrix addition:

$$\begin{bmatrix} 2 & 1 & 1 \\ 2 & 1 & 1 \\ 2 & 2 & 1 \end{bmatrix} + \begin{bmatrix} 1 & -1 & -1 \\ 1 & -1 & -1 \\ 1 & 1 & -1 \end{bmatrix} = \begin{bmatrix} 3 & 0 & 0 \\ 3 & 0 & 0 \\ 3 & 3 & 0 \end{bmatrix}.$$

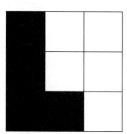

Figure 8.9 Changing contrast: the letter L

The picture corresponding to the matrix sum to the right of the equal sign is shown in Figure 8.9.

 Check Point 8 Change the contrast of the letter L in Figure 8.7 by making the L light gray and the background black. Use matrix addition to accomplish this.

Images of Space

Photographs sent back from space use matrices with thousands of pixels. Each pixel is assigned a number from 0 to 63 representing its color—0 for pure white and 63 for pure black. In the image of Saturn shown here, matrix operations provide false colors that emphasize the banding of the planet's upper atmosphere.

We have seen how functions can be transformed using translations, reflections, stretching, and shrinking. In a similar way, matrix operations are used to transform and manipulate computer graphics.

EXAMPLE 9   Transformations of an Image

The quadrilateral in Figure 8.10 can be represented by the matrix

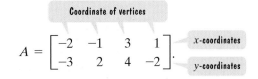

$$A = \begin{bmatrix} -2 & -1 & 3 & 1 \\ -3 & 2 & 4 & -2 \end{bmatrix}.$$

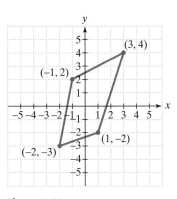

Figure 8.10

Each column in the matrix gives the coordinates of a vertex, or corner, of the quadrilateral. Use matrix operations to perfom the following transformations:

a. Move the quadrilateral 4 units to the right and 1 unit down.

b. Shrink the quadrilateral to half its perimeter.

c. Let $B = \begin{bmatrix} -1 & 0 \\ 0 & 1 \end{bmatrix}$. Find BA. What effect does this have on the quadrilateral in Figure 8.10?

Solution

a. We translate the quadrilateral 4 units right and 1 unit down by adding 4 to each x-coordinate and subtracting 1 from each y-coordinate. This is accomplished using the following matrix addition:

$$\begin{bmatrix} -2 & -1 & 3 & 1 \\ -3 & 2 & 4 & -2 \end{bmatrix} + \begin{bmatrix} 4 & 4 & 4 & 4 \\ -1 & -1 & -1 & -1 \end{bmatrix} = \begin{bmatrix} 2 & 3 & 7 & 5 \\ -4 & 1 & 3 & -3 \end{bmatrix}.$$

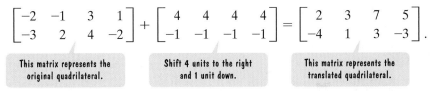

This matrix represents the original quadrilateral.   Shift 4 units to the right and 1 unit down.   This matrix represents the translated quadrilateral.

Each column in the matrix on the right gives the coordinates of a vertex of the translated quadrilateral. The original quadrilateral and the translated image are shown in Figure 8.11.

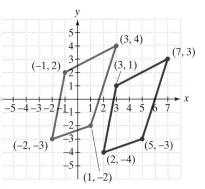

Figure 8.11 Shifting the quadrilateral 4 units right and 1 unit down

b. We shrink the quadrilateral in Figure 8.10 to half its perimeter by multiplying each x-coordinate and each y-coordinate by $\frac{1}{2}$. This is accomplished using the following scalar multiplication:

$$\frac{1}{2}\begin{bmatrix} -2 & -1 & 3 & 1 \\ -3 & 2 & 4 & -2 \end{bmatrix} = \begin{bmatrix} -1 & -\frac{1}{2} & \frac{3}{2} & \frac{1}{2} \\ -\frac{3}{2} & 1 & 2 & -1 \end{bmatrix}.$$

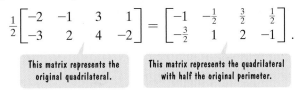

This matrix represents the original quadrilateral.   This matrix represents the quadrilateral with half the original perimeter.

Each column in the matrix on the right gives the coordinates of a vertex of the reduced quadrilateral. The original quadrilateral and the reduced image are shown in Figure 8.12.

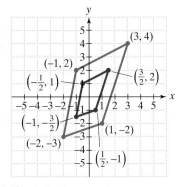

Figure 8.12 Shrinking the quadrilateral to half the original perimeter

c. We begin by finding BA. Keep in mind that A represents the original quadrilateral.

$$BA = \begin{bmatrix} -1 & 0 \\ 0 & 1 \end{bmatrix}\begin{bmatrix} -2 & -1 & 3 & 1 \\ -3 & 2 & 4 & -2 \end{bmatrix}$$

$$= \begin{bmatrix} (-1)(-2) + 0(-3) & (-1)(-1) + 0(2) & (-1)(3) + 0(4) & (-1)(1) + 0(-2) \\ 0(-2) + 1(-3) & 0(-1) + 1(2) & 0(3) + 1(4) & 0(1) + 1(-2) \end{bmatrix}$$

$$= \begin{bmatrix} 2 & 1 & -3 & -1 \\ -3 & 2 & 4 & -2 \end{bmatrix}$$

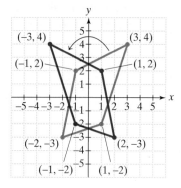

Figure 8.13

Each column in the matrix multiplication gives the coordinates of a vertex of the transformed image. The original quadrilateral and this transformed image are shown in Figure 8.13. Notice that each x-coordinate on the original blue image is replaced with its opposite on the transformed red image. We can conclude that

multiplication with $\begin{bmatrix} -1 & 0 \\ 0 & 1 \end{bmatrix}$ reflected the blue quadrilateral about the y-axis.

Check Point 9 Consider the triangle represented by the matrix

$$A = \begin{bmatrix} 0 & 3 & 4 \\ 0 & 5 & 2 \end{bmatrix}.$$

Use matrix operations to perform the following transformations:

a. Move the triangle 3 units to the left and 1 unit down.

b. Enlarge the triangle to twice its original perimeter.

Illustrate your results in parts (a) and (b) by showing the original triangle and the transformed image in a rectangular coordinate system.

c. Let $B = \begin{bmatrix} 1 & 0 \\ 0 & -1 \end{bmatrix}$. Find BA. What effect does this have on the original triangle?

EXERCISE SET 8.3

Practice Exercises

In Exercises 1–4,

a. *Give the order of each matrix.*

b. *If $A = [a_{ij}]$, identify a_{32}, and a_{23}, or explain why identification is not possible.*

1. $\begin{bmatrix} 4 & -7 & 5 \\ -6 & 8 & -1 \end{bmatrix}$

2. $\begin{bmatrix} -6 & 4 & -1 \\ -9 & 0 & \frac{1}{2} \end{bmatrix}$

3. $\begin{bmatrix} 1 & -5 & \pi & e \\ 0 & 7 & -6 & -\pi \\ -2 & \frac{1}{2} & 11 & -\frac{1}{5} \end{bmatrix}$

4. $\begin{bmatrix} -4 & 1 & 3 & -5 \\ 2 & -1 & \pi & 0 \\ 1 & 0 & -e & \frac{1}{5} \end{bmatrix}$

In Exercises 5–8, find values for the variables so that the matrices in each exercise are equal.

5. $\begin{bmatrix} x \\ 4 \end{bmatrix} = \begin{bmatrix} 6 \\ y \end{bmatrix}$

6. $\begin{bmatrix} x \\ 7 \end{bmatrix} = \begin{bmatrix} 11 \\ y \end{bmatrix}$

7. $\begin{bmatrix} x & 2y \\ z & 9 \end{bmatrix} = \begin{bmatrix} 4 & 12 \\ 3 & 9 \end{bmatrix}$

8. $\begin{bmatrix} x & y+3 \\ 2z & 8 \end{bmatrix} = \begin{bmatrix} 12 & 5 \\ 6 & 8 \end{bmatrix}$

In Exercises 9–16, find the following matrices:

a. $A + B$ **b.** $A - B$
c. $-4A$ **d.** $3A + 2B.$

9. $A = \begin{bmatrix} 4 & 1 \\ 3 & 2 \end{bmatrix}$, $B = \begin{bmatrix} 5 & 9 \\ 0 & 7 \end{bmatrix}$

10. $A = \begin{bmatrix} -2 & 3 \\ 0 & 1 \end{bmatrix}$, $B = \begin{bmatrix} 8 & 1 \\ 5 & 4 \end{bmatrix}$

11. $A = \begin{bmatrix} 1 & 3 \\ 3 & 4 \\ 5 & 6 \end{bmatrix}$, $B = \begin{bmatrix} 2 & -1 \\ 3 & -2 \\ 0 & 1 \end{bmatrix}$

12. $A = \begin{bmatrix} 3 & 1 & 1 \\ -1 & 2 & 5 \end{bmatrix}$, $B = \begin{bmatrix} 2 & -3 & 6 \\ -3 & 1 & -4 \end{bmatrix}$

13. $A = \begin{bmatrix} 2 \\ -4 \\ 1 \end{bmatrix}$, $B = \begin{bmatrix} -5 \\ 3 \\ -1 \end{bmatrix}$

14. $A = \begin{bmatrix} 6 & 2 & -3 \end{bmatrix}$, $B = \begin{bmatrix} 4 & -2 & 3 \end{bmatrix}$

15. $A = \begin{bmatrix} 2 & -10 & -2 \\ 14 & 12 & 10 \\ 4 & -2 & 2 \end{bmatrix}$, $B = \begin{bmatrix} 6 & 10 & -2 \\ 0 & -12 & -4 \\ -5 & 2 & -2 \end{bmatrix}$

16. $A = \begin{bmatrix} 6 & -3 & 5 \\ 6 & 0 & -2 \\ -4 & 2 & -1 \end{bmatrix}$, $B = \begin{bmatrix} -3 & 5 & 1 \\ -1 & 2 & -6 \\ 2 & 0 & 4 \end{bmatrix}$

In Exercises 17–26, let

$$A = \begin{bmatrix} -3 & -7 \\ 2 & -9 \\ 5 & 0 \end{bmatrix} \text{ and } B = \begin{bmatrix} -5 & -1 \\ 0 & 0 \\ 3 & -4 \end{bmatrix}.$$

Solve each matrix equation for X.

17. $X - A = B$

18. $X - B = A$

19. $2X + A = B$

20. $3X + A = B$

1. $3X + 2A = B$

22. $2X + 5A = B$

3. $B - X = 4A$

24. $A - X = 4B$

5. $4A + 3B = -2X$

26. $4B + 3A = -2X$

n Exercises 27–36, find (if possible) the following matrices:

   a. AB **b.** BA.

7. $A = \begin{bmatrix} 1 & 3 \\ 5 & 3 \end{bmatrix}$, $B = \begin{bmatrix} 3 & -2 \\ -1 & 6 \end{bmatrix}$

8. $A = \begin{bmatrix} 3 & -2 \\ 1 & 5 \end{bmatrix}$, $B = \begin{bmatrix} 0 & 0 \\ 5 & -6 \end{bmatrix}$

9. $A = \begin{bmatrix} 1 & 2 & 3 & 4 \end{bmatrix}$, $B = \begin{bmatrix} 1 \\ 2 \\ 3 \\ 4 \end{bmatrix}$

0. $A = \begin{bmatrix} -1 \\ -2 \\ -3 \end{bmatrix}$, $B = \begin{bmatrix} 1 & 2 & 3 \end{bmatrix}$

1. $A = \begin{bmatrix} 1 & -1 & 4 \\ 4 & -1 & 3 \\ 2 & 0 & -2 \end{bmatrix}$, $B = \begin{bmatrix} 1 & 1 & 0 \\ 1 & 2 & 4 \\ 1 & -1 & 3 \end{bmatrix}$

2. $A = \begin{bmatrix} 1 & -1 & 1 \\ 5 & 0 & -2 \\ 3 & -2 & 2 \end{bmatrix}$, $B = \begin{bmatrix} 1 & 1 & 0 \\ 1 & -4 & 5 \\ 3 & -1 & 2 \end{bmatrix}$

3. $A = \begin{bmatrix} 4 & 2 \\ 6 & 1 \\ 3 & 5 \end{bmatrix}$, $B = \begin{bmatrix} 2 & 3 & 4 \\ -1 & -2 & 0 \end{bmatrix}$

4. $A = \begin{bmatrix} 2 & 4 \\ 3 & 1 \\ 4 & 2 \end{bmatrix}$, $B = \begin{bmatrix} 3 & 2 & 0 \\ -1 & -3 & 5 \end{bmatrix}$

5. $A = \begin{bmatrix} 2 & -3 & 1 & -1 \\ 1 & 1 & -2 & 1 \end{bmatrix}$, $B = \begin{bmatrix} 1 & 2 \\ -1 & 1 \\ 5 & 4 \\ 10 & 5 \end{bmatrix}$

6. $A = \begin{bmatrix} 2 & -1 & 3 & 2 \\ 1 & 0 & -2 & 1 \end{bmatrix}$, $B = \begin{bmatrix} -1 & 2 \\ 1 & 1 \\ 3 & -4 \\ 6 & 5 \end{bmatrix}$

n Exercises 37–44, perform the indicated matrix operations given hat A, B, and C are defined as follows. If an operation is not defined, state the reason.

$$A = \begin{bmatrix} 4 & 0 \\ -3 & 5 \\ 0 & 1 \end{bmatrix} \quad B = \begin{bmatrix} 5 & 1 \\ -2 & -2 \end{bmatrix} \quad C = \begin{bmatrix} 1 & -1 \\ -1 & 1 \end{bmatrix}$$

37. $4B - 3C$

38. $5C - 2B$

39. $BC + CB$

40. $A(B + C)$

41. $A - C$

42. $B - A$

43. $A(BC)$

44. $A(CB)$

Practice Plus

In Exercises 45–50, let

$$A = \begin{bmatrix} 1 & 0 \\ 0 & 1 \end{bmatrix}, \quad B = \begin{bmatrix} 1 & 0 \\ 0 & -1 \end{bmatrix}, \quad C = \begin{bmatrix} -1 & 0 \\ 0 & 1 \end{bmatrix},$$

$$D = \begin{bmatrix} -1 & 0 \\ 0 & -1 \end{bmatrix}.$$

45. Find the product of the sum of A and B and the difference between C and D.

46. Find the product of the difference between A and B and the sum of C and D.

47. Use any three of the matrices to verify a distributive property.

48. Use any three of the matrices to verify an associative property.

In Exercises 49–50, suppose that the vertices of a computer graphic are points, (x, y), represented by the matrix

$$Z = \begin{bmatrix} x \\ y \end{bmatrix}.$$

49. Find BZ and explain why this reflects the graphic about the x-axis.

50. Find CZ and explain why this reflects the graphic about the y-axis.

Application Exercises

The + sign in the figure is shown using 9 pixels in a 3×3 grid. The color levels are given to the right of the figure. Each color is represented by a specific number: 0, 1, 2, or 3. Use this information to solve Exercises 51–52.

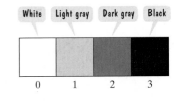

51. a. Find a matrix that represents a digital photograph of the + sign.

   b. Adjust the contrast by changing the black to dark gray and the light gray to white. Use matrix addition to accomplish this.

   c. Adjust the contrast by changing the black to light gray and the light gray to dark gray. Use matrix addition to accomplish this.

52. a. Find a matrix that represents a digital photograph of the + sign.

   b. Adjust the contrast by changing the black to dark gray and the light gray to black. Use matrix addition to accomplish this.

   c. Adjust the contrast by leaving the black alone and changing the light gray to white. Use matrix addition to accomplish this.

The figure shows the letter L in a rectangular coordinate system.

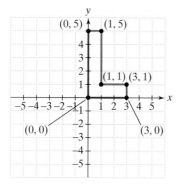

The figure can be represented by the matrix

$$B = \begin{bmatrix} 0 & 3 & 3 & 1 & 1 & 0 \\ 0 & 0 & 1 & 1 & 5 & 5 \end{bmatrix}.$$

Each column in the matrix describes a point on the letter. The order of the columns shows the direction in which a pencil must move to draw the letter. The L is completed by connecting the last point in the matrix, $(0, 5)$, to the starting point, $(0, 0)$. Use these ideas to solve Exercises 53–60.

53. Use matrix operations to move the L 2 units to the left and 3 units down. Then graph the letter and its transformation in a rectangular coordinate system.

54. Use matrix operations to move the L 2 units to the right and 3 units down. Then graph the letter and its transformation in a rectangular coordinate system.

55. Reduce the L to half its perimeter and move the reduced image 1 unit up. Then graph the letter and its transformation.

56. Reduce the L to half its perimeter and move the reduced image 2 units up. Then graph the letter and its transformation.

57. a. If $A = \begin{bmatrix} 1 & 0 \\ 0 & -1 \end{bmatrix}$, find AB.

b. Graph the object represented by matrix AB. What effect does the matrix multiplication have on the letter L represented by matrix B?

58. a. If $A = \begin{bmatrix} -1 & 0 \\ 0 & 1 \end{bmatrix}$, find AB.

b. Graph the object represented by matrix AB. What effect does the matrix multiplication have on the letter L represented by matrix B?

59. a. If $A = \begin{bmatrix} 0 & -1 \\ 1 & 0 \end{bmatrix}$, find AB.

b. Graph the object represented by matrix AB. What effect does the matrix multiplication have on the letter L represented by matrix B?

60. a. If $A = \begin{bmatrix} 2 & 0 \\ 0 & 1 \end{bmatrix}$, find AB.

b. Graph the object represented by matrix AB. What effect does the matrix multiplication have on the letter L represented by matrix B?

61. The graphs show how Republicans and Democrats identifie themselves in 2000 and 2004.

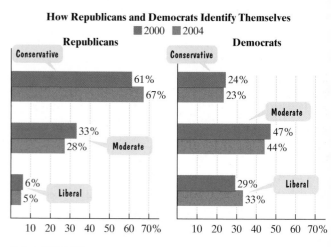

Source: Gallup Polls

a. Use a 3×2 matrix to represent the information for 2000 Entries in the matrix should be percents that are organized as follows:

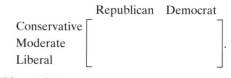

Call this matrix A.

b. Use a 3×2 matrix to represent the information given for 2004. Call this matrix B.

c. Find $B - A$. What does this matrix represent?

62. The graphs show the average weights of U.S. children, ages 10 and 15, in 1966 and 2002.

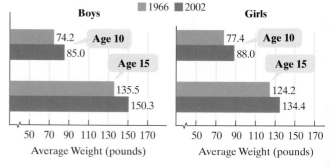

Source: National Center for Health Statistics

a. Use a 2×2 matrix to organize the data for 1966 as follows

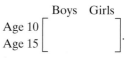

Call this matrix A.

b. Use a 2×2 matrix to organize the information given for 2002. Call this matrix B.

c. Find $B - A$. What does this matrix represent?

The solution is given by $X = A^{-1}B$. Consequently, we must find A^{-1}. We found the inverse of matrix A in Example 4. Using this result,

$$X = A^{-1}B = \begin{bmatrix} 3 & -3 & 1 \\ -2 & 2 & -1 \\ -4 & 5 & -2 \end{bmatrix}\begin{bmatrix} 2 \\ 2 \\ \frac{1}{2} \end{bmatrix} = \begin{bmatrix} 3\cdot2 + (-3)\cdot2 + 1\cdot\frac{1}{2} \\ -2\cdot2 + 2\cdot2 + (-1)\cdot\frac{1}{2} \\ -4\cdot2 + 5\cdot2 + (-2)\cdot\frac{1}{2} \end{bmatrix} = \begin{bmatrix} \frac{1}{2} \\ -\frac{1}{2} \\ 1 \end{bmatrix}.$$

Thus, $x = \frac{1}{2}$, $y = -\frac{1}{2}$, and $z = 1$. The solution set is $\left\{\left(\frac{1}{2}, -\frac{1}{2}, 1\right)\right\}$.

Technology

We can use a graphing utility to solve a linear system with a unique solution by entering the elements in A, the coefficient matrix, and B, the column matrix. Then find the product of A^{-1} and B. The screen on the right verifies our solution in Example 5.

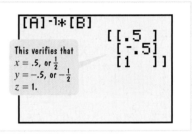

This verifies that
$x = .5$, or $\frac{1}{2}$
$y = -.5$, or $-\frac{1}{2}$
$z = 1$.

Check Point 5 Solve the system by using A^{-1}, the inverse of the coefficient matrix that you found in Check Point 4:

$$\begin{aligned} x \quad\quad + 2z &= 6 \\ -x + 2y + 3z &= -5 \\ x - y \quad\quad &= 6. \end{aligned}$$

③ Encode and decode messages.

Applications of Matrix Inverses to Coding

A **cryptogram** is a message written so that no one other than the intended recipient can understand it. To encode a message, we begin by assigning a number to each letter in the alphabet: $A = 1, B = 2, C = 3, \ldots, Z = 26$, and a space $= 0$. For example, the numerical equivalent of the word MATH is 13, 1, 20, 8. The numerical equivalent of the message is then converted into a matrix. Finally, an invertible matrix can be used to convert the message into code. The multiplicative inverse of this matrix can be used to decode the message.

Encoding a Word or Message

1. Express the word or message numerically.

2. List the numbers in step 1 by columns and form a square matrix. If you do not have enough numbers to form a square matrix, put zeros in any remaining spaces in the last column.

3. Select any square invertible matrix, called the **coding matrix**, the same size as the matrix in step 2. Multiply the coding matrix by the square matrix that expresses the message numerically. The resulting matrix is the **coded matrix**.

4. Use the numbers, by columns, from the coded matrix in step 3 to write the encoded message.

EXAMPLE 6   Encoding a Word

Use matrices to encode the word MATH.

Solution

Step 1   Express the word numerically. As shown previously, the numerical equivalent of MATH is 13, 1, 20, 8.

Step 2   List the numbers in step 1 by columns and form a square matrix. The 2×2 matrix for the numerical equivalent of MATH, 13, 1, 20, 8, is

$$\begin{bmatrix} 13 & 20 \\ 1 & 8 \end{bmatrix}.$$

Step 3   Multiply the matrix in step 2 by a square invertible matrix. We will use $\begin{bmatrix} -2 & -3 \\ 3 & 4 \end{bmatrix}$ as the coding matrix.

$$\begin{bmatrix} -2 & -3 \\ 3 & 4 \end{bmatrix} \begin{bmatrix} 13 & 20 \\ 1 & 8 \end{bmatrix} = \begin{bmatrix} -2(13) - 3(1) & -2(20) - 3(8) \\ 3(13) + 4(1) & 3(20) + 4(8) \end{bmatrix}$$

Coding matrix   Numerical representation of MATH

$$= \begin{bmatrix} -29 & -64 \\ 43 & 92 \end{bmatrix}$$

Coded matrix

Step 4   Use the numbers, by columns, from the coded matrix in step 3 to write the encoded message. The encoded message is $-29, 43, -64, 92$.

Check Point 6   Use the coding matrix in Example 6, $\begin{bmatrix} -2 & -3 \\ 3 & 4 \end{bmatrix}$, to encode the word BASE

The inverse of a coding matrix can be used to decode a word or message that was encoded.

Decoding a Word or Message That Was Encoded

1. Find the multiplicative inverse of the coding matrix.
2. Multiply the multiplicative inverse of the coding matrix and the coded matrix.
3. Express the numbers, by columns, from the matrix in step 2 as letters.

EXAMPLE 7   Decoding a Word

Decode $-29, 43, -64, 92$ from Example 6.

Solution

Step 1   Find the inverse of the coding matrix. The coding matrix in Example 6 was $\begin{bmatrix} -2 & -3 \\ 3 & 4 \end{bmatrix}$. We use the formula for the multiplicative inverse of a 2×2 matrix to find the multiplicative inverse of this matrix. It is $\begin{bmatrix} 4 & 3 \\ -3 & -2 \end{bmatrix}$.

Step 2   Multiply the multiplicative inverse of the coding matrix and the coded matrix.

$$\begin{bmatrix} 4 & 3 \\ -3 & -2 \end{bmatrix} \begin{bmatrix} -29 & -64 \\ 43 & 92 \end{bmatrix} = \begin{bmatrix} 4(-29) + 3(43) & 4(-64) + 3(92) \\ -3(-29) - 2(43) & -3(-64) - 2(92) \end{bmatrix}$$

Multiplicative inverse of the coding matrix   Coded matrix

$$= \begin{bmatrix} 13 & 20 \\ 1 & 8 \end{bmatrix}$$

Step 3   Express the numbers, by columns, from the matrix in step 2 as letters. The numbers are 13, 1, 20, and 8. Using letters, the decoded message is MATH.

Check Point 7 Decode the word that you encoded in Check Point 6.

Decoding is simple for an authorized receiver who knows the coding matrix. Because any invertible matrix can be used for the coding matrix, decoding a cryptogram for an unauthorized receiver who does not know this matrix is extremely difficult.

EXERCISE SET 8.4

Practice Exercises

In Exercises 1–12, find the products AB and BA to determine whether B is the multiplicative inverse of A.

1. $A = \begin{bmatrix} 4 & -3 \\ -5 & 4 \end{bmatrix}$, $B = \begin{bmatrix} 4 & 3 \\ 5 & 4 \end{bmatrix}$

2. $A = \begin{bmatrix} -2 & -1 \\ -1 & 1 \end{bmatrix}$, $B = \begin{bmatrix} 1 & 1 \\ 1 & 2 \end{bmatrix}$

3. $A = \begin{bmatrix} -4 & 0 \\ 1 & 3 \end{bmatrix}$, $B = \begin{bmatrix} -2 & 4 \\ 0 & 1 \end{bmatrix}$

4. $A = \begin{bmatrix} -2 & 4 \\ 1 & -2 \end{bmatrix}$, $B = \begin{bmatrix} 1 & 2 \\ -1 & -2 \end{bmatrix}$

5. $A = \begin{bmatrix} -2 & 1 \\ \frac{3}{2} & -\frac{1}{2} \end{bmatrix}$, $B = \begin{bmatrix} 1 & 2 \\ 3 & 4 \end{bmatrix}$

6. $A = \begin{bmatrix} 4 & 5 \\ 2 & 3 \end{bmatrix}$, $B = \begin{bmatrix} \frac{3}{2} & -\frac{5}{2} \\ -1 & 2 \end{bmatrix}$

7. $A = \begin{bmatrix} 0 & 1 & 0 \\ 0 & 0 & 1 \\ 1 & 0 & 0 \end{bmatrix}$, $B = \begin{bmatrix} 0 & 0 & 1 \\ 1 & 0 & 0 \\ 0 & 1 & 0 \end{bmatrix}$

8. $A = \begin{bmatrix} -2 & 1 & -1 \\ -5 & 2 & -1 \\ 3 & -1 & 1 \end{bmatrix}$, $B = \begin{bmatrix} 1 & 0 & 1 \\ 2 & 1 & 3 \\ -1 & 1 & 1 \end{bmatrix}$

9. $A = \begin{bmatrix} 1 & 2 & 3 \\ 1 & 3 & 4 \\ 1 & 4 & 3 \end{bmatrix}$, $B = \begin{bmatrix} \frac{7}{2} & -3 & \frac{1}{2} \\ -\frac{1}{2} & 0 & \frac{1}{2} \\ -\frac{1}{2} & 1 & -\frac{1}{2} \end{bmatrix}$

10. $A = \begin{bmatrix} 0 & 2 & 0 \\ 3 & 3 & 2 \\ 2 & 5 & 1 \end{bmatrix}$, $B = \begin{bmatrix} -3.5 & -1 & 2 \\ 0.5 & 0 & 0 \\ 4.5 & 2 & -3 \end{bmatrix}$

11. $A = \begin{bmatrix} 0 & 0 & -2 & 1 \\ -1 & 0 & 1 & 1 \\ 0 & 1 & -1 & 0 \\ 1 & 0 & 0 & -1 \end{bmatrix}$, $B = \begin{bmatrix} 1 & 2 & 0 & 3 \\ 0 & 1 & 1 & 1 \\ 0 & 1 & 0 & 1 \\ 1 & 2 & 0 & 2 \end{bmatrix}$

12. $A = \begin{bmatrix} 1 & -2 & 1 & 0 \\ 0 & 1 & -2 & 1 \\ 0 & 0 & 1 & -2 \\ 0 & 0 & 0 & 1 \end{bmatrix}$, $B = \begin{bmatrix} 1 & 2 & 3 & 4 \\ 0 & 1 & 2 & 3 \\ 0 & 0 & 1 & 2 \\ 0 & 0 & 0 & 1 \end{bmatrix}$

In Exercises 13–18, use the fact that if $A = \begin{bmatrix} a & b \\ c & d \end{bmatrix}$, then

$A^{-1} = \dfrac{1}{ad - bc} \begin{bmatrix} d & -b \\ -c & a \end{bmatrix}$ *to find the inverse of each matrix, if possible. Check that $AA^{-1} = I_2$ and $A^{-1}A = I_2$.*

13. $A = \begin{bmatrix} 2 & 3 \\ -1 & 2 \end{bmatrix}$

14. $A = \begin{bmatrix} 0 & 3 \\ 4 & -2 \end{bmatrix}$

15. $A = \begin{bmatrix} 3 & -1 \\ -4 & 2 \end{bmatrix}$

16. $A = \begin{bmatrix} 2 & -6 \\ 1 & -2 \end{bmatrix}$

17. $A = \begin{bmatrix} 10 & -2 \\ -5 & 1 \end{bmatrix}$

18. $A = \begin{bmatrix} 6 & -3 \\ -2 & 1 \end{bmatrix}$

In Exercises 19–28, find A^{-1} by forming $[A|I]$ and then using row operations to obtain $[I|B]$, where $A^{-1} = [B]$. Check that $AA^{-1} = I$ and $A^{-1}A = I$.

19. $A = \begin{bmatrix} 2 & 0 & 0 \\ 0 & 4 & 0 \\ 0 & 0 & 6 \end{bmatrix}$

20. $A = \begin{bmatrix} 3 & 0 & 0 \\ 0 & 6 & 0 \\ 0 & 0 & 9 \end{bmatrix}$

21. $A = \begin{bmatrix} 1 & 2 & -1 \\ -2 & 0 & 1 \\ 1 & -1 & 0 \end{bmatrix}$

22. $A = \begin{bmatrix} 1 & -1 & 1 \\ 0 & 2 & -1 \\ 2 & 3 & 0 \end{bmatrix}$

23. $A = \begin{bmatrix} 2 & 2 & -1 \\ 0 & 3 & -1 \\ -1 & -2 & 1 \end{bmatrix}$

24. $A = \begin{bmatrix} 2 & 4 & -4 \\ 1 & 3 & -4 \\ 2 & 4 & -3 \end{bmatrix}$

25. $A = \begin{bmatrix} 5 & 0 & 2 \\ 2 & 2 & 1 \\ -3 & 1 & -1 \end{bmatrix}$

26. $A = \begin{bmatrix} 3 & 2 & 6 \\ 1 & 1 & 2 \\ 2 & 2 & 5 \end{bmatrix}$

27. $A = \begin{bmatrix} 1 & 0 & 0 & 0 \\ 0 & -1 & 0 & 0 \\ 0 & 0 & 3 & 0 \\ 1 & 0 & 0 & 1 \end{bmatrix}$

28. $A = \begin{bmatrix} 2 & 0 & 0 & 1 \\ 0 & 1 & 0 & 0 \\ 0 & 0 & -1 & 0 \\ 0 & 0 & 0 & 2 \end{bmatrix}$

In Exercises 29–32, write each linear system as a matrix equation in the form $AX = B$, where A is the coefficient matrix and B is the constant matrix.

29. $6x + 5y = 13$
    $5x + 4y = 10$

30. $7x + 5y = 23$
    $3x + 2y = 10$

31. $x + 3y + 4z = -3$
    $x + 2y + 3z = -2$
    $x + 4y + 3z = -6$

32. $x + 4y - z = 3$
    $x + 3y - 2z = 5$
    $2x + 7y - 5z = 12$

In Exercises 33–36, write each matrix equation as a system of linear equations without matrices.

33. $\begin{bmatrix} 4 & -7 \\ 2 & -3 \end{bmatrix} \begin{bmatrix} x \\ y \end{bmatrix} = \begin{bmatrix} -3 \\ 1 \end{bmatrix}$

34. $\begin{bmatrix} 3 & 0 \\ -3 & 1 \end{bmatrix} \begin{bmatrix} x \\ y \end{bmatrix} = \begin{bmatrix} 6 \\ -7 \end{bmatrix}$

35. $\begin{bmatrix} 2 & 0 & -1 \\ 0 & 3 & 0 \\ 1 & 1 & 0 \end{bmatrix} \begin{bmatrix} x \\ y \\ z \end{bmatrix} = \begin{bmatrix} 6 \\ 9 \\ 5 \end{bmatrix}$

36. $\begin{bmatrix} -1 & 0 & 1 \\ 0 & -1 & 0 \\ 0 & 1 & 1 \end{bmatrix} \begin{bmatrix} x \\ y \\ z \end{bmatrix} = \begin{bmatrix} -4 \\ 2 \\ 4 \end{bmatrix}$

In Exercises 37–42,

    a. *Write each linear system as a matrix equation in the form* $AX = B$.

    b. *Solve the system using the inverse that is given for the coefficient matrix.*

37. $\begin{aligned} 2x + 6y + 6z &= 8 \\ 2x + 7y + 6z &= 10 \\ 2x + 7y + 7z &= 9 \end{aligned}$

The inverse of
$\begin{bmatrix} 2 & 6 & 6 \\ 2 & 7 & 6 \\ 2 & 7 & 7 \end{bmatrix}$ is $\begin{bmatrix} \frac{7}{2} & 0 & -3 \\ -1 & 1 & 0 \\ 0 & -1 & 1 \end{bmatrix}$.

38. $\begin{aligned} x + 2y + 5z &= 2 \\ 2x + 3y + 8z &= 3 \\ -x + y + 2z &= 3 \end{aligned}$

The inverse of
$\begin{bmatrix} 1 & 2 & 5 \\ 2 & 3 & 8 \\ -1 & 1 & 2 \end{bmatrix}$ is $\begin{bmatrix} 2 & -1 & -1 \\ 12 & -7 & -2 \\ -5 & 3 & 1 \end{bmatrix}$.

39. $\begin{aligned} x - y + z &= 8 \\ 2y - z &= -7 \\ 2x + 3y &= 1 \end{aligned}$

The inverse of
$\begin{bmatrix} 1 & -1 & 1 \\ 0 & 2 & -1 \\ 2 & 3 & 0 \end{bmatrix}$ is $\begin{bmatrix} 3 & 3 & -1 \\ -2 & -2 & 1 \\ -4 & -5 & 2 \end{bmatrix}$.

40. $\begin{aligned} x - 6y + 3z &= 11 \\ 2x - 7y + 3z &= 14 \\ 4x - 12y + 5z &= 25 \end{aligned}$

The inverse of
$\begin{bmatrix} 1 & -6 & 3 \\ 2 & -7 & 3 \\ 4 & -12 & 5 \end{bmatrix}$ is $\begin{bmatrix} 1 & -6 & 3 \\ 2 & -7 & 3 \\ 4 & -12 & 5 \end{bmatrix}$.

41. $\begin{aligned} w - x + 2y &= -3 \\ x - y + z &= 4 \\ -w + x - y + 2z &= 2 \\ -x + y - 2z &= -4 \end{aligned}$

The inverse of
$\begin{bmatrix} 1 & -1 & 2 & 0 \\ 0 & 1 & -1 & 1 \\ -1 & 1 & -1 & 2 \\ 0 & -1 & 1 & -2 \end{bmatrix}$ is $\begin{bmatrix} 0 & 0 & -1 & -1 \\ 1 & 4 & 1 & 3 \\ 1 & 2 & 1 & 2 \\ 0 & -1 & 0 & -1 \end{bmatrix}$.

42. $\begin{aligned} 2w + y + z &= 6 \\ 3w + z &= 9 \\ -w + x - 2y + z &= 4 \\ 4w - x + y &= 6 \end{aligned}$

The inverse of
$\begin{bmatrix} 2 & 0 & 1 & 1 \\ 3 & 0 & 0 & 1 \\ -1 & 1 & -2 & 1 \\ 4 & -1 & 1 & 0 \end{bmatrix}$ is $\begin{bmatrix} -1 & 2 & -1 & -1 \\ -4 & 9 & -5 & -6 \\ 0 & 1 & -1 & -1 \\ 3 & -5 & 3 & 3 \end{bmatrix}$.

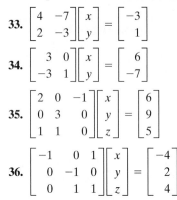

Practice Plus

In Exercises 43–44, find A^{-1} and check.

43. $A = \begin{bmatrix} e^x & e^{3x} \\ -e^{3x} & e^{5x} \end{bmatrix}$ **44.** $A = \begin{bmatrix} e^{2x} & -e^x \\ e^{3x} & e^{2x} \end{bmatrix}$

In Exercises 45–46, if I is the multiplicative identity matrix of order 2, find $(I - A)^{-1}$ for the given matrix A.

45. $\begin{bmatrix} 8 & -5 \\ -3 & 2 \end{bmatrix}$ **46.** $\begin{bmatrix} 7 & -5 \\ -4 & 3 \end{bmatrix}$

In Exercises 47–48, find $(AB)^{-1}$, $A^{-1}B^{-1}$, and $B^{-1}A^{-1}$. What do you observe?

47. $A = \begin{bmatrix} 2 & 1 \\ 3 & 1 \end{bmatrix}$ $B = \begin{bmatrix} 4 & 7 \\ 1 & 2 \end{bmatrix}$

48. $A = \begin{bmatrix} 2 & -9 \\ 1 & -4 \end{bmatrix}$ $B = \begin{bmatrix} 9 & 5 \\ 7 & 4 \end{bmatrix}$

49. Prove the following statement:

If $A = \begin{bmatrix} a & 0 & 0 \\ 0 & b & 0 \\ 0 & 0 & c \end{bmatrix}$, $a \neq 0, b \neq 0, c \neq 0$,

then $A^{-1} = \begin{bmatrix} \frac{1}{a} & 0 & 0 \\ 0 & \frac{1}{b} & 0 \\ 0 & 0 & \frac{1}{c} \end{bmatrix}$.

50. Prove the following statement:

If $A = \begin{bmatrix} a & b \\ c & d \end{bmatrix}$ and $ad - bc \neq 0$,

then $A^{-1} = \dfrac{1}{ad - bc} \begin{bmatrix} d & -b \\ -c & a \end{bmatrix}$.

    (*Hint:* Use the method of Example 2 on page 820.)

Application Exercises

In Exercises 51–52, use the coding matrix

$$A = \begin{bmatrix} 4 & -1 \\ -3 & 1 \end{bmatrix} \text{ and its inverse } A^{-1} = \begin{bmatrix} 1 & 1 \\ 3 & 4 \end{bmatrix}$$

to encode and then decode the given message.

51. HELP   **52.** LOVE

In Exercises 53–54, use the coding matrix

$$A = \begin{bmatrix} 1 & -1 & 0 \\ 3 & 0 & 2 \\ -1 & 0 & -1 \end{bmatrix} \text{ and its inverse}$$

$$A^{-1} = \begin{bmatrix} 0 & 1 & 2 \\ -1 & 1 & 2 \\ 0 & -1 & -3 \end{bmatrix} \text{ to write a cryptogram for each}$$

message. Check your result by decoding the cryptogram.

53. $\begin{array}{ccccccccc} \text{S} & \text{E} & \text{N} & \text{D} & \underline{\quad} & \text{C} & \text{A} & \text{S} & \text{H} \\ 19 & 5 & 14 & 4 & 0 & 3 & 1 & 19 & 8 \end{array}$

    Use $\begin{bmatrix} 19 & 4 & 1 \\ 5 & 0 & 19 \\ 14 & 3 & 8 \end{bmatrix}$.

64. S T A Y _ W E L L
19 20 1 25 $\overline{0}$ 23 5 12 12

Use $\begin{bmatrix} 19 & 25 & 5 \\ 20 & 0 & 12 \\ 1 & 23 & 12 \end{bmatrix}$.

Writing in Mathematics

65. What is the multiplicative identity matrix?

66. If you are given two matrices, A and B, explain how to determine if B is the multiplicative inverse of A.

67. Explain why a matrix that does not have the same number of rows and columns cannot have a multiplicative inverse.

68. Explain how to find the multiplicative inverse for a 2×2 invertible matrix.

69. Explain how to find the multiplicative inverse for a 3×3 invertible matrix.

70. Explain how to write a linear system of three equations in three variables as a matrix equation.

71. Explain how to solve the matrix equation $AX = B$.

72. What is a cryptogram?

73. It's January 1, and you've written down your major goal for the year. You do not want those closest to you to see what you've written in case you do not accomplish your objective. Consequently, you decide to use a coding matrix to encode your goal. Explain how this can be accomplished.

74. A year has passed since Exercise 63. (Time flies when you're solving exercises in algebra books.) It's been a terrific year and so many wonderful things have happened that you can't remember your goal from a year ago. You consult your personal journal and you find the encoded message and the coding matrix. How can you use these to find your original goal?

Technology Exercises

In Exercises 65–70, use a graphing utility to find the multiplicative inverse of each matrix. Check that the displayed inverse is correct.

65. $\begin{bmatrix} 3 & -1 \\ -2 & 1 \end{bmatrix}$

66. $\begin{bmatrix} -4 & 1 \\ 6 & -2 \end{bmatrix}$

67. $\begin{bmatrix} -2 & 1 & -1 \\ -5 & 2 & -1 \\ 3 & -1 & 1 \end{bmatrix}$

68. $\begin{bmatrix} 1 & 1 & -1 \\ -3 & 2 & -1 \\ 3 & -3 & 2 \end{bmatrix}$

69. $\begin{bmatrix} 7 & -3 & 0 & 2 \\ -2 & 1 & 0 & -1 \\ 4 & 0 & 1 & -2 \\ -1 & 1 & 0 & -1 \end{bmatrix}$

70. $\begin{bmatrix} 1 & 2 & 0 & 0 \\ 0 & 0 & 1 & 0 \\ 1 & 3 & 0 & 1 \\ 4 & 0 & 0 & 2 \end{bmatrix}$

In Exercises 71–76, write each system in the form $AX = B$. Then solve the system by entering A and B into your graphing utility and computing $A^{-1}B$.

71. $\begin{aligned} x - y + z &= -6 \\ 4x + 2y + z &= 9 \\ 4x - 2y + z &= -3 \end{aligned}$

72. $\begin{aligned} y + 2z &= 0 \\ -x + y &= 1 \\ 2x - y + z &= -1 \end{aligned}$

73. $\begin{aligned} 3x - 2y + z &= -2 \\ 4x - 5y + 3z &= -9 \\ 2x - y + 5z &= -5 \end{aligned}$

74. $\begin{aligned} x - y &= 1 \\ 6x + y + 20z &= 14 \\ y + 3z &= 1 \end{aligned}$

75. $\begin{aligned} v \quad\;\; -3x \quad\;\;\; + z &= -3 \\ w \quad\;\; + y \quad\;\;\;\; &= -1 \\ x \quad\;\;\; + z &= 7 \\ v + w - x + 4y \quad\;\;\; &= -8 \\ v + w + x + y + z &= 8 \end{aligned}$

76. $\begin{aligned} w + x + y + z &= 4 \\ w + 3x - 2y + 2z &= 7 \\ 2w + 2x + y + z &= 3 \\ w - x + 2y + 3z &= 5 \end{aligned}$

In Exercises 77–78, use a coding matrix A of your choice. Use a graphing utility to find the multiplicative inverse of your coding matrix. Write a cryptogram for each message. Check your result by decoding the cryptogram. Use your graphing utility to perform all necessary matrix multiplications.

77. A R R I V E D _ S A F E L Y
1 18 18 9 22 5 4 $\overline{0}$ 19 1 6 5 12 25

78. A R T _ E N R I C H E S
1 18 20 $\overline{0}$ 5 14 18 9 3 8 5 19

Critical Thinking Exercises

79. Which one of the following is true?

  a. Some nonsquare matrices have inverses.

  b. All square 2×2 matrices have inverses because there is a formula for finding these inverses.

  c. Two 2×2 invertible matrices can have a matrix sum that is not invertible.

  d. To solve the matrix equation $AX = B$ for X, multiply A and the inverse of B.

80. Which one of the following is true?

  a. $(AB)^{-1} = A^{-1}B^{-1}$, assuming A, B, and AB are invertible.

  b. $(A + B)^{-1} = A^{-1} + B^{-1}$, assuming A, B, and $A + B$ are invertible.

  c. $\begin{bmatrix} 1 & -3 \\ -1 & 3 \end{bmatrix}$ is an invertible matrix.

  d. None of the above is true.

81. Give an example of a 2×2 matrix that is its own inverse.

82. If $A = \begin{bmatrix} 3 & 5 \\ 2 & 4 \end{bmatrix}$, find $(A^{-1})^{-1}$.

83. Find values of a for which the following matrix is not invertible:

$$\begin{bmatrix} 1 & a + 1 \\ a - 2 & 4 \end{bmatrix}.$$

Group Exercise

84. Each person in the group should work with one partner. Send a coded word or message to each other by giving your partner the coded matrix and the coding matrix that you selected. Once messages are sent, each person should decode the message received.

Objectives

❶ Evaluate a second-order determinant.

❷ Solve a system of linear equations in two variables using Cramer's rule.

❸ Evaluate a third-order determinant.

❹ Solve a system of linear equations in three variables using Cramer's rule.

❺ Use determinants to identify inconsistent systems and systems with dependent equations.

❻ Evaluate higher-order determinants.

A portion of Charles Babbage's unrealized Difference Engine

As cyberspace absorbs more and more of our work, play, shopping, and socializing, where will it all end? Which activities will still be offline in 2025?

Our technologically transformed lives can be traced back to the English inventor Charles Babbage (1792–1871). Babbage knew of a method for solving linear systems called *Cramer's rule*, in honor of the Swiss geometer Gabriel Cramer (1704–1752). Cramer's rule was simple, but involved numerous multiplications for large systems. Babbage designed a machine, called the "difference engine," that consisted of toothed wheels on shafts for performing these multiplications. Despite the fact that only one-seventh of the functions ever worked, Babbage's invention demonstrated how complex calculations could be handled mechanically. In 1944, scientists at IBM used the lessons of the difference engine to create the world's first computer.

Those who invented computers hoped to relegate the drudgery of repeated computation to a machine. In this section, we look at a method for solving linear systems that played a critical role in this process. The method uses real numbers, called *determinants,* that are associated with arrays of numbers. As with matrix methods, solutions are obtained by writing down the coefficients and constants of a linear system and performing operations with them.

❶ Evaluate a second-order determinant.

The Determinant of a 2 × 2 Matrix

Associated with every square matrix is a real number, called its **determinant**. The determinant for a 2 × 2 square matrix is defined as follows:

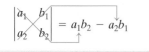
Definition of the Determinant of a 2 × 2 Matrix

The determinant of the matrix $\begin{bmatrix} a_1 & b_1 \\ a_2 & b_2 \end{bmatrix}$ is denoted by $\begin{vmatrix} a_1 & b_1 \\ a_2 & b_2 \end{vmatrix}$ and is defined by

$$\begin{vmatrix} a_1 & b_1 \\ a_2 & b_2 \end{vmatrix} = a_1 b_2 - a_2 b_1.$$

We also say that the **value** of the **second-order determinant** $\begin{vmatrix} a_1 & b_1 \\ a_2 & b_2 \end{vmatrix}$ is $a_1 b_2 - a_2 b_1$.

Example 1 illustrates that the determinant of a matrix may be positive or negative. The determinant can also have 0 as its value.

EXAMPLE 1   Evaluating the Determinant of a 2 × 2 Matrix

Evaluate the determinant of each of the following matrices:

$$\textbf{a.} \quad \begin{bmatrix} 5 & 6 \\ 7 & 3 \end{bmatrix} \quad \textbf{b.} \quad \begin{bmatrix} 2 & 4 \\ -3 & -5 \end{bmatrix}.$$

Solution   We multiply and subtract as indicated.

Discovery

Write and then evaluate three determinants, one whose value is positive, one whose value is negative, and one whose value is 0.

$$\textbf{a.} \quad \begin{vmatrix} 5 & 6 \\ 7 & 3 \end{vmatrix} = 5 \cdot 3 - 7 \cdot 6 = 15 - 42 = -27$$

The value of the second-order determinant is -27.

$$\textbf{b.} \quad \begin{vmatrix} 2 & 4 \\ -3 & -5 \end{vmatrix} = 2(-5) - (-3)(4) = -10 + 12 = 2$$

The value of the second-order determinant is 2.

Check Point 1   Evaluate the determinant of each of the following matrices:

$$\textbf{a.} \quad \begin{bmatrix} 10 & 9 \\ 6 & 5 \end{bmatrix} \quad \textbf{b.} \quad \begin{bmatrix} 4 & 3 \\ -5 & -8 \end{bmatrix}.$$

2 Solve a system of linear equations in two variables using Cramer's rule.

Solving Systems of Linear Equations in Two Variables Using Determinants

Determinants can be used to solve a linear system in two variables. In general, such a system appears as

$$a_1 x + b_1 y = c_1$$
$$a_2 x + b_2 y = c_2.$$

Let's first solve this system for x using the addition method. We can solve for x by eliminating y from the equations. Multiply the first equation by b_2 and the second equation by $-b_1$. Then add the two equations:

$$a_1 x + b_1 y = c_1 \quad \xrightarrow{\text{Multiply by } b_2.} \quad a_1 b_2 x + b_1 b_2 y = c_1 b_2$$
$$a_2 x + b_2 y = c_2 \quad \xrightarrow{\text{Multiply by } -b_1.} \quad \underline{-a_2 b_1 x - b_1 b_2 y = -c_2 b_1}$$
$$\text{Add:} \quad (a_1 b_2 - a_2 b_1) x = c_1 b_2 - c_2 b_1$$
$$x = \frac{c_1 b_2 - c_2 b_1}{a_1 b_2 - a_2 b_1}$$

Because

$$\begin{vmatrix} c_1 & b_1 \\ c_2 & b_2 \end{vmatrix} = c_1 b_2 - c_2 b_1 \quad \text{and} \quad \begin{vmatrix} a_1 & b_1 \\ a_2 & b_2 \end{vmatrix} = a_1 b_2 - a_2 b_1,$$

we can express our answer for x as the quotient of two determinants:

$$x = \frac{\begin{vmatrix} c_1 & b_1 \\ c_2 & b_2 \end{vmatrix}}{\begin{vmatrix} a_1 & b_1 \\ a_2 & b_2 \end{vmatrix}}.$$

Similarly, we could use the addition method to solve our system for y, again expressing y as the quotient of two determinants. This method of using determinants to solve the linear system, called **Cramer's rule**, is summarized in the box at the top of the next page.

Solving a Linear System in Two Variables Using Determinants

Cramer's Rule

If

$$a_1 x + b_1 y = c_1$$
$$a_2 x + b_2 y = c_2$$

then

$$x = \dfrac{\begin{vmatrix} c_1 & b_1 \\ c_2 & b_2 \end{vmatrix}}{\begin{vmatrix} a_1 & b_1 \\ a_2 & b_2 \end{vmatrix}} \quad \text{and} \quad y = \dfrac{\begin{vmatrix} a_1 & c_1 \\ a_2 & c_2 \end{vmatrix}}{\begin{vmatrix} a_1 & b_1 \\ a_2 & b_2 \end{vmatrix}}$$

where

$$\begin{vmatrix} a_1 & b_1 \\ a_2 & b_2 \end{vmatrix} \neq 0.$$

Here are some helpful tips when solving

$$a_1 x + b_1 y = c_1$$
$$a_2 x + b_2 y = c_2$$

using determinants:

1. Three different determinants are used to find x and y. The determinants in the denominators for x and y are identical. The determinants in the numerators for x and y differ. In abbreviated notation, we write

$$x = \frac{D_x}{D} \quad \text{and} \quad y = \frac{D_y}{D}, \text{ where } D \neq 0.$$

2. The elements of D, the determinant in the denominator, are the coefficients of the variables in the system.

$$D = \begin{vmatrix} a_1 & b_1 \\ a_2 & b_2 \end{vmatrix}$$

3. D_x, the determinant in the numerator of x, is obtained by replacing the x-coefficients, in D, a_1 and a_2, with the constants on the right sides of the equations, c_1 and c_2.

$$D = \begin{vmatrix} a_1 & b_1 \\ a_2 & b_2 \end{vmatrix} \quad \text{and} \quad D_x = \begin{vmatrix} c_1 & b_1 \\ c_2 & b_2 \end{vmatrix} \qquad \text{Replace the column with } a_1 \text{ and } a_2 \text{ with the constants } c_1 \text{ and } c_2 \text{ to get } D_x.$$

4. D_y, the determinant in the numerator for y, is obtained by replacing the y-coefficients, in D, b_1 and b_2, with the constants on the right sides of the equations, c_1 and c_2.

$$D = \begin{vmatrix} a_1 & b_1 \\ a_2 & b_2 \end{vmatrix} \quad \text{and} \quad D_y = \begin{vmatrix} a_1 & c_1 \\ a_2 & c_2 \end{vmatrix} \qquad \text{Replace the column with } b_1 \text{ and } b_2 \text{ with the constants } c_1 \text{ and } c_2 \text{ to get } D_y.$$

EXAMPLE 2 **Using Cramer's Rule to Solve a Linear System**

Use Cramer's rule to solve the system:

$$5x - 4y = 2$$
$$6x - 5y = 1.$$

Solution Because

$$x = \frac{D_x}{D} \quad \text{and} \quad y = \frac{D_y}{D},$$

we will set up and evaluate the three determinants $D, D_x,$ and D_y.

1. D, the determinant in both denominators, consists of the x- and y-coefficients.

$$D = \begin{vmatrix} 5 & -4 \\ 6 & -5 \end{vmatrix} = (5)(-5) - (6)(-4) = -25 + 24 = -1$$

Because this determinant is not zero, we continue to use Cramer's rule to solve the system.

2. D_x, the determinant in the numerator for x, is obtained by replacing the x-coefficients in D, 5 and 6, by the constants on the right sides of the equations, 2 and 1.

$$D_x = \begin{vmatrix} 2 & -4 \\ 1 & -5 \end{vmatrix} = (2)(-5) - (1)(-4) = -10 + 4 = -6$$

3. D_y, the determinant in the numerator for y, is obtained by replacing the y-coefficients in D, -4 and -5, by the constants on the right sides of the equations, 2 and 1.

$$D_y = \begin{vmatrix} 5 & 2 \\ 6 & 1 \end{vmatrix} = (5)(1) - (6)(2) = 5 - 12 = -7$$

4. Thus,

$$x = \frac{D_x}{D} = \frac{-6}{-1} = 6 \quad \text{and} \quad y = \frac{D_y}{D} = \frac{-7}{-1} = 7.$$

As always, the solution $(6, 7)$ can be checked by substituting these values into the original equations. The solution set is $\{(6, 7)\}$.

Check Point 2 Use Cramer's rule to solve the system:
$$5x + 4y = 12$$
$$3x - 6y = 24.$$

③ Evaluate a third-order determinant.

The Determinant of a 3 × 3 Matrix

Associated with every square matrix is a real number called its determinant. The determinant for a 3 × 3 matrix is defined as follows:

Definition of a Third-Order Determinant

$$\begin{vmatrix} a_1 & b_1 & c_1 \\ a_2 & b_2 & c_2 \\ a_3 & b_3 & c_3 \end{vmatrix} = a_1b_2c_3 + b_1c_2a_3 + c_1a_2b_3 - a_3b_2c_1 - b_3c_2a_1 - c_3a_2b_1$$

The six terms and the three factors in each term in this complicated evaluation formula can be rearranged, and then we can apply the distributive property. We obtain

$$a_1b_2c_3 - a_1b_3c_2 - a_2b_1c_3 + a_2b_3c_1 + a_3b_1c_2 - a_3b_2c_1$$
$$= a_1(b_2c_3 - b_3c_2) - a_2(b_1c_3 - b_3c_1) + a_3(b_1c_2 - b_2c_1)$$
$$= a_1\begin{vmatrix} b_2 & c_2 \\ b_3 & c_3 \end{vmatrix} - a_2\begin{vmatrix} b_1 & c_1 \\ b_3 & c_3 \end{vmatrix} + a_3\begin{vmatrix} b_1 & c_1 \\ b_2 & c_2 \end{vmatrix}.$$

You can evaluate each of the second-order determinants and obtain the three expressions in parentheses in the second step.

In summary, we now have arranged the definition of a third-order determinar as follows:

Definition of the Determinant of a 3 × 3 Matrix

A third-order determinant is defined by

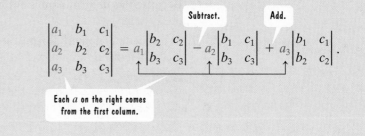

$$\begin{vmatrix} a_1 & b_1 & c_1 \\ a_2 & b_2 & c_2 \\ a_3 & b_3 & c_3 \end{vmatrix} = a_1 \begin{vmatrix} b_2 & c_2 \\ b_3 & c_3 \end{vmatrix} - a_2 \begin{vmatrix} b_1 & c_1 \\ b_3 & c_3 \end{vmatrix} + a_3 \begin{vmatrix} b_1 & c_1 \\ b_2 & c_2 \end{vmatrix}.$$

Each a on the right comes from the first column.

Here are some tips that may be helpful when evaluating the determinant of 3 × 3 matrix:

Evaluating the Determinant of a 3 × 3 Matrix

1. Each of the three terms in the definition contains two factors—a numerical factor and a second-order determinant.

2. The numerical factor in each term is an element from the first column of the third-order determinant.

3. The minus sign precedes the second term.

4. The second-order determinant that appears in each term is obtained by crossing out the row and the column containing the numerical factor.

$$a_1 \begin{vmatrix} b_2 & c_2 \\ b_3 & c_3 \end{vmatrix} - a_2 \begin{vmatrix} b_1 & c_1 \\ b_3 & c_3 \end{vmatrix} + a_3 \begin{vmatrix} b_1 & c_1 \\ b_2 & c_2 \end{vmatrix}$$

$$\begin{vmatrix} a_1 & b_1 & c_1 \\ a_2 & b_2 & c_2 \\ a_3 & b_3 & c_3 \end{vmatrix} \quad \begin{vmatrix} a_1 & b_1 & c_1 \\ a_2 & b_2 & c_2 \\ a_3 & b_3 & c_3 \end{vmatrix} \quad \begin{vmatrix} a_1 & b_1 & c_1 \\ a_2 & b_2 & c_2 \\ a_3 & b_3 & c_3 \end{vmatrix}$$

The **minor** of an element is the determinant that remains after deleting the row and column of that element. For this reason, we call this method **expansion by minors**.

EXAMPLE 3   Evaluating the Determinant of a 3 × 3 Matrix

Evaluate the determinant of the following matrix:

$$\begin{bmatrix} 4 & 1 & 0 \\ -9 & 3 & 4 \\ -3 & 8 & 1 \end{bmatrix}.$$

Solution We know that each of the three terms in the determinant contains numerical factor and a second-order determinant. The numerical factors are from the first column of the determinant of the given matrix. They are highlighted in th following matrix:

$$\begin{vmatrix} 4 & 1 & 0 \\ -9 & 3 & 4 \\ -3 & 8 & 1 \end{vmatrix}.$$

We find the minor for each numerical factor by deleting the row and column of that element:

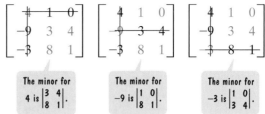

The minor for
4 is $\begin{vmatrix} 3 & 4 \\ 8 & 1 \end{vmatrix}$.

The minor for
-9 is $\begin{vmatrix} 1 & 0 \\ 8 & 1 \end{vmatrix}$.

The minor for
-3 is $\begin{vmatrix} 1 & 0 \\ 3 & 4 \end{vmatrix}$.

Now we have three numerical factors, 4, -9, and -3, and three second-order determinants. We multiply each numerical factor by its second-order determinant to find the three terms of the third-order determinant:

$$4\begin{vmatrix} 3 & 4 \\ 8 & 1 \end{vmatrix}, \quad -9\begin{vmatrix} 1 & 0 \\ 8 & 1 \end{vmatrix}, \quad -3\begin{vmatrix} 1 & 0 \\ 3 & 4 \end{vmatrix}.$$

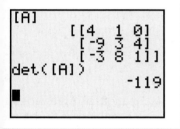
Based on the preceding definition, we subtract the second term from the first term and add the third term:

Don't forget to supply the minus sign.

$$\begin{vmatrix} 4 & 1 & 0 \\ -9 & 3 & 4 \\ -3 & 8 & 1 \end{vmatrix} = 4\begin{vmatrix} 3 & 4 \\ 8 & 1 \end{vmatrix} - (-9)\begin{vmatrix} 1 & 0 \\ 8 & 1 \end{vmatrix} - 3\begin{vmatrix} 1 & 0 \\ 3 & 4 \end{vmatrix} \qquad \text{Begin by evaluating the three second-order determinants.}$$

$$= 4(3 \cdot 1 - 8 \cdot 4) + 9(1 \cdot 1 - 8 \cdot 0) - 3(1 \cdot 4 - 3 \cdot 0)$$

$$= 4(3 - 32) + 9(1 - 0) - 3(4 - 0) \qquad \text{Multiply within parentheses.}$$

$$= 4(-29) + 9(1) - 3(4) \qquad \text{Subtract within parentheses.}$$

$$= -116 + 9 - 12 \qquad \text{Multiply.}$$

$$= -119 \qquad \text{Add and subtract as indicated.}$$

Check Point 3 Evaluate the determinant of the following matrix:

$$\begin{bmatrix} 2 & 1 & 7 \\ -5 & 6 & 0 \\ -4 & 3 & 1 \end{bmatrix}.$$

The six terms in the definition of a third-order determinant can be rearranged and factored in a variety of ways. Thus, it is possible to expand a determinant by minors about any row or any column. *Minus signs must be supplied preceding any element appearing in a position where the sum of its row and its column is an odd number.* For example, expanding about the elements in column 2 gives us

$$\begin{vmatrix} a_1 & b_1 & c_1 \\ a_2 & b_2 & c_2 \\ a_3 & b_3 & c_3 \end{vmatrix} = -b_1\begin{vmatrix} a_2 & c_2 \\ a_3 & c_3 \end{vmatrix} + b_2\begin{vmatrix} a_2 & c_1 \\ a_3 & c_3 \end{vmatrix} - b_3\begin{vmatrix} a_1 & c_1 \\ a_2 & c_2 \end{vmatrix}.$$

Minus sign is supplied because b_1 appears in row 1 and column 2; $1 + 2 = 3$, an odd number.

Minus sign is supplied because b_3 appears in row 3 and column 2; $3 + 2 = 5$, an odd number.

Expanding by minors about column 3, we obtain

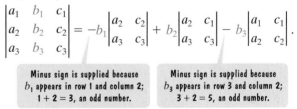

Minus sign must be supplied because c_2 appears in row 2 and column 3; $2 + 3 = 5$, an odd number.

When evaluating a 3×3 determinant using expansion by minors, you can expand about any row or column. To simplify the arithmetic, if a row or column contains one or more 0s, expand about that row or column.

EXAMPLE 4   Evaluating a Third-Order Determinant

Evaluate:

$$\begin{vmatrix} 9 & 5 & 0 \\ -2 & -3 & 0 \\ 1 & 4 & 2 \end{vmatrix}.$$

Solution   Note that the last column has two 0s. We will expand the determinant about the elements in that column.

$$\begin{vmatrix} 9 & 5 & 0 \\ -2 & -3 & 0 \\ 1 & 4 & 2 \end{vmatrix} = 0 \begin{vmatrix} -2 & -3 \\ 1 & 4 \end{vmatrix} - 0 \begin{vmatrix} 9 & 5 \\ 1 & 4 \end{vmatrix} + 2 \begin{vmatrix} 9 & 5 \\ -2 & -3 \end{vmatrix}$$

$$= 0 - 0 + 2[9(-3) - (-2) \cdot 5] \quad \text{Evaluate the second-order}$$
$$= 2(-27 + 10) \qquad\qquad\qquad \text{determinant whose numerical}$$
$$= 2(-17) \qquad\qquad\qquad\qquad \text{factor is not 0.}$$
$$= -34$$

Check Point 4   Evaluate:

$$\begin{vmatrix} 6 & 4 & 0 \\ -3 & -5 & 3 \\ 1 & 2 & 0 \end{vmatrix}.$$

④ Solve a system of linear equations in three variables using Cramer's rule.

Solving Systems of Linear Equations in Three Variables Using Determinants

Cramer's rule can be applied to solving systems of linear equations in three variables. The determinants in the numerator and denominator of all variables are third-order determinants.

Solving Three Equations in Three Variables Using Determinants

Cramer's Rule

If

$$a_1 x + b_1 y + c_1 z = d_1$$
$$a_2 x + b_2 y + c_2 z = d_2$$
$$a_3 x + b_3 y + c_3 z = d_3$$

then

$$x = \frac{D_x}{D}, y = \frac{D_y}{D}, \text{ and } z = \frac{D_z}{D}.$$

These four third-order determinants are given by

$$D = \begin{vmatrix} a_1 & b_1 & c_1 \\ a_2 & b_2 & c_2 \\ a_3 & b_3 & c_3 \end{vmatrix} \quad \text{These are the coefficients of the variables } x, y \text{ and } z.\ D \neq 0$$

$$D_x = \begin{vmatrix} d_1 & b_1 & c_1 \\ d_2 & b_2 & c_2 \\ d_3 & b_3 & c_3 \end{vmatrix} \quad \text{Replace } x\text{-coefficients in } D \text{ with the constants on the right of the three equations.}$$

$$D_y = \begin{vmatrix} a_1 & d_1 & c_1 \\ a_2 & d_2 & c_2 \\ a_3 & d_3 & c_3 \end{vmatrix} \quad \text{Replace } y\text{-coefficients in } D \text{ with the constants on the right of the three equations.}$$

$$D_z = \begin{vmatrix} a_1 & b_1 & d_1 \\ a_2 & b_2 & d_2 \\ a_3 & b_3 & d_3 \end{vmatrix}. \quad \text{Replace } z\text{-coefficients in } D \text{ with the constants on the right of the three equations.}$$

EXAMPLE 5   Using Cramer's Rule
to Solve a Linear System in Three Variables

Use Cramer's rule to solve:

$$\begin{aligned} x + 2y - z &= -4 \\ x + 4y - 2z &= -6 \\ 2x + 3y + z &= 3. \end{aligned}$$

Solution   Because

$$x = \frac{D_x}{D}, \quad y = \frac{D_y}{D}, \quad \text{and} \quad z = \frac{D_z}{D},$$

we need to set up and evaluate four determinants.

Step 1   Set up the determinants.

1. D, the determinant in all three denominators, consists of the x-, y-, and z-coefficients.

$$D = \begin{vmatrix} 1 & 2 & -1 \\ 1 & 4 & -2 \\ 2 & 3 & 1 \end{vmatrix}$$

2. D_x, the determinant in the numerator for x, is obtained by replacing the x-coefficients in D, 1, 1, and 2, with the constants on the right sides of the equations, -4, -6, and 3.

$$D_x = \begin{vmatrix} -4 & 2 & -1 \\ -6 & 4 & -2 \\ 3 & 3 & 1 \end{vmatrix}$$

3. D_y, the determinant in the numerator for y, is obtained by replacing the y-coefficients in D, 2, 4, and 3, with the constants on the right sides of the equations, -4, -6, and 3.

$$D_y = \begin{vmatrix} 1 & -4 & -1 \\ 1 & -6 & -2 \\ 2 & 3 & 1 \end{vmatrix}$$

4. D_z, the determinant in the numerator for z, is obtained by replacing the z-coefficients in D, -1, -2, and 1, with the constants on the right sides of the equations, -4, -6, and 3.

$$D_z = \begin{vmatrix} 1 & 2 & -4 \\ 1 & 4 & -6 \\ 2 & 3 & 3 \end{vmatrix}$$

Step 2   Evaluate the four determinants.

$$D = \begin{vmatrix} 1 & 2 & -1 \\ 1 & 4 & -2 \\ 2 & 3 & 1 \end{vmatrix} = 1 \begin{vmatrix} 4 & -2 \\ 3 & 1 \end{vmatrix} - 1 \begin{vmatrix} 2 & -1 \\ 3 & 1 \end{vmatrix} + 2 \begin{vmatrix} 2 & -1 \\ 4 & -2 \end{vmatrix}$$

> ## Study Tip
>
> To find D_x, D_y, and D_z, you'll need to apply the evaluation process for a 3×3 determinant three times. The values of D_x, D_y, and D_z cannot be obtained from the numbers that occur in the computation of D.

$$= 1(4 + 6) - 1(2 + 3) + 2(-4 + 4)$$
$$= 1(10) - 1(5) + 2(0) = 5$$

Using the same technique to evaluate each determinant, we obtain

$$D_x = -10, \quad D_y = 5, \quad \text{and} \quad D_z = 20.$$

Step 3 Substitute these four values and solve the system.

$$x = \frac{D_x}{D} = \frac{-10}{5} = -2$$

$$y = \frac{D_y}{D} = \frac{5}{5} = 1$$

$$z = \frac{D_z}{D} = \frac{20}{5} = 4$$

The solution $(-2, 1, 4)$ can be checked by substitution into the original three equations. The solution set is $\{(-2, 1, 4)\}$.

Check Point **5** Use Cramer's rule to solve the system:

$$3x - 2y + z = 16$$
$$2x + 3y - z = -9$$
$$x + 4y + 3z = 2.$$

Use determinants to identify inconsistent systems and systems with dependent equations.

Cramer's Rule with Inconsistent and Dependent Systems

If D, the determinant in the denominator, is 0, the variables described by the quotient of determinants are not real numbers. However, when $D = 0$, this indicates that the system is either inconsistent or contains dependent equations. This gives rise to the following two situations:

Discovery

Write a system of two equations that is inconsistent. Now use determinants and the result boxed on the right to verify that this is truly an inconsistent system. Repeat the same process for a system with two dependent equations.

> **Determinants: Inconsistent and Dependent Systems**
>
> 1. If $D = 0$ and at least one of the determinants in the numerator is not 0, then the system is inconsistent. The solution set is $\varnothing$.
> 2. If $D = 0$ and all the determinants in the numerators are 0, then the equations in the system are dependent. The system has infinitely many solutions.

Although we have focused on applying determinants to solve linear systems, they have other applications, some of which we consider in the exercise set that follows Example 6.

Evaluate higher-order determinants.

The Determinant of Any $n \times n$ Matrix

The determinant of a matrix with n rows and n columns is said to be an **nth-order determinant**. The value of an nth-order determinant $(n > 2)$ can be found in terms of determinants of order $n - 1$. For example, we found the value of a third-order determinant in terms of determinants of order 2.

We can generalize this idea for fourth-order determinants and higher. We have seen that the **minor** of the element a_{ij} is the determinant obtained by deleting the ith row and the jth column in the given array of numbers. The **cofactor** of the element a_{ij} is $(-1)^{i+j}$ times the minor of the a_{ij}th entry. If the sum of the row and column $(i + j)$ is even, the cofactor is the same as the minor. If the sum of the row and column $(i + j)$ is odd, the cofactor is the opposite of the minor.

Let's see what this means in the case of a fourth-order determinant.

EXAMPLE 6 Evaluating the Determinant of a 4 × 4 Matrix

Evaluate the determinant of the following matrix:

$$A = \begin{bmatrix} 1 & -2 & 3 & 0 \\ -1 & 1 & 0 & 2 \\ 0 & 2 & 0 & -3 \\ 2 & 3 & -4 & 1 \end{bmatrix}.$$

Solution

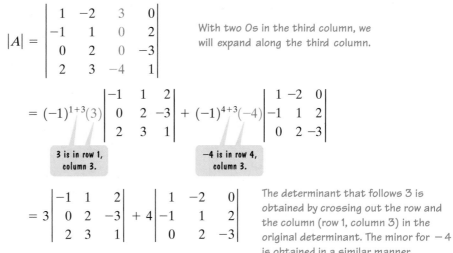

$$|A| = \begin{vmatrix} 1 & -2 & 3 & 0 \\ -1 & 1 & 0 & 2 \\ 0 & 2 & 0 & -3 \\ 2 & 3 & -4 & 1 \end{vmatrix}$$

With two 0s in the third column, we will expand along the third column.

$$= (-1)^{1+3}(3) \begin{vmatrix} -1 & 1 & 2 \\ 0 & 2 & -3 \\ 2 & 3 & 1 \end{vmatrix} + (-1)^{4+3}(-4) \begin{vmatrix} 1 & -2 & 0 \\ -1 & 1 & 2 \\ 0 & 2 & -3 \end{vmatrix}$$

3 is in row 1, column 3.

−4 is in row 4, column 3.

$$= 3 \begin{vmatrix} -1 & 1 & 2 \\ 0 & 2 & -3 \\ 2 & 3 & 1 \end{vmatrix} + 4 \begin{vmatrix} 1 & -2 & 0 \\ -1 & 1 & 2 \\ 0 & 2 & -3 \end{vmatrix}$$

The determinant that follows 3 is obtained by crossing out the row and the column (row 1, column 3) in the original determinant. The minor for −4 is obtained in a similar manner.

Evaluate the two third-order determinants to get

$$|A| = 3(-25) + 4(-1) = -79.$$

Check Point 6 Evaluate the determinant of the following matrix:

$$A = \begin{bmatrix} 0 & 4 & 0 & -3 \\ -1 & 1 & 5 & 2 \\ 1 & -2 & 0 & 6 \\ 3 & 0 & 0 & 1 \end{bmatrix}.$$

If a linear system has n equations, Cramer's rule requires you to compute $n + 1$ determinants of nth order. The excessive number of calculations required to perform Cramer's rule for systems with four or more equations makes it an inefficient method for solving large systems.

EXERCISE SET 8.5

Practice Exercises

Evaluate each determinant in Exercises 1–10.

1. $\begin{vmatrix} 5 & 7 \\ 2 & 3 \end{vmatrix}$

2. $\begin{vmatrix} 4 & 8 \\ 5 & 6 \end{vmatrix}$

3. $\begin{vmatrix} -4 & 1 \\ 5 & 6 \end{vmatrix}$

4. $\begin{vmatrix} 7 & 9 \\ -2 & -5 \end{vmatrix}$

5. $\begin{vmatrix} -7 & 14 \\ 2 & -4 \end{vmatrix}$

6. $\begin{vmatrix} 1 & -3 \\ -8 & 2 \end{vmatrix}$

7. $\begin{vmatrix} -5 & -1 \\ -2 & -7 \end{vmatrix}$

8. $\begin{vmatrix} \frac{1}{5} & \frac{1}{6} \\ -6 & 5 \end{vmatrix}$

9. $\begin{vmatrix} \frac{1}{2} & \frac{1}{2} \\ \frac{1}{8} & -\frac{3}{4} \end{vmatrix}$

10. $\begin{vmatrix} \frac{2}{3} & \frac{1}{3} \\ -\frac{1}{2} & \frac{3}{4} \end{vmatrix}$

For Exercises 11–26, use Cramer's rule to solve each system or to determine that the system is inconsistent or contains dependent equations.

11. $x + y = 7$
    $x - y = 3$

12. $2x + y = 3$
    $x - y = 3$

13. $12x + 3y = 15$
    $2x - 3y = 13$

14. $x - 2y = 5$
    $5x - y = -2$

15. $4x - 5y = 17$
    $2x + 3y = 3$

16. $3x + 2y = 2$
    $2x + 2y = 3$

17. $x + 2y = 3$
    $5x + 10y = 15$

18. $2x - 9y = 5$
    $3x - 3y = 11$

19. $3x - 4y = 4$
    $2x + 2y = 12$

20. $3x = 7y + 1$
    $2x = 3y - 1$

21. $2x = 3y + 2$
    $5x = 51 - 4y$

22. $y = -4x + 2$
    $2x = 3y + 8$

23. $3x = 2 - 3y$
    $2y = 3 - 2x$

24. $x + 2y - 3 = 0$
    $12 = 8y + 4x$

25. $4y = 16 - 3x$
    $6x = 32 - 8y$

26. $2x = 7 + 3y$
    $4x - 6y = 3$

Evaluate each determinant in Exercises 27–32.

27. $\begin{vmatrix} 3 & 0 & 0 \\ 2 & 1 & -5 \\ 2 & 5 & -1 \end{vmatrix}$

28. $\begin{vmatrix} 4 & 0 & 0 \\ 3 & -1 & 4 \\ 2 & -3 & 5 \end{vmatrix}$

29. $\begin{vmatrix} 3 & 1 & 0 \\ -3 & 4 & 0 \\ -1 & 3 & -5 \end{vmatrix}$

30. $\begin{vmatrix} 2 & -4 & 2 \\ -1 & 0 & 5 \\ 3 & 0 & 4 \end{vmatrix}$

31. $\begin{vmatrix} 1 & 1 & 1 \\ 2 & 2 & 2 \\ -3 & 4 & -5 \end{vmatrix}$

32. $\begin{vmatrix} 1 & 2 & 3 \\ 2 & 2 & -3 \\ 3 & 2 & 1 \end{vmatrix}$

In Exercises 33–40, use Cramer's rule to solve each system.

33. $x + y + z = 0$
    $2x - y + z = -1$
    $-x + 3y - z = -8$

34. $x - y + 2z = 3$
    $2x + 3y + z = 9$
    $-x - y + 3z = 11$

35. $4x - 5y - 6z = -1$
    $x - 2y - 5z = -12$
    $2x - y = 7$

36. $x - 3y + z = -2$
    $x + 2y = 8$
    $2x - y = 1$

37. $x + y + z = 4$
    $x - 2y + z = 7$
    $x + 3y + 2z = 4$

38. $2x + 2y + 3z = 10$
    $4x - y + z = -5$
    $5x - 2y + 6z = 1$

39. $x + 2z = 4$
    $2y - z = 5$
    $2x + 3y = 13$

40. $3x + 2z = 4$
    $5x - y = -4$
    $4y + 3z = 22$

Evaluate each determinant in Exercises 41–44.

41. $\begin{vmatrix} 4 & 2 & 8 & -7 \\ -2 & 0 & 4 & 1 \\ 5 & 0 & 0 & 5 \\ 4 & 0 & 0 & -1 \end{vmatrix}$

42. $\begin{vmatrix} 3 & -1 & 1 & 2 \\ -2 & 0 & 0 & 0 \\ 2 & -1 & -2 & 3 \\ 1 & 4 & 2 & 3 \end{vmatrix}$

43. $\begin{vmatrix} -2 & -3 & 3 & 5 \\ 1 & -4 & 0 & 0 \\ 1 & 2 & 2 & -3 \\ 2 & 0 & 1 & 1 \end{vmatrix}$

44. $\begin{vmatrix} 1 & -3 & 2 & 0 \\ -3 & -1 & 0 & -2 \\ 2 & 1 & 3 & 1 \\ 2 & 0 & -2 & 0 \end{vmatrix}$

Practice Plus

In Exercises 45–46, evaluate each determinant.

45. $\begin{vmatrix} \begin{vmatrix} 3 & 1 \\ -2 & 3 \end{vmatrix} & \begin{vmatrix} 7 & 0 \\ 1 & 5 \end{vmatrix} \\ \begin{vmatrix} 3 & 0 \\ 0 & 7 \end{vmatrix} & \begin{vmatrix} 9 & -6 \\ 3 & 5 \end{vmatrix} \end{vmatrix}$

46. $\begin{vmatrix} \begin{vmatrix} 5 & 0 \\ 4 & -3 \end{vmatrix} & \begin{vmatrix} -1 & 0 \\ 0 & -1 \end{vmatrix} \\ \begin{vmatrix} 7 & -5 \\ 4 & 6 \end{vmatrix} & \begin{vmatrix} 4 & 1 \\ -3 & 5 \end{vmatrix} \end{vmatrix}$

In Exercises 47–48, write the system of linear equations for which Cramer's rule yields the given determinants.

47. $D = \begin{vmatrix} 2 & -4 \\ 3 & 5 \end{vmatrix}$, $D_x = \begin{vmatrix} 8 & -4 \\ -10 & 5 \end{vmatrix}$

48. $D = \begin{vmatrix} 2 & -3 \\ 5 & 6 \end{vmatrix}$, $D_x = \begin{vmatrix} 8 & -3 \\ 11 & 6 \end{vmatrix}$

In Exercises 49–52, solve each equation for x.

49. $\begin{vmatrix} -2 & x \\ 4 & 6 \end{vmatrix} = 32$

50. $\begin{vmatrix} x + 3 & -6 \\ x - 2 & -4 \end{vmatrix} = 28$

51. $\begin{vmatrix} 1 & x & -2 \\ 3 & 1 & 1 \\ 0 & -2 & 2 \end{vmatrix} = -8$

52. $\begin{vmatrix} 2 & x & 1 \\ -3 & 1 & 0 \\ 2 & 1 & 4 \end{vmatrix} = 39$

Application Exercises

Determinants are used to find the area of a triangle whose vertices are given by three points in a rectangular coordinate system. The area of a triangle with vertices (x_1, y_1), (x_2, y_2), and (x_3, y_3) is

$$\text{Area} = \pm \frac{1}{2} \begin{vmatrix} x_1 & y_1 & 1 \\ x_2 & y_2 & 1 \\ x_3 & y_3 & 1 \end{vmatrix}$$

where the $\pm$ symbol indicates that the appropriate sign should be chosen to yield a positive area. Use this information to work Exercises 53–54.

53. Use determinants to find the area of the triangle whose vertices are $(3, -5)$, $(2, 6)$, and $(-3, 5)$.

54. Use determinants to find the area of the triangle whose vertices are $(1, 1)$, $(-2, -3)$, and $(11, -3)$.

Determinants are used to show that three points lie on the same line (are collinear). If

$$\begin{vmatrix} x_1 & y_1 & 1 \\ x_2 & y_2 & 1 \\ x_3 & y_3 & 1 \end{vmatrix} = 0,$$

then the points (x_1, y_1), (x_2, y_2), and (x_3, y_3) are collinear. If the determinant does not equal 0, then the points are not collinear. Use this information to work Exercises 55–56.

55. Are the points $(3, -1)$, $(0, -3)$, and $(12, 5)$ collinear?

56. Are the points $(-4, -6)$, $(1, 0)$, and $(11, 12)$ collinear?

Determinants are used to write an equation of a line passing through two points. An equation of the line passing through the distinct points (x_1, y_1) and (x_2, y_2) is given by

$$\begin{vmatrix} x & y & 1 \\ x_1 & y_1 & 1 \\ x_2 & y_2 & 1 \end{vmatrix} = 0.$$

Use this information to work Exercises 57–58.

57. Use the determinant to write an equation of the line passing through $(3, -5)$ and $(-2, 6)$. Then expand the determinant, expressing the line's equation in slope-intercept form.

58. Use the determinant to write an equation of the line passing through $(-1, 3)$ and $(2, 4)$. Then expand the determinant, expressing the line's equation in slope-intercept form.

Writing in Mathematics

9. Explain how to evaluate a second-order determinant.
0. Describe the determinants D_x and D_y in terms of the coefficients and constants in a system of two equations in two variables.
1. Explain how to evaluate a third-order determinant.
2. When expanding a determinant by minors, when is it necessary to supply minus signs?
3. Without going into too much detail, describe how to solve a linear system in three variables using Cramer's rule.
4. In applying Cramer's rule, what does it mean if $D = 0$?
5. The process of solving a linear system in three variables using Cramer's rule can involve tedious computation. Is there a way of speeding up this process, perhaps using Cramer's rule to find the value for only one of the variables? Describe how this process might work, presenting a specific example with your description. Remember that your goal is still to find the value for each variable in the system.
6. If you could use only one method to solve linear systems in three variables, which method would you select? Explain why this is so.

Technology Exercises

7. Use the feature of your graphing utility that evaluates the determinant of a square matrix to verify any five of the determinants that you evaluated by hand in Exercises 1–10, 27–32, or 41–44.

n Exercises 68–69, use a graphing utility to evaluate the eterminant for the given matrix.

8. $\begin{bmatrix} 3 & -2 & -1 & 4 \\ -5 & 1 & 2 & 7 \\ 2 & 4 & 5 & 0 \\ -1 & 3 & -6 & 5 \end{bmatrix}$

69. $\begin{bmatrix} 8 & 2 & 6 & -1 & 0 \\ 2 & 0 & -3 & 4 & 7 \\ 2 & 1 & -3 & 6 & -5 \\ -1 & 2 & 1 & 5 & -1 \\ 4 & 5 & -2 & 3 & -8 \end{bmatrix}$

0. What is the fastest method for solving a linear system with your graphing utility?

Critical Thinking Exercises

71. **a.** Evaluate: $\begin{vmatrix} a & a \\ 0 & a \end{vmatrix}$.

   b. Evaluate: $\begin{vmatrix} a & a & a \\ 0 & a & a \\ 0 & 0 & a \end{vmatrix}$.

   c. Evaluate: $\begin{vmatrix} a & a & a & a \\ 0 & a & a & a \\ 0 & 0 & a & a \\ 0 & 0 & 0 & a \end{vmatrix}$.

   d. Describe the pattern in the given determinants.

   e. Describe the pattern in the evaluations.

72. Evaluate: $\begin{vmatrix} 2 & 0 & 0 & 0 & 0 \\ 0 & 3 & 0 & 0 & 0 \\ 0 & 0 & 2 & 0 & 0 \\ 0 & 0 & 0 & 1 & 0 \\ 0 & 0 & 0 & 0 & 4 \end{vmatrix}$.

73. What happens to the value of a second-order determinant if the two columns are interchanged?

74. Consider the system

$$a_1 x + b_1 y = c_1$$
$$a_2 x + b_2 y = c_2.$$

   Use Cramer's rule to prove that if the first equation of the system is replaced by the sum of the two equations, the resulting system has the same solution as the original system.

75. Show that the equation of a line through (x_1, y_1) and (x_2, y_2) is given by the determinant equation in Exercises 57–58.

Group Exercise

76. We have seen that determinants can be used to solve linear equations, give areas of triangles in rectangular coordinates, and determine equations of lines. Not impressed with these applications? Members of the group should research an application of determinants that they find intriguing. The group should then present a seminar to the class about this application.

Chapter 8
Summary, Review, and Test

Summary

DEFINITIONS AND CONCEPTS	EXAMPLES
3.1 Matrix Solutions to Linear Systems	
a. Matrix row operations are described in the box on page 784.	Ex. 2, p. 784
b. To solve a linear system using Gaussian elimination, begin with the system's augmented matrix. Use matrix row operations to get 1s down the main diagonal from upper left to lower right, and 0s below the 1s. Such a matrix is in row-echelon form. Details are in the box on page 785.	Ex. 3, p. 785; Ex. 4, p. 787
c. To solve a linear system using Gauss-Jordan elimination, use the procedure of Gaussian elimination, but obtain 0s above and below the 1s in the main diagonal from upper left to lower right. Such a matrix is in reduced row-echelon form. Details are in the box on page 790.	Ex. 5, p. 790

DEFINITIONS AND CONCEPTS	EXAMPLES

8.2 Inconsistent and Dependent Systems and Their Applications

a. If Gaussian elimination results in a matrix with a row containing all 0s to the left of the vertical line and a nonzero number to the right, the system has no solution (is inconsistent).

Ex. 1, p. 794

b. If Gaussian elimination results in a matrix with a row with all 0s, the system has an infinite number of solutions (contains dependent equations).

Ex. 2, p. 796

c. In nonsquare systems, the number of variables differs from the number of equations.

Ex. 3, p. 798

8.3 Matrix Operations and Their Applications

a. A matrix of order $m \times n$ has m rows and n columns. Two matrices are equal if and only if they have the same order and corresponding elements are equal.

Ex. 1, p. 803

b. Matrix Addition and Subtraction: Matrices of the same order are added or subtracted by adding or subtracting corresponding elements. Properties of matrix addition are given in the box on page 805.

Ex. 2, p. 804

c. Scalar Multiplication: If A is a matrix and c is a scalar, then cA is the matrix formed by multiplying each element in A by c. Properties of scalar multiplication are given in the box on page 806.

Ex. 3, p. 806;
Ex. 4, p. 807

d. Matrix Multiplication: The product of an $m \times n$ matrix A and an $n \times p$ matrix B is an $m \times p$ matrix AB. The element in the ith row and jth column of AB is found by multiplying each element in the ith row of A by the corresponding element in the jth column of B and adding the products. Matrix multiplication is not commutative: $AB \neq BA$. Properties of matrix multiplication are given in the box on page 811.

Ex. 5, p. 808;
Ex. 6, p. 809;
Ex. 7, p. 810

8.4 Multiplicative Inverses of Matrices and Matrix Equations

a. The multiplicative identity matrix I_n is an $n \times n$ matrix with 1s down the main diagonal from upper left to lower right and 0s elsewhere.

b. Let A be an $n \times n$ square matrix. If there is a square matrix A^{-1} such that $AA^{-1} = I_n$ and $A^{-1}A = I_n$, then A^{-1} is the multiplicative inverse of A.

Ex. 1, p. 819

c. If a square matrix has a multiplicative inverse, it is invertible or nonsingular. Methods for finding multiplicative inverses for invertible matrices, including a formula for 2×2 matrices, are given in the box on page 825.

Ex. 2, p. 820;
Ex. 3, p. 822;
Ex. 4, p. 824

d. Linear systems can be represented by matrix equations of the form $AX = B$ in which A is the coefficient matrix and B is the constant matrix. If $AX = B$ has a unique solution, then $X = A^{-1}B$.

Ex. 5, p. 826

8.5 Determinants and Cramer's Rule

a. Value of a Second-Order Determinant:

$$\begin{vmatrix} a_1 & b_1 \\ a_2 & b_2 \end{vmatrix} = a_1 b_2 - a_2 b_1$$

Ex. 1, p. 833

b. Cramer's rule for solving systems of linear equations in two variables uses three second-order determinants and is stated in the box on page 834.

Ex. 2, p. 834

c. To evaluate an nth-order determinant, where $n > 2$,

  1. Select a row or column about which to expand.

  2. Multiply each element a_{ij} in the row or column by $(-1)^{i+j}$ times the determinant obtained by deleting the ith row and the jth column in the given array of numbers.

  3. The value of the determinant is the sum of the products found in step 2.

Ex. 3, p. 836;
Ex. 4, p. 838;
Ex. 6, p. 840

d. Cramer's rule for solving systems of linear equations in three variables uses four third-order determinants and is stated in the box on page 838.

Ex. 5, p. 839

e. Cramer's rule with inconsistent and dependent systems is summarized by the two situations in the box on page 840.

Review Exercises

8.1

In Exercises 1–2, write the system of linear equations represented by the augmented matrix. Use x, y, z, and, if necessary, w, x, y, and , for the variables. Once the system is written, use back-substitution to find its solution.

1. $\begin{bmatrix} 1 & 1 & 3 & | & 12 \\ 0 & 1 & -2 & | & -4 \\ 0 & 0 & 1 & | & 3 \end{bmatrix}$

2. $\begin{bmatrix} 1 & 0 & -2 & 2 & | & 1 \\ 0 & 1 & 1 & -1 & | & 0 \\ 0 & 0 & 1 & -\frac{7}{3} & | & -\frac{1}{3} \\ 0 & 0 & 0 & 1 & | & 1 \end{bmatrix}$

In Exercises 3–4, perform each matrix row operation and write the new matrix.

3. $\begin{bmatrix} 1 & 2 & 2 & | & 2 \\ 0 & 1 & -1 & | & 2 \\ 0 & 5 & 4 & | & 1 \end{bmatrix}$ $-5R_2 + R_3$

4. $\begin{bmatrix} 2 & -2 & 1 & | & -1 \\ 1 & 2 & -1 & | & 2 \\ 6 & 4 & 3 & | & 5 \end{bmatrix}$ $\frac{1}{2}R_1$

In Exercises 5–7, solve each system of equations using matrices. Use Gaussian elimination with back-substitution or Gauss-Jordan elimination.

5. $x + 2y + 3z = -5$
   $2x + y + z = 1$
   $x + y - z = 8$

6. $x - 2y + z = 0$
   $y - 3z = -1$
   $2y + 5z = -2$

7. $3x_1 + 5x_2 - 8x_3 + 5x_4 = -8$
   $x_1 + 2x_2 - 3x_3 + x_4 = -7$
   $2x_1 + 3x_2 - 7x_3 + 3x_4 = -11$
   $4x_1 + 8x_2 - 10x_3 + 7x_4 = -10$

8. The table shows the pollutants in the air in a city on a typical summer day.

x (Hours after 6 A.M.)	y (Amount of Pollutants in the Air, in parts per million)
2	98
4	138
10	162

   a. Use the function $y = ax^2 + bx + c$ to model the data. Use either Gaussian elimination with back-substitution or Gauss-Jordan elimination to find the values for a, b, and c.

   b. Use the function to find the time of day at which the city's air pollution level is at a maximum. What is the maximum level?

8.2

In Exercises 9–12, use Gaussian elimination to find the complete solution to each system, or show that none exists.

9. $2x - 3y + z = 1$
   $x - 2y + 3z = 2$
   $3x - 4y - z = 1$

10. $x - 3y + z = 1$
    $-2x + y + 3z = -7$
    $x - 4y + 2z = 0$

11. $x_1 + 4x_2 + 3x_3 - 6x_4 = 5$
    $x_1 + 3x_2 + x_3 - 4x_4 = 3$
    $2x_1 + 8x_2 + 7x_3 - 5x_4 = 11$
    $2x_1 + 5x_2 - 6x_4 = 4$

12. $2x + 3y - 5z = 15$
    $x + 2y - z = 4$

13. The figure shows the intersections of three one-way streets. The numbers given represent traffic flow, in cars per hour, at a peak period (from 4 P.M. to 6 P.M.).

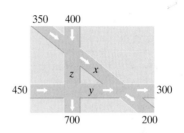

   a. Use the idea that the number of cars entering each intersection per hour must equal the number of cars leaving per hour to set up a system of linear equations involving x, y, and z.

   b. Use Gaussian elimination to solve the system.

   c. If construction limits the value of z to 400, how many cars per hour must pass between the other intersections to keep traffic flowing?

8.3

14. Find values for x, y, and z so that the following matrices are equal:

$$\begin{bmatrix} 2x & y + 7 \\ z & 4 \end{bmatrix} = \begin{bmatrix} -10 & 13 \\ 6 & 4 \end{bmatrix}.$$

In Exercises 15–28, perform the indicated matrix operations given that A, B, C, and D are defined as follows. If an operation is not defined, state the reason.

$$A = \begin{bmatrix} 2 & -1 & 2 \\ 5 & 3 & -1 \end{bmatrix} \quad B = \begin{bmatrix} 0 & -2 \\ 3 & 2 \\ 1 & -5 \end{bmatrix}$$

$$C = \begin{bmatrix} 1 & 2 & 3 \\ -1 & 1 & 2 \\ -1 & 2 & 1 \end{bmatrix} \quad D = \begin{bmatrix} -2 & 3 & 1 \\ 3 & -2 & 4 \end{bmatrix}$$

15. $A + D$ 16. $2B$ 17. $D - A$

18. $B + C$ 19. $3A + 2D$ 20. $-2A + 4D$

21. $-5(A + D)$ 22. AB 23. BA

24. BD 25. DB 26. $AB - BA$

27. $(A - D)C$ 28. $B(AC)$

29. Solve for X in the matrix equation

$$3X + A = B$$

where $A = \begin{bmatrix} 4 & 6 \\ -5 & 0 \end{bmatrix}$ and $B = \begin{bmatrix} -2 & -12 \\ 4 & 1 \end{bmatrix}.$

In Exercises 30–31, use nine pixels in a 3 × 3 grid and the color levels shown.

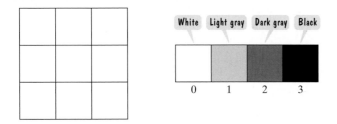

White   Light gray   Dark gray   Black

0   1   2   3

30. Write a 3 × 3 matrix that represents a digital photograph of the letter T in dark gray on a light gray background.

31. Find a matrix B so that $A + B$ increases the contrast of the letter T by changing the dark gray to black and the light gray to white.

The figure shows a right triangle in a rectangular coordinate system.

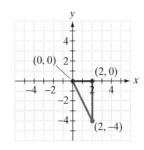

The figure can be represented by the matrix

$$B = \begin{bmatrix} 0 & 2 & 2 \\ 0 & 0 & -4 \end{bmatrix}.$$

Use the triangle and the matrix that represents it to solve Exercises 32–37.

32. Use matrix operations to move the triangle 2 units to the left and 1 unit up. Then graph the triangle and its transformation in a rectangular coordinate system.

33. Use matrix operations to reduce the triangle to half its perimeter and move the reduced image 2 units down. Then graph the triangle and its transformation in a rectangular coordinate system.

In Exercises 34–37, find AB and graph the resulting image. What effect does the multiplication have on the triangle represented by matrix B?

34. $A = \begin{bmatrix} 1 & 0 \\ 0 & -1 \end{bmatrix}$ **35.** $A = \begin{bmatrix} -1 & 0 \\ 0 & 1 \end{bmatrix}$

36. $A = \begin{bmatrix} 0 & -1 \\ 1 & 0 \end{bmatrix}$ **37.** $A = \begin{bmatrix} 2 & 0 \\ 0 & 1 \end{bmatrix}$

8.4

In Exercises 38–39, find the products AB and BA to determine whether B is the multiplicative inverse of A.

38. $A = \begin{bmatrix} 2 & 7 \\ 1 & 4 \end{bmatrix}$, $B = \begin{bmatrix} 4 & -7 \\ -1 & 3 \end{bmatrix}$

39. $A = \begin{bmatrix} 1 & 0 & 0 \\ 0 & 2 & -7 \\ 0 & -1 & 4 \end{bmatrix}$, $B = \begin{bmatrix} 1 & 0 & 0 \\ 0 & 4 & 7 \\ 0 & 1 & 2 \end{bmatrix}$

In Exercises 40–43, find A^{-1}. Check that $AA^{-1} = I$ and $A^{-1}A = I$.

40. $A = \begin{bmatrix} 1 & -1 \\ -2 & 3 \end{bmatrix}$ **41.** $A = \begin{bmatrix} 0 & 1 \\ 5 & 3 \end{bmatrix}$

42. $A = \begin{bmatrix} 1 & 0 & -2 \\ 2 & 1 & 0 \\ 1 & 0 & -3 \end{bmatrix}$ **43.** $A = \begin{bmatrix} 1 & 3 & -2 \\ 4 & 13 & -7 \\ 5 & 16 & -8 \end{bmatrix}$

In Exercises 44–45,

    a. *Write each linear system as a matrix equation in the form $AX = B$.*

    b. *Solve the system using the inverse that is given for th coefficient matrix.*

44. $\begin{aligned} x + y + 2z &= 7 \\ y + 3z &= -2 \\ 3x \quad\ -2z &= 0 \end{aligned}$
The inverse of $\begin{bmatrix} 1 & 1 & 2 \\ 0 & 1 & 3 \\ 3 & 0 & -2 \end{bmatrix}$ is $\begin{bmatrix} -2 & 2 & 1 \\ 9 & -8 & -3 \\ -3 & 3 & 1 \end{bmatrix}$.

45. $\begin{aligned} x - y + 2z &= 12 \\ y - z &= -5 \\ x \quad\ + 2z &= 10 \end{aligned}$
The inverse of $\begin{bmatrix} 1 & -1 & 2 \\ 0 & 1 & -1 \\ 1 & 0 & 2 \end{bmatrix}$ is $\begin{bmatrix} 2 & 2 & -1 \\ -1 & 0 & 1 \\ -1 & -1 & 1 \end{bmatrix}$.

46. Use the coding matrix $A = \begin{bmatrix} 3 & 2 \\ 4 & 3 \end{bmatrix}$ and its invers $A^{-1} = \begin{bmatrix} 3 & -2 \\ -4 & 3 \end{bmatrix}$ to encode and then decode the word RULE.

8.5

In Exercises 47–52, evaluate each determinant.

47. $\begin{vmatrix} 3 & 2 \\ -1 & 5 \end{vmatrix}$ **48.** $\begin{vmatrix} -2 & -3 \\ -4 & -8 \end{vmatrix}$

49. $\begin{vmatrix} 2 & 4 & -3 \\ 1 & -1 & 5 \\ -2 & 4 & 0 \end{vmatrix}$ **50.** $\begin{vmatrix} 4 & 7 & 0 \\ -5 & 6 & 0 \\ 3 & 2 & -4 \end{vmatrix}$

51. $\begin{vmatrix} 1 & 1 & 0 & 2 \\ 0 & 3 & 2 & 1 \\ 0 & -2 & 4 & 0 \\ 0 & 3 & 0 & 1 \end{vmatrix}$ **52.** $\begin{vmatrix} 2 & 2 & 2 & 2 \\ 0 & 2 & 2 & 2 \\ 0 & 0 & 2 & 2 \\ 0 & 0 & 0 & 2 \end{vmatrix}$

In Exercises 53–56, use Cramer's rule to solve each system.

53. $\begin{aligned} x - 2y &= 8 \\ 3x + 2y &= -1 \end{aligned}$ **54.** $\begin{aligned} 7x + 2y &= 0 \\ 2x + y &= -3 \end{aligned}$

55. $\begin{aligned} x + 2y + 2z &= 5 \\ 2x + 4y + 7z &= 19 \\ -2x - 5y - 2z &= 8 \end{aligned}$ **56.** $\begin{aligned} 2x + y \quad\ &= -4 \\ y - 2z &= 0 \\ 3x \quad\ - 2z &= -11 \end{aligned}$

7. Use the quadratic function $y = ax^2 + bx + c$ to model the following data:

x (Age of a Driver)	y (Average Number of Automobile Accidents per Day in the United States)
20	400
40	150
60	400

Use Cramer's rule to determine values for a, b, and c. Then use the model to write a statement about the average number of automobile accidents in which 30-year-olds and 50-year-olds are involved daily.

Chapter 8 Test

In Exercises 1–2, solve each system of equations using matrices.

1.
$$x + 2y - z = -3$$
$$2x - 4y + z = -7$$
$$-2x + 2y - 3z = 4$$

2.
$$x - 2y + z = 2$$
$$2x - y - z = 1$$

In Exercises 3–6, let

$$A = \begin{bmatrix} 3 & 1 \\ 1 & 0 \\ 2 & 1 \end{bmatrix}, \quad B = \begin{bmatrix} 1 & -1 \\ 2 & 1 \end{bmatrix}, \quad \text{and} \quad C = \begin{bmatrix} 1 & 2 \\ -1 & 3 \end{bmatrix}.$$

Carry out the indicated operations.

3. $2B + 3C$ **4.** AB

5. C^{-1} **6.** $BC - 3B$

7. If $A = \begin{bmatrix} 1 & 2 & 2 \\ 2 & 3 & 3 \\ 1 & -1 & -2 \end{bmatrix}$ and $B = \begin{bmatrix} -3 & 2 & 0 \\ 7 & -4 & 1 \\ -5 & 3 & -1 \end{bmatrix}$, show that B is the inverse of A.

8. Consider the system
$$3x + 5y = 9$$
$$2x - 3y = -13.$$

a. Express the system in the form $AX = B$, where A, X, and B are appropriate matrices.

b. Find A^{-1}, the inverse of the coefficient matrix.

c. Use A^{-1} to solve the given system.

9. Evaluate: $\begin{vmatrix} 4 & -1 & 3 \\ 0 & 5 & -1 \\ 5 & 2 & 4 \end{vmatrix}$.

10. Solve for x only using Cramer's rule:
$$3x + y - 2z = -3$$
$$2x + 7y + 3z = 9$$
$$4x - 3y - z = 7.$$

Cumulative Review Exercises (Chapters P–8)

Solve each equation or inequality in Exercises 1–6.

1. $2x^2 = 4 - x$ **2.** $5x + 8 \le 7(1 + x)$

3. $x^3 + x^2 - 4x - 4 \ge 0$

4. $3x^3 + 8x^2 - 15x + 4 = 0$

5. $e^{2x} - 14e^x + 45 = 0$

6. $\log_3 x + \log_3(x + 2) = 1$

7. Use matrices to solve this system:
$$x - y + z = 17$$
$$2x + 3y + z = 8$$
$$-4x + y + 5z = -2.$$

8. Solve for y using Cramer's rule:
$$x - 2y + z = 7$$
$$2x + y - z = 0$$
$$3x + 2y - 2z = -2.$$

9. If $f(x) = \sqrt{4x - 7}$, find $f^{-1}(x)$.

10. Graph: $f(x) = \dfrac{x}{x^2 - 16}$.

11. Use the graph of $f(x) = 4x^4 - 4x^3 - 25x^2 + x + 6$ shown in the figure to factor the polynomial completely.

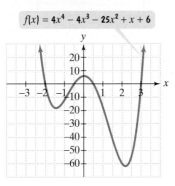

$f(x) = 4x^4 - 4x^3 - 25x^2 + x + 6$

12. Graph $y = \log_2 x$ and $y = \log_2(x + 1)$ in the same rectangular coordinate system.

13. Use the exponential decay model $A = A_0 e^{kt}$ to solve this problem. A radioactive substance has a half-life of 40 days. There are initially 900 grams of the substance.

a. Find the decay model for this substance. Round k to the nearest thousandth.

b. How much of the substance will remain after 10 days? Round to the nearest hundredth of a gram.

14. Multiply the matrices: $\begin{bmatrix} 1 & -1 & 0 \\ 2 & 1 & 3 \end{bmatrix} \begin{bmatrix} 4 & -1 \\ 2 & 0 \\ 1 & 1 \end{bmatrix}$.

15. Find the partial fraction decomposition of

$$\frac{3x^2 + 17x - 38}{(x - 3)(x - 2)(x + 2)}.$$

In Exercises 16–19, graph each equation, function, or inequality in a rectangular coordinate system.

16. $y = -\frac{2}{3}x - 1$

17. $3x - 5y < 15$

18. $f(x) = x^2 - 2x - 3$

19. $(x - 1)^2 + (y + 1)^2 = 9$

20. Use synthetic division to divide $x^3 - 6x + 4$ by $x - 2$.

21. Graph: $y = 2 \sin 2\pi x, \quad 0 \le x \le 2$.

22. Find the exact value of $\cos\left[\tan^{-1}\left(-\frac{4}{3}\right)\right]$.

23. Verify the identity: $\dfrac{\cos 2x}{\cos x - \sin x} = \cos x + \sin x$.

24. Solve on the interval $[0, 2\pi)$: $\cos^2 x + \sin x + 1 = 0$.

25. If $\mathbf{v} = -6\mathbf{i} + 5\mathbf{j}$ and $\mathbf{w} = -7\mathbf{i} + 3\mathbf{j}$, find $4\mathbf{w} - 5\mathbf{v}$.

Conic Sections and Analytic Geometry

F ROM RIPPLES IN WATER TO THE path on which humanity journeys through space, certain curves occur naturally throughout the universe. Over 2000 years ago, the ancient Greeks studied these curves, called conic sections, without regard to their immediate usefulness simply because the study elicited ideas that were exciting, challenging, and interesting. The ancient Greeks could not have imagined the applications of these curves in the twenty-first century. Overwhelmed by the choices on satellite television? Blame it on a conic section! In this chapter, we use the rectangular coordinate system to study the conic sections and the mathematics behind their surprising applications.

ONE MINUTE YOU'RE IN CLASS, enjoying the lecture. Then a sharp pain radiates down your side. The next minute you're being diagnosed with, of all things, a kidney stone. It took your cousin six weeks to recover from kidney stone surgery, but your doctor assures you there is nothing to worry about. A new procedure, based on a curve that looks like the cross section of a football, will dissolve the stone painlessly and let you return to class in a day or two. How can this be?

This problem appears in Section 9.1 in the discussion on applications of ellipses.

SECTION 9.1 *The Ellipse*

Objectives

1 Graph ellipses centered at the origin.

2 Write equations of ellipses in standard form.

3 Graph ellipses not centered at the origin.

4 Solve applied problems involving ellipses.

You took on a summer job driving a truck, delivering books that were ordered online. You're an avid reader, so just being around books sounded appealing. However, now you're feeling a bit shaky driving the truck for the first time. It's 10 feet wide and 9 feet high; compared to your compact car, it feels like you're behind the wheel of a tank. Up ahead you see a sign at the semielliptical entrance to a tunnel: Caution! Tunnel is 1... Feet High at Center Peak. Then you see another sign: Caution! Tunnel Is 40 Fee... Wide. Will your truck clear the opening of the tunnel's archway?

Mathematics is present in the movements of planets, bridge and tunnel construction, navigational systems used to keep track of a ship's location, manufacture of lenses for telescopes, and even in a procedure for disintegrating kidney stones. The mathematics behind these applications involves conic sections. **Conic sections** are curves that result from the intersection of a right circular cone and a plane. Figure 9.1 illustrates the four conic sections: the circle, the ellipse, the parabola, and the hyperbola.

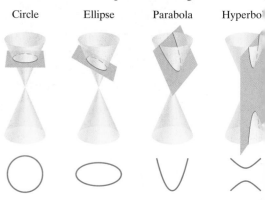

Figure 9.1 Obtaining the conic sections by intersecting a plane and a cone

In this section, we study the symmetric oval-shaped curve known as the ellipse. We will use a geometric definition for an ellipse to derive its equation. With this equation, we will determine if your delivery truck will clear the tunnel's entrance.

Definition of an Ellipse

Figure 9.2 illustrates how to draw an ellipse. Place pins at two fixed points, each of which is called a focus (plural: foci). If the ends of a fixed length of string are fastened to the pins and we draw the string taut with a pencil, the path traced by the pencil will be an ellipse. Notice that the sum of the distances of the pencil point from the foci remains constant because the length of the string is fixed. This procedure for drawing an ellipse illustrates its geometric definition.

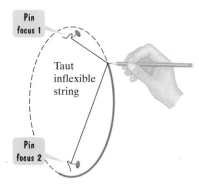

Figure 9.2 Drawing an ellipse

Definition of an Ellipse

An **ellipse** is the set of all points, P, in a plane the sum of whose distances from two fixed points, F_1 and F_2, is constant (see Figure 9.3). These two fixed points are called the **foci** (plural of **focus**). The midpoint of the segment connecting the foci is the **center** of the ellipse.

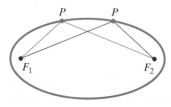

Figure 9.3

Figure 9.4 illustrates that an ellipse can be elongated in any direction. In this section, we will limit our discussion to ellipses that are elongated horizontally or vertically. The line through the foci intersects the ellipse at two points, called the **vertices** (singular: **vertex**). The line segment that joins the vertices is the **major axis**. Notice that the midpoint of the major axis is the center of the ellipse. The line segment whose endpoints are on the ellipse and that is perpendicular to the major axis at the center is called the **minor axis** of the ellipse.

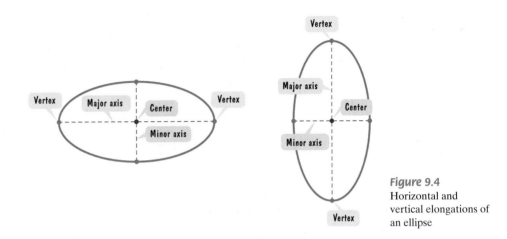

Figure 9.4
Horizontal and vertical elongations of an ellipse

Standard Form of the Equation of an Ellipse

The rectangular coordinate system gives us a unique way of describing an ellipse. It enables us to translate an ellipse's geometric definition into an algebraic equation.

We start with Figure 9.5 to obtain an ellipse's equation. We've placed an ellipse that is elongated horizontally into a rectangular coordinate system. The foci are on the x-axis at $(-c, 0)$ and $(c, 0)$, as in Figure 9.5. In this way, the center of the ellipse is at the origin. We let (x, y) represent the coordinates of any point on the ellipse.

What does the definition of an ellipse tell us about the point (x, y) in Figure 9.5? For any point (x, y) on the ellipse, the sum of the distances to the two foci, $d_1 + d_2$, must be constant. As we shall see, it is convenient to denote this constant by $2a$. Thus, the point (x, y) is on the ellipse if and only if

$$d_1 + d_2 = 2a.$$

$$\sqrt{(x + c)^2 + y^2} + \sqrt{(x - c)^2 + y^2} = 2a \qquad \text{Use the distance formula.}$$

After eliminating radicals and simplifying, we obtain

$$(a^2 - c^2)x^2 + a^2y^2 = a^2(a^2 - c^2).$$

Look at the triangle in Figure 9.5. Notice that the distance from F_1 to F_2 is $2c$. Because the length of any side of a triangle is less than the sum of the lengths of the other two sides, $2c < d_1 + d_2$. Equivalently, $2c < 2a$ and $c < a$. Consequently, $a^2 - c^2 > 0$. For convenience, let $b^2 = a^2 - c^2$. Substituting b^2 for $a^2 - c^2$ in the preceding equation, we obtain

$$b^2x^2 + a^2y^2 = a^2b^2$$

$$\frac{b^2x^2}{a^2b^2} + \frac{a^2y^2}{a^2b^2} = \frac{a^2b^2}{a^2b^2} \qquad \text{Divide both sides by } a^2b^2.$$

$$\frac{x^2}{a^2} + \frac{y^2}{b^2} = 1. \qquad \text{Simplify.}$$

This last equation is the **standard form of the equation of an ellipse centered at the origin**. There are two such equations, one for a horizontal major axis and one for a vertical major axis.

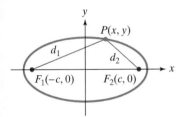

Figure 9.5

Study Tip

The algebraic details behind eliminating the radicals and obtaining the equation shown can be found in the appendix. There you will find a step-by-step derivation of the ellipse's equation.

Standard Forms of the Equations of an Ellipse

The **standard form of the equation of an ellipse** with center at the origin, and major and minor axes of lengths $2a$ and $2b$ (where a and b are positive, and $a^2 > b^2$) is

$$\frac{x^2}{a^2} + \frac{y^2}{b^2} = 1 \qquad \text{or} \qquad \frac{x^2}{b^2} + \frac{y^2}{a^2} = 1.$$

Figure 9.6 illustrates that the vertices are on the major axis, a units from the center. The foci are on the major axis, c units from the center. For both equations, $b^2 = a^2 - c^2$. Equivalently, $c^2 = a^2 - b^2$.

Study Tip

The form $c^2 = a^2 - b^2$ is the one you should remember. When finding the foci, this form is easy to manipulate.

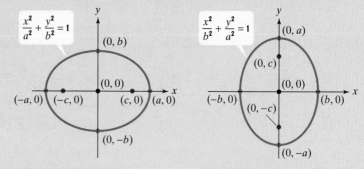

Figure 9.6(a) Major axis is horizontal with length $2a$.

Figure 9.6(b) Major axis is vertical with length $2a$.

The intercepts shown in Figure 9.6 can be obtained algebraically. Let's do this fo

$$\frac{x^2}{a^2} + \frac{y^2}{b^2} = 1.$$

x-intercepts: Set $y = 0$.

$$\frac{x^2}{a^2} = 1$$
$$x^2 = a^2$$
$$x = \pm a$$

> x-intercepts are $-a$ and a. The graph passes through $(-a, 0)$ and $(a, 0)$, which are the vertices.

y-intercepts: Set $x = 0$.

$$\frac{y^2}{b^2} = 1$$
$$y^2 = b^2$$
$$y = \pm b$$

> y-intercepts are $-b$ and b. The graph passes through $(0, -b)$ and $(0, b)$.

Using the Standard Form of the Equation of an Ellipse

We can use the standard form of an ellipse's equation to graph the ellipse. Althoug the definition of the ellipse is given in terms of its foci, the foci are not part of the graph. A complete graph of an ellipse can be obtained without graphing the foci.

① Graph ellipses centered at the origin.

EXAMPLE 1   Graphing an Ellipse Centered at the Origin

Graph and locate the foci: $\dfrac{x^2}{9} + \dfrac{y^2}{4} = 1$.

Solution   The given equation is the standard form of an ellipse's equation wit $a^2 = 9$ and $b^2 = 4$.

$$\frac{x^2}{9} + \frac{y^2}{4} = 1$$

> $a^2 = 9$. This is the larger of the two denominators.

> $b^2 = 4$. This is the smaller of the two denominators.

Technology

We graph $\dfrac{x^2}{9} + \dfrac{y^2}{4} = 1$ with a graphing utility by solving for y.

$$\frac{y^2}{4} = 1 - \frac{x^2}{9}$$

$$y^2 = 4\left(1 - \frac{x^2}{9}\right)$$

$$y = \pm 2\sqrt{1 - \frac{x^2}{9}}$$

Notice that the square root property requires us to define two functions. Enter

$$y_1 = 2 \boxed{\sqrt{}} \left(1 \boxed{-} x \boxed{\wedge} 2 \boxed{\div} 9 \right)$$

and

$$y_2 = -y_1.$$

To see the true shape of the ellipse, use the

$\boxed{\text{ZOOM SQUARE}}$ feature so that one unit on the y-axis is the same length as one unit on the x-axis.

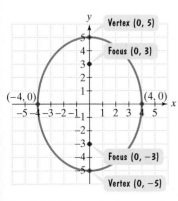

$[-5, 5, 1]$ by $[-3, 3, 1]$

Because the denominator of the x^2-term is greater than the denominator of the y^2-term, the major axis is horizontal. Based on the standard form of the equation, we know the vertices are $(-a, 0)$ and $(a, 0)$. Because $a^2 = 9$, $a = 3$. Thus, the vertices are $(-3, 0)$ and $(3, 0)$, shown in Figure 9.7.

Now let us find the endpoints of the vertical minor axis. According to the standard form of the equation, these endpoints are $(0, -b)$ and $(0, b)$. Because $b^2 = 4$, $b = 2$. Thus, the endpoints of the minor axis are $(0, -2)$ and $(0, 2)$. They are shown in Figure 9.7.

Finally, we find the foci, which are located at $(-c, 0)$ and $(c, 0)$. We can use the formula $c^2 = a^2 - b^2$ to do so. We know that $a^2 = 9$ and $b^2 = 4$. Thus,

$$c^2 = a^2 - b^2 = 9 - 4 = 5.$$

Because $c^2 = 5$, $c = \sqrt{5}$. The foci, $(-c, 0)$ and $(c, 0)$, are located at $\left(-\sqrt{5}, 0\right)$ and $\left(\sqrt{5}, 0\right)$. They are shown in Figure 9.7.

You can sketch the ellipse in Figure 9.7 by locating endpoints on the major and minor axes.

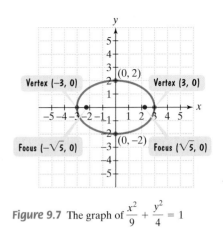

Figure 9.7 The graph of $\dfrac{x^2}{9} + \dfrac{y^2}{4} = 1$

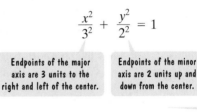

$$\frac{x^2}{3^2} + \frac{y^2}{2^2} = 1$$

Endpoints of the major axis are 3 units to the right and left of the center.   Endpoints of the minor axis are 2 units up and down from the center.

Check Point 1 Graph and locate the foci: $\dfrac{x^2}{36} + \dfrac{y^2}{9} = 1$.

EXAMPLE 2   Graphing an Ellipse Centered at the Origin

Graph and locate the foci: $25x^2 + 16y^2 = 400$.

Solution We begin by expressing the equation in standard form. Because we want 1 on the right side, we divide both sides by 400.

$$\frac{25x^2}{400} + \frac{16y^2}{400} = \frac{400}{400}$$

$$\frac{x^2}{16} + \frac{y^2}{25} = 1$$

$b^2 = 16$. This is the smaller of the two denominators. $a^2 = 25$. This is the larger of the two denominators.

The equation is the standard form of an ellipse's equation with $a^2 = 25$ and $b^2 = 16$. Because the denominator of the y^2-term is greater than the denominator of the x^2-term, the major axis is vertical. Based on the standard form of the equation, we know the vertices are $(0, -a)$ and $(0, a)$. Because $a^2 = 25$, $a = 5$. Thus, the vertices are $(0, -5)$ and $(0, 5)$, shown in Figure 9.8.

Now let us find the endpoints of the horizontal minor axis. According to the standard form of the equation, these endpoints are $(-b, 0)$ and $(b, 0)$. Because $b^2 = 16$, $b = 4$. Thus, the endpoints of the minor axis are $(-4, 0)$ and $(4, 0)$. They are shown in Figure 9.8.

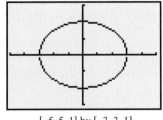

Figure 9.8 The graph of $25x^2 + 16y^2 = 400$, or $\dfrac{x^2}{16} + \dfrac{y^2}{25} = 1$

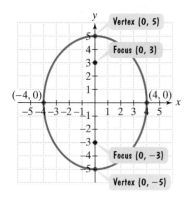

Figure 9.8 (repeated) The graph of $\dfrac{x^2}{16} + \dfrac{y^2}{25} = 1$

Finally, we find the foci, which are located at $(0, -c)$ and $(0, c)$. We can use the formula $c^2 = a^2 - b^2$ to do so. We know that $a^2 = 25$ and $b^2 = 16$. Thus,

$$c^2 = a^2 - b^2 = 25 - 16 = 9.$$

Because $c^2 = 9$, $c = 3$. The foci, $(0, -c)$ and $(0, c)$, are located at $(0, -3)$ and $(0, 3)$. They are shown in Figure 9.8.

You can sketch the ellipse in Figure 9.8 by locating endpoints on the major and minor axes.

$$\frac{x^2}{4^2} + \frac{y^2}{5^2} = 1$$

Endpoints of the minor axis are 4 units to the right and left of the center.   Endpoints of the major axis are 5 units up and down from the center.

 Check Point 2   Graph and locate the foci: $16x^2 + 9y^2 = 144$.

In Examples 1 and 2, we used the equation of an ellipse to find its foci and vertices. In the next example, we reverse this procedure.

② Write equations of ellipses in standard form.

EXAMPLE 3   Finding the Equation of an Ellipse from Its Foci and Vertices

Find the standard form of the equation of an ellipse with foci at $(-1, 0)$ and $(1, 0)$ and vertices $(-2, 0)$ and $(2, 0)$.

Solution   Because the foci are located at $(-1, 0)$ and $(1, 0)$, on the x-axis, the major axis is horizontal. The center of the ellipse is midway between the foci, located at $(0, 0)$. Thus, the form of the equation is

$$\frac{x^2}{a^2} + \frac{y^2}{b^2} = 1.$$

We need to determine the values for a^2 and b^2. The distance from the center, $(0, 0)$, to either vertex, $(-2, 0)$ or $(2, 0)$, is 2. Thus, $a = 2$.

$$\frac{x^2}{2^2} + \frac{y^2}{b^2} = 1 \qquad \text{or} \qquad \frac{x^2}{4} + \frac{y^2}{b^2} = 1$$

We must still find b^2. The distance from the center, $(0, 0)$, to either focus, $(-1, 0)$ or $(1, 0)$, is 1, so $c = 1$. Using $c^2 = a^2 - b^2$, we have

$$1^2 = 2^2 - b^2$$

and

$$b^2 = 2^2 - 1^2 = 4 - 1 = 3.$$

Substituting 3 for b^2 in $\dfrac{x^2}{4} + \dfrac{y^2}{b^2} = 1$ gives us the standard form of the ellipse's equation. The equation is

$$\frac{x^2}{4} + \frac{y^2}{3} = 1.$$

 Check Point 3   Find the standard form of the equation of an ellipse with foci at $(-2, 0)$ and $(2, 0)$ and vertices $(-3, 0)$ and $(3, 0)$.

③ Graph ellipses not centered at the origin.

Translations of Ellipses

Horizontal and vertical translations can be used to graph ellipses that are not centered at the origin. Figure 9.9 illustrates that the graphs of

$$\frac{(x-h)^2}{a^2} + \frac{(y-k)^2}{b^2} = 1 \quad \text{and} \quad \frac{x^2}{a^2} + \frac{y^2}{b^2} = 1$$

have the same size and shape. However, the graph of the first equation is centered at (h, k) rather than at the origin.

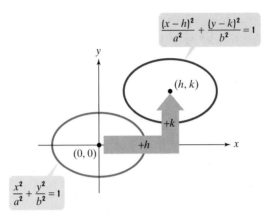

Figure 9.9 Translating an ellipse's graph

Table 9.1 gives the standard forms of equations of ellipses centered at (h, k) and shows their graphs.

Table 9.1   Standard Forms of Equations of Ellipses Centered at (h, k)

Equation	Center	Major Axis	Vertices	Graph
$\dfrac{(x-h)^2}{a^2} + \dfrac{(y-k)^2}{b^2} = 1$ *Endpoints of major axis are a units right and a units left of center.* $a^2 > b^2$ Foci are c units right and c units left of center, where $c^2 = a^2 - b^2$.	(h, k)	Parallel to the x-axis, horizontal	$(h - a, k)$ $(h + a, k)$	Vertex $(h + a, k)$; Focus $(h - c, k)$; Major axis; Vertex $(h - a, k)$; (h, k); Focus $(h + c, k)$
$\dfrac{(x-h)^2}{b^2} + \dfrac{(y-k)^2}{a^2} = 1$ $a^2 > b^2$ *Endpoints of the major axis are a units above and a units below the center.* Foci are c units above and c units below the center, where $c^2 = a^2 - b^2$.	(h, k)	Parallel to the y-axis, vertical	$(h, k - a)$ $(h, k + a)$	Vertex $(h, k + a)$; Focus $(h, k + c)$; (h, k); Vertex $(h, k - a)$; Focus $(h, k - c)$; Major axis

EXAMPLE 4   Graphing an Ellipse Centered at (h, k)

Graph: $\dfrac{(x-1)^2}{4} + \dfrac{(y+2)^2}{9} = 1$. Where are the foci located?

Solution   To graph the ellipse, we need to know its center, (h, k). In the standard forms of equations centered at (h, k), h is the number subtracted from x and k is the number subtracted from y.

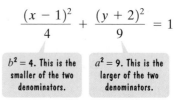

> This is $(x - h)^2$, with $h = 1$.   This is $(y - k)^2$, with $k = -2$.

$$\frac{(x-1)^2}{4} + \frac{(y-(-2))^2}{9} = 1$$

We see that $h = 1$ and $k = -2$. Thus, the center of the ellipse, (h, k), is $(1, -2)$. We can graph the ellipse by locating endpoints on the major and minor axes. To do this, we must identify a^2 and b^2.

$$\frac{(x-1)^2}{4} + \frac{(y+2)^2}{9} = 1$$

> $b^2 = 4$. This is the smaller of the two denominators. $a^2 = 9$. This is the larger of the two denominators.

The larger number is under the expression involving y. This means that the major axis is vertical and parallel to the y-axis.

We can sketch the ellipse by locating endpoints on the major and minor axes.

$$\frac{(x-1)^2}{2^2} + \frac{(y+2)^2}{3^2} = 1$$

> Endpoints of the minor axis are 2 units to the right and left of the center.   Endpoints of the major axis (the vertices) are 3 units up and down from the center.

We categorize the observations in the voice balloons as follows:

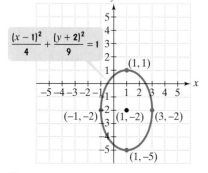

$\dfrac{(x-1)^2}{4} + \dfrac{(y+2)^2}{9} = 1$

Figure 9.10 The graph of an ellipse centered at $(1, -2)$

For a Vertical Major Axis with Center $(1, -2)$		
Vertices	**Endpoints of Minor Axis**	
$(1, -2 + 3) = (1, 1)$	$(1 + 2, -2) = (3, -2)$	
$(1, -2 - 3) = (1, -5)$	$(1 - 2, -2) = (-1, -2)$	

> 3 units above and below center

> 2 units right and left of center

Using the center and these four points, we can sketch the ellipse shown in Figure 9.10. With $c^2 = a^2 - b^2$, we have $c^2 = 9 - 4 = 5$. So the foci are located $\sqrt{5}$ units above and below the center, at $(1, -2 + \sqrt{5})$ and $(1, -2 - \sqrt{5})$.

Check Point 4   Graph: $\dfrac{(x+1)^2}{9} + \dfrac{(y-2)^2}{4} = 1$. Where are the foci located?

In some cases, it is necessary to convert the equation of an ellipse to standard form by completing the square on x and y. For example, suppose that we wish to graph the ellipse whose equation is

$$9x^2 + 4y^2 - 18x + 16y - 11 = 0.$$

Because we plan to complete the square on both x and y, we need to rearrange terms so that

- x-terms are arranged in descending order.
- y-terms are arranged in descending order.
- the constant term appears on the right.

$$9x^2 + 4y^2 - 18x + 16y - 11 = 0$$ This is the given equation.

$$(9x^2 - 18x) + (4y^2 + 16y) = 11$$ Group terms and add 11 to both sides.

$$9(x^2 - 2x + \square) + 4(y^2 + 4y + \square) = 11$$ To complete the square, coefficients of x^2 and y^2 must be 1. Factor out 9 and 4, respectively.

We added 9 · 1, or 9, to the left side.

We also added 4 · 4, or 16, to the left side.

$$9(x^2 - 2x + 1) + 4(y^2 + 4y + 4) = 11 + 9 + 16$$ Complete each square by adding the square of half the coefficient of x and y, respectively.

9 and 16, added on the left side, must also be added on the right side.

$$9(x - 1)^2 + 4(y + 2)^2 = 36$$ Factor.

$$\frac{9(x - 1)^2}{36} + \frac{4(y + 2)^2}{36} = \frac{36}{36}$$ Divide both sides by 36.

$$\frac{(x - 1)^2}{4} + \frac{(y + 2)^2}{9} = 1$$ Simplify.

The equation is now in standard form. This is precisely the form of the equation that we graphed in Example 4.

④ Solve applied problems involving ellipses.

Applications

Ellipses have many applications. German scientist Johannes Kepler (1571–1630) showed that the planets in our solar system move in elliptical orbits, with the sun at a focus. Earth satellites also travel in elliptical orbits, with Earth at a focus.

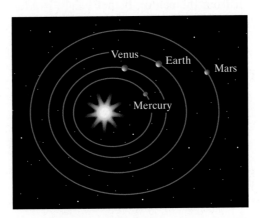

Planets move in elliptical orbits.

One intriguing aspect of the ellipse is that a ray of light or a sound wave emanating from one focus will be reflected from the ellipse to exactly the other focus. A whispering gallery is an elliptical room with an elliptical, dome-shaped ceiling. People standing at the foci can whisper and hear each other quite clearly, while persons in other locations in the room cannot hear them. Statuary Hall in the U.S. Capitol Building is elliptical. President John Quincy Adams, while a member of the House of Representatives, was aware of this acoustical phenomenon. He situated his desk at a focal point of the elliptical ceiling, easily eavesdropping on the private conversations of other House members located near the other focus.

The elliptical reflection principle is used in a procedure for disintegrating kidney stones. The patient is placed within a device that is elliptical in shape. The patient is placed so the kidney is centered at one focus, while ultrasound waves from the other focus hit the walls and are reflected to the kidney stone. The convergence of the ultrasound waves at the kidney stone causes vibrations that shatter it into fragments. The small pieces can then be passed painlessly through the patient's system. The patient recovers in days, as opposed to up to six weeks if surgery is used instead.

Whispering in an elliptical dome

Disintegrating kidney stones

Ellipses are often used for supporting arches of bridges and in tunnel construction. This application forms the basis of our next example.

EXAMPLE 5   An Application Involving an Ellipse

A semielliptical archway over a one-way road has a height of 10 feet and a width of 40 feet (see Figure 9.11). Your truck has a width of 10 feet and a height of 9 feet. Will your truck clear the opening of the archway?

Solution   Because your truck's width is 10 feet, to determine the clearance, we must find the height of the archway 5 feet from the center. If that height is 9 feet or less, the truck will not clear the opening.

Figure 9.11 A semi-elliptical archway

In Figure 9.12, we've constructed a coordinate system with the *x*-axis on the ground and the origin at the center of the archway. Also shown is the truck, whose height is 9 feet.

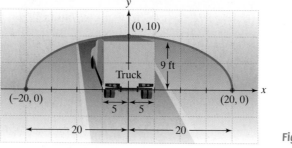

Figure 9.12

Using the equation $\dfrac{x^2}{a^2} + \dfrac{y^2}{b^2} = 1$, we can express the equation of the blue archway in Figure 9.12 as $\dfrac{x^2}{20^2} + \dfrac{y^2}{10^2} = 1$, or $\dfrac{x^2}{400} + \dfrac{y^2}{100} = 1$.

As shown in Figure 9.12, the edge of the 10-foot-wide truck corresponds to *x* = 5. We find the height of the archway 5 feet from the center by substituting 5 for *x* and solving for *y*.

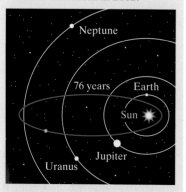

$$\frac{5^2}{400} + \frac{y^2}{100} = 1 \qquad \text{Substitute 5 for } x \text{ in } \frac{x^2}{400} + \frac{y^2}{100} = 1.$$

$$\frac{25}{400} + \frac{y^2}{100} = 1 \qquad \text{Square 5.}$$

$$400\left(\frac{25}{400} + \frac{y^2}{100}\right) = 400(1) \qquad \text{Clear fractions by multiplying both sides by 400.}$$

$$25 + 4y^2 = 400 \qquad \text{Use the distributive property and simplify.}$$

$$4y^2 = 375 \qquad \text{Subtract 25 from both sides.}$$

$$y^2 = \frac{375}{4} \qquad \text{Divide both sides by 4.}$$

$$y = \sqrt{\frac{375}{4}} \qquad \begin{array}{l}\text{Take only the positive square root. The archway is}\\ \text{above the } x\text{-axis, so } y \text{ is nonnegative.}\end{array}$$

$$\approx 9.68 \qquad \text{Use a calculator.}$$

Thus, the height of the archway 5 feet from the center is approximately 9.68 feet. Because your truck's height is 9 feet, there is enough room for the truck to clear the archway.

Check Point 5 Will a truck that is 12 feet wide and has a height of 9 feet clear the opening of the archway described in Example 5?

EXERCISE SET 9.1

Practice Exercises

In Exercises 1–18, graph each ellipse and locate the foci.

1. $\dfrac{x^2}{16} + \dfrac{y^2}{4} = 1$

2. $\dfrac{x^2}{25} + \dfrac{y^2}{16} = 1$

3. $\dfrac{x^2}{9} + \dfrac{y^2}{36} = 1$

4. $\dfrac{x^2}{16} + \dfrac{y^2}{49} = 1$

5. $\dfrac{x^2}{25} + \dfrac{y^2}{64} = 1$

6. $\dfrac{x^2}{49} + \dfrac{y^2}{36} = 1$

7. $\dfrac{x^2}{49} + \dfrac{y^2}{81} = 1$

8. $\dfrac{x^2}{64} + \dfrac{y^2}{100} = 1$

9. $\dfrac{x^2}{\frac{9}{4}} + \dfrac{y^2}{\frac{25}{4}} = 1$

10. $\dfrac{x^2}{\frac{81}{4}} + \dfrac{y^2}{\frac{25}{16}} = 1$

11. $x^2 = 1 - 4y^2$

12. $y^2 = 1 - 4x^2$

13. $25x^2 + 4y^2 = 100$

14. $9x^2 + 4y^2 = 36$

15. $4x^2 + 16y^2 = 64$

16. $4x^2 + 25y^2 = 100$

17. $7x^2 = 35 - 5y^2$

18. $6x^2 = 30 - 5y^2$

In Exercises 19–24, find the standard form of the equation of each ellipse and give the location of its foci.

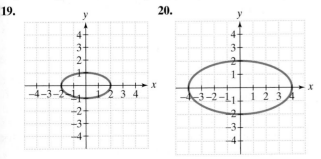

19.

20.

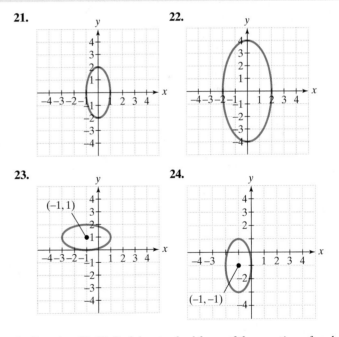

21.

22.

23.

24.

In Exercises 25–36, find the standard form of the equation of each ellipse satisfying the given conditions.

25. Foci: $(-5, 0), (5, 0)$; vertices: $(-8, 0), (8, 0)$

26. Foci: $(-2, 0), (2, 0)$; vertices: $(-6, 0), (6, 0)$

27. Foci: $(0, -4), (0, 4)$; vertices: $(0, -7), (0, 7)$

28. Foci: $(0, -3), (0, 3)$; vertices: $(0, -4), (0, 4)$

29. Foci: $(-2, 0), (2, 0)$; y-intercepts: -3 and 3

30. Foci: $(0, -2), (0, 2)$; x-intercepts: -2 and 2

31. Major axis horizontal with length 8; length of minor axis = 4; center: $(0, 0)$

32. Major axis horizontal with length 12; length of minor axis = 6; center: $(0, 0)$

33. Major axis vertical with length 10; length of minor axis = 4; center: $(-2, 3)$

34. Major axis vertical with length 20; length of minor axis = 10; center: $(2, -3)$

35. Endpoints of major axis: $(7, 9)$ and $(7, 3)$
Endpoints of minor axis: $(5, 6)$ and $(9, 6)$

36. Endpoints of major axis: $(2, 2)$ and $(8, 2)$
Endpoints of minor axis: $(5, 3)$ and $(5, 1)$

In Exercises 37–50, graph each ellipse and give the location of its foci.

37. $\dfrac{(x-2)^2}{9} + \dfrac{(y-1)^2}{4} = 1$

38. $\dfrac{(x-1)^2}{16} + \dfrac{(y+2)^2}{9} = 1$

39. $(x+3)^2 + 4(y-2)^2 = 16$

40. $(x-3)^2 + 9(y+2)^2 = 18$

41. $\dfrac{(x-4)^2}{9} + \dfrac{(y+2)^2}{25} = 1$

42. $\dfrac{(x-3)^2}{9} + \dfrac{(y+1)^2}{16} = 1$

43. $\dfrac{x^2}{25} + \dfrac{(y-2)^2}{36} = 1$

44. $\dfrac{(x-4)^2}{4} + \dfrac{y^2}{25} = 1$

45. $\dfrac{(x+3)^2}{9} + (y-2)^2 = 1$

46. $\dfrac{(x+2)^2}{16} + (y-3)^2 = 1$

47. $\dfrac{(x-1)^2}{2} + \dfrac{(y+3)^2}{5} = 1$

48. $\dfrac{(x+1)^2}{2} + \dfrac{(y-3)^2}{5} = 1$

49. $9(x-1)^2 + 4(y+3)^2 = 36$

50. $36(x+4)^2 + (y+3)^2 = 36$

In Exercises 51–56, convert each equation to standard form by completing the square on x and y. Then graph the ellipse and give the location of its foci.

51. $9x^2 + 25y^2 - 36x + 50y - 164 = 0$

52. $4x^2 + 9y^2 - 32x + 36y + 64 = 0$

53. $9x^2 + 16y^2 - 18x + 64y - 71 = 0$

54. $x^2 + 4y^2 + 10x - 8y + 13 = 0$

55. $4x^2 + y^2 + 16x - 6y - 39 = 0$

56. $4x^2 + 25y^2 - 24x + 100y + 36 = 0$

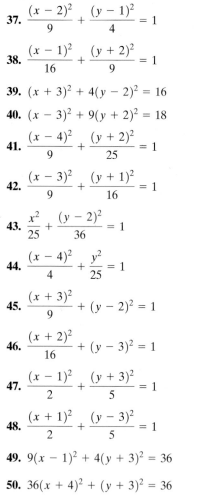

Practice Plus

In Exercises 57–62, find the solution set for each system by graphing both of the system's equations in the same rectangular coordinate system and finding points of intersection. Check all solutions in both equations.

57. $x^2 + y^2 = 1$
$x^2 + 9y^2 = 9$

58. $x^2 + y^2 = 25$
$25x^2 + y^2 = 25$

59. $\dfrac{x^2}{25} + \dfrac{y^2}{9} = 1$
$y = 3$

60. $\dfrac{x^2}{4} + \dfrac{y^2}{36} = 1$
$x = -2$

61. $4x^2 + y^2 = 4$
$2x - y = 2$

62. $4x^2 + y^2 = 4$
$x + y = 3$

In Exercises 63–64, graph each semiellipse.

63. $y = -\sqrt{16 - 4x^2}$

64. $y = -\sqrt{4 - 4x^2}$

Application Exercises

65. Will a truck that is 8 feet wide carrying a load that reaches 7 feet above the ground clear the semielliptical arch on the one-way road that passes under the bridge shown in the figure?

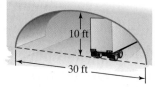

66. A semielliptic archway has a height of 20 feet and a width of 50 feet, as shown in the figure. Can a truck 14 feet high and 10 feet wide drive under the archway without going into the other lane?

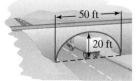

67. The elliptical ceiling in Statuary Hall in the U.S. Capitol Building is 96 feet long and 23 feet tall.

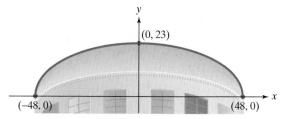

a. Using the rectangular coordinate system in the figure shown, write the standard form of the equation of the elliptical ceiling.

b. John Quincy Adams discovered that he could overhear the conversations of opposing party leaders near the left side of the chamber if he situated his desk at the focus at the right side of the chamber. How far from the center of the ellipse along the major axis did Adams situate his desk? (Round to the nearest foot.)

68. If an elliptical whispering room has a height of 30 feet and a width of 100 feet, where should two people stand if they would like to whisper back and forth and be heard?

Writing in Mathematics

69. What is an ellipse?

70. Describe how to graph $\frac{x^2}{25} + \frac{y^2}{16} = 1$.

71. Describe how to locate the foci for $\frac{x^2}{25} + \frac{y^2}{16} = 1$.

72. Describe one similarity and one difference between the graphs of $\frac{x^2}{25} + \frac{y^2}{16} = 1$ and $\frac{x^2}{16} + \frac{y^2}{25} = 1$.

73. Describe one similarity and one difference between the graphs of $\frac{x^2}{25} + \frac{y^2}{16} = 1$ and $\frac{(x-1)^2}{25} + \frac{(y-1)^2}{16} = 1$.

74. An elliptipool is an elliptical pool table with only one pocket. A pool shark places a ball on the table, hits it in what appears to be a random direction, and yet it bounces off the edge, falling directly into the pocket. Explain why this happens.

Technology Exercises

75. Use a graphing utility to graph any five of the ellipses that you graphed by hand in Exercises 1–18.

76. Use a graphing utility to graph any three of the ellipses that you graphed by hand in Exercises 37–50. First solve the given equation for y by using the square root property. Enter each of the two resulting equations to produce each half of the ellipse.

77. Use a graphing utility to graph any one of the ellipses that you graphed by hand in Exercises 51–56. Write the equation as a quadratic equation in y and use the quadratic formula to solve for y. Enter each of the two resulting equations to produce each half of the ellipse.

78. Write an equation for the path of each of the following elliptical orbits. Then use a graphing utility to graph the two ellipses in the same viewing rectangle. Can you see why early astronomers had difficulty detecting that these orbits are ellipses rather than circles?

Earth's orbit: Length of major axis: 186 million miles

Length of minor axis: 185.8 million miles

Mars's orbit: Length of major axis: 283.5 million miles

Length of minor axis: 278.5 million miles

Critical Thinking Exercises

79. Find the standard form of the equation of an ellipse with vertices at $(0, -6)$ and $(0, 6)$, passing through $(2, -4)$.

80. An Earth satellite has an elliptical orbit described by

$$\frac{x^2}{(5000)^2} + \frac{y^2}{(4750)^2} = 1.$$

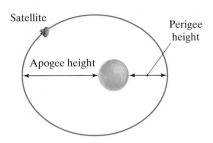

(All units are in miles.) The coordinates of the center of Earth are $(16, 0)$.

a. The perigee of the satellite's orbit is the point that is nearest Earth's center. If the radius of Earth is approximately 4000 miles, find the distance of the perigee above Earth's surface.

b. The apogee of the satellite's orbit is the point that is the greatest distance from Earth's center. Find the distance of the apogee above Earth's surface.

81. The equation of the red ellipse in the figure shown is

$$\frac{x^2}{25} + \frac{y^2}{9} = 1.$$

Write the equation for each circle shown in the figure.

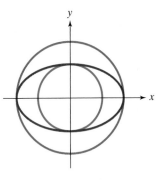

82. What happens to the shape of the graph of $\frac{x^2}{a^2} + \frac{y^2}{b^2} = 1$ as $\frac{c}{a} \to 0$?

SECTION 9.2 *The Hyperbola*

Objectives

❶ Locate a hyperbola's vertices and foci.

❷ Write equations of hyperbolas in standard form.

❸ Graph hyperbolas centered at the origin.

❹ Graph hyperbolas not centered at the origin.

❺ Solve applied problems involving hyperbolas.

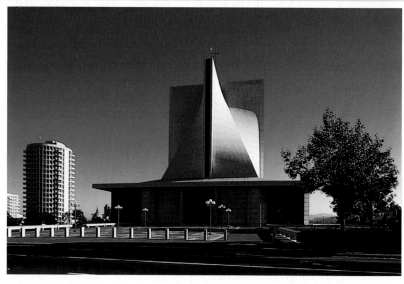

St. Mary's Cathedral

Conic sections are often used to create unusual architectural designs. The top of St. Mary's Cathedral in San Francisco is a 2135-cubic-foot dome with walls rising 200 feet above the floor and supported by four massive concrete pylons that extend 94 feet into the ground. Cross sections of the roof are parabolas and hyperbolas. In this section, we study the curve with two parts known as the hyperbola.

Definition of a Hyperbola

Figure 9.13 shows a cylindrical lampshade casting two shadows on a wall. These shadows indicate the distinguishing feature of hyperbolas: Their graphs contain two disjoint parts, called **branches**. Although each branch might look like a parabola, its shape is actually quite different.

The definition of a hyperbola is similar to that of an ellipse. For an ellipse, the *sum* of the distances to the foci is a constant. By contrast, for a hyperbola the *difference* of the distances to the foci is a constant.

Figure 9.13 Casting hyperbolic shadows

> ### Definition of a Hyperbola
>
> A **hyperbola** is the set of points in a plane the difference of whose distances from two fixed points, called foci, is constant.

Figure 9.14 illustrates the two branches of a hyperbola. The line through the foci intersects the hyperbola at two points, called the **vertices**. The line segment that joins the vertices is the **transverse axis**. The midpoint of the transverse axis is the **center** of the hyperbola. Notice that the center lies midway between the vertices, as well as midway between the foci.

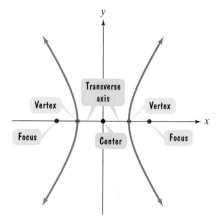

Figure 9.14 The two branches of a hyperbola

Standard Form of the Equation of a Hyperbola

The rectangular coordinate system enables us to translate a hyperbola's geometric definition into an algebraic equation. Figure 9.15 is our starting point for obtaining an equation. We place the foci, F_1 and F_2, on the x-axis at the points $(-c, 0)$ and $(c, 0)$. Note that the center of this hyperbola is at the origin. We let (x, y) represent the coordinates of any point, P, on the hyperbola.

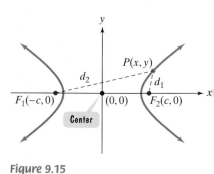

Figure 9.15

What does the definition of a hyperbola tell us about the point (x, y) in Figure 9.15? For any point (x, y) on the hyperbola, the absolute value of the difference of the distances from the two foci, $|d_2 - d_1|$, must be constant. We denote this constant by $2a$, just as we did for the ellipse. Thus, the point (x, y) is on the hyperbola if and only if

$$|d_2 - d_1| = 2a.$$

$$\left| \sqrt{(x + c)^2 + (y - 0)^2} - \sqrt{(x - c)^2 + (y - 0)^2} \right| = 2a \qquad \text{Use the distance formula.}$$

After eliminating radicals and simplifying, we obtain

$$(c^2 - a^2)x^2 - a^2 y^2 = a^2(c^2 - a^2).$$

For convenience, let $b^2 = c^2 - a^2$. Substituting b^2 for $c^2 - a^2$ in the preceding equation, we obtain

$$b^2 x^2 - a^2 y^2 = a^2 b^2.$$

$$\frac{b^2 x^2}{a^2 b^2} - \frac{a^2 y^2}{a^2 b^2} = \frac{a^2 b^2}{a^2 b^2} \qquad \text{Divide both sides by } a^2 b^2.$$

$$\frac{x^2}{a^2} - \frac{y^2}{b^2} = 1 \qquad \text{Simplify.}$$

This last equation is called the **standard form of the equation of a hyperbola centered at the origin**. There are two such equations. The first is for a hyperbola in which the transverse axis lies on the x-axis. The second is for a hyperbola in which the transverse axis lies on the y-axis.

> ### Standard Forms of the Equations of a Hyperbola
>
> The **standard form of the equation of a hyperbola** with center at the origin is
>
> $$\frac{x^2}{a^2} - \frac{y^2}{b^2} = 1 \qquad \text{or} \qquad \frac{y^2}{a^2} - \frac{x^2}{b^2} = 1.$$

Study Tip

The form $c^2 = a^2 + b^2$ is the one you should remember. When finding the foci, this form is easy to manipulate.

Figure 9.16(a) illustrates that for the equation on the left, the transverse axis lies on the x-axis. Figure 9.16(b) illustrates that for the equation on the right, the transverse axis lies on the y-axis. The vertices are a units from the center and the foci are c units from the center. For both equations, $b^2 = c^2 - a^2$. Equivalently, $c^2 = a^2 + b^2$.

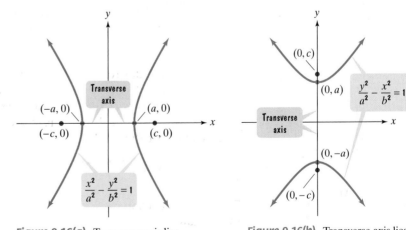

Figure 9.16(a) Transverse axis lies on the x-axis.

Figure 9.16(b) Transverse axis lies on the y-axis.

Study Tip

When the x^2-term is preceded by a plus sign, the transverse axis is horizontal. When the y^2-term is preceded by a plus sign, the transverse axis is vertical.

Locate a hyperbola's vertices and foci.

Study Tip

Notice the sign difference between the following equations:
Finding an ellipse's foci:
$$c^2 = a^2 - b^2$$
Finding a hyperbola's foci:
$$c^2 = a^2 + b^2.$$

Using the Standard Form of the Equation of a Hyperbola

We can use the standard form of the equation of a hyperbola to find its vertices and locate its foci. Because the vertices are a units from the center, begin by identifying a^2 in the equation. In the standard form of a hyperbola's equation, a^2 **is the number under the variable whose term is preceded by a plus sign** (+). If the x^2-term is preceded by a plus sign, the transverse axis lies along the x-axis. Thus, the vertices are a units to the left and right of the origin. If the y^2-term is preceded by a plus sign, the transverse axis lies along the y-axis. Thus, the vertices are a units above and below the origin.

We know that the foci are c units from the center. The substitution that is used to derive the hyperbola's equation, $c^2 = a^2 + b^2$, is needed to locate the foci when a^2 and b^2 are known.

EXAMPLE 1   Finding Vertices and Foci from a Hyperbola's Equation

Find the vertices and locate the foci for each of the following hyperbolas with the given equation:

a. $\dfrac{x^2}{16} - \dfrac{y^2}{9} = 1$ **b.** $\dfrac{y^2}{9} - \dfrac{x^2}{16} = 1$.

Solution   Both equations are in standard form. We begin by identifying a^2 and b^2 in each equation.

a. The first equation is in the form $\dfrac{x^2}{a^2} - \dfrac{y^2}{b^2} = 1$.

$$\dfrac{x^2}{16} - \dfrac{y^2}{9} = 1$$

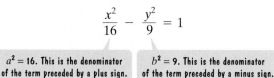

$a^2 = 16$. This is the denominator of the term preceded by a plus sign. $b^2 = 9$. This is the denominator of the term preceded by a minus sign.

Because the x^2-term is preceded by a plus sign, the transverse axis lies along the x-axis. Thus, the vertices are a units to the *left* and *right* of the origin. Based on the standard form of the equation, we know the vertices are $(-a, 0)$ and $(a, 0)$. Because $a^2 = 16$, $a = 4$. Thus, the vertices are $(-4, 0)$ and $(4, 0)$, shown in Figure 9.17.

We use $c^2 = a^2 + b^2$ to find the foci, which are located at $(-c, 0)$ and $(c, 0)$. We know that $a^2 = 16$ and $b^2 = 9$; we need to find c^2 in order to find c.

$$c^2 = a^2 + b^2 = 16 + 9 = 25$$

Because $c^2 = 25$, $c = 5$. The foci are located at $(-5, 0)$ and $(5, 0)$. They are shown in Figure 9.17.

b. The second given equation is in the form $\dfrac{y^2}{a^2} - \dfrac{x^2}{b^2} = 1$.

$$\dfrac{y^2}{9} - \dfrac{x^2}{16} = 1$$

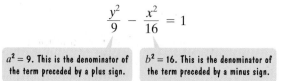

$a^2 = 9$. This is the denominator of the term preceded by a plus sign. $b^2 = 16$. This is the denominator of the term preceded by a minus sign.

Because the y^2-term is preceded by a plus sign, the transverse axis lies along the y-axis. Thus, the vertices are a units *above* and *below* the origin. Based on the standard form of the equation, we know the vertices are $(0, -a)$ and $(0, a)$. Because $a^2 = 9$, $a = 3$. Thus, the vertices are $(0, -3)$ and $(0, 3)$, shown in Figure 9.18.

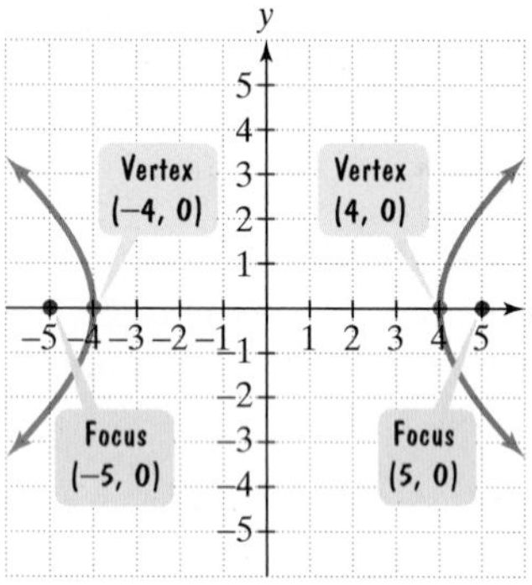

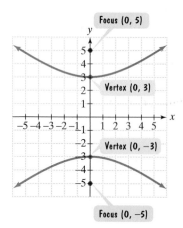

Figure 9.17 The graph of $\dfrac{x^2}{16} - \dfrac{y^2}{9} = 1$

Figure 9.18 The graph of $\dfrac{y^2}{9} - \dfrac{x^2}{16} = 1$

We use $c^2 = a^2 + b^2$ to find the foci, which are located at $(0, -c)$ and $(0, c)$.

$$c^2 = a^2 + b^2 = 9 + 16 = 25$$

Because $c^2 = 25$, $c = 5$. The foci are located at $(0, -5)$ and $(0, 5)$. They are shown in Figure 9.18.

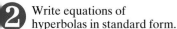 **Check Point 1** Find the vertices and locate the foci for each of the following hyperbolas with the given equation:

$$\textbf{a. } \frac{x^2}{25} - \frac{y^2}{16} = 1 \qquad \textbf{b. } \frac{y^2}{25} - \frac{x^2}{16} = 1.$$

In Example 1, we used equations of hyperbolas to find their foci and vertices. In the next example, we reverse this procedure.

② Write equations of hyperbolas in standard form.

EXAMPLE 2 Finding the Equation of a Hyperbola from Its Foci and Vertices

Find the standard form of the equation of a hyperbola with foci at $(0, -3)$ and $(0, 3)$ and vertices $(0, -2)$ and $(0, 2)$, shown in Figure 9.19.

Solution Because the foci are located at $(0, -3)$ and $(0, 3)$, on the y-axis, the transverse axis lies on the y-axis. The center of the hyperbola is midway between the foci, located at $(0, 0)$. Thus, the form of the equation is

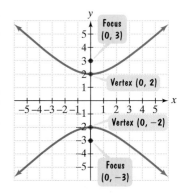

$$\frac{y^2}{a^2} - \frac{x^2}{b^2} = 1.$$

Figure 9.19

We need to determine the values for a^2 and b^2. The distance from the center, $(0, 0)$, to either vertex, $(0, -2)$ or $(0, 2)$, is 2, so $a = 2$.

$$\frac{y^2}{2^2} - \frac{x^2}{b^2} = 1 \qquad \text{or} \qquad \frac{y^2}{4} - \frac{x^2}{b^2} = 1$$

We must still find b^2. The distance from the center, $(0, 0)$, to either focus, $(0, -3)$ or $(0, 3)$, is 3. Thus, $c = 3$. Using $c^2 = a^2 + b^2$, we have

$$3^2 = 2^2 + b^2$$

and

$$b^2 = 3^2 - 2^2 = 9 - 4 = 5.$$

Substituting 5 for b^2 in $\dfrac{y^2}{4} - \dfrac{x^2}{b^2} = 1$ gives us the standard form of the hyperbola's equation. The equation is

$$\frac{y^2}{4} - \frac{x^2}{5} = 1.$$

 Check Point 2 Find the standard form of the equation of a hyperbola with foci at $(0, -5)$ and $(0, 5)$ and vertices $(0, -3)$ and $(0, 3)$.

The Asymptotes of a Hyperbola

As x and y get larger, the two branches of the graph of a hyperbola approach a pair of intersecting straight lines, called **asymptotes**. The asymptotes pass through the center of the hyperbola and are helpful in graphing hyperbolas.

Figure 9.20 shows the asymptotes for the graphs of hyperbolas centered at the origin. The asymptotes pass through the corners of a rectangle. Note that the dimensions of this rectangle are $2a$ by $2b$. The line segment of length $2b$ is the **conjugate axis** of the hyperbola and is perpendicular to the transverse axis through the center of the hyperbola.

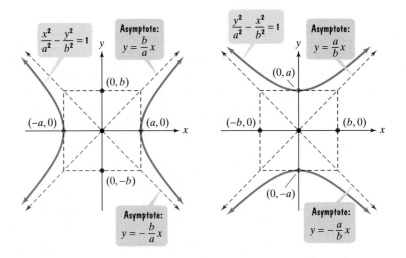

Figure 9.20 Asymptotes of a hyperbola

The Asymptotes of a Hyperbola Centered at the Origin

The hyperbola $\dfrac{x^2}{a^2} - \dfrac{y^2}{b^2} = 1$ has a horizontal transverse axis and two asymptotes

$$y = \frac{b}{a}x \qquad \text{and} \qquad y = -\frac{b}{a}x.$$

The hyperbola $\dfrac{y^2}{a^2} - \dfrac{x^2}{b^2} = 1$ has a vertical transverse axis and two asymptotes

$$y = \frac{a}{b}x \qquad \text{and} \qquad y = -\frac{a}{b}x.$$

Why are $y = \pm\dfrac{b}{a}x$ the asymptotes for a hyperbola whose transverse axis is horizontal? The proof can be found in the appendix.

③ Graph hyperbolas centered at the origin.

Graphing Hyperbolas Centered at the Origin

Hyperbolas are graphed using vertices and asymptotes.

Graphing Hyperbolas

1. Locate the vertices.
2. Use dashed lines to draw the rectangle centered at the origin with sides parallel to the axes, crossing one axis at $\pm a$ and the other at $\pm b$.
3. Use dashed lines to draw the diagonals of this rectangle and extend them to obtain the asymptotes.
4. Draw the two branches of the hyperbola by starting at each vertex and approaching the asymptotes.

EXAMPLE 3   Graphing a Hyperbola

Graph and locate the foci: $\dfrac{x^2}{25} - \dfrac{y^2}{16} = 1$. What are the equations of the asymptotes?

Solution

Step 1   Locate the vertices. The given equation is in the form $\dfrac{x^2}{a^2} - \dfrac{y^2}{b^2} = 1$, with $a^2 = 25$ and $b^2 = 16$.

$$\dfrac{x^2}{25} - \dfrac{y^2}{16} = 1$$

$$a^2 = 25 \qquad\qquad b^2 = 16$$

Based on the standard form of the equation with the transverse axis on the x-axis, we know that the vertices are $(-a, 0)$ and $(a, 0)$. Because $a^2 = 25$, $a = 5$. Thus, the vertices are $(-5, 0)$ and $(5, 0)$, shown in Figure 9.21.

Step 2   Draw a rectangle. Because $a^2 = 25$ and $b^2 = 16$, $a = 5$ and $b = 4$. We construct a rectangle to find the asymptotes, using -5 and 5 on the x-axis (the vertices are located here) and -4 and 4 on the y-axis. The rectangle passes through these four points, shown using dashed lines in Figure 9.21.

Step 3   Draw extended diagonals for the rectangle to obtain the asymptotes. We draw dashed lines through the opposite corners of the rectangle, shown in Figure 9.21, to obtain the graph of the asymptotes. Based on the standard form of the hyperbola's equation, the equations for these asymptotes are

$$y = \pm \dfrac{b}{a}x \qquad \text{or} \qquad y = \pm \dfrac{4}{5}x.$$

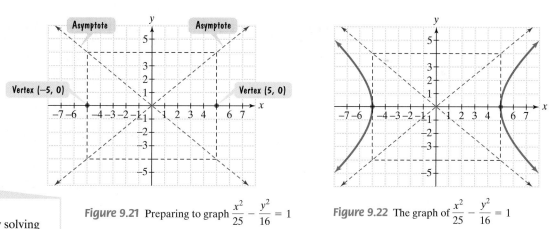

Figure 9.21 Preparing to graph $\dfrac{x^2}{25} - \dfrac{y^2}{16} = 1$

Figure 9.22 The graph of $\dfrac{x^2}{25} - \dfrac{y^2}{16} = 1$

Technology

Graph $\dfrac{x^2}{25} - \dfrac{y^2}{16} = 1$ by solving for y:

$$y_1 = \dfrac{\sqrt{16x^2 - 400}}{5}$$

$$y_2 = -\dfrac{\sqrt{16x^2 - 400}}{5} = -y_1.$$

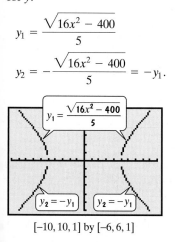

$y_1 = \dfrac{\sqrt{16x^2 - 400}}{5}$

$y_2 = -y_1$ $y_2 = -y_1$

$[-10, 10, 1]$ by $[-6, 6, 1]$

Step 4   Draw the two branches of the hyperbola by starting at each vertex and approaching the asymptotes. The hyperbola is shown in Figure 9.22.

We now consider the foci, located at $(-c, 0)$ and $(c, 0)$. We find c using $c^2 = a^2 + b^2$.

$$c^2 = 25 + 16 = 41$$

Because $c^2 = 41$, $c = \sqrt{41}$. The foci are located at $\left(-\sqrt{41}, 0\right)$ and $\left(\sqrt{41}, 0\right)$, approximately $(-6.4, 0)$ and $(6.4, 0)$.

Check Point 3   Graph and locate the foci: $\dfrac{x^2}{36} - \dfrac{y^2}{9} = 1$. What are the equations of the asymptotes?

EXAMPLE 4   Graphing a Hyperbola

Graph and locate the foci: $9y^2 - 4x^2 = 36$. What are the equations of the asymptotes?

Solution   We begin by writing the equation in standard form. The right side should be 1, so we divide both sides by 36.

$$\frac{9y^2}{36} - \frac{4x^2}{36} = \frac{36}{36}$$

$$\frac{y^2}{4} - \frac{x^2}{9} = 1 \qquad \text{Simplify. The right side is now 1.}$$

Now we are ready to use our four-step procedure for graphing hyperbolas.

Step 1 Locate the vertices. The equation that we obtained is in the form $\frac{y^2}{a^2} - \frac{x^2}{b^2} = 1$, with $a^2 = 4$ and $b^2 = 9$.

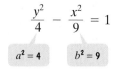

$$\frac{y^2}{4} - \frac{x^2}{9} = 1$$

$a^2 = 4 \qquad b^2 = 9$

Based on the standard form of the equation with the transverse axis on the y-axis, we know that the vertices are $(0, -a)$ and $(0, a)$. Because $a^2 = 4$, $a = 2$. Thus, the vertices are $(0, -2)$ and $(0, 2)$, shown in Figure 9.23.

Step 2   Draw a rectangle. Because $a^2 = 4$ and $b^2 = 9$, $a = 2$ and $b = 3$. We construct a rectangle to find the asymptotes, using -2 and 2 on the y-axis (the vertices are located here) and -3 and 3 on the x-axis. The rectangle passes through these four points, shown using dashed lines in Figure 9.23.

Step 3   Draw extended diagonals of the rectangle to obtain the asymptotes. We draw dashed lines through the opposite corners of the rectangle, shown in Figure 9.23, to obtain the graph of the asymptotes. Based on the standard form of the hyperbola's equation, the equations of these asymptotes are

$$y = \pm \frac{a}{b} x \qquad \text{or} \qquad y = \pm \frac{2}{3} x.$$

Step 4   Draw the two branches of the hyperbola by starting at each vertex and approaching the asymptotes. The hyperbola is shown in Figure 9.24.

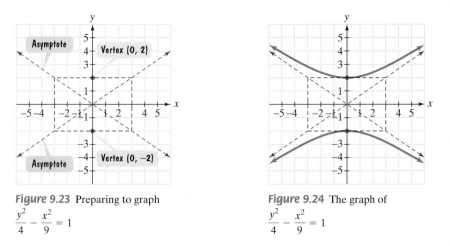

Figure 9.23 Preparing to graph $\frac{y^2}{4} - \frac{x^2}{9} = 1$

Figure 9.24 The graph of $\frac{y^2}{4} - \frac{x^2}{9} = 1$

We now consider the foci, located at $(0, -c)$ and $(0, c)$. We find c using $c^2 = a^2 + b^2$.

$$c^2 = 4 + 9 = 13$$

Because $c^2 = 13$, $c = \sqrt{13}$. The foci are located at $\left(0, -\sqrt{13}\right)$ and $\left(0, \sqrt{13}\right)$ approximately $(0, -3.6)$ and $(0, 3.6)$.

Check Point **4** Graph and locate the foci: $y^2 - 4x^2 = 4$. What are the equations of the asymptotes?

④ Graph hyperbolas not centered at the origin.

Translations of Hyperbolas

The graph of a hyperbola can be centered at (h, k), rather than at the origin. Horizontal and vertical translations are accomplished by replacing x with $x - h$ and y with $y - k$ in the standard form of the hyperbola's equation.

Table 9.2 gives the standard forms of equations of hyperbolas centered at (h, k) and shows their graphs.

Table 9.2   Standard Forms of Equations of Hyperbolas Centered at (h, k)

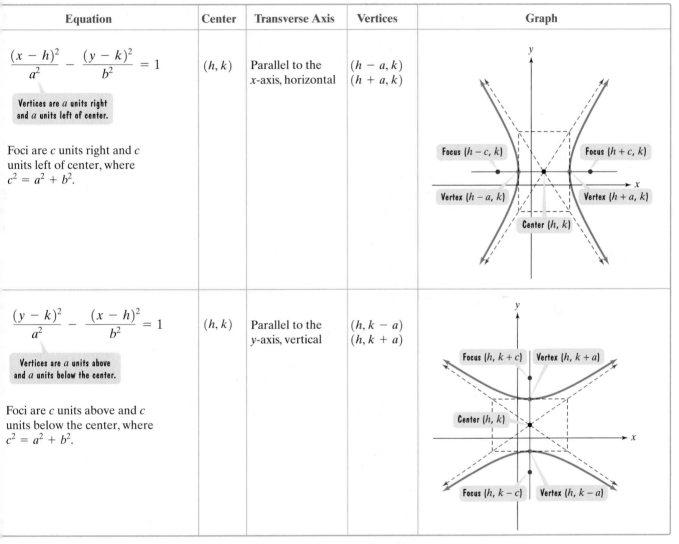

Equation	Center	Transverse Axis	Vertices	Graph
$\dfrac{(x - h)^2}{a^2} - \dfrac{(y - k)^2}{b^2} = 1$ Vertices are a units right and a units left of center. Foci are c units right and c units left of center, where $c^2 = a^2 + b^2$.	(h, k)	Parallel to the x-axis, horizontal	$(h - a, k)$ $(h + a, k)$	
$\dfrac{(y - k)^2}{a^2} - \dfrac{(x - h)^2}{b^2} = 1$ Vertices are a units above and a units below the center. Foci are c units above and c units below the center, where $c^2 = a^2 + b^2$.	(h, k)	Parallel to the y-axis, vertical	$(h, k - a)$ $(h, k + a)$	

EXAMPLE 5   Graphing a Hyperbola Centered at (h, k)

Graph: $\dfrac{(x - 2)^2}{16} - \dfrac{(y - 3)^2}{9} = 1$. Where are the foci located? What are the equations of the asymptotes?

Solution In order to graph the hyperbola, we need to know its center, (h, k). In the standard forms of equations centered at (h, k), h is the number subtracted from x and k is the number subtracted from y.

This is $(x - h)^2$, with $h = 2$.

This is $(y - k)^2$, with $k = 3$.

$$\frac{(x - 2)^2}{16} - \frac{(y - 3)^2}{9} = 1$$

We see that $h = 2$ and $k = 3$. Thus, the center of the hyperbola, (h, k), is $(2, 3)$. We can graph the hyperbola by using vertices, asymptotes, and our four-step graphing procedure.

Step 1 Locate the vertices. To do this, we must identify a^2.

Based on the standard form of the equation with a horizontal transverse axis, the vertices are a units to the left and right of the center. Because $a^2 = 16$, $a = 4$. This means that the vertices are 4 units to the left and right of the center, $(2, 3)$. Four units to the left of $(2, 3)$ puts one vertex at $(2 - 4, 3)$, or $(-2, 3)$. Four units to the right of $(2, 3)$ puts the other vertex at $(2 + 4, 3)$, or $(6, 3)$. The vertices are shown in Figure 9.25.

Figure 9.25 Locating a hyperbola's center and vertices

Step 2 Draw a rectangle. Because $a^2 = 16$ and $b^2 = 9$, $a = 4$ and $b = 3$. The rectangle passes through points that are 4 units to the right and left of the center (the vertices are located here) and 3 units above and below the center. The rectangle is shown using dashed lines in Figure 9.26.

Step 3 Draw extended diagonals of the rectangle to obtain the asymptotes. We draw dashed lines through the opposite corners of the rectangle, shown in Figure 9.26, to obtain the graph of the asymptotes. The equations of the asymptotes of the

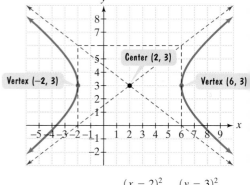

Figure 9.26 The graph of $\dfrac{(x - 2)^2}{16} - \dfrac{(y - 3)^2}{9} = 1$

unshifted hyperbola $\dfrac{x^2}{16} - \dfrac{y^2}{9} = 1$ are $y = \pm \dfrac{b}{a}x$, or $y = \pm \dfrac{3}{4}x$. Thus, the asymptotes for the hyperbola that is shifted two units to the right and three units up, namely

$$\frac{(x - 2)^2}{16} - \frac{(y - 3)^2}{9} = 1$$

have equations that can be expressed as

$$y - 3 = \pm \frac{3}{4}(x - 2).$$

Step 4 Draw the two branches of the hyperbola by starting at each vertex and approaching the asymptotes. The hyperbola is shown in Figure 9.26.

We now consider the foci, located c units to the right and left of the center. We find c using $c^2 = a^2 + b^2$.

$$c^2 = 16 + 9 = 25$$

Because $c^2 = 25$, $c = 5$. This means that the foci are 5 units to the left and right of the center, $(2, 3)$. Five units to the left of $(2, 3)$ puts one focus at $(2 - 5, 3)$, or $(-3, 3)$. Five units to the right of $(2, 3)$ puts the other focus at $(2 + 5, 3)$, or $(7, 3)$.

Study Tip

You can also use the point-slope form of a line's equation

$$y - y_1 = m(x - x_1)$$

to find the equations of the asymptotes. The center of the hyperbola, (h, k), is a point on each asymptote, so $x_1 = h$ and $y_1 = k$. The slopes, m, are $\pm \dfrac{b}{a}$ for a horizontal transverse axis and $\pm \dfrac{a}{b}$ for a vertical transverse axis.

Check Point 5 Graph: $\dfrac{(x - 3)^2}{4} - \dfrac{(y - 1)^2}{1} = 1$. Where are the foci located? What are the equations of the asymptotes?

In our next example, it is necessary to convert the equation of a hyperbola to standard form by completing the square on x and y.

EXAMPLE 6   Graphing a Hyperbola Centered at (h, k)

Graph: $4x^2 - 24x - 25y^2 + 250y - 489 = 0$. Where are the foci located? What are the equations of the asymptotes?

Solution   We begin by completing the square on x and y.

$$4x^2 - 24x - 25y^2 + 250y - 489 = 0$$ This is the given equation.

$$(4x^2 - 24x) + (-25y^2 + 250y) = 489$$ Group terms and add 489 to both sides.

$$4(x^2 - 6x + \square) - 25(y^2 - 10y + \square) = 489$$ Factor out 4 and -25, respectively, so coefficients of x^2 and y^2 are 1.

$$4(x^2 - 6x + 9) - 25(y^2 - 10y + 25) = 489 + 36 + (-625)$$ Complete each square by adding the square of half the coefficient of x and y, respectively.

We added 4 · 9, or 36, to the left side.   We added −25 · 25, or −625, to the left side.   Add 36 + (−625) to the right side.

$$4(x - 3)^2 - 25(y - 5)^2 = -100$$ Factor.

$$\frac{4(x - 3)^2}{-100} - \frac{25(y - 5)^2}{-100} = \frac{-100}{-100}$$ Divide both sides by -100.

$$\frac{(x - 3)^2}{-25} + \frac{(y - 5)^2}{4} = 1$$ Simplify.

This is $(y - k)^2$, with $k = 5$. $$\frac{(y - 5)^2}{4} - \frac{(x - 3)^2}{25} = 1$$ This is $(x - h)^2$, with $h = 3$.   Write the equation in standard form, $\frac{(y - k)^2}{a^2} - \frac{(x - h)^2}{b^2} = 1$.

We see that $h = 3$ and $k = 5$. Thus, the center of the hyperbola, (h, k), is $(3, 5)$. Because the x^2-term is being subtracted, the transverse axis is vertical and the hyperbola opens upward and downward.

We use our four-step procedure to obtain the graph of

$$\frac{(y - 5)^2}{4} - \frac{(x - 3)^2}{25} = 1.$$

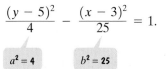

$a^2 = 4$ $b^2 = 25$

Step 1   Locate the vertices. Based on the standard form of the equation with a vertical transverse axis, the vertices are a units above and below the center. Because $a^2 = 4$, $a = 2$. This means that the vertices are 2 units above and below the center, $(3, 5)$. This puts the vertices at $(3, 7)$ and $(3, 3)$, shown in Figure 9.27.

Step 2   Draw a rectangle. Because $a^2 = 4$ and $b^2 = 25$, $a = 2$ and $b = 5$. The rectangle passes through points that are 2 units above and below the center (the vertices are located here) and 5 units to the right and left of the center. The rectangle is shown using dashed lines in Figure 9.27.

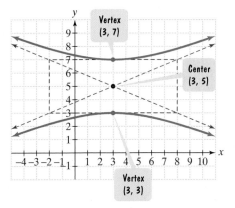

Figure 9.27 The graph of
$\frac{(y - 5)^2}{4} - \frac{(x - 3)^2}{25} = 1$

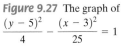

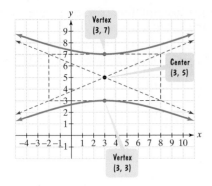

Figure 9.27 (repeated) The graph of
$$\frac{(y - 5)^2}{4} - \frac{(x - 3)^2}{25} = 1$$

Step 3 Draw extended diagonals of the rectangle to obtain the asymptotes. We draw dashed lines through the opposite corners of the rectangle, shown in Figure 9.27, to obtain the graph of the asymptotes. The equations of the asymptotes of the unshifted hyperbola $\frac{y^2}{4} - \frac{x^2}{25} = 1$ are $y = \pm\frac{a}{b}x$, or $y = \pm\frac{2}{5}x$. Thus, the asymptotes for the hyperbola that is shifted three units to the right and five units up, namely

$$\frac{(y - 5)^2}{4} - \frac{(x - 3)^2}{25} = 1$$

have equations that can be expressed as

$$y - 5 = \pm\frac{2}{5}(x - 3).$$

Step 4 Draw the two branches of the hyperbola by starting at each vertex and approaching the asymptotes. The hyperbola is shown in Figure 9.27.

We now consider the foci, located c units above and below the center, $(3, 5)$. We find c using $c^2 = a^2 + b^2$.

$$c^2 = 4 + 25 = 29.$$

Because $c^2 = 29$, $c = \sqrt{29}$. The foci are located at $(3, 5 + \sqrt{29})$ and $(3, 5 - \sqrt{29})$.

Check Point 6 Graph: $4x^2 - 24x - 9y^2 - 90y - 153 = 0$. Where are the foci located? What are the equations of the asymptotes?

⑤ Solve applied problems involving hyperbolas.

Applications

Hyperbolas have many applications. When a jet flies at a speed greater than the speed of sound, the shock wave that is created is heard as a sonic boom. The wave has the shape of a cone. The shape formed as the cone hits the ground is one branch of a hyperbola.

Halley's Comet, a permanent part of our solar system, travels around the sun in an elliptical orbit. Other comets pass through the solar system only once, following a hyperbolic path with the sun as a focus.

Hyperbolas are of practical importance in fields ranging from architecture to navigation. Cooling towers used in the design for nuclear power plants have cross sections that are both ellipses and hyperbolas. Three-dimensional solids whose cross sections are hyperbolas are used in some rather unique architectural creations, including the TWA building at Kennedy Airport in New York City and the St. Louis Science Center Planetarium.

The hyperbolic shape of a sonic boom

EXAMPLE 7   An Application Involving Hyperbolas

An explosion is recorded by two microphones that are 2 miles apart. Microphone M_1 received the sound 4 seconds before microphone M_2. Assuming sound travels at 1100 feet per second, determine the possible locations of the explosion relative to the location of the microphones.

Solution We begin by putting the microphones in a coordinate system. Because 1 mile = 5280 feet, we place M_1 5280 feet on a horizontal axis to the right of the origin and M_2 5280 feet on a horizontal axis to the left of the origin. Figure 9.28 illustrates that the two microphones are 2 miles apart.

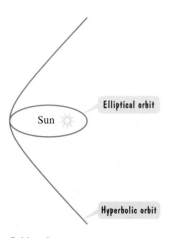

Orbits of comets

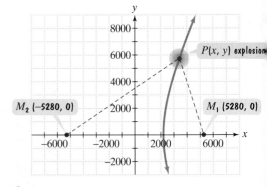

Figure 9.28 Locating an explosion on the branch of a hyperbola

We know that M_2 received the sound 4 seconds after M_1. Because sound travels at 1100 feet per second, the difference between the distance from P to M_1 and the distance from P to M_2 is 4400 feet. The set of all points P (or locations of the explosion) satisfying these conditions fits the definition of a hyperbola, with microphones M_1 and M_2 at the foci.

$$\frac{x^2}{a^2} - \frac{y^2}{b^2} = 1$$ Use the standard form of the hyperbola's equation. $P(x, y)$, the explosion point, lies on this hyperbola. We must find a^2 and b^2.

The difference between the distances, represented by $2a$ in the derivation of the hyperbola's equation, is 4400 feet. Thus, $2a = 4400$ and $a = 2200$.

$$\frac{x^2}{(2200)^2} - \frac{y^2}{b^2} = 1$$ Substitute 2200 for a.

We must still find b^2. We know that $a = 2200$. The distance from the center, $(0, 0)$, to either focus, $(-5280, 0)$ or $(5280, 0)$, is 5280. Thus, $c = 5280$. Using $c^2 = a^2 + b^2$, we have

$$5280^2 = 2200^2 + b^2$$

and

$$b^2 = 5280^2 - 2200^2 = 23{,}038{,}400.$$

The equation of the hyperbola with a microphone at each focus is

$$\frac{x^2}{4{,}840{,}000} - \frac{y^2}{23{,}038{,}400} = 1$$ Substitute 23,038,400 for b^2.

We can conclude that the explosion occurred somewhere on the right branch (the branch closer to M_1) of the hyperbola given by this equation.

In Example 7, we determined that the explosion occurred somewhere along one branch of a hyperbola, but not exactly where on the hyperbola. If, however, we had received the sound from another pair of microphones, we could locate the sound along a branch of another hyperbola. The exact location of the explosion would be the point where the two hyperbolas intersect.

Check Point 7   Rework Example 7 assuming microphone M_1 receives the sound 3 seconds before microphone M_2.

EXERCISE SET 9.2

Practice Exercises

In Exercises 1–4, find the vertices and locate the foci of each hyperbola with the given equation. Then match each equation to one of the graphs that are shown and labeled (a)–(d).

1. $\dfrac{x^2}{4} - \dfrac{y^2}{1} = 1$ **2.** $\dfrac{x^2}{1} - \dfrac{y^2}{4} = 1$

3. $\dfrac{y^2}{4} - \dfrac{x^2}{1} = 1$ **4.** $\dfrac{y^2}{1} - \dfrac{x^2}{4} = 1$

c.   **d.**
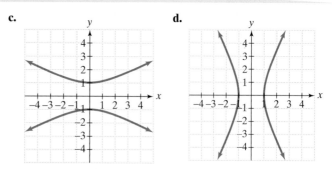

a.   **b.**
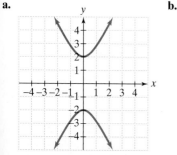
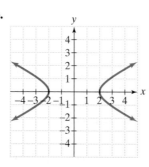

In Exercises 5–12, find the standard form of the equation of each hyperbola satisfying the given conditions.

5. Foci: $(0, -3), (0, 3)$; vertices: $(0, -1), (0, 1)$
6. Foci: $(0, -6), (0, 6)$; vertices: $(0, -2), (0, 2)$
7. Foci: $(-4, 0), (4, 0)$; vertices: $(-3, 0), (3, 0)$
8. Foci: $(-7, 0), (7, 0)$; vertices: $(-5, 0), (5, 0)$
9. Endpoints of transverse axis: $(0, -6), (0, 6)$; asymptote: $y = 2x$
10. Endpoints of transverse axis: $(-4, 0), (4, 0)$; asymptote: $y = 2x$

11. Center: $(4, -2)$; Focus: $(7, -2)$; vertex: $(6, -2)$

12. Center: $(-2, 1)$; Focus: $(-2, 6)$; vertex: $(-2, 4)$

In Exercises 13–26, use vertices and asymptotes to graph each hyperbola. Locate the foci and find the equations of the asymptotes.

13. $\dfrac{x^2}{9} - \dfrac{y^2}{25} = 1$

14. $\dfrac{x^2}{16} - \dfrac{y^2}{25} = 1$

15. $\dfrac{x^2}{100} - \dfrac{y^2}{64} = 1$

16. $\dfrac{x^2}{144} - \dfrac{y^2}{81} = 1$

17. $\dfrac{y^2}{16} - \dfrac{x^2}{36} = 1$

18. $\dfrac{y^2}{25} - \dfrac{x^2}{64} = 1$

19. $4y^2 - x^2 = 1$

20. $9y^2 - x^2 = 1$

21. $9x^2 - 4y^2 = 36$

22. $4x^2 - 25y^2 = 100$

23. $9y^2 - 25x^2 = 225$

24. $16y^2 - 9x^2 = 144$

25. $y = \pm\sqrt{x^2 - 2}$

26. $y = \pm\sqrt{x^2 - 3}$

In Exercises 27–32, find the standard form of the equation of each hyperbola.

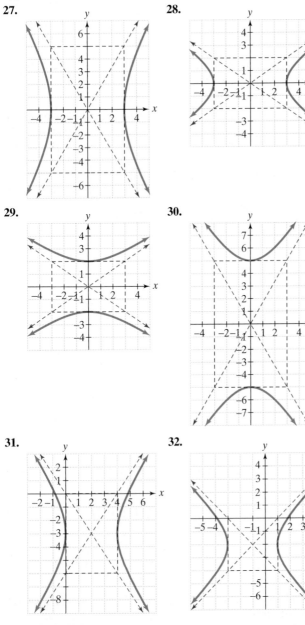

27.

28.

29.

30.

31.

32.

In Exercises 33–42, use the center, vertices, and asymptotes to graph each hyperbola. Locate the foci and find the equations of the asymptotes.

33. $\dfrac{(x + 4)^2}{9} - \dfrac{(y + 3)^2}{16} = 1$

34. $\dfrac{(x + 2)^2}{9} - \dfrac{(y - 1)^2}{25} = 1$

35. $\dfrac{(x + 3)^2}{25} - \dfrac{y^2}{16} = 1$

36. $\dfrac{(x + 2)^2}{9} - \dfrac{y^2}{25} = 1$

37. $\dfrac{(y + 2)^2}{4} - \dfrac{(x - 1)^2}{16} = 1$

38. $\dfrac{(y - 2)^2}{36} - \dfrac{(x + 1)^2}{49} = 1$

39. $(x - 3)^2 - 4(y + 3)^2 = 4$

40. $(x + 3)^2 - 9(y - 4)^2 = 9$

41. $(x - 1)^2 - (y - 2)^2 = 3$

42. $(y - 2)^2 - (x + 3)^2 = 5$

In Exercises 43–50, convert each equation to standard form by completing the square on x and y. Then graph the hyperbola. Locate the foci and find the equations of the asymptotes.

43. $x^2 - y^2 - 2x - 4y - 4 = 0$

44. $4x^2 - y^2 + 32x + 6y + 39 = 0$

45. $16x^2 - y^2 + 64x - 2y + 67 = 0$

46. $9y^2 - 4x^2 - 18y + 24x - 63 = 0$

47. $4x^2 - 9y^2 - 16x + 54y - 101 = 0$

48. $4x^2 - 9y^2 + 8x - 18y - 6 = 0$

49. $4x^2 - 25y^2 - 32x + 164 = 0$

50. $9x^2 - 16y^2 - 36x - 64y + 116 = 0$

Practice Plus

In Exercises 51–56, graph each relation. Use the relation's graph to determine its domain and range.

51. $\dfrac{x^2}{9} - \dfrac{y^2}{16} = 1$

52. $\dfrac{x^2}{25} - \dfrac{y^2}{4} = 1$

53. $\dfrac{x^2}{9} + \dfrac{y^2}{16} = 1$

54. $\dfrac{x^2}{25} + \dfrac{y^2}{4} = 1$

55. $\dfrac{y^2}{16} - \dfrac{x^2}{9} = 1$

56. $\dfrac{y^2}{4} - \dfrac{x^2}{25} = 1$

In Exercises 57–60, find the solution set for each system by graphing both of the system's equations in the same rectangular coordinate system and finding points of intersection. Check all solutions in both equations.

57. $x^2 - y^2 = 4$
$x^2 + y^2 = 4$

58. $x^2 - y^2 = 9$
$x^2 + y^2 = 9$

59. $9x^2 + y^2 = 9$
$y^2 - 9x^2 = 9$

60. $4x^2 + y^2 = 4$
$y^2 - 4x^2 = 4$

Application Exercises

61. An explosion is recorded by two microphones that are 1 mil apart. Microphone M_1 received the sound 2 seconds befor microphone M_2. Assuming sound travels at 1100 feet per sec ond, determine the possible locations of the explosio relative to the location of the microphones.

62. Radio towers A and B, 200 kilometers apart, are situated alon the coast, with A located due west of B. Simultaneous radi signals are sent from each tower to a ship, with the signal from B received 500 microseconds before the signal from A.

a. Assuming that the radio signals travel 300 meters per microsecond, determine the equation of the hyperbola on which the ship is located.

b. If the ship lies due north of tower B, how far out at sea is it?

63. An architect designs two houses that are shaped and positioned like a part of the branches of the hyperbola whose equation is $625y^2 - 400x^2 = 250{,}000$, where x and y are in yards. How far apart are the houses at their closest point?

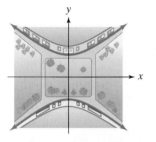

64. Scattering experiments, in which moving particles are deflected by various forces, led to the concept of the nucleus of an atom. In 1911, the physicist Ernest Rutherford (1871– 1937) discovered that when alpha particles are directed toward the nuclei of gold atoms, they are eventually deflected along hyperbolic paths, illustrated in the figure. If a particle gets as close as 3 units to the nucleus along a hyperbolic path with an asymptote given by $y = \frac{1}{2}x$, what is the equation of its path?

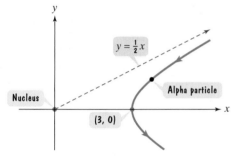

![pencil icon] **Writing in Mathematics**

65. What is a hyperbola?

66. Describe how to graph $\dfrac{x^2}{9} - \dfrac{y^2}{1} = 1$.

67. Describe how to locate the foci of the graph of $\dfrac{x^2}{9} - \dfrac{y^2}{1} = 1$.

68. Describe one similarity and one difference between the graphs of $\dfrac{x^2}{9} - \dfrac{y^2}{1} = 1$ and $\dfrac{y^2}{9} - \dfrac{x^2}{1} = 1$.

69. Describe one similarity and one difference between the graphs of $\dfrac{x^2}{9} - \dfrac{y^2}{1} = 1$ and $\dfrac{(x-3)^2}{9} - \dfrac{(y+3)^2}{1} = 1$.

70. How can you distinguish an ellipse from a hyperbola by looking at their equations?

71. In 1992, a NASA team began a project called Spaceguard Survey, calling for an international watch for comets that might collide with Earth. Why is it more difficult to detect a possible "doomsday comet" with a hyperbolic orbit than one with an elliptical orbit?

 Technology Exercises

72. Use a graphing utility to graph any five of the hyperbolas that you graphed by hand in Exercises 13–26.

73. Use a graphing utility to graph any three of the hyperbolas that you graphed by hand in Exercises 33–42. First solve the given equation for y by using the square root property. Enter each of the two resulting equations to produce each branch of the hyperbola.

74. Use a graphing utility to graph any one of the hyperbolas that you graphed by hand in Exercises 43–50. Write the equation as a quadratic equation in y and use the quadratic formula to solve for y. Enter each of the two resulting equations to produce each branch of the hyperbola.

75. Use a graphing utility to graph $\dfrac{x^2}{4} - \dfrac{y^2}{9} = 0$. Is the graph a hyperbola? In general, what is the graph of $\dfrac{x^2}{a^2} - \dfrac{y^2}{b^2} = 0$?

76. Graph $\dfrac{x^2}{a^2} - \dfrac{y^2}{b^2} = 1$ and $\dfrac{x^2}{a^2} - \dfrac{y^2}{b^2} = -1$ in the same viewing rectangle for values of a^2 and b^2 of your choice. Describe the relationship between the two graphs.

77. Write $4x^2 - 6xy + 2y^2 - 3x + 10y - 6 = 0$ as a quadratic equation in y and then use the quadratic formula to express y in terms of x. Graph the resulting two equations using a graphing utility in a $[-50, 70, 10]$ by $[-30, 50, 10]$ viewing rectangle. What effect does the xy-term have on the graph of the resulting hyperbola? What problems would you encounter if you attempted to write the given equation in standard form by completing the square?

78. Graph $\dfrac{x^2}{16} - \dfrac{y^2}{9} = 1$ and $\dfrac{x|x|}{16} - \dfrac{y|y|}{9} = 1$ in the same viewing rectangle. Explain why the graphs are not the same.

![lightbulb icon] **Critical Thinking Exercises**

79. Which one of the following is true?

a. If one branch of a hyperbola is removed from a graph, then the branch that remains must define y as a function of x.

b. All points on the asymptotes of a hyperbola also satisfy the hyperbola's equation.

c. The graph of $\dfrac{x^2}{9} - \dfrac{y^2}{4} = 1$ does not intersect the line $y = -\dfrac{2}{3}x$.

d. Two different hyperbolas can never share the same asymptotes.

80. What happens to the shape of the graph of $\dfrac{x^2}{a^2} - \dfrac{y^2}{b^2} = 1$ as $\dfrac{c}{a} \to \infty$?

81. Find the standard form of the equation of the hyperbola with vertices $(5, -6)$ and $(5, 6)$, passing through $(0, 9)$.

82. Find the equation of a hyperbola whose asymptotes are perpendicular.

SECTION 9.3 The Parabola

Objectives

1 Graph parabolas with vertices at the origin.

2 Write equations of parabolas in standard form.

3 Graph parabolas with vertices not at the origin.

4 Solve applied problems involving parabolas.

At first glance, this image looks like columns of smoke rising from a fire into a starry sky. Those are, indeed, stars in the background, but you are not looking at ordinary smoke columns. These stand almost 6 trillion miles high and are 7000 light-years from Earth—more than 400 million times as far away as the sun.

This NASA photograph is one of a series of stunning images captured from the end of the universe by the Hubble Space Telescope. The image shows infant star system the size of our solar system emerging from the gas and dust that shrouded their creation. Using a parabolic mirror that is 94.5 inches in diameter, the Hubble has provided answers to many of the profound mysteries of the cosmos: How big and how old is the universe? How did the galaxies come to exist? Do other Earth-like planets orbit other sun-like stars? In this section, we study parabolas and their applications, including parabolic shapes that gather distant rays of light and focus them into spectacular images.

Definition of a Parabola

In Chapter 2, we studied parabolas, viewing them as graphs of quadratic functions in the form

$$y = a(x - h)^2 + k \quad \text{or} \quad y = ax^2 + bx + c.$$

Study Tip

Here is a summary of what you should already know about graphing parabolas.

Graphing $y = a(x - h)^2 + k$ and $y = ax^2 + bx + c$

1. If $a > 0$, the graph opens upward. If $a < 0$, the graph opens downward.

2. The vertex of $y = a(x - h)^2 + k$ is (h, k).

3. The x-coordinate of the vertex of $y = ax^2 + bx + c$ is $x = -\dfrac{b}{2a}$.

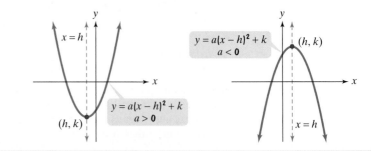

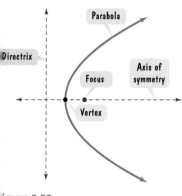

Parabola

Directrix

Focus

Axis of symmetry

Vertex

Figure 9.29

Parabolas can be given a geometric definition that enables us to include graphs that open to the left or to the right, as well as those that open obliquely. The definitions of ellipses and hyperbolas involved two fixed points, the foci. By contrast, the definition of a parabola is based on one point and a line.

Definition of a Parabola

A **parabola** is the set of all points in a plane that are equidistant from a fixed line, the **directrix**, and a fixed point, the **focus**, that is not on the line (see Figure 9.29).

In Figure 9.29, find the line passing through the focus and perpendicular to the directrix. This is the **axis of symmetry** of the parabola. The point of intersection of the parabola with its axis of symmetry is called the **vertex**. Notice that the vertex is midway between the focus and the directrix.

Standard Form of the Equation of a Parabola

The rectangular coordinate system enables us to translate a parabola's geometric definition into an algebraic equation. Figure 9.30 is our starting point for obtaining an equation. We place the focus on the x-axis at the point $(p, 0)$. The directrix has an equation given by $x = -p$. The vertex, located midway between the focus and the directrix, is at the origin.

What does the definition of a parabola tell us about the point (x, y) in Figure 9.30? For any point (x, y) on the parabola, the distance d_1 to the directrix is equal to the distance d_2 to the focus. Thus, the point (x, y) is on the parabola if and only if

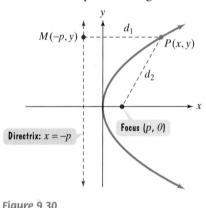

$M(-p, y)$ d_1 $P(x, y)$

d_2

Focus $(p, 0)$

Directrix: $x = -p$

Figure 9.30

$$d_1 = d_2.$$

$$\sqrt{(x + p)^2 + (y - y)^2} = \sqrt{(x - p)^2 + (y - 0)^2} \qquad \text{Use the distance formula.}$$

$$(x + p)^2 = (x - p)^2 + y^2 \qquad \text{Square both sides of the equation.}$$

$$x^2 + 2px + p^2 = x^2 - 2px + p^2 + y^2 \qquad \text{Square } x + p \text{ and } x - p.$$

$$2px = -2px + y^2 \qquad \text{Subtract } x^2 + p^2 \text{ from both sides of the equation.}$$

$$y^2 = 4px \qquad \text{Solve for } y^2.$$

This last equation is called the **standard form of the equation of a parabola with its vertex at the origin**. There are two such equations, one for a focus on the x-axis and one for a focus on the y-axis.

Standard Forms of the Equations of a Parabola

The **standard form of the equation of a parabola** with vertex at the origin is

$$y^2 = 4px \qquad \text{or} \qquad x^2 = 4py.$$

Figure 9.31(a) on the next page illustrates that for the equation on the left, the focus is on the x-axis, which is the axis of symmetry. Figure 9.31(b) on the next page illustrates that for the equation on the right, the focus is on the y-axis, which is the axis of symmetry.

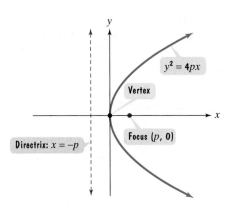

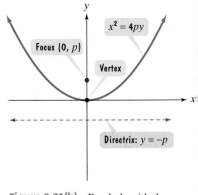

Figure 9.31(a) Parabola with the x-axis as the axis of symmetry. If $p > 0$, the graph opens to the right. If $p < 0$, the graph opens to the left.

Figure 9.31(b) Parabola with the y-axis as the axis of symmetry. If $p > 0$, the graph opens upward. If $p < 0$, the graph opens downward.

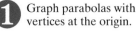 **①** Graph parabolas with vertices at the origin.

Using the Standard Form of the Equation of a Parabola

We can use the standard form of the equation of a parabola to find its focus and directrix. Observing the graph's symmetry from its equation is helpful in locating the focus.

$$y^2 = 4px \qquad\qquad x^2 = 4py$$

> The equation does not change if y is replaced with $-y$. There is x-axis symmetry and the focus is on the x-axis at $(p, 0)$.

> The equation does not change if x is replaced with $-x$. There is y-axis symmetry and the focus is on the y-axis at $(0, p)$.

Although the definition of a parabola is given in terms of its focus and its directrix, the focus and directrix are not part of the graph. The vertex, located at the origin, is a point on the graph of $y^2 = 4px$ and $x^2 = 4py$. Example 1 illustrates how you can find two additional points on the parabola.

EXAMPLE 1   Finding the Focus and Directrix of a Parabola

Find the focus and directrix of the parabola given by $y^2 = 12x$. Then graph the parabola.

Solution   The given equation is in the standard form $y^2 = 4px$, so $4p = 12$.

> No change if y is replaced with $-y$. The parabola has x-axis symmetry.

$$y^2 = 12x$$

> This is $4p$.

We can find both the focus and the directrix by finding p.

$$4p = 12$$
$$p = 3 \qquad \text{Divide both sides by 4.}$$

Because p is positive, the parabola, with its x-axis symmetry, opens to the right. The focus is 3 units to the right of the vertex, $(0, 0)$.

$$\text{Focus:}\quad (p, 0) = (3, 0)$$
$$\text{Directrix:}\quad x = -p; x = -3$$

③ Graph parabolas with vertices not at the origin.

Translations of Parabolas

The graph of a parabola can have its vertex at (h, k), rather than at the origin. Horizontal and vertical translations are accomplished by replacing x with $x - h$ and y with $y - k$ in the standard form of the parabola's equation.

Table 9.3 gives the standard forms of equations of parabolas with vertex at (h, k). Figure 9.36 shows their graphs.

Table 9.3 Standard Forms of Equations of Parabolas with Vertex at (h, k)

Equation	Vertex	Axis of Symmetry	Focus	Directrix	Description
$(y - k)^2 = 4p(x - h)$	(h, k)	Horizontal	$(h + p, k)$	$x = h - p$	If $p > 0$, opens to the right. If $p < 0$, opens to the left.
$(x - h)^2 = 4p(y - k)$	(h, k)	Vertical	$(h, k + p)$	$y = k - p$	If $p > 0$, opens upward. If $p < 0$, opens downward.

Study Tip

If y is the squared term, there is horizontal symmetry and the parabola's equation is not a function. If x is the squared term, there is vertical symmetry and the parabola's equation is a function. Continue to think of p as the directed distance from the vertex, (h, k), to the focus.

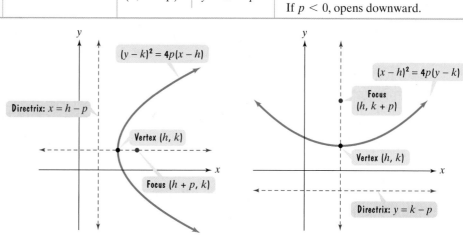

Figure 9.36 Graphs of parabolas with vertex at (h, k) and $p > 0$

The two parabolas shown in Figure 9.36 illustrate standard forms of equations for $p > 0$. If $p < 0$, a parabola with a horizontal axis of symmetry will open to the left and the focus will lie to the left of the directrix. If $p < 0$, a parabola with a vertical axis of symmetry will open downward and the focus will lie below the directrix.

EXAMPLE 4   Graphing a Parabola with Vertex at (h, k)

Find the vertex, focus, and directrix of the parabola given by

$$(x - 3)^2 = 8(y + 1).$$

Then graph the parabola.

Solution In order to find the focus and directrix, we need to know the vertex. In the standard forms of equations with vertex at (h, k), h is the number subtracted from x and k is the number subtracted from y.

$$(x - 3)^2 = 8(y - (-1))$$

This is $(x - h)^2$, with $h = 3$.   This is $y - k$, with $k = -1$.

We see that $h = 3$ and $k = -1$. Thus, the vertex of the parabola is $(h, k) = (3, -1)$. Now that we have the vertex, we can find both the focus and directrix by finding p.

$$(x - 3)^2 = 8(y + 1)$$

This is $4p$.

The equation is in the standard form $(x - h)^2 = 4p(y - k)$. Because x is the squared term, there is vertical symmetry and the parabola's equation is a function.

Because $4p = 8$, $p = 2$. Based on the standard form of the equation, the axis of symmetry is vertical. With a positive value for p and a vertical axis of symmetry, the parabola opens upward. Because $p = 2$, the focus is located 2 units above the vertex, $(3, -1)$. Likewise, the directrix is located 2 units below the vertex.

Focus:

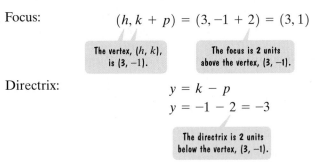

$$(h, k + p) = (3, -1 + 2) = (3, 1)$$

> The vertex, (h, k), is $(3, -1)$.

> The focus is 2 units above the vertex, $(3, -1)$.

Directrix:

$$y = k - p$$
$$y = -1 - 2 = -3$$

> The directrix is 2 units below the vertex, $(3, -1)$.

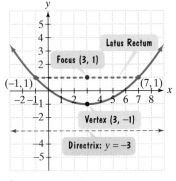

Figure 9.37 The graph of $(x - 3)^2 = 8(y + 1)$

Thus, the focus is $(3, 1)$ and the directrix is $y = -3$. They are shown in Figure 9.37. To graph the parabola, we will use the vertex, $(3, -1)$, and the two endpoints of the latus rectum. The length of the latus rectum is

$$|4p| = |4 \cdot 2| = |8| = 8.$$

Because the graph has vertical symmetry, the latus rectum extends 4 units to the left and 4 units to the right of the focus, $(3, 1)$. The endpoints of the latus rectum are $(3 - 4, 1)$, or $(-1, 1)$, and $(3 + 4, 1)$, or $(7, 1)$. Passing a smooth curve through the vertex and these two points, we sketch the parabola, shown in Figure 9.37.

Technology

Graph $(x - 3)^2 = 8(y + 1)$ by first solving for y:

$$\tfrac{1}{8}(x - 3)^2 = y + 1$$
$$y = \tfrac{1}{8}(x - 3)^2 - 1.$$

The graph passes the vertical line test. Because $(x - 3)^2 = 8(y + 1)$ is a function, you were familiar with the parabola's alternate algebraic form, $y = \tfrac{1}{8}(x - 3)^2 - 1$, in Chapter 2. The form is $y = a(x - h)^2 + k$ with $a = \tfrac{1}{8}$, $h = 3$, and $k = -1$.

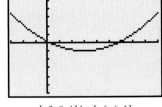

$[-3, 9, 1]$ by $[-6, 6, 1]$

Check Point 4 Find the vertex, focus, and directrix of the parabola given by $(x - 2)^2 = 4(y + 1)$. Then graph the parabola.

In some cases, we need to convert the equation of a parabola to standard form by completing the square on x or y, whichever variable is squared. Let's see how this is done.

EXAMPLE 5 Graphing a Parabola with Vertex at (h, k)

Find the vertex, focus, and directrix of the parabola given by

$$y^2 + 2y + 12x - 23 = 0.$$

Then graph the parabola.

Solution We convert the given equation to standard form by completing the square on the variable y. We isolate the terms involving y on the left side.

$$y^2 + 2y + 12x - 23 = 0 \qquad \text{This is the given equation.}$$
$$y^2 + 2y = -12x + 23 \qquad \text{Isolate the terms involving } y.$$
$$y^2 + 2y + 1 = -12x + 23 + 1 \qquad \text{Complete the square by adding the square of half the coefficient of } y.$$
$$(y + 1)^2 = -12x + 24 \qquad \text{Factor.}$$

Technology

Graph $y^2 + 2y + 12x - 23 = 0$ by solving the equation for y.

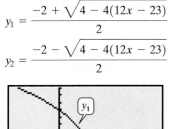

$$y^2 + 2y + (12x - 23) = 0$$

$a = 1$ $b = 2$ $c = 12x - 23$

Use the quadratic formula to solve for y and enter the resulting equations.

$$y_1 = \frac{-2 + \sqrt{4 - 4(12x - 23)}}{2}$$

$$y_2 = \frac{-2 - \sqrt{4 - 4(12x - 23)}}{2}$$

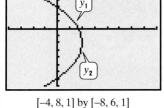

[-4, 8, 1] by [-8, 6, 1]

To express the equation $(y + 1)^2 = -12x + 24$ in the standard form $(y - k)^2 = 4p(x - h)$, we factor -12 on the right. The standard form of the parabola's equation is

$$(y + 1)^2 = -12(x - 2).$$

We use this form to identify the vertex, (h, k), and the value for p needed to locate the focus and the directrix.

$$[(y - (-1)]^2 = -12(x - 2)$$

This is $(y - k)^2$, with $k = -1$.   This is $4p$.   This is $x - h$, with $h = 2$.

The equation is in the standard form $(y - k)^2 = 4p(x - h)$. Because y is the squared term, there is horizontal symmetry and the parabola's equation is not a function.

We see that $h = 2$ and $k = -1$. Thus, the vertex of the parabola is $(h, k) = (2, -1)$. Because $4p = -12$, $p = -3$. Based on the standard form of the equation, the axis of symmetry is horizontal. With a negative value for p and a horizontal axis of symmetry, the parabola opens to the left. Because $p = -3$, the focus is located 3 units to the left of the vertex, $(2, -1)$. Likewise, the directrix is located 3 units to the right of the vertex.

Focus: $$(h + p, k) = (2 + (-3), -1) = (-1, -1)$$

The vertex, (h, k), is $(2, -1)$.   The focus is 3 units to the left of the vertex, $(2, -1)$.

Directrix: $$x = h - p$$
$$x = 2 - (-3) = 5$$

The directrix is 3 units to the right of the vertex, $(2, -1)$.

Thus, the focus is $(-1, -1)$ and the directrix is $x = 5$. They are shown in Figure 9.38.

To graph the parabola, we will use the vertex, $(2, -1)$, and the two endpoints of the latus rectum. The length of the latus rectum is

$$|4p| = |4(-3)| = |-12| = 12.$$

Because the graph has horizontal symmetry, the latus rectum extends 6 units above and 6 units below the focus, $(-1, -1)$. The endpoints of the latus rectum are $(-1, -1 + 6)$, or $(-1, 5)$, and $(-1, -1 - 6)$, or $(-1, -7)$. Passing a smooth curve through the vertex and these two points, we sketch the parabola shown in Figure 9.38.

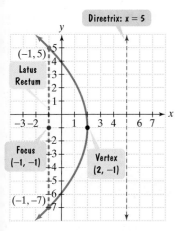

Figure 9.38 The graph of $y^2 + 2y + 12x - 23 = 0$, or $(y + 1)^2 = -12(x - 2)$

4 Solve applied problems involving parabolas.

Check Point 5 Find the vertex, focus, and directrix of the parabola given by $y^2 + 2y + 4x - 7 = 0$. Then graph the parabola.

Applications

Parabolas have many applications. Cables hung between structures to form suspension bridges form parabolas. Arches constructed of steel and concrete, whose main purpose is strength, are usually parabolic in shape.

Suspension bridge

Arch bridge

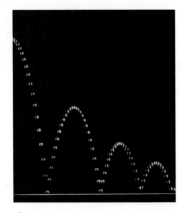

Figure 9.39 Multiflash photo showing the parabolic path of a ball thrown into the air

The Hubble Space Telescope

The Hubble Space Telescope

For decades, astronomers hoped to create an observatory above the atmosphere that would provide an unobscured view of the universe. This vision was realized with the 1990 launching of the Hubble Space Telescope. The telescope initially had blurred vision due to problems with its parabolic mirror. The mirror had been ground two millionths of a meter smaller than design specifications. In 1993, astronauts from the Space Shuttle *Endeavor* equipped the telescope with optics to correct the blurred vision. "A small change for a mirror, a giant leap for astronomy," Christopher J. Burrows of the Space Telescope Science Institute said when clear images from the ends of the universe were presented to the public after the repair mission. Although these images have helped unravel some of the universe's deepest mysteries, by mid-2005, the uncertain fate of the Hubble Space Telescope sparked debate in the U.S. Congress and the scientific community due to the growing costs of keeping it among the stars.

We have seen that comets in our solar system travel in orbits that are ellipse and hyperbolas. Some comets follow parabolic paths. Only comets with elliptica orbits, such as Halley's Comet, return to our part of the galaxy.

You throw a ball directly upward. As illustrated in Figure 9.39, the height o such a projectile as a function of time is parabolic.

If a parabola is rotated about its axis of symmetry, a parabolic surface i formed. Figure 9.40(a) shows how a parabolic surface can be used to reflect light Light originates at the focus. Note how the light is reflected by the parabolic surface so that the outgoing light is parallel to the axis of symmetry. The reflective proper ties of parabolic surfaces are used in the design of searchlights [see Figure 9.40(b)] automobile headlights, and parabolic microphones.

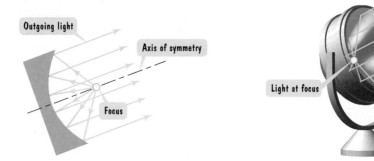

Figure 9.40(a) Parabolic surface reflecting light

Figure 9.40(b) Light from the focus is reflected parallel to the axis of symmetry.

Figure 9.41(a) shows how a parabolic surface can be used to reflect *incomin* light. Note that light rays strike the surface and are reflected *to the focus*. This prin ciple is used in the design of reflecting telescopes, radar, and television satellit dishes. Reflecting telescopes magnify the light from distant stars by reflecting th light from these bodies to the focus of a parabolic mirror [see Figure 9.41(b)].

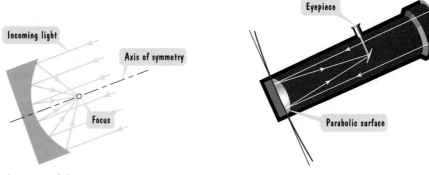

Figure 9.41(a) Parabolic surface reflecting incoming light

Figure 9.41(b) Incoming light rays are reflected to the focus.

EXAMPLE 6   Using the Reflection Property of Parabolas

An engineer is designing a flashlight using a parabolic reflecting mirror and a light source, shown in Figure 9.42. The casting has a diameter of 4 inches and a depth of 2 inches. What is the equation of the parabola used to shape the mirror? At what point should the light source be placed relative to the mirror's vertex?

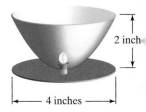

Figure 9.42 Designing a flashlight

4. In Exercise 63, if the diameter of the dish is halved and the depth stays the same, how far from the base of the smaller dish should the receiver be placed?

5. The towers of the Golden Gate Bridge connecting San Francisco to Marin County are 1280 meters apart and rise 160 meters above the road. The cable between the towers has the shape of a parabola and the cable just touches the sides of the road midway between the towers. What is the height of the cable 200 meters from a tower? Round to the nearest meter.

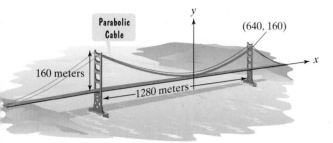

6. The towers of a suspension bridge are 800 feet apart and rise 160 feet above the road. The cable between the towers has the shape of a parabola and the cable just touches the sides of the road midway between the towers. What is the height of the cable 100 feet from a tower?

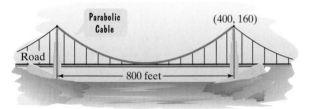

7. The parabolic arch shown in the figure is 50 feet above the water at the center and 200 feet wide at the base. Will a boat that is 30 feet tall clear the arch 30 feet from the center?

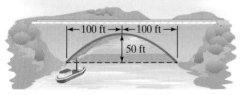

8. A satellite dish in the shape of a parabolic surface has a diameter of 20 feet. If the receiver is to be placed 6 feet from the base, how deep should the dish be?

Writing in Mathematics

9. What is a parabola?

0. Explain how to use $y^2 = 8x$ to find the parabola's focus and directrix.

1. If you are given the standard form of the equation of a parabola with vertex at the origin, explain how to determine if the parabola opens to the right, left, upward, or downward.

2. Describe one similarity and one difference between the graphs of $y^2 = 4x$ and $(y - 1)^2 = 4(x - 1)$.

73. How can you distinguish parabolas from other conic sections by looking at their equations?

74. Look at the satellite dish shown in Exercise 63. Why must the receiver for a shallow dish be farther from the base of the dish than for a deeper dish of the same diameter?

Technology Exercises

75. Use a graphing utility to graph any five of the parabolas that you graphed by hand in Exercises 5–16.

76. Use a graphing utility to graph any three of the parabolas that you graphed by hand in Exercises 35–42. First solve the given equation for y, possibly using the square root property. Enter each of the two resulting equations to produce the complete graph.

Use a graphing utility to graph the parabolas in Exercises 77–78. Write the given equation as a quadratic equation in y and use the quadratic formula to solve for y. Enter each of the equations to produce the complete graph.

77. $y^2 + 2y - 6x + 13 = 0$

78. $y^2 + 10y - x + 25 = 0$

In Exercises 79–80, write each equation as a quadratic equation in y and then use the quadratic formula to express y in terms of x. Graph the resulting two equations using a graphing utility. What effect does the xy-term have on the graph of the resulting parabola?

79. $16x^2 - 24xy + 9y^2 - 60x - 80y + 100 = 0$

80. $x^2 + 2\sqrt{3}xy + 3y^2 + 8\sqrt{3}x - 8y + 32 = 0$

Critical Thinking Exercises

81. Which one of the following is true?

    a. The parabola whose equation is $x = 2y - y^2 + 5$ opens to the right.

    b. If the parabola whose equation is $x = ay^2 + by + c$ has its vertex at $(3, 2)$ and $a > 0$, then it has no y-intercepts.

    c. Some parabolas that open to the right have equations that define y as a function of x.

    d. The graph of $x = a(y - k) + h$ is a parabola with vertex at (h, k).

82. Find the focus and directrix of a parabola whose equation is of the form $Ax^2 + Ey = 0$, $A \neq 0$, $E \neq 0$.

83. Write the standard form of the equation of a parabola whose points are equidistant from $y = 4$ and $(-1, 0)$.

Group Exercise

84. Consult the research department of your library or the Internet to find an example of architecture that incorporates one or more conic sections in its design. Share this example with other group members. Explain precisely how conic sections are used. Do conic sections enhance the appeal of the architecture? In what ways?

CHAPTER 9
MID-CHAPTER CHECK POINT

What You Know: We learned that the four conic sections are the circle, the ellipse, the hyperbola, and the parabola. Prior to this chapter, we graphed circles with center (h, k) and radius r:

$$(x - h)^2 + (y - k)^2 = r^2.$$

In this chapter, you learned to graph ellipses centered at the origin and ellipses centered at (h, k):

$$\frac{(x - h)^2}{a^2} + \frac{(y - k)^2}{b^2} = 1 \quad \text{or}$$

$$\frac{(x - h)^2}{b^2} + \frac{(y - k)^2}{a^2} = 1, a^2 > b^2.$$

We saw that the larger denominator (a^2) determines whether the major axis is horizontal or vertical. We used vertices and asymptotes to graph hyperbolas centered at the origin and hyperbolas centered at (h, k):

$$\frac{(x - h)^2}{a^2} - \frac{(y - k)^2}{b^2} = 1 \quad \text{or} \quad \frac{(y - k)^2}{a^2} - \frac{(x - h)^2}{b^2} = 1.$$

We used $c^2 = a^2 - b^2$ to locate the foci of an ellipse. We used $c^2 = a^2 + b^2$ to locate the foci of a hyperbola. Finally, we used the vertex and the latus rectum to graph parabolas with vertices at the origin and parabolas with vertices at (h, k):

$$(y - k)^2 = 4p(x - h) \quad \text{or} \quad (x - h)^2 = 4p(y - k).$$

In Exercises 1–5, graph each ellipse. Give the location of the foci.

1. $\dfrac{x^2}{25} + \dfrac{y^2}{4} = 1$

2. $9x^2 + 4y^2 = 36$

3. $\dfrac{(x - 2)^2}{16} + \dfrac{(y + 1)^2}{25} = 1$

4. $\dfrac{(x + 2)^2}{25} + \dfrac{(y - 1)^2}{16} = 1$

5. $x^2 + 9y^2 - 4x + 54y + 49 = 0$

In Exercises 6–11, graph each hyperbola. Give the location of the foci and the equations of the asymptotes.

6. $\dfrac{x^2}{9} - y^2 = 1$

7. $\dfrac{y^2}{9} - x^2 = 1$

8. $y^2 - 4x^2 = 16$

9. $4x^2 - 49y^2 = 196$

10. $\dfrac{(x - 2)^2}{9} - \dfrac{(y + 2)^2}{16} = 1$

11. $4x^2 - y^2 + 8x + 6y + 11 = 0$

In Exercises 12–13, graph each parabola. Give the location of the focus and the directrix.

12. $(x - 2)^2 = -12(y + 1)$ 13. $y^2 - 2x - 2y - 5 = 0$

In Exercises 14–21, graph each equation.

14. $x^2 + y^2 = 4$

15. $x + y = 4$

16. $x^2 - y^2 = 4$

17. $x^2 + 4y^2 = 4$

18. $(x + 1)^2 + (y - 1)^2 = 4$

19. $x^2 + 4(y - 1)^2 = 4$

20. $(x - 1)^2 - (y - 1)^2 = 4$

21. $(y + 1)^2 = 4(x - 1)$

In Exercises 22–27, find the standard form of the equation of the conic section satisfying the given conditions.

22. Ellipse; Foci: $(-4, 0)$, $(4, 0)$; Vertices: $(-5, 0)$, $(5, 0)$

23. Ellipse; Endpoints of major axis: $(-8, 2)$, $(10, 2)$
    Foci: $(-4, 2)$, $(6, 2)$

24. Hyperbola; Foci: $(0, -3)$, $(0, 3)$; Vertices: $(0, -2)$, $(0, 2)$

25. Hyperbola; Foci: $(-4, 5)$, $(2, 5)$; Vertices: $(-3, 5)$, $(1, 5)$

26. Parabola; Focus: $(4, 5)$; Directrix: $y = -1$

27. Parabola; Focus: $(-2, 6)$; Directrix: $x = 8$

28. A semielliptical archway over a one-way road has a height of 10 feet and a width of 30 feet. A truck has a width of 10 feet and a height of 9.5 feet. Will this truck clear the opening of the archway?

29. A lithotriper is used to disentegrate kidney stones. The patient is placed within an elliptical device with the kidney centered at one focus, while ultrasound waves from the other focus hit the walls and are reflected to the kidney stone, shattering the stone. Suppose that the length of the major axis of the ellipse is 40 centimeters and the length of the minor axis is 20 centimeters. How far from the kidney stone should the electrode that sends the ultrasound waves be placed in order to shatter the stone?

30. An explosion is recorded by two forest rangers, one at a primary station and the other at an outpost 6 kilometers away. The ranger at the primary station hears the explosion 6 seconds before the ranger at the outpost.

    a. Assuming sound travels at 0.35 kilometer per second, write an equation in standard form that gives all the possible locations of the explosion. Use a coordinate system with the two ranger stations on the x-axis and the midpoint between the stations at the origin.

    b. Graph the equation that gives the possible locations of the explosion. Show the locations of the ranger stations in your drawing.

31. A domed ceiling is a parabolic surface. Ten meters down from the top of the dome, the ceiling is 15 meters wide. For the best lighting on the floor, a light source should be placed at the focus of the parabolic surface. How far from the top of the dome, to the nearest tenth of a meter, should the light source be placed?

SECTION 9.4 Rotation of Axes

Objectives

1. Identify conics without completing the square.
2. Use rotation of axes formulas.
3. Write equations of rotated conics in standard form.
4. Identify conics without rotating axes.

Richard E. Prince "The Cone of Apollonius" (detail), fiberglass, steel, paint, graphite, $51 \times 18 \times 14$ in. Private collection, Vancouver. Photo courtesy of Equinox Gallery, Vancouver, Canada.

To recognize a conic section, you often need to pay close attention to its graph. Graphs powerfully enhance our understanding of algebra and trigonometry. However, it is not possible for people who are blind—or sometimes, visually impaired—to see a graph. Creating informative materials for the blind and visually impaired is a challenge for instructors and mathematicians. Many people who are visually impaired "see" a graph by touching a three-dimensional representation of that graph, perhaps while it is described verbally.

Is it possible to identify conic sections in nonvisual ways? The answer is yes, and the methods for doing so are related to the coefficients in their equations. As we present these methods, think about how you learn them. How would your approach to studying mathematics change if we removed all graphs and replaced them with verbal descriptions?

 Identify conics without completing the square.

Identifying Conic Sections without Completing the Square

Conic sections can be represented both geometrically (as intersecting planes and cones) and algebraically. The equations of the conic sections we have considered in the first three sections of this chapter can be expressed in the form

$$Ax^2 + Cy^2 + Dx + Ey + F = 0,$$

in which A and C are not both zero. You can use A and C, the coefficients of x^2 and y^2, respectively, to identify a conic section without completing the square.

Identifying a Conic Section without Completing the Square

A nondegenerate conic section of the form

$$Ax^2 + Cy^2 + Dx + Ey + F = 0,$$

in which A and C are not both zero, is

- a circle if $A = C$,
- a parabola if $AC = 0$,
- an ellipse if $A \neq C$ and $AC > 0$, and
- a hyperbola if $AC < 0$.

EXAMPLE 1 Identifying a Conic Section without Completing the Square

Identify the graph of each of the following nondegenerate conic sections:

a. $4x^2 - 25y^2 - 24x + 250y - 489 = 0$

b. $x^2 + y^2 + 6x - 2y + 6 = 0$

c. $y^2 + 12x + 2y - 23 = 0$

d. $9x^2 + 25y^2 - 54x + 50y - 119 = 0.$

Solution   We use A, the coefficient of x^2, and C, the coefficient of y^2, to identif
each conic section.

a. $4x^2 - 25y^2 - 24x + 250y - 489 = 0$

$A = 4$ $C = -25$

$AC = 4(-25) = -100 < 0$

Because $AC < 0$, the graph of the equation is a hyperbola.

b. $x^2 + y^2 + 6x - 2y + 6 = 0$

$A = 1$ $C = 1$

Because $A = C$, the graph of the equation is a circle.

c. We can write $y^2 + 12x + 2y - 23 = 0$ as

$$0x^2 + y^2 + 12x + 2y - 23 = 0.$$

$A = 0$ $C = 1$

$AC = 0(1) = 0$

Because $AC = 0$, the graph of the equation is a parabola.

d. $9x^2 + 25y^2 - 54x + 50y - 119 = 0$

$A = 9$ $C = 25$

$AC = 9(25) = 225 > 0.$

Because $AC > 0$ and $A \neq C$, the graph of the equation is an ellipse.

Check Point 1   Identify the graph of each of the following nondegenerate conic section:

a. $3x^2 + 2y^2 + 12x - 4y + 2 = 0$
b. $x^2 + y^2 - 6x + y + 3 = 0$
c. $y^2 - 12x - 4y + 52 = 0$
d. $9x^2 - 16y^2 - 90x + 64y + 17 = 0.$

② Use rotation of axes formulas.

Rotation of Axes

Figure 9.45 shows the graph of

$$7x^2 - 6\sqrt{3}xy + 13y^2 - 16 = 0.$$

The graph looks like an ellipse, although its major axis neither lies along the x-axis no
is parallel to the x-axis. Do you notice anything unusual about the equation?
contains an xy-term. However, look at what happens if we rotate the x- and y-axe
through an angle of $30°$. In the rotated $x'y'$-system, the major axis of the ellipse lie
along the x'-axis. We can write the equation of the ellipse in this rotated $x'y'$-system a

$$\frac{x'^2}{4} + \frac{y'^2}{1} = 1.$$

Observe that there is no $x'y'$-term in the equation.

Except for degenerate cases, the **general second-degree equation**

$$Ax^2 + Bxy + Cy^2 + Dx + Ey + F = 0$$

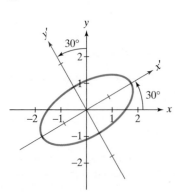

Figure 9.45 The graph of
$7x^2 - 6\sqrt{3}xy + 13y^2 - 16 = 0$,
a rotated ellipse

represents one of the conic sections. However, due to the xy-term in the equation, these conic sections are rotated in such a way that their axes are no longer parallel to the x- and y-axes. To reduce these equations to forms of the conic sections with which you are already familiar, we use a procedure called **rotation of axes**.

Suppose that the x- and y-axes are rotated through a positive angle θ, resulting in a new $x'y'$ coordinate system. This system is shown in Figure 9.46(a). The origin in the $x'y'$-system is the same as the origin in the xy-system. Point P in Figure 9.46(b) has coordinates (x, y) relative to the xy-system and coordinates (x', y') relative to the $x'y'$-system. Our goal is to obtain formulas relating the old and new coordinates. Thus, we need to express x and y in terms of x', y', and θ.

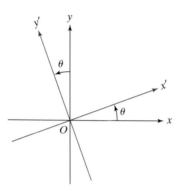

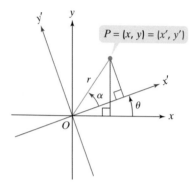

(a) Rotating the x- and y-axes through a positive angle θ

(b) Describing point P relative to the xy-system and the rotated $x'y'$-system

Figure 9.46 Rotating axes

Look at Figure 9.46(b). Notice that

$r = $ the distance from the origin O to point P.

$\alpha = $ the angle from the positive x'-axis to the ray from O through P.

Using the definitions of sine and cosine, we obtain

$$\cos \alpha = \frac{x'}{r} : x' = r \cos \alpha$$

This is from the right triangle with a leg along the x'-axis.

$$\sin \alpha = \frac{y'}{r} : y' = r \sin \alpha$$

$$\cos (\theta + \alpha) = \frac{x}{r} : x = r \cos (\theta + \alpha)$$

This is from the taller right triangle with a leg along the x-axis.

$$\sin (\theta + \alpha) = \frac{y}{r} : y = r \sin (\theta + \alpha).$$

Thus,

$x = r \cos(\theta + \alpha)$ This is the third of the preceding equations.

$\quad = r(\cos \theta \cos \alpha - \sin \theta \sin \alpha)$ Use the formula for the cosine of the sum of two angles.

$\quad = (r \cos \alpha) \cos \theta - (r \sin \alpha) \sin \theta$ Apply the distributive property and rearrange factors.

$\quad = x' \cos \theta - y' \sin \theta.$ Use the first and second of the preceding equations: $x' = r \cos \alpha$ and $y' = r \sin \alpha$.

Similarly,

$$y = r \sin(\theta + \alpha) = r(\sin \theta \cos \alpha + \cos \theta \sin \alpha) = x' \sin \theta + y' \cos \theta.$$

Rotation of Axes Formulas

Suppose an xy-coordinate system and an $x'y'$-coordinate system have the same origin and θ is the angle from the positive x-axis to the positive x'-axis. If the coordinates of point P are (x, y) in the xy-system and (x', y') in the rotated $x'y'$-system, then

$$x = x' \cos \theta - y' \sin \theta$$
$$y = x' \sin \theta + y' \cos \theta.$$

EXAMPLE 2 Rotating Axes

Write the equation $xy = 1$ in terms of a rotated $x'y'$-system if the angle of rotatic from the x-axis to the x'-axis is $45°$. Express the equation in standard form. Use th rotated system to graph $xy = 1$.

Solution With $\theta = 45°$, the rotation formulas for x and y are

$$x = x' \cos \theta - y' \sin \theta = x' \cos 45° - y' \sin 45°$$

$$= x'\left(\frac{\sqrt{2}}{2}\right) - y'\left(\frac{\sqrt{2}}{2}\right) = \frac{\sqrt{2}}{2}(x' - y')$$

$$y = x' \sin \theta + y' \cos \theta = x' \sin 45° + y' \cos 45°$$

$$= x'\left(\frac{\sqrt{2}}{2}\right) + y'\left(\frac{\sqrt{2}}{2}\right) = \frac{\sqrt{2}}{2}(x' + y').$$

Now substitute these expressions for x and y in the given equation, $xy = 1$.

$xy = 1$	This is the given equation.
$\left[\dfrac{\sqrt{2}}{2}(x' - y')\right]\left[\dfrac{\sqrt{2}}{2}(x' + y')\right] = 1$	Substitute the expressions for x and y from the rotation formulas.
$\dfrac{2}{4}(x' - y')(x' + y') = 1$	Multiply: $\dfrac{\sqrt{2}}{2} \cdot \dfrac{\sqrt{2}}{2} = \dfrac{2}{4}$.
$\dfrac{1}{2}(x'^2 - y'^2) = 1$	Reduce $\dfrac{2}{4}$ and multiply the binomials.
$\dfrac{x'^2}{2} - \dfrac{y'^2}{2} = 1$	Write the equation in standard form: $\dfrac{x^2}{a^2} - \dfrac{y^2}{b^2} = 1.$

$\boxed{a^2 = 2}$ $\boxed{b^2 = 2}$

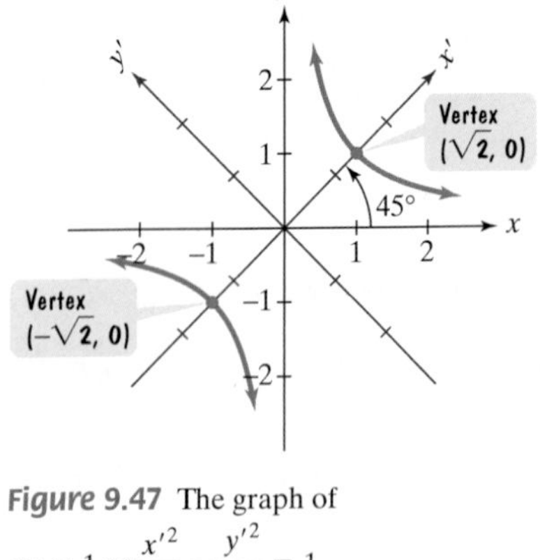

This equation expresses $xy = 1$ in terms of the rotated $x'y'$-system. Can you se that this is the standard form of the equation of a hyperbola? The hyperbola's cente is at $(0, 0)$, with the transverse axis on the x'-axis. The vertices are $(-a, 0)$ an $(a, 0)$. Because $a^2 = 2$, the vertices are $\left(-\sqrt{2}, 0\right)$ and $\left(\sqrt{2}, 0\right)$, located on th x'-axis. Based on the standard form of the hyperbola's equation, the equations fo the asymptotes are

$$y' = \pm\frac{b}{a}x' \quad \text{or} \quad y' = \pm\frac{\sqrt{2}}{\sqrt{2}}x'.$$

The equations of the asymptotes can be simplified to $y' = x'$ and $y' = -x'$, whic correspond to the original x- and y-axes. The graph of the hyperbola is shown i Figure 9.47.

Vertex $(\sqrt{2}, 0)$

Vertex $(-\sqrt{2}, 0)$

Figure 9.47 The graph of $xy = 1$ or $\dfrac{x'^2}{2} - \dfrac{y'^2}{2} = 1$

Check Point 2 Write the equation $xy = 2$ in terms of a rotated $x'y'$-system if the angle c rotation from the x-axis to the x'-axis is $45°$. Express the equation i standard form. Use the rotated system to graph $xy = 2$.

③ Write equations of rotated conics in standard form.

Using Rotations to Transform Equations with xy-Terms to Standard Equations of Conic Sections

We have noted that the appearance of the term Bxy $(B \neq 0)$ in the general second-degree equation indicates that the graph of the conic section has been rotated. A rotation of axes through an appropriate angle can transform the equation to one of the standard forms of the conic sections in x' and y' in which no $x'y'$-term appears.

Amount of Rotation Formula

The general second-degree equation

$$Ax^2 + Bxy + Cy^2 + Dx + Ey + F = 0, B \neq 0$$

can be rewritten as an equation in x' and y' without an $x'y'$-term by rotating the axes through angle θ, where

$$\cot 2\theta = \frac{A - C}{B}.$$

Before we learn to apply this formula, let's see how it can be derived. We begin with the general second-degree equation

$$Ax^2 + Bxy + Cy^2 + Dx + Ey + F = 0, B \neq 0.$$

Then we rotate the axes through an angle θ. In terms of the rotated $x'y'$-system, the general second-degree equation can be written as

$$A(x' \cos \theta - y' \sin \theta)^2 + B(x' \cos \theta - y' \sin \theta)(x' \sin \theta + y' \cos \theta)$$
$$+ C(x' \sin \theta + y' \cos \theta)^2 + D(x' \cos \theta - y' \sin \theta)$$
$$+ E(x' \sin \theta + y' \cos \theta) + F = 0.$$

After a lot of simplifying that involves expanding and collecting like terms, you will obtain the following equation:

> We want a rotation that results in no $x'y'$-term.

$$(A \cos^2 \theta + B \sin \theta \cos \theta + C \sin^2 \theta)x'^2 + [B(\cos^2 \theta - \sin^2 \theta) + 2(C - A)(\sin \theta \cos \theta)]x'y'$$
$$+ (A \sin^2 \theta - B \sin \theta \cos \theta + C \cos^2 \theta)y'^2$$
$$+ (D \cos \theta + E \sin \theta)x'$$
$$+ (-D \sin \theta + E \cos \theta)y' + F = 0.$$

If this looks somewhat ghastly, take a deep breath and focus only on the $x'y'$-term. We want to choose θ so that the coefficient of this term is zero. This will give the required rotation that results in no $x'y'$-term.

$B(\cos^2 \theta - \sin^2 \theta) + 2(C - A) \sin \theta \cos \theta = 0$	Set the coefficient of the $x'y'$-term equal to 0.
$B \cos 2\theta + (C - A) \sin 2\theta = 0$	Use the double-angle formulas: $\cos 2\theta = \cos^2 \theta - \sin^2 \theta$ and $\sin 2\theta = 2 \sin \theta \cos \theta$.
$B \cos 2\theta = -(C - A) \sin 2\theta$	Subtract $(C - A) \sin 2\theta$ from both sides.
$B \cos 2\theta = (A - C) \sin 2\theta$	Simplify
$\dfrac{B \cos 2\theta}{B \sin 2\theta} = \dfrac{(A - C) \sin 2\theta}{B \sin 2\theta}$	Divide both sides by $B \sin 2\theta$.
$\dfrac{\cos 2\theta}{\sin 2\theta} = \dfrac{A - C}{B}$	Simplify.
$\cot 2\theta = \dfrac{A - C}{B}$	Apply a quotient identity: $\cot 2\theta = \dfrac{\cos 2\theta}{\sin 2\theta}$.

If $\cot 2\theta$ is positive, we will select θ so that $0° < \theta < 45°$. If $\cot 2\theta$ is negative, we will select θ so that $45° < \theta < 90°$. Thus θ, the angle of rotation, is always an acute angle.

Here is a step-by-step procedure for writing the equation of a rotated conic section in standard form:

Writing the Equation of a Rotated Conic in Standard Form

1. Use the given equation

$$Ax^2 + Bxy + Cy^2 + Dx + Ey + F = 0, B \neq 0$$

to find $\cot 2\theta$.

$$\cot 2\theta = \frac{A - C}{B}$$

2. Use the expression for $\cot 2\theta$ to determine θ, the angle of rotation.
3. Substitute θ in the rotation formulas

$$x = x' \cos \theta - y' \sin \theta \quad \text{and} \quad y = x' \sin \theta + y' \cos \theta$$

and simplify.
4. Substitute the expressions for x and y from the rotation formulas in the given equation and simplify. The resulting equation should have no $x'y'$-term.
5. Write the equation involving x' and y' in standard form.

Using the equation in step 5, you can graph the conic section in the rotated $x'y'$-system.

EXAMPLE 3 Writing the Equation of a Rotated Conic Section in Standard Form

Rewrite the equation

$$7x^2 - 6\sqrt{3}xy + 13y^2 - 16 = 0$$

in a rotated $x'y'$-system without an $x'y'$-term. Express the equation in the standard form of a conic section. Graph the conic section in the rotated system.

Solution

Step 1   Use the given equation to find $\cot 2\theta$. We need to identify the constants A, B, and C in the given equation.

$$7x^2 - 6\sqrt{3}\, xy + 13y^2 - 16 = 0$$

| A is the coefficient of the x^2-term: $A = 7$. | B is the coefficient of the xy-term: $B = -6\sqrt{3}$. | C is the coefficient of the y^2-term: $C = 13$. |

The appropriate angle θ through which to rotate the axes satisfies the equation

$$\cot 2\theta = \frac{A - C}{B} = \frac{7 - 13}{-6\sqrt{3}} = \frac{-6}{-6\sqrt{3}} = \frac{1}{\sqrt{3}} \text{ or } \frac{\sqrt{3}}{3}.$$

Step 2   Use the expression for $\cot 2\theta$ to determine the angle of rotation. We have $\cot 2\theta = \frac{\sqrt{3}}{3}$. Based on our knowledge of exact values for trigonometric functions, we conclude that $2\theta = 60°$. Thus, $\theta = 30°$.

Step 3   Substitute θ in the rotation formulas $x = x' \cos \theta - y' \sin \theta$ and $y = x' \sin \theta + y' \cos \theta$ and simplify. Substituting 30° for θ,

$$x = x' \cos 30° - y' \sin 30° = x'\left(\frac{\sqrt{3}}{2}\right) - y'\left(\frac{1}{2}\right) = \frac{\sqrt{3}x' - y'}{2}$$

$$y = x' \sin 30° + y' \cos 30° = x'\left(\frac{1}{2}\right) + y'\left(\frac{\sqrt{3}}{2}\right) = \frac{x' + \sqrt{3}y'}{2}.$$

Step 4   Substitute the expressions for x and y from the rotation formulas in the given equation and simplify.

$$7x^2 - 6\sqrt{3}xy + 13y^2 - 16 = 0 \qquad \text{This is the given equation.}$$

$$7\left(\frac{\sqrt{3}x' - y'}{2}\right)^2 - 6\sqrt{3}\left(\frac{\sqrt{3}x' - y'}{2}\right)\left(\frac{x' + \sqrt{3}y'}{2}\right)$$

$$+ 13\left(\frac{x' + \sqrt{3}y'}{2}\right)^2 - 16 = 0 \qquad \begin{array}{l}\text{Substitute the expressions for } x \text{ and} \\ y \text{ from the rotation formulas.}\end{array}$$

$$7\left(\frac{3x'^2 - 2\sqrt{3}x'y' + y'^2}{4}\right) - 6\sqrt{3}\left(\frac{\sqrt{3}x'^2 + 3x'y' - x'y' - \sqrt{3}y'^2}{4}\right)$$

$$+ 13\left(\frac{x'^2 + 2\sqrt{3}x'y' + 3y'^2}{4}\right) - 16 = 0 \qquad \text{Square and multiply.}$$

$$7\left(3x'^2 - 2\sqrt{3}x'y' + y'^2\right) - 6\sqrt{3}\left(\sqrt{3}x'^2 + 2x'y' - \sqrt{3}y'^2\right)$$

$$+ 13\left(x'^2 + 2\sqrt{3}x'y' + 3y'^2\right) - 64 = 0 \qquad \text{Multiply both sides by 4.}$$

$$21x'^2 - 14\sqrt{3}x'y' + 7y'^2 - 18x'^2 - 12\sqrt{3}x'y' + 18y'^2$$

$$+ 13x'^2 + 26\sqrt{3}x'y' + 39y'^2 - 64 = 0 \qquad \text{Distribute throughout parentheses.}$$

$$21x'^2 - 18x'^2 + 13x'^2 - 14\sqrt{3}x'y' - 12\sqrt{3}x'y' + 26\sqrt{3}x'y'$$

$$+ 7y'^2 + 18y'^2 + 39y'^2 - 64 = 0 \qquad \text{Rearrange terms.}$$

$$16x'^2 + 64y'^2 - 64 = 0 \qquad \text{Combine like terms.}$$

Do you see how we "lost" the $x'y'$-term in the last equation?

$$-14\sqrt{3}x'y' - 12\sqrt{3}x'y' + 26\sqrt{3}x'y' = -26\sqrt{3}x'y' + 26\sqrt{3}x'y' = 0x'y' = 0$$

Step 5   Write the equation involving x' and y' in standard form. We can express $16x'^2 + 64y'^2 - 64 = 0$, an equation of an ellipse, in the standard form $\frac{x^2}{a^2} + \frac{y^2}{b^2} = 1$.

$$16x'^2 + 64y'^2 - 64 = 0 \qquad \begin{array}{l}\text{This equation describes the ellipse relative to a} \\ \text{system rotated through 30°.}\end{array}$$

$$16x'^2 + 64y'^2 = 64 \qquad \text{Add 64 to both sides.}$$

$$\frac{16x'^2}{64} + \frac{64y'^2}{64} = \frac{64}{64} \qquad \text{Divide both sides by 64.}$$

$$\frac{x'^2}{4} + \frac{y'^2}{1} = 1 \qquad \text{Simplify.}$$

The last equation is the standard form of the equation of an ellipse. The major axis is on the x'-axis and the vertices are $(-2, 0)$ and $(2, 0)$. The minor axis is on the y'-axis with endpoints $(0, -1)$ and $(0, 1)$. The graph of the ellipse is shown in Figure 9.45. Does this graph look familiar? It should—you saw it earlier in this section on page 890.

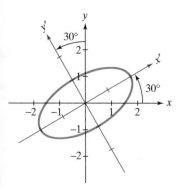

Figure 9.45 (repeated) The graph of $7x^2 - 6\sqrt{3}xy + 13y^2 - 16 = 0$ or $\frac{x'^2}{4} - \frac{y'^2}{1} = 1$, a rotated ellipse

Rewrite the equation

$$2x^2 + \sqrt{3}xy + y^2 - 2 = 0$$

in a rotated $x'y'$-system without an $x'y'$-term. Express the equation in the standard form of a conic section. Graph the conic section in the rotated system.

Technology

In order to graph a general second-degree equation in the form

$$Ax^2 + Bxy + Cy^2 + Dx + Ey + F = 0$$

using a graphing utility, it is necessary to solve for y. Rewrite the equation as a quadratic equation in y.

$$Cy^2 + (Bx + E)y + (Ax^2 + Dx + F) = 0$$

By applying the quadratic formula, the graph of this equation can be obtained by entering

$$y_1 = \frac{-(Bx + E) + \sqrt{(Bx + E)^2 - 4C(Ax^2 + Dx + F)}}{2C}$$

and

$$y_2 = \frac{-(Bx + E) - \sqrt{(Bx + E)^2 - 4C(Ax^2 + Dx + F)}}{2C}.$$

The graph of

$$7x^2 - 6\sqrt{3}xy + 13y^2 - 16 = 0$$

is shown on the right in a $[-2, 2, 1]$ by $[-2, 2, 1]$ viewing rectangle. The graph was obtained by entering the equations for y_1 and y_2 shown above with

$$A = 7, B = -6\sqrt{3}, C = 13, D = 0, E = 0,$$

and $F = -16.$

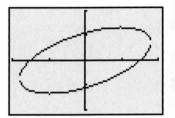

In Example 3 and Check Point 3, we found θ, the angle of rotation, directly because we recognized $\dfrac{\sqrt{3}}{3}$ as the value of $\cot 60°$. What do we do if $\cot 2\theta$ is not the cotangent of one of the familiar angles? We use $\cot 2\theta$ to find $\sin \theta$ and $\cos \theta$ a follows:

- Use a sketch of $\cot 2\theta$ to find $\cos 2\theta$.
- Find $\sin \theta$ and $\cos \theta$ using the identities

$$\sin \theta = \sqrt{\frac{1 - \cos 2\theta}{2}} \qquad \text{and} \qquad \cos \theta = \sqrt{\frac{1 + \cos 2\theta}{2}}.$$

> Because θ is an acute angle, the positive square roots are appropriate.

The resulting values for $\sin \theta$ and $\cos \theta$ are used to write the rotation formulas tha give an equation with no $x'y'$-term.

EXAMPLE 4   Graphing the Equation of a Rotated Conic

Graph relative to a rotated $x'y'$-system in which the equation has no $x'y'$-term:

$$16x^2 - 24xy + 9y^2 + 110x - 20y + 100 = 0.$$

Solution

Step 1   Use the given equation to find $\cot 2\theta$. With $A = 16$, $B = -24$, and $C = 9$ we have

$$\cot 2\theta = \frac{A - C}{B} = \frac{16 - 9}{-24} = -\frac{7}{24}.$$

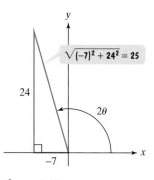

Figure 9.48 Using cot 2θ to find cos 2θ

Step 2   Use the expression for cot 2θ to determine sin θ and cos θ. A rough sketch showing cot 2θ is given in Figure 9.48. Because θ is always acute and cot 2θ is negative, 2θ is in quadrant II. The third side of the triangle is found using $r = \sqrt{x^2 + y^2}$. Thus, $r = \sqrt{(-7)^2 + 24^2} = \sqrt{625} = 25$. By the definition of the cosine function,

$$\cos 2\theta = \frac{x}{r} = \frac{-7}{25} = -\frac{7}{25}.$$

Now we use identities to find values for sin θ and cos θ.

$$\sin \theta = \sqrt{\frac{1 - \cos 2\theta}{2}} = \sqrt{\frac{1 - \left(-\dfrac{7}{25}\right)}{2}}$$

$$= \sqrt{\frac{\dfrac{25}{25} + \dfrac{7}{25}}{2}} = \sqrt{\frac{\dfrac{32}{25}}{2}} = \sqrt{\frac{32}{50}} = \sqrt{\frac{16}{25}} = \frac{4}{5}$$

$$\cos \theta = \sqrt{\frac{1 + \cos 2\theta}{2}} = \sqrt{\frac{1 + \left(-\dfrac{7}{25}\right)}{2}}$$

$$= \sqrt{\frac{\dfrac{25}{25} - \dfrac{7}{25}}{2}} = \sqrt{\frac{\dfrac{18}{25}}{2}} = \sqrt{\frac{18}{50}} = \sqrt{\frac{9}{25}} = \frac{3}{5}$$

Step 3   Substitute sin θ and cos θ in the rotation formulas

$$x = x' \cos \theta - y' \sin \theta \quad \text{and} \quad y = x' \sin \theta + y' \cos \theta$$

and simplify. Substituting $\frac{4}{5}$ for sin θ and $\frac{3}{5}$ for cos θ,

$$x = x'\left(\frac{3}{5}\right) - y'\left(\frac{4}{5}\right) = \frac{3x' - 4y'}{5}$$

$$y = x'\left(\frac{4}{5}\right) + y'\left(\frac{3}{5}\right) = \frac{4x' + 3y'}{5}.$$

Step 4   Substitute the expressions for x and y from the rotation formulas in the given equation and simplify.

$$16x^2 - 24xy + 9y^2 + 110x - 20y + 100 = 0 \qquad \text{This is the given equation.}$$

$$16\left(\frac{3x' - 4y'}{5}\right)^2 - 24\left(\frac{3x' - 4y'}{5}\right)\left(\frac{4x' + 3y'}{5}\right) + 9\left(\frac{4x' + 3y'}{5}\right)^2 \qquad \text{Substitute the expressions for } x \text{ and } y \text{ from the rotation formulas.}$$

$$+ 110\left(\frac{3x' - 4y'}{5}\right) - 20\left(\frac{4x' + 3y'}{5}\right) + 100 = 0$$

Take a few minutes to expand, multiply both sides of the equation by 25, and combine like terms. The resulting equation

$$y'^2 + 2x' - 4y' + 4 = 0$$

has no $x'y'$-term.

Step 5   Write the equation involving x' and y' in standard form. With only one variable that is squared, we have the equation of a parabola. We need to write the equation in the standard form $(y - k)^2 = 4p(x - h)$.

$$y'^2 + 2x' - 4y' + 4 = 0 \qquad \text{This is the equation without an } x'y'\text{-term.}$$

$$y'^2 - 4y' = -2x' - 4 \qquad \text{Isolate the terms involving } y'.$$

$$y'^2 - 4y' + 4 = -2x' - 4 + 4 \qquad \text{Complete the square by adding the square of half the coefficient of } y'.$$

$$(y' - 2)^2 = -2x' \qquad \text{Factor.}$$

The standard form of the parabola's equation in the rotated $x'y'$-system is

$$(y' - 2)^2 = -2x'.$$

This is $(y' - k)^2$, with $k = 2$.   This is $4p$.   This is $x' - h$, with $h = 0$.

We see that $h = 0$ and $k = 2$. Thus, the vertex of the parabola in the $x'y'$-system $(h, k) = (0, 2)$.

We can use the $x'y'$-system to graph the parabola. Using a calculator to solv

$\sin \theta = \dfrac{4}{5}$, we find that $\theta = \sin^{-1} \dfrac{4}{5} \approx 53°$. Rotate the axes through approximatel

$53°$. With $4p = -2$ and $p = -\dfrac{1}{2}$, the parabola's focus is $\dfrac{1}{2}$ unit to the left of th

vertex, $(0, 2)$. Thus, the focus in the $x'y'$-system is $\left(-\dfrac{1}{2}, 2\right)$.

To graph the parabola, we use the vertex, $(0, 2)$, and the two endpoints of th latus rectum.

$$\text{length of latus rectum} = |4p| = |-2| = 2$$

The latus rectum extends 1 unit above and 1 unit below the focus, $\left(-\dfrac{1}{2}, 2\right)$

Thus, the endpoints of the latus rectum in the $x'y'$-system are $\left(-\dfrac{1}{2}, 3\right)$ and $\left(-\dfrac{1}{2}, 1\right)$ Using the rotated system, pass a smooth curve through the vertex and the tw endpoints of the latus rectum. The graph of the parabola is shown in Figure 9.49.

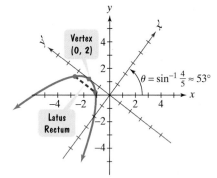

Figure 9.49 The graph of $(y' - 2)^2 = -2x'$ in a rotated $x'y'$-system

Check Point 4 Graph relative to a rotated $x'y'$-system in which the equation has n $x'y'$-term:

$$4x^2 - 4xy + y^2 - 8\sqrt{5}x - 16\sqrt{5}y = 0.$$

④ Identify conics without rotating axes.

Identifying Conic Sections without Rotating Axes

We now know that the general second-degree equation

$$Ax^2 + Bxy + Cy^2 + Dx + Ey + F = 0, B \neq 0$$

can be rewritten as

$$A'x'^2 + C'y'^2 + D'x' + E'y' + F' = 0$$

in a rotated $x'y'$-system. A relationship between the coefficients of the tw equations is given by

$$B^2 - 4AC = -4A'C'.$$

We also know that A' and C' can be used to identify the graph of the rotate equation. Thus, $B^2 - 4AC$ can also be used to identify the graph of the genera second-degree equation.

Identifying a Conic Section without a Rotation of Axes

A nondegenerate conic section of the form

$$Ax^2 + Bxy + Cy^2 + Dx + Ey + F = 0$$

is

- a parabola if $B^2 - 4AC = 0$,
- an ellipse or a circle if $B^2 - 4AC < 0$, and
- a hyperbola if $B^2 - 4AC > 0$.

Technology

The graph of
$$11x^2 + 10\sqrt{3}xy + y^2 - 4 = 0$$
is shown in a $\left[-1, 1, \frac{1}{4}\right]$ by $\left[-1, 1, \frac{1}{4}\right]$ viewing rectangle. The graph verifies that the equation represents a rotated hyperbola.

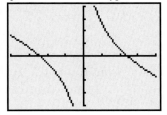

EXAMPLE 5   Identifying a Conic Section without Rotating Axes

Identify the graph of

$$11x^2 + 10\sqrt{3}xy + y^2 - 4 = 0.$$

Solution   We use A, B, and C to identify the conic section.

$$11x^2 + 10\sqrt{3}xy + y^2 - 4 = 0$$

$$A = 11 \qquad B = 10\sqrt{3} \qquad C = 1$$

$$B^2 - 4AC = \left(10\sqrt{3}\right)^2 - 4(11)(1) = 100 \cdot 3 - 44 = 256 > 0$$

Because $B^2 - 4AC > 0$, the graph of the equation is a hyperbola.

Check Point 5   Identify the graph of $3x^2 - 2\sqrt{3}xy + y^2 + 2x + 2\sqrt{3}y = 0$.

EXERCISE SET 9.4

Practice Exercises

In Exercises 1–8, identify each equation without completing the square.

1. $y^2 - 4x + 2y + 21 = 0$

2. $y^2 - 4x - 4y = 0$

3. $4x^2 - 9y^2 - 8x - 36y - 68 = 0$

4. $9x^2 + 25y^2 - 54x - 200y + 256 = 0$

5. $4x^2 + 4y^2 + 12x + 4y + 1 = 0$

6. $9x^2 + 4y^2 - 36x + 8y + 31 = 0$

7. $100x^2 - 7y^2 + 90y - 368 = 0$

8. $y^2 + 8x + 6y + 25 = 0$

In Exercises 9–14, write each equation in terms of a rotated $x'y'$-system using θ, the angle of rotation. Write the equation involving x' and y' in standard form.

9. $xy = -1; \theta = 45°$

10. $xy = -4; \theta = 45°$

11. $x^2 - 4xy + y^2 - 3 = 0; \theta = 45°$

12. $13x^2 - 10xy + 13y^2 - 72 = 0; \theta = 45°$

13. $23x^2 + 26\sqrt{3}xy - 3y^2 - 144 = 0; \theta = 30°$

14. $13x^2 - 6\sqrt{3}xy + 7y^2 - 16 = 0; \theta = 60°$

In Exercises 15–26, write the appropriate rotation formulas so that in a rotated system the equation has no $x'y'$-term.

15. $x^2 + xy + y^2 - 10 = 0$

16. $x^2 + 4xy + y^2 - 3 = 0$

17. $3x^2 - 10xy + 3y^2 - 32 = 0$

18. $5x^2 - 8xy + 5y^2 - 9 = 0$

19. $11x^2 + 10\sqrt{3}xy + y^2 - 4 = 0$

20. $7x^2 - 6\sqrt{3}xy + 13y^2 - 16 = 0$

21. $10x^2 + 24xy + 17y^2 - 9 = 0$

22. $32x^2 - 48xy + 18y^2 - 15x - 20y = 0$

23. $x^2 + 4xy - 2y^2 - 1 = 0$

24. $3xy - 4y^2 + 18 = 0$

25. $34x^2 - 24xy + 41y^2 - 25 = 0$

26. $6x^2 - 6xy + 14y^2 - 45 = 0$

In Exercises 27–38,

    a. *Rewrite the equation in a rotated $x'y'$-system without an $x'y'$ term. Use the appropriate rotation formulas from Exercises 15–26.*

    b. *Express the equation involving x' and y' in the standard form of a conic section.*

    c. *Use the rotated system to graph the equation.*

27. $x^2 + xy + y^2 - 10 = 0$

28. $x^2 + 4xy + y^2 - 3 = 0$

29. $3x^2 - 10xy + 3y^2 - 32 = 0$

30. $5x^2 - 8xy + 5y^2 - 9 = 0$

31. $11x^2 + 10\sqrt{3}xy + y^2 - 4 = 0$

32. $7x^2 - 6\sqrt{3}xy + 13y^2 - 16 = 0$

33. $10x^2 + 24xy + 17y^2 - 9 = 0$

34. $32x^2 - 48xy + 18y^2 - 15x - 20y = 0$

35. $x^2 + 4xy - 2y^2 - 1 = 0$

36. $3xy - 4y^2 + 18 = 0$

37. $34x^2 - 24xy + 41y^2 - 25 = 0$

38. $6x^2 - 6xy + 14y^2 - 45 = 0$

In Exercises 39–44, identify each equation without applying a rotation of axes.

39. $5x^2 - 2xy + 5y^2 - 12 = 0$

40. $10x^2 + 24xy + 17y^2 - 9 = 0$

41. $24x^2 + 16\sqrt{3}xy + 8y^2 - x + \sqrt{3}y - 8 = 0$

42. $3x^2 - 2\sqrt{3}xy + y^2 + 2x + 2\sqrt{3}y = 0$

43. $23x^2 + 26\sqrt{3}xy - 3y^2 - 144 = 0$

44. $4xy + 3y^2 + 4x + 6y - 1 = 0$

Practice Plus

In Exercises 45–48,

- *If the graph of the equation is an ellipse, find the coordinates of the vertices on the minor axis.*
- *If the graph of the equation is a hyperbola, find the equations of the asymptotes.*
- *If the graph of the equation is a parabola, find the coordinates of the vertex.*

Express answers relative to an $x'y'$-system in which the given equation has no $x'y'$-term. Assume that the $x'y'$-system has the same origin as the xy-system.

45. $5x^2 - 6xy + 5y^2 - 8 = 0$

46. $2x^2 - 4xy + 5y^2 - 36 = 0$

47. $x^2 - 4xy + 4y^2 + 5\sqrt{5}y - 10 = 0$

48. $x^2 + 4xy - 2y^2 - 6 = 0$

Writing in Mathematics

49. Explain how to identify the graph of

$$Ax^2 + Cy^2 + Dx + Ey + F = 0.$$

50. If there is a 60° angle from the positive x-axis to the positive x'-axis, explain how to obtain the rotation formulas for x and y.

51. How do you obtain the angle of rotation so that a general second-degree equation has no $x'y'$-term in a rotated $x'y'$-system?

52. What is the most time-consuming part in using a graphing utility to graph a general second-degree equation with an xy-term?

53. Explain how to identify the graph of

$$Ax^2 + Bxy + Cy^2 + Dx + Ey + F = 0.$$

Technology Exercises

In Exercises 54–60, use a graphing utility to graph each equation.

54. $x^2 + 4xy + y^2 - 3 = 0$

55. $7x^2 + 8xy + y^2 - 1 = 0$

56. $3x^2 + 4xy + 6y^2 - 7 = 0$

57. $3x^2 - 6xy + 3y^2 + 10x - 8y - 2 = 0$

58. $9x^2 + 24xy + 16y^2 + 90x - 130y = 0$

59. $x^2 + 4xy + 4y^2 + 10\sqrt{5}x - 9 = 0$

60. $7x^2 + 6xy + 2.5y^2 - 14x + 4y + 9 = 0$

Critical Thinking Exercises

61. Explain the relationship between the graph of $3x^2 - 2xy + 3y^2 + 2 = 0$ and the sound made by one hand clapping. Begin by following the directions for Exercise 27–38. (You will first need to write rotation formulas that eliminate the $x'y'$-term.)

62. What happens to the equation $x^2 + y^2 = r^2$ in a rotated $x'y'$-system?

In Exercises 63–64, let

$$Ax^2 + Bxy + Cy^2 + Dx + Ey + F = 0$$

be an equation of a conic section in an xy-coordinate system. Let $A'x'^2 + B'x'y' + C'y'^2 + D'x' + E'y' + F' = 0$ be the equation of the conic section in the rotated $x'y'$-coordinate system. Use the coefficients A', B', and C', shown in the equation with the voice balloon pointing to B' on page 907, to prove the following relationships.

63. $A' + C' = A + C$

64. $B'^2 - 4A'C' = B^2 - 4AC$

Group Exercise

65. Many public and private organizations and schools provide educational materials and information for the blind and visually impaired. Using your library, resources on the World Wide Web, or local organizations, investigate how your group or college could make a contribution to enhance the study of mathematics for the blind and visually impaired. In relation to conic sections, group members should discuss how to create graphs in tactile, or touchable, form that show blind students the visual structure of the conics, including asymptotes, intercepts, end behavior, and rotations.

Objectives

1. Use point plotting to graph plane curves described by parametric equations.

2. Eliminate the parameter.

3. Find parametric equations for functions.

4. Understand the advantages of parametric representations.

What a baseball game! You got to see the great Derek Jeter of the New York Yankees blast a powerful homer. In less than eight seconds, the parabolic path of his home run took the ball a horizontal distance of over 1000 feet. Is there a way to model this path that gives both the ball's location and the time that it is in each of its positions? In this section, we look at ways of describing curves that reveal the where and the when of motion.

Plane Curves and Parametric Equations

You throw a ball from a height of 6 feet, with an initial velocity of 90 feet per second and at an angle of $40°$ with the horizontal. After t seconds, the location of the ball can be described by

$$x = (90 \cos 40°)t \quad \text{and} \quad y = 6 + (90 \sin 40°)t - 16t^2.$$

> This is the ball's horizontal distance, in feet.

> This is the ball's vertical height, in feet.

Because we can use these equations to calculate the location of the ball at any time t, we can describe the path of the ball. For example, to determine the location when $t = 1$ second, substitute 1 for t in each equation:

$$x = (90 \cos 40°)t = (90 \cos 40°)(1) \approx 68.9 \text{ feet}$$

$$y = 6 + (90 \sin 40°)t - 16t^2 = 6 + (90 \sin 40°)(1) - 16(1)^2 \approx 47.9 \text{ feet}.$$

This tells us that after one second, the ball has traveled a horizontal distance of approximately 68.9 feet, and the height of the ball is approximately 47.9 feet. Figure 9.50 displays this information and the results for calculations corresponding to $t = 2$ seconds and $t = 3$ seconds.

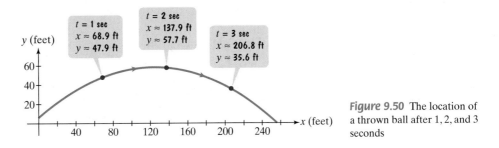

Figure 9.50 The location of a thrown ball after 1, 2, and 3 seconds

The voice balloons in Figure 9.50 tell where the ball is located and when the ball is at a given point (x, y) on its path. The variable t, called a **parameter**, gives the various times for the ball's location. The equations that describe where the ball i located express both x and y as functions of t, and are called **parametric equations**

$$x = (90 \cos 40°)t \qquad y = 6 + (90 \sin 40°)t - 16t^2$$

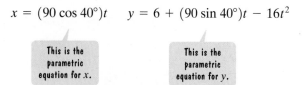

This is the parametric equation for x.

This is the parametric equation for y.

The collection of points (x, y) in Figure 9.50 on the previous page is called a **plan** **curve**.

Plane Curves and Parametric Equations

Suppose that t is a number in an interval I. A **plane curve** is the set of ordered pairs (x, y), where

$$x = f(t), \quad y = g(t) \quad \text{for } t \text{ in interval } I.$$

The variable t is called a **parameter**, and the equations $x = f(t)$ and $y = g(t)$ are called **parametric equations** for the curve.

Use point plotting to graph plane curves described by parametric equations.

Graphing Plane Curves

Graphing a plane curve represented by parametric equations involves plotting point in the rectangular coordinate system and connecting them with a smooth curve.

Graphing a Plane Curve Described by Parametric Equations

1. Select some values of t on the given interval.
2. For each value of t, use the given parametric equations to compute x and y.
3. Plot the points (x, y) in the order of increasing t and connect them with a smooth curve.

Turn back a page and take a second look at Figure 9.50. Do you notice arrow along the curve? These arrows show the direction, or **orientation**, along the curve a t increases. After graphing a plane curve described by parametric equations, use arrows between the points to show the orientation of the curve corresponding t increasing values of t.

EXAMPLE 1   Graphing a Curve Defined by Parametric Equations

Graph the plane curve defined by the parametric equations:

$$x = t^2 - 1, \qquad y = 2t, \qquad -2 \le t \le 2.$$

Solution

Step 1   Select some values of t on the given interval. We will select integral value of t on the interval $-2 \le t \le 2$. Let $t = -2, -1, 0, 1,$ and 2.

Step 2   For each value of t, use the given parametric equations to compute x and y We organize our work in a table. The first column lists the choices for the paramete t. The next two columns show the corresponding values for x and y. The last colum lists the ordered pair (x, y).

t	$x = t^2 - 1$	$y = 2t$	(x, y)
-2	$(-2)^2 - 1 = 4 - 1 = 3$	$2(-2) = -4$	$(3, -4)$
-1	$(-1)^2 - 1 = 1 - 1 = 0$	$2(-1) = -2$	$(0, -2)$
0	$0^2 - 1 = -1$	$2(0) = 0$	$(-1, 0)$
1	$1^2 - 1 = 0$	$2(1) = 2$	$(0, 2)$
2	$2^2 - 1 = 4 - 1 = 3$	$2(2) = 4$	$(3, 4)$

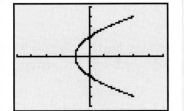

Figure 9.51 The plane curve defined by $x = t^2 - 1$, $y = 2t$, $-2 \le t \le 2$

② Eliminate the parameter.

Technology

A graphing utility can be used to obtain a plane curve represented by parametric equations. Set the mode to parametric and enter the equations. You must enter the minimum and maximum values for t, and an increment setting for t ($t$step). The setting $t$step determines the number of points the graphing utility will plot.

Shown below is the plane curve for

$$x = t^2 - 1$$
$$y = 2t$$

in a $[-5, 5, 1]$ by $[-5, 5, 1]$ viewing rectangle with $t$min $= -2$, $t$max $= 2$, and $t$step $= 0.01$.

Step 3 Plot the points (x, y) in the order of increasing t and connect them with a smooth curve. The plane curve defined by the parametric equations on the given interval is shown in Figure 9.51. The arrows show the direction, or orientation, along the curve as t varies from -2 to 2.

Check Point 1 Graph the plane curve defined by the parametric equations:

$$x = t^2 + 1, \qquad y = 3t, \qquad -2 \le t \le 2.$$

Eliminating the Parameter

The graph in Figure 9.51 shows the plane curve for $x = t^2 - 1$, $y = 2t$, $-2 \le t \le 2$. Even if we examine the parametric equations carefully, we may not be able to tell that the corresponding plane curve is a parabola. By **eliminating the parameter**, we can write one equation in x and y that is equivalent to the two parametric equations. The voice balloons illustrate this process.

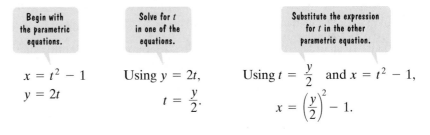

The rectangular equation (the equation in x and y), $x = \dfrac{y^2}{4} - 1$, can be written as $y^2 = 4(x + 1)$. This is the standard form of the equation of a parabola with vertex at $(-1, 0)$ and axis of symmetry along the x-axis. Because the parameter t is restricted to the interval $[-2, 2]$, the plane curve in the technology box on the left shows only a part of the parabola.

Our discussion illustrates a second method for graphing a plane curve described by parametric equations. Eliminate the parameter t and graph the resulting rectangular equation in x and y. However, **you may need to change the domain of the rectangular equation to be consistent with the domain for the parametric equation in x.** This situation is illustrated in Example 2.

EXAMPLE 2 Finding and Graphing the Rectangular Equation of a Curve Defined Parametrically

Sketch the plane curve represented by the parametric equations

$$x = \sqrt{t} \quad \text{and} \quad y = \tfrac{1}{2}t + 1$$

by eliminating the parameter.

Solution We eliminate the parameter t and then graph the resulting rectangular equation.

Begin with the parametric equations.	Solve for t in one of the equations.	Substitute the expression for t in the other parametric equation.
$x = \sqrt{t}$ $y = \dfrac{1}{2}t + 1$	Using $x = \sqrt{t}$ and squaring both sides, $t = x^2$.	Using $t = x^2$ and $y = \dfrac{1}{2}t + 1$, $y = \dfrac{1}{2}x^2 + 1$.

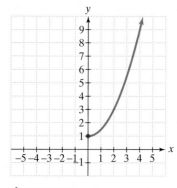

Figure 9.52 The plane curve for $x = \sqrt{t}$ and $y = \frac{1}{2}t + 1$, or $y = \frac{1}{2}x^2 + 1, x \geq 0$

Because t is not limited to a closed interval, you might be tempted to graph the entire U-shaped parabola whose equation is $y = \frac{1}{2}x^2 + 1$. However, take a second look at the parametric equation for x:

$$x = \sqrt{t}.$$

This equation is defined only when $t \geq 0$. Thus, x is nonnegative. The plane curve is the parabola given by $y = \frac{1}{2}x^2 + 1$ with the domain restricted to $x \geq 0$. The plane curve is shown in Figure 9.52.

Check Point 2 Sketch the plane curve represented by the parametric equations

$$x = \sqrt{t} \quad \text{and} \quad y = 2t - 1$$

by eliminating the parameter.

Eliminating the parameter is not always a simple matter. In some cases, it may not be possible. When this occurs, you can use point plotting to obtain a plane curve.

Trigonometric identities can be helpful in eliminating the parameter. For example, consider the plane curve defined by the parametric equations

$$x = \sin t, \quad y = \cos t, \quad 0 \leq t < 2\pi.$$

We use the trigonometric identity $\sin^2 t + \cos^2 t = 1$ to eliminate the parameter. Square each side of each parametric equation and then add.

$$
\begin{aligned}
x^2 &= \sin^2 t \\
y^2 &= \cos^2 t \\
\hline
x^2 + y^2 &= \sin^2 t + \cos^2 t
\end{aligned}
$$

> This is the sum of the two equations above the horizontal lines.

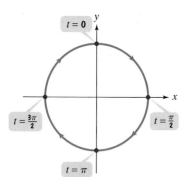

Figure 9.53 The plane curve defined by $x = \sin t, y = \cos t, 0 \leq t < 2\pi$

Using a Pythagorean identity, we write this equation as $x^2 + y^2 = 1$. The plane curve is a circle with center $(0, 0)$ and radius 1. It is shown in Figure 9.53.

EXAMPLE 3 Finding and Graphing the Rectangular Equation of a Curve Defined Parametrically

Sketch the plane curve represented by the parametric equations

$$x = 5 \cos t, \quad y = 2 \sin t, \quad 0 \leq t \leq \pi$$

by eliminating the parameter.

Solution We eliminate the parameter using the identity $\cos^2 t + \sin^2 t = 1$. To apply the identity, divide the parametric equation for x by 5 and the parametric equation for y by 2.

$$\frac{x}{5} = \cos t \quad \text{and} \quad \frac{y}{2} = \sin t$$

Square and add these two equations.

$$
\begin{aligned}
\frac{x^2}{25} &= \cos^2 t \\
\frac{y^2}{4} &= \sin^2 t \\
\hline
\frac{x^2}{25} + \frac{y^2}{4} &= \cos^2 t + \sin^2 t
\end{aligned}
$$

> This is the sum of the two equations above the horizontal lines.

Using a Pythagorean identity, we write this equation as

$$\frac{x^2}{25} + \frac{y^2}{4} = 1.$$

This rectangular equation is the standard form of the equation for an ellipse centered at $(0, 0)$.

$$\frac{x^2}{25} + \frac{y^2}{4} = 1$$

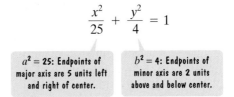

$a^2 = 25$: Endpoints of major axis are **5** units left and right of center.

$b^2 = 4$: Endpoints of minor axis are **2** units above and below center.

The ellipse is shown in Figure 9.54(a). However, this is not the plane curve. Because t is restricted to the interval $[0, \pi]$, the plane curve is only a portion of the ellipse. Use the starting and ending values for t, 0 and π, respectively, and a value of t in the interval $(0, \pi)$ to find which portion to include.

Begin at $t = 0$.

$x = 5 \cos t = 5 \cos 0 = 5 \cdot 1 = 5$

$y = 2 \sin t = 2 \sin 0 = 2 \cdot 0 = 0$

Increase to $t = \frac{\pi}{2}$.

$x = 5 \cos t = 5 \cos \frac{\pi}{2} = 5 \cdot 0 = 0$

$y = 2 \sin t = 2 \sin \frac{\pi}{2} = 2 \cdot 1 = 2$

End at $t = \pi$.

$x = 5 \cos t = 5 \cos \pi = 5(-1) = -5$

$y = 2 \sin t = 2 \sin \pi = 2(0) = 0$

Points on the plane curve include $(5, 0)$, which is the starting point, $(0, 2)$, and $(-5, 0)$, which is the ending point. The plane curve is the top half of the ellipse, shown in Figure 9.54(b).

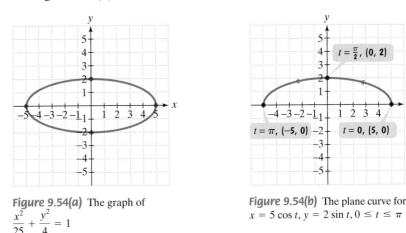

Figure 9.54(a) The graph of $\frac{x^2}{25} + \frac{y^2}{4} = 1$

Figure 9.54(b) The plane curve for $x = 5 \cos t,\ y = 2 \sin t,\ 0 \le t \le \pi$

Check Point 3 Sketch the plane curve represented by the parametric equations

$$x = 6 \cos t,\ y = 4 \sin t,\ \pi \le t \le 2\pi$$

by eliminating the parameter.

 Find parametric equations for functions.

Finding Parametric Equations

Infinitely many pairs of parametric equations can represent the same plane curve. If the plane curve is defined by the function $y = f(x)$, here is a procedure for finding a set of parametric equations:

Parametric Equations for the Function $y = f(x)$

One set of parametric equations for the plane curve defined by $y = f(x)$ is

$$x = t \quad \text{and} \quad y = f(t),$$

in which t is in the domain of f.

EXAMPLE 4   Finding Parametric Equations

Find a set of parametric equations for the parabola whose equation is $y = 9 - x^2$.

Solution   Let $x = t$. Parametric equations for $y = f(x)$ are $x = t$ and $y = f(t)$. Thus, parametric equations for $y = 9 - x^2$ are

$$x = t \quad \text{and} \quad y = 9 - t^2.$$

Check Point 4   Find a set of parametric equations for the parabola whose equation is $y = x^2 - 25$.

You can write other sets of parametric equations for $y = 9 - x^2$ by starting with a different parametric equation for x. Here are three more sets of parametric equations for

$$y = 9 - x^2:$$

- If $x = t^3$, $y = 9 - (t^3)^2 = 9 - t^6$.

  Parametric equations are $x = t^3$ and $y = 9 - t^6$.

- If $x = t + 1$, $y = 9 - (t + 1)^2 = 9 - (t^2 + 2t + 1) = 8 - t^2 - 2t$.

  Parametric equations are $x = t + 1$ and $y = 8 - t^2 - 2t$.

- If $x = \dfrac{t}{2}$, $y = 9 - \left(\dfrac{t}{2}\right)^2 = 9 - \dfrac{t^2}{4}$.

  Parametric equations are $x = \dfrac{t}{2}$ and $y = 9 - \dfrac{t^2}{4}$.

Can you start with any choice for the parametric equation for x? The answer is no. **The substitution for x must be a function that allows x to take on all the values in the domain of the given rectangular equation.** For example, the domain of the function $y = 9 - x^2$ is the set of all real numbers. If you incorrectly let $x = t^2$, these values of x exclude negative numbers that are included in $y = 9 - x^2$. The parametric equations

$$x = t^2 \quad \text{and} \quad y = 9 - (t^2)^2 = 9 - t^4$$

do not represent $y = 9 - x^2$ because only points for which $x \geq 0$ are obtained.

④ Understand the advantages of parametric representations.

Technology

The ellipse shown was obtained using the parametric mode and the radian mode of a graphing utility.

$$x(t) = 2 + 3 \cos t$$
$$y(t) = 3 + 2 \sin t$$

We used a $[-2, 6, 1]$ by $[-1, 6, 1]$ viewing rectangle with $t\text{min} = 0$, $t\text{max} = 6.2$, and $t\text{step} = 0.1$.

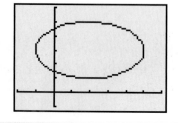

Advantages of Parametric Equations over Rectangular Equations

We opened this section with parametric equations that described the horizontal distance and the vertical height of your thrown baseball after t seconds. Parametric equations are frequently used to represent the path of a moving object. If t represents time, parametric equations give the location of a moving object and tell when the object is located at each of its positions. Rectangular equations tell where the moving object is located but do not reveal when the object is in a particular position.

When using technology to obtain graphs, parametric equations that represent relations that are not functions are often easier to use than their corresponding rectangular equations. It is far easier to enter the equation of an ellipse given by the parametric equations

$$x = 2 + 3 \cos t \quad \text{and} \quad y = 3 + 2 \sin t$$

than to use the rectangular equivalent

$$\frac{(x - 2)^2}{9} + \frac{(y - 3)^2}{4} = 1.$$

The rectangular equation must first be solved for y and then entered as two separate equations before a graphing utility reveals the ellipse.

A curve that is used in physics for much of the theory of light is called a **cycloid**. The path of a fixed point on the circumference of a circle as it rolls along a line is a cycloid. A point on the rim of a bicycle wheel traces out a cycloid curve, shown in Figure 9.55. If the radius of the circle is a, the parametric equations of the cycloid are

$$x = a(t - \sin t) \quad \text{and} \quad y = a(1 - \cos t).$$

It is an extremely complicated task to represent the cycloid in rectangular form.

Cycloids are used to solve problems that involve the "shortest time." For example, Figure 9.56 shows a bead sliding down a wire. The shape of the wire a bead could slide down so that the distance between two points is traveled in the shortest time is an inverted cycloid.

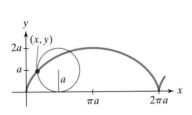

Figure 9.55 The curve traced by a fixed point on the circumference of a circle rolling along a straight line is a cycloid.

Figure 9.56

Rolling

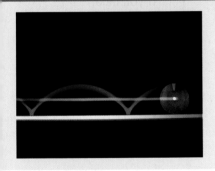

Linear functions and cycloids are used to describe rolling motion. The light at the rolling circle's center shows that it moves linearly. By contrast, the light at the circle's edge has rotational motion and traces out a cycloid. A number of sites on the Internet illustrate rotational motion and show how the cycloid is created.

EXERCISE SET 9.5

Practice Exercises

In Exercises 1–8, parametric equations and a value for the parameter t are given. Find the coordinates of the point on the plane curve described by the parametric equations corresponding to the given value of t.

1. $x = 3 - 5t, y = 4 + 2t; t = 1$

2. $x = 7 - 4t, y = 5 + 6t; t = 1$

3. $x = t^2 + 1, y = 5 - t^3; t = 2$

4. $x = t^2 + 3, y = 6 - t^3; t = 2$

5. $x = 4 + 2 \cos t, y = 3 + 5 \sin t; t = \dfrac{\pi}{2}$

6. $x = 2 + 3 \cos t, y = 4 + 2 \sin t; t = \pi$

7. $x = (60 \cos 30°)t, y = 5 + (60 \sin 30°)t - 16t^2; t = 2$

8. $x = (80 \cos 45°)t, y = 6 + (80 \sin 45°)t - 16t^2; t = 2$

In Exercises 9–20, use point plotting to graph the plane curve described by the given parametric equations. Use arrows to show the orientation of the curve corresponding to increasing values of t.

9. $x = t + 2, y = t^2; -2 \le t \le 2$

10. $x = t - 1, y = t^2; -2 \le t \le 2$

11. $x = t - 2, y = 2t + 1; -2 \le t \le 3$

12. $x = t - 3, y = 2t + 2; -2 \le t \le 3$

13. $x = t + 1, y = \sqrt{t}; t \ge 0$ **14.** $x = \sqrt{t}, y = t - 1; t \ge 0$

15. $x = \cos t, y = \sin t; 0 \le t < 2\pi$

16. $x = -\sin t, y = -\cos t; 0 \le t < 2\pi$

17. $x = t^2, y = t^3; -\infty < t < \infty$

18. $x = t^2 + 1, y = t^3 - 1; -\infty < t < \infty$
19. $x = 2t, y = |t - 1|; -\infty < t < \infty$
20. $x = |t + 1|, y = t - 2; -\infty < t < \infty$

In Exercises 21–40, eliminate the parameter t. Then use the rectangular equation to sketch the plane curve represented by the given parametric equations. Use arrows to show the orientation of the curve corresponding to increasing values of t. (If an interval for t is not specified, assume that $-\infty < t < \infty$.)

21. $x = t, y = 2t$ 22. $x = t, y = -2t$
23. $x = 2t - 4, y = 4t^2$ 24. $x = t - 2, y = t^2$
25. $x = \sqrt{t}, y = t - 1$ 26. $x = \sqrt{t}, y = t + 1$
27. $x = 2 \sin t, y = 2 \cos t; 0 \le t < 2\pi$
28. $x = 3 \sin t, y = 3 \cos t; 0 \le t < 2\pi$
29. $x = 1 + 3 \cos t, y = 2 + 3 \sin t; 0 \le t < 2\pi$
30. $x = -1 + 2 \cos t, y = 1 + 2 \sin t; 0 \le t < 2\pi$
31. $x = 2 \cos t, y = 3 \sin t; 0 \le t < 2\pi$
32. $x = 3 \cos t, y = 5 \sin t; 0 \le t < 2\pi$
33. $x = 1 + 3 \cos t, y = -1 + 2 \sin t; 0 \le t \le \pi$
34. $x = 2 + 4 \cos t, y = -1 + 3 \sin t; 0 \le t \le \pi$
35. $x = \sec t, y = \tan t$ 36. $x = 5 \sec t, y = 3 \tan t$
37. $x = t^2 + 2, y = t^2 - 2$ 38. $x = \sqrt{t} + 2, y = \sqrt{t} - 2$
39. $x = 2^t, y = 2^{-t}; t \ge 0$ 40. $x = e^t, y = e^{-t}; t \ge 0$

In Exercises 41–43, eliminate the parameter. Write the resulting equation in standard form.

41. A circle: $x = h + r \cos t, y = k + r \sin t$
42. An ellipse: $x = h + a \cos t, y = k + b \sin t$
43. A hyperbola: $x = h + a \sec t, y = k + b \tan t$
44. The parametric equations of the line through (x_1, y_1) and (x_2, y_2) are

$$x = x_1 + t(x_2 - x_1) \quad \text{and} \quad y = y_1 + t(y_2 - y_1).$$

Eliminate the parameter and write the resulting equation in point-slope form.

In Exercises 45–52, use your answers from Exercises 41–44 and the parametric equations given in Exercises 41–44 to find a set of parametric equations for the conic section or the line.

45. Circle: Center: $(3, 5)$; Radius: 6
46. Circle: Center: $(4, 6)$; Radius: 9
47. Ellipse: Center: $(-2, 3)$; Vertices: 5 units to the left and right of the center; Endpoints of Minor Axis: 2 units above and below the center
48. Ellipse: Center: $(4, -1)$; Vertices: 5 units above and below the center; Endpoints of Minor Axis: 3 units to the left and right of the center
49. Hyperbola: Vertices: $(4, 0)$ and $(-4, 0)$; Foci: $(6, 0)$ and $(-6, 0)$
50. Hyperbola: Vertices: $(0, 4)$ and $(0, -4)$; Foci: $(0, 5)$ and $(0, -5)$
51. Line: Passes through $(-2, 4)$ and $(1, 7)$
52. Line: Passes through $(3, -1)$ and $(9, 12)$

In Exercises 53–56, find two different sets of parametric equations for each rectangular equation.

53. $y = 4x - 3$ 54. $y = 2x - 5$
55. $y = x^2 + 4$ 56. $y = x^2 - 3$

In Exercises 57–58, the parametric equations of four plane curves are given. Graph each plane curve and determine how they differ from each other.

57. **a.** $x = t$ and $y = t^2 - 4$
    b. $x = t^2$ and $y = t^4 - 4$
    c. $x = \cos t$ and $y = \cos^2 t - 4$
    d. $x = e^t$ and $y = e^{2t} - 4$

58. **a.** $x = t, y = \sqrt{4 - t^2}; -2 \le t \le 2$
    b. $x = \sqrt{4 - t^2}, y = t; -2 \le t \le 2$
    c. $x = 2 \sin t, y = 2 \cos t; 0 \le t < 2\pi$
    d. $x = 2 \cos t, y = 2 \sin t; 0 \le t < 2\pi$

Practice Plus

In Exercises 59–62, sketch the plane curve represented by the given parametric equations. Then use interval notation to give each relation's domain and range.

59. $x = 4 \cos t + 2, y = 4 \cos t - 1$
60. $x = 2 \sin t - 3, y = 2 \sin t + 1$
61. $x = t^2 + t + 1, y = 2t$ 62. $x = t^2 - t + 6, y = 3t$

In Exercises 63–68, sketch the function represented by the given parametric equations. Then use the graph to determine each of the following:

a. *intervals, if any, on which the function is increasing and intervals, if any, on which the function is decreasing.*

b. *the number, if any, at which the function has a maximum and this maximum value, or the number, if any, at which the function has a minimum and this minimum value.*

63. $x = 2^t, y = t$ 64. $x = e^t, y = t$
65. $x = \dfrac{t}{2}, y = 2t^2 - 8t + 3$ 66. $x = \dfrac{t}{2}, y = -2t^2 + 8t - 1$
67. $x = 2(t - \sin t), y = 2(1 - \cos t); 0 \le t \le 2\pi$
68. $x = 3(t - \sin t), y = 3(1 - \cos t); 0 \le t \le 2\pi$

Application Exercises

The path of a projectile that is launched h feet above the ground with an initial velocity of v_0 feet per second and at an angle θ with the horizontal is given by the parametric equations

$$x = (v_0 \cos \theta)t \quad \text{and} \quad y = h + (v_0 \sin \theta)t - 16t^2,$$

where t is the time, in seconds, after the projectile was launched. The parametric equation for x gives the projectile's horizontal distance, in feet. The parametric equation for y gives the projectile's height, in feet. Use these parametric equations to solve Exercises 69–70.

69. The figure shows the path for a baseball hit by Derek Jeter. The ball was hit with an initial velocity of 180 feet per second at an angle of 40° to the horizontal. The ball was hit at a height 3 feet off the ground.

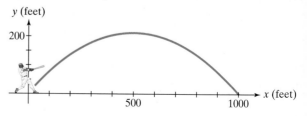

a. Find the parametric equations that describe the position of the ball as a function of time.

b. Describe the ball's position after 1, 2, and 3 seconds. Round to the nearest tenth of a foot. Locate your solutions on the plane curve.

c. How long, to the nearest tenth of a second, is the ball in flight? What is the total horizontal distance that it travels before it lands? Is your answer consistent with the figure shown?

d. You meet Derek Jeter and he asks you to tell him something interesting about the path of the baseball that he hit. Use the graph to respond to his request. Then verify your observation algebraically.

70. The figure shows the path for a baseball that was hit with an initial velocity of 150 feet per second at an angle of 35° to the horizontal. The ball was hit at a height of 3 feet off the ground.

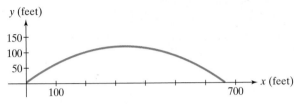

a. Find the parametric equations that describe the position of the ball as a function of time.

b. Describe the ball's position after 1, 2, and 3 seconds. Round to the nearest tenth of a foot. Locate your solutions on the plane curve.

c. How long is the ball in flight? (Round to the nearest tenth of a second.) What is the total horizontal distance that it travels, to the nearest tenth of a foot, before it lands? Is your answer consistent with the figure shown?

d. Use the graph to describe something about the path of the baseball that might be of interest to the player who hit the ball. Then verify your observation algebraically.

Writing in Mathematics

71. What are plane curves and parametric equations?

72. How is point plotting used to graph a plane curve described by parametric equations? Give an example with your description.

73. What is the significance of arrows along a plane curve?

74. What does it mean to eliminate the parameter? What useful information can be obtained by doing this?

75. Explain how the rectangular equation $y = 5x$ can have infinitely many sets of parametric equations.

76. Discuss how the parametric equations for the path of a projectile (see Exercises 69–70) and the ability to obtain plane curves with a graphing utility can be used by a baseball coach to analyze performances of team players.

Technology Exercises

77. Use a graphing utility in a parametric mode to verify any five of your hand-drawn graphs in Exercises 9–40.

In Exercises 78–82, use a graphing utility to obtain the plane curve represented by the given parametric equations.

78. Cycloid: $x = 3(t - \sin t)$, $y = 3(1 - \cos t); [0, 60, 5] \times [0, 8, 1], 0 \le t < 6\pi$

79. Cycloid: $x = 2(t - \sin t)$, $y = 2(1 - \cos t); [0, 60, 5] \times [0, 8, 1], 0 \le t < 6\pi$

80. Witch of Agnesi: $x = 2 \cot t, y = 2 \sin^2 t$; $[-6, 6, 1] \times [-4, 4, 1], 0 \le t < 2\pi$

81. Hypocycloid: $x = 4 \cos^3 t, y = 4 \sin^3 t$; $[-5, 5, 1] \times [-5, 5, 1], 0 \le t < 2\pi$

82. Lissajous Curve: $x = 2 \cos t, y = \sin 2t$; $[-3, 3, 1] \times [-2, 2, 1], 0 \le t < 2\pi$

Use the equations for the path of a projectile given prior to Exercises 69–70 to solve Exercises 83–85.

In Exercises 83–84, use a graphing utility to obtain the path of a projectile launched from the ground ($h = 0$) at the specified values of θ and v_0. In each exercise, use the graph to determine the maximum height and the time at which the projectile reaches its maximum height. Also use the graph to determine the range of the projectile and the time it hits the ground. Round all answers to the nearest tenth.

83. $\theta = 55°, v_0 = 200$ feet per second

84. $\theta = 35°, v_0 = 300$ feet per second

85. A baseball player throws a ball with an initial velocity of 140 feet per second at an angle of 22° to the horizontal. The ball leaves the player's hand at a height of 5 feet.

a. Write the parametric equations that describe the ball's position as a function of time.

b. Use a graphing utility to obtain the path of the baseball.

c. Find the ball's maximum height and the time at which it reaches this height. Round all answers to the nearest tenth.

d. How long is the ball in the air?

e. How far does the ball travel?

Critical Thinking Exercises

86. Eliminate the parameter: $x = \cos^3 t$ and $y = \sin^3 t$.

87. The plane curve described by the parametric equations $x = 3 \cos t$ and $y = 3 \sin t$, $0 \le t < 2\pi$, has a counterclockwise orientation. Alter one or both parametric equations so that you obtain the same plane curve with the opposite orientation.

88. The figure shows a circle of radius a rolling along a horizontal line. Point P traces out a cycloid. Angle t, in radians, is the angle through which the circle has rolled. C is the center of the circle.

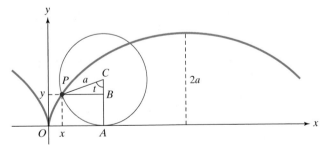

Use the suggestions in parts (a) and (b) to prove that the parametric equations of the cycloid are $x = a(t - \sin t)$ and $y = a(1 - \cos t)$.

a. Derive the parametric equation for x using the figure and
$$x = OA - xA.$$

b. Derive the parametric equation for y using the figure and
$$y = AC - BC.$$

Objectives

① Define conics in terms of a focus and a directrix.

② Graph the polar equations of conics.

John Glenn made the first U.S.-manned flight around Earth on *Friendship 7.*

On the morning of February 20, 1962, millions of Americans collectively held their breath as the world's newest pioneer swept across the threshold of one of our last frontiers. Roughly one hundred miles above Earth, astronaut John Glenn sat comfortably in the weightless environment of a $9\frac{1}{2}$-by-6-foot space capsule that offered the leg room of a Volkswagen "Beetle" and the aesthetics of a garbage can. Glenn became the first American to orbit Earth in a three-orbit mission that lasted slightly under 5 hours.

In this section, you will see how John Glenn's historic orbit can be described using conic sections in polar coordinates. To obtain this model, we begin with a definition that permits a unified approach to the conic sections.

① Define conics in terms of a focus and a directrix.

The Focus-Directrix Definitions of the Conic Sections

The definition of a parabola is given in terms of a fixed point, the focus, and a fixed line, the directrix. By contrast, the definitions of an ellipse and a hyperbola are given in terms of two fixed points, the foci. It is possible to define each of these conic sections in terms of a point and a line. Figure 9.57 shows a conic section in the polar coordinate system. The fixed point, the focus, is at the pole. The fixed line, the directrix, is perpendicular to the polar axis.

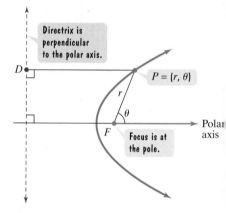

Figure 9.57 A conic in the polar coordinate system

Focus-Directrix Definitions of the Conic Sections

Let *F* be a fixed point, the focus, and let *D* be a fixed line, the directrix, in a plane (Figure 9.58). A **conic section**, or **conic**, is the set of all points *P* in the plane such that

$$\frac{PF}{PD} = e,$$

where *e* is a fixed positive number, called the **eccentricity**.

    If $e = 1$, the conic is a parabola.

    If $e < 1$, the conic is an ellipse.

    If $e > 1$, the conic is a hyperbola.

Figure 9.58 illustrates the eccentricity for each type of conic. Notice that if $e = 1$, the definition of the parabola is the same as the focus-directrix definition with which you are familiar.

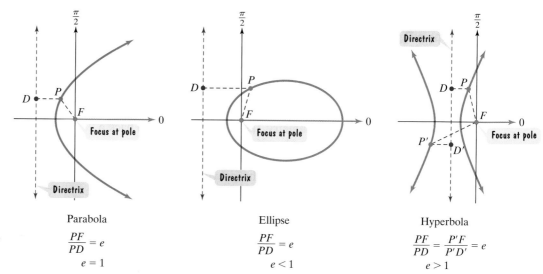

Figure 9.58 The eccentricity for each conic

Parabola
$$\frac{PF}{PD} = e$$
$$e = 1$$

Ellipse
$$\frac{PF}{PD} = e$$
$$e < 1$$

Hyperbola
$$\frac{PF}{PD} = \frac{P'F}{P'D'} = e$$
$$e > 1$$

② Graph the polar equations of conics.

Polar Equations of Conics

By locating a focus at the pole, all conics can be represented by similar equations in the polar coordinate system. In each of these equations,

- (r, θ) is a point on the graph of the conic.
- e is the eccentricity. (Remember that $e > 0$.)
- p is the distance between the focus (located at the pole) and the directrix.

Standard Forms of the Polar Equations of Conics

Let the pole be a focus of a conic section of eccentricity e with the directrix p units from the focus. The equation of the conic is given by one of the four equations listed.

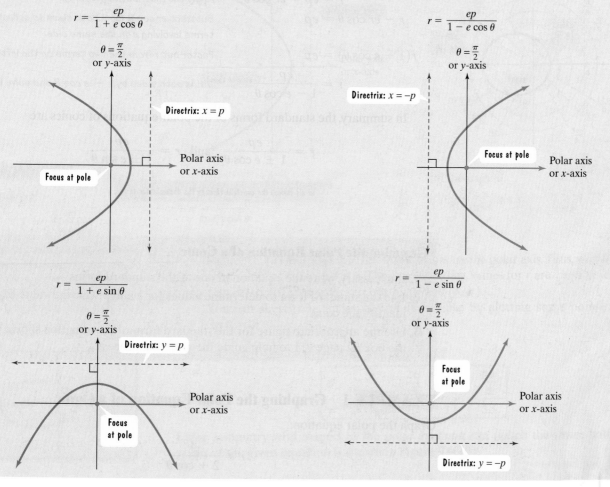

$$r = \frac{ep}{1 + e \cos \theta}$$

$$r = \frac{ep}{1 - e \cos \theta}$$

$$r = \frac{ep}{1 + e \sin \theta}$$

$$r = \frac{ep}{1 - e \sin \theta}$$

Check Point **1** Use the three steps shown in the box on page 912 to graph the polar equation

$$r = \frac{4}{2 - \cos\theta}.$$

EXAMPLE 2   Graphing the Polar Equation of a Conic

Graph the polar equation:

$$r = \frac{12}{3 + 3\sin\theta}.$$

Solution

Step 1   Write the equation in one of the standard forms. The equation is not in standard form because the constant term in the denominator is not 1. Divide the numerator and denominator by 3 to write the standard form.

$$r = \frac{4}{1 + 1\sin\theta}$$ This equation is in the form $r = \dfrac{ep}{1 + e\sin\theta}$.

$ep = 4$

$e = 1$

Step 2   Use the standard form to find e and p, and identify the conic. The voice balloons show that

$$e = 1 \quad \text{and} \quad ep = 1p = 4.$$

Thus, $e = 1$ and $p = 4$. Because $e = 1$, the conic is a parabola.

Step 3   Use the figure for the equation's standard form to guide the graphing process. Figure 9.61(a) indicates that we have symmetry with respect to $\theta = \dfrac{\pi}{2}$. The focus is at the pole and, with $p = 4$, the directrix is $y = 4$, located four units above the pole.

Figure 9.61(a) indicates that the vertex is on the line of $\theta = \dfrac{\pi}{2}$, or the y-axis. Thus, we find the vertex by selecting $\dfrac{\pi}{2}$ for θ. The corresponding value for r is 2.

Figure 9.61(b) shows the vertex, $\left(2, \dfrac{\pi}{2}\right)$.

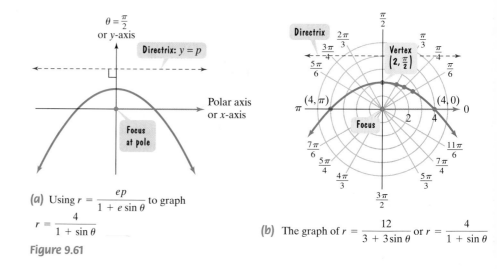

(a) Using $r = \dfrac{ep}{1 + e\sin\theta}$ to graph

$r = \dfrac{4}{1 + \sin\theta}$

(b) The graph of $r = \dfrac{12}{3 + 3\sin\theta}$ or $r = \dfrac{4}{1 + \sin\theta}$

Figure 9.61

To find where the parabola crosses the polar axis, select $\theta = 0$ and $\theta = \pi$. The corresponding values for r are 4 and 4, respectively. Figure 9.61(b) shows the points $(4, 0)$ and $(4, \pi)$ on the polar axis.

Technology

The graph of

$$r = \frac{12}{3 + 3 \sin \theta}$$

was obtained using

$$[-5, 5, 1] \text{ by } [-5, 5, 1]$$

with

$$\theta\text{min} = 0, \quad \theta\text{max} = 2\pi,$$
$$\theta\text{step} = \frac{\pi}{48}.$$

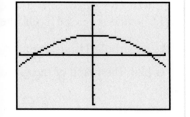

You can sketch the right half of the parabola by plotting some points from $\theta = 0$ to $\theta = \frac{\pi}{2}$.

$$r = \frac{12}{3 + 3 \sin \theta}$$

θ	$\frac{\pi}{6}$	$\frac{\pi}{4}$	$\frac{\pi}{3}$
r	2.7	2.3	2.1

Using symmetry with respect to $\theta = \frac{\pi}{2}$, you can sketch the left half. The graph of the given equation is shown in Figure 9.61(b).

Check Point 2 Use the three steps shown in the box on page 912 to graph the polar equation:

$$r = \frac{8}{4 + 4 \sin \theta}.$$

EXAMPLE 3 Graphing the Polar Equation of a Conic

Graph the polar equation:

$$r = \frac{9}{3 - 6 \cos \theta}.$$

Solution

Step 1 Write the equation in one of the standard forms. We can obtain a constant term of 1 in the denominator by dividing each term by 3.

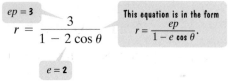

$$r = \frac{3}{1 - 2 \cos \theta}$$

$ep = 3$

$e = 2$

This equation is in the form $r = \frac{ep}{1 - e \cos \theta}$.

Step 2 Use the standard form to find e and p, and identify the conic. The voice balloons show that

$$e = 2 \quad \text{and} \quad ep = 2p = 3.$$

Thus, $e = 2$ and $p = \frac{3}{2}$. Because $e = 2 > 1$, the conic is a hyperbola.

Step 3 Use the figure for the equation's standard form to guide the graphing process. Figure 9.62(a) indicates that we have symmetry with respect to the polar axis. One focus is at the pole and, with $p = \frac{3}{2}$, a directrix is $x = -\frac{3}{2}$, located 1.5 units to the left of the pole.

Figure 9.62(a) indicates that the transverse axis is horizontal and the vertices lie on the polar axis. Thus, we find the vertices by selecting 0 and π for θ. Figure 9.62(b) shows the vertices, $(-3, 0)$ and $(1, \pi)$.

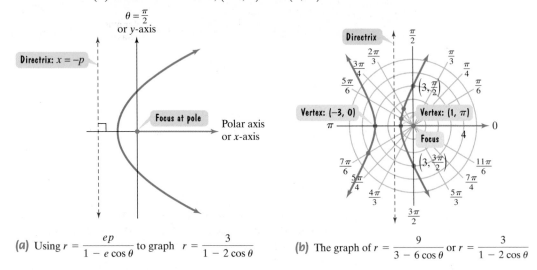

Figure 9.62 **(a)** Using $r = \frac{ep}{1 - e \cos \theta}$ to graph $r = \frac{3}{1 - 2 \cos \theta}$ **(b)** The graph of $r = \frac{9}{3 - 6 \cos \theta}$ or $r = \frac{3}{1 - 2 \cos \theta}$

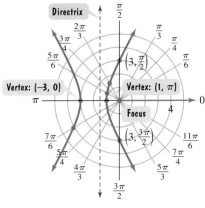

Figure 9.62(b) (repeated) The graph of
$r = \dfrac{9}{3 - 6 \cos \theta}$ or $r = \dfrac{3}{1 - 2 \cos \theta}$

To find where the hyperbola crosses the line $\theta = \dfrac{\pi}{2}$, select $\dfrac{\pi}{2}$ and $\dfrac{3\pi}{2}$ for θ.

Figure 9.62(b) shows the points $\left(3, \dfrac{\pi}{2}\right)$ and $\left(3, \dfrac{3\pi}{2}\right)$ on the graph.

We sketch the upper half of the hyperbola by plotting some points from $\theta = 0$ to $\theta = \pi$.

$$r = \frac{3}{1 - 2 \cos \theta}$$

θ	$\dfrac{\pi}{6}$	$\dfrac{2\pi}{3}$	$\dfrac{5\pi}{6}$
r	-4.1	1.5	1.1

Figure 9.62(b) shows the points $\left(\dfrac{\pi}{6}, -4.1\right)$, $\left(\dfrac{2\pi}{3}, 1.5\right)$, and $\left(\dfrac{5\pi}{6}, 1.1\right)$ on the graph. Observe that $\left(\dfrac{\pi}{6}, -4.1\right)$ is on the lower half of the hyperbola. Using symmetry with respect to the polar axis, we sketch the entire lower half. The graph of the given equation is shown in Figure 9.62(b).

Check Point 3 Use the three steps shown in the box on page 912 to graph the polar equation:

$$r = \frac{9}{3 - 9 \cos \theta}.$$

Modeling Planetary Motion

Polish astronomer Nicolaus Copernicus (1473–1543) was correct in stating that planets in our solar system revolve around the sun and not Earth. However, he incorrectly believed that celestial orbits move in perfect circles, calling his system "the ballet of the planets."

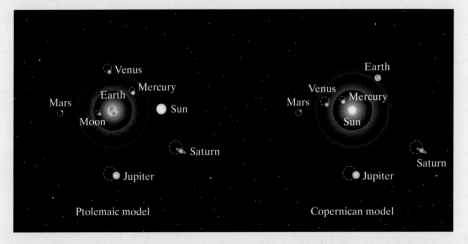

Table 9.4 indicates that the planets in our solar system have orbits with eccentricities that are much closer to 0 than to 1. Most of these orbits are almost circular, which made it difficult for early astronomers to detect that they are actually ellipses.

German scientist and mathematician Johannes Kepler (1571–1630) discovered that planets move in elliptical orbits with the sun at one focus. The polar equation for these orbits is

$$r = \frac{(1 - e^2)a}{1 - e \cos \theta},$$

Table 9.4 Eccentricities of Planetary Orbits

Mercury	0.2056	Saturn	0.0543
Venus	0.0068	Uranus	0.0460
Earth	0.0167	Neptune	0.0082
Mars	0.0934	Pluto	0.2481
Jupiter	0.0484		

where the length of the orbit's major axis is $2a$. Describing planetary orbits, Kepler wrote, "The heavenly motions are nothing but a continuous song for several voices, to be perceived by the intellect, not by the ear."

EXERCISE SET 9.6

Practice Exercises

In Exercises 1–8,

a. *Identify the conic section that each polar equation represents.*

b. *Describe the location of a directrix from the focus located at the pole.*

1. $r = \dfrac{3}{1 + \sin \theta}$

2. $r = \dfrac{3}{1 + \cos \theta}$

3. $r = \dfrac{6}{3 - 2 \cos \theta}$

4. $r = \dfrac{6}{3 + 2 \cos \theta}$

5. $r = \dfrac{8}{2 + 2 \sin \theta}$

6. $r = \dfrac{8}{2 - 2 \sin \theta}$

7. $r = \dfrac{12}{2 - 4 \cos \theta}$

8. $r = \dfrac{12}{2 + 4 \cos \theta}$

In Exercises 9–20, use the three steps shown in the box on page 912 to graph each polar equation.

9. $r = \dfrac{1}{1 + \sin \theta}$

10. $r = \dfrac{1}{1 + \cos \theta}$

11. $r = \dfrac{2}{1 - \cos \theta}$

12. $r = \dfrac{2}{1 - \sin \theta}$

13. $r = \dfrac{12}{5 + 3 \cos \theta}$

14. $r = \dfrac{12}{5 - 3 \cos \theta}$

15. $r = \dfrac{6}{2 - 2 \sin \theta}$

16. $r = \dfrac{6}{2 + 2 \sin \theta}$

17. $r = \dfrac{8}{2 - 4 \cos \theta}$

18. $r = \dfrac{8}{2 + 4 \cos \theta}$

19. $r = \dfrac{12}{3 - 6 \cos \theta}$

20. $r = \dfrac{12}{3 - 3 \cos \theta}$

Practice Plus

In Exercises 21–28, describe a viewing rectangle, or window, such as $[-30, 30, 3]$ by $[-8, 4, 1]$, that shows a complete graph of each polar equation and minimizes unused portions of the screen.

21. $r = \dfrac{15}{3 - 2 \cos \theta}$

22. $r = \dfrac{16}{5 - 3 \cos \theta}$

23. $r = \dfrac{8}{1 - \cos \theta}$

24. $r = \dfrac{8}{1 + \cos \theta}$

25. $r = \dfrac{4}{1 + 3 \cos \theta}$

26. $r = \dfrac{16}{3 + 5 \cos \theta}$

27. $r = \dfrac{4}{5 + 5 \sin \theta}$

28. $r = \dfrac{2}{3 + 3 \sin \theta}$

Application Exercises

Halley's comet has an elliptical orbit with the sun at one focus. Its orbit, shown in the figure at the top of the next column, is given approximately by

$$r = \dfrac{1.069}{1 + 0.967 \sin \theta}.$$

In the formula, r is measured in astronomical units. (One astronomical unit is the average distance from Earth to the sun, approximately 93 million miles.) Use the given formula and the

figure to solve Exercises 29–30. Round to the nearest hundredth of an astronomical unit and the nearest million miles.

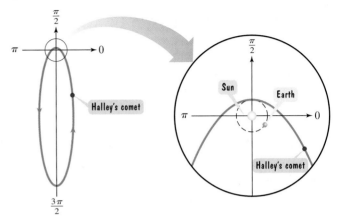

29. Find the distance from Halley's comet to the sun at its shortest distance from the sun.

30. Find the distance from Halley's comet to the sun at its greatest distance from the sun.

On February 20, 1962, John Glenn made the first U.S.-manned flight around the Earth for three orbits on Friendship 7. With Earth at one focus, the orbit of Friendship 7 is given approximately by

$$r = \dfrac{4090.76}{1 - 0.0076 \cos \theta},$$

where r is measured in miles from Earth's center. Use the formula and the figure shown to solve Exercises 31–32.

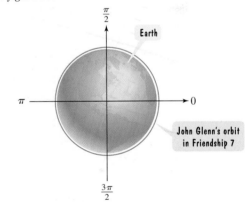

31. How far from Earth's center was John Glenn at his greatest distance from the planet? Round to the nearest mile. If the radius of Earth is 3960 miles, how far was he from Earth's surface at this point on the flight?

32. How far from Earth's center was John Glenn at his closest distance from the planet? Round to the nearest mile. If the radius of Earth is 3960 miles, how far was he from Earth's surface at this point on the flight?

Writing in Mathematics

33. How are the conics described in terms of a fixed point and a fixed line?

34. If all conics are defined in terms of a fixed point and a fixed line, how can you tell one kind of conic from another?

35. If you are given the standard form of the polar equation of a conic, how do you determine its eccentricity?

36. If you are given the standard form of the polar equation of a conic, how do you determine the location of a directrix from the focus at the pole?

37. Describe a strategy for graphing $r = \dfrac{1}{1 + \sin \theta}$.

38. You meet John Glenn and he asks you to tell him something of interest about the elliptical orbit of his first space voyage in 1962. Describe how to use the polar equation for orbits in the essay on page 916, the equation for his 1962 journey in Exercises 31–32, and a graphing utility to provide him with an interesting visual analysis.

Technology Exercises

Use the polar mode of a graphing utility with angle measure in radians to solve Exercises 39–42. Unless otherwise indicated, use θmin $= 0$, θmax $= 2\pi$, and θ step $= \dfrac{\pi}{48}$. If you are not satisfied with the quality of the graph, experiment with smaller values for θ step.

39. Use a graphing utility to verify any five of your hand-drawn graphs in Exercises 9–20.

In Exercises 40–42, identify the conic that each polar equation represents. Then use a graphing utility to graph the equation.

40. $r = \dfrac{16}{4 - 3 \cos \theta}$ **41.** $r = \dfrac{12}{4 + 5 \sin \theta}$ **42.** $r = \dfrac{18}{6 - 6 \cos \theta}$

In Exercises 43–44, use a graphing utility to graph the equation. Then answer the given question.

43. $r = \dfrac{4}{1 - \sin\left(\theta - \dfrac{\pi}{4}\right)}$; How does the graph differ from the

graph of $r = \dfrac{4}{1 - \sin \theta}$?

44. $r = \dfrac{3}{2 + 6 \cos\left(\theta + \dfrac{\pi}{3}\right)}$; How does the graph differ from the

graph of $r = \dfrac{3}{2 + 6 \cos \theta}$?

45. Use the polar equation for planetary orbits,

$$r = \frac{(1 - e^2)a}{1 - e \cos \theta},$$

to find the polar equation of the orbit for Mercury and Earth.

Mercury: $e = 0.2056$ and $a = 36.0 \times 10^6$ miles

Earth: $e = 0.0167$ and $a = 92.96 \times 10^6$ miles

Use a graphing utility to graph both orbits in the same viewing rectangle. What do you see about the orbits from their graphs that is not obvious from their equations?

Critical Thinking Exercises

46. Identify the conic and graph the equation:

$$r = \frac{4 \sec \theta}{2 \sec \theta - 1}.$$

In Exercises 47–48, write a polar equation of the conic that is named and described.

47. Ellipse: a focus at the pole; vertex: $(4, 0)$; $e = \frac{1}{2}$

48. Hyperbola: a focus at the pole; directrix: $x = -1$; $e = \frac{3}{2}$

49. Identify the conic and write its equation in rectangular coordinates: $r = \dfrac{1}{2 - 2 \cos \theta}$.

50. Prove that the polar equation of a planet's elliptical orbit is

$$r = \frac{(1 - e^2)a}{1 - e \cos \theta},$$

where e is the eccentricity and $2a$ is the length of the major axis.

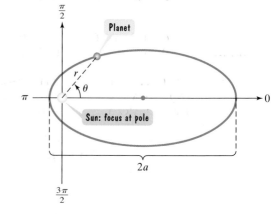

Chapter 9
Summary, Review, and Test

Summary

DEFINITIONS AND CONCEPTS	EXAMPLES

9.1 The Ellipse

a. An ellipse is the set of all points in a plane the sum of whose distances from two fixed points, the foci, is constant.

b. Standard forms of the equations of an ellipse with center at the origin are $\dfrac{x^2}{a^2} + \dfrac{y^2}{b^2} = 1$ [foci: $(-c, 0), (c, 0)$] and $\dfrac{x^2}{b^2} + \dfrac{y^2}{a^2} = 1$ [foci: $(0, -c), (0, c)$], where $c^2 = a^2 - b^2$ and $a^2 > b^2$. See the box on page 852 and Figure 9.6.   Ex. 1, p. 852; Ex. 2, p. 853; Ex. 3, p. 854

DEFINITIONS AND CONCEPTS	**EXAMPLES**

c. Standard forms of the equations of an ellipse centered at (h, k) are $\dfrac{(x - h)^2}{a^2} + \dfrac{(y - k)^2}{b^2} = 1$ and $\dfrac{(x - h)^2}{b^2} + \dfrac{(y - k)^2}{a^2} = 1, a^2 > b^2$. See Table 9.1 on page 855.

Ex. 4, p. 856

9.2 The Hyperbola

a. A hyperbola is the set of all points in a plane the difference of whose distances from two fixed points, the foci, is constant.

b. Standard forms of the equations of a hyperbola with center at the origin are $\dfrac{x^2}{a^2} - \dfrac{y^2}{b^2} = 1$ [foci: $(-c, 0), (c, 0)$] and $\dfrac{y^2}{a^2} - \dfrac{x^2}{b^2} = 1$ [foci: $(0, -c), (0, c)$], where $c^2 = a^2 + b^2$. See the box on page 863 and Figure 9.16.

Ex. 1, p. 864;
Ex. 2, p. 865

c. Asymptotes for $\dfrac{x^2}{a^2} - \dfrac{y^2}{b^2} = 1$ are $y = \pm \dfrac{b}{a}x$. Asymptotes for $\dfrac{y^2}{a^2} - \dfrac{x^2}{b^2} = 1$ are $y = \pm \dfrac{a}{b}x$.

d. A procedure for graphing hyperbolas is given in the box on page 866.

Ex. 3, p. 867;
Ex. 4, p. 868

e. Standard forms of the equations of a hyperbola centered at (h, k) are $\dfrac{(x - h)^2}{a^2} - \dfrac{(y - k)^2}{b^2} = 1$ and $\dfrac{(y - k)^2}{a^2} - \dfrac{(x - h)^2}{b^2} = 1$. See Table 9.2 on page 869.

Ex. 5, p. 869;
Ex. 6, p. 871

9.3 The Parabola

a. A parabola is the set of all points in a plane that are equidistant from a fixed line, the directrix, and a fixed point, the focus.

b. Standard forms of the equations of parabolas with vertex at the origin are $y^2 = 4px$ [focus: $(p, 0)$] and $x^2 = 4py$ [focus: $(0, p)$]. See the box on page 877 and Figure 9.31 on page 878.

Ex. 1, p. 878;
Ex. 3, p. 880

c. A parabola's latus rectum is a line segment that passes through its focus, is parallel to its directrix, and has its endpoints on the parabola. The length of the latus rectum for $y^2 = 4px$ and $x^2 = 4py$ is $|4p|$. A parabola can be graphed using the vertex and endpoints of the latus rectum.

Ex. 2, p. 879

d. Standard forms of the equations of a parabola with vertex at (h, k) are $(y - k)^2 = 4p(x - h)$ and $(x - h)^2 = 4p(y - k)$. See Table 9.3 on page 881 and Figure 9.36.

Ex. 4, p. 881;
Ex. 5, p. 882

9.4 Rotation of Axes

a. A nondegenerate conic section with equation of the form $Ax^2 + Cy^2 + Dx + Ey + F = 0$ in which A and C are not both zero is **1.** a circle if $A = C$; **2.** a parabola if $AC = 0$; **3.** an ellipse if $A \neq C$ and $AC > 0$; **4.** a hyperbola if $AC < 0$.

Ex. 1, p. 889

b. Rotation of Axes Formulas
θ is the angle from the positive x-axis to the positive x'-axis.
$$x = x' \cos \theta - y' \sin \theta \quad \text{and} \quad y = x' \sin \theta + y' \cos \theta$$

Ex. 2, p. 892

c. Amount of Rotation Formula
The general second-degree equation
$$Ax^2 + Bxy + Cy^2 + Dx + Ey + F = 0$$
can be rewritten in x' and y' without an $x'y'$-term by rotating the axes through angle θ, where
$\cot 2\theta = \dfrac{A - C}{B}$ and θ is an acute angle.

d. If 2θ in $\cot 2\theta$ is one of the familiar angles such as $30°$, $45°$, or $60°$, write the equation of a rotated conic in standard form using the five-step procedure in the box on page 894.

Ex. 3, p. 894

DEFINITIONS AND CONCEPTS   **EXAMPLES**

e. If $\cot 2\theta$ is not the cotangent of one of the more familiar angles, use a sketch of $\cot 2\theta$ to find $\cos 2\theta$. Then use   Ex. 4, p. 896

$$\sin \theta = \sqrt{\frac{1 - \cos 2\theta}{2}} \quad \text{and} \quad \cos \theta = \sqrt{\frac{1 + \cos 2\theta}{2}}$$

to find values for $\sin \theta$ and $\cos \theta$ in the rotation formulas.

f. A nondegenerate conic section of the form   Ex. 5, p. 899

$$Ax^2 + Bxy + Cy^2 + Dx + Ey + F = 0$$

is **1.** a parabola if $B^2 - 4AC = 0$; **2.** an ellipse or a circle if $B^2 - 4AC < 0$; **3.** a hyperbola if $B^2 - 4AC > 0$.

9.5 *Parametric Equations*

a. The relationship between the parametric equations $x = f(t)$ and $y = g(t)$ and plane curves is described in the first box on page 902.

b. Point plotting can be used to graph a plane curve described by parametric equations. See the second box on   Ex. 1, p. 902
page 902.

c. Plane curves can be sketched by eliminating the parameter t and graphing the resulting rectangular equa-   Ex. 2, p. 903;
tion. It is sometimes necessary to change the domain of the rectangular equation to be consistent with the   Ex. 3, p. 904
domain for the parametric equation in x.

d. Infinitely many pairs of parametric equations can represent the same plane curve. One pair for $y = f(x)$ is   Ex. 4, p. 906
$x = t$ and $y = f(t)$, in which t is in the domain of f.

9.6 *Conic Sections in Polar Coordinates*

a. The focus-directrix definitions of the conic sections are given in the box on page 910. For all points on a
conic, the ratio of the distance from a fixed point (focus) and the distance from a fixed line (directrix) is
constant and is called its eccentricity. If $e = 1$, the conic is a parabola. If $e < 1$, the conic is an ellipse. If
$e > 1$, the conic is a hyperbola.

b. Standard forms of the polar equations of conics are   Ex. 1, p. 912;
Ex. 2, p. 914;
$$r = \frac{ep}{1 \pm e \cos \theta} \quad \text{and} \quad r = \frac{ep}{1 \pm e \sin \theta},$$ Ex. 3, p. 915

in which (r, θ) is a point on the conic's graph, e is the eccentricity, and p is the distance between the focus
(located at the pole) and the directrix. Details are shown in the box on page 911.

c. A procedure for graphing the polar equation of a conic is given in the box on page 912.

Review Exercises

9.1

In Exercises 1–8, graph each ellipse and locate the foci.

1. $\dfrac{x^2}{36} + \dfrac{y^2}{25} = 1$ **2.** $\dfrac{y^2}{25} + \dfrac{x^2}{16} = 1$

3. $4x^2 + y^2 = 16$ **4.** $4x^2 + 9y^2 = 36$

5. $\dfrac{(x - 1)^2}{16} + \dfrac{(y + 2)^2}{9} = 1$ **6.** $\dfrac{(x + 1)^2}{9} + \dfrac{(y - 2)^2}{16} = 1$

7. $4x^2 + 9y^2 + 24x - 36y + 36 = 0$

8. $9x^2 + 4y^2 - 18x + 8y - 23 = 0$

In Exercises 9–11, find the standard form of the equation of each ellipse satisfying the given conditions.

9. Foci: $(-4, 0)$, $(4, 0)$; Vertices: $(-5, 0)$, $(5, 0)$

10. Foci: $(0, -3)$, $(0, 3)$; Vertices: $(0, -6)$, $(0, 6)$

11. Major axis horizontal with length 12; length of minor axis $= 4$; center: $(-3, 5)$.

12. A semielliptical arch supports a bridge that spans a river 20 yards wide. The center of the arch is 6 yards above the river's center. Write an equation for the ellipse so that the x-axis coincides with the water level and the y-axis passes through the center of the arch.

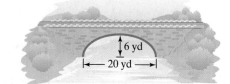

3. A semielliptic archway has a height of 15 feet at the center and a width of 50 feet, as shown in the figure. The 50-foot width consists of a two-lane road. Can a truck that is 12 feet high and 14 feet wide drive under the archway without going into the other lane?

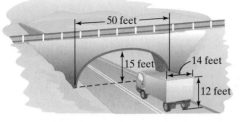

4. An elliptical pool table has a ball placed at each focus. If one ball is hit toward the side of the table, explain what will occur.

9.2

In Exercises 15–22, graph each hyperbola. Locate the foci and find the equations of the asymptotes.

15. $\dfrac{x^2}{16} - y^2 = 1$

16. $\dfrac{y^2}{16} - x^2 = 1$

17. $9x^2 - 16y^2 = 144$

18. $4y^2 - x^2 = 16$

19. $\dfrac{(x-2)^2}{25} - \dfrac{(y+3)^2}{16} = 1$

20. $\dfrac{(y+2)^2}{25} - \dfrac{(x-3)^2}{16} = 1$

21. $y^2 - 4y - 4x^2 + 8x - 4 = 0$

22. $x^2 - y^2 - 2x - 2y - 1 = 0$

In Exercises 23–24, find the standard form of the equation of each hyperbola satisfying the given conditions.

23. Foci: $(0, -4), (0, 4)$; Vertices: $(0, -2), (0, 2)$

24. Foci: $(-8, 0), (8, 0)$; Vertices: $(-3, 0), (3, 0)$

25. Explain why it is not possible for a hyperbola to have foci at $(0, -2)$ and $(0, 2)$ and vertices at $(0, -3)$ and $(0, 3)$.

26. Radio tower M_2 is located 200 miles due west of radio tower M_1. The situation is illustrated in the figure shown, where a coordinate system has been superimposed. Simultaneous radio signals are sent from each tower to a ship, with the signal from M_2 received 500 microseconds before the signal from M_1. Assuming that radio signals travel at 0.186 mile per microsecond, determine the equation of the hyperbola on which the ship is located.

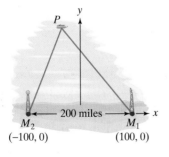

9.3

In Exercises 27–33, find the vertex, focus, and directrix of each parabola with the given equation. Then graph the parabola.

27. $y^2 = 8x$

28. $x^2 + 16y = 0$

29. $(y-2)^2 = -16x$

30. $(x-4)^2 = 4(y+1)$

31. $x^2 + 4y = 4$

32. $y^2 - 4x - 10y + 21 = 0$

33. $x^2 - 4x - 2y = 0$

In Exercises 34–35, find the standard form of the equation of each parabola satisfying the given conditions.

34. Focus: $(12, 0)$; Directrix: $x = -12$

35. Focus: $(0, -11)$; Directrix: $y = 11$

36. An engineer is designing headlight units for automobiles. The unit has a parabolic surface with a diameter of 12 inches and a depth of 3 inches. The situation is illustrated in the figure, where a coordinate system has been superimposed. What is the equation of the parabola in this system? Where should the light source be placed? Describe this placement relative to the vertex.

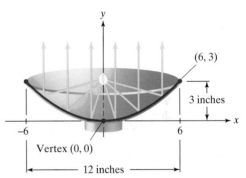

37. The George Washington Bridge spans the Hudson River from New York to New Jersey. Its two towers are 3500 feet apart and rise 316 feet above the road. As shown in the figure, the cable between the towers has the shape of a parabola and the cable just touches the sides of the road midway between the towers. What is the height of the cable 1000 feet from a tower?

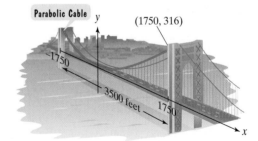

38. The giant satellite dish in the figure shown is in the shape of a parabolic surface. Signals strike the surface and are reflected to the focus, where the receiver is located. The diameter of the dish is 300 feet and its depth is 44 feet. How far, to the nearest foot, from the base of the dish should the receiver be placed?

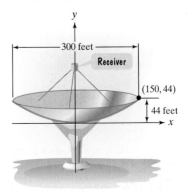

9.4

In Exercises 39–46, identify the conic represented by each equation without completing the square or using a rotation of axes.

39. $y^2 + 4x + 2y - 15 = 0$

40. $x^2 + 16y^2 - 160y + 384 = 0$

41. $16x^2 + 64x + 9y^2 - 54y + 1 = 0$

42. $4x^2 - 9y^2 - 8x + 12y - 144 = 0$

43. $5x^2 + 2\sqrt{3}xy + 3y^2 - 18 = 0$

44. $5x^2 - 8xy + 7y^2 - 9\sqrt{5}x - 9 = 0$

45. $x^2 + 6xy + 9y^2 - 2y = 0$

46. $x^2 - 2xy + 3y^2 + 2x + 4y - 1 = 0$

In Exercises 47–51,

a. *Rewrite the equation in a rotated $x'y'$-system without an $x'y'$-term.*

b. *Express the equation involving x' and y' in the standard form of a conic section.*

c. *Use the rotated system to graph the equation.*

47. $xy - 4 = 0$

48. $x^2 + xy + y^2 - 1 = 0$

49. $4x^2 + 10xy + 4y^2 - 9 = 0$

50. $6x^2 - 6xy + 14y^2 - 45 = 0$

51. $x^2 + 2\sqrt{3}xy + 3y^2 - 12\sqrt{3}x + 12y = 0$

9.5

In Exercises 52–57, eliminate the parameter and graph the plane curve represented by the parametric equations. Use arrows to show the orientation of each plane curve.

52. $x = 2t - 1, y = 1 - t; -\infty < t < \infty$

53. $x = t^2, y = t - 1; -1 \le t \le 3$

54. $x = 4t^2, y = t + 1; -\infty < t < \infty$

55. $x = 4 \sin t, y = 3 \cos t; 0 \le t < \pi$

56. $x = 3 + 2 \cos t, y = 1 + 2 \sin t; 0 \le t < 2\pi$

57. $x = 3 \sec t, y = 3 \tan t; 0 \le t \le \dfrac{\pi}{4}$

58. Find two different sets of parametric equations for $y = x^2 + 6$.

59. The path of a projectile that is launched h feet above the ground with an initial velocity of v_0 feet per second and at an angle θ with the horizontal is given by the parametric equations

$$x = (v_0 \cos \theta)t \quad \text{and} \quad y = h + (v_0 \sin \theta)t - 16t^2,$$

where t is the time, in seconds, after the projectile was launched. A football player throws a football with an initial velocity of 100 feet per second at an angle of 40° to the horizontal. The ball leaves the player's hand at a height of 6 feet.

a. Find the parametric equations that describe the position of the ball as a function of time.

b. Describe the ball's position after 1, 2, and 3 seconds. Round to the nearest tenth of a foot.

c. How long, to the nearest tenth of a second, is the ball in flight? What is the total horizontal distance that it travels before it lands?

d. Graph the parametric equations in part (a) using a graphing utility. Use the graph to determine when the ball is at its maximum height. What is its maximum height? Round all answers to the nearest tenth.

9.6

In Exercises 60–65,

a. *If necessary, write the equation in one of the standard forms for a conic in polar coordinates.*

b. *Determine values for e and p. Use the value of e to identify the conic section.*

c. *Graph the given polar equation.*

60. $r = \dfrac{4}{1 - \sin \theta}$

61. $r = \dfrac{6}{1 + \cos \theta}$

62. $r = \dfrac{6}{2 + \sin \theta}$

63. $r = \dfrac{2}{3 - 2 \cos \theta}$

64. $r = \dfrac{6}{3 + 6 \sin \theta}$

65. $r = \dfrac{8}{4 + 16 \cos \theta}$

Chapter 9 Test

In Exercises 1–5, graph the conic section with the given equation. For ellipses, find the foci. For hyperbolas, find the foci and give the equations of the asymptotes. For parabolas, find the vertex, focus, and directrix.

1. $9x^2 - 4y^2 = 36$ **2.** $x^2 = -8y$

3. $\dfrac{(x + 2)^2}{25} + \dfrac{(y - 5)^2}{9} = 1$

4. $4x^2 - y^2 + 8x + 2y + 7 = 0$

5. $(x + 5)^2 = 8(y - 1)$

In Exercises 6–8, find the standard form of the equation of the conic section satisfying the given conditions.

6. Ellipse; Foci: $(-7, 0), (7, 0)$; Vertices: $(-10, 0), (10, 0)$

7. Hyperbola; Foci: $(0, -10), (0, 10)$; Vertices: $(0, -7), (0, 7)$

8. Parabola; Focus: $(50, 0)$; Directrix: $x = -50$

9. A sound whispered at one focus of a whispering gallery can be heard at the other focus. The figure at the top of the next page shows a whispering gallery whose cross section is a semielliptical arch with a height of 24 feet and a width of 80 feet. How far from the room's center should two people stand so that they can whisper back and forth and be heard?

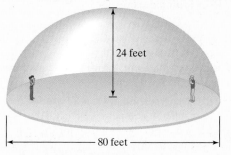

0. An engineer is designing headlight units for cars. The unit shown in the figure below has a parabolic surface with a diameter of 6 inches and a depth of 3 inches.

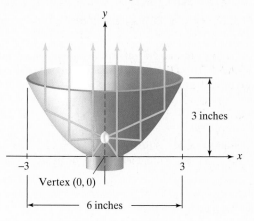

a. Using the coordinate system that has been positioned on the unit, find the parabola's equation.

b. If the light source is located at the focus, describe its placement relative to the vertex.

In Exercises 11–12, identify each equation without completing the square or using a rotation of axes.

11. $x^2 + 9y^2 + 10x - 18y + 25 = 0$

12. $x^2 + y^2 + xy + 3x - y - 3 = 0$

13. For the equation

$$7x^2 - 6\sqrt{3}xy + 13y^2 - 16 = 0,$$

determine what angle of rotation would eliminate the $x'y'$-term in a rotated $x'y'$-system.

In Exercises 14–15, eliminate the parameter and graph the plane curve represented by the parametric equations. Use arrows to show the orientation of each plane curve.

14. $x = t^2, y = t - 1; -\infty < t < \infty$

15. $x = 1 + 3 \sin t, y = 2 \cos t; 0 \le t < 2\pi$

In Exercises 16–17, identify the conic section and graph the polar equation.

16. $r = \dfrac{2}{1 - \cos \theta}$ **17.** $r = \dfrac{4}{2 + \sin \theta}$

Cumulative Review Exercises (Chapters P–9)

Solve each equation or inequality in Exercises 1–7.

1. $2(x - 3) + 5x = 8(x - 1)$

2. $-3(2x - 4) > 2(6x - 12)$

3. $x - 5 = \sqrt{x + 7}$ **4.** $(x - 2)^2 = 20$

5. $|2x - 1| \ge 7$ **6.** $3x^3 + 4x^2 - 7x + 2 = 0$

7. $\log_2(x + 1) + \log_2(x - 1) = 3$

Solve each system in Exercises 8–10.

8. $3x + 4y = 2$ **9.** $2x^2 - y^2 = -8$
   $2x + 5y = -1$ $x - y = 6$

0. (Use matrices.)

$$\begin{aligned} x - y + z &= 17 \\ -4x + y + 5z &= -2 \\ 2x + 3y + z &= 8 \end{aligned}$$

In Exercises 11–13, graph each equation, function, or system in a rectangular coordinate system.

1. $f(x) = (x - 1)^2 - 4$ **12.** $\dfrac{x^2}{9} + \dfrac{y^2}{4} = 1$

13. $5x + y \le 10$

$$y \ge \frac{1}{4}x + 2$$

14. a. List all possible rational roots of

$$32x^3 - 52x^2 + 17x + 3 = 0.$$

b. The graph of $f(x) = 32x^3 - 52x^2 + 17x + 3$ is shown in a $[-1, 3, 1]$ by $[-2, 6, 1]$ viewing rectangle. Use the graph of f and synthetic division to solve the equation in part (a).

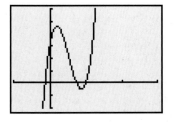

15. The figure shows the graph of $y = f(x)$ and its two vertical asymptotes.

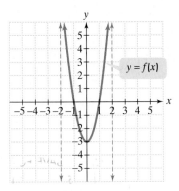

a. Find the domain and the range of f.

b. What is the relative minimum and where does it occur?

c. Find the interval on which f is increasing.

d. Find $f(-1) - f(0)$.

e. Find $(f \circ f)(1)$.

f. Use arrow notation to complete this statement:

$$f(x) \to \infty \text{ as } \underline{\hspace{1cm}} \text{ or as } \underline{\hspace{1cm}}.$$

g. Graph $g(x) = f(x - 2) + 1$.

h. Graph $h(x) = -f(2x)$.

16. If $f(x) = x^2 - 4$ and $g(x) = x + 2$, find $(g \circ f)(x)$.

17. Expand using logarithmic properties. Where possible evaluate logarithmic expressions.

$$\log_5\left(\frac{x^3\sqrt{y}}{125}\right)$$

18. Write the slope-intercept form of the equation of the line passing through $(1, -4)$ and $(-5, 8)$.

19. Rent-a-Truck charges a daily rental rate for a truck of $3 plus $0.16 a mile. A competing agency, Ace Truck Rentals charges $25 a day plus $0.24 a mile for the same truck. How many miles must be driven in a day to make the daily cost of both agencies the same? What will be the cost?

20. The local cable television company offers two deals. Basic cable service with one movie channel costs $35 per month. Basic service with two movie channels cost $45 per month. Find the charge for the basic cable service and the charge for each movie channel.

21. Verify the identity: $\dfrac{\csc \theta - \sin \theta}{\sin \theta} = \cot^2 \theta$.

22. Graph one complete cycle of $y = 2 \cos(2x + \pi)$.

23. If $\mathbf{v} = 3\mathbf{i} - 6\mathbf{j}$ and $\mathbf{w} = \mathbf{i} + \mathbf{j}$, find $(\mathbf{v} \cdot \mathbf{w})\mathbf{w}$.

24. Solve for θ: $\sin 2\theta = \sin \theta, 0 \leq \theta < 2\pi$.

25. In oblique triangle ABC, $A = 64°$, $B = 72°$, and $a = 13.6$. Solve the triangle. Round lengths to the nearest tenth.

Sequences, Induction, and Probability

WE OFTEN SAVE FOR THE FUTURE *by investing small amounts at periodic intervals. To understand how our savings accumulate, we need to understand properties of lists of numbers that are related to each other by a rule. Such lists are called sequences. Learning about properties of sequences will show you how to make your financial goals a reality. Your knowledge of sequences will enable you to inform your college roommate of the best of the three appealing offers described below.*

SOMETHING INCREDIBLE HAS HAPPENED. Your college roommate, a gifted athlete, has been given a six-year contract with a professional baseball team. He will be playing against the likes of Barry Bonds and Manny Ramirez. Management offers him three options. One is a beginning salary of $1,700,000 with annual increases of $70,000 per year starting in the second year. A second option is $1,700,000 the first year with an annual increase of 2% per year beginning in the second year. The third offer involves less money the first year— $1,500,000—but there is an annual increase of 9% yearly after that. Which option offers the most money over the six-year contract?

This problem appears as Exercise 67 in Exercise Set 10.3 and as the group project on page 958.

Objectives

❶ Find particular terms of a sequence from the general term.

❷ Use recursion formulas.

❸ Use factorial notation.

❹ Use summation notation.

Fibonacci Numbers on the Piano Keyboard

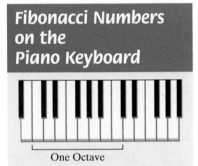

One Octave

Numbers in the Fibonacci sequence can be found in an octave on the piano keyboard. The octave contains 2 black keys in one cluster and 3 black keys in another cluster, for a total of 5 black keys. It also has 8 white keys, for a total of 13 keys. The numbers 2, 3, 5, 8, and 13 are the third through seventh terms of the Fibonacci sequence.

Sequences

Many creations in nature involve intricate mathematical designs, including a variety of spirals. For example, the arrangement of the individual florets in the head of a sunflower forms spirals. In some species, there are 21 spirals in the clockwise direction and 34 in the counterclockwise direction. The precise numbers depend on the species of sunflower: 21 and 34, or 34 and 55, or 55 and 89, or even 89 and 144.

This observation becomes even more interesting when we consider a sequence of numbers investigated by Leonardo of Pisa, also known as Fibonacci, an Italian mathematician of the thirteenth century. The **Fibonacci sequence** of numbers is an infinite sequence that begins as follows:

$$1, 1, 2, 3, 5, 8, 13, 21, 34, 55, 89, 144, 233, \dots.$$

The first two terms are 1. Every term thereafter is the sum of the two preceding terms. For example, the third term, 2, is the sum of the first and second terms: $1 + 1 = 2$. The fourth term, 3, is the sum of the second and third terms: $1 + 2 = 3$, and so on. Did you know that the number of spirals in a daisy or a sunflower, 21 and 34, are two Fibonacci numbers? The number of spirals in a pine cone, 8 and 13, and pineapple, 8 and 13, are also Fibonacci numbers.

We can think of the Fibonacci sequence as a function. The terms of the sequence

$$1, 1, 2, 3, 5, 8, 13, 21, 34, 55, 89, 144, 233, \dots$$

are the range values for a function whose domain is the set of positive integers.

Domain:	1,	2,	3,	4,	5,	6,	7,	...
	↓	↓	↓	↓	↓	↓	↓	
Range:	1,	1,	2,	3,	5,	8,	13,	...

Thus, $f(1) = 1, f(2) = 1, f(3) = 2, f(4) = 3, f(5) = 5, f(6) = 8, f(7) = 13$, and so on.

The letter a with a subscript is used to represent function values of a sequence rather than the usual function notation. The subscripts make up the domain of the sequence and they identify the location of a term. Thus, a_1 represents the first term of the sequence, a_2 represents the second term, a_3 the third term, and so on. This notation is shown for the first six terms of the Fibonacci sequence:

$$1, \quad 1, \quad 2, \quad 3, \quad 5, \quad 8.$$

The notation a_n represents the *n*th term, or **general term**, of a sequence. The entire sequence is represented by $\{a_n\}$.

Definition of a Sequence

An **infinite sequence** $\{a_n\}$ is a function whose domain is the set of positive integers. The function values, or **terms**, of the sequence are represented by

$$a_1, a_2, a_3, a_4, \ldots, a_n, \ldots.$$

Sequences whose domains consist only of the first *n* positive integers are called **finite sequences**.

---

① Find particular terms of a sequence from the general term.

EXAMPLE 1   Writing Terms of a Sequence from the General Term

Write the first four terms of the sequence whose *n*th term, or general term, is given:

a. $a_n = 3n + 4$ **b.** $a_n = \dfrac{(-1)^n}{3^n - 1}$.

Solution

a. We need to find the first four terms of the sequence whose general term is $a_n = 3n + 4$. To do so, we replace *n* in the formula with 1, 2, 3, and 4.

| a_1, 1st term | $3 \cdot 1 + 4 = 3 + 4 = 7$ | a_2, 2nd term | $3 \cdot 2 + 4 = 6 + 4 = 10$ |
| a_3, 3rd term | $3 \cdot 3 + 4 = 9 + 4 = 13$ | a_4, 4th term | $3 \cdot 4 + 4 = 12 + 4 = 16$ |

The first four terms are 7, 10, 13, and 16. The sequence defined by $a_n = 3n + 4$ can be written as

$$7, 10, 13, 16, \ldots, 3n + 4, \ldots.$$

b. We need to find the first four terms of the sequence whose general term is $a_n = \dfrac{(-1)^n}{3^n - 1}$. To do so, we replace each occurrence of *n* in the formula with 1, 2, 3, and 4.

| a_1, 1st term | $\dfrac{(-1)^1}{3^1 - 1} = \dfrac{-1}{3 - 1} = -\dfrac{1}{2}$ | a_2, 2nd term | $\dfrac{(-1)^2}{3^2 - 1} = \dfrac{1}{9 - 1} = \dfrac{1}{8}$ |
| a_3, 3rd term | $\dfrac{(-1)^3}{3^3 - 1} = \dfrac{-1}{27 - 1} = -\dfrac{1}{26}$ | a_4, 4th term | $\dfrac{(-1)^4}{3^4 - 1} = \dfrac{1}{81 - 1} = \dfrac{1}{80}$ |

The first four terms are $-\dfrac{1}{2}, \dfrac{1}{8}, -\dfrac{1}{26}$, and $\dfrac{1}{80}$. The sequence defined by $\dfrac{(-1)^n}{3^n - 1}$ can be written as

$$-\frac{1}{2}, \frac{1}{8}, -\frac{1}{26}, \frac{1}{80}, \ldots, \frac{(-1)^n}{3^n - 1}, \ldots.$$

Check Point 1   Write the first four terms of the sequence whose *n*th term, or general term, is given:

a. $a_n = 2n + 5$ **b.** $a_n = \dfrac{(-1)^n}{2^n + 1}$.

Although sequences are usually named with the letter *a*, any lowercase letter can be used. For example, the first four terms of the sequence $\{b_n\} = \left\{\left(\frac{1}{2}\right)^n\right\}$ are $b_1 = \frac{1}{2}, b_2 = \frac{1}{4}, b_3 = \frac{1}{8}$, and $b_4 = \frac{1}{16}$.

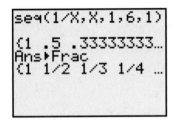
Because a sequence is a function whose domain is the set of positive integers, the **graph of a sequence** is a set of discrete points. For example, consider the sequence whose general term is $a_n = \frac{1}{n}$. How does the graph of this sequence differ from the graph of the function $f(x) = \frac{1}{x}$? The graph of $f(x) = \frac{1}{x}$ is shown in Figure 10.1(a) for positive values of x. To obtain the graph of the sequence $\{a_n\} = \left\{\frac{1}{n}\right\}$, remove all the points from the graph of f except those whose x-coordinates are positive integers. Thus, we remove all points except $(1, 1), \left(2, \frac{1}{2}\right), \left(3, \frac{1}{3}\right), \left(4, \frac{1}{4}\right)$, and so on. The remaining points are the graph of the sequence $\{a_n\} = \left\{\frac{1}{n}\right\}$, shown in Figure 10.1(b). Notice that the horizontal axis is labeled n and the vertical axis is labeled a_n.

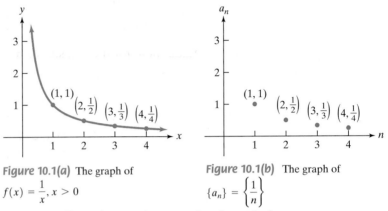

Figure 10.1(a) The graph of $f(x) = \frac{1}{x}, x > 0$

Figure 10.1(b) The graph of $\{a_n\} = \left\{\frac{1}{n}\right\}$

Comparing a continuous graph to the graph of a sequence

Recursion Formulas

In Example 1, the formulas used for the nth term of a sequence expressed the term as a function of n, the number of the term. Sequences can also be defined using **recursion formulas**. A recursion formula defines the nth term of a sequence as a function of the previous term. Our next example illustrates that if the first term of a sequence is known, then the recursion formula can be used to determine the remaining terms.

② Use recursion formulas.

EXAMPLE 2   Using a Recursion Formula

Find the first four terms of the sequence in which $a_1 = 5$ and $a_n = 3a_{n-1} + 2$ for $n \geq 2$.

Solution   Let's be sure we understand what is given.

$$a_1 = 5 \quad \text{and} \quad a_n = 3a_{n-1} + 2$$

The first term is 5.   Each term after the first   is   3 times the previous term   plus 2.

Now let's write the first four terms of this sequence.

$a_1 = 5$	This is the given first term.
$a_2 = 3a_1 + 2$	Use $a_n = 3a_{n-1} + 2$, with $n = 2$. Thus, $a_2 = 3a_{2-1} + 2 = 3a_1 + 2$.
$\quad = 3(5) + 2 = 17$	Substitute 5 for a_1.
$a_3 = 3a_2 + 2$	Again use $a_n = 3a_{n-1} + 2$, with $n = 3$.
$\quad = 3(17) + 2 = 53$	Substitute 17 for a_2.
$a_4 = 3a_3 + 2$	Notice that a_4 is defined in terms of a_3. We used $a_n = 3a_{n-1} + 2$, with $n = 4$.
$\quad = 3(53) + 2 = 161$	Use the value of a_3, the third term, obtained above.

The first four terms are $5, 17, 53$, and 161.

Check Point 2   Find the first four terms of the sequence in which $a_1 = 3$ and $a_n = 2a_{n-1} + 5$ for $n \geq 2$.

③ Use factorial notation.

Factorial Notation

Products of consecutive positive integers occur quite often in sequences. These products can be expressed in a special notation, called **factorial notation**.

Factorials from 0 through 20

0!	1
1!	1
2!	2
3!	6
4!	24
5!	120
6!	720
7!	5040
8!	40,320
9!	362,880
10!	3,628,800
11!	39,916,800
12!	479,001,600
13!	6,227,020,800
14!	87,178,291,200
15!	1,307,674,368,000
16!	20,922,789,888,000
17!	355,687,428,096,000
18!	6,402,373,705,728,000
19!	121,645,100,408,832,000
20!	2,432,902,008,176,640,000

As n increases, $n!$ grows very rapidly. Factorial growth is more explosive than exponential growth discussed in Chapter 3.

> ### Factorial Notation
>
> If n is a positive integer, the notation $n!$ (read "n factorial") is the product of all positive integers from n down through 1.
> $$n! = n(n-1)(n-2)\cdots(3)(2)(1)$$
> 0! (zero factorial), by definition, is 1.
> $$0! = 1$$

The values of $n!$ for the first six positive integers are

$$1! = 1$$
$$2! = 2 \cdot 1 = 2$$
$$3! = 3 \cdot 2 \cdot 1 = 6$$
$$4! = 4 \cdot 3 \cdot 2 \cdot 1 = 24$$
$$5! = 5 \cdot 4 \cdot 3 \cdot 2 \cdot 1 = 120$$
$$6! = 6 \cdot 5 \cdot 4 \cdot 3 \cdot 2 \cdot 1 = 720.$$

Factorials affect only the number or variable that they follow unless grouping symbols appear. For example,

$$2 \cdot 3! = 2(3 \cdot 2 \cdot 1) = 2 \cdot 6 = 12$$

whereas

$$(2 \cdot 3)! = 6! = 6 \cdot 5 \cdot 4 \cdot 3 \cdot 2 \cdot 1 = 720.$$

In this sense, factorials are similar to exponents.

Technology

Most calculators have factorial keys. To find 5!, most calculators use one of the following:

Many Scientific Calculators

$5 \boxed{x!}$

Many Graphing Calculators

$5 \boxed{!}$ $\boxed{\text{ENTER}}$.

Because $n!$ becomes quite large as n increases, your calculator will display these larger values in scientific notation.

EXAMPLE 3   Finding Terms of a Sequence Involving Factorials

Write the first four terms of the sequence whose nth term is

$$a_n = \frac{2^n}{(n-1)!}.$$

Solution   We need to find the first four terms of the sequence. To do so, we replace each n in $\dfrac{2^n}{(n-1)!}$ with $1, 2, 3,$ and 4.

a_1, 1st term
$$\frac{2^1}{(1-1)!} = \frac{2}{0!} = \frac{2}{1} = 2$$

a_2, 2nd term
$$\frac{2^2}{(2-1)!} = \frac{4}{1!} = \frac{4}{1} = 4$$

a_3, 3rd term
$$\frac{2^3}{(3-1)!} = \frac{8}{2!} = \frac{8}{2 \cdot 1} = 4$$

a_4, 4th term
$$\frac{2^4}{(4-1)!} = \frac{16}{3!} = \frac{16}{3 \cdot 2 \cdot 1} = \frac{16}{6} = \frac{8}{3}$$

The first four terms are $2, 4, 4,$ and $\frac{8}{3}$.

 Check Point **3**   Write the first four terms of the sequence whose nth term is

$$a_n = \frac{20}{(n+1)!}.$$

When evaluating fractions with factorials in the numerator and the denominator, try to reduce the fraction before performing the multiplications. For example, consider $\dfrac{26!}{21!}$. Rather than write out 26! as the product of all integers from 26 down to 1, we can express 26! as

$$26! = 26 \cdot 25 \cdot 24 \cdot 23 \cdot 22 \cdot 21!.$$

In this way, we can divide both the numerator and the denominator by the common factor, 21!.

$$\frac{26!}{21!} = \frac{26 \cdot 25 \cdot 24 \cdot 23 \cdot 22 \cdot \cancel{21!}}{\cancel{21!}} = 26 \cdot 25 \cdot 24 \cdot 23 \cdot 22 = 7,893,600$$

EXAMPLE 4   Evaluating Fractions with Factorials

Evaluate each factorial expression:

a. $\dfrac{10!}{2!8!}$ **b.** $\dfrac{(n+1)!}{n!}$.

Solution

a. $\dfrac{10!}{2!8!} = \dfrac{10 \cdot 9 \cdot \cancel{8!}}{2 \cdot 1 \cdot \cancel{8!}} = \dfrac{90}{2} = 45$

b. $\dfrac{(n+1)!}{n!} = \dfrac{(n+1) \cdot \cancel{n!}}{\cancel{n!}} = n + 1$

Check Point 4 Evaluate each factorial expression:

a. $\dfrac{14!}{2!12!}$ **b.** $\dfrac{n!}{(n-1)!}$.

4️⃣ Use summation notation.

Summation Notation

It is sometimes useful to find the sum of the first n terms of a sequence. For example, consider the cost of raising a child born in the United States in 2002 to a middle-income ($39,700–$66,900 per year) family, shown in Table 10.1.

Table 10.1   The Cost of Raising a Child Born in the U.S. in 2002 to a Middle-Income Family

Year	2002	2003	2004	2005	2006	2007	2008	2009	2010
Average Cost	$9230	$9530	$9830	$10,420	$10,750	$11,100	$11,440	$11,810	$12,180
	Child is under 1.	Child is 1.	Child is 2.	Child is 3.	Child is 4.	Child is 5.	Child is 6.	Child is 7.	Child is 8.

Year	2011	2012	2013	2014	2015	2016	2017	2018	2019
Average Cost	$12,440	$12,840	$13,250	$14,750	$15,230	$15,710	$16,520	$17,050	$17,600
	Child is 9.	Child is 10.	Child is 11.	Child is 12.	Child is 13.	Child is 14.	Child is 15.	Child is 16.	Child is 17.

Source: U.S. Department of Agriculture

We can let a_n represent the cost of raising a child in year n, where $n = 1$ corresponds to 2002, $n = 2$ to 2003, $n = 3$ to 2004, and so on. The terms of the finite sequence in Table 10.1 are given as follows:

9230,   9530,   9830,   10,420,   10,750,   11,100,   11,440,   11,810,   12,180,

a_1 a_2 a_3 a_4 a_5 a_6 a_7 a_8 a_9

12,440,   12,840,   13,250,   14,750,   15,230,   15,710,   16,520,   17,050,   17,600.

a_{10} a_{11} a_{12} a_{13} a_{14} a_{15} a_{16} a_{17} a_{18}

Why might we want to add the terms of this sequence? We do this to find the total cost of raising a child born in 2002 from birth through age 17. Thus,

$$a_1 + a_2 + a_3 + a_4 + a_5 + a_6 + a_7 + a_8 + a_9 + a_{10} + a_{11} + a_{12} + a_{13} + a_{14} + a_{15} + a_{16} + a_{17} + a_{18}$$
$$= 9230 + 9530 + 9830 + 10{,}420 + 10{,}750 + 11{,}100 + 11{,}440 + 11{,}810 + 12{,}180$$
$$+ 12{,}440 + 12{,}840 + 13{,}250 + 14{,}750 + 15{,}230 + 15{,}710 + 16{,}520 + 17{,}050 + 17{,}600$$
$$= 231{,}680.$$

We see that the total cost of raising a child born in 2002 from birth through age 17 is $231,680.

There is a compact notation for expressing the sum of the first n terms of a sequence. For example, rather than write

$$a_1 + a_2 + a_3 + a_4 + a_5 + a_6 + a_7 + a_8 + a_9 + a_{10} + a_{11} + a_{12} + a_{13} + a_{14} + a_{15} + a_{16} + a_{17} + a_{18},$$

we can use *summation notation* to express the sum as

$$\sum_{i=1}^{18} a_i.$$

We read this expression as "the sum as i goes from 1 to 18 of a_i." The letter i is called the *index of summation* and is not related to the use of i to represent $\sqrt{-1}$.

You can think of the symbol Σ (the uppercase Greek letter sigma) as an instruction to add up the terms of a sequence.

Summation Notation

The sum of the first n terms of a sequence is represented by the **summation notation**

$$\sum_{i=1}^{n} a_i = a_1 + a_2 + a_3 + a_4 + \cdots + a_n,$$

where i is the **index of summation**, n is the **upper limit of summation**, and 1 is the **lower limit of summation**.

Any letter can be used for the index of summation. The letters i, j, and k are used commonly. Furthermore, the lower limit of summation can be an integer other than 1.

When we write out a sum that is given in summation notation, we are **expanding the summation notation**. Example 5 shows how to do this.

Technology

Graphing utilities can calculate the sum of a sequence. For example, to find the sum of the sequence in Example 5(a), enter

$\boxed{\text{SUM}}$ $\boxed{\text{SEQ}}$ $(x^2 + 1, x, 1, 6, 1)$.

Then press $\boxed{\text{ENTER}}$; 97 should be displayed. Use this capability to verify Example 5(b).

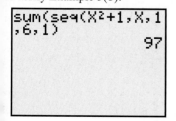

```
sum(seq(X²+1,X,1
,6,1)
               97
```

EXAMPLE 5   Using Summation Notation

Expand and evaluate the sum:

a. $\displaystyle\sum_{i=1}^{6} (i^2 + 1)$ **b.** $\displaystyle\sum_{k=4}^{7} [(-2)^k - 5]$ **c.** $\displaystyle\sum_{i=1}^{5} 3.$

Solution

a. To find $\displaystyle\sum_{i=1}^{6} (i^2 + 1)$, we must replace i in the expression $i^2 + 1$ with all consecutive integers from 1 to 6, inclusive. Then we add.

$$\sum_{i=1}^{6} (i^2 + 1) = (1^2 + 1) + (2^2 + 1) + (3^2 + 1) + (4^2 + 1)$$
$$+ (5^2 + 1) + (6^2 + 1)$$
$$= 2 + 5 + 10 + 17 + 26 + 37$$
$$= 97$$

b. The index of summation in $\displaystyle\sum_{k=4}^{7}[(-2)^k - 5]$ is k. First we evaluate $(-2)^k - \;$ for all consecutive integers from 4 through 7, inclusive. Then we add.

$$\sum_{k=4}^{7}[(-2)^k - 5] = [(-2)^4 - 5] + [(-2)^5 - 5]$$
$$+ [(-2)^6 - 5] + [(-2)^7 - 5]$$
$$= (16 - 5) + (-32 - 5) + (64 - 5) + (-128 - 5)$$
$$= 11 + (-37) + 59 + (-133)$$
$$= -100$$

c. To find $\displaystyle\sum_{i=1}^{5} 3$, we observe that every term of the sum is 3. The notation $i = $ through 5 indicates that we must add the first five terms of a sequence in which every term is 3.

$$\sum_{i=1}^{5} 3 = 3 + 3 + 3 + 3 + 3 = 15$$

Check Point 5 Expand and evaluate the sum:

a. $\displaystyle\sum_{i=1}^{6} 2i^2$ **b.** $\displaystyle\sum_{k=3}^{5} (2^k - 3)$ **c.** $\displaystyle\sum_{i=1}^{5} 4.$

For a given sum, we can vary the upper and lower limits of summation, as well as the letter used for the index of summation. By doing so, we can produce different looking summation notations for the same sum. For example, the sum of the squares of the first four positive integers, $1^2 + 2^2 + 3^2 + 4^2$, can be expressed in a number of equivalent ways:

$$\sum_{i=1}^{4} i^2 = 1^2 + 2^2 + 3^2 + 4^2 = 30$$

$$\sum_{i=0}^{3} (i + 1)^2 = (0 + 1)^2 + (1 + 1)^2 + (2 + 1)^2 + (3 + 1)^2$$
$$= 1^2 + 2^2 + 3^2 + 4^2 = 30$$

$$\sum_{k=2}^{5} (k - 1)^2 = (2 - 1)^2 + (3 - 1)^2 + (4 - 1)^2 + (5 - 1)^2$$
$$= 1^2 + 2^2 + 3^2 + 4^2 = 30.$$

EXAMPLE 6 Writing Sums in Summation Notation

Express each sum using summation notation:

a. $1^3 + 2^3 + 3^3 + \cdots + 7^3$ **b.** $1 + \dfrac{1}{3} + \dfrac{1}{9} + \dfrac{1}{27} + \cdots + \dfrac{1}{3^{n-1}}.$

Solution In each case, we will use 1 as the lower limit of summation and i for the index of summation.

a. The sum $1^3 + 2^3 + 3^3 + \cdots + 7^3$ has seven terms, each of the form i^3, starting at $i = 1$ and ending at $i = 7$. Thus,

$$1^3 + 2^3 + 3^3 + \cdots + 7^3 = \sum_{i=1}^{7} i^3.$$

b. The sum

$$1 + \frac{1}{3} + \frac{1}{9} + \frac{1}{27} + \cdots + \frac{1}{3^{n-1}}$$

has n terms, each of the form $\dfrac{1}{3^{i-1}}$, starting at $i = 1$ and ending at $i = n$. Thus,

$$1 + \frac{1}{3} + \frac{1}{9} + \frac{1}{27} + \cdots + \frac{1}{3^{n-1}} = \sum_{i=1}^{n} \frac{1}{3^{i-1}}.$$

Check Point 6 Express each sum using summation notation:

a. $1^2 + 2^2 + 3^2 + \cdots + 9^2$ **b.** $1 + \dfrac{1}{2} + \dfrac{1}{4} + \dfrac{1}{8} + \cdots + \dfrac{1}{2^{n-1}}.$

Table 10.2 contains some important properties of sums expressed in summation notation.

Table 10.2   Properties of Sums

Property	Example
1. $\displaystyle\sum_{i=1}^{n} ca_i = c \sum_{i=1}^{n} a_i$, c any real number	$\displaystyle\sum_{i=1}^{4} 3i^2 = 3 \cdot 1^2 + 3 \cdot 2^2 + 3 \cdot 3^2 + 3 \cdot 4^2$ $3\displaystyle\sum_{i=1}^{4} i^2 = 3(1^2 + 2^2 + 3^2 + 4^2) = 3 \cdot 1^2 + 3 \cdot 2^2 + 3 \cdot 3^2 + 3 \cdot 4^2$ Conclusion: $\displaystyle\sum_{i=1}^{4} 3i^2 = 3\sum_{i=1}^{4} i^2$
2. $\displaystyle\sum_{i=1}^{n} (a_i + b_i) = \sum_{i=1}^{n} a_i + \sum_{i=1}^{n} b_i$	$\displaystyle\sum_{i=1}^{4} (i + i^2) = (1 + 1^2) + (2 + 2^2) + (3 + 3^2) + (4 + 4^2)$ $\displaystyle\sum_{i=1}^{4} i + \sum_{i=1}^{4} i^2 = (1 + 2 + 3 + 4) + (1^2 + 2^2 + 3^2 + 4^2)$ $\qquad\qquad\qquad\quad = (1 + 1^2) + (2 + 2^2) + (3 + 3^2) + (4 + 4^2)$ Conclusion: $\displaystyle\sum_{i=1}^{4} (i + i^2) = \sum_{i=1}^{4} i + \sum_{i=1}^{4} i^2$
3. $\displaystyle\sum_{i=1}^{n} (a_i - b_i) = \sum_{i=1}^{n} a_i - \sum_{i=1}^{n} b_i$	$\displaystyle\sum_{i=3}^{5} (i^2 - i^3) = (3^2 - 3^3) + (4^2 - 4^3) + (5^2 - 5^3)$ $\displaystyle\sum_{i=3}^{5} i^2 - \sum_{i=3}^{5} i^3 = (3^2 + 4^2 + 5^2) - (3^3 + 4^3 + 5^3)$ $\qquad\qquad\qquad\quad = (3^2 - 3^3) + (4^2 - 4^3) + (5^2 - 5^3)$ Conclusion: $\displaystyle\sum_{i=3}^{5} (i^2 - i^3) = \sum_{i=3}^{5} i^2 - \sum_{i=3}^{5} i^3$

EXERCISE SET 10.1

Practice Exercises

In Exercises 1–12, write the first four terms of each sequence whose general term is given.

1. $a_n = 3n + 2$

2. $a_n = 4n - 1$

3. $a_n = 3^n$

4. $a_n = \left(\dfrac{1}{3}\right)^n$

5. $a_n = (-3)^n$

6. $a_n = \left(-\dfrac{1}{3}\right)^n$

7. $a_n = (-1)^n(n + 3)$

8. $a_n = (-1)^{n+1}(n + 4)$

9. $a_n = \dfrac{2n}{n + 4}$

10. $a_n = \dfrac{3n}{n + 5}$

11. $a_n = \dfrac{(-1)^{n+1}}{2^n - 1}$

12. $a_n = \dfrac{(-1)^{n+1}}{2^n + 1}$

The sequences in Exercises 13–18 are defined using recursion formulas. Write the first four terms of each sequence.

13. $a_1 = 7$ and $a_n = a_{n-1} + 5$ for $n \geq 2$

14. $a_1 = 12$ and $a_n = a_{n-1} + 4$ for $n \geq 2$

15. $a_1 = 3$ and $a_n = 4a_{n-1}$ for $n \geq 2$

16. $a_1 = 2$ and $a_n = 5a_{n-1}$ for $n \geq 2$

17. $a_1 = 4$ and $a_n = 2a_{n-1} + 3$ for $n \geq 2$

18. $a_1 = 5$ and $a_n = 3a_{n-1} - 1$ for $n \geq 2$

In Exercises 19–22, the general term of a sequence is given and involves a factorial. Write the first four terms of each sequence.

19. $a_n = \dfrac{n^2}{n!}$

20. $a_n = \dfrac{(n + 1)!}{n^2}$

21. $a_n = 2(n + 1)!$

22. $a_n = -2(n - 1)!$

In Exercises 23–28, evaluate each factorial expression.

23. $\dfrac{17!}{15!}$

24. $\dfrac{18!}{16!}$

25. $\dfrac{16!}{2!14!}$

26. $\dfrac{20!}{2!18!}$

27. $\dfrac{(n + 2)!}{n!}$

28. $\dfrac{(2n + 1)!}{(2n)!}$

In Exercises 29–42, find each indicated sum.

29. $\sum\limits_{i=1}^{6} 5i$

30. $\sum\limits_{i=1}^{6} 7i$

31. $\sum\limits_{i=1}^{4} 2i^2$

32. $\sum\limits_{i=1}^{5} i^3$

33. $\sum\limits_{k=1}^{5} k(k + 4)$

34. $\sum\limits_{k=1}^{4} (k - 3)(k + 2)$

35. $\sum\limits_{i=1}^{4} \left(-\dfrac{1}{2}\right)^i$

36. $\sum\limits_{i=2}^{4} \left(-\dfrac{1}{3}\right)^i$

37. $\sum\limits_{i=5}^{9} 11$

38. $\sum\limits_{i=3}^{7} 12$

39. $\sum\limits_{i=0}^{4} \dfrac{(-1)^i}{i!}$

40. $\sum\limits_{i=0}^{4} \dfrac{(-1)^{i+1}}{(i + 1)!}$

41. $\sum\limits_{i=1}^{5} \dfrac{i!}{(i - 1)!}$

42. $\sum\limits_{i=1}^{5} \dfrac{(i + 2)!}{i!}$

In Exercises 43–54, express each sum using summation notation. Use 1 as the lower limit of summation and i for the index of summation.

43. $1^2 + 2^2 + 3^2 + \cdots + 15^2$

44. $1^4 + 2^4 + 3^4 + \cdots + 12^4$

45. $2 + 2^2 + 2^3 + \cdots + 2^{11}$

46. $5 + 5^2 + 5^3 + \cdots + 5^{12}$

47. $1 + 2 + 3 + \cdots + 30$

48. $1 + 2 + 3 + \cdots + 40$

49. $\dfrac{1}{2} + \dfrac{2}{3} + \dfrac{3}{4} + \cdots + \dfrac{14}{14 + 1}$

50. $\dfrac{1}{3} + \dfrac{2}{4} + \dfrac{3}{5} + \cdots + \dfrac{16}{16 + 2}$

51. $4 + \dfrac{4^2}{2} + \dfrac{4^3}{3} + \cdots + \dfrac{4^n}{n}$

52. $\dfrac{1}{9} + \dfrac{2}{9^2} + \dfrac{3}{9^3} + \cdots + \dfrac{n}{9^n}$

53. $1 + 3 + 5 + \cdots + (2n - 1)$

54. $a + ar + ar^2 + \cdots + ar^{n-1}$

In Exercises 55–60, express each sum using summation notation. Use a lower limit of summation of your choice and k for the index of summation.

55. $5 + 7 + 9 + 11 + \cdots + 31$

56. $6 + 8 + 10 + 12 + \cdots + 32$

57. $a + ar + ar^2 + \cdots + ar^{12}$

58. $a + ar + ar^2 + \cdots + ar^{14}$

59. $a + (a + d) + (a + 2d) + \cdots + (a + nd)$

60. $(a + d) + (a + d^2) + \cdots + (a + d^n)$

Practice Plus

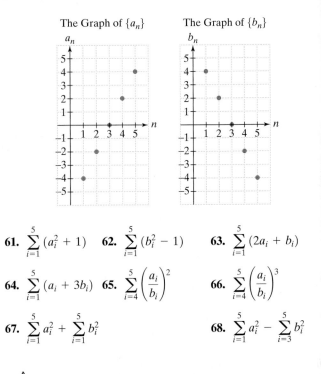

In Exercises 61–68, use the graphs of $\{a_n\}$ and $\{b_n\}$ to find each indicated sum.

The Graph of $\{a_n\}$ The Graph of $\{b_n\}$

61. $\sum\limits_{i=1}^{5} (a_i^2 + 1)$

62. $\sum\limits_{i=1}^{5} (b_i^2 - 1)$

63. $\sum\limits_{i=1}^{5} (2a_i + b_i)$

64. $\sum\limits_{i=1}^{5} (a_i + 3b_i)$

65. $\sum\limits_{i=4}^{5} \left(\dfrac{a_i}{b_i}\right)^2$

66. $\sum\limits_{i=4}^{5} \left(\dfrac{a_i}{b_i}\right)^3$

67. $\sum\limits_{i=1}^{5} a_i^2 + \sum\limits_{i=1}^{5} b_i^2$

68. $\sum\limits_{i=1}^{5} a_i^2 - \sum\limits_{i=3}^{5} b_i^2$

Application Exercises

69. The bar graph shows the number of people in the United States who lived below the poverty level from 1995 through 2002. Let a_n represent the number of people, in millions, living below the poverty level in year n, where $n = 1$ corresponds to 1995, $n = 2$ to 1996, and so on.

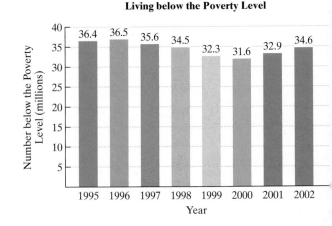

Number of People in the U.S. Living below the Poverty Level

Source: Bureau of the Census

a. Find $\sum\limits_{i=1}^{8} a_i$.

b. Find $\dfrac{\sum\limits_{i=1}^{8} a_i}{8}$. What does this number represent?

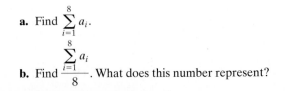

70. The bar graph shows the number of flu vaccine doses, in millions, that were available and distributed in the United States from 1999 through 2004. Let a_n represent the available doses, in millions, and let d_n represent the distributed doses, in millions, in year n, where $n = 1$ corresponds to 1999, $n = 2$ to 2000, and so on.

Flu Vaccine Doses in the U.S.

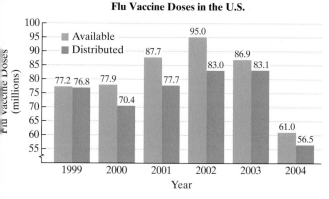

Source: Centers for Disease Control

Find $\sum\limits_{i=1}^{6} (a_i - d_i)$. What does this number represent?

The graph shows the millions of welfare recipients in the United States who received cash assistance from 1994 through 2003. In Exercises 71–72, consider a sequence whose general term, a_n, represents the millions of Americans receiving cash assistance n years after 1993.

Welfare Recipients in the U.S.

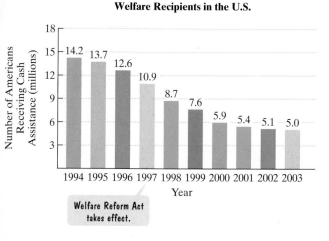

Source: Thomas R. Dye, *Politics in America*, Prentice Hall

71. a. Use the numbers given in the graph to find and interpret $\dfrac{1}{10}\sum\limits_{i=1}^{10} a_i$.

 b. The finite sequence whose general term is $a_n = -1.18n + 15.41$, where $n = 1, 2, 3, \ldots, 10$, models the millions of Americans receiving cash assistance, a_n, n years after 1993. Use the model to find $\dfrac{1}{10}\sum\limits_{i=1}^{10} a_i$. Does this seem reasonable in terms of the actual sum in part (a), or has model breakdown occurred?

72. a. Use the numbers given in the graph to find and interpret $\dfrac{1}{10}\sum\limits_{i=1}^{10} a_i$.

 b. The finite sequence whose general term is $a_n = 0.07n^2 - 1.98n + 17.01$, where $n = 1, 2, 3, \ldots, 10$, models the millions of Americans receiving cash assistance, a_n, n years after 1993. Use the model to find $\dfrac{1}{10}\sum\limits_{i=1}^{10} a_i$. Does this seem reasonable in terms of the actual sum in part (a), or has model breakdown occurred?

73. A deposit of $6000 is made in an account that earns 6% interest compounded quarterly. The balance in the account after n quarters is given by the sequence
$$a_n = 6000\left(1 + \frac{0.06}{4}\right)^n, \qquad n = 1, 2, 3, \ldots.$$
Find the balance in the account after five years. Round to the nearest cent.

74. A deposit of $10,000 is made in an account that earns 8% interest compounded quarterly. The balance in the account after n quarters is given by the sequence
$$a_n = 10{,}000\left(1 + \frac{0.08}{4}\right)^n, \qquad n = 1, 2, 3, \ldots.$$
Find the balance in the account after six years. Round to the nearest cent.

Writing in Mathematics

75. What is a sequence? Give an example with your description.

76. Explain how to write terms of a sequence if the formula for the general term is given.

77. What does the graph of a sequence look like? How is it obtained?

78. What is a recursion formula?

79. Explain how to find $n!$ if n is a positive integer.

80. Explain the best way to evaluate $\dfrac{900!}{899!}$ without a calculator.

81. What is the meaning of the symbol Σ? Give an example with your description.

82. You buy a new car for $24,000. At the end of n years, the value of your car is given by the sequence
$$a_n = 24{,}000\left(\frac{3}{4}\right)^n, \qquad n = 1, 2, 3, \ldots.$$
Find a_5 and write a sentence explaining what this value represents. Describe the nth term of the sequence in terms of the value of your car at the end of each year.

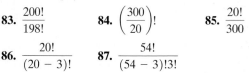

Technology Exercises

In Exercises 83–87, use a calculator's factorial key to evaluate each expression.

83. $\dfrac{200!}{198!}$

84. $\left(\dfrac{300}{20}\right)!$

85. $\dfrac{20!}{300}$

86. $\dfrac{20!}{(20-3)!}$

87. $\dfrac{54!}{(54-3)!3!}$

88. Use the SEQ (sequence) capability of a graphing utility to verify the terms of the sequences you obtained for any five sequences from Exercises 1–12 or 19–22.

89. Use the SUM SEQ (sum of the sequence) capability of a graphing utility to verify any five of the sums you obtained in Exercises 29–42.

90. As n increases, the terms of the sequence

$$a_n = \left(1 + \frac{1}{n}\right)^n$$

get closer and closer to the number e (where $e \approx 2.7183$). Use a calculator to find $a_{10}, a_{100}, a_{1000}, a_{10,000},$ and $a_{100,000}$, comparing these terms to your calculator's decimal approximation for e.

Many graphing utilities have a sequence-graphing mode that plots the terms of a sequence as points on a rectangular coordinate system. Consult your manual; if your graphing utility has this capability, use it to graph each of the sequences in Exercises 91–94. What appears to be happening to the terms of each sequence as n gets larger?

91. $a_n = \dfrac{n}{n + 1}$ $n{:}[0, 10, 1]$ by $a_n{:}[0, 1, 0.1]$

92. $a_n = \dfrac{100}{n}$ $n{:}[0, 1000, 100]$ by $a_n{:}[0, 1, 0.1]$

93. $a_n = \dfrac{2n^2 + 5n - 7}{n^3}$ $n{:}[0, 10, 1]$ by $a_n{:}[0, 2, 0.2]$

94. $a_n = \dfrac{3n^4 + n - 1}{5n^4 + 2n^2 + 1}$ $n{:}[0, 10, 1]$ by $a_n{:}[0, 1, 0.1]$

Critical Thinking Exercises

95. Which one of the following is true?

a. $\dfrac{n!}{(n - 1)!} = \dfrac{1}{n - 1}$

b. The Fibonacci sequence 1, 1, 2, 3, 5, 8, 13, 21, 34, 55, 89, 144, ... can be defined recursively using $a_0 = 1, a_1 = 1,$ $a_n = a_{n-2} + a_{n-1}$, where $n \geq 2$.

c. $\displaystyle\sum_{i=1}^{2} (-1)^i 2^i = 0$

d. $\displaystyle\sum_{i=1}^{2} a_i b_i = \sum_{i=1}^{2} a_i \sum_{i=1}^{2} b_i$

96. Write the first five terms of the sequence whose first term is ▪ and whose general term is

$$a_n = \begin{cases} \dfrac{a_{n-1}}{2} & \text{if } a_{n-1} \text{ is even} \\ 3a_{n-1} + 5 & \text{if } a_{n-1} \text{ is odd} \end{cases}$$

for $n \geq 2$.

Group Exercise

97. Enough curiosities involving the Fibonacci sequence exist to warrant a flourishing Fibonacci Association, which publishes a quarterly journal. Do some research on the Fibonacci sequence by consulting the Internet or the research department of your library, and find one property that interests you. After doing this research, get together with your group to share these intriguing properties.

SECTION 10.2   *Arithmetic Sequences*

Objectives

❶ Find the common difference for an arithmetic sequence.

❷ Write terms of an arithmetic sequence.

❸ Use the formula for the general term of an arithmetic sequence.

❹ Use the formula for the sum of the first n terms of an arithmetic sequence.

Your grandmother and her financial counselor are looking at options in case an adult residential facility is needed in the future. The good news is that your grandmother's total assets are $400,000. The bad news is that yearly adult residential community costs average $58,730, increasing by $1800 each year. In this section, we will see how sequences can be used to describe your grandmother's situation and help her to identify realistic options.

Find the common difference for an arithmetic sequence.

Arithmetic Sequences

The bar graph in Figure 10.2 shows annual salaries, rounded to the nearest thousand dollars, of U.S. senators from 2000 to 2005. The graph illustrates that each year salaries increased by $4 thousand. The sequence of annual salaries

142, 146, 150, 154, 158, 162, ...

shows that each term after the first, 142, differs from the preceding term by a constant amount, namely 4. This sequence is an example of an *arithmetic sequence*.

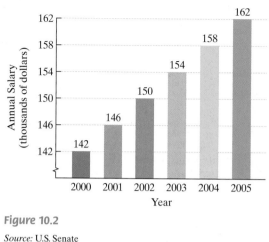

Annual Salaries of U.S. Senators

Figure 10.2

Source: U.S. Senate

Definition of an Arithmetic Sequence

An **arithmetic sequence** is a sequence in which each term after the first differs from the preceding term by a constant amount. The difference between consecutive terms is called the **common difference** of the sequence.

The common difference, d, is found by subtracting any term from the term that directly follows it. In the following examples, the common difference is found by subtracting the first term from the second term, $a_2 - a_1$.

Arithmetic Sequence	**Common Difference**
$142, 146, 150, 154, 158, \ldots$	$d = 146 - 142 = 4$
$-5, -2, 1, 4, 7, \ldots$	$d = -2 - (-5) = -2 + 5 = 3$
$8, 3, -2, -7, -12, \ldots$	$d = 3 - 8 = -5$

Figure 10.3 shows the graphs of the last two arithmetic sequences in our list. The common difference for the increasing sequence in Figure 10.3(a) is 3. The common difference for the decreasing sequence in Figure 10.3(b) is -5.

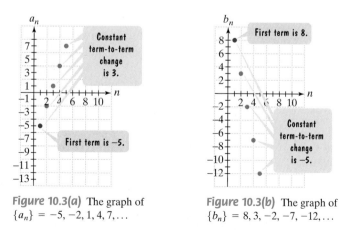

Figure 10.3(a) The graph of $\{a_n\} = -5, -2, 1, 4, 7, \ldots$

Figure 10.3(b) The graph of $\{b_n\} = 8, 3, -2, -7, -12, \ldots$

The graph of each arithmetic sequence in Figure 10.3 forms a set of discrete points lying on a straight line. This illustrates that **an arithmetic sequence is a linear function whose domain is the set of positive integers**.

If the first term of an arithmetic sequence is a_1, each term after the first i[s] obtained by adding d, the common difference, to the previous term. This can b[e] expressed recursively as follows:

$$a_n = a_{n-1} + d.$$

> Add d to the term in any position to get the next term.

To use this recursion formula, we must be given the first term.

 Write terms of an arithmetic sequence.

EXAMPLE 1 Writing the Terms of an Arithmetic Sequence Using the First Term and the Common Difference

Figure 10.4 shows the percentage of men and women in the U.S. labor force fo[r] five-year periods starting with 1980. The recursion formula $a_n = a_{n-1} - 0.67$ mod els the percentage of men working in the U.S. labor force, a_n, for each five-yea[r] period starting with 1980. Thus, $n = 1$ corresponds to 1980, $n = 2$ to 1985, $n = 3$ t[o] 1990, and so on. In 1980, 77.4% of U.S. men were working in the labor force. Fin[d] the first five terms of this arithmetic sequence in which $a_1 = 77.4$ an[d] $a_n = a_{n-1} - 0.67$.

Solution The recursion formula $a_1 = 77.4$ and $a_n = a_{n-1} - 0.67$ indicates tha[t] each term after the first, 77.4, is obtained by adding -0.67 to the previous term[.] Thus, during each five-year period, the percentage of men in the labor forc[e] decreased by 0.67%.

$a_1 = 77.4$ This is given.

$a_2 = a_1 - 0.67 = 77.4 - 0.67 = 76.73$ Use $a_n = a_{n-1} - 0.67$ with $n = 2$.

$a_3 = a_2 - 0.67 = 76.73 - 0.67 = 76.06$ Use $a_n = a_{n-1} - 0.67$ with $n = 3$.

$a_4 = a_3 - 0.67 = 76.06 - 0.67 = 75.39$ Use $a_n = a_{n-1} - 0.67$ with $n = 4$.

$a_5 = a_4 - 0.67 = 75.39 - 0.67 = 74.72$ Use $a_n = a_{n-1} - 0.67$ with $n = 5$.

The first five terms are

$$77.4, 76.73, 76.06, 75.39, \text{ and } 74.72.$$

These numbers represent the percentage of men working in the U.S. labor force i[n] 1980, 1985, 1990, 1995, and 2000, respectively, as given by the model.

Men in the U.S. Labor Force

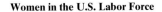

Women in the U.S. Labor Force

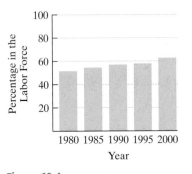

Figure 10.4

Source: U.S. Department of Labor

> **Check Point 1** The recursion formula $a_n = a_{n-1} + 2.18$ models the percentage of wome[n] working in the U.S. labor force, a_n, for each five-year period starting wi[th] 1980. In 1980, 51.5% of U.S. women were working in the labor force. Fin[d] the first five terms of the arithmetic sequence in which $a_1 = 51.5$ an[d] $a_n = a_{n-1} + 2.18$.

❸ Use the formula for the general term of an arithmetic sequence.

The General Term of an Arithmetic Sequence

Consider an arithmetic sequence whose first term is a_1 and whose common differenc[e] is d. We are looking for a formula for the general term, a_n. Let's begin by writing th[e] first six terms. The first term is a_1. The second term is $a_1 + d$. The third term [is] $a_1 + d + d$, or $a_1 + 2d$. Thus, we start with a_1 and add d to each successive ter[m.] The first six terms are

$$a_1, \quad a_1 + d, \quad a_1 + 2d, \quad a_1 + 3d, \quad a_1 + 4d, \quad a_1 + 5d.$$

| a_1, first term | a_2, second term | a_3, third term | a_4, fourth term | a_5, fifth term | a_6, sixth term |

Compare the coefficient of d and the subscript of a denoting the term number. Can you see that the coefficient of d is 1 less than the subscript of a denoting the term number?

$$a_3: \text{third term} = a_1 + 2d \qquad a_4: \text{fourth term} = a_1 + 3d$$

2 is one less than 3.   3 is one less than 4.

Thus, the formula for the nth term is

$$a_n: n\text{th term} = a_1 + (n-1)d.$$

$n-1$ is one less than n.

General Term of an Arithmetic Sequence

The nth term (the general term) of an arithmetic sequence with first term a_1 and common difference d is

$$a_n = a_1 + (n-1)d.$$

EXAMPLE 2   Using the Formula for the General Term of an Arithmetic Sequence

Find the eighth term of the arithmetic sequence whose first term is 4 and whose common difference is -7.

Solution   To find the eighth term, a_8, we replace n in the formula with 8, a_1 with 4, and d with -7.

$$a_n = a_1 + (n-1)d$$
$$a_8 = 4 + (8-1)(-7) = 4 + 7(-7) = 4 + (-49) = -45$$

The eighth term is -45. We can check this result by writing the first eight terms of the sequence:

$$4, -3, -10, -17, -24, -31, -38, -45.$$

Check Point 2   Find the ninth term of the arithmetic sequence whose first term is 6 and whose common difference is -5.

EXAMPLE 3   Using an Arithmetic Sequence to Model Teachers' Earnings

According to the National Education Association, teachers in the United States earned an average of $30,532 in 1990. This amount has increased by approximately $1472 per year.

a. Write a formula for the nth term of the arithmetic sequence that describes teachers' average earnings n years after 1989.

b. How much will U.S. teachers earn, on average, by the year 2010?

Solution

a. We can express teachers' earnings by the following arithmetic sequence:

$$30,532, \qquad 32,004, \qquad 33,476, \qquad 34,948, \dots.$$

a_1: earnings in a_2: earnings in a_3: earnings in a_4: earnings in
1990, 1 year   1991, 2 years   1992, 3 years   1993, 4 years
after 1989   after 1989   after 1989   after 1989

In the sequence $30{,}532, 32{,}004, 33{,}476, \ldots, a_1$, the first term, represents the amount teachers earned in 1990. Each subsequent year this amount increase by \$1472, so $d = 1472$. We use the formula for the general term of an arithmetic sequence to write the nth term of the sequence that describes teacher earnings n years after 1989.

$$a_n = a_1 + (n - 1)d \qquad \text{This is the formula for the general term of an arithmetic sequence.}$$

$$a_n = 30{,}532 + (n - 1)1472 \qquad a_1 = 30{,}532 \text{ and } d = 1472.$$

$$a_n = 30{,}532 + 1472n - 1472 \qquad \text{Distribute 1472 to each term in parentheses.}$$

$$a_n = 1472n + 29{,}060 \qquad \text{Simplify.}$$

Thus, teachers' earnings n years after 1989 can be described by $a_n = 1472n + 29{,}060$.

b. Now we need to find teachers' earnings in 2010. The year 2010 is 21 years after 1989: That is, $2010 - 1989 = 21$. Thus, $n = 21$. We substitute 21 for n in $a_n = 1472n + 29{,}060$.

$$a_{21} = 1472 \cdot 21 + 29{,}060 = 59{,}972$$

The 21st term of the sequence is 59,972. Therefore, U.S. teachers are predicted to earn an average of \$59,972 by the year 2010.

 Check Point 3 According to the U.S. Census Bureau, new one-family houses sold for an average of \$159,000 in 1995. This average sales price has increased by approximately \$9700 per year.

a. Write a formula for the nth term of the arithmetic sequence that describes the average cost of new one-family houses n years after 1995.

b. How much will new one-family houses cost, on average, by the year 2010?

④ Use the formula for the sum of the first n terms of an arithmetic sequence.

The Sum of the First n Terms of an Arithmetic Sequence

The sum of the first n terms of an arithmetic sequence, denoted by S_n, and called the **nth partial sum**, can be found without having to add up all the terms. Let

$$S_n = a_1 + a_2 + a_3 + \cdots + a_n$$

be the sum of the first n terms of an arithmetic sequence. Because d is the common difference between terms, S_n can be written forward and backward as follows.

Forward: Start with the first term, a_1. Keep adding d.

Backward: Start with the last term, a_n. Keep subtracting d.

$$\begin{aligned} S_n &= a_1 && + (a_1 + d) && + (a_1 + 2d) && + \cdots + a_n \\ S_n &= a_n && + (a_n - d) && + (a_n - 2d) && + \cdots + a_1 \\ \hline 2S_n &= (a_1 + a_n) && + (a_1 + a_n) && + (a_1 + a_n) && + \cdots + (a_1 + a_n) \end{aligned}$$

Add the two equations

Because there are n sums of $(a_1 + a_n)$ on the right side, we can express this side as $n(a_1 + a_n)$. Thus, the last equation can be written as follows:

$$2S_n = n(a_1 + a_n)$$

$$S_n = \frac{n}{2}(a_1 + a_n). \qquad \text{Solve for } S_n, \text{ dividing both sides by 2.}$$

We have proved the following result:

> **The Sum of the First n Terms of an Arithmetic Sequence**
>
> The sum, S_n, of the first n terms of an arithmetic sequence is given by
>
> $$S_n = \frac{n}{2}(a_1 + a_n),$$
>
> in which a_1 is the first term and a_n is the nth term.

To find the sum of the terms of an arithmetic sequence using $S_n = \frac{n}{2}(a_1 + a_n)$, we need to know the first term, a_1, the last term, a_n, and the number of terms, n. The following examples illustrate how to use this formula.

EXAMPLE 4   Finding the Sum of n Terms of an Arithmetic Sequence

Find the sum of the first 100 terms of the arithmetic sequence: $1, 3, 5, 7, \ldots$.

Solution   By finding the sum of the first 100 terms of $1, 3, 5, 7, \ldots$, we are finding the sum of the first 100 odd numbers. To find the sum of the first 100 terms, S_{100}, we replace n in the formula with 100.

$$S_n = \frac{n}{2}(a_1 + a_n)$$

$$S_{100} = \frac{100}{2}(a_1 + a_{100})$$

The first term, a_1, is 1.   We must find a_{100}, the 100th term.

We use the formula for the general term of a sequence to find a_{100}. The common difference, d, of $1, 3, 5, 7, \ldots$, is 2.

$$a_n = a_1 + (n-1)d$$

This is the formula for the nth term of an arithmetic sequence. Use it to find the 100th term.

$$a_{100} = 1 + (100 - 1) \cdot 2$$

Substitute 100 for n, 2 for d, and 1 (the first term) for a_1.

$$= 1 + 99 \cdot 2$$

$$= 1 + 198 = 199$$

Now we are ready to find the sum of the 100 terms $1, 3, 5, 7, \ldots, 199$.

$$S_n = \frac{n}{2}(a_1 + a_n)$$

Use the formula for the sum of the first n terms of an arithmetic sequence. Let $n = 100$, $a_1 = 1$, and $a_{100} = 199$.

$$S_{100} = \frac{100}{2}(1 + 199) = 50(200) = 10{,}000$$

The sum of the first 100 odd numbers is 10,000. Equivalently, the 100th partial sum of the sequence $1, 3, 5, 7, \ldots$ is 10,000.

Check Point 4   Find the sum of the first 15 terms of the arithmetic sequence: $3, 6, 9, 12, \ldots$.

EXAMPLE 5   Using S_n to Evaluate a Summation

Find the following sum: $\displaystyle\sum_{i=1}^{25}(5i-9)$.

Solution

$$\sum_{i=1}^{25}(5i-9) = (5\cdot 1 - 9) + (5\cdot 2 - 9) + (5\cdot 3 - 9) + \cdots + (5\cdot 25 - 9)$$

$$= \underset{-4}{\quad} \quad + \underset{+1}{\quad} \quad + \underset{+6}{\quad} \quad + \cdots + 116$$

By evaluating the first three terms and the last term, we see that $a_1 = -4$; d, the common difference, is $1 - (-4)$, or 5; and a_{25}, the last term, is 116.

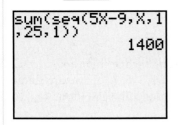

$$S_n = \frac{n}{2}(a_1 + a_n) \qquad \text{Use the formula for the sum of the first } n \text{ terms of an arithmetic sequence. Let } n = 25, a_1 = -4, \text{ and } a_{25} = 116.$$

$$S_{25} = \frac{25}{2}(-4 + 116) = \frac{25}{2}(112) = 1400$$

Thus,

$$\sum_{i=1}^{25}(5i-9) = 1400.$$

Check Point 5   Find the following sum: $\displaystyle\sum_{i=1}^{30}(6i-11)$.

EXAMPLE 6   Modeling Total Residential Community Costs over a Six-Year Period

Your grandmother has assets of \$400,000. One option that she is considering involves an adult residential community for a six-year period beginning in 2006. The model

$$a_n = 1800n + 58{,}730$$

describes yearly adult residential community costs n years after 2005. Does your grandmother have enough to pay for the facility?

Solution   We must find the sum of an arithmetic sequence. The first term of the sequence corresponds to the facility's costs in the year 2006. The last term corresponds to costs in the year 2011. Because the model describes costs n years after 2005, $n = 1$ describes the year 2006 and $n = 6$ describes the year 2011.

$$a_n = 1800n + 58{,}730 \qquad \text{This is the given formula for the general term of the sequence.}$$

$$a_1 = 1800\cdot 1 + 58{,}730 = 60{,}530 \qquad \text{Find } a_1 \text{ by replacing } n \text{ with 1.}$$

$$a_6 = 1800\cdot 6 + 58{,}730 = 69{,}530 \qquad \text{Find } a_6 \text{ by replacing } n \text{ with 6.}$$

The first year the facility will cost \$60,530. By year six, the facility will cost \$69,530. Now we must find the sum of the costs for all six years. We focus on the sum of the first six terms of the arithmetic sequence

$$60{,}530, \ 62{,}330, \ \ldots \ , \ 69{,}530.$$

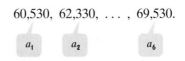

We find this sum using the formula for the sum of the first n terms of an arithmetic sequence. We are adding 6 terms: $n = 6$. The first term is 60,530: $a_1 = 60{,}530$. The last term—that is, the sixth term—is 69,530: $a_6 = 69{,}530$.

$$S_n = \frac{n}{2}(a_1 + a_n)$$

$$S_6 = \frac{6}{2}(60{,}530 + 69{,}530) = 3(130{,}060) = 390{,}180$$

The total adult residential community costs for your grandmother are predicted to be $390,180. Because your grandmother's assets are $400,000, she has enough to pay for the facility for the six-year period.

Check Point 6 In Example 6, how much would it cost for the adult residential community for a ten-year period beginning in 2006?

EXERCISE SET 10.2

Practice Exercises

In Exercises 1–14, write the first six terms of each arithmetic sequence.

1. $a_1 = 200, d = 20$ **2.** $a_1 = 300, d = 50$

3. $a_1 = -7, d = 4$ **4.** $a_1 = -8, d = 5$

5. $a_1 = 300, d = -90$ **6.** $a_1 = 200, d = -60$

7. $a_1 = \frac{5}{2}, d = -\frac{1}{2}$ **8.** $a_1 = \frac{3}{4}, d = -\frac{1}{4}$

9. $a_n = a_{n-1} + 6, a_1 = -9$ **10.** $a_n = a_{n-1} + 4, a_1 = -7$

11. $a_n = a_{n-1} - 10, a_1 = 30$ **12.** $a_n = a_{n-1} - 20, a_1 = 50$

13. $a_n = a_{n-1} - 0.4, a_1 = 1.6$

14. $a_n = a_{n-1} - 0.3, a_1 = -1.7$

In Exercises 15–22, find the indicated term of the arithmetic sequence with first term, a_1, and common difference, d.

15. Find a_6 when $a_1 = 13, d = 4$.

16. Find a_{16} when $a_1 = 9, d = 2$.

17. Find a_{50} when $a_1 = 7, d = 5$.

18. Find a_{60} when $a_1 = 8, d = 6$.

19. Find a_{200} when $a_1 = -40, d = 5$.

20. Find a_{150} when $a_1 = -60, d = 5$.

21. Find a_{60} when $a_1 = 35, d = -3$.

22. Find a_{70} when $a_1 = -32, d = 4$.

In Exercises 23–34, write a formula for the general term (the nth term) of each arithmetic sequence. Do not use a recursion formula. Then use the formula for a_n to find a_{20}, the 20th term of the sequence.

23. $1, 5, 9, 13, \ldots$ **24.** $2, 7, 12, 17, \ldots$

25. $7, 3, -1, -5, \ldots$ **26.** $6, 1, -4, -9, \ldots$

27. $a_1 = 9, d = 2$ **28.** $a_1 = 6, d = 3$

29. $a_1 = -20, d = -4$ **30.** $a_1 = -70, d = -5$

31. $a_n = a_{n-1} + 3, a_1 = 4$ **32.** $a_n = a_{n-1} + 5, a_1 = 6$

33. $a_n = a_{n-1} - 10, a_1 = 30$ **34.** $a_n = a_{n-1} - 12, a_1 = 24$

35. Find the sum of the first 20 terms of the arithmetic sequence: $4, 10, 16, 22, \ldots$.

36. Find the sum of the first 25 terms of the arithmetic sequence: $7, 19, 31, 43, \ldots$.

37. Find the sum of the first 50 terms of the arithmetic sequence: $-10, -6, -2, 2, \ldots$.

38. Find the sum of the first 50 terms of the arithmetic sequence: $-15, -9, -3, 3, \ldots$.

39. Find $1 + 2 + 3 + 4 + \cdots + 100$, the sum of the first 100 natural numbers.

40. Find $2 + 4 + 6 + 8 + \cdots + 200$, the sum of the first 100 positive even integers.

41. Find the sum of the first 60 positive even integers.

42. Find the sum of the first 80 positive even integers.

43. Find the sum of the even integers between 21 and 45.

44. Find the sum of the odd integers between 30 and 54.

For Exercises 45–50, write out the first three terms and the last term. Then use the formula for the sum of the first n terms of an arithmetic sequence to find the indicated sum.

45. $\sum_{i=1}^{17} (5i + 3)$ **46.** $\sum_{i=1}^{20} (6i - 4)$ **47.** $\sum_{i=1}^{30} (-3i + 5)$

48. $\sum_{i=1}^{40} (-2i + 6)$ **49.** $\sum_{i=1}^{100} 4i$ **50.** $\sum_{i=1}^{50} (-4i)$

Practice Plus

Use the graphs of the arithmetic sequences $\{a_n\}$ and $\{b_n\}$ to solve Exercises 51–58.

51. Find $a_{14} + b_{12}$.   **52.** Find $a_{16} + b_{18}$.

53. If $\{a_n\}$ is a finite sequence whose last term is -83, how many terms does $\{a_n\}$ contain?

54. If $\{b_n\}$ is a finite sequence whose last term is 93, how many terms does $\{b_n\}$ contain?

(Continue using the graphs at the bottom of page 943 to solve Exercises 55–58.)

55. Find the difference between the sum of the first 14 terms of $\{b_n\}$ and the sum of the first 14 terms of $\{a_n\}$.

56. Find the difference between the sum of the first 15 terms of $\{b_n\}$ and the sum of the first 15 terms of $\{a_n\}$.

57. Write a linear function $f(x) = mx + b$, whose domain is the set of positive integers, that represents $\{a_n\}$.

58. Write a linear function $g(x) = mx + b$, whose domain is the set of positive integers, that represents $\{b_n\}$.

Use a system of two equations in two variables, a_1 and d, to solve Exercises 59–60.

59. Write a formula for the general term (the nth term) of the arithmetic sequence whose second term, a_2, is 4 and whose sixth term, a_6, is 16.

60. Write a formula for the general term (the nth term) of the arithmetic sequence whose third term, a_3, is 7 and whose eighth term, a_8, is 17.

Application Exercises

The bar graphs show changes that have taken place in the United States from 1970 to 2002 or 2003. Exercises 61–63 involve developing arithmetic sequences that model the data.

Changing Times in the U.S.

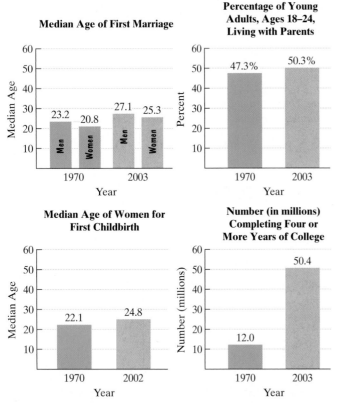

Source: U.S. Census Bureau

61. In 1970, the median age of first marriage for U.S. men was 23.2. On average, this age has increased by approximately 0.12 per year.

a. Write a formula for the nth term of the arithmetic sequence that describes the median age of first marriage for U.S. men n years after 1969.

b. What will be the median age of first marriage for U.S. men in 2009?

62. In 1970, the median age of women for first childbirth wa 22.1. On average, this age has increased by approximatel 0.08 per year.

a. Write a formula for the nth term of the arithmeti sequence that describes the median age for first childbirt for U.S. women n years after 1969.

b. What will be the median age for first childbirth for U.S women in 2009?

63. Repeat Exercise 61 or 62 for another one of the change from 1970 to 2003. Develop a formula for the nth term of th arithmetic sequence that describes the changing phenome non n years after 1969. Then make a prediction about wha might occur in 2009.

The bar graph shows the average cost of tuition, fees, and room ana board at public and private colleges in the United States for four academic years. Use this information to solve Exercises 64–66.

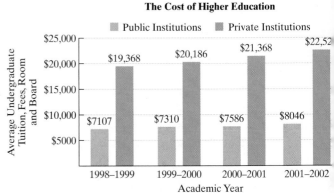

Source: U.S. Department of Education

64. a. Use the numbers shown in the bar graph to find the tot cost of higher education at a private college for a fou year period, beginning with the 1998–1999 academic yea and ending with the 2001–2002 academic year.

b. The model

$$a_n = 1064n + 18{,}201$$

describes the cost of higher education, a_n, at a private co lege in academic year n, where $n = 1$ corresponds 1998–1999, $n = 2$ to 1999–2000, and so on. Use this mode and the formula for S_n to find the total cost of a high education at a private college for a four-year period, begi ning with the 1998–1999 academic year and ending wit the 2001–2002 academic year. How well does the mode describe the actual sum that you obtained in part (a)?

65. a. Use the numbers shown in the bar graph to find the tot cost of higher education at a public college for a four-ye period, beginning with the 1998–1999 academic year an ending with the 2001–2002 academic year.

b. The model

$$a_n = 309n + 6739$$

describes the cost of higher education, a_n, at a publi college in academic year n, where $n = 1$ corresponds 1998–1999, $n = 2$ to 1999–2000, and so on. Use this mode and the formula for S_n to find the total cost of a high education at a public college for a four-year period, begi ning with the 1998–1999 academic year and ending wit the 2001–2002 academic year. How well does the mod describe the actual sum that you obtained in part (a)?

5. Use one of the models in Exercises 64–65 and the formula for S_n to find the total cost of your undergraduate education. How well does the model describe your anticipated costs?

7. A company offers a starting yearly salary of $33,000 with raises of $2500 per year. Find the total salary over a ten-year period.

8. You are considering two job offers. Company A will start you at $19,000 a year and guarantee a raise of $2600 per year. Company B will start you at a higher salary, $27,000 a year, but will only guarantee a raise of $1200 per year. Find the total salary that each company will pay over a ten-year period. Which company pays the greater total amount?

9. A theater has 30 seats in the first row, 32 seats in the second row, increasing by 2 seats per row for a total of 26 rows. How many seats are there in the theater?

9. A section in a stadium has 20 seats in the first row, 23 seats in the second row, increasing by 3 seats each row for a total of 38 rows. How many seats are in this section of the stadium?

Writing in Mathematics

1. What is an arithmetic sequence? Give an example with your explanation.

2. What is the common difference in an arithmetic sequence?

3. Explain how to find the general term of an arithmetic sequence.

4. Explain how to find the sum of the first n terms of an arithmetic sequence without having to add up all the terms.

Technology Exercises

75. Use the SEQ (sequence) capability of a graphing utility and the formula you obtained for a_n to verify the value you found for a_{20} in any five exercises from Exercises 23–34.

76. Use the capability of a graphing utility to calculate the sum of a sequence to verify any five of your answers to Exercises 45–50.

Critical Thinking Exercises

77. Give examples of two different arithmetic sequences whose fourth term, a_4, is 10.

78. In the sequence $21,700, 23,172, 24,644, 26,116, \ldots,$ which term is 314,628?

79. A *degree-day* is a unit used to measure the fuel requirements of buildings. By definition, each degree that the average daily temperature is below 65°F is 1 degree-day. For example, a temperature of 42°F constitutes 23 degree-days. If the average temperature on January 1 was 42°F and fell 2°F for each subsequent day up to and including January 10, how many degree-days are included from January 1 to January 10?

80. Show that the sum of the first n positive odd integers,

$$1 + 3 + 5 + \cdots + (2n - 1),$$

is n^2.

SECTION 10.3   *Geometric Sequences and Series*

Objectives

● Find the common ratio of a geometric sequence.

● Write terms of a geometric sequence.

● Use the formula for the general term of a geometric sequence.

● Use the formula for the sum of the first n terms of a geometric sequence.

● Find the value of an annuity.

● Use the formula for the sum of an infinite geometric series.

Here we are at the closing moments of a job interview. You're shaking hands with the manager. You managed to answer all the tough questions without losing your poise, and now you've been offered a job. As a matter of fact, your qualifications are so terrific that you've been offered two jobs—one just the day before, with a rival company in the same field! One company offers $30,000 the first year, with increases of 6% per year for four years after that. The other offers $32,000 the first year, with annual increases of 3% per year after that. Over a five-year period, which is the better offer?

If salary raises amount to a certain percent each year, the yearly salaries over time form a geometric sequence. In this section, we investigate geometric sequences and their properties. After studying the section, you will be in a position to decide which job offer to accept: You will know which company will pay you more over five years.

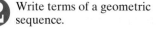

① Find the common ratio of a geometric sequence.

Geometric Sequences

Figure 10.5 shows a sequence in which the number of squares is increasing. From left to right, the number of squares is 1, 5, 25, 125, and 625. In this sequence, each term after the first, 1, is obtained by multiplying the preceding term by a constant amount, namely 5. This sequence of increasing numbers of squares is an example of a *geometric sequence.*

Figure 10.5 A geometric sequence of squares

Definition of a Geometric Sequence

A **geometric sequence** is a sequence in which each term after the first is obtained by multiplying the preceding term by a fixed nonzero constant. The amount by which we multiply each time is called the **common ratio** of the sequence.

The common ratio, r, is found by dividing any term after the first term by the term that directly precedes it. In the following examples, the common ratio is found by dividing the second term by the first term, $\dfrac{a_2}{a_1}$.

Geometric sequence	Common ratio
$1, 5, 25, 125, 625, \ldots$	$r = \dfrac{5}{1} = 5$
$4, 8, 16, 32, 64, \ldots$	$r = \dfrac{8}{4} = 2$
$6, -12, 24, -48, 96, \ldots$	$r = \dfrac{-12}{6} = -2$
$9, -3, 1, -\dfrac{1}{3}, \dfrac{1}{9}, \ldots$	$r = \dfrac{-3}{9} = -\dfrac{1}{3}$

Figure 10.6 shows a partial graph of the first geometric sequence in our list. The graph forms a set of discrete points lying on the exponential function $f(x) = 5^{x-1}$. This illustrates that **a geometric sequence with a positive common ratio other than 1 is an exponential function whose domain is the set of positive integers**.

② Write terms of a geometric sequence.

How do we write out the terms of a geometric sequence when the first term and the common ratio are known? We multiply the first term by the common ratio to get the second term, multiply the second term by the common ratio to get the third term, and so on.

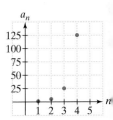

Figure 10.6 The graph $\{a_n\} = 1, 5, 25, 125, \ldots$

Check Point **3** Write the general term for the geometric sequence

$$3, 6, 12, 24, 48, \ldots.$$

Then use the formula for the general term to find the eighth term.

The Sum of the First n Terms of a Geometric Sequence

④ Use the formula for the sum of the first n terms of a geometric sequence.

The sum of the first n terms of a geometric sequence, denoted by S_n, and called the **nth partial sum**, can be found without having to add up all the terms. Recall that the first n terms of a geometric sequence are

$$a_1, a_1r, a_1r^2, \ldots, a_1r^{n-2}, a_1r^{n-1}.$$

We proceed as follows:

$$S_n = a_1 + a_1r + a_1r^2 + \cdots + a_1r^{n-2} + a_1r^{n-1} \qquad \text{S_n is the sum of the first n terms of the sequence.}$$

$$rS_n = a_1r + a_1r^2 + a_1r^3 + \cdots + a_1r^{n-1} + a_1r^n \qquad \text{Multiply both sides of the equation by r.}$$

$$S_n - rS_n = a_1 - a_1r^n \qquad \text{Subtract the second equation from the first equation.}$$

$$S_n(1 - r) = a_1(1 - r^n) \qquad \text{Factor out S_n on the left and a_1 on the right.}$$

$$S_n = \frac{a_1(1 - r^n)}{1 - r}. \qquad \text{Solve for S_n by dividing both sides by $1 - r$ (assuming that $r \neq 1$).}$$

We have proved the following result:

The Sum of the First n Terms of a Geometric Sequence

The sum, S_n, of the first n terms of a geometric sequence is given by

$$S_n = \frac{a_1(1 - r^n)}{1 - r}$$

in which a_1 is the first term and r is the common ratio ($r \neq 1$).

To find the sum of the terms of a geometric sequence, we need to know the first term, a_1, the common ratio, r, and the number of terms, n. The following examples illustrate how to use this formula.

EXAMPLE 4   Finding the Sum of the First n Terms of a Geometric Sequence

Find the sum of the first 18 terms of the geometric sequence: $2, -8, 32, -128, \ldots.$

Solution   To find the sum of the first 18 terms, S_{18}, we replace n in the formula with 18.

$$S_n = \frac{a_1(1 - r^n)}{1 - r}$$

$$S_{18} = \frac{a_1(1 - r^{18})}{1 - r}$$

The first term, a_1, is 2.   We must find r, the common ratio.

We can find the common ratio by dividing the second term of $2, -8, 32, -128, \ldots$ by the first term.

$$r = \frac{a_2}{a_1} = \frac{-8}{2} = -4$$

Now we are ready to find the sum of the first 18 terms of $2, -8, 32, -128, \ldots$.

$$S_n = \frac{a_1(1 - r^n)}{1 - r}$$ Use the formula for the sum of the first n terms of a geometric sequence.

$$S_{18} = \frac{2[1 - (-4)^{18}]}{1 - (-4)}$$ a_1 (the first term) $= 2$, $r = -4$, and $n = 18$ because we want the sum of the first 18 terms.

$$= -27{,}487{,}790{,}694$$ Use a calculator.

The sum of the first 18 terms is $-27{,}487{,}790{,}694$. Equivalently, this number is the 18th partial sum of the sequence $2, -8, 32, -128, \ldots$.

Check Point 4 Find the sum of the first nine terms of the geometric sequence $2, -6, 18, -54, \ldots$.

EXAMPLE 5 Using S_n to Evaluate a Summation

Find the following sum: $\displaystyle\sum_{i=1}^{10} 6 \cdot 2^i$.

Solution Let's write out a few terms in the sum.

$$\sum_{i=1}^{10} 6 \cdot 2^i = 6 \cdot 2 + 6 \cdot 2^2 + 6 \cdot 2^3 + \cdots + 6 \cdot 2^{10}$$

Do you see that each term after the first is obtained by multiplying the preceding term by 2? To find the sum of the 10 terms ($n = 10$), we need to know the first term, a_1, and the common ratio, r. The first term is $6 \cdot 2$ or 12: $a_1 = 12$. The common ratio is 2.

$$S_n = \frac{a_1(1 - r^n)}{1 - r}$$ Use the formula for the sum of the first n terms of a geometric sequence.

$$S_{10} = \frac{12(1 - 2^{10})}{1 - 2}$$ a_1 (the first term) $= 12$, $r = 2$, and $n = 10$ because we are adding ten terms.

$$= 12{,}276$$ Use a calculator.

Thus,

$$\sum_{i=1}^{10} 6 \cdot 2^i = 12{,}276.$$

Technology

To find

$$\sum_{i=1}^{10} 6 \cdot 2^i$$

on a graphing utility, enter
SUM SEQ $(6 \times 2^x, x, 1, 10, 1)$.
Then press ENTER.

```
sum(seq(6*2^X,X,
1,10,1))
        12276
```

Check Point 5 Find the following sum: $\displaystyle\sum_{i=1}^{8} 2 \cdot 3^i$.

Some of the exercises in the previous exercise set involved situations in which salaries increased by a fixed amount each year. A more realistic situation is one in which salary raises increase by a certain percent each year. Example 6 shows how such a situation can be described using a geometric sequence.

EXAMPLE 6   Computing a Lifetime Salary

A union contract specifies that each worker will receive a 5% pay increase each year for the next 30 years. One worker is paid $20,000 the first year. What is this person's total lifetime salary over a 30-year period?

Solution   The salary for the first year is $20,000. With a 5% raise, the second-year salary is computed as follows:

$$\text{Salary for year 2} = 20{,}000 + 20{,}000(0.05) = 20{,}000(1 + 0.05) = 20{,}000(1.05).$$

Each year, the salary is 1.05 times what it was in the previous year. Thus, the salary for year 3 is 1.05 times $20{,}000(1.05)$, or $20{,}000(1.05)^2$. The salaries for the first five years are given in the table.

Yearly Salaries					
Year 1	Year 2	Year 3	Year 4	Year 5	$\dots$
20,000	20,000(1.05)	$20{,}000(1.05)^2$	$20{,}000(1.05)^3$	$20{,}000(1.05)^4$	$\dots$

The numbers in the bottom row form a geometric sequence with $a_1 = 20{,}000$ and $r = 1.05$. To find the total salary over 30 years, we use the formula for the sum of the first n terms of a geometric sequence, with $n = 30$.

$$S_n = \frac{a_1(1 - r^n)}{1 - r}$$

$$S_{30} = \frac{20{,}000[1 - (1.05)^{30}]}{1 - 1.05}$$

Total salary over 30 years

$$= \frac{20{,}000[1 - (1.05)^{30}]}{-0.05}$$

$$\approx 1{,}328{,}777 \qquad \text{Use a calculator.}$$

The total salary over the 30-year period is approximately $1,328,777.

Check Point 6   A job pays a salary of $30,000 the first year. During the next 29 years, the salary increases by 6% each year. What is the total lifetime salary over the 30-year period?

 Find the value of an annuity.

Annuities

The compound interest formula

$$A = P(1 + r)^t$$

gives the future value, A, after t years, when a fixed amount of money, P, the principal, is deposited in an account that pays an annual interest rate r (in decimal form) compounded once a year. However, money is often invested in small amounts at periodic intervals. For example, to save for retirement, you might decide to place $1000 into an Individual Retirement Account (IRA) at the end of each year until you retire. An **annuity** is a sequence of equal payments made at equal time periods. An IRA is an example of an annuity.

Suppose P dollars is deposited into an account at the end of each year. The account pays an annual interest rate, r, compounded annually. At the end of the first year, the account contains P dollars. At the end of the second year, P dollars is

deposited again. At the time of this deposit, the first deposit has received interes earned during the second year. The **value of the annuity** is the sum of all deposit made plus all interest paid. Thus, the value of the annuity after two years is

$$P + P(1 + r).$$

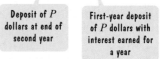

The value of the annuity after three years is

$$P \quad + \quad P(1 + r) \quad + \quad P(1 + r)^2.$$

The value of the annuity after t years is

$$P + P(1 + r) + P(1 + r)^2 + P(1 + r)^3 + \cdots + P(1 + r)^{t-1}.$$

| Deposit of P dollars at end of year t | | First-year deposit of P dollars with interest earned over $t - 1$ years |

This is the sum of the terms of a geometric sequence with first term P and commor ratio $1 + r$. We use the formula

$$S_n = \frac{a_1(1 - r^n)}{1 - r}$$

to find the sum of the terms:

$$S_t = \frac{P[1 - (1 + r)^t]}{1 - (1 + r)} = \frac{P[1 - (1 + r)^t]}{-r} = \frac{P[(1 + r)^t - 1]}{r}.$$

This formula gives the value of an annuity after t years if interest is compounded once a year. We can adjust the formula to find the value of an annuity if equal pay ments are made at the end of each of n yearly compounding periods.

Value of an Annuity: Interest Compounded n Times per Year

If P is the deposit made at the end of each compounding period for an annuity at r percent annual interest compounded n times per year, the value, A, of the annuity after t years is

$$A = \frac{P\left[\left(1 + \dfrac{r}{n}\right)^{nt} - 1\right]}{\dfrac{r}{n}}.$$

EXAMPLE 7 Determining the Value of an Annuity

To save for retirement, you decide to deposit $1000 into an IRA at the end of eacl year for the next 30 years. If the interest rate is 10% per year compounded annually find the value of the IRA after 30 years.

Solution   The annuity involves 30 year-end deposits of $P = \$1000$. The interes rate is 10%: $r = 0.10$. Because the deposits are made once a year and the interest i compounded once a year, $n = 1$. The number of years is 30: $t = 30$. We replace th variables in the formula for the value of an annuity with these numbers.

$$A = \dfrac{P\left[\left(1 + \dfrac{r}{n}\right)^{nt} - 1\right]}{\dfrac{r}{n}}$$

$$A = \dfrac{1000\left[\left(1 + \dfrac{0.10}{1}\right)^{1\cdot30} - 1\right]}{\dfrac{0.10}{1}} \approx 164{,}494$$

The value of the IRA at the end of 30 years is approximately $164,494.

Check Point 7 If $3000 is deposited into an IRA at the end of each year for 40 years and the interest rate is 10% per year compounded annually, find the value of the IRA after 40 years.

Geometric Series

⑥ Use the formula for the sum of an infinite geometric series.

An infinite sum of the form

$$a_1 + a_1r + a_1r^2 + a_1r^3 + \cdots + a_1r^{n-1} + \cdots$$

with first term a_1 and common ratio r is called an **infinite geometric series**. How can we determine which infinite geometric series have sums and which do not? We look at what happens to r^n as n gets larger in the formula for the sum of the first n terms of this series, namely

$$S_n = \dfrac{a_1(1 - r^n)}{1 - r}.$$

If r is any number between -1 and 1, that is, $-1 < r < 1$, the term r^n approaches 0 as n gets larger. For example, consider what happens to r^n for $r = \frac{1}{2}$:

$$\left(\dfrac{1}{2}\right)^1 = \dfrac{1}{2} \quad \left(\dfrac{1}{2}\right)^2 = \dfrac{1}{4} \quad \left(\dfrac{1}{2}\right)^3 = \dfrac{1}{8} \quad \left(\dfrac{1}{2}\right)^4 = \dfrac{1}{16} \quad \left(\dfrac{1}{2}\right)^5 = \dfrac{1}{32} \quad \left(\dfrac{1}{2}\right)^6 = \dfrac{1}{64}.$$

These numbers are approaching 0 as n gets larger.

Take another look at the formula for the sum of the first n terms of a geometric sequence.

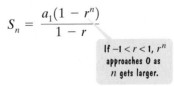

$$S_n = \dfrac{a_1(1 - r^n)}{1 - r}$$

If $-1 < r < 1$, r^n approaches 0 as n gets larger.

Let us replace r^n with 0 in the formula for S_n. This change gives us a formula for the sum of an infinite geometric series with a common ratio between -1 and 1.

The Sum of an Infinite Geometric Series

If $-1 < r < 1$ (equivalently, $|r| < 1$), then the sum of the infinite geometric series

$$a_1 + a_1r + a_1r^2 + a_1r^3 + \cdots$$

in which a_1 is the first term and r is the common ratio is given by

$$S = \dfrac{a_1}{1 - r}.$$

If $|r| \geq 1$, the infinite series does not have a sum.

To use the formula for the sum of an infinite geometric series, we need t
know the first term and the common ratio. For example, consider

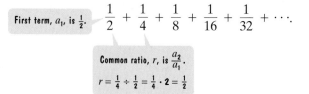

First term, a_1, is $\frac{1}{2}$.

$$\frac{1}{2} + \frac{1}{4} + \frac{1}{8} + \frac{1}{16} + \frac{1}{32} + \cdots.$$

Common ratio, r, is $\frac{a_2}{a_1}$.

$r = \frac{1}{4} \div \frac{1}{2} = \frac{1}{4} \cdot 2 = \frac{1}{2}$

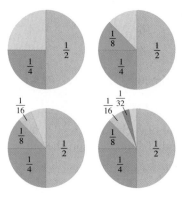

Figure 10.7 The sum
$\frac{1}{2} + \frac{1}{4} + \frac{1}{8} + \frac{1}{16} + \frac{1}{32} + \cdots$ is
approaching 1.

With $r = \dfrac{1}{2}$, the condition that $|r| < 1$ is met, so the infinite geometric serie

has a sum given by $S = \dfrac{a_1}{1-r}$. The sum of the series is found as follows:

$$\frac{1}{2} + \frac{1}{4} + \frac{1}{8} + \frac{1}{16} + \frac{1}{32} + \cdots = \frac{a_1}{1-r} = \frac{\frac{1}{2}}{1-\frac{1}{2}} = \frac{\frac{1}{2}}{\frac{1}{2}} = 1.$$

Thus, the sum of the infinite geometric series is 1. Notice how this is illustrated i
Figure 10.7. As more terms are included, the sum is approaching the area of on
complete circle.

EXAMPLE 8   Finding the Sum of an Infinite Geometric Series

Find the sum of the infinite geometric series: $\frac{3}{8} - \frac{3}{16} + \frac{3}{32} - \frac{3}{64} + \cdots$.

Solution   Before finding the sum, we must find the common ratio.

$$r = \frac{a_2}{a_1} = \frac{-\dfrac{3}{16}}{\dfrac{3}{8}} = -\frac{3}{16} \cdot \frac{8}{3} = -\frac{1}{2}$$

Because $r = -\frac{1}{2}$, the condition that $|r| < 1$ is met. Thus, the infinite geometric serie
has a sum.

$$S = \frac{a_1}{1-r}$$ This is the formula for the sum of an infinite
geometric series. Let $a_1 = \dfrac{3}{8}$ and $r = -\dfrac{1}{2}$.

$$= \frac{\dfrac{3}{8}}{1 - \left(-\dfrac{1}{2}\right)} = \frac{\dfrac{3}{8}}{\dfrac{3}{2}} = \frac{3}{8} \cdot \frac{2}{3} = \frac{1}{4}$$

Thus, the sum of $\frac{3}{8} - \frac{3}{16} + \frac{3}{32} - \frac{3}{64} + \cdots$ is $\frac{1}{4}$. Put in an informal way, as we continu
to add more and more terms, the sum is approximately $\frac{1}{4}$.

Check Point 8   Find the sum of the infinite geometric series: $3 + 2 + \frac{4}{3} + \frac{8}{9} + \cdots$.

We can use the formula for the sum of an infinite geometric series to express
repeating decimal as a fraction in lowest terms.

EXAMPLE 9   Writing a Repeating Decimal as a Fraction

Express $0.\overline{7}$ as a fraction in lowest terms.

Solution

$$0.\overline{7} = 0.7777\ldots = \frac{7}{10} + \frac{7}{100} + \frac{7}{1000} + \frac{7}{10,000} + \cdots$$

Observe that $0.\overline{7}$ is an infinite geometric series with first term $\frac{7}{10}$ and common ratio $\frac{1}{10}$. Because $r = \frac{1}{10}$, the condition that $|r| < 1$ is met. Thus, we can use our formula to find the sum. Therefore,

$$0.\overline{7} = \frac{a_1}{1 - r} = \frac{\frac{7}{10}}{1 - \frac{1}{10}} = \frac{\frac{7}{10}}{\frac{9}{10}} = \frac{7}{10} \cdot \frac{10}{9} = \frac{7}{9}.$$

An equivalent fraction for $0.\overline{7}$ is $\frac{7}{9}$.

Check Point **9** Express $0.\overline{9}$ as a fraction in lowest terms.

Infinite geometric series have many applications, as illustrated in Example 10.

EXAMPLE 10   Tax Rebates and the Multiplier Effect

A tax rebate that returns a certain amount of money to taxpayers can have a total effect on the economy that is many times this amount. In economics, this phenomenon is called the **multiplier effect**. Suppose, for example, that the government reduces taxes so that each consumer has $2000 more income. The government assumes that each person will spend 70% of this (= $1400). The individuals and businesses receiving this $1400 in turn spend 70% of it (= $980), creating extra income for other people to spend, and so on. Determine the total amount spent on consumer goods from the initial $2000 tax rebate.

$1400

70% is spent.

$980

70% is spent.

$686

Solution   The total amount spent is given by the infinite geometric series

$$1400 + 980 + 686 + \cdots.$$

70% of 1400   70% of 980

The first term is 1400: $a_1 = 1400$. The common ratio is 70%, or 0.7: $r = 0.7$. Because $r = 0.7$, the condition that $|r| < 1$ is met. Thus, we can use our formula to find the sum. Therefore,

$$1400 + 980 + 686 + \cdots = \frac{a_1}{1 - r} = \frac{1400}{1 - 0.7} \approx 4667.$$

This means that the total amount spent on consumer goods from the initial $2000 rebate is approximately $4667.

Check Point **10** Rework Example 10 and determine the total amount spent on consumer goods with a $1000 tax rebate and 80% spending down the line.

EXERCISE SET 10.3

Practice Exercises

In Exercises 1–8, write the first five terms of each geometric sequence.

1. $a_1 = 5, \quad r = 3$

2. $a_1 = 4, \quad r = 3$

3. $a_1 = 20, \quad r = \frac{1}{2}$

4. $a_1 = 24, \quad r = \frac{1}{3}$

5. $a_n = -4a_{n-1}, \quad a_1 = 10$

6. $a_n = -3a_{n-1}, \quad a_1 = 10$

7. $a_n = -5a_{n-1}, \quad a_1 = -6$

8. $a_n = -6a_{n-1}, \quad a_1 = -2$

In Exercises 9–16, use the formula for the general term (the nth term) of a geometric sequence to find the indicated term of each sequence with the given first term, a_1, and common ratio, r.

9. Find a_8 when $a_1 = 6, r = 2$.

10. Find a_8 when $a_1 = 5, r = 3$.

11. Find a_{12} when $a_1 = 5, r = -2$.

12. Find a_{12} when $a_1 = 4, r = -2$.

13. Find a_{40} when $a_1 = 1000, r = -\frac{1}{2}$.

14. Find a_{30} when $a_1 = 8000, r = -\frac{1}{2}$.

15. Find a_8 when $a_1 = 1,000,000, r = 0.1$.

16. Find a_8 when $a_1 = 40,000, r = 0.1$.

In Exercises 17–24, write a formula for the general term (the nth term) of each geometric sequence. Then use the formula for a_n to find a_7, the seventh term of the sequence.

17. $3, 12, 48, 192, \ldots$

18. $3, 15, 75, 375, \ldots$

19. $18, 6, 2, \frac{2}{3}, \ldots$

20. $12, 6, 3, \frac{3}{2}, \ldots$

21. $1.5, -3, 6, -12, \ldots$

22. $5, -1, \frac{1}{5}, -\frac{1}{25}, \ldots$

23. $0.0004, -0.004, 0.04, -0.4, \ldots$

24. $0.0007, -0.007, 0.07, -0.7, \ldots$

Use the formula for the sum of the first n terms of a geometric sequence to solve Exercises 25–30.

25. Find the sum of the first 12 terms of the geometric sequence: $2, 6, 18, 54, \ldots$.

26. Find the sum of the first 12 terms of the geometric sequence: $3, 6, 12, 24, \ldots$.

27. Find the sum of the first 11 terms of the geometric sequence: $3, -6, 12, -24, \ldots$.

28. Find the sum of the first 11 terms of the geometric sequence: $4, -12, 36, -108, \ldots$.

29. Find the sum of the first 14 terms of the geometric sequence: $-\frac{3}{2}, 3, -6, 12, \ldots$.

30. Find the sum of the first 14 terms of the geometric sequence: $-\frac{1}{24}, \frac{1}{12}, -\frac{1}{6}, \frac{1}{3}, \ldots$.

In Exercises 31–36, find the indicated sum. Use the formula for the sum of the first n terms of a geometric sequence.

31. $\displaystyle\sum_{i=1}^{8} 3^i$

32. $\displaystyle\sum_{i=1}^{6} 4^i$

33. $\displaystyle\sum_{i=1}^{10} 5 \cdot 2^i$

34. $\displaystyle\sum_{i=1}^{7} 4(-3)^i$

35. $\displaystyle\sum_{i=1}^{6} \left(\frac{1}{2}\right)^{i+1}$

36. $\displaystyle\sum_{i=1}^{6} \left(\frac{1}{3}\right)^{i+1}$

In Exercises 37–44, find the sum of each infinite geometric series.

37. $1 + \dfrac{1}{3} + \dfrac{1}{9} + \dfrac{1}{27} + \cdots$

38. $1 + \dfrac{1}{4} + \dfrac{1}{16} + \dfrac{1}{64} + \cdots$

39. $3 + \dfrac{3}{4} + \dfrac{3}{4^2} + \dfrac{3}{4^3} + \cdots$

40. $5 + \dfrac{5}{6} + \dfrac{5}{6^2} + \dfrac{5}{6^3} + \cdots$

41. $1 - \dfrac{1}{2} + \dfrac{1}{4} - \dfrac{1}{8} + \cdots$

42. $3 - 1 + \dfrac{1}{3} - \dfrac{1}{9} + \cdots$

43. $\displaystyle\sum_{i=1}^{\infty} 8(-0.3)^{i-1}$

44. $\displaystyle\sum_{i=1}^{\infty} 12(-0.7)^{i-1}$

In Exercises 45–50, express each repeating decimal as a fraction in lowest terms.

45. $0.\overline{5} = \dfrac{5}{10} + \dfrac{5}{100} + \dfrac{5}{1000} + \dfrac{5}{10,000} + \cdots$

46. $0.\overline{1} = \dfrac{1}{10} + \dfrac{1}{100} + \dfrac{1}{1000} + \dfrac{1}{10,000} + \cdots$

47. $0.\overline{47} = \dfrac{47}{100} + \dfrac{47}{10,000} + \dfrac{47}{1,000,000} + \cdots$

48. $0.\overline{83} = \dfrac{83}{100} + \dfrac{83}{10,000} + \dfrac{83}{1,000,000} + \cdots$

49. $0.\overline{257}$

50. $0.\overline{529}$

In Exercises 51–56, the general term of a sequence is given. Determine whether the sequence is arithmetic, geometric, or neither. If the sequence is arithmetic, find the common difference; if it is geometric, find the common ratio.

51. $a_n = n + 5$

52. $a_n = n - 3$

53. $a_n = 2^n$

54. $a_n = \left(\frac{1}{2}\right)^n$

55. $a_n = n^2 + 5$

56. $a_n = n^2 - 3$

Practice Plus

In Exercises 57–62, let
$$\{a_n\} = -5, 10, -20, 40, \ldots,$$
$$\{b_n\} = 10, -5, -20, -35, \ldots,$$
and
$$\{c_n\} = -2, 1, -\tfrac{1}{2}, \tfrac{1}{4}, \ldots.$$

57. Find $a_{10} + b_{10}$.

58. Find $a_{11} + b_{11}$.

59. Find the difference between the sum of the first 10 terms of $\{a_n\}$ and the sum of the first 10 terms of $\{b_n\}$.

60. Find the difference between the sum of the first 11 terms of $\{a_n\}$ and the sum of the first 11 terms of $\{b_n\}$.

61. Find the product of the sum of the first 6 terms of $\{a_n\}$ and the sum of the infinite series containing all the terms of $\{c_n\}$.

62. Find the product of the sum of the first 9 terms of $\{a_n\}$ and the sum of the infinite series containing all the terms of $\{c_n\}$.

In Exercises 63–64, find a_2 and a_3 for each geometric sequence.

63. $8, a_2, a_3, 27$

64. $2, a_2, a_3, -54$

Application Exercises

Use the formula for the general term (the nth term) of a geometric sequence to solve Exercises 65–68.

In Exercises 65–66, suppose you save \$1 the first day of a month, \$2 the second day, \$4 the third day, and so on. That is, each day you save twice as much as you did the day before.

65. What will you put aside for savings on the fifteenth day of the month?

66. What will you put aside for savings on the thirtieth day of the month?

67. A professional baseball player signs a contract with a beginning salary of \$3,000,000 for the first year and an annual increase of 4% per year beginning in the second year. That is, beginning in year 2, the athlete's salary will be 1.04 times what it was in the previous year. What is the athlete's salary for year 7 of the contract? Round to the nearest dollar.

68. You are offered a job that pays \$30,000 for the first year with an annual increase of 5% per year beginning in the second year. That is, beginning in year 2, your salary will be 1.05 times what it was in the previous year. What can you expect to earn in your sixth year on the job?

69. The population of California from 1990 through 1997 is shown in the following table.

Year	1990	1991	1992	1993
Population in millions	29.76	30.15	30.54	30.94

Year	1994	1995	1996	1997
Population in millions	31.34	31.75	32.16	32.58

a. Divide the population for each year by the population in the preceding year. Round to three decimal places and show that the population of California is increasing geometrically.

b. Write the general term of the geometric sequence describing population growth for California n years after 1989.

c. Use your model from part (b) to estimate California's population, in millions, for the year 2000. According to the U.S. Census Bureau, California's population in 2000 was 33.87 million. How well does your geometric sequence model the actual population?

70. The population of Texas from 1990 through 1997 is shown in the following table.

Year	1990	1991	1992	1993
Population in millions	16.99	17.35	17.71	18.08

Year	1994	1995	1996	1997
Population in millions	18.46	18.85	19.25	19.65

a. Divide the population for each year by the population in the preceding year. Round to three decimal places and show that the population of Texas is increasing geometrically.

b. Write the general term of the geometric sequence describing population growth for Texas n years after 1989.

c. Use your model from part (b) to estimate Texas's population in millions for the year 2000. According to the U.S. Census Bureau, Texas's population in 2000 was 20.85 million. How well does your geometric sequence model the actual population?

Use the formula for the sum of the first n terms of a geometric sequence to solve Exercises 71–76.

In Exercises 71–72, you save $1 the first day of a month, $2 the second day, $4 the third day, continuing to double your savings each day.

71. What will your total savings be for the first 15 days?

72. What will your total savings be for the first 30 days?

73. A job pays a salary of $24,000 the first year. During the next 19 years, the salary increases by 5% each year. What is the total lifetime salary over the 20-year period? Round to the nearest dollar.

74. You are investigating two employment opportunities. Company A offers $30,000 the first year. During the next four years, the salary is guaranteed to increase by 6% per year. Company B offers $32,000 the first year, with guaranteed annual increases of 3% per year after that. Which company offers the better total salary for a five-year contract? By how much? Round to the nearest dollar.

75. A pendulum swings through an arc of 20 inches. On each successive swing, the length of the arc is 90% of the previous length.

$$20, \quad 0.9(20), \quad 0.9^2(20), \quad 0.9^3(20),\dots$$

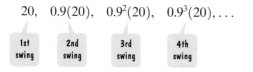

After 10 swings, what is the total length of the distance the pendulum has swung?

76. A pendulum swings through an arc of 16 inches. On each successive swing, the length of the arc is 96% of the previous length.

$$16, \quad 0.96(16), \quad (0.96)^2(16), \quad (0.96)^3(16),\dots$$

After 10 swings, what is the total length of the distance the pendulum has swung?

Use the formula for the value of an annuity to solve Exercises 77–80. Round answers to the nearest dollar.

77. To save for retirement, you decide to deposit $2500 into an IRA at the end of each year for the next 40 years. If the interest rate is 9% per year compounded annually, find the value of the IRA after 40 years.

78. You decide to deposit $100 at the end of each month into an account paying 8% interest compounded monthly to save for your child's education. How much will you save over 16 years?

79. You contribute $600 at the end of each quarter to a Tax Sheltered Annuity (TSA) paying 8% annual interest compounded quarterly. Find the value of the TSA after 18 years.

80. To save for a new home, you invest $500 per month in a mutual fund with an annual rate of return of 10% compounded monthly. How much will you have saved after four years?

Use the formula for the sum of an infinite geometric series to solve Exercises 81–83.

81. A new factory in a small town has an annual payroll of $6 million. It is expected that 60% of this money will be spent in the town by factory personnel. The people in the town who receive this money are expected to spend 60% of what they receive in the town, and so on. What is the total of all this spending, called the *total economic impact* of the factory, on the town each year?

82. How much additional spending will be generated by a $10 billion tax rebate if 60% of all income is spent?

83. If the shading process shown in the figure is continued indefinitely, what fractional part of the largest square will eventually be shaded?

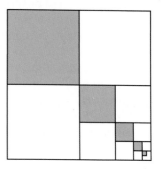

Writing in Mathematics

84. What is a geometric sequence? Give an example with your explanation.

85. What is the common ratio in a geometric sequence?

86. Explain how to find the general term of a geometric sequence.

87. Explain how to find the sum of the first n terms of a geometric sequence without having to add up all the terms.

88. What is an annuity?

89. What is the difference between a geometric sequence and an infinite geometric series?

90. How do you determine if an infinite geometric series has a sum? Explain how to find the sum of such an infinite geometric series.

91. Would you rather have $10,000,000 and a brand new BMW, or 1¢ today, 2¢ tomorrow, 4¢ on day 3, 8¢ on day 4, 16¢ on day 5, and so on, for 30 days? Explain.

92. For the first 30 days of a flu outbreak, the number of students on your campus who become ill is increasing. Which is worse: The number of students with the flu is increasing arithmetically or is increasing geometrically? Explain your answer.

Technology Exercises

93. Use the $\boxed{\text{SEQ}}$ (sequence) capability of a graphing utility and the formula you obtained for a_n to verify the value you found for a_7 in any three exercises from Exercises 17–24.

94. Use the capability of a graphing utility to calculate the sum of a sequence to verify any three of your answers to Exercises 31–36.

In Exercises 95–96, use a graphing utility to graph the function. Determine the horizontal asymptote for the graph of f and discuss its relationship to the sum of the given series.

95. Function   **Series**

$$f(x) = \frac{2\left[1 - \left(\frac{1}{3}\right)^x\right]}{1 - \frac{1}{3}} \qquad 2 + 2\left(\frac{1}{3}\right) + 2\left(\frac{1}{3}\right)^2 + 2\left(\frac{1}{3}\right)^3 + \cdots$$

96. Function   **Series**

$$f(x) = \frac{4[1 - (0.6)^x]}{1 - 0.6} \qquad 4 + 4(0.6) + 4(0.6)^2 + 4(0.6)^3 + \cdots$$

Critical Thinking Exercises

97. Which one of the following is true?
   a. The sequence $2, 6, 24, 120, \ldots$ is an example of a geometric sequence.
   b. The sum of the geometric series $\frac{1}{2} + \frac{1}{4} + \frac{1}{8} + \cdots + \frac{1}{512}$ can only be estimated without knowing precisely which terms occur between $\frac{1}{8}$ and $\frac{1}{512}$.
   c. $10 - 5 + \frac{5}{2} - \frac{5}{4} + \cdots = \dfrac{10}{1 - \frac{1}{2}}$
   d. If the nth term of a geometric sequence is $a_n = 3(0.5)^{n-1}$, the common ratio is $\frac{1}{2}$.

98. In a pest-eradication program, sterilized male flies are released into the general population each day. Ninety percent of those flies will survive a given day. How many flies should be released each day if the long-range goal of the program is to keep 20,000 sterilized flies in the population?

99. You are now 25 years old and would like to retire at age 55 with a retirement fund of $1,000,000. How much should you deposit at the end of each month for the next 30 years in an IRA paying 10% annual interest compounded monthly to achieve your goal? Round to the nearest dollar.

Group Exercise

100. Group members serve as a financial team analyzing the three options given to the professional baseball player described in the chapter opener on page 925. As a group, determine which option provides the most amount of money over the six-year contract and which provides the least. Describe one advantage and one disadvantage to each option.

CHAPTER 10
MID-CHAPTER CHECK POINT

What You Know: We learned that a sequence is a function whose domain is the set of positive integers. In an arithmetic sequence, each term after the first differs from the preceding term by a constant, the common difference, d. In a geometric sequence, each term after the first is obtained by multiplying the preceding term by a nonzero constant, the common ratio, r. We found the general term of arithmetic sequences $[a_n = a_1 + (n-1)d]$ and geometric sequences $[a_n = a_1 r^{n-1}]$ and used these formulas to find particular terms. We determined the sum of the first n terms of arithmetic sequences $\left[S_n = \frac{n}{2}(a_1 + a_n)\right]$ and geometric sequences $\left[S_n = \frac{a_1(1 - r^n)}{1 - r}\right]$. Finally, we determined the sum of an infinite geometric series,

$$a_1 + a_1 r + a_1 r^2 + a_1 r^3 + \cdots, \text{ if } -1 < r < 1 \left(S = \frac{a_1}{1 - r}\right).$$

In Exercises 1–4, write the first five terms of each sequence. Assume that d represents the common difference of an arithmetic sequence and r represents the common ratio of a geometric sequence.

1. $a_n = (-1)^{n+1}\dfrac{n}{(n - 1)!}$ **2.** $a_1 = 5, d = -3$

3. $a_1 = 5, r = -3$ **4.** $a_1 = 3, a_n = -a_{n-1} + 4$

In Exercises 5–7, write a formula for the general term (the nth term) of each sequence. Then use the formula to find the indicated term.

5. $2, 6, 10, 14, \ldots; a_{20}$ **6.** $3, 6, 12, 24, \ldots; a_{10}$

7. $\dfrac{3}{2}, 1, \dfrac{1}{2}, 0, \ldots; a_{30}$

8. Find the sum of the first ten terms of the sequence:
$$5, 10, 20, 40, \ldots.$$

9. Find the sum of the first 50 terms of the sequence:
$$-2, 0, 2, 4, \ldots.$$

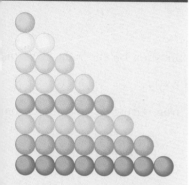

Finding the sum of consecutive positive integers leads to **triangular numbers** of the form $\dfrac{n(n + 1)}{2}$.

$\dfrac{n(n + 1)}{2}$ $\dfrac{n(n + 1)}{2}$

$n = 1:$ $n = 2:$
1   3

$\dfrac{n(n + 1)}{2}$ $\dfrac{n(n + 1)}{2}$

$n = 3:$ $n = 4:$
6   10

Solution

Step 1   Show that S_1 is true. Statement S_1 is

$$1 = \frac{1(1 + 1)}{2}.$$

Simplifying on the right, we obtain $1 = 1$. This true statement shows that S_1 is true.

Step 2   Show that if S_k is true, then S_{k+1} is true. Using S_k and S_{k+1} from Example 1(a), show that the truth of S_k,

$$1 + 2 + 3 + \cdots + k = \frac{k(k + 1)}{2},$$

implies the truth of S_{k+1},

$$1 + 2 + 3 + \cdots + (k + 1) = \frac{(k + 1)(k + 2)}{2}.$$

We will work with S_k. Because we assume that S_k is true, we add the next consecutive integer after k—namely, $k + 1$—to both sides.

$$1 + 2 + 3 + \cdots + k = \frac{k(k + 1)}{2} \qquad \text{This is } S_k, \text{ which we assume is true.}$$

$$1 + 2 + 3 + \cdots + k + (k + 1) = \frac{k(k + 1)}{2} + (k + 1) \qquad \text{Add } k + 1 \text{ to both sides of the equation.}$$

We do not have to write this k because k is understood to be the integer that precedes $k + 1$.

$$1 + 2 + 3 + \cdots + (k + 1) = \frac{k(k + 1)}{2} + \frac{2(k + 1)}{2} \qquad \text{Write the right side with a common denominator of 2.}$$

$$1 + 2 + 3 + \cdots + (k + 1) = \frac{(k + 1)}{2}(k + 2) \qquad \text{Factor out the common factor } \frac{k + 1}{2} \text{ on the right.}$$

$$1 + 2 + 3 + \cdots + (k + 1) = \frac{(k + 1)(k + 2)}{2} \qquad \text{This final result is the statement } S_{k+1}.$$

We have shown that if we assume that S_k is true and we add $k + 1$ to both sides of S_k, then S_{k+1} is also true. By the principle of mathematical induction, the statement S_n, namely,

$$1 + 2 + 3 + \cdots + n = \frac{n(n + 1)}{2}$$

is true for every positive integer n.

Check Point 2   Use mathematical induction to prove that

$$2 + 4 + 6 + \cdots + 2n = n(n + 1)$$

for all positive integers n.

EXAMPLE 3   Proving a Formula by Mathematical Induction

Use mathematical induction to prove that

$$1^2 + 2^2 + 3^2 + \cdots + n^2 = \frac{n(n + 1)(2n + 1)}{6}$$

for all positive integers n.

$S_n : 1^2 + 2^2 + 3^2 + \cdots + n^2$

$$= \frac{n(n + 1)(2n + 1)}{6}$$

The given statement (repeated)

Solution

Step 1 Show that S_1 is true. Statement S_1 is

$$1^2 = \frac{1(1 + 1)(2 \cdot 1 + 1)}{6}.$$

Simplifying, we obtain $1 = \dfrac{1 \cdot 2 \cdot 3}{6}$. Further simplification on the right gives the statement $1 = 1$. This true statement shows that S_1 is true.

Step 2 Show that if S_k is true, then S_{k+1} is true. Using S_k and S_{k+1} from Example 1(b), show that the truth of

$$S_k : 1^2 + 2^2 + 3^2 + \cdots + k^2 = \frac{k(k + 1)(2k + 1)}{6}$$

implies the truth of

$$S_{k+1} : 1^2 + 2^2 + 3^2 + \cdots + (k + 1)^2 = \frac{(k + 1)(k + 2)(2k + 3)}{6}.$$

We will work with S_k. Because we assume that S_k is true, we add the square of the next consecutive integer after k—namely, $(k + 1)^2$—to both sides of the equation.

$$1^2 + 2^2 + 3^2 + \cdots + k^2 = \frac{k(k + 1)(2k + 1)}{6}$$

This is S_k, assumed to be true. We must work with this and show S_{k+1} is true.

$$1^2 + 2^2 + 3^2 + \cdots + k^2 + (k + 1)^2 = \frac{k(k + 1)(2k + 1)}{6} + (k + 1)^2$$

Add $(k + 1)^2$ to both sides.

$$1^2 + 2^2 + 3^2 + \cdots + (k + 1)^2 = \frac{k(k + 1)(2k + 1)}{6} + \frac{6(k + 1)^2}{6}$$

It is not necessary to write k^2 on the left. Express the right side with the least common denominator, 6.

$$= \frac{(k + 1)}{6}[k(2k + 1) + 6(k + 1)]$$

Factor out the common factor $\dfrac{k + 1}{6}$.

$$= \frac{(k + 1)}{6}(2k^2 + 7k + 6)$$

Multiply and combine like terms.

$$= \frac{(k + 1)}{6}(k + 2)(2k + 3)$$

Factor $2k^2 + 7k + 6$.

$$= \frac{(k + 1)(k + 2)(2k + 3)}{6}$$

This final statement is S_{k+1}.

We have shown that if we assume that S_k is true, and we add $(k + 1)^2$ to both sides of S_k, then S_{k+1} is also true. By the principle of mathematical induction, the statement S_n, namely,

$$1^2 + 2^2 + 3^2 + \cdots + n^2 = \frac{n(n + 1)(2n + 1)}{6}$$

is true for every positive integer n.

Check Point 3 Use mathematical induction to prove that

$$1^3 + 2^3 + 3^3 + \cdots + n^3 = \frac{n^2(n + 1)^2}{4}$$

for all positive integers n.

Example 4 illustrates how mathematical induction can be used to prove statements about positive integers that do not involve sums.

EXAMPLE 4   Using the Principle of Mathematical Induction

Prove that 2 is a factor of $n^2 + 5n$ for all positive integers n.

Solution

Step 1   Show that S_1 is true. Statement S_1 reads

$$2 \text{ is a factor of } 1^2 + 5 \cdot 1.$$

Simplifying the arithmetic, the statement reads

$$2 \text{ is a factor of } 6.$$

This statement is true: that is, $6 = 2 \cdot 3$. This shows that S_1 is true.

Step 2   Show that if S_k is true, then S_{k+1} is true. Let's write S_k and S_{k+1}:

$$S_k: \quad 2 \text{ is a factor of } k^2 + 5k.$$
$$S_{k+1}: \quad 2 \text{ is a factor of } (k+1)^2 + 5(k+1).$$

We can rewrite statement S_{k+1} by simplifying the algebraic expression in the statement as follows:

$$(k+1)^2 + 5(k+1) = k^2 + 2k + 1 + 5k + 5 = k^2 + 7k + 6.$$

Use the formula $(A + B)^2 = A^2 + 2AB + B^2.$

Statement S_{k+1} now reads

$$2 \text{ is a factor of } k^2 + 7k + 6.$$

We need to use statement S_k—that is, 2 is a factor of $k^2 + 5k$—to prove statement S_{k+1}. We do this as follows:

$$k^2 + 7k + 6 = (k^2 + 5k) + (2k + 6) = (k^2 + 5k) + 2(k + 3).$$

We assume that 2 is a factor of $k^2 + 5k$ because we assume S_k is true.

Factoring the last two terms shows that 2 is a factor of $2k + 6$.

The voice balloons show that 2 is a factor of $k^2 + 5k$ and of $2(k + 3)$. Thus, if S_k is true, 2 is a factor of the sum $(k^2 + 5k) + 2(k + 3)$, or of $k^2 + 7k + 6$. This is precisely statement S_{k+1}. We have shown that if we assume that S_k is true, then S_{k+1} is also true. By the principle of mathematical induction, the statement S_n, namely 2 is a factor of $n^2 + 5n$, is true for every positive integer n.

Check Point 4   Prove that 2 is a factor of $n^2 + n$ for all positive integers n.

EXERCISE SET 10.4

Practice Exercises

In Exercises 1–4, a statement S_n about the positive integers is given. Write statements S_1, S_2, and S_3, and show that each of these statements is true.

1. S_n: $1 + 3 + 5 + \cdots + (2n - 1) = n^2$

2. S_n: $3 + 4 + 5 + \cdots + (n + 2) = \dfrac{n(n + 5)}{2}$

3. S_n: 2 is a factor of $n^2 - n$.

4. S_n: 3 is a factor of $n^3 - n$.

In Exercises 5–10, a statement S_n about the positive integers is given. Write statements S_k and S_{k+1}, simplifying statement S_{k+1} completely.

5. S_n: $4 + 8 + 12 + \cdots + 4n = 2n(n + 1)$

6. S_n: $3 + 4 + 5 + \cdots + (n + 2) = \dfrac{n(n + 5)}{2}$

7. S_n: $3 + 7 + 11 + \cdots + (4n - 1) = n(2n + 1)$

8. $S_n: 2 + 7 + 12 + \cdots + (5n - 3) = \dfrac{n(5n - 1)}{2}$

9. $S_n:$ 2 is a factor of $n^2 - n + 2$.

10. $S_n:$ 2 is a factor of $n^2 - n$.

In Exercises 11–24, use mathematical induction to prove that each statement is true for every positive integer n.

11. $4 + 8 + 12 + \cdots + 4n = 2n(n + 1)$

12. $3 + 4 + 5 + \cdots + (n + 2) = \dfrac{n(n + 5)}{2}$

13. $1 + 3 + 5 + \cdots + (2n - 1) = n^2$

14. $3 + 6 + 9 + \cdots + 3n = \dfrac{3n(n + 1)}{2}$

15. $3 + 7 + 11 + \cdots + (4n - 1) = n(2n + 1)$

16. $2 + 7 + 12 + \cdots + (5n - 3) = \dfrac{n(5n - 1)}{2}$

17. $1 + 2 + 2^2 + \cdots + 2^{n-1} = 2^n - 1$

18. $1 + 3 + 3^2 + \cdots + 3^{n-1} = \dfrac{3^n - 1}{2}$

19. $2 + 4 + 8 + \cdots + 2^n = 2^{n+1} - 2$

20. $\dfrac{1}{2} + \dfrac{1}{4} + \dfrac{1}{8} + \cdots + \dfrac{1}{2^n} = 1 - \dfrac{1}{2^n}$

21. $1 \cdot 2 + 2 \cdot 3 + 3 \cdot 4 + \cdots + n(n + 1) = \dfrac{n(n + 1)(n + 2)}{3}$

22. $1 \cdot 3 + 2 \cdot 4 + 3 \cdot 5 + \cdots + n(n + 2) = \dfrac{n(n + 1)(2n + 7)}{6}$

23. $\dfrac{1}{1 \cdot 2} + \dfrac{1}{2 \cdot 3} + \dfrac{1}{3 \cdot 4} + \cdots + \dfrac{1}{n(n + 1)} = \dfrac{n}{n + 1}$

24. $\dfrac{1}{2 \cdot 3} + \dfrac{1}{3 \cdot 4} + \dfrac{1}{4 \cdot 5} + \cdots + \dfrac{1}{(n + 1)(n + 2)} = \dfrac{n}{2n + 4}$

Practice Plus

In Exercises 25–34, use mathematical induction to prove that each statement is true for every positive integer n.

25. 2 is a factor of $n^2 - n$.

26. 2 is a factor of $n^2 + 3n$.

27. 6 is a factor of $n(n + 1)(n + 2)$.

28. 3 is a factor of $n(n + 1)(n - 1)$.

29. $\displaystyle\sum_{i=1}^{n} 5 \cdot 6^i = 6(6^n - 1)$

30. $\displaystyle\sum_{i=1}^{n} 7 \cdot 8^i = 8(8^n - 1)$

31. $n + 2 > n$

32. If $0 < x < 1$, then $0 < x^n < 1$.

33. $(ab)^n = a^n b^n$

34. $\left(\dfrac{a}{b}\right)^n = \dfrac{a^n}{b^n}$

Writing in Mathematics

35. Explain how to use mathematical induction to prove that a statement is true for every positive integer n.

36. Consider the statement S_n given by

$$n^2 - n + 41 \text{ is prime.}$$

Although $S_1, S_2, \ldots, S_{40}$ are true, S_{41} is false. Verify that S_{41} is false. Then describe how this is illustrated by the dominoes in the figure. What does this tell you about a pattern, or formula that seems to work for several values of n?

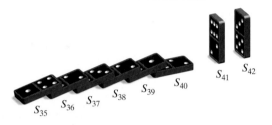

S_{35} S_{36} S_{37} S_{38} S_{39} S_{40} S_{41} S_{42}

Critical Thinking Exercises

Some statements are false for the first few positive integers, but true for some positive integer on. In these instances, you can prove S_n for $n \geq k$ by showing that S_k is true and that S_k implies S_{k+1}. Use this extended principle of mathematical induction to prove that each statement in Exercises 37–38 is true.

37. Prove that $n^2 > 2n + 1$ for $n \geq 3$. Show that the formula is true for $n = 3$ and then use step 2 of mathematical induction.

38. Prove that $2^n > n^2$ for $n \geq 5$. Show that the formula is true for $n = 5$ and then use step 2 of mathematical induction.

In Exercises 39–40, find S_1 through S_5 and then use the pattern to make a conjecture about S_n. Prove the conjectured formula for S_n by mathematical induction.

39. $S_n: \dfrac{1}{4} + \dfrac{1}{12} + \dfrac{1}{24} + \cdots + \dfrac{1}{2n(n + 1)} = ?$

40. $S_n: \left(1 - \dfrac{1}{2}\right)\left(1 - \dfrac{1}{3}\right)\left(1 - \dfrac{1}{4}\right) \cdots \left(1 - \dfrac{1}{n + 1}\right) = ?$

Group Exercise

41. Fermat's most notorious theorem, described in the section opener on page 959, baffled the greatest minds for more than three centuries. In 1994, after ten years of work, Princeton University's Andrew Wiles proved Fermat's Last Theorem. *People* magazine put him on its list of "the 25 most intriguing people of the year," the Gap asked him to model jeans, and Barbara Walters chased him for an interview. "Who's Barbara Walters?" asked the bookish Wiles, who had somehow gone through life without a television.

Using the 1993 PBS documentary "Solving Fermat: Andrew Wiles" or information about Andrew Wiles on the Internet, research and present a group seminar on what Wiles did to prove Fermat's Last Theorem, problems along the way, and the role of mathematical induction in the proof.

SECTION 10.5   *The Binomial Theorem*

Objectives

① Evaluate a binomial coefficient.

② Expand a binomial raised to a power.

③ Find a particular term in a binomial expansion.

Galaxies are groupings of billions of stars bound together by gravity. Some galaxies, such as the Centaurus galaxy shown here, are elliptical in shape.

Is mathematics discovered or invented? For example, planets revolve in elliptical orbits. Does that mean that the ellipse is out there, waiting for the mind to discover it? Or do people create the definition of an ellipse just as they compose a song? And is it possible for the same mathematics to be discovered/invented by independent researchers separated by time, place, and culture? This is precisely what occurred when mathematicians attempted to find efficient methods for raising binomials to higher and higher powers, such as

$$(x + 2)^3, (x + 2)^4, (x + 2)^5, (x + 2)^6,$$

and so on. In this section, we study higher powers of binomials and a method first discovered/invented by great minds in Eastern and Western cultures working independently.

① Evaluate a binomial coefficient.

Binomial Coefficients

Before turning to powers of binomials, we introduce a special notation that uses factorials.

Definition of a Binomial Coefficient $\binom{n}{r}$

For nonnegative integers n and r, with $n \geq r$, the expression $\binom{n}{r}$ (read "n above r") is called a **binomial coefficient** and is defined by

$$\binom{n}{r} = \frac{n!}{r!(n-r)!}.$$

The symbol $_nC_r$ is often used in place of $\binom{n}{r}$ to denote binomial coefficients.

Technology

Graphing utilities can compute binomial coefficients. For example, to find $\binom{6}{2}$, many utilities require the sequence

6 $\boxed{_nC_r}$ 2 $\boxed{\text{ENTER}}$.

The graphing utility will display 15. Consult your manual and verify the other evaluations in Example 1.

EXAMPLE 1   Evaluating Binomial Coefficients

Evaluate: **a.** $\binom{6}{2}$ **b.** $\binom{3}{0}$ **c.** $\binom{9}{3}$ **d.** $\binom{4}{4}$.

Solution   In each case, we apply the definition of the binomial coefficient.

a. $\binom{6}{2} = \frac{6!}{2!(6-2)!} = \frac{6!}{2!4!} = \frac{6 \cdot 5 \cdot 4!}{2 \cdot 1 \cdot 4!} = 15$

b. $\dbinom{3}{0} = \dfrac{3!}{0!(3-0)!} = \dfrac{3!}{0!3!} = \dfrac{1}{1} = 1$

> Remember that $0! = 1$.

c. $\dbinom{9}{3} = \dfrac{9!}{3!(9-3)!} = \dfrac{9!}{3!6!} = \dfrac{9 \cdot 8 \cdot 7 \cdot \cancel{6!}}{3 \cdot 2 \cdot 1 \cdot \cancel{6!}} = 84$

d. $\dbinom{4}{4} = \dfrac{4!}{4!(4-4)!} = \dfrac{\cancel{4!}}{\cancel{4!}0!} = \dfrac{1}{1} = 1$

Check Point 1 Evaluate:

a. $\dbinom{6}{3}$ **b.** $\dbinom{6}{0}$ **c.** $\dbinom{8}{2}$ **d.** $\dbinom{3}{3}$.

② Expand a binomial raised to a power.

The Binomial Theorem

When we write out the *binomial expression* $(a + b)^n$, where n is a positive integer, a number of patterns begin to appear.

$$(a + b)^1 = a + b$$
$$(a + b)^2 = a^2 + 2ab + b^2$$
$$(a + b)^3 = a^3 + 3a^2b + 3ab^2 + b^3$$
$$(a + b)^4 = a^4 + 4a^3b + 6a^2b^2 + 4ab^3 + b^4$$
$$(a + b)^5 = a^5 + 5a^4b + 10a^3b^2 + 10a^2b^3 + 5ab^4 + b^5$$

Each expanded form of the binomial expression is a polynomial. Observe the following patterns:

1. The first term in the expansion of $(a + b)^n$ is a^n. The exponents on a decrease by 1 in each successive term.

2. The exponents on b in the expansion of $(a + b)^n$ increase by 1 in each successive term. In the first term, the exponent on b is 0. (Because $b^0 = 1$, b is not shown in the first term.) The last term is b^n.

3. The sum of the exponents on the variables in any term in the expansion of $(a + b)^n$ is equal to n.

4. The number of terms in the polynomial expansion is one greater than the power of the binomial, n. There are $n + 1$ terms in the expanded form of $(a + b)^n$.

Using these observations, the variable parts of the expansion of $(a + b)^6$ are

$$a^6, \quad a^5b, \quad a^4b^2, \quad a^3b^3, \quad a^2b^4, \quad ab^5, \quad b^6.$$

The first term is a^6, with the exponents on a decreasing by 1 in each successive term. The exponents on b increase from 0 to 6, with the last term being b^6. The sum of the exponents in each term is equal to 6.

We can generalize from these observations to obtain the variable parts of the expansion of $(a + b)^n$. They are

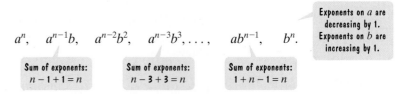

$$a^n, \quad a^{n-1}b, \quad a^{n-2}b^2, \quad a^{n-3}b^3, \dots, \quad ab^{n-1}, \quad b^n.$$

> Exponents on a are decreasing by 1. Exponents on b are increasing by 1.

> Sum of exponents:
> $n - 1 + 1 = n$

> Sum of exponents:
> $n - 3 + 3 = n$

> Sum of exponents:
> $1 + n - 1 = n$

If we use binomial coefficients and the pattern for the variable part of each term, a formula called the **Binomial Theorem** can be used to expand any positive integral power of a binomial.

EXAMPLE 4   Finding a Single Term of a Binomial Expansion

Find the fourth term in the expansion of $(3x + 2y)^7$.

Solution   The fourth term in the expansion of $(3x + 2y)^7$ contains $(2y)^3$. To find the fourth term, first note that $4 = 3 + 1$. Equivalently, the fourth term of $(3x + 2y)^7$ is the $(3 + 1)$st term. Thus, $r = 3, a = 3x, b = 2y,$ and $n = 7$. The fourth term is

$$\binom{7}{3}(3x)^{7-3}(2y)^3 = \binom{7}{3}(3x)^4(2y)^3 = \frac{7!}{3!(7-3)!}(3x)^4(2y)^3.$$

Use the formula for the $(r+1)$st term of $(a+b)^n$: $\binom{n}{r}a^{n-r}b^r$.

We use $\binom{n}{r} = \frac{n!}{r!(n-r)!}$ to evaluate $\binom{7}{3}$.

Now we need to evaluate the factorial expression and raise $3x$ and $2y$ to the indicated powers. We obtain

$$\frac{7!}{3!4!}(81x^4)(8y^3) = \frac{7 \cdot 6 \cdot 5 \cdot 4!}{3 \cdot 2 \cdot 1 \cdot 4!}(81x^4)(8y^3) = 35(81x^4)(8y^3) = 22{,}680x^4y^3.$$

The fourth term of $(3x + 2y)^7$ is $22{,}680x^4y^3$.

Check Point 4   Find the fifth term in the expansion of $(2x + y)^9$.

The Universality of Mathematics

Pascal's triangle is an array of numbers showing coefficients of the terms in the expansions of $(a + b)^n$. Although credited to French mathematician Blaise Pascal (1623–1662), the triangular array of numbers appeared in a Chinese document printed in 1303. The Binomial Theorem was known in Eastern cultures prior to its discovery in Europe. The same mathematics is often discovered/invented by independent researchers separated by time, place, and culture.

Binomial Expansions

$(a + b)^0 = 1$
$(a + b)^1 = a + b$
$(a + b)^2 = a^2 + 2ab + b^2$
$(a + b)^3 = a^3 + 3a^2b + 3ab^2 + b^3$
$(a + b)^4 = a^4 + 4a^3b + 6a^2b^2 + 4ab^3 + b^4$
$(a + b)^5 = a^5 + 5a^4b + 10a^3b^2 + 10a^2b^3 + 5ab^4 + b^5$

Pascal's Triangle
Coefficients in the Expansions

```
            1
          1   1
         1   2   1
        1   3   3   1
      1   4   6   4   1
    1   5  10  10   5   1
  1   6  15  20  15   6   1
 1  7  21  35  35  21   7  1
1  8  28  56  70  56  28  8  1
```

Chinese Document: 1303

EXERCISE SET 10.5

Practice Exercises

In Exercises 1–8, evaluate the given binomial coefficient.

1. $\binom{8}{3}$ **2.** $\binom{7}{2}$ **3.** $\binom{12}{1}$

4. $\binom{11}{1}$ **5.** $\binom{6}{6}$ **6.** $\binom{15}{2}$

7. $\binom{100}{2}$ **8.** $\binom{100}{98}$

In Exercises 9–30, use the Binomial Theorem to expand each binomial and express the result in simplified form.

9. $(x + 2)^3$ **10.** $(x + 4)^3$
11. $(3x + y)^3$ **12.** $(x + 3y)^3$
13. $(5x - 1)^3$ **14.** $(4x - 1)^3$
15. $(2x + 1)^4$ **16.** $(3x + 1)^4$
17. $(x^2 + 2y)^4$ **18.** $(x^2 + y)^4$
19. $(y - 3)^4$ **20.** $(y - 4)^4$
21. $(2x^3 - 1)^4$ **22.** $(2x^5 - 1)^4$
23. $(c + 2)^5$ **24.** $(c + 3)^5$
25. $(x - 1)^5$ **26.** $(x - 2)^5$
27. $(3x - y)^5$ **28.** $(x - 3y)^5$
29. $(2a + b)^6$ **30.** $(a + 2b)^6$

In Exercises 31–38, write the first three terms in each binomial expansion, expressing the result in simplified form.

31. $(x + 2)^8$ **32.** $(x + 3)^8$
33. $(x - 2y)^{10}$ **34.** $(x - 2y)^9$
35. $(x^2 + 1)^{16}$ **36.** $(x^2 + 1)^{17}$
37. $(y^3 - 1)^{20}$ **38.** $(y^3 - 1)^{21}$

In Exercises 39–48, find the term indicated in each expansion.

39. $(2x + y)^6$; third term   **40.** $(x + 2y)^6$; third term
41. $(x - 1)^9$; fifth term   **42.** $(x - 1)^{10}$; fifth term
43. $(x^2 + y^3)^8$; sixth term   **44.** $(x^3 + y^2)^8$; sixth term
45. $\left(x - \frac{1}{2}\right)^9$; fourth term   **46.** $\left(x + \frac{1}{2}\right)^8$; fourth term
47. $(x^2 + y)^{22}$; the term containing y^{14}
48. $(x + 2y)^{10}$; the term containing y^6

Practice Plus

In Exercises 49–52, use the Binomial Theorem to expand each expression and write the result in simplified form.

49. $(x^3 + x^{-2})^4$ **50.** $(x^2 + x^{-3})^4$

51. $\left(x^{\frac{1}{3}} - x^{-\frac{1}{3}}\right)^3$ **52.** $\left(x^{\frac{2}{3}} - \frac{1}{\sqrt[3]{x}}\right)^3$

In Exercises 53–54, find $\dfrac{f(x + h) - f(x)}{h}$ and simplify.

53. $f(x) = x^4 + 7$ **54.** $f(x) = x^5 + 8$

55. Find the middle term in the expansion of $\left(\dfrac{3}{x} + \dfrac{x}{3}\right)^{10}$.

56. Find the middle term in the expansion of $\left(\dfrac{1}{x} - x^2\right)^{12}$.

Application Exercises

Bariatrics is the field of medicine that deals with the overweight. Bariatric surgery closes off a large part of the stomach. As a result, patients eat less and have a diminished appetite. Celebrities like pop singer Carnie Wilson and the Today show's weatherman Al Roker have become no-longer-larger-than-life walking billboards for the operation. The figure shows the number of bariatric surgeries from 2000 through 2005.

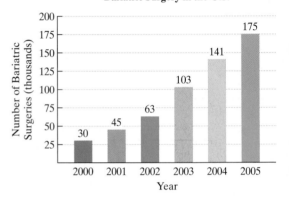

Bariatric Surgery in the U.S.

Source: American Society for Bariatric Surgery

The function

$$f(x) = -x^3 + 11x^2 + x + 31$$

models the number of bariatric surgeries, $f(x)$, in thousands, x years after 2000. Use this function to solve Exercises 57–58.

57. a. How can we adjust the function so that $x = 0$ correspond to 2003 rather than 2000? We shift the graph of f 3 unit to the left. We obtain $g(x) = f(x + 3)$. Use the Binomia Theorem to express g in descending powers of x.

b. Find $f(3)$ and $g(0)$. How well do these function value model the number shown in the bar graph?

58. a. How can we adjust the function so that $x = 0$ correspond to 2002 rather than 2000? We shift the graph of f 2 unit to the left. We obtain $g(x) = f(x + 2)$. Use the Binomia Theorem to express g in descending powers of x.

b. Find $f(4)$ and $g(2)$. How well do these function value model the number shown in the bar graph?

Writing in Mathematics

59. Explain how to evaluate $\binom{n}{r}$. Provide an example with you explanation.

60. Describe the pattern on the exponents on a in the expansio of $(a + b)^n$.

61. Describe the pattern on the exponents on b in the expansio of $(a + b)^n$.

62. What is true about the sum of the exponents on a and b i any term in the expansion of $(a + b)^n$?

73. How do you determine how many terms there are in a binomial expansion?

74. Explain how to use the Binomial Theorem to expand a binomial. Provide an example with your explanation.

75. Explain how to find a particular term in a binomial expansion without having to write out the entire expansion.

76. Describe how you would use mathematical induction to prove

$$(a + b)^n = \binom{n}{0}a^n + \binom{n}{1}a^{n-1}b + \binom{n}{2}a^{n-2}b^2$$

$$+ \cdots + \binom{n}{n-1}ab^{n-1} + \binom{n}{n}b^n.$$

What happens when $n = 1$? Write the statement that we assume to be true. Write the statement that we must prove. What must be done to the left side of the assumed statement to make it look like the left side of the statement that must be proved? (More detail on the actual proof is found in Exercise 78.)

Technology Exercises

67. Use the $\boxed{_nC_r}$ key on a graphing utility to verify your answers in Exercises 1–8.

In Exercises 68–69, graph each of the functions in the same viewing rectangle. Describe how the graphs illustrate the Binomial Theorem.

68. $f_1(x) = (x + 2)^3$
$f_2(x) = x^3$
$f_3(x) = x^3 + 6x^2$
$f_4(x) = x^3 + 6x^2 + 12x$
$f_5(x) = x^3 + 6x^2 + 12x + 8$
Use a $[-10, 10, 1]$ by $[-30, 30, 10]$ viewing rectangle.

69. $f_1(x) = (x + 1)^4$
$f_2(x) = x^4$
$f_3(x) = x^4 + 4x^3$
$f_4(x) = x^4 + 4x^3 + 6x^2$
$f_5(x) = x^4 + 4x^3 + 6x^2 + 4x$
$f_6(x) = x^4 + 4x^3 + 6x^2 + 4x + 1$
Use a $[-5, 5, 1]$ by $[-30, 30, 10]$ viewing rectangle.

In Exercises 70–72, use the Binomial Theorem to find a polynomial expansion for each function. Then use a graphing utility and an approach similar to the one in Exercises 68 and 69 to verify the expansion.

70. $f_1(x) = (x - 1)^3$ **71.** $f_1(x) = (x - 2)^4$

72. $f_1(x) = (x + 2)^6$

Critical Thinking Exercises

73. Which one of the following is true?

a. The binomial expansion for $(a + b)^n$ contains n terms.

b. The Binomial Theorem can be written in condensed form as $(a + b)^n = \sum_{r=0}^{n} \binom{n}{r}a^{n-r}b^r$.

c. The sum of the binomial coefficients in $(a + b)^n$ cannot be 2^n.

d. There are no values of a and b such that $(a + b)^4 = a^4 + b^4$.

74. Use the Binomial Theorem to expand and then simplify the result: $(x^2 + x + 1)^3$.

Hint: Write $x^2 + x + 1$ as $x^2 + (x + 1)$.

75. Find the term in the expansion of $(x^2 + y^2)^5$ containing x^4 as a factor.

76. Prove that

$$\binom{n}{r} = \binom{n}{n - r}.$$

77. Show that

$$\binom{n}{r} + \binom{n}{r + 1} = \binom{n + 1}{r + 1}.$$

Hints:

$$(n - r)! = (n - r)(n - r - 1)!$$
$$(r + 1)! = (r + 1)r!$$

78. Follow the outline below and use mathematical induction to prove the Binomial Theorem:

$$(a + b)^n = \binom{n}{0}a^n + \binom{n}{1}a^{n-1}b + \binom{n}{2}a^{n-2}b^2$$

$$+ \cdots + \binom{n}{n-1}ab^{n-1} + \binom{n}{n}b^n.$$

a. Verify the formula for $n = 1$.

b. Replace n with k and write the statement that is assumed true. Replace n with $k + 1$ and write the statement that must be proved.

c. Multiply both sides of the statement assumed to be true by $a + b$. Add exponents on the left. On the right, distribute a and b, respectively.

d. Collect like terms on the right. At this point, you should have

$$(a + b)^{k+1} = \binom{k}{0}a^{k+1} + \left[\binom{k}{0} + \binom{k}{1}\right]a^k b$$

$$+ \left[\binom{k}{1} + \binom{k}{2}\right]a^{k-1}b^2 + \left[\binom{k}{2} + \binom{k}{3}\right]a^{k-2}b^3$$

$$+ \cdots + \left[\binom{k}{k-1} + \binom{k}{k}\right]ab^k + \binom{k}{k}b^{k+1}$$

e. Use the result of Exercise 77 to add the binomial sums in brackets. For example, because $\binom{n}{r} + \binom{n}{r + 1}$

$$= \binom{n + 1}{r + 1}, \text{ then } \binom{k}{0} + \binom{k}{1} = \binom{k + 1}{1} \text{ and}$$

$$\binom{k}{1} + \binom{k}{2} = \binom{k + 1}{2}.$$

f. Because $\binom{k}{0} = \binom{k + 1}{0}$ (why?) and $\binom{k}{k} = \binom{k + 1}{k + 1}$ (why?), substitute these results and the results from part (e) into the equation in part (d). This should give the statement that we were required to prove in the second step of the mathematical induction process.

SECTION 10.6 *Counting Principles, Permutations, and Combinations*

Objectives

Objectives

1 Use the Fundamental Counting Principle.

2 Use the permutations formula.

3 Distinguish between permutation problems and combination problems.

4 Use the combinations formula.

Have you ever imagined what your life would be like if you won the lottery? What changes would you make? Before you fantasize about becoming a person of leisure with a staff of obedient elves, think about this: The probability of winning top prize in the lottery is about the same as the probability of being struck by lightning. There are millions of possible number combinations in lottery games and only one way of winning the grand prize. Determining the probability of winning involves calculating the chance of getting the winning combination from all possible outcomes. In this section, we begin preparing for the surprising world of probability by looking at methods for counting possible outcomes.

The Fundamental Counting Principle

 Use the Fundamental Counting Principle.

It's early morning, you're groggy, and you have to select something to wear for your 8 A.M. class. (What *were* you thinking of when you signed up for a class at that hour?!) Fortunately, your "lecture wardrobe" is rather limited—just two pairs of jeans to choose from (one blue, one black), three T-shirts to choose from (one beige, one yellow, and one blue), and two pairs of sneakers to select from (one black pair, one red pair). Your possible outfits are shown in Figure 10.9.

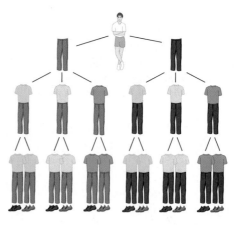

Figure 10.9 Selecting a wardrobe

The **tree diagram**, so named because of its branches, shows that you can form 12 outfits from your two pairs of jeans, three T-shirts, and two pairs of sneakers. Notice that the number of outfits can be obtained by multiplying the number of choices for jeans, 2, the number of choices for the T-shirts, 3, and the number of choices for the sneakers, 2:

$$2 \cdot 3 \cdot 2 = 12.$$

We can generalize this idea to any two or more groups of items—not just jeans, T-shirts, and sneakers—with the **Fundamental Counting Principle**:

> ### The Fundamental Counting Principle
> The number of ways in which a series of successive things can occur is found by multiplying the number of ways in which each thing can occur.

For example, if you own 30 pairs of jeans, 20 T-shirts, and 12 pairs of sneakers, you have

$$30 \cdot 20 \cdot 12 = 7200$$

choices for your wardrobe!

EXAMPLE 1   Options in Planning a Course Schedule

Next semester you are planning to take three courses—math, English, and humanities. Based on time blocks and highly recommended professors, there are 8 sections of math, 5 of English, and 4 of humanities that you find suitable. Assuming no scheduling conflicts, how many different three-course schedules are possible?

Solution   This situation involves making choices with three groups of items.

Math	English	Humanities
8 choices	*5 choices*	*4 choices*

We use the Fundamental Counting Principle to find the number of three-course schedules. Multiply the number of choices for each of the three groups:

$$8 \cdot 5 \cdot 4 = 160.$$

Thus, there are 160 different three-course schedules.

Check Point 1   A pizza can be ordered with three choices of size (small, medium, or large), four choices of crust (thin, thick, crispy, or regular), and six choices of toppings (ground beef, sausage, pepperoni, bacon, mushrooms, or onions). How many different one-topping pizzas can be ordered?

EXAMPLE 2   A Multiple-Choice Test

You are taking a multiple-choice test that has ten questions. Each of the questions has four answer choices, with one correct answer per question. If you select one of these four choices for each question and leave nothing blank, in how many ways can you answer the questions?

Solution   This situation involves making choices with ten questions.

Question 1	Question 2	Question 3	· · ·	Question 9	Question 10
4 choices	*4 choices*	*4 choices*		*4 choices*	*4 choices*

We use the Fundamental Counting Principle to determine the number of ways that you can answer the questions on the test. Multiply the number of choices, 4, for each of the ten questions.

$$4 \cdot 4 \cdot 4 \cdot 4 \cdot 4 \cdot 4 \cdot 4 \cdot 4 \cdot 4 \cdot 4 = 4^{10} = 1,048,576$$

Thus, you can answer the questions in 1,048,576 different ways.

e number of possible ways of play-
g the first four moves on each side
a game of chess is 318,979,564,000.

Are you surprised that there are over one million ways of answering a ten question multiple-choice test? Of course, there is only one way to answer the test and receive a perfect score. The probability of guessing your way into a perfect score involves calculating the chance of getting a perfect score, just one way, from all 1,048,576 possible outcomes. In short, prepare for the test and do not rely on guessing.

 Check Point 2 You are taking a multiple-choice test that has six questions. Each of the questions has three answer choices, with one correct answer per question. If you select one of these three choices for each question and leave nothing blank, in how many ways can you answer the questions?

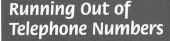
Running Out of Telephone Numbers

By the year 2020, portable telephones used for business and pleasure will all be videophones. At that time, the U.S. population is expected to be 323 million. Faxes, beepers, cell phones, computer phone lines, and business lines may result in certain areas running out of phone numbers. Solution: Add more digits!

EXAMPLE 3 Telephone Numbers in the United States

Telephone numbers in the United States begin with three-digit area codes followed by seven-digit local telephone numbers. Area codes and local telephone numbers cannot begin with 0 or 1. How many different telephone numbers are possible?

Solution This situation involves making choices with ten groups of items.

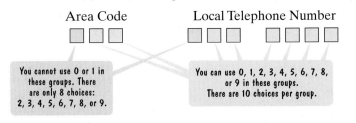

Here are the numbers of choices for each of the ten groups of items:

Area Code			Local Telephone Number						
8	10	10	8	10	10	10	10	10	10

We use the Fundamental Counting Principle to determine the number of different telephone numbers that are possible. The total number of telephone numbers possible is

$$8 \cdot 10 \cdot 10 \cdot 8 \cdot 10 \cdot 10 \cdot 10 \cdot 10 \cdot 10 \cdot 10 = 6,400,000,000.$$

There are six billion four hundred million different telephone numbers that are possible.

 Check Point 3 License plates in a particular state display two letters followed by three numbers, such as AT-887 or BB-013. How many different license plates can be manufactured?

 Use the permutations formula.

Permutations

You are the coach of a little league baseball team. There are 13 players on the team (and lots of parents hovering in the background, dreaming of stardom for their little "Barry Bonds"). You need to choose a batting order having 9 players. The order makes a difference, because, for instance, if bases are loaded and "Little Barry" is fourth or fifth at bat, his possible home run will drive in three additional runs. How many batting orders can you form?

You can choose any of 13 players for the first person at bat. Then you will have 12 players from which to choose the second batter, then 11 from which to choose the third batter, and so on. The situation can be shown as follows:

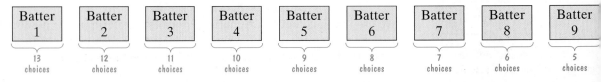

We use the Fundamental Counting Principle to find the number of batting orders. The total number of batting orders is

$$13 \cdot 12 \cdot 11 \cdot 10 \cdot 9 \cdot 8 \cdot 7 \cdot 6 \cdot 5 = 259,459,200.$$

Nearly 260 million batting orders are possible for your 13-player little league team. Each batting order is called a *permutation* of 13 players taken 9 at a time. The number of permutations of 13 players taken 9 at a time is 259,459,200.

A **permutation** is an ordered arrangement of items that occurs when

- No item is used more than once. (Each of the 9 players in the batting order bats exactly once.)
- The order of arrangement makes a difference.

We can obtain a formula for finding the number of permutations of 13 players taken 9 at a time by rewriting our computation:

$$13 \cdot 12 \cdot 11 \cdot 10 \cdot 9 \cdot 8 \cdot 7 \cdot 6 \cdot 5$$

$$= \frac{13 \cdot 12 \cdot 11 \cdot 10 \cdot 9 \cdot 8 \cdot 7 \cdot 6 \cdot 5 \cdot \boxed{4 \cdot 3 \cdot 2 \cdot 1}}{\boxed{4 \cdot 3 \cdot 2 \cdot 1}} = \frac{13!}{4!} = \frac{13!}{(13-9)!}.$$

Thus, the number of permutations of 13 things taken 9 at a time is $\frac{13!}{(13-9)!}$. The special notation $_{13}P_9$ is used to replace the phrase "the number of permutations of 13 things taken 9 at a time." Using this new notation, we can write

$$_{13}P_9 = \frac{13!}{(13-9)!}.$$

The numerator of this expression is the number of items, 13 team members, expressed as a factorial: 13!. The denominator is also a factorial. It is the factorial of the difference between the number of items, 13, and the number of items in each permutation, 9 batters: $(13-9)!$.

The notation $_nP_r$ means the **number of permutations of n things taken r at a time**. We can generalize from the situation in which 9 batters were taken from 13 players. By generalizing, we obtain the following formula for the number of permutations if r items are taken from n items.

Permutations of n Things Taken r at a Time

The number of possible permutations if r items are taken from n items is

$$_nP_r = \frac{n!}{(n-r)!}.$$

Study Tip

Because all permutation problems are also Fundamental Counting problems, they can be solved using the formula for $_nP_r$ or using the Fundamental Counting Principle.

Technology

Graphing utilities have a menu item for calculating permutations, usually labeled $\boxed{_nP_r}$. For example, to find $_{20}P_3$, the keystrokes are

20 $\boxed{_nP_r}$ 3 $\boxed{\text{ENTER}}$.

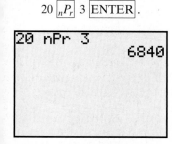

If you are using a scientific calculator, check your manual for the location of the menu item for calculating permutations and the required keystrokes.

EXAMPLE 4   Using the Formula for Permutations

You and 19 of your friends have decided to form an Internet marketing consulting firm. The group needs to choose three officers—a CEO, an operating manager, and a treasurer. In how many ways can those offices be filled?

Solution   Your group is choosing $r = 3$ officers from a group of $n = 20$ people (you and 19 friends). The order in which the officers are chosen matters because the CEO, the operating manager, and the treasurer each have different responsibilities. Thus, we are looking for the number of permutations of 20 things taken 3 at a time. We use the formula

$$_nP_r = \frac{n!}{(n-r)!}$$

with $n = 20$ and $r = 3$.

$$_{20}P_3 = \frac{20!}{(20-3)!} = \frac{20!}{17!} = \frac{20 \cdot 19 \cdot 18 \cdot 17!}{17!} = \frac{20 \cdot 19 \cdot 18 \cdot \cancel{17!}}{\cancel{17!}} = 20 \cdot 19 \cdot 18 = 6840$$

Thus, there are 6840 different ways of filling the three offices.

 Check Point 4   A corporation has seven members on its board of directors. In how many different ways can it elect a president, vice-president, secretary, and treasurer?

EXAMPLE 5   Using the Formula for Permutations

You need to arrange seven of your favorite books along a small shelf. How many different ways can you arrange the books, assuming that the order of the books makes a difference to you?

Solution   Because you are using all seven of your books in every possible arrangement, you are arranging $r = 7$ books from a group of $n = 7$ books. Thus, we are looking for the number of permutations of 7 things taken 7 at a time. We use the formula

$$_nP_r = \frac{n!}{(n-r)!}$$

with $n = 7$ and $r = 7$.

$$_7P_7 = \frac{7!}{(7-7)!} = \frac{7!}{0!} = \frac{7!}{1} = 5040$$

Thus, you can arrange the books in 5040 ways. There are 5040 different possible permutations.

How to Pass the Time for $2\frac{1}{2}$ Million Years

If you were to arrange 15 different books on a shelf and it took you one minute for each permutation, the entire task would take 2,487,965 years.

Source: Isaac Asimov's *Book of Facts.*

Check Point 5   In how many ways can 6 books be lined up along a shelf?

Combinations

As the twentieth century drew to a close, *Time* magazine presented a series of special issues on the most influential people of the century. In their issue on heroes and icons (June 14, 1999), they discussed a number of people whose careers became more profitable after their tragic deaths, including Marilyn Monroe, James Dean, Jim Morrison, Kurt Cobain, and Selena.

Imagine that you ask your friends the following question: "Of these five people, which three would you select to be included in a documentary featuring the best of their work?" You are not asking your friends to rank their three favorite artists in any kind of order—they should merely select the three to be included in the documentary.

One friend answers, "Jim Morrison, Kurt Cobain, and Selena." Another responds, "Selena, Kurt Cobain, and Jim Morrison." These two people have the same artists in their group of selections, even if they are named in a different order. We are interested *in which artists are named, not the order in which they are named* for the documentary. Because the items are taken without regard to order, this is not a permutation problem. No ranking of any sort is involved.

③ Distinguish between permutation problems and combination problems.

Marilyn Monroe, actress (1927–1962)

James Dean, actor (1931–1955)

Jim Morrison, musician and lead singer of the Doors (1943–1971)

Kurt Cobain, musician and front man for Nirvana (1967–1994)

Selena, musician of Tejano music (1971–1995)

Later on, you ask your roommate which three artists she would select for the documentary. She names Marilyn Monroe, James Dean, and Selena. Her selection is different from those of your two other friends because different entertainers are cited.

Mathematicians describe the group of artists given by your roommate as a *combination*. A **combination** of items occurs when

- The items are selected from the same group (the five stars who died young and tragically).
- No item is used more than once. (You may adore Selena, but your three selections cannot be Selena, Selena, and Selena.)
- The order of items makes no difference. (Morrison, Cobain, Selena is the same group in the documentary as Selena, Cobain, Morrison.)

Do you see the difference between a permutation and a combination? A permutation is an ordered arrangement of a given group of items. A combination is a group of items taken without regard to their order. **Permutation** problems involve situations in which **order matters**. **Combination** problems involve situations in which the **order** of the items **makes no difference**.

EXAMPLE 6   Distinguishing between Permutations and Combinations

For each of the following problems, determine whether the problem is one involving permutations or combinations. (It is not necessary to solve the problem.)

a. Six students are running for student government president, vice-president, and treasurer. The student with the greatest number of votes becomes the president, the second highest vote-getter becomes vice-president, and the student who gets the third largest number of votes will be treasurer. How many different outcomes are possible for these three positions?

b. Six people are on the board of supervisors for your neighborhood park. A three-person committee is needed to study the possibility of expanding the park. How many different committees could be formed from the six people?

c. Baskin-Robbins offers 31 different flavors of ice cream. One of their items is a bowl consisting of three scoops of ice cream, each a different flavor. How many such bowls are possible?

Solution

a. Students are choosing three student government officers from six candidates. The order in which the officers are chosen makes a difference because each of the offices (president, vice-president, treasurer) is different. Order matters. This is a problem involving permutations.

b. A three-person committee is to be formed from the six-person board of supervisors. The order in which the three people are selected does not matter because they are not filling different roles on the committee. Because order makes no difference, this is a problem involving combinations.

c. A three-scoop bowl of three different flavors is to be formed from Baskin-Robbin's 31 flavors. The order in which the three scoops of ice cream are put into the bowl is irrelevant. A bowl with chocolate, vanilla, and strawberry is exactly the same as a bowl with vanilla, strawberry, and chocolate. Different orderings do not change things, and so this is a problem involving combinations.

 Check Point 6   For each of the following problems, explain if the problem is one involving permutations or combinations. (It is not necessary to solve the problem.)

a. How many ways can you select 6 free videos from a list of 200 videos?

b. In a race in which there are 50 runners and no ties, in how many ways can the first three finishers come in?

4 Use the combinations formula.

A Formula for Combinations

We have seen that the notation $_nP_r$ means the number of permutations of n things taken r at a time. Similarly, the notation $_nC_r$ **means the number of combinations of** things taken r at a time.

We can develop a formula for $_nC_r$ by comparing permutations and combinations. Consider the letters A, B, C, and D. The number of permutations of these four letters taken three at a time is

$$_4P_3 = \frac{4!}{(4-3)!} = \frac{4!}{1!} = \frac{4 \cdot 3 \cdot 2 \cdot 1}{1} = 24.$$

Here are the 24 permutations:

ABC,	ABD,	ACD,	BCD,
ACB,	ADB,	ADC,	BDC,
BAC,	BAD,	CAD,	CBD,
BCA,	BDA,	CDA,	CDB,
CAB,	DAB,	DAC,	DBC,
CBA,	DBA,	DCA,	DCB.

This column contains only one combination, ABC.	This column contains only one combination, ABD.	This column contains only one combination, ACD.	This column contains only one combination, BCD.

Because the order of items makes no difference in determining combinations, each column of six permutations represents one combination. There is a total of four combinations:

$$\text{ABC,} \quad \text{ABD,} \quad \text{ACD,} \quad \text{BCD.}$$

Thus, $_4C_3 = 4$: The number of combinations of 4 things taken 3 at a time is 4. With 2 permutations and only four combinations, there are 6, or 3!, times as man permutations as there are combinations.

In general, there are $r!$ times as many permutations of n things taken r at time as there are combinations of n things taken r at a time. Thus, we find the number of combinations of n things taken r at a time by dividing the number of permutations of n things taken r at a time by $r!$.

$$_nC_r = \frac{_nP_r}{r!} = \frac{\dfrac{n!}{(n-r)!}}{r!} = \frac{n!}{(n-r)!r!}$$

Study Tip

The number of combinations if r items are taken from n items cannot be found using the Fundamental Counting Principle and requires the use of the formula shown on the right.

Combinations of n Things Taken r at a Time

The number of possible combinations if r items are taken from n items is

$$_nC_r = \frac{n!}{(n-r)!r!}.$$

Notice that the formula for $_nC_r$ is the same as the formula for the binomial coefficient $\binom{n}{r}$.

EXAMPLE 7   Using the Formula for Combinations

A three-person committee is needed to study ways of improving public transportation. How many committees could be formed from the eight people on the board of supervisors?

Solution   The order in which the three people are selected does not matter. This is a problem of selecting $r = 3$ people from a group of $n = 8$ people. We are looking for the number of combinations of eight things taken three at a time. We use the formula

$$_nC_r = \frac{n!}{(n-r)!r!}$$

with $n = 8$ and $r = 3$.

$$_8C_3 = \frac{8!}{(8-3)!3!} = \frac{8!}{5!3!} = \frac{8 \cdot 7 \cdot 6 \cdot 5!}{5! \cdot 3 \cdot 2 \cdot 1} = \frac{8 \cdot 7 \cdot 6 \cdot 5!}{5! \cdot 3 \cdot 2 \cdot 1} = 56$$

Thus, 56 committees of three people each can be formed from the eight people on the board of supervisors.

Check Point 7   From a group of 10 physicians, in how many ways can four people be selected to attend a conference on acupuncture?

EXAMPLE 8   Using the Formula for Combinations

In poker, a person is dealt 5 cards from a standard 52-card deck. The order in which you are dealt the 5 cards does not matter. How many different 5-card poker hands are possible?

Solution   Because the order in which the 5 cards are dealt does not matter, this is a problem involving combinations. We are looking for the number of combinations of $n = 52$ cards drawn $r = 5$ at a time. We use the formula

$$_nC_r = \frac{n!}{(n-r)!r!}$$

with $n = 52$ and $r = 5$.

$$_{52}C_5 = \frac{52!}{(52-5)!5!} = \frac{52!}{47!5!} = \frac{52 \cdot 51 \cdot 50 \cdot 49 \cdot 48 \cdot 47!}{47! \cdot 5 \cdot 4 \cdot 3 \cdot 2 \cdot 1} = 2{,}598{,}960$$

Thus, there are 2,598,960 different 5-card poker hands possible. It surprises many people that more than 2.5 million 5-card hands can be dealt from a mere 52 cards.

Figure 10.10 A royal flush

If you are a card player, it does not get any better than to be dealt the 5-card poker hand shown in Figure 10.10. This hand is called a *royal flush*. It consists of an ace, king, queen, jack, and 10, all of the same suit: all hearts, all diamonds, all clubs, or all spades. The probability of being dealt a royal flush involves calculating the number of ways of being dealt such a hand: just 4 of all 2,598,960 possible hands. In the next section, we move from counting possibilities to computing probabilities.

Check Point 8   How many different 4-card hands can be dealt from a deck that has 16 different cards?

EXERCISE SET 10.6

 Practice Exercises

In Exercises 1–8, use the formula for $_nP_r$ to evaluate each expression.

1. $_9P_4$

2. $_7P_3$

3. $_8P_5$

4. $_{10}P_4$

5. $_6P_6$

6. $_9P_9$

7. $_8P_0$

8. $_6P_0$

In Exercises 9–16, use the formula for $_nC_r$ to evaluate each expression.

9. $_9C_5$

10. $_{10}C_6$

11. $_{11}C_4$

12. $_{12}C_5$

13. $_7C_7$

14. $_4C_4$

15. $_5C_0$

16. $_6C_0$

In Exercises 17–20, does the problem involve permutations or combinations? Explain your answer. (It is not necessary to solve the problem.)

17. A medical researcher needs 6 people to test the effectiveness of an experimental drug. If 13 people have volunteered for the test, in how many ways can 6 people be selected?

18. Fifty people purchase raffle tickets. Three winning tickets are selected at random. If first prize is $1000, second prize is $500, and third prize is $100, in how many different ways can the prizes be awarded?

19. How many different four-letter passwords can be formed from the letters A, B, C, D, E, F, and G if no repetition of letters is allowed?

20. Fifty people purchase raffle tickets. Three winning tickets are selected at random. If each prize is $500, in how many different ways can the prizes be awarded?

Practice Plus

In Exercises 21–28, evaluate each expression.

21. $\dfrac{_7P_3}{3!} - {_7C_3}$

22. $\dfrac{_{20}P_2}{2!} - {_{20}C_2}$

23. $1 - \dfrac{_3P_2}{_4P_3}$

24. $1 - \dfrac{_5P_3}{_{10}P_4}$

25. $\dfrac{_7C_3}{_5C_4} - \dfrac{98!}{96!}$

26. $\dfrac{_{10}C_3}{_6C_4} - \dfrac{46!}{44!}$

27. $\dfrac{_4C_2 \cdot {_6C_1}}{_{18}C_3}$

28. $\dfrac{_5C_1 \cdot {_7C_2}}{_{12}C_3}$

Application Exercises

Use the Fundamental Counting Principle to solve Exercises 29–40.

29. The model of the car you are thinking of buying is available in nine different colors and three different styles (hatchback, sedan, or station wagon). In how many ways can you order the car?

30. A popular brand of pen is available in three colors (red, green, or blue) and four writing tips (bold, medium, fine, or micro). How many different choices of pens do you have with this brand?

31. An ice cream store sells two drinks (sodas or milk shakes), in four sizes (small, medium, large, or jumbo), and five flavors (vanilla, strawberry, chocolate, coffee, or pistachio). In how many ways can a customer order a drink?

32. A restaurant offers the following lunch menu.

Main Course	Vegetables	Beverages	Desserts
Ham	Potatoes	Coffee	Cake
Chicken	Peas	Tea	Pie
Fish	Green beans	Milk	Ice cream
Beef		Soda	

If one item is selected from each of the four groups, in how many ways can a meal be ordered? Describe two such orders.

33. You are taking a multiple-choice test that has five questions. Each of the questions has three answer choices, with one correct answer per question. If you select one of these three choices for each question and leave nothing blank, in how many ways can you answer the questions?

34. You are taking a multiple-choice test that has eight questions. Each of the questions has three answer choices, with one correct answer per question. If you select one of these three choices for each question and leave nothing blank, in how many ways can you answer the questions?

35. In the original plan for area codes in 1945, the first digit could be any number from 2 through 9, the second digit was either 0 or 1, and the third digit could be any number except 0. With this plan, how many different area codes were possible?

36. How many different four-letter radio station call letters can be formed if the first letter must be W or K?

37. Six performers are to present their comedy acts on a weekend evening at a comedy club. One of the performers insists on being the last stand-up comic of the evening. If this performer's request is granted, how many different ways are there to schedule the appearances?

38. Five singers are to perform at a night club. One of the singers insists on being the last performer of the evening. If this singer's request is granted, how many different ways are there to schedule the appearances?

39. In the *Cambridge Encyclopedia of Language* (Cambridge University Press, 1987), author David Crystal presents five sentences that make a reasonable paragraph regardless of their order. The sentences are as follows:

- Mark had told him about the foxes.
- John looked out the window.
- Could it be a fox?
- However, nobody had seen one for months.
- He thought he saw a shape in the bushes.

How many different five-sentence paragraphs can be formed if the paragraph begins with "He thought he saw a shape in the bushes" and ends with "John looked out of the window"?

40. A television programmer is arranging the order that five movies will be seen between the hours of 6 P.M. and 4 A.M. Two of the movies have a G rating and they are to be shown in the first two time blocks. One of the movies is rated NC-17 and is to be shown in the last of the time blocks, from 2 A.M. until 4 A.M. Given these restrictions, in how many ways can the five movies be arranged during the indicated time blocks?

Use the formula for $_nP_r$ to solve Exercises 41–48.

41. A club with ten members is to choose three officers—president, vice-president, and secretary-treasurer. If each office is to be held by one person and no person can hold more than one office, in how many ways can those offices be filled?

42. A corporation has ten members on its board of directors. In how many different ways can it elect a president, vice-president, secretary, and treasurer?

43. For a segment of a radio show, a disc jockey can play 7 songs. If there are 13 songs to select from, in how many ways can the program for this segment be arranged?

44. Suppose you are asked to list, in order of preference, the three best movies you have seen this year. If you saw 20 movies during the year, in how many ways can the three best be chosen and ranked?

45. In a race in which six automobiles are entered and there are no ties, in how many ways can the first three finishers come in?

46. In a production of *West Side Story*, eight actors are considered for the male roles of Tony, Riff, and Bernardo. In how many ways can the director cast the male roles?

47. Nine bands have volunteered to perform at a benefit concert, but there is only enough time for five of the bands to play. How many lineups are possible?

48. How many arrangements can be made using four of the letters of the word COMBINE if no letter is to be used more than once?

se the formula for $_nC_r$ to solve Exercises 49–56.

9. An election ballot asks voters to select three city commissioners from a group of six candidates. In how many ways can this be done?

0. A four-person committee is to be elected from an organization's membership of 11 people. How many different committees are possible?

1. Of 12 possible books, you plan to take 4 with you on vacation. How many different collections of 4 books can you take?

2. There are 14 standbys who hope to get seats on a flight, but only 6 seats are available on the plane. How many different ways can the 6 people be selected?

3. You volunteer to help drive children at a charity event to the zoo, but you can fit only 8 of the 17 children present in your van. How many different groups of 8 children can you drive?

4. Of the 100 people in the U.S. Senate, 18 serve on the Foreign Relations Committee. How many ways are there to select Senate members for this committee (assuming party affiliation is not a factor in selection)?

5. To win at LOTTO in the state of Florida, one must correctly select 6 numbers from a collection of 53 numbers (1 through 53). The order in which the selection is made does not matter. How many different selections are possible?

6. To win in the New York State lottery, one must correctly select 6 numbers from 59 numbers. The order in which the selection is made does not matter. How many different selections are possible?

Exercises 57–66, solve by the method of your choice.

7. In a race in which six automobiles are entered and there are no ties, in how many ways can the first four finishers come in?

8. A book club offers a choice of 8 books from a list of 40. In how many ways can a member make a selection?

9. A medical researcher needs 6 people to test the effectiveness of an experimental drug. If 13 people have volunteered for the test, in how many ways can 6 people be selected?

0. Fifty people purchase raffle tickets. Three winning tickets are selected at random. If first prize is $1000, second prize is $500, and third prize is $100, in how many different ways can the prizes be awarded?

1. From a club of 20 people, in how many ways can a group of three members be selected to attend a conference?

2. Fifty people purchase raffle tickets. Three winning tickets are selected at random. If each prize is $500, in how many different ways can the prizes be awarded?

3. How many different four-letter passwords can be formed from the letters A, B, C, D, E, F, and G if no repetition of letters is allowed?

4. Nine comedy acts will perform over two evenings. Five of the acts will perform on the first evening and the order in which the acts perform is important. How many ways can the schedule for the first evening be made?

5. Using 15 flavors of ice cream, how many cones with three different flavors can you create if it is important to you which flavor goes on the top, middle, and bottom?

6. Baskin-Robbins offers 31 different flavors of ice cream. One of their items is a bowl consisting of three scoops of ice cream, each a different flavor. How many such bowls are possible?

Writing in Mathematics

67. Explain the Fundamental Counting Principle.

68. Write an original problem that can be solved using the Fundamental Counting Principle. Then solve the problem.

69. What is a permutation?

70. Describe what $_nP_r$ represents.

71. Write a word problem that can be solved by evaluating $_7P_3$.

72. What is a combination?

73. Explain how to distinguish between permutation and combination problems.

74. Write a word problem that can be solved by evaluating $_7C_3$.

Technology Exercises

75. Use a graphing utility with an $\boxed{_nP_r}$ menu item to verify your answers in Exercises 1–8.

76. Use a graphing utility with an $\boxed{_nC_r}$ menu item to verify your answers in Exercises 9–16.

Critical Thinking Exercises

77. Which one of the following is true?
   a. The number of ways to choose four questions out of ten questions on an essay test is $_{10}P_4$.
   b. If $r > 1$, $_nP_r$ is less than $_nC_r$.
   c. $_7P_3 = 3!_7C_3$
   d. The number of ways to pick a winner and first runner-up in a piano recital with 20 contestants is $_{20}C_2$.

78. Five men and five women line up at a checkout counter in a store. In how many ways can they line up if the first person in line is a woman and the people in line alternate woman, man, woman, man, and so on?

79. How many four-digit odd numbers less than 6000 can be formed using the digits 2, 4, 6, 7, 8, and 9?

80. A mathematics exam consists of 10 multiple-choice questions and 5 open-ended problems in which all work must be shown. If an examinee must answer 8 of the multiple-choice questions and 3 of the open-ended problems, in how many ways can the questions and problems be chosen?

Group Exercise

81. The group should select real-world situations where the Fundamental Counting Principle can be applied. These could involve the number of possible student ID numbers on your campus, the number of possible phone numbers in your community, the number of meal options at a local restaurant, the number of ways a person in the group can select outfits for class, the number of ways a condominium can be purchased in a nearby community, and so on. Once situations have been selected, group members should determine in how many ways each part of the task can be done. Group members will need to obtain menus, find out about telephone-digit requirements in the community, count shirts, pants, shoes in closets, visit condominium sales offices, and so on. Once the group reassembles, apply the Fundamental Counting Principle to determine the number of available options in each situation. Because these numbers may be quite large, use a calculator.

SECTION 10.7 *Probability*

Objectives

❶ Compute empirical probability.

❷ Compute theoretical probability.

❸ Find the probability that an event will not occur.

❹ Find the probability of one event or a second event occurring.

❺ Find the probability of one event and a second event occurring.

Table 10.3 The Hours of Slee Americans Get on a Typical Night

Hours of Sleep	Number of Americans, in millions
4 or less	11.36
5	25.56
6	71.00
7	85.20
8	76.68
9	8.52
10 or more	5.68

Total: 284.00

Source: Discovery Health Media

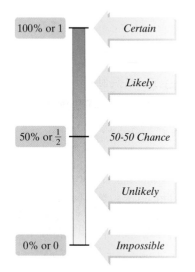

Possible Values for Probabilities

How many hours of sleep do you typically get each night? Table 10.3 indicates th. 71 million out of 284 million Americans are getting six hours of sleep on a typic night. The *probability* of an American getting six hours of sleep on a typical night is $\frac{71}{28}$ This fraction can be reduced to $\frac{1}{4}$, or expressed as 0.25, or 25%. Thus, 25% of Ame icans get six hours of sleep each night.

We find a probability by dividing one number by another. Probabilities a assigned to an *event*, such as getting six hours of sleep on a typical night. Events th are certain to occur are assigned probabilities of 1, or 100%. For example, the pro ability that a given individual will eventually die is 1. Regrettably, taxes and dea are always certain! By contrast, if an event cannot occur, its probability is 0. F example, the probability that Elvis will return from the dead and serenade us wi one final reprise of "Heartbreak Hotel" is 0.

Probabilities of events are expressed as numbers ranging from 0 to 1, or 0% 100%. The closer the probability of a given event is to 1, the more likely it is that t event will occur. The closer the probability of a given event is to 0, the less likely it that the event will occur.

Empirical Probability

Empirical probability applies to situations in which we observe how frequently an eve occurs. We use the following formula to compute the empirical probability of an ever

❶ Compute empirical probability.

Computing Empirical Probability

The **empirical probability** of event E, denoted by $P(E)$, is

$$P(E) = \frac{\text{observed number of times } E \text{ occurs}}{\text{total number of observed occurrences}}.$$

EXAMPLE 1 Empirical Probabilities with Real-World Data

When women turn 40, their gynecologists typically remind them that it is time undergo mammography screening for breast cancer. The data in Table 10.4 at t top of the next page are based on 100,000 U.S. women, ages 40 to 50, who participa ed in mammography screening.

  a. Use Table 10.4 to find the probability that a woman aged 40 to 50 has brea cancer.

  b. Among women without breast cancer, find the probability of a positi mammogram.

  c. Among women with positive mammograms, find the probability of not havi breast cancer.

Table 10.4   Mammography Screening on 100,000 U.S. Women, Ages 40 to 50

	Breast Cancer	**No Breast Cancer**
Positive Mammogram	720	6944
Negative Mammogram	80	92,256

720 + 6944 = 7664 women have positive mammograms.

80 + 92,256 = 92,336 women have negative mammograms.

720 + 80 = 800 women have breast cancer.

6944 + 92,256 = 99,200 women do not have breast cancer.

Source: Gerd Gigerenzer, *Calculated Risks*, Simon and Schuster, 2002

Solution

a. We begin with the probability that a woman aged 40 to 50 has breast cancer. The probability of having breast cancer is the number of women with breast cancer divided by the total number of women.

$$P(\text{breast cancer}) = \frac{\text{number of women with breast cancer}}{\text{total number of women}}$$

$$= \frac{800}{100,000} = \frac{1}{125} = 0.008$$

The empirical probability that a woman aged 40 to 50 has breast cancer is $\frac{1}{125}$, or 0.008.

b. Now, we find the probability of a positive mammogram among women without breast cancer. Thus, we restrict the data to women without breast cancer:

	No Breast Cancer
Positive Mammogram	6944
Negative Mammogram	92,256

Within the restricted data, the probability of a positive mammogram is the number of women with positive mammograms divided by the total number of women.

$$P(\text{positive mammogram}) = \frac{\text{number of women with positive mammograms}}{\text{total number of women in the restricted data}}$$

$$= \frac{6944}{6944 + 92,256} = \frac{6944}{99,200} = 0.07$$

This is the total number of women without breast cancer.

Among women without breast cancer, the empirical probability of a positive mammogram is $\frac{6944}{99,200}$, or 0.07.

c. Now, we find the probability of not having breast cancer among women with positive mammograms. Thus, we restrict the data to women with positive mammograms:

	Breast Cancer	**No Breast Cancer**
Positive Mammogram	720	6944

Within the restricted data, the probability of not having breast cancer is the number of women with no breast cancer divided by the total number of women.

$$P(\text{no breast cancer}) = \frac{\text{number of women with no breast cancer}}{\text{total number of women in the restricted data}}$$

$$= \frac{6944}{720 + 6944} = \frac{6944}{7664} \approx 0.906$$

This is the total number of women with positive mammograms.

Among women with positive mammograms, the probability of not havin$\overline{}$ breast cancer is $\frac{6944}{7664}$, or approximately 0.906.

 Check Point 1 Use the data in Table 10.4 on page 985 to answer this exercise. Expres$\overline{}$ probabilities as fractions and as decimals to three decimal places.

a. Find the probability that a woman aged 40 to 50 has a positiv$\overline{}$ mammogram.
b. Among women with breast cancer, find the probability of a positiv$\overline{}$ mammogram.
c. Among women with positive mammograms, find the probability ($\overline{}$ having breast cancer.

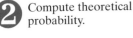 **2** Compute theoretical probability.

Theoretical Probability

You toss a coin. Although it is equally likely to land either heads up, denoted by H, $\overline{}$ tails up, denoted by T, the actual outcome is uncertain. Any occurrence for which the ou$\overline{}$ come is uncertain is called an **experiment**. Thus, tossing a coin is an example of a$\overline{}$ experiment. The set of all possible outcomes of an experiment is the **sample space** of th$\overline{}$ experiment, denoted by S. The sample space for the coin-tossing experiment is

$$S = \{H, T\}.$$

Lands heads up   Lands tails up

We can define an event more formally using these concepts. An **event**, denoted by $\overline{}$ is any subcollection, or subset, of a sample space. For example, the subset $E = \{T$ is the event of landing tails up when a coin is tossed.

Theoretical probability applies to situations like this, in which the samp$\overline{}$ space only contains equally likely outcomes, all of which are known. To calculate th$\overline{}$ theoretical probability of an event, we divide the number of outcomes resulting $\overline{}$ the event by the number of outcomes in the sample space.

Computing Theoretical Probability

If an event E has $n(E)$ equally likely outcomes and its sample space S has $n(S)$ equally likely outcomes, the **theoretical probability** of event E, denoted by $P(E)$, is

$$P(E) = \frac{\text{number of outcomes in event } E}{\text{number of outcomes in sample space } S} = \frac{n(E)}{n(S)}.$$

The sum of the theoretical probabilities of all possible outcomes in the sample space is 1.

How can we use this formula to compute the probability of a coin landing tai$\overline{}$ up? We use the following sets:

$$E = \{T\} \qquad S = \{H, T\}.$$

This is the event of landing tails up.   This is the sample space with all equally likely outcomes.

The probability of a coin landing tails up is

$$P(E) = \frac{n(E)}{n(S)} = \frac{1}{2}.$$

Theoretical probability applies to many games of chance, including di$\overline{}$ rolling, lotteries, card games, and roulette. The next example deals with the exper$\overline{}$ ment of rolling a die. Figure 10.11 illustrates that when a die is rolled, there are s$\overline{}$ equally likely outcomes. The sample space can be shown as

$$S = \{1, 2, 3, 4, 5, 6\}.$$

Figure 10.11 Outcomes when a die is rolled

EXAMPLE 2   Computing Theoretical Probability

A die is rolled. Find the probability of getting a number less than 5.

Solution   The sample space of equally likely outcomes is $S = \{1, 2, 3, 4, 5, 6\}$. There are six outcomes in the sample space, so $n(S) = 6$.

We are interested in the probability of getting a number less than 5. The event of getting a number less than 5 can be represented by

$$E = \{1, 2, 3, 4\}.$$

There are four outcomes in this event, so $n(E) = 4$.

The probability of rolling a number less than 5 is

$$P(E) = \frac{n(E)}{n(S)} = \frac{4}{6} = \frac{2}{3}.$$

Check Point 2   A die is rolled. Find the probability of getting a number greater than 4.

EXAMPLE 3   Computing Theoretical Probability

Two ordinary six-sided dice are rolled. What is the probability of getting a sum of 8?

Solution   Each die has six equally likely outcomes. By the Fundamental Counting Principle, there are $6 \cdot 6$, or 36, equally likely outcomes in the sample space. That is, $n(S) = 36$. The 36 outcomes are shown below as ordered pairs. The five ways of rolling a sum of 8 appear in the green highlighted diagonal.

		Second Die				
	·	:·	·:·	::	·::·	:::
·	(1,1)	(1,2)	(1,3)	(1,4)	(1,5)	(1,6)
:·	(2,1)	(2,2)	(2,3)	(2,4)	(2,5)	(2,6)
·:·	(3,1)	(3,2)	(3,3)	(3,4)	(3,5)	(3,6)
::	(4,1)	(4,2)	(4,3)	(4,4)	(4,5)	(4,6)
·::·	(5,1)	(5,2)	(5,3)	(5,4)	(5,5)	(5,6)
:::	(6,1)	(6,2)	(6,3)	(6,4)	(6,5)	(6,6)

First Die

$$S = \{(1, 1), (1, 2), (1, 3), (1, 4),$$
$$(1, 5), (1, 6), (2, 1), (2, 2),$$
$$(2, 3), (2, 4), (2, 5), (2, 6),$$
$$(3, 1), (3, 2), (3, 3), (3, 4),$$
$$(3, 5), (3, 6), (4, 1), (4, 2),$$
$$(4, 3), (4, 4), (4, 5), (4, 6),$$
$$(5, 1), (5, 2), (5, 3), (5, 4),$$
$$(5, 5), (5, 6), (6, 1), (6, 2),$$
$$(6, 3), (6, 4), (6, 5), (6, 6)\}$$

The phrase "getting a sum of 8" describes the event

$$E = \{(6, 2), (5, 3), (4, 4), (3, 5), (2, 6)\}.$$

This event has 5 outcomes, so $n(E) = 5$. Thus, the probability of getting a sum of 8 is

$$P(E) = \frac{n(E)}{n(S)} = \frac{5}{36}.$$

Check Point 3   What is the probability of getting a sum of 5 when two six-sided dice are rolled?

Computing Theoretical Probability without Listing an Event and the Sample Space

In some situations, we can compute theoretical probability without having to write out each event and each sample space. For example, suppose you are dealt one card from a standard 52-card deck, illustrated in Figure 10.12. The deck has four suits: Hearts and diamonds are red, and clubs and spades are black. Each suit has 13 different face values—A(ace), 2, 3, 4, 5, 6, 7, 8, 9, 10, J(jack), Q(queen), and K(king). Jacks, queens, and kings are called **picture cards** or **face cards**.

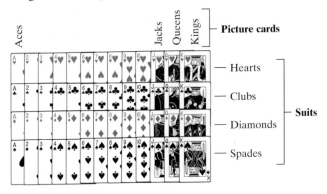

Figure 10.12 A standard 52-card bridge deck

EXAMPLE 4   Probability and a Deck of 52 Cards

You are dealt one card from a standard 52-card deck. Find the probability of being dealt a heart.

Solution   Let E be the event of being dealt a heart. Because there are 13 hearts in the deck, the event of being dealt a heart can occur in 13 ways. The number of outcomes in event E is 13: $n(E) = 13$. With 52 cards in the deck, the total number of possible ways of being dealt a single card is 52. The number of outcomes in the sample space is 52: $n(S) = 52$. The probability of being dealt a heart is

$$P(E) = \frac{n(E)}{n(S)} = \frac{13}{52} = \frac{1}{4}.$$

Check Point 4   If you are dealt one card from a standard 52-card deck, find the probability of being dealt a king.

If your state has a lottery drawing each week, the probability that someone will win the top prize is relatively high. If there is no winner this week, it is virtually certain that eventually someone will be graced with millions of dollars. So, why are you so unlucky compared to this undisclosed someone? In Example 5, we provide an answer to this question, using the counting principles discussed in Section 8.6.

EXAMPLE 5   Probability and Combinations: Winning the Lottery

Florida's lottery game, LOTTO, is set up so that each player chooses six different numbers from 1 to 53. If the six numbers chosen match the six numbers drawn randomly, the player wins (or shares) the top cash prize. (As of this writing, the top cash prize has ranged from $7 million to $106.5 million.) With one LOTTO ticket, what is the probability of winning this prize?

Solution   Because the order of the six numbers does not matter, this is a situation involving combinations. Let E be the event of winning the lottery with one ticket. With one LOTTO ticket, there is only one way of winning. Thus, $n(E) = $

State lotteries keep 50 cents on the dollar, resulting in $10 billion a year for public funding.

© Damon Higgins/The Palm Beach Post

The sample space is the set of all possible six-number combinations. We can use the combinations formula

$$_nC_r = \frac{n!}{(n-r)!r!}$$

to find the total number of possible combinations. We are selecting $r = 6$ numbers from a collection of $n = 53$ numbers.

$$_{53}C_6 = \frac{53!}{(53-6)!6!} = \frac{53!}{47!6!} = \frac{53 \cdot 52 \cdot 51 \cdot 50 \cdot 49 \cdot 48 \cdot 47!}{47! \cdot 6 \cdot 5 \cdot 4 \cdot 3 \cdot 2 \cdot 1} = 22,957,480$$

There are nearly 23 million number combinations possible in LOTTO. If a person buys one LOTTO ticket, the probability of winning is

$$P(E) = \frac{n(E)}{n(S)} = \frac{1}{22,957,480} \approx 0.0000000436.$$

The probability of winning the top prize with one LOTTO ticket is $\frac{1}{22,957,480}$, or about 1 in 23 million.

Comparing the Probability of Dying to the Probability of Winning Florida's LOTTO

As a healthy nonsmoking 30-year-old, your probability of dying this year is approximately 0.001. Divide this probability by the probability of winning LOTTO with one ticket:

$$\frac{0.001}{0.0000000436} \approx 22,936.$$

A healthy 30-year-old is nearly 23,000 times more likely to die this year than to win Florida's lottery.

In 2003, Americans spent nearly 19 billion dollars on lotteries set up by revenue-hungry states. If a person buys, say 5000 different tickets in Florida's LOTTO, that person has selected 5000 different combinations of the six numbers. The probability of winning is

$$\frac{5000}{22,957,480} \approx 0.000218.$$

The chances of winning top prize are about 218 in a million. At $1 per LOTTO ticket, it is highly probable that our LOTTO player will be $5000 poorer.

Check Point 5 People lose interest when they do not win at games of chance, including Florida's LOTTO. With drawings twice weekly instead of once, the game described in Example 5 was brought in to bring back lost players and increase ticket sales. The original LOTTO was set up so that each player chose six different numbers from 1 to 49, rather than from 1 to 53, with a lottery drawing only once a week. With one LOTTO ticket, what was the probability of winning the top cash prize in Florida's original LOTTO? Express the answer as a fraction and as a decimal correct to ten places.

3 Find the probability that an event will not occur.

Probability of an Event Not Occurring

If we know $P(E)$, the probability of an event E, we can determine the probability that the event will not occur, denoted by $P(\text{not } E)$. Because the sum of the probabilities of all possible outcomes in any situation is 1,

$$P(E) + P(\text{not } E) = 1.$$

We now solve this equation for $P(\text{not } E)$, the probability that event E will not occur, by subtracting $P(E)$ from both sides. The resulting formula is given in the following box:

The Probability of an Event Not Occurring
The probability that an event E will not occur is equal to 1 minus the probability that it will occur.

$$P(\text{not } E) = 1 - P(E)$$

EXAMPLE 6   The Probability of an Event Not Occurring

The graph in Figure 10.13 shows the distribution, by branch and gender, of the 1.4 million, or 1430 thousand, active-duty personnel in the U.S. military in 200. Numbers are given in thousands and rounded to the nearest ten thousand. If on person is randomly selected from the U.S. military and the distribution is the sam as it was in 2003, find the probability that this person is not in the Army.

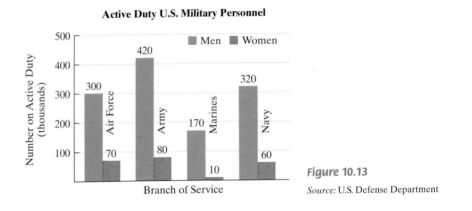

Active Duty U.S. Military Personnel

Figure 10.13

Source: U.S. Defense Department

Solution   We begin by finding the probability that the selected person *is* in the Arm

$$P(\text{Army}) = \frac{\text{number of people in the Army}}{\text{total number of people in the U.S. military}}$$

$$= \frac{420 + 80}{1430}$$

> The graph shows 420 thousand men and 80 thousand women in the Army.

> This number was given, but can be obtained by adding the eight numbers above the bars.

$$= \frac{500}{1430} = \frac{50}{143}$$

Thus,

$$P(\text{not in Army}) = 1 - P(\text{Army}) = 1 - \frac{50}{143} = \frac{143}{143} - \frac{50}{143} = \frac{93}{143}.$$

The probability that a person selected from the U.S. military is not in the Army is $\frac{93}{14}$

Check Point 6   Use the graph in Figure 10.13. If one person is randomly selected from th U.S. military, find the probability that this person is not in the Marines.

④ Find the probability of one event or a second event occurring.

Or Probabilities with Mutually Exclusive Events

Suppose that you randomly select one card from a deck of 52 cards. Let *A* be th event of selecting a king and let *B* be the event of selecting a queen. Only one car is selected, so it is impossible to get both a king and a queen. The events of selectin a king and a queen cannot occur simultaneously. They are called *mutually exclusiv events.* If it is impossible for any two events, *A* and *B*, to occur simultaneously, the are said to be **mutually exclusive**. If *A* and *B* are mutually exclusive events, th probability that either *A* or *B* will occur is determined by adding their individua probabilities.

Or **Probabilities with Mutually Exclusive Events**

If A and B are mutually exclusive events, then

$$P(A \text{ or } B) = P(A) + P(B).$$

Using set notation, $P(A \cup B) = P(A) + P(B)$.

EXAMPLE 7   The Probability of Either of Two Mutually Exclusive Events Occurring

If one card is randomly selected from a deck of cards, what is the probability of selecting a king or a queen?

Solution   We find the probability that either of these mutually exclusive events will occur by adding their individual probabilities.

$$P(\text{king or queen}) = P(\text{king}) + P(\text{queen}) = \frac{4}{52} + \frac{4}{52} = \frac{8}{52} = \frac{2}{13}$$

The probability of selecting a king or a queen is $\frac{2}{13}$.

Check Point 7   If you roll a single, six-sided die, what is the probability of getting either a 4 or a 5?

Or **Probabilities with Events That Are Not Mutually Exclusive**

Consider the deck of 52 cards shown in Figure 10.14. Suppose that these cards are shuffled and you randomly select one card from the deck. What is the probability of selecting a diamond or a picture card (jack, queen, king)? Begin by adding their individual probabilities.

$$P(\text{diamond}) + P(\text{picture card}) = \frac{13}{52} + \frac{12}{52}$$

There are 13 diamonds in the deck of 52 cards.   There are 12 picture cards in the deck of 52 cards.

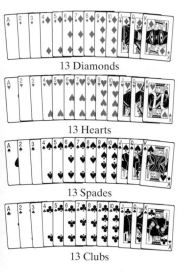

Figure 10.14 A deck of 52 cards

13 Diamonds

13 Hearts

13 Spades

13 Clubs

However, this sum is not the probability of selecting a diamond or a picture card. The problem is that there are three cards that are *simultaneously* diamonds and picture cards, shown in Figure 10.15. The events of selecting a diamond and selecting a picture card are not mutually exclusive. It is possible to select a card that is both a diamond and a picture card.

Figure 10.15 Three diamonds are picture cards.

The situation is illustrated in the diagram in Figure 10.16. Why can't we find the probability of selecting a diamond or a picture card by adding their individual probabilities? The diagram shows that three of the cards, the three diamonds that are picture cards, get counted twice when we add the individual probabilities. First the three cards get counted as diamonds and then they get counted as picture cards. In order to avoid the error of counting the three cards twice, we need to subtract the probability of getting a diamond and a picture card, $\frac{3}{52}$, as follows:

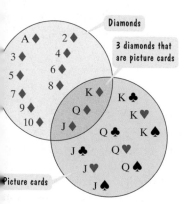

Diamonds

3 diamonds that are picture cards

Picture cards

Figure 10.16

$P(\text{diamond or picture card})$

$$= P(\text{diamond}) + P(\text{picture card}) - P(\text{diamond and picture card})$$

$$= \frac{13}{52} + \frac{12}{52} - \frac{3}{52} = \frac{13 + 12 - 3}{52} = \frac{22}{52} = \frac{11}{26}.$$

Thus, the probability of selecting a diamond or a picture card is $\frac{11}{26}$.

In general, if A and B are events that are not mutually exclusive, the probabi ity that A or B will occur is determined by adding their individual probabilities an then subtracting the probability that A and B occur simultaneously.

> ### *Or* Probabilities with Events That Are Not Mutually Exclusive
>
> If A and B are not mutually exclusive events, then
> $$P(A \text{ or } B) = P(A) + P(B) - P(A \text{ and } B).$$
> Using set notation,
> $$P(A \cup B) = P(A) + P(B) - P(A \cap B).$$

EXAMPLE 8 An *Or* Probability with Events That Are Not Mutually Exclusive

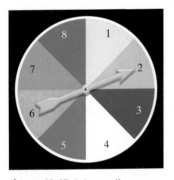

Figure 10.17 It is equally probable that the pointer will land on any one of the eight regions.

Figure 10.17 illustrates a spinner. It is equally probable that the pointer will land o any one of the eight regions, numbered 1 through 8. If the pointer lands on borderline, spin again. Find the probability that the pointer will stop on an eve number or a number greater than 5.

Solution   It is possible for the pointer to land on a number that is both even an greater than 5. Two of the numbers, 6 and 8, are even and greater than 5. Thes events are not mutually exclusive. The probability of landing on a number that i even or greater than 5 is calculated as follows:

$$P\left(\begin{array}{c}\text{even or}\\ \text{greater than 5}\end{array}\right) = P(\text{even}) + P(\text{greater than 5}) - P\left(\begin{array}{c}\text{even and}\\ \text{greater than 5}\end{array}\right)$$

$$= \quad \frac{4}{8} \quad + \quad \frac{3}{8} \quad - \quad \frac{2}{8}$$

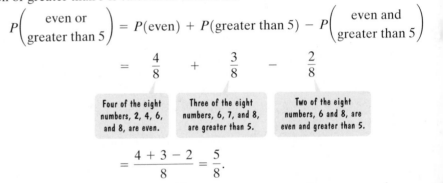

Four of the eight numbers, 2, 4, 6, and 8, are even.	Three of the eight numbers, 6, 7, and 8, are greater than 5.	Two of the eight numbers, 6 and 8, are even and greater than 5.

$$= \frac{4 + 3 - 2}{8} = \frac{5}{8}.$$

The probability that the pointer will stop on an even number or a number greate than 5 is $\frac{5}{8}$.

Check Point 8   Use Figure 10.17 to find the probability that the pointer will stop on a odd number or a number less than 5.

EXAMPLE 9   An *Or* Probability with Real-World Data

Earlier in this section, we saw a graph showing the distribution, by branch an gender, of active-duty personnel in the U.S. military. The data are shown again i Table 10.5. If one person is randomly selected from the U.S. military, find th probability that this person is in the Army or is a woman.

Table 10.5   Active-Duty U.S. Military Personnel, in Thousands

	Air Force	Army	Marine Corps	Navy	Total	
Male	300	420	170	320	1210	Total male: 300 + 420 + 170 + 320 = 1210
Female	70	80	10	60	220	
Total	370	500	180	380	1430	Total female: 70 + 80 + 10 + 60 = 220

Total Air Force   Total Army   Total Marines   Total Navy   Total on active duty

Source: U.S. Defense Department

Solution   It is possible to select a person who is both in the Army and is a woman. Thus, these events are not mutually exclusive.

$$P(\text{Army or woman}) = P(\text{Army}) + P(\text{woman}) - P(\text{Army and woman})$$

$$= \frac{500}{1430} + \frac{220}{1430} - \frac{80}{1430}$$

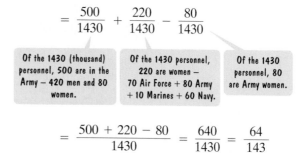

| Of the 1430 (thousand) personnel, 500 are in the Army – 420 men and 80 women. | Of the 1430 personnel, 220 are women – 70 Air Force + 80 Army + 10 Marines + 60 Navy. | Of the 1430 personnel, 80 are Army women. |

$$= \frac{500 + 220 - 80}{1430} = \frac{640}{1430} = \frac{64}{143}$$

The probability that a person selected from the U.S. military is in the Army or is a woman is $\frac{64}{143}$.

Check Point 9   Use Table 10.5. If one person is randomly selected from the U.S. military, find the probability that this person is in the Navy or is a man.

And Probabilities with Independent Events

Suppose that you toss a fair coin two times in succession. The outcome of the first toss, heads or tails, does not affect what happens when you toss the coin a second time. For example, the occurrence of tails on the first toss does not make tails more likely or less likely to occur on the second toss. The repeated toss of a coin produces *independent events* because the outcome of one toss does not influence the outcome of others. Two events are **independent events** if the occurrence of either of them has no effect on the probability of the other.

If two events are independent, we can calculate the probability of the first occurring and the second occurring by multiplying their probabilities.

> **And Probabilities with Independent Events**
> If A and B are independent events, then
> $$P(A \text{ and } B) = P(A) \cdot P(B).$$

EXAMPLE 10   Independent Events on a Roulette Wheel

Figure 10.18 shows a U.S. roulette wheel that has 38 numbered slots (1 through 36, 0, and 00). Of the 38 compartments, 18 are black, 18 are red, and two are green. A play has the dealer spin the wheel and a small ball in opposite directions. As the ball slows to a stop, it can land with equal probability on any one of the 38 numbered slots. Find the probability of red occurring on two consecutive plays.

Solution   The wheel has 38 equally likely outcomes and 18 are red. Thus, the probability of red occurring on a play is $\frac{18}{38}$, or $\frac{9}{19}$. The result that occurs on each play is independent of all previous results. Thus,

$$P(\text{red and red}) = P(\text{red}) \cdot P(\text{red}) = \frac{9}{19} \cdot \frac{9}{19} = \frac{81}{361} \approx 0.224.$$

The probability of red occurring on two consecutive plays is $\frac{81}{361}$.

Figure 10.18 A U.S. roulette wheel

Some roulette players incorrectly believe that if red occurs on two consecutiv
plays, then another color is "due." Because the events are independent, the ou
comes of previous spins have no effect on any other spins.

Check Point 10 Find the probability of green occurring on two consecutive plays on
roulette wheel.

The *and* rule for independent events can be extended to cover three or mor
events. Thus, if A, B, and C are independent events, then

$$P(A \text{ and } B \text{ and } C) = P(A) \cdot P(B) \cdot P(C).$$

EXAMPLE 11   Independent Events in a Family

The picture in the margin shows a family that has had nine girls in a row. Find th
probability of this occurrence.

Solution   If two or more events are independent, we can find the probability o
them all occurring by multiplying their probabilities. The probability of a baby girl i
$\frac{1}{2}$, so the probability of nine girls in a row is $\frac{1}{2}$ used as a factor nine times.

$$P(\text{nine girls in a row}) = \frac{1}{2} \cdot \frac{1}{2} \cdot \frac{1}{2} \cdot \frac{1}{2} \cdot \frac{1}{2} \cdot \frac{1}{2} \cdot \frac{1}{2} \cdot \frac{1}{2} \cdot \frac{1}{2}$$

$$= \left(\frac{1}{2}\right)^9 = \frac{1}{512}$$

The probability of a run of nine girls in a row is $\frac{1}{512}$. (If another child is born int
the family, this event is independent of the other nine, and the probability of a gi
is still $\frac{1}{2}$.)

Check Point 11 Find the probability of a family having four boys in a row.

EXERCISE SET 10.7

Practice and Application Exercises

The table shows the breakdown of the 89 thousand single parents on active duty in the U.S. military in 2002. All numbers are
in thousands and rounded to the nearest thousand. Use the data in the table to solve Exercises 1–10.

Single Parents on Active Duty in the U.S. Military, in Thousands

	Army	Navy	Marine Corps	Air Force	Total
Male	26	23	5	12	66
Female	10	6	1	6	23
Total	36	29	6	18	89

Total male:
26 + 23 + 5 + 12 = 66

Total female:
10 + 6 + 1 + 6 = 23

Total Army   Total Navy   Total Marines   Total Air Force   Total on active duty

Source: U.S. Defense Department

*Find the probability that a randomly selected single parent in the
U.S. military is*

1. female.

2. male.

3. in the Army.

4. in the Navy.

5. a woman in the Air Force.

6. a man in the Marine Corps.

7. Among single parents in the Air Force, find the probability o
selecting a woman.

8. Among single parents in the Marine Corps, find the probabilit
of selecting a man.

9. Among the female single parents in the military, find th
probability of selecting a woman in the Air Force.

10. Among the male single parents in the military, find th
probability of selecting a man in the Marine Corps.

In Exercises 11–16, a die is rolled. Find the probability of getting

1. a 4.

12. a 5.

3. an odd number.

14. a number greater than 3.

5. a number greater than 4.

16. a number greater than 7.

In Exercises 17–20, you are dealt one card from a standard 52-card deck. Find the probability of being dealt

7. a queen.

18. a diamond.

9. a picture card.

0. a card greater than 3 and less than 7.

In Exercises 21–22, a fair coin is tossed two times in succession. The sample space of equally likely outcomes is {HH, HT, TH, TT}. Find the probability of getting

1. two heads.

2. the same outcome on each toss.

In Exercises 23–24, you select a family with three children. If M represents a male child and F a female child, the sample space of equally likely outcomes is {MMM, MMF, MFM, MFF, FMM, FMF, FFM, FFF}. Find the probability of selecting a family with

3. at least one male child.

4. at least two female children.

In Exercises 25–26, a single die is rolled twice. The 36 equally likely outcomes are shown as follows:

	Second Roll					
First Roll	(1,1) (1,2) (1,3) (1,4) (1,5) (1,6)
	(2,1) (2,2) (2,3) (2,4) (2,5) (2,6)
	(3,1) (3,2) (3,3) (3,4) (3,5) (3,6)
	(4,1) (4,2) (4,3) (4,4) (4,5) (4,6)
	(5,1) (5,2) (5,3) (5,4) (5,5) (5,6)
	(6,1) (6,2) (6,3) (6,4) (6,5) (6,6)

Find the probability of getting

5. two numbers whose sum is 4.

6. two numbers whose sum is 6.

7. To play the California lottery, a person has to correctly select 6 out of 51 numbers, paying $1 for each six-number selection. If you pick six numbers that are the same as the ones drawn by the lottery, you win mountains of money. What is the probability that a person with one combination of six numbers will win? What is the probability of winning if 100 different lottery tickets are purchased?

8. A state lottery is designed so that a player chooses six numbers from 1 to 30 on one lottery ticket. What is the probability that a player with one lottery ticket will win? What is the probability of winning if 100 different lottery tickets are purchased?

Exercises 29–30 involve a deck of 52 cards. If necessary, refer to the picture of a deck of cards, Figure 10.12 on page 988.

9. A poker hand consists of five cards.

   a. Find the total number of possible five-card poker hands.

   b. A diamond flush is a five-card hand consisting of all diamonds. Find the number of possible diamond flushes.

   c. Find the probability of being dealt a diamond flush.

30. If you are dealt 3 cards from a shuffled deck of 52 cards, find the probability that all 3 cards are picture cards.

The table shows the educational attainment of the U.S. population, ages 25 and over. Use the data in the table, expressed in millions, to solve Exercises 31–36.

Educational Attainment, in Millions, of the U.S. Population, Ages 25 and Over

	Less Than 4 Years High School	4 Years High School Only	Some College (Less than 4 years)	4 Years College (or More)	Total
Male	14	25	20	23	82
Female	15	31	24	22	92
Total	29	56	44	45	174

Source: U.S. Census Bureau

Find the probability that a randomly selected American, aged 25 or over

31. has not completed four years (or more) of college.

32. has not completed four years of high school.

33. has completed four years of high school only or less than four years of college.

34. has completed less than four years of high school or four years of high school only.

35. has completed four years of high school only or is a man.

36. has completed four years of high school only or is a woman.

In Exercises 37–42, you are dealt one card from a 52-card deck. Find the probability that

37. you are not dealt a king.

38. you are not dealt a picture card.

39. you are dealt a 2 or a 3.

40. you are dealt a red 7 or a black 8.

41. you are dealt a 7 or a red card.

42. you are dealt a 5 or a black card.

In Exercises 43–44, it is equally probable that the pointer on the spinner shown will land on any one of the eight regions, numbered 1 through 8. If the pointer lands on a borderline, spin again.

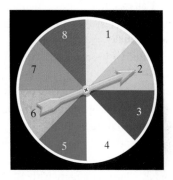

Find the probability that the pointer will stop on

43. an odd number or a number less than 6.

44. an odd number or a number greater than 3.

Use this information to solve Exercises 45–46. The mathematics department of a college has 8 male professors, 11 female professors, 14 male teaching assistants, and 7 female teaching assistants. If a person is selected at random from the group, find the probability that the selected person is

45. a professor or a male.

46. a professor or a female.

In Exercises 47–50, a single die is rolled twice. Find the probability of rolling

47. a 2 the first time and a 3 the second time.

48. a 5 the first time and a 1 the second time.

49. an even number the first time and a number greater than 2 the second time.

50. an odd number the first time and a number less than 3 the second time.

51. If you toss a fair coin six times, what is the probability of getting all heads?

52. If you toss a fair coin seven times, what is the probability of getting all tails?

53. The probability that South Florida will be hit by a major hurricane (category 4 or 5) in any single year is $\frac{1}{16}$.
(*Source*: National Hurricane Center)

   a. What is the probability that South Florida will be hit by a major hurricane two years in a row?

   b. What is the probability that South Florida will be hit by a major hurricane in three consecutive years?

   c. What is the probability that South Florida will not be hit by a major hurricane in the next ten years?

   d. What is the probability that South Florida will be hit by a major hurricane at least once in the next ten years?

Writing in Mathematics

54. Describe the difference between theoretical probability and empirical probability.

55. Give an example of an event whose probability must be determined empirically rather than theoretically.

56. Write a probability word problem whose answer is one of the following fractions: $\frac{1}{6}$ or $\frac{1}{4}$ or $\frac{1}{3}$.

57. Explain how to find the probability of an event not occurring. Give an example.

58. What are mutually exclusive events? Give an example of two events that are mutually exclusive.

59. Explain how to find *or* probabilities with mutually exclusive events. Give an example.

60. Give an example of two events that are not mutually exclusive.

61. Explain how to find *or* probabilities with events that are not mutually exclusive. Give an example.

62. Explain how to find *and* probabilities with independent events. Give an example.

63. The president of a large company with 10,000 employees is considering mandatory cocaine testing for every employee. The test that would be used is 90% accurate, meaning that it will detect 90% of the cocaine users who are tested, and that 90% of the nonusers will test negative. This also means that the test gives 10% false positive. Suppose that 1% of th employees actually use cocaine. Find the probability tha someone who tests positive for cocaine use is, indeed, a use

Hint: Find the following probability fraction:

the number of employees who test positive
and are cocaine users
_____ .
the number of employees who test positive

This fraction is given by

90% of 1% of 10,000
_____ .
the number who test positive who actually use
cocaine plus the number who test positive
who do not use cocaine.

What does this probability indicate in terms of the percent age of employees who test positive who are not actuall users? Discuss these numbers in terms of the issue of manda tory drug testing. Write a paper either in favor of or agains mandatory drug testing, incorporating the actual percentag accuracy for such tests.

Critical Thinking Exercises

64. The target in the figure shown contains four squares. If a dar thrown at random hits the target, find the probability that i will land in a yellow region.

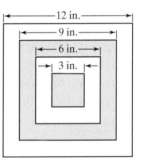

65. Suppose that it is a week in which the cash prize in Florida LOTTO is promised to exceed $50 million. If a perso purchases 22,957,480 tickets in LOTTO at $1 per ticket (a possible combinations), isn't this a guarantee of winning th lottery? Because the probability in this situation is 1, what wrong with doing this?

66. Some three-digit numbers, such as 101 and 313, read th same forward and backward. If you select a number from al three-digit numbers, find the probability that it will read th same forward and backward.

67. In a class of 50 students, 29 are Democrats, 11 are busines majors, and 5 of the business majors are Democrats. If on student is randomly selected from the class, find th probability of choosing

   a. a Democrat who is not a business major.

   b. a student who is neither a Democrat nor a business majo

68. On New Year's Eve, the probability of a person driving whil intoxicated or having a driving accident is 0.35. If the proba bility of driving while intoxicated is 0.32 and the probabilit of having a driving accident is 0.09, find the probability of person having a driving accident while intoxicated.

a. If two people are selected at random, the probability that they do not have the same birthday (day and month) is $\frac{365}{365} \cdot \frac{364}{365}$. Explain why this is so. (Ignore leap years and assume 365 days in a year.)

b. If three people are selected at random, find the probability that they all have different birthdays.

c. If three people are selected at random, find the probability that at least two of them have the same birthday.

d. If 20 people are selected at random, find the probability that at least 2 of them have the same birthday.

e. How large a group is needed to give a 0.5 chance of at least two people having the same birthday?

Group Exercise

70. Research and present a group report on state lotteries. Include answers to some or all of the following questions: Which states do not have lotteries? Why not? How much is spent per capita on lotteries? What are some of the lottery games? What is the probability of winning top prize in these games? What income groups spend the greatest amount of money on lotteries? If your state has a lottery, what does it do with the money it makes? Is the way the money is spent what was promised when the lottery first began?

Chapter 10
Summary, Review, and Test

Summary

DEFINITIONS AND CONCEPTS	EXAMPLES
10.1 Sequences and Summation Notation	
a. An infinite sequence $\{a_n\}$ is a function whose domain is the set of positive integers. The function values, or terms, are represented by $$a_1, a_2, a_3, a_4, \ldots, a_n, \ldots.$$	Ex. 1, p. 927
b. Sequences can be defined using recursion formulas that define the nth term as a function of the previous term.	Ex. 2, p. 928
c. Factorial Notation: $$n! = n(n-1)(n-2)\cdots(3)(2)(1) \quad \text{and} \quad 0! = 1$$	Ex. 3, p. 929; Ex. 4, p. 930
d. Summation Notation: $$\sum_{i=1}^{n} a_i = a_1 + a_2 + a_3 + a_4 + \cdots + a_n$$	Ex. 5, p. 931; Ex. 6, p. 932
10.2 Arithmetic Sequences	List of arithmetic sequences and common differences, p. 937;
a. In an arithmetic sequence, each term after the first differs from the preceding term by a constant, the common difference. Subtract any term from the term that directly follows to find the common difference.	Ex. 1, p. 938
b. General term or nth term: $a_n = a_1 + (n-1)d$. The first term is a_1 and the common difference is d.	Ex. 2, p. 939; Ex. 3, p. 939
c. Sum of the first n terms: $S_n = \dfrac{n}{2}(a_1 + a_n)$	Ex. 4, p. 941; Ex. 5, p. 942; Ex. 6, p. 942
10.3 Geometric Sequences and Series	List of geometric sequences and common ratios, p. 946;
a. In a geometric sequence, each term after the first is obtained by multiplying the preceding term by a nonzero constant, the common ratio. Divide any term after the first by the term that directly precedes it to find the common ratio.	Ex. 1, p. 947
b. General term or nth term: $a_n = a_1 r^{n-1}$. The first term is a_1 and the common ratio is r.	Ex. 2, p. 947; Ex. 3, p. 948
c. Sum of the first n terms: $S_n = \dfrac{a_1(1 - r^n)}{1 - r}, r \neq 1$	Ex. 4, p. 949; Ex. 5, p. 950; Ex. 6, p. 951

DEFINITIONS AND CONCEPTS	EXAMPLES
d. An annuity is a sequence of equal payments made at equal time periods. The value of an annuity, A, is the sum of all deposits made plus all interest paid, given by $$A = \frac{P\left[\left(1 + \frac{r}{n}\right)^{nt} - 1\right]}{\frac{r}{n}}.$$ The deposit made at the end of each period is P, the annual interest rate is r, compounded n times per year, and t is the number of years deposits have been made.	Ex. 7, p. 952
e. The sum of the infinite geometric series $a_1 + a_1 r + a_1 r^2 + a_1 r^3 + \cdots$ is $S = \frac{a_1}{1 - r}$; $|r| < 1$. If $|r| \geq 1$, the infinite series does not have a sum.	Ex. 8, p. 954; Ex. 9, p. 954; Ex. 10, p. 955

10.4 Mathematical Induction

To prove that S_n is true for all positive integers n,	Ex. 2, p. 962;
1. Show that S_1 is true.	Ex. 3, p. 963;
2. Show that if S_k is assumed true, then S_{k+1} is also true, for every positive integer k.	Ex. 4, p. 965

10.5 The Binomial Theorem

a. Binomial coefficient: $\binom{n}{r} = \frac{n!}{r!(n - r)!}$	Ex. 1, p. 967
b. Binomial Theorem: $$(a + b)^n = \binom{n}{0}a^n + \binom{n}{1}a^{n-1}b + \binom{n}{2}a^{n-2}b^2 + \cdots + \binom{n}{n}b^n$$	Ex. 2, p. 969; Ex. 3, p. 969
c. The $(r + 1)$st term in the expansion of $(a + b)^n$ is $$\binom{n}{r}a^{n-r}b^r.$$	Ex. 4, p. 971

10.6 Counting Principles, Permutations, and Combinations

a. The Fundamental Counting Principle: The number of ways in which a series of successive things can occur is found by multiplying the number of ways in which each thing can occur.	Ex. 1, p. 975; Ex. 2, p. 975; Ex. 3, p. 976
b. A permutation from a group of items occurs when no item is used more than once and the order of arrangement makes a difference.	
c. Permutations Formula: The number of possible permutations if r items are taken from n items is $$_nP_r = \frac{n!}{(n - r)!}.$$	Ex. 4, p. 977; Ex. 5, p. 978
d. A combination from a group of items occurs when no item is used more than once and the order of items makes no difference.	Ex. 6, p. 979
e. Combinations Formula: The number of possible combinations if r items are taken from n items is $$_nC_r = \frac{n!}{(n - r)!r!}.$$	Ex. 7, p. 980; Ex. 8, p. 981

10.7 Probability

a. Empirical probability applies to situations in which we observe the frequency of the occurrence of an event. The empirical probability of event E is $$P(E) = \frac{\text{observed number of times } E \text{ occurs}}{\text{total number of observed occurrences}}.$$	Ex. 1, p. 984
b. Theoretical probability applies to situations in which the sample space of all equally likely outcomes is known. The theoretical probability of event E is $$P(E) = \frac{\text{number of outcomes in event } E}{\text{number of outcomes in sample space } S} = \frac{n(E)}{n(S)}.$$	Ex. 2, p. 987; Ex. 3, p. 987; Ex. 4, p. 988; Ex. 5, p. 988

DEFINITIONS AND CONCEPTS

c. Probability of an event not occurring: $P(\text{not } E) = 1 - P(E)$.   Ex. 6. p. 990

d. If it is impossible for events A and B to occur simultaneously, the events are mutually exclusive.

e. If A and B are mutually exclusive events, then $P(A \text{ or } B) = P(A) + P(B)$.   Ex. 7, p. 991

f. If A and B are not mutually exclusive events, then   Ex. 8, p. 992;
$$P(A \text{ or } B) = P(A) + P(B) - P(A \text{ and } B).$$ Ex. 9, p. 992

g. Two events are independent if the occurrence of either of them has no effect on the probability of the other.

h. If A and B are independent events, then   Ex. 10, p. 993
$$P(A \text{ and } B) = P(A) \cdot P(B).$$

i. The probability of a succession of independent events is the product of each of their probabilities.   Ex. 11, p. 994

Review Exercises

10.1

In Exercises 1–6, write the first four terms of each sequence whose general term is given.

1. $a_n = 7n - 4$

2. $a_n = (-1)^n \dfrac{n + 2}{n + 1}$

3. $a_n = \dfrac{1}{(n - 1)!}$

4. $a_n = \dfrac{(-1)^{n+1}}{2^n}$

5. $a_1 = 9$ and $a_n = \dfrac{2}{3a_{n-1}}$ for $n \ge 2$

6. $a_1 = 4$ and $a_n = 2a_{n-1} + 3$ for $n \ge 2$

7. Evaluate: $\dfrac{40!}{4!38!}$.

In Exercises 8–9, find each indicated sum.

8. $\displaystyle\sum_{i=1}^{5} (2i^2 - 3)$

9. $\displaystyle\sum_{i=0}^{4} (-1)^{i+1} i!$

In Exercises 10–11, express each sum using summation notation. Use i for the index of summation.

10. $\dfrac{1}{3} + \dfrac{2}{4} + \dfrac{3}{5} + \cdots + \dfrac{15}{17}$

11. $4^3 + 5^3 + 6^3 + \cdots + 13^3$

10.2

In Exercises 12–15, write the first six terms of each arithmetic sequence.

12. $a_1 = 7, d = 4$

13. $a_1 = -4, d = -5$

14. $a_1 = \frac{3}{2}, d = -\frac{1}{2}$

15. $a_{n+1} = a_n + 5, a_1 = -2$

In Exercises 16–18, find the indicated term of the arithmetic sequence with first term, a_1, and common difference, d.

16. Find a_6 when $a_1 = 5, d = 3$.

17. Find a_{12} when $a_1 = -8, d = -2$.

18. Find a_{14} when $a_1 = 14, d = -4$.

In Exercises 19–21, write a formula for the general term (the nth term) of each arithmetic sequence. Do not use a recursion formula. Then use the formula for a_n to find a_{20}, the 20th term of the sequence.

19. $-7, -3, 1, 5, \ldots$

20. $a_1 = 200, d = -20$

21. $a_n = a_{n-1} - 5, a_1 = 3$

22. Find the sum of the first 22 terms of the arithmetic sequence: $5, 12, 19, 26, \ldots$.

23. Find the sum of the first 15 terms of the arithmetic sequence: $-6, -3, 0, 3, \ldots$.

24. Find $3 + 6 + 9 + \cdots + 300$, the sum of the first 100 positive multiples of 3.

In Exercises 25–27, use the formula for the sum of the first n terms of an arithmetic sequence to find the indicated sum.

25. $\displaystyle\sum_{i=1}^{16} (3i + 2)$

26. $\displaystyle\sum_{i=1}^{25} (-2i + 6)$

27. $\displaystyle\sum_{i=1}^{30} (-5i)$

28. The graph shows the percentage of in-home dinners in the United States having various items as a side dish from 1993 through 2004.

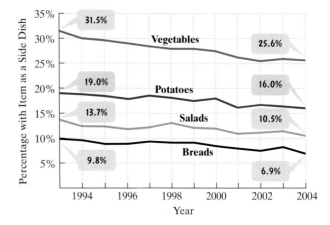

Percentage of U.S. In-Home Dinners Having Various Items as a Side Dish

Source: The NPD Group

In 1993, 31.5% of home dinners had vegetables as a side dish. On average, this decreased by approximately 0.54% per year since then.

(Be sure to refer to the graph and the information given below it on page 999.)

a. Write a formula for the nth term of the arithmetic sequence that describes the percentage of dinners that included vegetables n years after 1992.

b. Use the model to predict the percentage of dinners that will include vegetables by the year 2010.

c. Repeat parts (a) and (b) for the change in another one of the items from 1993 to 2004.

29. A company offers a starting salary of $31,500 with raises of $2300 per year. Find the total salary over a ten-year period.

30. A theater has 25 seats in the first row and 35 rows in all. Each successive row contains one additional seat. How many seats are in the theater?

10.3

In Exercises 31–34, write the first five terms of each geometric sequence.

31. $a_1 = 3, r = 2$

32. $a_1 = \frac{1}{2}, r = \frac{1}{2}$

33. $a_1 = 16, r = -\frac{1}{2}$

34. $a_n = -5a_{n-1}, a_1 = -1$

In Exercises 35–37, use the formula for the general term (the nth term) of a geometric sequence to find the indicated term of each sequence.

35. Find a_7 when $a_1 = 2, r = 3$.

36. Find a_6 when $a_1 = 16, r = \frac{1}{2}$.

37. Find a_5 when $a_1 = -3, r = 2$.

In Exercises 38–40, write a formula for the general term (the nth term) of each geometric sequence. Then use the formula for a_n to find a_8, the eighth term of the sequence.

38. $1, 2, 4, 8, \ldots$

39. $100, 10, 1, \frac{1}{10}, \ldots$

40. $12, -4, \frac{4}{3}, -\frac{4}{9}, \ldots$

41. Find the sum of the first 15 terms of the geometric sequence: $5, -15, 45, -135, \ldots$.

42. Find the sum of the first 7 terms of the geometric sequence: $8, 4, 2, 1, \ldots$.

In Exercises 43–45, use the formula for the sum of the first n terms of a geometric sequence to find the indicated sum.

43. $\sum_{i=1}^{6} 5^i$

44. $\sum_{i=1}^{7} 3(-2)^i$

45. $\sum_{i=1}^{5} 2\left(\frac{1}{4}\right)^{i-1}$

In Exercises 46–49, find the sum of each infinite geometric series.

46. $9 + 3 + 1 + \frac{1}{3} + \cdots$

47. $2 - 1 + \frac{1}{2} - \frac{1}{4} + \cdots$

48. $-6 + 4 - \frac{8}{3} + \frac{16}{9} - \cdots$

49. $\sum_{i=1}^{\infty} 5(0.8)^i$

In Exercises 50–51, express each repeating decimal as a fraction in lowest terms.

50. $0.\overline{6}$

51. $0.\overline{47}$

52. Projections for the U.S. population, ages 85 and older, are shown in the following table.

Year	2000	2010	2020	2030	2040
Projected Population in millions	4.2	5.9	8.3	11.6	16.2

Actual 2000 population

Source: U.S. Census Bureau

a. Show that the U.S. population, ages 85 and older, is projected to increase geometrically.

b. Write the general term of the geometric sequence describing the U.S. population ages 85 and older, in millions, n decades after 2000.

c. Use the model in part (b) to project the U.S. population ages 85 and older, in 2080.

53. A job pays $32,000 for the first year with an annual increase of 6% per year beginning in the second year. What is the salary in the sixth year? What is the total salary paid over this six-year period? Round answers to the nearest dollar.

54. You decide to deposit $200 at the end of each month into an account paying 10% interest compounded monthly to save for your child's education. How much will you save over 18 years?

55. A factory in an isolated town has an annual payroll of $4 million. It is estimated that 70% of this money is spent within the town, that people in the town receiving this money will again spend 70% of what they receive in the town, and so on. What is the total of all this spending in the town each year?

10.4

In Exercises 56–60, use mathematical induction to prove that each statement is true for every positive integer n.

56. $5 + 10 + 15 + \cdots + 5n = \dfrac{5n(n+1)}{2}$

57. $1 + 4 + 4^2 + \cdots + 4^{n-1} = \dfrac{4^n - 1}{3}$

58. $2 + 6 + 10 + \cdots + (4n - 2) = 2n^2$

59. $1 \cdot 3 + 2 \cdot 4 + 3 \cdot 5 + \cdots + n(n+2) = \dfrac{n(n+1)(2n+7)}{6}$

60. 2 is a factor of $n^2 + 5n$.

10.5

In Exercises 61–62, evaluate the given binomial coefficient.

61. $\dbinom{11}{8}$

62. $\dbinom{90}{2}$

In Exercises 63–66, use the Binomial Theorem to expand each binomial and express the result in simplified form.

63. $(2x + 1)^3$

64. $(x^2 - 1)^4$

65. $(x + 2y)^5$

66. $(x - 2)^6$

Exercises 67–68, write the first three terms in each binomial expansion, expressing the result in simplified form.

67. $(x^2 + 3)^8$ **68.** $(x - 3)^9$

Exercises 69–70, find the term indicated in each expansion.

69. $(x + 2)^5$; fourth term   **70.** $(2x - 3)^6$; fifth term

10.6

Exercises 71–74, evaluate each expression.

71. $_8P_3$ **72.** $_9P_5$

73. $_8C_3$ **74.** $_{13}C_{11}$

Exercises 75–81, solve by the method of your choice.

75. A popular brand of pen comes in red, green, blue, or black ink. The writing tip can be chosen from extra bold, bold, regular, fine, or micro. How many different choices of pens do you have with this brand?

76. A stock can go up, go down, or stay unchanged. How many possibilities are there if you own five stocks?

77. A club with 15 members is to choose four officers—president, vice-president, secretary, and treasurer. In how many ways can these offices be filled?

78. How many different ways can a director select 4 actors from a group of 20 actors to attend a workshop on performing in rock musicals?

79. From the 20 CDs that you've bought during the past year, you plan to take 3 with you on vacation. How many different sets of three CDs can you take?

80. How many different ways can a director select from 20 male actors and cast the roles of Mark, Roger, Angel, and Collins in the musical *Rent*?

81. In how many ways can five airplanes line up for departure on a runway?

10.7

The table shows differences in political ideology, by education, for a random sample of U.S. voters. (The ratios for each group's ideologies are sourced from voting patterns in the 2000 U.S. election. The frequencies shown are hypothetical.) Use the data to solve Exercises 82–87. Express probabilities as simplified fractions.

	Liberal	Moderate	Conservative
High School only	7	35	13
College	10	15	20

Find the probability that a randomly selected person from this group

82. is liberal.

83. is not conservative.

84. is moderate or conservative.

85. is conservative or attended college.

86. Among people with a conservative ideology, find the probability of selecting a person who attended high school only.

87. Among people who attended college, find the probability of selecting a person with a liberal ideology.

In Exercises 88–89, a die is rolled. Find the probability of

88. getting a number less than 5.

89. getting a number less than 3 or greater than 4.

In Exercises 90–91, you are dealt one card from a 52-card deck. Find the probability of

90. getting an ace or a king.

91. getting a queen or a red card.

In Exercises 92–94, it is equally probable that the pointer on the spinner shown will land on any one of the six regions, numbered 1 through 6, and colored as shown. If the pointer lands on a borderline, spin again. Find the probability of

92. not stopping on yellow.

93. stopping on red or a number greater than 3.

94. stopping on green on the first spin and stopping on a number less than 4 on the second spin.

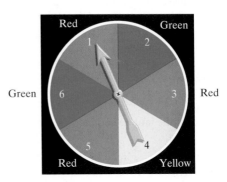

95. A lottery game is set up so that each player chooses five different numbers from 1 to 20. If the five numbers match the five numbers drawn in the lottery, the player wins (or shares) the top cash prize. What is the probability of winning the prize

  a. with one lottery ticket?

  b. with 100 different lottery tickets?

96. What is the probability of a family having five boys born in a row?

97. The probability of a flood in any given year in a region prone to floods is 0.2.

  a. What is the probability of a flood two years in a row?

  b. What is the probability of a flood for three consecutive years?

  c. What is the probability of no flooding for four consecutive years?

Chapter 10 Test

1. Write the first five terms of the sequence whose general term is $a_n = \dfrac{(-1)^{n+1}}{n^2}$.

In Exercises 2–4, find each indicated sum.

2. $\displaystyle\sum_{i=1}^{5}(i^2 + 10)$

3. $\displaystyle\sum_{i=1}^{20}(3i - 4)$

4. $\displaystyle\sum_{i=1}^{15}(-2)^i$

In Exercises 5–7, evaluate each expression.

5. $\dbinom{9}{2}$

6. $_{10}P_3$

7. $_{10}C_3$

8. Express the sum using summation notation. Use i for the index of summation.

$$\frac{2}{3} + \frac{3}{4} + \frac{4}{5} + \cdots + \frac{21}{22}$$

In Exercises 9–10, write a formula for the general term (the nth term) of each sequence. Do not use a recursion formula. Then use the formula to find the twelfth term of the sequence.

9. $4, 9, 14, 19, \ldots$

10. $16, 4, 1, \frac{1}{4}, \ldots$

In Exercises 11–12, use a formula to find the sum of the first ten terms of each sequence.

11. $7, -14, 28, -56, \ldots$

12. $-7, -14, -21, -28, \ldots$

13. Find the sum of the infinite geometric series:

$$4 + \frac{4}{2} + \frac{4}{2^2} + \frac{4}{2^3} + \cdots.$$

14. Express $0.\overline{73}$ in fractional notation.

15. A job pays $30,000 for the first year with an annual increase of 4% per year beginning in the second year. What is the total salary paid over an eight-year period? Round to the nearest dollar.

16. Use mathematical induction to prove that for every positive integer n,

$$1 + 4 + 7 + \cdots + (3n - 2) = \frac{n(3n - 1)}{2}.$$

17. Use the Binomial Theorem to expand and simplify: $(x^2 - 1)^5$.

18. Use the Binomial Theorem to write the first three terms in the expansion and simplify: $(x + y^2)^8$.

19. A human resource manager has 11 applicants to fill three different positions. Assuming that all applicants are equally qualified for any of the three positions, in how many ways can this be done?

20. From the ten books that you've recently bought but not read, you plan to take four with you on vacation. How many different sets of four books can you take?

21. How many seven-digit local telephone numbers can be formed if the first three digits are 279?

A class is collecting data on eye color and gender. They organize the data they collected into the table shown. Numbers in the table represent the number of students from the class that belong to each of the categories. Use the data to solve Exercises 22–25. Express probabilities as simplified fractions.

	Brown	Blue	Green
Male	22	18	10
Female	18	20	12

Find the probability that a randomly selected student from this class

22. does not have brown eyes.

23. has brown eyes or blue eyes.

24. is female or has green eyes.

25. Among the students with blue eyes, find the probability of selecting a male.

26. A lottery game is set up so that each player chooses six different numbers from 1 to 15. If the six numbers match the six numbers drawn in the lottery, the player wins (or shares) the top cash prize. What is the probability of winning the prize with 50 different lottery tickets?

27. One card is randomly selected from a deck of 52 cards. Find the probability of selecting a black card or a picture card.

28. A group of students consists of 10 male freshmen, 15 female freshmen, 20 male sophomores, and 5 female sophomores. If one person is randomly selected from the group, find the probability of selecting a freshman or a female.

29. A quiz consisting of four multiple-choice questions has four available options (a, b, c, or d) for each question. If a person guesses at every question, what is the probability of answering all questions correctly?

30. If the spinner shown is spun twice, find the probability that the pointer lands on red on the first spin and blue on the second spin.

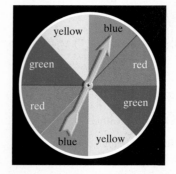

Cumulative Review Exercises (Chapters P–10)

The figure shows the graph of $y = f(x)$ and its vertical asymptote. Use the graph to solve Exercises 1–9.

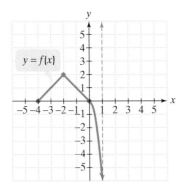

1. Find the domain and the range of f.

2. Does f have a relative maximum or a relative minimum? What is this relative maximum or minimum and where does it occur?

3. Find the interval on which f is decreasing.

4. Is f even, odd, or neither?

5. For what value(s) of x is $f(x) = 1$?

6. Find $(f \circ f)(-4)$.

7. Use arrow notation to complete this statement:

$$f(x) \to -\infty \quad \text{as} \quad \underline{\hspace{2cm}}.$$

8. Graph $g(x) = f(x - 2) + 1$.

9. Graph $h(x) = -f(2x)$.

In Exercises 10–22, solve each equation, inequality, or system of equations.

10. $-2(x - 5) + 10 = 3(x + 2)$

11. $3x^2 - 6x + 2 = 0$

12. $\log_2 x + \log_2(2x - 3) = 1$

13. $x^{\frac{1}{2}} - 6x^{\frac{1}{4}} + 8 = 0$

14. $e^{2x} - 6e^x + 8 = 0$

15. $|2x + 1| \le 1$

16. $6x^2 - 6 < 5x$

17. $\dfrac{x - 1}{x + 3} \le 0$

18. $30e^{0.7x} = 240$

19. $2x^3 + 3x^2 - 8x + 3 = 0$

20. $4x^2 + 3y^2 = 48$
$3x^2 + 2y^2 = 35$

21. (Use matrices.)

$x - 2y + \ z = \ \ \ 16$
$2x - \ y - \ z = \ \ \ 14$
$3x + 5y - 4z = -10$

22. $x - y = 1$
$x^2 - x - y = 1$

In Exercises 23–29, graph each equation, function, or system in a rectangular coordinate system. If two functions are indicated, graph both in the same system.

23. $100x^2 + y^2 = 25$

24. $4x^2 - 9y^2 - 16x + 54y - 29 = 0$

25. $f(x) = \dfrac{x^2 - 1}{x - 2}$

26. $2x - y \ge 4$
$x \le 2$

27. $f(x) = x^2 - 4x - 5$

28. $f(x) = \sqrt[3]{x + 4}$ and f^{-1}

29. $f(x) = \log_2 x$ and $g(x) = -\log_2(x + 1)$

In Exercises 30–31, let $f(x) = -x^2 - 2x + 1$ and $g(x) = x - 1$.

30. Find $(f \circ g)(x)$ and $(g \circ f)(x)$.

31. Find $\dfrac{f(x + h) - f(x)}{h}$ and simplify.

32. If $A = \begin{bmatrix} 4 & 2 \\ 1 & -1 \\ 0 & 5 \end{bmatrix}$ and $B = \begin{bmatrix} 2 & 4 \\ 3 & 1 \end{bmatrix}$, find $AB - 4A$.

33. Find the partial fraction decomposition for

$$\frac{2x^2 - 10x + 2}{(x - 2)(x^2 + 2x + 2)}.$$

34. Expand and simplify: $(x^3 + 2y)^5$.

35. Use the formula for the sum of the first n terms of an arithmetic sequence to find $\displaystyle\sum_{i=1}^{50}(4i - 25)$.

In Exercises 36–37, write the linear function in slope-intercept form satisfying the given conditions.

36. Graph of f passes through $(6, 3)$ and $(-2, 1)$.

37. Graph of g passes through $(0, -2)$ and is perpendicular to the line whose equation is $x - 5y - 20 = 0$.

38. For a summer sales job, you are choosing between two pay arrangements: a weekly salary of $200 plus 5% commission on sales, or a straight 15% commission. For how many dollars of sales will the earnings be the same regardless of the pay arrangement?

39. You have 900 feet of fencing to enclose a rectangular plot that borders on a river. If you do not fence the side along the river, find the length and width of the plot that will maximize the area. What is the largest area that can be enclosed?

40. If 10 pens and 12 pads cost $42, and 5 of the same pens and 10 of the same pads cost $29, find the cost of a pen and a pad.

41. A ball is thrown vertically upward from the top of a 96-foot tall building with an initial velocity of 80 feet per second. The height of the ball above ground, $s(t)$, in feet, after t seconds is modeled by the position function

$$s(t) = -16t^2 + 80t + 96.$$

  a. After how many seconds will the ball strike the ground?

  b. When does the ball reach its maximum height? What is the maximum height?

42. The current, I, in amperes, flowing in an electrical circuit varies inversely as the resistance, R, in ohms, in the circuit. When the resistance of an electric percolator is 22 ohms, it draws 5 amperes of current. How much current is needed when the resistance is 10 ohms?

43. The bar graph shows the decline in the number of U.S. farms growing tobacco from 1982 through 2002. Develop a linear function that models the data and then use the function to make a prediction about what might occur in the future.

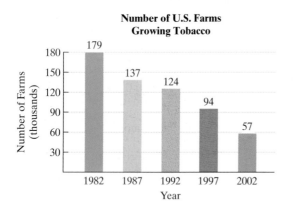

Number of U.S. Farms Growing Tobacco

Source: U.S. Bureau of Agriculture

44. An object moves in simple harmonic motion described by $d = 10 \sin \frac{3\pi}{4} t$, where t is measured in seconds and d in inches. Find **a.** the maximum displacement; **b.** the frequency and **c.** the time required for one oscillation.

Verify each identity in Exercises 45–46.

45. $\tan x + \dfrac{1}{\tan x} = \dfrac{1}{\sin x \cos x}$

46. $\dfrac{1 - \tan^2 x}{1 + \tan^2 x} = \cos 2x$

47. Graph one period: $y = -2 \cos(3x - \pi)$.

In Exercises 48–49, solve each equation on the interval $[0, 2\pi)$.

48. $4 \cos^2 x = 3$

49. $2 \sin^2 x + 3 \cos x - 3 = 0$

50. Find the exact value of $\cot \left[\cos^{-1} \left(-\frac{5}{6} \right) \right]$.

51. Graph the polar equation: $r = 1 + 2 \cos \theta$.

52. In oblique triangle ABC, $A = 34°$, $a = 22$, and $b = 32$. Solve the triangle(s). Round lengths to the nearest tenth and angle measures to the nearest degree.

53. Use the parametric equations

$$x = \sin t, \quad y = 1 + \cos^2 t, \quad -\frac{\pi}{2} < t < \frac{\pi}{2}$$

and eliminate the parameter. Graph the plane curve represented by the parametric equations. Use arrows to show the orientation of the curve.

Introduction to Calculus

T AKE A RAPID SEQUENCE OF still photographs of a moving scene and project them onto a screen at thirty shots a second or faster. Our eyes see the result as continuous motion. The small difference between one frame and the next cannot be detected by the human visual system. The idea of calculus likewise regards continuous motion as made up of a sequence of still configurations. In this chapter, you will see how calculus masters the mystery of movement by "freezing the frame" instant by instant. You will learn to use mathematics in a way that is similar to making a movie.

IN THE DRAMATIC ARTS, OURS IS THE ERA of the movies. As individuals and as a nation, we've grown up with them. Our images of love, war, family, country—even of things that terrify us—owe much to what we've seen on screen.

Using mathematics in a way that is similar to making a movie is introduced in the Section 11.4 opener and developed throughout the section.

SECTION 11.1 Finding Limits Using Tables and Graphs

Objectives

1. Understand limit notation.
2. Find limits using tables.
3. Find limits using graphs.
4. Find one-sided limits and use them to determine if a limit exists.

Motion and change are the very essence of life. Moving air brushes against our faces, rain falls on our heads, birds fly past us, plants spring from the earth, grow, and then die, and rocks thrown upward reach a maximum height before falling to the ground. The tools of algebra and trigonometry are essentially static; numbers, points, lines, equations, functions, and graphs do not incorporate motion. The development of calculus in the middle of the seventeenth century provided a way to use these static tools to analyze motion and change. It took nearly two thousand years of effort for humankind to achieve this feat, made possible by a revolutionary concept called *limits*. The invention of limits marked a turning point in human history, having as dramatic an impact on our lives as the invention of the wheel and the printing press. In this section, we introduce this bold and dramatic style of thinking about mathematics.

1 Understand limit notation.

An Introduction to Limits

Suppose that you and a friend are walking along the graph of the function

$$f(x) = \frac{x^2 - 4}{x - 2}.$$

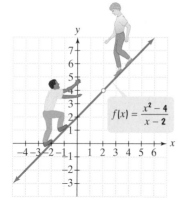

Figure 11.1 illustrates that you are walking uphill and your friend is walking downhill. Because 2 is not in the domain of the function, there is a hole in the graph at $x = 2$. Warning signs along the graph might be appropriate: Caution: $f(2)$ is undefined! If you or your friend reach 2, you will fall through the hole and splatter onto the x-axis.

Obviously, there is a problem at $x = 2$. But what happens along the graph of $f(x) = \frac{x^2 - 4}{x - 2}$ as you and your friend walk very, very close to $x = 2$? What function value, $f(x)$, will each of you approach? One way to answer this question is to construct a table of function values to analyze numerically the behavior of f as x gets closer and closer to 2. Remember that you are walking uphill, approaching 2 from the left side of 2. Your friend is walking downhill, approaching 2 from the right side of 2. Thus, we must include values of x that are less than 2 and values of x that are greater than 2.

Figure 11.1 Walking along the graph of f, very close to 2

In Table 11.1 at the top of the next page, we choose values of x close to 2. As x approaches 2 from the left, we arbitrarily start with $x = 1.99$. Then we select two additional values of x that are closer to 2, but still less than 2. We choose 1.999 and 1.9999. As x approaches 2 from the right, we arbitrarily start with $x = 2.01$. Then we select two additional values of x that are closer to 2, but still greater than 2. We choose 2.001 and 2.0001. Finally, evaluate f at each chosen value of x to obtain Table 11.1.

Technology

A graphing utility with a TABLE feature can be used to generate the entries in Table 11.1. In TBLSET, change Auto to Ask for Indpnt, the independent variable. Here is a typical screen that verifies Table 11.1.

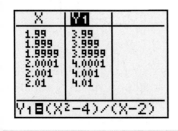

Table 11.1

		x approaches 2 from the left.				x approaches 2 from the right.		
x		1.99	1.999	1.9999	2	2.0001	2.001	2.01
$f(x) = \dfrac{x^2 - 4}{x - 2}$		3.99	3.999	3.9999	Undefined	4.0001	4.001	4.01
			$f(x)$ gets closer to 4.			$f(x)$ gets closer to 4.		

From Table 11.1, it appears that as x gets closer to 2, the values of $f(x) = \dfrac{x^2 - 4}{x - 2}$ get closer to 4. We say that

"The limit of $\dfrac{x^2 - 4}{x - 2}$ as x approaches 2 equals the number 4."

We can express this sentence in a mathematical notation called **limit notation**. We use an arrow for the word *approaches*. Likewise, we use *lim* as shorthand for the word *limit*. Thus, the limit notation for the English sentence in quotations is

$$\lim_{x \to 2} \frac{x^2 - 4}{x - 2} = 4.$$ *The limit of $\dfrac{x^2 - 4}{x - 2}$ as x approaches 2 equals the number 4.*

Calculus is the study of limits and their applications. Concepts that you will encounter in calculus are limits.

Limit Notation and Its Description

Suppose that f is a function defined on some open interval containing the number a. The function f may or may not be defined at a. The **limit notation**

$$\lim_{x \to a} f(x) = L$$

is read "the limit of $f(x)$ as x approaches a equals the number L." This means that as x gets closer to a, but remains unequal to a, the corresponding values of $f(x)$ get closer to L.

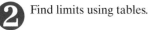 Find limits using tables.

Finding Limits Using Tables

To find $\lim_{x \to a} f(x)$, use a graphing utility with a TABLE feature or create a table by hand. Approach a from the left, choosing values of x that are close to a, but still less than a. Then approach a from the right, choosing values of x that are close to a, but still greater than a. Evaluate f at each chosen value of x to obtain the desired table.

Choose values of x so that the table makes it obvious what the corresponding values of $f(x)$ are getting close to. If the values of $f(x)$ are getting close to the number L, we infer that

$$\lim_{x \to a} f(x) = L.$$

EXAMPLE 1   Finding a Limit Using a Table

Find: $\lim\limits_{x \to 4} 3x^2$.

Solution   As x gets closer to 4, but remains unequal to 4, we must find the number that the corresponding values of $3x^2$ get closer to. The voice balloons shown below indicate that in this limit problem, $f(x) = 3x^2$ and $a = 4$.

$$\lim_{x \to 4} 3x^2$$

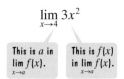

This is a in
$\lim\limits_{x \to a} f(x)$.

This is $f(x)$
in $\lim\limits_{x \to a} f(x)$.

In making a table, we choose values of x close to 4. As x approaches 4 from the left, we arbitrarily start with $x = 3.99$. Then we select two additional values of x that are closer to 4, but still less than 4. We choose 3.999 and 3.9999. As x approaches 4 from the right, we arbitrarily start with $x = 4.01$. Then we select two additional numbers that are closer to 4, but still greater than 4. We choose 4.001 and 4.0001. Finally, we evaluate f at each chosen value of x to obtain Table 11.2. The values of $f(x)$ in the table are rounded to four decimal places.

Table 11.2

	x approaches 4 from the left.					x approaches 4 from the right.		
x	3.99	3.999	3.9999	$\longrightarrow$	$\longleftarrow$	4.0001	4.001	4.01
$f(x) = 3x^2$	47.7603	47.9760	47.9976	$\longrightarrow$	$\longleftarrow$	48.0024	48.0240	48.2403

$f(x)$ gets closer to 48. $f(x)$ gets closer to 48.

From Table 11.2, it appears that as x gets closer to 4, the values of $3x^2$ get closer to 48. We infer that

$$\lim_{x \to 4} 3x^2 = 48.$$

Check Point 1   Find: $\lim\limits_{x \to 3} 4x^2$.

EXAMPLE 2   Finding a Limit Using a Table

Find: $\lim\limits_{x \to 0} \dfrac{\sin x}{x}$.

Solution   As x gets closer to 0, but remains unequal to 0, we must find the number that the corresponding values of $\dfrac{\sin x}{x}$ get closer to. The voice balloons shown below indicate that in this limit problem, $f(x) = \dfrac{\sin x}{x}$ and $a = 0$.

$$\lim_{x \to 0} \frac{\sin x}{x}$$

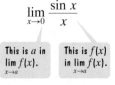

This is a in
$\lim\limits_{x \to a} f(x)$.

This is $f(x)$
in $\lim\limits_{x \to a} f(x)$.

Because division by 0 is undefined, the domain of $f(x) = \dfrac{\sin x}{x}$ is $\{x | x \neq 0\}$. Thus, f is not defined at 0. However, in this limit problem, we do not care what is happening at $x = 0$. We are interested in the behavior of the function as x gets close to 0. Table 11.3 shows the values of $f(x)$, rounded to five decimal places, as x approaches 0 from the left and from the right. Values of x in the table are measured in radians.

Technology

The graph of $f(x) = \dfrac{\sin x}{x}$ illustrates that as x gets closer to 0, the values of $f(x)$ are approaching 1. This supports our inference that

$$\lim_{x \to 0} \frac{\sin x}{x} = 1.$$

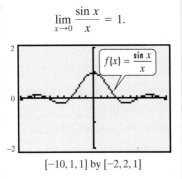

$[-10, 1, 1]$ by $[-2, 2, 1]$

Table 11.3

	x approaches 0 from the left.				x approaches 0 from the right.		
x	-0.03	-0.02	-0.01	$\longrightarrow \longleftarrow$	0.01	0.02	0.03
$f(x) = \dfrac{\sin x}{x}$	0.99985	0.99993	0.99998	$\longrightarrow \longleftarrow$	0.99998	0.99993	0.99985

$f(x)$ gets closer to 1.	$f(x)$ gets closer to 1.

From Table 11.3, it appears that as x gets closer to 0, the values of $\dfrac{\sin x}{x}$ get closer to 1. We infer that

$$\lim_{x \to 0} \frac{\sin x}{x} = 1.$$

Check Point 2 Find: $\displaystyle\lim_{x \to 0} \frac{\cos x - 1}{x}$.

Finding Limits Using Graphs

3 Find limits using graphs.

The limit statement

$$\lim_{x \to a} f(x) = L$$

is illustrated in Figure 11.2. In the three graphs, the number that x is approaching, a, is shown on the x-axis. The limit, L, is shown on the y-axis. Take a few minutes to examine the graphs. Can you see that as x approaches a along the x-axis, $f(x)$ approaches L along the y-axis? In each graph, as x gets closer to a, the values of $f(x)$ get closer to L.

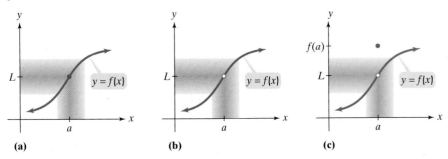

(a)   (b)   (c)

Figure 11.2 In each graph, as x gets closer to a, the values of f get closer to L: $\displaystyle\lim_{x \to a} f(x) = L$.

In Figure 11.2(a), as x approaches a, $f(x)$ approaches L. At a, the value of the function is L: $f(a) = L$. In Figure 11.2(b), as x approaches a, $f(x)$ approaches L. This is true although f is not defined at a, shown by the hole in the graph. In Figure 11.2(c), we again see that as x approaches a, $f(x)$ approaches L. Notice, however, that the value of the function at a, $f(a)$, shown by the blue dot, is not equal to the limit: $f(a) \neq L$. What you get as you approach a is not the same as what you get at a.

Example 3 illustrates that the graph of a function can sometimes be helpful in finding limits.

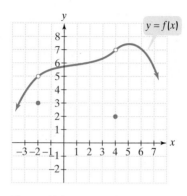

Figure 11.3

EXAMPLE 3   Finding a Limit Using a Graph

Use the graph in Figure 11.3 to find each of the following:

    a. $\lim\limits_{x \to 4} f(x)$ **b.** $f(4)$.

Solution

a. To find $\lim\limits_{x \to 4} f(x)$, examine the graph of f *near* $x = 4$. As x gets closer to 4, the values of $f(x)$ get closer to the y-coordinate of the point shown by the open dot on the right. The y-coordinate of this point is 7. Thus, as x gets closer to 4, the values of $f(x)$ get closer to 7. We conclude from the graph that

$$\lim\limits_{x \to 4} f(x) = 7.$$

b. To find $f(4)$, examine the graph of f *at* $x = 4$. At $x = 4$, the open dot is no included in the graph of f. The graph of f at 4 is shown by the closed dot with coordinates $(4, 2)$. Thus, $f(4) = 2$.

In Example 3, notice that the value of f at 4 has nothing to do with the conclusion that $\lim\limits_{x \to 4} f(x) = 7$. Regardless of how f is defined at 4, it is still true that $\lim\limits_{x \to 4} f(x) = 7$. Furthermore, if f were undefined at 4, the limit of $f(x)$ as $x \to 4$ would still equal 7.

Check Point 3 Use the graph in Figure 11.3 to find each of the following:

    a. $\lim\limits_{x \to -2} f(x)$ **b.** $f(-2)$.

EXAMPLE 4   Finding a Limit by Graphing a Function

Graph the function

$$f(x) = \begin{cases} 2x - 4 & \text{if } x \neq 3 \\ -5 & \text{if } x = 3. \end{cases}$$

Use the graph to find $\lim\limits_{x \to 3} f(x)$.

Solution This piecewise function is defined by two equations. Graph the piece defined by the linear function, $f(x) = 2x - 4$, using the y-intercept, -4, and the slope, 2. Because $x = 3$ is not included, show an open dot on the line corresponding to $x = 3$. This open dot, with coordinates $(3, 2)$, is shown in Figure 11.4.

Now we complete the graph using $f(x) = -5$ if $x = 3$. This part of the function is graphed as the point $(3, -5)$, shown as a closed blue dot in Figure 11.4.

To find $\lim\limits_{x \to 3} f(x)$, examine the graph of f near $x = 3$. As x gets closer to 3, the values of $f(x)$ get closer to the y-coordinate of the point shown by the open dot. The y-coordinate of this point is 2. We conclude from the graph that

$$\lim\limits_{x \to 3} f(x) = 2.$$

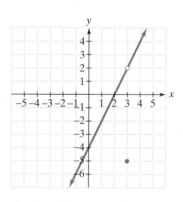

Figure 11.4 As x gets closer to 3, what number are the function values getting closer to?

Check Point 4 Graph the function

$$f(x) = \begin{cases} 3x - 2 & \text{if } x \neq 2 \\ 1 & \text{if } x = 2. \end{cases}$$

Use the graph to find $\lim\limits_{x \to 2} f(x)$.

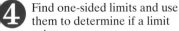

Find one-sided limits and use them to determine if a limit exists.

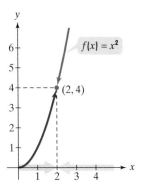

Figure 11.5 As x approaches 2 from the left (red arrow) or from the right (blue arrow), values of $f(x)$ get closer to 4.

One-Sided Limits

The graph in Figure 11.5 shows a portion of the graph of the function $f(x) = x^2$. The graph illustrates that

$$\lim_{x \to 2} x^2 = 4.$$

As x gets closer to 2, but remains unequal to 2, the corresponding values of $f(x)$ get closer to 4. The values of x near 2 fall into two categories: those that lie to the left of 2, shown by the red arrow on the x-axis, and those that lie to the right of 2, shown by the blue arrow on the x-axis.

The values of x can get closer to 2 in two ways. The values of x can approach 2 from the left, through numbers that are less than 2. Table 11.4 shows some values of x and the corresponding values of $f(x)$ rounded to four decimal places. The red portion of the graph in Figure 11.5 shows that as x approaches 2 from the left of 2, $f(x)$ approaches 4.

Table 11.4

x	1.99	1.999	1.9999 →
$f(x) = x^2$	3.9601	3.9960	3.9996 →

We say that "the limit of x^2 as x approaches 2 from the left equals 4." The mathematical notation for this English sentence is

$$\lim_{x \to 2^-} x^2 = 4.$$

The notation $x \to 2^-$ indicates that x is less than 2 and is approaching 2 from the left.

The values of x can also approach 2 from the right, through numbers that are greater than 2. Table 11.5 shows some values of x and the corresponding values of $f(x)$ rounded to four decimal places. The blue portion of the graph in Figure 11.5 shows that as x approaches 2 from the right of 2, $f(x)$ approaches 4.

Table 11.5

x	← 2.0001	2.001	2.01
$f(x) = x^2$	← 4.0004	4.0040	4.0401

We say that "the limit of x^2 as x approaches 2 from the right equals 4." The mathematical notation for this English sentence is

$$\lim_{x \to 2^+} x^2 = 4.$$

The notation $x \to 2^+$ indicates that x is greater than 2 and is approaching 2 from the right.

In general, if x approaches a from one side, we have a **one-sided limit**.

One-Sided Limits

Left-Hand Limit The limit notation

$$\lim_{x \to a^-} f(x) = L$$

is read "the limit of $f(x)$ as x approaches a from the left equals L" and is called the **left-hand limit**. This means that as x gets closer to a, but remains less than a, the corresponding values of $f(x)$ get closer to L.

Right-Hand Limit The limit notation

$$\lim_{x \to a^+} f(x) = L$$

is read "the limit of $f(x)$ as x approaches a from the right equals L" and is called the **right-hand limit**. This means that as x gets closer to a, but remains greater than a, the corresponding values of $f(x)$ get closer to L.

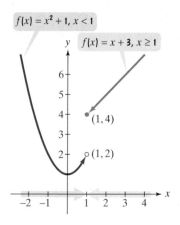

$f(x) = x^2 + 1, x < 1$

$f(x) = x + 3, x \geq 1$

(1, 4)

(1, 2)

Figure 11.6 As x approaches 1 from the left (red arrow) and from the right (blue arrow), values of $f(x)$ do not get closer to a single number.

A function's graph can be helpful in finding one-sided limits. For example Figure 11.6 shows the graph of the piecewise function

$$f(x) = \begin{cases} x^2 + 1 & \text{if } x < 1 \\ x + 3 & \text{if } x \geq 1. \end{cases}$$

The red portion of the graph, part of a parabola, illustrates that as x approaches 1 from the left, the corresponding values of $f(x)$ get closer to 2. The left-hand limit is 2:

$$\lim_{x \to 1^-} f(x) = 2.$$

The blue portion of the graph, part of a line, illustrates that as x approaches 1 from the right, the corresponding values of $f(x)$ get closer to 4. The right-hand limit is 4:

$$\lim_{x \to 1^+} f(x) = 4.$$

Because $\lim_{x \to 1^-} f(x) = 2$ and $\lim_{x \to 1^+} f(x) = 4$, there is no single number that the values of $f(x)$ are close to when x is close to 1. In this case, we say that **f has no limit as x approaches 1** or that $\lim_{x \to 1} f(x)$ **does not exist**.

In general, a function f has a limit as x approaches a if and only if the left-hand limit equals the right-hand limit.

> **Equal and Unequal One-Sided Limits**
>
> • One-sided limits can be used to show that a function has a limit as x approaches a.
>
> $$\lim_{x \to a} f(x) = L \text{ if and only if both}$$
>
> $$\lim_{x \to a^-} f(x) = L \quad \text{and} \quad \lim_{x \to a^+} f(x) = L.$$
>
> • One-sided limits can be used to show that a function has no limit as x approaches a.
>
> If $\lim_{x \to a^-} f(x) = L \quad \text{and} \quad \lim_{x \to a^+} f(x) = M$, where $L \neq M$,
>
> $$\lim_{x \to a} f(x) \text{ does not exist.}$$

Study Tip

The word *from* is helpful in distinguishing left- and right-hand limits. A left-hand limit means you approach the given x-value *from* the left. It does not mean that you approach toward the left on the graph. A right-hand limit means you approach the given x-value *from* the right. It does not mean you approach toward the right on the graph.

EXAMPLE 5   Finding One-Sided Limits Using a Graph

Use the graph of the piecewise function f in Figure 11.7 to find each of the following, or state that a limit or function value does not exist:

a. $\lim_{x \to -2^-} f(x)$ **b.** $\lim_{x \to -2^+} f(x)$ **c.** $\lim_{x \to -2} f(x)$ **d.** $f(-2)$.

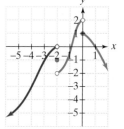

Figure 11.7

Solution

a. To find $\lim_{x \to -2^-} f(x)$, examine the portion of the graph shown in red that is near, but to the left of $x = -2$. As x approaches -2 from the left, the values of $f(x)$ get close to the y-coordinate of the point shown by the red open dot. This point, $(-2, 0)$, has a y-coordinate of 0. Thus,

$$\lim_{x \to -2^-} f(x) = 0.$$

b. To find $\lim_{x \to -2^+} f(x)$, examine the portion of the graph shown in blue that is near, but to the right of $x = -2$. As x approaches -2 from the right, the values of $f(x)$ get close to the y-coordinate of the point shown by the blue open dot. This point, $(-2, -2)$, has a y-coordinate of -2. Thus,

$$\lim_{x \to -2^+} f(x) = -2.$$

c. We found that $\lim_{x \to -2^-} f(x) = 0$ and $\lim_{x \to -2^+} f(x) = -2$.

> The limit as x approaches -2 from the left equals 0.

> The limit as x approaches -2 from the right equals -2.

Because the left- and right-hand limits are unequal, $\lim_{x \to -2} f(x)$ does not exist.

d. To find $f(-2)$, examine the graph of f at $x = -2$. The graph of f at -2 is shown by the blue closed dot with coordinates $(-2, -1)$. Thus, $f(-2) = -1$.

Check Point 5 Use the graph of the piecewise function f in Figure 11.7 to find each of the following, or state that a limit or function value does not exist:

a. $\lim_{x \to 0^-} f(x)$ **b.** $\lim_{x \to 0^+} f(x)$ **c.** $\lim_{x \to 0} f(x)$ **d.** $f(0)$.

EXERCISE SET 11.1

Practice Exercises

In Exercises 1–4, use each table to find the indicated limit.

1. $\lim_{x \to 2} x^2$

x	1.99	1.999	1.9999 →	← 2.0001	2.001	2.01
$f(x) = x^2$	3.960	3.996	3.9996 →	← 4.0004	4.004	4.040

2. $\lim_{x \to 3} 5x^2$

x	2.99	2.999	2.9999 →	← 3.0001	3.001	3.01
$f(x) = 5x^2$	44.970	44.970	44.997 →	← 45.003	45.03	45.301

3. $\lim_{x \to 0} \dfrac{\sin 3x}{x}$

x	−0.03	−0.02	−0.01 →	← 0.01	0.02	0.03
$f(x) = \dfrac{\sin 3x}{x}$	2.9960	2.9982	2.9996 →	← 2.9996	2.9982	2.996

4. $\lim_{x \to 0} \dfrac{\sin 4x}{\sin 2x}$

x	−0.03	−0.02	−0.01 →	← 0.01	0.02	0.03
$f(x) = \dfrac{\sin 4x}{\sin 2x}$	1.9964	1.9984	1.9996 →	← 1.9996	1.9984	1.9964

In Exercises 5–18, construct a table to find the indicated limit.

5. $\lim_{x \to 2} 5x^2$ **6.** $\lim_{x \to 2}(x^2 - 1)$

7. $\lim_{x \to 3} \dfrac{1}{x - 2}$ **8.** $\lim_{x \to 4} \dfrac{1}{x - 3}$

9. $\lim_{x \to 0} \dfrac{x}{x^2 + 1}$ **10.** $\lim_{x \to 0} \dfrac{x + 1}{x^2 + 1}$

11. $\lim_{x \to -2} \dfrac{x^3 + 8}{x + 2}$ **12.** $\lim_{x \to -5} \dfrac{x^2 - 25}{x + 5}$

13. $\lim_{x \to 0} \dfrac{2x^2 + x}{\sin x}$ **14.** $\lim_{x \to 0} \dfrac{\sin x^2}{x}$

15. $\lim_{x \to 0} \dfrac{\tan x}{x}$ **16.** $\lim_{x \to 0} \dfrac{x^2}{\sec x - 1}$

17. $\lim_{x \to 0} f(x)$, where $f(x) = \begin{cases} x + 1 & \text{if } x < 0 \\ 2x + 1 & \text{if } x \geq 0 \end{cases}$

18. $\lim_{x \to 0} f(x)$, where $f(x) = \begin{cases} x + 2 & \text{if } x < 0 \\ 3x + 2 & \text{if } x \geq 0 \end{cases}$

In Exercises 19–22, use the graph of f to find the indicated limit and function value.

19.

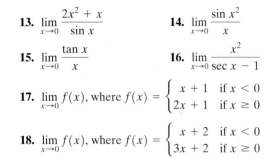

a. $\lim_{x \to 3} f(x)$ **b.** $f(3)$

20.

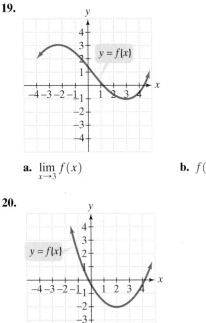

a. $\lim_{x \to 2} f(x)$ **b.** $f(2)$

21.

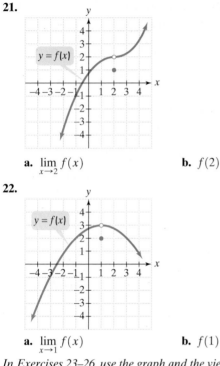

a. $\lim\limits_{x \to 2} f(x)$ **b.** $f(2)$

22.

$y = f(x)$

a. $\lim\limits_{x \to 1} f(x)$ **b.** $f(1)$

In Exercises 23–26, use the graph and the viewing rectangle shown below the graph to find the indicated limit.

23. $\lim\limits_{x \to 2} (1 - x^2)$

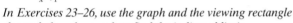

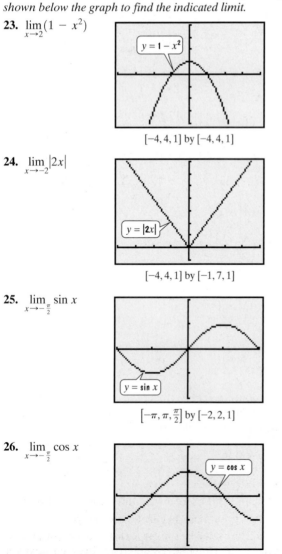

$y = 1 - x^2$

$[-4, 4, 1]$ by $[-4, 4, 1]$

24. $\lim\limits_{x \to -2} |2x|$

$y = |2x|$

$[-4, 4, 1]$ by $[-1, 7, 1]$

25. $\lim\limits_{x \to -\frac{\pi}{2}} \sin x$

$y = \sin x$

$\left[-\pi, \pi, \frac{\pi}{2}\right]$ by $[-2, 2, 1]$

26. $\lim\limits_{x \to -\frac{\pi}{2}} \cos x$

$y = \cos x$

$\left[-\pi, \pi, \frac{\pi}{2}\right]$ by $[-2, 2, 1]$

In Exercises 27–32, the graph of a function f is given. Use the graph to find the indicated limits and function values, or state that a limit or function value does not exist.

27.

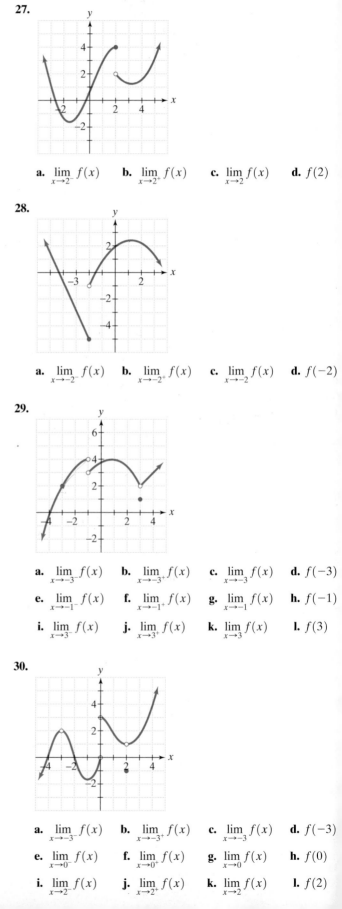

a. $\lim\limits_{x \to 2^-} f(x)$ **b.** $\lim\limits_{x \to 2^+} f(x)$ **c.** $\lim\limits_{x \to 2} f(x)$ **d.** $f(2)$

28.

a. $\lim\limits_{x \to -2^-} f(x)$ **b.** $\lim\limits_{x \to -2^+} f(x)$ **c.** $\lim\limits_{x \to -2} f(x)$ **d.** $f(-2)$

29.

a. $\lim\limits_{x \to -3^-} f(x)$ **b.** $\lim\limits_{x \to -3^+} f(x)$ **c.** $\lim\limits_{x \to -3} f(x)$ **d.** $f(-3)$

e. $\lim\limits_{x \to -1^-} f(x)$ **f.** $\lim\limits_{x \to -1^+} f(x)$ **g.** $\lim\limits_{x \to -1} f(x)$ **h.** $f(-1)$

i. $\lim\limits_{x \to 3^-} f(x)$ **j.** $\lim\limits_{x \to 3^+} f(x)$ **k.** $\lim\limits_{x \to 3} f(x)$ **l.** $f(3)$

30.

a. $\lim\limits_{x \to -3^-} f(x)$ **b.** $\lim\limits_{x \to -3^+} f(x)$ **c.** $\lim\limits_{x \to -3} f(x)$ **d.** $f(-3)$

e. $\lim\limits_{x \to 0^-} f(x)$ **f.** $\lim\limits_{x \to 0^+} f(x)$ **g.** $\lim\limits_{x \to 0} f(x)$ **h.** $f(0)$

i. $\lim\limits_{x \to 2^-} f(x)$ **j.** $\lim\limits_{x \to 2^+} f(x)$ **k.** $\lim\limits_{x \to 2} f(x)$ **l.** $f(2)$

31.

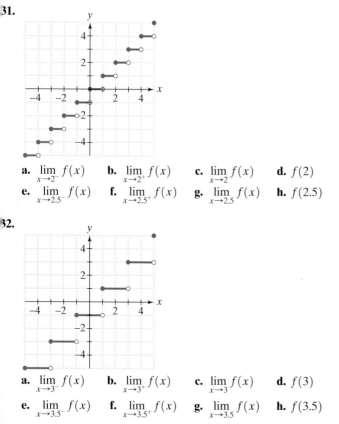

a. $\lim_{x \to 2^-} f(x)$ **b.** $\lim_{x \to 2^+} f(x)$ **c.** $\lim_{x \to 2} f(x)$ **d.** $f(2)$

e. $\lim_{x \to 2.5^-} f(x)$ **f.** $\lim_{x \to 2.5^+} f(x)$ **g.** $\lim_{x \to 2.5} f(x)$ **h.** $f(2.5)$

32.

a. $\lim_{x \to 3^-} f(x)$ **b.** $\lim_{x \to 3^+} f(x)$ **c.** $\lim_{x \to 3} f(x)$ **d.** $f(3)$

e. $\lim_{x \to 3.5^-} f(x)$ **f.** $\lim_{x \to 3.5^+} f(x)$ **g.** $\lim_{x \to 3.5} f(x)$ **h.** $f(3.5)$

In Exercises 33–54, graph each function. Then use your graph to find the indicated limit, or state that the limit does not exist.

33. $f(x) = 2x + 1, \lim_{x \to 3} f(x)$ **34.** $f(x) = 2x - 1, \lim_{x \to 3} f(x)$

35. $f(x) = 4 - x^2, \lim_{x \to -3} f(x)$ **36.** $f(x) = 9 - x^2, \lim_{x \to -2} f(x)$

37. $f(x) = |x + 1|, \lim_{x \to -1} f(x)$ **38.** $f(x) = |x + 2|, \lim_{x \to -2} f(x)$

39. $f(x) = \dfrac{1}{x}, \lim_{x \to -1} f(x)$ **40.** $f(x) = \dfrac{1}{x^2}, \lim_{x \to -1} f(x)$

41. $f(x) = \dfrac{x^2 - 1}{x - 1}, \lim_{x \to 1} f(x)$ **42.** $f(x) = \dfrac{x^2 - 4}{x - 2}, \lim_{x \to 2} f(x)$

43. $f(x) = e^x, \lim_{x \to 0} f(x)$ **44.** $f(x) = \ln x, \lim_{x \to 1} f(x)$

45. $f(x) = \sin x, \lim_{x \to \pi} f(x)$ **46.** $f(x) = \cos x, \lim_{x \to \pi} f(x)$

47. $f(x) = \begin{cases} x + 1 & \text{if } x \neq 2 \\ 5 & \text{if } x = 2, \lim_{x \to 2} f(x) \end{cases}$

48. $f(x) = \begin{cases} x - 1 & \text{if } x \neq 3 \\ 4 & \text{if } x = 3, \lim_{x \to 3} f(x) \end{cases}$

49. $f(x) = \begin{cases} x + 3 & \text{if } x < 0 \\ 4 & \text{if } x \geq 0, \lim_{x \to 0} f(x) \end{cases}$

50. $f(x) = \begin{cases} x + 4 & \text{if } x < 0 \\ 5 & \text{if } x \geq 0, \lim_{x \to 0} f(x) \end{cases}$

51. $f(x) = \begin{cases} 2x & \text{if } x < 1 \\ x + 1 & \text{if } x \geq 1, \lim_{x \to 1} f(x) \end{cases}$

52. $f(x) = \begin{cases} 3x & \text{if } x < 1 \\ x + 2 & \text{if } x \geq 1, \lim_{x \to 1} f(x) \end{cases}$

53. $f(x) = \begin{cases} x + 1 & \text{if } x < 0 \\ \sin x & \text{if } x \geq 0, \lim_{x \to 0} f(x) \end{cases}$

54. $f(x) = \begin{cases} x & \text{if } x < 0 \\ \cos x & \text{if } x \geq 0, \lim_{x \to 0} f(x) \end{cases}$

Practice Plus

In Exercises 55–56, use the equations for the functions f and g to graph the function $y = (f \circ g)(x)$. Then use the graph of $f \circ g$ to find the indicated limit.

55. $f(x) = x^2 - 5, g(x) = \sqrt{x}; \lim_{x \to 2}(f \circ g)(x)$

56. $f(x) = x^2 + 3, g(x) = \sqrt{x}; \lim_{x \to 1}(f \circ g)(x)$

In Exercises 57–58, use the equation for the function f to find and graph the function $y = f^{-1}(x)$. Then use the graph of f^{-1} to find the indicated limit.

57. $f(x) = x^3 - 2; \lim_{x \to 6} f^{-1}(x)$

58. $f(x) = x^3 - 4; \lim_{x \to 4} f^{-1}(x)$

In Exercises 59–66, use the graph of $y = f(x)$ to graph each function g. Then use the graph of g to find the indicated limit.

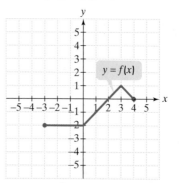

$y = f(x)$

59. $g(x) = f(x) + 2; \lim_{x \to 3} g(x)$ **60.** $g(x) = f(x) - 2; \lim_{x \to 3} g(x)$

61. $g(x) = f(x + 3); \lim_{x \to -1} g(x)$ **62.** $g(x) = f(x + 2); \lim_{x \to -2} g(x)$

63. $g(x) = -f(x); \lim_{x \to -3^+} g(x)$ **64.** $g(x) = -2f(x); \lim_{x \to -3^+} g(x)$

65. $g(x) = f(2x); \lim_{x \to 1} g(x)$ **66.** $g(x) = f\left(\tfrac{1}{2}x\right); \lim_{x \to 1} g(x)$

Application Exercises

67. You are approaching a fan located at 3 on the *x*-axis.

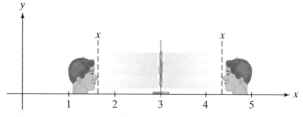

The function f describes the breeze that you feel, $f(x)$, in miles per hour, when your nose is at position *x* on the *x*-axis. Use the values in the table to solve this exercise.

x	2.9	2.99	2.999 → ← 3.001	3.01	3.1
$f(x)$	7.7	7.92	7.991 → ← 7.991	7.92	7.7

a. Find $\lim_{x \to 3} f(x)$. Describe what this means in terms of the location of your nose and the breeze that you feel.

b. Would it be a good idea to move closer so that you actually reach $x = 3$? Describe the difference between what you feel for $\lim_{x \to 3} f(x)$ and $f(3)$.

68. You are riding along an expressway traveling x miles per hour. The function $f(x) = 0.015x^2 + x + 10$ describes the recommended safe distance, $f(x)$, in feet, between your car and other cars on the expressway. Use the values in the table below to find $\lim\limits_{x \to 60} f(x)$. Describe what this means in terms of your car's speed and the recommended safe distance.

x	59.9	59.99	59.999 → ← 60.001	60.01	60.1
$f(x) = 0.015x^2 + x + 10$	123.72	123.972	123.997 → ← 124.003	124.028	124.28

Functions can be used to model changes in intellectual abilities over one's life span. The graphs of f and g show mean scores on standardized tests measuring spatial orientation and verbal ability, respectively, as a function of age. Use the graphs of f and g to solve Exercises 69–70.

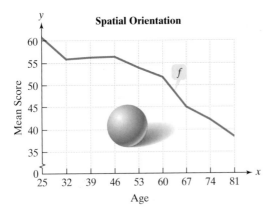

Spatial Orientation

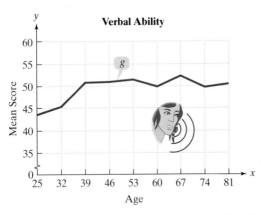

Verbal Ability

Source: Wade and Tavris, *Psychology Sixth Edition*, Prentice Hall, 2000

69. What mean score in spatial orientation is associated with a person whose age is close to 67? Use limit notation to express the answer.

70. What mean score in verbal ability is associated with a person whose age is close to 60? Use limit notation to express the answer.

71. You rent a car from a company that charges $20 per day plus $0.10 per mile. The car is driven 200 miles in the first day. The figure at the top of the next column shows the graph of the cost, $f(x)$, in dollars, as a function of the miles, x, that you drive the car.

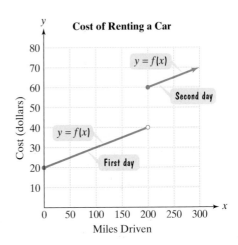

Cost of Renting a Car

a. Find $\lim\limits_{x \to 100} f(x)$. Interpret the limit, referring to miles driven and cost.

b. For the first day only, what is the rental cost approaching as the mileage gets closer to 200?

c. What is the cost to rent the car at the start of the second day?

72. You are building a greenhouse next to your house, as shown in the figure. Because the house will be used for one side of the enclosure, only three sides will need to be enclosed. You have 60 feet of fiberglass to enclose the three walls.

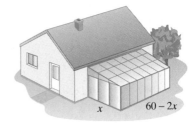

The function $f(x) = x(60 - 2x)$ describes the area of the greenhouse that you can enclose, $f(x)$, in square feet, if the width of the greenhouse is x feet.

a. Use the table shown to find $\lim\limits_{x \to 15} f(x)$.

X	Y1	
14.7	449.82	
14.8	449.92	
14.9	449.98	
15		
15.1	449.98	
15.2	449.92	
15.3	449.82	
Y1⊟X*(60-2X)		

b. Use the graph shown to find $\lim_{x \to 15} f(x)$. Do you get the same limit as you did in part (a)? What information about the limit is shown by the graph that might not be obvious from the table?

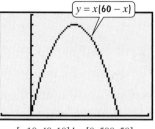

$y = x(60 - x)$

$[-10, 40, 10]$ by $[0, 500, 50]$

Writing in Mathematics

73. Explain how to read $\lim_{x \to a} f(x) = L$.

74. What does the limit notation $\lim_{x \to a} f(x) = L$ mean?

75. Without showing the details, explain how to use a table to find $\lim_{x \to 4} x^2$.

76. Explain how a graph can be used to find a limit.

77. When we find $\lim_{x \to a} f(x)$, we do not care about the value of the function at $x = a$. Explain why this is so.

78. Explain how to read $\lim_{x \to a^-} f(x) = L$.

79. What does the limit notation $\lim_{x \to a^-} f(x) = L$ mean?

80. Explain how to read $\lim_{x \to a^+} f(x) = L$.

81. What does the limit notation $\lim_{x \to a^+} f(x) = L$ mean?

82. What does it mean if the limits in Exercises 79 and 81 are not both equal to the same number L?

Technology Exercises

83. Use the $\boxed{\text{TABLE}}$ feature of your graphing utility to verify any five of the limits that you found in Exercises 5–16.

84. Use the $\boxed{\text{ZOOM IN}}$ feature of your graphing utility to verify any five of the limits that you found in Exercises 33–46. Zoom in on the graph of the given function, f, near $x = a$ to verify each limit.

In Exercises 85–88, estimate $\lim_{x \to a} f(x)$ by using the $\boxed{\text{TABLE}}$ feature of your graphing utility to create a table of values. Then use the $\boxed{\text{ZOOM IN}}$ feature to zoom in on the graph of f near $x = a$ to justify or improve your estimate.

85. $\lim_{x \to 0} \dfrac{2^x - 1}{x}$

86. $\lim_{x \to 4} \dfrac{\ln x - \ln 4}{x - 4}$

87. $\lim_{x \to 1} \dfrac{x^{3/2} - 1}{x - 1}$

88. $\lim_{x \to 0} \dfrac{x^2}{1 - \cos 2x}$

Critical Thinking Exercises

89. Give an example of a function that is not defined at 2 for which $\lim_{x \to 2} f(x) = 5$.

90. Consider the function $f(x) = 3x + 2$. As x approaches 1, $f(x)$ approaches 5: $\lim_{x \to 1} f(x) = 5$. Find the values of x such that $f(x)$ is within 0.1 of 5 by solving
$$|f(x) - 5| < 0.1.$$
Then find the values of x such that $f(x)$ is within 0.01 of 5.

91. Find an estimate of $3^\pi (\pi \approx 3.14159265)$ by taking a sequence of rational numbers, $x_1, x_2, x_3, \dots$ that approaches π. Obtain your estimate by evaluating $3^{x_1}, 3^{x_2}, 3^{x_3}, \dots$.

SECTION 11.2 *Finding Limits Using Properties of Limits*

Objectives

❶ Find limits of constant functions and the identity function.

❷ Find limits using properties of limits.

❸ Find one-sided limits using properties of limits.

❹ Find limits of fractional expressions in which the limit of the denominator is zero.

Isaac Newton

Gottfried Leibniz

Calculus was invented independently by British mathematician Isaac Newton (1642–1727) and German mathematician Gottfried Leibniz (1646–1716). Although Newton stated that limits were the basic concept in calculus, neither he nor Leibniz was able to express the idea of a limit in a precise mathematical fashion. In essence, Newton and Leibniz developed calculus into a powerful tool even though they could not fully understand why the tool worked.

A great triumph of calculus came with the work of German mathematician Karl Weierstrass (1815–1897). Weierstrass provided a precise definition of $\lim_{x \to a} f(x) = L$, placing calculus on a sound footing almost two hundred years after its invention. The properties of limits presented in this section are theorems that you will prove in calculus using this definition. In this section, you will learn to apply these properties to find limits.

Limits Involving Constant Functions and the Identity Function

① Find limits of constant functions and the identity function.

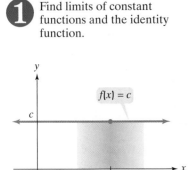

We frequently encounter the constant function, $f(x) = c$, and the identity function, $f(x) = x$. Figure 11.8 shows the graph of the constant function. The graph is a horizontal line. What does this mean about the limit as x approaches a? Regardless of how close x is to a, the corresponding value of $f(x)$ is c. Thus, if $f(x) = c$, then $\lim_{x \to a} f(x) = c$.

> **Limit of a Constant Function**
>
> For the constant function $f(x) = c$,
> $$\lim_{x \to a} f(x) = \lim_{x \to a} c = c,$$
> where a is any number. In words, regardless of what number x is approaching, the limit of any constant is that constant.

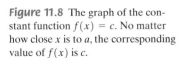

Figure 11.8 The graph of the constant function $f(x) = c$. No matter how close x is to a, the corresponding value of $f(x)$ is c.

EXAMPLE 1   Finding Limits of Constant Functions

Find the following limits:

    a. $\lim_{x \to 4} 7$ **b.** $\lim_{x \to 0} (-5)$.

Solution   Regardless of what number x is approaching, the limit of any constant is that constant: $\lim_{x \to a} c = c$. Using this formula, we find the given limits.

    a. $\lim_{x \to 4} 7 = 7$ **b.** $\lim_{x \to 0} (-5) = -5$

Check Point 1   Find the following limits:

    a. $\lim_{x \to 8} 11$ **b.** $\lim_{x \to 0} (-9)$.

The graph of the identity function, $f(x) = x$, is shown in Figure 11.9. Each input for this function is associated with an identical output. What does this mean about the limit as x approaches a? For any value of a, as x gets closer to a, the corresponding value of $f(x)$ is just as close to a. Thus, if $f(x) = x$, then $\lim_{x \to a} f(x) = a$.

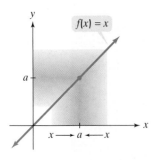

Figure 11.9 The graph of the identity function $f(x) = x$. No matter how close x is to a, the corresponding value of $f(x)$ is just as close to a.

> **Limit of the Identity Function**
>
> For the identity function $f(x) = x$,
> $$\lim_{x \to a} f(x) = \lim_{x \to a} x = a,$$
> where a is any number. In words, the limit of x as x approaches any number is that number.

EXAMPLE 2   Finding Limits of the Identity Function

Find the following limits:

    a. $\lim_{x \to 7} x$ **b.** $\lim_{x \to -\pi} x$.

Solution   We use the formula $\lim_{x \to a} x = a$. The number that x is approaching is also the limit.

    a. $\lim_{x \to 7} x = 7$ **b.** $\lim_{x \to -\pi} x = -\pi$

Check Point 2 Find the following limits:

a. $\lim\limits_{x \to 19} x$ b. $\lim\limits_{x \to -\sqrt{2}} x.$

2 Find limits using properties of limits.

Properties of Limits

How do we find the limit of a sum, such as

$$\lim_{x \to 5}(x + 7)?$$

We find the limit of each function in the sum:

$$\lim_{x \to 5} x = 5 \quad \text{and} \quad \lim_{x \to 5} 7 = 7.$$

Use the formula
$\lim\limits_{x \to a} x = a.$

Use the formula
$\lim\limits_{x \to a} c = c.$

Then we add each of these limits. Thus,

$$\lim_{x \to 5}(x + 7) = \lim_{x \to 5} x + \lim_{x \to 5} 7 = 5 + 7 = 12.$$

This is an application of a limit property involving the limit of a sum.

The Limit of a Sum

If $\lim\limits_{x \to a} f(x) = L$ and $\lim\limits_{x \to a} g(x) = M$, then

$$\lim_{x \to a}[f(x) + g(x)] = \lim_{x \to a} f(x) + \lim_{x \to a} g(x) = L + M.$$

In words, the limit of the sum of two functions equals the sum of their limits.

EXAMPLE 3 Finding the Limit of a Sum

Find: $\lim\limits_{x \to -4}(x + 9)$.

Solution The two functions in this limit problem are $f(x) = x$ and $g(x) = 9$. We seek the limit of the sum of these functions.

$$\begin{aligned} \lim_{x \to -4}(x + 9) &= \lim_{x \to -4} x + \lim_{x \to -4} 9 \quad &\text{The limit of a sum is the sum of the limits.}\\ &= -4 + 9 \quad &\lim_{x \to a} x = a \text{ and } \lim_{x \to a} c = c.\\ &= 5 \end{aligned}$$

Check Point 3 Find: $\lim\limits_{x \to -3}(x + 16)$.

In calculus, you will prove the following property involving the limit of the difference of two functions:

The Limit of a Difference

If $\lim\limits_{x \to a} f(x) = L$ and $\lim\limits_{x \to a} g(x) = M$, then

$$\lim_{x \to a}[f(x) - g(x)] = \lim_{x \to a} f(x) - \lim_{x \to a} g(x) = L - M.$$

In words, the limit of the difference of two functions equals the difference of their limits.

EXAMPLE 4 Finding the Limit of a Difference

Find: $\lim\limits_{x \to 5}(12 - x)$.

Solution The two functions in this limit problem are $f(x) = 12$ and $g(x) = x$. We seek the limit of the difference of these functions.

$$\lim\limits_{x \to 5}(12 - x) = \lim\limits_{x \to 5} 12 - \lim\limits_{x \to 5} x \qquad \text{The limit of a difference is the difference of the limits.}$$

$$= 12 - 5 \qquad \lim\limits_{x \to a} c = c \text{ and } \lim\limits_{x \to a} x = a.$$

$$= 7$$

Check Point 4 Find: $\lim\limits_{x \to 14}(19 - x)$.

Now we consider a property that will enable you to find the limit of the product of two functions.

> ### The Limit of a Product
>
> If $\lim\limits_{x \to a} f(x) = L$ and $\lim\limits_{x \to a} g(x) = M$, then
>
> $$\lim\limits_{x \to a}[f(x) \cdot g(x)] = \lim\limits_{x \to a} f(x) \cdot \lim\limits_{x \to a} g(x) = LM.$$
>
> In words, the limit of the product of two functions equals the product of their limits.

EXAMPLE 5 Finding the Limit of a Product

Find: $\lim\limits_{x \to 5}(-6x)$.

Solution The two functions in this limit problem are $f(x) = -6$ and $g(x) = x$. We seek the limit of the product of these functions.

$$\lim\limits_{x \to 5}(-6x) = \lim\limits_{x \to 5}(-6) \cdot \lim\limits_{x \to 5} x \qquad \text{The limit of a product is the product of the limits.}$$

$$= -6 \cdot 5 \qquad \lim\limits_{x \to a} c = c \text{ and } \lim\limits_{x \to a} x = a.$$

$$= -30$$

Check Point 5 Find: $\lim\limits_{x \to 7}(-10x)$.

EXAMPLE 6 Finding Limits Using Properties of Limits

Find the following limits:

a. $\lim\limits_{x \to -3}(7x - 4)$ **b.** $\lim\limits_{x \to 5} 6x^2$.

Solution

a. $\lim\limits_{x \to -3}(7x - 4) = \lim\limits_{x \to -3}(7x) - \lim\limits_{x \to -3} 4$ The limit of a difference is the difference of the limits.

$$= \lim\limits_{x \to -3} 7 \cdot \lim\limits_{x \to -3} x - \lim\limits_{x \to -3} 4 \qquad \text{The limit of a product is the product of the limits.}$$

$$= 7(-3) - 4 \qquad \lim\limits_{x \to a} c = c \text{ and } \lim\limits_{x \to a} x = a.$$

$$= -21 - 4$$

$$= -25$$

b. $\displaystyle\lim_{x\to5} 6x^2 = \lim_{x\to5} 6 \cdot \lim_{x\to5} x^2$ *The limit of a product is the product of the limits.*

$\displaystyle = 6 \cdot \lim_{x\to5}(x \cdot x)$ $\displaystyle\lim_{x\to a} c = c$

$\displaystyle = 6 \cdot \lim_{x\to5} x \cdot \lim_{x\to5} x$ *The limit of a product is the product of the limits.*

$= 6 \cdot 5 \cdot 5$ $\displaystyle\lim_{x\to a} x = a$

$= 150$

Check Point 6   Find the following limits:

a. $\displaystyle\lim_{x\to-5}(3x - 7)$ **b.** $\displaystyle\lim_{x\to3} 8x^2.$

The procedure used in Example 6(b) can be used to find the limit of any monomial function in the form $f(x) = b_n x^n$, where n is a positive integer and b_n is a constant.

$\displaystyle\lim_{x\to a} b_n x^n = \lim_{x\to a} b_n \cdot \lim_{x\to a} x^n$ *The limit of a product is the product of the limits.*

$\displaystyle = b_n \cdot \lim_{x\to a} (\underbrace{x \cdot x \cdot x \cdot \cdots \cdot x})$ $\displaystyle\lim_{x\to a} c = c$

By definition, x^n contains n factors of x.

$\displaystyle = b_n \cdot \underbrace{\lim_{x\to a} x \cdot \lim_{x\to a} x \cdot \lim_{x\to a} x \cdot \cdots \cdot \lim_{x\to a} x}$ *The limit of the product containing n factors is the product of the limits.*

There are n factors of $\displaystyle\lim_{x\to a} x$.

$= b_n \cdot \underbrace{a \cdot a \cdot a \cdot \cdots \cdot a}$ $\displaystyle\lim_{x\to a} x = a$

There are n factors of a.

$= b_n a^n$

This is the monomial function $f(x) = b_n x^n$ evaluated at a.

> **Limit of a Monomial**
>
> If n is a positive integer and b_n is a constant, then
> $$\lim_{x\to a} b_n x^n = b_n a^n$$
> for any number a. In words, the limit of a monomial as x approaches a is the monomial evaluated at a.

EXAMPLE 7   Finding the Limit of a Monomial

Find: $\displaystyle\lim_{x\to2}(-6x^4).$

Solution   The limit of the monomial $-6x^4$ as x approaches 2 is the monomial evaluated at 2. Thus, we find the limit by substituting 2 for x.

$$\lim_{x\to2}(-6x^4) = -6 \cdot 2^4 = -6 \cdot 16 = -96$$

Check Point 7   Find: $\displaystyle\lim_{x\to2}(-7x^3).$

How do we find the limit of a polynomial function

$$f(x) = b_n x^n + b_{n-1} x^{n-1} + \cdots + b_1 x + b_0$$

as x approaches a? A polynomial is a sum of monomials. Thus, the limit of a polynomial is the sum of the limits of its monomials.

$$\lim_{x \to a} f(x) = \lim_{x \to a} (b_n x^n + b_{n-1} x^{n-1} + \cdots + b_1 x + b_0) \qquad \textit{f is a polynomial function.}$$

$$= \lim_{x \to a} b_n x^n + \lim_{x \to a} b_{n-1} x^{n-1} + \cdots + \lim_{x \to a} b_1 x + \lim_{x \to a} b_0 \qquad \textit{The limit of a sum is the sum of the limits.}$$

$$= b_n a^n + b_{n-1} a^{n-1} + \cdots + b_1 a + b_0 \qquad \textit{Find limits by evaluating monomials at a. Find the last limit in the sum using } \lim_{x \to a} c = c \textit{ with } c = b_0.$$

This is the polynomial function
$$f(x) = b_n x^n + b_{n-1} x^{n-1} + \cdots + b_1 x + b_0$$
evaluated at a.

$$= f(a)$$

> ## Limit of a Polynomial
>
> If f is a polynomial function, then
>
> $$\lim_{x \to a} f(x) = f(a)$$
>
> for any number a. In words, the limit of a polynomial as x approaches a is the polynomial evaluated at a.

EXAMPLE 8 Finding the Limit of a Polynomial

Find: $\lim_{x \to 3} (4x^3 + 2x^2 - 6x + 5)$.

Solution The limit of the polynomial $4x^3 + 2x^2 - 6x + 5$ as x approaches 3 is the polynomial evaluated at 3. Thus, we find the limit by substituting 3 for x.

$$\lim_{x \to 3} (4x^3 + 2x^2 - 6x + 5)$$
$$= 4 \cdot 3^3 + 2 \cdot 3^2 - 6 \cdot 3 + 5$$
$$= 4 \cdot 27 + 2 \cdot 9 - 6 \cdot 3 + 5$$
$$= 108 + 18 - 18 + 5$$
$$= 113$$

Check Point 8 Find: $\lim_{x \to 2} (7x^3 + 3x^2 - 5x + 3)$.

A linear function, $f(x) = mx + b$, is a polynomial function of degree one. This means that the limit of a linear function as x approaches a is the linear function evaluated at a:

$$\lim_{x \to a} (mx + b) = ma + b.$$

For example,

$$\lim_{x \to 4} (3x - 7) = 3 \cdot 4 - 7 = 12 - 7 = 5.$$

The next limit property involves the limit of a polynomial to a power, such as

$$\lim_{x \to 2} (x^2 + 2x - 3)^4.$$

To find such a limit, first find $\lim_{x \to 2} (x^2 + 2x - 3)$:

$$\lim_{x \to 2} (x^2 + 2x - 3) = 2^2 + 2 \cdot 2 - 3 = 4 + 4 - 3 = 5.$$

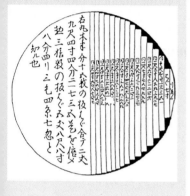

The limit that we seek is found by taking this limit, 5, and raising it to the fourth power. Thus,

$$\lim_{x \to 2}(x^2 + 2x - 3)^4 = \left[\lim_{x \to 2}(x^2 + 2x - 3)\right]^4 = 5^4 = 625.$$

The Limit of a Power

If $\lim_{x \to a} f(x) = L$ and n is a positive integer, then

$$\lim_{x \to a}[f(x)]^n = \left[\lim_{x \to a} f(x)\right]^n = L^n.$$

In words, the limit of a function to a power is found by taking the limit of the function and then raising this limit to the power.

EXAMPLE 9   Finding the Limit of a Power

Find: $\lim_{x \to 5}(2x - 7)^3$.

Solution   The limit of the linear function $f(x) = 2x - 7$ as x approaches 5 is the linear function evaluated at 5. Because this function is raised to the third power, the limit that we seek is the limit of the linear function raised to the third power.

$$\lim_{x \to 5}(2x - 7)^3 = \left[\lim_{x \to 5}(2x - 7)\right]^3 = (2 \cdot 5 - 7)^3 = 3^3 = 27$$

Check Point 9   Find: $\lim_{x \to 4}(3x - 5)^3$.

How do we find the limit of a root? Recall that if $\sqrt[n]{L}$ represents a real number and $n \geq 2$, then

$$\sqrt[n]{L} = L^{1/n}.$$

Because a root is a power, we find the limit of a root using a similar procedure for finding the limit of a power.

The Limit of a Root

If $\lim_{x \to a} f(x) = L$ and n is a positive integer greater than or equal to 2, then

$$\lim_{x \to a}\sqrt[n]{f(x)} = \sqrt[n]{\lim_{x \to a} f(x)} = \sqrt[n]{L}$$

provided that all roots represent real numbers. In words, the limit of the nth root of a function is found by taking the limit of the function and then taking the nth root of this limit.

EXAMPLE 10   Finding the Limit of a Root

Find: $\lim_{x \to -2}\sqrt{4x^2 + 5}$.

Solution   The limit of the quadratic (polynomial) function $f(x) = 4x^2 + 5$ as x approaches -2 is the function evaluated at -2. Because we have the square root of this function, the limit that we seek is the square root of the limit of the quadratic function.

$$\lim_{x \to -2}\sqrt{4x^2 + 5} = \sqrt{\lim_{x \to -2}(4x^2 + 5)} = \sqrt{4(-2)^2 + 5} = \sqrt{16 + 5} = \sqrt{21}$$

Check Point 10   Find: $\lim_{x \to -1}\sqrt{6x^2 - 4}$.

We have considered limits of sums, differences, products, and roots. We conclude with a property that deals with the limit of a quotient.

The Limit of a Quotient

If $\lim\limits_{x \to a} f(x) = L$ and $\lim\limits_{x \to a} g(x) = M$, $M \neq 0$, then

$$\lim_{x \to a} \frac{f(x)}{g(x)} = \frac{\lim\limits_{x \to a} f(x)}{\lim\limits_{x \to a} g(x)} = \frac{L}{M}, M \neq 0.$$

In words, the limit of the quotient of two functions equals the quotient of their limits, **provided that the limit of the denominator is not zero**.

Before possibly applying the quotient property, begin by finding the limit of the denominator. If this limit is not zero, you can apply the quotient property. If this limit is zero, the quotient property cannot be used.

EXAMPLE 11 Finding the Limit of a Quotient

Find: $\lim\limits_{x \to 1} \dfrac{x^3 - 3x^2 + 7}{2x - 5}$.

Solution The two functions in this limit problem are $f(x) = x^3 - 3x^2 + 7$ and $g(x) = 2x - 5$. We seek the limit of the quotient of these functions. Can we use the quotient property for limits? We answer the question by finding the limit of the denominator, $g(x)$.

$$\lim_{x \to 1} (2x - 5) = 2 \cdot 1 - 5 = -3$$

Because the limit of the denominator is not zero, we can apply the quotient property for limits. The limit of the quotient is the quotient of the limits.

$$\lim_{x \to 1} \frac{x^3 - 3x^2 + 7}{2x - 5} = \frac{\lim\limits_{x \to 1}(x^3 - 3x^2 + 7)}{\lim\limits_{x \to 1}(2x - 5)} = \frac{1^3 - 3 \cdot 1^2 + 7}{2 \cdot 1 - 5} = \frac{5}{-3} = -\frac{5}{3}$$

Check Point 11 Find: $\lim\limits_{x \to 2} \dfrac{x^2 - 4x + 1}{3x - 5}$.

We've considered a number of limit properties. Let's take a moment to summarize these properties.

Properties of Limits

Formulas for Finding Limits

1. $\lim\limits_{x \to a} c = c$ **2.** $\lim\limits_{x \to a} x = a$

3. If f is a polynomial (linear, quadratic, cubic, etc.) function, $\lim\limits_{x \to a} f(x) = f(a)$.

Limits of Sums, Differences, Products, Powers, Roots, and Quotients

If $\lim\limits_{x \to a} f(x) = L$ and $\lim\limits_{x \to a} g(x) = M$, then

4. $\lim\limits_{x \to a}[f(x) + g(x)] = \lim\limits_{x \to a} f(x) + \lim\limits_{x \to a} g(x) = L + M.$

5. $\lim\limits_{x \to a}[f(x) - g(x)] = \lim\limits_{x \to a} f(x) - \lim\limits_{x \to a} g(x) = L - M.$

6. $\lim\limits_{x \to a}[f(x) \cdot g(x)] = \lim\limits_{x \to a} f(x) \cdot \lim\limits_{x \to a} g(x) = LM.$

7. $\lim\limits_{x \to a}[f(x)]^n = [\lim\limits_{x \to a} f(x)]^n = L^n$, where $n \geq 2$ is an integer.

8. $\lim\limits_{x \to a} \sqrt[n]{f(x)} = \sqrt[n]{\lim\limits_{x \to a} f(x)} = \sqrt[n]{L}$, where $n \geq 2$ is an integer and all roots represent real numbers.

9. $\lim\limits_{x \to a} \dfrac{f(x)}{g(x)} = \dfrac{\lim\limits_{x \to a} f(x)}{\lim\limits_{x \to a} g(x)} = \dfrac{L}{M}, M \neq 0.$

③ Find one-sided limits using properties of limits.

Properties of Limits and Piecewise Functions

In Section 11.1, we used graphs of piecewise functions to find one-sided limits. We can now find such limits by applying properties of limits to the appropriate part of a piecewise function's equation.

EXAMPLE 12   Using Limit Properties to Find One-Sided Limits

Consider the piecewise function defined by

$$f(x) = \begin{cases} x^2 + 5 & \text{if } x < 2 \\ 3x + 1 & \text{if } x \geq 2. \end{cases}$$

Find each of the following limits, or state that the limit does not exist:

a. $\lim\limits_{x \to 2^-} f(x)$ **b.** $\lim\limits_{x \to 2^+} f(x)$ **c.** $\lim\limits_{x \to 2} f(x)$.

Solution

a. To find $\lim\limits_{x \to 2^-} f(x)$, we look at values of $f(x)$ when x is close to 2, but less than 2. Because x is less than 2, we use the first line of the piecewise function's equation, $f(x) = x^2 + 5$.

$$\lim\limits_{x \to 2^-} f(x) = \lim\limits_{x \to 2^-} (x^2 + 5) = 2^2 + 5 = 9$$

b. To find $\lim\limits_{x \to 2^+} f(x)$, we look at values of $f(x)$ when x is close to 2, but greater than 2. Because x is greater than 2, we use the second line of the piecewise function's equation, $f(x) = 3x + 1$.

$$\lim\limits_{x \to 2^+} f(x) = \lim\limits_{x \to 2^+} (3x + 1) = 3 \cdot 2 + 1 = 7$$

c. We found that $\lim\limits_{x \to 2^-} f(x) = 9$ and $\lim\limits_{x \to 2^+} f(x) = 7$.

> The limit as x approaches 2 from the left equals 9.

> The limit as x approaches 2 from the right equals 7.

These one-sided limits are illustrated in Figure 11.10. Because the left- and right-hand limits are unequal, $\lim\limits_{x \to 2} f(x)$ does not exist.

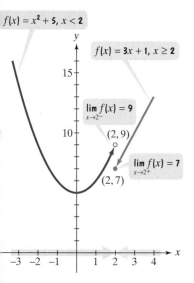

$f(x) = x^2 + 5, x < 2$

$f(x) = 3x + 1, x \geq 2$

$\lim\limits_{x \to 2^-} f(x) = 9$

$(2, 9)$

$\lim\limits_{x \to 2^+} f(x) = 7$

$(2, 7)$

Figure 11.10

Check Point 12   Consider the piecewise function defined by

$$f(x) = \begin{cases} -1 & \text{if } x < 1 \\ \sqrt[3]{2x - 1} & \text{if } x \geq 1. \end{cases}$$

Find each of the following limits, or state that the limit does not exist:

a. $\lim\limits_{x \to 1^-} f(x)$ **b.** $\lim\limits_{x \to 1^+} f(x)$ **c.** $\lim\limits_{x \to 1} f(x)$.

④ Find limits of fractional expressions in which the limit of the denominator is zero.

Strategies for Finding Limits When the Limit of the Denominator is Zero

When taking the limit of a fractional expression in which the limit of the denominator is zero, the quotient property for limits cannot be used. In such cases, it is necessary to rewrite the expression before the limit can be found. Factoring is one technique that can be used to rewrite an expression.

EXAMPLE 13   Using Factoring to Find a Limit

Find: $\lim\limits_{x \to 3} \dfrac{x^2 - x - 6}{x - 3}$.

Solution   The limit of the denominator is zero:

$$\lim\limits_{x \to 3} (x - 3) = 3 - 3 = 0.$$

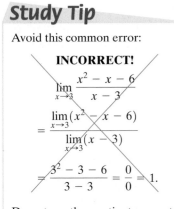
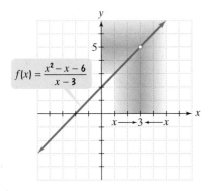

Figure 11.11 As x approaches 3, values of f get closer to 5:

$$\lim_{x \to 3} \frac{x^2 - x - 6}{x - 3} = 5.$$

Thus, the quotient property for limits cannot be used. Instead, try simplifying the expression using factoring:

$$\frac{x^2 - x - 6}{x - 3} = \frac{(x - 3)(x + 2)}{x - 3}.$$

We seek the limit of this expression as x approaches 3. Because x is close to 3, but not equal to 3, the common factor in the numerator and denominator, $x - 3$, is not equal to zero. With $x - 3 \neq 0$, we can divide the numerator and denominator by $x - 3$. Cancel the common factor, $x - 3$, and then take the limit.

$$\lim_{x \to 3} \frac{x^2 - x - 6}{x - 3} = \lim_{x \to 3} \frac{(x - 3)(x + 2)}{x - 3} = \lim_{x \to 3}(x + 2) = 3 + 2 = 5$$

The graph of $f(x) = \dfrac{x^2 - x - 6}{x - 3}$ is shown in Figure 11.11. The hole in the graph at $x = 3$ shows that $f(3)$ is undefined. However, as x approaches 3, the graph shows that the values of f get closer to 5. This verifies the limit that we found in Example 13:

$$\lim_{x \to 3} \frac{x^2 - x - 6}{x - 3} = 5.$$

Check Point 13 Find: $\displaystyle\lim_{x \to 1} \frac{x^2 + 2x - 3}{x - 1}$.

Rationalizing the numerator or denominator of a fractional expression is another technique that can be used to find a limit when the limit of the denominator is zero.

EXAMPLE 14 Rationalizing a Numerator to Find a Limit

Find: $\displaystyle\lim_{x \to 0} \frac{\sqrt{4 + x} - 2}{x}$.

Solution As x approaches 0, the denominator of the expression approaches zero. Thus, the quotient property for limits cannot be used. Instead, try rewriting the expression by rationalizing the numerator. If we multiply the numerator and denominator by $\sqrt{4 + x} + 2$, the numerator will not contain radicals.

$$\lim_{x \to 0} \frac{\sqrt{4 + x} - 2}{x}$$

$$= \lim_{x \to 0} \frac{\sqrt{4 + x} - 2}{x} \cdot \frac{\sqrt{4 + x} + 2}{\sqrt{4 + x} + 2} \qquad \text{Rationalize the numerator.}$$

$$= \lim_{x \to 0} \frac{\left(\sqrt{4 + x}\right)^2 - 2^2}{x\left(\sqrt{4 + x} + 2\right)} \qquad \left(\sqrt{a} - \sqrt{b}\right)\left(\sqrt{a} + \sqrt{b}\right) = \left(\sqrt{a}\right)^2 - \left(\sqrt{b}\right)^2$$

$$= \lim_{x \to 0} \frac{4 + x - 4}{x\left(\sqrt{4 + x} + 2\right)} \qquad \left(\sqrt{4 + x}\right)^2 = 4 + x$$

$$= \lim_{x \to 0} \frac{x}{x\left(\sqrt{4 + x} + 2\right)} \qquad \text{Simplify: } 4 + x - 4 = x.$$

$$= \lim_{x \to 0} \frac{1}{\sqrt{4 + x} + 2} \qquad \begin{array}{l}\text{Divide both the numerator and denominator by } x. \\ \text{This is permitted because } x \text{ approaches 0, but is} \\ \text{not equal to 0.}\end{array}$$

$$= \frac{\lim_{x \to 0} 1}{\sqrt{\lim_{x \to 0}(4 + x)} + \lim_{x \to 0} 2} \qquad \text{Use limit properties.}$$

$$= \frac{1}{\sqrt{4 + 0} + 2} \qquad \text{Take the limits.}$$

$$= \frac{1}{2 + 2} = \frac{1}{4} \qquad \text{Simplify.}$$

Check Point 14 Find: $\lim\limits_{x\to 0} \dfrac{\sqrt{9+x}-3}{x}$.

EXERCISE SET 11.2

Practice Exercises

In Exercises 1–42, use properties of limits to find the indicated limit. It may be necessary to rewrite an expression before limit properties can be applied.

1. $\lim\limits_{x\to 2} 8$

2. $\lim\limits_{x\to 3}(-6)$

3. $\lim\limits_{x\to 2} x$

4. $\lim\limits_{x\to 3} x$

5. $\lim\limits_{x\to 6}(3x-4)$

6. $\lim\limits_{x\to 7}(4x-3)$

7. $\lim\limits_{x\to -2} 7x^2$

8. $\lim\limits_{x\to -3} 5x^2$

9. $\lim\limits_{x\to 5}(x^2-3x-4)$

10. $\lim\limits_{x\to 6}(x^2-4x-7)$

11. $\lim\limits_{x\to 2}(5x-8)^3$

12. $\lim\limits_{x\to 4}(6x-21)^3$

13. $\lim\limits_{x\to 1}(2x^2-3x+5)^2$

14. $\lim\limits_{x\to 2}(2x^2+3x-1)^2$

15. $\lim\limits_{x\to -4}\sqrt{x^2+9}$

16. $\lim\limits_{x\to -1}\sqrt{5x^2+4}$

17. $\lim\limits_{x\to 5}\dfrac{x}{x+1}$

18. $\lim\limits_{x\to 2}\dfrac{3x}{x-4}$

19. $\lim\limits_{x\to 2}\dfrac{x^2-1}{x-1}$

20. $\lim\limits_{x\to 3}\dfrac{x^2-4}{x-2}$

21. $\lim\limits_{x\to 1}\dfrac{x^2-1}{x-1}$

22. $\lim\limits_{x\to 2}\dfrac{x^2-4}{x-2}$

23. $\lim\limits_{x\to 2}\dfrac{2x-4}{x-2}$

24. $\lim\limits_{x\to 3}\dfrac{4x-12}{x-3}$

25. $\lim\limits_{x\to 1}\dfrac{x^2+2x-3}{x^2-1}$

26. $\lim\limits_{x\to 3}\dfrac{x^2-x-6}{x^2-9}$

27. $\lim\limits_{x\to 2}\dfrac{x^3-2x^2+4x-8}{x^4-2x^3+x-2}$

28. $\lim\limits_{x\to -1}\dfrac{x^3+2x^2+x}{x^4+x^3+2x+2}$

29. $\lim\limits_{x\to 0}\dfrac{\sqrt{1+x}-1}{x}$

30. $\lim\limits_{x\to 0}\dfrac{\sqrt{16+x}-4}{x}$

31. $\lim\limits_{x\to 2}(x+1)^2(3x-1)^3$

32. $\lim\limits_{x\to -1}(x+2)^3(3x+2)$

33. $\lim\limits_{x\to 4}\dfrac{\sqrt{x}-2}{x-4}$

34. $\lim\limits_{x\to 9}\dfrac{\sqrt{x}-3}{x-9}$

35. $\lim\limits_{x\to 2}\dfrac{\frac{1}{x}-\frac{1}{2}}{x-2}$

36. $\lim\limits_{x\to 3}\dfrac{\frac{1}{x}-\frac{1}{3}}{x-3}$

37. $\lim\limits_{x\to 4}\dfrac{\sqrt{x}+5}{x-5}$

38. $\lim\limits_{x\to 9}\dfrac{\sqrt{x}+10}{x-10}$

39. $\lim\limits_{x\to 0}\dfrac{\frac{1}{x+3}-\frac{1}{3}}{x}$

40. $\lim\limits_{x\to 0}\dfrac{\frac{1}{x+4}-\frac{1}{4}}{x}$

41. $\lim\limits_{x\to 2}\dfrac{x^2-4}{x^3-8}$

42. $\lim\limits_{x\to 1}\dfrac{x^2-1}{x^3-1}$

In Exercises 43–50, a piecewise function is given. Use properties of limits to find the indicated limit, or state that the limit does not exist.

43. $f(x) = \begin{cases} x+5 & \text{if } x < 1 \\ x+7 & \text{if } x \geq 1 \end{cases}$

   a. $\lim\limits_{x\to 1^-} f(x)$ **b.** $\lim\limits_{x\to 1^+} f(x)$ **c.** $\lim\limits_{x\to 1} f(x)$

44. $f(x) = \begin{cases} x+6 & \text{if } x < 1 \\ x+9 & \text{if } x \geq 1 \end{cases}$

   a. $\lim\limits_{x\to 1^-} f(x)$ **b.** $\lim\limits_{x\to 1^+} f(x)$ **c.** $\lim\limits_{x\to 1} f(x)$

45. $f(x) = \begin{cases} x^2+5 & \text{if } x < 2 \\ x^3+1 & \text{if } x \geq 2 \end{cases}$

   a. $\lim\limits_{x\to 2^-} f(x)$ **b.** $\lim\limits_{x\to 2^+} f(x)$ **c.** $\lim\limits_{x\to 2} f(x)$

46. $f(x) = \begin{cases} x^2+6 & \text{if } x < 2 \\ x^3+2 & \text{if } x \geq 2 \end{cases}$

   a. $\lim\limits_{x\to 2^-} f(x)$ **b.** $\lim\limits_{x\to 2^+} f(x)$ **c.** $\lim\limits_{x\to 2} f(x)$

47. $f(x) = \begin{cases} \dfrac{x^2-9}{x-3} & \text{if } x \neq 3 \\ 5 & \text{if } x = 3 \end{cases}$

   a. $\lim\limits_{x\to 3^-} f(x)$ **b.** $\lim\limits_{x\to 3^+} f(x)$ **c.** $\lim\limits_{x\to 3} f(x)$

48. $f(x) = \begin{cases} \dfrac{x^2-16}{x-4} & \text{if } x \neq 4 \\ 7 & \text{if } x = 4 \end{cases}$

   a. $\lim\limits_{x\to 4^-} f(x)$ **b.** $\lim\limits_{x\to 4^+} f(x)$ **c.** $\lim\limits_{x\to 4} f(x)$

49. $f(x) = \begin{cases} 1-x & \text{if } x < 1 \\ 2 & \text{if } x = 1 \\ x^2-1 & \text{if } x > 1 \end{cases}$

   a. $\lim\limits_{x\to 1^-} f(x)$ **b.** $\lim\limits_{x\to 1^+} f(x)$ **c.** $\lim\limits_{x\to 1} f(x)$

50. $f(x) = \begin{cases} 4 - x & \text{if } x < 1 \\ 2 & \text{if } x = 1 \\ x^2 + 2 & \text{if } x > 1 \end{cases}$

  a. $\lim\limits_{x \to 1^-} f(x)$ **b.** $\lim\limits_{x \to 1^+} f(x)$ **c.** $\lim\limits_{x \to 1} f(x)$

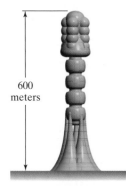

Practice Plus

51. Let $f(x) = x^3 - x^2 + 5x - 1$ and $g(x) = 2$.
Find $\lim\limits_{x \to 3}(f \circ g)(x)$ and $\lim\limits_{x \to 3}(g \circ f)(x)$.

52. Let $f(x) = x^3 + x^2 - 6x - 1$ and $g(x) = 3$.
Find $\lim\limits_{x \to 4}(f \circ g)(x)$ and $\lim\limits_{x \to 4}(g \circ f)(x)$.

53. Let $f(x) = \dfrac{2}{x}$ and $g(x) = \dfrac{3}{x - 1}$.
Find $\lim\limits_{x \to 1}(f \circ g)(x)$ and $\lim\limits_{x \to 1}(g \circ f)(x)$.

54. Let $f(x) = \dfrac{4}{x - 1}$ and $g(x) = \dfrac{1}{x + 2}$.
Find $\lim\limits_{x \to 1}(f \circ g)(x)$ and $\lim\limits_{x \to 1}(g \circ f)(x)$.

55. Let $f(x) = x^2 + 4, x \ge 0$. Find $\lim\limits_{x \to 8} f^{-1}(x)$.

56. Let $f(x) = x^2 + 9, x \ge 0$. Find $\lim\limits_{x \to 25} f^{-1}(x)$.

57. Let $f(x) = \dfrac{2x + 1}{x - 1}$. Find $\lim\limits_{x \to 4} f^{-1}(x)$.

58. Let $f(x) = \dfrac{2x + 3}{x + 4}$. Find $\lim\limits_{x \to 5} f^{-1}(x)$.

Application Exercises

In Albert Einstein's special theory of relativity, time slows down and length in the direction of motion decreases from the point of view of an observer watching an object moving at velocities approaching the speed of light. (The speed of light is approximately 186,000 miles per second. At this speed, a beam of light can travel around the world about seven times in a single second.) Einstein's theory, verified with experiments in atomic physics, forms the basis of Exercises 59–60.

600 meters — Starship at rest

84 meters — Change in starship's length when moving at 99% of light's speed when viewed by an observer

59. The formula

$$L = L_0 \sqrt{1 - \frac{v^2}{c^2}}$$

expresses the length, L, of a starship moving at velocity v with respect to an observer on Earth, where L_0 is the length of the starship at rest and c is the speed of light.

  a. Find $\lim\limits_{v \to c^-} L$.

b. If a starship is traveling at velocities approaching the speed of light, what does the limit in part (a) indicate about its length from the perspective of a stationary viewer on Earth?

c. Explain why a left-hand limit is used in part (a).

60. The formula

$$R_a = R_f \sqrt{1 - \left(\frac{v}{c}\right)^2}$$

expresses the aging rate of an astronaut, R_a, relative to the aging rate of a friend on Earth, R_f, where v is the astronaut's velocity and c is the speed of light.

  a. Find $\lim\limits_{v \to c^-} R_a$.

  b. If you are traveling in a starship at velocities approaching the speed of light, what does the limit in part (a) indicate about your aging rate relative to a friend on Earth?

  c. Explain why a left-hand limit is used in part (a).

Writing in Mathematics

61. Explain how to find the limit of a constant. Then express your written explanation using limit notation.

62. Explain how to find the limit of the identity function $f(x) = x$. Then express your written explanation using limit notation.

63. Explain how to find the limit of a sum. Then express your written explanation using limit notation.

64. Explain how to find the limit of a difference. Then express your written explanation using limit notation.

65. Explain how to find the limit of a product. Then express your written explanation using limit notation.

66. Describe how to find the limit of a polynomial function. Provide an example with your description.

67. Explain how to find the following limit: $\lim\limits_{x \to 2}(3x^2 - 10)^3$. Then use limit notation to write the limit property that supports your explanation.

68. Explain how to find the following limit: $\lim\limits_{x \to 2} \sqrt{5x - 6}$. Then use limit notation to write the limit property that supports your explanation.

69. Explain how to find the limit of a quotient if the limit of the denominator is not zero. Then express your written explanation using limit notation.

70. Write an example involving the limit of a quotient in which the quotient property for limits cannot be applied. Explain why the property cannot be applied to your limit problem.

71. Explain why

$$\lim\limits_{x \to 4} \frac{(x + 4)(x - 4)}{x - 4}$$

can be found by first dividing the numerator and the denominator of the expression by $x - 4$. Division by zero is undefined. How can we be sure that we are not dividing the numerator and the denominator by zero?

Technology Exercises

72. Use the TABLE feature of your graphing utility to verify any five of the limits that you found in Exercises 1–42.

73. Use the ZOOM IN feature of your graphing utility to verify any five of the limits that you found in Exercises 1–42. Zoom in on the graph of the given function, f, near $x = a$ to verify each limit.

Critical Thinking Exercises

In Exercises 74–75, find the indicated limit.

74. $\lim\limits_{x \to 0} x\left(1 - \dfrac{1}{x}\right)$ **75.** $\lim\limits_{x \to 4}\left(\dfrac{1}{x} - \dfrac{1}{4}\right)\left(\dfrac{1}{x - 4}\right)$

In Exercises 76–77, find $\lim\limits_{h \to 0} \dfrac{f(a + h) - f(a)}{h}$.

76. $f(x) = x^2 + 2x - 3, a = 1$

77. $f(x) = \sqrt{x}, a = 1$

In Exercises 78–79, use properties of limits and the following limits

$$\lim\limits_{x \to 0} \dfrac{\sin x}{x} = 1, \quad \lim\limits_{x \to 0} \dfrac{\cos x - 1}{x} = 0,$$

$$\lim\limits_{x \to 0} \sin x = 0, \quad \lim\limits_{x \to 0} \cos x = 1$$

to find the indicated limit.

78. $\lim\limits_{x \to 0} \dfrac{\tan x}{x}$ **79.** $\lim\limits_{x \to 0} \dfrac{2\sin x + \cos x - 1}{3x}$

Group Exercises

80. Here is a list of ten common errors involving algebra, trigonometry, and limits that students frequently make in calculus. Group members should examine each error and describe the mistake. Where possible, correct each error. Finally, group members should offer suggestions for avoiding each error.

a. $(x + h)^3 - x^3 = x^3 + h^3 - x^3 = h^3$

b. $\dfrac{1}{a + b} = \dfrac{1}{a} + \dfrac{1}{b}$

c. $\dfrac{1}{a + b} = \dfrac{1}{a} + b$

d $\sqrt{x + h} - \sqrt{x} = \sqrt{x} + \sqrt{h} - \sqrt{x} = \sqrt{h}$

e. $\dfrac{\sin 2x}{x} = \sin 2$

f. $\dfrac{a + bx}{a} = 1 + bx$

g. $\lim\limits_{x \to 1} \dfrac{x^3 - 1}{x - 1} = \dfrac{1^3 - 1}{1 - 1} = \dfrac{0}{0} = 1$

h. $\sin(x + h) - \sin x = \sin x + \sin h - \sin x = \sin h$

i. $ax = bx$, so $a = b$

j. To find $\lim\limits_{x \to 4} \dfrac{x^2 - 9}{x - 3}$, it is necessary to rewrite $\dfrac{x^2 - 9}{x - 3}$ by factoring $x^2 - 9$.

81. Research and present a group report about the history of the feud between Newton and Leibniz over who invented calculus. What other interests did these men have in addition to mathematics? What practical problems led them to the invention of calculus? What were their personalities like? Whose version established the notation and rules of calculus that we use today?

SECTION 11.3   *Limits and Continuity*

Objectives

1 Determine whether a function is continuous at a number.

2 Determine for what numbers a function is discontinuous.

Why you should not ski down discontinuous slopes

In everyday speech, a continuous process is one that goes on without interruption and without abrupt changes. In mathematics, a continuous function has much the same meaning. The graph of a continuous function does not have interrupting breaks, such as holes, gaps, or jumps. Thus, the graph of a continuous function can be drawn without lifting a pencil off the paper. In this section, you will learn how limits can be used to describe continuity.

Determine whether a function is continuous at a number.

Limits and Continuity

Figure 11.12 shows three graphs that cannot be drawn without lifting a pencil from the paper. In each case, there appears to be an interruption of the graph of f at $x = a$

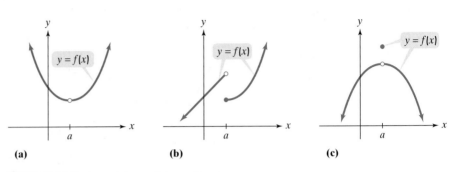

(a) (b) (c)

Figure 11.12 Each graph has an interruption at $x = a$.

Examine Figure 11.12(a). The interruption occurs because of the open dot at $x = a$. This shows that $f(a)$ is not defined.

Now, examine Figure 11.12(b). The closed blue dot at $x = a$ shows that $f(a)$ is defined. However, there is still a jump at a. As x approaches a from the left, the values of f get closer to the y-coordinate of the point shown by the open dot. By contrast, as x approaches a from the right, the values of f get closer to the y-coordinate of the point shown by the closed dot. There is no single limit as x approaches a. The jump in the graph reflects the fact that $\lim\limits_{x \to a} f(x)$ does not exist.

Finally, examine Figure 11.12(c). The closed blue dot at $x = a$ shows that $f(a)$ is defined. Furthermore, as x approaches a from the left or from the right, the values of f get closer to the y-coordinate of the point shown by the open dot. Thus, $\lim\limits_{x \to a} f(x)$ exists. However, there is still an interruption at a. Do you see why? The limit as x approaches a, $\lim\limits_{x \to a} f(x)$, is the y-coordinate of the open dot. By contrast, the value of the function at a, $f(a)$, is the y-coordinate of the closed dot. The jump in the graph reflects the fact that $\lim\limits_{x \to a} f(x)$ and $f(a)$ are not equal.

We now provide a precise definition of what it means for a function to be continuous at a number. Notice how each part of this definition avoids the interruptions that occurred in Figure 11.12.

Definition of a Function Continuous at a Number

A function f **is continuous at a** when three conditions are satisfied.

1. f is defined at a; that is, a is in the domain of f, so that $f(a)$ is a real number.
2. $\lim\limits_{x \to a} f(x)$ exists.
3. $\lim\limits_{x \to a} f(x) = f(a)$

If f is not continuous at a, we say that f is **discontinuous at a**. Each of the functions whose graph is shown in Figure 11.12 is discontinuous at a.

EXAMPLE 1 Determining Whether a Function Is Continuous at a Number

Determine whether the function

$$f(x) = \frac{2x + 1}{2x^2 - x - 1}$$

is continuous: **a.** at 2; **b.** at 1.

Solution   According to the definition, three conditions must be satisfied to have continuity at a.

a. To determine whether the function is continuous at 2, we check the conditions for continuity with $a = 2$.

Condition 1 f is defined at a.   Is $f(2)$ defined?

$$f(2) = \frac{2 \cdot 2 + 1}{2 \cdot 2^2 - 2 - 1} = \frac{4 + 1}{8 - 2 - 1} = \frac{5}{5} = 1$$

Because $f(2)$ is a real number, 1, $f(2)$ is defined.

Condition 2 $\lim\limits_{x \to a} f(x)$ exists.   Does $\lim\limits_{x \to 2} f(x)$ exist?

$$\lim_{x \to 2} f(x) = \lim_{x \to 2} \frac{2x + 1}{2x^2 - x - 1} = \frac{\lim\limits_{x \to 2}(2x + 1)}{\lim\limits_{x \to 2}(2x^2 - x - 1)}$$

$$= \frac{2 \cdot 2 + 1}{2 \cdot 2^2 - 2 - 1} = \frac{4 + 1}{8 - 2 - 1} = \frac{5}{5} = 1$$

Using properties of limits, we see that $\lim\limits_{x \to 2} f(x)$ exists.

Condition 3 $\lim\limits_{x \to a} f(x) = f(a)$ Does $\lim\limits_{x \to 2} f(x) = f(2)$? We found that $\lim\limits_{x \to 2} f(x) = 1$ and $f(2) = 1$. Thus, as x gets closer to 2, the corresponding values of $f(x)$ get closer to the function value at 2: $\lim\limits_{x \to 2} f(x) = f(2)$.

Because the three conditions are satisfied, we conclude that f is continuous at 2.

b. To determine whether the function is continuous at 1, we check the conditions for continuity with $a = 1$.

Condition 1 f is defined at a.   Is $f(1)$ defined? Factor the denominator of the function's equation:

$$f(x) = \frac{2x + 1}{(x - 1)(2x + 1)}$$

> Denominator is zero at $x = 1$.   Denominator is zero at $x = -\frac{1}{2}$.

Because division by zero is undefined, the domain of f is $\left\{ x \mid x \neq 1, x \neq -\frac{1}{2} \right\}$. Thus, f is not defined at 1.

Because one of the three conditions is not satisfied, we conclude that f is not continuous at 1. Equivalently, we can say that f is discontinuous at 1.

The graph of $f(x) = \dfrac{2x + 1}{2x^2 - x - 1}$ is shown in Figure 11.13. The graph verifies our work in Example 1. Can you see that f is continuous at 2? By contrast, it is not continuous at 1, where the graph has a vertical asymptote.

The graph in Figure 11.13 reveals the discontinuity at $-\frac{1}{2}$. The open dot indicates that $f\left(-\frac{1}{2}\right)$ is not defined. Can you see what is happening as x approaches $-\frac{1}{2}$?

$$\lim_{x \to -\frac{1}{2}} \frac{2x + 1}{2x^2 - x - 1} = \lim_{x \to -\frac{1}{2}} \frac{2x + 1}{(x - 1)(2x + 1)} = \lim_{x \to -\frac{1}{2}} \frac{1}{x - 1} = \frac{1}{-\frac{1}{2} - 1} = -\frac{2}{3}$$

As x gets closer to $-\frac{1}{2}$, the graph of f gets closer to $-\frac{2}{3}$. Because f is not defined at $-\frac{1}{2}$, the graph has a hole at $\left(-\frac{1}{2}, -\frac{2}{3}\right)$. This is shown by the open dot in Figure 11.13.

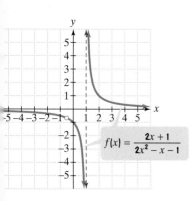

$$f(x) = \frac{2x + 1}{2x^2 - x - 1}$$

Figure 11.13 f is continuous at 2. It is not continuous at 1 or at $-\frac{1}{2}$.

Check Point 1   Determine whether the function

$$f(x) = \frac{x - 2}{x^2 - 4}$$

is continuous: **a.** at 1; **b.** at 2.

Determine for what numbers a function is discontinuous.

Determining Where Functions Are Discontinuous

We have seen that the limit of a polynomial function as x approaches a is the polynomial function evaluated at a. Thus, if f is a polynomial function, then $\lim_{x \to a} f(x) = f(a)$ for any number a. This means that **a polynomial function is continuous at every number**.

Many of the functions discussed throughout this book are continuous at every number in their domain. For example, rational functions are continuous at every number, except any at which they are not defined. At numbers that are not in the domain of a rational function, a hole in the graph or an asymptote appears. Exponential, logarithmic, sine, and cosine functions are continuous at every number in their domain. Like rational functions, the tangent, cotangent, secant, and cosecant functions are continuous at every number, except any at which they are not defined. At numbers that are not in the domain of these trigonometric functions, an asymptote occurs.

Study Tip

Most functions are always continuous at every number in their domain, including polynomial, rational, radical, exponential, logarithmic, and trigonometric functions. Most of the discontinuities you will encounter in calculus will be due to jumps in piecewise functions.

Example 2 illustrates how to determine where a piecewise function is discontinuous.

EXAMPLE 2   Determining Where a Piecewise Function Is Discontinuous

Determine for what numbers x, if any, the following function is discontinuous:

$$f(x) = \begin{cases} x + 2 & \text{if } x \leq 0 \\ 2 & \text{if } 0 < x \leq 1. \\ x^2 + 2 & \text{if } x > 1 \end{cases}$$

Solution   First, let's determine whether each of the three pieces of f is continuous. The first piece, $f(x) = x + 2$, is a linear function; it is continuous at every number x. The second piece, $f(x) = 2$, a constant function, is continuous at every number x. And the third piece, $f(x) = x^2 + 2$, a polynomial function, is also continuous at every number x. Thus, these three functions, a linear function, a constant function, and a polynomial function, can be graphed without lifting a pencil from the paper. However, the pieces change at $x = 0$ and at $x = 1$. Is it necessary to lift a pencil from the paper when graphing f at these values? It appears that we must investigate continuity at 0 and at 1.

To determine whether the function is continuous at 0, we check the conditions for continuity with $a = 0$.

Condition 1 f is defined at a.   Is $f(0)$ defined? Because $a = 0$, we use the first line of the piecewise function, where $x \leq 0$.

$$f(x) = x + 2 \qquad \text{\small This is the function's equation for } x \leq 0, \text{ which includes } x = 0.$$

$$f(0) = 0 + 2 \qquad \text{\small Replace } x \text{ with } 0.$$

$$= 2$$

Because $f(0)$ is a real number, 2, $f(0)$ is defined.

Condition 2 $\lim_{x \to a} f(x)$ exists.   Does $\lim_{x \to 0} f(x)$ exist? To answer this question, we look at the values of $f(x)$ when x is close to 0. Let us investigate the left- and right-hand limits. If these limits are equal, then $\lim_{x \to 0} f(x)$ exists. To find $\lim_{x \to 0^-} f(x)$

the left-hand limit, we look at the values of $f(x)$ when x is close to 0, but less than 0. Because x is less than 0, we use the first line of the piecewise function, $f(x) = x + 2$ if $x \leq 0$. Thus,

$$\lim_{x \to 0^-} f(x) = \lim_{x \to 0^-} (x + 2) = 0 + 2 = 2.$$

To find $\lim_{x \to 0^+} f(x)$, the right-hand limit, we look at the values of $f(x)$ when x is close to 0, but greater than 0. Because x is greater than 0, we use the second line of the piecewise function, $f(x) = 2$ if $0 < x \leq 1$. Thus,

$$\lim_{x \to 0^+} f(x) = \lim_{x \to 0^+} 2 = 2.$$

Because the left- and right-hand limits are both equal to 2, $\lim_{x \to 0} f(x) = 2$. Thus, we see that $\lim_{x \to 0} f(x)$ exists.

Condition 3 $\lim_{x \to a} f(x) = f(a)$ Does $\lim_{x \to 0} f(x) = f(0)$? We found that $\lim_{x \to 0} f(x) = 2$ and $f(0) = 2$. This means that as x gets closer to 0, the corresponding values of $f(x)$ get closer to the function value at 0: $\lim_{x \to 0} f(x) = f(0)$.

Because the three conditions are satisfied, we conclude that f is continuous at 0.

Now we must determine whether the function is continuous at 1, the other value of x where the pieces change. We check the conditions for continuity with $a = 1$.

Condition 1 f **is defined at a.**   Is $f(1)$ defined? Because $a = 1$, we use the second line of the piecewise function, where $0 < x \leq 1$.

$$f(x) = 2 \quad \text{This is the function's equation for } 0 < x \leq 1, \text{ which includes } x = 1.$$

$$f(1) = 2 \quad \text{Replace } x \text{ with 1.}$$

Because $f(1)$ is a real number, 2, $f(1)$ is defined.

Condition 2 $\lim_{x \to a} f(x)$ **exists.**   Does $\lim_{x \to 1} f(x)$ exist? We investigate left- and right-hand limits as x approaches 1. To find $\lim_{x \to 1^-} f(x)$, the left-hand limit, we look at values of $f(x)$ when x is close to 1, but less than 1. Thus, we use the second line of the piecewise function, $f(x) = 2$ if $0 < x \leq 1$. The left-hand limit is

$$\lim_{x \to 1^-} f(x) = \lim_{x \to 1^-} 2 = 2.$$

To find $\lim_{x \to 1^+} f(x)$, the right-hand limit, we look at values of $f(x)$ when x is close to 1, but greater than 1. Thus, we use the third line of the piecewise function, $f(x) = x^2 + 2$ if $x > 1$. The right-hand limit is

$$\lim_{x \to 1^+} f(x) = \lim_{x \to 1^+} (x^2 + 2) = 1^2 + 2 = 3.$$

The left- and right-hand limits are not equal: $\lim_{x \to 1^-} f(x) = 2$ and $\lim_{x \to 1^+} f(x) = 3$. This means that $\lim_{x \to 1} f(x)$ does not exist.

Because one of the three conditions is not satisfied, we conclude that f is not continuous at 1.

In summary, the given function is discontinuous at 1 only. The graph of f, shown in Figure 11.14, illustrates this conclusion.

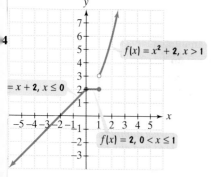

Figure 11.14 This piecewise function is continuous at 0, where pieces change, and discontinuous at 1, where pieces change.

Check Point 2   Determine for what numbers x, if any, the following function is discontinuous:

$$f(x) = \begin{cases} 2x & \text{if } x \leq 0 \\ x^2 + 1 & \text{if } 0 < x \leq 2. \\ 7 - x & \text{if } x > 2 \end{cases}$$

Critical Thinking Exercises

54. Define $f(x) = \dfrac{x^2 - 81}{x - 9}$ at $x = 9$ so that the function becomes continuous at 9.

55. Is it possible to define $f(x) = \dfrac{1}{x - 9}$ at $x = 9$ so that the function becomes continuous at 9? How does this discontinuity differ from the discontinuity in Exercise 54?

56. For the function

$$f(x) = \begin{cases} x^2 & \text{if } x < 1 \\ Ax - 3 & \text{if } x \geq 1 \end{cases}$$

find A so that the function is continuous at 1.

 **Group Exercise**

57. In this exercise, the group will define three piecewise functions. Each function should have three pieces and two values of x at which the pieces change.

  a. Define and graph a piecewise function that is continuous at both values of x where the pieces change.

  b. Define and graph a piecewise function that is continuous at one value of x where the pieces change and discontinuous at the other value of x where the pieces change.

  c. Define and graph a piecewise function that is discontinuous at both values of x where the pieces change.

At the end of the activity, group members should turn in the functions and their graphs. Do not use any of the piecewise functions or graphs that appear anywhere in this book.

CHAPTER 11
MID-CHAPTER CHECK POINT

What You Know: We learned that $\lim\limits_{x \to a} f(x) = L$ means that as x gets closer to a, but remains unequal to a, the corresponding values of $f(x)$ get closer to L. We found limits using tables, graphs, and properties of limits. The quotient property for limits did not apply to fractional expressions in which the limit of the denominator is zero. In these cases, rewriting the expression using factoring or rationalizing the numerator or denominator was helpful before finding the limit. We saw that if the left-hand limit, $\lim\limits_{x \to a^-} f(x)$, ($x$ approaches a from the left) is not equal to the right-hand limit, $\lim\limits_{x \to a^+} f(x)$, ($x$ approaches from the right), then $\lim\limits_{x \to a} f(x)$ does not exist. Finally, we defined continuity in terms of limits. A function f is continuous at a when f is defined at a, $\lim\limits_{x \to a} f(x)$ exists, and $\lim\limits_{x \to a} f(x) = f(a)$. If f is not continuous at a, we say that is discontinuous at a.

In Exercises 1–7, use the graphs of f and g to find the indicated limit or function value, or state that the limit or function value does not exist.

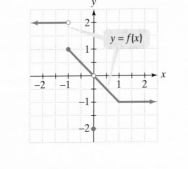

1. $\lim\limits_{x \to -1^-} f(x)$

2. $\lim\limits_{x \to -1^+} f(x)$

3. $\lim\limits_{x \to -1} f(x)$

4. $\lim\limits_{x \to 1}[f(x) + g(x)]$

5. $\lim\limits_{x \to 0}[f(x) - g(x)]$

6. $(f - g)(0)$

7. $\lim\limits_{x \to 1} \sqrt{10 + f(x)}$

8. Use the graph of f shown above to determine for what numbers the function is discontinuous. Then use the definition of continuity to verify each discontinuity.

In Exercises 9–11, use the table to find the indicated limit.

x	-0.03	-0.02	-0.01	-0.007	0.007	0.01	0.02	0.03
$f(x) = \dfrac{\sin x}{2x^2 - x}$	-0.9433	-0.9615	-0.9804	-0.9862	-1.014	-1.02	-1.042	-1.064
$g(x) = \dfrac{e^x - \tan x}{\cos^2 x}$	1.0014	1.0006	1.0002	1.0001	1.0001	1.0001	1.0006	1.0013

9. $\lim_{x \to 0} f(x)$ **10.** $\lim_{x \to 0} g(x)$ **11.** $\lim_{x \to 0} \dfrac{4g(x)}{[f(x)]^2}$

In Exercises 12–17, find the limits.

12. $\lim_{x \to -2} (x^3 - x + 5)$ **13.** $\lim_{x \to 3} \sqrt{x^2 - 3x + 4}$

14. $\lim_{x \to 5} \dfrac{2x^2 - x + 4}{x - 1}$ **15.** $\lim_{x \to 5} \dfrac{2x^2 - 7x - 15}{x - 5}$

16. $\lim_{x \to 0} \dfrac{\sqrt{x^2 + 9} - 3}{x^2}$ **17.** $\lim_{x \to 0} \dfrac{\frac{1}{x + 10} - \frac{1}{x}}{x}$

In Exercises 18–19, a piecewise function is given. Use the function to find the indicated limit, or state that the limit does not exist.

18. $f(x) = \begin{cases} 9 - 2x & \text{if } x < 4 \\ \sqrt{x - 4} & \text{if } x \ge 4 \end{cases}$

    a. $\lim_{x \to 4^-} f(x)$ **b.** $\lim_{x \to 4^+} f(x)$ **c.** $\lim_{x \to 4} f(x)$

19. $f(x) = \begin{cases} \dfrac{x^4 - 16}{x - 2} & \text{if } x \ne 2 \\ 32 & \text{if } x = 2 \end{cases}$

    a. $\lim_{x \to 2^-} f(x)$ **b.** $\lim_{x \to 2^+} f(x)$ **c.** $\lim_{x \to 2} f(x)$

In Exercises 20–21, use the definition of continuity to determine whether f is continuous at a.

20. $f(x) = \begin{cases} \sqrt{3 - x} & \text{if } x \le 3 \\ x^2 - 3x & \text{if } x > 3 \end{cases}$

    $a = 3$

21. $f(x) = \begin{cases} \dfrac{(x + 3)^2 - 9}{x} & \text{if } x \ne 0 \\ 6 & \text{if } x = 0 \end{cases}$

    $a = 0$

22. Determine for what numbers, if any, the following function is discontinuous:

$$f(x) = \begin{cases} \dfrac{x^2 - 1}{x + 1} & \text{if } x < -1 \\ 2x & \text{if } -1 \le x \le 5. \\ 3x - 4 & \text{if } x > 5 \end{cases}$$

SECTION 11.4   *Introduction to Derivatives*

Objectives

- Find slopes and equations of tangent lines.
- Find the derivative of a function.
- Find average and instantaneous rates of change.
- Find instantaneous velocity.

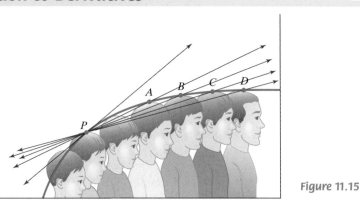

Figure 11.15

Can this possibly be your little cousin, whom you haven't seen in five years? He's changed from a kid to an attractive young adult. You know that things change over time and that most changes occur at uneven rates, but this is a radical transformation. What does calculus have to say about this?

In this section, we will see how calculus allows motion and change to be analyzed by "freezing the frame" of a continuous changing process, instant by instant. For example, Figure 11.15 shows your cousin's changing height over intervals of time. Over the period of time from P to D, his average rate of growth is his change in height—that is, his height at time D minus his height at time P—divided by the change in time from P to D.

The lines PD, PC, PB, and PA shown in Figure 11.15 have slopes that show your cousin's average growth rates for successively shorter periods of time. Calculus makes these time frames so small that their limit approaches a single point—that is, a single instant in time. This point is shown as point P in Figure 11.15. The slope of the line that touches the graph at P gives your cousin's growth rate at one instant in time, P.

Keep this informal discussion of your cousin and his growth rate in mind as you read this section. We begin with the calculus that describes the slope of the line that touches the graph in Figure 11.15 at P.

Find slopes and equations of tangent lines.

Slopes and Equations of Tangent Lines

In Chapter 1, we saw that if the graph of a function is not a straight line, the **averag** **rate of change** between any two points is the slope of the line containing the tw points. We called this line a **secant line**.

Figure 11.16 shows the graph of your cousin's height, in inches, as a function o his age, in years. Two points on the graph are labeled: $(13, 57)$ and $(18, 76)$. At ag 13, your cousin was 57 inches tall, and at age 18, he was 76 inches tall. The slope o the secant line containing these two point is

$$\frac{76 - 57}{18 - 13} = \frac{19}{5} = 3\frac{4}{5}.$$

Slope is the change in the y-coordinates divided by the change in the x-coordinates.

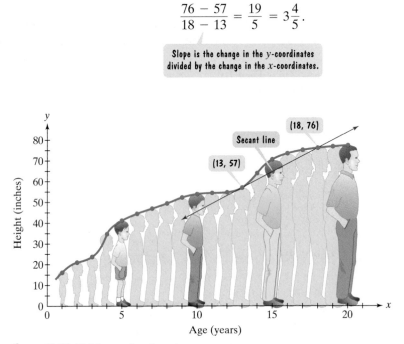

Figure 11.16 Height as a function of age

Your cousin's average rate of change, or average growth rate, from 13 to 18 was 3 inches per year.

How can we find your cousin's growth rate at the instant when he was 13? W can find this *instantaneous rate of change* by repeating the computation of slop from 13 to 17, then from 13 to 16, then from 13 to 15, again from 13 to 14, again from 13 to $13\frac{1}{2}$, and once again from 13 to 13.01. What limit is approached by these com putations as the shrinking interval of time gets closer and closer to the instant whe your cousin was 13?

We answer these questions by considering the graph of any function f, shown i Figure 11.17. We need to find the slope, or steepness, of this curve at the point $P = (a, f(a))$. This slope will reveal the function's instantaneous rate of change at a. We begin by choosing a second point, Q, whose x-coordinate is $a + h$, where $h \neq 0$. The point $Q = (a + h, f(a + h))$ is shown in Figure 11.17.

How do we find the aver- age rate of change of f between points P and Q? We find the slope of the secant line, the line containing P and Q.

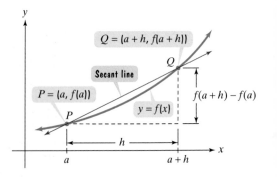

Figure 11.17 Finding the average rate of change, or the slope of the secant line

Slope of secant line

$$= \frac{f(a + h) - f(a)}{a + h - a}$$ *Slope is the change in y-coordinates, f(a + h) − f(a),*
divided by the change in x-coordinates, (a + h) − a.

$$= \frac{f(a + h) - f(a)}{h}$$ *Simplify.*

Do you recognize this expression as the difference quotient presented in Chapter 1? We will make use of this expression and our understanding of limits to find the slope of a graph at a specific point.

What happens if the distance labeled h in Figure 11.17 approaches 0? The value of the x-coordinate of point Q, $a + h$, will get closer and closer to a. Can you see that a is the x-coordinate of point P? Thus, as h approaches 0, point Q approaches point P. Examine Figure 11.18 to see how we visualize the changing position of point Q.

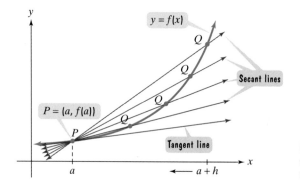

Figure 11.18 As Q approaches P, the succession of secant lines approaches the tangent line.

Figure 11.18 also shows how the secant line between points P and Q changes as h approaches 0. Note how the position of the secant line changes as the position of Q changes. The secant line between point P and point Q approaches the red line that touches the graph of f at point P. This limiting position of the secant line is called the **tangent line** to the graph of f at the point $P = (a, f(a))$.

According to our earlier derivation, the slope of each secant line in Figure 11.18 is

$$\frac{f(a + h) - f(a)}{h}.$$ *This difference quotient is also the average rate of*
change of f from x₁ = a to x₂ = a + h.

As h approaches 0, this slope approaches the slope of the tangent line to the curve at $(a, f(a))$. Thus, the slope of the tangent line to the curve at $(a, f(a))$ is

$$\lim_{h \to 0} \frac{f(a + h) - f(a)}{h}.$$

This limit also represents the **instantaneous rate of change of f with respect to x at a.**

Slope of the Tangent Line to a Curve at a Point

The **slope of the tangent line** to the graph of a function $y = f(x)$ at $(a, f(a))$ is given by

$$m_{\tan} = \lim_{h \to 0} \frac{f(a + h) - f(a)}{h}$$

provided that this limit exists. This limit also describes

- the **slope of the graph** of f at $(a, f(a))$.
- the **instantaneous rate of change** of f with respect to x at a.

EXAMPLE 1   Finding the Slope of the Tangent Line

Find the slope of the tangent line to the graph of $f(x) = x^2 + x$ at $(2, 6)$.

Solution   The slope of the tangent line at $(a, f(a))$ is

$$m_{\tan} = \lim_{h \to 0} \frac{f(a + h) - f(a)}{h}.$$

We use this formula to find the slope of the tangent line at the given point. Because we are finding the slope of the tangent line at $(2, 6)$, we know that $a = 2$.

$$m_{\tan} = \lim_{h \to 0} \frac{f(2 + h) - f(2)}{h}$$

Because $a = 2$, substitute 2 into the formula for each occurrence of a.

$$= \lim_{h \to 0} \frac{[(2 + h)^2 + (2 + h)] - [2^2 + 2]}{h}$$

To find $f(2 + h)$, replace x in $f(x) = x^2 + x$ with $2 + h$. To find $f(2)$, replace x with 2.

$$= \lim_{h \to 0} \frac{[4 + 4h + h^2 + 2 + h] - 6}{h}$$

Square $2 + h$ using $(A + B)^2 = A^2 + 2AB + B^2$.

$$= \lim_{h \to 0} \frac{h^2 + 5h}{h}$$

Combine like terms in the numerator.

$$= \lim_{h \to 0} \frac{h(h + 5)}{h}$$

Factor the numerator.

$$= \lim_{h \to 0} (h + 5)$$

Divide both the numerator and denominator by h. This is permitted because h approaches 0, but $h \neq 0$.

$$= 0 + 5$$

Use limit properties.

$$= 5$$

Thus, the slope of the tangent line to the graph of $f(x) = x^2 + x$ at $(2, 6)$ is 5. This is shown in Figure 11.19. We also say that the slope of the graph of $f(x) = x^2 + x$ a $(2, 6)$ is 5.

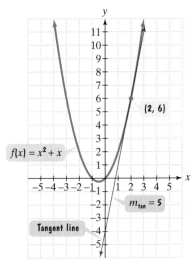

$f(x) = x^2 + x$

$(2, 6)$

$m_{\tan} = 5$

Tangent line

Figure 11.19

Check Point **1**   Find the slope of the tangent line to the graph of $f(x) = x^2 - x$ at $(4, 12)$

In Example 1, we found the *slope* of the tangent line shown in Figure 11.19. We can find an *equation* of this line using the point-slope form of the equation of a line

$$y - y_1 = m(x - x_1).$$

The tangent line passes through $(2, 6)$: $x_1 = 2$ and $y_1 = 6$. The slope of the tangent line is 5: $m = 5$. The point-slope equation of the tangent line is

$$y - 6 = 5(x - 2).$$

We can solve for y and express the equation of the tangent line in slope-intercept form: $y = mx + b$. The slope-intercept equation of the tangent line is

$$y - 6 = 5x - 10$$ Apply the distributive property.

$$y = 5x - 4.$$ Add 6 to both sides and write in slope-intercept form.

Technology

Graphing utilities with a
DRAW TANGENT feature
will draw tangent lines to curves
and display their slope-intercept
equations. Figure 11.20 shows the
tangent line to the graph of
$y = x^2 + x$ at the point whose
x-coordinate is 2. Also displayed
is the slope-intercept equation of
the tangent line, $y = 5x - 4$.

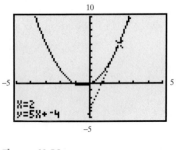

Figure 11.20

EXAMPLE 2   Finding the Slope-Intercept Equation of the Tangent Line

Find the slope-intercept equation of the tangent line to the graph of $f(x) = \sqrt{x}$ at $(4, 2)$.

$$= \lim_{h \to 0} \frac{h(2x + h + 3)}{h}$$

Factor the numerator.

$$= \lim_{h \to 0} (2x + h + 3)$$

Divide the numerator and the denominator by h.

$$= 2x + 0 + 3$$

Use limit properties. As h approaches 0, only the term containing h is affected.

$$= 2x + 3$$

The derivative of $f(x) = x^2 + 3x$ is

$$f'(x) = 2x + 3.$$

b. The derivative gives the slope of the tangent line at any point. Thus, to find the slope of the tangent line to $f(x) = x^2 + 3x$ at $x = -2$, evaluate the derivative at -2. Similarly, to find the slope of the tangent line at $x = -\frac{3}{2}$, evaluate the derivative at $-\frac{3}{2}$.

$$f'(x) = 2x + 3$$

$$f'(-2) = 2(-2) + 3 = -4 + 3 = -1$$

$$f'\left(-\tfrac{3}{2}\right) = 2\left(-\tfrac{3}{2}\right) + 3 = -3 + 3 = 0$$

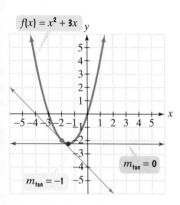

$f(x) = x^2 + 3x$

$m_{tan} = 0$

$m_{tan} = -1$

Figure 11.23 Two tangent lines to $f(x) = x^2 + 3x$ and their slopes

Figure 11.23 shows the graph of $f(x) = x^2 + 3x$ and tangent lines at $x = -2$ and $x = -\frac{3}{2}$. The slope of the decreasing green tangent line at $x = -2$ is -1. The slope of the horizontal red tangent line at $x = -\frac{3}{2}$ is 0.

 Check Point 3   **a.** Find the derivative of $f(x) = x^2 - 5x$ at x. That is, find $f'(x)$.
   b. Find the slope of the tangent line to $f(x) = x^2 - 5x$ at $x = -1$ and $x = 3$.

 Find average and instantaneous rates of change.

Applications of the Derivative

Many applications of the derivative involve analyzing change by determining a function's instantaneous rate of change at any moment. How do we use the derivative of a function to reveal such a change? We know that the derivative of f at x is defined by

$$f'(x) = \lim_{h \to 0} \frac{f(x + h) - f(x)}{h}.$$

Thus, the derivative of f at a is

$$f'(a) = \lim_{h \to 0} \frac{f(a + h) - f(a)}{h}.$$

Do you recognize this limit? It describes the instantaneous rate of change of f with respect to x at a.

Average and Instantaneous Rates of Change

Average Rate of Change The **average rate of change** of f from $x = a$ to $x = a + h$ is given by the difference quotient

$$\frac{f(a + h) - f(a)}{h}.$$

Instantaneous Rate of Change The **instantaneous rate of change** of f with respect to x at a is the derivative of f at a:

$$f'(a) = \lim_{h \to 0} \frac{f(a + h) - f(a)}{h}.$$

EXAMPLE 4   Finding Average and Instantaneous Rates of Change

The function $f(x) = x^3$ describes the volume of a cube, $f(x)$, in cubic inches, whose length, width, and height each measure x inches. If x is changing,

    a. Find the average rate of change of the volume with respect to x as x changes from 5 inches to 5.1 inches and from 5 inches to 5.01 inches.

    b. Find the instantaneous rate of change of the volume with respect to x at the moment when $x = 5$ inches.

Solution

    a. As x changes from 5 to 5.1, $a = 5$ and $h = 0.1$. The average rate of change of the volume with respect to x as x changes from 5 to 5.1 is determined as follows.

$$\frac{f(a + h) - f(a)}{h}$$
    The difference quotient gives the average rate of change from a to $a + h$.

$$= \frac{f(5 + 0.1) - f(5)}{0.1}$$
    This is the average rate of change from 5 to 5.1.

$$= \frac{f(5.1) - f(5)}{0.1}$$
    Simplify.

$$= \frac{5.1^3 - 5^3}{0.1}$$
    Use $f(x) = x^3$ and substitute 5.1 and 5, respectively, for x.

$$= 76.51$$

The average rate of change in the volume is 76.51 cubic inches per inch as x changes from 5 to 5.1 inches.

    As x changes from 5 to 5.01, $a = 5$ and $h = 0.01$. The average rate of change of the volume with respect to x as x changes from 5 to 5.01 is determined as follows.

$$\frac{f(a + h) - f(a)}{h}$$
    The difference quotient gives the average rate of change from a to $a + h$.

$$= \frac{f(5 + 0.01) - f(5)}{0.01}$$
    This is the average rate of change from 5 to 5.01.

$$= \frac{f(5.01) - f(5)}{0.01}$$
    Simplify.

$$= \frac{5.01^3 - 5^3}{0.01}$$
    Use $f(x) = x^3$ and substitute 5.01 and 5, respectively, for x.

$$= 75.1501$$

The average rate of change in the volume is 75.1501 cubic inches per inch as x changes from 5 to 5.01 inches.

    b. Instantaneous rates of change are given by the derivative. The derivative of f at a, $f'(a)$, is the instantaneous rate of change of f with respect to x at a. We must find the instantaneous rate of change of the volume with respect to x at the moment when $x = 5$ inches. This means that we must find $f'(5)$. We find $f'(5)$ by first finding $f'(x)$, the derivative, and then evaluating f' at 5.

$$f'(x) = \lim_{h \to 0} \frac{f(x + h) - f(x)}{h}$$

Use the definition of the derivative.

$$= \lim_{h \to 0} \frac{(x + h)^3 - x^3}{h}$$

To find $f(x + h)$, replace x in $f(x) = x^3$ with $x + h$.

$$= \lim_{h \to 0} \frac{(x^3 + 3x^2h + 3xh^2 + h^3) - x^3}{h}$$

Use the Binomial Theorem to cube $x + h$.

$$= \lim_{h \to 0} \frac{3x^2h + 3xh^2 + h^3}{h}$$

Simplify the numerator: $x^3 - x^3 = 0$.

$$= \lim_{h \to 0} \frac{h(3x^2 + 3xh + h^2)}{h}$$

Factor the numerator.

$$= \lim_{h \to 0} (3x^2 + 3xh + h^2)$$

Divide the numerator and the denominator by h.

$$= 3x^2 + 3x \cdot 0 + 0^2$$

Use limit properties. As h approaches 0, only terms containing h are affected.

$$= 3x^2$$

Technology
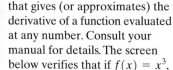

Graphing utilities have a feature that gives (or approximates) the derivative of a function evaluated at any number. Consult your manual for details. The screen below verifies that if $f(x) = x^3$, then $f'(5) = 75$.

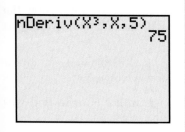

The derivative of $f(x) = x^3$ is $f'(x) = 3x^2$. To find the instantaneous change of f at 5, evaluate the derivative at 5.

$$f'(x) = 3x^2$$
$$f'(5) = 3 \cdot 5^2 = 75$$

The instantaneous rate of change of the volume with respect to x at the moment when $x = 5$ inches is 75 cubic inches per inch. Notice how the average rates of change that we computed in part (a), 76.51 and 75.1501, are approaching the instantaneous rate of change, 75.

 Check Point 4 Use the function in Example 4, $f(x) = x^3$, to find each of the following:

a. the average rate of change of the volume with respect to x as x changes from 4 inches to 4.1 inches and from 4 inches to 4.01 inches.

b. the instantaneous rate of change of the volume with respect to x at the moment when $x = 4$ inches.

 Find instantaneous velocity.

The ideas of calculus are frequently applied to position functions that express an object's position, $s(t)$, in terms of time, t. In the time interval from $t = a$ to $t = a + h$, the change in the object's position is

$$s(a + h) - s(a).$$

The **average velocity** over this time interval is

$$\frac{s(a + h) - s(a)}{h}.$$

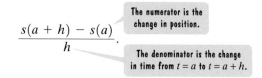

Now suppose that we compute the average velocities over shorter and shorter time intervals $[a, a + h]$. This means that we let h approach 0. As in our previous discussion, we define the *instantaneous velocity* at time $t = a$ to be the limit of these average velocities. This limit is the derivative of s at a.

Instantaneous Velocity

Suppose that a function expresses an object's position, $s(t)$, in terms of time, t. The **instantaneous velocity** of the object at time $t = a$ is

$$s'(a) = \lim_{h \to 0} \frac{s(a + h) - s(a)}{h}.$$

Instantaneous velocity at time a is also called **velocity** at time a.

EXAMPLE 5 Finding Instantaneous Velocity

A ball is thrown straight up from a rooftop 160 feet high with an initial velocity of 48 feet per second. The function

$$s(t) = -16t^2 + 48t + 160$$

describes the ball's height above the ground, $s(t)$, in feet, t seconds after it is thrown. The ball misses the rooftop on its way down and eventually strikes the ground.

a. What is the instantaneous velocity of the ball 2 seconds after it is thrown?

b. What is the instantaneous velocity of the ball when it hits the ground?

Solution Instantaneous velocity is given by the derivative of a function that expresses an object's position, $s(t)$, in terms of time, t. The instantaneous velocity of the ball at a seconds is $s'(a)$.

$$s'(a) = \lim_{h \to 0} \frac{s(a + h) - s(a)}{h}$$
   This derivative describes instantaneous velocity at time a.

To find $s(a + h)$, replace t in $s(t) = -16t^2 + 48t + 160$ with $a + h$. To find $s(a)$ replace t with a. Thus,

$$s'(a) = \lim_{h \to 0} \frac{-16(a + h)^2 + 48(a + h) + 160 - (-16a^2 + 48a + 160)}{h}$$

Take a few minutes to simplify the numerator of the difference quotient and factor out h. You should obtain

$$s'(a) = \lim_{h \to 0} \frac{h(-32a - 16h + 48)}{h} = -32a - 16 \cdot 0 + 48 = -32a + 48.$$

The instantaneous velocity of the ball at a seconds is

$$s'(a) = -32a + 48.$$

a. The instantaneous velocity of the ball at 2 seconds is found by replacing a with 2.

$$s'(2) = -32 \cdot 2 + 48 = -64 + 48 = -16$$

Two seconds after the ball is thrown, its instantaneous velocity is -16 feet per second. The negative sign indicates that the ball is moving downward when $t = 2$ seconds.

b. To find the instantaneous velocity of the ball when it hits the ground, we need to know how many seconds elapse between the time the ball is thrown from the rooftop and the time it hits the ground. The ball hits the ground when $s(t)$, its height above the ground, is 0. Thus, we set $s(t)$ equal to 0.

$$-16t^2 + 48t + 160 = 0 \qquad \text{Set } s(t) = 0.$$
$$-16(t^2 - 3t - 10) = 0 \qquad \text{Factor out } -16.$$
$$-16(t - 5)(t + 2) = 0 \qquad \text{Factor completely.}$$
$$t - 5 = 0 \qquad t + 2 = 0 \qquad \text{Set each variable factor equal to 0.}$$
$$t = 5 \qquad t = -2 \qquad \text{Solve for } t.$$

Roller Coasters and Derivatives

Roller coaster rides give you the opportunity to spend a few hair-raising minutes plunging hundreds of feet, accelerating to 80 miles an hour in seven seconds, and enduring vertical loops that turn you upside-down. By finding a function that models your distance above the ground at every moment of the ride and taking its derivative, you can determine when the instantaneous velocity is the greatest. As you experience the glorious agony of the roller coaster, this is your moment of peak terror.

7. a. How well does the function model the data shown for 2003?

b. Find $f'(x)$ for the piece of the function given by a linear function.

c. At what rate, in millions of audits per year, was the number of IRS audits of individual taxpayers increasing in 2003?

8. a. How well does the function model the data shown for 1997?

b. Find $f'(x)$ for the piece of the function given by a quadratic function.

c. At what rate, in millions of audits per year, was the number of IRS audits of individual taxpayers decreasing in 1997? Round to the nearest tenth of a million.

Writing in Mathematics

49. Explain how the tangent line to the graph of a function at point P is related to the secant lines between points P and Q on the function's graph.

50. Explain what we mean by the slope of the graph of a function at a point.

51. Explain how to find the slope of $f(x) = x^2$ at $(2, 4)$.

52. Explain how to write an equation of the tangent line to the graph of $f(x) = x^2$ at $(2, 4)$.

53. If you are given $y = f(x)$, the equation of function f, describe how to find $f'(x)$.

54. Explain how to use the derivative to compute the slopes of various tangent lines to the graph of a function.

55. Explain how the instantaneous rate of change of a function at a point is related to its average rates of change.

56. If a function expresses an object's distance in terms of time, how do you find the instantaneous velocity of the object at any time during its motion?

57. Use the concept of an interval of time to describe how calculus views a particular instant of time.

58. You are about to take a great picture of fog rolling into San Francisco from the middle of the Golden Gate Bridge, 400 feet above the water. Whoops! You accidently lean too far over the safety rail and drop your camera. Your friend quips, "Well at least you know calculus; you can figure out the velocity with which the camera is going to hit the water." If the camera's height, $s(t)$, in feet, over the water after t seconds is $s(t) = 400 - 16t^2$, describe how to determine the camera's velocity at the instant of its demise.

59. A calculus professor introduced the derivative by saying that it could be summed up in one word: *slope*. Explain what this means.

60. For an unusual introduction to calculus by a poetic, quirky, and funny writer who loves the subject, read *A Tour of the Calculus* by David Berlinski (Vintage Books, 1995). Write a report describing two new things that you learned from the book about algebra, trigonometry, limits, or derivatives.

Technology Exercises

61. Use the DRAW TANGENT feature of a graphing utility to graph the functions and tangent lines for any five exercises from Exercises 1–14. Use the equation that is displayed on the screen to verify the slope-intercept equation of the tangent line that you found in each exercise.

62. Without using the DRAW TANGENT feature of a graphing utility, graph the function and the tangent line whose equation you found for any five exercises from Exercises 1–14. Does the line appear to be tangent to the graph of f at the point on f that is given in the exercise?

63. Use the feature on a graphing utility that gives the derivative of a function evaluated at any number to verify part (b) for any five of your answers in Exercises 15–28.

In Exercises 64–67, find, or approximate to two decimal places, the derivative of each function at the given number using a graphing utility.

64. $f(x) = x^4 - x^3 + x^2 - x + 1$ at 1

65. $f(x) = \dfrac{x}{x - 3}$ at 6

66. $f(x) = x^2 \cos x$ at $\dfrac{\pi}{4}$

67. $f(x) = e^x \sin x$ at 2

Critical Thinking Exercises

In Exercises 68–73, graphs of functions are shown in $[-5, 5, 1]$ by $[-5, 5, 1]$ viewing rectangles. Match each function with the graph of its derivative. Graphs of derivatives are labeled (a)–(f) and are shown on the next page in $[-5, 5, 1]$ by $[-5, 5, 1]$ viewing rectangles.

68.

69.

70.

71.

72.

73.

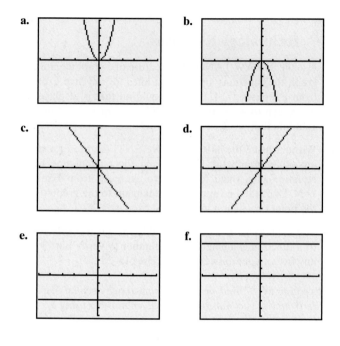

a.

b.

c.

d.

e.

f.

74. A ball is thrown straight up from a rooftop 96 feet high with an initial velocity of 80 feet per second. The function

$$s(t) = -16t^2 + 80t + 96$$

describes the ball's height above the ground, $s(t)$, in feet t seconds after it is thrown. The ball misses the rooftop on its way down and eventually strikes the ground. What is its instantaneous velocity as it passes the rooftop on the way down?

75. Show that the rate of change of the area of a circle with respect to its radius is equal to the circumference of the circle.

76. Show that the x-coordinate of the vertex of the parabola whose equation is $y = ax^2 + bx + c$ occurs when the derivative of the function is zero.

77. For any positive integer n, prove that if $f(x) = x^n$, then $f'(x) = nx^{n-1}$.

Chapter 11
Summary, Review, and Test

Summary

DEFINITIONS AND CONCEPTS	EXAMPLES

11.1 Finding Limits Using Tables and Graphs

a. Limit Notation and Its Description

$\lim\limits_{x \to a} f(x) = L$ is read "the limit of $f(x)$ as x approaches a equals the number L." This means that as x gets closer to a, but remains unequal to a, the corresponding values of $f(x)$ get closer to L.

b. Limits can be found using tables.

Ex. 1, p. 1008;
Ex. 2, p. 1008

c. Limits can be found using graphs.

Ex. 3, p. 1010;
Ex. 4, p. 1010

d. Left-Hand Limit

$\lim\limits_{x \to a^-} f(x) = L$ is read "the limit of $f(x)$ as x approaches a from the left equals L." This means that as x gets closer to a, but remains less than a, the corresponding values of $f(x)$ get closer to L.

e. Right-Hand Limit

$\lim\limits_{x \to a^+} f(x) = L$ is read "the limit of $f(x)$ as x approaches a from the right equals L." This means that as x gets closer to a, but remains greater than a, the corresponding values of $f(x)$ get closer to L.

f. If $\lim\limits_{x \to a^-} f(x) \neq \lim\limits_{x \to a^+} f(x)$, then $\lim\limits_{x \to a} f(x)$ does not exist.

Ex. 5, p. 1012

DEFINITIONS AND CONCEPTS   **EXAMPLES**

11.2 Finding Limits Using Properties of Limits

a. Properties of limits are given in the box on page 1024.

Ex. 1–Ex. 11, pp. 1018–1024

b. Properties of limits can be used to find one-sided limits.

Ex. 12, p. 1025

c. When taking the limit of a fractional expression in which the limit of the denominator is zero, the quotient property for limits cannot be used. Rewriting the expression using factoring or rationalizing the numerator or denominator may be helpful before the limit is found.

Ex. 13, p. 1025;
Ex. 14, p. 1026

11.3 Limits and Continuity

a. A function f is continuous at a when f is defined at a, $\lim_{x \to a} f(x)$ exists, and $\lim_{x \to a} f(x) = f(a)$.

If f is not continuous at a, we say that f is discontinuous at a.

Ex. 1, p. 1030;

Ex. 2, p. 1032

11.4 Introduction to Derivatives

a. The slope of the tangent line to the graph of a function $y = f(x)$ at $(a, f(a))$ is given by

$$m_{\tan} = \lim_{h \to 0} \frac{f(a + h) - f(a)}{h}$$

provided that this limit exists. The limit also describes the slope of the graph of f at $(a, f(a))$.

Ex. 1, p. 1040;
Ex. 2, p. 1040

b. The Derivative of a Function
The derivative of f at x is given by

$$f'(x) = \lim_{h \to 0} \frac{f(x + h) - f(x)}{h}$$

provided that this limit exists. The derivative gives the slope of f for any value of x

Ex. 3, p. 1042.

c. The derivative of f at a, $f'(a) = \lim_{h \to 0} \dfrac{f(a + h) - f(a)}{h}$, gives the instantaneous rate of change of f with respect to x at a. Expressions for average and instantaneous rates of change are given in the box on page 1043.

Ex. 4, p. 1044

d. If a function expresses an object's position, $s(t)$, in terms of time, t, the instantaneous velocity of the object at time $t = a$ is

$$s'(a) = \lim_{h \to 0} \frac{s(a + h) - s(a)}{h}.$$

Ex. 5, p. 1046

Review Exercises

11.1

In Exercises 1–3, construct a table to find the indicated limit.

1. $\lim\limits_{x \to 1} \dfrac{x^3 - 1}{x - 1}$

2. $\lim\limits_{x \to 0} \dfrac{\sqrt{x + 1} - 1}{x}$

3. $\lim\limits_{x \to 0} \dfrac{\sin 2x}{x}$

In Exercises 4–8, use the graph of f to find the indicated limit or function value.

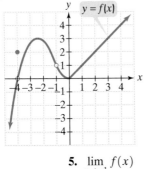

4. $\lim\limits_{x \to -4} f(x)$

5. $\lim\limits_{x \to -1} f(x)$

6. $\lim\limits_{x \to 3} f(x)$

7. $f(-4)$

8. $f(3)$

In Exercises 9–23, use the graph of function f to find the indicated limit or function value, or state that the limit or function value does not exist.

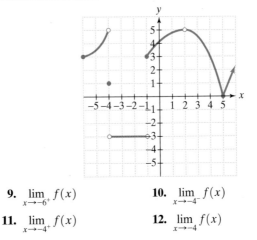

9. $\lim\limits_{x \to -6^+} f(x)$

10. $\lim\limits_{x \to -4^-} f(x)$

11. $\lim\limits_{x \to -4^+} f(x)$

12. $\lim\limits_{x \to -4} f(x)$

13. $f(-4)$

14. $\lim\limits_{x \to -1^+} f(x)$

15. $\lim\limits_{x \to -1^-} f(x)$

16. $\lim\limits_{x \to -1} f(x)$

17. $f(-1)$

18. $f(2)$

19. $\lim\limits_{x \to 2^-} f(x)$

20. $\lim\limits_{x \to 2^+} f(x)$

21. $\lim\limits_{x \to 2} f(x)$

22. $\lim\limits_{x \to 5} f(x)$

23. $f(5)$

In Exercises 24–26, graph each function. Then use your graph to find the indicated limit.

24. $f(x) = \dfrac{x^2 - 9}{x - 3}, \quad \lim\limits_{x \to 3} f(x)$

25. $f(x) = \sin x, \quad \lim\limits_{x \to \frac{3\pi}{2}} f(x)$

26. $f(x) = \begin{cases} 1 - x & \text{if } x < 0 \\ \cos x & \text{if } x \ge 0 \end{cases}, \lim\limits_{x \to 0} f(x)$

11.2

In Exercises 27–37, find the limit.

27. $\lim\limits_{x \to 4} (2x^2 - 5x + 3)$

28. $\lim\limits_{x \to -1} (-2x^3 - x + 5)$

29. $\lim\limits_{x \to -3} (x^2 + 1)^3$

30. $\lim\limits_{x \to 4} \sqrt{x^2 + 9}$

31. $\lim\limits_{x \to 5} \dfrac{11x - 3}{x^2 + 1}$

32. $\lim\limits_{x \to -4} \dfrac{x^2 - 16}{x + 4}$

33. $\lim\limits_{x \to 7} \dfrac{5x - 35}{x - 7}$

34. $\lim\limits_{x \to 0} \dfrac{\sqrt{x + 100} - 10}{x}$

35. $\lim\limits_{x \to -1} \dfrac{x^2 - 1}{x^2 + x}$

36. $\lim\limits_{x \to 100} \dfrac{\sqrt{x} - 10}{x - 100}$

37. $\lim\limits_{x \to 0} \dfrac{\dfrac{1}{x + 5} - \dfrac{1}{5}}{x}$

In Exercises 38–40, a piecewise function is given. Use properties of limits to find the indicated limit, or state that the limit does not exist.

38. $f(x) = \begin{cases} x^2 + 1 & \text{if } x < 2 \\ 3x + 1 & \text{if } x \ge 2 \end{cases}$

  a. $\lim\limits_{x \to 2^-} f(x)$ **b.** $\lim\limits_{x \to 2^+} f(x)$ **c.** $\lim\limits_{x \to 2} f(x)$

39. $f(x) = \begin{cases} \sqrt[3]{x^2 + 7} & \text{if } x < 1 \\ 4x & \text{if } x \ge 1 \end{cases}$

  a. $\lim\limits_{x \to 1^-} f(x)$ **b.** $\lim\limits_{x \to 1^+} f(x)$ **c.** $\lim\limits_{x \to 1} f(x)$

40. $f(x) = \begin{cases} \dfrac{x^2 - 25}{x + 5} & \text{if } x \ne -5 \\ 13 & \text{if } x = -5 \end{cases}$

  a. $\lim\limits_{x \to -5^-} f(x)$ **b.** $\lim\limits_{x \to -5^+} f(x)$ **c.** $\lim\limits_{x \to -5} f(x)$

11.3

In Exercises 41–45, use the definition of continuity to determine whether f is continuous at a.

41. $f(x) = 3x^2 - 2x + 1$

  $a = 4$

42. $f(x) = \dfrac{x^2 - 9}{x + 3}$

  $a = -3$

43. $f(x) = \begin{cases} \dfrac{x^2 + 5x}{x^2 - 5x} & \text{if } x \ne 0 \\ -2 & \text{if } x = 0 \end{cases}$

  $a = 0$

44. $f(x) = \begin{cases} \dfrac{x^2 + x}{x^2 - 3x - 4} & \text{if } x \ne -1 \\ \frac{1}{5} & \text{if } x = -1 \end{cases}$

  $a = -1$

45. $f(x) = \begin{cases} 3x & \text{if } x < 2 \\ 5 & \text{if } x = 2 \\ x + 4 & \text{if } x > 2 \end{cases}$

  $a = 2$

In Exercises 46–51, determine for what numbers, if any, the given function is discontinuous.

46. $f(x) = x^3 + 5x^2 - 1$

47. $f(x) = \dfrac{x - 1}{(x - 1)(x + 3)}$

48. $f(x) = \begin{cases} -1 & \text{if } x < 0 \\ 1 & \text{if } x \ge 0 \end{cases}$

49. $f(x) = \begin{cases} 4x & \text{if } x < 5 \\ x^2 - 5 & \text{if } x \ge 5 \end{cases}$

50. $f(x) = \begin{cases} \dfrac{x^2 - 4}{x + 2} & \text{if } x \ne -2 \\ 4 & \text{if } x = -2 \end{cases}$

51. $f(x) = \begin{cases} \dfrac{x^2 - 121}{x - 11} & \text{if } x \ne 11 \\ 20 & \text{if } x = 11 \end{cases}$

11.4

In Exercises 52–53,

  a. *Find the slope of the tangent line to the graph of f at the given point.*

  b. *Find the slope-intercept equation of the tangent line to the graph of f at the given point.*

52. $f(x) = 2x^2 + 5x$ at $(1, 7)$

53. $f(x) = x^2 - 7x - 4$ at $(-1, 4)$

In Exercises 54–57,

  a. *Find $f'(x)$.*

  b. *Find the slope of the tangent line to the graph of f at each of the two values of x given to the right of the function.*

54. $f(x) = 3x^2 + 12x - 1; \ x = -2, x = 1$

55. $f(x) = 2x^3 - x; \ x = -1, x = 1$

56. $f(x) = \dfrac{1}{x}; \ x = -2, x = 2$

57. $f(x) = \sqrt{x}; \ x = 36, x = 81$

5. The function $f(x) = 5x^2$ describes the volume of a rectangular box, $f(x)$, in cubic inches, whose square base has sides that each measure x inches and whose height is 5 inches. If x is changing,

   a. Find the average rate of change of the volume with respect to x as x changes from 2 inches to 2.1 inches and from 2 inches to 2.01 inches.

   b. Find the instantaneous rate of change of the volume with respect to x at the moment when $x = 2$ inches.

59. The function $f(x) = \frac{4}{3}\pi x^3$ describes the volume, $f(x)$, of a sphere of radius x inches. If the radius is changing, find the instantaneous rate of change of the volume with respect to the radius when the radius is 5 inches. Express the answer in terms of π.

60. A baseball is thrown straight upward from a height of 5 feet with an initial velocity of 80 feet per second. The function

$$s(t) = -16t^2 + 80t + 5$$

describes the ball's height above the ground, $s(t)$, in feet, t seconds after it is thrown.

   a. What is the instantaneous velocity of the ball 2 seconds after it is thrown? 4 seconds after it is thrown?

   b. The ball reaches its maximum height above the ground when the instantaneous velocity is zero. After how many seconds does the ball reach its maximum height? What is the maximum height?

Chapter 11 Test

1. Construct a table to find $\lim\limits_{x \to 9} \dfrac{9 - x}{3 - \sqrt{x}}$.

10. $\lim\limits_{x \to 9} \dfrac{\sqrt{x} - 3}{x - 9}$

In Exercises 2–7, use the graph of function f to find the indicated limit or function value, or state that the limit or function value does not exist.

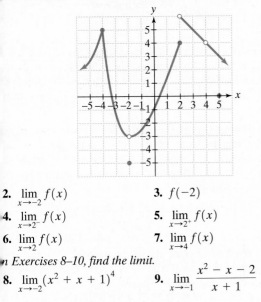

2. $\lim\limits_{x \to -2} f(x)$

3. $f(-2)$

4. $\lim\limits_{x \to 2^-} f(x)$

5. $\lim\limits_{x \to 2^+} f(x)$

6. $\lim\limits_{x \to 2} f(x)$

7. $\lim\limits_{x \to 4} f(x)$

In Exercises 8–10, find the limit.

8. $\lim\limits_{x \to -2} (x^2 + x + 1)^4$

9. $\lim\limits_{x \to -1} \dfrac{x^2 - x - 2}{x + 1}$

In Exercises 11–12, determine whether f is continuous at a.

11. $f(x) = \begin{cases} \dfrac{x^2 - 1}{x + 1} & \text{if } x \neq -1 \\ 6 & \text{if } x = -1 \end{cases}$

$a = -1$

12. $f(x) = \begin{cases} 2 - x & \text{if } x \leq 2 \\ x^2 - 2x & \text{if } x > 2 \end{cases}$

$a = 2$

In Exercises 13–14, find $f'(x)$.

13. $f(x) = x^2 - 5x + 1$

14. $f(x) = \dfrac{10}{x}$

15. Find the slope-intercept equation of the tangent line to the graph of $f(x) = x^2$ at $(-3, 9)$.

16. A ball is thrown straight upward. The function

$$s(t) = -16t^2 + 72t$$

describes the ball's height above the ground, $s(t)$, in feet, t seconds after it is thrown. What is the instantaneous velocity of the ball 3 seconds after it is thrown?

Cumulative Review Exercises (Chapters P–11)

Solve each equation or inequality in Exercises 1–5.

1. $\dfrac{1}{x + 2} > \dfrac{3}{x + 1}$

2. $2x^3 + 11x^2 - 7x - 6 = 0$

3. $|2x + 4| > 3$

4. $\cos^2 x + \sin x + 1 = 0, 0 \leq x < 2\pi$

5. $\log_4(x^2 - 9) - \log_4(x + 3) = 3$

In Exercises 6–15, graph each equation, function, or system in a rectangular coordinate system.

6. $f(x) = x^3 + x^2 - 12x$

7. $f(x) = \dfrac{2x^2 - 5x + 2}{x^2 - 4}$

8. $f(x) = \begin{cases} -x + 1 & \text{if } -1 \leq x < 1 \\ 2 & \text{if } x = 1 \\ x^2 & \text{if } x > 1 \end{cases}$

9. $y = 2\sin\left(2x + \dfrac{\pi}{2}\right)$ (Graph one period.)

10. $y = \frac{1}{2}\sec 2\pi x, \quad 0 \leq x \leq 2$

11. $x - 2y \leq 4$

$x \geq 2$

12. $x^2 - 4y^2 - 4x - 24y - 48 = 0$

13. $f(x) = \sqrt{x}, g(x) = \sqrt{x - 2} + 1$ (Graph f and g in the same rectangular coordinate system.)

14. $x = 3 \sin t$, $y = 4 \cos t + 2$; $0 \le t \le 2\pi$

15. $2x^2 + 5xy + 2y^2 - \frac{9}{2} = 0$

16. Find $f'(x)$ if $f(x) = -2x^2 + 7x - 1$.

17. Find $f^{-1}(x)$ if $f(x) = 7x - 1$.

18. Find the limit: $\lim\limits_{x \to -3} \dfrac{x^2 + x - 6}{x^2 + 2x - 3}$.

19. Expand and simplify: $(x^2 - 3y)^4$.

20. Write the slope-intercept form of the equation of the line passing through the point $(2, -3)$ and parallel to the line whose equation is $2x + y - 6 = 0$.

21. Find the dot product $\mathbf{v} \cdot \mathbf{w}$ and the angle between $\mathbf{v}$ and $\mathbf{w}$:

$$\mathbf{v} = -2\mathbf{i} + \mathbf{j}, \quad \mathbf{w} = 4\mathbf{i} - 3\mathbf{j}$$

22. Find the partial fraction decomposition for

$$\frac{1}{x(x^2 + x + 1)}.$$

Verify each identity in Exercises 23–24.

23. $\tan \theta + \cot \theta = \sec \theta \csc \theta$

24. $\tan(\theta + \pi) = \tan \theta$

25. If $A = \begin{bmatrix} 2 & 1 & 3 \\ 1 & -1 & 0 \end{bmatrix}$ and $B = \begin{bmatrix} 1 & 0 \\ 3 & 2 \\ 2 & 1 \end{bmatrix}$, find BA.

26. Graph the polar equation: $r = 4 \sin \theta$.

27. Express $h(x) = (x^2 - 3x + 7)^9$ as a composition of two functions f and g such that $h(x) = (f \circ g)(x)$.

28. Solve using matrices:

$$2x - y - 2z = -1$$
$$x - 2y - z = 1$$
$$x + y + z = 4.$$

29. Use the formula for the sum of the first n terms of a geometric sequence to find $\sum\limits_{i=1}^{6} 4(-2)^i$.

30. Use DeMoivre's Theorem to find

$$\left[\sqrt{2}(\cos 15° + i \sin 15°) \right]^4.$$

Write the answer in rectangular form.

31. A bank loaned out $120,000, part of it at 8% per year and the rest at 18% per year.

   a. Express the interest, I, on the two loans as a function of the amount loaned at 8%, x.

   b. If the interest received totaled $10,000, how much was loaned at each rate?

32. A machine produces open boxes using square sheets of metal. The machine cuts equal-sized squares measuring 9 centimeters on a side from each corner. Then the machine shapes the metal into an open box by turning up the sides. If each box must have a volume of 225 cubic centimeters, what should be the dimensions of the piece of sheet metal?

33. You have 200 feet of fencing to enclose a small rectangular garden with one side against a barn. If you do not fence the side along the barn, find the length and width of the garden that will maximize its area. What is the largest area that can be enclosed?

34. Use Newton's Law of Cooling, $T = C + (T_0 - C)e^{kt}$, t solve this exercise. You remove a pie that has a temperatur of 375°F from the oven. You leave the pie in a room whos temperature is 72°F. After 60 minutes, the temperature of th pie is 75°F.

   a. Write a model for the temperature of the pie, T, after minutes.

   b. When will the temperature of the pie be 250°F?

35. You just purchased a rectangular waterfront lot along river's edge. The area of the lot is 60,000 square feet. T create a sense of privacy, you decide to fence along thre sides, excluding the side that fronts the river. An expensiv fencing along the lot's front length costs $25 per foot. An in expensive fencing along the two side widths costs only $5 pe foot. Express the total cost, C, of fencing along the three side as a function of the lot's length, x.

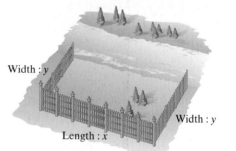

Width : y

Width : y

Length : x

36. Two ships leave a harbor at the same time. One ship travels a a bearing of N42°E for 23 miles. The other ship travels at bearing of N38°W for 72 miles. After both ships are anchored, how far apart are they? Round to the nearest tent of a mile.

37. At a fixed temperature, the volume of a given mass of ga varies inversely as the pressure applied to the gas. A certai mass of gas has a volume of 40 cubic inches when th pressure is 22 pounds. What is the volume of the gas when th pressure is 30 pounds?

38. A ball is thrown straight upward. The function

$$s(t) = -16t^2 + 40t$$

describes the ball's height above the ground, $s(t)$, in feet, seconds after it is thrown. What is the instantaneous velocity of the ball 2 seconds after it is thrown?

39. The figure shows an open box with a square base. The box i to have a volume of 4 cubic feet. Express the surface area o the box, A, as a function of the length of a side of its squar base, x.

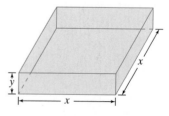

40. The function $f(x) = -2.32x^2 + 76.58x - 559.87$ models the percentage of U.S. students, $f(x)$, who are x years old who say their school is not drug free, where $12 \le x \le 17$. At wha age do 70% of U.S. students say that their school is not drug free? Round to the nearest tenth of a year.

Appendix
Where Did That Come From? Selected Proofs

SECTION 3.3 **Properties of Logarithms**

The Product Rule
Let b, M, and N be positive real numbers with $b \neq 1$.

$$\log_b(MN) = \log_b M + \log_b N$$

Proof

We begin by letting $\log_b M = R$ and $\log_b N = S$.
Now we write each logarithm in exponential form.

$$\log_b M = R \quad \text{means} \quad b^R = M.$$
$$\log_b N = S \quad \text{means} \quad b^S = N.$$

By substituting and using a property of exponents, we see that

$$MN = b^R b^S = b^{R+S}.$$

Now we change $MN = b^{R+S}$ to logarithmic form.

$$MN = b^{R+S} \quad \text{means} \quad \log_b(MN) = R + S.$$

Finally, substituting $\log_b M$ for R and $\log_b N$ for S gives us

$$\log_b(MN) = \log_b M + \log_b N,$$

the property that we wanted to prove.

The quotient and power rules for logarithms are proved using similar procedures.

The Change-of-Base Property
For any logarithmic bases a and b, and any positive number M,

$$\log_b M = \frac{\log_a M}{\log_a b}.$$

Proof

To prove the change-of-base property, we let x equal the logarithm on the left side:

$$\log_b M = x.$$

Now we rewrite this logarithm in exponential form.

$$\log_b M = x \quad \text{means} \quad b^x = M.$$

Because b^x and M are equal, the logarithms with base a for each of these expressions must be equal. This means that

$$\log_a b^x = \log_a M$$

$$x \log_a b = \log_a M \qquad \text{\textit{Apply the power rule for logarithms on the left side.}}$$

$$x = \frac{\log_a M}{\log_a b} \qquad \text{\textit{Solve for x by dividing both sides by } \log_a b.}$$

In our first step we let x equal $\log_b M$. Replacing x on the left side by $\log_b M$ gives us

$$\log_b M = \frac{\log_a M}{\log_a b},$$

which is the change-of-base property.

SECTION 6.2 · *The Law of Cosines*

Heron's Formula for the Area of a Triangle

The area of a triangle with sides $a, b,$ and c is

$$\text{Area} = \sqrt{s(s - a)(s - b)(s - c)},$$

where s is one-half its perimeter: $s = \frac{1}{2}(a + b + c)$.

Proof

The proof of Heron's formula begins with a half-angle formula and the Law of Cosines.

$$\cos \frac{C}{2} = \sqrt{\frac{1 + \cos C}{2}} = \sqrt{\frac{1 + \dfrac{a^2 + b^2 - c^2}{2ab}}{2}}$$

This is the Law of Cosines $c^2 = a^2 + b^2 - 2ab \cos C$ solved for $\cos C$.

$$= \sqrt{\frac{a^2 + 2ab + b^2 - c^2}{4ab}} = \sqrt{\frac{(a + b)^2 - c^2}{4ab}} = \sqrt{\frac{(a + b + c)(a + b - c)}{4ab}}$$

Multiply the numerator and denominator of the radicand by $2ab$.

Factor $a^2 + 2ab + b^2$.

Factor the numerator as the difference of two squares.

We now introduce the expression for one-half the perimeter: $s = \frac{1}{2}(a + b + c)$. We replace $a + b + c$ in the numerator by $2s$. We also find an expression for $a + b - c$ as follows:

$$a + b - c = a + b + c - 2c = 2s - 2c = 2(s - c).$$

Thus,

$$\cos \frac{C}{2} = \sqrt{\frac{(a + b + c)(a + b - c)}{4ab}} = \sqrt{\frac{2s \cdot 2(s - c)}{4ab}} = \sqrt{\frac{s(s - c)}{ab}}.$$

In a similar manner, we obtain

$$\sin \frac{C}{2} = \sqrt{\frac{1 - \cos C}{2}} = \sqrt{\frac{(s - a)(s - b)}{ab}}.$$

From our work in Section 6.1, we know that the area of a triangle is one-half the product of the length of two sides times the sine of their included angle.

$$\text{Area} = \frac{1}{2}ab \sin C$$

$$= \frac{1}{2}ab \cdot 2 \sin \frac{C}{2} \cos \frac{C}{2} \qquad \qquad \sin C = \sin 2\frac{C}{2} = 2 \sin \frac{C}{2} \cos \frac{C}{2}$$

$$= ab \sqrt{\frac{(s-a)(s-b)}{ab}} \sqrt{\frac{s(s-c)}{ab}} \qquad \text{Use the expressions for } \sin \frac{C}{2} \text{ and } \cos \frac{C}{2}$$
$$\text{on page A2.}$$

$$= ab \frac{\sqrt{s(s-a)(s-b)(s-c)}}{\sqrt{a^2 b^2}} \qquad \qquad \text{Multiply the radicands.}$$

$$= \sqrt{s(s-a)(s-b)(s-c)} \qquad \qquad \text{Simplify: } \frac{ab}{\sqrt{a^2 b^2}} = \frac{ab}{ab} = 1.$$

SECTION 6.5   *Complex Numbers in Polar Form; DeMoivre's Theorem*

The Quotient of Two Complex Numbers in Polar Form

Let $z_1 = r_1(\cos \theta_1 + i \sin \theta_1)$ and $z_2 = r_2(\cos \theta_2 + i \sin \theta_2)$ be two complex numbers in polar form. Their quotient, $\dfrac{z_1}{z_2}$, is

$$\frac{z_1}{z_2} = \frac{r_1}{r_2}[\cos(\theta_1 - \theta_2) + i \sin(\theta_1 - \theta_2)].$$

Proof

We begin by multiplying the numerator and denominator of the quotient, $\dfrac{z_1}{z_2}$, by the conjugate of the denominator. Then we simplify the quotient using the difference formulas for sine and cosine.

$$\frac{z_1}{z_2} = \frac{r_1(\cos \theta_1 + i \sin \theta_1)}{r_2(\cos \theta_2 + i \sin \theta_2)} \qquad \qquad \text{This is the given quotient.}$$

$$= \frac{r_1(\cos \theta_1 + i \sin \theta_1)(\cos \theta_2 - i \sin \theta_2)}{r_2(\cos \theta_2 + i \sin \theta_2)(\cos \theta_2 - i \sin \theta_2)} \qquad \text{Multiply the numerator and denominator by the}$$
$$\text{conjugate of the denominator. Recall that the}$$
$$\text{conjugate of } a + bi \text{ is } a - bi.$$

$$= \frac{r_1(\cos \theta_1 + i \sin \theta_1)(\cos \theta_2 - i \sin \theta_2)}{r_2(\cos^2 \theta_2 + \sin^2 \theta_2)} \qquad \text{Multiply the conjugates in the denominator.}$$

$$= \frac{r_1(\cos \theta_1 + i \sin \theta_1)(\cos \theta_2 - i \sin \theta_2)}{r_2} \qquad \text{Use a Pythagorean identity:}$$
$$\cos^2 \theta_2 + \sin^2 \theta_2 = 1.$$

$$= \frac{r_1}{r_2}(\cos \theta_1 \cos \theta_2 - i \cos \theta_1 \sin \theta_2 + i \sin \theta_1 \cos \theta_2 - i^2 \sin \theta_1 \sin \theta_2) \qquad \text{Use the FOIL method.}$$

$$= \frac{r_1}{r_2}[\cos \theta_1 \cos \theta_2 + i(\sin \theta_1 \cos \theta_2 - \cos \theta_1 \sin \theta_2) - i^2 \sin \theta_1 \sin \theta_2] \qquad \text{Factor } i \text{ from the second and third terms.}$$

$$= \frac{r_1}{r_2}[\cos \theta_1 \cos \theta_2 + i(\sin \theta_1 \cos \theta_2 - \cos \theta_1 \sin \theta_2) - (-1)\sin \theta_1 \sin \theta_2] \qquad i^2 = -1.$$

$$= \frac{r_1}{r_2}\left[\cos \theta_1 \cos \theta_2 + \sin \theta_1 \sin \theta_2 + i(\sin \theta_1 \cos \theta_2 - \cos \theta_1 \sin \theta_2)\right] \qquad \text{Rearrange terms.}$$

This is cos $(\theta_1 - \theta_2)$.   This is sin $(\theta_1 - \theta_2)$.

$$= \frac{r_1}{r_2}[\cos(\theta_1 - \theta_2) + i \sin(\theta_1 - \theta_2)]$$

SECTION 6.7 *The Dot Product*

Properties of the Dot Product

If **u**, **v**, and **w** are vectors, and c is a scalar, then

1. $\mathbf{u} \cdot \mathbf{v} = \mathbf{v} \cdot \mathbf{u}$
2. $\mathbf{u} \cdot (\mathbf{v} + \mathbf{w}) = \mathbf{u} \cdot \mathbf{v} + \mathbf{u} \cdot \mathbf{w}$
3. $\mathbf{0} \cdot \mathbf{v} = 0$
4. $\mathbf{v} \cdot \mathbf{v} = \|\mathbf{v}\|^2$
5. $(c\mathbf{u}) \cdot \mathbf{v} = c(\mathbf{u} \cdot \mathbf{v}) = \mathbf{u} \cdot (c\mathbf{v})$

Proof

To prove the second property, let

$$\mathbf{u} = u_1\mathbf{i} + u_2\mathbf{j}, \quad \mathbf{v} = v_1\mathbf{i} + v_2\mathbf{j}, \quad \text{and} \quad \mathbf{w} = w_1\mathbf{i} + w_2\mathbf{j}.$$

Then,

$$\mathbf{u} \cdot (\mathbf{v} + \mathbf{w}) = (u_1\mathbf{i} + u_2\mathbf{j}) \cdot [(v_1\mathbf{i} + v_2\mathbf{j}) + (w_1\mathbf{i} + w_2\mathbf{j})] \quad \text{These are the giver vectors.}$$

$$= (u_1\mathbf{i} + u_2\mathbf{j}) \cdot [(v_1 + w_1)\mathbf{i} + (v_2 + w_2)\mathbf{j}] \quad \text{Add horizontal components and add vertical components.}$$

$$= u_1(v_1 + w_1) + u_2(v_2 + w_2) \quad \text{Multiply horizontal components and multiply vertical components.}$$

$$= u_1v_1 + u_1w_1 + u_2v_2 + u_2w_2 \quad \text{Use the distributive property.}$$

$$= u_1v_1 + u_2v_2 + u_1w_1 + u_2w_2 \quad \text{Rearrange terms.}$$

> This is the dot product of **u** and **v**.

> This is the dot product of **u** and **w**.

$$= \mathbf{u} \cdot \mathbf{v} + \mathbf{u} \cdot \mathbf{w}.$$

To prove the third property, let

$$\mathbf{0} = 0\mathbf{i} + 0\mathbf{j} \quad \text{and} \quad \mathbf{v} = v_1\mathbf{i} + v_2\mathbf{j}.$$

Then,

$$\mathbf{0} \cdot \mathbf{v} = (0\mathbf{i} + 0\mathbf{j}) \cdot (v_1\mathbf{i} + v_2\mathbf{j}) \quad \text{These are the given vectors.}$$

$$= 0 \cdot v_1 + 0 \cdot v_2 \quad \text{Multiply horizontal components and multiply vertical components.}$$

$$= 0 + 0$$

$$= 0.$$

To prove the first part of the fifth property, let

$$\mathbf{u} = u_1\mathbf{i} + u_2\mathbf{j} \quad \text{and} \quad \mathbf{v} = v_1\mathbf{i} + v_2\mathbf{j}.$$

Then,

$$(c\mathbf{u}) \cdot \mathbf{v} = [c(u_1\mathbf{i} + u_2\mathbf{j})] \cdot (v_1\mathbf{i} + v_2\mathbf{j}) \quad \text{These are the given vectors.}$$

$$= (cu_1\mathbf{i} + cu_2\mathbf{j}) \cdot (v_1\mathbf{i} + v_2\mathbf{j}) \quad \text{Multiply each component of } u_1\mathbf{i} + u_2\mathbf{j} \text{ by } c.$$

$$= cu_1v_1 + cu_2v_2 \quad \text{Multiply horizontal components and multiply vertical components.}$$

$$= c(u_1v_1 + u_2v_2) \quad \text{Factor out } c \text{ from both terms.}$$

> This is the dot product of **u** and **v**.

$$= c(\mathbf{u} \cdot \mathbf{v})$$

SECTION 9.1 *The Ellipse*

The Standard Form of the Equation of an Ellipse with a Horizontal Major Axis Centered at the Origin

Proof

Refer to Figure A.1.

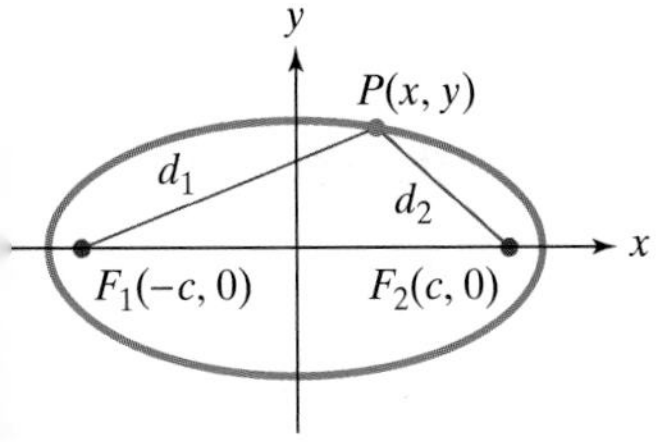

Figure A.1

$d_1 + d_2 = 2a$	The sum of the distances from P to the foci equals a constant, $2a$.
$\sqrt{(x + c)^2 + y^2} + \sqrt{(x - c)^2 + y^2} = 2a$	Use the distance formula.
$\sqrt{(x + c)^2 + y^2} = 2a - \sqrt{(x - c)^2 + y^2}$	Isolate a radical.
$(x + c)^2 + y^2 = 4a^2 - 4a\sqrt{(x - c)^2 + y^2}$ $+ (x - c)^2 + y^2$	Square both sides.
$x^2 + 2cx + c^2 + y^2 = 4a^2 - 4a\sqrt{(x - c)^2 + y^2}$ $+ x^2 - 2cx + c^2 + y^2$	Square $x + c$ and $x - c$.
$4cx - 4a^2 = -4a\sqrt{(x - c)^2 + y^2}$	Simplify and isolate the radical.
$cx - a^2 = -a\sqrt{(x - c)^2 + y^2}$	Divide both sides by 4.
$(cx - a^2)^2 = a^2[(x - c)^2 + y^2]$	Square both sides.
$c^2x^2 - 2a^2cx + a^4 = a^2(x^2 - 2cx + c^2 + y^2)$	Square $cx - a^2$ and $x - c$.
$c^2x^2 - 2a^2cx + a^4 = a^2x^2 - 2a^2cx + a^2c^2 + a^2y^2$	Use the distributive property.
$c^2x^2 + a^4 = a^2x^2 + a^2c^2 + a^2y^2$	Add $2a^2cx$ to both sides.
$c^2x^2 - a^2x^2 - a^2y^2 = a^2c^2 - a^4$	Rearrange the terms.
$(c^2 - a^2)x^2 - a^2y^2 = a^2(c^2 - a^2)$	Factor out x^2 and a^2, respectively.
$(a^2 - c^2)x^2 + a^2y^2 = a^2(a^2 - c^2)$	Multiply both sides by -1.

Refer to the discussion on page 851 and let $b^2 = a^2 - c^2$ in the preceding equation.

$b^2x^2 + a^2y^2 = a^2b^2$	
$\dfrac{x^2}{a^2} + \dfrac{y^2}{b^2} = 1$	Divide both sides by a^2b^2.

SECTION 9.2 *The Hyperbola*

The Asymptotes of a Hyperbola Centered at the Origin

The hyperbola

$$\frac{x^2}{a^2} - \frac{y^2}{b^2} = 1$$

with a horizontal transverse axis has the two asymptotes

$$y = \frac{b}{a}x \quad \text{and} \quad y = -\frac{b}{a}x.$$

Proof

Begin by solving the hyperbola's equation for y.

$$\frac{x^2}{a^2} - \frac{y^2}{b^2} = 1 \qquad \text{This is the standard form of the equation of a hyperbola.}$$

$$\frac{y^2}{b^2} = \frac{x^2}{a^2} - 1 \qquad \text{We isolate the term involving } y^2 \text{ to solve for } y.$$

$$y^2 = \frac{b^2 x^2}{a^2} - b^2 \qquad \text{Multiply both sides by } b^2.$$

$$y^2 = \frac{b^2 x^2}{a^2}\left(1 - \frac{a^2}{x^2}\right) \qquad \text{Factor out } \frac{b^2 x^2}{a^2} \text{ on the right. Verify that this result is correct by multiplying using the distributive property and obtaining the previous step.}$$

$$y = \pm\sqrt{\frac{b^2 x^2}{a^2}\left(1 - \frac{a^2}{x^2}\right)} \qquad \text{Solve for } y \text{ using the square root property: If } u^2 = d, \text{ then } u = \pm\sqrt{d}.$$

$$y = \pm\frac{b}{a}x\sqrt{1 - \frac{a^2}{x^2}} \qquad \text{Simplify.}$$

As $|x| \to \infty$, the value of $\frac{a^2}{x^2}$ approaches 0. Consequently, the value of y can be approximated by

$$y = \pm\frac{b}{a}x.$$

This means that the lines whose equations are $y = \frac{b}{a}x$ and $y = -\frac{b}{a}x$ are asymptotes for the graph of the hyperbola.

Answers to Selected Exercises

CHAPTER P

Section P.1

Check Point Exercises

1. 608   **2.** 2311; The formula models the data quite well.   **3.** $\{3, 7\}$ **4.** $\{3, 4, 5, 6, 7, 8, 9\}$ **5. a.** $\sqrt{9}$ **b.** $0, \sqrt{9}$ **c.** $-9, 0, \sqrt{9}$
d. $-9, -1.3, 0, 0.\overline{3}, \sqrt{9}$ **e.** $\frac{\pi}{2}, \sqrt{10}$ **f.** $-9, -1.3, 0, 0.\overline{3}, \frac{\pi}{2}, \sqrt{9}, \sqrt{10}$ **6. a.** $\sqrt{2} - 1$ **b.** $\pi - 3$ **c.** 1   **7.** 9   **8.** $38x^2 + 23x$
9. $42 - 4x$

Exercise Set P.1

1. 57   **3.** 10   **5.** 88   **7.** 10   **9.** 44   **11.** 46   **13.** 10   **15.** -8 **17.** 10°C   **19.** 60 ft   **21.** $\{2, 4\}$ **23.** $\{s, e, t\}$ **25.** $\varnothing$
27. $\varnothing$ **29.** $\{1, 2, 3, 4, 5\}$ **31.** $\{1, 2, 3, 4, 5, 6, 7, 8, 10\}$ **33.** $\{a, e, i, o, u\}$ **35. a.** $\sqrt{100}$ **b.** $0, \sqrt{100}$ **c.** $-9, 0, \sqrt{100}$
d. $-9, -\frac{4}{5}, 0, 0.25, 9.2, \sqrt{100}$ **e.** $\sqrt{3}$ **f.** $-9, -\frac{4}{5}, 0, 0.25, \sqrt{3}, 9.2, \sqrt{100}$ **37. a.** $\sqrt{64}$ **b.** $0, \sqrt{64}$ **c.** $-11, 0, \sqrt{64}$
d. $-11, -\frac{5}{6}, 0, 0.75, \sqrt{64}$ **e.** $\sqrt{5}, \pi$ **f.** $-11, -\frac{5}{6}, 0, 0.75, \sqrt{5}, \pi, \sqrt{64}$ **39.** 0   **41.** Answers may vary.   **43.** true   **45.** true
47. true   **49.** true   **51.** 300   **53.** $12 - \pi$ **55.** $5 - \sqrt{2}$ **57.** -1 **59.** 4   **61.** 3   **63.** 7   **65.** -1 **67.** $|17 - 2|; 15$
69. $|5 - (-2)|; 7$ **71.** $|-4 - (-19)|; 15$ **73.** $|-1.4 - (-3.6)|; 2.2$ **75.** commutative property of addition
77. associative property of addition   **79.** commutative property of addition   **81.** distributive property of multiplication over addition
83. inverse property of multiplication   **85.** $15x + 16$ **87.** $27x - 10$ **89.** $29y - 29$ **91.** $8y - 12$ **93.** $16y - 25$ **95.** $12x^2 + 11$
97. $14x$ **99.** $-2x + 3y + 6$ **101.** x **103.** $>$ **105.** $=$ **107.** $<$ **109.** $=$ **111.** $x - (x + 4); -4$ **113.** $6(-5x); -30x$
115. $5x - 2x; 3x$ **117.** $8x - (3x + 6); 5x - 6$ **119.** 313; very well   **121.** 522   **123.** Model 3   **125.** Model 3   **127. a.** $1200 - 0.07x$
b. \$780   **141.** d   **143.** $>$ **145. a.** \$50.50, \$5.50, \$1.00   **b.** no

Section P.2

Check Point Exercises

1. a. $16x^{12}y^{24}$ **b.** $-18x^3y^8$ **c.** $\frac{5y^6}{x^4}$ **d.** $\frac{y^8}{25x^2}$ **2. a.** $-2,600,000,000$ **b.** 0.000003017   **3. a.** 5.21×10^9 **b.** -6.893×10^{-8}
4. 4.1×10^9 **5. a.** 3.55×10^{-1} **b.** 4×10^8 **6.** \$7,014   **7.** $2.5344 \times 10^3 = 2534.4$

Exercise Set P.2

1. 50   **3.** 64   **5.** -64 **7.** 1   **9.** -1 **11.** $\frac{1}{64}$ **13.** 32   **15.** 64   **17.** 16   **19.** $\frac{1}{9}$ **21.** $\frac{1}{16}$ **23.** $\frac{y}{x^2}$ **25.** y^5 **27.** x^{10}
29. x^5 **31.** x^{21} **33.** $\frac{1}{x^{15}}$ **35.** x^7 **37.** x^{21} **39.** $64x^6$ **41.** $-\frac{64}{x^3}$ **43.** $9x^4y^{10}$ **45.** $6x^{11}$ **47.** $18x^9y^5$ **49.** $4x^{16}$
51. $-5a^{11}b$ **53.** $\frac{2}{b^7}$ **55.** $\frac{1}{16x^6}$ **57.** $\frac{3y^{14}}{4x^4}$ **59.** $\frac{y^2}{25x^6}$ **61.** $-\frac{27\,b^{15}}{a^{18}}$ **63.** 1   **65.** 380   **67.** 0.0006   **69.** $-7,160,000$ **71.** 0.79
73. -0.00415 **75.** $-60,000,100,000$ **77.** 3.2×10^4 **79.** 6.38×10^{17} **81.** -5.716×10^3 **83.** 2.7×10^{-3} **85.** -5.04×10^{-9}
87. 6.3×10^7 **89.** 6.4×10^4 **91.** 1.22×10^{-11} **93.** 2.67×10^{13} **95.** 2.1×10^3 **97.** 4×10^5 **99.** 2×10^{-8} **101.** 5×10^3
103. 4×10^{15} **105.** 9×10^{-3} **107.** 1   **109.** $\frac{y}{16x^8z^6}$ **111.** $\frac{1}{x^{12}y^{16}z^{20}}$ **113.** $\frac{x^{18}y^6}{4}$ **115.** 6.26×10^7 people
117. 9.63×10^7 people   **119.** approximately 68 hot dogs per person   **121.** $2.5 \times 10^2 = 250$ chickens   **123.** approximately \$23,448
133. b   **135.** $A = C + D$

Section P.3

Check Point Exercises

1. a. 9   **b.** -3 **c.** $\frac{1}{5}$ **d.** 10   **e.** 14   **2. a.** $5\sqrt{3}$ **b.** $5x\sqrt{2}$ **3. a.** $\frac{5}{4}$ **b.** $5x\sqrt{3}$ **4. a.** $17\sqrt{13}$ **b.** $-19\sqrt{17x}$
5. a. $17\sqrt{3}$ **b.** $10\sqrt{2x}$ **6. a.** $\frac{5\sqrt{3}}{3}$ **b.** $\sqrt{3}$ **7.** $\frac{32 - 8\sqrt{5}}{11}$ **8. a.** $2\sqrt[3]{5}$ **b.** $2\sqrt[5]{2}$ **c.** $\frac{5}{3}$ **9.** $5\sqrt[3]{3}$ **10. a.** 5   **b.** 2
c. -3 **d.** -2 **e.** $\frac{1}{3}$ **11. a.** 81   **b.** 8   **c.** $\frac{1}{4}$ **12. a.** $10x^4$ **b.** $4x^{5/2}$ **13.** $\sqrt{x}$

Exercise Set P.3

1. 6   **3.** -6 **5.** not a real number   **7.** 3   **9.** 1   **11.** 13   **13.** $5\sqrt{2}$ **15.** $3|x|\sqrt{5}$ **17.** $2x\sqrt{3}$ **19.** $x\sqrt{x}$ **21.** $2x\sqrt{3x}$
23. $\frac{1}{9}$ **25.** $\frac{7}{4}$ **27.** $4x$ **29.** $5x\sqrt{2x}$ **31.** $2x^2\sqrt{5}$ **33.** $13\sqrt{3}$ **35.** $-2\sqrt{17x}$ **37.** $5\sqrt{2}$ **39.** $3\sqrt{2x}$ **41.** $34\sqrt{2}$
43. $20\sqrt{2} - 5\sqrt{3}$ **45.** $\frac{\sqrt{7}}{7}$ **47.** $\frac{\sqrt{10}}{5}$ **49.** $\frac{13(3 - \sqrt{11})}{-2}$ **51.** $7(\sqrt{5} + 2)$ **53.** $3(\sqrt{5} - \sqrt{3})$ **55.** 5   **57.** -2

59. not a real number   **61.** 3   **63.** -3 **65.** $-\dfrac{1}{2}$ **67.** $2\sqrt[3]{4}$ **69.** $x\sqrt[3]{x}$ **71.** $3\sqrt[3]{2}$ **73.** $2x$ **75.** $7\sqrt[5]{2}$ **77.** $13\sqrt[3]{2}$ **79.** $-y\sqrt[3]{2x}$

81. $\sqrt{2}+2$ **83.** 6   **85.** 2   **87.** 25   **89.** $\dfrac{1}{16}$ **91.** $14x^{7/12}$ **93.** $4x^{1/4}$ **95.** x^2 **97.** $5x^2|y|^3$ **99.** $27y^{2/3}$ **101.** $\sqrt{5}$ **103.** x^2

105. $\sqrt[3]{x^2}$ **107.** $\sqrt[6]{x^2y}$ **109.** 3   **111.** $\dfrac{x^2}{7y^{3/2}}$ **113.** $\dfrac{x^3}{y^2}$ **115.** $6\sqrt{3}$ miles; 10.4 miles   **117.** 70 mph; He was speeding.

119. $\dfrac{7\sqrt{2\cdot2\cdot3}}{6}=\dfrac{7\cdot2\sqrt{3}}{6}=\dfrac{14\sqrt{3}}{6}=\dfrac{7\sqrt{3}}{3}=\dfrac{7}{3}\sqrt{3}$ **121. a.** $C=35.74+0.6215t-35.74v^{4/25}+0.4275tv^{4/25}$ **b.** $8°F$

123. $P=18\sqrt{5}$ ft ; $A=100$ sq ft   **133.** d   **135.** Let $\square=25$ and $\square=14$.   **137. a.** $>$ **b.** $>$

Section P.4

Check Point Exercises

1. a. $-x^3+x^2-8x-20$ **b.** $20x^3-11x^2-2x-8$ **2.** $15x^3-31x^2+30x-8$ **3.** $28x^2-41x+15$ **4. a.** $21x^2-25xy+6y^2$
b. $4x^2+16xy+16y^2$ **5. a.** $9x^2+12x+4-25y^2$ **b.** $4x^2+4xy+y^2+12x+6y+9$

Exercise Set P.4

1. yes; $3x^2+2x-5$ **3.** no   **5.** 2   **7.** 4   **9.** $11x^3+7x^2-12x-4$; 3   **11.** $12x^3+4x^2+12x-14$; 3   **13.** $6x^2-6x+2$; 2
15. x^3+1 **17.** $2x^3-9x^2+19x-15$ **19.** $x^2+10x+21$ **21.** $x^2-2x-15$ **23.** $6x^2+13x+5$ **25.** $10x^2-9x-9$
27. $15x^4-47x^2+28$ **29.** $8x^5-40x^3+3x^2-15$ **31.** x^2-9 **33.** $9x^2-4$ **35.** $25-49x^2$ **37.** $16x^4-25x^2$ **39.** $1-y^{10}$
41. x^2+4x+4 **43.** $4x^2+12x+9$ **45.** x^2-6x+9 **47.** $16x^4-8x^2+1$ **49.** $4x^2-28x+49$ **51.** x^3+3x^2+3x+1
53. $8x^3+36x^2+54x+27$ **55.** $x^3-9x^2+27x-27$ **57.** $27x^3-108x^2+144x-64$ **59.** $7x^2+38xy+15y^2$
61. $2x^2+xy-21y^2$ **63.** $15x^2y^2+xy-2$ **65.** $49x^2+70xy+25y^2$ **67.** $x^4y^4-6x^2y^2+9$ **69.** x^3-y^3 **71.** $9x^2-25y^2$
73. $x^2+2xy+y^2-9$ **75.** $9x^2+42x+49-25y^2$ **77.** $25y^2-4x^2-12x-9$ **79.** $x^2+2xy+y^2+2x+2y+1$
81. $4x^2+4xy+y^2+4x+2y+1$ **83.** $48xy$ **85.** $-9x^2+3x+9$ **87.** $16x^4-625$ **89.** $4x^2-28x+49$
91. Model 4; $0.01x^3+0.09x^2+1.1x+5.64$ **93.** Model 1   **95.** very well   **97.** $4x^3-36x^2+80x$ **99.** $6x+22$
109. $V=x^3+7x^2-3x$ **111.** $6y^n-13$

Section P.5

Check Point Exercises

1. a. $2x^2(5x-2)$ **b.** $(x-7)(2x+3)$ **2.** $(x+5)(x^2-2)$ **3. a.** $(x+8)(x+5)$ **b.** $(x-7)(x+2)$ **4.** $(3x-1)(2x+7)$
5. $(3x-y)(x-4y)$ **6. a.** $(x+9)(x-9)$ **b.** $(6x+5)(6x-5)$ **7.** $(9x^2+4)(3x+2)(3x-2)$ **8. a.** $(x+7)^2$ **b.** $(4x-7)^2$
9. a. $(x+1)(x^2-x+1)$ **b.** $(5x-2)(25x^2+10x+4)$ **10.** $3x(x-5)^2$ **11.** $(x+10+6a)(x+10-6a)$ **12.** $\dfrac{2x-1}{(x-1)^{1/2}}$

Exercise Set P.5

1. $9(2x+3)$ **3.** $3x(x+2)$ **5.** $9x^2(x^2-2x+3)$ **7.** $(x+5)(x+3)$ **9.** $(x-3)(x^2+12)$ **11.** $(x-2)(x^2+5)$ **13.** $(x-1)(x^2+2)$
15. $(3x-2)(x^2-2)$ **17.** $(x+2)(x+3)$ **19.** $(x-5)(x+3)$ **21.** $(x-5)(x-3)$ **23.** $(3x+2)(x-1)$ **25.** $(3x-28)(x+1)$
27. $(2x-1)(3x-4)$ **29.** $(2x+3)(2x+5)$ **31.** $(3x-2)(3x-1)$ **33.** $(5x+8)(4x-1)$ **35.** $(2x+y)(x+y)$
37. $(3x+2y)(2x-3y)$ **39.** $(x+10)(x-10)$ **41.** $(6x+7)(6x-7)$ **43.** $(3x+5y)(3x-5y)$ **45.** $(x^2+4)(x+2)(x-2)$
47. $(4x^2+9)(2x+3)(2x-3)$ **49.** $(x+1)^2$ **51.** $(x-7)^2$ **53.** $(2x+1)^2$ **55.** $(3x-1)^2$ **57.** $(x+3)(x^2-3x+9)$
59. $(x-4)(x^2+4x+16)$ **61.** $(2x-1)(4x^2+2x+1)$ **63.** $(4x+3)(16x^2-12x+9)$ **65.** $3x(x+1)(x-1)$ **67.** $4(x+2)(x-3)$
69. $2(x^2+9)(x+3)(x-3)$ **71.** $(x-3)(x+3)(x+2)$ **73.** $2(x-8)(x+7)$ **75.** $x(x-2)(x+2)$ **77.** prime   **79.** $(x-2)(x+2)^2$
81. $y(y^2+9)(y+3)(y-3)$ **83.** $5y^2(2y+3)(2y-3)$ **85.** $(x-6+7y)(x-6-7y)$ **87.** $(x+y)(3b+4)(3b-4)$
89. $(y-2)(x+4)(x-4)$ **91.** $2x(x+6+2a)(x+6-2a)$ **93.** $x^{1/2}(x-1)$ **95.** $\dfrac{4(1+2x)}{x^{2/3}}$ **97.** $-(x+3)^{1/2}(x+2)$

99. $\dfrac{x+4}{(x+5)^{3/2}}$ **101.** $-\dfrac{4(4x-1)^{1/2}(x-1)}{3}$ **103.** $(x+1)(5x-6)(2x+1)$ **105.** $(x^2+6)(6x^2-1)$ **107.** $y(y^2+1)(y^4-y^2+1)$

109. $(x+2y)(x-2y)(x+y)(x-y)$ **111.** $(x-y)^2(x-y+2)(x-y-2)$ **113.** $(2x-y^2)(x-3y^2)$
115. a. $(x-0.4x)(1-0.4)=(0.6x)(0.6)=0.36x$ **b.** no; 36%   **117. a.** $9x^2-16$ **b.** $(3x+4)(3x-4)$ **119. a.** $x(x+y)-y(x+y)$
b. $(x+y)(x-y)$ **121.** $4a^3-4ab^2=4a(a+b)(a-b)$ **131.** $(x^n+4)(x^n+2)$ **133.** $(x-y)^3(x+y)$ **135.** $b=8,-8,16,-16$

Mid-Chapter P Check Point

1. $12x^2-x-35$ **2.** $-x+12$ **3.** $10\sqrt{6}$ **4.** $3\sqrt{3}$ **5.** $x+45$ **6.** $6x^2-48x+9$ **7.** $\dfrac{x^2}{y^3}$ **8.** $\dfrac{3}{4}$ **9.** $-x^2+5x-6$
10. $2x^3-11x^2+17x-5$ **11.** $-x^6+2x^3$ **12.** $18a^2-11ab-10b^2$ **13.** $\{a,c,d,e,f,h\}$ **14.** $\{c,d\}$ **15.** $5x^2y^3+2xy-y^2$
16. $-\dfrac{12y^{15}}{x^3}$ **17.** $\dfrac{6y^3}{x^7}$ **18.** $|\sqrt[3]{x}|$ **19.** $16y^2-9x^2-12x-4$ **20.** $x^2-4xy+4y^2-2x+4y+1$ **21.** 1.2×10^{-2} **22.** $2\sqrt[3]{2}$
23. x^6-4 **24.** x^4+4x^2+4 **25.** $10\sqrt{3}$ **26.** $\dfrac{77+11\sqrt{3}}{46}$ **27.** $\dfrac{11\sqrt{3}}{3}$ **28.** $(7x-1)(x-3)$ **29.** prime   **30.** $(x^2+3)(x+5)$
31. $(3x-7y)(x+y)$ **32.** $y(4-y)(16+4y+y^2)$ **33.** $2x(5x+1)^2$ **34.** $(x-3-7y)(x-3+7y)$ **35.** $\dfrac{(1-x)^2}{x^{3/2}}$ **36.** $\dfrac{(x-3)(x+3)}{(x^2+1)^{1/2}}$
37. $-11,-\dfrac{3}{7},0,0.45,\sqrt{25}$ **38.** $\sqrt{13}-2$ **39.** $-x^3$ **40.** $\$3.48\times10^{10}$ **41.** 4 times   **42. a.** Model 3   **b.** \$1001 million, or \$1,001,000,000

Section P.6

Check Point Exercises

1. **a.** -5 **b.** $6, -6$ **2. a.** $x^2, x \neq -3$ **b.** $\dfrac{x-1}{x+1}, x \neq -1$ **3.** $\dfrac{x-3}{(x-2)(x+3)}, x \neq 2, x \neq -2, x \neq -3$

4. $\dfrac{3(x-1)}{x(x+2)}, x \neq 1, x \neq 0, x \neq -2$ **5.** $-2, x \neq -1$ **6.** $\dfrac{2(4x+1)}{(x+1)(x-1)}, x \neq 1, x \neq -1$ **7.** $(x-3)(x-3)(x+3)$

8. $\dfrac{-x^2+11x-20}{2(x-5)^2}, x \neq 5$ **9.** $\dfrac{2(2-3x)}{4+3x}, x \neq 0, x \neq -\dfrac{4}{3}$ **10.** $-\dfrac{1}{x(x+7)}, x \neq 0, x \neq -7$ **11.** $\dfrac{x+1}{x^{3/2}}$ **12.** $\dfrac{1}{\sqrt{x+3}+\sqrt{x}}$

Exercise Set P.6

1. 3 **3.** $5, -5$ **5.** $-1, -10$ **7.** $\dfrac{3}{x-3}, x \neq 3$ **9.** $\dfrac{x-6}{4}, x \neq 6$ **11.** $\dfrac{y+9}{y-1}, y \neq 1, 2$ **13.** $\dfrac{x+6}{x-6}, x \neq 6, -6$ **15.** $\dfrac{1}{3}, x \neq 2, -3$

17. $\dfrac{(x-3)(x+3)}{x(x+4)}, x \neq 0, -4, 3$ **19.** $\dfrac{x-1}{x+2}, x \neq -2, -1, 2, 3$ **21.** $\dfrac{x^2+2x+4}{3x}, x \neq -2, 0, 2$ **23.** $\dfrac{7}{9}, x \neq -1$ **25.** $\dfrac{(x-2)^2}{x}, x \neq 0, -2, 2$

27. $\dfrac{2(x+3)}{3}, x \neq 3, -3$ **29.** $\dfrac{x-5}{2}, x \neq 1, -5$ **31.** $\dfrac{(x+2)(x+4)}{x-5}, x \neq -6, -3, -1, 3, 5$ **33.** $2, x \neq -\dfrac{5}{6}$ **35.** $\dfrac{2x-1}{x+3}, x \neq 0, -3$

37. $3, x \neq 2$ **39.** $\dfrac{3}{x-3}, x \neq 3, -4$ **41.** $\dfrac{9x+39}{(x+4)(x+5)}, x \neq -4, -5$ **43.** $-\dfrac{3}{x(x+1)}, x \neq -1, 0$ **45.** $\dfrac{3x^2+4}{(x+2)(x-2)}, x \neq -2, 2$

47. $\dfrac{2x^2+50}{(x-5)(x+5)}, x \neq -5, 5$ **49.** $\dfrac{4x+16}{(x+3)^2}, x \neq -3$ **51.** $\dfrac{x^2-x}{(x+5)(x-2)(x+3)}, x \neq -5, 2, -3$ **53.** $\dfrac{x-1}{x+2}, x \neq -2, -1$ **55.** $\dfrac{1}{3}, x \neq 3$

57. $\dfrac{x+1}{3x-1}, x \neq 0, \dfrac{1}{3}$ **59.** $\dfrac{1}{xy}, x \neq 0, y \neq 0, x \neq -y$ **61.** $\dfrac{x}{x+3}, x \neq -2, -3$ **63.** $-\dfrac{x-14}{7}, x \neq -2, 2$ **65.** $\dfrac{x-3}{x+2}, x \neq -2, -1, 3$

67. $-\dfrac{2x+h}{x^2(x+h)^2}, x \neq 0, h \neq 0, x \neq -h$ **69.** $1-\dfrac{1}{3x}; x > 0$ **71.** $-\dfrac{2}{x^2\sqrt{x^2+2}}$ **73.** $\dfrac{\sqrt{x}-\sqrt{x+h}}{h\sqrt{x}\sqrt{x+h}}; h \neq 0$ **75.** $\dfrac{1}{\sqrt{x+5}+\sqrt{x}}$

77. $\dfrac{1}{(x+y)(\sqrt{x}-\sqrt{y})}$ **79.** $\dfrac{x^2+5x+8}{(x+2)(x+1)}$ **81.** 2 **83.** $\dfrac{1}{y(y+5)}$ **85.** $\dfrac{2d}{a^2+ab+b^2}$ **87. a.** $86.67, 520, 1170;$

It costs $86,670,000 to inoculate 40% of the population against this strain of flu, $520,000,000 to inoculate 80% of the population, and $1,170,000,000

to inoculate 90% of the population. **b.** $x = 100$ **c.** increases rapidly; impossible to inoculate 100% of the population.

89. **a.** $\dfrac{-0.3t+14}{3.6t+260}$ **b.** $0.04; 4000$ **c.** fairly well **91.** $\dfrac{4x^2+14x}{(x+3)(x+4)}$ **93.** $\dfrac{R_1 R_2 R_3}{R_2 R_3 + R_1 R_3 + R_1 R_2}; \dfrac{24}{11}$ ohms **107.** e

109. $\dfrac{x-1}{x+3}$ **111.** It cubes x.

Section P.7

Check Point Exercises

1. $\{5\}$ **2.** $\{1\}$ **3.** $\{7\}$ **4.** $\varnothing$ **5.** $q = \dfrac{pf}{p-f}$ **6.** $\{-2, 3\}$ **7. a.** $\{0, 3\}$ **b.** $\left\{\dfrac{1}{2}, -1\right\}$ **8. a.** $\{-\sqrt{7}, \sqrt{7}\}$

9. $\{-5+\sqrt{11}, -5-\sqrt{11}\}$ **9.** $\{-2 \pm \sqrt{5}\}$ **10.** $\left\{\dfrac{-1+\sqrt{3}}{2}, \dfrac{-1-\sqrt{3}}{2}\right\}$ **11.** -56; no real solutions **12.** $\{6\}$

Exercise Set P.7

1. $\{11\}$ **3.** $\{7\}$ **5.** $\{13\}$ **7.** $\{2\}$ **9.** $\{9\}$ **11.** $\left\{\dfrac{33}{2}\right\}$ **13.** $\{-12\}$ **15.** $\left\{\dfrac{46}{5}\right\}$ **17. a.** 1 **b.** $\{3\}$ **19. a.** -1 **b.** $\varnothing$

21. **a.** 1 **b.** $\{2\}$ **23. a.** $-1, 1$ **b.** $\{-3\}$ **25. a.** $-2, 4$ **b.** $\varnothing$ **27.** $p = \dfrac{I}{rt}$ **29.** $p = \dfrac{T-D}{m}$ **31.** $a = \dfrac{2A}{h} - b$ **33.** $r = \dfrac{S-P}{Pt}$

35. $S = \dfrac{F}{B} + V$ **37.** $I = \dfrac{E}{R+r}$ **39.** $f = \dfrac{pq}{p+q}$ **41.** $f_1 = -\dfrac{ff_2}{f-f_2}$ or $f_1 = \dfrac{ff_2}{f_2-f}$ **43.** $\{-5, 9\}$ **45.** $\{-2, 3\}$ **47.** $\left\{-\dfrac{5}{3}, 3\right\}$

49. $\left\{-\dfrac{4}{5}, 4\right\}$ **51.** $\varnothing$ **53.** $\left\{\dfrac{1}{2}\right\}$ **55.** $\{-2, 5\}$ **57.** $\{3, 5\}$ **59.** $\{0, 4\}$ **61.** $\{\pm 3\}$ **63.** $\{\pm \sqrt{10}\}$ **65.** $\{4 \pm \sqrt{5}\}$ **67.** $\{-7, 1\}$

69. $\{1+\sqrt{3}, 1-\sqrt{3}\}$ **71.** $\{3+2\sqrt{5}, 3-2\sqrt{5}\}$ **73.** $\{-2+\sqrt{3}, -2-\sqrt{3}\}$ **75.** $\{-5, -3\}$ **77.** $\left\{\dfrac{-5+\sqrt{13}}{2}, \dfrac{-5-\sqrt{13}}{2}\right\}$

79. $\left\{\dfrac{3+\sqrt{57}}{6}, \dfrac{3-\sqrt{57}}{6}\right\}$ **81.** $\left\{\dfrac{1+\sqrt{29}}{4}, \dfrac{1-\sqrt{29}}{4}\right\}$ **83.** 36; 2 unequal real solutions **85.** 97; 2 unequal real solutions

87. 0; 1 real solution **89.** 37; 2 unequal real solutions **91.** $\left\{-\dfrac{1}{2}, 1\right\}$ **93.** $\left\{\dfrac{1}{5}, 2\right\}$ **95.** $\{-2\sqrt{5}, 2\sqrt{5}\}$ **97.** $\{1+\sqrt{2}, 1-\sqrt{2}\}$

99. $\left\{\dfrac{-11+\sqrt{33}}{4}, \dfrac{-11-\sqrt{33}}{4}\right\}$ **101.** $\left\{0, \dfrac{8}{3}\right\}$ **103.** $\{2\}$ **105.** $\{-2, 2\}$ **107.** $\{2 \pm \sqrt{2}\}$ **109.** $\left\{0, \dfrac{7}{2}\right\}$ **111.** $\{2+\sqrt{10}, 2-\sqrt{10}\}$

113. $\{-5, -1\}$ **115.** $\{6\}$ **117.** $\{6\}$ **119.** $\{-6\}$ **121.** $\{10\}$ **123.** $\{-5\}$ **125.** $\{-2\}$ **127.** $\{-3, 1\}$ **129.** $\{-8, -6, 4, 6\}$

131. $\left\{\dfrac{-1+\sqrt{21}}{2}\right\}$ **133.** $\{8\}$ **135.** $\dfrac{-2-\sqrt{22}}{2}$ and $\dfrac{-2+\sqrt{22}}{2}$ **137.** 5.5; point $(5.5, 3.5)$ on high-humor graph **139.** 125 liters

141. 33-year-olds and 58-year-olds; The formula models the actual data well. **143.** High: $66.0°$; Low: $61.5°$ **145.** Using H: 2014; Using L: 2010

161. c **163.** $x^2 - 2x - 15 = 0$ **165.** $t = \dfrac{v_0 \pm \sqrt{v_0^2 - 64s}}{32}$

Section P.8

Check Point Exercises

1. basketball: 1.6 million; bicycle riding: 1.3 million; football: 1 million   **2.** 2011   **3.** $1200   **4.** 50 ft by 94 ft   **5.** 2 ft
6. 120 yd   **7.** 5 people

Exercise Set P.8

1. births: 375 thousand; deaths: 146 thousand   **3.** 2050   **5.** 2008   **7.** 2014; 22,300 students   **9.** $420   **11.** $150   **13.** $36,000
15. $467.20   **17.** 50 yd by 100 yd   **19.** 36 ft by 78 ft   **21.** 2 in.   **23.** length: 9 ft; width: 6 ft   **25.** 5 in.   **27.** 5 m   **29.** 3 ft
31. 13.2 ft   **33.** 13 ft   **35.** 21.9 yd   **37.** 8 people   **39.** car: 50 miles per hour; bus: 30 miles per hour   **41.** 6 miles per hour
43. 11 hr   **45.** 5 ft 7 in.   **47.** 10   **55.** 3 miles, 4 miles, 5 miles   **57.** Coburn = 60 years old; woman = 20 years old
59. $4000 for the mother; $8000 for the boy; $2000 for the girl

Section P.9

Check Point Exercises

1. a. $\{x | -2 \le x < 5\}$ **b.** $\{x | 1 \le x \le 3.5\}$ **c.** $\{x | x < -1\}$ **2. a.** $(2, 3]$ **b.** $[1, 6)$

3. $[-1, \infty)$ or $\{x | x \ge -1\}$ **4.** $\{x | x < 4\}$ or $(-\infty, 4)$ **5.** $[-1, 4)$ or $\{x | -1 \le x < 4\}$

6. $(-3, 7)$ or $\{x | -3 < x < 7\}$ **7.** $\left\{x | -\frac{11}{5} \le x \le 3\right\}$ or $\left[-\frac{11}{5}, 3\right]$ **8.** $\{x | x < -4 \text{ or } x > 8\}$ or $(-\infty, -4) \cup (8, \infty)$

9. more than 720 mi per week

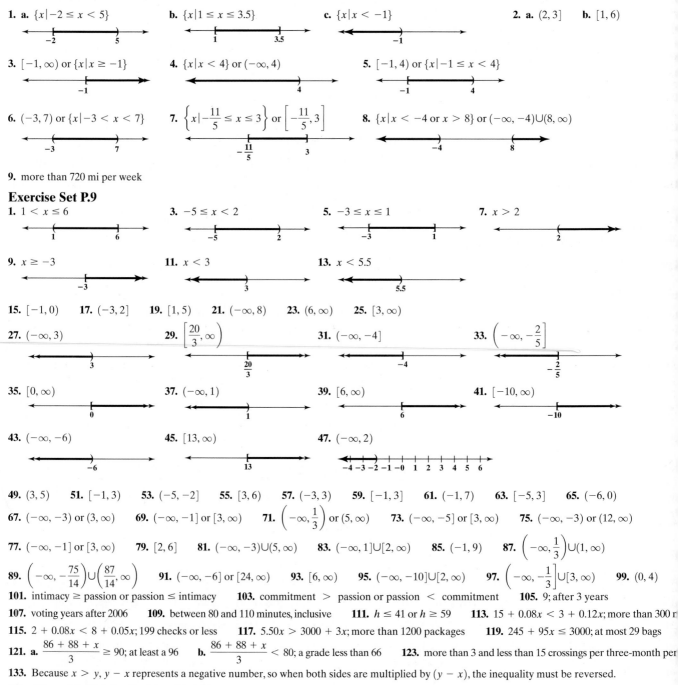

Exercise Set P.9

1. $1 < x \le 6$ **3.** $-5 \le x < 2$ **5.** $-3 \le x \le 1$ **7.** $x > 2$

9. $x \ge -3$ **11.** $x < 3$ **13.** $x < 5.5$

15. $[-1, 0)$ **17.** $(-3, 2]$ **19.** $[1, 5)$ **21.** $(-\infty, 8)$ **23.** $(6, \infty)$ **25.** $[3, \infty)$

27. $(-\infty, 3)$ **29.** $\left[\frac{20}{3}, \infty\right)$ **31.** $(-\infty, -4]$ **33.** $\left(-\infty, -\frac{2}{5}\right]$

35. $[0, \infty)$ **37.** $(-\infty, 1)$ **39.** $[6, \infty)$ **41.** $[-10, \infty)$

43. $(-\infty, -6)$ **45.** $[13, \infty)$ **47.** $(-\infty, 2)$

49. $(3, 5)$ **51.** $[-1, 3)$ **53.** $(-5, -2]$ **55.** $[3, 6)$ **57.** $(-3, 3)$ **59.** $[-1, 3]$ **61.** $(-1, 7)$ **63.** $[-5, 3]$ **65.** $(-6, 0)$
67. $(-\infty, -3)$ or $(3, \infty)$ **69.** $(-\infty, -1]$ or $[3, \infty)$ **71.** $\left(-\infty, \frac{1}{3}\right)$ or $(5, \infty)$ **73.** $(-\infty, -5]$ or $[3, \infty)$ **75.** $(-\infty, -3)$ or $(12, \infty)$
77. $(-\infty, -1]$ or $[3, \infty)$ **79.** $[2, 6]$ **81.** $(-\infty, -3) \cup (5, \infty)$ **83.** $(-\infty, 1] \cup [2, \infty)$ **85.** $(-1, 9)$ **87.** $\left(-\infty, \frac{1}{3}\right) \cup (1, \infty)$
89. $\left(-\infty, -\frac{75}{14}\right) \cup \left(\frac{87}{14}, \infty\right)$ **91.** $(-\infty, -6]$ or $[24, \infty)$ **93.** $[6, \infty)$ **95.** $(-\infty, -10] \cup [2, \infty)$ **97.** $\left(-\infty, -\frac{1}{3}\right] \cup [3, \infty)$ **99.** $(0, 4)$
101. intimacy $\ge$ passion or passion $\le$ intimacy   **103.** commitment $>$ passion or passion $<$ commitment   **105.** 9; after 3 years
107. voting years after 2006   **109.** between 80 and 110 minutes, inclusive   **111.** $h \le 41$ or $h \ge 59$ **113.** $15 + 0.08x < 3 + 0.12x$; more than 300 r
115. $2 + 0.08x < 8 + 0.05x$; 199 checks or less   **117.** $5.50x > 3000 + 3x$; more than 1200 packages   **119.** $245 + 95x \le 3000$; at most 29 bags
121. a. $\frac{86 + 88 + x}{3} \ge 90$; at least a 96   **b.** $\frac{86 + 88 + x}{3} < 80$; a grade less than 66   **123.** more than 3 and less than 15 crossings per three-month per
133. Because $x > y$, $y - x$ represents a negative number, so when both sides are multiplied by $(y - x)$, the inequality must be reversed.
135. Albany: Model 2; San Francisco: Model 1

Chapter P Review Exercises

1. 51   **2.** 16   **3.** 124 ft   **4.** $\{a, c\}$ **5.** $\{a, b, c, d, e\}$ **6.** $\{a, b, c, d, f, g\}$ **7.** $\{a\}$ **8. a.** $\sqrt{81}$ **b.** $0, \sqrt{81}$
c. $-17, 0, \sqrt{81}$ **d.** $-17, -\frac{9}{13}, 0, 0.75, \sqrt{81}$ **e.** $\sqrt{2}, \pi$ **f.** $-17, -\frac{9}{13}, 0, 0.75, \sqrt{2}, \pi, \sqrt{81}$ **9.** 103   **10.** $\sqrt{2} - 1$
11. $\sqrt{17} - 3$ **12.** $|4 - (-17)|; 21$ **13.** commutative property of addition   **14.** associative property of multiplication
15. distributive property of multiplication over addition   **16.** commutative property of multiplication   **17.** commutative property
of multiplication   **18.** commutative property of addition   **19.** $17x - 15$ **20.** $2x$ **21.** $5y - 17$ **22.** $10x$
23. $E = 0.04x^2 + 9.2x + 169$ **24.** -108 **25.** $\frac{5}{16}$ **26.** $\frac{1}{25}$ **27.** $\frac{1}{27}$ **28.** $-8x^{12}y^9$ **29.** $\frac{10}{x^8}$ **30.** $\frac{1}{16x^{12}}$ **31.** $\frac{y^8}{4x^{10}}$ **32.** 37,400
33. 0.0000745   **34.** 3.59×10^6 **35.** 7.25×10^{-3} **36.** 390,000   **37.** 0.023   **38.** 10^3 or 1000 yr   **39.** $\$4.35 \times 10^{10}$ **40.** $10\sqrt{3}$
41. $2|x|\sqrt{3}$ **42.** $2x\sqrt{5}$ **43.** $r\sqrt{r}$ **44.** $\frac{11}{2}$ **45.** $4x\sqrt{3}$ **46.** $20\sqrt{5}$ **47.** $16\sqrt{2}$ **48.** $24\sqrt{2} - 8\sqrt{3}$ **49.** $6\sqrt{5}$ **50.** $\frac{\sqrt{6}}{3}$
51. $\frac{5(6 - \sqrt{3})}{33}$ **52.** $7(\sqrt{7} + \sqrt{5})$ **53.** 5   **54.** -2 **55.** not a real number   **56.** 5   **57.** $3\sqrt[3]{3}$ **58.** $y\sqrt[3]{y^2}$ **59.** $2\sqrt[5]{5}$
60. $13\sqrt[3]{2}$ **61.** $x\sqrt[4]{2}$ **62.** 4   **63.** $\frac{1}{5}$ **64.** 5   **65.** $\frac{1}{3}$ **66.** 16   **67.** $\frac{1}{81}$ **68.** $20x^{11/12}$ **69.** $3x^{1/4}$ **70.** $25x^4$ **71.** $\sqrt{y}$
72. $8x^3 + 10x^2 - 20x - 4$; degree 3   **73.** $8x^4 - 5x^3 + 6$; degree 4   **74.** $12x^3 + x^2 - 21x + 10$ **75.** $6x^2 - 7x - 5$ **76.** $16x^2 - 25$
77. $4x^2 + 20x + 25$ **78.** $9x^2 - 24x + 16$ **79.** $8x^3 + 12x^2 + 6x + 1$ **80.** $125x^3 - 150x^2 + 60x - 8$ **81.** $3x^2 + 16xy - 35y^2$
82. $9x^2 - 30xy + 25y^2$ **83.** $9x^4 + 12x^2y + 4y^2$ **84.** $49x^2 - 16y^2$ **85.** $a^3 - b^3$ **86.** $25y^2 - 4x^2 - 8x - 1$
87. $x^2 + 4xy + 4y^2 + 8x + 16y + 16$ **88.** $3x^2(5x + 1)$ **89.** $(x - 4)(x - 7)$ **90.** $(3x + 1)(5x - 2)$ **91.** $(8 - x)(8 + x)$
92. prime   **93.** $3x^2(x - 5)(x + 2)$ **94.** $4x^3(5x^4 - 9)$ **95.** $(x + 3)(x - 3)^2$ **96.** $(4x - 5)^2$ **97.** $(x^2 + 4)(x + 2)(x - 2)$
98. $(y - 2)(y^2 + 2y + 4)$ **99.** $(x + 4)(x^2 - 4x + 16)$ **100.** $3x^2(x - 2)(x + 2)$ **101.** $(3x - 5)(9x^2 + 15x + 25)$
102. $x(x - 1)(x + 1)(x^2 + 1)$ **103.** $(x^2 - 2)(x + 5)$ **104.** $(x + 9 + y)(x + 9 - y)$ **105.** $\frac{16(1 + 2x)}{x^{3/4}}$
106. $(x + 2)(x - 2)(x^2 + 3)^{1/2}(-x^4 + x^2 + 13)$ **107.** $\frac{6(2x + 1)}{x^{3/2}}$ **108.** $x^2, x \neq -2$ **109.** $\frac{x - 3}{x - 6}, x \neq -6, 6$ **110.** $\frac{x}{x + 2}, x \neq -2$
111. $\frac{(x + 3)^3}{(x - 2)^2(x + 2)}, x \neq 2, -2$ **112.** $\frac{2}{x(x + 1)}, x \neq 0, 1, -1, -\frac{1}{3}$ **113.** $\frac{x + 3}{x - 4}, x \neq -3, 4, 2, 8$ **114.** $\frac{1}{x - 3}, x \neq 3, -3$
115. $\frac{4x(x - 1)}{(x + 2)(x - 2)}, x \neq 2, -2$ **116.** $\frac{2x^2 - 3}{(x - 3)(x + 3)(x - 2)}, x \neq 3, -3, 2$ **117.** $\frac{11x^2 - x - 11}{(2x - 1)(x + 3)(3x + 2)}, x \neq \frac{1}{2}, -3, -\frac{2}{3}$ **118.** $\frac{3}{x}, x \neq 0, 2$
119. $\frac{3x}{x - 4}, x \neq 0, 4, -4$ **120.** $\frac{3x + 8}{3x + 10}, x \neq -3, -\frac{10}{3}$ **121.** $\frac{25\sqrt{25 - x^2}}{(5 - x)^2(5 + x)^2}$ **122.** $\{-13\}$; conditional equation
123. $\{-3\}$; conditional equation   **124.** $\{-1\}$; conditional equation   **125.** all real numbers except 1 and -1
126. $\{7\}$ **127.** $\{-2, 1\}$ **128.** $\left\{\frac{1}{2}, 5\right\}$ **129.** $\left\{-2, \frac{10}{3}\right\}$ **130.** $\left\{\frac{7 + \sqrt{37}}{6}, \frac{7 - \sqrt{37}}{6}\right\}$ **131.** $\{-3, 3\}$ **132.** $\{3 \pm 2\sqrt{6}\}$
133. $\{4\}$ **134.** $\{2\}$ **135.** $\{2\}$ **136.** $g = \frac{s - vt}{t^2}$ **137.** $P = \frac{A}{1 + rt}$ **138.** no real solutions   **139.** one repeated real solution
140. Chicken Caesar: 495; Express Taco: 620; Mandarin Chicken: 590   **141.** 2008   **142.** $60   **143.** $10,000 in sales
144. 44 yd by 126 yd   **145.** 2019; 32,100   **146.** length = 5 yd; width = 3 yd   **147.** approximately 134 m   **148.** 2 in.   **149.** 10 people
150. $\{x | -3 \leq x < 5\}$ **151.** $\{x | x > -2\}$ **152.** $\{x | x \leq 0\}$ **153.** $[-1, 1]$ **154.** $(-2, 3)$
155. $[1, 3)$ **156.** $(0, 4)$

157. $[-2, \infty)$ **158.** $\left[\frac{3}{5}, \infty\right)$ **159.** $\left(-\infty, -\frac{21}{2}\right)$ **160.** $(-3, \infty)$

161. $(-\infty, -2]$ **162.** $(2, 3]$ **163.** $[-9, 6]$ **164.** $(-\infty, -6)$ or $(0, \infty)$

165. $(-\infty, -3]$ or $[-2, \infty)$ **166.** $(-\infty, -5]\cup[1, \infty)$;

167. no more than 80 miles per day   **168.** $[49\%, 99\%)$

Chapter P Test

1. $6x^2 - 27x$ **2.** $-6x + 17$ **3.** $\{5\}$ **4.** $\{1, 2, 5, a\}$ **5.** $\frac{5y^8}{x^6}$ **6.** $3r\sqrt{2}$ **7.** $11\sqrt{2}$ **8.** $\frac{3(5 - \sqrt{2})}{23}$
9. $2x\sqrt[3]{2x}$ **10.** $\frac{x + 3}{x - 2}, x \neq 2, 1$ **11.** 2.5×10^1 **12.** $2x^3 - 13x^2 + 26x - 15$ **13.** $25x^2 + 30xy + 9y^2$ **14.** $\frac{2(x + 3)}{x + 1}, x \neq 3, -1, -4, -3$
15. $\frac{x^2 + 2x + 15}{(x + 3)(x - 3)}, x \neq 3, -3$ **16.** $\frac{11}{(x - 3)(x - 4)}, x \neq 3, 4$ **17.** $\frac{2x}{(x + 2)(x + 1)}$ **18.** $\frac{10x}{\sqrt{(x^2 + 5)^3}}$ **19.** $(x - 3)(x - 6)$

20. $(x^2 + 3)(x + 2)$ **21.** $(5x - 3)(5x + 3)$ **22.** $(6x - 7)^2$ **23.** $(y - 5)(y^2 + 5y + 25)$ **24.** $(x + 5 + 3y)(x + 5 - 3y)$ **25.** $\dfrac{2x + 3}{(x + 3)^{3/5}}$

26. $-7, -\dfrac{4}{5}, 0, 0.25, \sqrt{4}, \dfrac{22}{7}$ **27.** commutative property of addition   **28.** distributive property of multiplication over addition   **29.** 7.6×10^{-4}

30. $\dfrac{1}{243}$ **31.** 1.26×10^{10} **32. a.** men: Model 2; women: Model 1   **b.** 74; fairly well   **c.** 2050   **33.** $\{-1\}$ **34.** $\{-6\}$ **35.** $\{5\}$

36. $\left\{-\dfrac{1}{2}, 2\right\}$ **37.** $\left\{\dfrac{1 - 5\sqrt{3}}{3}, \dfrac{1 + 5\sqrt{3}}{3}\right\}$ **38.** $\{1 - \sqrt{5}, 1 + \sqrt{5}\}$ **39.** $\{7\}$ **40.** $\{2\}$ **41.** $\{6, 12\}$ **42.** $\left\{\dfrac{1}{2}, 3\right\}$ **43.** $\{4\}$

44. $(-\infty, 12]$ **45.** $\left[\dfrac{21}{8}, \infty\right)$ **46.** $\left[-7, \dfrac{13}{2}\right)$ **47.** $\left(-\infty, -\dfrac{5}{3}\right]$ or $\left[\dfrac{1}{3}, \infty\right)$

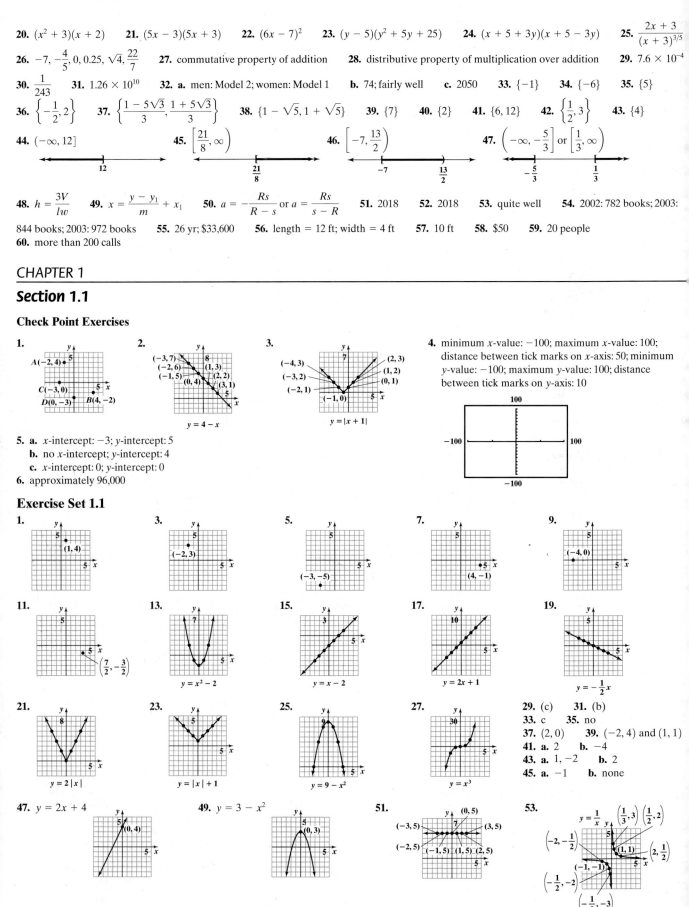

48. $h = \dfrac{3V}{lw}$ **49.** $x = \dfrac{y - y_1}{m} + x_1$ **50.** $a = -\dfrac{Rs}{R - s}$ or $a = \dfrac{Rs}{s - R}$ **51.** 2018   **52.** 2018   **53.** quite well   **54.** 2002: 782 books; 2003:

844 books; 2003: 972 books   **55.** 26 yr; $33,600   **56.** length = 12 ft; width = 4 ft   **57.** 10 ft   **58.** $50   **59.** 20 people
60. more than 200 calls

CHAPTER 1

Section 1.1

Check Point Exercises

1.   **2.**   **3.**   **4.** minimum x-value: -100; maximum x-value: 100; distance between tick marks on x-axis: 50; minimum y-value: -100; maximum y-value: 100; distance between tick marks on y-axis: 10

5. a. x-intercept: -3; y-intercept: 5
   b. no x-intercept; y-intercept: 4
   c. x-intercept: 0; y-intercept: 0
6. approximately 96,000

Exercise Set 1.1

1.   **3.**   **5.**   **7.**   **9.**

11.   **13.**   **15.**   **17.**   **19.**

21.   **23.**   **25.**   **27.**   **29.** (c)   **31.** (b)
33. c   **35.** no
37. $(2, 0)$ **39.** $(-2, 4)$ and $(1, 1)$
41. a. 2   **b.** -4
43. a. $1, -2$ **b.** 2
45. a. -1 **b.** none

47. $y = 2x + 4$ **49.** $y = 3 - x^2$ **51.**   **53.**

5. 65   **57.** 1989–1993   **59.** 1977   **61.** 135 beats/min; $(40, 135)$ on blue graph   **63. a.** 36 cm   **b.** 44.7 cm   **c.** 46.9 cm
, describes healthy children   **71.**

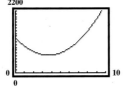

73. a   **75.** b   **77.** b

ection 1.2

heck Point Exercises

1. domain: $\{5, 10, 15, 20, 25\}$; range: $\{12.8, 16.2, 18.9, 20.7, 21.8\}$ **2. a.** not a function   **b.** function   **3. a.** $y = 6 - 2x$; function
b. $y = \pm\sqrt{1 - x^2}$; not a function   **4. a.** 42   **b.** $x^2 + 6x + 15$ **c.** $x^2 + 2x + 7$

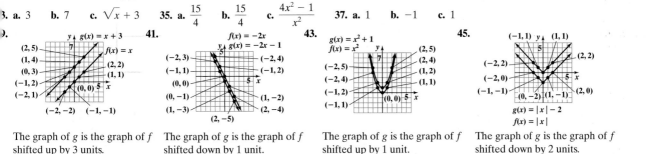

5. ; The graph of g is the graph of f shifted down by 3 units.
6. a. function   **b.** function   **c.** not a function   **7. a.** $f(10) \approx 16$ **b.** $x \approx 8$
8. a. Domain = $\{x|-2 \le x \le 1\}$; Range = $\{y|0 \le y \le 3\}$
b. Domain = $\{x|-2 < x \le 1\}$; Range = $\{y|-1 \le y < 2\}$
c. Domain = $\{x|-3 \le x < 0\}$; Range = $\{-3, -2, -1\}$

xercise Set 1.2

1. function; $\{1, 3, 5\}$; $\{2, 4, 5\}$ **3.** not a function; $\{3, 4\}$; $\{4, 5\}$ **5.** function; $\{3, 4, 5, 7\}$; $\{-2, 1, 9\}$ **7.** function; $\{-3, -2, -1, 0\}$; $\{-3, -2, -1, 0\}$
9. not a function; $\{1\}$; $\{4, 5, 6\}$ **11.** y is a function of x.   **13.** y is a function of x.   **15.** y is not a function of x.   **17.** y is not a function of x.
19. y is a function of x.   **21.** y is a function of x.   **23.** y is a function of x.   **25.** y is a function of x.   **27. a.** 29   **b.** $4x + 9$ **c.** $-4x + 5$
29. a. 2   **b.** $x^2 + 12x + 38$ **c.** $x^2 - 2x + 3$ **31. a.** 13   **b.** 1   **c.** $x^4 - x^2 + 1$ **d.** $81a^4 - 9a^2 + 1$
33. a. 3   **b.** 7   **c.** $\sqrt{x} + 3$ **35. a.** $\dfrac{15}{4}$ **b.** $\dfrac{15}{4}$ **c.** $\dfrac{4x^2 - 1}{x^2}$ **37. a.** 1   **b.** -1 **c.** 1
39. **41.** **43.** **45.**

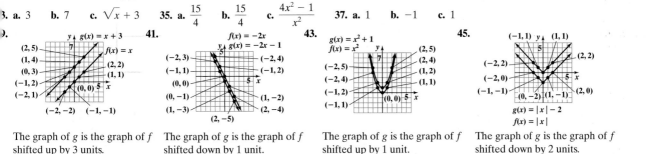

The graph of g is the graph of f The graph of g is the graph of f The graph of g is the graph of f The graph of g is the graph of f
shifted up by 3 units.   shifted down by 1 unit.   shifted up by 1 unit.   shifted down by 2 units.

47. **49.** **51.** **53.**

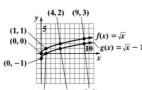
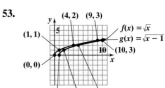

The graph of g is the graph of f The graph of g is the graph of f The graph of g is the graph of f The graph of g is the graph of f
shifted up by 2 units.   shifted up by 2 units.   shifted down by 1 unit.   shifted to the right by 1 unit.

55. function   **57.** function   **59.** not a function   **61.** function   **63.** function   **65.** -4 **67.** 4   **69.** 0   **71.** 2   **73.** 2   **75.** -2
77. a. $(-\infty, \infty)$ **b.** $[-4, \infty)$ **c.** -3 and 1   **d.** -3 **e.** $f(-2) = -3$ and $f(2) = 5$ **79. a.** $(-\infty, \infty)$ **b.** $[1, \infty)$ **c.** none
d. 1   **e.** $f(-1) = 2$ and $f(3) = 4$ **81. a.** $[0, 5)$ **b.** $[-1, 5)$ **c.** 2   **d.** -1 **e.** $f(3) = 1$ **83. a.** $[0, \infty)$ **b.** $[1, \infty)$
c. none   **d.** 1   **e.** $f(4) = 3$ **85. a.** $[-2, 6]$ **b.** $[-2, 6]$ **c.** 4   **d.** 4   **e.** $f(-1) = 5$ **87. a.** $(-\infty, \infty)$ **b.** $(-\infty, -2]$
c. none   **d.** -2 **e.** $f(-4) = -5$ and $f(4) = -2$ **89. a.** $(-\infty, \infty)$ **b.** $(0, \infty)$ **c.** none   **d.** 1.5   **e.** $f(4) = 6$
91. a. $\{-5, -2, 0, 1, 3\}$ **b.** $\{2\}$ **c.** none   **d.** 2   **e.** $f(-5) + f(3) = 4$ **93.** $-2; 10$ **95.** -38 **97.** $-2x^3 - 2x$
99. a. $\{(\text{U.S.}, 80\%), (\text{Japan}, 64\%), (\text{France}, 64\%), (\text{Germany}, 61\%), (\text{England}, 59\%), (\text{China}, 47\%)\}$ **b.** Yes; Each country corresponds to a
unique percent.   **c.** $\{(80\%, \text{U.S.}), (64\%, \text{Japan}), (64\%, \text{France}), (61\%, \text{Germany}), (59\% \text{England}), (47\%, \text{China})\}$ **d.** No; 64% in the domain
corresponds to two members of the range, Japan and France.   **101.** 5.22; In 2000, there were 5.22 million women enrolled in U.S.
colleges; $(2000, 5.22)$.   **103.** 1.4; In 2004, there were 1.4 million more women than men enrolled in U.S. colleges.   **105. a.** 73%   **b.** 72.8%
73.3%   **107.** $C = 100,000 + 100x$, where x is the number of bicycles produced; $C(90) = 109,000$; It cost \$109,000 to produce 90 bicycles.
109. $T = \dfrac{40}{x} + \dfrac{40}{x + 30}$, where x is the rate on the outgoing trip; $T(30) = 2$; It takes 2 hours, traveling 30 mph outgoing and 60 mph returning.
121. c   **123.** Answers will vary; an example is $\{(1, 1), (2, 1)\}$.

Section 1.3

Check Point Exercises

1. a. $-2x^2 - 4xh - 2h^2 + x + h + 5$ **b.** $-4x - 2h + 1$ **2. a.** 20; With 40 calling minutes, the cost is \$20; (40, 20).   **b.** 28; With 80 calling minutes, the cost is \$28; (80, 28).   **3.** increasing on $(-\infty, -1)$, decreasing on $(-1, 1)$, increasing on $(1, \infty)$ **4. a.** even   **b.** odd   **c.** neither

Exercise Set 1.3

1. $4, h \neq 0$ **3.** $3, h \neq 0$ **5.** $2x + h, h \neq 0$ **7.** $2x + h - 4, h \neq 0$ **9.** $4x + 2h + 1, h \neq 0$ **11.** $-2x - h + 2, h \neq 0$
13. $-4x - 2h + 5, h \neq 0$ **15.** $-4x - 2h - 1, h \neq 0$ **17.** $0, h \neq 0$ **19.** $-\dfrac{1}{x(x+h)}, h \neq 0$ **21.** $\dfrac{1}{\sqrt{x+h}+\sqrt{x}}, h \neq 0$
23. a. -1 **b.** 7   **c.** 19   **25. a.** 3   **b.** 3   **c.** 0   **27. a.** 8   **b.** 3   **c.** 6   **29. a.** increasing: $(-1, \infty)$ **b.** decreasing: $(-\infty, -1)$
c. constant: none   **31. a.** increasing: $(0, \infty)$ **b.** decreasing: none   **c.** constant: none   **33. a.** increasing: none   **b.** decreasing: $(-2, 6)$
c. constant: none   **35. a.** increasing: $(-\infty, -1)$ **b.** decreasing: none   **c.** constant: $(-1, \infty)$ **37. a.** increasing: $(-\infty, 0)$ or $(1.5, 3)$
b. decreasing: $(0, 1.5)$ or $(3, \infty)$ **c.** constant: none   **39. a.** increasing: $(-2, 4)$ **b.** decreasing: none   **c.** constant: $(-\infty, -2)$ or $(4, \infty)$
41. a. $0; f(0) = 4$ **b.** $-3, 3; f(-3) = f(3) = 0$ **43. a.** $-2; f(-2) = 21$ **b.** $1; f(1) = -6$ **45.** odd   **47.** neither   **49.** even
51. even   **53.** even   **55.** odd   **57.** even   **59.** odd   **61. a.** $(-\infty, \infty)$ **b.** $[-4, \infty)$ **c.** 1 and 7   **d.** 4   **e.** $(4, \infty)$ **f.** $(0, 4)$
g. $(-\infty, 0)$ **h.** 4   **i.** -4 **j.** 4   **k.** 2 and 6   **l.** neither   **63. a.** $(-\infty, 3]$ **b.** $(-\infty, 4]$ **c.** -3 and 3   **d.** 3   **e.** $(-\infty, 1)$
f. $(1, 3)$ **g.** $(-\infty, -3]$ **h.** A relative maximum of 4 occurs at 1.   **i.** 1   **j.** positive   **65.** $f(1.06) = 1$ **67.** $f\left(\dfrac{1}{3}\right) = 0$

69. $f(-2.3) = -3$ **71.** -18

73. $0.30t - 6$ **75.** $C(t) = \begin{cases} 50 & \text{if } 0 \le t \le 400 \\ 50 + 0.30(t - 400) & \text{if } t > 400 \end{cases}$

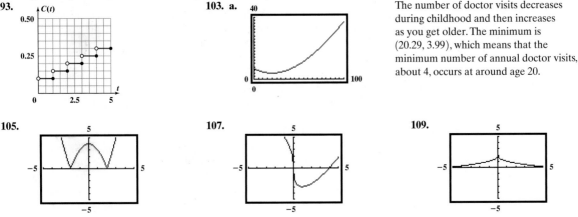

77. $f(60) \approx 3.1$; In 1960, Jewish Americans made up about 3.1% of the U.S. population.   **79.** $x \approx 19$ and $x \approx 64$; In 1919 and 1964, Jewish Americans made up about 3% of the U.S. population.   **81.** 1940; 3.7%   **83.** Each year corresponds to only one percentage.
85. increasing: $(45, 74)$; decreasing: $(16, 45)$; The number of accidents occurring per 50,000 miles driven increases with age starting at age 45, while it decreases with age starting at age 16.   **87.** Answers will vary; an example is 16 and 74 years old. For those ages, the number of accidents is 526.4 per 50 million miles.   **89.** 1989 cigarettes per adult; quite well   **91.** 1960; 4100 cigarettes per adult; The function estimates 3797 cigarettes per adult, which models the graph reasonably well.

93.

103. a.

The number of doctor visits decreases during childhood and then increases as you get older. The minimum is $(20.29, 3.99)$, which means that the minimum number of annual doctor visits, about 4, occurs at around age 20.

105.   **107.**   **109.**

Increasing: $(-2, 0)$ or $(2, \infty)$ Increasing: $(1, \infty)$ Increasing: $(-\infty, 0)$
Decreasing: $(-\infty, -2)$ or $(0, 2)$ Decreasing: $(-\infty, 1)$ Decreasing: $(0, \infty)$
113. a. h is even if both f and g are even or if both f and g are odd.
b. h is odd if f is odd and g is even or if f is even and g is odd.

Section 1.4

Check Point Exercises

1. a. 6   **b.** $-\dfrac{7}{5}$ **2.** $y + 5 = 6(x - 2); y = 6x - 17$ **3.** $y + 1 = -5(x + 2); y = -5x - 11$

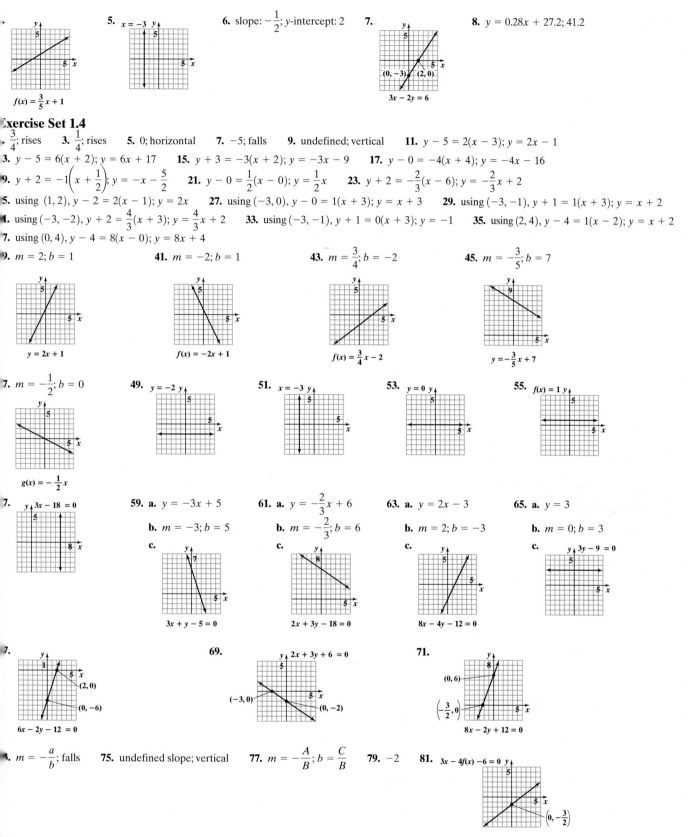

5. $x = -3$ **6.** slope: $-\dfrac{1}{2}$; y-intercept: 2   **7.**   **8.** $y = 0.28x + 27.2$; 41.2

$f(x) = \dfrac{3}{5}x + 1$ $3x - 2y = 6$ $(0, -3)$ $(2, 0)$

Exercise Set 1.4

1. $\dfrac{3}{4}$; rises   **3.** $\dfrac{1}{4}$; rises   **5.** 0; horizontal   **7.** -5; falls   **9.** undefined; vertical   **11.** $y - 5 = 2(x - 3)$; $y = 2x - 1$

13. $y - 5 = 6(x + 2)$; $y = 6x + 17$ **15.** $y + 3 = -3(x + 2)$; $y = -3x - 9$ **17.** $y - 0 = -4(x + 4)$; $y = -4x - 16$

19. $y + 2 = -1\left(x + \dfrac{1}{2}\right)$; $y = -x - \dfrac{5}{2}$ **21.** $y - 0 = \dfrac{1}{2}(x - 0)$; $y = \dfrac{1}{2}x$ **23.** $y + 2 = -\dfrac{2}{3}(x - 6)$; $y = -\dfrac{2}{3}x + 2$

25. using $(1, 2)$, $y - 2 = 2(x - 1)$; $y = 2x$ **27.** using $(-3, 0)$, $y - 0 = 1(x + 3)$; $y = x + 3$ **29.** using $(-3, -1)$, $y + 1 = 1(x + 3)$; $y = x + 2$

31. using $(-3, -2)$, $y + 2 = \dfrac{4}{3}(x + 3)$; $y = \dfrac{4}{3}x + 2$ **33.** using $(-3, -1)$, $y + 1 = 0(x + 3)$; $y = -1$ **35.** using $(2, 4)$, $y - 4 = 1(x - 2)$; $y = x + 2$

37. using $(0, 4)$, $y - 4 = 8(x - 0)$; $y = 8x + 4$

39. $m = 2$; $b = 1$ **41.** $m = -2$; $b = 1$ **43.** $m = \dfrac{3}{4}$; $b = -2$ **45.** $m = -\dfrac{3}{5}$; $b = 7$

$y = 2x + 1$ $f(x) = -2x + 1$ $f(x) = \dfrac{3}{4}x - 2$ $y = -\dfrac{3}{5}x + 7$

47. $m = -\dfrac{1}{2}$; $b = 0$ **49.** $y = -2$ **51.** $x = -3$ **53.** $y = 0$ **55.** $f(x) = 1$

$g(x) = -\dfrac{1}{2}x$

57. $3x - 18 = 0$ **59. a.** $y = -3x + 5$ **61. a.** $y = -\dfrac{2}{3}x + 6$ **63. a.** $y = 2x - 3$ **65. a.** $y = 3$

b. $m = -3$; $b = 5$ **b.** $m = -\dfrac{2}{3}$; $b = 6$ **b.** $m = 2$; $b = -3$ **b.** $m = 0$; $b = 3$

c.   **c.**   **c.**   **c.** $3y - 9 = 0$

$3x + y - 5 = 0$ $2x + 3y - 18 = 0$ $8x - 4y - 12 = 0$

67. $6x - 2y - 12 = 0$ $(2, 0)$ $(0, -6)$ **69.** $2x + 3y + 6 = 0$ $(-3, 0)$ $(0, -2)$ **71.** $8x - 2y + 12 = 0$ $(0, 6)$ $\left(-\dfrac{3}{2}, 0\right)$

73. $m = -\dfrac{a}{b}$; falls   **75.** undefined slope; vertical   **77.** $m = -\dfrac{A}{B}$; $b = \dfrac{C}{B}$ **79.** -2 **81.** $3x - 4f(x) - 6 = 0$ $\left(0, -\dfrac{3}{2}\right)$

83. 5   **85.** m_1, m_3, m_2, m_4 **87. a.** $y - 16 = -0.55(x - 10)$ or $y - 12.7 = -0.55(x - 16)$ **b.** $f(x) = -0.55x + 21.5$ **c.** 10.5%

89. a.

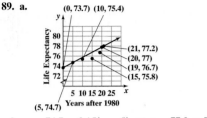

b. $y - 74.7 = 0.15(x - 5)$ or $y - 77.0 = 0.15(x - 20)$;
$y = 0.15x + 73.95$ or $y = 0.15x + 74$
c. $E(x) = 0.15x + 73.95$ or $E(x) = 0.15x + 74$; 80 years
91. $y = -2.3x + 255$, where x is the percentage of adult females who are literate and y is under-five mortality per thousand; For each percent increase in adult female literacy, under-five mortality decreases by 2.3 per thousand.
101. $m = -3$ **103.** $m = \dfrac{3}{4}$

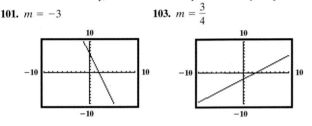

105. c   **107.** coefficient of x: 1; coefficient of y: -2 **109.** $E = 2.4M - 20$

Section 1.5

Check Point Exercises

1. $y - 5 = 3(x + 2)$; $y = 3x + 11$ or $f(x) = 3x + 11$ **2. a.** 3   **b.** $3x - y = 0$ **3.** $\dfrac{2}{15} \approx 0.13$; The number of U.S. men living alone is projected to increase by 0.13 million each year.   **4. a.** 1   **b.** 7   **c.** 4   **5.** 0.01 mg per 100 ml per hr
6. a. 12 ft/sec   **b.** 10 ft/sec   **c.** 8.04 ft/sec

Exercise Set 1.5

1. $y - 2 = 2(x - 4)$; $y = 2x - 6$ or $f(x) = 2x - 6$ **3.** $y - 4 = -\dfrac{1}{2}(x - 2)$; $y = -\dfrac{1}{2}x + 5$ or $f(x) = -\dfrac{1}{2}x + 5$

5. $y + 10 = -4(x + 8)$; $y = -4x - 42$ **7.** $y + 3 = -5(x - 2)$; $y = -5x + 7$ **9.** $y - 2 = \dfrac{2}{3}(x + 2)$; $2x - 3y + 10 = 0$

11. $y + 7 = -2(x - 4)$; $2x + y - 1 = 0$ **13.** 3   **15.** 10   **17.** $\dfrac{1}{5}$ **19. a.** 70 ft/sec   **b.** 65 ft/sec   **c.** 60.1 ft/sec   **d.** 60.01 ft/sec

21. $f(x) = 5$ **23.** $f(x) = -\dfrac{1}{2}x + 1$ **25.** $f(x) = -\dfrac{2}{3}x - 2$ **27.** $m = 0.01$; The temperature of Earth is increasing by 0.01°F per year.

29. $m = -0.52$; The percentage of U.S. adults who smoke cigarettes is decreasing by 0.52% each year.   **31.** $f(x) = 13x + 222$

33. $f(x) = -2.40x + 52.40$ **35.** $m = -1.22$; average decrease of 1.22% of total sales per year

43. a. The product of their slopes is -1.   **45.**   **47.** $-\dfrac{3}{7}$

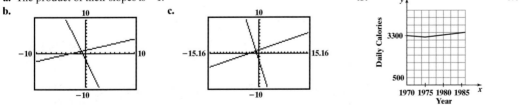

Mid-Chapter 1 Check Point

1. not a function; Domain: $\{1, 2\}$; Range: $\{-6, 4, 6\}$ **2.** function; Domain: $\{0, 2, 3\}$; Range: $\{1, 4\}$ **3.** function; Domain: $\{x|-2 \le x < 2\}$;
Range: $\{y|0 \le y \le 3\}$ **4.** not a function; Domain: $\{x|-3 < x \le 4\}$; Range: $\{y|-1 \le y \le 2\}$ **5.** not a function; Domain: $\{-2, -1, 0, 1, 2\}$;
Range: $\{-2, -1, 1, 3\}$ **6.** function; Domain: $\{x|x \le 1\}$; Range: $\{y|y \ge -1\}$ **7.** y is a function of x **8.** y is not a function of x
9. No vertical line intersects the graph in more than one point.   **10.** $(-\infty, \infty)$ **11.** $(-\infty, 4]$ **12.** -6 and 2   **13.** 3   **14.** $(-\infty, -2)$
15. $(-2, \infty)$ **16.** -2 **17.** 4   **18.** 3   **19.** -7 and 3   **20.** -6 and 2   **21.** $(-6, 2)$ **22.** negative   **23.** neither   **24.** -1

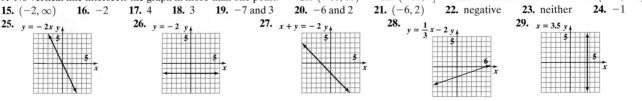

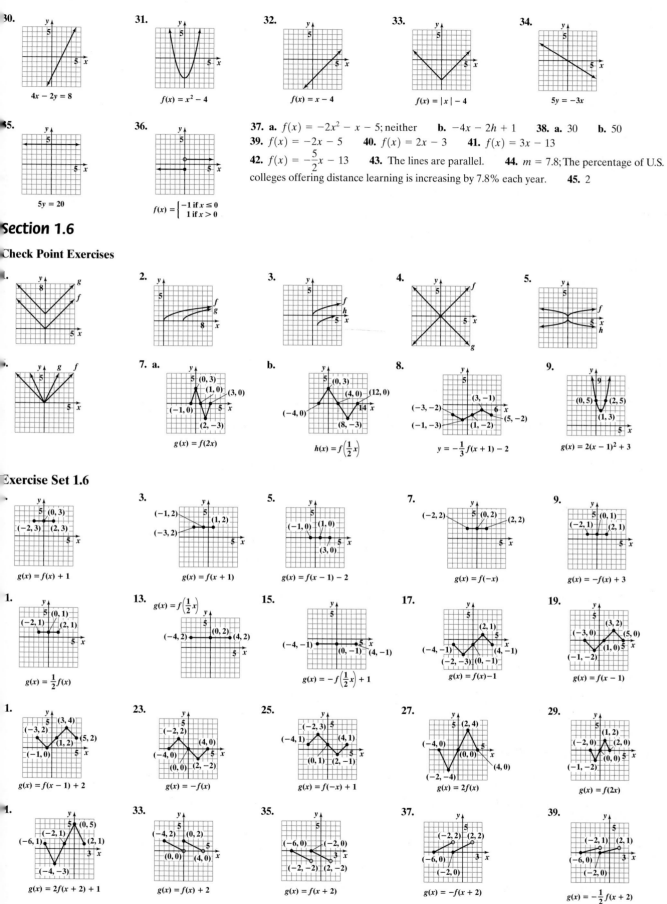

30. $4x - 2y = 8$

31. $f(x) = x^2 - 4$

32. $f(x) = x - 4$

33. $f(x) = |x| - 4$

34. $5y = -3x$

35. $5y = 20$

36. $f(x) = \begin{cases} -1 \text{ if } x \le 0 \\ 1 \text{ if } x > 0 \end{cases}$

37. a. $f(x) = -2x^2 - x - 5$; neither   **b.** $-4x - 2h + 1$ **38. a.** 30   **b.** 50
39. $f(x) = -2x - 5$ **40.** $f(x) = 2x - 3$ **41.** $f(x) = 3x - 13$
42. $f(x) = -\dfrac{5}{2}x - 13$ **43.** The lines are parallel.   **44.** $m = 7.8$; The percentage of U.S. colleges offering distance learning is increasing by 7.8% each year.   **45.** 2

Section 1.6

Check Point Exercises

1.

2.

3.

4.

5.

6.

7. a. $g(x) = f(2x)$

b. $h(x) = f\left(\dfrac{1}{2}x\right)$

8. $y = -\dfrac{1}{3}f(x + 1) - 2$

9. $g(x) = 2(x - 1)^2 + 3$

Exercise Set 1.6

1. $g(x) = f(x) + 1$

3. $g(x) = f(x + 1)$

5. $g(x) = f(x - 1) - 2$

7. $g(x) = f(-x)$

9. $g(x) = -f(x) + 3$

11. $g(x) = \dfrac{1}{2}f(x)$

13. $g(x) = f\left(\dfrac{1}{2}x\right)$

15. $g(x) = -f\left(\dfrac{1}{2}x\right) + 1$

17. $g(x) = f(x) - 1$

19. $g(x) = f(x - 1)$

21. $g(x) = f(x - 1) + 2$

23. $g(x) = -f(x)$

25. $g(x) = f(-x) + 1$

27. $g(x) = 2f(x)$

29. $g(x) = f(2x)$

31. $g(x) = 2f(x + 2) + 1$

33. $g(x) = f(x) + 2$

35. $g(x) = f(x + 2)$

37. $g(x) = -f(x + 2)$

39. $g(x) = -\dfrac{1}{2}f(x + 2)$

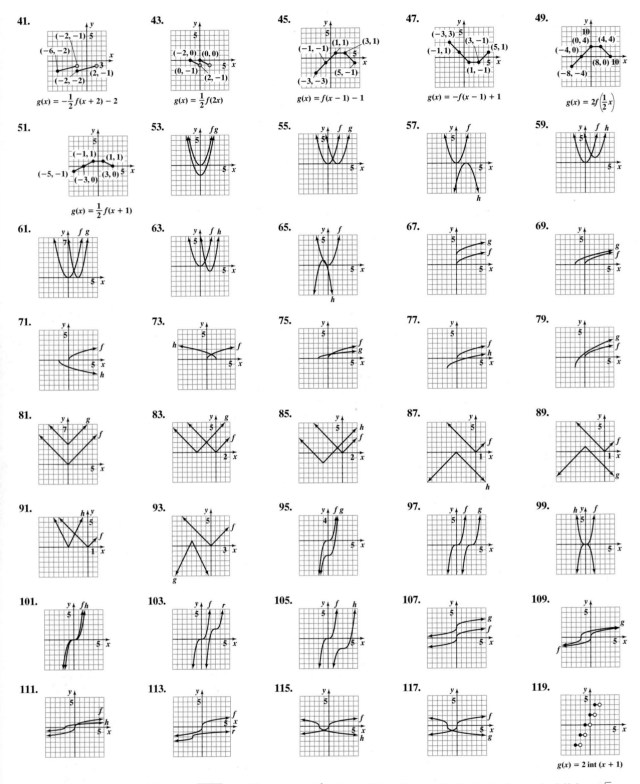

41. $g(x) = -\frac{1}{2}f(x+2) - 2$

43. $g(x) = \frac{1}{2}f(2x)$

45. $g(x) = f(x-1) - 1$

47. $g(x) = -f(x-1) + 1$

49. $g(x) = 2f\left(\frac{1}{2}x\right)$

51. $g(x) = \frac{1}{2}f(x+1)$

119. $g(x) = 2 \text{ int } (x+1)$

121.

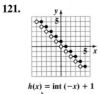

$h(x) = \text{int } (-x) + 1$

123. $y = \sqrt{x-2}$ **125.** $y = (x+1)^2 - 4$ **127. a.** First, vertically stretch the graph of $f(x) = \sqrt{x}$ by the factor 2.9; then, shift the result up 20.1 units. **b.** 40.2 in.; very well **c.** 0.9 in. per month **d.** 0.2 in. per month; This is a much smaller rate of change; The graph is not as steep between 50 and 60 as it is between 0 and 10.

35. a.

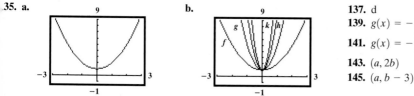

b.

137. d
139. $g(x) = -|x - 5| + 1$
141. $g(x) = -\frac{1}{4}\sqrt{16 - x^2} - 1$
143. $(a, 2b)$
145. $(a, b - 3)$

Section 1.7

Check Point Exercises

1. a. $(-\infty, \infty)$ **b.** $(-\infty, -7)\cup(-7, 7)\cup(7, \infty)$ **c.** $[3, \infty)$ **2. a.** $(f + g)(x) = x^2 + x - 6$ **b.** $(f - g)(x) = -x^2 + x - 4$

$(fg)(x) = x^3 - 5x^2 - x + 5$ **d.** $\left(\dfrac{f}{g}\right)(x) = \dfrac{x - 5}{x^2 - 1}, x \neq \pm 1$ **3. a.** $(f + g)(x) = \sqrt{x - 3} + \sqrt{x + 1}$ **b.** $[3, \infty)$

4. a. $(f \circ g)(x) = 10x^2 - 5x + 1$ **b.** $(g \circ f)(x) = 50x^2 + 115x + 65$ **5. a.** $(f \circ g)(x) = \dfrac{4x}{1 + 2x}$ **b.** $\left(-\infty, -\dfrac{1}{2}\right)\cup\left(-\dfrac{1}{2}, 0\right)\cup(0, \infty)$

6. if $f(x) = \sqrt{x}$ and $g(x) = x^2 + 5$, then $h(x) = (f \circ g)(x)$

Exercise Set 1.7

1. $(-\infty, \infty)$ **3.** $(-\infty, 4)\cup(4, \infty)$ **5.** $(-\infty, \infty)$ **7.** $(-\infty, -3)\cup(-3, 5)\cup(5, \infty)$ **9.** $(-\infty, -7)\cup(-7, 9)\cup(9, \infty)$
11. $(-\infty, -1)\cup(-1, 1)\cup(1, \infty)$ **13.** $(-\infty, 0)\cup(0, 3)\cup(3, \infty)$ **15.** $(-\infty, 1)\cup(1, 3)\cup(3, \infty)$ **17.** $[3, \infty)$ **19.** $(3, \infty)$ **21.** $[-7, \infty)$
23. $(-\infty, 12]$ **25.** $[2, \infty)$ **27.** $[2, 5)\cup(5, \infty)$ **29.** $(-\infty, -2)\cup(-2, 2)\cup(2, 5)\cup(5, \infty)$ **31.** $(f + g)(x) = 3x + 2$;

Domain: $(-\infty, \infty)$; $(f - g)(x) = x + 4$; Domain: $(-\infty, \infty)$; $(fg)(x) = 2x^2 + x - 3$; Domain: $(-\infty, \infty)$; $\left(\dfrac{f}{g}\right)(x) = \dfrac{2x + 3}{x - 1}$; Domain: $(-\infty, 1)\cup(1, \infty)$

33. $(f + g)(x) = 3x^2 + x - 5$; Domain: $(-\infty, \infty)$; $(f - g)(x) = -3x^2 + x - 5$; Domain: $(-\infty, \infty)$; $(fg)(x) = 3x^3 - 15x^2$;

Domain: $(-\infty, \infty)$; $\left(\dfrac{f}{g}\right)(x) = \dfrac{x - 5}{3x^2}$; Domain: $(-\infty, 0)\cup(0, \infty)$ **35.** $(f + g)(x) = 2x^2 - 2$; Domain: $(-\infty, \infty)$; $(f - g)(x) = 2x^2 - 2x - 4$;

Domain: $(-\infty, \infty)$; $(fg)(x) = 2x^3 + x^2 - 4x - 3$; Domain: $(-\infty, \infty)$; $\left(\dfrac{f}{g}\right)(x) = 2x - 3$; Domain: $(-\infty, -1)\cup(-1, \infty)$

37. $(f + g)(x) = 2x - 12$; Domain: $(-\infty, \infty)$; $(f - g)(x) = -2x^2 - 2x + 18$; Domain: $(-\infty, \infty)$; $(fg)(x) = -x^4 - 2x^3 + 18x^2 + 6x - 45$;

Domain: $(-\infty, \infty)$; $\left(\dfrac{f}{g}\right)(x) = \dfrac{3 - x^2}{x^2 + 2x - 15}$; Domain: $(-\infty, -5) \cup (-5, 3)\cup(3, \infty)$ **39.** $(f + g)(x) = \sqrt{x} + x - 4$;

Domain: $[0, \infty)$; $(f - g)(x) = \sqrt{x} - x + 4$; Domain: $[0, \infty)$; $(fg)(x) = \sqrt{x}(x - 4)$; Domain: $[0, \infty)$; $\left(\dfrac{f}{g}\right)(x) = \dfrac{\sqrt{x}}{x - 4}$; Domain: $[0, 4)\cup(4, \infty)$

41. $(f + g)(x) = \dfrac{2x + 2}{x}$; Domain: $(-\infty, 0)\cup(0, \infty)$; $(f - g)(x) = 2$; Domain: $(-\infty, 0)\cup(0, \infty)$; $(fg)(x) = \dfrac{2x + 1}{x^2}$;

Domain: $(-\infty, 0)\cup(0, \infty)$; $\left(\dfrac{f}{g}\right)(x) = 2x + 1$; Domain: $(-\infty, 0)\cup(0, \infty)$ **43.** $(f + g)(x) = \dfrac{9x - 1}{x^2 - 9}$;

Domain: $(-\infty, -3)\cup(-3, 3)\cup(3, \infty)$; $(f - g)(x) = \dfrac{x + 3}{x^2 - 9} = \dfrac{1}{x - 3}$; Domain: $(-\infty, -3)\cup(-3, 3)\cup(3, \infty)$; $(fg)(x) = \dfrac{20x^2 - 6x - 2}{(x^2 - 9)^2}$;

Domain: $(-\infty, -3)\cup(-3, 3)\cup(3, \infty)$; $\left(\dfrac{f}{g}\right)(x) = \dfrac{5x + 1}{4x - 2}$; Domain: $(-\infty, -3)\cup\left(-3, \dfrac{1}{2}\right)\cup\left(\dfrac{1}{2}, 3\right)\cup(3, \infty)$

45. $(f + g)(x) = \sqrt{x + 4} + \sqrt{x - 1}$; Domain: $[1, \infty)$; $(f - g)(x) = \sqrt{x + 4} - \sqrt{x - 1}$; Domain: $[1, \infty)$; $(fg)(x) = \sqrt{x^2 + 3x - 4}$;

Domain: $[1, \infty)$; $\left(\dfrac{f}{g}\right)(x) = \dfrac{\sqrt{x + 4}}{\sqrt{x - 1}}$; Domain: $(1, \infty)$ **47.** $(f + g)(x) = \sqrt{x - 2} + \sqrt{2 - x}$; Domain: $\{2\}$; $(f - g)(x) = \sqrt{x - 2} - \sqrt{2 - x}$;

Domain: $\{2\}$; $(fg)(x) = \sqrt{x - 2} \cdot \sqrt{2 - x}$; Domain: $\{2\}$; $\left(\dfrac{f}{g}\right)(x) = \dfrac{\sqrt{x - 2}}{\sqrt{2 - x}}$; Domain: $\varnothing$ **49. a.** $(f \circ g)(x) = 2x + 14$

b. $(g \circ f)(x) = 2x + 7$ **c.** $(f \circ g)(2) = 18$ **51. a.** $(f \circ g)(x) = 2x + 5$ **b.** $(g \circ f)(x) = 2x + 9$ **c.** $(f \circ g)(2) = 9$

53. a. $(f \circ g)(x) = 20x^2 - 11$ **b.** $(g \circ f)(x) = 80x^2 - 120x + 43$ **c.** $(f \circ g)(2) = 69$ **55. a.** $(f \circ g)(x) = x^4 - 4x^2 + 6$

b. $(g \circ f)(x) = x^4 + 4x^2 + 2$ **c.** $(f \circ g)(2) = 6$ **57. a.** $(f \circ g)(x) = -2x^2 - x - 1$ **b.** $(g \circ f)(x) = 2x^2 - 17x + 41$

c. -11 **59. a.** $(f \circ g)(x) = \sqrt{x - 1}$ **b.** $(g \circ f)(x) = \sqrt{x} - 1$ **c.** $(f \circ g)(2) = 1$ **61. a.** $(f \circ g)(x) = x$

b. $(g \circ f)(x) = x$ **c.** $(f \circ g)(2) = 2$ **63. a.** $(f \circ g)(x) = x$ **b.** $(g \circ f)(x) = x$ **c.** 2

65. a. $(f \circ g)(x) = \dfrac{2x}{1 + 3x}$ **b.** $\left(-\infty, -\dfrac{1}{3}\right)\cup\left(-\dfrac{1}{3}, 0\right)\cup(0, \infty)$ **67. a.** $(f \circ g)(x) = \dfrac{4}{4 + x}$ **b.** $(-\infty, -4)\cup(-4, 0)\cup(0, \infty)$

69. a. $(f \circ g)(x) = \sqrt{x - 2}$ **b.** $[2, \infty)$ **71. a.** $(f \circ g)(x) = 5 - x$ **b.** $(-\infty, 1]$ **73.** $f(x) = x^4, g(x) = 3x - 1$

75. $f(x) = \sqrt[3]{x}, g(x) = x^2 - 9$ **77.** $f(x) = |x|, g(x) = 2x - 5$ **79.** $f(x) = \dfrac{1}{x}, g(x) = 2x - 3$ **81.** 5   **83.** -1 **85.** $\{x|-4 \leq x \leq 3\}$

87.

89. 1   **91.** −6   **93.** 1 and 2   **95.** $\{x|x = 0, 1, 2, \dots, 8\}$

97. a. $(B - D)(x) = 8244x + 1,569,712$; change in U.S. population

   b. 1,635,664; In 2003, the U.S. population increased by 1,635,664.   **c.** 1,670,000; fairly well

99. $f + g$ represents the total world population in year x.

101. $(f + g)(2000) \approx 6$ billion people

103. $(R - C)(20,000) = -200,000$; The company lost $200,000 since costs exceeded revenues; $(R - C)(30,000) = 0$; The company broke even since revenues equaled cost; $(R - C)(40,000) = 200,000$; The company made a profit of $200,000.

105. a. f gives the price of the computer after a $400 discount. g gives the price of the computer after a 25% discount.

   b. $(f \circ g)(x) = 0.75x - 400$; This models the price of a computer after first a 25% discount and then a $400 discount.

   c. $(g \circ f)(x) = 0.75(x - 400)$; This models the price of a computer after first a $400 discount and then a 25% discount.

   d. The function $f \circ g$ models the greater discount, since the 25% discount is taken on the regular price first.

113.

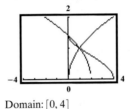

The per capita costs are increasing over time.

115.

Domain: $[0, 4]$

117. Assume f and g are even; then $f(-x) = f(x)$ and $g(-x) = g(x)$. $(fg)(-x) = f(-x)g(-x) = f(x)g(x) = (fg)(x)$, so fg is even.

119.

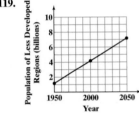

Section 1.8

Check Point Exercises

1. $f(g(x)) = 4\left(\dfrac{x + 7}{4}\right) - 7 = x + 7 - 7 = x$; $g(f(x)) = \dfrac{(4x - 7) + 7}{4} = \dfrac{4x}{4} = x$ **2.** $f^{-1}(x) = \dfrac{x - 7}{2}$ **3.** $f^{-1}(x) = \sqrt[3]{\dfrac{x + 1}{4}}$

4. $f^{-1}(x) = \dfrac{3}{x + 1}$ **5.** (b) and (c)

6.

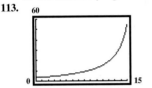

7. $f^{-1}(x) = \sqrt{x - 1}$

Exercise Set 1.8

1. $f(g(x)) = x$; $g(f(x)) = x$; f and g are inverses.   **3.** $f(g(x)) = x$; $g(f(x)) = x$; f and g are inverses.

5. $f(g(x)) = \dfrac{5x - 56}{9}$; $g(f(x)) = \dfrac{5x - 4}{9}$; f and g are not inverses.   **7.** $f(g(x)) = x$; $g(f(x)) = x$; f and g are inverses.

9. $f(g(x)) = x$; $g(f(x)) = x$; f and g are inverses.   **11.** $f^{-1}(x) = x - 3$ **13.** $f^{-1}(x) = \dfrac{x}{2}$ **15.** $f^{-1}(x) = \dfrac{x - 3}{2}$ **17.** $f^{-1}(x) = \sqrt[3]{x - 2}$

19. $f^{-1}(x) = \sqrt[3]{x} - 2$ **21.** $f^{-1}(x) = \dfrac{1}{x}$ **23.** $f^{-1}(x) = x^2, x \geq 0$ **25.** $f^{-1}(x) = \dfrac{7}{x + 3}$ **27.** $f^{-1}(x) = \dfrac{3x + 1}{x - 2}; x \neq 2$

29. The function is not one-to-one, so it does not have an inverse function.

31. The function is not one-to-one, so it does not have an inverse function.

33. The function is one-to-one, so it does have an inverse function.

35.

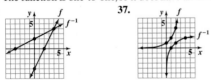

37.

9. a. $f^{-1}(x) = \dfrac{x+1}{2}$

b.
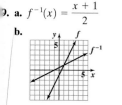

c. Domain of f = Range of $f^{-1} = (-\infty, \infty)$;
Range of f = Domain of $f^{-1} = (-\infty, \infty)$

41. a. $f^{-1}(x) = \sqrt{x} + 4$

b.
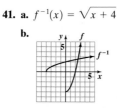

c. Domain of f = Range of $f^{-1} = [0, \infty)$;
Range of f = Domain of $f^{-1} = [-4, \infty)$

43. a. $f^{-1}(x) = -\sqrt{x} + 1$

b.

c. Domain of f = Range of $f^{-1} = (-\infty, 1]$;
Range of f = Domain of $f^{-1} = [0, \infty)$

45. a. $f^{-1}(x) = \sqrt[3]{x} + 1$

b.
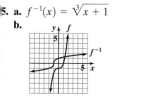

c. Domain of f = Range of $f^{-1} = (-\infty, \infty)$;
Range of f = Domain of $f^{-1} = (-\infty, \infty)$

47. a. $f^{-1}(x) = \sqrt[3]{x} - 2$

b.

c. Domain of f = Range of $f^{-1} = (-\infty, \infty)$;
Range of f = Domain of $f^{-1} = (-\infty, \infty)$

49. a. $f^{-1}(x) = x^2 + 1, x \geq 0$

b.

c. Domain of f = Range of $f^{-1} = [1, \infty)$;
Range of f = Domain of $f^{-1} = [0, \infty)$

51. a. $f^{-1}(x) = (x-1)^3$

b.

c. Domain of f = Range of $f^{-1} = (-\infty, \infty)$;
Range of f = Domain of $f^{-1} = (-\infty, \infty)$

53. 5 **55.** 1 **57.** 2 **59.** −7 **61.** 3 **63.** 11 **65. a.** $f = \{(\text{Zambia}, -7.3), (\text{Columbia}, -4.5), (\text{Poland}, -2.8), (\text{Italy}, -2.8),$ (United States, −1.9))} **b.** $f^{-1} = \{(-7.3, \text{Zambia}), (-4.5, \text{Columbia}), (-2.8, \text{Poland}), (-2.8, \text{Italy}), (-1.9, \text{United States})\}$; No; One member of the domain, −2.8, corresponds to more than one member of the range, Poland and Italy. **67. a.** f is a one-to-one function. **b.** $f^{-1}(0.25)$ is the number of people in a room for a 25% probability of two people sharing a birthday. $f^{-1}(0.5)$ is the number of people in a room for a 50% probability of two people sharing a birthday. $f^{-1}(0.7)$ is the number of people in a room for a 70% probability of two people sharing a birthday.

69. $f(g(x)) = \dfrac{9}{5}\left[\dfrac{5}{9}(x-32)\right] + 32 = x$ and $g(f(x)) = \dfrac{5}{9}\left[\left(\dfrac{9}{5}x + 32\right) - 32\right] = x$

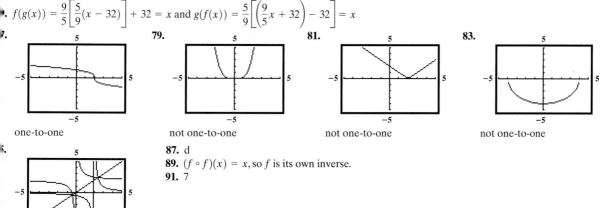

77. one-to-one **79.** not one-to-one **81.** not one-to-one **83.** not one-to-one

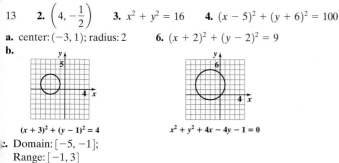

87. d
89. $(f \circ f)(x) = x$, so f is its own inverse.
91. 7

f and g are inverses.

ection 1.9

heck Point Exercises

1. 13 **2.** $\left(4, -\dfrac{1}{2}\right)$ **3.** $x^2 + y^2 = 16$ **4.** $(x-5)^2 + (y+6)^2 = 100$

5. a. center: $(-3, 1)$; radius: 2 **6.** $(x+2)^2 + (y-2)^2 = 9$

b.

$(x+3)^2 + (y-1)^2 = 4$

$x^2 + y^2 + 4x - 4y - 1 = 0$

c. Domain: $[-5, -1]$;
Range: $[-1, 3]$

Exercise Set 1.9

1. 13   **3.** $2\sqrt{29} \approx 10.77$ **5.** 5   **7.** $\sqrt{29} \approx 5.39$ **9.** $4\sqrt{2} \approx 5.66$ **11.** $2\sqrt{5} \approx 4.47$ **13.** $2\sqrt{2} \approx 2.83$ **15.** $\sqrt{93} \approx 9.64$

17. $\sqrt{5} \approx 2.24$ **19.** $(4, 6)$ **21.** $(-4, -5)$ **23.** $\left(\dfrac{3}{2}, -6\right)$ **25.** $(-3, -2)$ **27.** $(1, 5\sqrt{5})$ **29.** $(2\sqrt{2}, 0)$ **31.** $x^2 + y^2 = 49$

33. $(x - 3)^2 + (y - 2)^2 = 25$ **35.** $(x + 1)^2 + (y - 4)^2 = 4$ **37.** $(x + 3)^2 + (y + 1)^2 = 3$ **39.** $(x + 4)^2 + (y - 0)^2 = 100$

41. center: $(0, 0)$ **43.** center: $(3, 1)$ **45.** center: $(-3, 2)$ **47.** center: $(-2, -2)$
radius: 4   radius: 6   radius: 2   radius: 2
Domain: $[-4, 4]$;   Domain: $[-3, 9]$;   Domain: $[-5, -1]$;   Domain: $[-4, 0]$;
Range: $[-4, 4]$ Range: $[-5, 7]$ Range: $[0, 4]$ Range: $[-4, 0]$

49. $(x + 3)^2 + (y + 1)^2 = 4$ **51.** $(x - 5)^2 + (y - 3)^2 = 64$ **53.** $(x + 4)^2 + (y - 1)^2 = 25$ **55.** $(x - 1)^2 + (y - 0)^2 = 16$
center: $(-3, -1)$ center: $(5, 3)$ center: $(-4, 1)$ center: $(1, 0)$
radius: 2   radius: 8   radius: 5   radius: 4

57. $\left(x - \dfrac{1}{2}\right)^2 + (y + 1)^2 = \dfrac{1}{4}$ **59.** $\left(x + \dfrac{3}{2}\right)^2 + (y - 1)^2 = \dfrac{17}{4}$ **61. a.** $(5, 10)$ **b.** $\sqrt{5}$ **c.** $(x - 5)^2 + (y - 10)^2 = 5$
center: $\left(\dfrac{1}{2}, -1\right)$ center: $\left(-\dfrac{3}{2}, 1\right)$
radius: $\dfrac{1}{2}$ radius: $\dfrac{\sqrt{17}}{2}$

63. $\{(0, -4), (4, 0)\}$ **65.** $\{(0, -3), (2, -1)\}$ **67.** 0.5hr; 30 min   **69.** $(x + 2.4)^2 + (y + 2.7)^2 = 900$

77.   **79.**   **81.** $2\sqrt{2} + 3\sqrt{2} = 5\sqrt{2}$ **83.** 11π

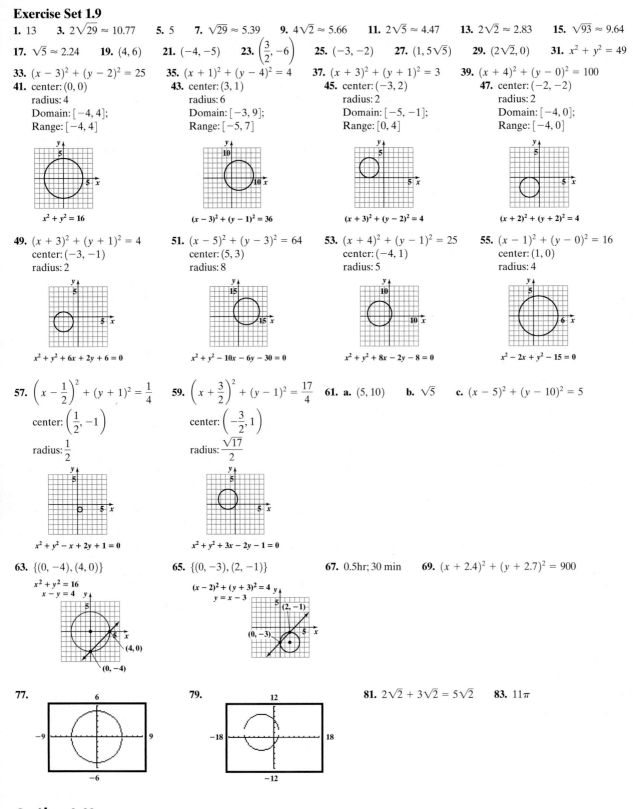

Section 1.10

Check Point Exercises

1. a. $f(x) = 0.08x + 15$ **b.** $g(x) = 0.12x + 3$ **c.** 300 min   **2. a.** $N(x) = -100x + 18{,}000$ **b.** $R(x) = -100x^2 + 18{,}000x$

3. a. $V(x) = x(15 - 2x)(8 - 2x)$ **b.** $\{x \mid 0 < x < 4\}$ or $(0, 4)$ **4.** $x(100 - x) = 100x - x^2$ ft^2 **5.** $A(r) = 2\pi r^2 + \dfrac{2000}{r}$

6. $I(x) = 0.07x + 0.09(25{,}000 - x)$ **7.** $d = \sqrt{x^6 + x^2}$